# Occupational Ergonomics

## Theory and Applications

Second Edition

# Occupational Ergonomics

## Theory and Applications

Second Edition

edited by

**Amit Bhattacharya**

University of Cincinnati Medical College
Cincinnati, Ohio
Co-Founder, OsteoDynamics, Inc.
Cincinnati, Ohio

**James D. McGlothlin**

Purdue University
West Lafayette, Indiana

CRC Press is an imprint of the
Taylor & Francis Group, an **informa** business

CRC Press
Taylor & Francis Group
6000 Broken Sound Parkway NW, Suite 300
Boca Raton, FL 33487-2742

First issued in paperback 2017



ISBN-13: 978-1-4398-1934-0 (hbk)
ISBN-13: 978-1-138-07471-2 (pbk)



**Visit the Taylor & Francis Web site at**
**http://www.taylorandfrancis.com**

**and the CRC Press Web site at**
**http://www.crcpress.com**

# Contents

## Part II Application of Ergonomic Principles



## Part III Medical Surveillance for Ergonomics Programs

## PART IV Ergonomic Case Studies



## PART V Physical Agents in Workplace

## Part VI Current Topics



## Part VII International Perspective on Ergonomics

# Preface

Since the publication of the first edition in 1996, significant new advances have been made in the field of ergonomics. These advances include the impact of aging and obesity in the workplace, the role of ergonomics in promoting healthy workplaces and healthy lifestyles, the role of ergonomic science in the design of consumer products, and much more. Therefore, we saw a critical need to update the first edition, which has had more than a dozen printings and can be found in libraries all over the world.

We hope you agree that the second edition fills a critical gap between existing ergonomic books by providing a comprehensive approach that encompasses the principles of ergonomics from theory to practice. It also blends medical and engineering applications to solve musculoskeletal, safety, and health problems in a variety of traditional and emerging industries ranging from the office to the operating room to operations engineering. The scope and contents of this comprehensive book are such that it is suitable for graduate (and senior-level undergraduate) students as well as a reference text for occupational safety and health professionals.

This book is designed to address both the fundamentals of ergonomics (Part I) as well as practical applications (Part II) of those fundamentals in solving ergonomic problems to provide an insight into methods used to assess ergonomic risks and their control. Part III addresses the issue of connecting physical risk factors to the development of adverse health outcomes, thus providing a framework for the development of medical surveillance programs designed to monitor/control ergonomic-related challenges in the workplace. In Part IV, the reader is given the opportunity to review ergonomic case studies from a variety of industries such as health care, agriculture and construction. Part V consists of new material covering workplace-related physical agents such as workplace noise, workplace nonionizing radiation, heat, and vibration. As the field of ergonomic is growing rapidly, we have included emerging topics in Part VI, such as economics of ergonomics, application of research methods for solving "real-world" ergonomic problems, healthy workplace and lifestyle, cumulative trauma disorders (CTDs), NIOSH's Applications Manual for the Revised NIOSH Lifting Equation, and the impact of the Americans with Disabilities Act (ADA) on ergonomics. Finally, as the field of ergonomics has had a significant impact at the international level, Part VII covers the current trends in this field from around the world. The end of the book has appendices, which include several useful ergonomic tables, list of vendors of ergonomic tools, software and video training material, ergonomic checklists, homework problems, and other background material. The benefits to the potential users of this book, we hope, will be numerous, including learning, understanding, and applying the fundamentals of ergonomics from theory to practice, yet appreciating the expanding applications for occupational safety and health students as well as professionals. Further, this book contains several ergonomic solutions from the very basic to systems design to prevention through design (PtD). We are very excited and proud to offer you the second edition of this book, and we are indebted to our chapter authors for their diligence in bringing you the very best the field of ergonomics has to offer.

The second edition of this book could not have been possible without the many talented administrators, staff, and students who helped bring this book to fruition. In the following, we have listed some of those who helped toward this end. There may be many more whom we may have inadvertently missed. We will always be grateful for all their help. Special thanks are to Yvonne Nash for her word processing and indexing assistance; to Mike Johnson for his help in editing and indexing our many figures and photographs, and Mark Sharp for his IT support; to Dr. McGlothlin's many students—Stephanie Snack, Maggie Cappel, Sandra Cole, and Michael White—for their organization and proofreading of the new chapters. We would also like to thank the many contributors to our appendices, including Dr. Vincent Duffy and his graduate student H. Onan Demirel. Above all, we are especially grateful for the patience and understanding of our wives, Prakriti Bhattacharya and Nancy McGlothlin.

# Editors

**Amit Bhattacharya** received his PhD in biomedical/mechanical engineering and his MS in fluid mechanics/heat transfer, from the University of Kentucky in Lexington, Kentucky, United States. He is a professor of environmental health and also a professor of biomedical engineering and mechanical engineering at the University of Cincinnati Medical College, Cincinnati, Ohio, United States. He is the founding director of the Biomechanics-Ergonomics Research Laboratories of the Department of Environmental Health. He is also the founding director of the Occupational Ergonomics/Safety graduate education program and the Pilot Research Training program sponsored by the National Institute for Occupational Safety and Health (NIOSH) and housed in the Department of Environmental Health within the College of Medicine at the University of Cincinnati. He has been teaching and conducting ergonomics/biomechanics research for more than 30 years. Dr. Bhattacharya has been active as an ergonomics/biomechanics consultant to various private industries as well as governmental agencies such as NIOSH and NIH. He serves on the editorial board of the *Journal of Occupational Ergonomics* and as an ad hoc reviewer for several peer-reviewed journals such as *Human Factors, Clinical Biomechanics, Gait and Posture, Neurotoxicology,* and the *International Journal of Audiology and Environmental Health Perspectives.* Dr. Bhattacharya has made significant contributions in the areas of biomechanics of slips/falls in the workplace, heat stress, occupational biomechanics of repetitive trauma, workstation design, physiological/biomechanical effects of external vibration on animals and humans, therapeutic aspects of whole-body vibration, development of countermeasures for cardiovascular deconditioning resulting from weightlessness, and the development of noninvasive, sensitive techniques for the quantification of postural imbalance as an indicator of neurotoxicity and identification of preclinical biomechanical parameters of osteoarthritis and osteoporosis. Dr. Bhattacharya's current research activities include (1) impact of environmental toxicants (e.g., Pb, Mn, pesticide) on the human neuromuscular system and susceptibility of developing degenerative skeletal disorder (e.g., osteoporosis); (2) design, development, and application of nano-sensors/BIOMEMS technology for early detection of neurodegenerative (e.g., Parkinson's disease) and degenerative skeletal disorders; and (3) use of wearable and ingestible sensors for real-time assessment of heat stress among firefighters during live firefighting incidents. A recently completed NIH-supported study by his group developed a novel *noninvasive* tool to quantitate bone's shock absorption (BSA) property as an aggregate response of bone's structural integrity for early detection of fracture in osteoporosis. With support from BIOSTART organization (Ohio's Edison incubator program for life science and health) and the University of Cincinnati, Dr. Bhattacharya cofounded a start-up company (OsteoDynamics Inc.) for commercialization of BSA system for early detection of skeletal disorders. In the "Cincinnati Innovates-2010" event, OsteoDynamics Inc. received the First Place Award for Legal and Patent Awards sponsored by Taft Law Firm of Ohio. In 2011, Dr. Bhattacharya received the finalist award from *Business Courier Journal's* annual Health Care Heroes for innovation that makes an *impact on health care* in the Greater Cincinnati region. He is a fellow of the Biomedical Engineering Society and a charter member of the National Academy of Inventors. He is a also full member of the Human Factors and Ergonomics Society and the American Industrial Hygiene Association.

**James D. McGlothlin** received his BA (1975) in psychology, his MPH (1977) in epidemiology, and his MS (1977) in environmental and industrial health, all from the University of Hawaii, Honolulu. He received his PhD (1988) in industrial health from the Rackham School of Graduate Studies, with a specialty in ergonomics from the University of Michigan, Ann Arbor. Dr. McGlothlin is an associate professor of health sciences in the College of Health and Human Sciences at Purdue University. His research interests include ergonomics, exposure assessment, occupational hygiene, engineering controls, and epidemiology. He is the director of the Graduate Program in Occupational and Environmental Sciences in the School of Health Sciences at Purdue University and codirector of the Center for Virtual Reality of Healthcare Center Design. He is one of the co-principle investigators for the Regenstrief Center for Healthcare Engineering (RCHE). He has served on the University Senate and on the Senate Advisory Committee to President Cordova and Provost Sands through 2011. Prior to his appointment at Purdue University (January 4, 1999), Dr. McGlothlin was a senior researcher in ergonomics and occupational hygiene engineering controls with the Centers for Disease Control and Prevention, National Institute for Occupational Safety and Health (CDC/NIOSH). The author of more than 150 scientific, technical, and government reports, he has served as a course director in ergonomics at Northwestern University School of Engineering, Evanston, Illinois, and at the University of Cincinnati School of Medicine, Ohio. He is currently an adjunct professor in the College of Public Health at the University of Iowa, and he has courtesy appointments at Purdue's School of Industrial Technology and Purdue Calumet. He has received several awards for his research and service to the U.S. Public Health Service, including the Surgeon General's Exemplary Service Medal, the Outstanding Service Medal, and the Stanley J. Kessel Award for Outstanding Health Services Professional of the Year. At Purdue University, his awards include the Focus Award for outstanding contributions to the furthering of Purdue University's commitment to disability accessibility and diversity. He is recognized as a faculty scholar for outstanding academic distinction as well as a RCHE scholar. Dr. McGlothlin became a fellow of the American Industrial Hygiene Association in March 2006. He has three patents to his credit, with one pending on wireless real-time video exposure monitoring systems. He also serves on several national and international professional committees and currently serves on the editorial boards of the *Occupational and Environmental Health Journal* and the *Occupational Hazards Journal*. Dr. McGlothlin is a certified professional ergonomist. You can find out more about him by visiting his website at www.DrMcGlothlin.com.

# Contributors

**James F. Annis**
Anthropology Research Project, Inc.
Yellow Springs, Ohio

**Margaret R. Atterbury**
City of Cincinnati
Department of Health
Millvale Health Center
Cincinnati, Ohio

**Angus Bagchee**
Department of Environmental Health
University of Cincinnati
Cincinnati, Ohio

**David J. Beebe**
Department of Biomedical Engineering
University of Wisconsin-Madison
Madison, Wisconsin

**Thomas E. Bernard**
College of Public Health
University of South Florida
Tampa, Florida

**Amit Bhattacharya**
Biomechanics-Ergonomics Research Labs
Department of Environmental Health
University of Cincinnati Medical College
Cincinnati, Ohio

**Balmatee Bidassie**
Purdue University
West Lafayette, Indiana

**Donald S. Bloswick**
Department of Mechanical Engineering
and
Rocky Mountain Center for Occupational and Environmental Health
University of Utah
Salt Lake City, Utah

**Robert E. Breslin**
Breslin Vocational Consultation, Ltd.
Cincinnati, Ohio

**Don R. Brown**
Department of Mechanical Engineering
and
Rocky Mountain Center for Occupational and Environmental Health
University of Utah
Salt Lake City, Utah

**Peter M. Budnick**
Ergoweb, Inc.
Park City, Utah

**John Cardarelli II**
U.S. Public Health Services
Washington, District of Columbia

**Claire C. Caruso**
Division of Applied Research and Technology
National Institute for Occupational Safety and Health
Cincinnati, Ohio

**Sharon S. Chiou**
Division of Safety Research
National Institute for Occupational Safety and Health
Centers for Disease Control and Prevention
Atlanta, Georgia

**David R. Clark**
Industrial and Manufacturing Engineering Department
GMI Engineering & Management Institute
Kettering University
Flint, Michigan

**Alexander L. Cohen**
Occupational Human Factors
Cincinnati, Ohio

**Sandra S. Cole**
Purdue University
West Lafayette, Indiana

**Robert G. Cutlip**
Health Effects Laboratory Division
National Institute for Occupational Safety and Health
Centers for Disease Control and Prevention
Atlanta, Georgia

**Elizabeth Dalsey**
Office of Research and Technology Transfer
National Institute for Occupational Safety and Health
Cincinnati, Ohio

**Kermit G. Davis**
Low Back Biomechanics and Workplace Stress Laboratory
Department of Environmental Health
University of Cincinnati
Cincinnati, Ohio

**H. Onan Demirel**
Department of Industrial Engineering
Purdue University
West Lafayette, Indiana

**Lennart Dimberg**
Volvo Aero Corporation
Trollhättan, Sweden

**Vincent Duffy**
Department of Industrial Engineering
Purdue University
West Lafayette, Indiana

**Andris Freivalds**
Department of Industrial & Manufacturing Systems Engineering
The Pennsylvania State University
University Park, Pennsylvania

**Emil Golias**
Occupational Safety and Health Administration
Salt Lake City, Utah

**Kevin P. Granata**
Virginia Polytechnic Institute and State University
Blacksburg, Virginia

**Katharyn A. Grant**
National Institute for Occupational Safety and Health
Cincinnati, Ohio

**Michael J. Griffin**
Human Factors Research Unit
Institute of Sound and Vibration Research
University of Southampton
Southampton, England

**Sergey A. Grinshpun**
Department of Environmental Health
College of Medicine
University of Cincinnati
Cincinnati, Ohio

**John Howard**
National Institute for Occupational Safety and Health
Washington, District of Columbia

**Hongwei Hsiao**
Division of Safety Research
National Institute for Occupational Safety and Health
Morgantown, West Virginia

**Steve Hudock**
Division of Applied Research and Technology
National Institute for Occupational Safety and Health
Centers for Disease Control and Prevention
Cincinnati, Ohio

**Heidi Hudson**
Division of Applied Research and Technology
National Institute for Occupational Safety and Health
Cincinnati, Ohio

**R. DeLon Hull**
Office of Research and Technology Transfer
National Institute for Occupational Safety and Health
Cincinnati, Ohio

**Janice Huy**
Office of Research and Technology Transfer
National Institute for Occupational Safety and Health
Cincinnati, Ohio

**Glenn Jimmerson**
Ford Motor Company
Dearborn, Michigan

**Bradley S. Joseph**
Ford Motor Company
Dearborn, Michigan

**Myung-Chul Jung**
Department of Industrial and Information Systems Engineering
Ajou University
Suwon, Korea

**Paul Keane**
Division of Safety Research
National Institute for Occupational Safety and Health
Morgantown, West Virginia

**Jay Kim**
School of Dynamic Systems—Mechanical Engineering
College of Engineering and Applied Science
University of Cincinnati
Cincinnati, Ohio

**Yong-Ku Kong**
Department of Systems Management Engineering
Sungkyunkwan University
Suwon, Korea

**Susan E. Kotowski**
Department of Rehabilitation Sciences
College of Allied Health Sciences
University of Cincinnati
Cincinnati, Ohio

**Sheila Krawczyk**
Liberty Mutual Research Center for Safety and Health
Hopkinton, Massachusetts

**Kyung-Suk Lee**
Agricultural Safety Engineering Division
National Academy of Agricultural Science
Suwon, Korea

**Soo-Jin Lee**
Department of Occupational and Environmental Medicine
Hanyang University
Seoul, Korea

**Grace Kawas Lemasters**
Department of Environmental Health
University of Cincinnati Medical School
Cincinnati, Ohio

**He Lihua**
School of Public Health
Peking University
Beijing, People's Republic of China

**Arthur R. Longmate**
Johnson & Johnson
New Brunswick, New Jersey

**Brian D. Lowe**
Division of Applied Research and Technology
National Institute for Occupational Safety and Health
Cincinnati, Ohio

**Jean Mangharam**
Worksafe Western Australia
West Perth, Western Australia

**William S. Marras**
Biodynamics Laboratory
Department of Integrated Systems Engineering
Baker Systems Engineering
The Ohio State University
Columbus, Ohio

**John T. McConville**
Anthropology Research Project, Inc.
Yellow Springs, Ohio

**James D. McGlothlin**
Purdue University
West Lafayette, Indiana

**James J. Montgomery**
Montgomery, Rennie & Jonson
Cincinnati, Ohio

**William J. Murphy**
Center for Health-Related Aerosol Studies
University of Cincinnati
Cincinnati, Ohio

**Jerry A. Olsheski**
School of Applied Behavioral Sciences
Ohio University
Athens, Ohio

**Mary O'Reilly**
School of Public Health
State University of New York
New York, New York

**Christopher S. Pan**
Division of Safety Research
National Institute for Occupational Safety and Health
Morgantown, West Virginia

**Robert G. Radwin**
Department of Biomedical Engineering
University of Wisconsin-Madison
Madison, Wisconsin

**Tapas K. Ray**
Department of Health and Human Services
National Institute for Occupational Safety and Health
Centers for Disease Control and Prevention
Cincinnati, Ohio

**Roderick J. Reasor (deceased)**
Virginia Polytechnic and State University
Blacksburg, Virginia

**Mark S. Redfern**
Department of Otolaryngology
and
Departments of Industrial Engineering
University of Pittsburgh
Pittsburgh, Pennsylvania

**Timothy P. Rhoades**
Applied Safety and Ergonomics, Inc.
Ann Arbor, Michigan

**Norka Saldaña**
Johnson & Johnson Shared Services
Caguas, Puerto Rico

**Rabindra Nath Sen**
Ergonomics Laboratory
Department of Physiology
University of Calcutta
Calcutta, India

**Ronald L. Stanevich**
Division of Safety Research
National Institute for Occupational Safety and Health
Morgantown, West Virginia

**Shiro Tanaka**
Division of Surveillance
National Institute for Occupational Safety and Health
Cincinnati, Ohio

**Bingshi Wang**
Ethicon Endo-Surgery, Inc.
Cincinnati, Ohio

**Eric Min-yang Wang**
Department of Industrial Engineering
National Tsing-Hua University
Taiwan, Republic of China

**Sheng Wang**
School of Public Health
Peking University
Beijing, People's Republic of China

**Thomas R. Waters**
Applied Psychology and Ergonomics Branch
National Institute for Occupational Safety and Health
Cincinnati, Ohio

**John G. Webster**
Department of Biomedical Engineering
University of Wisconsin-Madison
Madison, Wisconsin

**Michael White**
Purdue University
West Lafayette, Indiana

**Rick Wickstrom**
WorkAbility Wellness Center
WorkAbility Systems, Inc.
West Chester, Ohio

**Jeffrey C. Woldstad**
University of Nebraska-Lincoln
Lincoln, Nebraska

**Thomas Y. Yen**
Department of Biomedical Engineering
University of Wisconsin-Madison
Madison, Wisconsin

**Chi-Yuang Yu**
Department of Industrial Engineering
National Tsing-Hua University
Taiwan, Republic of China

# I

# Principles of Ergonomics

CHAPTER 1

# Anthropometry

James F. Annis and John T. McConville

## CONTENTS

## 1.1 INTRODUCTION

Anthropometry involves the systematic measurement of the physical properties of the human body, primarily dimensional descriptors of body size and shape. Anthropologists have been measuring humans for hundreds of years, but for only the last 50 years or so have the dimensions been used in an organized fashion to improve the design and sizing of the things we use in everyday life. Often the problem with the application of anthropometry to a design problem will be the lack of certain necessary measurements or the need to accommodate a wide range in size and shape variability into a single, often inflexible design. Applied anthropometry—that is, the use of anthropometric data in the design and construction of a wide variety of items from clothing to spacecraft—is a relatively new discipline whose practitioners are still learning to cope with the exponential character of technology and its impact on the kinds of information needed to describe the physical and biological characteristics of our species. It grew out of physical anthropology, which traditionally studied body size and function with the goal of resolving our ancestry and identifying the existing varieties of *Homo sapiens*. Today, we are still trying to learn what to measure to satisfy an even younger discipline called ergonomics.

The original impetus for the development of applied anthropometries, and perhaps ultimately ergonomics, was the need to improve the effectiveness and efficiency of equipment used in combat during World War II. The military tie was so strong in the United States that it has been recognized only relatively recently that civilian industry could also benefit from the proper use of anthropometry in the design of products and workstations. Today industry has embraced the concept so eagerly that the word *ergonomics*, which derives from the Greek *ergon* (work) and *nomos* (natural laws of), has become popularly used in advertisements. A concern for ergonomics currently spawns industrial action committees, comprising members of both management and labor, whose purpose is to improve the human–machine interface to achieve a healthier, safer, and more efficient workforce. Anthropometric data are a necessary and basic tool for attaining this end as well as improving the design of a wide variety of products.

## 1.2 BACKGROUND AND SIGNIFICANCE TO OCCUPATIONAL ERGONOMICS

There appears to be two major divisions of ergonomics. The first deals with the worker, the machine the worker uses, and the environment in which the worker operates. The objective of this branch of ergonomics is to create the best possible situation on the job relative to the welfare of the worker's physical and mental health, the efficiency of production, and the quality of the product produced. Second, there are the characteristics of the manufactured product(s) that interact with the human user. In reviewing the literature we have found no less than 20 different definitions of ergonomics. One of the most inclusive examples is as follows:

> The ability to apply information regarding human character, capacities, and limitations to the design of human tasks, machines, machine systems, living spaces, and environment so that people can live, work, and play safely, comfortably, and efficiently.

All of the definitions of ergonomics mention work or the workplace, but, more important from our perspective, they also mention the human operator in the equation. The ways in which work is related to the individual vary greatly throughout industry, yet it is very difficult to think of any work situation in which the application of anthropometry or anthropometric principles could not make the work environment healthier, safer, and more efficient. At the same time, ergonomists and designers are not trying to make their products either cheaper or more expensive but rather better suited to the limitations of the human user. Anthropometry cannot be separated from these ergonomic processes, because they cannot be carried out without the knowledge of human dimensionalities.

The relationship of anthropometry to occupational ergonomics is both straightforward and complex. All the tools used in manufacturing, all the workspaces in which the manufacturing is done, and virtually all of the items produced by the manufacturing process interact with the human body or human body space. In the most automated of manufacturing environments, humans must still make and repair the machines and robots, and the products that come off the assembly line must be designed for human users. Software is still written by humans and entered onto the computer disk or tape through a keyboard operated by the human hand. It is hard to think of an exception. Ultimately there is a human in the loop; hence, human dimensions are likely to be needed for some time to come.

Obviously, physical anthropologists are prepared to include almost any measurement that describes the shape, size, or function of the body or its parts as components that make up the area of anthropometry. The data discussed most frequently in this chapter are those used as dimensional descriptors of general body size, those used for specific types of working postures, and those describing reach capabilities. Data for related areas of investigation, such as mass distribution and segmental moments of inertia, range of joint motion (ROJM), strength, and biomechanical aspects of the human body, are discussed elsewhere in this book. In one way or another all of these data are at some point in the design process useful to the occupational ergonomist.

## 1.3 RELEVANT CONCEPTS AND TERMINOLOGY

### 1.3.1 Anatomic Concepts

In performing anthropometric measurements, some knowledge of human anatomy is essential, because almost all measurements are defined in terms of some body part or some specific location on a specified part. Subjects being measured are directed to assume specific predefined positions. The standard reference point is the *anatomical position*, in which the person stands erect with arms at the sides and the palms of the hands facing forward. From this posture the descriptive terms that define the body's principal axes and the resultant planes are derived. From these planes, too, the basic terms used to describe the relative position or location of relevant points on the body structure are developed. The most commonly used terms are shown diagrammatically in Figure 1.1. Typically, the principal axes, $X$, (front-to-back), $Y$ (side-to-side), and $Z$ (head-to-foot), divide the body into three planes: sagittal, which divides the body into right and left parts ($XZ$); coronal, which divides the body front to back ($YZ$); and transverse, which divides the body cross-sectionally ($XY$).

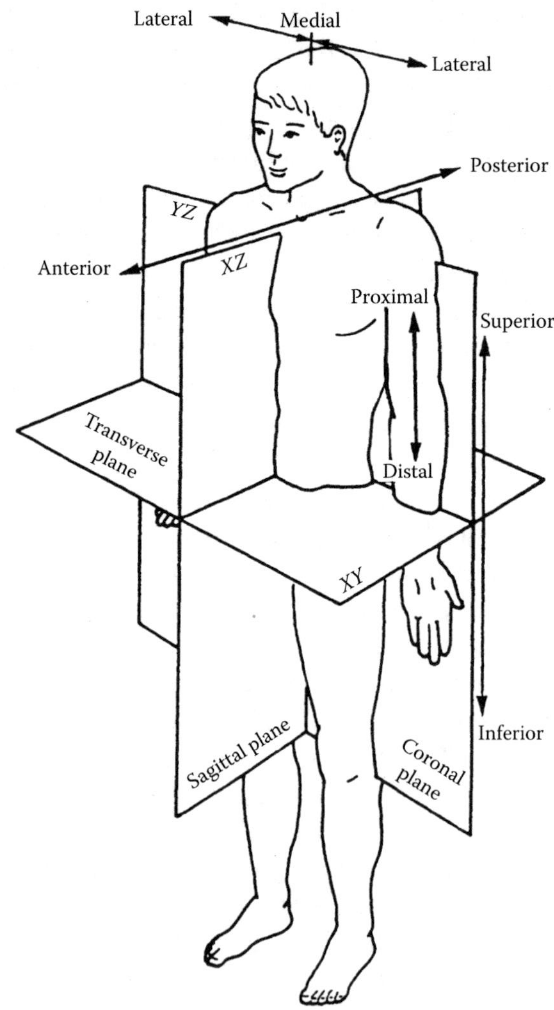


FIGURE 1.1 Terminology used to define position and location on the body.

The relative positions of particular structures or features in or on the body are defined as follows:

*Anterior/posterior*: Structures nearer the front or ventral side are *anterior* (+*X*) to those located nearer to the back or dorsal surface, which are *posterior* (–*X*).

*Medial/lateral*: Structures located nearer to the center of the body or to the midsagittal plane relative to others are *medial* to those located away from the central body on the left (+*Y*) or right (–*Y*) side, which are *lateral*.

*Superior/inferior*: Structures located nearer the head are located superior (+*Z*) to those below, which are termed *inferior* (–*Z*). For example, the heart is located superior to the kidneys.

*Proximal/distal*: On the limbs, parts that are near the trunk are *proximal* whereas those farther from the body central are *distal*. A finger is distal to the elbow.

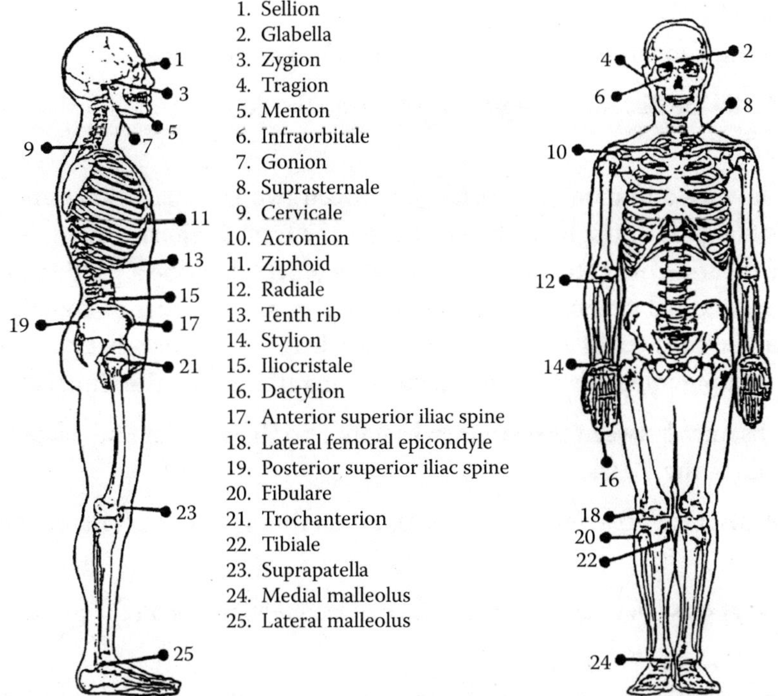


FIGURE 1.2 Selected skeletal landmarks used to define traditional anthropometric measurements.

Frequently the points between which a given dimension is measured are actually drawn on the subject's skin by the anthropometrist. Such points are called *landmarks*. Some landmarks are simply determined by a certain feature found in the topography of the body surface, for example, the tip of a finger; others must be palpated and marked in relation to skeletal architecture. Twenty-four of the twenty-five landmarks shown in Figure 1.2 are of the latter variety.

In a recent survey of U.S. Army personnel (ANSUR) [1], nearly 100 landmarks were used to define the basic group of 132 dimensions measured. Approximately 70 of the landmarks were actually drawn on the subjects' skin by landmarking specialists (e.g., landmarks shown on Figure 1.2), and another 30 landmarks were located by observation by the measurer, for example, distal tip of thumb, inferior tip of earlobe.

### 1.3.2 Types of Measurements

Most commonly, anthropometry refers to traditional dimensional descriptors of body size. Except for weight, these measurements basically provide the straight-line or curvilinear distance between two points obtained under static prescribed conditions. Basic categories of static anthropometric dimensions include lengths, depths, breadths, and distances that are basic descriptors of body size. Surface contour measurements, such as arcs and circumferences, are more complex because they contain elements of three-dimensional (3D) shape in one plane. In most major surveys, the use of other types of anthropometric measurements is limited. Simple forms of reach, for example, functional leg length, overhead reach, grip reach, and thumb tip reach, may be measured.

#### *1.3.2.1 Static Measurements*

Traditional static dimensions may be briefly described as follows:

*Height*: Typically the distance along the *Z* axis from the floor or seated surface to a specific point on the body.

*Length*: Usually used to name the distance between two landmarks that are found on a single segmental part of the body. In some cases the term *distance* is used. Some lengths or distances describe the entire segment, whereas others describe a portion of a segment along the longitudinal axis of the part. Some lengths are contours or complex distances measured on the body surface, for example, sleeve length. Distances between linkage centers used in computer models and drafting board manikins are called link lengths.

*Depth*: The distance between two landmarks found on the anterior and posterior surfaces of the body along the *X* axis.

*Breadth*: The distance between two points found on the right and left (lateral) sides of the body.

*Arc*: Curvilinear or surface contour distance between two points on the body, often on the head or face.

*Circumference*: Closed curvilinear contour that provides the distance completely around the body part. In most cases the circumference is located perpendicular to the longitudinal axis of the body part.

*Reach*: Specialized arm–hand distance in a particular posture or condition.

In most cases, dimension names include the previous designations, but in some cases no such descriptive label is used, for example, span, stature, or scye. The idea of the defining labels used to describe a particular type or class of measurements begs the question of standardization of measurement definitions, a matter over which anthropologists have often been unable to agree. One text that deals with standards was published in 1988 [2]. This book concentrates on nutritional/health assessment measurements, but a number of applied dimensions are discussed. At a conference of practicing anthropologists held at Wright–Patterson Air Force Base in the late 1960s, agreement on a list of 29 dimensions fundamental to applied users was barely achieved [3]. It would be helpful to the user if dimensions with the same name were always measured in the same way. This is very often not the case. Dimensions with the same name from any two surveys may have well been measured differently; hence, the difference observed between the two values may in part be due to procedures used and not to population or sampling differences.

#### *1.3.2.2 Dynamic Measurements*

Dynamic measurements such as isometric strength and ROJM have traditionally been measured in separate surveys. The measurement of ROJM is probably the simplest of such measurements in anthropometries. Traditionally, planar ROJM measurements are made using a goniometer, which in its simplest form is nothing more than a 180° or 360° protractor with extended arms, one of which is movable to track the segment. A related measurement of this type is the reach envelope, which at some point in the measuring process always involves movement of the arm-shoulder complex and sometimes movement of the trunk or torso. In most cases, however, these measurements are obtained statically, that is, they are expressed as the change or difference in location across the movement, that is, begin-end delta.

Too few truly dynamic anthropometric measurements are made on humans, and most of those that may be thought to qualify as dynamic contain static elements. For example, isotonic strength testing in which muscle lengths change may be said to be dynamic. Recently, isokinetic devices, in which the speed of the movement produced is controlled, have been developed for this type of testing. Strength and biomechanics are discussed elsewhere in this book.

## 1.3.3 Equipment

### *1.3.3.1 Traditional Static Measurements*

The basic tools of the anthropometrist are the anthropometer, a variety of calipers, and a tape measure. The most commonly used instruments are shown in Figure 1.3.

Typically, anthropometers are precision instruments made up of four interconnecting sections of tubular metal that are engraved in millimeter (mm) intervals. Current models are square in cross-section and are capable of measuring stature or other heights from the floor and seated surfaces as well as straight lengths and distances up to 210 cm when completely assembled. The heights (starting with 0 mm from the floor or seated surface) are read using a movable slide housing that contains an adjustable perpendicular blade, which is placed in alignment with, or lightly on, the desired measuring point. The slide housing contains a window with centerline that enables the user to read the distance to the nearest 0.5 mm on the engraved scale. Typically, only the nearest whole millimeter is recorded.

The upper two sections of the anthropometer may be used as a beam caliper, as they are equipped with a millimeter scale on the side opposite the main scale that starts with

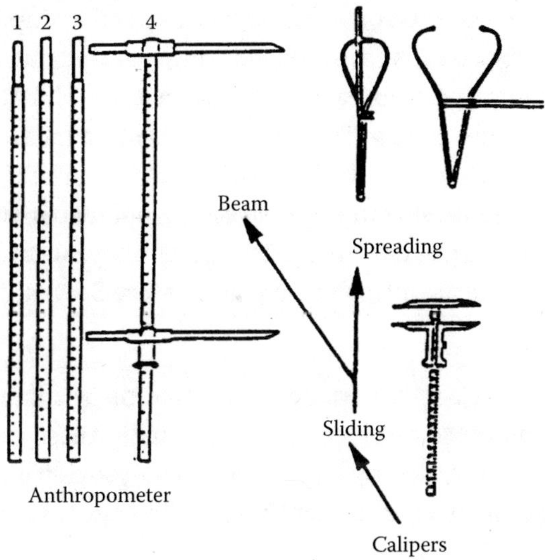


FIGURE 1.3 Basic equipment used to perform traditional anthropometric measurements.

0 mm at the top fixture (see Figure 1.3). The beam caliper, which is capable of measuring distances of up to 95.0 cm, is used for measuring whole body depths and breadths as well as many straight linear distances between landmarked points. Anthropometers may be purchased individually or as part of a set that includes a sliding and a spreading caliper and a tape measure.* These tools have made up the professional anthropometrist's basic measuring kit over a long time period; however, they are slowly being displaced by more modern automated systems. A wide variety of special application calipers and other instruments are also available for anthropometric measuring, although most of these will not be needed by the industrial ergonomist. Other, less complex items requiring a minimum of equipment or shop skills may be homemade. Of the latter, the most frequently used are special tables for seated measurements with adjustable buttock plates and foot rests. Many of the seated workspace dimensions are obtained using such tables. Static measuring of various regions of the body such as the head, hands, or feet frequently requires the use of stabilizing or referencing surfaces to help control the repeatability of the measurement. A foot box is often used, for example, in measuring various foot lengths and breadths. To obtain head and face dimensions referenced to fixed surfaces, a device called a headboard has been used in a number of military surveys [4,5]. A subject places the top (vertex) and back (occiput) of the head firmly against the two perpendicular surfaces, and the anthropometrist measures the distance from the two reference surfaces to defined landmark locations using a depth/height gauge. An automated form of this device was developed for use in the 1988 survey of U.S. Army personnel [6,7]. The automated headboard provided 3D coordinates for 26 selected landmark locations on over 8000 soldiers.

An industrial ergonomist may find it necessary to perform measurements of employees on the job. In this case an effort should be made to incorporate defined and reproducible controls into the measuring methods. A number of simple devices can be designed and built by users to improve reliability and accuracy or to provide special dimensional values to suit a specific need. If the anthropometrist wishes to compare any data collected to those existing in some database, however, care should be exercised to duplicate the original procedure. In some cases this caveat should extend to the equipment used. The final report that describes the methods and summary statistics for the ANSUR survey [1] contains a description of a number of devices used for special application measurements as well as the procedures used to measure over 132 body dimensions. The ANSUR database represents the largest, most recent, and most comprehensive anthropometric survey of Americans in existence. Thus, we recommend that the reader follow the procedures developed for this survey until comparable data are available on civilians (see Section 1.3.5).

#### *1.3.3.2 Looking Ahead*

To date, most anthropometric data have been obtained using manually operated instruments or devices such as those discussed earlier. Usually these measurements are performed one at a time and recorded by hand. These data provide a two-dimensional (2D) description of the body worthy of an earlier time. But biologists have for centuries wanted

* The best anthropometer currently on the market is the GPM, manufactured in Switzerland and imported into this country by Seritex, Inc., 450 Barell Avenue, Carlstadt, NJ 07072. A catalog is available.

to be able to quantitatively describe the 3D form of the human body. Early in this century, researchers began exploring techniques to obtain large quantities of digital information about the body surface size and structure through the use of stereophotogrammetric cameras [8]. A slow and tedious digitization process was required to obtain the 3D data from the resulting photographs. Today, the use of lasers, video, cameras, and other devices in combination with graphics software makes it possible to rapidly collect large quantities of high-density digital information on limited areas of the body. In the near future, it will likely be possible to scan the entire body and generate 3D coordinates on up to 500,000 points on the body surface in just a few seconds. Before the use of true 3D shape becomes routine, work remains to be done on methods used to summarize the massive quantity of 3D coordinates and improve the ability of optoelectronic devices to resolve the many cracks and crevasses on the surface of the body. This is one of the reasons that traditional anthropometry will continue to be needed for some time to come.

With the onset of the ability to collect high-density digital information in three dimensions, we now can talk about body shape in quantitative ways, for both surface contours and internal organs. Scientists and physicians can now examine 3D full-color images of a person's internal structures using a variety of computerized scanning techniques. It will be some time before people outside the medical field will know exactly how to use it on the job, but currently such information is invaluable in the early detection or confirmation of cumulative trauma disorders and other afflictions that result from poor ergonomics in the workplace. When high-density 3D coordinate data for the whole body surface become state-of-the-art, simple point-to-point measurement may be viewed as rudimentary. More important, the engineer, the ergonomist, and the designer will have at their fingertips any human dimension they may need to resolve a given design problem.

### 1.3.4 Statistical Considerations

Fortunately, values for most traditional static anthropometric dimensions are normally distributed. This characteristic permits the use of a number of simplifications that the reader may find helpful. The classic shape of the normal distribution curve is shown in Figure 1.4 in conjunction with increments of standard deviation (SD) relative to the mean ($\bar{x}$). The mean ±1 SD can quickly be seen to include 68% of the population sampled, while the mean ±2 SD and the mean ±3 SD include the variability exhibited by approximately 95% and 99.8%, respectively, of the population. The relationships between the mean ± increments of the SD and selected percentile levels are given in Table 1.1. As can be seen, the ubiquitous 5th to 95th percentile range is approximately equivalent to the mean ±1.65 SD.

Two fundamental questions to be resolved in the application of anthropometric data to design are (1) what database should be selected? and (2) what statistical value(s) should be used? The theoretical answer to the first question is simple: The database of choice is the most recent survey containing the desired dimensions obtained on a large sample of the target user population. In practical fact, this desirable database usually does not exist. Hence, the database will probably have to be selected from the most appropriate of those available.

The choice of a statistical approach depends on the problem to be solved. If the design requires only a single dimension, the designer could attempt to accommodate the entire

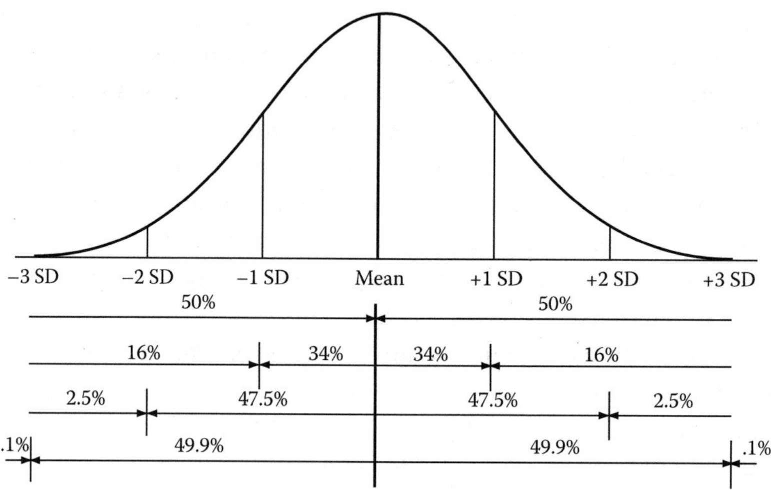


FIGURE 1.4 The normal distribution curve and relative proportions of the population represented by multiples of the standard deviation.

TABLE 1.1 Using the Mean and Standard Deviation to Estimate Percentile Values for Normally Distributed Data

| Percentile Value | Formula |
|---|---|
| 99.5 | Mean + (2.58 × SD) |
| 99 | Mean + (2.32 × SD) |
| 97.5 | Mean + (1.95 × SD) |
| 97 | Mean + (1.88 × SD) |
| 95 | Mean + (1.65 × SD) |
| 90 | Mean + (1.28 × SD) |
| 80 | Mean + (0.84 × SD) |
| 75 | Mean + (0.67 × SD) |
| 70 | Mean + (0.52 × SD) |
| 50 | Mean |
| 30 | Mean − (0.52 × SD) |
| 25 | Mean − (0.67 × SD) |
| 20 | Mean − (0.84 × SD) |
| 10 | Mean − (1.28 × SD) |
| 5 | Mean − (1.65 × SD) |
| 3 | Mean − (1.88 × SD) |
| 2.5 | Mean − (1.95 × SD) |
| 1 | Mean − (2.32 × SD) |
| 0.5 | Mean − (2.58 × SD) |

range of variability in the user population for that variable. Such simple design problems are in reality rare. Usually, a number of body dimensions are required, and it is here that the difficulty begins.

Probably the most common error in thinking about the use of anthropometry in sizing and design is that if an individual is small (or large) for a given dimension, then that person is small (or large) for all other dimensions. This is seldom, if ever, true—a fact that can be demonstrated in at least two ways.

One common misconception is that the mean, which lies near the center of the most densely populated portion of the distribution, is the best value to use to satisfy size requirements of the largest number of people. This is true only if the problem is univariate and if the design will include no adjustability. The "average person" has long been dear to designers' hearts, but, as has been demonstrated by Daniels [9], this individual probably does not exist. In 1952, Daniels analyzed a sample of 4063 U.S. Air Force flyers to determine how many men with average stature (mean ± 0.3 SD) would also be of average size for a successively inputted series of clothing design dimensions for which the same criterion for average was used. A total of 1055 men met the average criterion for stature, but not a single individual was found who also had average values for all 10 clothing dimensions. We recently conducted a similar analysis on male and female subsamples taken from the ANSUR survey [10]. To reduce overall variability, samples of 2074 white males and 1438 white females were used in the analysis. The results of the analysis are shown in Table 1.2.

As can be seen in Table 1.2, no men remained in the sample after screening for eight dimensions, and only one woman of the 309 with average stature was average for all 10 dimensions measured. The rate of dropout would vary with a different list of dimensions, of course, but in all cases a very small group of individuals would remain after a few rounds of screening. As can be seen from Table 1.2, for example, only 5%–6% of the original subjects remained after only the second round of selection. The "average person" is a statistical concept; such a person does not in fact exist.

A second misconception involves use of percentile values in designs. Probably the most frequently specified design limits are the 5th and 95th percentiles. These percentiles are

TABLE 1.2 The "Average"[a] Man and Woman (Values in mm)

| | Men | | | Women | | |
|---|---|---|---|---|---|---|
| **Dimension** | **Mean ± 0.3 SD** | ***n*[b]** | **Percent[c]** | **Mean ± 0.3 SD** | ***n*[b]** | **Percent[c]** |
| Stature | 1745–1784 | 486 | 23.4 | 1614–1652 | 309 | 21.5 |
| Chest circumference | 978–1018 | 115 | 5.5 | 891–930 | 86 | 6.0 |
| Sleeve length | 876–898 | 60 | 2.9 | 789–809 | 35 | 2.4 |
| Crotch height | 823–849 | 28 | 1.4 | 749–773 | 21 | 1.5 |
| Vert. trunk circ. | 1624–1668 | 18 | 0.9 | 1527–1566 | 14 | 1.0 |
| Hip circumference | 968–1004 | 10 | 0.5 | 951–987 | 6 | 0.4 |
| Neck circumference | 374–385 | 3 | <0.1 | 310–319 | 3 | 0.2 |
| Waist circumference | 849–899 | 1 | 0.1 | 771–821 | 2 | 0.1 |
| Thigh circumference | 580–608 | 0 | 0.0 | 563–590 | 2 | 0.1 |
| Crotch length | 632–656 | 0 | 0.0 | 603–636 | 1 | <0.1 |

*Source:* Annis, J.F. and McConville, J.T., Application of anthropometric data in sizing and design, in *Advances in Industrial Ergonomics and Safety*, Vol. 2, B. Das, Ed., Taylor & Francis, Philadelphia, PA, 1990.

[a] Average defined as the mean ± 0.3 SD.

[b] Number of original sample remaining.

[c] Percentage of original sample remaining.

TABLE 1.3 5th and 95th Percentile Models: Males and Females[a] (Values in cm)

| | 5th Percentile | | 95th Percentile | |
|---|---|---|---|---|
| **Segment** | **Male** | **Female** | **Male** | **Female** |
| Floor to lateral malleolus | 5.84 | 5.23 | 7.64 | 6.97 |
| Lateral malleolus to lateral femoral epicondyle | 39.63 | 36.28 | 47.62 | 44.05 |
| Lateral femoral epicondyle to greater trochanter | 38.67 | 36.27 | 46.90 | 43.97 |
| Greater trochanter to iliac crest | 11.97 | 10.22 | 17.08 | 15.24 |
| Iliac crest to tenth rib | 2.39 | 3.22 | 7.52 | 8.10 |
| 10th to 7th | 36.21 | 33.02 | 43.43 | 39.79 |
| Cervical | | | | |
| Seventh cervical to tragus of ear | 8.67 | 8.04 | 12.35 | 11.56 |
| Tragus of ear to top of head | 12.18 | 11.42 | 14.04 | 13.25 |
| Sum for segments | 155.56 | 143.70 | 196.58 | 182.93 |
| ANSUR stature | 164.69 | 152.78 | 186.65 | 173.73 |
| Difference | −9.13 | −9.08 | +9.93 | +9.20 |

[a] The data presented here were derived from ANSUR [1] and originally presented in Ref. [10]. *N* for the females = 2208, *N* for the males = 1774.

often used to designate the smallest and largest individuals for whom an item or workspace will be designed, but just as with "average," a person with a 5th percentile value for one dimension will be 5th percentile for very few others. Thus, if a group of 5th percentile (or 95th percentile) dimensions are used in a design, one can be quite sure that only a very few, if any, individuals will be totally accommodated.

Another pitfall encountered in the use of percentiles for design purposes is that, except for the 50th percentile, they are not additive. This was demonstrated by McConville and Churchill [11] and by Robinette and Churchill [12], who examined the effect of combining percentile body segments. When seven 5th and 95th percentile height segments were each added together, the total was significantly below the 5th percentile stature and well above the 95th percentile stature, respectively.

A similar analysis of the more recent ANSUR data reveals similar results [10]. As can be seen in Table 1.3, the segmental totals are over 9 cm (approximately 3.5 in.) different from the direct percentile value. The segments are depicted by number in Figure 1.5. Certainly the cumulative error is significant, even for the most lenient design requirements. As with the earlier example, the error is negative (sum of segments is less than directly computed percentiles) at the 5th percentile level and equivalently positive at the 95th percentile level. Although the cumulative errors are large, the error associated with each segment may result in a poor design when minimum or maximum values used in conjunction with preferred clearances are mixed.

### 1.3.5 Sources of Data

Most engineers, ergonomists, and designers have a favorite source for the anthropometric data they use in solving design problems. The large number of textbooks, handbooks, and data collections that include some anthropometric data on American adults would

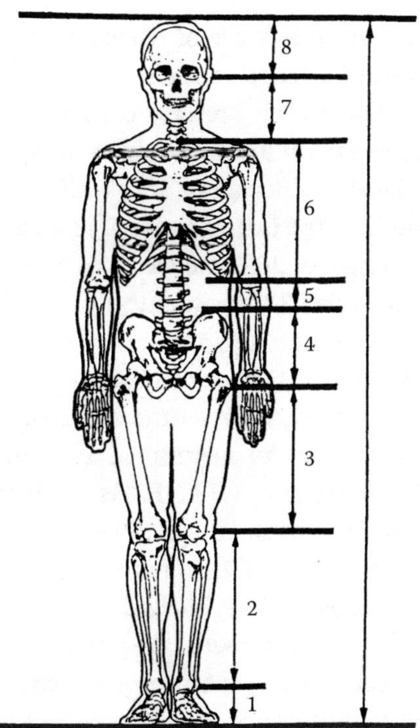


FIGURE 1.5 Segmental heights used in Table 1.3.

lead one to assume that there are ample data to solve almost any design or sizing problem. Unfortunately, this is far from the truth. For the most part, available data are simple univariate summary statistics that suffer from the drawbacks discussed earlier. Often, the authors say little about the source of the data, offering little description of the makeup and size of the sample represented.

The lack of a current and comprehensive database representing the U.S. civilian population is a serious shortcoming. So far as we know, only one nationwide survey of U.S. civilian men and women is available that includes applied workspace dimensions, and its data are now over 30 years old.

Fortunately, the status of anthropometry on U.S. military personnel is much better. In fact, data tables presented in many textbooks and handbooks in this country were compiled from U.S. military samples. Data that appear in popular texts authored by Europeans are also occasionally based on U.S. military data [13,14]. For the immediate future at least, users may be forced to employ military data for civilian problem solving.

#### *1.3.5.1 Civilian Data*

In 1939–1940, more than 10,000 women and more than 100,000 children were measured in major surveys conducted by the Department of Agriculture to obtain dimensional data relevant to clothing sizing and pattern design [15,16]. These surveys were well planned and executed, with excellent data analysis and reporting, but they are clearly dated for purposes of describing the current civilian population. In addition, few of the clothing dimensions selected for measurement in the surveys are of relevance to ergonomists. The data from these surveys formed the basis of a series of recommended sizing standards for women's and children's clothing. These standards are now considered defunct. No such clothing survey has ever been conducted for adult civilian males in the United States.

The U.S. Department of Health, Education, and Welfare initiated a cycle of national health surveys (HES) in 1960 [17]. The first of the series included 18 anthropometric measurements on a nationwide sample of 3091 men and 3581 women between the ages of 18 and 79. The measurements included 12 ergonomically useful dimensions, such as knee height, sitting height, and buttock–knee length that are relevant in seated workspace layouts. Although these surveys continue to be carried out at 10 year intervals, the most recent anthropometric data collected include only stature and weight plus a series of skinfolds, girths, and bony breadths to be used for nutritional assessment. The currently ongoing health and nutrition (examination) survey (HANES) is the third in the series (HANES III).

There have been quite a number of small specialized anthropometric surveys on U.S. civilians. Some results are proprietary and thus not widely available. Nonproprietary data may be found in journals and occasionally in books, but because the samples are small they are typically composed of some selected subset of the population. One example can be found in the excellent two-volume set on work ergonomics produced by the human factors group at Kodak [18]. Some of the anthropometry presented was collected on Kodak employees. Care should be exercised in using such data, however, because local population demographics or employee selection processes may be reflected in them. For example, selection criteria can have an effect on the correlation coefficient ($r$) of stature and weight. The $r$ value for a sample of airline stewardesses measured in 1971 [19] is 0.729, whereas for U.S. army women measured in 1988 [1], $r = 0.529$, and for U.S. civilian women as measured in the HES [17], $r = 0.205$. The data show that although both stewardesses and army women are subsets of the general female civilian population, they are clearly selectively sized subsets of that population.

#### *1.3.5.2 Military Data*

The anthropometric database for U.S. military populations is far more current and comprehensive than that for the U.S. civilian population. Since the end of World War II, each of the military services has conducted one or more major body size surveys of their populations. Such surveys are designed to obtain body size information for a wide variety of needs—sizing and design of clothing and personal protective equipment, workspace layout, and modeling, among others. They usually include some 70–150 body measurements made on relatively large samples of 500 or more. Thus, they are an excellent source of anthropometric design data for the ergonomist if they are used with knowledge of their limitations.

The chief problem associated with the use of military data in civilian applications is fairly obvious. Military samples are truncated at the ends of the distribution for most variables, and very few military personnel are over 50 years of age. Military selection criteria limit stature to approximately 155–203 cm for men and 145–203 cm for women. And, because of existing requirements for fitness levels in the military as well as height and weight limits, the physique of military personnel reflects less body fat and a greater lean body mass on the average than will be found in the civilian population. The pros and cons of applying military anthropometry to the design problems of a civilian workforce have been discussed by a number of authors [20,21]. Despite the potential problems, military data are often the only source for body measurements of interest.

The majority of the military data have been published as technical reports prepared by the various sponsoring agencies and are available through the National Technical Information Service (NTIS). These reports vary in completeness from a presentation of simple univariate statistical summaries accompanied by a description of the survey methodology and measurement techniques used [22] to detailed data analyses that supplement summary statistics with correlation matrices, simple and multiple regressions, selected bivariate frequency tables, and subset data analysis [5]. ANSUR, the most recent military survey [1], is modeled on the Clauser et al. report [5] and is reasonably comprehensive. It is a particularly useful survey in that both army men and women were measured in the same survey, which was conducted by a measurement team trained to perform the measurements in as nearly as possible exactly the same way on both sexes. In addition, considerable effort was put into the sampling strategy so that special databases could be constructed to match the demographics (age and racial/ethnic proportions) of the U.S. Army for some time to come [23].

In the 1970s, raw data tapes of several military surveys and a few civilian surveys that were suitable for mainframe computers [24] were made available through NTIS to interested parties so that researchers could obtain the complete data sets to conduct their own analysis. Under contract to the U.S. Air Force [25,26] and through arrangement with other agencies, Anthropology Research Project, Inc. (ARP) has collected raw data files in its data bank for some 50 surveys—military and civilian, American and foreign, male and female. A summary listing of these surveys by name, number of subjects, and number of variables is provided in Table 1.A.1. This database and associated analysis software are available to support a variety of ergonomic design functions including sizing of personal items and workstation design.

Hundreds of different anthropometric dimensions have been measured over the years. Of these, nearly 300 are listed in the *NASA Anthropometric Source Book*, Vol. II [27]. This data book provides the mean, SD, coefficient of variation, and percentile breakouts for 295 traditional static dimensions. The data derive from 61 foreign and domestic military and civilian surveys of males and females. Although somewhat outdated, this work is still one of the most comprehensive presentations of anthropometric data available.

In sum, there is a fairly impressive array of anthropometric data available on the U.S. military and a rather more limited amount on civilians. Because data will not be available for every sample of interest or for every dimension required, the user must be knowledgeable about the strengths and limitations inherent in the available data and use them appropriately.

#### *1.3.5.3 Future Needs*

At no time in history has the workforce been so diverse, and going on into the twenty first century we expect that trend to continue. As diversity due to age, sex, race, and ethnicity increases, so too does the complexity of industrial and military design problems. A given job may be performed in the morning by an athletic 20 year old man, 6 ft 8 in. tall, weighing 220 lb and in the afternoon by a 50 year old woman, 5 ft tall, weighing 100 lb. Such problems already confront today's ergonomists; in the near future they will become routine.

Fortunately, the advent of scanning techniques combined with computer graphics technology has expanded our ability to rapidly describe the 3D shape of the body in digital form. So far as is known, a full body surface scanner is not yet available. However, a system that uses a low-power He-Ne laser combined with a color video camera to digitize over 150,000 coordinates in 3D space in about 17 s is in use in our laboratory at the present time. This scanner is capable of digitizing a volume as large as 14 in. on a side. Currently, scan data on the heads of over 1000 children and youths are being analyzed. Ultimately such instrumentation will enable the derivation of custom dimensions upon demand, both simple point-to-point descriptors and sufficient 3D coordinates to define the human shape.

### 1.3.6 Applications for Anthropometry

To ensure that the 1988 ANSUR survey would include dimensions that could be used for a variety of purposes for some time to come, over 360 dimensions that had been previously measured in 34 domestic and foreign surveys were examined for their potential utility [28]. Thirteen categories of measurements based upon area of application were identified:

1. Basic body size descriptors
2. Key dimensions for specifying and directing the sizing and design of personal items
3. Clothing and personal equipment pattern and/or construction dimensions
4. Dimensions used to develop manikins or test dummies
5. Dimensions useful for the design and sizing of load-carrying equipment
6. Head and face dimensions used primarily in the design of headgear, optical, and auditory devices
7. Hand, finger, wrist, and forearm measurements used in the design and sizing of gloves and in the construction of hand forms
8. Foot and ankle dimensions needed for the design and sizing of shoes and boots
9. Dimensions central to the design and layout of single-person and multiperson workstations
10. Dimensions used to design and specify anthropometrically compatible aircraft and other vehicles
11. Dimensions used for developing the link or skeletal system for most 3D kinematic anthropomorphic analogs
12. Dimensions useful for the development of 3D anthropomorphic dummies, 3D computer models of humans, and evaluation of workstations
13. Dimensions recommended for use in standards

These categories are not mutually exclusive, and in many instances dimensions appear in two or more categories.

Only 30–40 dimensions common to workspace or workstation design are likely to be of interest to the practicing ergonomist. Approximately this number of traditional dimensions have been selected, principally from the HES [17] and the 1988 ANSUR [1] surveys, for discussion in this chapter. These dimensions are illustrated in Figures 1.A.1 through 1.A.4. For detailed descriptions of the measurement procedures used in the two surveys, the reader is referred to the referenced publications.

## 1.4 ANTHROPOMETRIC CRITERIA FOR ERGONOMIC APPLICATION

Three major factors related to body size variability that the ergonomist must consider in specifying the dimensions of a workstation or design are sex, age, and race or ethnicity. It is likely that in many cases all three areas of concern will apply to the problem at hand. To examine the range of anthropometric variability within these subgroups as briefly and efficiently as possible, the presentation here is restricted principally to those dimensions for adults most relevant to workspace sizing and design.

### 1.4.1 General Body Size Variability

Probably the most frequently used body size descriptors are stature and weight. Not only are they more readily available than most other body size data, but, taken together, these two variables have a higher correlation with a larger number of dimensions than any other pair of measurements [15]. The range of variability in stature and weight for both sexes is shown in Figures 1.A.5 [1,17,22,29–31] and 1.A.6 [1,5,17,19,29,32] for a number of military and civilian anthropometric surveys going back to the 1960s.

When military and civilian samples such as those presented in Figures 1.A.5 and 1.A.6 are examined for the range of the 1st to 99th percentile values in stature and weight, the differences are surprisingly small. A notable exception is the much wider distribution of civilian women's weight values, which reflect the many heavier women found in the civilian population. The data also reveal the selective hiring requirements for such special populations as female flight attendants [19] and law enforcement officers [30]. As expected, the distributions in the nationwide civilian surveys, that is, HANES [29] and HES [17], are wider than in the smaller surveys. There is no clear evidence for secular change (people getting larger over time) in either variable when data from the older surveys are compared to those in the more recent ANSUR survey [1]. Evidence of body size difference for other dimensions will be found in association with much of the discussion to follow.

### 1.4.2 Sexual Variation

Women today occupy many of the workplaces that had been until recently the sole province of men. This phenomenon presents a wide range of new challenges to designers and ergonomists, who must now be concerned about a wider range of users. Size and shape differences between the sexes have been, until recently, poorly understood or ignored in design. Historically, the assumption has been that the 50th percentile female is essentially equivalent to the 5th percentile male. Although this is a reasonable assumption for height and weight, it does not necessarily hold true for many other variables. A study conducted by Robinette et al. [33] demonstrated that women are not simply scaled-down versions of men.

A more recent analysis of ANSUR [1] data, performed at ARP, compared differences in the mean values for 130 dimensions for samples that were closely matched in stature and weight (and age). The data for 30 dimensions selected to cover the range of differences observed in the analysis are presented in Table 1.A.2. The female mean as a percentage of the male mean was found to range from a low of 86.8% (neck circumference) to a high of 110.2% (hip breadth). Basically the data reflect differences in proportion caused by male muscularity and female pelvic structure as well as variations in the amount and location of body fat. ANSUR data are particularly useful for this type of analysis, because both sexes were measured within the same time period by the same measuring team and using identical techniques. Table 1.A.3 shows a comparison of female and male proportions for civilians taken from the HES survey [17]. Like the military samples of Table 1.A.2, the two samples were matched on the basis of ±2.5 cm in stature and ±5 kg in weight. When men and women are nearly the same in height and weight, women are slightly larger on the average for a number of important workspace dimensions, particularly hip-related dimensions that are important in seated workspace/workstation design.

### 1.4.3 Age Variation

It is common knowledge that adult body size and shape alters with aging, but there are few anthropometric data to help the ergonomist quantify these changes [34]. No large sample of civilians has been specifically measured to track this phenomenon. Ideally, such a study should be longitudinal—that is, measurements should be repeated periodically on the same individuals over their adult lifetime. Some short-term longitudinal data that include dimensions that would be useful to the ergonomist have been collected [35–39], usually over 5–10 year periods. The largest, longest, ongoing longitudinal study, covering over 60 years, is the Fels Institute study of human growth [40], and unfortunately it does not include applied dimensions.

By far the most anthropometric data available in the United States are cross-sectional, that is, measured on a sample of subjects at one point in time. Although not an ideal source, cross-sectional data from a large random sample containing subjects of a wide range of ages can nevertheless reflect differences in body size and shape between persons of different ages in the population at one point in time. These are the people the ergonomist/designer must fit into workspaces.

Such data may be examined by breaking out groups of subjects from contiguous age brackets. This approach is illustrated in Tables 1.A.4 and 1.A.5, which give the data for men and women from the HES survey [17] ranging from less than 24 to more than 75 years of age. The data were analyzed in terms of changes in the mean values observed for 6 age brackets referenced to the youngest group, those of ≤24 years of age. The greatest change observed for either sex occurs in the apparent loss in stature and associated components of standing body height. Also of some significance is the increase in seat breadth and elbow–elbow breadth, particularly among women. The rather large increase in the distance between women's elbows with age is believed to reflect the change occurring in the architecture of the shoulder girdle, for example, a forward rotation of the glenoid cavity of the scapula.

Military data are basically limited in demonstrating age-related changes in body size, because the oldest individuals on active duty are typically in their early fifties. ANSUR [1] data were examined for the same basic dimensions listed in Tables 1.A.4 and 1.A.5 for the HES samples. Briefly, soldiers show the same increases in mass-related dimensions through their mid-to-late forties. Changes in other applied dimensions prior to the beginning of the sixth decade appeared not to be significant for design purposes.

### 1.4.4 Racial/Ethnic Variation

A major component in the diversity to be found in the workplace of the twenty first century will be due to racial/ethnic variation. Anthropologists have never fully agreed on a universal definition of race, but it is herein reserved to identify the major classic subdivisions: white, black, and Asian. Ethnic difference or ethnicity refers to those populations that can be identified in association with a geographic or national boundary or a social or religious population comprising individuals belonging to one of the races or to an admixture of races. In the United States, for example, Hispanics constitute a large and growing population that, strictly speaking, is an ethnic group. Depending on the demographic distribution of racial/ethnic groups sampled in a given survey, the label "Other" has traditionally been used to designate individuals in a total sample who have not been sampled in sufficient numbers to make up a statistically useful subsample. Considering the major surveys used in the chapter, the "Other" group for the HES database [17] is believed to be principally composed of Hispanics, Asian Americans, and Native Americans, whereas the "Other" subsample from ANSUR [1] is known to include mainly the latter two groups.

The means and SDs for the dimensions measured in the HES survey were broken out by the available racial/ethnic groups and are presented in Tables 1.A.6 and 1.A.7 for men and women, respectively. The same values for similar dimensions measured on the men and women of the ANSUR survey are given in Tables 1.A.8 and 1.A.9. Although comparisons between the two databases, even for dimensions with the same name, should be viewed with caution because differences in measuring techniques may confound the results, the data reveal some instructive points. Black/white mean values for buttock–knee length, for example, differ by 1–1.5 cm for men and 2 cm for women for both civilian and military personnel. For both of these groups, the difference in sitting height as a proportion of stature can be explained by the length of the legs. In short, African Americans of a given stature have longer legs than white Americans. Hispanic values given with the ANSUR data in Tables 1.A.8 and 1.A.9 reflect overall smaller body sizes than either the black or white sample. These and other findings of racial/ethnic body size variation in the U.S. population should be taken into consideration in plotting effective designs for both civilian and military workspaces.

### 1.4.5 Specialized Data

Although traditional anthropometry has many uses, its applications often baffle engineers and ergonomists faced with real-life design problems. Traditional anthropometry had its origins in the laboratory, where lightly clothed subjects stand or sit immobile in the strictly controlled postures required to obtain reliable measurement results. In the workplace, people rarely stand with their heads in the Frankfort plane* or sit at display consoles with hips and knees at perfect 90° angles. They slump, stoop, or stretch to reach controls. They may be required to wear bulky clothes or protective headgear on the job, which further complicates the problems. For these and other reasons the engineer/designer may not find the anthropometry needed to solve a particular problem.

Some data for a number of specialized measurements of use to the ergonomist or engineer are available, particularly for military personnel. The following sections provide some data and suggested sources of information that can supplement the more traditional dimensional variables.

#### *1.4.5.1 Clothed Anthropometry*

Commonly, anthropometric dimensions are measured on the bare skin or over minimal clothing. Most ergonomists and engineers are concerned with dimensions on individuals arrayed in possible combinations of clothing. Many jobs require, at least occasionally, that the worker wear restrictive multilayered clothing. The effect of such heavy clothing on the ability to perform required tasks was probably first recognized by the military. As a result, almost all available, clothed anthropometric data were obtained from military ensembles. The first data of this type were published in 1946 [41]. Summaries of subsequent studies may be found in a variety of handbooks and government technical reports, but most of the studies were conducted on fairly small samples of subjects.

Results of a typical clothed anthropometry study are summarized in Table 1.A.10 [42]. The four ensembles tested in this study are listed at the bottom of the table. Ensemble 1 (ground soldier—hot weather) may be similar to normal clothing worn in the work environment, whereas Ensemble 4 (extremely cold weather, chemical protection added) is something like what the handler of hazardous material might wear. In general, body size can be seen to increase while overhead reach and mobility decrease from Ensemble 1 to Ensemble 4, as chemical protective clothing and/or cold weather layers were added. Forearm–forearm breadth and hip breadth, for example, increase by two-thirds, while overhead reach capability is reduced by about 20% in Ensemble 4. As can be seen, increases in body depths and breadths can be quite large, and mobility is substantially reduced in overhead reach. For example, Forearm–forearm breadth, hip breadth, and sitting were increased by approximately 66.7%, and the overhead reaching capability was reduced by about 20% in Ensemble 4.

ARP recently completed a pilot study of five different U.S. Army ensembles on a sample of five civilian male and five civilian female subjects [43]. The data are currently being analyzed; however, indications are that the range of variability is not substantially different from that presented in Table 1.A.10, except that the replicability of the measurements was found to be very labile even under strict controls for many measurements. A companion

* The Frankfort plane is an arbitrary reference position of the head used for anthropometry in which a landmark just above the tragus of the right ear (tragion) is horizontally aligned with a landmark on the lowest border of both right and left orbital ridges (infraorbitale).

study [44] on the effects of the same ensembles on more than 36 planar joint motions was undertaken at the same time. Although different subjects were used in the studies, the data provide evidence of the combined effects of complex clothing ensembles on body size and mobility. Although the number of subjects was small in both studies, the data were collected under what were perhaps the best documented and best controlled studies of this type in recent times. The full technical reports of these studies should be available through the NTIS in 1995.

#### *1.4.5.2 Working Postures*

Recognizing the limitations of measurement data collected from subjects sitting and standing in stylized postures, anthropologists at the Aeromedical Laboratory at Wright–Patterson Air Force Base in the 1950s collected data on men positioned in a few common working postures [45]. Investigators from the same laboratory later repeated and expanded upon the earlier study [46]. Thirteen of twenty-six dimensions were measured, and the remainder were derived photogrammetrically. All were related to common working postures. Data for the 13 directly measured dimensions and the age, stature, and weight of the sample are summarized in Table 1.A.11, and the associated measurements are illustrated in Figure 1.A.7a and b.

Using the mean stature as a reference, interesting relationships can be computed. For example, a man on his knees with his torso erect may require overhead clearance averaging more than 74% of his stature. A standing man should be able to grasp an object that is at a height equal to approximately 122% of his stature. An average-sized man lying on his back may be able to grip an object about 750 mm (29.5 in.) above the floor or at a distance of about 42% of his stature. The same man lying on the floor will have a horizontal length about 25 mm (1 in.) more than his stature.

These data are presented principally because of their unusual nature and because no comparable data have been collected in the intervening years. The measurements should be repeated and expanded upon using a current workforce population.

#### *1.4.5.3 Reach*

Traditional anthropometry provides little help for the ergonomist who must position equipment or controls within the reach of a worker. Like many of the special applied anthropometric measurements, interest in reach and reach envelope measurements was spurred by problems encountered in designing complex aircraft cockpits. One of the first to investigate reach envelopes was Dempster [47]. Some 20 years later, Kennedy [48] conducted a 3D reach study on a relatively small sample of civilian men ($n = 20$) and women ($n = 30$) who were believed to be the representative of the USAF population. Examples of these data are shown in Figures 1.6 and 1.7.

The measurements were made on seated subjects in seats designed to simulate those found in an aircraft cockpit. The seat was unpadded, with a pan angle of 6° and a back angle of 103°. Investigators used a special reach apparatus that permitted reaches to be controlled in defined positions over most of the sphere that encompasses the seated worker. The reach task was to grasp a small knob between the thumb and index finger and push it

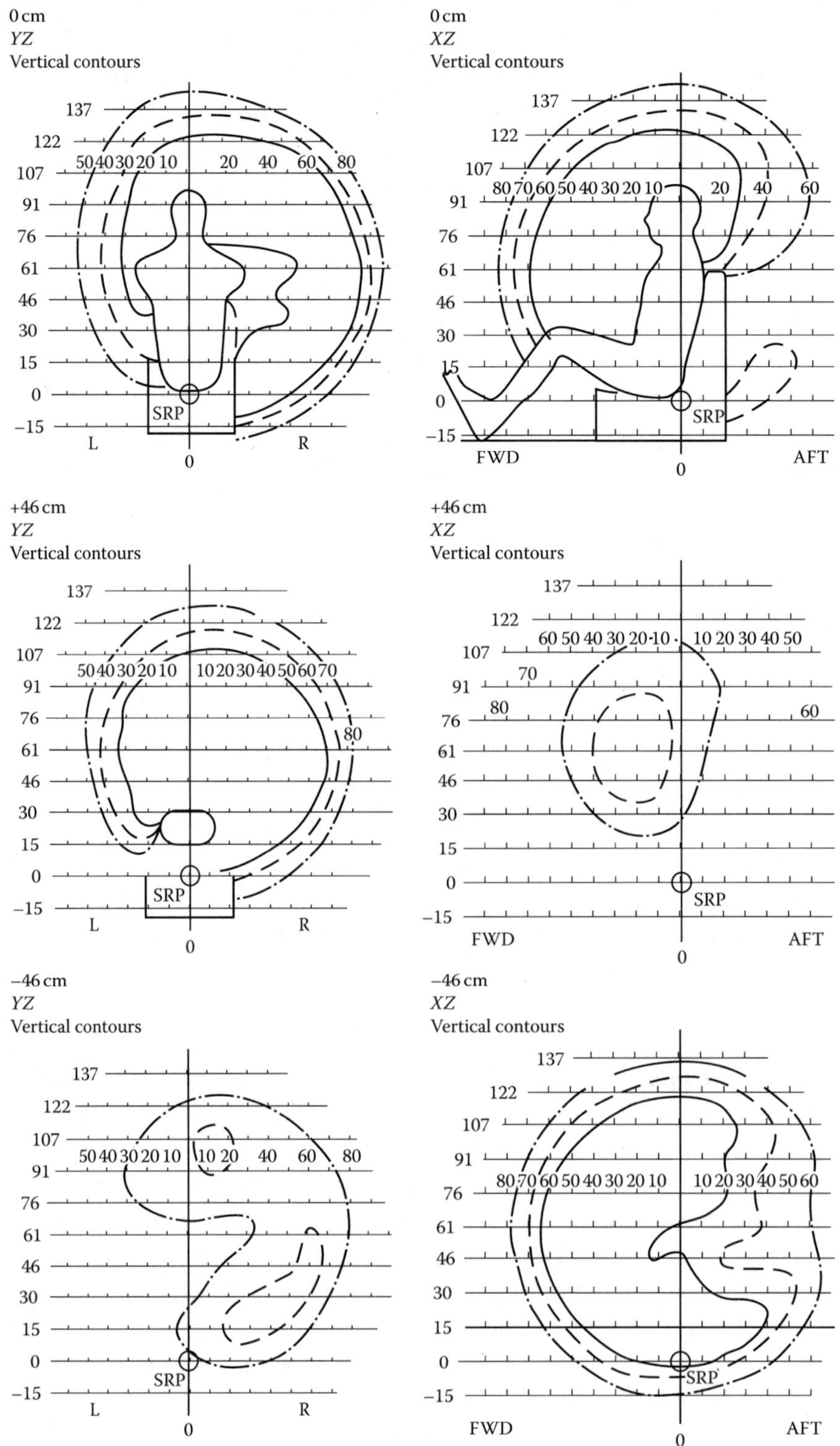


FIGURE 1.6 Selected vertical reach envelopes for seated men in the coronal plane (*YZ*) and sagittal plane (*XZ*) relative to the seat reference point (SRP).

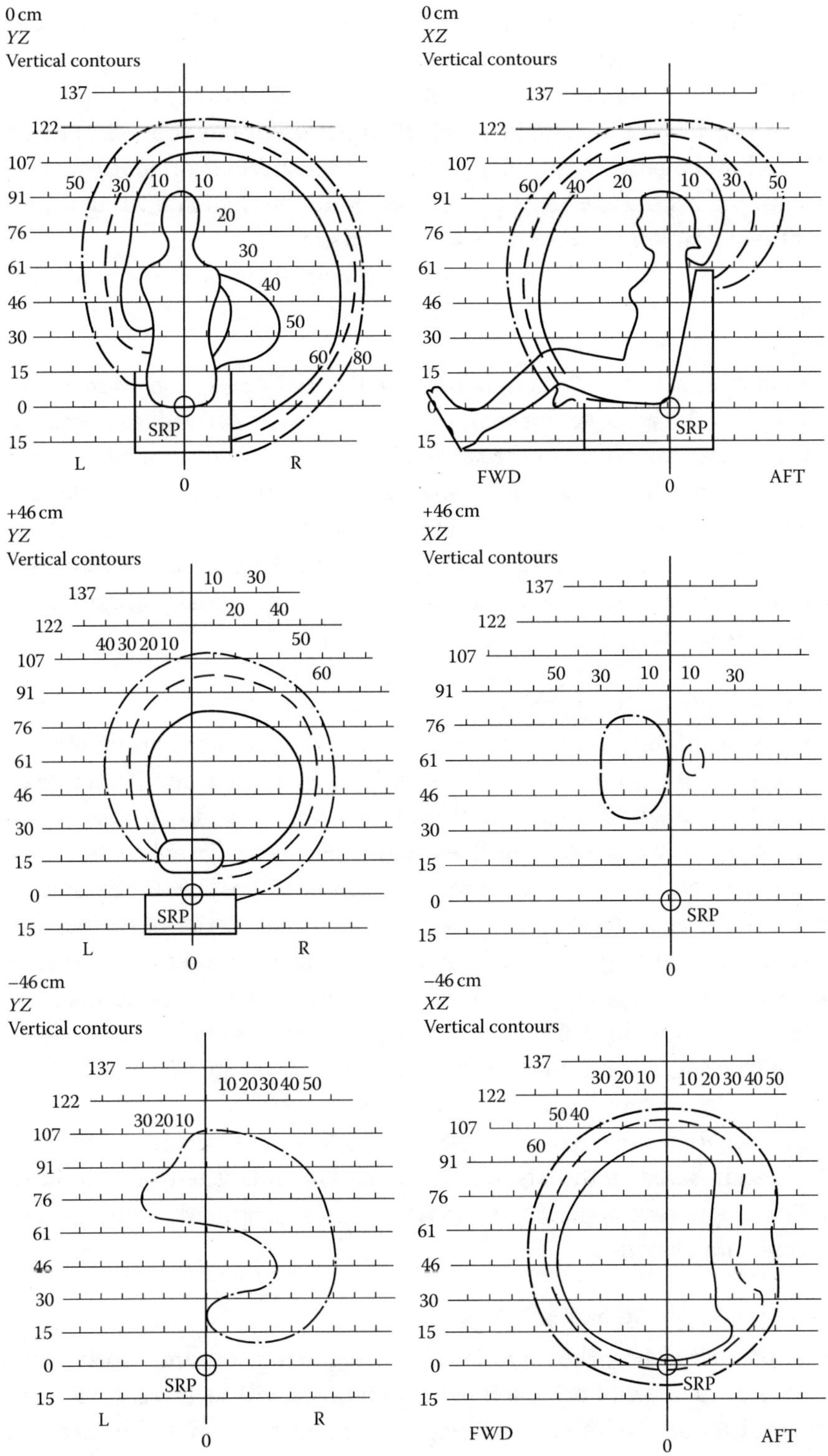


FIGURE 1.7 Selected vertical reach envelopes for seated women in the coronal plane (*YZ*) and sagittal plane (*XZ*) relative to the SRP.

away along the test azimuth until the arm was extended to the fullest extent possible with the shoulders still in contact with the seat back. The push rod could be positioned at 15° intervals relative to the seat reference point (SRP). The illustrations at top left and right in Figures 1.6 and 1.7 give the 5th, 50th, and 95th percentile envelopes in the midcoronal (0 cm, *YZ*) and midsaggital (0 cm, *XZ*) planes. The illustrations provide the envelopes at 46 cm in front and behind the SRP (*YZ*), and 46 cm to the right and left of the SRP (*XZ*). The effect of arm position on reach is clearly demonstrated.

Many additional reach envelopes, including horizontal views, are provided in this and other publications [49–53]. In many cases the data are from the same source, and they are almost entirely military.

More traditional 2D reach data obtained from ANSUR men and women [1] are provided in Table 1.A.12. The values for some of the reaches were derived mathematically from other measurements completed in the survey. The reaches are illustrated in Figures 1.A.3 and 1.A.4. In general, the female reaches average more than 90% of the male values, with functional leg length, wrist-thumbtip length, and wrist-center of grip having the nearest equivalence to male values. Men retain about 68% of their overhead reach when seated, compared to 64% in women. These and other evidence of sexually dimorphic traits are reflected in the reach data.

#### *1.4.5.4 Linkage and Computer Models*

Drafting board manikins and the majority of current computer man-models are based on the concept of the anthropometric links outlined by Braune and Fischer [54] in their classic biomechanical study. This concept was later refined and expanded by Dempster [47], who defined the link as a straight-line distance between adjacent centers of joint rotation on a segment of the body. Although it has not been possible to measure the internal links as defined by Dempster, it is possible to measure a series of surface dimensions that approximate, to varying degrees, the internal link lengths. These anthropometric link lengths are summarized for the ANSUR [1] males and females in Table 1.A.13 and illustrated in Figure 1.8. The links are either directly measured, as in biacromial breadth and acromiale–radiale length, or derived by subtraction, where thigh link length, for example, is trochanteric height minus lateral femoral epicondyle height.

Regression equations for predicting link lengths from stature and weight are given in Table 1.A.14, again based on the ANSUR [1] survey data. It is reiterated that these anthropometric link lengths are approximations from surface landmarks and not the internal center-to-center link lengths.

### 1.4.6 Anthropometry and Seated Work

As automation encroaches, the number of jobs requiring standing work postures will decline and the attention of occupational ergonomists will be increasingly focused on the seated workstation. With each passing year, more workers will be seated at some form of video display terminal (VDT) or at least in front of a control panel, which may include video displays. All too frequently, seated postures assumed over extended periods lead to severe discomfort and a variety of job-related disorders. The hand, wrist,

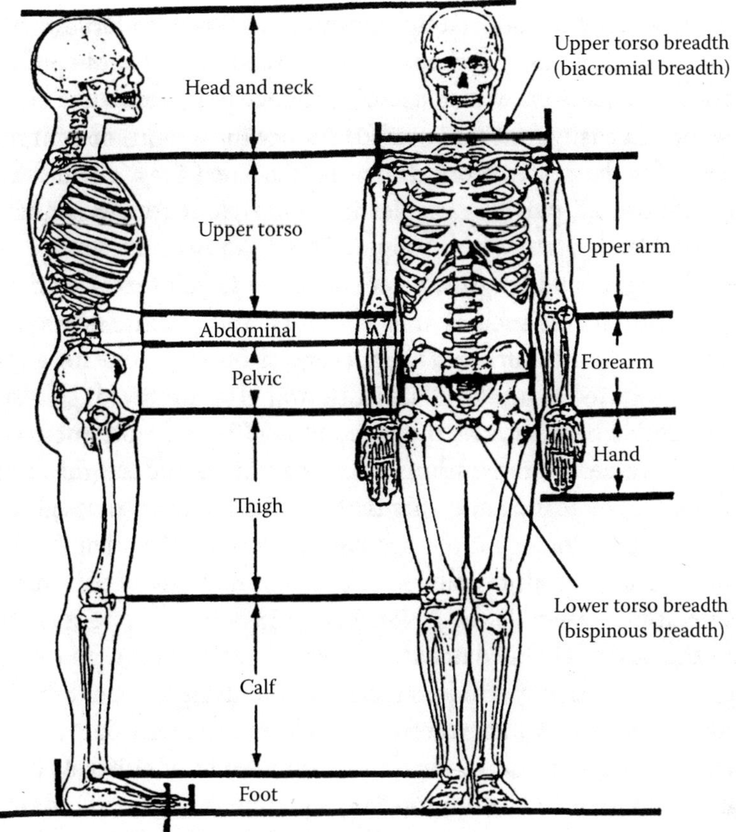


FIGURE 1.8 Links and related landmark centers on the skeleton.

arm, and back disorders that sometimes result from either stasis or repetitive movements associated with postures used in badly fitted workstations are discussed elsewhere in this text.

A critique of the current recommended standards for VDT design [55] is beyond the scope of this chapter. It suffices to say that the univariate statistics used as a basis for most such standards are not adequate to the task of designing workstations for an increasingly diverse population of users. Among available workspace dimensions, the importance of popliteal height (PH) and buttock–popliteal length to the chair are clear, but the relationship of thigh clearance height and elbow rest height to the preferred work surface height is considerably more difficult to deal with. Statistically speaking, this is because the correlation of these two dimensions with other workspace dimensions is quite poor. It is very easy to forget that the work surface has some thickness, that handbook values for PH and sitting knee height frequently do not include shoes, and that the buttock–knee length does not take much clothing into account.

With the previous caveats in mind, let us reexamine some concepts introduced earlier in this chapter and then look briefly at a more promising way of selecting dimensional values for seated work designs. Faulty thinking related to the "average man" concept and the cumulative error introduced by summing percentile values for body segments were discussed in Section 1.3.4. Perhaps the best way to avoid these pitfalls is through the use of regression equations, which usually produce less cumulative error. Frequently one may know or can easily find statistics for stature (or weight) of a target population, but no data exist for the seated work dimensions needed for a given user population. As noted, the best overall pair of variables in a multiple regression equation with two independent variables is stature and weight. This is so because stature relates well to many segmental heights and lengths, whereas weight is volumetric and correlates well with complex measurements such as arcs, contours, and circumferences. Tables 1.A.15 and 1.A.16 give a series of stature and weight regression equations for a group of workspace dimensions selected from the ANSUR [1] and HES surveys [17], respectively. The equations are included here because they are not readily available but are invaluable in estimating reliable values for dimensions when the stature and weight ranges desired in a design are known. The principal errors that may be incurred through their use arise from the fact that regression equations are population-specific. That is, in this case, the value computed for a given dimension will be accurately predicted only to the extent that target populations resemble the ANSUR or HES databases. Because these two surveys represent the best available databases covering both civilian and military populations, the regressed values may still be superior to inputting a series of percentile values or taking a value for a given dimension from a table of univariate data from an unknown source. Any value within the range for stature and weight of either ANSUR [1] or HES [17] data can be inputted in the appropriate equation, and the standard error of the estimate ($SE_{est}$) can be used in the calculations much like the SD is used with the mean. That is, by adding and subtracting increments of the $SE_{est}$ to/from the predicted value, various levels of confidence limits can be established for the variable. If one wanted to establish the range of adjustability for the seat pan of a chair, the height and weight for a very small person and that for a very large person could be inputted into a regression equation for predicting PH. For example, the lower limit could be established by a person who is 1524 mm (60 in.) tall and weights 45.4 kg (100 lb), and the upper limit might be a person who is 1980 mm (78 in.) tall and weighs 95 kg (210 lb). By inputting these values into the PH equations (Table 1.A.16) for HES women and men, the following results are obtained:

$$\text{PH(women)} = (0.301 \times 1524) - (0.347 \times 45.4) - 62.7 = 380.3 \text{ mm}$$

$$\text{PH(men)} = (0.321 \times 1980) - (0.372 \times 95.0) - 88.6 = 511.6 \text{ mm}$$

As noted previously, the predicted range of 380.3–511.6 mm (15–20 in.) can now be expanded by using $SE_{est}$ to ensure inclusion of a high percentage of the population.

The value can also be adjusted to account for various heel heights. If designing for a population with characteristics of age and weight per height that are similar to those of the military, the ANSUR regressions can be used.

### 1.4.7 General Guidelines

The compilation of general guidelines to use when applying anthropometric data to ergonomic design problems can be dangerous. Nevertheless, the statements hereafter offer an approach that should be of relatively universal usefulness.

1. Identify the specific dimension(s) you need for the problem at hand. If more than one major design dimension is involved, you have a multivariate problem to solve.
2. Determine the maximum error the design can accept for the dimension(s) identified. If the allowable error is ±10% or greater, or if multiple dimensions are not involved, the database selection is less critical.
3. Identify the demographics of your user population with regard to such variables as age range, sex, and race/ethnicity. Determine whether there are any special anthropometric characteristics of the target population that could affect your sizing or design options.
4. Seek out the availability of relatively recent anthropometry on the user population (the database should not be older than one generation). If satisfactory data cannot be found, and if it is not feasible to create the needed database, as is likely the case, use the best available database that includes the dimension(s) you require. For example, as mentioned earlier, the most comprehensive listing of foreign and domestic databases currently available in Volume 2 of the *NASA Anthropometric Source Book* [27]. Use the HES data [17] if the design cannot tolerate the shortcomings of a military database. The best current source of military data is ANSUR [1].
5. If the problem is multivariate (most are) and the allowable error is small, try to locate multiple regression equations for the selected database that permit the use of independent input variables (a) for which you know the target population range and (b) that have the best correlation with your list of design variables. Remember that regression equations are population specific, so some error will be incurred if the user population is very different.
6. In the final analysis, if data on the required dimensions do not exist, collect data on at least 30 individuals representative of the target population. Be consistent with your procedures, record the method used, and add to the database should the design effort prove to be successful.

## 1.A APPENDIX: ERGONOMIC TABLES AND FIGURES

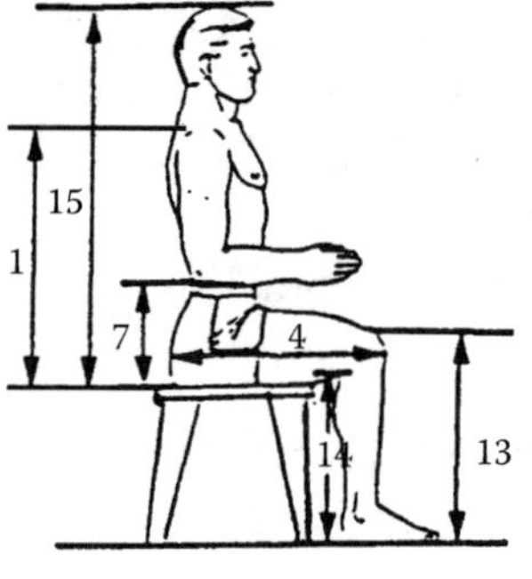


1. Acromial height, sitting
2. Biacromial breadth
3. Bideltoid breadth
4. Buttock–knee length
5. Buttock–popliteal length
6. Chest circumference
7. Elbow-rest height
8. Elbow–elbow height
9. Eye height, sitting
10. Forearm–forearm breadth
11. Hip breadth, sitting
12. Hip circumference
13. Knee height, sitting
14. Popliteal height
15. Sitting height
16. Stature
17. Thigh clearance height
18. Waist circumference
19. Waist height, sitting

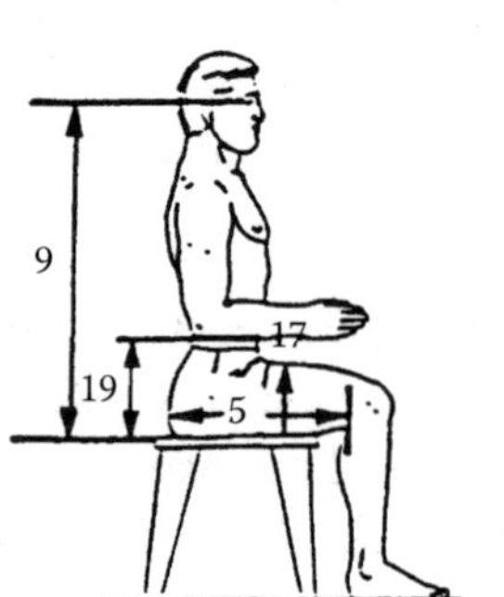


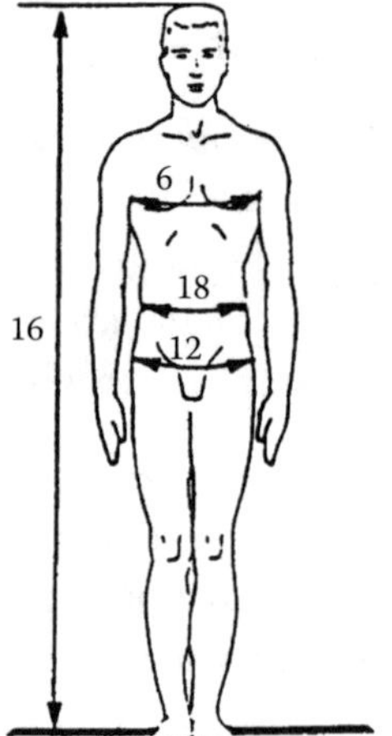


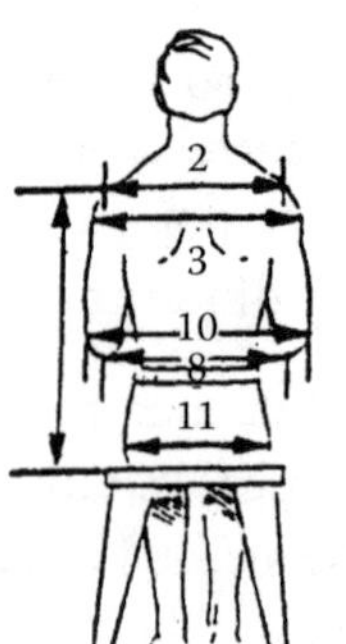


FIGURE 1.A.1 Selected traditional dimensions useful in workspace design and evaluation.

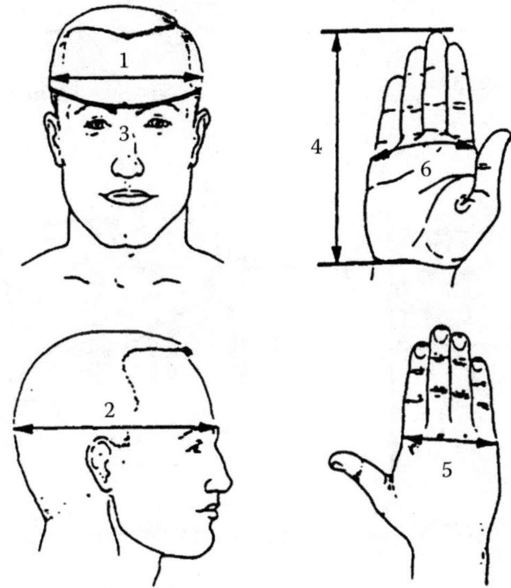


1. Head breadth
2. Head length
3. Head circumference
4. Hand length
5. Hand breadth
6. Hand circumference

FIGURE 1.A.2 Basic head and hand dimensions.

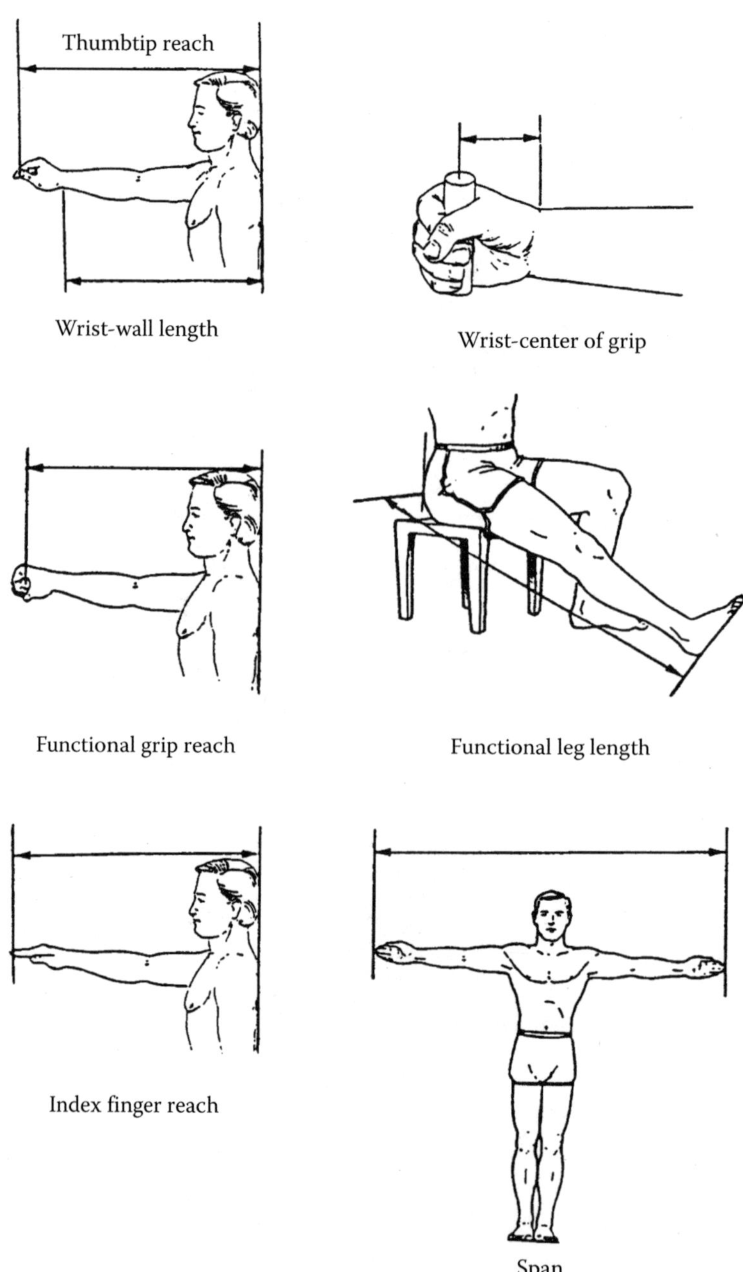


FIGURE 1.A.3 Selected reaches from the wall and other functional dimensions.

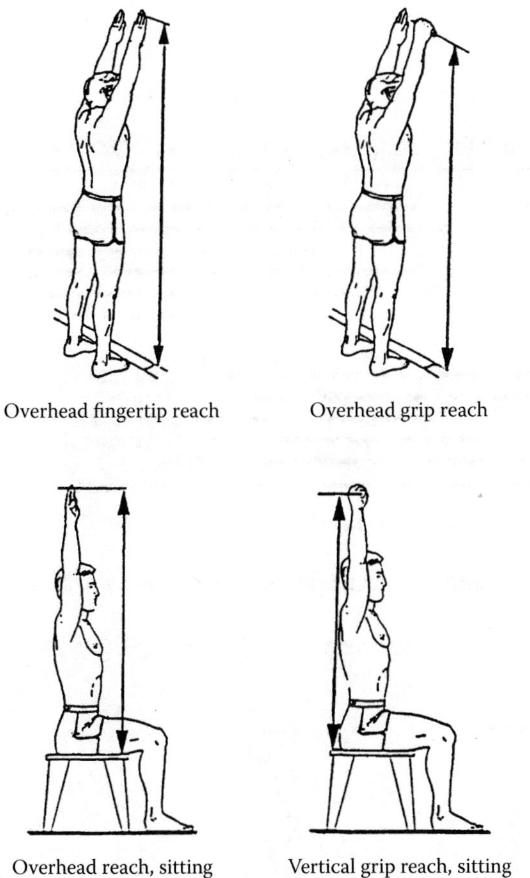


FIGURE 1.A.4 Selected overhead reach dimensions.

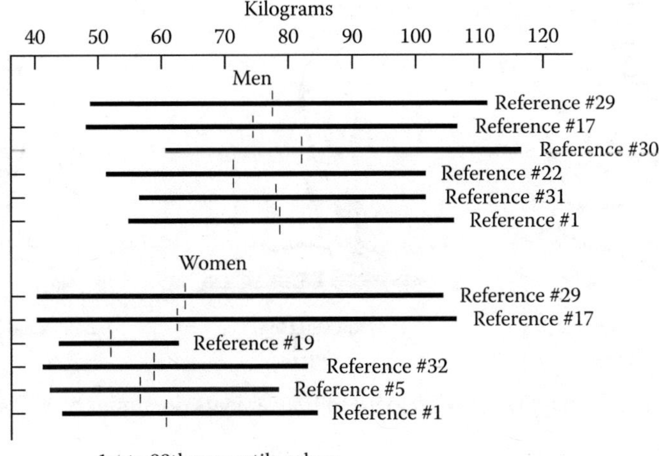


FIGURE 1.A.5 Range of weight for men and women in selected U.S. anthropometric surveys.

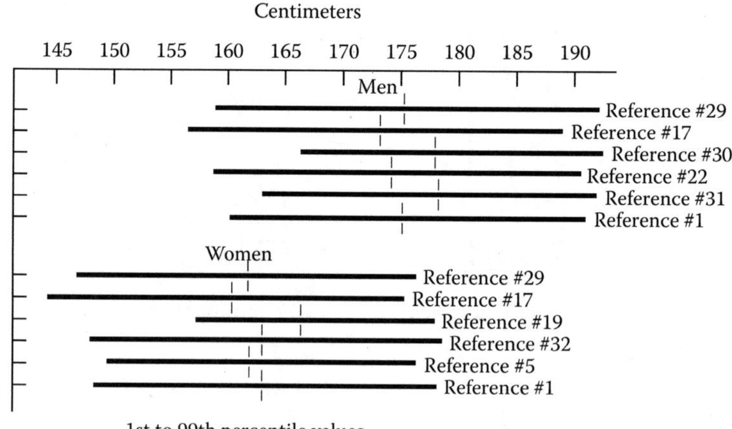


FIGURE 1.A.6 Ranges of stature for men and women in selected U.S. anthropometric surveys.

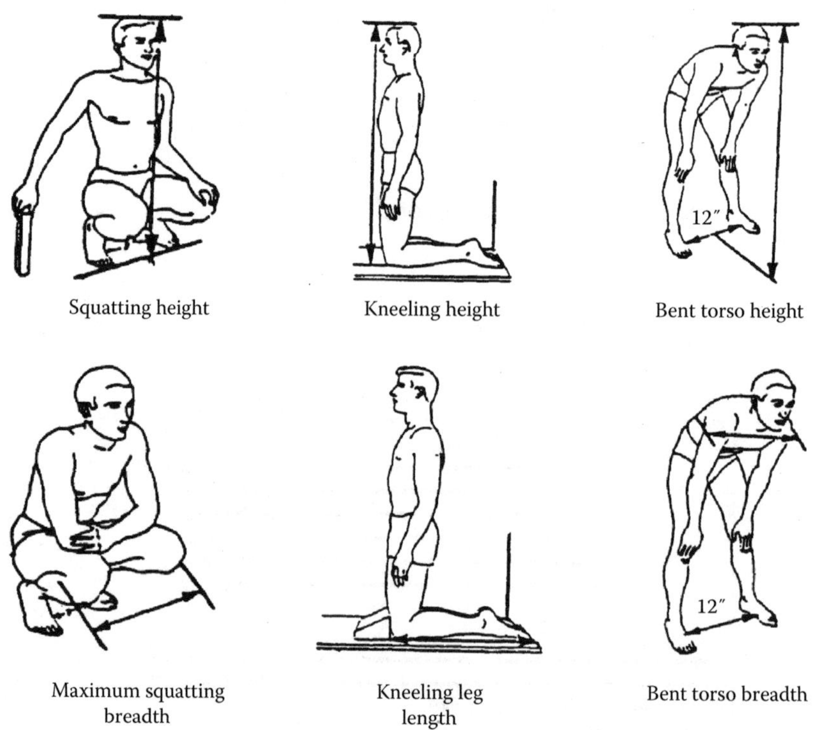


FIGURE 1.A.7 Visual description of workspace measurements documented in Table 1.A.14.

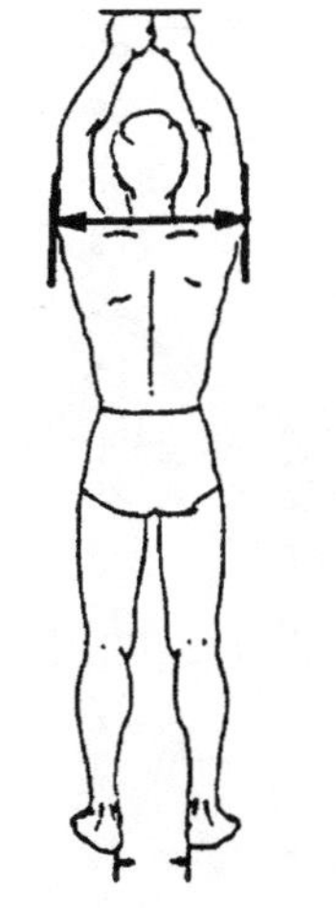

Overhead reach breadth

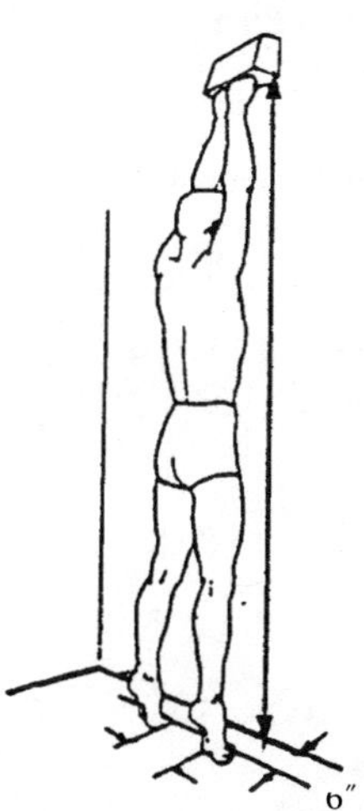


Maximum overhead reach height

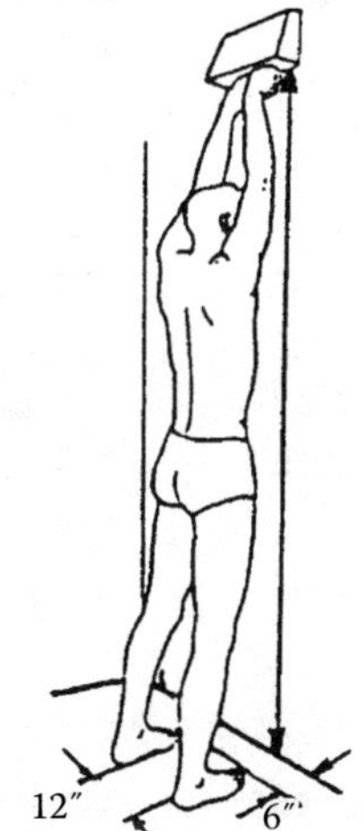


Overhead reach height, fist

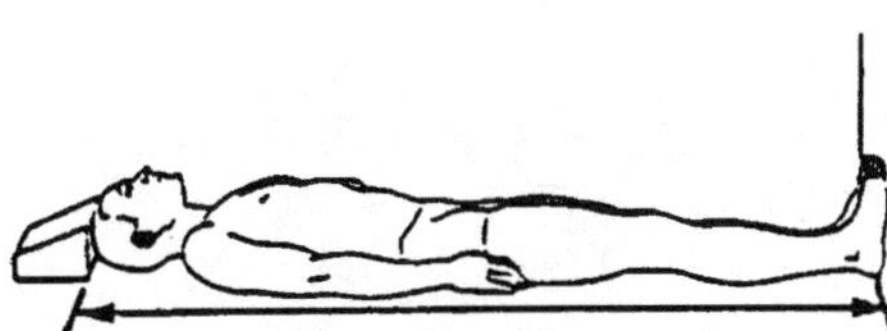

Horizontal length

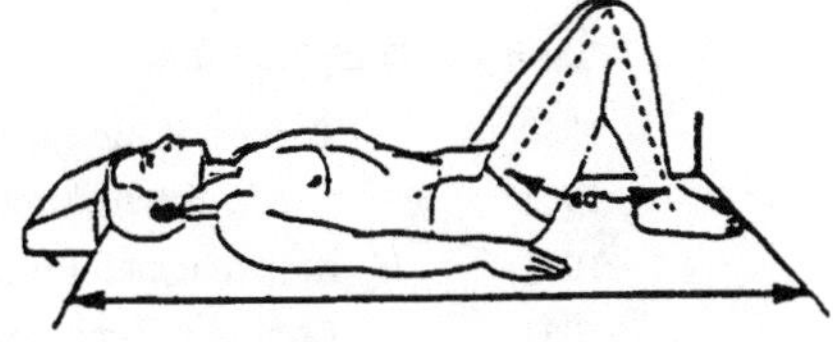

Horizontal length, knee bent

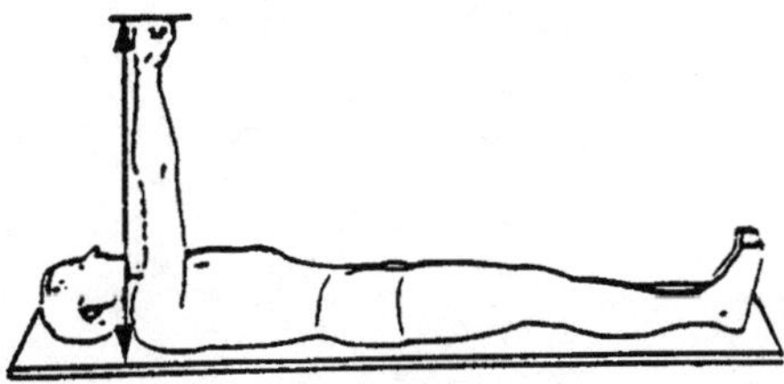

Arm reach, supine

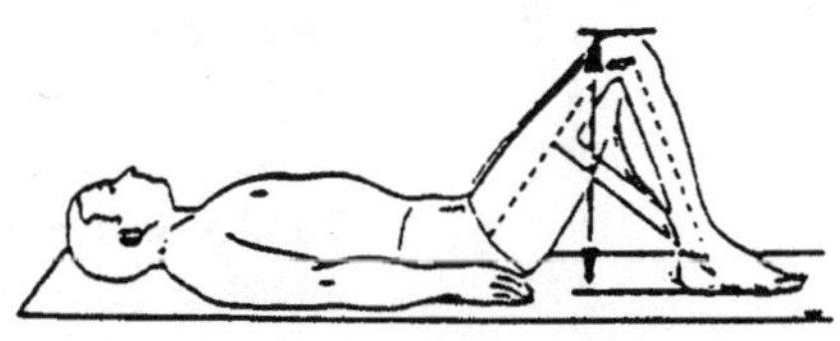

Bent knee height, supine

FIGURE 1.A.7 (continued)

TABLE 1.A.1 Databank Holdings

| Survey Date | Surveyed Population | Sample Size | Variables |
|---|---|---|---|
| *U.S. military populations* | | | |
| Males | | | |
| 1950 | U.S. Air Force pilots | 4,000 | 146 |
| 1959 | U.S. Army aviators | 500 | 46 |
| 1964 | U.S. Navy aviators | 1,529 | 98 |
| 1965 | U.S. Air Force ground personnel | 3,869 | 161 |
| 1966 | U.S. Army ground personnel | 6,682 | 73 |
| 1966 | U.S. Navy enlisted | 4,095 | 73 |
| 1966 | U.S. Marines enlisted | 2,008 | 73 |
| 1967 | U.S. Air Force flyers | 2,420 | 189 |
| 1970 | U.S. Army flyers | 1,482 | 88 |
| 1981 | U.S. Navy flyers | 1,087 | 108 |
| 1988 | U.S. Army personnel | 5,500 | 225 |
| Females | | | |
| 1946 | U.S. Women's Army Corps | 7,563 | 65 |
| 1968 | U.S. Air Force women | 1,905 | 139 |
| 1977 | U.S. Army women | 1,331 | 151 |
| 1981 | U.S. Navy flyers | 351 | 112 |
| 1988 | U.S. Army personnel | 3,500 | 225 |
| *Foreign military populations* | | | |
| 1960 | Turkish armed forces | 912 | 151 |
| 1961 | Greek armed forces | 1,071 | 151 |
| 1961 | Italian armed forces | 1,342 | 151 |
| 1961 | Korean military flyers | 264 | 132 |
| 1964 | Vietnamese military forces | 2,129 | 51 |
| 1967 | German Air Force | 1,466 | 152 |
| 1969 | Iranian military | 9,414 | 74 |
| 1970 | Latin-American armed forces | 1,985 | 76 |
| 1970 | Royal Air Force aircrew | 2,000 | 64 |
| 1972 | Royal Air Force head study | 500 | 46 |
| 1972 | Royal Australian Air Force | 482 | 18 |
| 1973 | French military flyers | 174 | 118 |
| 1974 | Royal New Zealand Air Force aircrew | 238 | 63 |
| 1974 | Canadian military forces | 565 | 33 |
| 1975 | British Army survey | 1,537 | 61 |
| 1975 | English guardsmen | 100 | 61 |
| 1976 | English transport corpsmen | 161 | 61 |
| 1976 | United Kingdom gurkhas | 36 | 61 |
| 1976 | Hong Kong Chinese military | 73 | 47 |

TABLE 1.A.1 (continued) Databank Holdings

| Survey Date | Surveyed Population | Sample Size | Variables |
|---|---|---|---|
| 1977 | Australian personnel | 2,945 | 32 |
| 1981 | Israeli aircrew men | 360 | 63 |
| 1985 | Dutch military | 1,010 | 40 |
| 1985 | Canadian aircrew | 519 | 32 |
| *U.S. civilian populations* | | | |
| Adult males | | | |
| 1961 | Air traffic controllers | 678 | 65 |
| 1962 | Health examination survey (HES) | 3,091 | 18 |
| 1962/1981 | Matched health examination survey (HES) (ages 18–65) | 2,761 | 70 |
| 1974 | Law enforcement officers | 2,989 | 23 |
| 1975 | Health and nutrition examination survey (HANES) (ages 18–74) | 6,563 | 11 |
| 1980 | Health and nutrition examination survey (HANES II) (ages 18–75) | 5,921 | 13 |
| 1981 | U.S. miners | 270 | 44 |
| 1984 | Hispanic health and nutrition examination Survey (HHANES) (ages 18–65) | 1,619 | 21 |
| Adult females | | | |
| 1962 | Health examination survey (HES) | 3,581 | 18 |
| 1971 | Airline stewardesses | 423 | 73 |
| 1975 | Health and nutrition examination survey (HANES) (ages 18–74) | 10,123 | 11 |
| 1980 | Health and nutrition examination survey (HANES II) (ages 18–75) | 6,598 | 13 |
| 1981 | U.S. miners | 86 | 44 |
| 1984 | Hispanic health and nutrition examination survey (HHANES) (ages 18–65) | 2,037 | 21 |
| *Children and youths* | | | |
| Males | | | |
| 1975 | Health and nutrition examination survey (HANES) (ages 1–18) | 3,571 | 13 |
| Females | | | |
| 1975 | Health and nutrition examination survey (HANES) (ages 1–18) | 3,533 | 13 |

TABLE 1.A.2 Comparison of Dimensional Values of Military Men and Women Matched for Stature and Weight[a] (Weight in kg, Age in Years, All Others in cm)

| | Men (*n* = 91) | | Women (*n* = 153) | | Percent |
|---|---|---|---|---|---|
| **Dimension** | **Mean** | **SD** | **Mean** | **SD** | **(F/M × 100)** |
| Age | 25.6 | 0.67 | 25.7 | 0.6 | — |
| Stature | 168 | 1.5 | 167.2 | 1.3 | — |
| Weight | 65.3 | 2.67 | 64.7 | 2.8 | — |
| Neck circumference | 36.5 | 1.28 | 31.7 | 1.1 | 86.8 |
| Biceps circumference | 31.6 | 1.36 | 28.2 | 1.5 | 89.2 |
| Chest below bust circum. | 87.2 | 3.15 | 78.4 | 3.1 | 89.9 |
| Scye circumference | 41.9 | 1.59 | 37.9 | 1.4 | 90.5 |
| Shoulder circumference | 111.6 | 3.17 | 103.8 | 3 | 93 |
| Bideltoid breadth | 46.5 | 1.48 | 43.7 | 1.5 | 93.8 |
| Axillary arm circumference | 31.3 | 1.24 | 29.4 | 1.4 | 93.9 |
| Bimalleolar breadth | 6.9 | 0.29 | 6.5 | 0.3 | 94.2 |
| Chest breadth | 30 | 1.4 | 28.5 | 1.4 | 95 |
| Waist-to-hip length | 17.1 | 1.61 | 16.4 | 2 | 95.9 |
| Biacromial breadth | 38.4 | 1.37 | 36.9 | 1.3 | 96.1 |
| Radiale–stylion length | 25.2 | 0.91 | 24.4 | 0.8 | 96.8 |
| Sleeve outseam | 57 | 1.78 | 55.5 | 1.6 | 97.4 |
| Shoulder–elbow length | 35.1 | 1.06 | 34.3 | 1 | 97.7 |
| Head circumference | 55.9 | 1.22 | 54.6 | 1.2 | 97.7 |
| Hand length | 18.4 | 0.64 | 18.1 | 0.6 | 98.4 |
| Acromion–radiale length | 32.3 | 0.91 | 31.8 | 1 | 98.5 |
| Knee height | 47.2 | 1.26 | 46.7 | 1.3 | 98.9 |
| Chest height | 121.3 | 1.89 | 120.3 | 2.3 | 99.2 |
| Crotch height | 79 | 2.22 | 78.4 | 1.8 | 99.2 |
| Cervicale height | 194.5 | 1.41 | 194.4 | 1.5 | 99.9 |
| Acromial height | 136.8 | 1.8 | 136.8 | 1.8 | 100 |
| Shoulder length | 14.6 | 1.06 | 14.6 | 1 | 100 |
| Waist height, navel | 100.3 | 2.27 | 100.7 | 2 | 100.4 |
| Calf circumference | 35.6 | 1.47 | 35.9 | 1.6 | 100.8 |
| Buttock height | 83.6 | 2.1 | 84.8 | 2 | 101.4 |
| Waist circumference | 79 | 4.01 | 80.4 | 5.5 | 101.8 |
| Waist breadth | 28.7 | 1.29 | 29.8 | 1.9 | 103.8 |
| Buttock circumference | 91.8 | 2.37 | 99.1 | 3.6 | 108 |
| Hip breadth | 32.2 | 0.93 | 35.5 | 1.5 | 110.2 |

*Source:* Gordon, C.C., Bradtmiller, B., Clauser, C.E., Churchill, T., McConville, J.T., Tebbetts, I., and Walker, R.A., 1987–1988 Anthropometric survey of U.S. army personnel: Methods and summary statistics, Techn. Rep. NATICK/TR-89-044, U.S. Army Natick Res. Dev. Engineering Center, Natick, MA, 1989.

[a] Individuals matched within +5 kg in weight and +2.5 cm in stature.

TABLE 1.A.3 Comparison of Dimensional Values of Civilian Men and Women Matched for Stature and Weight (Weight in kg, Age in Years, All Others in cm)

| | Men ($n$ = 198) | | Women ($n$ = 154) | | Percent |
|---|---|---|---|---|---|
| **Dimension** | **Mean** | **SD** | **Mean** | **SD** | **(F/M × 100)** |
| Age | 46.1 | 15.5 | 41.1 | 14.3 | — |
| Stature | 167.1 | 1.45 | 166.4 | 1.46 | — |
| Weight | 69.4 | 2.88 | 68.6 | 2.66 | — |
| Biacromial breadth | 38.7 | 1.81 | 36.1 | 1.63 | 93.3 |
| Buttock–knee length | 56.9 | 1.90 | 59.1 | 2.04 | 103.9 |
| Buttock–popliteal length | 47.4 | 2.47 | 50.2 | 2.48 | 105.9 |
| Chest girth[a] | 96.9 | 4.05 | 89.6 | 3.96 | 92.5 |
| Elbow–elbow breadth | 40.9 | 2.56 | 39.5 | 2.81 | 96.6 |
| Elbow-rest height | 23.7 | 2.93 | 23.5 | 2.83 | 99.2 |
| Knee height | 51.9 | 1.88 | 51.8 | 1.80 | 99.8 |
| Popliteal height | 42.0 | 1.78 | 41.4 | 1.94 | 98.6 |
| Right arm girth[a] | 30.4 | 1.86 | 29.5 | 1.79 | 97.0 |
| Seat breadth | 34.3 | 1.73 | 39.5 | 2.81 | 115.2 |
| Sitting height | 84.2 | 2.50 | 84.5 | 2.56 | 100.4 |
| Sitting height erect | 88.2 | 2.50 | 87.4 | 2.41 | 99.1 |
| Thigh clearance height | 14.0 | 1.11 | 14.2 | 1.30 | 101.4 |
| Waist girth[a] | 87.0 | 7.04 | 78.2 | 5.35 | 89.9 |

*Source:* Stoudt, H., et al., *Weight, Height and Selected Body Dimensions of Adults, United States 1960–1962*, Public Health Service, Publ. No. 1000, Ser. 11, No. 8, U.S. Gov. Printing Office, Washington, DC, 1965.

[a] Original name.

TABLE 1.A.4 Mean Changes in Workspace Dimensions with Aging Civilian Men (Values in cm)

| | Age Group | | | | | | |
|---|---|---|---|---|---|---|---|
| **Dimension** | **≤24** | **25–34** | **35–44** | **45–54** | **55–64** | **65–74** | **≥75** |
| Biacromial breadth | 40.0 | +0.1 | −0.1 | −0.4 | −0.8 | −1.3 | −2.6 |
| Buttock–knee length | 58.9 | +1.0 | +0.2 | +0.7 | −0.3 | −0.6 | −1.1 |
| Buttock–popliteal length | 49.2 | +0.6 | +0.0 | −0.1 | −0.2 | −0.4 | −0.8 |
| Elbow–elbow breadth | 40.3 | +1.4 | +2.4 | +2.9 | +2.8 | +2.7 | +1.7 |
| Elbow-rest height | 24.8 | +0.1 | +0.1 | −0.3 | −0.8 | −1.7 | −2.9 |
| Knee height | 54.2 | +0.6 | 0.0 | −0.2 | −0.6 | −1.0 | −1.8 |
| Popliteal height | 44.3 | +0.4 | −0.5 | −0.8 | −1.0 | −1.3 | −2.1 |
| Seat breadth | 34.8 | +0.8 | +1.0 | +1.2 | +0.9 | +0.5 | +0.1 |
| Sitting height | 91.4 | +0.6 | +0.1 | −0.4 | −1.7 | −3.0 | −4.6 |
| Stature | 174.3 | +1.6 | −0.4 | −1.1 | −2.8 | −4.3 | −7.1 |
| Thigh clearance height | 14.4 | +0.3 | +0.1 | −0.1 | −0.5 | −0.8 | −1.1 |

*Source:* Stoudt, H. et al., *Weight, Height and Selected Body Dimensions of Adults, United States 1960–1962*, Public Health Service, Publ. No. 1000, Ser. 11, No. 8, U.S. Gov. Printing Office, Washington, DC, 1965.

TABLE 1.A.5 Mean Change of Workspace Dimensions with Aging in Civilian Women (Values in cm)

| | Age Group | | | | | | |
|---|---|---|---|---|---|---|---|
| **Dimension** | **≤24** | **25–34** | **35–44** | **45–54** | **55–64** | **65–74** | **≥75** |
| Biacromial breadth | 35.3 | +0.3 | +0.5 | +0.3 | 0.0 | −0.5 | −0.8 |
| Butt–knee length | 56.6 | +0.3 | +0.6 | +0.0 | 0.0 | −0.2 | −0.7 |
| Buttock–popliteal length | 47.8 | +0.2 | +0.2 | +0.2 | +0.2 | 0.0 | −0.6 |
| Elbow–elbow breadth | 35.6 | +1.2 | +3.3 | +4.5 | +6.1 | +6.1 | 4.5 |
| Elbow-rest Height | 22.9 | +0.7 | +1.0 | +0.5 | −0.3 | −1.6 | −2.1 |
| Knee height | 50.0 | 0.0 | 0.0 | −0.5 | −0.7 | −1.0 | −0.7 |
| Popliteal height | 40.6 | −0.7 | −1.0 | −1.2 | −1.5 | −1.7 | −1.2 |
| Seat breadth | 35.1 | +1.0 | +2.0 | +2.2 | +2.5 | +2.2 | +1.0 |
| Sitting height | 85.3 | +0.3 | +0.3 | −0.5 | −1.5 | −3.8 | −4.8 |
| Stature | 162.1 | −0.3 | −0.8 | −2.3 | −3.6 | −5.9 | −6.9 |
| Thigh clearance height | 13.5 | +0.2 | +0.5 | +0.5 | +0.2 | 0.0 | −0.3 |

*Source:* Stoudt, H. et al., *Weight, Height and Selected Body Dimensions of Adults, United States 1960–1962*, Public Health Service, Publ. No. 1000, Ser. 11, No. 8, U.S. Gov. Printing Office, Washington, DC, 1965.

TABLE 1.A.6 Selected Workspace Dimensions for U.S. Civilian Men by Race/Ethnicity (Weight in kg, All Others in cm)

| | White ($n$ = 2669) | | Black ($n$ = 358) | | Other ($n$ = 64) | |
|---|---|---|---|---|---|---|
| **Dimension** | **Mean** | **SD** | **Mean** | **SD** | **Mean** | **SD** |
| Biacromial breadth | 39.6 | 2.10 | 39.9 | 2.17 | 39.5 | 2.45 |
| Buttock–knee length | 59.1 | 2.89 | 60.1 | 2.87 | 56.5 | 2.44 |
| Buttock–popliteal length | 49.2 | 3.02 | 50.8 | 2.93 | 47.7 | 3.44 |
| Chest girth[a] | 99.7 | 8.19 | 96.3 | 8.76 | 96.7 | 7.75 |
| Elbow–elbow breadth | 42.2 | 4.62 | 40.5 | 4.54 | 40.5 | 4.64 |
| Elbow-rest height | 24.4 | 2.83 | 21.7 | 3.05 | 25.2 | 3.12 |
| Knee height | 54.1 | 2.86 | 55.1 | 2.99 | 52.1 | 2.27 |
| Popliteal height | 43.9 | 2.64 | 45.0 | 2.62 | 42.7 | 24.4 |
| Right arm girth[a] | 30.7 | 3.18 | 31.1 | 3.56 | 28.9 | 3.03 |
| Seat breadth | 35.6 | 2.74 | 34.0 | 2.79 | 34.3 | 2.40 |
| Sitting height | 86.8 | 3.60 | 84.0 | 3.30 | 84.5 | 3.39 |
| Sitting height, erect | 90.9 | 3.54 | 87.7 | 3.44 | 88.7 | 3.23 |
| Stature | 173.5 | 6.93 | 172.2 | 6.46 | 168.7 | 5.30 |
| Thigh clearance height | 14.3 | 1.68 | 14.3 | 1.77 | 13.5 | 1.50 |
| Waist girth[a] | 89.2 | 11.39 | 85.4 | 11.69 | 86.5 | 10.62 |
| Weight | 75.3 | 12.54 | 73.1 | 13.20 | 69.0 | 10.35 |

*Source:* Stoudt, H. et al., *Weight, Height and Selected Body Dimensions of Adults, United States 1960–1962*, Public Health Service, Publ. No. 1000, Ser. 11, No. 8, U.S. Gov. Printing Office, Washington, DC, 1965.

[a] Original name.

TABLE 1.A.7 Selected Workspace Dimensions of U.S. Civilian Women by Race/Ethnicity (Weight in kg, All Others in cm)

| | **White ($n$ = 3051)** | | **Black ($n$ = 469)** | | **Other ($n$ = 62)** | |
|---|---|---|---|---|---|---|
| **Dimension** | **Mean** | **SD** | **Mean** | **SD** | **Mean** | **SD** |
| Biacromial breadth | 35.3 | 1.88 | 36.3 | 1.94 | 34.8 | 2.47 |
| Buttock–knee length | 56.6 | 2.95 | 58.6 | 3.43 | 54.9 | 2.99 |
| Buttock–popliteal length | 47.7 | 2.91 | 49.7 | 3.40 | 46.6 | 3.37 |
| Chest girth[a] | 87.8 | 7.84 | 89.5 | 9.47 | 90.9 | 9.00 |
| Elbow–elbow breadth | 38.8 | 5.23 | 39.7 | 5.99 | 39.8 | 5.61 |
| Elbow-rest height | 23.4 | 2.82 | 21.2 | 2.79 | 22.4 | 3.60 |
| Knee height | 49.5 | 2.63 | 50.8 | 2.95 | 48.8 | 2.50 |
| Popliteal height | 39.6 | 2.52 | 40.8 | 2.89 | 39.0 | 2.00 |
| Right arm girth[a] | 28.5 | 4.06 | 29.7 | 5.08 | 28.4 | 4.24 |
| Seat breadth | 36.7 | 3.63 | 36.5 | 4.30 | 36.0 | 3.85 |
| Sitting height | 82.2 | 3.72 | 79.4 | 3.66 | 79.7 | 3.42 |
| Sitting height, erect | 85.1 | 3.51 | 82.6 | 3.62 | 82.5 | 3.21 |
| Stature | 160.4 | 6.51 | 160.3 | 6.92 | 156.5 | 5.67 |
| Thigh clearance height | 13.7 | 1.77 | 14.1 | 2.18 | 12.8 | 1.76 |
| Waist girth[a] | 76.0 | 11.70 | 80.4 | 13.48 | 81.5 | 13.05 |
| Weight | 63.1 | 13.15 | 67.9 | 16.95 | 62.0 | 13.91 |

*Source:* Stoudt, H. et al., *Weight, Height and Selected Body Dimensions of Adults, United States 1960–1962*, Public Health Service, Publ. No. 1000, Ser. 11, No. 8, U.S. Gov. Printing Office, Washington, DC, 1965.

[a] Original name.

TABLE 1.A.8 Selected Workspace Dimensions for Military Men by Race/Ethnicity (Values in cm)

| | Black ($n$ = 1465) | | White ($n$ = 1979) | | Hispanic ($n$ = 1075) | | Other ($n$ = 985) | |
|---|---|---|---|---|---|---|---|---|
| **Dimension** | **Mean** | **SD** | **Mean** | **SD** | **Mean** | **SD** | **Mean** | **SD** |
| Acromial height, sitting | 57.8 | 2.82 | 60.7 | 2.73 | 58.8 | 2.58 | 59.2 | 2.80 |
| Bideltoid breadth | 49.2 | 2.62 | 49.2 | 2.52 | 48.8 | 2.46 | 48.9 | 2.76 |
| Buttock–knee length | 63.0 | 2.97 | 61.5 | 2.76 | 59.7 | 2.72 | 59.7 | 3.21 |
| Buttock–popliteal length | 51.3 | 2.63 | 49.9 | 2.48 | 48.3 | 2.42 | 48.3 | 2.82 |
| Elbow-rest height | 20.7 | 2.48 | 24.0 | 2.34 | 23.2 | 2.31 | 23.6 | 2.49 |
| Eye height, sitting | 77.0 | 3.23 | 80.4 | 3.15 | 77.4 | 3.06 | 78.3 | 3.21 |
| Forearm–forearm breadth | 54.1 | 4.33 | 54.9 | 4.20 | 54.5 | 4.26 | 54.0 | 4.48 |
| Hand circumference | 21.6 | 1.00 | 21.4 | 0.91 | 20.9 | 0.91 | 21.0 | 1.04 |
| Hip breadth, sitting | 36.0 | 2.64 | 37.0 | 2.41 | 36.3 | 2.40 | 36.1 | 2.53 |
| Knee height, sitting | 57.0 | 2.76 | 55.8 | 2.65 | 54.3 | 2.56 | 54.1 | 2.98 |
| Popliteal height | 44.4 | 2.54 | 43.3 | 2.36 | 42.0 | 2.33 | 41.8 | 2.67 |
| Sitting height | 89.0 | 3.34 | 92.6 | 3.22 | 89.6 | 3.15 | 90.6 | 3.35 |
| Span | 186.4 | 8.31 | 181.7 | 7.56 | 177.3 | 7.65 | 177.8 | 8.60 |
| Stature | 175.6 | 6.79 | 176.5 | 6.50 | 170.9 | 6.40 | 171.7 | 7.07 |
| Thigh clearance height | 17.1 | 1.34 | 16.7 | 1.17 | 16.6 | 1.20 | 16.6 | 1.31 |
| Waist depth | 22.1 | 2.45 | 22.7 | 2.55 | 22.5 | 2.56 | 22.1 | 2.55 |
| Waist height, sitting | 22.8 | 1.51 | 23.9 | 1.47 | 22.9 | 1.47 | 23.3 | 1.46 |

*Source:* Gordon, C.C. et al., 1987–1988 Anthropometric survey of U.S. army personnel: Methods and summary statistics, Techn. Rep. NATICK/TR-89-044, U.S. Army Natick Res. Dev. Engineering Center, Natick, MA, 1989.

TABLE 1.A.9 Selected Workspace Dimensions for Military Women by Race/Ethnicity (Values in cm)

| | Black (*n* = 1360) | | White (*n* = 1387) | | Hispanic (*n* = 337) | | Others (*n* =403) | |
|---|---|---|---|---|---|---|---|---|
| **Dimension** | **Mean** | **SD** | **Mean** | **SD** | **Mean** | **SD** | **Mean** | **SD** |
| Acromial height, sitting | 54.3 | 2.64 | 56.7 | 2.61 | 54.7 | 2.42 | 55.1 | 2.85 |
| Bideltoid breadth | 43.4 | 2.27 | 43.1 | 2.24 | 43.1 | 2.07 | 43.4 | 2.42 |
| Buttock–knee length | 60.1 | 2.79 | 58.1 | 2.70 | 56.9 | 2.66 | 57.1 | 3.36 |
| Buttock–popliteal length | 49.3 | 2.53 | 47.4 | 2.41 | 46.4 | 2.39 | 46.6 | 3.00 |
| Elbow-rest height | 20.4 | 2.34 | 23.4 | 2.19 | 22.3 | 2.11 | 22.5 | 2.48 |
| Eye height, sitting | 72.4 | 2.97 | 75.1 | 3.10 | 72.6 | 2.93 | 73.1 | 3.14 |
| Forearm–forearm breadth | 46.5 | 3.38 | 47.0 | 3.48 | 46.6 | 3.28 | 46.7 | 3.66 |
| Hand circumference | 18.8 | 0.84 | 18.5 | 8.11 | 18.2 | 0.75 | 18.4 | 0.92 |
| Hip breadth, sitting | 38.1 | 2.70 | 38.8 | 2.80 | 38.1 | 2.48 | 37.8 | 2.94 |
| Knee height, sitting | 52.5 | 2.53 | 51.0 | 2.45 | 49.8 | 2.34 | 50.1 | 3.01 |
| Popliteal height | 39.7 | 2.37 | 38.6 | 2.23 | 37.5 | 2.06 | 37.8 | 2.53 |
| Sitting height | 83.6 | 3.08 | 86.6 | 3.21 | 84.0 | 3.01 | 84.7 | 3.29 |
| Span | 170.9 | 7.60 | 164.6 | 7.28 | 162.5 | 7.22 | 163.9 | 8.77 |
| Stature | 163.0 | 6.22 | 163.3 | 6.34 | 158.7 | 5.92 | 160.0 | 6.93 |
| Thigh clearance height | 16.2 | 1.21 | 15.7 | 11.60 | 15.5 | 1.08 | 15.5 | 1.26 |
| Waist depth | 20.5 | 2.44 | 20.1 | 2.47 | 20.6 | 2.30 | 20.2 | 2.52 |
| Waist height, sitting | 22.1 | 1.42 | 23.3 | 1.37 | 22.4 | 1.40 | 22.4 | 1.44 |

*Source:* Gordon, C.C. et al., 1987–1988 Anthropometric survey of U.S. army personnel: Methods and summary statistics, Techn. Rep. NATICK/TR-89-044, U.S. Army Natick Res. Dev. Engineering Center, Natick, MA, 1989.

TABLE 1.A.10 The Effect of Various Military Ensembles[a] on Selected Workspace Dimensions—Mean Change from Nude Value (Values in cm)

| Dimension | Ensemble 1 | Ensemble 2 | Ensemble 3 | Ensemble 4 |
|---|---|---|---|---|
| Stature | 4.7 | 6.5 | 7.5 | 8.9 |
| Chest depth | 6.5 | 9.1 | 12.6 | 15.6 |
| Thumbtip reach | 1.7 | 0.8 | 6.9 | 4.3 |
| Overhead reach, sitting | −6.6 | −9.1 | −21.1 | −24.1 |
| Sitting height | 1.9 | 2.7 | 4.6 | 6.0 |
| Eye height, sitting | −1.2 | −1.6 | −0.5 | −1.9 |
| Knee height, sitting | 3.4 | 4.3 | 4.1 | 5.8 |
| Popliteal height, sitting | −2.5 | −1.7 | −0.6 | −1.6 |
| Buttock–knee length | 4.7 | 3.9 | 6.7 | 8.5 |
| Buttock–popliteal length | 3.1 | 1.0 | 4.7 | 2.2 |
| Shoulder breadth, sitting | 1.5 | 6.6 | 12.5 | 15.0 |
| Forearm–forearm breadth, sitting | 13.0 | 20.2 | 27.0 | 31.7 |
| Hip breadth, sitting | 14.3 | 16.7 | 18.2 | 22.6 |
| Head length | 9.0 | 9.0 | 12.3 | 15.3 |
| Head breadth | 8.8 | 8.8 | 13.7 | 14.8 |
| Foot length | 3.4 | 8.4 | 6.4 | 11.7 |
| Foot breadth | 1.1 | 5.8 | 3.0 | 6.3 |
| Heel breadth | 1.7 | 4.3 | 3.6 | 3.7 |
| Hand length | — | 0.0 | 2.4 | 1.4 |
| Hand breadth | — | 0.8 | — | — |

*Source:* Johnson, R.F., Anthropometry of the clothed U.S. army ground troop and combat vehicle crewman, Techn. Rep. NATICK/TR-84/034, U.S. Army Natick Res. Dev. Electronics Center, Natick, MA, 1984.

[a] Ensemble 1: Ground soldier—hot weather.
Ensemble 2: Ground soldier—hot weather, plus chemical protection.
Ensemble 3: Ground soldier—extremely cold weather.
Ensemble 4: Ground soldier—extremely cold weather, plus chemical protection.

TABLE 1.A.11 Selected Body Dimensions of Men in Common Working Positions (Values in mm, All Others Indicated)

| | | | | Percentile | | |
|---|---|---|---|---|---|---|
| **Dimension** | **Mean** | **SD** | **Range** | **5th** | **50th** | **95th** |
| Overhead reach breadth | 377 | 17.3 | 340–422 | 345 | 378 | 404 |
| Maximum overhead reach height | 2240 | 85.9 | 1913–2410 | 2116 | 2235 | 2377 |
| Overhead reach height, fist | 2141 | 82.9 | 1836–2314 | 1996 | 2146 | 2225 |
| Horizontal length | 1778 | 59.7 | 1582–1895 | 1676 | 1773 | 1877 |
| Horizontal length, knee bent | 1484 | 53.6 | 1349–1621 | 1400 | 1466 | 1575 |
| Arm reach, supine | 747 | 32.6 | 630–836 | 671 | 742 | 818 |
| Bent knee height, supine | 502 | 22.9 | 432–559 | 462 | 500 | 536 |
| Squatting height | 1117 | 49.3 | 1034–1224 | 1036 | 1107 | 1194 |
| Kneeling height | 1304 | 45.0 | 1199–1392 | 1224 | 1295 | 1382 |
| Bent torso height | 1309 | 70.1 | 1120–1443 | 1176 | 1321 | 1420 |
| Maximum squatting breadth | 562 | 54.1 | 447–716 | 478 | 559 | 653 |
| Kneeling leg length | 672 | 33.5 | 569–744 | 617 | 673 | 729 |
| Bent torso breadth | 448 | 22.4 | 401–518 | 414 | 445 | 485 |
| Age (year) | 2.9 | 9.0 | 18–61 | — | — | — |
| Stature | 1752 | 57.9 | 1575–1859 | — | — | — |
| Weight (kg) | 73 | 9.6 | 154.9–104.8 | — | — | — |

*Source:* Kennedy, K.W., Reach capability of men and women: A three-dimensional analysis, Techn. Rep. AMRL-TR-77-50, Aerospace Med. Res. Lab., Wright-Patterson Air Force Base, OH, 1978.

TABLE 1.A.12 Selected Reach and Related Length Data from Military Subjects (Values in cm)

| | | | | Percentile | | | | |
|---|---|---|---|---|---|---|---|---|
| **Dimension** | **Sex** | **Mean** | **SD** | **1st** | **5th** | **50th** | **95th** | **99th** |
| Finger tip reach | M | 87.5 | 4.21 | 78.7 | 80.9 | 87.3 | 94.6 | 98.0 |
| | F | 80.0 | 3.94 | 71.8 | 73.8 | 79.9 | 86.8 | 89.6 |
| Functional grip reach | M | 75.1 | 3.68 | 67.3 | 69.3 | 75.0 | 81.3 | 84.0 |
| | F | 68.6 | 3.39 | 61.5 | 63.2 | 68.5 | 74.4 | 77.0 |
| Functional leg length | M | 108.2 | 5.10 | 96.9 | 100.2 | 108.0 | 116.9 | 120.3 |
| | F | 101.2 | 4.91 | 89.8 | 93.3 | 101.1 | 109.4 | 112.8 |
| Index finger reach | M | 86.2 | 4.14 | 77.4 | 79.6 | 86.1 | 93.2 | 96.5 |
| | F | 78.9 | 3.87 | 70.8 | 72.8 | 78.8 | 85.5 | 88.3 |
| Overhead fingertip reach | M | 223.1 | 9.75 | 200.6 | 207.3 | 223.0 | 239.3 | 245.0 |
| | F | 206.2 | 9.24 | 185.0 | 191.5 | 205.9 | 221.7 | 227.8 |
| Overhead fingertip reach, sitting | M | 143.4 | 5.90 | 129.3 | 133.8 | 143.3 | 153.2 | 156.7 |
| | F | 132.7 | 5.59 | 119.7 | 123.3 | 132.7 | 141.8 | 145.4 |
| Vertical grip reach | M | 210.7 | 9.24 | 189.3 | 195.8 | 210.6 | 226.0 | 231.3 |
| | F | 134.7 | 8.71 | 174.6 | 180.8 | 194.5 | 209.4 | 215.1 |
| Vertical grip reach, sitting | M | 131.0 | 5.45 | 117.8 | 122.1 | 130.9 | 14.01 | 143.2 |
| | F | 121.2 | 5.13 | 109.2 | 112.7 | 121.3 | 129.6 | 132.9 |
| Wrist-center of grip | M | 7.0 | .49 | 6.0 | 6.2 | 6.9 | 7.8 | 8.2 |
| | F | 6.6 | .49 | 5.6 | 5.9 | 6.6 | 7.5 | 7.9 |
| Wrist-wall length | M | 74.8 | 3.73 | 66.4 | 68.8 | 74.8 | 81.1 | 84.1 |
| | F | 67.9 | 3.43 | 60.6 | 62.5 | 67.9 | 73.7 | 76.2 |

*Source:* Gordon, C.C. et al., 1987–1988 Anthropometric survey of U.S. army personnel: Methods and summary statistics, Techn. Rep. NATICK/TR-89-044, U.S. Army Natick Res. Dev. Engineering Center, Natick, MA, 1989.

TABLE 1.A.13 Link Lengths for Military Men and Women (Values in cm)

| | Males | | Females | | |
|---|---|---|---|---|---|
| Link | Mean | SD | Mean | SD | Derivation |
| Head | 13.1 | — | 11.4 | — | Vertex–tragion |
| Neck | 10.6 | 1.12 | 9.8 | 1.07 | Tragion–cervicale |
| Upper torso (vertical)[a] | 39.8 | 2.18 | 36.4 | 2.06 | Cervicale–10th rib |
| Upper torso (horizontal)[b] | 39.7 | 1.80 | 36.3 | 1.74 | Biacromial breadth |
| Abdominal | 4.8 | 1.57 | 5.6 | 1.47 | 10th Rib–iliocristale |
| Lower torso (horizontal) | 23.1 | 1.98 | 22.0 | 2.05 | Bispinous breadth |
| Pelvic | 14.5 | 1.56 | 12.7 | 1.52 | Iliocristale–trochanterion |
| Thigh | 42.7 | 2.51 | 4.01 | 2.35 | Trochanterion–lateral femoral epicondyle |
| Calf | 43.5 | 2.45 | 40.0 | 2.36 | Lateral femoral epicondyle–lateral malleolus |
| Foot (vertical) | 6.7 | 0.55 | 6.1 | 0.53 | Lateral malleolus height |
| Foot (horizontal) | 27.0 | 1.31 | 24.4 | 1.22 | Foot length |
| Upper arm | 34.1 | 1.72 | 31.2 | 1.67 | Acromion–radiale |
| Forearm | 27.0 | 1.57 | 24.3 | 1.55 | Radiale–stylion |
| Hand | 19.4 | 0.98 | 18.1 | 0.97 | Hand length |

*Source:* Gordon, C.C. et al., 1987–1988 Anthropometric survey of U.S. army personnel: Methods and summary statistics, Techn. Rep. NATICK/TR-89-044, U.S. Army Natick Res. Dev. Engineering Center, Natick, MA, 1989.

[a] Also called thorax link.

[b] For clavicle link, divide biacromial breadth by 2.

TABLE 1.A.14 Stature (*S*) and Weight (*W*) Regression Equations for Selected Link Lengths on ANSUR Males (M) and Females (F) (Weight in kg, Stature in cm)

| **Dependent Variable** | | | **Slope Stature** | | **Slope Weight** | | **Intercept** | $SE_{est}$ | $R^2$ |
|---|---|---|---|---|---|---|---|---|---|
| Lateral malleolus height | (M) | = | 0.032 × *S* | + | 0.006 × *W* | + | 6.2 | 4.8 | 0.218 |
| | (F) | | 0.033 × *S* | + | —[a] | + | 6.4 | 4.9 | 0.159 |
| Foot length | (M) | = | 0.117 × *S* | + | 0.022 × *W* | + | 46.3 | 91.3 | 0.514 |
| | (F) | = | 0.110 × *S* | + | 0.028 × *W* | + | 47.9 | 8.8 | 0.478 |
| Ball of foot length | (M) | = | 0.084 × *S* | + | 0.017 × W | + | 35.3 | 7.9 | 0.426 |
| | (F) | = | 0.084 × *S* | + | 0.028 × *W* | + | 29.0 | 70.2 | 0.463 |
| Lateral femoral | (M) | = | 0.342 × *S* | + | —[a] | – | 99.3 | 13.0 | 0.757 |
| Epicondyle height | (F) | = | 0.332 × *S* | + | —[a] | – | 76.8 | 13.1 | 0.720 |
| Trochanterion height | (M) | = | 0.648 × *S* | – | 0.034 × *W* | – | 182.8 | 23.8 | 0.751 |
| | (F) | = | 0.601 × *S* | + | —[a] | – | 117.5 | 24.2 | 0.714 |
| Illiocristale height | (M) | = | 0.722 × *S* | – | 0.016 × *W* | – | 181.8 | 20.1 | 0.847 |
| | (F) | = | 0.694 × *S* | + | —[a] | – | 141.6 | 19.5 | 0.837 |
| Waist height (Omph) | (M) | = | 0.736 × *S* | – | 0.040 × *W* | – | 200.5 | 20.1 | 0.844 |
| | (F) | = | 0.736 × *S* | – | 0.045 × *W* | – | 188.7 | 19.1 | 0.847 |
| Tenth rib height | (M) | = | 0.698 × *S* | + | 0.018 × *W* | – | 118.5 | 18.8 | 0.865 |
| | (F) | = | 0.714 × *S* | + | —[a] | – | 118.7 | 17.6 | 0.869 |
| Cervicale height | (M) | = | 0.900 × *S* | + | 0.028 × *W* | – | 82.8 | 10.6 | 0.972 |
| | (F) | = | 0.805 × *S* | + | 0.017 × *W* | – | 77.7 | 10.2 | 0.970 |
| Suprasternale height | (M) | = | 0.843 × *S* | + | 0.030 × *W* | – | 64.8 | 11.4 | 0.963 |
| | (F) | = | 0.849 × *S* | + | 0.023 × *W* | – | 68.0 | 10.1 | 0.967 |
| Biacromial breadth | (M) | = | 0.083 × *S* | + | 0.053 × *W* | + | 210.0 | 14.9 | 0.312 |
| | (F) | = | 0.104 × *S* | + | 0.050 × *W* | + | 162.8 | 14.6 | 0.296 |
| Acroraion–radiale length | (M) | = | 0.192 × *S* | + | 0.015 × *W* | – | 8.5 | 10.2 | 0.647 |
| | (F) | = | 0.195 × *S* | + | 0.012 × *W* | – | 13.3 | 10.5 | 0.604 |
| Radiale–stylion length | (M) | = | 0.156 × *S* | + | 0.008 × *W* | – | 10.3 | 11.2 | 0.485 |
| | (F) | = | 0.352 × *S* | + | —[a] | – | 20.3 | 11.5 | 0.444 |
| Hand length | (M) | = | 0.084 × *S* | + | 0.012 × *W* | + | 36.6 | 7.3 | 0.435 |
| | (F) | = | 0.086 × *S* | + | 0.015 × *W* | + | 30.6 | 7.4 | 0.416 |
| Wrist-index finger length | (M) | | 0.081 × *S* | + | 0.010 × *W* | + | 29.9 | 6.7 | 0.457 |
| | (F) | = | 0.085 × *S* | + | 0.012 × *W* | + | 23.6 | 6.5 | 0.453 |
| Wrist-thumbtip length | (M) | = | 0.055 × *S* | + | 0.007 × *W* | + | 22.7 | 5.3 | 0.373 |
| | (F) | = | 0.056 × *S* | + | 0.009 × *W* | + | 21.4 | 5.3 | 0.357 |

*Source:* Cheverud, J. et al., 1988 Anthropometric survey of U.S. army personnel: Correlation coefficients and regression equations, Techn. Rep. NATICK/TR-90/036, U.S. Army Natick Res. Dev. Eng. Center, Natick, MA, 1990.

[a] This independent variable does not add significantly.

TABLE 1.A.15 Stature (*S*) and Weight (*W*) Regression Equations for Selected Workspace Dimensions on ANSUR Males (M) and Females (F) Soldiers (Weight in kg, Stature in cm)

| Dependent Variable | | | Slope Stature | | Slope Weight | | Intercept | $SE_{est}$ | $R^2$ |
|---|---|---|---|---|---|---|---|---|---|
| Acromial height, sitting | (M) | = | 0.240 × *S* | + | 0.057 × *W* | + | 132.5 | 21.7 | 0.463 |
| | (F) | = | 0.269 × *S* | + | 0.044 × *W* | + | 89.6 | 21.2 | 0.454 |
| Bideltoid breadth | (M) | = | −0.039 × *S* | + | 2.009 × *W* | + | 395.8 | 13.8 | 0.716 |
| | (F) | = | −0.026 × *S* | + | 2.030 × *W* | + | 331.9 | 13.1 | 0.662 |
| Buttock–knee length | (M) | = | 0.287 × *S* | + | 0.079 × *W* | + | 50.6 | 16.2 | 0.705 |
| | (F) | = | 0.254 × *S* | + | 1.033 × *W* | + | 92.2 | 17.4 | 0.654 |
| Buttock–popliteal length | (M) | = | 0.274 × *S* | + | 0.038 × *W* | − | 9.7 | 16.5 | 0.616 |
| | (F) | = | 0.263 × *S* | + | 0.060 × *W* | + | 15.6 | 17.6 | 0.555 |
| Elbow-rest height | (M) | = | —[a] | + | 0.060 × *W* | + | 183.6 | 26.4 | 0.060 |
| | (F) | = | 0.047 × *S* | + | 0.042 × *W* | | 117.7 | 26.2 | 0.044 |
| Eye height, sitting | (M) | = | 0.359 × *S* | + | 0.014 × *W* | + | 149.7 | 23.5 | 0.529 |
| | (F) | = | 0.391 × *S* | + | —[a] | + | 101.6 | 22.1 | 0.560 |
| Forearm–forearm breadth | (M) | = | −0.172 × S | + | 3.051 × *W* | + | 572.6 | 27.2 | 0.611 |
| | (F) | = | −0.118 × *S* | + | 3.069 × *W* | + | 432.5 | 21.0 | 0.633 |
| Hand circumference | (M) | = | 0.029 × *S* | + | 0.043 × *W* | + | 128.8 | 7.5 | 0.393 |
| | (F) | = | 0.003 × *S* | + | 0.041 × *W* | + | 106.7 | 6.9 | 0.333 |
| Hip breadth, sitting | (M) | = | −0.026 × *S* | + | 2.006 × *W* | + | 250.8 | 12.2 | 0.764 |
| | (F) | = | −0.052 × *S* | + | 2.085 × *W* | + | 292.6 | 15.8 | 0.664 |
| Knee height, sitting | (M) | = | 0.343 × *S* | + | 0.030 × *W* | − | 66.7 | 12.7 | 0.793 |
| | (F) | = | 0.332 × *S* | + | 0.033 × *W* | − | 45.8 | 13.4 | 0.742 |
| Popliteal height | (M) | = | 0.354 × *S* | − | 0.040 × *W* | − | 155.7 | 12.5 | 0.748 |
| | (F) | = | 0.352 × *S* | − | 0.073 × *W* | − | 138.9 | 13.0 | 0.699 |
| Sitting height | (M) | = | 0.377 × *S* | + | 0.019 × *W* | + | 236.7 | 23.8 | 0.551 |
| | (F) | = | 0.414 × *S* | + | —[a] | + | 176.7 | 22.9 | 0.570 |
| Span | (M) | = | 0.951 × *S* | + | 0.055 × *W* | + | 110.9 | 47.2 | 0.688 |
| | (F) | = | 0.986 × *S* | + | 0.030 × *W* | + | 47.4 | 50.1 | 0.620 |
| Thigh clearance | (M) | = | −0.046 × *S* | + | 1.007 × *W* | + | 164.0 | 6.8 | 0.707 |
| | (F) | = | −0.041 × *S* | + | 1.027 × *W* | + | 147.5 | 7.5 | 0.618 |
| Waist depth | (M) | | −0.121 × *S* | + | 2.022 × *W* | | 263.5 | 14.0 | 0.699 |
| | (F) | = | −0.148 × *S* | + | 2.078 × *W* | + | 273.5 | 15.0 | 0.638 |
| Waist circumference | (M) | = | −0.347 × *S* | + | 7.075 × *W* | + | 863.1 | 41.3 | 0.771 |
| | (F) | = | −0.424 × *S* | + | 9.028 × *W* | + | 907.1 | 48.2 | 0.661 |

*Source:* From Cheverud, J. et al., 1988 Anthropometric survey of U.S. army personnel: Correlation coefficients and regression equations, Techn. Rep. NATICK/TR-90/036, U.S. Army Natick Res. Dev. Eng. Center, Natick, MA, 1990.

[a] This independent variable does not add significantly.

TABLE 1.A.16 Stature (*S*) and Weight (*W*) Regression Equations for the HES Males (Weight in kg, Stature in mm)

| Dependent Variable | | | Slope Stature | | Slope Weight | | Intercept | $SE_{est}$ | $R^2$ |
|---|---|---|---|---|---|---|---|---|---|
| Biacromial breadth | (M) | = | 0.069 × *S* | + | 0.648 × *W* | + | 228.2 | 18.1 | 0.268 |
| | (F) | = | 0.097 × *S* | + | 0.538 × *W* | + | 164.8 | 16.1 | 0.309 |
| Right arm girth[a] | (M) | = | −0.109 × *S* | + | 2.412 × *W* | + | 314.3 | 15.6 | 0.768 |
| | (F) | = | −0.134 × *S* | + | 2.853 × *W* | + | 319.5 | 17.3 | 0.832 |
| Chest girth[a] | (M) | = | −0.217 × *S* | + | 6.300 × *W* | + | 896.1 | 36.3 | 0.810 |
| | (F) | = | −0.184 × *S* | + | 5.384 × *W* | + | 832.0 | 35.9 | 0.804 |
| Waist girth[a] | (M) | = | −0.511 × *S* | + | 8.622 × *W* | + | 1125.6 | 55.6 | 0.765 |
| | (F) | | −0.450 × *S* | + | 7.901 × *W* | + | 985.4 | 55.8 | 0.786 |
| Sitting height | (M) | = | 0.335 × *S* | + | 0.607 × *W* | + | 237.4 | 24.8 | 0.540 |
| | (F) | = | 0.416 × *S* | + | 0.096 × *W* | + | 144.5 | 26.7 | 0.519 |
| Sitting height, erect | (M) | = | 0.379 × *S* | + | 0.385 × *W* | + | 219.5 | 23.3 | 0.597 |
| | (F) | = | 0.416 × *S* | + | 0.149 × *W* | + | 171.1 | 23.3 | 0.586 |
| Knee height, sitting | (M) | = | 0.318 × *S* | + | 0.243 × *W* | − | 28.2 | 17.1 | 0.649 |
| | (F) | = | 0.309 × *S* | + | 0.347 × *W* | − | 19.8 | 16.1 | 0.646 |
| Popliteal height | (M) | | 0.321 × *S* | − | 0.372 × *W* | − | 88.6 | 16.8 | 0.604 |
| | (F) | = | 0.301 × *S* | − | 0.347 × *W* | − | 62.7 | 17.4 | 0.554 |
| Thigh clearance | (M) | = | −0.020 × *S* | + | 0.973 × *W* | + | 105.6 | 12.1 | 0.489 |
| | (F) | = | −0.002 × *S* | + | 0.938 × *W* | + | 80.9 | 13.1 | 0.494 |
| Buttock–knee length | (M) | = | 0.260 × *S* | | 0.734 × *W* | + | 85.7 | 17.8 | 0.628 |
| | (F) | = | 0.240 × *S* | + | 1.189 × *W* | + | 108.5 | 18.5 | 0.646 |
| Buttock–popliteal length | (M) | = | 0.232 × *S* | + | 0.438 × *W* | + | 58.8 | 24.2 | 0.378 |
| | (F) | | 0.203 × *S* | + | 0.930 × *W* | + | 94.7 | 23.0 | 0.439 |
| Seat breadth | (M) | = | −0.005 × *S* | + | 1.816 × *W* | + | 226.7 | 16.2 | 0.666 |
| | (F) | = | −0.009 × *S* | + | 2.195 × *W* | + | 241.9 | 21.8 | 0.657 |
| Elbow–elbow breadth | (M) | | −0.189 × *S* | + | 3.383 × *W* | + | 494.6 | 24.5 | 0.721 |
| | (F) | = | −0.185 × *S* | + | 3.483 × *W* | + | 464.7 | 25.3 | 0.776 |
| Elbow-rest height | (M) | = | 0.042 × *S* | + | 0.646 × *W* | + | 119.5 | 28.4 | 0.104 |
| | (F) | = | 0.084 × *S* | + | 0.463 × *W* | + | 66.3 | 27.6 | 0.101 |

*Source:* From Rodgers, S.H., (Ed.), *Ergonomic Design for People at Work*, Vol. 1 and 2, Van Nostrand Reinhold, New York, 1986.

[a] Original name.

## REFERENCES


1. C. C. Gordon, B. Bradtmiller, C. E. Clauser, T. Churchill, J. T. McConville, I. Tebbetts, and R. A. Walker, *1987–1988 Anthropometric Survey of U.S. Army Personnel: Methods and Summary Statistics*, Techn. Rep. NATICK/TR-89-044, U.S. Army Natick Research, Development and Engineering Center, Natick, MA, 1989.
2. T. G. Lohman, A. F. Roche, and R. Martorell, *Anthropometric Standardization Reference Manual*, Human Kinetics Books, Champaign, IL, 1988.
3. H. T. E. Hertzberg, The conference on standardization of anthropometric techniques and terminology, *Am. J. Phys. Anthropol. 28*(1):1–16 (1968).
4. H. T. E. Hertzberg, E. Churchill, C. W. Dupertuis, R. M. White, and A. Damon, *Anthropometric Survey of Turkey, Greece and Italy*, Macmillan, New York, 1963.
5. C. E. Clauser, P. E. Tucker, J. T. McConville, E. Churchill, L. L. Laubach, and J. A. Reardon, *Anthropometry of Air Force Women*, Techn. Rep. AMRL-TR-70-5 (AD 743 113), Aerospace Medical Research Laboratory, Wright-Patterson Air Force Base, OH, 1972.
6. J. F. Annis and C. C. Gordon, *The Development and Validation of an Automated Headboard Device for Measurement of Three-Dimensional Coordinates for the Head and Face*, Techn. Rep. NATICK/TR-88/048 (AD A201-186), U. S. Army Natick Research, Development and Engineering Center, Natick, MA, 1988.
7. J. F. Annis, An automated device used to develop a new 3-D data base for head and face anthropometry, in *Advances in Industrial Ergonomics and Safety*, Vol. 1, A. Mital, Ed., Taylor & Francis, Philadelphia, PA, 1989.
8. R. E. Herron, Biostereometric measurement of body form, *Yearb. Phys. Anthropol. 16*:80–121 (1972).
9. G. S. Daniels, *The "Average Man"?* Techn. Note WCRD 53–7, Wright Air Dev. Center, Wright-Patterson Air Force Base, OH, 1952.
10. J. F. Annis and J. T. McConville, Application of anthropometric data in sizing and design, in *Advances in Industrial Ergonomics and Safety*, Vol. II, B. Das, Ed., Taylor & Francis, Philadelphia, PA, 1990.
11. J. T. McConville and E. Churchill, *Statistical Concepts in Design*, AMRL-TR-76-29, Aerospace-Medical Res. Lab., Wright-Patterson Air Force Base, OH, 1976.
12. K. Robinette and T. Churchill, *Design Criteria for Characterizing Individuals in the Extreme Upper and Lower Body Size Ranges*, AMRL-TR-79-33 (AD A072 353), Aerospace Medical Res. Lab., Wright-Patterson Air Force Base, OH, 1979.
13. S. Pheasant, *Bodyspace Anthropometry, Ergonomics, and Design*, Taylor & Francis, London, U.K., 1986.
14. E. Grandjean, *Fitting the Task to the Man*, 4th edn., Taylor & Francis, London, U.K., 1988.
15. R. O'Brien and W. C. Shelton, *Women's Measurements for Garment and Pattern Construction*, Misc. Publ. No. 454, U.S. Gov. Printing Office, Washington, DC, 1941.
16. R. O'Brien, M. A. Girshick, and E. P. Hunt, *Body Measurements of American Boys and Girls for Garment Pattern Construction*, Misc. Publ. 366, U.S. Dept. of Agr., U.S. Gov. Printing Office, Washington, DC, 1941.
17. H. Stoudt, A. Damon, R. MacFarland, and J. Roberts, *Weight, Height and Selected Body Dimensions of Adults, United States 1960–1962*, Public Health Service Publ. No. 1000, Ser. 11, No. 8, U.S. Gov. Printing Office, Washington, DC, 1965.
18. S. H. Rodgers (Ed.), *Ergonomic Design for People at Work*, Vol. 1 and 2, Van Nostrand Reinhold, New York, 1986.
19. C. C. Snow, H. M. Reynolds, and M. A. Allgood, *Anthropometry of Airline Stewardesses*, Dept. Transportation Rep. No. FAA-AM-75-2, FAA Office of Aviation Medicine, Civil Aeromed. Inst., Oklahoma City, OK, 1975.
20. J. T. McConville and C. E. Clauser, Anthropometric resources vs. civilian needs, Techn. Rep. AMRL-TR-78-111 (AD A061-390), Aerospace Med. Res. Lab., Wright-Patterson Air Force Base, OH, 1978.

21. W. S. Marras and J. Y. Kim, Anthropometry of industrial populations, *Ergonomics 36*(4): 371–378 (1993).
22. R. M. White and E. Churchill, The body size of soldiers: U.S. army anthropometry—1966, Techn. Rep. 72-51-CE (AD 743 465), U.S. Army Natick Lab., Natick, MA, 1971.
23. B. Bradtmiller, J. Ratnaparkhi, and I. Tebbetts, Demographic and anthropometric assessment of U.S. army anthropometric data base, Techn. Rep. NATICK/TR-86/004 (AD A164-637), U.S. Army Natick Res. Dev. Eng. Center, Natick, MA, 1985.
24. E. Churchill, P. Kikta, and T. Churchill, The AMRL anthropometric data bank library, Vols. 1–5, Techn. Rep. AMRL-TR-77-1 (AD A047 314), Aerospace Med. Res. Lab., Wright-Patterson Air Force Base, OH, 1977.
25. J. T. McConville, E. Churchill, C. E. Clauser, and M. Alexander, The aerospace medical research laboratory's anthropometric data bank: A resource for designers, Preprints 1977 Sci. Program, Aerospace Med. Assoc., 1977 Annu. Sci. Meeting, Las Vegas, NV, 1977.
26. J. C. Robinson, K. M. Robinette, and G. F. Zehner, User's guide to accessing the anthropometric data base at the center for anthropometric research data, 2nd edn., Techn. Rep. AL-TR-1992-0036, Armstrong Lab., U.S. Air Force Systems Command, Wright-Patterson Air Force Base, OH, 1992.
27. E. Churchill, J. T. McConville, L. L. Laubach, T. Churchill, P. Erskine, and K. Downing, *Anthropometric Source Book*, Vol. II, A handbook of anthropometric data, NASA Ref. Publ. 1024 (NTIS No. N79-13711/3/XPS), National Aeronautics and Space Administration, Washington, DC, 1978.
28. C. E. Clauser, J. T. McConville, C. C. Gordon, and I. O. Tebbetts, Selection of dimensions for an anthropometric data base, Vol. 1, Rationale, summary and conclusions, Techn. Rep. NATICK/TR-86/053 (AD A179 566), U.S. Army Natick Res. Dev. Eng. Center, Natick, MA, 1986.
29. A. Engle, R. S. Murphy, K. Maurer, and E. Collins, *Plan and Operation of the HANES I Augmentation Survey of Adults 25 to 74, United States 1974–1975*, DHEW Publ. PHS 78-1314, Ser. 1, No. 14, U.S. Public Health Service, NCHS, Hyattsville, MD, 1978.
30. J. I. Martin, R. Sabeh, L. L. Driver, T. D. Lowe, R. W. Hintz, and P. A. C. Peters, Anthropometry of law enforcement officers, Rep. No. NELC, Techn. Docu. 442, TD 442, Law Enforcement Standards Laboratory, Nat. Bureau of Standards, Naval Electron. Lab. Center, San Diego, CA, 1975.
31. H. G. Grunhofer and G. Kroh (Eds.), A review of anthropometric data of german air force and United States air force personnel, Techn. Rep. AGARD-AG-205, Advisory Group for Aerospace Res. Dev., Neuilly sur Seine, France, 1975.
32. L. L. Laubach, J. T. McConville, E. Churchill, and R. M. White, Anthropometry of women of the U.S. army—1977, Rep. No. 1—Method and survey plan, Techn. Rep. NATICK/TR-77/021 (AD A043 715), U.S. Army Natick Res. Dev. Command, Natick, MA, 1977.
33. K. Robinette, T. Churchill, and J. T. McConville, A comparison of male and female body sizes and proportions, Techn. Rep. AMRL-TR-79-69 (AD A074 807), Aerospace Med. Res. Labor., Wright-Patterson Air Force Base, OH, 1979.
34. J. F. Annis, H. W. Case, C. E. Clauser, and B. Bradtmiller, Anthropometry of an aging work force, *Exp. Aging Res. 17*(3):157–176 (1991).
35. J. S. Friedlaender, P. T. Costa, Jr., R. Bosse, E. Ellis, J. G. Rhoads, and H. W. Stoudt, Longitudinal physique changes among healthy white veterans at Boston, *Hum. Biol. 49*(4):541–558 (1977).
36. A. Damon and H. W. Stoudt, The functional anthropometry of old men, *Hum. Factors 5*:485–491 (1963).
37. B. Bell, C. L. Rose, and A. Damon, The Normative Aging Study (NAS), the Veterans administration longitudinal study of healthy aging, *Gerontologist 6*:179–184 (1966).
38. W. C. Chumlea, P. J. Garry, W. C. Hunt, and R. L. Rhyne, Distributions of serial changes in stature and weight in a healthy elderly population, *Hum. Biol. 60*(6):917–925 (1988).

39. N. W. Shock, R. C. Greulich, R. Andres, D. Arenberg, P. T. Costa, Jr., E. G. Lakatta, and J. D. Tobin, *Normal Human Aging: The Baltimore Longitudinal Study of Aging*, Vols. 1 and 2, (NIH Publ. AM-84-2450). U.S. Govt., Printing Office, Washington, DC, 1984.
40. A. F. Roche, *Growth, Maturation and Body Composition: The Fels Longitudinal Study 1929–1991*, Cambridge University Press, Cambridge, 1992.
41. F. E. Randall, A. Damon, R. S. Benton, and D. I. Patt, Human body size in military aircraft and personal equipment, Techn. Rep. AAF-TR-5501 (ATI 25 419), Air Materiel Command, Wright Field, Dayton, OH, 1946.
42. R. F. Johnson, Anthropometry of the clothed U.S. army ground troop and combat vehicle crewman, Techn. Rep. NATICK/TR-84/034, U.S. Army Natick Res. Dev. Electronics Center, Natick, MA, 1984.
43. S. P. Paquette, H. W. Case, J. F. Annis, T. L. Mayfield, S. E. Kristensen, and D. N. Mountjoy, The effect of multilayered military ensembles on body size, NATICK TR/-, U.S. Army Natick Res. Dev. Eng. Center, Natick, MA, 1993.
44. S. P. Paquette and R. A. Maulucci, The effects of clothing on body size and range of joint motion, Techn. Rep. NATICK/TR- Natick Res. Dev. Eng. Center, Natick, MA, 1993.
45. H. T. E. Hertzberg, I. Emanuel, and M. Alexander, The anthropometry of working positions, WADC Techn. Rep. 54-520, Wright Air Dev. Center, Wright-Patterson Air Force Base, OH, 1956.
46. M. Alexander and C. E. Clauser, Anthropometry of common working positions, Techn. Rep. AMRL-TR-65-73, Aerospace Med. Res. Lab., Wright-Patterson Air Force Base, OH, 1965.
47. W. T. Dempster, The anthropometry of body action, *Ann. NY. Acad. Sci. 63*(4):559–585 (1955).
48. K. W. Kennedy, Reach capability of men and women: A three-dimensional analysis, Techn. Rep. AMRL-TR-77-50, Aerospace Med. Res. Lab., Wright-Patterson Air Force Base, OH, 1978.
49. H. W. Stoudt, Arm-leg reach and workspace layout, *in Anthropometric Source Book*, Vol. I, A handbook of anthropometric data, NASA Ref. Publ. 1024 (NTIS No. N79-13711/3/XPS), National Aeronautics and Space Administration, Washington, DC, 1978, Chapter 5.
50. W. E. Woodson, B. Tillman, and P. Tillman, *Human Factors Design Handbook*, 2nd edn., McGraw-Hill, New York, 1992.
51. R. D. Huchingson, *New Horizons for Human Factors Design*, McGraw-Hill, New York, 1981.
52. A. Damon, H. W. Stoudt, and R. A. McFarland, *The Human Body in Equipment Design*, Harvard University Press, Cambridge, MA, 1966.
53. J. S. Roebuck, Jr., K. H. E. Kroemer, and W. G. Thomson, *Engineering Anthropometry Methods*, Wiley, New York, 1975.
54. W. Braune and O. Fischer, *The Center of Gravity of the Human Body as Related to the German Infantryman*, Leipzig, Germany, 1889, (ATI 138 452; available from Defense Documentation Center).
55. The Human Factors Society, *American National Standard for Human Factors Engineering of Visual Display Terminal Workstations*, ANSI/HFS Standard 100–1988, Human Factors Society, Santa Monica, CA, 1988.
56. J. Cheverud, C. C. Gordon, R. A. Walker, C. Jacquish, L. Kohn, A. Moore, and N. Yamashita, 1988 Anthropometric survey of U.S. army personnel: Correlation coefficients and regression equations, Techn. Rep. NATICK/TR-90/036, U.S. Army Natick Res. Dev. Eng. Center, Natick, MA, 1990.

CHAPTER 2

# Skeletal Muscle Physiology and Its Application to Occupational Ergonomics

Robert G. Cutlip and Sharon S. Chiou

CONTENTS

## 2.1 INTRODUCTION

In the United States, work-related musculoskeletal disorders account for approximately 38% of cases involving days away from work [1], thus making it an enormous economic and health care burden. A large component in musculoskeletal disorders is acute and chronic contraction-induced skeletal muscle injury [2]. In order to address this issue, there have been extensive studies to-date on acute contraction-induced muscle injury using both animals and humans. Occupationally related musculoskeletal disorders have been associated with exposure to excessive physical loads, repetitive movements, awkward postures, and vibration [3]. A number of different tissues, including skeletal muscles, can be injured by exposure to these various factors [3]. To understand how exposure to these factors results in muscle injury, it is necessary to understand the biological and physiological mechanisms that allow skeletal muscles to generate movement, maintain posture, and support loads. The goal of this chapter is to provide a basic description of skeletal muscle physiology, injury mechanics, and motor control with application to occupational musculoskeletal disorders.

## 2.2 MUSCLE PHYSIOLOGY AND ANATOMY

Individual skeletal muscles are comprised of bundles of muscle cells or myofibers (Figure 2.1). Each myofiber is surrounded by a collagenous basement membrane (basal lamina) in addition to a cellular membrane called the sarcolemma. Myofibers are similar to other cells in the body, but they have a couple of unique features. First, myofibers contain a modified endoplasmic reticulum called a sarcoplasmic reticulum (SR). The SR functions as a protein processing and distribution organelle, and it regulates the levels of free intracellular calcium ($Ca^{2+}$) within the myofiber. Second, most of the intracellular space within the myofiber comprises the contractile elements or myofibrils (80% of a muscle's volume) (Figures 2.1 and 2.2). Each myofibril comprises thick and thin filaments. Each thick filament (12–18 nm diameter) is composed of several hundred myosin proteins. Within the thick filaments, each myosin protein has a projection or a globular head. These globular heads have binding sites that can interact with and form crossbridges with the thin filaments, and an ATPase binding site. Thin filaments (5–8 nm diameter) are made

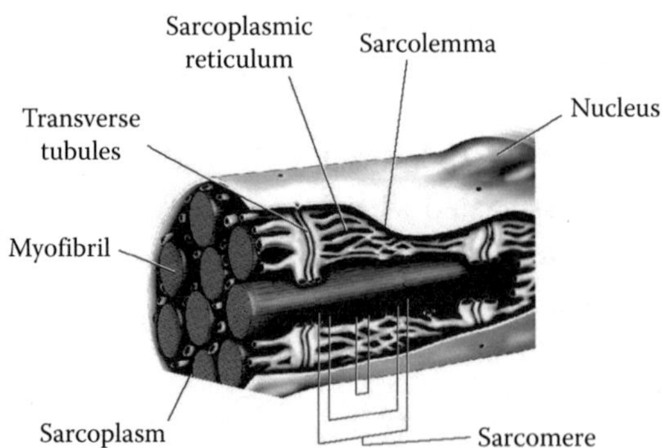


FIGURE 2.1 Organization of a skeletal muscle cell showing the sarcolemma, sarcoplasm, sarcomere, and transverse T-tubules.

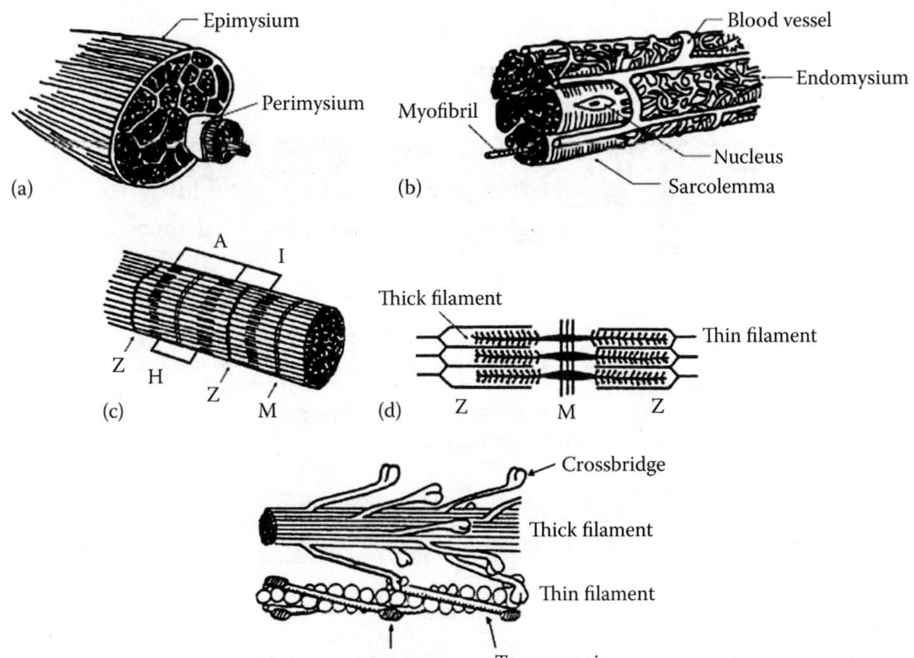


FIGURE 2.2 Skeletal muscle organization from (a) epimysium and perimysium, (b) the endomysium and myofibrils, (c) the myofilaments, (d) the crossbridge, (e) the contractile and regulatory proteins.

of actin molecules that are organized in two strands twisted together to form a helix, that are covered by threadlike tropomyosin molecules and spherical troponin molecules. Thin and thick filaments are organized in a specific pattern which is repeated down the length of the muscle. It is this patterning that gives skeletal muscle its striated appearance [4]. Myofibers are wrapped in a connective tissue sheath called the endomysium, bundles of myofibers are wrapped in an additional sheath (the perimysium), and whole muscle is wrapped in the epimysium (Figure 2.2). Each repeated segment of thin and thick fibers forms a sarcomere [4]. The sarcomere is defined as the area between the Z-disk or Z-line (Figure 2.3). Each sarcomere comprises dark areas (A bands), that contain the thick filaments, and light areas (I bands), that contain the ends of the thin filaments that do not overlap with the thick filaments (Figures 2.3 and 2.4). At each end of the sarcomere is a three-dimensional (3D) structure referred to as a Z-disk or Z-line. Muscle contractions are produced when crossbridges are formed between overlapping thin and thick fibers in the sarcomere, making the sarcomere the smallest contractile unit in the myofiber. The protein titin attaches the A-band to the Z-disk (Figures 2.3 and 2.4).

## 2.3 SOMATIC NERVOUS SYSTEM

Skeletal muscle is innervated by motor neurons that control activation and subsequent contraction. The somatic nervous system which innervates skeletal muscle has nerve fibers that are continuous from their origin in the cell bodies on the spinal cord to their termination

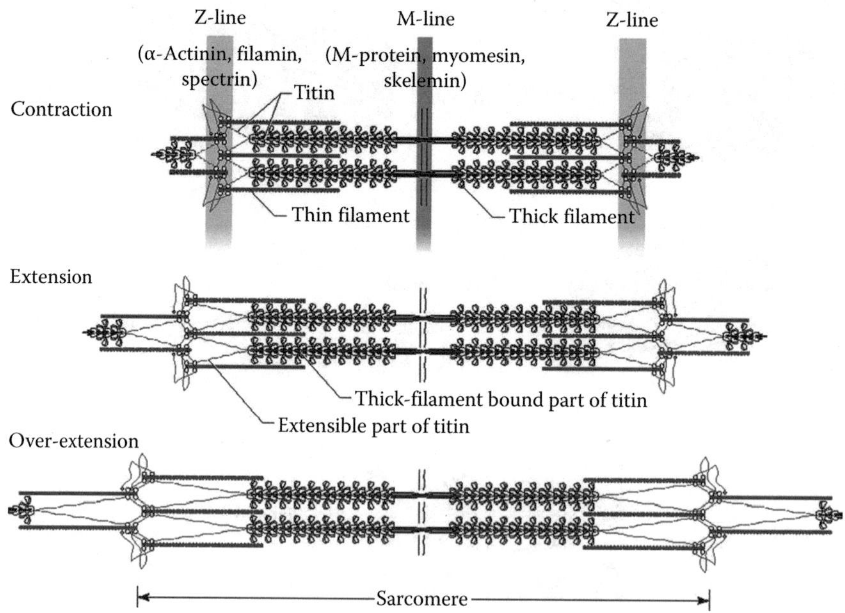


FIGURE 2.3 The crossbridge, showing the Z-line, M-line, and thick and thin filaments.

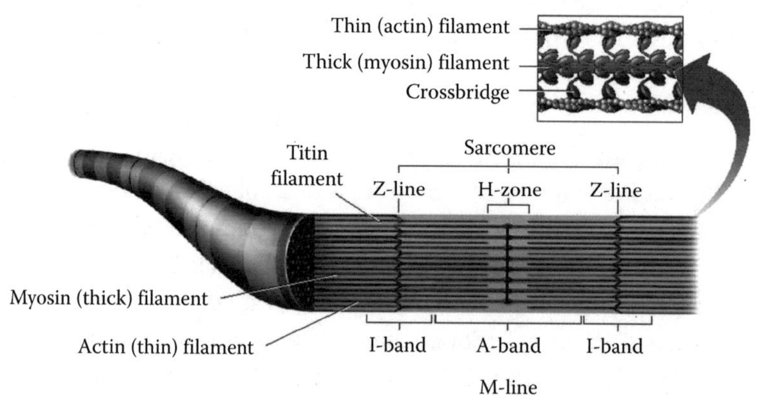


FIGURE 2.4 The crossbridge, showing the I-band, A-band, titin filaments, and Z-line.

on skeletal muscle at the myofibrils (Figure 2.5). This is in contrast to the autonomic nervous system, which contains two neuron chains of preganglionic and postganglionic fibers used to innervate cardiac and smooth muscles. To initiate a muscle contraction, myofibers must receive stimulation from motor neurons located in the ventral horn of the spinal cord. This stimulation is based on total synaptic input at the cell body from inhibitory postsynaptic potentials (IPSPs) and excitatory postsynaptic potentials (EPSPs) produced by presynaptic inputs originating from diverse sites in the brain. Presynaptic inputs can be influenced by many peripheral sensory receptors in the ascending pathways that can influence descending pathway output via the motor neurons. These inputs can either be excitatory or inhibitory. Areas of the brain that control skeletal muscle activity are motor regions of the cortex, basal nuclei, brain stem, and cerebellum. When summed at the cell

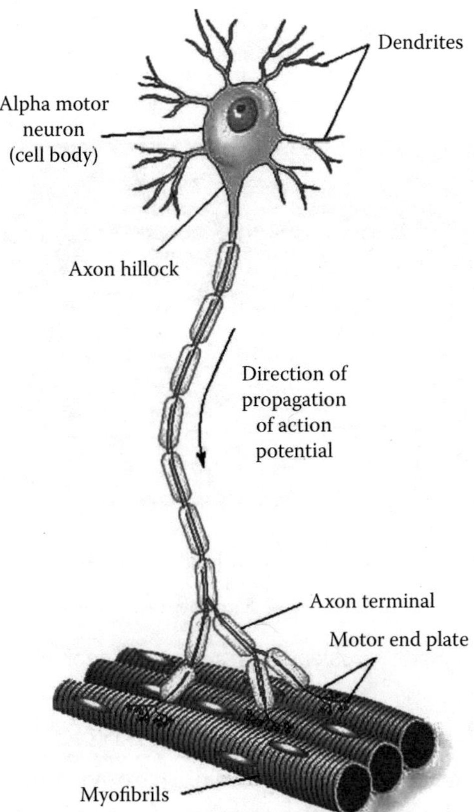


FIGURE 2.5 The motor unit that consists of the alpha motor neuron, motor end plate, and the myofibrils it innervates.

body on the spinal cord, if EPSP's are more dominant, an action potential is then initiated. Alpha motor neurons are considered the final common pathway where skeletal muscle activity can be governed only by input from these neurons; thus, it is considered the final common pathway (Figure 2.5). Other parts of the nervous system can only affect skeletal muscle activity by influencing the motor neurons in the somatic nervous system. While skeletal muscle is under voluntary control, lower brain centers control involuntary activity that assists in bringing about the voluntary action that is typically controlled by motor regions in the higher brain centers. Indeed, much of skeletal muscle control is involuntary.

When a motor neuron is activated, an action potential is propagated down the nerve axon and terminates on the neuromuscular junction (NMJ). The NMJ is a space where the action potential cannot cross from the nerve axon to the muscle fibers it innervates. Thus, a chemical messenger is used to transmit the signal from the nerve axon to the muscle fibers. As the signal is transmitted down the nerve axon, voltage-gated channels open to release calcium into the terminal button of the NMJ. This facilitates the release of the chemical messenger acetylcholine (ACH) that crosses the space to the motor end plate. This causes an ionic shift which results in the propagation of the action potential down the basement membrane of the muscle fiber and then down the transverse tubule (T-tubules) of the muscle cell (Figure 2.6).

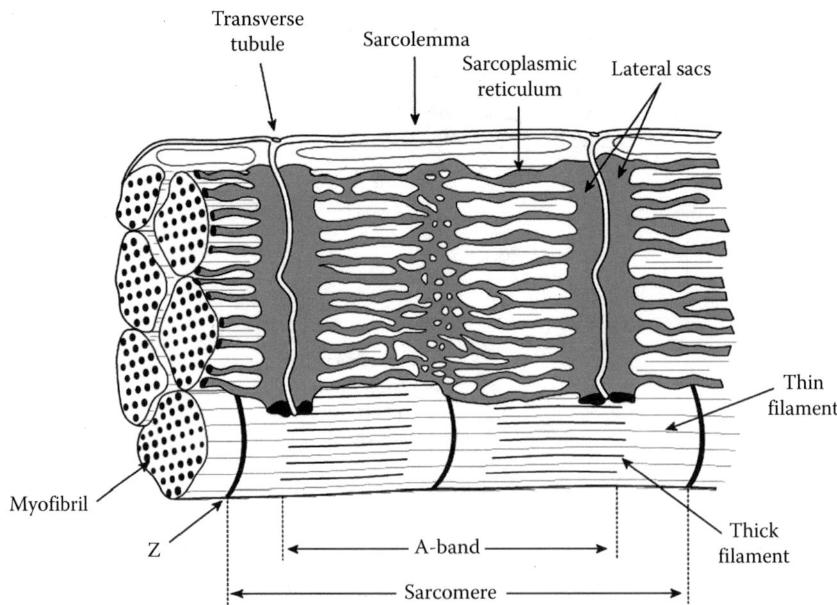


FIGURE 2.6 The muscle cell, with sarcolemma, transverse T-tubule, SR, and myofibrils.

## 2.4 MUSCLE CONTRACTION AND FORCE GENERATION

The action potential activates the voltage-gated dihydropyridine receptors in the T-tubule. This change in the T-tubules triggers the opening of calcium release channels (ryanodine receptors) on the SR. $Ca^{2+}$ leaves the SR lateral sacs through the ryanodine receptors, enters the cytoplasm, and binds to troponin, one of the proteins on the thin filaments (Figure 2.6). Troponin has three polypeptide units: one binds to tropomyosin, one binds to actin, and a third which binds to $Ca^{2+}$. Under resting conditions, tropomyosin is bound to actin and it blocks the myosin binding site on the actin protein, preventing the formation of crossbridges (Figure 2.7). However, when free $Ca^{2+}$ rises in the cytoplasm of a myofiber, it binds to troponin and tropomyosin is pulled away from the myosin binding site on actin, leaving it open for crossbridge formation. Once crossbridges are formed, the ATPase located on the myosin head increases its activity and hydrolyzes ATP. This causes the crossbridge to break, and $Ca^{2+}$ then dissociates from its binding site on tropomyosin. When $Ca^{2+}$ is removed, tropomyosin slides back into the blocking position and the muscle relaxes. Thus, troponin and tropomyosin are referred to as regulatory proteins in muscle contraction (Figure 2.7).

One of the main functions of skeletal muscle is to generate and transmit force. Force, or muscle tension, is directly related to the number of actin and myosin crossbridges that are formed and the frequency of stimulation. A single action potential results in a single muscle contraction referred to as "twitch". As the frequency of stimulation increases, the resultant twitch tension increases with increasing stimulation frequency until a force plateau results. Force is produced at each attached crossbridge, so the total force development is proportional to the number of attached crossbridges. The number of crossbridges that

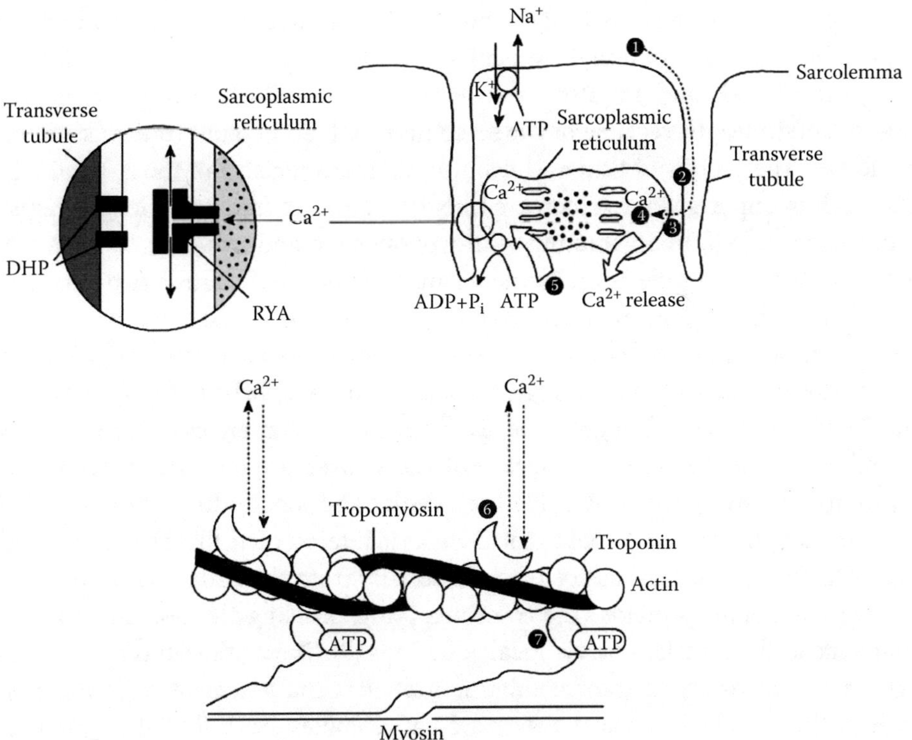


FIGURE 2.7 The NA-K pump on the cell membrane, and the binding of calcium to the tropomyosin molecule.

can be formed depends upon the degree of overlap between the thin and thick filaments. When a sarcomere is overstretched or compressed, the area over which thin and thick filaments overlap is reduced, and thus there is a decrease in the number of crossbridges that can be formed resulting in a reduction in force (Figures 2.3 and 2.8). Thus, maximal force is generated when sarcomeres are at a length that produces the optimal overlap between thin and thick fibers. Passive tension is due to connective tissue and increases dramatically as the length increases, usually outside of the normal physiological range. The total tension in the muscle is the sum of the active and passive tension (Figure 2.8).

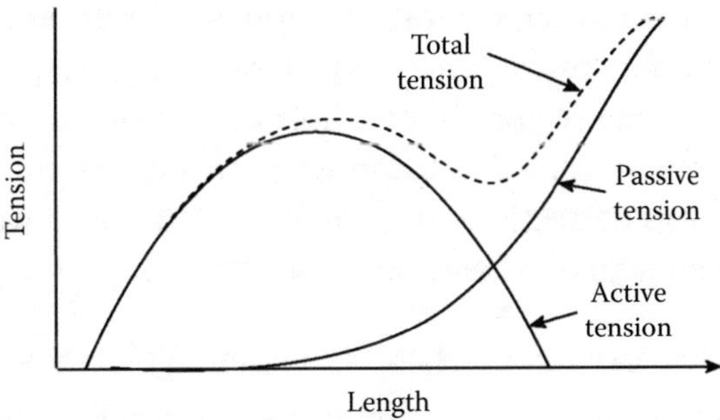


FIGURE 2.8 The length–tension relationship of skeletal muscle.

Force is generated at the crossbridges, but it is transmitted longitudinally and radially along myofibrils. The longitudinal transmission of force occurs down the thick myosin filament to the Z-disk, and on to the next serial set of myofibrils. Two proteins titin and nebulin, maintain length registry of the sarcomere and aid in axial transmission of contractile forces. The actions of titin and nebulin maintain registry of the A-band with the Z-disk which is important for sarcomere integrity. Nebulin maintains length registry of the thin filaments [5,6] by interacting with tropomyosin and troponin to form a lateral network with actin to regulate thin filament length. Titin functions as a two part spring to transmit force from the thick filaments to the corresponding Z-disk.

Radial forces are transmitted via lateral stabilization of adjacent myofibrils. The protein responsible for maintaining lateral registry of adjacent myofibrils at the Z line is desmin [7]. The Z-disk structure is thought to be 3D in nature and comprises the proteins desmin, actin, and α-actinin. The radial enclosure of these three proteins also extends longitudinally along the myofibrils to provide both radial and longitudinal stability [8]. These proteins are thought to be anchored to the Z-disk via intermediate filament-associated proteins (IFAP). The cytoskeletal lattice extends radially from the Z-disk to the sarcolemma via the transmembrane proteins. The transmembrane proteins are thought to anchor the myofilaments to the sarcolemma via focal adhesions [9]. These adhesions or "costameres" are made up of a variety of transmembrane proteins. The basement membrane is then attached to the sarcolemma via the dystroglycan complex [8,9]. Radial transmission of forces occurs through structural proteins located in and outside of the sarcomeric region via the intermediate filament network, and to the sarcolemma via the transmembrane proteins [8]. Capability of radial force transmission is necessary for redundancy in case of fiber injury. Thus, force can be transmitted in any direction in relation to the axis of the muscle fibers via endosarcomeric and exosarcomeric protein lattices.

### 2.4.1 Types of Muscle Contractions

There are three primary types of muscle contractions. These contraction types are distinguished by how the muscle length changes during the contraction [10,11]. Isometric contractions are defined as muscle activity where tension is generated without a change in length. This is also referred to as a static contraction where muscle is generating tension but does not result in a change in length, and thus, there is no segmental (about a single joint) or whole body motion. Shortening contractions (often referred to as concentric contractions [10] are defined as the muscle generating tension while getting shorter. Concentric contractions usually generate segmental or whole body motion. Lengthening (or eccentric) muscle contractions are defined as the muscle generating tension while the muscle is lengthening. Lengthening contractions are usually used to absorb work or energy, thereby applying braking to segmental or whole body motion.

### 2.4.2 Concentric Muscle Contractions and the Force–Velocity Relationship

During concentric muscle actions, the shortening velocity varies as a function of applied load to the muscle, thus shortening velocity progressively decreases as applied load is increased. The resultant relationship is hyperbolic (Figure 2.9). The maximum shortening velocity

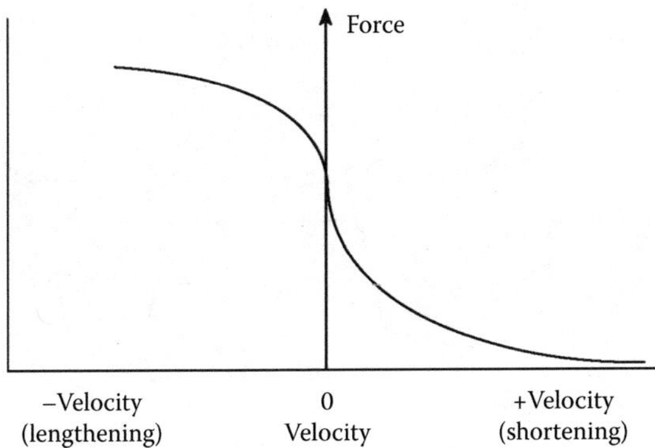


FIGURE 2.9 The force–velocity relationship of skeletal muscle.

occurs when there is no applied load, and the shortening velocity progressively approaches zero velocity at an applied load equivalent to the maximum isometric force of the muscle [12].

### 2.4.3 Eccentric Muscle Contractions

It is now well known that muscle can generate more tension during eccentric muscle actions than during concentric or isometric contractions. This was first reported in a study involving human muscles under volitional control [13]. It is also interesting that while muscles generate more tension during eccentric muscle actions than concentric muscle actions, EMG activity is less in muscles during stretch than during shortening at the same tension. During maximal effort, the EMG signature remains constant and force varies due to the length–tension relation of that specific muscle or muscle group; however, force during volitional eccentric activity never exceeded 140% of maximal shortening forces [14]. In animal studies that employ electrical stimulation to activate the muscle of interest, forces of 180% of maximum isometric force are typical [15]. High eccentric forces in humans with spastic paresis have been attained to levels similar as those seen in animal studies [16]. In addition, if muscles in humans are stimulated by external electrical stimulators, as in the case of spinal cord injured patients, the external forces generated during eccentric muscle actions are nearly 200% of the forces generated concentrically using the same electrical stimulation paradigm [17]. Thus, exogenous electrical stimulation overrides the inhibitory influences that moderate muscle output force. The force–velocity relationship during lengthening is near linear as the force does not increase as the lengthening velocity increases (Figure 2.9).

### 2.4.4 Stretch-Shortening Cycles

Stretch-shortening cycles (SSCs) are a type of muscle action that incorporates both concentric and eccentric muscle actions. Natural movement usually comprises SSCs. In most sports-related activities, it involves a prior stretch before shortening to enhance the shortening phase of the movement. Activities that typically use SSCs are jumping, walking, running, and movement in and around obstacles. In occupational-related activities, it is most related to reciprocal lifting and lowering activities and repetitive lift and carry tasks. It is an excellent model to study physiological muscle function [18]. It also allows for simultaneous study of concentric and eccentric muscle function and their synergism.

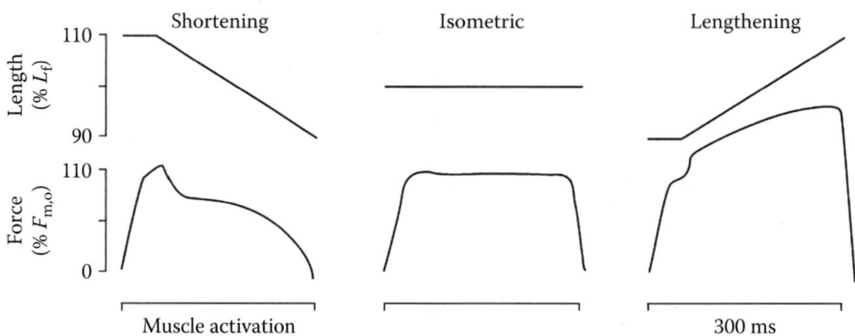


FIGURE 2.10 Length–tension relationship during shortening, isometric, and lengthening muscle actions.

### 2.4.5 Length–Tension Relationship

During isometric muscle contractions, it is well understood that force varies as a function of muscle length while the muscle is shortening or lengthening (Figure 2.10). Also, peak muscle tension reaches a plateau during maximal isometric contractions. It has been shown that muscle tension is lowest at very short and very long muscle lengths and develops higher tension in the intermediate lengths [19,20]. This is due to the degree of sarcomere overlap in the crossbridge. Thus, the length–tension curve has an ascending and descending limb as length increases (see Figure 2.8). The ascending limb, defined as the increase in force with increase in length, is due to more actin binding sites being available to bind with the myosin filaments. As tension plateaus, this is thought to be due to all the actin binding sites being bound to the myosin filaments. The descending limb, defined as the decrease in muscle tension with increasing length, is due to less actin binding sites being available as the actin filaments are pulled out of register with the myosin filaments. Thus, the length–tension relationship of muscle is due to myofilament overlap in the sarcomere (as shown in Figures 2.3 and 2.8).

## 2.5 MUSCULOTENDON ACTUATOR

Muscle and tendon have typically been studied in isolation although they function synergistically. Their integrative function has been defined as the musculotendon actuator [21]. While physiologists have long recognized that muscle and tendon act in a synergistic fashion, they have studied those tissues in isolation to better understand the function of muscle and tendon separately. The musculotendon actuators interact with body segments to produce movement and the dynamics of movement are dependent upon the contraction dynamics of the actuator. This system also functions as a feedback loop where the dynamics of body segments affect the force output of the actuator via the length and velocity of the actuator and the inherent compliance [21].

Tendon compliance affects the contraction dynamics of the muscle. In actuators with highly compliant tendons, a length change of the actuator would be mostly realized by the length change in the tendon, with very little concomitant length change in muscle. Compliance of an actuator is defined by the ratio of tendon slack length to muscle fiber length. Muscle length changes will be commensurate with length perturbations of the musculotendon actuator if the actuator is stiff. However in compliant actuators, muscle length changes will not follow exogenous length perturbations. This is quite relevant since changes in muscle length are rarely measured directly, indeed in human studies of muscle function the kinematics of the musculotendon are measured. The assertion that changes in musclotendon length are representative of muscle length changes may be incorrect, particularly for actuators with highly compliant tendons [21]. The active and passive force–length relation of muscles that have been published to-date may be erroneous due to the fact that muscle fibers are at different lengths in the active versus passive state even though the actuator is at the same length. This is due to differing amounts of tendon stretch which are caused by different forces exerted by the muscle fibers. Thus, the stretch of the tendon must be accounted for to accurately represent muscle stretch [22]. The muscle tendon interface (at the aponeurosis) exists in a state of dynamic equilibria, where force transients are equalized via stretch of the tendon and muscle activation and muscle length change. The dynamic equilibria also are governed by the response time of the tendon and muscle which are often different. Actuator compliance varies depending on the muscle group and animal species. In humans, actuator compliance appears to be highest on the plantar flexor group and lowest on proximal groups such as biceps and triceps [23–25]. In summary, one must think in terms of the musculotendon when investigating in vivo muscle function or reviewing scientific studies of in vivo function. While most studies refer specifically to muscle function, the measurements are typically made on the musculotendon group. Thus, one must be cognizant of the influence of tendon mechanics on muscle function and the musculotendon unit.

## 2.6 ENERGETICS OF MUSCLE CONTRACTIONS

Skeletal muscle is an electro-chemical-mechanical transducer. In other words, muscle activity is governed by electrical input from the somatic nervous system evoking chemical changes that provide energy for the contractile activity, and the subsequent mechanical output which results from the contractile activity. Muscle activity is fueled by several metabolic pathways; each pathway is most suited for a type of muscle contraction. Muscle fibers have these alternate pathways that generate adenosine triphosphate (ATP), the fuel for muscle contraction. Myosin ATPase splits ATP to provide energy for the power stroke of the crossbridge (Figure 2.11). To repeat the cycle, a fresh molecule of ATP binds to the myosin thick filament to allow detachment from the actin filament. This ATP is then split to recycle the bridge for the next stroke. The active resequestration of calcium back into the SR is also fueled by ATP. ATP is the only energy source used for muscle activity and must be supplied at a continuous rate for activity to be maintained. Limited supplies of ATP are located in the muscle, but three additional pathways provide supply of fresh ATP as needed during activity. The three pathways are (1) creatine phosphate pathway which donates high energy phosphates to ADP to form ATP, (2) oxidative phosphorylation (citric acid cycle and the electron transport chain), and (3) glycolysis. Creatine phosphate is the first storehouse tapped during muscle contractile activity. The reversible reaction is based on the donation of a high energy phosphate from creatine phosphate to ADP in the presence of the enzyme creatine kinase to form creatine and ATP. As energy reserves are built up in resting muscle, the high energy phosphate is transferred to creatine phosphate, thus

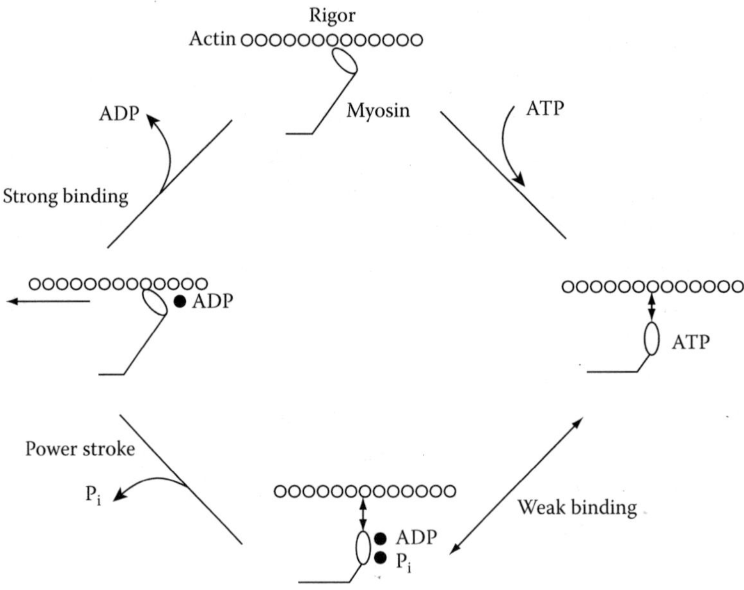


FIGURE 2.11 Crossbridge cycling and the role of ATP.

a rested muscle contains five times as much creatine phosphate as ATP. When the small pools of ATP are utilized at the onset of muscle contractile activity, the creatine phosphate pools can be rapidly converted to ATP to fuel the muscle activity. This energy pathway is mostly tapped during short, high intensity activity such as sprints, jumping, or lifting of objects for a short duration. Longer duration activities will use alternative energy pathways of oxidative phosphorylation and glycolysis to produce ATP. Oxidative phosphorylation provides a rich yield of 36 ATP molecules for each molecule of glucose. This pathway is slow, due to the number of intermediate steps. Sufficient oxygen is needed in the muscle mitochondria to fuel this pathway. Long-term, lower intensity activities such as repetitive motion tasks typical in manufacturing or manual materials handling, or walking would use this pathway provided that sufficient oxygen and nutrient supplies are available. This would be categorized as an aerobic pathway since it is done in the presence of oxygen. Oxygen is transported to the active muscles by the bloodstream. The flow of oxygen is potentiated by increased inotropic and chronotropic actions of the heart to pump more blood, vasodilation of the blood vessels to increase mass flow rate, and increased pulmonary activity to provide more oxygen to the bloodstream. Slow twitch muscle fibers have more myoglobin for storage of oxygen to assist in transfer of oxygen to active muscle fibers. Nutrient fuel can also be transported by the blood for use by active muscle or stored in glucose chains for later use.

Cardiovascular performance can limit the amount of oxygen that can be delivered to active muscle. During maximal activity, vasoconstriction occurs and the relatively slow oxidative phosphorylation pathway cannot keep up with the task demands for delivery of ATP to the muscle. When this occurs, muscles will rely on glycolysis to meet energy demands. Using this pathway, a glucose molecule is broken down into two ATP molecules and two pyruvic acid molecules. In the presence of oxygen, the pyruvic acid molecules can then enter the oxidative phosphorylation cycle to yield more nutrient fuel. However, the glycolytic pathway can proceed in the absence of oxygen to produce two ATP molecules for every glucose molecule. This is not as rich of ATP yield as produced by the oxidative phosphorylation pathway; however, this pathway is much faster and can proceed in the absence of oxygen, thus an anaerobic pathway. This pathway is best suited for short duration, intense physical activity. However, this pathway rapidly depletes the muscles' small supply of glycogen and in the absence of oxygen, the molecules of pyruvic acid are converted into lactic acid, thus promoting muscle soreness and metabolic acidosis leading to significant muscle fatigue. In most occupational tasks that require manual material handling, this pathway is probably not used since most tasks in the industrial setting are of lower intensity and longer duration.

## 2.7 MUSCLE FIBER TYPES

There are three types of skeletal muscle fibers that differ by their contractile speed and the ATP enzymatic processes used. There are slow oxidative fibers (Type I), fast oxidative fibers (Type IIa), and fast glycolytic fibers (Type IIb) (Figure 2.12). The faster fibers (Type IIa and IIb) are also referred to as fast myosin-ATPase for their ability to process ATP more rapidly. This also means that the crossbridge cycles more rapidly, resulting in a faster contractile velocity. Type IIa and IIb fibers are also referred to as fast-twitch-fatigue resistant and fast-twitch fatigable, respectively. Type I muscle fibers are slow twitch fibers that have a greater capacity to form ATP, are richer in mitochondria that contain enzymes in the oxidative phosphorylation pathway, and are richer in myoglobin due to the fiber's dependency on oxygen for contractile activity. This gives Type I muscles a red color. Type I muscles are typically used to maintain posture, and for long duration, low-intensity tasks that are typically found in many work environments. Type I muscles are better equipped for oxidative phosphorylation, a slow ATP synthesizing pathway. The rich yield of ATP in this pathway does not readily deplete energy stores, and also does not accumulate lactic acid since this pathway always operates in the presence of oxygen; thus, they are more fatigue resistant. Fast fibers have far fewer mitochondria than their slower twitch counterparts and have more stored glycogen and the enzymes necessary to break down glycogen into ATP.

| | **Type I** | **Type IIA** | **Type IIB** |
|---|---|---|---|
| Other names | Red, slow twitch (ST)<br>Slow oxidative (SO) | White, fast twitch (FT)<br>Fast oxidative glycolytic (FOG) | Fast glycolytic (FG) |
| Speed of contraction | Slow | Fast | Fast |
| Strength of contraction | Low | High | High |
| Fatigability | Fatigue resistant | Fatigable | Most fatigable |
| Aerobic capacity | High | Medium | Low |
| Anaerobic capacity | Low | Medium | High |
| Motor unit size | Small | Larger | Largest |
| Capillary density | High | High | Low |

FIGURE 2.12 The characteristics of muscle fiber types. (Adapted from Simon S.R. (ed.), *Orthopaedic Basic Science*, American Academy of Orthopaedic Surgeons, Rosemont, IL, 1994. With permission.)

They also have much less myoglobin than the Type I fibers, gives them a white color. They also have less capillary density since they do not rely on oxygenation for contraction. They have a higher myosin-ATPase content, and are typically larger in diameter than their Type I counterparts and have more myofilaments of actin and myosin to produce greater tension. Thus, the Type II glycolytic fibers are most suited for tasks that require high power output for a short duration, due to their dependence on glycogen to fuel the activity. Type IIa fibers are hybrids between the Type I and Type IIb fibers. They have both the high ATPase capacity of the Type IIb fibers and the higher oxidative capacity of the Type I fibers. Their contraction velocity is higher than the slow twitch Type I fibers and can maintain contraction for longer than the Type IIb fibers. However, their contractile activity needs more ATP than the oxidative phosphorylation pathway can provide, they do rely somewhat on the glycolytic pathway to meet the contractile demands, thus they fatigue more rapidly than their Type I counterparts.

Whole muscles contain a mixture of fiber types, dependent upon the intended use of the muscle. Muscles that have a large percentage of fast twitch glycolytic fibers are used for short duration, higher intensity tasks, while those that have a larger percentage of slow twitch fibers are used more for maintenance of posture. The fiber type composition can vary from muscle to muscle and person to person. For example, elite marathon runners have a higher population of slow twitch fibers than elite strength athletes. Strength athletes normally have a high population of Type II fibers to generate power for short durations.

## 2.8 MOTOR CONTROL

Motor control is the system that controls and coordinates human movements. It involves a dynamic process that regulates the ability of the human body to move, adapt, learn, use tools, conduct occupational tasks, and carry out many more activities. Motor control functions involve continuously processing sensory information related to body movements (e.g., force magnitude, directions, etc.) and environment (e.g., floor slipperiness, lighting, etc.), and initiating the commands necessary for successful task completion.

Factors affecting motor control are complex. The study of motor control is generally considered multidimensional; as such, it necessarily includes many dimensions of control, theoretical perspectives, approaches, and disciplinary methods. Such complexity of approaches reflects the fact that human activities—from the simple to the most complex—rely on physiological functions that cross numerous boundaries, and that draw on complex control functions within the human brain, nervous system, and musculoskeletal system, and that rely on the body's ability to dynamically integrate sensory information and control functions. Traditionally, research on motor control has focused on the function of the nervous system and how it controls body movements. More current research clearly establishes that motor controls arise from the interaction between neural and biomechanical dynamics [26].

### 2.8.1 Structure of the Motor Control System

The part of nervous system pertaining to motor control includes the brain, spinal cord, motor nerves, sensory nerves, and proprioceptors. The brain and the spinal cord comprise the central nervous system, which initiates and controls human movement.

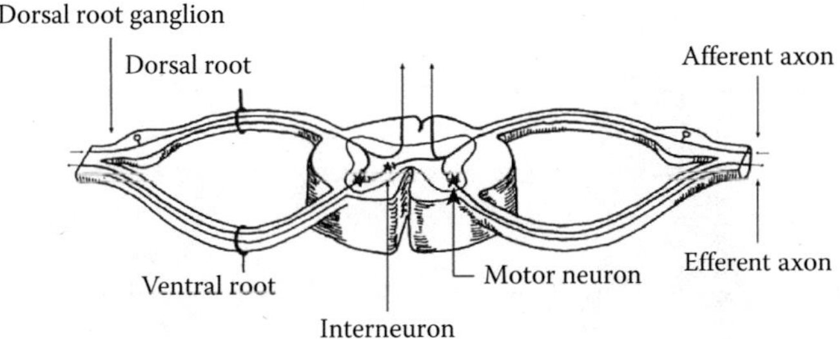


FIGURE 2.13 The afferent and efferent axons from the motor neuron and the ventral and dorsal root ganglia.

Human muscles are connected to the central nervous system by both motor and sensory nerves. The motor nerves exit the spinal cord on the ventral side or front side of the cord while the sensory nerves enter the spinal cord from the dorsal or back side (Figure 2.13). The motor nerves, also called efferent nerves, carry impulses outward from the central nervous system to the muscles. This pathway is termed the efferent or descending pathway. The sensory nerves, also known as the afferent nerves, transmit impulses inward from the muscles, joints, skin, and other sensory organs to the central nervous system. This pathway is termed afferent or ascending pathway.

Proprioceptors are sensory receptors located in the muscles, joints, and tendons that transform mechanical distortion of the muscle or joint into nerve impulses to provide the central nervous system with information on the changing conditions of the musculoskeletal system, such as muscle length and joint position. Proprioceptors include muscle spindles and Golgi tendon organs. The muscle spindles lie parallel with the muscle fibers in the belly of the muscle and are activated by the gamma motor neurons to stay in synchronization with the muscle fibers. They provide inputs about muscle length to the central nervous system. Golgi tendon organs monitor forces and tensions in the muscle; they lie at the muscle–tendon junctions (Figure 2.14). Another important proprioceptor is the semicircular canals of the inner ear that sense the position of the body. Collectively, these receptors provide continuous input to the nervous system to determine how many muscle fibers should be recruited to perform a task. In addition to proprioceptors, vision plays an important role in motor control. Research evidence suggests that vision is used to define the trajectory and kinematics of movements while proprioceptor is critical in the transmission of the plan into motor commands sent to the muscles [27,28]. The interactive process can explain how the human body maintains postural balance and how human gait is accomplished.

## 2.8.2 Control of Postural Balance

The maintenance of postural equilibrium is a complex process. In daily life or in an occupational setting, postural equilibrium is constantly challenged by perturbations arising from environmental changes, sudden movement of body segments, or task demands. These perturbations could be visual, vestibular, or mechanical changes resulting in conditions that may displace the body away from its equilibrium.

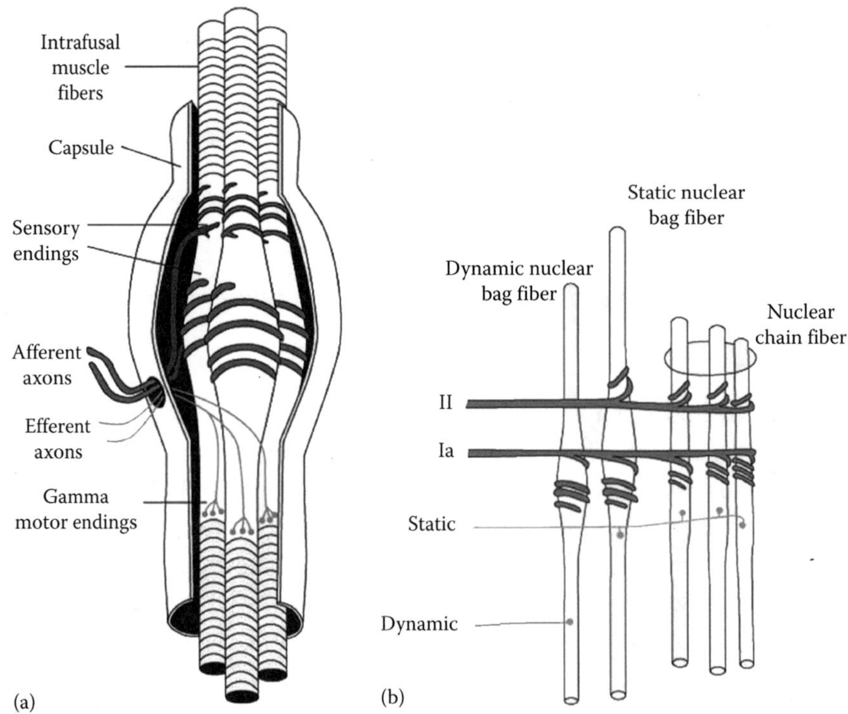


FIGURE 2.14 Intrafusal fibers complete with (a) gamma motor neurons and (b) sensory endings.

Biomechanically, the body's equilibrium is achieved when all forces acting on the body segments are balanced and the center of mass is well controlled within the base of support. The maintenance of postural balance is dependent on both sensory and motor processes. When the body's center of mass is shifted near the outer perimeter of the basal support due to perturbations, the instability is detected via afferent inputs from muscles, joints, vestibular system, and vision. Subsequently, motor processes coordinate the muscle actions into discrete synergies to minimize postural sway and keep the body's center of mass within the base of support [29,30]. If the static balance cannot be maintained, a rapid step or additional external support such as holding onto a guard rail is needed to reestablish the base of support.

Research on the human body's responses to postural perturbations provides important views of postural control mechanisms. In order to obtain a measure of the relative importance of the participating afferent inputs, experiments were designed to enhance, weaken, or exclude one or more afferent inputs [27,31–33]. For example, the role of the vestibular system becomes more important when the standing surface is uneven and the vision input is excluded. Despite the availability of multiple sensory inputs for healthy adults, the preferred sensory input for balance control is the proprioceptors at the feet [34]. Although the central nervous system generally relies on one sense at a time for balance, when one source is reduced, alternative sensory inputs are used for balance [35]. Postural balance is widely used as an indicator of susceptibility to loss of balance or fall. The control of postural stability plays an important role in fall prevention among elderly as well as among individuals performing workplace activities.

### 2.8.3 Control of Human Gait

Human gait is the most common human movements which occur subconsciously; however, the control mechanism behind such common movement is little understood by most people. Gait requires an integration of a complex neuromuscular-skeletal system as well as the coordination of muscles acting across many joints. This dynamic balance could be disturbed by traumatic injury, neurological damage, gradual degeneration, and even fatigue [36]. It can also be challenged by perturbations arising from environmental changes, that is, floor slipperiness and compliance, or job–task requirements, such as working at heights [37–40].

To achieve safe and efficient locomotion, previous studies indicated that major motor functional requirements needed to be met. Upright posture and total body balance must be maintained [41] and the upper body should be fully supported against the force of gravity during locomotion. Foot trajectory needs to be well controlled to achieve safe ground clearance [42]. In addition, sufficient mechanical energy needs to be generated by the body to maintain forward velocity during progression [43]. Moreover, the motor patterns at the hips, knees, and ankles have the major function of absorbing and generating energy [43] and the central nervous system must integrate and coordinate efferent commands with proprioceptive feedback and vestibular and visual inputs to generate the correct patterns of moment of force at each joint.

Previous studies on gait [38,44,45] have documented that people changed their gait as they approached and encountered slippery surfaces. Humans can adjust their gait to perceived changes indicated by sensory feedback, and can safely negotiate many different friction-surface levels. The typical protective gait strategy adopted in response to increased slipperiness includes shorter steps, and increased knee flexion to reduce vertical acceleration and forward velocity [46]. Gait changes are also observed in poorly lit environments. Subjects experiencing these conditions walked significantly slower, and exhibited decreased incoming velocity and heel contact angle [38]. It is the sudden and unanticipated changes in surface slipperiness or other environmental factors that cause most slips and falls.

### 2.8.4 Muscle Mechanics of Movement

The motor unit is the most fundamental element of the neuromuscular system; it consists of a group of muscle fibers that are innervated by the same motor neurons (Figure 2.15). When a single motor neuron is stimulated to exceed the threshold of contraction, all fibers of the motor unit will contract [47]. A large motor unit with a larger number of muscle fibers can exert greater force than a small motor unit. Other factors affecting the force output of a muscle are the muscle fiber length, muscle cross-section, and the rate at which the motor unit discharges the impulse. The force generated by a muscle is also determined by the number of motor units actively stimulated at the same time. Muscle recruitment is the mechanism of muscle force production and involves the order pattern in which pools of motor units are sequentially activated [48].

### 2.8.5 Motor Learning

There are two major types of motor abilities that are critical for our day-to-day activities—postural equilibrium control and voluntary motor control involving a particular movement

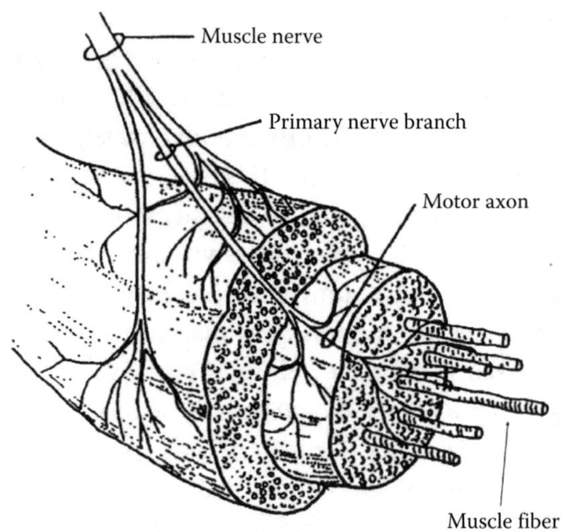


FIGURE 2.15 The muscle nerve, motor axons, and muscle fibers.

goal [49]. The latter is the basis for learning complex motor skills, such as sports routines and job tasks requiring specific motor skills. Motor learning is one of the most complex human behaviors, and enables the human body to adapt and learn in a continuously changing environment. The body relies on a series of sensory impulses followed by motor commands from the brain to accomplish a skilled task with efficiency and precision.

The initial stage of learning relies greatly on visual feedback and the requirement that the movements must be performed consciously. Eye–hand coordination is essential during this stage as many occupational tasks, such as reaching or lifting an object, are guided by visual input and feedback. During the second stage of learning, proprioceptive cues provide feedback to integrate the techniques [50]. When the third stage is reached, a stable movement pattern is formed. New pathways and junctions are established in the brain and the control of movements gradually becomes autonomous, unlike the conscious control needed in the early stage [51]. Muscle firing patterns used for the control of movement is sophisticated and is dependent on a number of variables. As one becomes more accomplished in a task, such as in martial arts training, ballet dancing, basketball, or manual materials handling, those firing patterns become more optimized for the task demands. However, muscle injury can occur when tasks demands exceed the tolerance of skeletal muscle, and those demands can comprise muscle forces, strain rates, repetitions, or work-rest cycles. While evidence indicates that more accomplished firing patterns can reduce muscle injury susceptibility, we must be cognizant of what factors contribute to muscle injury, and the mitigating factors as a foundation to design effective intervention strategies.

## 2.9 INJURY MECHANICS

We know that eccentric muscle actions are known to cause a greater amount of muscle damage than concentric or isometric contractions (Figure 2.16). This suggests that high load tensions in fibers may be more important than physiologic considerations in the etiology

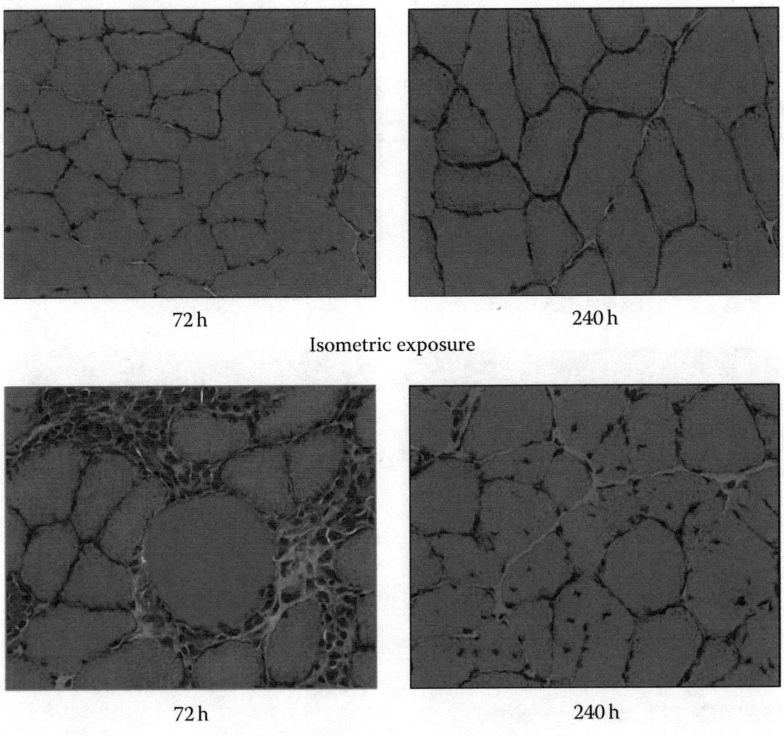


FIGURE 2.16 Histographs of tibialis anterior muscle sections exposed to isometric contractions and injurious SSCs 72 and 240 h after exposure.

of the injury process [52–54]. For decades, it has been hypothesized that high mechanical forces produced during eccentric muscle actions have been causal in the underlying etiology of muscle strain injuries [15,55]. This was thought to be due to high fiber stresses in the contractile apparatus due to high forces transmitted axially to the actin and myosin contractile proteins. Additionally, high mechanical forces produced during muscular contractions, particularly in eccentric exercise, where forces are distributed over relatively small cross sectional areas of muscles, cause disruption of contractile and intermediate filament proteins in skeletal muscle fibers and connective tissues [56,57]. A single exposure to damaging eccentric muscle actions results in loss of performance immediately after exposure and can last for up to 30 days [58]. Past investigations of eccentric contraction-induced muscle injury have indicated that mechanical factors such as peak force and average force [59], work during stretch [60], fiber length [59,60], strain [61], range of motion [62], number of repetitions [63] (Figure 2.17), and work-rest cycles [64] influence the amount of muscle damage. Change in maximum isometric force after injurious exposure has been shown to be the best indicator of the degree of muscle damage [65]. Interestingly, exposure to concentric (shortening) or isometric muscle actions does not normally produce muscle injury [15,66–68]. Recently, the ability to rapidly quantify both skeletal muscle degeneration and inflammation following an injurious exposure in the same tissue using a novel stereological technique has been shown to reveal further insight into the injury and repair process of skeletal muscle [63,67,69,70].

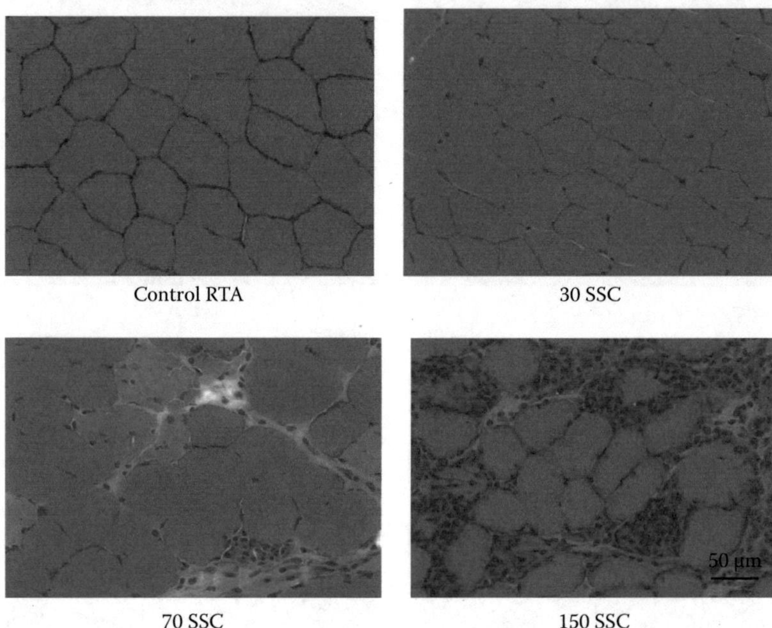


FIGURE 2.17 Histographs of tibialis anterior muscle sections exposed to either none, 30 SSCs, 70 SSCs, or 150 SSCs 72 h after exposure.

Fortunately, our understanding of the cellular and molecular mechanisms involved in mediating muscle damage and repair after strain, or low force, repetitive motion induced injuries is rapidly increasing [2,71,72]. Understanding how muscle tissue responds to these various stressors, and the time course of those responses, is necessary for defining approaches that can be taken to reduce the chances of obtaining an injury, or approaches to enhance the rate of recovery once an injury occurs. The cellular changes that occur as a result of acute muscle strain and repetitive motion are different. Strain injuries are associated with structural damage to myofibers, blood vessels and nerves, and cause edema and inflammation [73,74]. Repetitive motion damage can be associated with muscle ischemia, pain and mitochondrial dysfunction [75,76]. Details regarding the cellular responses to repetitive use and strain injuries are discussed hereafter.

### 2.9.1 Strain Injuries and Skeletal Muscle

Strain injuries are the result of three basic processes: (1) Initially, excessive forces coupled with lengthening result in structural damage to muscle cells (myofibers), including tearing of the cell membrane (i.e., the sarcolemma). (2) Structural damage to the sarcolemma results in an increase in intracellular calcium levels, modifications of myofiber proteins and lipids and the activation of intracellular pathways that regulate the injured muscle's response to damage. (3) Pro- and anti-inflammatory factors (i.e., cytokines and chemokines) are released by local tissues and infiltrating immune cells. Cytokines and chemokines cause inflammation and stimulate cellular pathways mediating muscle regeneration and repair. These three processes are involved in mediating the extent of physical damage, functional changes of the muscle, pain, and repair.

### 2.9.2 Muscle Injuries Associated with Repetitive Tasks

Repetitive tasks involve movements that typically require little force generation by a muscle. Instead, the muscle action may need to be maintained over long periods of time, or repeated over and over again during a work cycle [76,77]. Injuries caused by these types of activities are not usually associated with inflammation or large areas of myofiber degeneration, but instead are characterized by muscle pain and/or rapid fatigability of the muscle [78–82]. Repetitive task injuries are prevalent in people working with computers and other less physically demanding tasks. With the decrease in jobs that require high levels of physical exertion, understanding the mechanisms underlying this type of damage is crucial.

Injury due to overuse has been studied in people with trapezius myalgia. This disorder is often seen in workers whose job requires them to maintain stable upper body postures for extended periods of time such as computer and clerical work [78,80]. Trapezius mylagia is associated with the appearance of ragged Type I muscle fibers and with a decrease in muscle blood flow to the injured region. This ragged appearance of myofibers is an indicator of mitochondrial dysfunction in the cell [75,82,83]. The dysfunction and pain associated with trapezius myalgia and with other overuse injuries have been linked to changes in calcium regulation in the damaged area, changes in the pH of the intra extracellular fluid, and changes in the local concentrations of specific ions involved in mediating muscle activity [79,84,85]. These biochemical alterations may have profound effects on myofiber metabolism and on the activity of sensory pathways carrying pain information (reviewed in [76]). Early onset overuse related MSDs have been shown to have an inflammatory component, with increased levels of CRP, TNF-$\alpha$, IL-1$\beta$, and IL-6 in serum [86]. Mechanical tissue injury that is localized from repetitive tasks can result in both local and systemic inflammation followed by structural changes in the tissue [2]. Barbe and Barr developed an animal model of repetitive upper limb tasks to model repetitive use injuries [87]. Based on their results, they argue that inflammation reduces soft tissue tolerance to continued repetitive loading, and is related to pain and movement dysfunction [88]. Also, they established a causal relationship between repetitive tasks and the development of carpal tunnel syndrome with all the associated physiological changes such as reduced conduction velocity, motor function deficits, and sensory changes [89]. As the upper limb task demands increase, serum and tissue levels of cytokines and chemokines increase, along with deleterious changes in grip strength [90]. This is an important finding because it demonstrates a causal link between the task demand, changes in inflammatory mediator levels in the tissue and blood, and motor performance and strength. High force, moderate repetitive tasks of the upper extremities also change central nervous system sensitization by increasing levels of substance P and neurokinin at the dorsal horn of the spinal cord that resulted in reduced motor function, neuropathic pain, mechanical allodynia, and widespread inflammation [91].

### 2.9.3 Physical Damage to Tissue

Strain injuries are caused by exposure of muscle and other soft tissues, including vasculature, nerves and tendons to excessive strain or lengthening [92–94] Studies performed in both humans [73,95–98], and animals [55,99,100] have demonstrated that exposure to lengthening contractions results in physical damage to muscle tissue that can include shearing of myofibers, the loss of intermediate filaments and mitochondria, damage to the extracellular membrane, and disruptions in Z-line alignment (i.e., Z-line streaming). Immunostaining for structural proteins that maintain the integrity of the myofiber, such as desmin, titin, and fibronectin, have demonstrated that there are disruptions of the exo- and endosarcomeric membranes [66,94,101], and of the extracellular matrix [100,101] in strain-injured muscle tissue. In lengthening contraction-induced injuries, damage within the muscle is most often seen at the myotendinous junction and at specific sarcomeres [102–105]. In fact, it has been hypothesized that there is a population of sarcomeres that are weaker, and tear more easily under lengthening conditions [94,106,107]. The stretch-induced damage to the extracellular matrix, sarcomeres, and critical cell organelles are associated with reduced isometric muscle force [62,65,108,109], but not with pain. It has been hypothesized that these initial structural changes initiate a chain of events that may maintain an injury-induced force deficit, result in inflammation and pain, and stimulate pathways important for regulating muscle repair and regeneration.

### 2.9.4 Muscle Inflammation

The physical disruptions of muscle fibers along with increases in intracellular calcium are the initial effects of muscle injury. However, force deficits, muscle swelling, and soreness occur 1–7 days after the initial injury, and are associated with muscle inflammation [55,63,93,101,110]. Neutrophils, the first immune cells to enter damaged tissue, actually infiltrate muscle within 2h of the initial injury [111,112]. We are only beginning to understand the complex roles that various immune cell types play in the damage and repair processes. However, it has been demonstrated that neutrophils phagocytize degenerating fibers and debris produced by injury. In addition, neutrophils can participate in the production and release of free radicals from damaged tissue, which can exacerbate the damage [113]. Neutrophils can also produce proteases and a number of cytokines, including tumor necrosis factor-α (TNF-α; [114]) and IL1-β [115]. These cytokines might increase muscle catabolism and degradation and act to attract monocytes to the site of the injury [116–118].

Monocytes/macrophages, the other inflammatory cells commonly seen in injured muscle, can be found between 12h and 14 days after the initial muscle injury [119,120]. In rats, macrophages expressing specific cell-surface molecules, including ED1 and ED2, have been identified in damaged muscle tissues [120–122]. ED1 expressing macrophages infiltrate damaged and necrotic tissue and remove debris. These macrophages also express pro-inflammatory cytokines including TNF-α [123,124]. Aside from increasing muscle catabolism and promoting protease activity, TNF-α also activates the transcription factor, nuclear factor κB (NF-κB), to stimulate transcription of genes encoding for proteins that are part of the ubiquitin proteolytic pathway [125,126]. TNF-α may also stimulate the transcription of other pro-inflammatory cytokines including IL-1β and IL-6 and the chemokine monocyte chemoattractant protein (MCP-1; [127,128]). The increased production of cytokines and chemokines by inflammatory cells in damaged tissue enhance local pathways mediating tissue inflammation and may act to exacerbate damage during the first 5 days after muscle injury.

### 2.9.5 Muscle Regeneration and Repair

The infiltration of immune cells along with the release of pro-inflammatory cytokines appears to enhance muscle damage. During the acute phase of the injury, both pain and force deficits appear to be reduced by treating animals with nonsteroidal anti-inflammatory drugs (NSAIDs). For example, rabbits which were exposed to a session of repeated eccentric contractions and treated with the NSAID flurbiprofen, showed improved functional recovery during the first week after injury as compared to controls [129]. However, 4 weeks after the injury, the NSAID treated animals demonstrated reduced force generation. The authors hypothesized that treatment with NSAIDs may have interfered with or delayed the recovery process in these animals [129]. This finding was supported in more recent work by LaPointe and colleagues [121,122]. In humans with muscle damage caused by downhill walking, treatment with over the counter doses of acetaminophen or ibuprofen results in a decrease in pain, but these anti-inflammatory agents also decrease protein synthesis which may be needed for muscle repair [130–132]. These findings suggest that inhibition of the inflammatory response interferes with normal recovery of muscle after a strain-induced injury.

Inflammatory cells, particularly macrophages, may stimulate myofiber regeneration through a number of different mechanisms. ED2 expressing macrophages, also referred to as resident macrophages, are thought to play a role in muscle repair and myofiber regeneration [120–122,133]. These macrophages may stimulate growth and repair by releasing a number of factors that could stimulate the division, migration, and differentiation of muscle precursor cells. These factors include, fibroblast growth factor-2 (FGF-2; [134,135]), insulin-like growth factor-1 (IGF-1; [136–138] and hepatocyte growth factor (HGF; [134,139]). In addition, ED2 expressing macrophages may also release the anti-inflammatory cytokines IL-6 [140,141] and IL-15 [142–144]. Although inflammation causes pain and appears to exacerbate myofiber damage, the inflammatory process also appears to be necessary for complete repair of tissue and recovery of muscle function [145].

## 2.10 FUTURE DIRECTIONS

In this chapter, basic aspects of skeletal muscle physiology, motor control and learning, and more advanced concepts of muscle injury were discussed. The control mechanism for postural balance and human gait was also reviewed. There are many directions in which research on the mechanisms of motor control described in the chapter could be extended. Besides locomotion, other motor behaviors that are commonly seen in occupational settings could be further researched. For example, the control of reaching movement using different arm postures, body movement strategies during obstacle crossing, postural control strategies after slips and falls, and so forth. The contribution of these postures and strategies to common injury risks for overexertion, traumatic injuries, and musculoskeletal disorders, for example, is intuitively apparent, but not substantiated by sufficient research efforts. Further in-depth understanding in these areas will provide insights to improve job performance and to prevent such common occupational injuries as slips- and falls-related trauma. In addition, skeletal muscle injury mechanics is important to the field of ergonomics and the mitigation of occupational musculoskeletal disorders [146]. Recent work in muscle pathomechanics have allowed better understanding of the response of tissues to repeated mechanical exposures [70]. Efforts to incorporate epidemiological, biomechanical loading, soft tissue tolerance, and psychosocial findings into a systems-engineering approach should be pursued to establish a better understanding of the pathways of musculoskeletal injury and the resultant preventive strategies. Although we are beginning to understand how the major risk factors influence the load-tolerance relationship of human tissue, the mechanism of interaction of these risk factors in producing tissue damage is virtually unexplored [147]. Further work is needed to clarify the response of tissue to acute and repetitive loading and the pain pathways associated with these physiological responses. Much of our knowledge about tissue tolerance, especially that related to repetitive loading, has been gained from cadaver studies. Such data do not reflect the biological system's ability to physiologically adapt to exposures that occur during actual working conditions. Future research efforts should examine the in vivo tissue tolerance of healthy populations as well as compromised populations that are reflective of workplace populations [147]. Additionally, research involving the risk of further injury associated with return-to-work is sparse. Integration of epidemiological data, biomechanical exposures, soft tissue pathomechanics, and psychosocial data are needed to determine how the risk of injury is affected when an individual is working while recovering from a musculoskeletal disorder [147]. The findings from research on the pathomechanics of skeletal muscle injury are beneficial for incorporation into modern biomechanical and psychosocial models. In addition, biomarkers that are identified from pathomechanics studies could be beneficial for use in field studies to help determine jobs and populations at risk.

## REFERENCES


1. United States Department of Labour (USDL), Work-related musculoskeletal disorders, *B.o.L. Statistics* (ed.), Table 10a, 2007.
2. M.F. Barbe and A.E. Barr, Inflammation and the pathophysiology of work-related musculoskeletal disorders, *Brain Behav Immun*, 20: 423–429 (2006).
3. B.P. Bernard, V. Putz-Anderson, S.E. Burt, L.L. Cole, C. Fairfield-Estill, L.J. Fine, K.A. Grant, C. Gjessing, L. Jenkins, J.J. Hurrell, N. Nelson, D. Pfirman, R.L. Roberts, D. Stetson, M. Haring-Sweeney, and S. Tanaka, *Musculoskeletal Disorders and Workplace Factors: A Critical Review of Epidemiologic Evidence for Work-Related Musculoskeletal Disorders of the Neck, Upper Extremity, and Low Back*, US Department of Health and Human Services, National Institute for Occupational Safety and Health (DHHS–NIOSH), Cincinnati, OH, pp. 400, 1997.
4. L. Sherwood, *Human Physiology: From Cells to Systems*, Brooks/Cole, Pacific Grove, CA, 2001.
5. J. Wright, Q.Q. Huang, and K. Wang, Nebulin is a full-length template of actin filaments in the skeletal muscle sarcomere: An immunoelectron microscopic study of its orientation and span with site-specific monoclonal antibodies, *J Muscle Res Cell Motil*, 14: 476–483 (1993).
6. M.J. Chen, C.L. Shih, and K. Wang, Nebulin as an actin zipper. A two-module nebulin fragment promotes actin nucleation and stabilizes actin filaments, *J Biol Chem*, 268: 20327–20334 (1993).
7. E. Lazarides, Intermediate filaments as mechanical integrators of cellular space, *Nature*, 283: 249–256 (1980).
8. T.J. Patel and R.L. Lieber, Force transmission in skeletal muscle: From actomyosin to external tendons, *Exerc Sport Sci Rev*, 25: 321–363 (1997).

9. J.V. Pardo, J.D. Siliciano, and S.W. Craig, A vinculin-containing cortical lattice in skeletal muscle: Transverse lattice elements ("costameres") mark sites of attachment between myofibrils and sarcolemma, *Proc Natl Acad Sci USA*, 80: 1008–1012 (1983).
10. E. Asmussen, Positive and Negative Muscular Work, *ACTA Physiol Scand*, 28: 365–382 (1953).
11. P. Cavanagh, On "muscle action" vs "muscle contraction", *J Biomech*, 21: 69 (1988).
12. D.R. Wilkie, The relation between force and velocity in human muscle, *J Physiol*, 110: 249–280 (1950).
13. M. Singh and P.V. Karpovich, Isotonic and isometric forces of forearm flexors and extensors, *J Appl Physiol*, 21: 1435–1437 (1966).
14. P.V. Komi, Measurement of the force-velocity relationship in human muscle under concentric and eccentric contractions, *Medicine and Sport*, Vol. 8, pp. 224–229 (1973).
15. G.L. Warren, D.A. Hayes, D.A. Lowe, and R.B. Armstrong, Mechanical factors in the initiation of eccentric contraction-induced injury in rat soleus muscle, *J Physiol*, 464: 457–475 (1993).
16. E. Knutsson, Analysis of spastic paresis., *Proceedings of the 10th International Congress of the World Confederation for Physical Therapy*, Sydney, NSW, Australia, pp. 629–633, 1987.
17. R. Triolo, D. Robinson, E. Gardner, and R. Betz, The eccentric strength of electrically stimulated paralyzed muscle, *IEEE T Biomed Eng*, 651–652 (1987).
18. P.V. Komi, Stretch-shortening cycle: A powerful model to study normal and fatigued muscle, *J Biomech*, 33: 1197–1206 (2000).
19. A.M. Gordon, A.F. Huxley, and F.J. Julian, The variation in isometric tension with sarcomere length in vertebrate muscle fibres, *J Physiol*, 184: 170–192 (1966).
20. K.A. Edman, The velocity of unloaded shortening and its relation to sarcomere length and isometric force in vertebrate muscle fibres, *J Physiol*, 291: 143–159 (1979).
21. F.E. Zajac, Muscle and tendon: Properties, models, scaling, and application to biomechanics and motor control, *Crit Rev Biomed Eng*, 17: 359–411 (1989).
22. A.V. Hill, The heat of shortening and the dynamic constants of muscle, *Proc R Soc London Ser B*, 126: 136–195 (1938).
23. M.F. Bobbert, P.A. Huijing, and G.J. van Ingen Schenau, A model of the human triceps surae muscle-tendon complex applied to jumping, *J Biomech*, 19: 887–898 (1986).
24. T.L. Wickiewicz, R.R. Roy, P.L. Powell, and V.R. Edgerton, Muscle architecture of the human lower limb, *Clin Orthop*, 275–283 (1983).
25. P.A. Huijing, Architecture of the human gastrocnemius muscle and some functional consequences, *Acta Anat* (*Basel*), 123: 101–107 (1985).
26. R.D. Beer, Beyond control: The dynamics of brain-body-environment interaction in motor systems, *Adv Exp Med Biol*, 629: 7–24 (2009).
27. R.L. Sainburg, M.F. Ghilardi, H. Poizner, and C. Ghez, Control of limb dynamics in normal subjects and patients without proprioception, *J Neurophysiol*, 73: 820–835 (1995).
28. R.L. Sainburg, H. Poizner, and C. Ghez, Loss of proprioception produces deficits in interjoint coordination, *J Neurophysiol*, 70: 2136–2147 (1993).
29. L.M. Nasher, Adapting reflexes controlling the human posture, *Exp Brain Res*, 26: 59–72 (1976).
30. F.B. Horak, and L.M. Nashner, Central programming of postural movements: Adaptation to altered support-surface configurations, *J Neurophysiol*, 55: 1369–1381 (1986).
31. T. Sahlstrand, R. Ortengren, and A. Nachemson, Postural equilibrium in adolescent idiopathic scoliosis, *Acta Orthop Scand*, 49: 354–365 (1978).
32. J.P. Scholz, G. Schoner, W.L. Hsu, J.J. Jeka, F. Horak, and V. Martin, Motor equivalent control of the center of mass in response to support surface perturbation, *Exp Brain Res*, 180: 163–179 (1997).
33. A. Bhattacharya, R. Morgan, R. Shukla, H.K. Ramakrishanan, and L. Wang, Non-invasive estimation of afferent inputs for postural stability under low levels of alcohol, *Ann Biomed Eng*, 15: 533–550 (1987).

34. A. Shumway-Cook and F.B. Horak, Assessing the influence of sensory interaction of balance. Suggestion from the field, *Phys Ther*, 66: 1548–1550 (1986).
35. L.M. Nasher, Adaptation of human movement to altered environments, *Trends in Neuroscience*, 5: 351–361 (1982).
36. D.A. Winter, *Biomechanics and Motor Control of Human Movement*, University of Waterloo Press, Waterloo, ON, Canada, 1991.
37. T. Bhatt, J.D. Wening, and Y.C. Pai, Adaptive control of gait stability in reducing slip-related backward loss of balance, *Exp Brain Res*, 170: 61–73 (2006).
38. S. Chiou, A. Bhattacharya, and P.A. Succop, Evaluation of workers' perceived sense of slip and effect of prior knowledge of slipperiness during task performance on slippery surfaces, *AIHAJ*, 61: 492–500 (2000).
39. M.S. Redfern, R. Cham, K. Gielo-Perczak, R. Gronqvist, M. Hirvonen, H. Lanshammar, M. Marpet, C.Y. Pai, and C. Powers, Biomechanics of slips, *Ergonomics*, 44: 1138–1166 (2001).
40. B. Wang, A. Bhattacharya, A. Bagchee, and W. Wang, Kinematic methods for quantitating loss of balance while negotiating a curved path on a slippery surface, *J Test Eval*, 25(1): 135–142 (1997).
41. L.M. Nasher, Balance adjustments of human perturbed while walking, *J Neurophysiol*, 44: 650–664 (1980).
42. D.A. Winter, Foot trajectory in human gait: A precise and multifactorial motor control task, *Phys Ther*, 72: 45–53 (1992); discussion 54–56.
43. D.A. Winter, Energy generation and absorption at the ankle and knee during fast, natural, and slow cadences, *Clin Orthop Relat Res*, 147–154 (1983).
44. R. Cham and M.S. Redfern, Heel contact dynamics during slip events on level and inclined surfaces, *Safety Sci*, 40(7–8): 559–576 (2002).
45. P.F. Tang, M.H. Woollacott, and R.K. Chong, Control of reactive balance adjustments in perturbed human walking: Roles of proximal and distal postural muscle activity, *Exp Brain Res*, 119: 141–152 (1998).
46. M.G.A. Llewellyn and V.R. Nevola, Strategies for walking on low-friction surfaces, in: W.A.L.a.G. Havenith, (ed.), *Proceedings of the Fifth International Conference on Environmental Ergonomics*, Maastricht, the Netherlands, pp. 156–157,1992.
47. R.M. Enoka, Central modulation of motor unit activity, *Med Sci Sports Exerc*, 37: 2111–2112 (2005).
48. R.E. Burke, The control of muscle force: Motor unit recruitment and firing patterns, in: *Human Muscle Power*, N.L. Jones, N. McCartney, and A.J. McComas (eds.), Human Kinetics, Champaign, IL, pp. 97–109, 1986.
49. J. Massion, A. Alexandrov, and A. Frolov, Why and how are posture and movement coordinated? *Prog Brain Res*, 143: 13–27 (2004).
50. R. Barlett, *Sport Biomechanics: Reducing Injury and Improving Performance*, Taylor & Francis, New York, pp. 245–265, 2007.
51. K.H.E. Kroemer and E. Grandjean, *Fitting the Task to the Human*, Taylor & Francis Group, New York, pp. 147–155, 2005.
52. W.T. Stauber, Eccentric action of muscles: Physiology, injury, and adaptation, *Exerc Sport Sci Rev*, 17: 157–185 (1989).
53. W.T. Stauber, Measurement of muscle function in man, sports injuries, in: *International Perspectives in Physical Therapy*, Churchill-Livingstone, Philadelphia, PA, 1989.
54. R.B. Armstrong, Muscle damage and endurance events, *Sports Med*, 3: 370–381 (1986).
55. R.B. Armstrong, R.W. Ogilvie, and J.A. Schwane, Eccentric exercise-induced injury to rat skeletal muscle, *J Appl Physiol*, 54: 80–93 (1983).
56. R.B. Armstrong, Mechanisms of exercise-induced delayed onset muscular soreness: A brief review, *Med Sci Sports Exerc*, 16: 529–538 (1984).

57. R.B. Armstrong, G.L. Warren, and J.A. Warren, Mechanisms of exercise-induced muscle fibre injury, *Sports Med*, 12: 184–207 (1991).
58. G.L. Warren, C.P. Ingalls, S.J. Shah, and R.B. Armstrong, Uncoupling of in vivo torque production from EMG in mouse muscles injured by eccentric contractions, *J Physiol*, 515(Pt 2): 609–19 (1999).
59. L.E. Gosselin and H. Burton, Impact of initial muscle length on force deficit following lengthening contractions in mammalian skeletal muscle, *Muscle Nerve*, 25: 822–827 (2002).
60. K.D. Hunter and J.A. Faulkner, Pliometric contraction-induced injury of mouse skeletal muscle: Effect of initial length, *J Appl Physiol*, 82: 278–283 (1997).
61. R.L. Lieber and J. Friden, Muscle damage is not a function of muscle force but active muscle strain, *J Appl Physiol*, 74: 520–526 (1993).
62. R.G. Cutlip, K.B. Geronilla, B.A. Baker, M.L. Kashon, G.R. Miller, and A.W. Schopper, Impact of muscle length during stretch-shortening contractions on real-time and temporal muscle performance measures in rats in vivo, *J Appl Physiol*, 96: 507–516 (2004).
63. B.A. Baker, R.R. Mercer, K.B. Geronilla, M.L. Kashon, G.R. Miller, and R.G. Cutlip, Impact of repetition number on muscle performance and histological response, *Med Sci Sports Exerc*, 39: 1275–1281 (2007).
64. R.G. Cutlip, K.B. Geronilla, B.A. Baker, R.D. Chetlin, I. Hover, M.L. Kashon, and J.Z. Wu, Impact of stretch-shortening cycle rest interval on in vivo muscle performance, *Med Sci Sports Exerc*, 37: 1345–1355 (2005).
65. G.L. Warren, D.A. Lowe, and R.B. Armstrong, Measurement tools used in the study of eccentric contraction-induced injury, *Sports Med*, 27: 43–59 (1999).
66. R.L. Lieber, L.E. Thornell, and J. Friden, Muscle cytoskeletal disruption occurs within the first 15 min of cyclic eccentric contraction, *J Appl Physiol*, 80: 278–284 (1996).
67. B.A. Baker, K.M. Rao, R.R. Mercer, K.B. Geronilla, M.L. Kashon, G.R. Miller, and R.G. Cutlip, Quantitative histology and MGF gene expression in rats following SSC exercise in vivo, *Med Sci Sports Exerc*, 38: 463–471 (2006).
68. J.A. Faulkner, S.V. Brooks, and E. Zerba, Muscle atrophy and weakness with aging: Contraction-induced injury as an underlying mechanism, *J Gerontol A Biol Sci Med Sci*, 50 Spec No: 124–129 (1995).
69. B.A. Baker, R.R. Mercer, K.B. Geronilla, M.L. Kashon, G.R. Miller, and R.G. Cutlip, Stereological analysis of muscle morphology following exposure to repetitive stretch-shortening cycles in a rat model, *Appl Physiol Nutr Metab*, 31: 167–179 (2006).
70. R.G. Cutlip, B.A. Baker, K.B. Geronilla, R.R. Mercer, M.L. Kashon, G.R. Miller, Z.S. Murlasits, and S.E. Alway, Chronic exposure to stretch-shortening contractions results in skeletal muscle adaptation in young rats and maladaptation in old rats, *Appl Physiol Nutr Metab*, 31: 573–587 (2006).
71. P.M. Clarkson and M.J. Hubal, Exercise-induced muscle damage in humans, *Am J Phys Med Rehabil*, 81: S52–S69 (2002).
72. A.E. Barr and M.F. Barbe, Pathophysiological tissue changes associated with repetitive movement: A review of the evidence, *Phys Ther*, 82: 173–187 (2002).
73. J. Friden, M. Sjostrom, and B. Ekblom, A morphological study of delayed muscle soreness, *Experientia*, 37: 506–507 (1981).
74. W.T. Stauber, V.K. Fritz, D.W. Vogelbach, and B. Dahlmann, Characterization of muscles injured by forced lengthening. I. Cellular infiltrates, *Med Sci Sports Exerc*, 20: 345–353 (1988).
75. F. Kadi, G. Hagg, R. Hakansson, S. Holmner, G.S. Butler-Browne, and L.-E. Thornell, Structural changes in male trapezius muscle with work-related myalgia, *Acta Neurophathol*, 95: 352–360 (1998).
76. G. Sjogaard and K. Sogaard, Muscle injury in repetitive motion disorders, *Clin Orthop*, 21–31 (1998).

77. G. Sjogaard, U. Lundberg, and R. Kadefors, The role of muscle activity and mental load in the development of pain and degenerative processes at the muscle cell level during computer work, *Eur J Appl Physiol*, 83: 99–105 (2000).
78. K. Keller, J. Corbett, and D. Nichols, Repetitive strain injury in computer keyboard users: Pathomechanics and treatment principles in individual and group intervention, *J Hand Ther*, 11: 9–26 (1998).
79. G. Sjogaard, Muscle energy metabolism and electrolyte shifts during low-level prolonged static contraction in man, *Acta Physiol Scand*, 134: 181–187 (1988).
80. B.R. Jensen, B. Schibye, K. Sogaard, E.B. Simonsen, and G. Sjogaard, Shoulder muscle load and muscle fatigue among industrial sewing-machine operators, *Eur J Appl Physiol Occup Physiol*, 67: 467–475 (1993).
81. G. Sjogaard, and B.R. Jensen, Muscle Pathology with Overuse, in: D. Ranney, (ed.), *Chronic Musculoskeletal Injuries in the Workplace*, W.B. Saunders Company, Philadelphia, PA, pp. 17–40, 1997.
82. S.E. Larsson, L. Bodegard, K.G. Henriksson, and P.A. Oberg, Chronic trapezius myalgia. Morphology and blood flow studied in 17 patients, *Acta Orthop Scand*, 61: 394–398 (1990).
83. B. Larsson, J. Bjork, K.G. Henriksson, B. Gerdle, and R. Lindman, The prevalences of cytochrome c oxidase negative and superpositive fibres and ragged-red fibres in the trapezius muscle of female cleaners with and without myalgia and of female healthy controls, *Pain*, 84: 379–387 (2000).
84. N. Ortenblad, G. Sjogaard, and K. Madsen, Impaired sarcoplasmic reticulum Ca(2+) release rate after fatiguing stimulation in rat skeletal muscle, *J Appl Physiol*, 89: 210–217 (2000).
85. B.R. Jensen, G. Sjogaard, S. Bornmyr, M. Arborelius, and K. Jorgensen, Intramuscular laser-Doppler flowmetry in the supraspinatus muscle during isometric contractions, *Eur J Appl Physiol Occup Physiol*, 71: 373–378 (1995).
86. S.J. Carp, M.F. Barbe, K.A. Winter, M. Amin, and A.E. Barr, Inflammatory biomarkers increase with severity of upper-extremity overuse disorders, *Clin Sci* (*Lond*), 112: 305–314 (2007).
87. A.E. Barr, F.F. Safadi, R.P. Garvin, S.N. Popoff, and M.F. Barbe, Evidence of progressive tissue pathophysiology and motor behavior degradation in a rat model of work related musculoskeletal disease, *Proceedings of the International Ergonomics Association/Human Factors and Ergonomics Society 2000 Congress*, San Diego, CA, pp. 5/584–5/587, 2000.
88. A.E. Barr, and M.F. Barbe, Inflammation reduces physiological tissue tolerance in the development of work-related musculoskeletal disorders, *J Electromyogr Kinesiol*, 14: 77–85 (2004).
89. B.D. Clark, T.A. Al-Shatti, A.E. Barr, M. Amin, and M.F. Barbe, Performance of a high-repetition, high-force task induces carpal tunnel syndrome in rats, *J Orthop Sports Phys Ther*, 34: 244–253 (2004).
90. M.F. Barbe, M.B. Elliott, S.M. Abdelmagid, M. Amin, S.N. Popoff, F.F. Safadi, and A.E. Barr, Serum and tissue cytokines and chemokines increase with repetitive upper extremity tasks, *J Orthop Res*, 26: 1320–1326 (2008).
91. M.B. Elliott, A.E. Barr, B.D. Clark, M. Amin, S. Amin, and M.F. Barbe, High force reaching task induces widespread inflammation, increased spinal cord neurochemicals and neuropathic pain, *Neuroscience*, 158: 922–931 (2009).
92. R.L. Lieber, T.M. Woodburn, and J. Friden, Muscle damage induced by eccentric contractions of 25% strain, *J Appl Physiol*, 70: 2498–2507 (1991).
93. J. Friden, P.N. Sfakianos, and A.R. Hargens, Muscle soreness and intramuscular fluid pressure: Comparison between eccentric and concentric load, *J Appl Physiol*, 61: 2175–2179 (1986).
94. J. Friden, and R.L. Lieber, Segmental muscle fiber lesions after repetitive eccentric contractions, *Cell Tissue Res*, 293: 165–171 (1998).
95. J. Friden, J. Seger, M. Sjostrom, and B. Ekblom, Adaptive response in human skeletal muscle subjected to prolonged eccentric training, *Int J Sports Med*, 4: 177–183 (1983).

96. D.J. Newham, G. McPhail, K.R. Mills, and R.H. Edwards, Ultrastructural changes after concentric and eccentric contractions of human muscle, *J Neurol Sci*, 61: 109–122 (1983).
97. J. Friden, M. Sjostrom, and B. Ekblom, Myofibrillar damage following intense eccentric exercise in man, *Int J Sports Med*, 4: 170–176 (1983).
98. D.J. Newham, D.A. Jones, and R.H. Edwards, Large delayed plasma creatine kinase changes after stepping exercise, *Muscle Nerve*, 6: 380–385 (1983).
99. R.L. Lieber, and J. Friden, Selective damage of fast glycolytic muscle fibres with eccentric contraction of the rabbit tibialis anterior, *Acta Physiol Scand*, 133: 587–588 (1988).
100. W.T. Stauber, and C.A. Smith, Cellular responses in exertion-induced skeletal muscle injury, *Mol Cell Biochem*, 179: 189–196 (1998).
101. R.L. Lieber, M.C. Schmitz, D.K. Mishra, and J. Friden, Contractile and cellular remodeling in rabbit skeletal muscle after cyclic eccentric contractions, *J Appl Physiol*, 77: 1926–1934 (1994).
102. W.E. Garrett, Jr., Muscle strain injuries, *Am J Sports Med*, 24: S2–S8 (1996).
103. C.T. Hasselman, T.M. Best, A.V. Seaber, and W.E. Garrett, Jr., A threshold and continuum of injury during active stretch of rabbit skeletal muscle, *Am J Sports Med*, 23: 65–73 (1995).
104. T.J. Noonan, T.M. Best, A.V. Seaber, and W.E. Garrett, Jr., Identification of a threshold for skeletal muscle injury, *Am J Sports Med*, 22: 257–261 (1994).
105. W.T. Obremsky, A.V. Seaber, B.M. Ribbeck, and W.E. Garrett, Jr., Biomechanical and histologic assessment of a controlled muscle strain injury treated with piroxicam, *Am J Sports Med*, 22: 558–561 (1994).
106. D.L. Morgan, New insights into the behavior of muscle during active lengthening, *Biophys J*, 57: 209–221 (1990).
107. J.A. Talbot, and D.L. Morgan, Quantitative analysis of sarcomere non-uniformities in active muscle following a stretch, *J Muscle Res Cell Motil*, 17: 261–268 (1996).
108. K.B. Geronilla, G.R. Miller, K. Mowrey, M.L. Kashon, and R.G. Cutlip, The mechanical and histological response of rat skeletal muscle to oscillatory contractions, American college of sports medicine, *Medicine and Science in Sports and Exercise*, Medicine and Science in Sports and Exercise, Baltimore, MD, pp. 474, 2001.
109. G.L. Warren, D.A. Hayes, D.A. Lowe, J.H. Williams, and R.B. Armstrong, Eccentric contraction-induced injury in normal and hindlimb-suspended mouse soleus and EDL muscles, *J Appl Physiol*, 77: 1421–1430 (1994).
110. K.B. Geronilla, G.R. Miller, K.F. Mowrey, J.Z. Wu, M.L. Kashon, K. Brumbaugh, J. Reynolds, A. Hubbs, and R.G. Cutlip, Dynamic force responses of skeletal muscle during stretch-shortening cycles, *Eur J Appl Physiol*, 90: 144–153 (2003).
111. R.A. Fielding, T.J. Manfredi, W. Ding, M.A. Fiatarone, W.J. Evans, and J.G. Cannon, Acute phase response in exercise. III. Neutrophil and IL-1 beta accumulation in skeletal muscle, *Am J Physiol*, 265: R166–R172 (1993).
112. J.G. Tidball, Inflammatory cell response to acute muscle injury, *Med Sci Sports Exerc*, 27: 1022–1032 (1995).
113. T.M. Best, and K.D. Hunter, Muscle injury and repair, *Phys Med Rehabil Clin N Am*, 11: 251–266 (2000).
114. D.B. Dubravec, D.R. Spriggs, J.A. Mannick, and M.L. Rodrick, Circulating human peripheral blood granulocytes synthesize and secrete tumor necrosis factor alpha, *Proc Natl Acad Sci USA*, 87: 6758–6761 (1990).
115. K. Tiku, M.L. Tiku, S. Liu, and J.L. Skosey, Normal human neutrophils are a source of a specific interleukin 1 inhibitor, *J Immunol*, 136: 3686–3692 (1986).
116. P.O. Hasselgren, P. Pedersen, H.C. Sax, B.W. Warner, and J.E. Fischer, Current concepts of protein turnover and amino acid transport in liver and skeletal muscle during sepsis, *Arch Surg*, 123: 992–999 (1988).

117. L.L. Moldawer, G. Svaninger, J. Gelin, and K.G. Lundholm, Interleukin 1 and tumor necrosis factor do not regulate protein balance in skeletal muscle, *Am J Physiol*, 253: C766–C773 (1987).
118. B. Alvarez, L.S. Quinn, S. Busquets, F.J. Lopez-Soriano, and J.M. Argiles, Direct effects of tumor necrosis factor alpha (TNF-alpha) on murine skeletal muscle cell lines, Bimodal effects on protein metabolism, *Eur Cytokine Netw*, 12: 399–410 (2001).
119. J.M. Round, D.A. Jones, and G. Cambridge, Cellular infiltrates in human skeletal muscle: Exercise induced damage as a model for inflammatory muscle disease? *J Neurol Sci*, 82: 1–11 (1987).
120. B.A. St Pierre and J.G. Tidball, Macrophage activation and muscle remodeling at myotendinous junctions after modifications in muscle loading, *Am J Pathol*, 145: 1463–1471 (1994).
121. B.M. Lapointe, P. Fremont, and C.H. Cote, Adaptation to lengthening contractions is independent of voluntary muscle recruitment but relies on inflammation, *Am J Physiol Regul Integr Comp Physiol*, 282: R323–R329 (2002).
122. B.M. Lapointe, J. Frenette, and C.H. Cote, Lengthening contraction-induced inflammation is linked to secondary damage but devoid of neutrophil invasion, *J Appl Physiol*, 92: 1995–2004 (2002).
123. J.L. De Bleecker, V.I. Meire, W. Declercq, and E.H. Van Aken, Immunolocalization of tumor necrosis factor-alpha and its receptors in inflammatory myopathies, *Neuromuscul Disord*, 9: 239–246 (1999).
124. E. Zador, L. Mendler, V. Takacs, J. de Bleecker, and F. Wuytack, Regenerating soleus and extensor digitorum longus muscles of the rat show elevated levels of TNF-alpha and its receptors, TNFR-60 and TNFR-80, *Muscle Nerve*, 24: 1058–1067 (2001).
125. R. Medina, S.S. Wing, and A.L. Goldberg, Increase in levels of polyubiquitin and proteasome mRNA in skeletal muscle during starvation and denervation atrophy, *Biochem J*, 307(Pt 3): 631–637 (1995).
126. P.O. Hasselgren, Role of the ubiquitin-proteasome pathway in sepsis-induced muscle catabolism, *Mol Biol Rep*, 26: 71–76 (1999).
127. B.S. Wung, J.J. Cheng, H.J. Hsieh, Y.J. Shyy, and D.L. Wang, Cyclic strain-induced monocyte chemotactic protein-1 gene expression in endothelial cells involves reactive oxygen species activation of activator protein 1, *Circ Res*, 81: 1–7 (1997).
128. M. Baggiolini, B. Dewald, and B. Moser, Chemokines, in: J.I. Gallin and R. Snyderman, (eds.), *Inflammation, Basic Principles and Clinical Correlates*, Lippincott Williams & Wilkins, Philadelphia, PA, pp. 419–431, 1991.
129. D.K. Mishra, J. Friden, M.C. Schmitz, and R.L. Lieber, Anti-inflammatory medication after muscle injury, A treatment resulting in short-term improvement but subsequent loss of muscle function, *J Bone Joint Surg Am*, 77: 1510–1519 (1995).
130. T.A. Trappe, J.D. Fluckey, F. White, C.P. Lambert, and W.J. Evans, Skeletal muscle PGF(2)(alpha) and PGE(2) in response to eccentric resistance exercise: Influence of ibuprofen acetaminophen, *J Clin Endocrinol Metab*, 86: 5067–5070 (2001).
131. T.A. Trappe, F. White, C.P. Lambert, D. Cesar, M. Hellerstein, and W.J. Evans, Effect of ibuprofen and acetaminophen on postexercise muscle protein synthesis, *Am J Physiol Endocrinol Metab*, 282: E551–E556 (2002).
132. J.M. Peterson, T.A. Trappe, E. Mylona, F. White, C.P. Lambert, W.J. Evans, and F.X. Pizza, Ibuprofen and acetaminophen: Effect on muscle inflammation after eccentric exercise, *Med Sci Sports Exerc*, 35: 892–896 (2003).
133. M. Cantini, M.L. Massimino, A. Bruson, C. Catani, L.D. Libera, and U. Carraro, Macrophages regulate proliferation and differentiation of satellite cells, *Biochem Biophys Res Commun*, 202: 1688–1696 (1994).
134. S.M. Sheehan and R.E. Allen, Skeletal muscle satellite cell proliferation in response to members of the fibroblast growth factor family and hepatocyte growth factor, *J Cell Physiol*, 181: 499–506 (1999).

135. G. Fibbi, S. D'Alessio, M. Pucci, M. Cerletti, and M. Del Rosso, Growth factor-dependent proliferation and invasion of muscle satellite cells require the cell-associated fibrinolytic system, *Biol Chem*, 383: 127–136 (2002).
136. M. Hill, A. Wernig, and G. Goldspink, Muscle satellite (stem) cell activation during local tissue injury and repair, *J Anat*, 203: 89–99 (2003).
137. E.D. Rabinovsky, E. Gelir, S. Gelir, H. Lui, M. Kattash, F.J. DeMayo, S.M. Shenaq, and R.J. Schwartz, Targeted expression of IGF-1 transgene to skeletal muscle accelerates muscle and motor neuron regeneration, *Faseb J*, 17: 53–55 (2003).
138. G.R. Adams, Autocrine and/or paracrine insulin-like growth factor-I activity in skeletal muscle, *Clin Orthop*, S188–S196 (2002).
139. C. Zeng, J.E. Pesall, K.K. Gilkerson, and D.C. McFarland, The effect of hepatocyte growth factor on turkey satellite cell proliferation and differentiation, *Poult Sci*, 81: 1191–1198 (2002).
140. S. Gallucci, C. Provenzano, P. Mazzarelli, F. Scuderi, and E. Bartoccioni, Myoblasts produce IL-6 in response to inflammatory stimuli, *Int Immunol*, 10: 267–273 (1998).
141. M. Cantini, M.L. Massimino, E. Rapizzi, K. Rossini, C. Catani, L. Dalla Libera, and U. Carraro, Human satellite cell proliferation in vitro is regulated by autocrine secretion of IL-6 stimulated by a soluble factor(s) released by activated monocytes, *Biochem Biophys Res Commun*, 216: 49–53 (1995).
142. L.S. Quinn, K.L. Haugk, and K.H. Grabstein, Interleukin-15: A novel anabolic cytokine for skeletal muscle, *Endocrinology*, 136: 3669–3672 (1995).
143. L.S. Quinn, B.G. Anderson, R.H. Drivdahl, B. Alvarez, and J.M. Argiles, Overexpression of interleukin-15 induces skeletal muscle hypertrophy in vitro: Implications for treatment of muscle wasting disorders, *Exp Cell Res*, 280: 55–63 (2002).
144. N. Carbo, J. Lopez-Soriano, P. Costelli, S. Busquets, B. Alvarez, F.M. Baccino, L.S. Quinn, F.J. Lopez-Soriano, and J.M. Argiles, Interleukin-15 antagonizes muscle protein waste in tumour-bearing rats, *Br J Cancer*, 83: 526–531 (2000).
145. V. Prisk and J. Huard, Muscle injuries and repair: The role of prostaglandins and inflammation, *Histol Histopathol*, 18: 1243–1256 (2003).
146. R.G. Cutlip, Soft Tissue Pathomechanics, in: W.S.M.a.W. Karowski, (ed.), *The Occupational Ergonomics Handbook*, CRC Press, Boca Raton, FL, pp. 15-1–15-43, 2006.
147. W.S. Marras, R.G. Cutlip, S.E. Burt, and T.R. Waters, National occupational research agenda (NORA) future directions in occupational musculoskeletal disorder health research, *Appl Ergon*, 40: 15–22 (2009).

CHAPTER 3

# Physiological Aspects of Neuromuscular Function

Thomas R. Waters and Amit Bhattacharya

CONTENTS

## 3.1 INTRODUCTION

The mechanical motions of the body segments necessary to accomplish a task are skillfully controlled by a myriad of neuromuscular components and a series of well-orchestrated neural events encompassing both the central and peripheral nervous systems. Some of these issues are discussed in the following sections of this chapter.

The maintenance of ergonomically efficient posture requires the optimal orientation of various interconnected human body segments to produce minimal biomechanical loadings of the joints. The neural components of movements are the planning and programming units and the performance units. Planning and programming are the functions of the precortical centers (cerebral cortex, bilateral movement), the basal ganglia, the cerebellum, and the thalamus. The premotor and sensory regions provide the input for planning to the basal ganglia and the cerebellum. The performance units include the motor cortex and spinal cord, with the smoothing function performed by the cerebellum. In order for a person to perform a motor act smoothly, all the somatosensory systems have to work in harmony to help provide "accurate" information regarding the position of body segments, muscle tensions, and joint motions to the higher centers. For a detailed discussion of this topic, readers should refer to comprehensive texts of physiology and neuroanatomy [1–5].

## 3.2 BACKGROUND AND SIGNIFICANCE TO OCCUPATIONAL ERGONOMICS

A worker who performs a simple manual task, such as picking up a box from a table, does so without much difficulty. The underlying physiological chain of events required to perform this seemingly simple task, however, is very complex and involves a number of physiological systems including the muscular, sensory, and central and peripheral nervous systems. Prior to the actual event of reaching for the box, the worker uses his or her visual system and the higher brain centers to make a judgment regarding its weight, size, and location with respect to the orientation and location of his or her body. This information is synthesized by the brain, and a movement plan is developed that will generate a series of complex control commands. These control commands are relayed to the appropriate muscle groups (via motor neurons) to allow for smooth movement of the body segment and load unit.

At the muscle–bone level, when the appropriate commands are received by the muscle, the muscle contracts and applies a pulling force on the tendon attached to the bone segment. If the tensile force generated by the muscle is strong enough to create a moment that will overcome the moment created by the weight of the segment and any external load applied to the segment, then the segment and load will be moved.

To ensure smooth movement, a complex system of sensory elements monitors the condition of the musculoskeletal system. These sensors measure the joint forces, muscle forces, and muscle length and provide continuous feedback to the central nervous system, which then modifies the commands sent to the muscles. The sensors consist of various muscle stretch (change-of-length detector) and tension-monitoring receptors, which balance reciprocally acting pairs of muscles to control the position of a segment at any point in time (e.g., extensor versus flexor muscles around a joint). This positional control system is crucial to coordinated muscle movement. The sensory units are integrated into a highly evolved system of reflexes that provide a short-latency connection between the sensory and motor components of the system that result in stereotypical motor responses to sensory stimulus. This arrangement allows for a rapid response to perturbations in the status of the system (e.g., the knee jerk response to stretching the patella ligament of the knee by striking it with a hammer).

## 3.3 RELEVANT CONCEPTS AND TERMINOLOGY

### 3.3.1 Major Components of the Neuromuscular System

Table 3.1 lists the five main elements of the neuromuscular system and their purpose. As shown in Figure 3.1, the five component groups of the neuromuscular system are organized into a highly complex command and control system that integrates physiological and mechanical functions into a single system capable of performing work.

#### *3.3.1.1 Cerebral Cortex, Cerebellum, and Subcortical Centers*

In humans, the programming center for a movement is made up of the cerebral cortex of the brain and various components of the subcortical centers (basal ganglia, brainstem nuclei, and brainstem reticular formation). These systems send descending commands (efferent) to the motor neurons and finally to the muscles (Figure 3.2). During contraction of the muscles, various receptors in the muscles (muscle spindles), tendons (Golgi tendon organ), and joints send real-time information (afferent) about the status of the body segment movement to the higher centers for processing. The role of the cerebellum is to smooth the movement of the body segment. It receives information from afferent systems (vestibular, proprioception, and visual) as well as commands from the higher centers. Input signals from the cerebral cortex are conveyed via brainstem nuclei to the cerebellum and tell the brain what the muscles should be doing. Based on its knowledge of the status of various afferent systems (as it relates to motor coordination) and the nature of "expected" controlling commands from the higher centers, the cerebellum is capable of producing smooth movement. It is not clear how the cerebellum processes this information. However,

TABLE 3.1 Major Components of the Neuromuscular System

| Component | Purpose |
|---|---|
| Cerebral cortex, cerebellum, and subcortical centers | Central processing unit for coordination of muscular activity |
| Motor neurons | Neural circuitry |
| Muscle fibers | Force-generating neuromechanical actuator |
| Visual, vestibular, and somatic sensory receptors | Feedback sensors for detection of position, stretch, pressure, tension, etc. |
| Bones, joints, ligaments, and tendons | Mechanical support and linkage connectors |

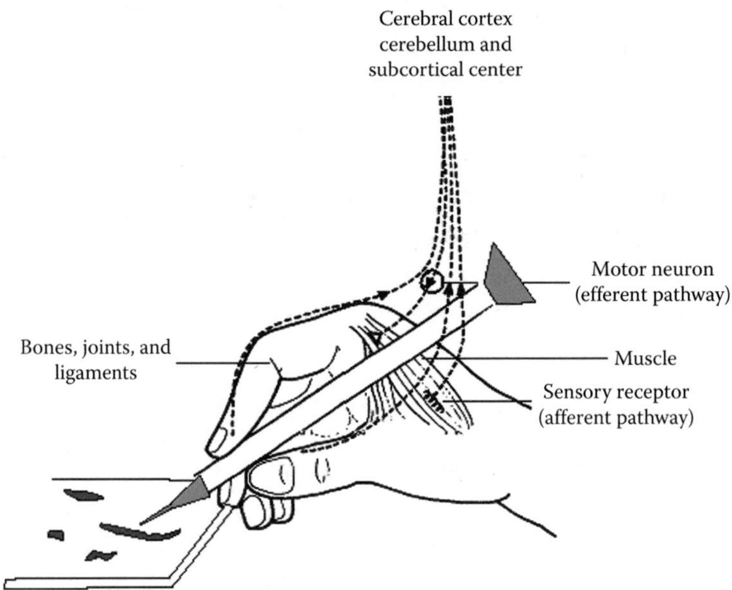


FIGURE 3.1 Components of the neuromuscular system showing the direction of flow of neural signals. A complex system of interneurons (not shown) also provides reflex pathways in which afferent sensory signals can directly affect motor signals without traveling to the higher centers of the brain. The final common pathway for motor function is the motor neuron, which connects directly to the muscle fibers. (Adapted from Astrand, P. and Rodahl, K., *Textbook of Work Physiology*, McGraw-Hill, New York, 1986, pp. 19, 115, 334.)

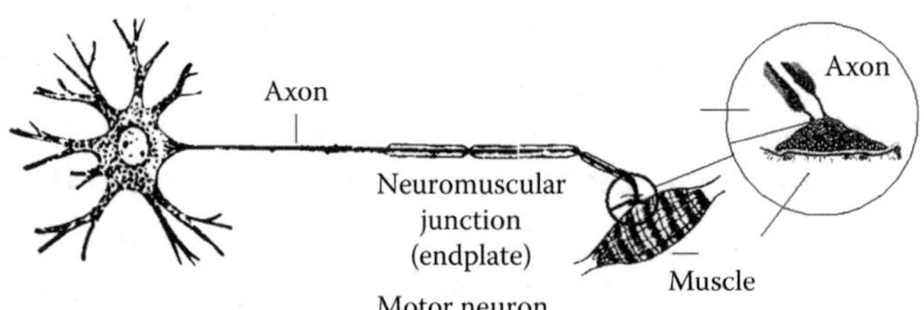


FIGURE 3.2 Illustration of a motor neuron showing the axon and the neuromuscular junction (endplate). When the cell body is stimulated, an action potential is transmitted to the endplate, where a neurotransmitter (acetylcholine) is released into the neuromuscular junction. The neurotransmitter causes the muscle fibers to contract.

although its exact mechanism of action is not clear, it certainly plays an important role in minimizing the error signal between what the muscles are doing and what they "should" be doing. The cerebellum does not initiate muscle contractions directly but controls the activity in the descending motor commands. People with cerebellum damage have jerky uncoordinated movements, poor balance, and unsteady gait [1].

#### *3.3.1.2 Motor Neurons*

The motor neuron, illustrated in Figure 3.2, is a nerve cell that links the central nervous system and the appropriate muscle fiber or fibers. The motor neuron consists of a cell body containing the cell nucleus, an elongated segment or axon, and the endplate or neuromuscular junction. When a command signal of sufficient strength is received at the body of the motor neuron, the cell membrane is depolarized (there is a drop in voltage due to a change in cell membrane permeability. This depolarization initiates an action potential along the nerve axon. The action potential is a propagating electric signal that travels toward the endplate. When the action potential reaches the endplate, acetylcholine, a chemical transmitter, is released into the neuromuscular junction. This chemical transmitter then depolarizes the endplate membrane of the muscle fiber to a threshold value (to +30 mV from a resting potential of −70 mV). This action electrically triggers the muscle fiber contraction.

Motor neurons and muscles are organized into motor units that consist of a single motor neuron, its axon, and all the muscle fibers innervated by it. The number of muscle fibers innervated by a single motor unit varies from a few (e.g., muscles that move fingers) to several hundred (e.g., back muscles). The number of muscle fibers innervated by one motor unit is dependent on the function of the muscle rather than its size. Muscles that cause large and strong body motions usually have more muscle fibers under the control of a motor unit than those required to perform fine precision movements.

#### *3.3.1.3 Muscles*

There are over 600 muscles in the human body accounting for about 45% of the total body weight. Muscles are composed of one of three kinds of fibers, depending upon the function of the muscle. The three types are skeletal, smooth, and cardiac. Skeletal muscle is connected to the bones of the body, and when contracted it causes the body segments to move. Smooth muscle is found in the stomach, intestinal tracts, and walls of blood vessels. Cardiac muscle is the contractile tissue found in the heart that pumps the blood for circulation. In this chapter, we focus our attention on the skeletal muscles.

A single skeletal muscle, as shown in Figure 3.3, consists of hundreds to tens of thousands of muscle fibers, totaling about a quarter of a billion in an average person. Each muscle fiber consists of a single cylindrical muscle cell with a diameter of 10–90 μm and a length of up to 30 cm. Muscle fibers are further divided into individual contractile

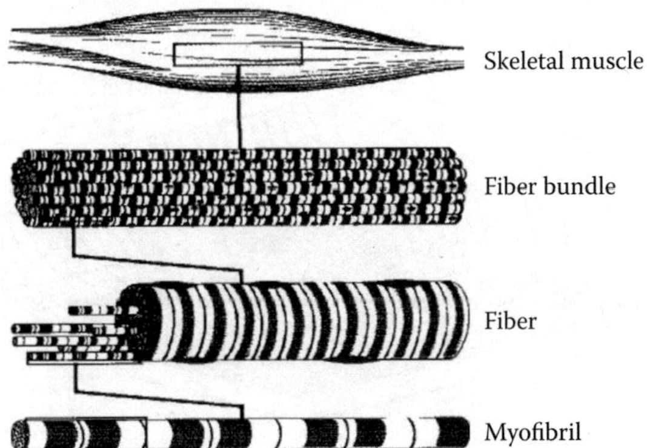


FIGURE 3.3 Illustration of a skeletal muscle showing the fiber composition. Each fiber is innervated by one motor neuron that controls the contraction of the muscle fiber. (Adapted from Dudel, J., in *Fundamentals of Neurophysiology*, Schmidt, R.F., Ed., Springer-Verlag, New York, 1978, Chapter 5, pp. 129–137.)

elements called myofibrils. Myofibrils are the basic contractile element of the muscle, and the forces and movement in a muscle cell are generated by special protein molecules and contractile proteins in the myofibrils. Each muscle fiber contains hundreds of myofibrils that actively contract in the presence of calcium, which is released when a contraction is initiated. Contraction is accomplished by a complex chemical process whereby adjacent filaments of molecular proteins, actin, and myosin, are pulled toward each other in a sliding motion that results in shortening of the muscle cell and the development of tensile force. Generally, muscle fibers are shorter than the muscle they make up, but some fibers run the entire length of the muscle.

The amount of tension or force generated by a contracting muscle is dependent on its precontraction length, the velocity of the contraction, and the direction of muscle movement during the contraction (i.e., whether the muscle is lengthening or shortening). When a muscle is shortening during a contraction, the activity is defined as *concentric*. Conversely, when a muscle is lengthening during a contraction, the activity is defined as *eccentric*. Figure 3.4 graphically illustrates the relationship between muscle force, length, velocity of contraction, and direction of movement.

For a concentric contraction, there is an optimal precontraction muscle length that will produce a maximum tension force when the muscle is stimulated. This length is called the *resting length* of the muscle. If the precontraction muscle length is at or below 60% of the resting length, the muscle will not produce any tension when stimulated. Therefore, the ability of a muscle to produce an optimal force is strongly dependent upon the position of the body segment to which the muscle is attached. For example, optimal biceps muscle tension is generated when the elbow joint angle (subtended between the forearm and the upper arm) is in the region of 90°–100°. Moreover, as the velocity of a concentric contraction increases, the force decreases.

Muscles are usually attached to bones in a paired arrangement, with the *agonist* muscles performing the main movement of the body segment and the *antagonist* muscles acting as the controller. *Fixator* muscles provide support to the proximal joints, and *synergist*

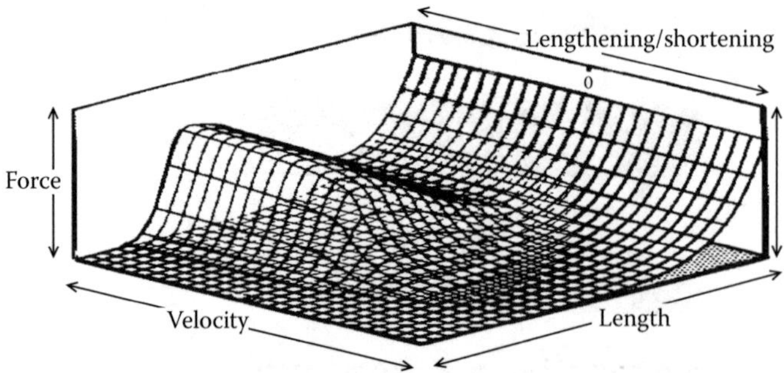


FIGURE 3.4 Graphical illustration of the effects of muscle length, contraction velocity, and direction of muscle movement on muscle force (assuming a fixed excitation level). The surface plot shows the magnitude of force developed as a function of muscle length, velocity, and movement direction. The ridge of the peak corresponds to the resting length of the muscle. (Adapted from Brooks, V.B., *The Neural Basis of Motor Control*, Oxford University Press, New York, 1986.)

*muscles* help prevent undesirable movements of other joints when the agonist muscle passes over more than one joint.

#### 3.3.1.4 Visual, Vestibular, and Sensory Receptors

In order to move a body segment(s) and/or maintain whole-body upright balance during static and dynamic conditions, the brain must receive feedback information regarding the position and movement of the body. This type of information is provided by a series of biological sensors located in the joints, muscles, and tendons and under the skin, as well as the visual and vestibular systems. The output from these sensory components is transmitted by afferent neurons to the spinal cord and brain, where they are processed and used to alter the motor signals. Feedback from these sensory receptors is essential for smooth, coordinated movement because the sensory input provides the cues necessary to alter the timing of the motor program that controls the motor function. For new, unlearned tasks, the visual and vestibular systems are especially important in generating the coordinated motor patterns for smooth motion. The flowchart shown in Figure 3.5 illustrates how the sensory system provides feedback information to alter the motor signals sent to the muscles.

The visual and vestibular systems provide information about the spatial orientation of the body and the movement of the head. The *visual system* provides information regarding the orientations (horizontal and vertical) of objects in three-dimensional (3-D) space. The *vestibular system,* illustrated in Figure 3.6, provides information regarding the position and movements of the head and their relationship to gravitational forces. The three orthogonally placed semicircular canals provide information regarding acceleration of the head in 3-D space. The utricle and the saccule are position sensors that provide information regarding the position of the head in space.

The actions of the vestibular system are not consciously felt unless one is required to perform a motor task under dim light and/or walk or stand on an uneven surface. The literature indicates that the role of the vestibular system for motor task performance is not as critical when the other afferents from the visual and proprioception systems are intact.

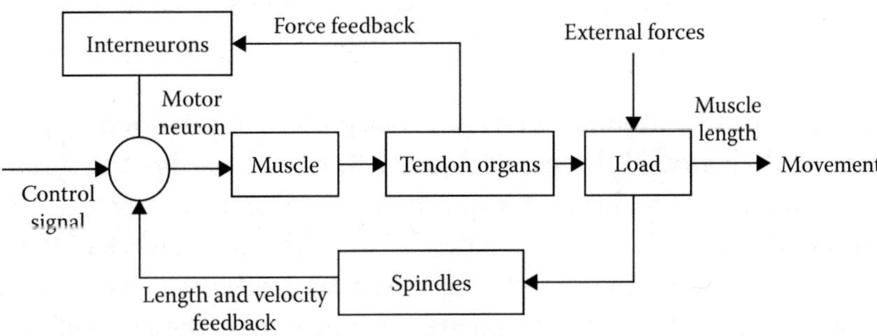


FIGURE 3.5 Flowchart of muscle control system. Feedback loops from tendon organs and muscle spindles provide sensory feedback for control of muscular function. Descending pathways from the brain provide control signals, where they are integrated with the sensory input and sent to the muscles via the motor neurons. (Adapted from Brooks, V.B., *The Neural Basis of Motor Control,* Oxford University Press, New York, 1986.)

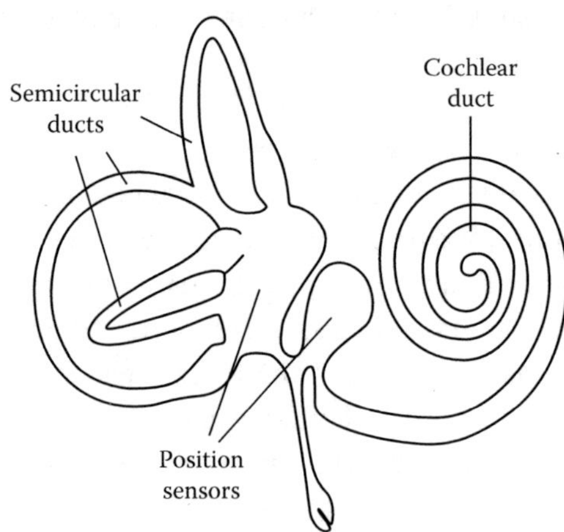


FIGURE 3.6 Vestibular sensory receptors. Semicircular ducts provide sensory input regarding angular acceleration. The position receptors provide information about the orientation of the head. (Adapted from Brooks, V.B., *The Neural Basis of Motor Control*, Oxford University Press, New York, 1986.)

Proprioceptive sensors provide information about the status of the muscles themselves, such as how hard the muscles are pulling and how fast they are being stretched. Muscle stretch receptors (muscle spindle) and the Golgi tendon organs (tension-monitoring receptors) transmit information regarding muscle length and tension to the controlling centers in the cortex and subcortical units as well as to the cerebellum.

Kinesthetic sensors measure joint movements, and somatic sensors in the skin provide information about temperature, pressure, and pain.

#### *3.3.1.5 Bones, Joints, Tendons, and Ligaments*

The bones, joints, tendons, and ligaments serve as an integrated support frame for the body and provide an attachment point for the muscles. The bones are connected by a complex arrangement of connective tissue to form a custom-fitted joint that provides a flexible yet stable connection with a wide range of motion. As the muscles contract, they transmit their forces to the bones through the tendons, thereby causing the bones to move around the joints.

There are three classes of joints: fibrous, cartilaginous, and synovial [4]. Most body movements take place around the synovial joints. There are six types of synovial joints: hinge, pivot, ellipsoid, ball-and-socket, plane, and saddle joints. Examples of these joints are found in the elbow, neck, wrist, shoulder, between the bony arches of the vertebrae, and at the base of the thumb. The type of motion and the maximum range of motion of the body segment possible at the synovial joint are dependent upon the shape of the articulating bones, the strength and the orientation of the surrounding ligaments, and the size and strength of the muscles.

The muscles are attached to bones by collagen fibers called tendons. Tendons have strong tensile characteristics but do not possess contracting properties. Tendons are instrumental in transmitting forces generated by the contracting muscles. For example, finger motions

TABLE 3.2 Segment Movements and Muscles Used in Making Those Movements

| Body Movement | Agonist | Antagonist |
|---|---|---|
| Elbow flexion | Biceps brachii | Triceps brachii |
| Shoulder abduction | Deltoid | Pectoralis |
| Spinal extension | Erector spinae | Rectus abdominus |
| Hip flexion | Psoas, iliacus, rectus femoris, pectineus, tensor fascia | Gluteus maximus and hamstring |

are caused by the force transmitted by the long tendons attached to the contracting muscles in the lower arm (between the elbow and the wrist).

Movements of the body segments around the articulating joint surfaces have been given specific names. In general, when the angle between two attached bone segments decreases during a movement, the movement is called *flexion,* and when the angle between the two attached bone segments increases, the movement is called *extension.* When the movement of a limb is away from the midline of the body, the movement is called *abduction,* and when the movement is toward the midline of the body, the movement is called *adduction. Pronation* and *supination* movement occur when a body segment undergoes a rotation about its long axis. Pronation or medial rotation is a movement toward the center of the body, whereas supination or lateral rotation defines a movement away from the center of the body. Table 3.2 provides some examples of body movements and the corresponding muscles producing those movements.

### 3.3.2 Motor Control

#### *3.3.2.1 Motor Programs*

Learned tasks, such as touching your nose, standing on one foot, or hitting a baseball, are controlled by overall motor plans that are stored in the central nervous system. These programs provide the proper sequence of motor activity needed to coordinate a specific planned motor function. It is important to note that the motor program must not only provide motor signals to initiate a specific movement but must also provide the motor signals needed to maintain posture. Complex, multijoint movement patterns are difficult to learn, because they are composed of numerous nested subprograms of varying complexity. On the other hand, some neural patterns are repeated so frequently that they may function without external timing cues or sensory input. These programs are sometimes referred to as central pattern generators. Motor activity is monitored by sensory receptors, and the status of the system is relayed to the central nervous system, where the motor program may be altered. Additional neural networks within the spinal cord may modify the motor signals through a feedback system of reflexes.

#### *3.3.2.2 Reflexes*

The neuromuscular system has the capability of rapidly adjusting motor function as a result of sensory input through highly developed neural circuits or reflex pathways. These reflex pathways consist of efferent motor neurons, afferent sensory neurons, and interneurons that link and integrate the signals between the sensory and motor neurons. A reflex can be thought of as a negative feedback control system that results in a stereotyped motor response to a sensory stimulus. A number of reflexes have been identified, such as the antigravity reflex for assistance in standing, the stretch reflex for maintaining constant muscle length (demonstrated by the knee tendon jerk test), the flexor and crossed-extensor reflexes that provide a matched response of leg withdrawal to painful stimuli and a resultant extension of the opposite leg, and the reciprocal inhibition reflex that turns the antagonist muscle off during a motor function. It is the network of interneurons in the spinal cord that integrates the control signals from the brain and the afferent signals from the sensory receptors to produce a flow of triggering signals to the muscles. The motor neurons, however, are the final common pathway for muscle activation.

#### *3.3.2.3 Static and Dynamic Contractions*

Muscles perform two basic types of tasks: dynamic body segment movement and the holding of body segments in static postures. Concentric contractions involve shortening of the muscles, causing a movement such as lifting a box. Eccentric contractions involve lengthening of an actively contracting muscle, such as the controlled lowering of a weight on a glass surface. In this case, the active muscle is controlling the rate and the motion under the action of the external force of gravity. A static contraction is performed when a posture is maintained without any movement. This type of contraction can cause rapid muscle fatigue due to poor blood circulation and the buildup of metabolites and waste products in the muscle. Most body movements or sustained postures require the use of more than one muscle.

#### *3.3.2.4 Recruitment and Rate Coding*

To allow for a wide range of muscle force levels, two physiological mechanisms are used to control the amount of force generated by a muscle: recruitment and rate coding.

*Recruitment* is defined as the number of motor units involved in the muscle contraction, and *rate coding* is defined as the frequency of discharge of the motor neurons. For slow contractions, such as postural control, the small, slow, fatigue-resistant motor units are recruited first, and the larger, fast, fatiguable motor units are recruited as the desired movement becomes faster or more powerful. Rate coding is used to modulate the force level by increasing the discharge rate as more force is needed. In general, for powerful contractions, all of the motor units are recruited before maximum discharge rates are reached.

### 3.3.3 Types of Movements

When the decision has been made to move one or more body segments, the neuromuscular system must choose which muscles to turn on, when they should be turned on, and how much force each muscle should develop. Since a wide range of activation sequences, muscle combinations, movement speeds, and motion trajectories could be used to achieve a desired movement, a predictable set of movement strategies must exist in order to optimize the efficiency of the movement. In general, complex movements can be resolved into simpler individual movements that are controlled by the agonist and antagonist muscles, which work together to achieve a specified movement. In a typical movement, for example, the agonist muscle contracts first to initiate the rotation of the segment about the joint at the proper velocity, and a short time later the antagonist muscle contracts to slow or stop the motion of the segment. In broader terms, the accuracy and speed of a movement dictate the selection and sequencing of the muscles, but postural constraints and system equilibrium must also be considered.

A number of goal-directed movement strategies have been proposed. Some of these strategies are based on the method of excitation control. For example, speed-dependent movements are modulated by the excitation amplitude, whereas speed-independent strategies are modulated by the excitation duration [8]. In some movements, coactivation of the agonist and antagonist muscles is needed to increase joint stiffness to stabilize the joint when it is exposed to high inertial loading. Significant coactivation has been shown in the spine for dynamic movements [9]. The stretch–shorten cycle has been proposed as another strategy for controlling movement. The stretch–shorten cycle is a linked eccentric–concentric contraction designed to use the additional force generated by a muscle when it is contracting eccentrically (lengthening), which is helpful in high-performance movements. Also, synergistic muscle action (cooperating agonists) may be used to assist in the development of the proper segment velocities.

### 3.3.4 Assessing Neuromuscular Function

#### *3.3.4.1 Electromyography*

When a muscle is activated, it creates an electric discharge (myoelectric signal) that can be measured directly from the muscle or through an electrode attached to the surface of the skin. Measurement and recording of these signals is called electromyography (EMG). These electric signals provide information about the intensity and duration of the contraction. Although the activity of individual motor units can be measured with needle and fine wire electrodes, surface electrodes are typically used to measure the activity of whole muscle groups. In this case, the measured signal is the summation of all the active motor units within the recording area of the electrodes. Under certain conditions, EMG activity can be used to measure the magnitude of the force being developed in a selected muscle group. Care must be taken, however, in determining muscle force from EMG activity, due to the length–tension and velocity–tension characteristics mentioned previously. For more information about making EMG measurements, the National Institute for Occupational Safety and Health has published a users' manual [10].

#### *3.3.4.2 Evoked Potentials*

Evoked potentials are electric signals initiated in a neural pathway by the application of an external stimulus (electric pulse) to a motor or sensory component to determine the functional status of the pathway of interest. Measurement of the amplitude or conduction velocity of an evoked potential along a neural pathway is useful in evaluating the function of the sensory or motor system. For example, visual or auditory evoked potentials can be used to assess the function of the optic or auditory nerve, and somatosensory evoked potentials can be used to assess the function of the various peripheral nerves. Multiple stimulations are required to measure evoked potentials from the skull because the signals must be averaged to remove the random electroencephalographic (EEG) activity of the brain.

#### 3.3.4.3 Postural Stability

When a person is standing upright, the neuromuscular system provides motor control to the musculature of the supporting limbs to maintain a stable posture. These fine motor contractions result in a small natural oscillation or swaying of the body. Techniques for measuring the extent of this natural swaying are discussed in Chapter 4. It has been shown that the natural swaying of the body is modified in clinical and neurological disorders, such as in humans exposed to neurotoxic industrial chemicals and by man-made drugs and physically fatiguing tasks [11–14]. Excessive body sway could very well interfere with the safe performance of tasks and may jeopardize safety in the workplace. Successful application of this method has been illustrated for workers exposed to chemicals that affect the neurological system and may impair their balance, making them sway more than members of an unexposed population [14].

#### 3.3.4.4 Muscular Strength Measurement

Strength is a measure of the maximum force that can be produced by a single muscle or by a series of muscles under prescribed conditions. Owing to the voluntary nature of the test, it is usually thought of as the maximum voluntary contraction level. Standardized tests have been developed for measuring static (isometric) strength as well as dynamic (isokinetic and isoinertial) strength. Recall that muscle force is a function of the resting length of the muscle, the velocity of contraction, and the direction of motion. Thus, muscle strength measures depend upon the position of the body and the type of motion occurring at the time of measurement. Maximum strength values have been published for a wide range of test conditions. For more information on assessment of strength, see the chapter on physical work capacity in this book or refer to studies reported in the literature [15,16].

#### 3.3.4.5 Kinesiology

In its broadest sense, the term *kinesiology* refers to the study of movement. In the context of this chapter, however, kinesiology is the measurement and analysis of human motion and the way in which it relates to the musculoskeletal system that generates those movements. Kinesiological measurements are important to understand musculoskeletal functions such as gait, posture, and static and dynamic muscle action. Kinesiology, which is closely related to biomechanics, is useful for identifying abnormalities in musculoskeletal function (see Chapter 4). For example, movement requirements can be analyzed and used to assess work demands in ergonomics, motion patterns can be compared to normative patterns to assist the clinician in diagnosing neuromuscular dysfunction, and kinematic measurements may be used to improve athletic mechanics.

Measurement techniques are available for either 2- or 3-D analysis. 2-D kinesiological measurements may be made with a simple goniometer (a device designed to measure the relative rotation of a given joint) or with a video camera and markers placed at the joints of interest. The linear translation, velocity, and acceleration of the given joint as well as the relative angle between the segments and the rotation velocity and acceleration of the segments can then be obtained from a frame-by-frame analysis of the videotape. 3-D measurements are usually complex and require sophisticated methods for acquiring and analyzing the kinesiological data. A full description of kinesiology is beyond the scope of this chapter. For more information, refer to studies reported in the literature [15,17,18].

#### *3.3.4.6 Tremor*

The muscular contraction needed to move a body part or maintain it in a fixed position is accompanied by small, arrhythmic, involuntary oscillations in the muscle forces that are not visible to the untrained eye. These muscle oscillations are referred to as *physiological tremor.* When muscles become fatigued or damaged, the physiological tremor increases, and large-amplitude monorhythmic oscillations are observed. Musculoskeletal function can thus be assessed by measuring the amplitude and frequency of these oscillations. Findley and Capildeo [19] have edited a book that describes methods of assessing movement disorders from tremor measurements. In addition, Galinsky et al. [20] describe a portable device that can be used to measure tremor in a field environment.

### 3.3.5 Metabolic Considerations

As mentioned previously, the basis of muscular contraction is the transformation of chemical energy derived from food taken into the body into useful mechanical energy in the form of muscular contractions. To achieve this transformation, high-energy phosphate compounds, such as adenosine triphosphate (ATP) and phosphocreatine (PCr), provide the chemical energy needed for muscular contractions. These high-energy-yielding compounds are crucial in both aerobic metabolism (metabolic processes that use oxygen) and anaerobic metabolism (metabolic processes that occur without oxygen).

#### *3.3.5.1 Functional Fiber Types*

Because a whole muscle is required to operate across a wide range of exertion conditions, it is composed of a mixture of different functional types of muscle fibers. Based on the amount of time it takes a muscle fiber to reach its peak tension, two types of muscle fibers have been identified: slow (Type I, or red) and fast (Type II, or white). The slow fibers, which take 80–100 ms to reach peak tension, have more myoglobin and rely on aerobic metabolism. They do not fatigue easily, and therefore a task can be maintained over long periods (e.g., endurance running activities). The fast fibers, which usually take about 40 ms to reach peak tension, have higher concentrations of glycolytic enzymes and glycogen and rely on anaerobic metabolism. The fast fibers fatigue easily, but they are best suited for strong and quick body movements (e.g., weight-lifting activities). The exact reasons behind the nature of the functional behavior of these fibers are still not clear. The literature suggests that the type of nerve fibers that innervate these muscle fibers dictate their functional behavior [2]. The slow muscle fibers are innervated by small-diameter, low-conduction velocity nerve fibers, whereas the fast muscle fibers are innervated by large-diameter, high-conduction velocity nerves. The slow fibers are always active at low levels of contraction, providing a sustained tonic muscle activity such as that required for maintaining posture of the body. The fast fibers are only active during strong movements of the body segments. Most body motions require a combination of slow and fast motor activities. For example, a hand–wrist manipulative task of using a screwdriver above head level may require slow unit activity of the shoulder muscles to stabilize the posture of the arm, but fast units of the hand wrist muscle are needed to turn the screwdriver. A muscle may consist of anywhere between 10% and 90% of any one type, but most muscles have a relatively even distribution (soleus muscle may range as high as 70% Type I).

#### *3.3.5.2 Aerobic Metabolism and Bloodflow*

Muscular exertions requiring aerobic metabolism need ample blood flow to carry oxygen to the tissues and to carry away metabolic by-products. Therefore, it is essential that blood flow be maintained to muscles with high workloads. When a muscle is contracted at high tonic levels, such as when a sustained static posture is required, the muscle contractions may inhibit adequate blood flow, thereby reducing the capability for aerobic metabolism. From an ergonomic perspective, it is important to limit static postures so that physiologic function is not compromised.

Maximum aerobic power, which is defined as the highest oxygen uptake an individual can attain during exercise, is a measure of the capability of the cardiovascular system to provide oxygen to the muscles for aerobic metabolism [3]. Maximum aerobic power has been measured as high as 7.4 L/min for a male and 4.5 L/min for a female cross-country skier [3]. The mean for an industrial population would be about 3.0 L/min for men and about 2.0 L/min for women [3,16]. Refer to Chapters 2 and 9 in this book for more information on maximum aerobic power, physical work capacity, and cardiovascular capacity.

#### *3.3.5.3 Anaerobic Metabolism*

During activities requiring strong muscular contractions, anaerobic metabolism plays a crucial role in providing energy to the muscles. Energy is provided to the muscles by the anaerobic breakdown of ATP, PCr, and glycogen in a lactic phase. It is difficult to measure an individual's capacity for anaerobic metabolism, but it is known that there is limited capability to sustain high workloads due to the limited supply of energy-yielding substrates. High sustained workloads will result in the buildup of high concentrations of lactate in the muscle, which is removed by the circulatory system. When a muscle has depleted its stores of anaerobic substrate and the workload is higher than about 50% of the maximum voluntary contraction, the muscle may begin to lose strength and become fatigued.

#### *3.3.5.4 Oxygen Debt*

During the recovery period following exercise, the amount of oxygen consumed in excess of the resting value is called *oxygen debt*. The higher the exercise level or workload, the higher is the level of oxygen debt incurred. In an exhaustive workload, the energy demand is not met adequately by aerobic metabolism; therefore, anaerobic energy production provides the necessary energy, causing lactic acid to build up. In other words, the more strenuous the workload, the longer it takes to achieve preexercise-level metabolism. Therefore, the concept of oxygen debt is critical in the design of a work (i.e., exercise) and rest regimen so that a task can be performed without experiencing fatigue. The literature provides recommendations of work–rest regimens using the principles of oxygen consumption and oxygen debt [3,16].

#### *3.3.5.5 Local Muscle Fatigue*

Static as well as dynamic muscular contractions can result in local muscle fatigue. Local muscle fatigue occurs when the endurance time for the muscle is exceeded. The endurance time for a muscle is dependent on the amount of force developed by the muscle as a percentage of the maximum force attainable by the muscle. For example, a muscle can sustain a force of about 15% of its maximum indefinitely without becoming fatigued, but it can sustain 50% of its maximum force for only about 1 min [3]. Similarly, a muscle can sustain a repetitive contraction rate of about 30 contractions/min if the force is about 60% of maximum but can sustain a rate of only about 10 contractions/min if the force is about 80% of maximum [3].

#### *3.3.5.6 Whole Body Fatigue*

When the metabolic demands of dynamic and sustained activity exceed the energy-producing capacity of a worker, muscle contraction is affected and whole body fatigue is usually experienced. Physiologists generally recommend that energy expenditure not exceed about 50% of maximum aerobic power for 1 h of work, about 40% for 2 h of work, and about 33% for 8 h of continuous work [21]. These values are designed to prevent fatigue, which is believed to increase a worker's risk of musculoskeletal injury. Intervals of heavy, continuous work should be separated by light duty jobs, so that recovery can occur. For more information on whole body fatigue, refer to Astrand and Rodahl [3] and McArdle et al. [22].

## REFERENCES


1. A. J. Vander, J. H. Sherman, and D. S. Luciano, *Human Physiology: The Mechanism of Body Function*, McGraw-Hill, New York, 1970.
2. A. C. Guyton, *Function of the Human Body*, Saunders, Philadelphia, PA, 1969.
3. P. Astrand and K. Rodahl, *Textbook of Work Physiology*, McGraw-Hill, New York, 1986, pp. 19, 115, 334.
4. B. Tyldesley and J. I. Grieve, *Muscles, Nerves and Movement: Kinesiology in Daily Living*, Blackwell Scientific, London, U.K., 1989.
5. A. M. Burt, *Textbook of Neuroanatomy*, Saunders, Philadelphia, PA, 1993.
6. J. Dudel, in *Fundamentals of Neurophysiology*, R. F. Schmidt, Ed., Springer-Verlag, New York, 1978, Chapter 5, pp. 129–137.
7. V. B. Brooks, *The Neural Basis of Motor Control*, Oxford University Press, New York, 1986.
8. R. M. Enoka, *Neuromechanical Basis of Kinesiology*, Human Kinetics, Champaign, IL, 1988.
9. W. S. Marras, Toward an understanding of dynamic variables in ergonomics, *Occup. Med.: State Rev.* *7*(4):655–677 (1992).
10. DHHS(NIOSH), *Selected Topics in Surface Electromyography for Use in the Occupational Setting: Expert Perspectives*, Department of Health and Human Services, National Institute for Occupational Safety and Health, Cincinnati, OH, 1992.
11. A. Bhattacharya, R. Shukla, K. Dietrich, and R. Bornschein, Effect of early lead exposure on the maturation of children's postural balance: A longitudinal study, *Neurotoxicol. Teratol.* *28*:376–385 (2006).
12. R. Seliga, A. Bhattacharya, P. Succop, R. Wickstrom, D. Smith, and K. Willeke, Effect of workload and respirator wear on postural stability, heart rate and perceived exertion, *Am. Ind. Hyg. Assoc. J.* *52*(10):417–422 (1991).

13. A. Bhattacharya, R. Shukla, E. Auyang, K. N. Dietrich, and R. L. Bornschein, Effect of succimer chelation therapy on postural balance and gait outcomes in children with early exposure to environmental lead, *Neurotoxicology 28*:686–695 (2007).
14. J. S. Standridge, A. Bhattacharya, P. S. Succop, C. Cox, and E. Haynes, Effect of chronic low level manganese exposure on postural balance: A pilot study of residents in southwest Ohio, *J. Occup. Environ. Med. 50*(12):1421–1429 (December 2008).
15. D. B. Chaffin, G. B. J. Andersson, and B. Martin, *Occupational Biomechanics*, 4th edn., Wiley, New York, 2006.
16. Eastman Kodak, *Kodak's Ergonomic Design for People at Work*, John Wiley & Sons Inc., Hoboken, NJ, 2003.
17. P. Allard, I. A. F. Stokes, and J. P. Blanchi (Eds.), *Three-Dimensional Analysis of Movement*, Human Kinetics, Champaign, IL, 1995.
18. W. J. Vincent, *Statistics in Kinesiology*, Human Kinetics, Champaign, IL, 1995.
19. L. J. Findley and R. Capildeo (Eds.), *Movement Disorders: Tremor*, Oxford Press, New York, 1984.
20. T. L. Galinsky, R. R. Rosa, and D. D. Wheeler, Assessing muscular fatigue with a portable tremor measurement system suitable for field use, *Behav. Res. Methods, Instrum. Comput. 22*(6):507–516 (1990).
21. T. Waters, V. Putz-Anderson, A. Garg, and L. J. Fine, Revised NIOSH equation for the design and evaluation of manual lifting tasks, *Ergonomics 36*(7):749–776 (1993).
22. W. D. McArdle, F. I. Katch, and V. L. Katch, *Exercise Physiology: Energy, Nutrition, and Human Performance,* Lea & Febiger, Philadelphia, PA, 1981.

CHAPTER 4

# Biomechanical Aspects of Body Movement

Angus Bagchee and Amit Bhattacharya

CONTENTS

## 4.1 INTRODUCTION

A significant number of the principles of ergonomics are derived from biomechanics, which can be defined as the systematic study of the human body as governed by the laws of physics. Laws of classical mechanics, in the form of fluid mechanics, statics and dynamics, and solid-state mechanics are applied to the scientific study of human posture and movement. Several definitions of the term *biomechanics* have been proposed by various authors. Fung [1] describes it as the mechanism of the living system. It is simply the application of the laws of mechanics to the biological system. The study of biomechanics dates back to the medieval ages, where the references to physical laws that govern human body movements can be found in the works of Leonardo da Vinci (1452–1519). Advances in the last two decades in the field of biomechanics have helped us understand several important physiological functions, like the role of the neuromuscular system in human movement, fluid mechanical principles governing the blood flow and respiratory systems, tissue biomechanics, and the finite-element modeling of bone structures.

The application of biomechanics to the occupational field, however, did not gain momentum until after World War II. Occupational biomechanics, a subfield of biomechanics, can be defined as the application of biomechanical principles to characterize and evaluate the effect of work task demand on the kinematic and kinetic responses of the workers [2,3].

## 4.2 BACKGROUND AND SIGNIFICANCE TO OCCUPATIONAL ERGONOMICS

With the advent of the industrial revolution, the emphasis shifted toward mass production. Occupational injuries and diseases were either neglected as an inevitable side effect of a mass production system or were attributed to several other factors not related to the workplace. In the 1700s, Ramazzini suggested that the cause of such occupational diseases was the use of excessive force and unnatural postures [4]. It was not until recently that the principles of biomechanics were applied to a systematic study of human performance in the work environment.

Applied biomechanics forms the primary structure of occupational ergonomics, dealing with characterization of the loading of the musculoskeletal system. From the simple lever system that helps in quantitating the loading of the musculoskeletal system due to manual lifting of a weight, to complex measurements of interdiscal pressure to estimate the loading of the spinal unit, biomechanics helps us explain and improve a number of ergonomic problems and is a rapidly expanding field of study. Occupational biomechanics is applied in the determination of the bone–muscle-joint loading of the worker due to his or her interaction with tools, equipment, and the workplace. It also provides scientific guidelines for developing new tools that will reduce musculoskeletal disorders and for developing and modifying workstations to reduce worker discomfort. Progress in computer technology has allowed the development of computer models for predicting musculoskeletal loading associated with the performance of certain tasks. It allows, through the use of these models, the development of safe weight-lifting limits, chair design, and workplace layouts to conform to the specific working population. Since occupational biomechanics combines engineering concepts and the laws of physics with medicine, a multidisciplinary approach is required with expertise from a number of other fields, including bioinstrumentation, kinesiology, physiology, engineering, occupational therapy, rehabilitation engineering, and several other allied fields. This chapter outlines the major concepts of biomechanics that are relevant to occupational ergonomics. Appropriate references are cited throughout the text for in-depth information on specific topics.

## 4.3 RELEVANT CONCEPTS AND TERMINOLOGY

### 4.3.1 Principles of Applied Mechanics

Mechanics is the branch of physics that deals with the motion and deformations resulting from forces and moments acting upon a body. Applied mechanics is limited to the application of these principles to rigid bodies, deformable bodies, or fluids. Though the human body segments are somewhat deformable, for simplification they are assumed to be rigid bodies acted upon by point forces and moments. Depending on the state of the object under the influence of disturbing forces, the branch of mechanics is further subdivided into *statics*, which deals with bodies that are in a state of static equilibrium ("at rest"), and *dynamics*, which deals with bodies that are in a state of motion due to the action of the acting forces. The principles of statics are useful in determining the loading on the musculoskeletal system and the forces acting at the different body segment joints and muscles, while the principles of dynamics are useful in determining the forces generated when the body is in motion, as in running or diving.

Newtonian mechanics is based on the three dimensions of length, width, and height, along with the measurement of time. Quantities that require only magnitude for their description are known as scalars. Some examples of scalar quantities are mass, time, temperature, and speed. Vector quantities require the following for their specification: (1) magnitude, (2) direction, and (3) point of application. Examples of vector quantities include velocity, force, and momentum. In a three-dimensional case, a velocity vector **v** may be expressed as $\mathbf{v} = a\mathbf{i} + b\mathbf{j} + c\mathbf{k}$, where **i**, **j**, and **k** are three orthogonal unit vectors representing the three-dimensional axis system (Figure 4.1). The quantities *a*, *b*, and *c* are magnitudes (real numbers). Although scalar quantities can be added and multiplied algebraically, vector quantities require special mathematical methods for performing these operations. Some of the mathematical operations with vector quantities are described next. Refer to Özkaya and Nordin [5] for a detailed treatment of the subject as applied to biomechanics. It is customary to use boldface letters or letters with a bar or arrowhead over them to signify vectors. Boldface letters—**u**, **v**, **w**, etc., are used here to denote vector quantities.

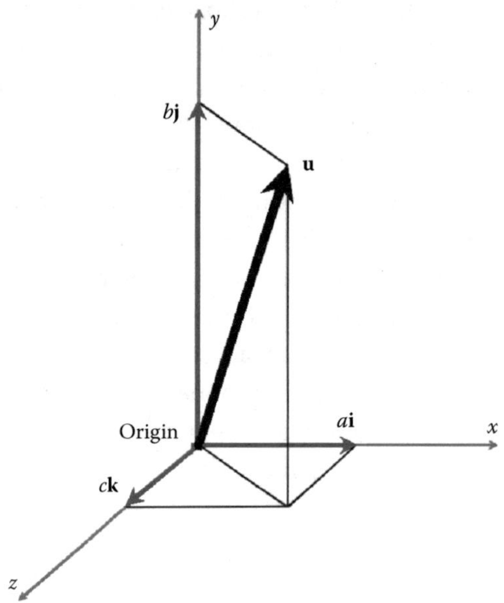


FIGURE 4.1 Representation of a vector quantity.

### 4.3.1.1 Vector Addition

Two vectors $\mathbf{u} = a_1\mathbf{i} + b_1\mathbf{j} + c_1\mathbf{k}$ and $\mathbf{v} = a_2\mathbf{i} + b_2\mathbf{j} + c_2\mathbf{k}$ can be added by adding the individual components along the **i**, **j**, and **k** directions as follows:

$$\mathbf{w} = \mathbf{u} + \mathbf{v} = (a_1 + a_2)\mathbf{i} + (b_1 + b_2)\mathbf{j} + (c_1 + c_2)\mathbf{k}.$$

### 4.3.1.2 Vector Product

Two different kinds of products can be formed with vectors. The *dot* or *scalar product*, which has a magnitude but no direction, is formed as follows:

$$\mathbf{w} = \mathbf{u} \cdot \mathbf{v} = |\mathbf{u}| \; |\mathbf{v}| \cos\theta = (a_1 \times a_2 + b_1 \times b_2 + c_1 \times c_2),$$

where θ is the angle between the two vectors **u** and **v**.

The *cross* or *vector product* between the two vectors, resulting in another vector with both magnitude and direction, is given by

$$\begin{aligned} \mathbf{w} = \mathbf{u} \times \mathbf{v} &= |\mathbf{u}| \; |\mathbf{v}| \sin\theta \\ &= (b_1c_2 - b_2c_1)\mathbf{i} + (a_1c_2 - a_2c_1)\mathbf{j} + (a_1b_2 - a_2b_1)\mathbf{k}. \end{aligned}$$

Figure 4.2 shows the vectors **u** and **v** and their cross product **w**. The direction of the resultant vector **w** is perpendicular to the plane formed by the vectors **u** and **v**, shown as plane P in Figure 4.2. If **u** and **v** lie in the plane of this page, **w** would then be sticking into or out of the page. The exact direction of the resultant vector **w** can also be given by the *right-hand rule*, whereby the direction of the vector **w** is along the thumb if the rest of the

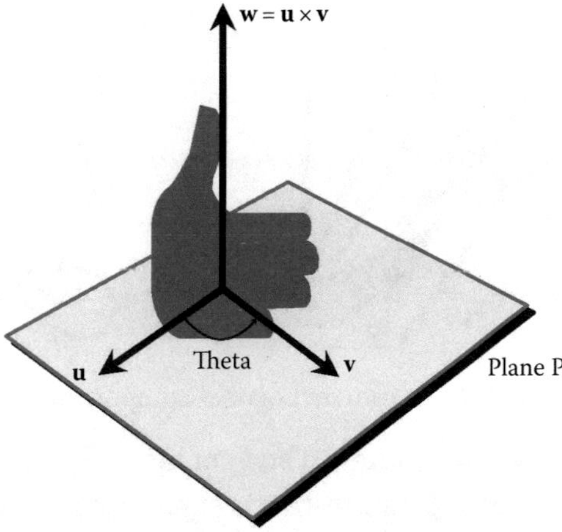


FIGURE 4.2 Right-hand rule used in calculating cross product.

fingers curl up from **u** toward **v** (see Figure 4.2). Note that the vector product is not commutative, that is, $\mathbf{u} \times \mathbf{v} \neq \mathbf{v} \times \mathbf{u}$.

#### *4.3.1.3 Resolving Vectors into Components*

Vector quantities can be resolved into their component vectors along any set of directions. Vectors are usually resolved into components along mutually perpendicular directions that form the *axis system*. An axis system is important to specify the location of any vector in space. In the previous example, $\mathbf{u} = a\mathbf{i} + b\mathbf{j} + c\mathbf{k}$ may be resolved into components along the orthogonal axis system formed by $x$, $y$, and $z$ (**i**, **j**, and **k** are unit vectors along $x$, $y$, and $z$), as shown in Figure 4.1. This kind of coordinate system is the most widely used and is known as a *rectangular* or *Cartesian coordinate system*. In a two-dimensional system, the $z$ axis is absent. For example, $\mathbf{q} = x\mathbf{i} + y\mathbf{j}$ represents a two-dimensional vector.

#### *4.3.1.4 Forces*

To understand the concepts of mechanics as applied to the human body, it is important to be familiar with several terms associated with forces and moments. *External* forces act on the outside of the object; examples of such forces are gravity, inertia, and ground reaction forces. *Internal* forces act within the structure of the body in reaction to the external forces and are called *stresses*. If we consider the forces in the body segments due to external forces like the gravitational pull of a load being carried in the hand, then the muscle forces generated can be assumed to be internal forces opposing the load. A force is a vector quantity that requires direction, magnitude, and point of application for it to be specified.

The application of forces on a rigid body may or may not cause the body to move or change direction. Figure 4.3 shows a force vector **F** acting at the forearm. The rotation of the arm occurs about the instantaneous center of rotation located at the elbow joint. This point can be considered as the origin, with the $x$ axis along the forearm and the $y$ axis orthogonal to it. The direction of the force vector can be specified with respect to this coordinate system. Such a coordinate system is termed the local or joint-based coordinate system. In a metric system of measurement, forces are expressed in newtons (N), where 1 N

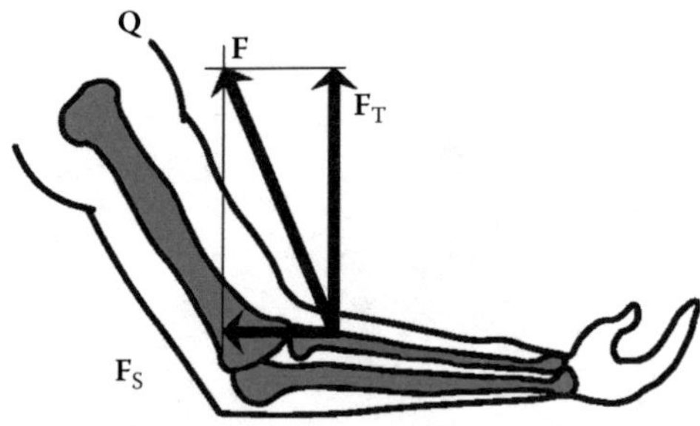


FIGURE 4.3 Resolution of muscle force into orthogonal components.

is defined as the force required to maintain a body of mass 1 kg at a constant acceleration of 1 m/s². Force in the U.S. customary unit system is expressed in pounds (lb).

### 4.3.1.5 Newton's Laws of Motion

The basis of mechanics is formed primarily by Newton's laws of motion (Newtonian mechanics). Briefly these laws can be described as follows:

*First Law*: A body tends to remain in its inertial state of rest or motion unless and until acted upon by an external disturbing force.

*Second Law*: The net rate of change of momentum of a body is equal to the external forces acting on the body. Mathematically, this can be stated as

$$\mathbf{F} = \frac{\Delta(m\mathbf{v})}{t},$$

where $\Delta(m\mathbf{v})$ denotes the change in the momentum of the body. For a constant mass, the equation becomes

$$\mathbf{F} = \frac{\Delta(m\mathbf{v})}{t} = m\mathbf{a}.$$

where

$\mathbf{F}$ denotes the net force (algebraic sum of all forces) acting on the rigid body
$\mathbf{a}$ is **the** acceleration achieved by the body as a result of the force
$m$ is the mass of the body

This forms the basis for the governing equations discussed later.

*Third Law*: For every action of force, there is always an equal and opposite reaction force produced. Thus, forces work in pairs. As an example, ground reaction forces are produced due to the action of body force on the ground.

### 4.3.1.6 Moments

Moments are formed by forces acting in a certain manner on a rigid body that causes the body to rotate about an axis. For example, a push applied to the edge of the door causes the door to rotate about its hinges. Similarly, torque is a moment that causes the body to twist as a result of a force. Moment is a vector quantity and is obtained by forming a cross product of the distance vector with the force vector, as follows:

$$\mathbf{M} = \mathbf{r} \times \mathbf{F} = |\mathbf{r}|\,|\mathbf{F}|\sin\theta,$$

where **r** is the distance vector from the axis of rotation to the force vector **F**. If $d$ is the perpendicular distance of the force vector **F** from the axis, the aforementioned expression simplifies to a moment vector with magnitude $M = dF$ and the direction given by the right-hand rule. In the metric system, moments are expressed in newton-meters (N-m); in the U.S. customary system it is expressed in pound-feet (lb-ft) or pound-inches (lb-in.).

#### *4.3.1.7 Governing Equations*

Newton's second law of motion states that the net resultant force on the body is equal to the rate of change of momentum for the body. Mathematically,

$$\Sigma\mathbf{F} = m\mathbf{a},$$

where

$\Sigma\mathbf{F}$ denotes the vector sum of all the forces acting on the rigid body
**a** denotes the resulting acceleration of the body due to the action of the forces

A similar expression can be stated for the moments acting on the body:

$$\Sigma\mathbf{M} = \boldsymbol{I\alpha},$$

where

$\Sigma\mathbf{M}$ denotes the vector sum of all external moments acting on the rigid body
$\boldsymbol{\alpha}$ is the resulting angular acceleration
$\boldsymbol{I}$ is the moment of inertia of the rigid body about the axis of rotation

Moment of inertia is the resistance to change in angular velocity, $\boldsymbol{I} = \Sigma m\mathbf{r}^2$.

#### *4.3.1.8 Equilibrium*

Equilibrium is a special state of the body in which the body either is at rest (static equilibrium) or is moving with a constant velocity (dynamic equilibrium). In the latter case, a constant velocity signifies zero acceleration ($\mathbf{a} = 0$; $\boldsymbol{\alpha} = 0$). In cases of either static or dynamic equilibrium, the net acceleration of the body is zero, and the previous governing equations can be written as

$$\Sigma\mathbf{F} = 0 \quad \text{and} \quad \Sigma\mathbf{M} = 0,$$

where the linear and angular acceleration of the body are both zero. Though a true condition of equilibrium is not always achieved, many biomechanical systems can be approximated to be in a state of equilibrium to simplify the evaluation of forces and moments. Care should be taken, however, to understand the factors and events that may contribute to violate the assumption of equilibrium and place the calculated forces and moments in error.

For example, a running activity is highly dynamic, and thus principles of static equilibrium may not be applicable in calculating the joint forces and moments for such an activity.

### 4.3.2 Statics

Statics is concerned with bodies that are in a state of rest or under a static equilibrium. The forces acting upon the body in a static state are balanced; in other words, the net resultant of the forces acting on the body is zero. The principles of statics can be applied to stationary or near-stationary body segments.

As discussed earlier, forces acting on a body can be expressed as vectors. These forces acting at various segments of the human body are both two- and three-dimensional in nature. In most cases, the degrees of freedom for these segments are limited to a single plane. For example, the elbow joint is formed by the upper and lower arm, which move relative to each other in a single plane when performing simple activities such as lifting. Thus, in most cases, the forces can be approximated to be two-dimensional. Proper placement of the local coordinate system can yield values of forces and moments that are biomechanically useful and easier to interpret. Forces acting on the body segment, being vector quantities, can be resolved into components along the axes, as illustrated in the following example. The reader is encouraged to consult further references for a detailed treatment of vectors [6].

**Example Problem 1**

Figure 4.3 shows the force **F** generated by the biceps brachii of the forearm. The magnitude of the biceps force vector **F** is 40 N, and it is making an angle of 60° with the forearm. We wish to calculate the force component along the forearm (stabilizing force), $\mathbf{F}_S$, and the force component perpendicular to the forearm, $\mathbf{F}_T$ (tangential, so called because it is acting at a tangent to the circular path in which the forearm will rotate due to the force).

**Solution**

The forces $\mathbf{F}_T$ and $\mathbf{F}_S$ can be calculated by resolving the force **F** into its components (see Figure 4.3) along the direction of the forearm ($\mathbf{F}_S$) and along a direction perpendicular to the forearm ($\mathbf{F}_T$), using the following set of equations [6]:

$$\mathbf{F}_T = |\mathbf{F}|\sin\theta = 40.0 \times \sin 60^\circ = 34.64 \text{ N}$$

$$\mathbf{F}_S = |\mathbf{F}|\cos\theta = 40.0 \times \cos 60^\circ = 20.0 \text{ N}.$$

The $\mathbf{F}_T$ component of the force **F** in the previous example is called the "rotational" force, for it has the effect of rotating the forearm about the elbow joint. The $\mathbf{F}_S$ component is the "stabilizing" component; it compresses or pulls apart the joint, depending on the direction of the force.

Application of force gives rise to moment or torque as described earlier and may cause the body segment to rotate. Thus, moment can be calculated by multiplying the force by the perpendicular distance of the force vector from the pivotal axis. In the previous example, if the distance of the point of application of the tangential force $\mathbf{F}_T$ from the elbow (pivotal point) is $d = 5$ cm, then the moment would be calculated as $\mathbf{M} = d\mathbf{F}_T = 34.64\text{N} \times 5\text{cm} = 173.2\text{N-cm} = 1.732\text{N-m}$.

Note that the direction of the moment vector **M** is given by the right-hand rule discussed earlier and would be along an axis perpendicular to the plane of the forces, coming out of the plane of the paper in the aforementioned example. Also note that the component $\mathbf{F}_S$ will not produce any moment as it passes through the pivotal axis, and the corresponding $d$ is zero.

In a condition of static equilibrium, the body remains stationary under the action of external forces ($\mathbf{F}_1$, $\mathbf{F}_2$, $\mathbf{F}_3$, $\mathbf{F}_4$, $\mathbf{F}_5$,…) and moments ($\mathbf{M}_1$, $\mathbf{M}_2$, $\mathbf{M}_3$, $\mathbf{M}_4$, $\mathbf{M}_5$,…). Using the equation stated earlier for equilibrium, the net algebraic sum of the forces and moments acting on the body at any time is zero and can be expressed by the following equations:

$$\Sigma\mathbf{F} = \mathbf{F}_1 + \mathbf{F}_2 + \mathbf{F}_3 + \mathbf{F}_4 + \cdots = 0$$

$$\Sigma\mathbf{M} = \mathbf{M}_1 + \mathbf{M}_2 + \mathbf{M}_3 + \mathbf{M}_4 + \cdots = 0.$$

This sum of vectors can be simplified for a two-dimensional case by equating all the components of forces and moments to zero:

$$\sum_{i=1}^{i=n} (\mathbf{F}_x)_i = 0, \quad \sum_{i=1}^{i=n} (\mathbf{F}_y)_i = 0, \quad \sum_{i=1}^{i=n} (\mathbf{M}_z)_i = 0.$$

Here, the forces $(\mathbf{F}_x)_i$ and $(\mathbf{F}_y)_i$ lie in a plane while $(\mathbf{M}_z)_i$ is along an axis perpendicular to the plane of the forces. Using these equations, any set of forces and moments can be added algebraically to obtain the unknown forces and moments in a system at equilibrium. We will be applying the concept of equilibrium when solving for forces in a lever system.

### 4.3.3 Lever System

A lever system operates on the application of the governing laws of mechanics discussed previously. It is the simplest form of machine system existing in nature. One example of this system is the seesaw system shown in Figure 4.4. The shortest distance of the load and effort forces from the instantaneous center of rotation (or fulcrum) are called the *load arm* and *effort arm*, respectively. Taking the moments about the fulcrum F, and applying the principles of static equilibrium, we have

$$\Sigma\mathbf{M}_z = 0$$

$$\text{Load} \times \text{Load arm} - \text{Effort} \times \text{Effort arm} = 0$$

or

$$\text{Load} \times \text{Load arm} = \text{Effort} \times \text{Effort arm}.$$

In terms of the quantities shown in Figure 4.4,

$$\mathbf{L}d_\text{L} = \mathbf{E}d_\text{E}$$

$$\mathbf{E} = \frac{d_\text{L}}{d_\text{E}}\mathbf{L}.$$

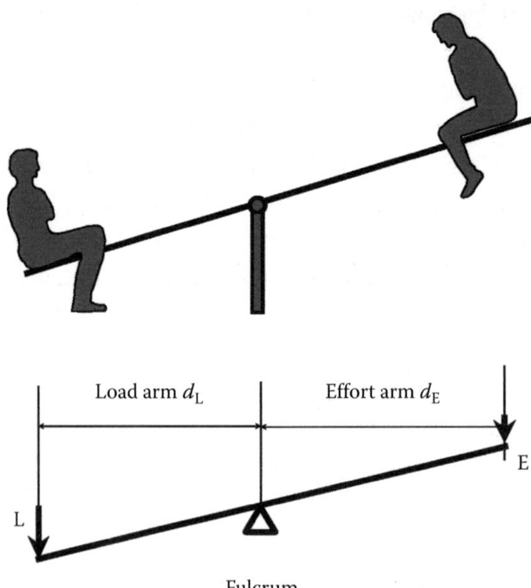


FIGURE 4.4 Lever system.

Thus,

$$\text{Effort} = \frac{\text{load arm}}{\text{effort arm}} \times \text{load}.$$

The amount of force, or effort, needed to balance the load can be varied by changing the ratio of the load arm to the effort arm. Hence, the placement of the fulcrum with respect to the load and the effort is critical in determining the value of the force needed to balance the load. *Mechanical advantage* is defined as the ratio of the load to the effort required to balance it. From the aforementioned equation, this is also equal to (effort arm)/(load arm). A mechanical advantage value greater than unity means that the effort required to balance the load is *smaller* than the load itself. Most of the lever systems found in the human body actually work at a mechanical disadvantage (mechanical advantage < 1.0), requiring greater effort than the balanced load. On the other hand, having a smaller effort arm is anatomically advantageous, because the muscle has to move very little to achieve a large movement of the load.

Several examples of lever systems can be found in the human body. There exist three classes of lever systems, based on the positioning of the load and effort with respect to the fulcrum. These are termed (1) first-class; (2) second-class; and (3) third-class lever systems. The system in which the fulcrum lies between the load and the effort, as shown in Figures 4.4 and 4.5a, is known as a first-class lever system. An example of this kind of lever system in the human body exists at the neck joint as shown in Figure 4.5a. A second-class lever system has the load situated between the effort and the fulcrum. An example of the second-class of lever system can be found in the human body at the ankle joint, where standing on the toes would require the forces through the Achilles tendons to balance the body weight

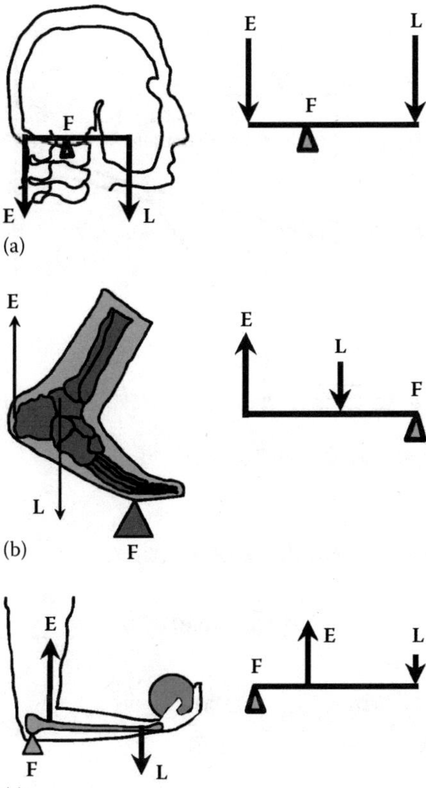


FIGURE 4.5 Three classes of lever systems. Examples and their schematic representation. (a) Lever system of first class (b) lever system of second class (c) lever system of third class.

with the fulcrum at the big toe, so as to produce a second-class lever system as shown in Figure 4.5b. A third-class lever system is the type most commonly found in the human body. Here, the effort lies between the load and the fulcrum. The effort required to counterbalance the load is always greater than the load. Figure 4.5c shows a third-class lever system. An example of the third class of lever system is found in the forearm, where the effort is exerted by the biceps brachii muscles as shown in Figure 4.5c.

The principles of a lever system are useful in determining the muscle forces required to perform several activities of daily life. They also provide better understanding of the importance of a reduced load arm in order to reduce the effort (muscle force) required for performing the task. Several examples are presented next to illustrate the principles of levers as they apply to the various body segments.

**Example Problem 2**

Calculate the lifting force required by the operator to lift the wheelbarrow shown in Figure 4.6. The load **L** of the wheelbarrow is 70 kg. The perpendicular distances of the load vector **L** and the effort vector **E** from the instantaneous center of rotation (fulcrum) are 30 and 120 cm, respectively. What class of lever system is illustrated in this example? Also, calculate the mechanical advantage of the system.

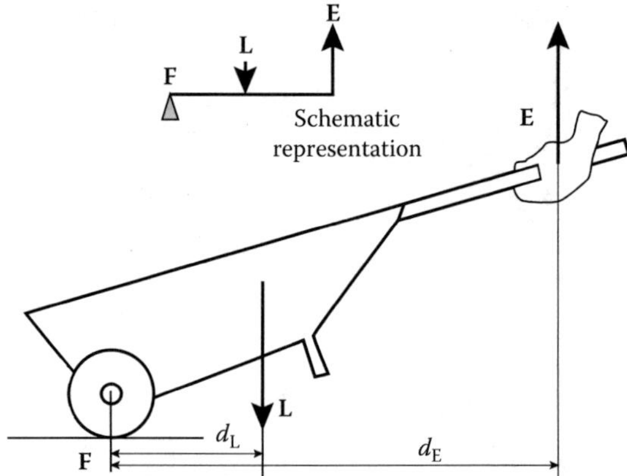


FIGURE 4.6 Lever system applied to a wheelbarrow.

**Solution**

Using the equation for a lever system described earlier:

$$\text{Effort} = \text{load arm}$$

$$\text{Effort} = \frac{\text{load arm}}{\text{effort arm}} \times \text{load} = \frac{30}{120} \times 70 = 17.5 \text{ kg } (=171.7 \text{ N}).$$

The mechanical advantage is defined as follows:

$$\text{Mechanical advantage (MA)} = \frac{\text{load}}{\text{effort}}.$$

Since load × load arm = effort × effort arm, we have

$$\text{MA} = \frac{\text{effort arm}}{\text{load arm}} = \frac{120}{30} = 4.0.$$

**Example Problem 3**

Figure 4.7 shows a worker performing a task on a table that requires bending of the torso, head, and arms. The balancing force can be assumed to be provided by the extensor muscles of the back, shown by the vector **E** acting at a distance of $d_e$ (=5 cm) from the instantaneous center of rotation (fulcrum) situated between the fifth lumbar and first sacral ($L_5S_1$) vertebrae. The approximate location of the gravitational forces (loads) due to the head, trunk, and arms from the fulcrum are also shown. The values of loads and their load arm distances for a 61 kg (600 N) man are given in Table 4.1.

Calculate the muscle force **E** required for this lever system. What is the total weight (or force) experienced by the $L_5S_1$ vertebrae? Assume that the component of **E** acting through $L_5S_1$ is given by **E** cos θ, where θ is the angle of inclination of the vector **E** to the vertical direction (assume θ = 30°).

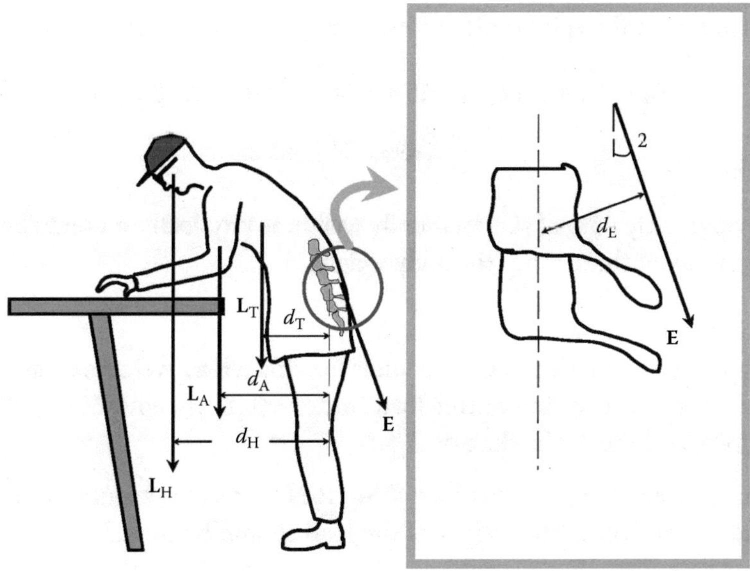


FIGURE 4.7 Calculation of back muscle forces.

TABLE 4.1

| Load | Load Arm (cm) |
|---|---|
| $\mathbf{L}_H$ = 8.4% of bw = 50.4 N | 55.0 |
| $\mathbf{L}_A$ = 5.1% of bw = 30.6 N | 30.0 |
| $\mathbf{L}_T$ = 50% of bw = 300.0 N | 13.0 |

*Note:* bw is the body weight (600 N).

**Solution**

Summing up the load moment about the fulcrum,

$$\mathbf{L}_H d_H + \mathbf{L}_A d_A + \mathbf{L}_T d_T = 50.4 \times 55 + 30.6 \times 30 + 300 \times 13$$
$$= 7590 \text{ N cm}.$$

Equating the net moment about the fulcrum to zero, we have

$$\mathbf{E} \times d_E = 7590.$$

Therefore,

$$\mathbf{E} = 7590/d_E = 7590/5 = 1518 \text{ N}.$$

In terms of body weight, this effort is 1518/600.0 = 2.53 bw.

The total loading on the spinal column is the net sum of forces acting on $L_5S_1$:

$$\mathbf{L}_H + \mathbf{L}_A + \mathbf{L}_T + \mathbf{E}\cos\theta = 50.4 + 30.6 + 300 + 1518 \times \cos 30^\circ$$

$$= 1695.6 \text{ N (2.83 times bw)}.$$

Thus, even though only 63% of the total body weight is physically present above $L_5S_1$, the actual loading is about three times the body weight.

**Example Problem 4**

Assume that, in the previous case, the person is holding a load weighing 1 kg (10 N) in his hands. Assume that the distance of this load from his $L_5S_1$ is about 60 cm. What are the total muscle force and the net loading on $L_5S_1$?

**Hint**: Add the moment due to the load weighing 10 N to the aforementioned equation and reevaluate the muscle force. The weight of the load should be included when calculating the net loading on $L_5S_1$.

**Answer**: 1638 N (2.73 bw); 1809.5 (3.0 bw).

**Example Problem 5**

Figure 4.8 shows three postures, commonly found at the workplace: (a) standing upright, (b) sitting upright, (c) sitting and performing a task at arms' reach. The gravitational force of the upper body is assumed to be acting along the line G. The distance of the line G

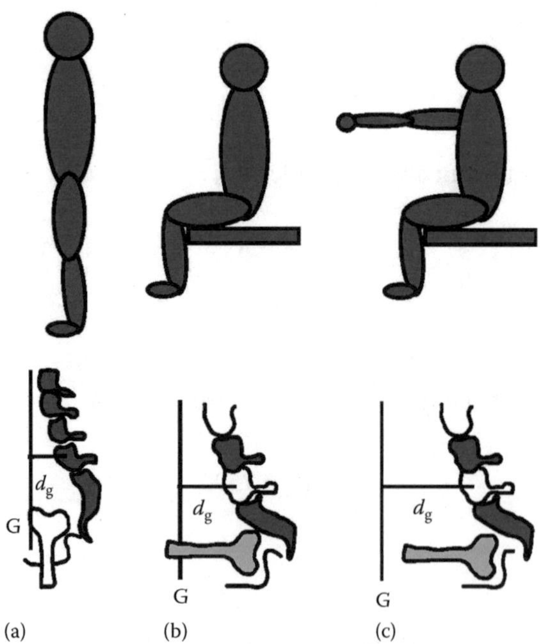


FIGURE 4.8 Influence of posture in the loading of the spinal column. (a) Standing (b) sitting (c) sitting with arms extended. (Modified from Eastman Kodak Co., *Ergonomics for People at Work*, Van Nostrand Rein-hold, New York, 1986.)

from the fifth lumbar and first sacral vertebrae ($L_5S_1$) is shown to be (a) 8 cm in a standing posture, (b) 8.5 cm in a seated posture, and (c) 10 cm in a seated posture with arms extended for stenographic work. The weight of the head, trunk, and arms constitute 62% of the body weight. The balancing moment is provided by the back extensor muscles. The distance of the line of back extensor muscle force **F** acting on the vertebrae ($d_F$) is assumed to be 4 cm. Calculate the resultant loading **R** on the $L_5S_1$ vertebrae for a woman weighing 600 N (61 kg). What are the biomechanical implications of these postures based on your calculations?

**Hint**: Use the following equation for calculating the net loading **R** on $L_5S_1$:

$$\mathbf{R} = \mathbf{G} + \mathbf{F}_m,$$

where

$\mathbf{F}_m$ is the back muscle force
**R** is the net loading on $L_5S_1$
**G** is the gravitational force

$$\mathbf{R} = \mathbf{G} + \frac{d_g}{d_F}\mathbf{G} = \frac{d_g + d_F}{d_F}\mathbf{G}$$

where

$d_g$ is the distance of the force **G**
$d_F$ is the distance of the back muscles from the instantaneous center of rotation at $L_5S_1$

**Answer**: (a) 1116 N; (b) 1164 N; (c) 1302 N.

### 4.3.4 Principles of Statics Applied to Specific Human Body Joints

In this section, several numerical examples of the application of principles of statics are presented for various body segments. The aim is to illustrate the application of simple principles of statics in evaluating loading of the musculoskeletal system. A comprehensive discussion of the biomechanics of all body segment movements can be found elsewhere [3,6–8] and is beyond the scope of this chapter.

#### *4.3.4.1 Wrist and Hand*

The wrist and hand are primary body segments for performing the activities of daily life. The movement is achieved through the articulation of several small bone structures in a highly complex interaction, generating forces and movements in three-dimensional planes. A detailed discussion of the bone articulation can be found in [3].

When gripping or supporting an object, forces are applied at the distal joints of the fingers. Forces are transmitted through a series of bone joint articulations of the hand. These joints are the distal interphalangeal joint (DIP) between the distal and proximal phalanges, the proximal interphalangeal joint (PIP) between the middle and proximal phalanges, the meta-carpophalangeal joint (MCP) between proximal carpal and the metacarpal bones,

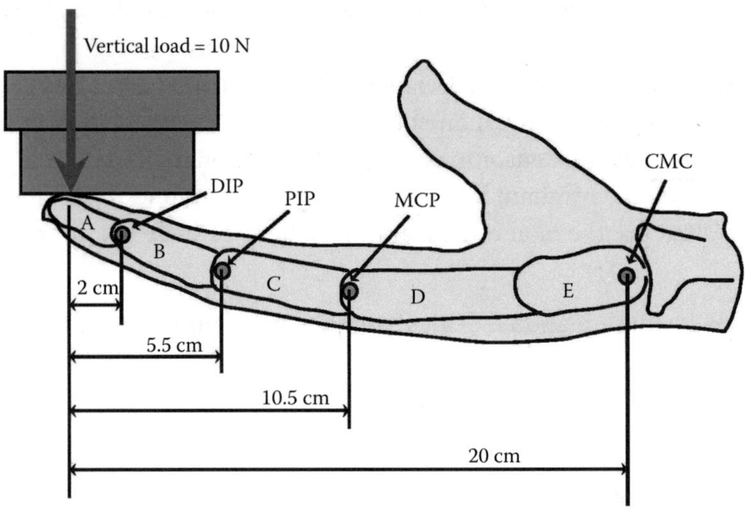


FIGURE 4.9 Forces acting on the hand. (Modified from Nordin, M. and Frankel, V.H., *Basic Biomechanics of the Musculoskeletal System*, Lea and Febiger, Philadelphia, PA, 1989.)

and the carpometacarpal joint (CMC) between metacarpal and carpal bone, as illustrated in Figure 4.9. Since each of these joints has ligamental tissues attached to the bones for supporting the structure, the external forces are balanced in part by each of the four major joints mentioned earlier and as illustrated in the following example.

**Example Problem 6**

Figure 4.9 shows the hand pressing on a surface with 1 kg force (≈10 N). Calculate the moment at each of the joints, using the distance of the joints and the effort arm tabulated in Table 4.2 (distance of the muscle force at each joint from its center of rotation, or fulcrum, shown in Figure 4.9).

**Solution**

The moment about each of the joints can be calculated by multiplying the distance of the load from the fulcrum and the load. The values can be tabulated as in Table 4.3.

**Example Problem 7**

Using the effort arm distances listed in Example Problem 6, calculate the muscle force that will be required, if it is assumed that each of the joints is acting alone to balance the moment due to the load.

TABLE 4.2

| Joint | Distance from Load (cm) | Effort Arm (cm) |
|---|---|---|
| DIP | 2.0 | 0.50 |
| PIP | 5.5 | 0.75 |
| MCP | 10.5 | 1.00 |
| CMC | 20.0 | 1.25 |

TABLE 4.3

| Joint | Moment (Load × Load Arm) |
|---|---|
| DIP | 20 N-cm (=0.2 N-m) |
| PIP | 55 N-cm (=0.55 N-m) |
| MCP | 105 N-cm (=1.05 N-m) |
| CMC | 200 N-cm (=2.0 N-m) |

TABLE 4.4

| Joint | Muscle Effort (Moment/Effort Arm) (N) |
|---|---|
| DIP | 40 |
| PIP | 73.3 |
| MCP | 105 |
| CMC | 160 |

**Solution**

The muscle force (effort) required at each joint can be calculated by using the principles of statics, assuming that the net moment about the particular joint is zero. Thus, if **E** is the muscle effort acting at the effort arm $d_F$, then $\mathbf{E} = \mathbf{M}/d_F$, where **M** is the moment calculated in the previous example. Table 4.4 lists the muscle forces.

As we can see in the earlier examples, the moment due to the external load increases as we go from fingertip to the wrist, due to an increase in the moment arm. The muscle force needed to balance the moment would thus be excessively high if only one of the joints were to act alone. In reality, the moment at each of the joints is counterbalanced by smaller moments produced by the flexor profundus tendons in each of the individual joints [3].

The grip force required to hold an object depends on the size and weight of the object and the type of grip. Figure 4.10 shows two different kinds of grips and the tangential forces produced in the synovium tendons of the hand. The tendon, in turn, wraps around the bones of the wrist to create a pulley system as depicted in Figure 4.11, where the person is gripping a tool while flexing his hand. This flexion produces a normal force on the tendon synovium itself that is given by [2]

$$\mathbf{F}_N = \frac{\mathbf{F}_T e^{\mu\phi}}{R},$$

where

$\mathbf{F}_N$ is the normal force on the tendons
$\mathbf{F}_T$ is the tangential force produce along the tendon
$R$ is the radius of the pulley system
$\phi$ is the angle subtended by the tendon on the pulley
$\mu$ is the coefficient of friction between the tendon and the pulley

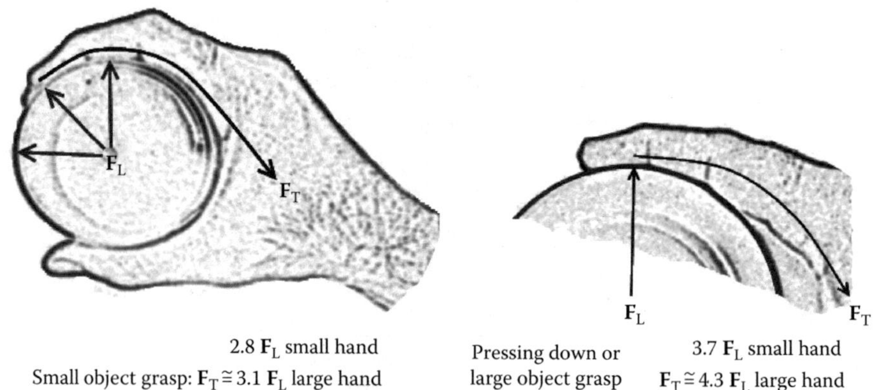


FIGURE 4.10 Forces experienced by wrist during gripping. (Adapted from Chaffin, D.B. and Andersson, G.B., *Occupational Biomechanics*, Wiley, New York, 1984.)

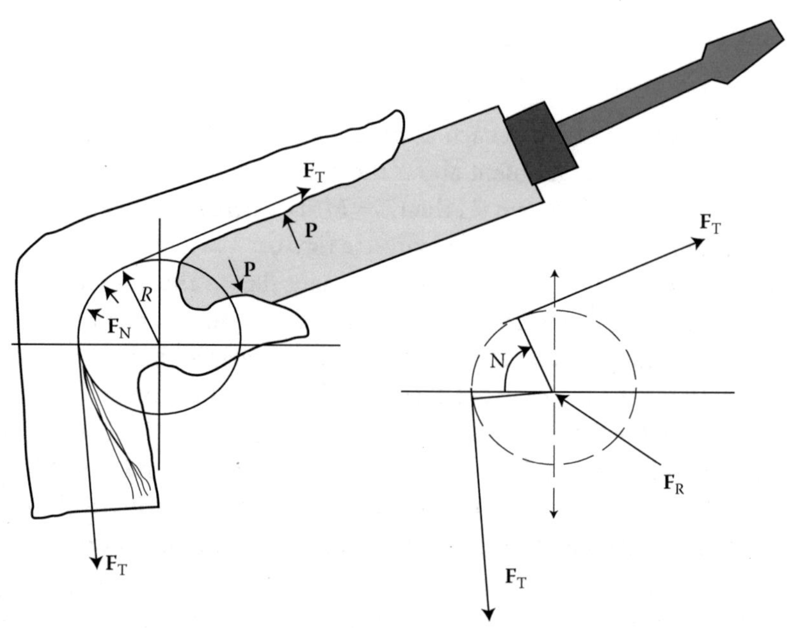


FIGURE 4.11 Forces acting on the wrist during flexion.

For the purpose of simplification, the value of μ can be assumed to be zero [6]. The previous equation is further simplified to

$$\mathbf{F}_N = \mathbf{F}_T / R.$$

The reaction force on the bones due to the tendons, $F_R$, can be calculated by resolving the forces, yielding

$$\mathbf{F}_R = 2\mathbf{F}_T \sin(\phi/2).$$

**Example Problem 8**

Figure 4.11 shows the forces acting on the hand of a person operating a screwdriver. Calculate the tendon force $\mathbf{F}_T$ and the reaction force $\mathbf{F}_R$. For the purpose of this example, the grip force **P** acting at the finger may be assumed to be 1 N. Assume a small hand in operation on a small grip size. Also assume that the subtended angle (wrist angle) is 30°.

**Solution**

Using the equation shown in Figure 4.10,

$$\mathbf{F}_T = 2.8 \quad \mathbf{P} = 2.8 \times 1 = 2.8 \text{ N}.$$

Using the equation for $\mathbf{F}_R$, we have

$$\mathbf{F}_R = 2 \quad \mathbf{F}_T \sin\phi/2 = 2 \times 2.8 \times \sin 30^\circ/2 = 1.45 \text{ N}.$$

**Example Problem 9**

Repeat the previous analysis for an extension of the wrist as shown in Figure 4.12, where the person is using a paint roller. Assume that the grip pressure force **P** required for holding is 5 N. Assume a wrist angle of 25° and a small grip size. Assume a factory worker with large hand size.

**Hint**: Use the methodology mentioned previously.

**Answers**: $\mathbf{F}_T$ = 15.5 N; $\mathbf{F}_R$ = 6.7 N.

Several studies have been conducted to investigate the range of motion of the hand and wrist and the analysis of the forces acting on the adjacent structures [2,3]. These values act

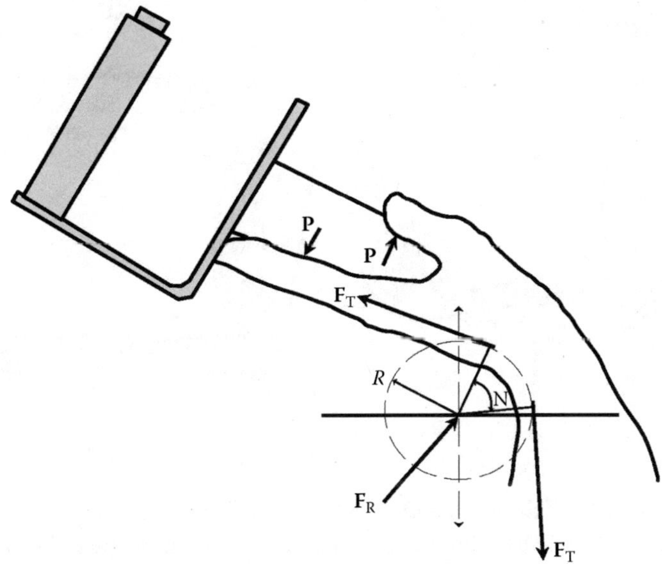


FIGURE 4.12 Forces acting on the wrist during extension.

as guidelines for determining untoward postures of the hand and the extent of forces that may be detrimental when applied at extreme angles of flexion and extension of the wrist. One of the most common musculoskeletal cumulative trauma disorders is the carpal tunnel syndrome, which causes inflammation of the nerves at the wrist joints due to excessive and repeated loading of the median nerve at the wrist joint structure. Additional issues related to carpal tunnel syndrome are given elsewhere in this book.

#### *4.3.4.2 Elbow*

Weight lifting is a common activity we perform every day. Most of the time we simply grab the weight in one hand and flex our arms to raise it up. Lifting can be performed using one or both hands. It can be pointed out here that one-arm lifting would load only one side of the body framework. This will produce an asymmetric loading of the spinal column and skeletal system as a whole. In arm lifting, the elbow joint acts as the instantaneous center of rotation (fulcrum).

As was described earlier, depending on the muscles involved in supporting the weight, the arm uses either a second- or third-class lever system. This is illustrated further in Figure 4.13. In Figure 4.13a, the principal muscle applying the force is the brachoradialis, attached to the radial bone near the wrist, as shown. The combined load of the arm and the weight is shown as **L**. Since the load held by the hand (point of application of the load) lies between the effort and fulcrum, it can be termed a second-class lever system. In Figure 4.13b, the muscle applying the force is the biceps brachii. This is a third-class lever system [7,9].

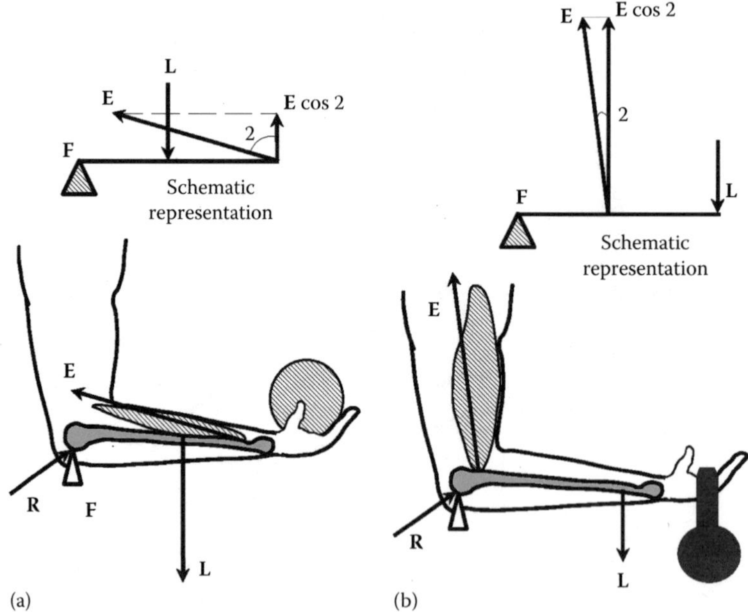


FIGURE 4.13 Application of different muscle forces may determine what lever system should be applied for calculating the resultant forces (a) the principal muscle applying the force is the brachoradialis, attached to the radial bone near the wrist (Class II Lever System) (b) the principal muscle applying the force is the biceps brachii (Class III Lever System).

**Example Problem 10**

Calculate the muscle force (effort) needed in Figure 4.13a and b if the value of the load **L** is 90 N and the load arm (perpendicular distance of the application of **L** from the fulcrum) is 18 cm. The effort arm (perpendicular distance of the force vector **E** from the fulcrum) can be assumed to be 4 cm for (a) and 2 cm for (b). Calculate the resultant force **R** on the fulcrum F, if it is given that **E** acts at an angle of 15° in Figure 4.13a and 60° in Figure 4.13b, measured from the left horizontal.

**Hint**: Calculate the value of **E** using the equation **E** = [(load arm)/(effort arm)] × load. Resultant force **R** can be calculated by resolving the forces **L** and **E** along the two axes.

**Answer**: (a) 405 N; (b) 810 N. Resultants: (a) 391.48 N at −2.2° to right horizontal; (b) 733.4 N at −56.5° to right horizontal.

*4.3.4.3 Shoulder*

Movement at the shoulder level is one of the most intricate to quantify. The motion around the shoulder joint involves a number of muscles acting together at several possible joints to achieve the wide range of movement of the arm with respect to the torso. The combined range of motion is achieved by four articulations: the glenohumeral, sternoclavicular, acromioclavicular, and scapulothoracic [3]. The near-global range of motion exceeds the individual range of motion at any of these articulations. The main range of motion occurs at the glenohumeral joint formed by the humeral head and the glenoid fossa of the scapula. It forms an open-ended ball-and-socket joint, formed by a pear-shaped depression at the glenoid fossa and the humeral head capable of performing rotational, translational, and rolling motion with respect to the fossa. Several muscles act at the four articulations, making it difficult to calculate the precise forces experienced by each muscle group. Simple assumptions can be made to calculate the major forces and reactions at these joints, as shown in the following example problem.

**Example Problem 11**

The main joint forces at the shoulder act at the glenohumeral joint. The major muscle group involved in the articulation is the deltoid muscle. Figure 4.14 shows a construction worker performing a ceiling operation, which requires him to operate a tool with stretched arms as shown. Assume the weight of the tool to be 0.5 kg (5 N), acting at a distance of 80 cm from the instantaneous center of rotation *O* at the shoulder. Assume that the weight of the arm is 0.05 times the body weight and acts at a distance of 35 cm from *O*. For a man weighing 100 kg (1000 N), calculate the effort **E** applied by the deltoid muscles with an effort arm of 3.5 cm.

**Solution**

Equating the moments about the point *O* to zero, we have

$$-(\mathbf{E} \times 3.5) + (0.05 \times 1000 \times 35) + (5 \times 80) = 0$$

$$\mathbf{E} = 2150 / 3.5 = 614.3\,\text{N}\,(0.6\,\text{bw}).$$

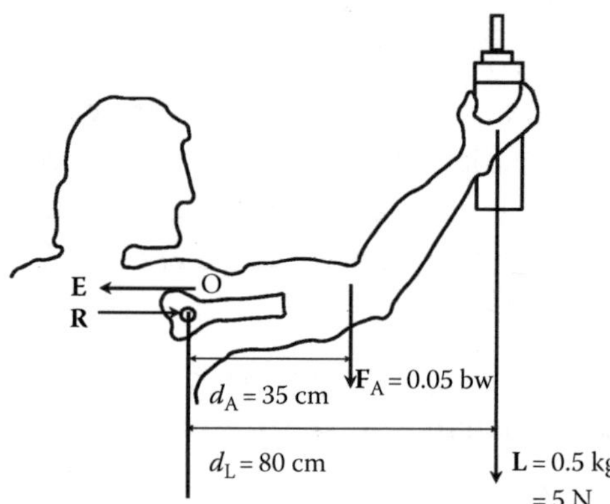


FIGURE 4.14 Forces at shoulder during an overhead task performance.

Since the only horizontal force acting at the joint is the muscle effort **E**, the horizontal reaction force **R** on the joint is equal to **E**, or 614.3 N.

**Example Problem 12**

Suppose the worker in the previous example has to apply an upward thrust to the tool for driving the nails into the roof, increasing the effective downward force to 50 N (instead of 5 N). Recalculate the deltoid force **E**.

**Hint**: Use the aforementioned equation for calculating **E** (=**R**).

**Answer**: **E** = 1642.9 N (1.64 times bw).

**Example Problem 13**

At a Stockholm construction site, several complaints were received concerning strains and aches in the upper extremities due to the excessive use of bolt guns (Hilti Model DX450). A bolt gun support was developed to eliminate the loading on the upper extremities by transferring the loading to the waist belt. Figure 4.15 shows the biomechanical configuration of the loading (left) without bolt-gun support, and (right) with bolt-gun support. Calculate the shoulder abduction force being provided by the deltoid muscles, which are assumed to be acting at a distance of 3 cm from the instantaneous center of rotation (fulcrum), to support the gun in Figure 4.15. The weight of the gun ($\mathbf{W}_H$) and the weight of the arm ($\mathbf{W}_A$) and their distances from the fulcrum *O* are shown in the figure for an 800 N worker.

**Hint**: Calculate the moments about *O* due to the arm and gun, and equate them to the moment created by the shoulder deltoid muscles.

**Answer**: 373.3 N.

### 4.3.4.4 Knee

The human knee acts as a pulley system as shown in Figure 4.16. The patellar tendon and the quadriceps each apply balancing forces to maintain the flexion of the knee. In a static case of a lifting task, forces generated in the knee can be calculated, as illustrated in the following example problem.

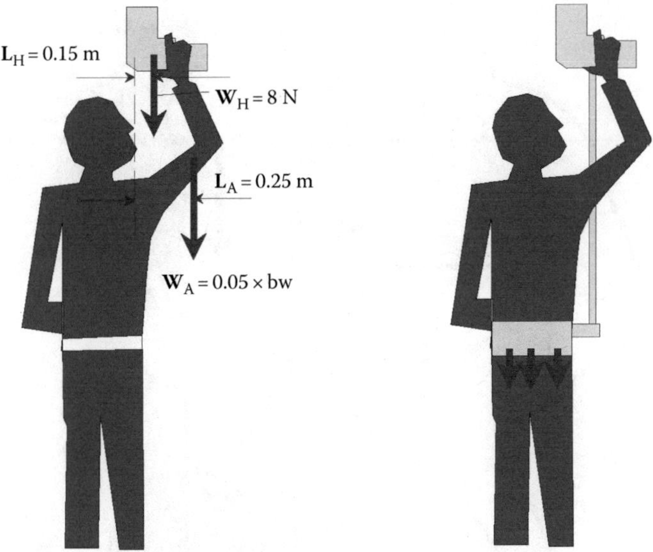


FIGURE 4.15 Modification of a heavy tool to reduce the loading of the shoulder.

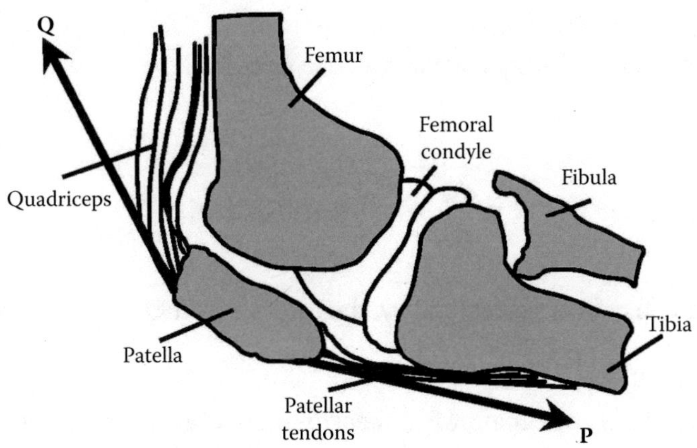


FIGURE 4.16 Anatomy of a knee joint.

**Example Problem 14**

Figure 4.17 shows a person lifting a weight. The superincumbent weight (above the knee joint) of the person with the package can be assumed to be 900 N. Also, assume that $d_w$, the distance of the line of the center of gravity from the instantaneous center of rotation (fulcrum), is 5 cm. The corresponding distance, $d_p$, of the patellar force vector **P** is 4 cm. Assuming a static posture as shown, calculate the patellar tendon forces.

**Solution**

Using the equilibrium of moments about the fulcrum,

$$\mathbf{W} \times d_W = \mathbf{P} \times d_P.$$

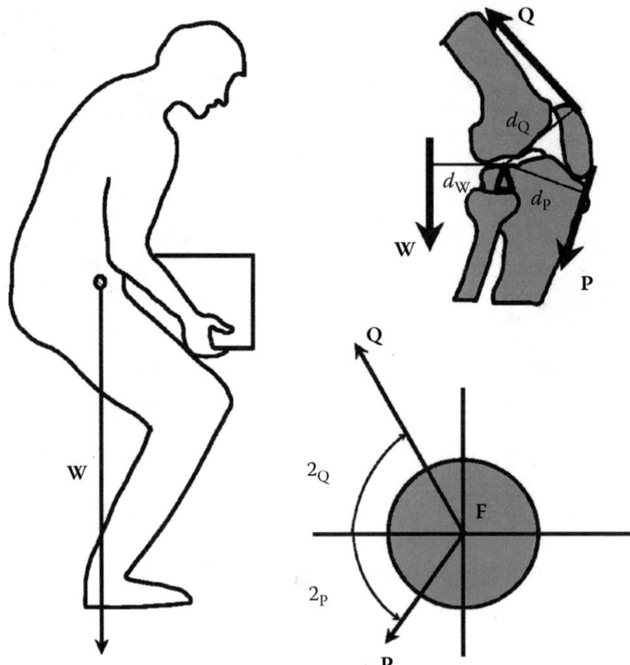


FIGURE 4.17 Forces at knee during manual materials handling.

Hence,

$$\mathbf{P} = \frac{d_W}{d_P}\mathbf{W} = \frac{5}{4}900 = 1125\text{ N}.$$

Assuming the patella to be a perfect pulley, the forces **P** and **Q** are equal in magnitude.

**Example Problem 15**

In the previous example, the anatomical measurements reveal the approximate angles of the force vectors **P** and **Q** with the left horizontal to be $\theta_P$ and $\theta_Q$, with the values of 60° and 50°, respectively (see Figure 4.17). Calculate the resultant force on the patella.

**Hint**: The resultant force on the patella can be calculated by resolving the vectors **P**, **Q**, and resultant **R** into *x* and *y* components (assume that the *x* axis is along the right horizontal).

**Answers**: 1290.5 N acting at an angle of 5° to the right horizontal.

The aforementioned examples assume that the knee is acting as a perfect pulley, with equal tension on the ligaments carrying the forces. However, this approximation is not valid for extreme angles of knee flexion [10].

### 4.3.5 Dynamics

Dynamics is concerned with bodies that are in motion. Apart from the system of forces acting on the body, there exist the inertial forces due to the motion of the individual segments of the body. The forces that cause the motion may vary with time, as do the position and orientation of the segments. Such time-dependent measurements are termed temporal quantities and play a vital role in dynamics. Dynamics can be further subdivided into kinetics, which is concerned with the study of forces and moments that cause the motion, and kinematics, which deals with the measurement of motion itself. Newton's second law of motion can then be applied to the body segments for evaluating the forces and moments. As mentioned in the laws governing the principles of mechanics, for a system of forces **F** and moments **M** acting on a body, the law can be mathematically stated as

$$\Sigma \mathbf{F} = m\mathbf{a}, \quad \Sigma \mathbf{M} = \boldsymbol{I}_o \boldsymbol{\alpha}.$$

Here

- $\mathbf{a}$ (in m/s$^2$) and $\boldsymbol{\alpha}$ (rad/s$^2$) denote the resulting linear and angular acceleration of the body
- $m$ (kg) is the mass of the body
- $\boldsymbol{I}_o$ (kg/m$^2$) is the moment of inertia of the body about the axis of rotation

Thus, both kinematic and kinetic measurements are required to fully describe the motion of the body and the forces that cause the motion.

In the case of a static loading, the body is assumed to be at rest. Hence, the inertial forces are absent in a static loading, and the right-hand sides of the previous equations are equal to zero (no linear or angular accelerations). Thus, static loading constitutes a special case of dynamic loading. Since, in a real-life scenario, the worker is involved in various activities that may include dynamic motion of the body segments, inertial forces play a very important role in determining the total musculoskeletal loading on the human body. The importance of dynamic motion of the segments on the loading of the body segment joints is further explained in another chapter of this book.

#### *4.3.5.1 Kinematics*

Kinematics describes the geometry of motion. It is concerned with the temporal change in the position of the body segments. Both linear and angular displacement of the body segments are important in characterizing motion. For simplification, we assume that the body segments behave as rigid links, and the displacement, velocity, and acceleration of a fixed point on this rigid link are evaluated.

If linear displacement is denoted as $x$, velocity is defined as the rate of change of $x$ in a particular direction. Acceleration is defined as the rate of change of velocity, or the rate of change of displacement. Mathematically,

$$\text{Velocity} = \mathbf{v} = \frac{d}{dt}(x)$$

$$\text{Acceleration} = \mathbf{a} = \frac{d}{dt}(\mathbf{v}) = \frac{d}{dt}\left(\frac{d}{dt}x\right) = \frac{d^2}{dt^2}(x).$$

Similarly, if angular displacement is defined as θ, the angular velocity and angular acceleration are defined as follows:

$$\text{Angular velocity} = \boldsymbol{\omega} = \frac{d}{dt}\theta$$

$$\text{Angular acceleration} = \boldsymbol{\alpha} = \frac{d}{dt}\left(\frac{d}{dt}\theta\right) = \frac{d^2}{dt^2}\theta.$$

Since linear and angular displacement, velocity, and acceleration are vector quantities, they require magnitude, direction, and point of application for their specification. These quantities must be specified with respect to a global reference axis system or a local joint-based axis system. Special passively illuminated markers can be placed at body segments and captured in a video image. This video can be digitized using specialized instrumentation to obtain the movement of each body segment in space. Figure 4.18a shows the digitized image of the lower limb of a child walking over an obstacle. Figure 4.18b shows how the angle between the ankle and lower leg (shin) changes during this walk [11]. Additional information needed to measure kinematic information is included in another chapter of this book.

### 4.3.5.2 Kinetics

Kinetics is concerned with the measurement of forces and moments that cause motion. For simplification, the human body is assumed to be made up of individual link segments,

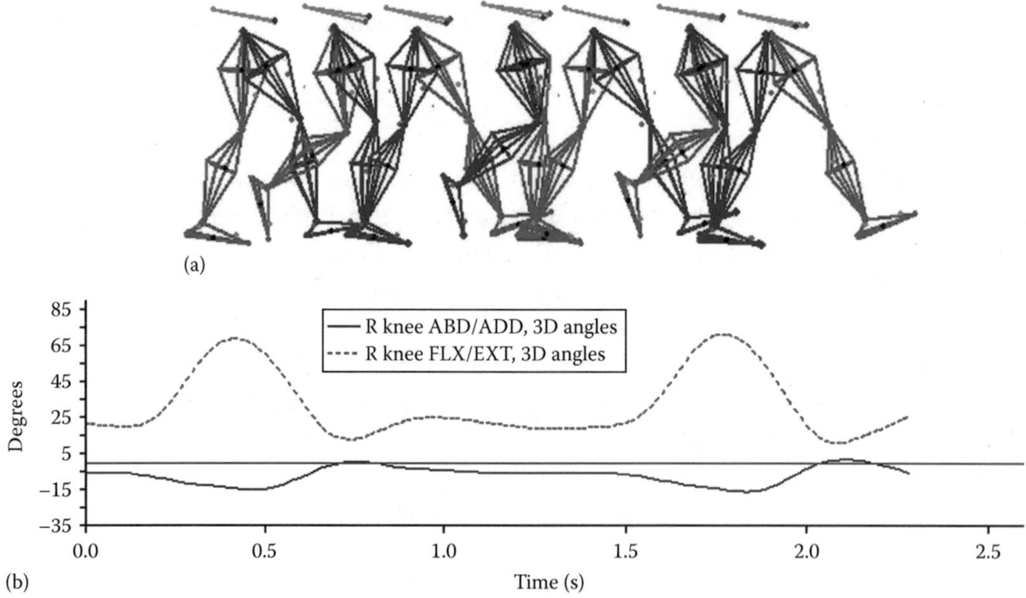


FIGURE 4.18 (a) Use of videography for digitization of human movement and (b) measurement of segmental displacement and angles.

each of which behaves as a rigid body, that interact with each other to transmit the forces, moments, and the motion caused by the action of these forces. The rigid link segments behave in accordance with Newton's laws of motion.

Application of Newton's second law of motion yields the set of equations for dynamic equilibrium stated at the beginning of this section. The calculation of ground reaction forces plays an important role in understanding the result of body movement during the performance of tasks. As we will see later in this chapter, the principles of dynamics can be applied to complex link-segment models of the human body in order to evaluate the joint forces and moments, which play an important part in understanding the functions and limitations that govern human movement.

### 4.3.6 Spinal Unit

#### *4.3.6.1 Structure*

The spinal unit forms the integral part of the framework of bone that provides structural support to the upper part of the body. The complex movement pattern allowed by the spinal column is key to the maintenance of this structural support, while allowing the necessary range of motion of the upper body. The spinal unit bears a major portion of the weight of the head, arm, and trunk (HAT). During simple activities like walking or standing erect to perform a task, the spinal column has to not only support the gravitational forces of the HAT but also support the forces and moments created due to motion of the upper body parts. Such forces may often exceed the limitations of biomechanical loading that can be endured by the spinal unit and can result in long-term or acute injury. Understanding the loading on the spinal column and its behavior under the action of these forces is important in deriving the criteria of safe manual materials handling practices.

The spinal column, also known as the vertebral column, is a complicated structure. The structure can be simplified by assuming it to consist of fairly rigid bone segments (vertebrae) connected by means of ligaments and intervertebral disks. The column consists of 33 vertebrae linked to one another through intricate connections of tissues, ligaments, and intervertebral disks to form a strong but flexible supporting column for the head and trunk. The vertebral column can be subdivided into several regions. The cervical region consists of seven vertebrae and is followed by the thoracic region, which contains 12 vertebrae. The lumbar region follows, with five vertebrae. This is followed by the five fused vertebrae of the sacral region and three to four fused coccygeal segments. Normally, the column appears symmetrical in the frontal plane and has characteristic curvatures in the sagittal plane. These curvatures provide additional natural shock absorbance and flexibility at the intervertebral joints. There exist anterior convexity in the cervical and lumbar regions and posterior convexity in the thoracic and sacral regions. The curvatures in the thoracic and sacral regions are primarily due to the wedge-shaped structure of the vertebrate unit, while the curves in the cervical and lumbar regions are due to wedge-shaped intervertebral disks. In the event of the loading of the complete spinal unit, the lumbar and cervical regions are flattened to a greater extent [12]. Deviations from these natural curvatures of the spine may generate excessive stresses in the intervertebral units and can cause permanent deformations, depending on the extent of the loading.

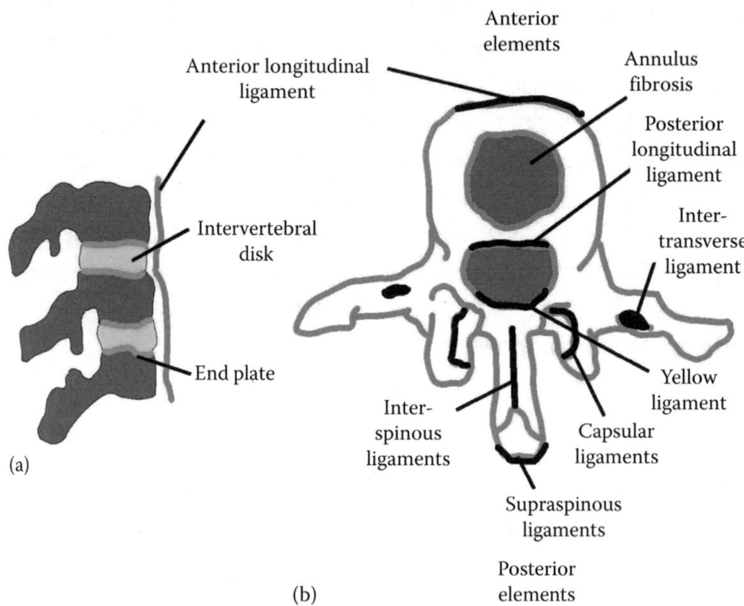


FIGURE 4.19 Elements of a spinal unit.

Integrated vertebral movement of the spine is the result of coordinated action of the soft tissues interlacing the vertebrae. In order to describe the biomechanics of spinal movement, it is important to understand the structure of the spinal units or vertebrae. Figure 4.19 shows side and top views of the vertebral units including the ligaments found in the region. Figure 4.19a shows vertebrae with the intervertebral disks between them. The structure and function of the intervertebral disk will be discussed later. Figure 4.19b shows the major ligaments associated with the vertebral units. These ligaments allow smooth movement of the spinal column within the feasible ranges of motion with minimum resistance and expenditure of energy and provide vital support in the event of a traumatic shock to the spinal column. Without the fine coordination of these soft tissues, smooth spinal motion would not be possible. Each of the components of the spinal column has its own physical properties that combine to provide a quantitative picture of the spinal column's functional capabilities.

#### *4.3.6.2 Intervertebral Disk*

The intervertebral disk is present between two vertebrae as discussed earlier and is shown in Figure 4.19a. It is subjected to a variety of loadings during the movement of the column. In fact, when a person is standing upright, the forces to which a disk is subjected are much greater than the weight of the portion of the body above it. This is due to the fact that the line through the center of gravity of the HAT lies a little in front of the spinal column, which creates moments about the instantaneous center of rotation situated somewhere in the middle of the vertebral unit. This is further illustrated in the following example problem.

**Example Problem 16**

Figure 4.20 shows a single vertebral unit under the action of the gravitational load **W** of the head and trunk acting at a distance $d_w$ (=6 cm) in front of the instantaneous axis of rotation

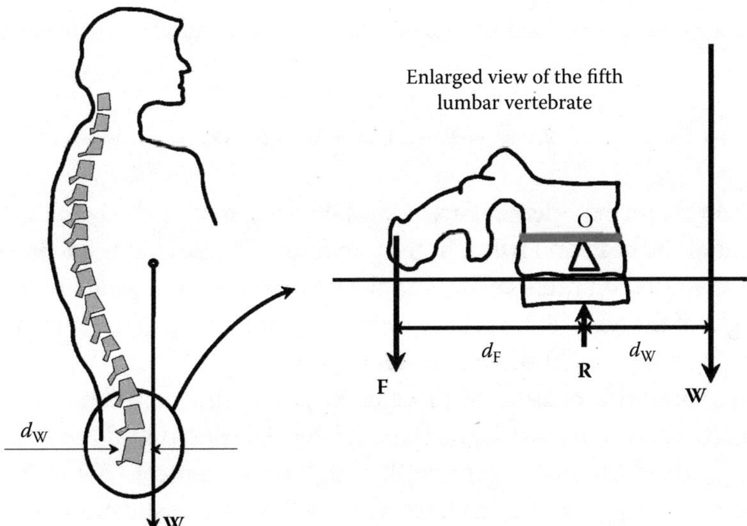


FIGURE 4.20 Forces experienced by the fifth lumbar vertebra due to the body weight. (Adapted from White, A.A. and Panjabi, M.M., *Clinical Biomechanics of the Spine*, Lippincott, Philadelphia, PA, 1978.)

(fulcrum) on the vertebrae. The muscle force exerted is shown as **F** and acts at a distance $d_F$. Calculate the ratio of the forces **F** and **W**. What is the combined compression force on the vertebral unit?

**Solution**

Using the laws of static equilibrium and applying the equation for a lever system, we have that the net moment about the instantaneous center of rotation (fulcrum) is zero:

$$\mathbf{W} \times d_W + (-\mathbf{F} \times d_F) = 0$$

$$\mathbf{W} \times d_W = \mathbf{F} \times d_F.$$

Therefore,

$$\frac{\mathbf{F}}{\mathbf{W}} = \frac{d_W}{d_F}.$$

For $d_w = 6\,\text{cm}$ and $d_F = 3\,\text{cm}$, we have

$$\frac{\mathbf{F}}{\mathbf{W}} = \frac{6}{3} = 2.0.$$

To calculate the compressive load on the vertebral unit, we add the forces experienced by the unit:

$$\mathbf{R} = \mathbf{F} + \mathbf{W} = 2.0\mathbf{W} + \mathbf{W} = 3.0\mathbf{W}.$$

Thus, the vertebral unit experiences three times the weight of the body mass above it.

If the weight of the head and trunk in the previous example equals 0.6 times the weight of the person, then the compressive force on the vertebral column would be $3.0 \times 0.6$ or 1.8 times the weight of the body. Studies have estimated the loading of the spine to be as much as three times the body weight when in a seated posture.

The intervertebral disk consists of three main parts. These are the nucleus pulposus, the annulus fibrosus, and the cartilaginous endplates. The nucleus pulposus is a gelatinous center with embedded fibrous content, with a high water content (70%–90%). The water content decreases with age, which in effect changes the viscoelastic properties of the disk. The nucleus pulposus is more dorsally placed in the lower back and has a larger swelling capacity in the cervical and lumbar regions. The annulus fibrosis forms the outer peripheral ring of the disk and consists of fiber running in a helical pattern with alternating layers of fibers criss-crossing each other. The fibers run at about 30° to the disk plane and cross each other in adjacent layers at an angle of about 120°. The fibers are attached directly to the vertebral units on the periphery and to the endplates in the interior of the structure. This allows relative movement of the different spinal substructures in a smoother manner. The cartilaginous endplates form the upper and lower surfaces of the intervertebral disks and are formed out of hyaline cartilage tissues.

Since the disk is the major load-bearing component of the spine, several studies have been performed to establish the behavior of the disk under various loading conditions. In general, the disk is subjected to loads (both compressive and tensile) and shear stresses or a combination of loading. The forces may be exerted suddenly (jerk loading) or over a longer period of time. The loading can be sustained, causing a creep failure, or cyclic, causing a fatigue failure. In cases of creep or fatigue failure, the level of force required to cause the failure is often much lower than that required for failure under normal conditions. The deformation curve for the disk under normal loading condition is sigmoidal, implying that the disk offers little resistance at low loads, but its stiffness increases dramatically with increase in the applied loading. This provides flexibility of movement at low-loading conditions and provides supportive action at higher loading conditions [12]. It has been demonstrated that the disk is very stable at high levels of compressive loading. This type of loading causes the disk material to bulge out, which in turn introduces compressive and shear stresses in the disk material. The degenerated disk behaves in a slightly different fashion. The load is transmitted from one endplate to the other primarily via the annulus. This causes the endplates to sustain the major portion of the loading at the disk's outer edges. At this outer layer of the disk, the tangential forces are small but the annulus fibers are exposed to twice as much stress. Unlike the nondegenerated disk, the axial stress becomes compressive.

Torsional forces introduce similar internal stresses in the disk structure. These generated stresses are not evenly distributed throughout the disk. The magnitude of these shear forces is usually higher at the periphery than in the central parts of the disk. Experiments indicate that a relatively high level of shear force is required to cause any failure in the disk structure in the horizontal plane [12].

Due to the fact that the instantaneous center of rotation (fulcrum) for a vertebral unit lies somewhere in the middle of the unit, both compressive and tensile loads are generated in different parts of the disk. Measurement of the tensile strength of the disk has been made along various directions to the plane of the disk [12]. The tensile strength is the highest along a 15° direction to the disk plane. The anterior regions are stronger than the posterior regions. The strength distribution provides most protection at areas prone to herniation under severe loading. Unequal distribution of the tensile strength along the cross section of the disk shows that the disk is highly anisotropic and is more adaptive in resisting loads in certain directions than in others. The disk has been found to be stiffer in compression than in tension. This has been attributed to the buildup of fluid pressure within the nucleus pulposus under compressive loading. The tensile stress can be divided into its normal and shear components along and perpendicular to the plane of the disk. Although the normal component can be resisted by the alternating fibers, there exists no provision for resisting shear stresses. Another factor contributing to the vulnerability of the disk to tensile loading is Poisson's effect, whereby the cross-sectional area is reduced in tensile loading whereas it is increased due to the bulging in compressive loading.

Experimental findings suggest that bending and torsional loads, and not compressive loads, are the most potentially damaging to the disk. Bending of the spinal column produces both compressive and shear forces on the material of the disk. The disk material is pushed out in the direction of the curve as shown in Figure 4.21. Excessive bending may cause this bulge to impinge against nerve roots, causing painful physiological conditions.

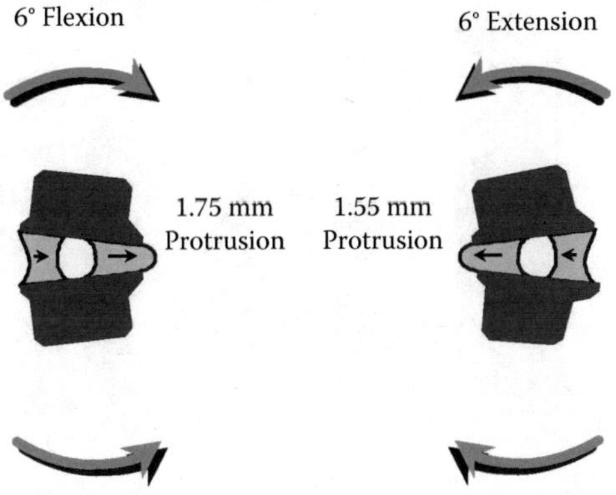


FIGURE 4.21 Protrusion of the discal material during flexion or extension of the spine. (Adapted from White, A.A. and Panjabi, M.M., *Clinical Biomechanics of the Spine*, Lippincott, Philadelphia, PA, 1978.)

The loading in occupational tasks is often a complex combination of the different loading conditions defined earlier. Experimental values of the stiffness coefficients of the intervertebral disk may be found in several references [12]. The stress values are important in determining safe limits of musculoskeletal loading and are used as guidelines for loading limits in manual materials handling.

#### 4.3.6.3 Intradiscal Pressure

Measurement of the fluid pressure in the intradiscal cavity is probably the most effective way of studying the loading characteristics *in vivo*. In this method, the intradiscal pressure of the nucleus pulposus acts as a load transducer and indicates the magnitude of axial loading on the spinal column. Figure 4.22 shows the level of intradiscal pressure for various lumbar support during seated postures. Figure 4.23 indicates the measured intradiscal pressure for various daily life postures. Increased pressure indicates a greater muscular effort in maintaining the posture and hence a larger stress on the spinal column. As is evident from Figure 4.23, a seated posture may actually produce higher disk pressures than a standing posture. This is attributed to the fact that the line of gravity for the upper body acts farther away from the instantaneous center of rotation.

#### 4.3.6.4 Biomechanical Aspect of Back Pain

Back injury and pain can have a variety of causes. The phenomenon of back pain is most prevalent in the occupational setting. It is the most frequently reported form of musculoskeletal disorder. Several physical, psychosomatic, biomechanical, and anatomical factors have been shown to cause back pain, alone or in combination with one another.

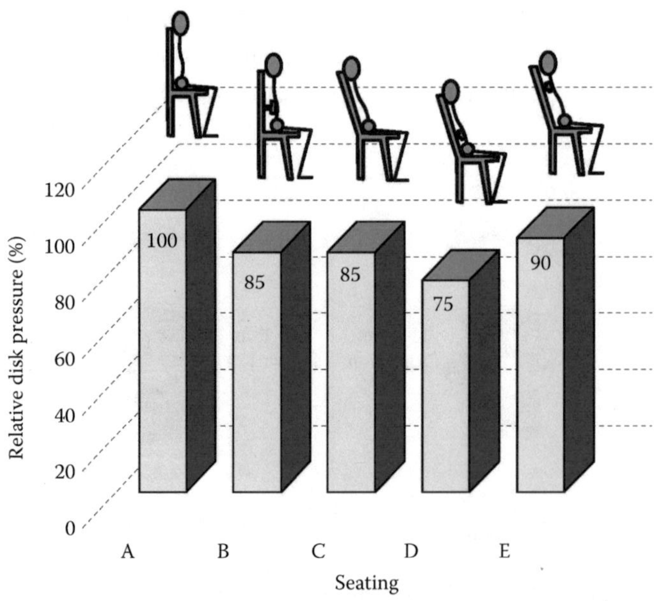


FIGURE 4.22 Relative disk pressure experienced during sitting with various inclinations of the back support and presence of a lumbar support. (Modified from Eastman Kodak Co., *Ergonomics for People at Work*, Van Nostrand Rein-hold, New York, 1986.)

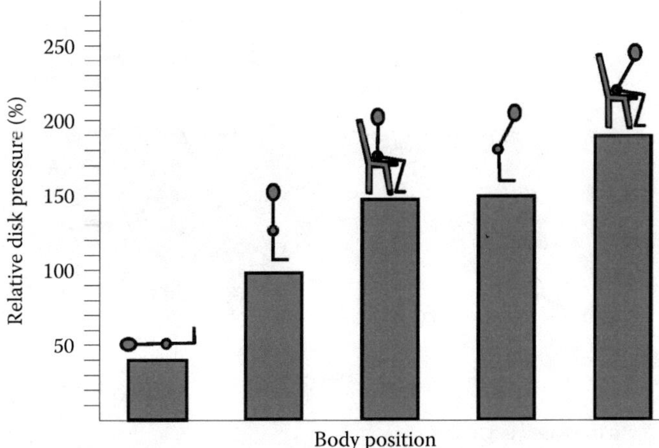


FIGURE 4.23 Relative disk pressure during various postures.

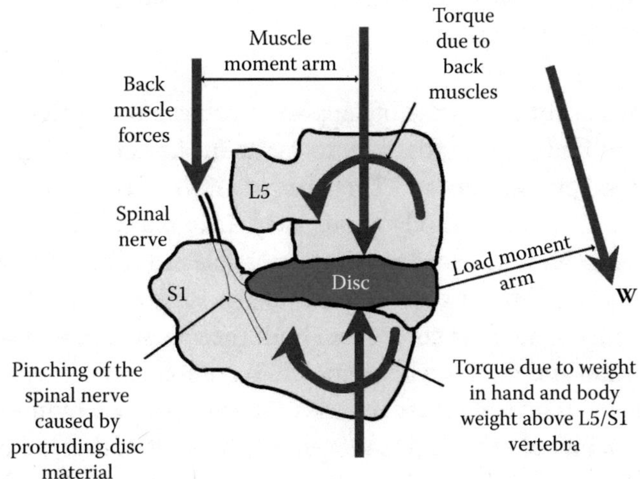


FIGURE 4.24 Pinching of the spinal nerve. (Adapted from Cailliet, R., *Low Back Pain Syndrome*, F. A. Davis Co., Philadelphia, PA, 1981.)

As the disk material does not have any nerve endings, simple compressive forces are not responsible for pain. During the application of compressive loads on the spinal column, the disk material can protrude outside and impinge upon the spinal nerves that surround the disk and the vertebral units. This phenomenon is depicted in Figure 4.24. In the static condition of the loading of the spine, an increase in the lumbosacral angle can cause an increase in the natural lumbar curvature (lordosis). This is termed *swayback*. About three-fourths of all back pain cases are due to excessive lordosis [13]. Pain is generated due to facet impingement and irritation, as depicted diagrammatically in Figure 4.25.

## 4.3.7 Center of Gravity

Upright stability in human beings is possible due to a complex mechanism of the neuromusculoskeletal system that acts in a systematic manner to control the orientation of

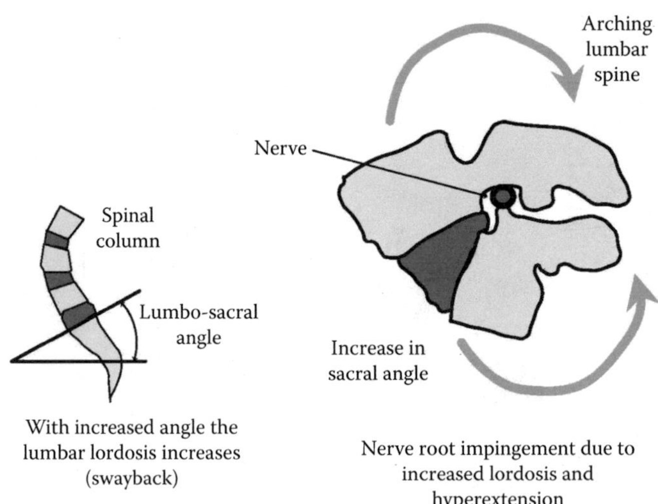


FIGURE 4.25 Backache caused by "swayback." (Adapted from Cailliet, R., *Low Back Pain Syndrome*, F. A. Davis Co., Philadelphia, PA, 1981.)

various body segments above its base of support to achieve the upright stance. To attain upright stability, the body constantly tries to maintain its center of gravity (COG) at a comfortable location between the feet. Three major joints of the body—ankle, knee, and hip—help move the COG of the body appreciably. Like many other body postures, the same location of the COG is achieved by several possible movement strategies of the three joints. The muscle groups moving these joints are in turn activated by motor neurons. Commands to the motor neurons come from both the spinal cord and subcortical levels, depending on the kind and duration of excitation. A combination of reflex action and voluntary movements helps activate these muscles. At least one pair of muscles are acting at each of these three joints to achieve fine movement. A result of that is a continuous effort to correct the positioning of the COG, creating a slight swaying of the body in both the anterior–posterior and medial–lateral directions.

The physics of upright stability dictates that the line of projection through the COG must lie within the base of support for the body to maintain its position in a static stability. Several factors, including muscle strength, age, reaction time, and environmental conditions, affect the actual limit within which the COG must be maintained at all times for stable upright support [14]. In reality, the limits within which the person's COG must be maintained for him or her to hold the upright posture without falling may actually be smaller than the outer periphery of the feet but depends heavily on the stance or support. This is illustrated in Figure 4.26, where the base of support is estimated for three different stances. Studies have indicated the presence of a closed-loop feedback system for maintaining the COG within the limits of basal support [15] or performing a movement strategy to modify or increase the base of support, either by taking a step (staggered feet stance) or by grabbing an external support.

In static postures, the projection of the COG can be estimated by the center of pressure (COP) measurement under the feet [16]. Under normal circumstances, the COP is continuously shifting as the body is trying to maintain balance, creating a bodily sway as discussed previously.

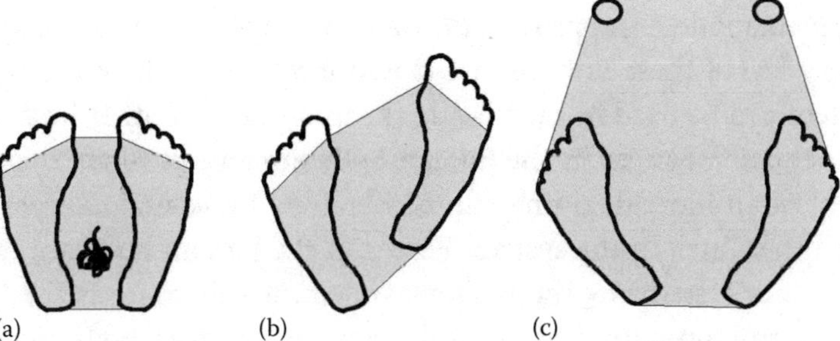


FIGURE 4.26 Base of support during various upright stances. (a) Footprints of a person with normal stance. Also shown in the figure is the superimposed trace of the movement of the center of pressure. (b) Footprints of a person with staggered feet stance. This stance produces a larger base of support. (c) Footprints of a person with walking aid (crutches). Use of the crutches increases the base of support considerably.

The locus of the COP can be experimentally determined by using force platforms [16,17]. As the subject stands on a force platform, the forces and moments generated by the plantar surface of the subject's feet are collected via an amplifier and analog-to-digital board interfaced to a personal computer. These forces and moments are further analyzed to determine the time-variant location of the COP. The trace of the locus of the COP is plotted for the entire test period. The length and spread of COP sway, with respect to the stability boundary described earlier, are important indirect measures of lack of stability. It forms a tool for evaluating workplaces and tasks that present a very high potential for loss of balance for the worker and possible fall or injury due to a momentary loss of balance [18].

### 4.3.8 Link-Segment Model

The skeletal system along with the muscles, jointly called the musculoskeletal system, is the primary source of human body movement. The skeletal system consists of the bones that form the hard interior structure of the body and provides support to many of the body organs. The bones are also responsible for articulation, and a combination of these articulations give rise to human movement. Movement of each individual bone is accomplished by the musculature attached to it. The musculature is in turn activated by the nervous system. As we saw in Chapter 3, the nervous system along with the musculoskeletal system forms a closed-loop feedback system to achieve complex movement patterns such as walking or playing a piano.

To apply the laws of physics and engineering in a simplified manner to describe human movement, the body is divided into segments [19]. This simplified approach assumes that the body segments are formed of rigid links joined to each other by pin joints that allow relative motion similar to the body joint that is being studied. So far, we have discussed the individual body segments (elbow, shoulder, wrist, lower extremities, etc.) and how each of these segments behaves when acted upon by external loading of the joints. In the workplace, a worker might be involved in a task that creates a simultaneous complex loading on several body segments. In order to evaluate the musculoskeletal loading characteristics of the entire body, it is essential to combine the loadings of the various body segments in a composite link-segment model.

In a link-segment modeling approach, the human body is modeled as a kinematic chain of linkages. Very few of these linkages are closed-loop, where the movement of any one link in the system can be used in calculating the movement of other links in the system. Most of the kinematic linkages in the human body are an open-loop system, where the same movement of an individual link can be obtained by several alternative movement patterns of the other links in the system. Forces in the human body are transmitted via these open-loop link systems by the nervous system, which controls the link segments. Because of the open-loop nature of the link segments, a given body movement can be achieved by the nervous system in several alternative movement patterns of the kinematic chain, though the end result is the same. It is important to understand this limitation of the mathematical link-segment model in reproducing human movement. Most of the time, a careful examination of the physically possible range of motion of these link segments can narrow down the choice of alternative movement patterns. Figure 4.27a shows the link segments that model the human leg. The upper leg (thigh, lower leg, and foot can be assumed to be three separate rigid body segments, attached to each other through pin joints. Each of these segments can be separately considered for analysis of the forces and moments exerted on it. Figure 4.27b and c shows the separated segments for ease of analysis. These are called *free-body diagrams.* In a state of either static or dynamic equilibrium, the net algebraic sum of the forces and moments acting on the segment must be equal to zero (in dynamic equilibrium, the body has a constant linear and/or angular velocity). Within each of these segments, there may be internal muscle forces acting to keep the segment in a state of equilibrium.

Several assumptions are made to simplify the analysis and apply the existing equation of motion to the body segments.

1. Each body segment has a fixed mass located at its COG.
2. Bone joints are approximated as pure hinge joints (with no loss of motion).
3. The location of the COG does not shift with motion.
4. Mass moment of inertia (denoted by ***I***) of each body segment about its COG is constant during the movement.
5. At each joint, a resultant force acts that is equal to the joint reaction force plus the result of all muscle forces (agonists and antagonists) acting across the joint.
6. Only the moment or torque generated by the resultant muscle force (not individual muscles) around the joint is used for analysis.
7. Each muscle crosses only one joint.

An axis system can be placed on the segmental joint (where it connects and interacts with other segments). Forces acting at the joint can be resolved into components acting along the $x$ and $y$ axes. The equation of motion can be formulated for the three body segments as shown next.

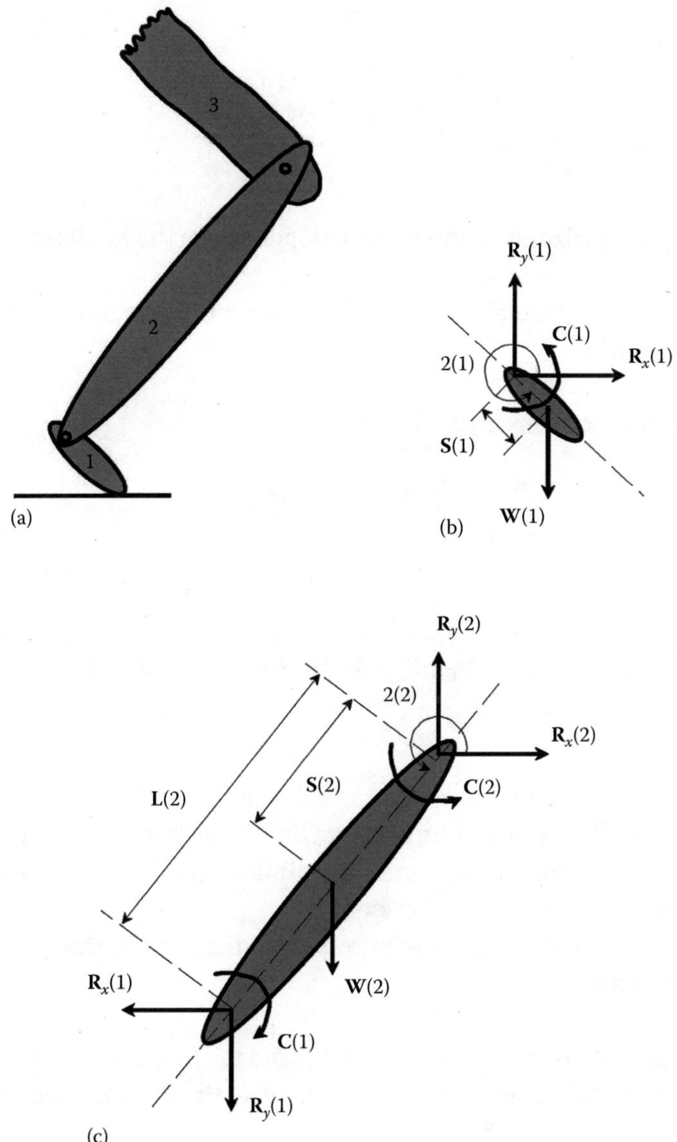


FIGURE 4.27 Line-segment modeling of the lower leg. (a) Lower leg and foot represented as body segments (b) segment 1 (foot) (c) segment 2 (lower leg). (Modified from Nordin, M. and Frankel, V.H., *Basic Biomechanics of the Musculoskeletal System*, Lea and Febiger, Philadelphia, PA, 1989.)

**Foot Segment**: From the free-body diagram shown in Figure 4.27b, summation of the horizontal and vertical forces yields

$$\Sigma \mathbf{F}_x = 0 \quad \text{and} \quad \Sigma \mathbf{F}_y = 0$$

$$\mathbf{R}_x(1) = \mathbf{M}(1) \times \mathbf{A}_x(1)$$

and

$$\mathbf{R}_y(1) - \mathbf{W}(1) = \mathbf{M}(1) \times \mathbf{A}_y(1),$$

where

**R**$_x$(1) and **R**$_y$(1) are horizontal and vertical components of the resultant muscle and reaction forces at the ankle joint

**A**$_x$(1) and **A**$_y$(1) are horizontal and vertical components of the linear acceleration of the center of mass (CM)

Summation of the moments about the CM yields

$$\Sigma \mathbf{M}_{\text{CM}} = \boldsymbol{I}(1)\boldsymbol{\alpha}(1)$$

or

$$\mathbf{C}(1) + \mathbf{R}_x(1)[\mathbf{S}(1)\sin\theta_1] - \mathbf{R}_y(1)[\mathbf{S}(1)\cos\theta_1] = \boldsymbol{I}(1)\boldsymbol{\alpha}(1),$$

where

**C**(1) is the resultant muscular torque at the ankle joint

***I***(1) is the moment of inertia of the foot about the COG of the foot (perpendicular to the sagittal plane, assuming that the motion is limited to the sagittal plane)

$\boldsymbol{\alpha}(1)$ is the angular acceleration of the foot

$\theta_1$ is the angle of the foot segment with respect to the right horizontal in the counterclockwise direction

**S**(1) is the distance between the proximal joint and the CM or COG of the foot segment Similar to segment 1 (foot segment), we can formulate the three equations for segment 2 (lower leg) shown in Figure 4.27c:

$$\mathbf{R}_x(2) - \mathbf{R}_x(1) = \mathbf{M}(2) \times \mathbf{A}_x(2)$$

$$-\mathbf{R}_y(1) + \mathbf{R}_y(2) - \mathbf{W}(2) = \mathbf{M}(2) \times \mathbf{A}_y(2)$$

$$-\mathbf{C}(1) + \mathbf{C}(2) + \mathbf{R}_x(1)\,[\mathbf{L}(2) - \mathbf{S}(2)]\sin\theta_2 - \mathbf{R}_y(1)\,[\mathbf{L}(2) - \mathbf{S}(2)]\cos\theta_2$$

$$-\mathbf{R}_y(2)\,\mathbf{S}(2)\cos\theta_2 + \mathbf{R}_x(2)\,\mathbf{S}(2)\sin\theta_2 = \boldsymbol{I}(2)\,\boldsymbol{\alpha}(2).$$

Similarly, the equation of motion can be formed for each of the connecting segments. Because of the number of variables involved, the equations become highly complex for a three-dimensional analysis. A generalized set of two-dimensional equations for static and dynamic cases is given as follows to illustrate the process.

**Generalized Equations for a Two-Dimensional Static Model**: Note: Here, $\mathbf{A}_x$, $\mathbf{A}_y$, and the horizontal force may be assumed to be zero:

$$\Sigma \mathbf{R}_j(y) = \mathbf{R}_{j-1}(y) + \mathbf{W}_j,$$

where

$\mathbf{R}_j(y)$ is the resultant muscular and reaction force at $j$th joint (vertical)
$\mathbf{R}_{j-1}(y)$ is the resultant muscular and reaction force at the previous $(j-1)$th joint
$\mathbf{W}_j$ is the weight of the $j$th body segment

For the torque at the joint,

$$\mathbf{C}_j = \mathbf{C}_{j-1} + \mathbf{L}_j \cos\theta_j \; \mathbf{W}_j + \mathbf{L}_{j-1} \cos\theta_j \, \mathbf{R}_{j-1}(y)$$

where

$\mathbf{C}_{j-1}$, $\mathbf{C}_j$ are the torques at the $(j-1)$th and $j$th joints
$\mathbf{L}_j$ is the distance from joint $j$ to the CM or COG of the link
$\mathbf{L}_{j-1}$ is the link length from joint $j$ to the adjacent joining joint $(j-1)$
$\theta_j$ are the angles of the links at each joint $j$ with respect to the right horizontal

**Generalized Equations for a Two-Dimensional Dynamic Model**: For the dynamic model, the orthogonal components of the linear acceleration, $\mathbf{A}_x$ and $\mathbf{A}_y$, need to be calculated for each joint. The equations are as follows:

$$\mathbf{A}_x(j) = 0$$

$$\mathbf{A}_x(j+1) = -\mathbf{L}_i[\dot{\theta}_j^2 \cos\theta_j + \ddot{\theta}_j \sin\theta_j] + \mathbf{A}_x(j)$$

$$\mathbf{A}_y(j+1) = -\mathbf{L}_i[\dot{\theta}_j^2 \cos\theta_j - \ddot{\theta}_j \sin\theta_j] + \mathbf{A}_y(j),$$

where

$\mathbf{L}_i$ is the length of segment $i$
$\theta_j$ is the angular position of joint $j$ with respect to the right horizontal, measured counterclockwise
$\dot{\theta}_j$ is the angular velocity of segment $i$ at joint $j$
$\ddot{\theta}_j$ is the angular acceleration of segment $i$ at joint $j$

As in the previous analysis, equations can be formed for each individual body segment, and the forces and moments at each individual joint can be calculated. The moment of inertia of the segment is calculated based on the fact that each body segment can be assumed to be a simple conical volume [20], as shown in Figure 4.28.

The aforementioned equations can be programmed into a computer for calculation of the joint forces and moments. The anthropometric measurements can be fed into the

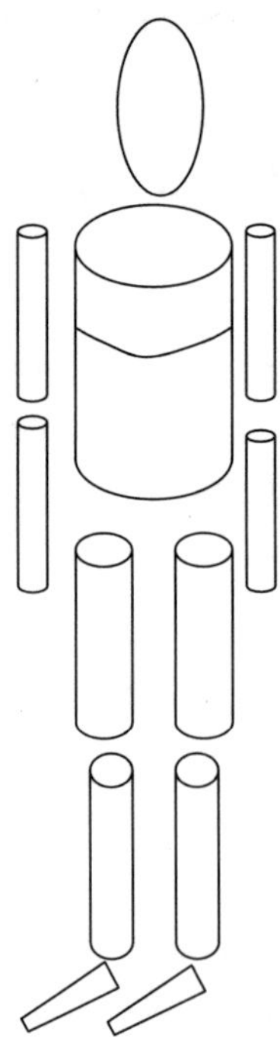

FIGURE 4.28 Three-dimensional modeling of a human body. (Adapted from Materials Handling Research Unit, University of Surrey (England) (Preparers), Force Limits in Manual Work, IPC Science and Technology Press, Surrey, England, 1980.)

model that includes the segmental length, weight, diameter/circumference, and segment orientation with respect to each other. The displacement, velocity, and acceleration of the segments can be obtained from kinematic data collected through videographic techniques. Several commercial packages are available for performing such sophisticated analyses. A simple application of the aforementioned link-segment modeling can be found in models that calculate the joint forces and moments as well as the loading on the $L_5S_1$ vertebral unit, for manual materials handling tasks [2,20].

## 4.4 BIOMECHANICAL CRITERIA FOR ERGONOMIC APPLICATIONS

Why is it essential to study the biomechanical criteria for ergonomic applications? This chapter has presented a number of biomechanical principles that can be applied to human body movement and form the basis for developing criteria to be used in applied ergonomics. For example, the limitation of the forces produced in the fingers at various flexions or extensions of the wrist may help us design hand tools that require optimum forces for a worker to operate them without excessive strain. Such interventions, on the basis of biomechanical criteria, reduce the risk of injury to the body due to impact loads or cumulative trauma by giving due consideration to the physical limitations of the human body. Determination of the criteria for cumulative trauma disorders due to incorrect postures, loading, and frequency of task performance have been based on the biomechanical analysis of the static as well as dynamic loading of the body segments. Such standards are already under preparation [21].

Several other important applications of biomechanical analysis may be cited. The lifting guidelines suggested by the National Institute of Occupational Safety and Health (NIOSH) use the biomechanical criteria of a maximum allowable compressive loading on the $L_5S_1$ vertebral units (3.4 kN, 770 lb) in setting the maximum load that can be safely lifted by a person. This recommendation includes several other biomechanical, physiological, and psychophysical factors for evaluating this loading of the human body that are based on the physical limitations due to the strength and posture of the body [22]. These include the frequency of lifting (which increases the dynamic loading as well as the metabolic demand), forward and upward reach (static loading), travel of the load, control at the endpoint of the loading, asymmetry of lift, coupling of the load, etc. Another application of biomechanical analysis is found in the design of equipments, tools, and workstations to optimally utilize the midrange of motion of the body segments, since it has been shown that forces and moments can be applied maximally at the midrange of motion of the segments. Existing databases have documented the maximum forces that can be applied at various limb positions and body postures. These databases may be used to control the amount of work output depending on the posture of the worker, without inflicting any form of harmful musculoskeletal loading. For example, the maximal two-handed upward vertical force (including lifting) that can be applied repetitively by a 41–50-year-old male in a seated position (without backrest) is 11 kg when the person has to stretch out his arms [23]. Any job requiring the person to perform such a task should be designed on the basis of this criterion for avoiding worker discomfort and possibility of injury. Such databases are limited in their scope of population size and type and need to be expanded to include the specific working population under consideration.

At present, lack of extensive data for all segments of the human population prevents us from more accurate analysis of safe limits of the loading of the body segments. With the help of further research, further advances in computer capabilities, and sophisticated equipment for measurement, better and more realistic biomechanical models can be generated to enable us to set forth safe and acceptable limits of physical work and human output.

## REFERENCES


1. Y. C. Fung, *Biomechanics*, Springer-Verlag, New York, 1981.
2. D. B. Chaffin and G. B. Andersson, *Occupational Biomechanics*, Wiley, New York, 1984.
3. M. Nordin and V. H. Frankel, *Basic Biomechanics of the Musculoskeletal System*, 2nd edn., Lea and Febiger, Philadelphia, PA, 1989.
4. E. R. Tichauer, *The Biomechanical Basis of Ergonomics*, Wiley-Interscience, New York, 1978.
5. N. Özkaya and M. Nordin, in *Fundamentals of Biomechanics: Equilibrium, Motion, and Deformation*, D. Leger, Ed., Van Nostrand Reinhold, New York, 1991.
6. B. LeVeau, *Biomechanics of Human Motion*, Saunders, Philadelphia, PA, 1976.
7. B. A. Gowitzke and M. Milner, *Scientific Basis of Human Movement*, 3rd edn., Williams and Wilkins, Baltimore, MD, 1988.
8. Eastman Kodak Co., Human Factors Section, *Ergonomics for People at Work*, Van Nostrand Rein-hold, New York, 1986.
9. B. Tyldesley and J. I. Grieve, *Muscles, Nerves, and Movement: Kinesiology in Daily Living*, Blackwell Scientific, London, U.K., 1989.
10. M. I. Ellis, B. B. Seedhom, V. Wright, and D. Dowson, An evaluation of the ratio between the tensions along the quadriceps tendon and the patellar ligament, *Eng. Med.* 9(4): 189–194 (1980).

11. Peak Performance Technologies, *User's Reference Manual,* Englewood, CO, 1994.
12. A. A. White and M. M. Panjabi, *Clinical Biomechanics of the Spine*, Lippincott, Philadelphia, PA, 1978.
13. R. Cailliet, *Low Back Pain Syndrome*, 3rd edn., F. A. Davis Co., Philadelphia, PA, 1981.
14. A. Bagchee and A. Bhattacharya, Method for estimating fall potential in clinical, occupational, and environmental exposure cases (Abstract), *J. Biomech. 26*(3): 305 (1992).
15. M. L. Root, W. P. Orien, and J. H. Weed, Normal and abnormal function of the foot, in *Clinical Biomechanics*, Vol. II, Clinical Biomechanics Corp., Los Angeles, CA, 1977.
16. T. Shimba, An estimation of center of gravity from force platform data, *J. Biomech. 15*: 533–550 (1987).
17. A. Bhattacharya, R. Morgan, R. Shukla et al., Non-invasive estimation of afferent inputs for postural stability under low levels of alcohol, *Ann. Biomed. Eng. 15*: 533–550 (1987).
18. A. Bhattacharya, The role of worker's balance in falls at the workplace, *Regional Conference on Ergonomics, Safety and Health in Construction,* July 19–20, 1994.
19. W. T. Dempster, *Space Requirements of the Seated Operator: Geometrical, Kinematic, and Mechanical Aspects of the Body with Special Reference to the Limbs*, Wright Air Development Center Tech. Rep. No. 55-159, Dayton, OH, Wright Patterson Air Force Base, Nat. Tech. Info. Service No. AD-087892, 1955.
20. E. P. Hanavan, *A Mathematical Model of the Human Body*, Wright-Patterson Air Force Base, OH, AMRL-TR 64-102, 1964.
21. ANSI, American National Standard Institute: Control of cumulative trauma disorders, Draft ANSI Z-365, 1993
22. T. R. Waters, V. Putz-Anderson, A. Garg et al., Revised NIOSH equation for the design and evaluation of manual lifting tasks, *Ergonomics 36*(7): 749–776 (1993).
23. Materials Handling Research Unit, University of Surrey (England) (Preparers), *Force Limits in Manual Work*, IPC Science and Technology Press, Surrey, England, 1980.

CHAPTER 5

# Biomechanical Models in Ergonomics

Kevin P. Granata*, Kermit G. Davis, and William S. Marras

CONTENTS



## 5.1 INTRODUCTION

Why should an ergonomist be interested in biomechanical models of the human body? In order to answer this question, one must consider what can be gained from biomechanical modeling. The goal of biomechanical modeling, in an ergonomics context, is to gain insight into the how the joints of the body are strained during work as well as to provide a framework for understanding how the musculoskeletal system behaves during work task performance. Biomechanical models provide qualitative and quantitative answers that explain how the body is loaded during work. Quantification is important because it allows one to address the issue of how much is too much loading on a joint. Through quantification of joint loading, one can not only assess the risk associated with existing jobs but also predict whether someone is likely to suffer an injury with a proposed design of a workplace. Qualitative analyses allow comparison of biomedical loads between two or more

* In memoriam, Dr. Kevin P. Granata was a victim of a senseless act of violence on the campus of Virginia Tech on Monday, April 16, 2007.

sets of conditions. Models of musculoskeletal stresses may allow ergonomists to determine which of several workstation designs will result in the least amount of structural loading on a specific joint. Thus, biomechanical models used in ergonomics provide both a means to quantitatively assess work and a means to predict whether a proposed workstation can be expected to lead to injuries.

Traditionally, our knowledge of joint biomechanics was limited in that we are able to provide only gross approximations and estimates of forces supplied by the muscles or imposed on the body. Although the complexity of joint motion and force development can be appreciated from studies of anatomy, physiology, and kinesiology, most quantitative descriptions of their mechanical behavior used for ergonomics purposes are still simplistic in comparison. Through the application of biomechanical modeling, we can test our hypotheses, simulate joint dynamics, and predict the loads developed within a joint. By developing validated biomechanical simulations, complex anatomical structures and motions can be not only described but also explained.

## 5.2 BACKGROUND AND SIGNIFICANCE TO OCCUPATIONAL ERGONOMICS

The objective of biomechanical analyses is to explain how the body works from a loading perspective. Application of these analyses also provides insight into how, why, and when a joint may fail. A biomechanical model is simply a mathematical representation of the musculoskeletal system and permits one to quantitatively describe the loading effects of a task. A biomechanical model that is capable of accurately simulating the forces developed within a joint may be employed to investigate the relation between joint load and exertion parameters such as joint angle, torsion, velocity, applied load, and impact. When these joint loads are compared to the structural strengths or tolerance limits of the joint, one may use this information to explain why an injury occurred during a particular exertion. When the probabilistic nature of joint tolerances is considered, these models could help explain why some individuals will suffer an occupationally related injury under certain conditions whereas others may not. In addition, common injuries that occur for unknown reasons may be examined to determine the causative factors, thereby indicating a possible means for injury prevention.

Models are used to help us organize and test our knowledge and understanding of how different workplace or loading factors are related to the probability of failure of a joint or structure. Perhaps one of the greatest values of biomechanical modeling is that it provides a mechanism to test hypotheses generated from epidemiological studies of injuries in the workplace. For example, studies [1] have shown that twisting motions in the workplace are a contributing factor in occupationally related low back disorder risk. Biomechanical models [2] can explain how spine loading patterns differ during twisting and bending motions. Another example is that psychosocial factors such as job stress and mental concentration demands have been linked to increased injuries by epidemiological studies [3,4], and loading mechanisms were identified by increased loading on the joints [4–6]. Thus, biomechanical models could provide a cross-validation for suspected risk factors for musculoskeletal injuries. In addition, once we understand the mechanism of occupationally related injury, we can use these models to determine how to change the work environment to minimize the risk of injury. One example was reported by Marras and associates [7] who found that lifting cases in a warehouse setting can produce significant loads on the spine but these loads can be reduced by either by lifting lighter weights or by lifting from higher lift origin height levels of a pallet. In this manner, biomechanical models can help us determine how much exposure to a risk factor is too much, so that low-risk work environments can be created.

Occupationally related low back pain provides a classic example of how biomechanical modeling can be employed to investigate why and how an injury occurred as well as to suggest possible solutions. The epidemiological literature demonstrates that, in general, occupations associated with manual materials handling industries are subject to high incidence rates of low back disorders (LBDs). The studies clearly indicate that lifting and parameters associated with lifting result in increased risk of injury. Biomechanical models of the trunk and low back have been developed to qualitatively and quantitatively describe trunk and spinal loads as a function of the lifted weight, the distance it is held from the body, and associated trunk moment, trunk posture, and motion [8–11]. These models have been successful in describing why people who perform certain lifting tasks are more susceptible to injury than those who perform different tasks. Furthermore, the models have provided estimates of the direction and magnitude of loads on the spine. These data can be employed to determine how the spine fails under typical lifting conditions. Although we are far from a complete understanding of the biomechanical cause of occupationally related low back pain, significant progress has been made in the recognition and explanation of many of the causative parameters. However, the complexity of the human body and specifically the lower back necessitates more complex models to further refine our knowledge about the biomechanical mechanisms of LBD, potentially integrating actual pain pathways into biomechanical models in the future [12].

Biomechanical models are developed to understand or interpret the functional behavior and joint loadings that occur during a task or exertion of interest. Therefore, the conditions under which the model is developed and operates must be realistic and accurate for the model to be useful for ergonomic purposes. For example, static models [13–15] that assume symmetric lines of action and neglect motion are not intended to represent the biomechanics of common, complex, dynamic motions such as many of those tasks observed in most workplaces. Similarly, analyses that neglect significant anatomic structures and physiologic principles can misrepresent the biomechanical loads imparted by and upon the joints. The fact that a computer model "works" does not guarantee that it is correct or useful. Inaccurate or unrealistic model output may lead to misdiagnosis of the joint load, potential, injury, and risk, thereby permitting conclusions that may exacerbate the problem.

Unfortunately, as a biomechanical model becomes more accurate and realistic, it often becomes more complex. Simplistic biomechanical models are often easy to use and can be employed by those who are not necessarily experts in ergonomics. Results from simple models are also easier to interpret than their complicated counterparts. These types of models may be somewhat useful for gross approximations of tissue load when no other means of assessment are available (due to a lack of resources such as time, lack of sophisticated equipment, and financial funds). These models are particularly attractive to nonsophisticated users. The danger with these simple models is that the user may not be aware of the model limitations or of assumptions inherent to the model and thus may misapply the model or use it to lead them to an incorrect solution to a manual materials handling problem. On the other hand, highly complex models representing realistic responses require significantly more input data to achieve those ends resulting in more accurate representation of the loading on the joints. Another factor to consider is complex models require more resources to develop, often taking years to be completely functional and accurate. In most cases, the most complex models must be used by experts and this often limits the use by individuals whose primary concern is the design of the workplace. Therefore, there is often a trade-off between criteria of accuracy and realism versus simplicity and ease of use as well as the resources required to develop and use the models. In general, for ergonomic purposes, a model need only be as complex as is necessary to accurately and reasonably describe the nature of loads occurring at a joint of interest due to a particular work task. Thus, user experience and ergonomics insight is an important element of proper model usage.

It is important to recognize that the fundamental goal of a biomechanical model is to quantitatively describe the resultant internal loads experienced by a joint as a consequence of an occupationally related task. As mentioned in an earlier chapter, both external and internal loads can be imposed on the body as a result of work. External loads are those associated with forces acting on the body from outside the body, such as the forces of gravity acting upon the body segments and a box that one is attempting to lift. Internal loads are associated with forces and moments within the body used to counteract the external load. These typically are generated by muscles and ligaments within the body. However, the distance between a joint's center of rotation and the external load is typically much greater than the distance from the joint's center of rotation to the muscle attachment. Therefore, muscles, tendons, and ligaments are at a mechanical disadvantage, and the internal forces are typically much larger, often orders of magnitude larger, than the applied external force. For example, holding a 10-lb weight (e.g., an external load) in front of one's body at arm's length could require internal forces within the back muscles in excess of 150 lb. Biomechanical models help us to place the contributions of both the internal and external forces in perspective relative to occupationally related risk of injury.

Biomechanical models used for ergonomic purposes can be found for almost any joint in the body but the two most often joints modeled are the back and the wrist. The anatomy of the back and that of the wrist are drastically different; therefore, these models have developed along different paths.

### 5.2.1 Models of the Trunk and Back

Several mathematical models of the lumbar spine have been developed to determine the loads on the spine during lifting exertions [8–16]. These models have been designed to enhance the knowledge of low back mechanics, improve the ability to diagnose the cause and risk of injury during manual materials handling tasks, and appraise recommendations designed to reduce lifting injuries.

One must recognize that loading of the spine can occur in several different ways. Compression, shear, and torsion loading can be imposed on the spine during the performance of a task. Realistically, most lifting tasks result in some combination of these loadings on the spine. In addition, loading can occur at various levels of the spine. The spine is a series of 33 vertebrae, and each vertebra could be loaded differently during an occupational task. Hence, the system has the potential to be represented in a very complex model. Most models used in ergonomics assume that the spine consists of a single rod or a single or double joint so the models can be solved. In recent years, Marras and colleagues [17] have developed a multilevel spine model capable of assessing the loading patterns throughout the lumbar spine as well as complex loading resulting from handling tasks such as pushing and pulling.

Model development continues to evolve quickly as a result of greater computer power and better imaging techniques. This evolution has allowed biomechanical models to be personalized by accounting for individual muscle recruitment patterns as well as individual muscle anatomy [17]. The personalization of models will ultimately provide better estimates of the spine loads and identification of individual specific tissue breakdown through better quantification of loads on discs as well as facet joints [17]. These personalized models will allow for estimation of complex loading specific to individual's anatomy and previous structural damage (e.g., cumulative trauma). As the imaging procedures provide better descriptions of the structures of the spine, more specificity of injuries will arise, which is a key to understanding such a complex system as the lower back.

Once estimates of loading are derived for a particular task, these loadings are typically compared to spine tolerances. Tolerances of the vertebral end plates have been determined in vitro by Evans and Lissner [18], Sonada [19], and Jagger et al. [20] for pure compressive loads. Brinckmann et al. [21] determined compressive tolerance of the spine as a function of combinations of loading level and frequency of loading for a population of cadaver specimens. Other researchers have identified much lower spine disc tolerance levels for shear loading [22]. Further, Gallagher and associates [23] have shown the tolerances levels of the disc are impacted by the complex loading.

#### *5.2.1.1 Biomechanical Risk Factors*

One of the major challenges for any biomechanical model of the spine is to determine which components of the workplace need to be considered in order to accurately account for the risk associated with the occupational demands of the work. As mentioned earlier, biomechanical models can be extremely complex and difficult to use if they comprehensively include all biomechanical components of work. Similarly, if the models are too simplistic they may not be very useful for the control of occupationally related LBDs. Thus, it is imperative to identify those components that are necessary to include in an accurate biomechanical model that can be applied to the workplace.

Marras and associates [1,24] performed an industrial in vivo study to assess the contribution of various biomechanical workplace factors to the risk of suffering a work-related LBD. Over 400 manual materials handling jobs were observed in 48 different industries. The medical records in these industries were examined so that specific jobs historically categorized as presenting low, medium, or high risk of occupationally related LBD could be identified. One hundred fourteen job-dependent characteristics were recorded for each job. Among these characteristics, workplace factors such as the weight of the object lifted, the distance of the object from the worker, work heights, and cycle times were recorded. In addition, workers wore a triaxial goniometer on the back (lumbar motion monitor [LMM]) that documented the three-dimensional (3D) angular position, velocity, and acceleration characteristics of the lumbar spine associated with the various jobs. A multiple logistic regression model indicated that a combination of five trunk motion and workplace factors was best associated with the risk of LBD. These factors were lifting frequency, load moment, trunk lateral velocity, trunk twisting velocity, and trunk sagittal angle. Increases in the magnitude of these factors significantly increased the risk of LBD. This study indicated that by quantitatively considering the combination of these five factors it is possible to predict the probability of membership in the high-risk category 10 times more effectively than chance. As this was one of the first field-based studies to comprehensively examine the biomechanical factors associated with occupationally related LBDs, these results provide valuable information as to which factors need to be included in biomechanical models of the trunk. The study suggests that if all five of the risk factors could be accounted for in a biomechanical model, one may gain an appreciation for how back injuries are associated with spine loading. A more recent field-study has extended the quantification of these physical exposure factors to dynamic moment exposure and has been able to further identify the nature of physical risk factor exposure [25]. Many other factors such as job stress, mental concentration, cold, and hot temperatures, and work organization factors may also contribute to low back injuries through biomechanical loading [5,12].

#### *5.2.1.2 Static Models*

Biomechanical models of lifting have progressed from static two-dimensional (2D) analyses to more recent attempts at understanding dynamic 3D stresses on the spine. The newer models endeavor to accurately and realistically represent the mechanical loading and behavior of the lower back while refraining from as much unnecessary complexity as possible.

A simple 2D lifting model can be developed by representing the body as a stick figure as shown in Figure 5.1. If the spine is assumed to be a rigid rod attached to the pelvis by a single hinge, all motion and trunk mechanics are centered about that point. The back muscles can be represented by a single tensioned cable strung between the upper spine and the posterior aspect of the pelvis (single equivalent muscle). When weight is placed in the stick man's hands, the only thing preventing the hinged spine from falling forward is the tension in the back muscle equivalent. The external moment tending to cause the stick man to bend forward can be described by the 2D vector product of the lifted external load and the moment arm from the back joint to the weight. Similarly, the internal, restorative moment that supports the trunk in an upright posture against the external load can be described by the vector product of the back muscle equivalent and the moment arm from the back joint to the muscle. If we assume that there is no motion and know the moment arm distances,

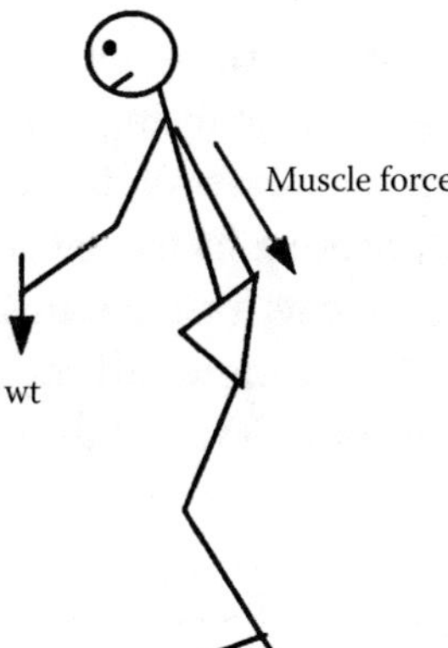


FIGURE 5.1 A rudimentary lifting model representing sagittally symmetric, 2D exertions. The spine is modeled as a rigid rod with a single hinge at its base. Back muscles are represented by a single tensioned cable. Although the model is neither accurate nor realistic, it may provide a "ballpark" estimate of spinal loading.

the laws of static equilibrium can be applied to determine the tensile force in the modeled back muscle required to hold a given external load, resulting in the estimation of the tensile force in the back muscle equivalent.

This simple analysis can be extended to determine the compressive load imposed on the spine by the stick man when he holds the weight. In this model, compression is simply represented as the sum of vertical forces. Clearly, the weight held by the stick man adds compression on the spine. Note also that the back muscle equivalent adds a significant vertical force that must be supported by the spine. Therefore, in this example, the compressive force is the sum of the weight in the hand, body segment weights, and, approximately, the back muscle equivalent tensile force.

The analysis described here is not realistic and has several limitations. It simply provides "ballpark" estimates of the loads that act on the spine. The biomechanical model is 2D, whereas the human body is 3D. Great numbers of muscles in the trunk have been neglected, each with unique endpoints, i.e., insertions and origins. Those muscles exist in 3D space and often wrap around the skeletal structures of the trunk. Hence, such a model is unable to accurately evaluate shear or torsional loads (two very important risk factors) applied to the spine. Representing the spine as a rigid rod with a frictionless hinge at its base also severely limits model realism. The spine is a multielement structure with complex geometry and passive nonlinear resistance to bending and twisting motions. In addition, no passive forces from the stretching of muscles and ligaments are included in such models. The modeled exertion was represented as a static, sagittally symmetric lift. Next, lifting velocity and acceleration influence trunk mechanics and spinal loads. Lifts are seldom performed directly in front of the body and never statically. Instead, most occur in combination with twisting and lateral bending and with the weight held in an asymmetric posture (e.g., off to one side) as well as require the object to move over a distance. Thus, when using such a model, one must consider how model predictions may be limited by inappropriate assumptions about the workplace conditions.

Clearly, the simple stick man model described here is useful for its designed purpose—to demonstrate the fundamental concepts of lifting models, but it is greatly limited in its usefulness for accurate biomechanical analyses of lifting exertion and concomitant spinal loads. One of the original, groundbreaking lifting models [8] is very similar to the stick man model described here. It has served as a simple analytical tool for job and workplace evaluation. To overcome its limitations, more advanced models have been developed. The single-extensor stick man model has served as the ancestral foundation for the more advanced models.

Models have also been developed that attempt to account for the 3D nature of trunk loading. Chaffin and Erig [25] extended the single-equivalent stick man representation to 3D space. Their model was able to accommodate asymmetric lifts but was still limited to static evaluations and was not sensitive to complex loadings of the spine due to the activities of the multiple internal muscle forces involved in a lift. Schultz and Andersson [13] developed a model that accounts for the 3D architecture of the trunk. As shown in Figure 5.2, this model assumes that the trunk can be viewed as a cross section through a level of the lumbar spine. Schultz and Andersson described the relationship between the 10 trunk muscles, intra-abdominal pressure (IAP), and 3D spine loading in six force and moment equations. Since there were 11 unknown muscle and IAP values and only six equations available to derive a solution, the model resulted in a statically indeterminate problem. This static model was solved either by accepting simplifying assumptions about the muscle activities (i.e., no coactivation) or through the application of linear programming techniques. However, because of the limited number of functional equations in the model, the linear programming approach was not able to predict coactivation of the trunk muscles that had been observed even under static conditions [26]. More recent models

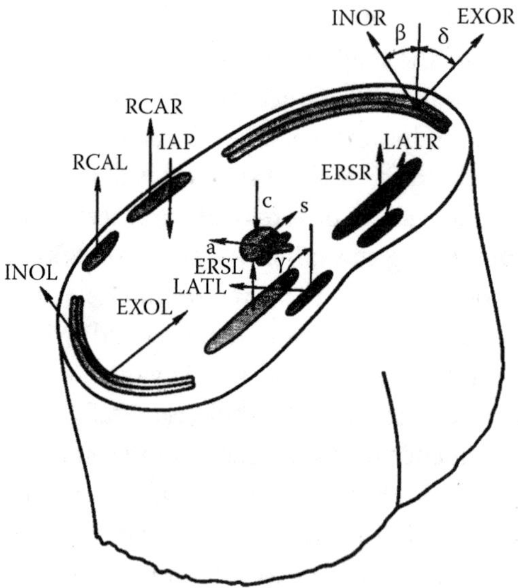


FIGURE 5.2 Cross-sectional view of the human trunk at the lumbosacral junction. (Adapted from Schultz, A. and Andersson, G., *Spine*, 6, 76–82, 1981.)

have attempted to include coactivation in the solution by using adjustments to the linear programming model [27] and double optimization functions [28].

Biomechanical models of spinal loading require accurate kinetic assessment of muscle forces and realistic representation of lifting kinematics. Static modeling may represent static exertions, however, lifting exertions, by definition, include a dynamic component.

#### *5.2.1.3 Lifting Dynamics*

The lifting models of Chaffin [8], Shultz and Andersson [13], and Gracovetsky and Farfan [29] attempted to estimate spinal load from equations of static equilibrium. Consequently, the models were representations of static posture lifting. Schultz and Andersson [13] indicated that their model of static lifting could be employed to model quasi-dynamic movement. They implied that motion can be represented as a series of static postures linked together in time to give the illusion of motion. This concept is similar to connecting a series of still photographs to produce a motion picture. However, the laws of physics demonstrate that forces associated with static exertions are not the same as those associated with dynamic exertions. Furthermore, physiology demonstrates that the contractile velocity of a muscle influences its ability to generate force. Static modeling cannot satisfy these conditions. The significance of considering these dynamic factors was confirmed via industrial biomechanical assessments [1,24].

Evidence has shown that static lifting models underpredict spinal loading during dynamic lifting exertions. Marras and Sommerich [11] concluded that "spine compression increases directly with [isokinetic] trunk velocity" during lifting exertions. Similar conclusions were reached by Granata and Marras [30], Frievalds et al. [31], McGill and Norman [14], and Goel et al. [32], who predicted a 30%–40% increase in dynamic lifting moment and spinal compression compared to static exertions.

In addition to the mechanical factors, lifting dynamics also influence muscle coactivity. Muscle coactivity is important because it is responsible for the complex loading of the spine. Because many of the trunk muscles are not oriented in a vertical direction, they can apply compression, shear, and torsional forces on the spine. The number of active trunk muscles and the magnitude of the force in each muscle are also significantly affected by lifting velocity. Measures of muscle coactivity [33,34] have demonstrated that the activity in muscles other than prime movers, i.e., synergistic and antagonistic muscles, increases with lifting velocity. The relation between dynamic motion and muscle coactivity directly affects the loads on the lumbar spine [35]. Therefore, dynamic modeling must be employed to accurately represent spinal loads during lifting exertions.

Quasi-dynamic models of isokinetic lifting have been developed [10,35,36] that incorporate the behavior of muscle coactivity patterns. The models are labeled quasi-dynamic because lifting exertions are represented as smooth, continuous trunk motions, typically isokinetic, but the influences of mass dynamics are not considered. McGill and Norman [14] concluded that isokinetic models overpredict peak dynamic lifting moments by an average of 25%. To accurately represent the biomechanics of lifting, fully dynamic modeling must be employed [37].

#### *5.2.1.4 Trunk Mechanics*

Both external and internal forces contribute to spinal loading. In fact, internal (i.e., muscle) forces typically contribute far more to spinal stress than external forces. Moment arm distances of the muscles are small compared to almost any external moment arm distance. Because muscles are at a severe mechanical disadvantage when attempting to offset external moments, extremely large forces are required to overcome this handicap. Estimates of spinal loading must therefore include both the external and the much larger internal (muscle) forces.

Chaffin [8] and Chaffin and Baker [38] developed an early low back lifting model to predict spinal loads under static 2D sagittally symmetric conditions. The model computed the force in a single extensor muscle required to offset reactive trunk moments generated by external forces. It did not account for the muscle synergy or flexor antagonism evident in multiple-muscle systems [26,39–41]. Hof and Van Den Berg [42] demonstrated that the forces exerted by synergistic and antagonistic muscles significantly affect the total joint torque. Hence, multiple-muscle models must be employed to accurately represent lifting biomechanics.

One of the first multiple-muscle models intended to overcome these limitations was the model of Schultz and Andersson [13]. This model used 10 muscle equivalents to predict static sagittally symmetric and asymmetric [27] spinal loading on the lumbar spine. Optimization techniques were employed to estimate muscle forces from an indeterminate system of equations. Gracovetsky and Farfan [21] implemented a similar analysis with a much more comprehensive anatomic model of the low back. In addition, Jager and Luttmann [43] and Goel et al. [32] used optimization procedures to predict loading during dynamic and asymmetric lifting exertions. However, optimization-based predictions of muscle coactivation do not agree well with observed muscle coactivation during dynamic lifting activities [34,44].

Deterministic models, including the single-equivalent-muscle models, by their nature cannot account for muscle force variability within and between subjects performing a lifting exertion. The models assume that internal forces can be completely described from the applied external forces and trunk kinematics. Muscle tensile forces in these models are determined from a system of equations representing lifting mechanics and possible constraints introduced by the optimization objective functions [45]. Variability within the system is not possible in deterministic modeling, because muscle kinetics are reduced to a single, "unique," optimum solution [46,47]. Therefore, deterministic modeling represents the optimal solution to the objective function of the model, but not necessarily the realistic muscle activity and spinal load. It is important to be able to account for this variability in muscle recruitment during an occupational activity. Variation in muscle force can significantly influence spinal loading. Therefore, to accurately model lifting mechanics, the variation in muscle coactivity must be considered.

There is little question that a significant amount of antagonistic coactivation exists and is the rule rather than the exception under typical manual materials handling conditions. This coactivity has been noted in response to several workplace factors in several studies. Marras et al. [26], Gracovetsky et al. [39], Zetterburg et al. [40], and McGill and Norman [41] measured antagonistic muscle activity and concluded that coactivity contributes to trunk stability. Rajulu [48] and Marras and Mirka [34,49] also measured significant antagonistic coactivity as a function of external load, trunk position, velocity, and acceleration.

Optimization methods have often been suggested as a means to overcome the statically indeterminate nature of the trunk's multiple-muscle system and permit analytic solutions to more complex biomechanical models. However, these optimization schemes typically predict unrealistic muscle coactivity levels, including an absence of antagonistic activity [35,45]. In an attempt to improve upon this situation, muscle coactivity has been predicted using nonlinear optimization techniques [32,50], but with limited success. The predicted muscle activity amplitudes do not agree with measured values.

It is extremely important to account for muscle coactivity because this coactivity significantly influences spinal load [35,42]. Trunk flexor muscles of the abdomen have a greater mechanical advantage than the extensor muscles of the back. Minimal abdominal activity requires substantial back muscle tension to offset the antagonistic moments. Therefore, muscle cocontraction influences the activity in the prime movers necessary to achieve static or dynamic equilibrium. If optimization techniques inaccurately predict coactivity, then muscle activity in the prime movers will also be incorrectly estimated, and subsequent predictions of spinal loads will be inaccurate. Furthermore, if the coactivity of all trunk muscle equivalents is not monitored, then damaging components of torsion and shear loads remain unpredicted [51,52]. The assumption of insignificant coactivity and antagonism is valid in a very limited and artificial set of circumstances and ensures underestimation of spinal loading. Therefore, ignoring muscle coactivity through deterministic optimization risks inaccurate estimation of spinal loading.

#### *5.2.1.5 Electromyography-Assisted Models*

To avoid estimation errors associated with deterministic prediction of muscle force, biomechanical models of lifting exertions have been developed that are assisted or guided by biological input. The most common biological input used in biomechanical modeling involves electromyographic (EMG) measures of muscle activity. Electromyography-assisted models employ measured myoelectric activity to determine individual muscle forces. Thus, EMG-assisted models avoid the problem of neglected muscle coactivity associated with statistically indeterminate models. Relative weighted EMG values recorded during lifting exertions are used to predict the tension developed in muscles and their resultant trunk moments and spinal loads. By their nature, EMG-assisted models accurately represent the neuromuscular control system of the trunk through direct measurement of muscle activity [50,53]. The advantages of EMG-assisted models are that (1) they are not limited by the constraints of an optimization objective function, (2) muscle coactivity and antagonistic forces are accurately represented via measurement, (3) coactive variability within and between subjects is directly measured, (4) physiologic coefficients predicted by the model can be used for instantaneous validity checking, and (5) accuracy can be documented via direct comparison of measured and predicted lifting moments.

McGill and Norman [15] developed a dynamic model using EMG measures to estimate force in 12 trunk muscles. Myoelectric activity was measured on one half of the body, and contralateral muscles were presumed to behave similarly to their measured counterparts. Therefore, the model represents sagittally symmetric lifting only and disregards muscle-induced lateral shear and torsion at the base of the spine. A more recent publication by McGill [54] employs bilateral placement of electrodes in an attempt to predict lateral bending moments. Muscle force $F$ is represented by

$$F = \text{gain} \times \frac{\text{EMG}(t)}{\text{EMG}_{\text{Max}}} \times \text{Area} \times F(\text{Vel}) \times F(L) \tag{5.1}$$

where

Area is the cross-sectional area of the muscle

gain represents the physiologic muscle force per unit area assigned to 35 or 50 N/cm$^2$

$F$(Vel) and $F(L)$ are modulation factors representing theoretical muscle force–velocity [55,56] and length–strength [56,57] relations, respectively

Muscle activity is represented by normalized EMG levels, i.e., EMG/EMG$_{\text{Max}}$, where EMG$_{\text{Max}}$ is a constant. Unfortunately, normalizing by a constant value of EMG$_{\text{Max}}$ may have limited applicability. Marras et al. [26,33] demonstrate that maximum EMG levels change as a function of trunk angle and subsequent muscle length. Thus, the normalization constant must be a function of trunk position. Muscle length directly influences the relation between measured EMG and joint torque [58], due in part to the physiologic length–strength relation of muscle [57]. Isokinetic velocity and acceleration also significantly affect measured EMG levels [34,44,49,59]. Therefore, dynamic effects on myoelectric activity and muscle force must be included in any EMG-assisted low back model.

An EMG-assisted model developed by Reilly and Marras [36] incorporated intramuscular EMG data from five left–right pairs of trunk muscles. Bilateral muscle measures allowed prediction of relative asymmetric moments, although experimental data were collected only in the sagittal plane. A similar model by Marras and Sommerich [10] was validated under asymmetric moments, although experimental data were collected only in the sagittal plane. A similar model by Marras and Sommerich [10] was validated under asymmetric exertions, i.e., in a vertical bending plane rotated from the sagittal. Myoelectric inputs were normalized as a function of both trunk bending angle and asymmetry and were modified for length and velocity artifact via empirically derived regression equations [10,34]. Time-dependent kinetic, kinematic, and EMG data were approximated by a three-segment, straight-line representation of the data profiles. Good correlations were achieved between predicted and measured trunk moments (Figure 5.3). However, approximating the dynamic data profiles by straight-line segments reduced the true power of the EMG-assisted model by artificially representing the measured muscle activity.

The Marras and Sommerich [10] model was the first EMG-assisted lifting model to allow validation of predicted results. Straight-line segment representations of the measured trunk extension moments were compared with predicted moments to test the model accuracy. A validity check was performed by requiring a physiologically reasonable gain, i.e., muscle force per unit area, predicted from each trial. Although the output gain factor

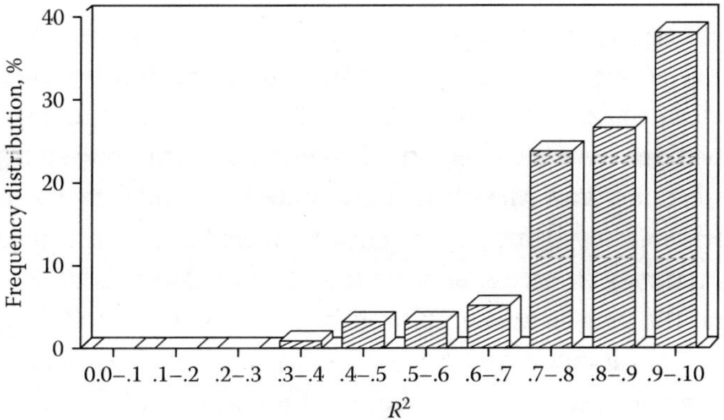


FIGURE 5.3 Association ($R^2$) between predicted and measured trunk torque (total of 98 trials). (From Marras, W.S. and Sommerich, C.M., *Hum. Factors*, 33, 123–137, 1991.)

was checked for validity, it was allowed to vary with each lifting task, implying that the subject's strength capacity changes with each exertion. Clearly, a subject's muscle strength per unit area does not change from trial to trial. A dynamic EMG-assisted lifting model proposed by Granata and Marras [16] treated gain as a subject-dependent input constant. This introduced possible intertask variability into the magnitude of predicted extension moment with each lifting task. The model sampled and processed dynamic data at each point in time (100 Hz) throughout small time windows, requiring no interpolation or linearity assumptions, although the entire lifting motion was not represented. As shown in Figure 5.4, the processed EMG represented dynamic changes in muscle forces and measured the coactivation in the multimuscle system. The model was tested under isometric,

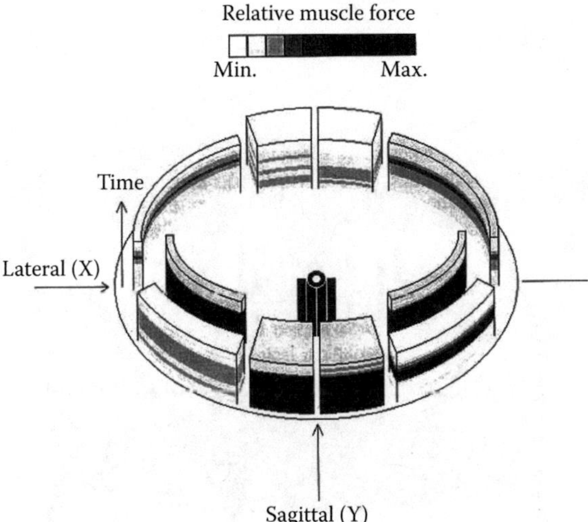


FIGURE 5.4 Relative muscle force as a function of time (vertical axis) in each of the 10 modeled trunk muscles viewed from the posterior aspect. Note the measured coactivity, especially in the antagonistic musculature. (Adapted from Marras, W.S. and Granata, K.P., *Spine*, 20, 1440–1451, 1995.)

isokinetic, and isoinertial conditions at three separate angles of asymmetry. Coactive bilateral EMG data were normalized as a function of trunk angle and asymmetry.

Models extending the concept of EMG-assisted modeling to trunk loading during axial twisting exertions have also been published. Pope et al. [60] endeavored to model static twisting exertions, and McGill [61] attempted to model dynamic twisting exertions. Both models suffered from physiologically unrealistic gain values. They required muscle force capacity far in excess of accepted values [62–64]. Marras and Granata [2] demonstrated that the muscle area employed in EMG-assisted modeling must represent the maximum area of the muscle as opposed to the area of the muscles that is found at the transverse plane defined by the lumbosacral junction. This modification permitted successful simulation of static and isokinetic torsional exertions.

In order to allow the biomechanical modeling of typical daily lifting activities, a model must be capable of simulating spinal loading during unconstrained, free-dynamic coupled motion. It is necessary that muscles represented by the model be permitted to change length, direction, and velocity so that appropriate modulation of the muscle myoelectric activity can occur. Recently, EMG-assisted models have been developed that successfully treat these three measures as dependent variables [2,37,56]. These models represent the trunk musculature as a series of force vectors as shown in Figure 5.5. These vectors can change in orientation and magnitude as the trunk moves through space and can thus more accurately account for the loading contributions of the trunk musculature. The model was further modified to account for differences in trunk musculature between males and females [65,66].

Interpreting EMG data collected during dynamic motions can be difficult because of the motion artifact related to physiologic and analytic phenomena [34,44,59,67–69]. Electromyography-assisted models must appropriately handle EMG dynamics. This has

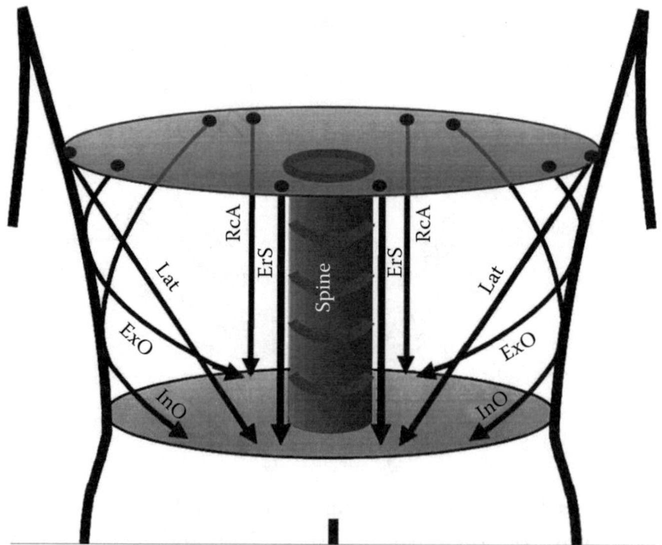


FIGURE 5.5 Vectors representing internal muscle forces from the dynamic model of Granata and Marras [30] and Marras and Granata [2].

been accomplished by modulating the EMG on either a purely theoretical [15,54] or purely empirical [10,16] basis without examining the relation between the two. Analysis of muscle length and velocity modulation factors appropriate for free-dynamic lifting exertions [70] demonstrated that empirically developed relations agree with the theoretical relations reported in the physiological literature. Basically, the modulation factors account for variation in muscle force generating capability that result from changes in muscle length and velocity. The muscle activity level for a given force will depend on the position of the muscle (length-strength relationship) where a muscle that is elongated or shortened from its resting length has capability to produce force [56]. Similarly, the rate of the change of muscle length (force-velocity relationship) also influences the force generating capability of the muscle. Under concentric muscle exertions—contraction of the muscle as muscle length shortens—faster muscle contractions result in a decreased force generating capability of a muscle due to inefficient coupling of the cross-bridges of the muscle filaments [56]. During eccentric exertions—muscle lengthening during contraction—there is no deficit due to faster muscle contractions and muscles are about 20%–50% stronger as compared to concentric muscle contractions [11,16,56]. This increased force output results from the stretching and eventual breaking of the cross-bridges between the muscle filaments [11,16,56].

As can be seen from this discussion, EMG-assisted models are the only models available at this time that have the potential to accurately assess the loads on the lumbar spine associated with the biomechanical risk factors identified via industrial biomechanical studies [1,24]. This is true because these models are the only ones that are capable of accurately assessing muscle coactivity. Granata and Marras [30] assessed the influence of coactivation upon spinal compression estimates. Their study concluded that during free-dynamic lifting tasks a single-equivalent-muscle model would underestimate the compression on the spine by 40% compared to a full 10-muscle, EMG-driven coactivation model. Hence it is extremely important to consider the activities of the complete set of trunk muscles when attempting to assess loads on the lumbar spine.

Data necessary as input to an EMG-assisted model are dynamic myoelectrical data and trunk kinematics. Dynamic kinetic data are also necessary if model performance validation and quality assurance measures are desired. Because EMG activity must be measured along with kinematic information in order to run an EMG-driven model, most occupational tasks of interest are simulated in the laboratory.

The major disadvantage of EMG-assisted models is that they require the measurement of myoelectric activity. This measurement can be tedious and laborious, making it difficult to assess tasks on a routine basis. Mirka and Marras [71] have developed a means to estimate expected EMG activity given the parameters of a lifting exertion. They developed a stochastic EMG model that would simulate the range of expected EMG activity from each of 10 trunk muscles given the load lifted, dynamic characteristics of the trunk, and position profile of the trunk. Figure 5.6a shows the mean and standard deviation of expected EMG activity in the erector spinae muscle for a given trunk lift. Mirka and Marras used these stochastically generated EMG activities as input to an EMG-assisted model. These estimated muscle activities agreed well with measured muscle activities during a validation study. By running the model over many trials the authors were able to predict the

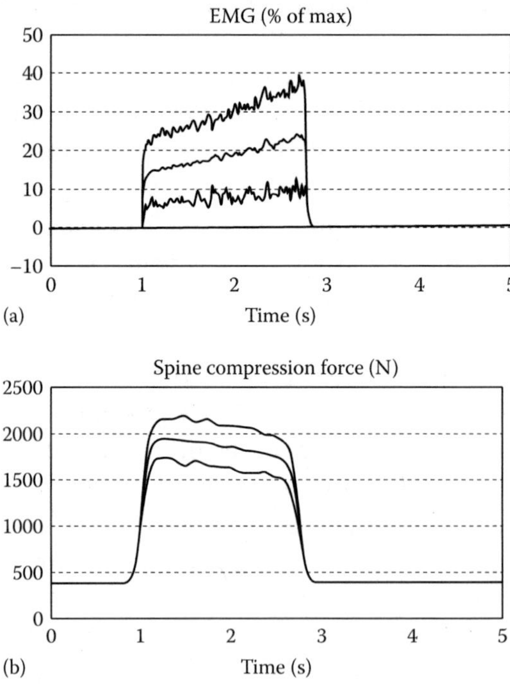


FIGURE 5.6 Mean and standard deviation of expected EMG activity in erector spinae (a) and spinal compression (b) muscle during an extension motion of the trunk.

expected loading on the spine over repeated lifting without measuring muscle activity. Figure 5.6b shows the expected range of spinal loads that would be expected via 50 simulations of a lifting activity using the stochastically generated EMG as input to the EMG-assisted model described earlier. Hence, future multiple-muscle models may be developed that can enjoy the advantages of EMG-assisted models without the need to actually measure muscle activity for a particular task.

In more recent years, a group lead by Marras, Karwowski and Davis developed a spine load model based on fuzzy logic and neural network techniques [72,73]. A complex engine capable of predicting the muscle activation patterns that result during 3D lifting was developed using fuzzy average with fuzzy cluster distribution techniques in combination with neural network structure. Muscle activity of the ten trunk muscles could now be estimated based on kinematic, kinetic, anthropometric, and work condition variables as inputs including physical and psychosocial characteristics and then feed into the EMG-assisted spine load model (as described earlier). This technique eliminated the need to collect the muscle activities in real time but accounts for the neuromuscular coactivity during dynamic lifting.

### 5.2.2 Models of the Hand and Wrist

As with the back, it is important that to understand how the structures of the hand and wrist are loaded during work so that a better understanding of how the risk of injury changes as working conditions change can bet developed. Over the past several decades, cumulative trauma injuries to the hands and wrist have continued to plague many different industries with hand intensive work. Illnesses such as carpal tunnel syndrome, tendinitis, and tenosynovitis have remained common in industry, whereas they were unheard of in most industries just a decade or two ago. Thus, it is imperative that we begin to understand those occupationally related factors that are associated with an increased risk of hand and wrist disorders, allowing those types of exposure to be controlled on the job.

Compared to the back, biomechanical modeling efforts have been sparse for the hand and wrist with even fewer biomechanical models being used for ergonomic assessment purposes. This dearth of models of the hand and wrist may be due, in part, to the high degree of complexity in the hand and wrist structure compared to the back. This complexity is a function of the elaborate anatomic construction of the wrist. The hand and wrist complex is unique in that most of the muscles that supply the internal force during gripping and finger action are located in the forearm rather than in the hand or wrist itself. Thus, internal forces must be transmitted from the forearm muscles through tendons to the fingers and hand joints. For that reason, biomechanical models have traditionally represented the wrist as a pulley system as shown in Figure 5.7. The modeling issues, therefore, have centered on including the appropriate anatomical representation of the hand and wrist and including enough information that solutions to the models could be found.

In an attempt to reduce the degree of complexity associated with biomechanical modeling of the hand and wrist, most of the existing models focus on specific segments of the hand (i.e., finger or wrist joint) and the forces exerted during specific postural conditions.

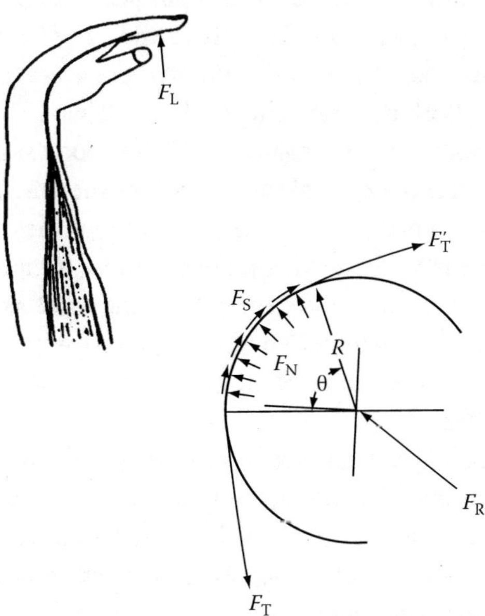


FIGURE 5.7 Armstrong and Chaffin's [78] biomechanical model of a flexor tendon wrapping around the flexor retinaculum, $F_R = 2F_T \sin(\theta/2)$, where $F_T$ is the tendon force and $F_R$ is the resultant reaction force exerted against the tendon. (Adapted from D. Chaffin and G. Andersson: *Occupational Biomechanics*. p. 219. 1984. Copyright Wiley-VCH Verlag GmbH & Co. KGaA. Reproduced with permission)

For the most part, the vast majority of these models were developed for the purpose of understanding the implications of hand injury or hand surgery; very few were developed specifically for ergonomic purposes. Few models have been concerned with the assessment of cumulative trauma disorders (CTDs), which have been identified as a major risk factor by ergonomists for many occupationally related wrist injuries. However, we can still gain an appreciation for the state of ergonomic biomechanical modeling through these injury-based models.

As was the case for the back, the early models of the hand and wrist attempted to evaluate the structure and loading in the hand and wrist while the wrist assumed static postures. Landsmeer [74] was one of the first to formulate a model of the structures in the fingers. Through anatomical investigations he was able to describe, with a series of three 2D models, how the tendons bridge a finger joint. His models consisted of two interphalangeal joints that formed a biarticulating system, which was traversed by both a flexor tendon and an extensor tendon. Thus, the model could permit both agonist and antagonist actions of the musculoskeletal system. This model was able to explain how changes in the state of equilibrium between the two tendons were able to generate variations in finger positions. Chao et al. [75] later developed a 3D model of a finger joint in selected isometric hand positions. The joint and tendon orientations were derived from biplanar x-ray analyses. Coordinate systems defined at each joint helped identify the constraint forces and moments. These authors formulated a pylon concept that explained the relationship between applied force, passive tendon force, active tendon force, and the finger joint. This formulation resulted in a statically indeterminate problem. EMG and physiological assessments were used to solve the problems. This study did provide an understanding of the functional anatomy of the hand. The model predicted the average forces in the finger tendons in selected positions. Both the Landsmeer [74] and Chao [75] models relied on simple static equilibrium conditions for their solutions. These models, although simple in their conception, required detailed anatomical input for their success. The complexity also limited their use to static hand postures. Another difficulty of these hand and wrist models was that there was no means to validate them. Furthermore, simplifying assumptions were used to solve the indeterminacy of the system, generating results that neglected coactivity. Analyses by Ferguson et al. [76] demonstrated that not only muscle coactivity in the hand and wrist are significant, but their neglect may cause joint and tendon loads to be underpredicted by as much as 30%.

The model of Chao et al. [75] was further developed by An and associates [77] to generate a normative model of the human hand that computed the tendon locations and excursions under various functional configurations of the hand. In this model the tendon orientations and their force and moment parameter contributions to the analytical model were added to the governing equilibrium equations.

Armstrong and Chaffin [78] developed a 2D static model of the wrist that calculated the resultant force exerted by a tendon on adjacent wrist structures. This model represented the extrinsic finger flexor tendons as frictionless pulley–belt mechanisms passing through the carpal tunnel. The model permitted one to quantitatively determine how wrist size and wrist position affect forces on the tendons and their adjacent structures. It assumed that a single equivalent tendon force produced a reaction force in the carpal tunnel during flexion or extension. It was intended to determine when loading of the carpal canal reached the point where the tendon sheaths would become inflamed and/or when significant pressure would develop around the median nerve. Goldstein et al. [79] showed that a sufficient magnitude of tensile force applied to the tendons can cause residual strain in these tendons. This was the first model intended for ergonomic purposes. As with previous finger models, their wrist model did not attempt to account for the dynamic nature of many occupational tasks. Marras and Schoenmarklin [80] and Schoenmarklin et al. [81] have identified rapid motion as a risk factor for CTDs under some work situations. Schoenmarklin and Marras [82] considered the effects of incorporating acceleration into the Armstrong–Chaff in [78] model. As shown in Figure 5.8, acceleration significantly increases the friction in the wrist. Thus, the model emphasizes the importance of including wrist motion in ergonomic models.

In order to deal with the problem of indeterminacy, researchers have attempted to find solutions to the models through optimization techniques. Penrod et al. [83] used the principle of minimal muscle effort to create a linear optimization program that calculated the contributions of tendons passing through the wrist to externally applied static moments. Chao and An [84] also used linear programming methods to solve the redundancy problem associated with hand modeling. Again, the appropriateness of these methods depends upon whether coactivation muscle is present in the task of interest.

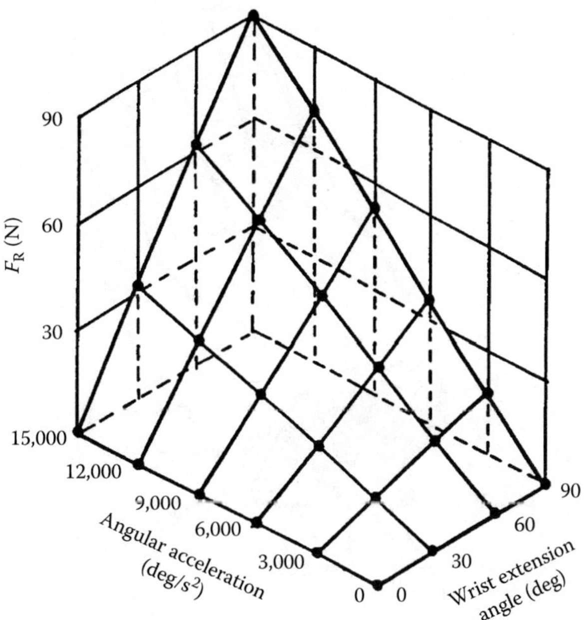


FIGURE 5.8 The resultant reaction force ($F_R$) exerted by the carpal bones or flexor retinaculum against a flexor tendon as a function of wrist angle and acceleration. (From Komi, P.V., *New Dev. Electromyogr. Clin. Neurophysiol.* 1, 596–606, 1973.)

Many of the more recent models have been concerned with more than simply predicting the forces occurring in the muscles and tendons of interest in the hand and wrist. Recently, ergonomic biomechanical models have attempted to predict hand postures when grasping handles in a power grip orientation. Buchholz and Armstrong [85] developed a 3D kinematic model of the hand that predicts joint angles of the fingers and the hand. They assumed that the hand and fingers are a series of ellipses that wrap around a handle. However, no estimates of internal loading are provided in this model. Moore et al. [86] developed a model of the wrist based upon the relationship between hand forces and tendon contributions described by An et al. [87]. This model computes the frictional work between tendons and their adjacent structures during power grasps.

In an attempt to consider the coactive nature of loading on the wrist, Schoenmarklin [88] developed an EMG-assisted model of the wrist. Figure 5.9 shows the anatomical relationship among the tendons passing through the wrist. This model monitors the wrist flexor and extensor muscles and conditions the EMG signal for velocity artifact and muscle length changes. By considering both of the flexor and extensor muscle forces, the model is able to estimate the amount of friction associated with dynamic wrist motion activities.

In conclusion, biomechanical modeling of the hand and wrist can be considered still in its infancy compared to that of other parts of the body. Progress is occurring slowly but steadily. Upper extremity models have the additional burden of attempting to describe

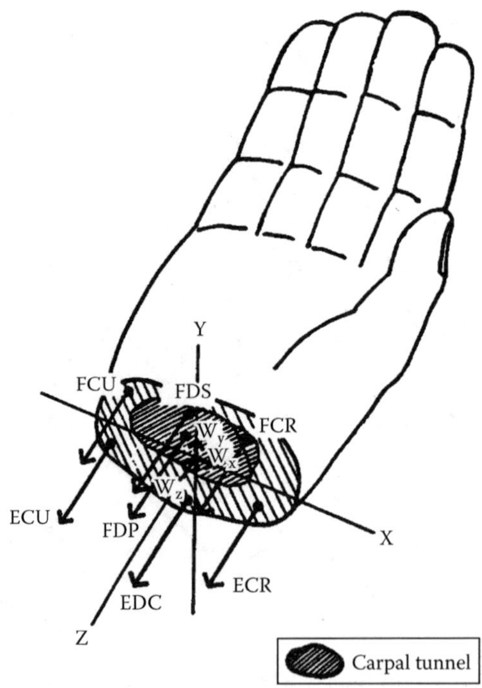


FIGURE 5.9 Cross-sectional view of the wrist from the EMG-driven model developed by Schoenmarklin [88].

phenomena that are not well understood such as tendon sheath inflammation. However, these models are crucial in the understanding of the CTD process and the proper design of the workplace.

## 5.3 CRITICAL REVIEW OF CURRENT STATUS

As is evident from this chapter, our knowledge of the human body and how its loading is associated with the risk of injury is still evolving. It is much better developed with respect to the back than with respect to the hand and wrist. We can use the development of the models in these fields as a lesson on how to approach industrial biomechanical problems in the workplace. For both the back and the hand, our representation of available knowledge began with models of the structures under static (isometric) conditions. These models were relatively straightforward in their development, and some could be validated under idealized, isometric conditions. However, there is only limited evidence that the application of such models has made a difference in industrial injury rates. Thus, although these models have been a useful starting point for the consolidation of knowledge of the biomechanics of a particular part of the body, their usefulness has been limited because many of them do not have the realism necessary to describe and understand the risk factors inherent in many industrial jobs.

In the case of both the back and the wrist, we have seen that once we begin to relax the assumption that the body is subject to isometric forces in the workplace, our ability to account for risk via modeling improves significantly [1,24,81]. This situation can be appreciated via a review of possible biomechanical variables that could be included in our models. Table 5.1 shows the trade-off between the degree of realism associated with various test conditions (under which biomechanical models have been developed) and the components of the physical sciences that are accounted for in the models [89]. This table shows that the physical conditions associated with isometric models (shown in the left-hand column) are the most limited in their ability to relate to realistic physical conditions. The only factors associated with force on a structure are the mass handled or force applied. The models do not have the ability to consider the effects of acceleration or jerk forces that could surely become significant when considering internal loadings of the body.

As we consider more of the physical factors associated with the variables in each column, more realistic conditions can be accounted for. For example, in Table 5.1 as we move through the columns from left to right, we find more factors such as acceleration and jerk forces included in a model based on a given technique of measurement, and a greater appreciation for the collection of variables influencing the loading on a particular joint can be gained. Once valid models have been developed that can account for and predict the variables that are listed as dependent variables in the free-dynamic (far right) column in this table, our knowledge of biomechanical loading in the body will be complete. However, it is expected that these models will be extremely complicated and difficult to apply. Therefore, the goal of current biomechanical modeling efforts should be to develop models that are not unnecessarily complex but include enough of the dependent variables that reasonable control of workplace risk can be attained.

TABLE 5.1 Independent and Dependent Variables in Several Techniques Used to Measure Motor Performance

| | Isometric (Static) | | Isokinetic | | Isoacceleration | | Isojerk | | Isoforce | | Isoinertial | | Free Dynamic | |
|---|---|---|---|---|---|---|---|---|---|---|---|---|---|---|
| **Variables** | **Indep.** | **Dep.** | **Indep.** | **Dep.** | **Indep.** | **Dep.** | **Indep.** | **Dep.** | **Indep.** | **Dep.** | **Indep.** | **Dep.** | **Indep.** | **Dep.** |
| Displacement, linear/angular | Const.* | | C | X | C | X | C | X | C | X | C | X | | X |
| Velocity, linear/angular | 0 | Const.* | | C | X | C | X | C | X | C | X | X | | |
| Acceleration, linear/angular | 0 | | 0 | | Const.* | | C | X | C | X | C | X | X | |
| Jerk, linear/angular | 0 | | 0 | | 0 | | Const. * | | C | X | C | X | X | |
| Force, torque | C | X | C | X | C | X | C | X | Const.* | C | X | | X | |
| Mass, moment of inertia | C | | C | | C | | C | | C | Const.* | | C | X | |
| Repetition | C | X | C | X | C | X | C | X | C | X | C | X | C | X |

*Source:* Marras, W.S., et al., *American Industrial Hygiene Association Ergonomics Guide,* American Industrial Hygiene Association, Fairfax, VA, 174-ER-93, 1993.

*Note:* C variable can be controlled, 0 variable is not present (zero), X can be dependent variable, * set to zero. The boxed constant variable provides the descriptive name.

## 5.4 FUTURE CONCERNS

Current biomechanical models consist of isolated models of a particular part of the body. As shown in this chapter, models useful for ergonomics purposes exist for only a few portions of the body. It is expected that our knowledge of joint loading for each particular part of the body will improve to the point where we can accurately assess the loads on a joint in isolation. However, there are two further challenges that must be addressed.

First, we must work toward developing a whole body model that can accurately reflect imposed loads on several joints of the body simultaneously. We know that the body acts in a coordinated fashion to accomplish many routine tasks. Thus, the load imposed upon one joint of the body can affect joints in other parts of the body. It is also evident that if one joint becomes compromised, the human can compensate by changing the loading sequence experienced by the other joints of the body. A real challenge for the future will be to develop isolated joint models so that they can accurately account for the chain of events that occurs throughout the body, thereby treating the body as a true system. Only then can we truly begin to understand the limitations that might be imposed upon the body through injury and appreciate the implications of legislative acts such as the Americans with Disabilities Act.

Finally, human loadings are known to be subject to temporal demands. Greater effort must be placed upon understanding how the imposed loads of a joint are altered by age, tissue degeneration, genetics, and other individual characteristics. We must also increase our understanding of how loading of a joint changes throughout the workshift. In addition, we must begin to develop models that can describe how tissues repair themselves and how this process changes as a function of age. Ultimately, models might be able to someday account for pain and the process of pain sensation which is a large component of whether discomfort becomes an occupational disorder or injury. Tissue breakdown often occurs when a tolerance is exceeded but the reporting of an injury on the job does not occur until the pain tolerance is exceeded. Only when these elements are in place can we begin to truly understand and quantify the risk associated with occupational demands.

## REFERENCES

1. W. S. Marras, S. A. Lavender, S. E. Leurgans, S. L. Rajulu, W. G. Allread, F. A. Fathallah, and S. A. Fersuson, The role of dynamic three-dimensional trunk motion in occupationally-related low back disorders: The effects of workplace factors, trunk position and trunk motion characteristics on risk of injury, *Spine 18*: 617–628 (1993).
2. W. S. Marras and K. P. Granata, A biomechanical assessment and model of axial twisting in the thoraco-lumbar spine, *Spine 20*: 1440–1451 (1995).
3. P. M. Bongers, C. R. de Winter, M. A. J. Kompier, and V. H. Wildebrandt, Psychosocial factors at work and musculoskeletal disease, *Scand. J. Work Environ. Health 19*: 297–312 (1993).
4. K. G. Davis and C. A. Heaney, The relationship between psychosocial work characteristics and low back pain: Underlying methodological issues, *Clin. Biomech. 15*: 389–406 (2000).
5. K. G. Davis, W. S. Marras, C. A. Heaney, T. R. Waters, and P. Gupta, The impact of mental processing and pacing on spine loading, *Spine 27*: 2645–2653 (2002).
6. W. S. Marras, K. G. Davis, C. A. Heaney, A. B. Maronitis, and W. G. Allread, The influence of psychosocial stress, gender, and personality on mechanical loading of the lumbar spine, *Spine 25*: 3045–3054 (2000).

7. W. S. Marras, K. G. Granata, K. G. Davis, W. G. Allread, and M. J. Jorgensen, Effects of box features on spinal loading during order selecting, *Ergonomics 42*: 980–996 (1999).
8. D. Chaffin, A computerized biomechanical model. Development and use in studying gross body actions, *J. Biomech. 2*: 429–441 (1969).
9. Department of Health and Human Services (DHHS), *Work Practices Guide for Manual Lifting*, National Institute for Occupational Safety and Health (NIOSH), Publ. No. 81–122, 1981.
10. W. S. Marras and C. M. Sommerich, A three-dimensional motion model of loads on the lumbar spine. I. Model structure, *Hum. Factors 33*: 123–137 (1991).
11. W. S. Marras and C. M. Sommerich, A three-dimensional motion model of loads on the lumbar spine. II. Model validation, *Hum. Factors 33*: 139–149 (1991).
12. K. G. Davis and M. J. Jorgensen, Biomechanical modeling for understanding of low back injuries: A systematic review, *Occup. Ergonomics, (Invited)*, *5*(1): 57–76 (2005).
13. A. Schultz and G. Andersson, Analysis of loads on the lumbar spine, *Spine 6*: 76–82 (1981).
14. S. M. McGill and R. W. Norman, Dynamically and statically determined low back moments during lifting, *J. Biomech. 8*: 877–885 (1985).
15. S. M. McGill and R. Norman, Partitioning the L4–L5 dynamic moment into disc, ligamentous, and muscular components during lifting, *Spine 11*: 666–678 (1986).
16. K. P. Granata and W. S. Marras, An EMG-assisted model of loads on the lumbar spine during asymmetric trunk extensions, *J. Biomech. 26*: 1429–1438 (1993).
17. G. G. Knapik and W. S. Marras, Spine loading at different lumbar levels during pushing and pulling, *Ergonomics 52*: 60–70 (2009).
18. F. G. Evans and H. R. Lissner, Biomechanical studies on the lumbar spine and pelvis, *J. Bone Joint Surg. 41A*: 218–290 (1959).
19. T. Sonada, Studies on the compression, tension, and torsion strength of the human vertebral column, *J. Kyoto Prefect Med. Univ. 71*: 659–702 (1962).
20. M. Jager, A. Luttmann, and W. Laurig, Lumbar load during one-handed bricklaying, *Int. J. Ind. Ergon. 8*: 261–277 (1991).
21. P. Brinckmann, M. Biggemann, and D. Hilwed, Fatigue fracture of human lumbar vertebrae, *Clin. Biomech. 3*: S1–S23 (1988).
22. S. M. McGill, The biomechanics of low back injury: Implications on current practice in industry and the clinic, *J. Biomech. 30*: 465–475 (1997).
23. S. Gallagher, W. S. Marras, A. S. Litsky, D. Burr, J. Landoll, and V. Matkovic, A comparison of fatigue failure responses of young and old lumbar motion segments in simulated flexed lifting, *Spine 32*: 1832–1839 (2007).
24. W. S. Marras, S. A. Lavender, S. E. Leurgans, F. A. Fathallah, S. A. Fersuson, W. G. Allread, and S. L. Rajulu, Biomechanical risk factors for occupationally-related low back disorders, *Ergonomics 28*: 377–410 (1995).
25. D. B. Chaffin and M. Erig, Three-dimensional biomechanic static strength prediction model sensitivity to postural and anthropometric inaccuracies, *IIE Trans. 23*: 215–227 (1991).
26. W. S. Marras, A. I. King, and R. L. Joynt, Measurements of loads on the lumbar spine under isometric and isokinetic conditions, *Spine 9*: 176–188 (1984).
27. A. Schultz, G. Anderson, R. Ortengren, K. Haderspeck, A. Nachemson, and S. Gotegorg, Loads on the lumbar spine, *J. Bone Joint Surg. 64*: 713–720 (1982).
28. J. C. Bean, D. B. Chaffin, and A. B. Schultz, Biomechanical model calculation of muscle forces: A double linear programming method, *J. Biomech. 21*: 59–66 (1988).
29. S. Gracovetsky and H. Farfan, The optimum spine, *Spine 11*: 543–573 (1986).
30. K. P. Granata and W. S. Marras, The influence of trunk muscle coactivity upon dynamic spinal loads, *Spine 20*: 913–919 (1995).
31. A. Freivalds, D. B. Chaffin, A. Garg, and K. S. Lee, A dynamic biomechanical evaluation of lifting maximum acceptable loads, *J. Biomech. 17*: 251–262 (1984).

32. V. K. Goel, J. S. Han, J. Y. Ahn, T. Cook, J. N. Weinstein, J. Winterbottom, D. McGowan, and D. Dawson, Loads on the human spine during dynamic lifting with knees straight, *Adv. Bioeng. 20*: 33–36 (1991).
33. W. S. Marras, P. E. Wongsam, and S. L. Rangarajulu, Trunk motion during lifting: The relative cost, *Int. J. Ind. Ergon. 1*: 103–113 (1986).
34. W. S. Marras and G. A. Mirka, A comprehensive evaluation of trunk response to asymmetric trunk motion, *Spine 17*: 318–326 (1992).
35. W. S. Marras, Predictions of forces acting upon the lumbar spine under isometric and isokinetic conditions: A model—Experimental comparison, *Int. J. Ind. Ergon. 3*: 19–27 (1988).
36. C. Reilly and W. S. Marras, Simulift: A simulation model of the human trunk motion, *Spine 14*: 5–11 (1989).
37. K. P. Granata and W. S. Marras, An EMG-assisted model of trunk loading during free-dynamic lifting exertions, *J. Biomech. 28*: 1309–1317 (1995).
38. D. Chaffin and W. Baker, A biomechanical model for analysis of symmetric sagittal plane lifting, *AIIE Trans. 2*: 16–27 (1970).
39. S. Gracovetsky, H. F. Farfan, and C. Helluer, The abdominal mechanism, *Spine 10*: 317–324 (1985).
40. C. Zetterberg, G. B. Andersson, and A. B. Schultz, The activity of individual trunk muscles during heavy physical loading, *Spine 12*: 1035–1040 (1987).
41. S. M. McGill and R. W. Norman, Potential of the lumbodorsal fascia forces to generate back extension moments during squat lifts, *J. Biomed. Eng. 10*: 312–318 (1988).
42. A. L. Hof and J. W. Van Den Berg, Linearity between the weighted sum of the EMG's of the human triceps surae and the total torque, *J. Biomech. 10*: 529–539 (1977).
43. M. Jager and A. Luttmann, Biomechanical analysis and assessment of lumbar stress during lifting using a dynamic 19-segment human model, *Ergonomics 32*: 93–112 (1989).
44. W. S. Marras and G. A. Mirka, Muscle activities during asymmetric trunk angular accelerations, *J. Orthop. Res. 8*: 824–832 (1990).
45. D. E. Hardt, Determining muscle forces in the leg during normal human walking. An application and evaluation of optimization methods, *J. Biomech. Eng. 100*: 72–78 (1978).
46. K. N. An, B. M. Kwak, E. Y. Chao, and B. F. Morrey, Determination of muscle and joint forces: A new technique to solve the indeterminate problem, *J. Biomech. Eng. 106*: 364–367 (1984).
47. D. R. Pederson, R. A. Brand, C. Chang, and J. S. Arora, Direct comparison of muscle force predictions using linear and nonlinear programming, *J. Biomech. Eng. 109*: 192–199 (1987).
48. S. Rajulu, Decomposition of electromyographic signals for biomechanical interpretation, PhD Dissertation, Ohio State University, Columbus, OH, 1990.
49. W. S. Marras and G. A. Mirka, Electromyographic studies of the lumbar trunk musculature during the generation of low-level trunk acceleration, *J. Orthop. Res. 11*: 811–817 (1993).
50. D. G. Thelen, A. B. Schultz, and J. A. Ashton-Miller, Prediction of dynamic isometric muscle forces from myoelectric signals during complex loadings of the trunk, *Adv. Bioeng. 20*: 505–508 (1991).
51. A. Shirazi-Adl, Stress in fibers of a lumbar disc, analysis of the role of lifting in producing disc prolapse, *Spine 14*: 96–103 (1989).
52. A. Shirazi-Adl, A. M. Ahmed, and S. C. Shrivastava, Mechanical response of the lumbar motion segment in axial torque alone and in combination with compression, *Spine 11*: 914–927 (1986).
53. W. S. Marras and C. Reilly, Networks: Internal trunk loading activities under controlled trunk motion conditions, *Spine 13*: 661–667 (1988).
54. S. M. McGill, A myoelectrically based dynamic three dimensional model to predict loads on lumbar spine tissues during lateral bending, *J. Biomech. 25*: 395–414 (1992).
55. A. V. Hill, The heat of shortening and the dynamic constants of muscle, *Proc. Roy. Soc. Biol. 126*: 136–195 (1938).

56. K. G. Davis, W. S. Marras, and T. R. Waters, Evaluation of spinal loading during lowering and lifting, *Clinical Biomech. 13*: 141–152 (1997).
57. A. M. Gordon, A. F. Huxley, and F. J. Julian, The variation in isometric tension with sacromere length in vertebrate muscle fibers, *J. Physiol. (Lond.) 184*: 170–192 (1966).
58. J. Vredenbregt and G. Rau, Surface electromyography in relation to force, muscle length and endurance, in *New Developments in Electromyography and Clinical Neurophysiology,* Vol. 1, J. E. Desmedt, Ed., Karger, Basel, Switzerland, 1973, pp. 607–622.
59. B. Bigland and O. C. J. Lippold, The relation between force velocity and integrated electrical activity in human muscles, *J. Physiol. 123*: 214–224 (1954).
60. M. H. Pope, G. B. J. Andersson, H. Broman, M. Svensson, and C. Zetterburg, Electromyographic studies of the lumbar trunk musculature during the development of axial torques, *J. Orthop. Res. 4*: 288–297 (1986).
61. S. M. McGill, Electromyographic activity of the abdominal and low back musculature during the generation of isometric and dynamic axial trunk torque: Implications for lumbar mechanics, *J. Orthop. Res. 9*: 91–103 (1991).
62. T. Weis-Fogh and R. M. Alexander, *Scale Effects in Animal Locomotion,* Academic Press, London, 1977, pp. 511–525.
63. S. M. McGill and R. W. Norman, Effects of an anatomically detailed erector spinae model on L4-S1 disc compression and shear, *J. Biomech. 20*: 591–600 (1987).
64. J. G. Reid and P. A. Costigan, Trunk muscle balance and muscular force, *Spine 12*: 783–786 (1987).
65. M. J. Jorgensen, W. S. Marras, K. P. Granata, and B. Wiand, MRI-derived moment-arms of the female and male spine loading muscles, *Clin. Biomech. 16*: 182–193 (2001).
66. W. S. Marras, M. J. Jorgensen, K. P. Granata, and B. Wiand, Female and male trunk geometry, size and prediction of the spine loading muscles derived from MRI, *Clin. Biomech. 16*: 38–46 (2000).
67. P. V. Komi, Measurement of force–velocity relationship in human muscle under concentric and eccentric contractions, *Med. Sport, Biomech. III 8*: 224–229 (1973).
68. P. V. Komi, Relationship between muscle tension, EMG and velocity of contraction under concentric and eccentric work, *New Dev. Electromyogr. Clin. Neurophysiol. 1*: 596–606 (1973).
69. W. S. Marras, Industrial electromyography, *Int. J. Ind. Ergon. 6*: 89–93 (1990).
70. K. P. Granata, An EMG-assisted model of trunk loading during free-dynamic lifting exertions, PhD Dissertation, Ohio State University, Columbus, OH, 1993.
71. G. A. Mirka and W. S. Marras, A stochastic model of trunk muscle coactivation during trunk bending, *Spine 18*: 1396–1409 (1993).
72. Y. Hou, J. M. Aurada, W. Karwowski, W. S. Marras, and K. G. Davis, Estimation of the dynamic spinal forces using a recurrent fuzzy neural network, *IEEE Trans. Systems Man Cybrenet. 37*: 100–109 (2007).
73. W. Karwowski, A. Gaweda, W. S. Marras, K. G. Davis, J. M. Zurada, and D. Rodrick, A fuzzy relational rule network modeling of electromyographical activity of trunk muscles in manual lifting based on trunk angles, moments, pelvic tilt and rotation angels, *IJIE, 36*: 847–859 (2006).
74. J. M. F. Landsmeer, The coordination of finger joint motions, *J. Bone Joint Surg. 45*: 1654–1662 (1963).
75. E. Y. Chao, J. D. Opgrande, and F. E. Axmear, Three-dimensional force analysis of finger joints in selected isometric hand functions, *J. Biomech. 9*: 387–396 (1976).
76. S. A. Ferguson, F. A. Fathallah, K. P. Granata, J. Y. Kim, and W. S. Marras, Coactivity effects upon carpal tunnel contact forces, *Proceedings of the Human Factors and Ergonomics Society Annual Meeting*, Seattle, WA, 1993, pp. 705–709.
77. K. N. An, E. Y. Chao, W. P. Cooney, and R. L. Linscheid, Normative model of the human hand for biomechanical analysis, *J. Biomed. Eng. 9*: 313–320 (1979).

78. T. J. Armstrong and D. B. Chaffin, Some biomechanical aspects of the carpal tunnel, *J. Biomech. 12*: 567–570 (1978).
79. S. A. Goldstein, T. J. Armstrong, D. B. Chaffin, and L. S. Mathews, Analysis of cumulative strain in tendons and tendon sheaths, *J. Biomech. 20*: 1–6 (1987).
80. W. S. Marras and R. W. Schoenmarklin, Wrist motions in industry, *Ergonomics 36*: 341–351 (1993).
81. R. W. Schoenmarklin, W. S. Marras, and S. E. Leurgans, Industrial wrist motions and incidence of hand/wrist cumulative trauma disorders, *Ergonomics 37*: 1449–1459 (1994).
82. R. W. Schoenmarklin and W. S. Marras, A dynamic biomechanical model of the wrist joint, *Proceedings of the Human Factors and Ergonomics Society 34th Annual Meeting*, Orlando, FL, pp. 805–809, 1990. Human Factors and Ergonomics Society.
83. D. D. Penrod, D. T. Davey, and D. P. Singh, An optimization approach to tendon force analysis, *J. Biomech. 7*: 123–129 (1974).
84. E. Y. Chao and K. N. An, Graphical interpretation of the solution to the redundant problem in biomechanics, *J. Biomech. Eng. 100*: 159–167 (1978).
85. B. Buschholz and T. J. Armstrong, A kinematic model of the human hand to evaluate its prehensile capabilities, *J. Biomech. 25*: 149–162 (1992).
86. A. Moore, R. Wells, and D. Ranney, Quantifying exposure in occupational manual tasks with cumulative trauma disorder potential, *Ergonomics 34*: 1433–1453 (1991).
87. K. N. An, E. Y. Chao, W. P. Cooney, and R. L. Linscheid, Forces in the normal and abnormal hand, *J. Orthop. Res. 3*: 202–211 (1985).
88. R. W. Schoenmarklin, Biomechanical analysis of wrist motion in highly repetitive, hand-intensive industrial jobs, PhD Dissertation, Ohio State University, Columbus, OH, 1991.
89. W. S. Marras, J. D. McGlothlin, D. R. McIntyre, M. Nordin, and K. H. E. Kroemer, Dynamic measures of low back performance, in *American Industrial Hygiene Association Ergonomics Guide*, 174-ER-93, 1993.
90. D. Chaffin and G. Andersson, *Occupational Biomechanics*, John Wiley & Sons, New York, p. 219, 1984.

CHAPTER 6

# Psychophysical Methodology and the Evaluation of Manual Materials Handling and Upper Extremity Intensive Work

Sheila Krawczyk

CONTENTS

## 6.1 INTRODUCTION: WHY PSYCHOPHYSICS?

Psychophysical methods are a consistent, reproducible, quick, inexpensive, and convenient way to assess the degree of physical strain on the human body. Psychophysical criteria have also been correlated with physiological criteria and some injury indices. Psychophysical methods utilize the results of the central nervous system integration of various information, including the many signals elicited from the peripheral working muscles and joints, and from the central cardiovascular and respiratory functions. All of these signals, perceptions, and experiences are combined and utilized by means of psychophysical methods.

Various physical stressors found in manual work, such as excessive forces, high rates of repetition, and awkward or sustained postures, have been associated with musculoskeletal injuries. There is an absence of quantitative dose–response data that examine the relationship between work parameters and morbidity patterns. Epidemiological studies that examine the relationship between work and morbidity patterns require a lot of time and resources for data collection [1]. Biomechanical methodologies cannot address issues of repetitive work and fatigue [2,3]. Separate physiological criteria are needed for whole body or lower body tasks and upper body work [4–12]. More information is needed about the relationship between risk factors, work parameters, and the development of musculoskeletal injuries so that work can be designed to reduce the risk of the development of musculoskeletal injuries.

Another approach is a psychophysical methodology, in which subjects or workers determine the relationship between work factors and the perception of physical stress, exertion, fatigue, and discomfort on the body. Since localized muscle fatigue may be an early symptom of some use-related musculoskeletal injuries [13–15], psychophysical methodologies may serve as a more sensitive indicator for the risk of the development of musculoskeletal injuries [16].

The psychophysical methodology is an approach that allows for the simultaneous evaluation of the combined effects of different physical stressors. Specifically, the effects of different work task parameters, as well as the combined effects, can be evaluated using psychophysical methods. Psychophysical methods have been used extensively to evaluate manual materials-handling tasks, upper extremity–intensive tasks, and other manual work. A major advantage of psychophysical studies is that the results can be readily applied as guidelines in the workplace. Psychophysical data can provide guidance in the analysis and design of repetitive manual work that is commonly found in manufacturing and production assembly facilities, warehouses, retail trades such as grocery and discount stores, and other workplaces.

## 6.2 BACKGROUND

Psychophysics is an old branch of psychology that studies the relationship between sensation and physical stimulus intensity. Weber's law, from the early 1800s, states that

$$\frac{\Delta I}{I} = k$$

where

$I$ is the intensity of a physical stimulus
$\Delta I$ is the increment of $I$ producing a just noticeable difference (jnd)
$k$ is the constant

In essence, this means that the percentage error stays constant. The constant $k$ is a function of the particular parameter being measured. For instance, the constant k is 3% for brightness, 7% for length, and about 2.5% (or 1/40) for weight. In practice, one cannot tell the difference between a 40 g weight and a 40.5 g weight. For one to be able to just notice the difference the second weight would have to weigh 41 g.

Fechner's law, which is based on Weber's law, states that sensation increases as the logarithm of physical stimulus intensity:

$$S = k \log I$$

where

$S$ is the strength of sensation
$k$ is the constant
$I$ is the intensity of a physical stimulus

For instance, in the judgment of loudness of sound, the sound energy increases logarithmically with respect to linear judgments, and that is why the decibel (dB) scale is used.

Over the years, it became clear that the logarithmic-linear relationship between physical stimulus intensity and sensation was an accurate description over only a limited range of physical stimulus intensity. In 1960, Stevens showed that a wider range of physical stimulus intensity could be accurately described if sensation were expressed as a power function of physical stimulus intensity [17].

## 6.3 RELEVANT CONCEPTS AND TERMINOLOGY

### 6.3.1 Psychophysical Methodology and Physiological Criteria

#### *6.3.1.1 Psychophysical Power Law*

The general form of Stevens' power law is

$$\psi = k(\phi - \phi_0)^n$$

where

$\psi$ is the psychological magnitude
$k$ is the constant
$\phi$ is the physical magnitude
$\phi_0$ is a constant value corresponding to the threshold of detection [17]

The value of $\phi_0$ is usually negligible, but its importance assumes larger proportions when subjective scales are extended downward to very low values [17]. For ranges of stimuli well above the minimum detectable level, the value of $\phi_0$ is usually negligible [17].

Thus, neglecting $\phi_0$, the psychophysical power law proposed by Stevens states that the psychological magnitude $\psi$ is related to the physical magnitude $\phi$ by

$$\psi = k\phi^n$$

Or, as stated in the terms used previously:

$$S = kI^n$$

where

$S$ is the strength of sensation
$k$ is the constant
$I$ is the intensity of a physical stimulus

The constant $k$ is a function of the particular unit of measurement. The exponent $n$ depends mainly on the modality tested. For perception of muscular force $n$ is about 1.6 [18], and for force of handgrip it is 1.7 [17]. The value of $n$ ranges from 0.33 for brightness to 3.5 for electric shock and is determined by magnitude estimation [17]. Using this method, the observer estimates the apparent strength or intensity of his or her subjective impressions relative to a standard. Magnitude estimation is the most useful of four principal methods to construct these ratio scales of apparent magnitude.

The power function is plotted as a straight line in log–log coordinates, with the slope of the line equal to the value of the exponent $n$. Accordingly, modern psychophysical theory [17] provides that the strength of a sensation is related by a power function to the intensity of its physical stimulus [16–26]. Most psychophysical relations may be described by this power function [17].

#### *6.3.1.2 Borg Scales*

The aforementioned ratio scaling methods are very good methods to describe how the subjective intensity varies with physical stimulus intensity. One major drawback of the aforementioned ratio scaling methods is that they do not provide absolute levels for inter-individual comparisons. Although general functions for a group of subjects can be determined, it is difficult to compare subjects with each other because subjects are asked only to make relative comparisons. To overcome the difficulties associated with the ratio scaling methods, a category scale was developed by Borg [18]. Borg's [18] first subjective rating scale had 21 grades. All of the odd scale values from 3 to 19 were anchored with verbal expressions, chosen so that the scale should receive a good interindividual reliability. The scale was presented to the subjects with equal distances between the figures and in the following terms: 3, Extremely light; 5, Very light; 7, Light; 9, Rather light; 11, Neither light nor laborious; 13, Rather laborious; 15, Laborious; 17, Very laborious; and 19, Extremely laborious.

Very high correlations have been obtained between heart rate and these ratings during work tests [18]. This indicates the differential value of the scale but not the general validity of the growth function [27]. Since perceived exertion determined by ratio scaling methods increased with an exponent of about 1.6, Borg [18] concluded that an integration of central factors (such as heart rates) and peripheral factors (such as blood lactates, with an exponent of about 2) would explain the psychophysical variation better than any single physiological variable [27].

Another category scale for ratings of perceived exertion (RPE), shown in Table 6.1, was constructed by Borg [28] to increase linearly with the exercise intensity for work on a cycle ergometer. Since oxygen consumption and heart rate increase linearly with workload, this would be a convenient means of constructing a scale, even if it did violate the true growth of the perceived intensities [27]. The values of the RPE scale, shown in Table 6.1, grow fairly linearly with workload. The correlation between the ratings and heart rate is also very high [28]. For middle-aged persons at workloads of medium intensity, heart rate should be fairly close to 10 times the RPE value.

The Borg RPE scale has been widely used to study the perception of exertion in laboratory, clinical, and occupational settings. One of the most common uses of the RPE scale is in the clinical diagnosis of patients with coronary and respiratory disturbances. The normal growth pattern and the level of exertion change dramatically in different clinical populations. The RPE scale is also used in rehabilitation, and for the prescription and regulation of exercise intensities, or as a means to evaluate certain training situations [16]. A similar use, of interest to ergonomists, is in the evaluation of different work tasks. Perceived exertion has been used in ergonomic evaluations of heavy aerobic work tasks. However, the value of subjective estimations may be especially evident in job situations where the work tasks consist of short-term static work, intermittent or varied work, or upper extremity intensive work for which valid physiological measurements are difficult to obtain. Borg [27] suggests

TABLE 6.1 The Borg Scale for Ratings of Perceived Exertion (RPE)

| | |
|---|---|
| 6 | |
| 7 | Very, very light |
| 8 | |
| 9 | Very light |
| 10 | |
| 11 | Fairly light |
| 12 | |
| 13 | Somewhat hard |
| 14 | |
| 15 | Hard |
| 16 | |
| 17 | Very hard |
| 18 | |
| 19 | Very, very hard |
| 20 | |

TABLE 6.2 The Borg Category Ratio (CR) Scale of Perceived Exertion, Which Was Constructed as a Category Scale with Ratio Properties

| | | |
|---|---|---|
| 0 | Nothing at all | |
| 0.5 | Very, very weak | (just noticeable) |
| 1 | Very weak | |
| 2 | Weak | (light) |
| 3 | Moderate | |
| 4 | Somewhat strong | |
| 5 | Strong | (heavy) |
| 6 | | |
| 7 | Very strong | |
| 8 | | |
| 9 | | |
| 10 | Very, very strong | (almost max) |
| • | Maximal | |

that the RPE scale is the best one for most applied studies of perceived exertion. The RPE scale is one of the most frequently used indices of physical stress [16–20,22–25,27–30].

To meet the twofold demands of ratio scaling and level estimations, Borg [27] developed the category ratio (CR) scale, shown in Table 6.2, so that perceptual ratings would increase as a positively accelerating function. This scale contains some of the category properties of the RPE scale, and also contains ratio properties. The verbal expressions are set so that perceptual intensity increases according to a power function. The number "10" is defined as the strongest effort and exertion a person has ever experienced [16]. Since a person may imagine an intensity that is even stronger, the "absolute" maximum is somewhat higher.

By anchoring the highest number at a well-defined perception, with some degree of "sameness" for different individuals, a good point of reference is obtained [16]. Thus, two individuals working at their respective maximal working capacities will be experiencing the same degree of perceived exertion, even though their physical outputs may be different. Similarly, two individuals working at 50% of their respective working capacities will experience the same amount of perceived exertion, even though their physical outputs may be different [31]. This scale gives psychophysical functions comparable to those obtained with magnitude estimation. Exponents of about 1.6 have been obtained for perceived exertion in cycle ergometer exercise. In addition, close correlation between scale ratings and both blood lactate and muscle lactate levels have been obtained [32]. Borg [27] suggests that this scale may be best suited for subjective symptoms, such as aches and pain.

#### *6.3.1.3 Visual Analog Scales*

The visual analog scales (VAS) [19,26,30–31,33–39] is another frequently used measurement of physical stress. One of the most common VAS consists of a horizontal 10 cm line with verbal anchors or descriptions at the endpoints. The subject is instructed to indicate the degree of perceived sensory intensity by placing a vertical line at the appropriate position along the horizontal continuum. The distance from the left end of the VAS to

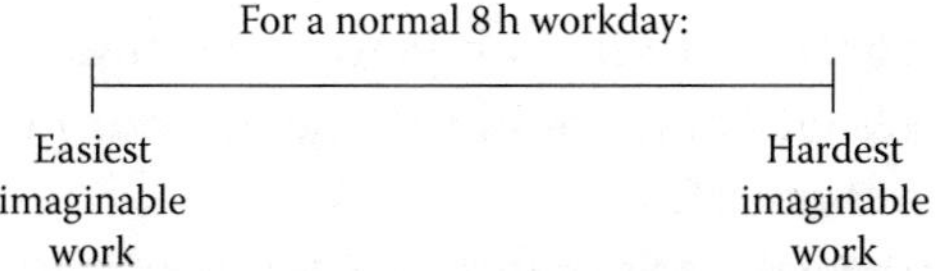


FIGURE 6.1 A 10 cm VAS with the verbal anchors "easiest imaginable work" and "hardest imaginable work" at the left and right endpoints, respectively, that has been used for the evaluation of manual work.

the subject's vertical line is measured in centimeters to obtain a value between 0 and 10. A visual analog scale that has been used for the evaluation of upper extremity intensive work [30,35–38] is shown in Figure 6.1.

Ulin et al. [30] used this scale and another VAS with the verbal anchors "very uncomfortable work" and "very comfortable work" on the left and right ends, respectively, and Borg's CR scale for university students in the assessment of screw-driving tasks at different heights. She found that the scales compared in sensitivity and that anyone could be used with reliable results. She suggests that the VAS may be preferred by subject populations that are not as verbally oriented as university students, or in production situations where workers do not have the time to read and consider all of the verbal anchor points of the Borg scale.

In an automobile assembly plant, Armstrong et al. [39] utilized 10 cm VAS for worker assessments of hand tool mass, tool handle circumference, horizontal work location, vertical work location, and overall ratings. For tool mass, the verbal anchors used were "too heavy," "just right," and "too light" at 0, 5, and 10 cm, respectively, and "very uncomfortable," "somewhat uncomfortable," and "very comfortable" at 0, 5, and 10 cm, respectively. For tool handle circumference, the verbal anchors were "too large," "just right," and "too small" at 0, 5, and 10 cm, respectively. For the horizontal and vertical work locations, the verbal anchors were "very uncomfortable," "somewhat uncomfortable," and "very comfortable" at 0, 5, and 10 cm, respectively. The overall ratings had the verbal anchors "very good," "fair," and "very poor" at 0, 5, and 10 cm, respectively. Males and females did not differ significantly in their assessments. The workers' subjective assessments using the 10 cm VAS showed strong correlations with the tool characteristics and the work tasks.

Borg [19] compared different rating methods, among them a 11 cm VAS with the verbal anchors "no exertion at all" and "maximal exertion" on the left and right endpoints of the scale, respectively. For short-time work on a bicycle ergometer, good correlations between heart rate and perceived exertion ratings were obtained independently of which scale was used. In addition, the correlation coefficients between heart rate and perceived exertion ratings were very similar for the different methods of the physical work tests. Similarly, Harms-Ringdahl et al. [33] found that there was no significant difference between the ratings made on a 10 cm VAS and Borg's category scale for ratings of perceived pain (BRPP), which is similar to the CR scale, for the assessment of pain in response to loading of soft tissue structures at the radial side of the elbow joint.

Neely et al. [31] compared a 10 cm VAS with the verbal anchors "MIN" and "MAX" on the left and right endpoints, respectively, with Borg's CR scale, heart rate, and blood lactate levels for leg exertion during an exercise test using a bicycle ergometer. As is usual in these kinds of tests, heart rate correlated well with power. The physiological measures of heart rate and blood lactate levels (an indicator of peripheral strain) were correlated with both the CR and VAS ratings of leg exertion.

Seymour et al. [34] evaluated VASs of different lengths (5, 10, 15, and 20 cm) with the verbal anchors "no pain" and "worst pain imaginable" on the left and right endpoints, respectively. Also evaluated were 10 cm VASs with different verbal anchors on the right endpoints ("troublesome," "miserable," "intense," "unbearable," and "worst pain imaginable"). For dental pain among males and females, high correlations were found between the scores on all of the scales. Scales of length 10 or 15 cm had the smallest measurement error. Considering the different verbal anchors, the scale with the verbal anchor "worst pain imaginable" was found to be the best choice for comparing present pain or worst pain between different groups. Using this scale, no significant difference was found between males and females, or between patients with different dental conditions. Seymour et al. [34] suggest that the use of a 10 cm VAS with the verbal anchor "worst pain imaginable" was the most suitable in the measurement of dental pain. Similarly, Price et al. [26] found that a 15 cm VAS was a valid and reliable measure for both the intensity and the unpleasantness of experimentally induced pain from noxious heat stimuli delivered to the forearm, or for chronic back and/or shoulder pain.

In conclusion, it appears that VAS with a length of approximately 10 cm and verbal anchors suited to the extremes of the intensity of the physical stress of concern may provide results that are well correlated with both physical measures (e.g., tool characteristics, work tasks) and physiological measures.

#### *6.3.1.4 Preferred Maximums*

In ergonomic assessments of physical work tasks, the various Borg scales and VAS have been used to rate the physical intensity of the work tasks. The subjects are given a task to perform and then are instructed to rate the work, utilizing the scales as if they were performing the task for a normal 8 h workday. Thus, a determination of the intensity of the work can be made to compare different work tasks and their parameters, such as weight, frequency, and so forth.

Another way that psychophysics has been applied in the study of work is in the determination of preferred maximums. The subject is given control of one of the task parameters or variables, usually the weight of the object being handled. All of the other task parameters or variables, such as frequency, size, distance, and so on are kept constant. Subjects are instructed to work as hard as they can without straining themselves or without becoming unusually tired, weakened, overheated, or out of breath [40,41]. Individual subjects integrate all of their sensory inputs, monitor their feelings of exertion and fatigue, and adjust the weight accordingly. In this way, a preferred maximum for the particular task is obtained.

This preferred maximum methodology has been utilized extensively in studies of manual materials handling [24,40–58], to determine maximum weights or forces, frequencies, and so on for various lifting, lowering, pushing, pulling, and carrying tasks. When weight is the variable that is adjusted, the resulting preferred maximum may be referred to as the maximum acceptable weight of lift (MAL) [42]. When weight and/or frequency are

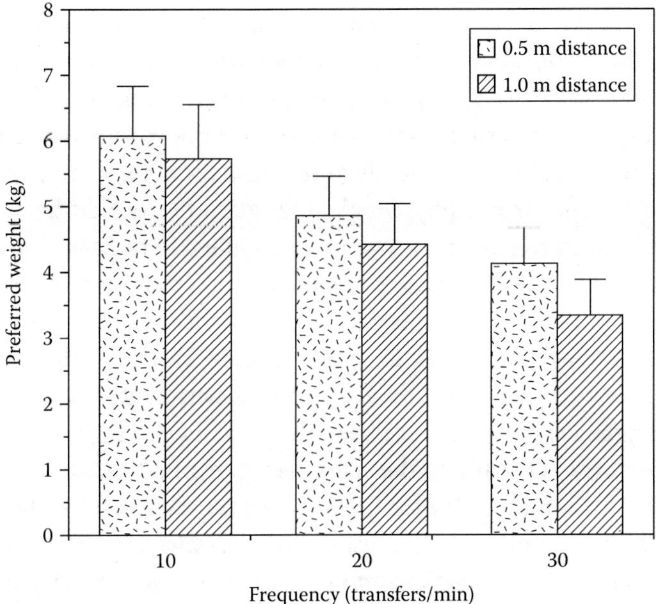


FIGURE 6.2 Mean preferred weights for each of the frequency and distance combinations for an upper extremity transfer task, over an 8 h workday, for 20 industrial subjects. 95% confidence intervals are shown by error bars.

the variables adjusted, the resulting preferred maximum workload may be calculated for comparison purposes as the maximum acceptable workload (MAWL), which is equal to the load in kg × frequency in actions/min × distance lifted in meters [43].

This preferred maximum methodology has also been utilized in upper extremity intensive work. Krawczyk et al. [35,37] had subjects adjust the weight for a repetitive upper extremity transfer task. The results are shown in Figure 6.2. As frequency and distance increased, the preferred weight decreased. Similarly, Kim and Fernandez [59] and Marley and Fernandez [60,61] had subjects determine the maximum acceptable frequency for handheld pneumatic drilling tasks. Likewise, Putz-Anderson and Galinsky [62] had subjects psychophysically determine maximum work durations for repetitive elevated arm movement tasks. Snook et al. [63] had subjects psychophysically determine maximum forces for wrist flexion and extension.

Researchers [42,44] have found significant differences in psychophysical studies between male industrial workers and students, and between female industrial workers and housewives, respectively. Thus, in psychophysical studies, it is especially important that the subject population be the same type as the target population in which the psychophysical results are to be applied. For instance, if the target population is industrial workers, then the experimental subject population should also be industrial workers.

#### *6.3.1.5 Lower Body versus Upper Body Work*

For whole body or lower body tasks, where the large muscle groups in the legs are used, perceived exertion ratings and physiological criteria, such as heart rate and oxygen consumption, are highly correlated [5,19,24,28,31,64–66]. Some physiological measures, such as heart rate, do not respond equivalently for whole body or lower body tasks and upper body work. Physiological methodologies where heart rate, measures of oxygen consumption, and energy expenditure criteria are monitored are not as sensitive to upper extremity work [4–12]. For instance, heart rate tends to increase more for arm work than for whole body or lower body tasks. Thus, physiological fatigue criteria that have been developed from leg or whole body exertions cannot validly be applied to tasks performed by the arms. Separate physiological fatigue criteria are needed for tasks that involve mainly arm work.

Snook and Irvine [4] found that there were no significant differences in heart rate for arm work in a laboratory setting and at a shoe factory. However, there was a significant heart rate difference between arm lifts and leg lifts performed in the laboratory. They concluded that fatigue criteria that have been developed from leg tasks cannot validly be applied to tasks performed mainly by the arms; energy expenditure rate guidelines for arm tasks should be lower than those for leg tasks. Based on their results, the mean heart rate should not exceed 99 beats/min for arm tasks and 112 beats/min for leg tasks.

Gamberale [5] examined the relationships between perceived exertion and physiological indicators of exertion during different exercises. Borg's RPE scale was used to determine perceived exertion at different workloads for lifting weights with the arms, working with a wheelbarrow, and also exercising on a bicycle ergometer. In addition, heart rate, oxygen uptake, and blood lactate concentration were measured. There was a linear relationship between RPE and heart rate independent of the kind of work producing the physical exertion. For all three exercises, a closer relationship to heart rate was obtained when the subjects rated their overall feeling of exertion, rather than the exertion on the arms or on the legs. However, the highest level of perceived exertion in relation to heart rate was found in the exercise of lifting weights with the arms. This exercise also yielded the highest level of blood lactate concentration in relation to oxygen uptake. The blood lactate concentrations in the exercise of lifting weights were higher than in the exercise of pushing the wheelbarrow, even though the oxygen uptake was lower. At a given level of oxygen uptake, the exercise of lifting weights was more anaerobic in character than the exercise of pushing the wheelbarrow and of working on the bicycle ergometer. These results suggest that the higher the blood lactate concentration an exercise produces as compared with oxygen uptake, the higher will be the level of the overall perception of exertion as compared with heart rate. In addition, the higher the blood lactate concentration an exercise produces as compared to oxygen uptake, the higher will be the perceived exertion on the most involved muscle groups as compared with the overall perceived exertion.

Similarly, Borg et al. [6] examined perceived exertion using the RPE and the CR scales related to heart rate and blood lactate concentration during arm and leg exercise. The arm and leg exercise was performed on bicycle ergometers, one of which was specially adapted for arm exercise. The responses obtained were at least twice as high for arm cranking as for cycling. The largest difference was found for blood lactate concentration and the smallest for RPE and heart rate. The incremental functions were similar in both exercises, with approximately linear increases in RPE and heart rate, and positively accelerating functions for CR and blood lactate concentration.

When perceived exertion on the CR scale was set as the dependent variable and a simple combination of heart rate and blood lactate was used as the independent variable, a linear relationship was obtained for both kinds of exercise, as had been previously found for cycling, running, and walking. Thus, for exercise of a steady-state type with increasing loads, the incremental curve for perceived exertion can be predicted from a simple combination of heart rate and blood lactate [6].

Mermier et al. [7] also compared upper and lower body activities and found significantly different results for upper body versus lower body tasks. Three tasks were used: exercising on a cycle ergometer, lifting (waist-to-shoulder height), and vacuuming. Their goal was to estimate ventilation by using ventilation-on-heart-rate regressions established during exercise testing to estimate ventilation in the field. For men and women, ventilation increased more steeply relative to heart rate for the exercises involving the upper body (lifting and vacuuming) compared with the lower body exercise (cycling). The regression coefficient describing the increase of ventilation with heart rate was approximately 30% greater with upper body exercise. The differences in the mean regressions for upper and lower body exercise tended to be greater in women than in men. However, these physiological criteria were consistent in that ventilation-on-heart-rate regression slopes derived from tests in which progressively increasing workloads were used were comparable to those obtained during variable and nonprogressive protocols.

From these studies, it can be concluded that the psychophysical indices are correlated with the physiological indices; however, the relationships are different for arm work than for whole body or lower body work. The physiological indices are clearly more sensitive for upper body or arm work than for whole body or lower body tasks. Thus, different physiological guidelines should be used for arm work compared with whole body or lower body work.

### 6.3.2 Psychophysical Consistency and Reproducibility

#### *6.3.2.1 Borg Scales and Visual Analog Scales*

The Borg scales, VASs, and variations of these scales have been shown to be consistent and reproducible for measurement and comparison of physical stimuli [4,34–38,40,45–53, 56,59,60,61,67] regardless of which scale was used [19,25,26,30–33]. Borg [19] compared four different rating methods: the RPE scale, graded from 6 to 20; the original Borg scale, graded from 1 to 21; a 9-point graded scale; and an 11 cm VAS. Good correlations between heart rates and perceived exertion ratings were obtained independently of which scale was used. The correlations between the different ratings were also satisfactory. Similarly, Arstila et al. [25] compared three variations of the Borg scale and found good correlation of the perceived exertion ratings with heart rate and excellent reproducibility. Krawczyk et al. [35,37] found that the VAS-perceived exertion ratings for an upper extremity transfer task were consistent throughout an 8 h workday, as shown in Figure 6.3. Harms-Ringdahl et al. [33] compared Borg's category scale for BRPP and a 10 cm VAS and found no significant difference between the two scales. There was also no significant difference between the first and second time a scale was used by the same subject. Ulin et al. [30] found equivalent results for Borg's CR scale and two different 10 cm VAS with different sets of verbal anchors in the evaluation of pneumatic screw-driving tasks. Seymour et al. [34]

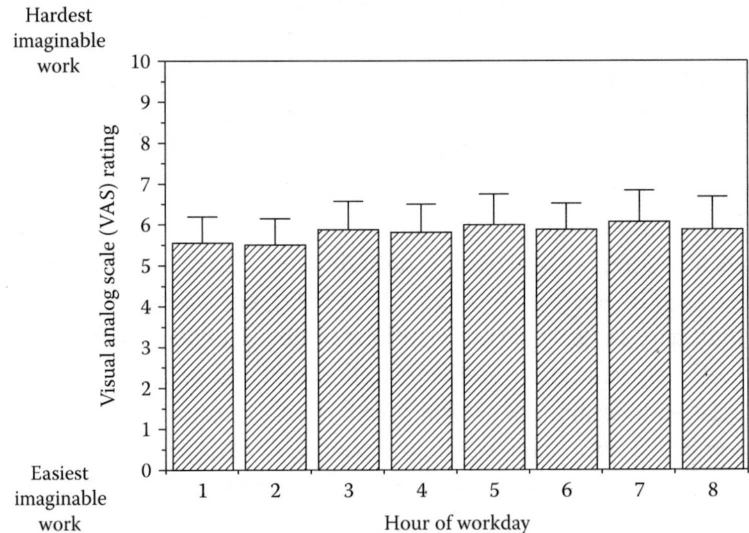


FIGURE 6.3 Overall mean VAS-perceived exertion ratings combined for all frequencies and distances at the end of each hour throughout an 8 h workday, for an upper extremity transfer task, for 20 industrial subjects. 95% confidence intervals are shown by error bars.

compared VAS of different lengths (5, 10, 15, and 20 cm) and 10 cm VAS with different verbal anchors on the right endpoints. High correlation was found between the scores on all of the scales. No significant difference was found between scores recorded by males and females or between those given by patients with two different dental conditions. Likewise, Price et al. [26] found that 15 cm VASs can be used as a valid and reliable measure for both the intensity and the unpleasantness of experimentally induced pain or chronic (back and/or shoulder) pain.

#### 6.3.2.2 *Manual Materials Handling*

Psychophysical methodologies have been shown to be consistent for manual materials-handling tasks. Ciriello et al. [45] found that subjects were able to determine their maximum acceptable weights and forces for various manual materials-handling tasks during the first 40 min of testing, and that they remained consistent throughout the 4 h duration of the test session. Ljungberg et al. [24] found that it took subjects only 5–10 min to select a weight for lifting tasks that remained consistent over 1 h. Snook and Irvine [40] found that subjects were able to determine their maximum acceptable frequency of lifting in 40 min.

Psychophysical methodologies have been shown to be reproducible for manual materials-handling tasks. Snook and Irvine [40] replicated an experiment three times and showed only insignificant differences among the three replications. Griffin et al. [46] found only a 7% decrease, which was not significant, between test and retest 1 week later for acceptable weights of lifting. Similarly, Foreman et al. [47] found a 7% decrease in the rating of acceptable dynamic lifting strength between 2 days, and the difference was not statistically significant. Fernandez et al. [48,49] concluded that the psychophysical approach was reproducible as subjects arrived at the same estimates of lifting capacity in repeated trials. Legg and Myles [52,53] found that maximum acceptable load for lifting determined two times a day for 5 days did not change significantly. When this load was used over an 8h workday, soldiers performed the lifting and lowering tasks without metabolic, cardiovascular, or subjective evidence of fatigue.

Conversely, Mital [54] found a larger decrease in the MAL throughout an 8h workday when compared to estimates made at 25min. Much of this decrease in the maximum acceptable weight took place by the completion of the second hour. After 2h, the slope of the decrease in acceptable weight had leveled off.

The differences in results from these manual materials-handling studies may be due to a frequency effect. Ciriello and Snook [55] reported the tendency of the psychophysical method to produce overestimates of maximum acceptable weights and forces for tasks with very high frequencies. This was verified with experiments of 4h duration by Karwowski and Yates [50,51] and 8h duration by Fernandez et al. [48,49]. These studies [48–51] found that selected lifting weights did not differ significantly with time at low frequencies, but that at high frequencies there were larger, sometimes significant decreases. Since Mital [54] combined all frequencies in his results, the decrease with time may have been due to this frequency effect.

#### *6.3.2.3 Upper Extremity Intensive Work*

Psychophysical methods have been shown to be consistent for upper extremity intensive work. Krawczyk et al. [35,37] found that subjects could accurately predict their perceived exertion for a normal 8h workday for a repetitive upper extremity transfer task after 1 h. After subjects performed the same task throughout a full 8h workday, the difference in the perceived exertion between the first 2h and the last 2h of performing the repetitive upper extremity transfer tasks was only 8%, as shown in Figure 6.3. Thus, the perceived exertion rating made in the first 2h, where subjects were supposed to imagine what it would be like to perform the task "for a normal 8h workday," closely agreed with what they thought 8h later, after they had actually performed the transfer task for 8h. Likewise, the difference in the maximum preferred weights between the first 2h and the last 2h of performing the repetitive upper extremity transfer tasks was only 5%, as shown in Figure 6.4.

Thus, perceived exertion and preferred weight were both consistent when determined at hourly intervals throughout an 8h workday. In addition, these results verified that a psychophysical determination for a "normal 8h workday" does not require a *full* 8h workday.

Other studies have also shown that perceived exertion is consistent over time. Krawczyk and Armstrong [35,36] found no significant differences in perceived exertion at 30, 60, 90, and 120min for a similar repetitive upper extremity transfer task performed over a 2h time period. Likewise, in another study, Krawczyk et al. [35,38] found no significant differences in perceived exertion determined after 30 and 60min for combinations of transferring and screw-driving tasks.

Psychophysical methods have been shown to be reproducible for upper extremity intensive work too. Ulin et al. [30] used Borg and two different VASs to rate perceived exertion for pneumatic screw-driving tasks and found equivalent results regardless of which scale

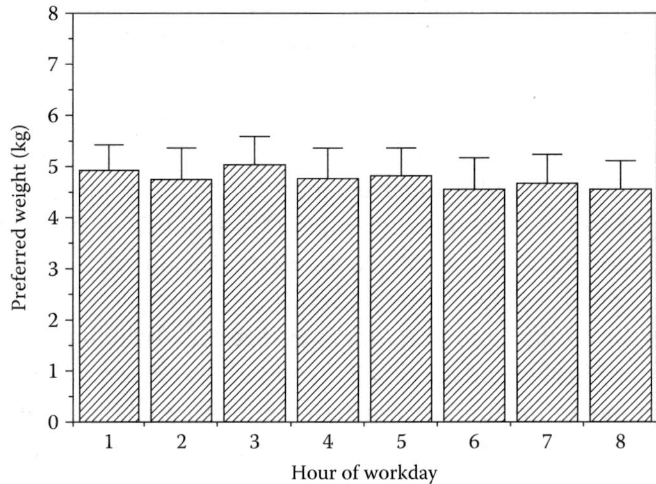


FIGURE 6.4 Overall mean preferred weights combined for all frequencies and distances at the end of each hour throughout an 8 h workday, for an upper extremity transfer task, for 20 industrial subjects. 95% confidence intervals are shown by error bars.

was used. Kim and Fernandez [59] and Marley and Fernandez [60,61] found no significant differences in the maximum acceptable frequency for three experimental replications of handheld pneumatic drilling tasks.

#### *6.3.2.4 Preferred Maximums: Initially Heavy versus Initially Light*

In the determination of maximum acceptable preferred weights, when subjects started with a heavy container weight, Krawczyk et al. [35,37] found that the resultant preferred weight for an upper extremity transfer task was greater than that determined when starting with a light weight. This is shown in Figure 6.4, where hours 1, 3, 5, and 7 were started with a heavy container weight and hours 2, 4, 6, and 8 were started with a lightweight. This difference was statistically significant. However, it is not of practical significance since this difference was only about 3%. The perceived exertions shown in Figure 6.3 also reflect this: hours 1, 3, 5, and 7 were slightly greater than hours 2, 4, 6, and 8. Legg and Myles [52,53] found this same trend for manual materials-handling tasks. The difference was 15%; however, it was not statistically significant. To eliminate the effect of the initial weight, the resultant preferred weights determined from an initially heavy container weight and an initially light container weight should be averaged together.

## 6.4 PSYCHOPHYSICAL CRITERIA FOR ERGONOMIC APPLICATION

### 6.4.1 Psychophysical Criteria and Injury Indices

Psychophysical measures of physical stress have been correlated with health outcomes, disability, and compensation [68,69]. Psychophysical criteria have been used in developing recommendations for permissible workloads. After numerous studies investigating different manual materials-handling tasks, Snook [56] concluded that designing the job to fit the worker, using psychophysically determined guidelines, can reduce up to one-third of industrial back injuries. Other studies [68–71] have shown that overexertion injuries would be reduced if manual materials-handling tasks were designed to match acceptable levels of perceived exertion. Snook [69] reviewed a number of studies, suggesting that the setting of maximum permissible workloads in industry would have a significant effect upon low-back disability and low-back compensation.

### 6.4.2 Psychophysical Studies in Manual Materials Handling and Upper Extremity Intensive Work

#### *6.4.2.1 Overview*

The psychophysical methodology is an approach that allows the evaluation of the combined effects of different physical stressors to be evaluated simultaneously. Psychophysical methodologies have been used extensively in the evaluation and design of manual materials-handling tasks [4,5,22,24,40–58,64–67,72–77] such as lifting, lowering, pushing, pulling, carrying, and also walking tasks. The National Institute for Occupational Safety and Health (NIOSH) [78–80] equation for the design and evaluation of manual lifting tasks utilizes biomechanical, physiological, and psychophysical criteria. Researchers at the Liberty Mutual Insurance Company [56,57] have composed extensive tables of psychophysically determined maximum acceptable weights and forces for manual materials-handling tasks to serve as guidelines that are consistent with worker capabilities and limitations. The variables in these studies included task frequency, distance, height and duration, object size and handles, extended horizontal reach, and combination tasks. The guidelines are intended to assist industry in the control of low back pain through reductions in initial episodes, length of disability, and recurrences [81].

Likewise, utilizing psychophysically determined guidelines for upper extremity intensive work may reduce the risk of the development of upper extremity cumulative trauma disorders (CTDs) or repetitive stress injuries (RSIs), such as carpal tunnel syndrome (CTS). Psychophysical methodologies have been used to assess upper extremity intensive work and different hand tools, to derive guidelines for the design of upper extremity work [4,10,11,30,35–39,59–62,82–95].

At present, the psychophysical methodology may be the most appropriate way to evaluate upper extremity intensive work. Unfortunately, there are no definitive data concerning what constitutes excessive quantities of the work-related risk factors: force, repetition, awkward postures, mechanical stresses, vibration, and cold temperature [96,97]. More information is needed about the relationship between risk factors, work parameters, and the development of CTDs so that work can be designed to reduce the risk of the development of CTDs. Epidemiological studies that examine the relationship between work and morbidity patterns require a lot of time and resources to collect these types of data [1]. Biomechanical methodologies cannot address repetitive work and fatigue issues [2,3]. As discussed earlier (Section 6.3.1.5), physiological methodologies, where heart rate, measures of oxygen consumption, and energy expenditure criteria are monitored, are not as sensitive for upper extremity work [4–12].

In the absence of validated epidemiological, biomechanical, or physiological methods of assessing upper extremity intensive work, an alternative approach is a psychophysical methodology, in which subjects or workers determine the relationship between work factors and the perception of physical stress, exertion, fatigue, and discomfort on the body. Since localized muscle fatigue may be an early symptom of use-related CTDs [13–15], psychophysical methodologies may serve as a more sensitive indicator for the risk of the development of CTDs [16].

#### *6.4.2.2 Task Variables*

6.4.2.2.1 Weight An increase in perceived exertion with increased weight has been found in manual materials-handling tasks (whole-body exertions) [5,22,24,52,64,73,98], as well as a decrease in working endurance time [76]. Similarly, an increase in perceived exertion with increased weight or force on the upper extremity has been found by researchers examining one-handed lifts [10,11,35–37], repetitive arm elevations [12], screw-driving [92,94], drilling [59], and gripping tasks [90]. Decreased psychophysically determined work durations [62] and increased electromyographic activities [99] have also been found in the upper extremities with increased weight or force.

When the variable adjusted by the subject is weight, how does the preferred maximum weight for an 8 h workday compare to maximum strength? Pytel and Kamon [58] found that for males and females, the maximum acceptable load selected for repetitive lifting (manual materials handling, whole-body exertions) was 22% of the experimental maximum load that an individual was able to lift once without risk of injury. In comparison, Krawczyk et al. [35,37] found that for males and females, the maximum preferred weights for a repetitive upper extremity transfer task ranged from 3.3 to 6.1 kg, as shown in Figure 6.2. The smallest preferred weight of 3.3 kg occurred at a repetition rate of 30 transfers/min and a transfer distance of 1.0 m, and the largest maximum preferred weight of 6.1 kg occurred at 10 transfers/min and 0.5 m distance. These maximum preferred weights were 16%–29% of the measured "upper extremity strength."

As frequency increases, preferred weight decreases in studies of maximum preferred weights or forces, when weight or force is the variable adjusted by the subject. This is shown in Figure 6.2. Numerous studies have shown a frequency and weight trade-off: As frequency increased, psychophysically determined acceptable weight or force decreased [35,37,41,45,48–51,55,56,66,67,73]; or, as weight or force increased, psychophysically determined acceptable frequency decreased [11,40,56,59,73,90,91].

Some upper extremity studies have shown that the effect of weight was greater than the effect of frequency [10,35,36,99]. For manual materials-handling tasks, the total psychophysically determined MAWL (load in kg × frequency in actions/min × distance lifted in meters) was affected more by weight than by frequency, which results in a higher workload performed at higher frequencies [24,43,56,74]. This result is in accordance with the frequency effect discussed earlier (Section 6.3.2.2) for manual materials-handling tasks: the tendency of the psychophysical method to produce overestimates of maximum acceptable weights and forces for tasks with high frequencies. Similarly, the results of an 8 h study [35,37] of preferred maximum weights for the upper extremity showed that subjects tolerated a higher perceived exertion rating with higher task frequency, as shown in Figure 6.5, even though they were always supposed to be working at their maximum level.

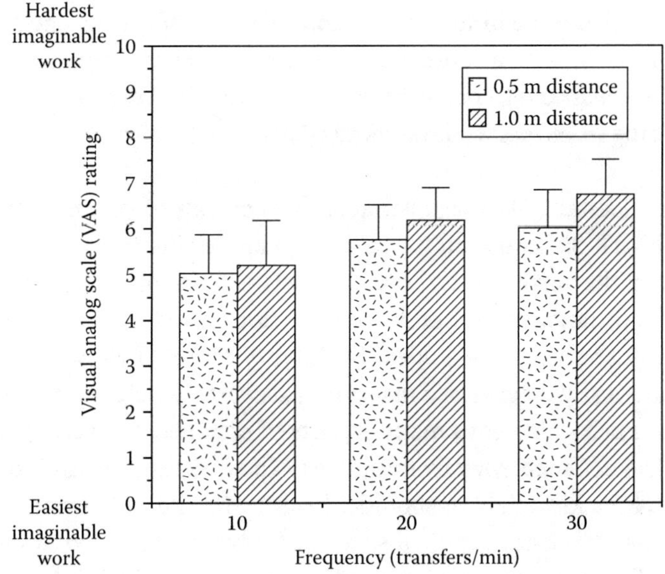


FIGURE 6.5 Mean VAS-perceived exertion ratings for each of the frequency and distance combinations for an upper extremity transfer task, over an 8 h workday, for 20 industrial subjects. 95% confidence intervals are shown by error bars.

6.4.2.2.2 Frequency An increase in perceived exertion with increased task frequency has been found in manual materials-handling tasks [64,66,73,75], as well as a decrease in working endurance time [75,76]. Similarly, an increase in perceived exertion with increased task frequency for upper extremity tasks has been found by researchers for one-handed lifts [10,35–37], as shown in Figure 6.5, and for screw-driving tasks [95]. Decreased psychophysically determined work durations [62] and increased electromyographic activities [99] have also been found in the upper extremities with increased task frequency.

Increasing frequency obviously does not increase the static strength required for the task. However, the rate of energy expenditure will increase as transfer frequency increases, and to a lesser degree as weight increases. Increases in the rate of energy expenditure have been shown to increase perceived exertion for whole-body exertions [5,19,24,64–66,72]. However, remember (Section 6.3.1.5) that physiological methodologies are not as sensitive for upper extremity work [4–12].

Energy expenditure rates do not appear to be the limiting factor in upper extremity intensive work. By process of elimination, this provides more credence that inadequate recovery time for the upper extremity, and the resulting localized discomfort and fatigue may be a limiting factor. For the upper extremity, a frequency threshold may exist [35,36]. Below this threshold, significant increases or decreases in perceived exertion do not occur, and above this threshold there is inadequate recovery time for the upper extremity, resulting in greater perceived exertion, localized discomfort and fatigue, and presumably greater risk of the development of CTDs. Indeed, morphological tissue changes (resembling peritendinitis crepitans) may occur in overexercised limbs and have been experimentally induced by overexercising the hind limb in rabbits [100]. Thus, time and load characteristics have been correlated with a subsequent cumulative trauma injury. Likewise, an accumulation of strain was found to occur in tendinous tissues of the upper extremity during physiological loading in human cadaver hands [101].

6.4.2.2.3 Distance Distance is a task variable that consists of yet another whole series of variables. The NIOSH [80] equation for the design and evaluation of manual lifting tasks has three distances: the vertical travel distance between the origin and the destination of the lift, the vertical distance of the hands from the floor, and the horizontal distance of the hands from midpoint between the ankles. The Liberty Mutual Insurance Company [56,57] tables of psychophysically determined maximum acceptable weights and forces for manual materials-handling tasks also have many aspects of distance: the vertical distance of lifting or lowering; the distance away from the body (box width); the distances of floor level to knuckle height, knuckle height to shoulder height, and shoulder height to arm reach for lifting and lowering; the distances of close to the body and extended horizontal reach for lift; the distance of pushing, pulling, or carrying; and the vertical distance from the floor to the hands for the pushing, pulling, and carrying tasks. Usually these distances are significant and should be considered, but for some tasks they may not be significant. This will depend on the distance considered and the task involved. For instance, no significant differences in maximum acceptable weight were observed among six different box sizes in the carrying tasks [57]. Thus, one should refer to the psychophysical tables [56,57] to determine what is appropriate for the application involved.

Similarly, there are many aspects of distance for upper extremity intensive work. For upper extremity work, the distance component may be significant in that it acutely affects posture. In a repetitive upper extremity transfer task [35,37], as distance increased, perceived exertion increased and preferred weight decreased for distances of 50 and 100 cm, as shown in Figures 6.5 and 6.2, respectively. The distance was lateral, producing a side-to-side motion in front of the subjects. The 50 cm distance neither challenged nor exceeded the reach envelope (where one can reach) for all subjects, while the 100 cm distance challenged the reach envelope. Thus, the different distances had some effect on the trunk posture and the resultant moments produced about the shoulder, elbow, and wrist. However, in a similar repetitive upper extremity transfer task, Krawczyk and Armstrong [35,36] found no significant distance effect on perceived exertion for distances of 25, 51, and 76 cm. Again, these transfer distances required a lateral displacement, producing a side-to-side transfer motion right in front of the subject. However, none of these distances challenged or exceeded the reach envelope for all subjects. Thus, postural changes were not required at the different distances.

In a similar study [11] of upper extremity one-handed lifts in the horizontal plane that considered distances measured from in front of the body, subjects stood in front of a 91 cm-high work table and were required to move dumbbells distances of 38 and 63 cm toward themselves. This task required some trunk flexion, and the distances were found to be significantly different for a 2.3 kg load, while no significant difference was found between the two distances for a 4.5 kg load. In another study where the distance parameter required postural changes, Ulin et al. [93] found that RPE increased with increasing horizontal distance away from the body for screw-driving tasks. In a checkstand configuration study [88], design variations (89 and 76 cm heights, and 8 and 23 cm distances), which ultimately affected the transfer distance and body posture, had significant effects on some comfort ratings. Likewise, postural discomfort has been shown to increase as a function of horizontal and vertical distance from the body for automobile assembly tasks [39]. Other upper extremity studies have found significant distance effects when the distances studied were vertical work heights [30,62].

Another factor which was shown importance is the direction of the transfer distance. Increased electromyographic activities [99] have been found in the upper extremity with different directions of movements from outward points within the reach (20°–230° measured from the frontal plane of the subjects) to a fixed point near the body for a constant 38 cm transfer distance.

#### 6.4.2.3 *Design Applicability*

Psychophysical methodologies are quick, relatively inexpensive, and convenient. Psychophysical criteria have also been shown to be consistent and reproducible, and well correlated with physiological criteria and some injury indices. In addition, a major advantage of psychophysical studies is that the results can be readily applied as guidelines in the workplace, such as in manufacturing and production assembly facilities, warehouses, and retail trades, for example, grocery and discount stores.

The results of preferred maximum studies can be directly applied, such as the Liberty Mutual Insurance Company [56,57] tables of psychophysically determined maximum acceptable weights and forces for manual materials-handling tasks, and the preferred weight results for the upper extremity transfer tasks shown in Figure 6.2. In this case, work tasks can be designed within these work parameters. The objective would be to modify the weight, frequency, distance, and other task parameters singularly or in combination to fit both the psychophysically determined acceptable task parameters (Liberty Mutual Insurance Company tables or Figure 6.2) and then given requirements for the particular work task. For instance, if for a given transfer task the frequency were fixed, say, at a particular assembly line speed, weight and/or distance could be modified appropriately. Conversely, if weight were held constant, frequency and/or distance could be modified. For example, Figure 6.2 shows that the average maximum preferred weight should be 6.1 kg for a one-handed upper extremity transfer task at a work pace of 10 transfers/min and a distance of 0.5 m. Conversely, for a one-handed upper extremity transfer task of 6.1 kg at a distance of 0.5 m, the frequency should be ≤10 transfers/min.

The NIOSH [80] equation for the design and evaluation of manual lifting tasks utilizes biomechanical, physiological, and psychophysical criteria, and thus provides another method to directly apply psychophysical criteria as workplace guidelines. For studies of perceived exertion, at least a relative rating with respect to different work tasks and sometimes an absolute measure with respect to heart rate (Borg RPE scale) can be obtained. In addition, Section 6.4.3 shows how discomfort analyses may be used to determine the effects of specific task and workstation attributes.

For both manual materials-handling tasks and upper extremity intensive work, weight or force has been shown to have the most significant effect [10,35,36,99]. Thus, weight and force may be the most important work parameter to consider in the analysis and design of work. For example, to decrease the weight for the worker, the following questions should be considered: Does the object to be lifted need to be that heavy? Could the object be made out of a lighter material? Could the object be made less bulky, so that it could be lifted closer to the body? Could it be lifted in parts? Does the object need to be manually lifted and/or transferred? Could the workstation be modified to partially or fully support the weight of the object? Could a mechanical assist (e.g., hoist, articulating arm) be used to transfer or support the object? Weight or force guidelines are shown in the Liberty Mutual Insurance Company tables and in Figure 6.2.

The second most important parameter to consider may be frequency or work pace [35,36]. As frequency increases, perceived exertion increases and preferred weight decreases, as shown in Figures 6.5 and 6.2, respectively. The Liberty Mutual Insurance Company tables and Figure 6.2 show some frequency guidelines. Recalling that there may be a frequency threshold for upper extremity transfer tasks [35,36], if the task frequency were $\leq 30$ transfers/min, a decrease in perceived exertion would occur if the frequency were decreased to 20 transfers/min. However, further decreases in frequency below 20 transfers/min may not have a significant effect on perceived exertion.

Distance may also be an appropriate parameter to consider in the analysis and design of work. As distance increases, perceived exertion increases and preferred weight decreases, as shown in Figures 6.5 and 6.2, respectively. The Liberty Mutual Insurance Company tables and Figure 6.2 give some distance guidelines. For the upper extremity, some studies have shown the importance of different distances, especially if the different distances produce postural changes, whereas lateral, side-to-side distances within the reach envelope may have only a minimal effect on perceived exertion [35,36]. Thus, decreases in distance within the reach envelope may not significantly decrease perceived exertion.

In conclusion, psychophysical data can provide guidance in the analysis and design of manual materials-handling tasks, upper extremity intensive tasks, and other manual work. Further studies of work-related musculoskeletal injuries and health surveillance will be required to determine the effect of decreased perceived exertion, discomfort, and fatigue, and to verify the effectiveness of psychophysically determined guidelines in reducing the risk of work-related musculoskeletal injuries. Until psychophysically determined guidelines are validated as effective in helping to reduce the risk of the development of work-related musculoskeletal injuries, they should be used in conjunction with both an active and passive surveillance program. If it is found that workers are adversely affected while using psychophysically determined guidelines, appropriate workplace and medical interventions should be implemented.

#### *6.4.2.4 Work Enlargement*

Psychophysical studies can be used to verify the positive effects of work enlargement. Krawczyk et al. [35,38] used a psychophysical methodology to examine combination tasks consisting of upper extremity transferring and screw-driving components. Five combination tasks were performed with different proportions of transferring and screw driving: 100% transfer, 75% transfer and 25% screw drive, 50% transfer and 50% screw drive, 25% transfer and 75% screw drive, and 100% screw drive. The left hand always performed the transferring component and the right hand always performed the screw-driving component of the combination tasks. The positive effect of using both extremities to perform the different components of the combination tasks was verified using VAS-perceived exertion ratings and body part discomfort analyses.

Overall VAS-perceived exertion decreased as the work was enlarged to utilize both upper extremities. Figure 6.6 shows that the lowest mean overall VAS-perceived exertion rating was for the 50% transferring and 50% screw-driving task. The workload of this task was evenly distributed between the left and right upper extremities, since the transferring was done with the left hand and the screw driving was done with the right hand. By allowing different extremities to perform the task components, the frequency for each extremity was effectively decreased. Consequently, this task allowed the maximum amount of physiological recovery time that could be provided simultaneously for both the left and right upper extremities. Most subjects reported a preference for the balanced nature of this combination task.

The overall VAS-perceived exertion increased as the combination task utilized more of one upper extremity than the other and involved more of either the transferring or screw-driving task components, as shown in Figure 6.6. Thus, the left upper extremity (transferring component) or the right upper extremity (screw-driving component) would become the limiting factor of the work. Even though the less varied tasks affected only one extremity, this limiting factor affected *overall* VAS ratings.

The highest overall perceived exertions were for the tasks that required only transferring or only screw driving, as shown in Figure 6.6. These tasks allowed the least amount of

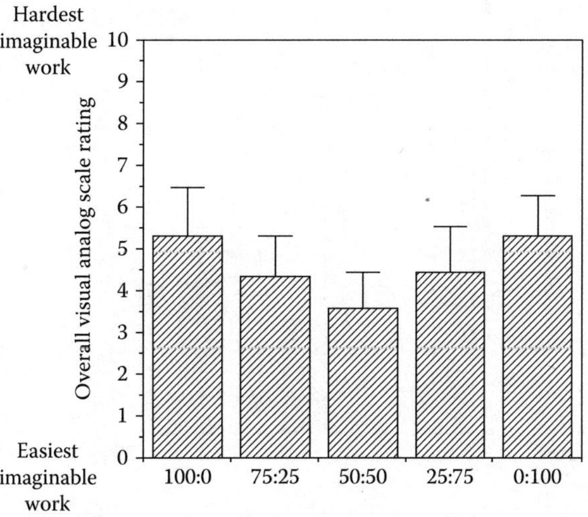


FIGURE 6.6 Mean overall VAS-perceived exertion ratings for an upper extremity combination task involving transferring (performed with the left hand) and screw-driving (performed with the right hand) components, for 24 industrial subjects. 95% confidence intervals are shown by error bars.

physiological recovery time for the upper extremity (left or right) that was responsible for performing the task. When there is not enough recovery time, the body part experiences fatigue and discomfort. Body part discomfort surveys showed a proportional increase in frequency and severity of discomfort (using a discomfort VAS, shown in Section 6.4.3, Figure 6.8) in the respective upper extremity that was required to perform a greater proportion of the combination tasks.

In addition to the overall VAS-perceived exertion, the subjects used separate VAS to rate the perceived exertion of the transferring and screw-driving components. As shown in Figure 6.7, as the transferring proportion of the task increased, the transferring perceived exertion increased. Likewise, as the screw-driving proportion of the task increased, the screw-driving perceived exertion increased. Psychophysical methods are usually used to combine the effects of physical stressors such as weight and frequency [4,5,22,24,35–37, 40–58,64–67,72–77]. However, in these upper extremity combination tasks that required two components (transferring and screw-driving), subjects were able to do the opposite: to separate the effects of physical stressors. Subjects were able to discriminate between the transferring and screw-driving parts of the tasks, and to give a psychophysical rating accordingly.

The body part discomfort surveys revealed that the less varied work, which utilized more of one upper extremity than the other and involved more of one particular task component (transferring or screw-driving), produced greater body part discomfort severity concentrated in fewer body parts. Conversely, the more varied work produced decreased body part discomfort severity, and the discomfort was more evenly distributed throughout the

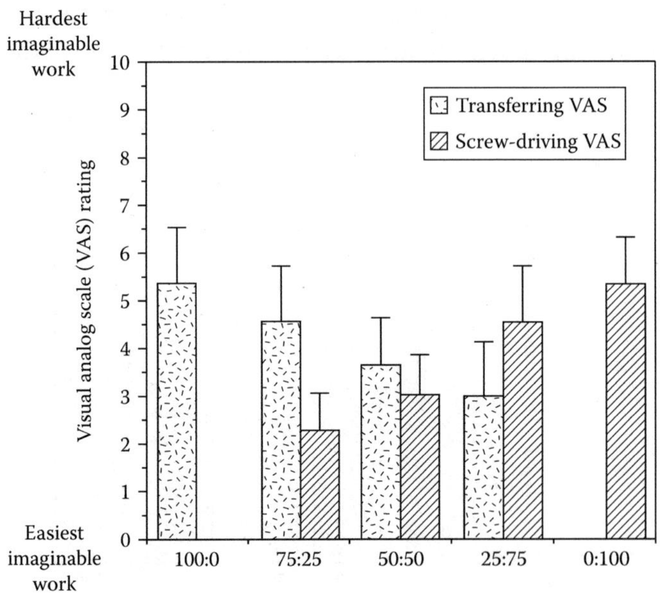


FIGURE 6.7 Mean transferring and screw-driving VAS-perceived exertion ratings for an upper extremity combination task involving transferring (performed with the left hand) and screw-driving (performed with the right hand) components, for 24 industrial subjects. 95% confidence intervals are shown by error bars.

different body parts. Similarly, Westgaard and Jansen [102] found that production workers had significantly higher symptom scores of self-reported musculoskeletal complaints for some body parts than did the group with more varied work tasks.

### 6.4.3 Discomfort Analyses

Psychophysical methodologies utilize an individual's perception of physical strain on the human body. From this perception, individuals may rate how hard they feel that they are working using Borg or VASs, or they may adjust a work variable to achieve a maximum level of normal workday exertion. When carried out over a period of time, this perception will incorporate feelings of exertion and also feelings of fatigue and discomfort.

Another way to assess individual perception of exertion over time is by tracking discomfort directly. Both overall and body part discomfort can be monitored for both frequency (how often) and severity (how much) [30,35–39,65,82–89,94,95,98,103–106,108–111]. The frequency of overall and body part discomfort can be determined simply by asking individuals if they have any discomfort, either overall or in particular body parts or regions of the body. Likewise, the severity of discomfort can be quantified using various rating scales for overall discomfort or discomfort in particular body parts or regions of the body. A VAS such as the one shown in Figure 6.1, with the verbal anchors "easiest imaginable work" and "hardest imaginable work" on the left and right endpoints, respectively, can be used to determine overall discomfort. As explained earlier, the results from the VAS shown in Figure 6.1 were comparable to the results from the VAS with "very uncomfortable work" and "very comfortable work" anchors [30]. Thus, it appears that as long as the two verbal anchors represent extremes of some work-related physical intensity, the results will be comparable. For body part discomfort, separate VASs can be used for each body part or region, with the anchors "no discomfort at all" and "worst imaginable discomfort" [35,38], as shown in Figure 6.8. A simple numerical scale may also be used to quantify discomfort severity. In this case, the individual picks a number to represent his or her discomfort, say, from 0 to 7 or from 0 to 9 [103], where the numbers are verbally anchored by comfort and discomfort descriptions. This can be done for overall discomfort and also separately for each body part or region. Specific types of discomfort, such as pain, soreness, stiffness, numbness, and tingling, may also be quantified [63]. The relative severity of body part discomfort may be determined by using a body part discomfort survey that shows a body part diagram and asks the individual to rate the different body parts from most discomfort to least discomfort [104,105].

Discomfort has been associated with poor work performance [83,84,104–106]. Corlett and Bishop [104,105] found that relieving the discomfort of workers increased work output and reduced production costs. Many upper extremity intensive tasks cause discomfort, fatigue, and pain [30,35–38,87–89,94,95,107]. Discomfort analyses have been shown to be beneficial in the evaluation of various work apparatus [39,65,94,108–110], tasks [35–38,82–86,95,98,103,111], and workstations [30,39,87–89,94,95,104–106]. For instance, for automobile assembly tasks, postural discomfort has been shown to increase as a function of horizontal and vertical distance from the body [39]. Since localized muscle fatigue may be an early symptom of use-related upper extremity CTD [13–15],

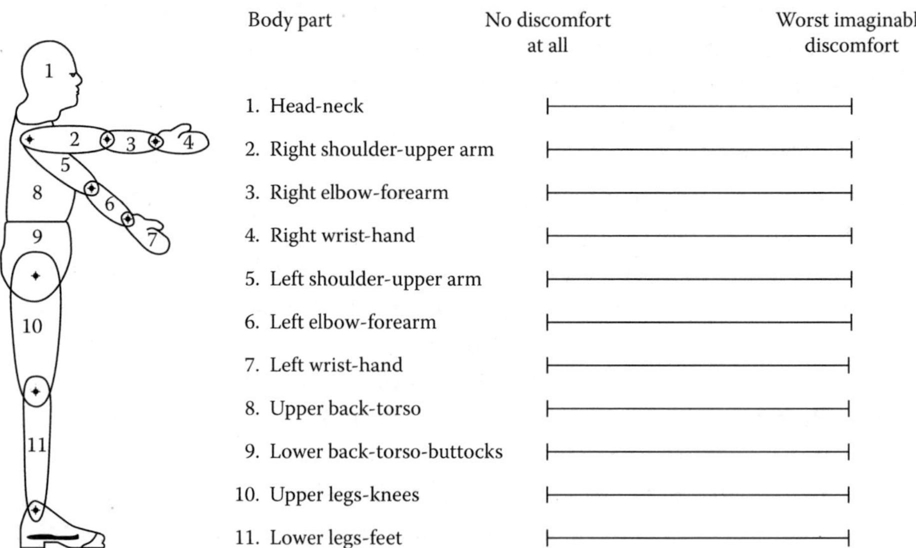


FIGURE 6.8 A body part discomfort survey with 10 cm VAS for each body part, with the verbal anchors "no discomfort at all" and "worst imaginable discomfort" on the left and right endpoints, respectively, that has been used for the evaluation of manual work. Using this body part discomfort survey, discomfort frequency and severity can be assessed for particular body parts or regions of the body.

discomfort analyses may serve as a more sensitive indicator for the risk of the development of some work-related musculoskeletal disorders.

Krawczyk et al. [35–38] used both VAS perceived exertion ratings and body part discomfort data to verify that the overall VAS-perceived exertion ratings were a reflection of the discomfort in the upper extremities due to the performance of upper extremity intensive work, and that perceived exertion ratings may be an appropriate measure to use for overall assessment of the tasks. The tasks were repetitive upper extremity transfer tasks and pneumatic screw-driving tasks. Hagberg [12] found that increases in ratings or perceived exertion with workload indicated an importance of local factors, that is, strain on muscles and tendons. Likewise, Wiker et al. [86,87] found that global reports of discomfort or fatigue were strongly related to the severity of symptoms experienced in specific muscle groups. Similarly, Yoshitake [103] found that general fatigue correlated well with symptoms of fatigue in some part of the body. Nevertheless, Kuorinka [82] concluded that local discomfort ratings were more sensitive to differences in work methods, and thus, more reliable than general discomfort ratings for ergonomic purposes.

Body part discomfort analyses may provide specific information about the relationship between task and workstation attributes (e.g., upper extremity intensive, height adjusted, standing) and their effects on the body. In a checkstand configuration study [88], subjects were found capable of distinguishing specific symptoms related to different checkstand designs.

Body part discomfort analyses may reveal additional aspects of a job that may be important for task analysis and redesign. For example, Krawczyk et al. [35–37] used body part discomfort analyses in two different studies of upper extremity transfer tasks and verified a positive effect of table height adjustment with respect to lower back, torso, and buttocks discomfort. In one study [37], where the height of the horizontal conveyor was adjusted so that the vertically oriented container handle was at elbow height for each subject, the reported lower back, torso, and buttocks discomfort was minimal, with a cumulative response rate of less than 14%. Whereas, in the other study [36], when the table height was fixed at 91 cm for each subject, the lower back, torso, and buttocks region was reported to be among most uncomfortable body regions. A full 75% rated this as one of the five more uncomfortable body regions (of a total of eleven). On average, the lower back, torso, and buttocks region was the fourth most uncomfortable body part—ranked just after the dominant upper extremity body parts and just ahead of the lower legs and feet. This was despite the fact that in the former study with minimal lower back, torso, and buttocks discomfort, the experimental workday was a full 8 h duration. And, in the latter study with much more lower back, torso, and buttocks discomfort, the experimental task was carried out for only 2 h each day, with lower weights and shorter distances too. These studies certainly suggest that the discomfort of the lower back can be decreased or minimized by adjusting the workstation height appropriately for each worker. These discomfort analyses show the relative importance of the effects of different workplace design attributes that may be addressed to minimize body part discomfort. Similar results have been shown in other upper extremity intensive work. Ayoub [87] found that all cashiers experienced some type of body pain. The areas that caused discomfort were the back, feet, legs, shoulders, neck, knees, hands, elbows, and wrists. Back pains occurred among 90% of the cashiers.

In Krawczyk et al.'s [35–38] studies of upper extremity intensive work, the subjects were standing while performing the work tasks. Even though the experimental tasks were mainly upper extremity intensive, the standing component taxed the lower extremities. This was evident since in addition to upper extremity discomfort, there was also a substantial number of discomfort responses in the lower legs and feet, and the number of discomfort responses for the lower legs and feet increased throughout the workday. Similarly, in a study of seven supermarkets, Ryan [89] found that the checkout department had the highest rates of discomfort or pain symptoms for almost all body areas. The lower back, lower limbs, and feet were the body areas with the highest rates of discomfort or pain symptoms. Ryan [89] found a positive and significant correlation between the proportion of time spent standing and symptoms in the lower limb and foot, especially in the checkout department, where 90% of the time was spent standing in one place. Rys and Konz [109] also found that body comfort decreases as the duration of the standing work increases, regardless of the type of floor. They found the greatest increases in discomfort in the lower body parts. None of the upper parts of the body had comfort significantly affected by floor surface.

In conclusion, discomfort analyses may serve as a sensitive indicator for the risk of the development of work-related musculoskeletal disorders. Overall and body part discomfort analyses may provide specific information about aspects of the work apparatus, tasks, and workstations that need to be addressed. Minimizing work-related discomfort may improve worker performance, increase output, decrease production costs [83,84,104–106], and increase worker satisfaction.

## 6.5 CONCLUSION

Psychophysical methods are a consistent, reproducible, quick, inexpensive, and convenient way to assess the degree of physical strain on the human body. Psychophysical criteria have been correlated with physiological criteria and some injury indices. Psychophysical methods utilize the results of the central nervous system integration of various information, including the many signals elicited from the peripheral working muscles and joints and from the central cardiovascular and respiratory functions. All of these signals, perceptions, and experiences are combined and utilized with psychophysical methods.

The effects of different work task parameters, as well as the combined effects, can be evaluated using psychophysical methods. A major advantage of psychophysical studies is that the results can be readily applied as guidelines in the workplace. Psychophysical data can provide guidance in the analysis and design of repetitive manual work tasks, such as those found in manufacturing and production assembly facilities, warehouses, retail trades (e.g., grocery and discount stores), and other workplaces. The objective should be to modify the weight, frequency, distance, and other task parameters singularly or in combination to fit both the psychophysically determined acceptable task parameters and the given requirements for a particular work task. In addition, body part discomfort analyses may contribute specific information about the relationship between task and workstation attributes (e.g., upper extremity intensive, height adjusted, standing) and their effects on the body.

Health surveillance is required to verify the effectiveness of psychophysically determined guidelines in reducing the risk of work-related musculoskeletal injuries. Further studies of work-related musculoskeletal injuries and health surveillance is required to determine the effect of decreased perceived exertion, discomfort, and fatigue, and to verify the effectiveness of psychophysically determined guidelines in reducing the risk of work-related musculoskeletal injuries. Until psychophysically determined guidelines are validated as effective in helping to reduce the risk of the development of work-related musculoskeletal injuries, they should be used in conjunction with both an active and passive surveillance program. If it is found that workers are adversely affected while using psychophysically determined guidelines, appropriate workplace and medical interventions should be implemented.

## REFERENCES

1. C. H. Hennekens and J. E. Buring. *Epidemiology in Medicine*. Little, Brown and Company, Boston, MA/Toronto, Ontario, Canada, 1987, pp. 258–271.
2. M. F. Tracy, Biomechanical methods in posture analysis, in *Evaluation of Human Work*, J. R. Wilson and E. Nigel Corlett, Eds., Taylor & Francis Group, New York, 1990, pp. 571–604.
3. T. J. Armstrong and D. B. Chaffin, Some biomechanical aspects of the carpal tunnel, *J. Biomech. 12*:567–570 (1979).
4. S. H. Snook and C. H. Irvine, Psychophysical studies of physiological fatigue criteria, *Hum. Factors 11*(3):291–300 (1969).
5. F. Gamberale, Perceived exertion, heart rate, oxygen uptake and blood lactate in different work operations, *Ergonomics 15*(5):545–554 (1972).
6. G. Borg, P. Hassmén, and M. Lagerström, Perceived exertion related to heart rate and blood lactate during arm and leg exercise, *Eur. J. Appl. Physiol. 65*:679–685 (1987).

7. C. M. Mermier, J. M. Samet, W. E. Lambert, and T. W. Chick, Evaluation of the relationship between heart rate and ventilation for epidemiologic studies, *Arch. Environ. Health* *48*(4):263–269 (1993).
8. R. B. Andrews, Estimation of values of energy expenditure rate from observed values of heart rate, *Hum. Factors* *9*(6):581–586 (1967).
9. E. Asmussen and I. Hemmingsen, Determination of maximum working capacity at different ages in work with the legs or with the arms, *Scand. J. Clin. Lab. Invest.* *10*:67–71 (1958).
10. A. Garg, Physiological responses to one-handed lift in the horizontal plane by female workers, *Am. Ind. Hyg. Assoc. J.* *44*(3):190–200 (1983).
11. A. Garg and U. Saxena, Maximum frequency acceptable to female workers for one-handed lifts in the horizontal plane, *Ergonomics* *25*(9):839–853 (1982).
12. M. Hagberg, Work load and fatigue in repetitive arm elevations, *Ergonomics* *24*(7):543–555 (1981).
13. K. N. Baidya and M. G. Stevenson, Local muscle fatigue in repetitive work, *Ergonomics* *31*(2):227–239 (1988).
14. R. H. T. Edwards, Muscle fatigue and pain, *Acta Med. Scand. Suppl.* *717*:179–188 (1986).
15. F. Valencia, Local muscle fatigue: A precursor to RSI? *Med. J. Ausl.* *145*:327–330 (1986).
16. G. A. V. Borg, Psychophysical scaling with applications in physical work and the perception of exertion, *Scand. J. Work Environ. Health* *16*(Suppl. 1):55–58 (1990).
17. S. S. Stevens, The psychophysics of sensory function, *Am. Sci.* *48*:226–253 (1960).
18. G. A. V. Borg, *Physical Performance and Perceived Exertion, Studia Psychologica et Paedagogica, Series altera, Investigationes XI,* CWK Gleerup, Lunds, Sweden, 1962.
19. G. A. V. Borg, Perceived exertion: A note on "history" and methods, *Med. Sci. Sports,* *5*(2):90–93 (1973).
20. H. Eisler, Subjective scale of force for a large muscle group, *J. Exp. Psychol.* *64*(3):253–257 (1962).
21. J. C. Stevens and W. S. Cain, Effort in isometric muscular contractions related to force level and duration, *Percept. Psychophy.* *8*(4):240–244 (1970).
22. C. E. Baxter, H. Stålhammar, and J. D. G. Troup, A psychophysical study of heaviness for box lifting and lowering, *Ergonomics* *29*(9):1055–1062 (1986).
23. F. I. Gamberale, I. Holmér, A. S. Kindblom, and A. Nordström, Magnitude perception of added inspiratory resistance during steady-state exercise, *Ergonomics* *21*(7):531–538 (1978).
24. A.-S. Ljungberg, F. Gamberale, and A. Kilbom, Horizontal lifting: Physiological and psychological responses, *Ergonomics* *25*(8):741–757 (1982).
25. M. Arstila, H. Wendelin, I. Vuori, and I. Välimäki, Comparison of two rating scales in the estimation of perceived exertion in a pulse–conducted exercise test, *Ergonomics* *17*(5):577–584 (1974).
26. D. D. Price, P. A. McGrath, A. Rafii, and B. Buckingham, The validation of visual analog scales as ratio scale measures for chronic and experimental pain, *Pain* *17*:45–56 (1983).
27. G. A. V. Borg, Psychophysical bases of perceived exertion, *Med. Sci. Sports Exerc.* *14*(5):377–381 (1982).
28. G. A. V. Borg, Perceived exertion as an indicator of somatic stress, *Scand. J. Rehabil. Med.* *2*(3):92–98 (1970).
29. B. J. Noble, Clinical application of perceived exertion, *Med. Sci. Sports Exerc.* *14*(5):406–411 (1982).
30. S. S. Ulin, C. M. Ways, T. J. Armstrong, and S. H. Snook, Perceived exertion and discomfort versus work height with a pistol-shaped screwdriver, *Am. Ind. Hyg. Assoc. J.* *51*(11):588–594 (1990).
31. G. Neely, G. Ljunggren, C. Sylven, and G. Borg, Comparison between the visual analogue scale (VAS) and the category ratio scale (CR-10) for the evaluation of leg exertion, *Int. J. Sports Med.* *13*:133–136 (1992).
32. B. J. Noble, G. A. V. Borg, I. Jacobs, R. Ceci, and P. Kaiser, A category-ratio perceived exertion scale: Relationship to blood and muscle lactates and heart rate, *Med. Sci. Sports Exerc.* *15*(6):523–528 (1983).

33. K. Harms-Ringdahl, A. M. Carlsson, J. Ekholm, A. Raustorp, T. Svensson, and H.-G. Toresson, Pain assessment with different intensity scales in response to loading of joint structures, *Pain 27*:401–411 (1986).
34. R. A. Seymour, J. M. Simpson, J. E. Charlton, and M. E. Phillips, An evaluation of length and end-phrase of visual analogue scales in dental pain, *Pain 21*:177–185 (1985).
35. S. Krawczyk, Psychophysical determination of work design guidelines for repetitive upper extremity transfer tasks over an eight hour workday, PhD dissertation, The University of Michigan, Ann Arbor, MI, 1993.
36. S. Krawczyk and T. J. Armstrong, Perceived exertion over time of hand transfer tasks: Weight, frequency, and distance, in *Designing for Everyone,* Y. Quéinnec and F. Daniellou, Eds., Taylor & Francis Group, London, U.K., 1991, pp. 167–169.
37. S. Krawczyk, T. J. Armstrong, and S. H. Snook, Preferred weights for hand transfer tasks for an eight hour workday, in *Arbete Och Hälsa Vetenskaplig Skriftserie, Proceedings of National Institute of Occupational Health International Scientific Conference on Prevention of Work-Related Musculoskeletal Disorders PREMUS,* M. Hagberg and Å. Kilbom, Eds., Stockholm, Sweden, 1992, pp. 157–159.
38. S. Krawczyk, T. J. Armstrong, and S. H. Snook, Psychophysical assessment of simulated assembly line work: Combinations of transferring and screw driving tasks, *Proceedings of the Human Factors and Ergonomics Society 37th Annual Meeting,* Seattle, Washington, DC, 1993, pp. 803–807.
39. T. J. Armstrong, L. Punnett, and P. Ketner, Subjective worker assessments of hand tools used in automobile assembly, *Am. Ind. Hyg. Assoc. J. 50*(12):639–645 (1989).
40. S. H. Snook and C. H. Irvine, Maximum frequency of lift acceptable to male industrial workers, *Am. Ind. Hyg. Assoc. J. 29*(November–December):531–536 (1968).
41. S. H. Snook, C. H. Irvine, and S. F. Bass, Maximum weights and work loads acceptable to male industrial workers, *Am. Ind. Hyg. Assoc. J. 31*:579–588 (1970).
42. A. Mital, Patterns of differences between the maximum weights of lift acceptable to experienced and inexperienced materials handlers, *Ergonomics 30*(8):1137–1147 (1987).
43. L. M. Nicholson and S. J. Legg, A psychophysical study of the effects of load and frequency upon selection of workload in repetitive lifting, *Ergonomics 29*(7):903–911 (1986).
44. S. H. Snook and V. M. Ciriello, Maximum weights and work loads acceptable to female workers, *J. Occup. Med. 16*(8):527–534 (1974).
45. V. M. Ciriello, S. H. Snook, A. C. Blick, and P. L. Wilkinson, The effects of task duration on psychophysical-determined maximum acceptable weights and forces, *Ergonomics 33*(2):187–200, (1990).
46. A. B. Griffin, J. D. G. Troup, and D. C. E. F. Lloyd, Tests of lifting and handling capacity: Their repeatability and relationship to back symptoms, *Ergonomics 27*(3):305–320, (1984).
47. T. K. Foreman, C. E. Baxter, and J. D. G. Troup, Ratings of acceptable load and maximal isometric lifting strengths: The effects of repetition, *Ergonomics 27*(12):1283–1288, (1984).
48. J. E. Fernandez and M. M. Ayoub, The psychophysical approach: The valid measure of lifting capacity, in *Trends in Ergonomics/Human Factors V,* F. Aghazadeh, Ed., North-Holland, Elsevier, Amsterdam, the Netherlands, 1988, pp. 837–845.
49. J. E. Fernandez, M. M. Ayoub, and J. L. Smith, Psychophysical lifting capacity over extended periods, *Ergonomics 34*(1):23–32 (1991).
50. W. Karwowski and J. W. Yates, The effect of time in the psychophysical study of the maximum acceptable amounts of liquid lifted by females, *Proceedings of the Human Factors Society 28th Annual Meeting,* San Antonio, TX, 1984, pp. 586–590.
51. W. Karwowski and J. W. Yates, Reliability of the psychophysical approach to manual lifting of liquids by females, *Ergonomics 29*(2):237–248 (1986).
52. S. J. Legg and W. S. Myles, Maximum acceptable repetitive lifting workloads for an 8-hour workday using psychophysical and subjective rating methods, *Ergonomics 24*(12):907–916 (1981).

53. S. J. Legg and W. S. Myles, Metabolic and cardiovascular cost, and perceived effort over an 8 hour day when lifting loads selected by the psychophysical method, *Ergonomics 28*(1):337–343 (1985).
54. A. Mital, The psychophysical approach in manual lifting: A verification study, *Hum. Factors 25*(5):485–491 (1983).
55. V. M. Ciriello and S. H. Snook, A study of size, distance, height, and frequency effects on manual handling tasks, *Hum. Factors 25*(5):473–483 (1983).
56. S. H. Snook, The design of manual handling tasks, *Ergonomics 21*(12):963–985 (1978).
57. S. H. Snook and V. M. Ciriello, The design of manual handling tasks: Revised tables of maximum acceptable weights and forces, *Ergonomics 34*(9):1197–1213 (1991).
58. J. L. Pytel and E. Kamon, Dynamic strength test as a predictor for maximal and acceptable lifting, *Ergonomics 24*(9):663–672 (1981).
59. C. H. Kim and J. E. Fernandez, Psychophysical frequency for a drilling task, *Int. J. Ind. Ergonomics 12*:209–218 (1993).
60. R. J. Marley and J. E. Fernandez, A psychophysical approach to establish maximum acceptable frequency for hand/wrist work, in *Advances in Industrial Ergonomics and Safety III*, W. Karwowski and J. W. Yates, Eds., Taylor & Francis, New York, 1991, pp. 75–82.
61. R. J. Marley and J. E. Fernandez, Psychophysical frequency and sustained exertion at varying wrist postures for a drilling task, *Ergonomics 38*(2):303–325 (1995).
62. V. Putz-Anderson and T. L. Galinsky, Psychophysically determined work durations for limiting shoulder girdle fatigue from elevated manual work, *Int. J. Ind. Ergon. 11*:19–28 (1993).
63. S. H. Snook, D. R. Vaillancourt, V. M. Ciriello, and B. S. Webster, Pyschophysical studies of repetitive wrist flexion and extension, *Ergonomics 38*(7):1488–1507 (1995).
64. S. S. Asfour, M. M. Ayoub, A. Mital, and N. J. Bethea, Perceived exertion of physical effort for various manual handling tasks, *Am. Ind. Hyg. Assoc. J. 44*(3):223–228 (1983).
65. C. G. Drury and J. M. Deeb, Handle positions and angles in a dynamic lifting task, part 2. Psychophysical measures and heart rate, *Ergonomics 29*(6):769–777 (1986).
66. A. Garg and J. Banaag, Maximum acceptable weights, heart rates and RPEs for one hour's repetitive asymmetric lifting, *Ergonomics 31*(1):77–96 (1988).
67. S. H. Snook, Psychophysical acceptability as a constraint in manual working capacity, *Ergonomics 28*(1):331–335 (1985).
68. G. D. Herrin, M. Jaraiedi, and C. K. Anderson, Prediction of overexertion injuries using biomechanical and psychophysical models, *Am. Ind. Hyg. Assoc. J. 47*(6):322–330 (1986).
69. S. H. Snook, Psychophysical considerations in permissible loads, *Ergonomics 28*(1):327–330 (1985).
70. S. H. Snook, R. A. Campanelli, and J. W. Hart, A study of three preventive approaches to low back injury, *J. Occup. Med. 20*(7):478–481 (1978).
71. D. H. Liles, S. Deivanayagam, M. M. Ayoub, and P. Mahajan, A job severity index for the evaluation and control of lifting injury, *Hum. Factors 26*(6):683–693 (1984).
72. M. G. Wardle and D. S. Gloss, A psychophysical approach to estimating endurance in performing physically demanding work, *Hum. Factors 29*(6):745–747 (1978).
73. F. Gamberale, A.-S. Ljungberg, G. Annwall, and Å. Kilbom, An experimental evaluation of psychophysical criteria for repetitive lifting work, *Appl. Ergon. 18*(4):311–321 (1987).
74. A. Garg and U. Saxena, Effects of lifting frequency and technique on physical fatigue with special reference to psychophysical methodology and metabolic rate, *Am. Ind. Hyg. Assoc. J. 40*(10):894–903 (1979).
75. A. M. Genaidy and S. Al-Rayes, A psychophysical approach to determine the frequency and duration of work-rest schedules for manual handling operations, *Ergonomics 36*(5):509–518 (1993).
76. A. M. Genaidy and S. S. Asfour, Effects of frequency and load of lift on endurance time, *Ergonomics 32*(1):51–57 (1989).
77. S. H. Snook and V. M. Ciriello, Maximum weights and work loads acceptable to female worker, *Occup. Health Nurs. 22*(8):11–20 (1974).

78. V. Putz-Anderson and T. Waters, Revisions in NIOSH guide to manual lifting, *Proceedings of a National Strategy for Occupational Musculoskeletal Injury Prevention—Implementation Issues and Research Needs*, The University of Michigan, Ann Arbor, MI, 1991.
79. National Institute for Occupational Safety and Health (NIOSH), scientific support documentation for the revised 1991 NIOSH lifting equation: Technical contract reports, *Psychophysical Basis for Manual Lifting Guidelines* by M. M. Ayoub (purchase order number 88–79313, 1988), U.S. Department of Health and Human Services, Public Health Service, Centers for Disease Control, NIOSH, Cincinnati, OH, 1989.
80. T. R. Waters, V. Putz-Anderson, A. Garg, and L. J. Fine, Revised NIOSH equation for the design and evaluation of manual lifting tasks, *Ergonomics 36*(7):749–776 (1993).
81. S. H. Snook, Approaches to the control of back pain in industry: Job design, job placement and education/training, *SPINE: State of the Art Reviews 2*(1):45–59 (1987).
82. I. Kuorinka, Subjective discomfort in a simulated repetitive task, *Ergonomics 26*(11):1089–1101 (1983).
83. R. W. Schoenmarklin and W. S. Marras, Effects of handle and work orientation on hammering: I. Wrist motion and hammering performance, *Hum. Factors 31*(4):397–411 (1989).
84. R. W. Schoenmarklin and W. S. Marras, Effects of handle and work orientation on hammering: II. Muscle fatigue and subjective ratings of body discomfort, *Hum. Factors 31*(4):413–420 (1989).
85. S. F. Wiker, D. B. Chaffin, and G. D. Langolf, Shoulder postural fatigue and discomfort: A preliminary finding of no relationship with isometric strength capability in a light-weight manual assembly task, *Int. J. Ind. Ergon. 5*:133–146 (1990).
86. S. F. Wiker, D. B. Chaffin, and G. D. Langolf, Shoulder posture and localized muscle fatigue and discomfort, *Ergonomics 32*(2):211–237 (1989).
87. M. A. Ayoub, Ergonomic deficiencies: I. Pain at work, *J. Occup. Med. 32*(1):52–57 (1990).
88. P. Harber, D. Bloswick, J. Luo, J. Beck, D. Greer, L. F. Peña, Work-related symptoms and checkstand configuration: An experimental study, *Am. Ind. Hyg. Assoc. J. 54*(7):371–375 (1993).
89. G. A. Ryan, The prevalence of musculo-skeletal symptoms in supermarket workers, *Ergonomics 32*(4):359–371 (1989).
90. J. B. Dahalan and J. E. Fernandez, Psychophysical frequency for a gripping task, *Int. J. Ind. Ergon. 12*:214–230 (1993).
91. J. E. Fernandez, J. B. Dahalan, M. G. Klein, and R. J. Marley, Using the psychophysical approach in hand-wrist work, *Proceedings of M. M. Ayoub Institute for Ergonomics Research, Occupational Ergonomics Symposium*, Lubbock, TX, April 1993, pp. 63–70.
92. R. Örtengren, T. Cederqvist, M. Lindberg, and B. Magnusson, Workload in lower arm and shoulder when using manual and powdered screwdrivers at different working heights, *Int. J. Ind. Ergon. 8*:225–235 (1991).
93. S. S. Ulin, T. J. Armstrong, S. H. Snook, and A. Franzblau, Effect of tool shape and work location on perceived exertion for work on horizontal surfaces, *Am. Ind. Hyg. Assoc. J. 54*(7):383–391 (1993).
94. S. S. Ulin, T. J. Armstrong, S. H. Snook, and W. M. Keyserling, Examination of the effect of tool mass and work posture on ratings of perceived exertion for a screw driving task, *Int. J. Ind. Ergon. 12*(1–2):105–115 (1993).
95. S. S. Ulin, T. J. Armstrong, S. H. Snook, and W. M. Keyserling, Perceived exertion and discomfort associated with driving screws at various work locations and at different work frequencies, *Ergonomics 36*(7):833–846 (1993).
96. K. H. E. Kroemer, Cumulative trauma disorders: Their recognition and ergonomics measures to avoid them, *Appl. Ergon. 20*(4):274–280 (1989).
97. T. J. Armstrong, R. G. Radwin, D. J. Hansen, and K. W. Kennedy, Repetitive trauma disorders: Job evaluation and design, *Hum. Factors 28*(3):325–336 (1986).
98. B. G. Coury and C. G. Drury, Optimum handle positions in a box-holding task, *Ergonomics 25*(7):645–662 (1982).

99. H. Strasser and E. Keller, Local muscular strain dependent on the direction of horizontal arm movements, *Ergonomics 32*(7):899–910 (1989).
100. O. Rais, Heparin treatment of peritenomyosis (peritendinitis) crepitans acuta, *Acta Chirurgica Scand. Suppl. 268*:1–88 (1961).
101. S. A. Goldstein, T. J. Armstrong, D. B. Chaffin, and L. S. Matthews, Analysis of cumulative strain in tendons and tendon sheaths, *J. Biomech. 20*(1):1–6 (1987).
102. R. H. Westgaard and T. Jansen, Individual and work related factors associated with symptoms of musculoskeletal complaints. II. Different risk factors among sewing machine operators, *Br. J. Ind. Med. 49*:154–162 (1992).
103. H. Yoshitake, Relations between the symptoms and feeling of fatigue, *Ergonomics 14*(1): 175–185 (1971).
104. E. N. Corlett and R. P. Bishop, A technique for assessing postural discomfort, *Ergonomics 19*(2):175–182 (1976).
105. E. N. Corlett and R. P. Bishop, The ergonomics of spot welders, *Appl. Ergon.* (March):23–32 (1978).
106. V. Bhatnager, C. G. Drury, and S. G. Schiro, Posture, postural discomfort, and performance, *Hum. Factors 27*(2):189–199 (1985).
107. L. Lannersten and K. Harms-Ringdahl, Neck and shoulder muscle activity during work with different cash register systems, *Ergonomics 33*(1):49–65 (1990).
108. C. G. Drury and B. G. Coury, A methodology for chair evaluation, *Appl. Ergon. 13*(3):195–202 (1982).
109. M. Rys and S. Konz, Floor mats, *Proceedings of the Human Factors Society 34th Annual Meeting*, Orlando, FL, 1990, pp. 575–579.
110. S. S. Ulin, S. H. Snook, T. J. Armstrong, and G. D. Herrin, Preferred tool shapes for various horizontal and vertical work locations, *Appl. Occup. Environ. Hyg. 7*(5):327–337 (1992).
111. J. M. Deeb, C. G. Drury, and K. L. Begbie, Handle positions in a holding task as a function of task height, *Ergonomics 28*(5):747–763 (1985).

CHAPTER 7

# Instrumentation for Occupational Ergonomics

Robert G. Radwin, Thomas Y. Yen,
David J. Beebe, and John G. Webster

CONTENTS

## 7.1 INTRODUCTION

Ergonomists are concerned with measurement of a variety of physical, physiological, biophysical, and environmental factors. These include physical phenomena such as human motion (e.g., joint angles, force, acceleration), physiological processes related to energy expenditure and metabolism (e.g., heart rate, oxygen consumption, body temperature), and biopotentials associated with muscle contraction (e.g., electrocardiograms, electromyograms (EMGs)) and environmental exposures (illumination, acoustic noise, whole body and hand-arm vibration). An understanding of the principles, operation, and use of biomedical instruments is necessary for acquiring accurate measurements of any of these phenomena.

Many measurements in industrial ergonomics involve the use of electronic or mechanical instruments. Too often a lack of the basic fundamentals of measurement, electronic principles, and sensors can lead to unwitting measurement errors. Ergonomics practitioners and researchers should therefore be familiar with how electronic instruments operate and be aware of possible sources of error that can affect measurement systems.

Figure 7.1 shows a block diagram of a generalized instrumentation system. Most instrument systems include (1) a sensor, (2) an amplifier, (3) a signal conditioning stage, and (4) a display and storage device. Sensors are transducers that measure physical, chemical, or thermal events as electric signals. Because these signals are often very small, they have to be amplified in order to be accurately measured. The signals usually contain noise not related to the activity being measured, and because these unrelated signals are amplified along with the desired signal, signal conditioning is necessary to remove them from the wanted signals. The display presents the measured information so an observer can access the data. The signal may also be stored for further processing or viewed at a later time. Often the storage, processing, and display device is a digital computer.

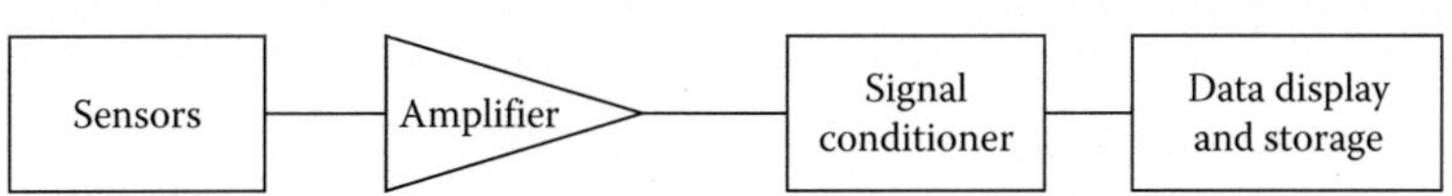


FIGURE 7.1 Block diagram of generalized instrumentation system.

## 7.2 RELEVANT CONCEPTS AND TERMINOLOGY

This section reviews the fundamentals of physics, electricity, and electronics that govern the operation and use of various instruments for measuring physical and physiological phenomena.

### 7.2.1 Measurement Principles

#### *7.2.1.1 Accuracy*

Accuracy is the difference between the quantity as measured and its true value. This difference may arise from a variety of sources. It can be a gross measurement error (attributed to misreading or improper use of the instrument), a systematic error (caused by a particular instrument's shortcomings or environmental factors), or a random error (due to random variations in the measurement itself). Random variations can often be reduced by making multiple measurements and taking an average. When the true value of a measurement is not known, error is sometimes determined by taking the standard deviation of a large number of repeated measurements. Accuracy is sometimes expressed in terms of percent error of full scale and usually varies over the range of the instrument.

#### *7.2.1.2 Precision*

Precision is the ability of an instrument to reproduce the same measurement over and over again. The precision of an instrument is often qualified by the number of significant figures the instrument is capable of resolving (see the following). It is important to realize that higher precision does not necessarily imply better accuracy.

#### *7.2.1.3 Resolution*

Resolution is the smallest change that an instrument is capable of detecting. Resolution is determined by the manufacturer based on the instrument's accuracy and precision. The resolution of an analog meter is limited by the smallest division on the meter scale. A digital instrument's resolution is the quantity that the least significant digit represents.

#### *7.2.1.4 Range*

Minimum range is limited by the minimum quantity the instrument can resolve. The maximum range is limited by the greatest input that will not damage the instrument or the greatest quantity that an instrument can measure within a certain accuracy.

#### *7.2.1.5 Calibration*

Calibration references a measurement against a more accurate reference standard, usually traceable to the National Bureau of Standards. Typically, the reference standard should be at least an order of magnitude more accurate than the instrument being calibrated. Since accuracy depends on the quantity measured, calibration should be performed over the complete range of the instrument.

#### *7.2.1.6 Reading an Analog Scale*

Errors can arise from improper readings off an analog display. An anthropometeric study found that considerable error was attributable to the precision used for reading the instrument rather than the actual precision limits of the instrument. Untrained subjects tended to measure body dimensions consistently to the nearest 1/4 in. (6 mm) or less, irrespective that the rule they were provided had 1/16 in. (1 mm) precision [1]. Measurements should be recorded from analog scales using all the digits that one is sure of plus one digit estimated nearest to the true value. For instance, if an analog scale is divided into tenths of a unit, the measurement may be recorded using all digits up to a tenth of a unit, plus a last digit that is the best estimation read by the observer.

### 7.2.2 Instrument Dynamics

#### *7.2.2.1 Response Characteristics*

Because physiological and physical measurements are seldom static, the dynamic features of instrumentation must be taken into consideration. System response characteristics should be considered for all sensors, amplifiers, signal conditioners, and display devices relative to the signal being measured. The response to a step input is often used for characterizing instruments. The specific response depends on the ordinary differential equations that describe the instrument input–output relationships. Figure 7.2 shows typical dynamic characteristics. Most instruments can be classified as zeroth-order, first-order, or second-order systems.

A zeroth-order system output is always proportional to its input. Consequently the step response for zeroth-order systems has no time delays, oscillations, or overshoot. Zeroth-order systems can be described by their static sensitivity, which is the ratio of the output signal divided by the input signal. Therefore the output is proportional to the input signal times the static sensitivity. A simple potentiometer is an example of a zeroth-order instrument.

The output of a first-order instrument requires a finite time to reach its final level for a step input, but no overshoot occurs. First-order systems are characterized by their static sensitivity and time constant. The time constant is defined as the time required for the system to reach 63% of its final value for a step input. The smaller the time constant, the faster the output rises. A mercury thermometer is an example of a first-order instrument.

For a step input, second-order systems take a finite time to reach the final value, may overshoot, and can yield oscillations. Many higher order instruments can be treated as second-order systems by making some simplifying assumptions. A simple spring scale is an example of a second-order instrument. Second-order systems can be modeled by a mass, spring, and viscous damping element, as shown in Figure 7.3a; and their characteristics may be described by their static sensitivity, natural frequency, and damping ratio. An excited second-order system oscillates or resonates as its natural frequency. Inputs that

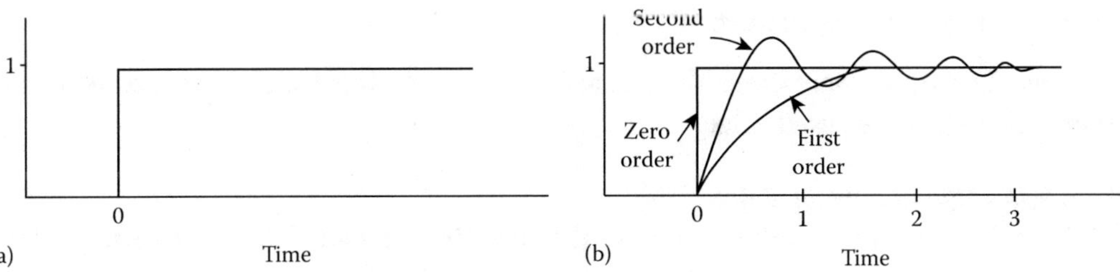


FIGURE 7.2 (a) Step input and (b) step response for zeroth-, first-, and second-order systems.

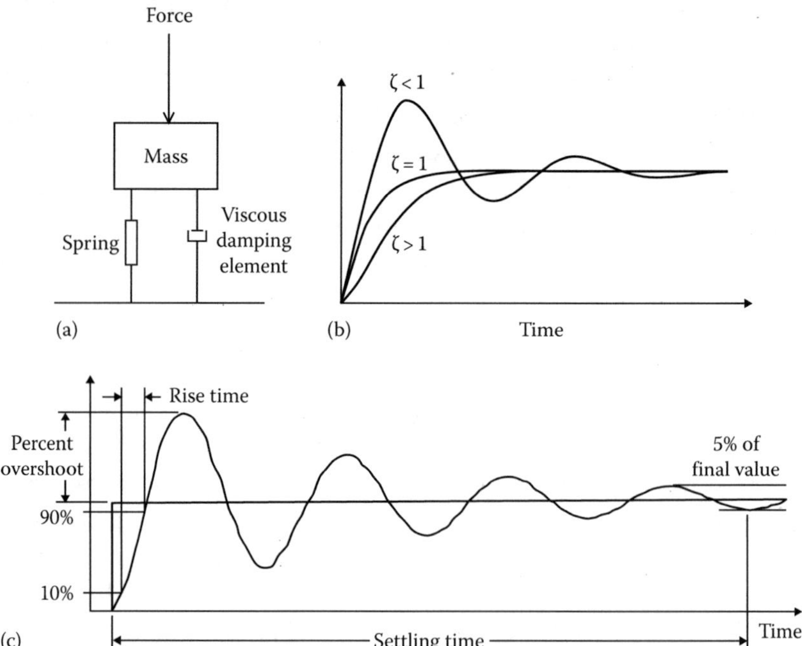


FIGURE 7.3 (a) Generalized second-order system model consisting of mass, spring, and viscous damping element. (b) Second-order system response based on damping ratio ζ. (c) Step response parameters describing a second-order signal.

are near or at the system's natural frequency should be avoided because they cause the instrument to oscillate uncontrollably. The damping ratio is a dimensionless quantity that is greater than 1 when the system is overdampened and less than 1 when the system is underdampened, in which case the output for a step input will overshoot the final value (see Figure 7.3b).

### *7.2.2.2 System Step Response*

Often the actual mass, spring, and viscous damping elements are not known. Second-order system characteristics are sometimes obtained empirically by their step response using the following parameters (see Figure 7.3c).

*Percent overshoot*: The amount by which the peak output exceeds the final value is expressed as a percentage of the final value. Percent overshoot is a convenient measure of relative stability.

*Settling time*: Settling time is the total time needed for the response to settle within a certain percentage (e.g., 5%) of the final value.

*Rise time*: Rise time is sometimes defined as the transition time for the response to change from 10% to 90% of its final value.

### *7.2.2.3 Instrument Specifications*

Additional instrument specifications describe operating characteristics that are important to consider when selecting an instrument for a particular application. Among others, the following specifications should be considered.

*Input and output range*: These ranges are usually specified as the minimum and maximum voltages for the input and output. This is particularly important to consider when interconnecting instruments or instrument components such as sensors and amplifiers.

*Overload range*: The maximum input that will not damage the instrument is the overload range. The overload range should be sufficiently large that any anticipated input does not exceed that level.

*Input and output impedance*: Impedance is the resistance to the flow of current. The most fundamental equation in electric circuits is Ohm's law, which states that voltage across an electrical conductor is directly proportional to the current passing through it such that $V = IR$, where $V$ is voltage (volts), $I$ is electric current (amperes), and the constant $R$ is electrical resistance in ohms (Ω). In direct current (dc) circuits, impedance corresponds to resistance. The impedance for direct current is not necessarily the same as for alternating current (ac). Impedance is a function of resistance, inductance, and capacitance for ac circuits and is expressed as ohms.

Resistors offer the same opposition to current flow for dc and ac circuits, and therefore have the same impedance. Inductors and capacitors build up voltage that opposes the flow of ac. This opposition to alternating current, called reactance, must be combined with resistance, resulting in a complex quantity that has properties of both magnitude and phase. Consequently, impedance is a function of ac frequency. The reactance produced by inductance is proportional to frequency, whereas the reactance produced by capacitance is inversely proportional to frequency. The dc impedance for a capacitor is infinite, and its ac impedance is $1/2\pi fC$, where f is frequency (Hz). The dc impedance for an inductor is zero, and its ac impedance is $2\pi fL$. A circuit's total impedance is determined by the complex combination of the resistance and reactance for all of the circuit components.

In order to transfer maximum electric power from one device to another, the two impedances must be matched. To prevent an output component from being loaded down by having the input component draw current and changing its output, the impedance of the output component should be much less (usually many orders of magnitude less) than the impedance of the input component.

*Frequency response*: The frequency response describes the characteristics of an instrument as a function of the input frequency. Frequency response is often presented as a plot of the frequencies that shows that input signals are attenuated, or amplified, or where resonances occur. Frequency is usually plotted on the $x$-axis using a logarithmic scale, and the response ratio of the output/input is plotted on a decibel scale such that

$$\text{Response}(\text{dB}) = 20\log_{10}\left(\frac{\text{output}}{\text{input}}\right)$$

It is often desirable for the input signal to be within a frequency where the response is flat so that the output signal is unaffected by the input signal frequency (does not change with frequency).

*Nonlinearity*: Nonlinearity is a measure of how much an instrument output varies from its input. It is sometimes expressed as the maximum deviation from a linear output in terms of a percentage of the reading or a percentage of full scale, whichever is greater.

*Repeatability*: Repeatability is a measure of the variation of the output for the same input over time.

*Signal-to-Noise Ratio*: The ratio between the peak or rms amplitude of the signal and the noise is usually expressed as a fraction (e.g., 1000/1) or in decibels as SNR = 20 $\log_{10}$ (signal/noise).

*Stability*: Stability is a measure of the output drift as a function of time, temperature, humidity, shock, or vibration.

## 7.3 SENSOR PRINCIPLES

Sensors can be electromechanical, electrochemical, or electrothermal. Electromechanical sensors measure phenomena associated with kinetics and kinematics, such as motion and force. Electrochemical sensors include electrodes for measuring gas concentrations or bioelectrical phenomena. Electrothermal sensors are used for measuring temperature.

### 7.3.1 Electromechanical Sensors

Electromechanical sensors convert mechanical inputs into electrical outputs. Mechanical parameters usually include displacement, velocity, acceleration, and force. Electromechanical sensors are based on resistive, piezoelectric, capacitive, and inductive transduction.

#### *7.3.1.1 Sensors Based on Electrical Resistance*

The resistance of a cylindrical electrical conductor is $R = \rho L/A$, where $L$ is its length, $A$ is its cross-sectional area, and $\rho$ is the resistivity of the particular material (Ω-m). The higher the resistivity of a material, the more the material acts as an insulator. Resistors are often made from carbon composites because of their very high resistivity, or from high-resistivity metal wire, which is usually wound in coils to increase the length. The resistivity of metals increases with increasing temperatures because of the increased number of collisions that electrons make, thus increasing their electrical resistance.

Mechanical actions usually affect sensors based on electrical resistance either by changing $L$ or $A$ (for metals) or by changing $\rho$ (for semiconductors), consequently changing the sensor's resistance and the voltage across it. The simplest displacement sensor is a variable resistor, or potentiometer. The electrical resistance is proportional to the position of the slider. Metal strain gauges are made from lengths of very fine wire (<25 μm diameter). When the wire is stretched, its resistance changes, mainly because of changes in its cross-sectional area $A$ and length $L$. Resistive sensors are sensitive to temperature changes; this must be taken into account when measurements are being made.

Silicon strain gauges are made of diffused resistors integrated into a silicon substrate. Silicon and metal strain gauges are used in a similar manner, and both provide a very linear response within the elastic limits of the material to which they are fastened. Silicon strain gauges exhibit even higher temperature effects than metal because deformation affects $\rho$. These temperature effects can often be controlled using special compensation circuits.

The gauge factor

$$\frac{\Delta R / R_0}{\Delta L / L_0}$$

specifies a strain gauge's sensitivity to mechanical deformation. In this proportion, $\Delta R/R_0$ is the fractional change in resistance due to strain and $\Delta L/L_0$ is the fractional change in strain gauge length. The gauge factor for metal strain gauges is typically between 2 and 5, whereas the gauge factor for silicon can be as high as 170. This mechanical deflection, usually a fraction of 1%, causes a small but measurable change in resistance.

Strain gauges are usually used in pairs. Two gauges can be placed on the top of a cantilever beam and two on the bottom. The folded cantilever beam permits four gauges to be placed on the top surface, which eases manufacture and increases performance [2]. When strain gauges are arranged in a Wheatstone bridge circuit, their sensitivity can be increased. The circuit in Figure 7.4 uses two strain gauges in adjacent legs of the bridge. The output voltage of the bridge, $V_o$, is given by the following equation:

$$V_o = V_e \left( \frac{R_2}{R_2 + R_4} - \frac{R_1}{R_1 + R_3} \right)$$

When the beam bends, gauge 1 on the top of the beam elongates and increases $\Delta R$, while gauge 2 on the bottom is compressed and decreases $\Delta R$. If two fixed resistors in the opposite two legs of the bridge have resistance identical to that of the undeformed strain gauges, $R_0$, they do not deform. This arrangement is known as a half-bridge circuit. If $\Delta R$ is much less than $R_0$, then

$$V_o \approx -\frac{V_e}{2R_0} \Delta R$$

and the output is proportional to the change in resistance. A full-bridge configuration in which all four legs contain strain gauges ($R_1$ and $R_3$ on top of the beam and $R_2$ and $R_4$ on the bottom) yields even higher sensitivity with an output given by $V_o \approx \Delta R/R_0$.

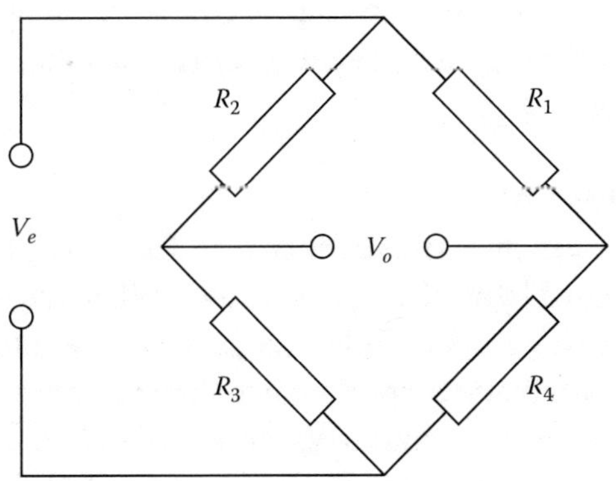


FIGURE 7.4 Wheatstone bridge circuit.

The bridge circuit configuration increases sensitivity by $2(1 + \nu)$ over that of a single strain gauge configuration, where $\nu$ is Poisson's ratio. Bridge circuits also provide temperature compensation, which prevents output shifts due to temperature effects on the transducers. When maximum output is desired for any strain gauge, a low-modulus material (such as aluminum) is often used to increase strain per unit force. When four strain gauges are used in a Wheatstone bridge, they yield an electrical output that is insensitive to bending stresses due to the force being applied off center or at an angle, and they can be temperature-compensated [2]. Strain gauges are often used in load cells and as force sensors in scales and pressure transducers.

#### *7.3.1.2 Piezoelectric and Piezoresistive Sensors*

Piezoelectric materials, such as quartz and barium titanate, produce a change in charge distribution when subjected to a mechanical stress. Piezoresistive sensors produce changes in resistance when stressed. Piezoelectric and piezoresistive load cells require minute deformations of their atomic structure within a block of crystalline material. Typical uses for piezoelectric and piezoresistive sensors include accelerometers and load cells.

Because piezoelectric sensors operate on changes in charge distribution, their most notable drawback is their inability to respond to static loads. Quartz is a naturally occurring piezoelectric material, and deformation of its crystalline structure changes its electrical characteristics such that the electric charge across its surface is altered. The charges collect on metal electrodes deposited onto the surface of piezoelectric material. Special amplifiers called charge amplifiers are used that output a voltage proportional to a charge at its input.

Piezoelectric force sensors operate on the principle that the charge across a piezoelectric material is proportional to its deformation, which is proportional to the applied force. Piezoelectric accelerometers operate the same way, except that a small mass $M$ is mounted on top of the piezoelectric material. When the accelerometer moves with an acceleration $a$, the mass exerts against the piezoelectric material a force $F = Ma$ that is proportional to its acceleration. Piezoelectric device sensitivity is expressed in terms of coulomb charge per acceleration or force unit. Small-mass devices have low charge sensitivity.

Piezoresistive sensors are not limited to measuring changing signals and are suitable for static loads. Piezoresistive sensors are usually configured in a Wheatstone bridge like strain gauges. Consequently they require an external excitation voltage and a sensitive instrumentation amplifier.

#### *7.3.1.3 Electrocapacitive Sensors*

Capacitors have the capacity to store an electric charge. The simplest capacitor consists of two parallel metal plates separated by an electrical insulator. When the voltage across a capacitor changes, opposite polarity charges accumulate on the plates, resulting in an electric current flowing through the capacitor that is proportional to the voltage change. The current through a capacitor is $I = C\, dV/dt$, where $C$ is capacitance measured in farads (F). The capacitance of a plate capacitor is described by the equation $C = \varepsilon A/x$, where $x$ is the plate separation distance, $\varepsilon$ is the dielectric constant of the insulator, and $A$ is the plate area. These plates are constructed such that when an applied force changes $x$, the capacitance $C$ changes. Applications for electrocapacitive sensors include their use in transducers for measuring minute displacement changes and for microphones.

#### 7.3.1.4 Inductive Transducers

An inductor consists of multiple turns of insulated wire wound as a coil. When current flows through the coil, a magnetic field is produced that is proportional to the current passing through the coil. If the current changes, the magnetic field changes, and a voltage is induced across the ends of the coil. The voltage across an inductor is $V = dI/dt$, where $L$ is inductance measured in henrys (H). The inductance of a coiled wire is proportional to the square of the number of turns of wire, $n$, such that $L = n^2G\mu$, where $G$ is a geometric form factor and $\mu$ is the effective permeability of the material the wire is looped around.

Inductors are often used as displacement sensors where changes in displacement result in changes in inductance. The most common sensor using this principle is the linear voltage displacement transducer (LVDT). LVDTs have better resolution than potentiometers and provide very high motion sensitivity at the expense of more complex signal-processing circuitry. Usually an ac signal is fed through a primary coil. The coil is wrapped around a movable core along with a secondary coil, so that a voltage is induced across its output. The amplitude of the ac voltage depends on the position of the core relative to the coils. The core can be moved in and out along the axis of the two coils. The ac-induced voltage is converted to direct current by using a rectifying circuit. LVDTs make good linear displacement transducers because of their high resolution, almost zero friction, and no wear.

### 7.3.2 Electrochemical Sensors

Electrodes convert ionic bioelectrical potentials into an electric current or voltage that can be amplified or measured. The most common types of electrochemical sensors used in ergonomics include electrodes for measuring oxygen concentrations in expired air and EMG and electrocardiogram electrodes.

#### 7.3.2.1 Biopotential Electrodes

Needle or wire electrodes are invasive and are used for measuring ionic activity in a localized region is tissue such as muscle or nerve. Surface electrodes are noninvasive and measure potentials externally at the surface of the skin, such as for EMG, ECG, and EEG measurements. Figures 7.5 and 7.6 show examples of both types of electrodes.

Biopotentials are usually measured as the potential difference between two electrodes. All electrodes involve an interface between a metal and an electrolyte. Ions of opposite charge migrate from the metal to the electrolytic solution and from the solution into the electrode, so that an electrical double layer is formed and a potential difference is developed across the interface. This electrode potential occurs independently of the specific bioelectrical activity present. When the potential difference between two electrodes is measured, the electrode potential cancels out. Small, random fluctuations in electrode potential sometimes do not cancel and cause drift in the output voltage.

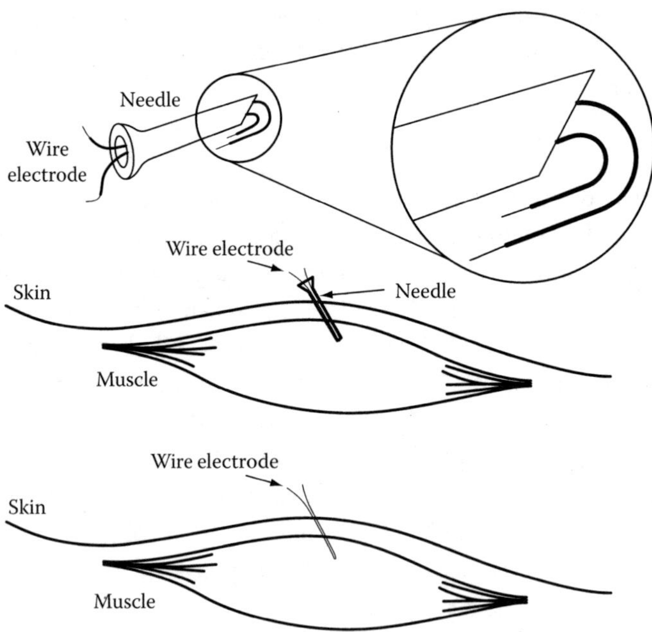


FIGURE 7.5 Needle/wire electrode shown placed in a muscle.

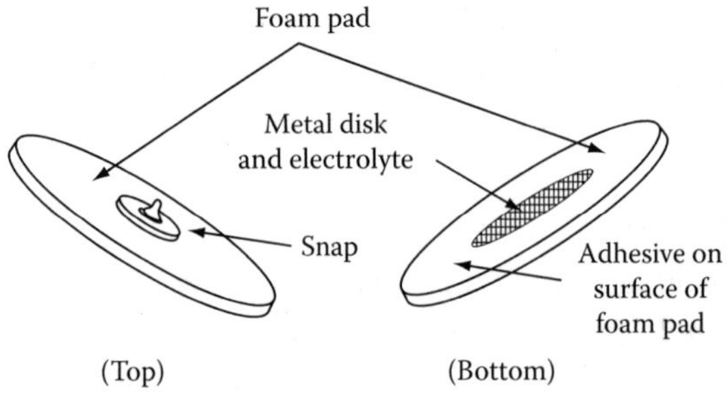


FIGURE 7.6 Typical foam pad surface electrodes.

Surface electrodes usually consist of a metal disk mounted in a plastic receptacle. Silver–silver chloride (Ag–AgCl) electrodes, consisting of a silver chloride–coated silver disk, are common, although other metals are sometimes used. The space between the disk and the skin is filled with an electrolyte gel or paste such as a mixture of sodium chloride and glycerol. Disposable electrodes are available that already contain the gel.

#### 7.3.2.2 Ion-Selective Sensors

The Nernst equation describes the relationship between ion concentrations and the voltage across a membrane $E_k$ such that

$$E_k = \frac{RT}{nF} \ln\left[\frac{\text{ion concentration inside}}{\text{ion concentration outside}}\right]$$

where

*R* is the universal gas constant
*T* is the temperature (K)
*n* is the ion valence
*F* is Faraday's constant

When an ion-selective membrane separates the solution from another (i.e., the ion concentrations are different), a voltage will appear across the membrane. This principle can be used for measuring the concentrations of different ions, depending on the composition of the membrane. pH electrodes use glass membranes to measure $H^+$ concentration. The principle is also used for $CO_2$ and $O_2$ concentration measurements.

### 7.3.3 Electrothermal Sensors

Although mercury column thermometers are widely used for measuring temperature and are reliable and inexpensive, and they have a high thermal capacity, so they may alter the temperature being measured. They also respond slowly, so rapid changes in temperature cannot be measured. Perhaps the greatest disadvantage is that they do not produce an electric signal that can be used for continuous measurements and recordings. A variety of electronic sensors are available for measuring temperatures that overcome these limitations.

Thermistors are ceramic materials that, unlike metals, decrease in resistance with increasing temperatures. Transistor-based temperature sensors use the temperature dependence of voltage across a diode junction. Thermocouples operate on the principle that when two dissimilar metals are brought together, a voltage is observed across the junction. The magnitude of that voltage is dependent on temperature.

### 7.3.4 Photosensors

Photosensors can detect radiant energy or light. Photoconductive cells, or photoresistors, contain elements that change conductivity as a function of incident electromagnetic radiation. Materials such as cadmium sulfide are photoconductive. Photosensors are often used for detecting when an incident beam of light is broken or for quantifying the amount of incident light such as used in a light meter.

## 7.4 AMPLIFICATION

Electric signals produced by most sensors are usually too small to measure directly (a body surface electrocardiogram signal measured at the electrode is usually less than 1 mV). In order for signals to be observed, processed, or recorded, they need to be amplified to a magnitude of 1 V or more. The characteristics of an amplifier describe how an input signal appears on its output.

### 7.4.1 Gain

The amplifier output voltage $V_o$ is the product of the input voltage $V_{\text{in}}$ multiplied by the amplitude factor (gain) *A* such that $V_o = AV_{\text{in}}$. An amplifier having a gain of 100,000 will amplify a 50 μV EMG signal to 0.5 V.

### 7.4.2 Impedance

Instrumentation amplifiers should have high input impedance (at least 10 MΩ) so they provide minimal loading of the signal being measured.

### 7.4.3 Frequency Response

Frequency response specifies the upper and lower frequency of the range in which a signal can be amplified without a loss in gain.

### 7.4.4 AC and DC Coupling

When a sensor is connected to an amplifier directly, it is said to be dc-coupled. That is, the complete signal is input to the amplifier. If a dc offset voltage is present, that offset will be amplified along with the ac signal. When a sensor is coupled through a series capacitor, it is said to be ac-coupled, and only the alternating signal is input to the amplifier because capacitors have infinite impedance to dc signals. Examples for both ac- and dc-coupled amplifier outputs are shown in Figure 7.7.

### 7.4.5 Saturation

The maximum output of any instrument is limited to its power supply voltage. When this maximum output is reached, no further increase is possible, regardless of what input is applied. An instrument in this state is said to be saturated. Just as a saturated sponge cannot hold any more water, a saturated instrument is unaffected by any further increase in the input signal. Figure 7.6 illustrates a saturated and an unsaturated amplifier.

### 7.4.6 Zero Suppression

Direct current amplifiers may have an offset adjustment to compensate for drift in the output.

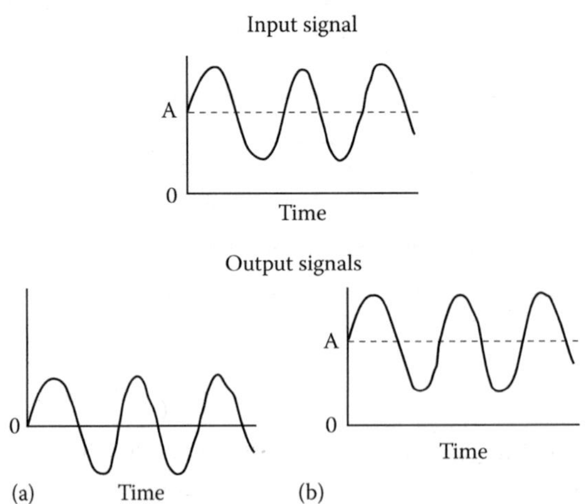


FIGURE 7.7 Signal output for (a) ac-coupled amplifier and (b) dc-coupled amplifier.

### 7.4.7 Differential Amplifiers

Differential amplifiers amplify the difference between the signals at the two inputs. If the same signal is sent to both inputs of a differential amplifier, the output is expected to be zero. This requires identical impedance characteristics at each input. In the real world this is usually not the case. The common mode rejection ratio (CMRR) is a measure of an amplifier's ability to reject common mode signals (i.e., signals that are equally applied to both inputs of a differential amplifier). The CMRR for a differential amplifier is determined by measuring the output voltage for the same voltage at both inputs of a difference amplifier. The common mode gain (CMG) is the ratio between the output voltage $V_o$ and the input voltage $V_i$ such that CMG = $V_o/V_i$. The CMRR is the ratio between the amplifier gain (differential gain) $A$ and the CMG such that CMRR = $A$/CMG. Since many unwanted or modifying inputs are common mode (the same signal appears on both amplifier inputs), a large CMRR is desirable. The CMRR should be greater than 10,000 for instrumentation amplifiers. Differential amplifiers are often used for biopotential recordings such as EMGs to eliminate electrical noise common to both inputs, such as line current interference.

The intrinsic mismatch between the inputs gives rise to a voltage between the inputs called the offset voltage. Although offset voltages are typically only a few millivolts, they can give rise to significant errors if the signal one is attempting to measure is also small. Many amplifiers have adjustments for eliminating offset voltages.

### 7.4.8 Temperature

The environment in which instruments are used can affect their performance. Most electric components are affected by temperature. The resistance of a conductor, for instance, increases with temperature. The offset voltage of an amplifier is usually temperature-dependent, giving rise to measurement errors if large temperature changes occur.

### 7.4.9 Noise

Any unwanted signal at an amplifier's input is amplified along with the desired signal. Noise can come from a variety of sources, including capacitive coupling between electrode leads and ac wiring, induced currents from ac magnetic fields, skin-electrode movement artifact, and popcorn noise from the amplifier itself. Fortunately, the effect of noise can often be minimized by taking simple corrective precautions. For example, the use of shielded cabling can eliminate capacitive coupling.

## 7.5 SIGNAL CONDITIONING

### 7.5.1 Signals

Bioelectric signals are sometimes simply recorded as time-varying signals. Often complex signals have to be reduced to a form related to the quantity being measured using circuitry for rectification, integration, differentiation, and root mean square (rms).

#### *7.5.1.1 Time-Variant Signals*

Signals measured from electronic sensors are usually represented as a voltage changing with time. These time signals are usually displayed on an oscilloscope or printed on a strip

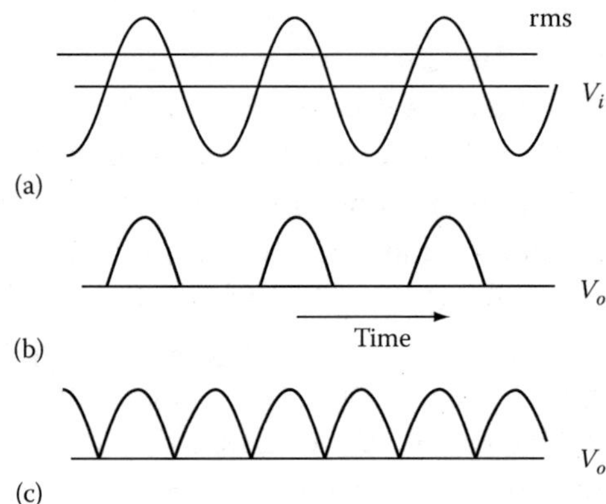


FIGURE 7.8 (a) Simple sinusoidal signal; (b) half-wave rectified sine wave; (c) full-wave rectified sine wave.

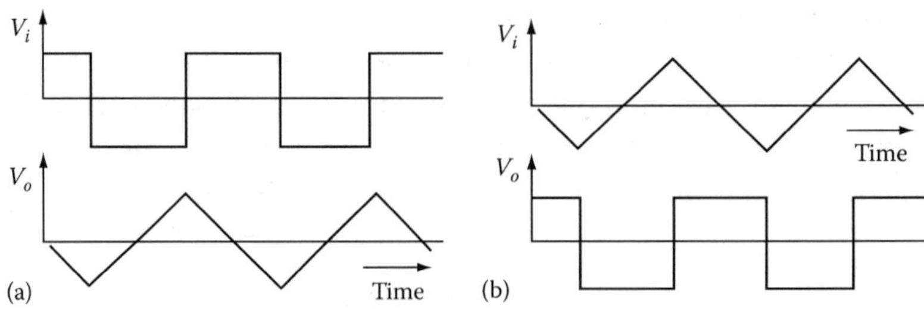


FIGURE 7.9 (a) Differentiation; (b) integration.

chart recorder, or they may be converted into digital form and displayed on a computer screen. A simple sinusoidal signal, as shown in Figure 7.8, is simply described by its amplitude and period. When the signal is more complex or random, time plots are inadequate for extracting useful information. Since physiological signals are often complex, signal-processing methods are needed for displaying and interpreting the information the signals contain. Common signal-conditioning functions include rectification, logarithm, integration, differentiation, and rms (see Figures 7.8 and 7.9).

#### *7.5.1.2 Rectification*

A half-wave rectifier has an output signal equivalent to the input signal when the input is a positive value and an output of zero when the input is negative. A full-wave rectifier is a circuit that produces the absolute value $|x|$ of the input signal. Figure 7.8b and c shows a half-wave and full-wave rectified output for a sine wave.

#### *7.5.1.3 Logarithmic Functions*

Logarithm circuits depend on the logarithmic relationship between a transistor base-emitter voltage and its collector current. Antilog circuits can be obtained by interchanging the circuitry. Log and antilog circuits may be used for multiplication, division, raising a voltage to a power, or taking a root.

#### 7.5.1.4 Differentiation

The output of a differentiator circuit is proportional to the derivative of the input signal (see Figure 7.9a). Differentiators may be used for transforming displacement data into velocity or acceleration. Differentiators tend to accentuate the high-frequency noise, which can often mask the desired output.

#### 7.5.1.5 Integration

An integrator outputs the integral of the input signal (see Figure 7.9b). Since differentiation tends to enhance noise in a signal, integration is often a more desirable approach. For example, velocity can be obtained by differentiating a displacement signal or integrating an acceleration signal. A less noisy signal would be obtained by integrating the output of an accelerometer. Often integration occurs over a specified time period. If, on the other hand, the signal being integrated has a low-frequency drift, differentiation may be more desirable because over integration time these drifts accumulate into large errors.

#### 7.5.1.6 Root Mean Square

The rms of a signal is the square root of the average of the square of the signal, such that

$$\text{Root mean square} = \left[ \frac{1}{T} \int_0^T f^2(t)\, dt \right]^{1/2}$$

For a simple sine wave,

$$f(x) = A \sin \frac{2\pi}{T} t, \quad \text{rms} = \left[ \frac{1}{T} \int_0^T A^2 \sin^2 \frac{2\pi t}{T}\, dt \right]^{1/2} = \frac{A}{\sqrt{2}} = .707\, A$$

Some rms detectors simply scale the peak amplitude of a sinusoidal signal by 70.7% for determining rms (see Figure 7.8a). Such circuitry would give false readings for any signal except a sine wave. A true rms detector contains circuitry for determining the rms of an arbitrary function $f(x)$ using circuitry for instantaneously squaring the signal, integrating over a defined time constant, and taking the square root. Some averaging rms detectors simply rectify and integrate the signal without squaring as an approximation. The accuracy of this approximation depends on the magnitude of the signal.

### 7.5.2 Filters

Electronic filters separate unwanted signals from sensor signals according to frequency. Three general types of filters are (1) low-pass, (2) high-pass, and (3) bandpass filters (see Figure 7.10). Low-pass filters allow low frequencies to pass unattenuated, while high-frequency signals are attenuated and vice versa for high-pass filters. Bandpass filters pass signals unattenuated within a specific frequency range.

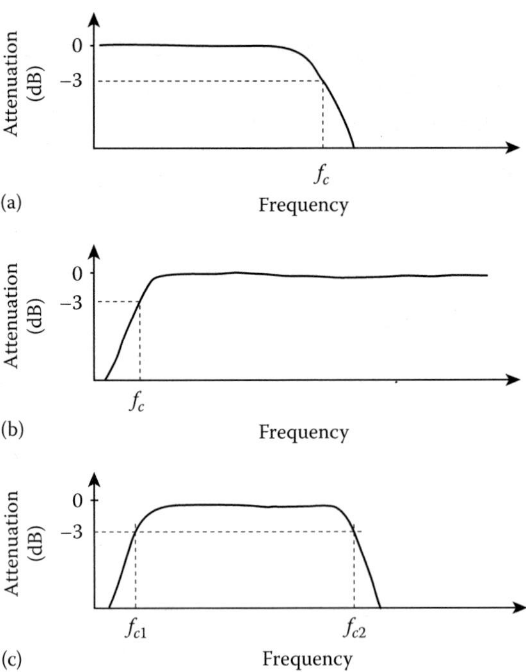


FIGURE 7.10 Three general types of filters: (a) Low-pass; (b) high-pass; (c) bandpass.

In practice, filters are not ideal. Signals at the unwanted frequencies are not completely eliminated; they are only reduced in amplitude. Often the signal-to-noise ratio is great enough that the noise is practically eliminated. Significant attenuation of unwanted signals is greatest when there is a large frequency difference between the signal and the noise. The corner frequency ($f_c$) is usually defined as the frequency at which the signal is attenuated by 71% or 3 dB [20 log(0.707)]. Frequencies inside the corner frequency are much less attenuated than 3 dB, where attenuation is insignificant. Attenuation beyond the corner frequency depends on the filter roll-off, which is specified as a slope in terms of either decibels per octave or decibels per decade. An octave is a frequency multiplied by 2, and a decade is a multiple of 10. Therefore, if a filter attenuates −60 dB/octave, then the signal is attenuated −60 dB for each doubling of frequency. The more negative the slope, the sharper the filter and the more noise is reduced beyond the corner frequency.

## 7.6 SIGNAL PROCESSING

### 7.6.1 Digital Circuits

Computers represent signals as binary digits containing 0's and 1's. The conversion of analog signals into digital representations accounts for the powerful manipulation and storage capabilities of the computer. Many instruments contain interface circuitry, so they can communicate data with a computer and permit the computer to remotely operate the instrument.

#### *7.6.1.1 Digital Devices*

Digital electronic devices use two different voltage levels to represent the two logic states of 0 and 1. Computers generally use 0 and +5 V for representing logic levels of 0 and 1, respectively.

This is the convention used for transistor-transistor logic (TTL), named for the type of circuitry. Some digital devices, such as modems and printers, use different voltage level conventions such as −12 V for logical 1 and +12 V for logical 0. Low-power devices now use 0 and +3 V conventions for 0 and 1. Although less common, some digital devices use current levels instead of voltage levels.

#### *7.6.1.2 Digital-to-Analog Conversion*

A digital-to-analog converter (DAC) in its simplest form is a set of resistors each having a different resistance, providing a binary-weighted scaling of input digital voltages. The output voltage is proportional to the current, which is determined by which of the resistors are connected to the reference voltage.

#### *7.6.1.3 Analog-to-Digital Conversion*

There are numerous types of analog-to-digital converter (ADC) circuits available, differing in conversion speed and complexity. Voltage-to-frequency converters provide an output train of pulses at a frequency proportional to input current or voltage. A counter accumulates the pulse count for a fixed period of time. High resolution and relatively good accuracy over a wide range of input voltage are possible, but conversions are slow. A simple counter, or servo converter, contains a DAC and a digital counter. When the analog output of the DAC is greater than or equal to the input analog voltage, the counter stops counting and the digitized value is the counter value. Because a simple up-counter converter takes a long time for each conversion, an up-down counter can be used to track the input level. This counting process is faster but still relatively slow, especially for rapidly changing signals. Successive approximation converters are the most popular type of ADC converter because of their conversion speed and simplicity. Successive approximation converters operate at a fixed conversion time independent of the amplitude of the input analog signal. The converter operates by approximating the analog input signal with a binary code and successively revises this code for each bit in the code until the best approximation is achieved.

ADC range and resolution are limited by the number of bits in the converter. Conversions involving more bits require more time and therefore need faster converters if they are to maintain a nominal conversion rate. Conventional 12-bit ADC converters are capable of operating at conversion speeds of better than 1000 samples per second. An $N$-bit ADC converter has a range of $2^N$ levels. Therefore the range for a 12-bit ADC is $2^{12} = 4096$ discrete levels. A converter's range is sometimes expressed as its dynamic range computed as 20 log $2^N$ (dB), or 20 log 4096 = 72 dB for a 12-bit converter. An ADC voltage resolution depends on its input voltage limits. For example, a 12-bit converter with a −5 to +5 V input range provides a resolution of (10 V)/(4096 bits) = 2.44 mV/bit.

### 7.6.2 Sampling Theorem

Analog signals must be sampled at a sufficiently high sampling rate in order to completely represent the data while not accumulating so many data that it is difficult to store and handle them. The Nyquist sampling theorem states that if a continuous band-limited signal contains no frequency component greater than $f_c$, then the original signal can be completely represented if it is sampled at a rate of at least $2f_c$ samples per second. The frequency $2f_c$. is known as the Nyquist frequency. Although completely represented, signals sampled just as the Nyquist frequency are usually not visually satisfactory. Oversampling, or sampling a signal at rates greater than the Nyquist frequency, is often used for sampling data that are intended for visual display.

If an analog signal is undersampled when digitized by using a sampling frequency less than the Nyquist frequency, aliasing can occur. Aliasing confounds the sampled signal with a signal different from the actual signal and results in unreliable data. This phenomenon is sometimes observed in the cinema when the wheels of a moving vehicle appear to be rotating more slowly than they should, or even rotating backward, because of the limited number of frames per second at which the event is captured relative to the rotating speed of the wheels. Aliasing is easily prevented by sampling using a sampling rate greater than the Nyquist frequency. Sometimes the signal of interest is much less than the total frequency content of the entire signal. Antialiasing filters are nothing more than low-pass filters that attenuate signals greater than the Nyquist frequency in order to prevent aliasing. These filters also help eliminate interference noise at frequencies above the Nyquist frequency that can corrupt digitized data and are recommended whenever analog data are digitized.

### 7.6.3 Digital Filters

Digital filters can implement all analog filter functions (i.e., low-pass, high-pass, bandpass, integration, differentiation) via software. Two advantages of digital filters are that the filters can be adjusted without the need for special hardware and filtering can be performed after the data have been sampled. This is very useful for extracting noise that was not evident while the data were being collected. Digital filters can be subdivided into two categories: (1) finite impulse response (FIR) filters and (2) infinite impulse response (IIR) filters. Digital filters are implemented using difference equations. The speed of the filter is dependent on the complexity and the number of coefficients in the difference equation. The design of digital filters often involves a trade-off between desired performance and speed. In other words, it is possible to design a digital filter with optimal performance, but often the resulting difference equation is computationally inefficient.

FIR digital filters have $M + 1$ coefficients and have the general equation $y_i = b_0x_i + b_1x_{i-1} + b_2x_{i-2} + \cdots + b_Mx_{i-M}$, where $x_i$ is the digitized signal input and $y_i$ is filtered output. For example, if a two-coefficient FIR filter has coefficients $b_0 = 1/2$ and $b_1 = -1/2$, then the filter response is $y_i = 1/2x_i + 1/2x_{i-1}$, which is simply the average of every input data point with the point preceding it. FIR filters are inherently stable, but sharp roll-offs are difficult to achieve with low-order (containing few coefficients) filters.

An IIR filter uses both input and output values and thus has feedback. The general IIR filter has $N + M + 1$ coefficients and has the equation $y_i = a_1y_{i-1} + a_2y_{i-2} + \cdots + a_Ny_{i-N} + b_0x_i + b_1x_{i-1} + b_2x_{i-2} + \cdots + b_Mx_{i-M}$. A simple IIR filter has the equation $y_i = (1 - a)y_{i-1} + ax_{i-1}$ where $a = 1 - e - T/t$, $T$ = sampling period, $t = 1/2\pi f_c$, and $f_c$ is the filter corner frequency. This digital filter behaves similarly to a single-pole Butterworth filter.

Integer filters are a special class of filters designed for maximum computational efficiency at the expense of very specific performance limitations. To achieve computational speed, integer filters use only integer coefficients. This allows the computer to perform multiplication via bit shift operations, which are much faster than standard multiply operations. The use of only integer coefficients results in a limited choice in the placement of poles during the filter design.

Standard design methodologies have been developed for many digital filters; they are discussed and applied by Tompkins [3]. It is relatively easy to implement digital filters without an in-depth knowledge of signal processing theory. However, one must be careful to consider the effects of phase, aliasing, and sampling rate when implementing them.

### 7.6.4 Spectrum Analysis

Spectrum analysis views a time-variant signal as a function of frequency. Real-time spectrum analyzers resolve the frequency content of complex signals using a network of narrowband filters tuned to different frequencies. Simple spectrum analysis may be performed using filter networks containing filters that are successively tuned to multiples of the previous filter. This is the kind of spectrum obtained using octave-band and one-third octave-band analyzers.

Digital computers and analog-to-digital converters make it possible for signals to be sampled, digitized, and mathematically transformed as a function of frequency using the Fourier transform. The fast Fourier transform (FFT) is an efficient algorithm commonly used for calculating Fourier transforms. The major limitation of FFT analyzers is that signals may be sampled for only a limited time window. The FFT algorithm assumes that the signal being transformed is periodic and therefore repeats indefinitely, which may not be appropriate in some cases.

Many FFT analyzers contain a fixed number of points resolution for the spectrum; 400 points is typical. The frequency resolution of an FFT analyzer is inversely proportional to the time window for a fixed number of points spectrum. Sample rates are therefore adjusted in proportion to the time window. Consequently a spectrum for a longer time window will contain a finer frequency resolution than one for a shorter time window, but it will have a smaller bandwidth. Consider a spectrum analyzer that has a 1 kHz sample rate for a 1 s time window and a frequency resolution limited to 1 Hz for a 500 Hz spectrum. If the time window were cut in half to 0.5 s, the sample rate would be doubled to 2 kHz, so the frequency resolution would be limited to 2 Hz for a 1000 Hz bandwidth. An increased spectrum resolution would also be obtainable if the number of points were increased without changing the sample rate.

Digital FFT spectrum analyzers often allow averaging repeated samples of the signal. Although FFT analyzers are extremely useful for ergonomics, extreme care must be taken to ensure that the signal sampled is representative of the actual signal under study and that the signal's spectrum does not vary over time. When a signal's statistical properties remain the same over time, the signal is said to be stationary.

### 7.6.5 Interfaces

Many modern electronic instruments are now being equipped with special circuitry for interfacing them with a computer. Such an interface allows a computer to read data directly from the instrument as well as externally control most of the instrument functions. Interfaces are circuits that convert data from an external instrument or circuit into a form suitable for a computer or convert digital data from the computer to a form accessible to an external device. A brief description of commonly available interfaces is provided.

#### *7.6.5.1 Serial Interfaces*

Serial interfaces transmit and receive digital data one bit at a time. Because data bits are sent one at a time, only two data lines are necessary; one line is used for transmission and the other is used for reception. This makes serial transmission cable small and inexpensive.

All serial interfaces operate in similar manner. Serial interface circuits perform two basic operations. First, they convert an 8-bit parallel data word from the computer into a serial data word to be sent to the serial device. Second, they convert the data from a serial device into an 8-bit parallel data word that is transferred to the computer. Because of the extensive use of asynchronous serial communication, a common device performing the parallel-to-serial and serial-to-parallel conversions was developed that is called a universal asynchronous receiver transmitter (UART). The rate at which serial transmission occurs is called the baud rate. This is the number of bits of information that are transmitted during 1 s. If $x$ bits are transmitted during the time interval $T_b$, then the baud rate is

$$\text{Baur rate (bit/s)} = x/T_b$$

For proper interpretation of the received data, the receiver must be operating at the same baud rate as the transmitter.

There are many serial interface standards describing the protocol used for communication between devices. Three common serial interface standards are presented.

A. RS-232 (TIA-232, EAI-232, ITU-T V.24) and RS-422 (TIA-422, EIA-422, ITU-T V.11)

A standard originally designed to interconnect data terminal equipment (DTE) such as teletypewriters and electronic terminals and data circuit terminating equipment (DCE) such as modem RS-232 is commonly used in computer serial ports.

Baud rates between 75 and 115,200 bit/s are possible although rates above 20,000 bit/s were beyond the scope of the standard. In the particular case of the IBM PC, baud rates were programmable allowing connection to devices requiring different baud rates. Since most devices do not have automatic baud rate detection, the baud rate (and all other parameters) must be manually set at both ends of the RS-232 connection. Connection distance with standard quality cable can reach to 10 m. By using high-quality low-capacitance cables, communication can be maintained over distances up to about 300 m.

Common problems were inconsistent pin assignment on connectors, and incorrect or missing control signals. A common deviation was the signal voltage from the standard +12 and −12 V to a reduced voltage, as low as +3 and −3 V. As personal computers became popular in the mid-1980s, signal voltages of +5 and −5 V were used and labeled as "RS-232 compatible."

The name of the standard has changed over time as the sponsoring organization changed. The Electronics Industries Association maintained the standard beginning in 1969. Since 1988, the Telecommunications Industry Association (TIA) has maintained the standard leading to the current name TIA-232. The International Telecommunication Union Standardization sector (ITU-T) coordinates telecommunication standards worldwide, provides the V.24 designation for the RS-232 standard.

The RS-422 standard provides more robust signal levels for RS-232 by using differential drivers. The standard describes signal degradation with respect to cable length. The maximum cable length is 1200 m. Maximum data rates are 10 Mbit/s at 12 m or 100 kbit/s at 1200 m. RS-232 extenders are commonly implemented using RS-422. An RS-232-compatible variant of RS-422 was widely used on Apple Macintosh computer during the 1990s.

B. Universal Serial Bus (USB)

The USB (v1.0) specification was introduced in 1996, which addressed many of the weaknesses of the previous serial communication standards. Connection of external devices to PCs was greatly simplified by replacing the multitude of connectors on computers. Interface usability and software configuration of all devices ("plug and play") was also greatly simplified. Greater signal bandwidth for external devices was possible. The original v1.0 specification had a theoretical maximum data transfer rate of 12 Mbit/s.

The USB Implementers Forum standardized v2.0 specification at the end of 2001, which developed higher data transfer rate of 480 Mbit/s. v3.0 specification release in November, 2008 provides for a maximum transfer up to 10 times faster than v2.0. At the time this chapter is written, v3.0 has yet to be standardized. All specifications after v1.0 have substantial backward compatibility allowing devices implementing previous version to be operable with current versions.

A USB system consists of a host, and many downstream USB ports connected in a tiered-star topology. Additional tiers, allowing branching into a tree structure with up to five tier levels were possible. A USB host may have multiple host controllers and each host controller may provide one or more USB ports. Up to 127 devices, including the hub devices, may be connected to a single host controller.

USB simplifies software configuration by using predefined class codes (i.e., keyboards, flash memory drives, printers, etc.) to identify a device's functionality and to install the required device driver. This enables different manufacturers to support devices with drivers for a given class code. USB connectors are made to ensure correct connection. Connectors cannot be plugged in upside down, and it is clear from the appearance and tactile sensation when the plug and socket are correctly mated. The connectors are also designed for electrically safe hot swapping (connection/disconnection without powering off the device).

Maximum cable length is 3 m for USB 1.1 and 5 m for USB 2.0. The data cables for USB 1.x and USB 2.x use a twisted pair to reduce noise and cross talk. USB 3.0 cables are more complex and employ shielding for some of the added data lines. Cable length beyond 5 m is possible by using low-resistance cable. USB 3.0 cable may be of any length as long as all requirements defined in the specification are met. However, maximum speed can be achieved across a maximum cable length of approximately 3 m.

The USB 1.x and 2.0 specifications provide 5 V from which connected USB devices may draw power. For each device, a 100 mA load is permitted for USB 2.0, and 150 mA in USB 3.0. A maximum of 500 mA can be drawn from a single port in USB 2.0, and 900 mA in USB 3.0.

C. IEEE 1394 High-Performance Serial Bus

The IEEE 1394 is a serial bus interface standard for high-speed communications and isochronous real-time data transfer, used by personal computers, digital audio and video, automotive, and aeronautics applications. The interface is also known by many brand names—FireWire (Apple), i.LINK (Sony), and Lynx (Texas Instruments). IEEE 1394 can be implemented using wireless, fiber optics, and coaxial versions. IEEE 1394 was initiated by Apple Computers in 1986 and developed by the IEEE P1394 Working Group. It was intended as a serial replacement for the parallel SCSI bus implementing all SCSI capabilities and high sustained data transfer rates. It became an IEEE standard in 1995 with its current version as IEEE Std. 1394–2008.

The standard is now used for connection of data storage devices and DV (digital video) cameras. It is also popular in industrial systems for machine vision and professional audio systems. It is preferred over USB for its greater effective transfer speed and power distribution capabilities. FireWire can connect up to 63 peripherals in a tree chain topology. It allows peer-to-peer device communication—such as communication between a scanner and a printer—not requiring a host, using system memory or the CPU, although multiple hosts per bus are supported. It is designed to support "Plug and play" and "hot swapping." The copper cable (the most common implementation) can be up to 4.5 m long. It can supply up to 45 W of power per port at up to 30 V, allowing moderate load devices to operate without a separate power supply.

The original standard (IEEE 1394–1995) commonly referred to as Firewire 400 supported half-duplex transfer rates of 100, 200, and 400 Mbit/s, respectively, called transfer mode S100, S200, and S400. In 2002, Firewire 800 (IEEE 1394b–2002) was introduced bringing the full-duplex transfer rate to near 800 Mbit/s. Although the standard is backward compatible with Firewire 400, the Firewire 800 connector was changed requiring new cables. The full IEEE 1394b specification supports data rates up to 3200 Mbit/s optical connections up to 100 m in length. Standard Category 5e unshielded twisted pair copper wire (networking) cable supports 100 m at S100.

IEEE 1394–2008 introduced transfer modes S1600 (1.6 Gbit/s) and S3200 (3.2 Gbit/s) using the same Firewire 800 connector but fully compatible with S400 and S800 devices. This is intended to compete with the forthcoming USB 3.0.

#### *7.6.5.2 Parallel Interfaces*

Parallel communication interfaces send and receive data on multiple signal lines at the same time. If a computer communicates eight bits of parallel data, eight separate lines are needed. Devices that operate using parallel data interfaces include printers and a variety of test and measurement instruments.

In the late 1960s, Hewlett-Parkard (HP) introduced an interface to interconnect their test and measurement instruments to controllers such as computers. The interface was called HP Interface Bus (HP-IB). The interface was easy to implement with the technology at the time, so many other manufacturers copied the HP-IB and called it General Purpose Interface Bus (GPIB). In 1975, the bus was standardized by the Institute of Electrical and Electronics Engineers as IEEE-488–1975 (now 488.1) The standard is based on 8-bit parallel bi-directional data transmission bus, and it is suitable only over short distances from the computer. Instruments using the IEEE-488 interface have a standard 24-pin connector. Up to 15 devices can share the data bus by daisy chaining connections. The slowest device determines the speed of the transaction. The maximum data rate is about 1 MB/s in the original standard, and about 8 MB/s with later extensions. Many instruments now have this interface as a standard feature or it is available as an option. Interface boards and software for the IEEE-488 interface are available for most microcomputers.

### 7.6.6 Virtual Instruments

Software like LabVIEW is capable of performing many of the functions of hard-wired instruments, such as generating signals, measuring voltages, processing, and displaying signals using a PC. This software allows the computer to control physical instruments through many of the interfaces mentioned earlier such as IEEE-488 and RS-232 interfaces and microcomputer data acquisition boards containing analog-to-digital converters, DAC, timing circuitry, and digital I/O ports. Virtual instruments are created using interactive software that provides a graphical user interface (GUI) for simulating the front panel of an instrument on the computer screen, analogous to the front panel of a physical instrument, and the instrument is operated from the keyboard or mouse. Users write programs using virtual instrument software instead of physically interconnecting electronic components. The software provides icons for analog and digital displays, control switches and knobs, and signal processing operations. Icons can be interconnected to create arbitrary new instruments at the desire of the computer operator. Physical instruments containing serial or parallel interfaces for communicating with the computer are represented by instrument driver icons, are accessible and controllable, and can be programmed remotely from the computer. Such software is a powerful tool for controlling hardware and processing and storing data while taking advantage of the flexibility of a computer. An example of an instrument implemented through software as a virtual instrument is shown in Figure 7.11a. The icon block diagram of the virtual instrument is shown in Figure 7.11b.

## 7.7 INSTRUMENTS IN INDUSTRIAL ERGONOMICS PRACTICE AND RESEARCH

Various instruments are used by ergonomics practitioners and researchers. Often a specific measurement need necessitates selecting an instrument. The following discussion describes some common instruments used in ergonomics and their principles of operation. An understanding of particular instrumentation benefits and limitations will result in more accurate measurements and selection of the instrument most suited for a specific need.

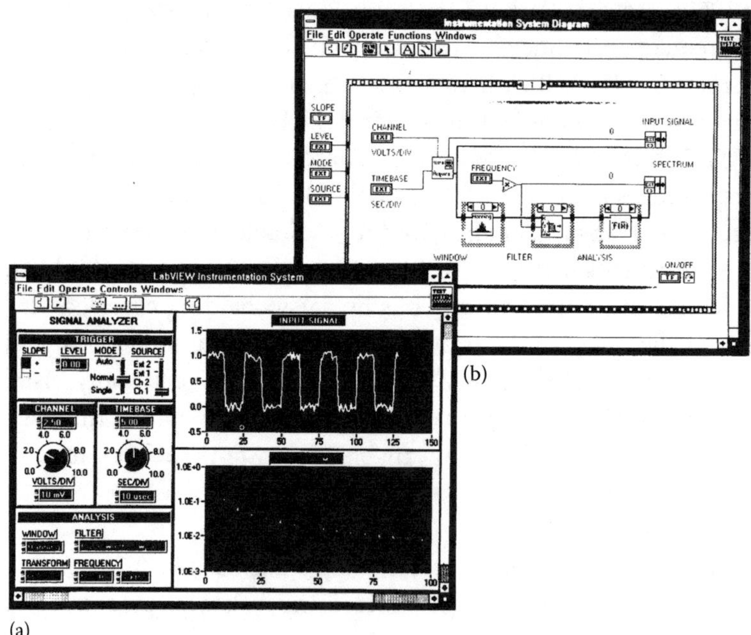

FIGURE 7.11 (a) A virtual instrument implemented with LabView software and (b) icon block diagram. (Courtesy of National Instruments.)

### 7.7.1 Human Kinematics

Kinematic measurements in ergonomics include linear and angular displacements, velocities, and acceleration. Movement is often measured relative to an anatomical landmark such as a body segment, joint, or other anatomical prominence, using a body landmark, or with respect to an external reference.

#### *7.7.1.1 Acceleration*

Acceleration is usually measured using devices called accelerometers. An accelerometer consists of a small mass and a sensor that measures the force associated with the acceleration of that mass. Because the mass is constant, the force sensor voltage is proportional to acceleration. Today's accelerometers are small enough and light enough to attach directly to the limbs for measuring body motions. Accelerometers are usually sensitive to movement in only one direction. Triaxial accelerometers can be used to measure movement in any direction by resolving the motion into three orthogonal components. These are simply three accelerometers mounted at right angles to each other.

#### *7.7.1.2 Velocity*

Velocity is seldom measured directly but can be calculated through integration of acceleration measurements or differentiation of displacement signals (see the following).

#### *7.7.1.3 Displacement*

Although accelerometers can be used for measuring displacement by integration, direct-displacement measurements available for displacement are more accurate and include optical methods, electroacoustic instruments, and electrogoniometers.

*Accelerometers for displacement measurements*: Double integration of the acceleration data will measure displacement but constant errors will compound in the integration. Displacement can also be calculated by integrating velocity data once. Piezoelectric accelerometers are usually satisfactory for measuring only changing acceleration signals and are not adequate for constant acceleration. Consequently, constant velocity and displacement cannot be measured using piezoelectric devices.

*Optical motion analysis*: Optical displacement measurement systems capture movements using stroboscopic photographs, cinematography, video, or some other optoelectrical device. In the simplest form, body segments can be illuminated using reflective markers located on the joints or limbs for producing bright points or lines. Movement can be measured simply by using multiple-exposure images from stroboscopic photographs and manually measuring displacement or inputting the data using a computer digitizing tablet.

Computerized motion analysis systems, such as Selspot, contain light-emitting diode (LED) markers affixed to anatomical landmarks that are flashed and picked up by a special camera that locates the x and y coordinates of the specific marker that is flashing and sends the information to a computer. Since the LED markers are illuminated sequentially, the computer is able to track a large amount of markers, typically 30 or more.

Video motion analysis uses a conventional video camera mounted in a fixed location for recording brightly illuminated reflective markers attached to anatomical landmarks. The video signal is recorded on tape for off-line analysis or directly digitized using video capture circuitry into a pixel image for a microcomputer. Either an observer or software is used for identifying the high-contrast illuminated markers, and locations of x and y coordinates of the markers are measured as pixels relative to a calibrated image. Since markers have to be tracked in space, these systems are limited in their number of markers. A commercially available system using this approach has been developed and sold by Peak Performance Technologies which is now part of Vicon Motion Systems.

Three-dimensional motion tracking is possible by using the aforementioned technologies with a minimum two cameras or LED sensors located at known positions. Additional cameras/sensors are needed for tracking markers when they may become hidden from the view of one camera/sensor. As many as five or more cameras/sensors may be used to improve accuracy and resolution for motion analysis. Many commercially available 2D and 3D optical motion tracking systems are produced and sold by Vicon Motion Systems (MX, Bonita, and Motus systems) and Northern Digital, Inc. (Optotrak and Polaris systems).

*Electrogoniometers*: Electrogoniometers are instruments that can directly measure the angular displacement of a joint. They are usually made from potentiometers, angle encoders, or strain gauges attached to body segments. Electrogoniometers range in complexity. The best instruments account for nonlinearities of joint rotation and have little cross talk between multiarticular joints. Electrogoniometer signals can be directly recorded or sampled by an ADC, and they do not have a problem with visual obstructions, but they do have some limitations. They may interfere with normal movement patterns by limiting the range motion or restricting movement. A commercially available lightweight strain gauge type of electrogoniometer is shown in Figure 7.12. Spectral analysis was used for

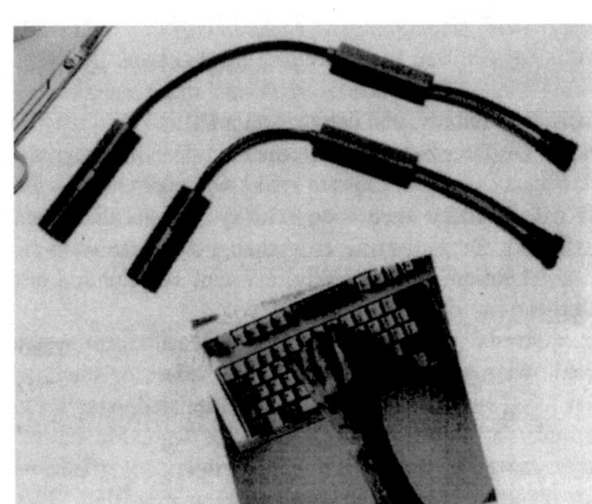

FIGURE 7.12 A lightweight strain gauge type of electrogoniometer shown mounted on the wrist. (Courtesy of Penny and Giles.)

quantifying postural repetitiveness, sustained posture, and postural deviation magnitude from continuously measured electrogoniometer angles [4,5].

### 7.7.2 Force Measurements

#### 7.7.2.1 Mechanical Force Transducers

Simple spring scales are useful for measuring static forces in the field, such as the weight of objects or constant loads. A push–pull spring force transducer with a variety of attachments is shown in Figure 7.13. But because spring scales behave as second-order systems with resonant frequencies often less than 1 Hz, they are unsuitable for measuring varying forces. To measure a force that changes with time, electronic load cells are used. Because springs can become permanently deformed when stretched beyond their elastic limits, their calibration should be periodically tested using known loads.

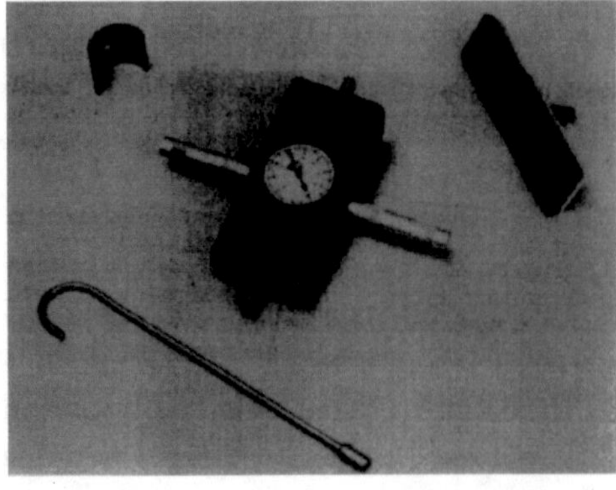

FIGURE 7.13 Spring-type force transducer with a variety of attachments. (From Eastman Kodak Company, *Ergonomic Design for People at Work*, Van Nostrand Reinhold, New York, 1986.)

#### 7.7.2.2 Strain Gauge Load Cells

A simple load cell consists of a metal beam with strain gauges bonded on the sides. Commercial load cells of various types are available for full-scale loads of a few grams up to hundreds of thousands of newtons, often in several degrees of accuracy and price ranges.

A difficulty with strain gauge load cells is that the measured force is limited to a single point on the beam or plate. Forces applied at points further away from where the strain gauge is mounted produce greater moments than forces applied at a point closer to the strain gauge. To get consistent and repeatable force measurements, the same point of force application must be maintained. Radwin et al. [6] built a strain gauge instrument that linearly summed forces applied at multiple locations along the length of a beam that was independent of the point of application. Such an instrument is useful for measuring forces that are summed, such as grip force produced by several fingers. This type of strain gauge instrument can be used as a handle for measuring hand force while operating tools as equipment (see Figure 7.14a and b).

#### 7.7.2.3 Piezoelectric Load Cells

Rapidly changing forces can be measured using piezoelectric force sensors. Because of their small size, many commercial piezo load cells have the necessary support electronic circuit built into their already small package. A single piezo load cell is useful over a very wide range of forces because the 1% nonlinearity applies to any calibration range. Piezoelectric load cells can measure forces as great as 16,000 N. Because piezoelectric load cells are very stiff, they are suitable for measuring isometric forces. Because they respond only to changing and impulsive forces, piezoelectric load cells are unsuitable for measuring steady-state forces.

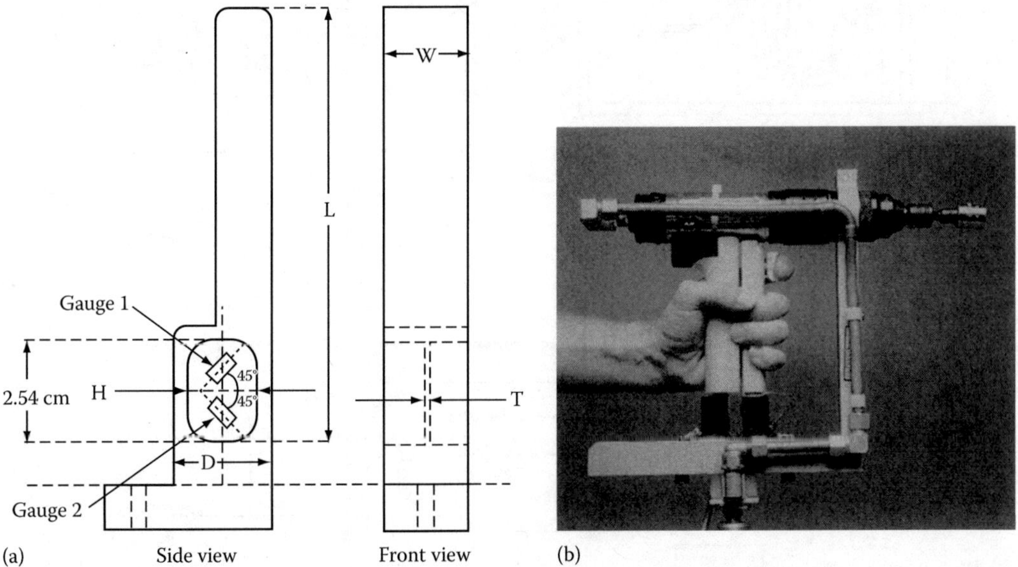


FIGURE 7.14 (a) Diagram of a shear stress strain gauge load cell. (b) The shear stress strain gauge load cell mounted on a pistol-grip pneumatic hand tool. (From Radwin, R.G. et al., *Appl. Ergon.*, 22, 339, 1991.)

### *7.7.2.4 Conductive Polymer Force Sensors*

A durable and thin conductive polymer sensor was found useful for measuring external forces when conventional force sensors are too large. Because of their small size and durability, these sensors can be attached to the fingers and hands [7]. The conductive polymer sensing elements are composed of two conducting interdigitated patterns deposited on a thermoplastic sheet facing another sheet that contains a conductive polyetherimide film. As applied force increases, the two layers compress together, increasing the contact area and decreasing its electrical resistance. A dome for distributing force over the active sensing area is necessary for these elements to operate as force sensors. Without the dome the measurements are erroneous. These sensors are very limited; their useful range is up to 30 N with an accuracy of 1 N; however, there are few alternatives available for directly measuring finger and hand forces. A schematic diagram of this type of force sensor is shown in Figure 7.15a. A force–time plot for force measurements from four fingers during the lifting of a box is shown in Figure 7.15b. Commercially available conductive polymer force sensors are produced and sold by Tekscan, Inc. (FlexiForce sensors) and Interlink Electronics (FSR sensors).

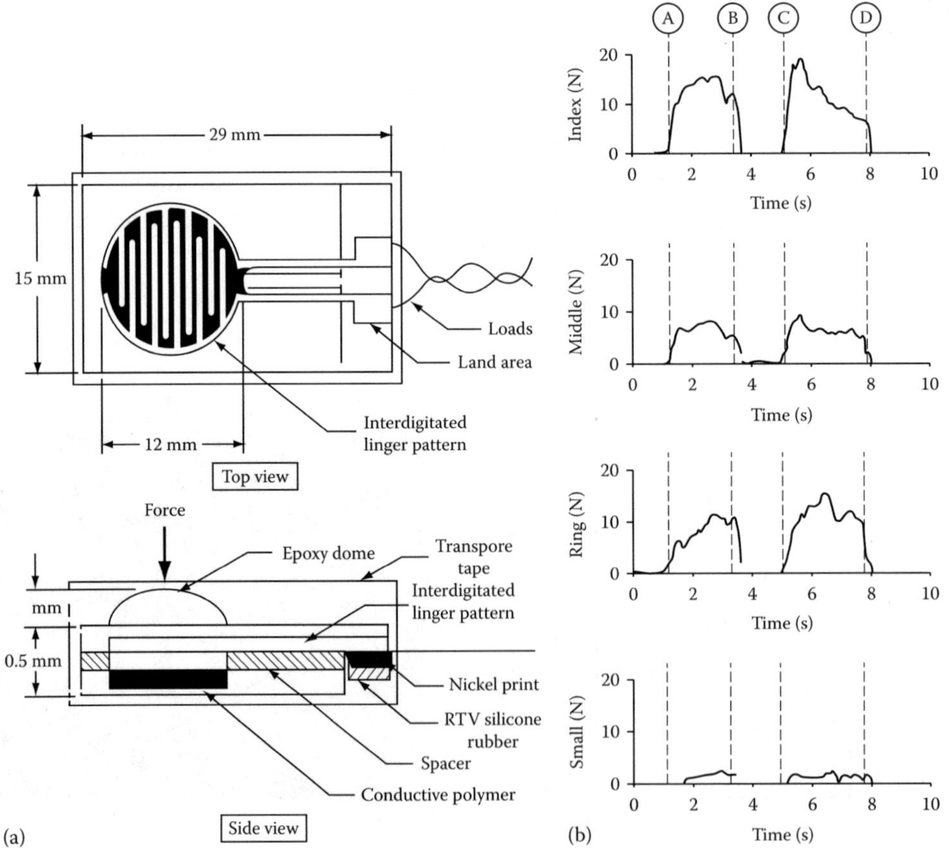


FIGURE 7.15 (a) Conductive polymer force sensor; (b) finger forces measured during a box-lifting task using the conductive polymer force sensors. The box was lifted from the floor at A, placed on a table at B, lifted from the table at C, and placed back on the floor at D. (From Jensen, T.R. et al., *J. Biomech.*, 24, 851, 1991.)

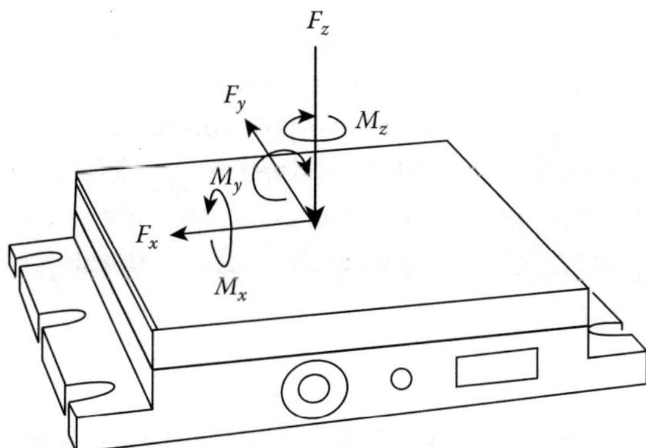


FIGURE 7.16 A force platform for measuring the three orthogonal force vectors ($F_x$, $F_y$, $F_z$) and three moments ($M_x$, $M_y$, $M_z$) about a single force vector applied to the surface.

#### *7.7.2.5 Force Platforms*

Force platforms are used for measuring forces in two or more directions, such as ground reaction force acting on the feet during standing or walking. Ground reaction force contains a vertical component plus two shear components acting along the surface [8]. This measurement is accomplished by using three or more strain gauge piezoelectric, piezoresistive, or capacitative force sensors that are arranged at right angles to each other. A common force plate configuration contains a flat plate supported by four triaxial force transducers (see Figure 7.16).

### 7.7.3 Electromyogram

EMGs are recorded using indwelling electrodes or surface electrodes. Concentric needle electrodes have a central insulated wire surrounded by a 200 μm-diameter cannula. The tip is beveled for insertion. Insulated wire electrodes are inserted through a carrier needle, which is then withdrawn. Both types permit superior localization near the tip, but they are invasive, making them unsuitable for field use. Surface electrodes are more practical for industrial ergonomics applications outside of the laboratory. A popular commercially available surface electrode is the Beckman silver–silver chloride miniature electrode. It is small (about 10 mm in diameter), light (250 mg), and reusable [9].

Being very small in amplitude, ranging from less than 50 μV up to 5 mV, EMG signals have to be amplified. A differential instrumentation amplifier with a gain greater than 1000 is commonly used. Since the input impedances for differential amplifiers used for measuring EMGs are never perfectly balanced, objectionable interference from power line sources is inevitable. This can sometimes be minimized by shielding leads and grounding each shield. Lowering skin impedance and using very high input impedance amplifiers is also helpful. Skin impedance is lowered by cleansing the skin with rubbing alcohol and using conductive electrode gel. The EMG amplitude is greatly affected by electrode placement and skin impedance.

The bandwidth for raw EMG signals is less than 500 Hz, with most of the power at frequencies less than 100 Hz. Integrated or rms EMGs are obtained by passing the EMG signal through a rectifier and an integrator circuit or rms detector. The signal is usually integrated over a specific time period, such as 100 ms, and then integration begins again. The time integral has a linear relationship with muscle tension under isometric conditions. Sometimes the EMG signal is rectified and low-pass filtered so that the output has a deliberate change with frequency. The resulting envelope of the EMG signal is related to muscle activity and has a much lower frequency content than the raw EMG (less than 10 Hz). This enables the signal to be recorded in conjunction with the development of force produced from the muscle activity.

### 7.7.4 Energy Expenditure

Heart rate is sometimes used for approximating energy expenditure. The respiratory demands of physical work, however, are best assessed by measuring the fractions of oxygen and carbon dioxide in expired air. Ventilation rate can be measured by exhaling through a gas meter that mechanically measures the volume of breath that accumulates over a fixed time period, typically 1 min. A face mask is usually worn as the subject performs manual tasks. Oxygen consumption and carbon dioxide production are measured from the collected air.

#### *7.7.4.1 Heart Rate*

Heart rate is usually measured using electrocardiograph (ECG) signals and a cardiotachometer. ECG electrodes are similar to EMG electrodes and amplified using similar amplifiers. Interference from contracting muscle EMG signals can sometimes be picked up by ECG electrodes. This problem is sometimes minimized by relocating the electrodes. Pulse rate monitors can continuously monitor heart rate using a photooptical pulse sensor clipped to the earlobe.

#### *7.7.4.2 Ventilation*

Rotating vane flowmeters contain a small turbine in the flow path air. The number of revolutions of the rotating turbine is proportional to the volume of air flowing past it. Rotation is measured using electromechanical sensors or the interruption of a light beam using a photosensor.

#### *7.7.4.3 Oxygen and Carbon Dioxide Concentrations*

The percentage of expired $CO_2$ can be measured by using an infrared detector or mass spectrometer. Mass spectrometers are instruments that produce a stream of ions from a substance and separate the ions into a spectrum according to their mass-to-charge ratios. The instrument determines the partial pressure or molar fraction of each type of ion present for $O_2$, $CO_2$, $N_2$, etc. Some systems can measure as many as eight component gases. Mass spectrometers are very versatile but quite expensive. Infrared analyzers are more cost-effective and are usually suitable for ergonomic applications. Most gases absorb varying levels of infrared light (3–30 μm wavelength) at distinct characteristic wavelengths. Hence the concentrations of components of a gas mixture can be measured by the transmission of infrared light. When an instrument measures transmission for a well-defined set of wavelengths selected for particular gases, such as $CO_2$, it is known as a nondispersive infrared (NDIR) analyzer.

The polarographic $P_{O_2}$ electrode is based on an oxidation reaction that occurs at the Ag–AgCl anode and a reduction reaction that occurs at the glass-coated Pt cathode [10]. A polarizing voltage sufficient to drive the reaction (600–800 mV) produces a current linearly proportional to the number of $O_2$ molecules in solution. A membrane permeable to $O_2$ and other gases separates the electrode from the surroundings. The electrode is zeroed using a $CO_2/N_2$ gas and calibrated using two gases containing known concentrations of $O_2$. The reaction partially consumes $O_2$ and is very sensitive to temperature. The amount of oxygen consumed is measured from the difference between the amounts of oxygen inspired and expired. Expired gas collected in a mixing chamber or in a sample bag is analyzed for $P_{O_2}$. Portable battery-powered measurement systems, such as the Oxylog, are available for measuring oxygen uptake of mobile subjects.

### 7.7.5 Vibration

Since vibration is oscillatory motion, it is most conveniently measured by using an accelerometer. Accelerometers should be selected carefully. It is important that the size and mass of an accelerometer be small enough not to interfere with the measurement by loading. Accelerometers weighing more than 15 g are unsuitable for vibration measurements made by mounting them on a human body. Accelerometer sensitivities are proportional to their mass; the smaller the accelerometer, the less sensitive it is. It is important that an accelerometer not be excited by frequencies near or at its resonant frequency. This would produce erroneously large measurements. Accelerometer resonant frequencies vary in inverse proportion to their mass. Accelerometers are also influenced by temperature changes, humidity, and other harsh environmental conditions.

#### *7.7.5.1 Whole-Body Vibration*

The frequencies of interest for whole-body vibration are dc to 80 Hz. Piezoresistive accelerometers are best suited for whole body vibration measurements because they operate at dc and low frequencies. Furthermore, piezoresistive accelerometers are self-calibrating because they can be oriented in the direction of gravity for an input equivalent to the acceleration of gravity and then turned 90° to gravity for removing gravity from the accelerometer.

Accelerometers are mounted on the body near bony eminences and surfaces. A triaxial seat disk accelerometer may be inserted between a vehicle seat and a passenger's buttocks. Vibration at the feet of a vehicle passenger can be measured by mounting an accelerometer directly to the floor.

#### *7.7.5.2 Hand-Arm Vibration*

Hand-arm vibration usually is transmitted through the handles of manually operated equipment such as powered hand tools. Piezoelectric accelerometers are often used for hand-arm vibration measurements. Triaxial accelerometers can be mounted directly on a tool handle for measuring vibration along orthogonal axes. Accelerometers are usually mounted directly to the handle using a hose clamp or similar strap or by welding a mounting block with a stud for the accelerometer.

#### 7.7.5.3 *Vibration Analysis*

Human vibration measurements are usually recorded and spectrum analysis is used for identifying the frequency characteristics of vibration exposure. Vibration exposure time is determined by measuring the duration of tool operation during a suitable sample period.

### 7.7.6 Temperature and Humidity

Electronic thermometers based on thermocouples and thermistors are commercially available and provide a calibrated temperature output. Rise time, size, long-term stability, and sensitivity should all be considered when choosing a temperature sensor. Thermal expansion thermometers, such as the mercury column variety, typically have a rise time of 10 s, whereas a thermistor may have a rise time of 1 s and a thermocouple may have a rise time on the order of 0.05 s.

Electronic sensors are capable of determining relative humidity. Resistive humidity sensors use materials whose resistance decreases when they absorb moisture. Capacitive humidity sensors use materials for which the dielectric constant (and thus capacitance) changes with moisture uptake.

### 7.7.7 Integration of Measurements and Video

The effective integration of measurements using multiple instruments with video of human activities can be overwhelming. A large amount of collected data requires an efficient way to extract meaningful information. Radwin and Yen [11] developed a method for integrating the data collected from transducers with the video image recorded using a digital camcorder. An example of this system is shown in Figure 7.17. Analog signals instruments (electrogoniometers, force sensors, EMGs, etc.) are digitized, coded, and recorded directly in synchronization with video images of the subject performing the operation. The data and video are analyzed using an interactive computer-controlled video-based analysis system.

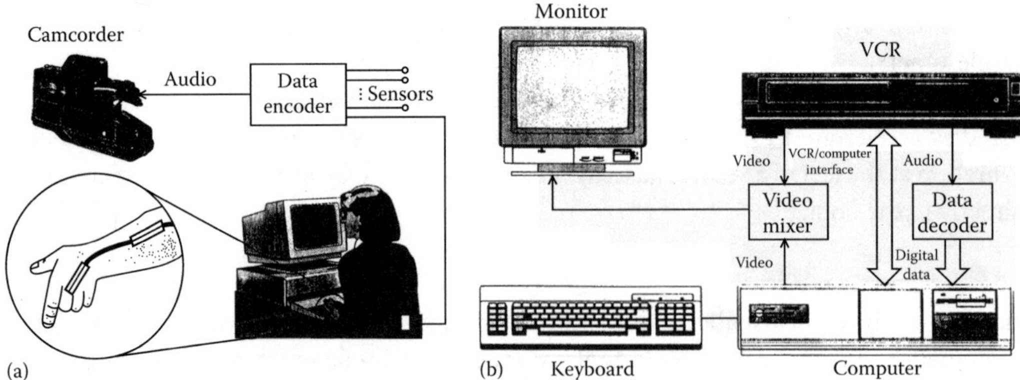


FIGURE 7.17 (a) Video-based data collection system and (b) interactive multimedia data extraction and analysis system.

The video can be viewed in real time, slow, faster, or frame by frame in either the forward or backward direction, allowing flexible breakpoint assignments.

Commercially available digital video technology and data acquisition instrumentation has allowed the development of the Multimedia Video Task Analysis™ (MVTA™) system based on the system described by Radwin and Yen [11]. MVTA contains unique software for time-based activity and event analysis with synchronized data collection. The hardware integrates computer and digital video technologies to eliminate much of the manual and tedious activities associated with conventional timing and video analysis. Digital video and sensor-based analog biomechanical data can be collected in direct synchronization. The video and biomechanical data can be displayed simultaneously in synchronization allowing easy exploration of relationships between the data and video (see Figure 7.18).

The software provides a convenient user interface for information entry, data management, and reporting. The computer keeps track of the current video frame and time code embedded in the video recording and therefore enables the computer program to maintain data linking each occurrence with the time it occurred. All timing measurements are automatically performed by the computer program freeing the analyst to concentrate on activity and event identification. Manual control of the video player is also eliminated, whereby the user needs only to select the event of interest and the computer can instruct the video player to locate the correct segment on the recording. A time study analysis report can be generated from the activities and events indentified.

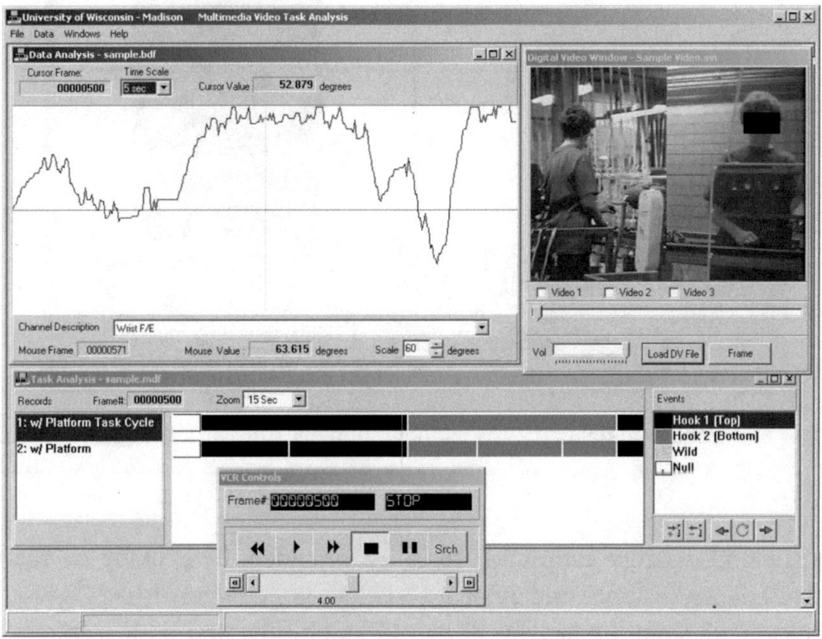


FIGURE 7.18 Sample task and data analysis windows from the MVTA™ software.

FIGURE 7.19 Angle estimation from video.

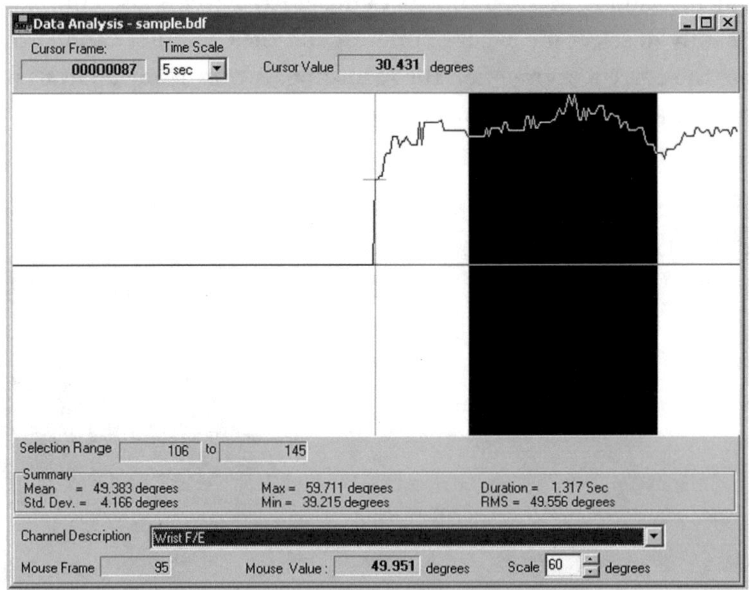


FIGURE 7.20 Biomechanical data analysis window showing summary statistics of highlight data segment.

Basic angle and distance estimation tools are provided. By clicking on the frame of the recorded images, distance and angle estimations are computed (see Figure 7.19). The biomechanical data from sensors are displayed graphically and basic statistics (average, standard deviation, maximum, minimum, rms, duration) can be calculated on all or any portion of the data, using a convenient user interface (see Figure 7.20). Digital filtering and other signal processing features are included.

## REFERENCES


1. J. C. Yoon and R. G. Radwin, The accuracy of consumer-made body measurements for women's mail-order clothing, *Hum. Factors 26*: 557–568 (1994).
2. E. O. Doebelin, *Measurement Systems: Application and Design,* 4th edn., McGraw-Hill, New York, 1990.
3. W. J. Tompkins, Ed., *Biomedical Digital Signal Processing: C-Language Examples and Laboratory Experiments for the IBM PC,* Prentice-Hall, Englewood Cliffs, NJ, 1993.
4. R. G. Radwin and M. L. Lin, An analytical method for characterizing repetitive motion and postural stress using spectral analysis, *Ergonomics 56*(4): 379–389 (1993).
5. R. G. Radwin, M. L. Lin, and T. Y. Yen, Exposure assessment of biomechanical stress in repetitive manual work using frequency-weighted filters, *Ergonomics 37*(12): 1984–1998 (1993).
6. R. G. Radwin, G. P. Masters, and F. W. Lupton, A linear force-summing hand dynamometer independent of point of application, *Appl. Ergon. 22*: 339–345 (1991).
7. T. R. Jensen, R. G. Radwin, and J. G. Webster, A conductive polymer sensor for measuring external finger forces, *J. Biomech. 24*: 851–858 (1991).
8. D. A. Winter, *Biomechanics and Motor Control of Human Movement,* Wiley, New York, 1990.
9. C. J. De Luca, Electromyography, in *Encyclopedia of Medical Devices and Instrumentation,* J. G. Webster, Ed., Wiley, New York, 1988, pp. 1111–1120.
10. J. G. Webster, Ed., *Medical Instrumentation: Application and Design,* 2nd edn., Houghton Mifflin, Boston, MA, 1992.
11. T. Y. Yen and R. G. Radwin, A video-based system for acquiring biomechanical data synchronized with arbitrary events and activities, *IEEE Trans. Biomed. Eng. 42*(9): 944–948 (1995).
12. Eastman Kodak Company, *Ergonomic Design for People at Work,* Vol. 2, Van Nostrand Reinhold, New York, 1986.

CHAPTER 8

# Worker Participation

## *Approaches and Issues*

Alexander L. Cohen

CONTENTS

## 8.1 INTRODUCTION

This chapter summarizes the literature on participative approaches in addressing workplace problems, with mention made of their application to workplace safety and health issues in general and ergonomic problems in particular. The material reproduces with few modifications a section of a recent National Institute for Occupational Safety and Health (NIOSH) report [1] describing team techniques to define ergonomic problems in meatpacking jobs and the merits of this approach in proposing effective control measures. The text is used to set the stage for three case studies that are described in the report in elaborating on the processes involved in these types of interventions. The results of these studies are discussed briefly later in the chapter.

The term *participatory approach* as used in the work setting has a number of meanings. In this chapter, its essential meaning is *worker involvement*. Hence, references to teams, groups, and committees formed to deal with work-related issues (ergonomic hazards in this instance) are assumed to include front-line employees or their representatives. Other members of such bodies may be supervisory or managerial persons, staff from other departments whose duties pertain to matters at issue, and outside consultants. Lawler [2] characterizes employee participation as the movement of decision making, information, knowledge and skill, and rewards to the lower levels of an organization. References to these and other elements are apparent in the various forms and levels of worker participation noted in the chapter.

## 8.2 BACKGROUND

### 8.2.1 Rationale for Worker Participation

A review of the industrial psychology and organizational behavior management literature makes clear the benefits that can accrue from worker involvement in dealing with organizational issues, along with some important qualifiers [2–4]. The results are summarized in the following paragraphs.

#### *8.2.1.1 Enhanced Worker Motivation and Job Satisfaction*

An employee's work motivation and job satisfaction are increased not only by added pay but also by the opportunities to provide input into decisions affecting work methods, everyday job routines, and performance goals. Having control over one's own work is especially satisfying and enhances commitment and quality effort. Positive results, though, are conditioned by a number of factors such as the following:

1. The perception that an important work performance matter is at issue, not some trivial concern (e.g., the color of the hallways).
2. The perception that the work is interesting and challenging. Worker participation to address repetitive, simplistic, standard tasks in and of itself would not be a good candidate unless the concern was to consider job redesign or other changes.
3. Workforce educational level and knowledge that indicate capabilities for offering meaningful input. Today's workforce, being better educated than their forebearers, has greater expectations about job roles and greater need for self-esteem.

#### *8.2.1.2 Added Problem-Solving Capabilities*

Employee involvement in decisions affecting their work situations can capitalize on their unique and relevant experience. Indeed, the person doing the job often has the best knowledge of the problem elements and insights into ways to improve the work. Effectiveness here can depend upon whether the individuals have the problem-solving skills needed to identify valid solutions and the ability to argue effectively for their adoption. Another factor is whether the issue is a local one in which the group can make decisions and take action or one having broader implications that require higher level review and approval. If the latter is the case, delays in responsiveness to recommendations can create cynical attitudes about the participative process.

#### *8.2.1.3 Greater Acceptance of Change*

There is evidence that participation in decision making regarding a major organizational change can lead to a significant reduction in resistance to that change. Creating better understanding of the needs for change through improved communications and the involvement of those affected in structuring the change in desirable ways can do much to gain employee commitment to a successful implementation. Lacking these efforts, change can be perceived as threatening job security or as having other negative consequences.

#### *8.2.1.4 Greater Knowledge of the Work and the Organization*

Taking part in problem solving in workplace conditions and in decision making in work design with those in one's own work group and/or with others from different units or areas will invariably increase the employee's knowledge of his or her own job if not of the overall company operation. An important payoff from such interaction can be improved communications and coordination among the members and their respective departments. However, for this to occur requires employee training in communication skills and motivation to use the knowledge of how other units function.

### 8.2.2 Forms and Levels of Participation*

Employee participation in work organizations can take a variety of forms. Among the shaping factors are (1) the nature of the issues requiring consideration, (2) whether the matters are broad based or specific to a local operation or group, (3) whether the needs for response or action are time limited or necessitate continuing efforts, (4) the abilities of the group most affected, and (5) the organization's prevailing practices for joint labor-management or participative approaches in resolving workplace issues. The degree or level of involvement may also vary. At one extreme may be simple consultations with individual workers or groups to obtain their reactions to ideas of superiors who will make the final decision. At the other extreme may be obtaining worker ideas along with those from management and other affected parties in addressing issues, with decisions based on consensus. The fact-finding report from the Commission on the Future of Worker-Management Relations [6] outlines the variety and scope of employee participation and labor-management cooperation in U.S. workplaces. In this section, common forms of worker participation found in industry are described as are different levels of sharing in decision making and other factors reflecting the degree of actual worker involvement.

#### *8.2.2.1 Quality Circles*

Quality circles are generally defined as small groups of worker volunteers from the same work area who, with their supervisor, agree to meet regularly to identify, analyze, and solve quality and related problems in their areas of responsibility [2,7]. They usually consist of 8–10 members who meet once a week during work hours. The volunteers typically receive training in some form of problem-solving techniques as part of this activity.

The use of quality circles is attributed to Deming's introduction of data-based quality control techniques in Japan to rebuild their industry after World War II [7]. Although originally intended as a program for troubleshooting by engineers, the movement quickly evolved to include line workers in accord with Deming's view that quality must concern every employee rather than be limited to the engineers or the quality control department. The success of Japanese industry in capturing large market shares for their products in the early 1970s led American businesses to emulate their techniques. In 1986, it was reported that more than 40% of U.S. companies employing more than 500 workers were using quality circles [8].

As Krigsman and O'Brien [7] note, quality circles in Japan were focused on performance data and quality control issues. Worker involvement was based on the underlying idea that workers ought to be responsible for the quality of their work and are in the best position to troubleshoot it. In the United States, quality circles became more of a participatory management technique intended not only to yield increased productivity and product quality

* The legality of management forming certain groups with employee participants to address productivity, quality, and safety matters has been questioned. The National Labor Relations Act forbids such actions, fearing domination of such groups by management. In response, some employers have gone to self-directed work teams, and others are keeping with the existing forms but including volunteer employees as members who represent themselves in such groups. The issue may be resolved through court tests or OSHA reform legislation, which may exempt well-intentioned groups such as those dealing with safety and health concerns. See LaBar [5] for further details on this subject.

but also to enhance employee motivation and job satisfaction. Although experiences in the United States tended to support these various outcomes, the results were not always up to expectations [9,10]. For example, Griffin [10], in his study of U.S. electronics plants, found that quality circles produced initial improvements in job satisfaction, organizational commitment, and performance measures but that over time and in the absence of other supportive measures, these indicators reverted back to their original levels. When asked about this end result, quality circle members in this study stated that they felt that management was no longer interested in their recommendations and that their supervisors asked fewer questions as to how the group was functioning and displayed less enthusiasm about evaluating suggestions. Without continued management support for this program, the early improvements could not be sustained. On this point, Lawler [2] and Griffin [10] view quality circles as a building block to other forms of worker participation that could ultimately create a more participative culture in an organization. Cascio [3] notes too that worker participation programs can die out eventually if the organization does not change in a manner consistent with the democratic values that characterize such practices.

Safety circles represent a variation on the quality circle form of worker involvement, the difference being that the thrust of the group effort is directed to identifying, analyzing, and solving safety and related health risk problems in their work area [11,12]. The National Safety Council [13] describes a step-by-step approach to establishing safety circles. Needs for management support and resources for implementing recommendations, decision-making authority to be invested in the group, training of members in safety subjects, and interpersonal relationships are duly noted.

#### *8.2.2.2 Labor-Management Committees*

Whereas quality circles are small in size, composed of volunteers from a single work area who are brought together to address problems specific to their job tasks, labor-management committees are more expansive, including elected or appointed members from different areas within an organization and charged with a broader agenda. Also, unlike quality circles whose members can actually implement solutions, most committees only recommend actions, and their recommendations are then forwarded to other parties for concurrence or coordination in determining how and when approved actions can be carried out.

Joint labor-management committees offer opportunities to identify areas of mutual concern and to engage in cooperative activities that can reduce the level of traditional adversarial behavior between the two parties [2,14]. Two areas, quality of working life (QWL) and occupational safety and health, have been the focus of much joint committee activity. QWL committees seek ways to improve working conditions so as to enhance worker job satisfaction and morale and therein increase company productivity. QWL efforts can encompass recommendations for making a more pleasant physical environment, furnishing educational opportunities during off-job hours and facilities for recreation. In some instances, collective bargaining agreements struck between unions and management have enabled QWL committees to also address certain aspects of job classification and work schedule issues. The reader is referred to Lawler [2] for more details and examples of QWL committee work.

Joint labor-management safety and health committees offer opportunities for cooperative problem solving with regard to hazard recognition and control concerns as well as for the recommendation of preventive measures [14]. The effectiveness of these groups is the topic of a later discussion.

The membership of joint labor-management QWL and safety and health committees includes representatives from affected groups. Worker participation may be attained through elected workers or local union leaders, with management represented by department heads or other key figures. The success of such groups in effecting actions depends upon their own decision-making authority or links to others who have that role. As already noted, the committees make recommendations whose implementation may take the form of establishing task forces, work teams whose job is to formulate and carry out specific plans. A by-product of the committee deliberations and follow-on actions by these groups is that information is shared widely in the organization and more channels are opened for communication. As a result, more employees can understand the business better and participate more effectively in problem-solving activities.

#### *8.2.2.3 Work Teams*

Work teams are referred to in the literature as "self-regulating" work groups in that they can make decisions in terms of setting production goals, selecting work methods and quality control procedures, and managing inventory [2]. In some cases, such groups may have even more autonomy, for example, determining pay rates and hiring/firing policies. Management maintains oversight of the group's practices and operations and has the right to challenge any of its decisions. Work teams include all of the employees working in a given area who, with a chosen lead worker or supervisor, are given the responsibility for producing an entire product or offering a complete service. Because of their broadened roles, work team members are cross-trained so that each can do the various tasks that fall within the domain of the team, and frequently workers rotate their work assignments. In addition to the extensive training that may be needed to perform these multiple job functions, work team members require instruction in interpersonal skills. As explained, these skills are necessary to ensure positive, effective interactions among the group members. Indeed, their varied responsibilities demand that work teams meet often to discuss and reach agreement on numerous matters. Experiences with work teams in mining and various manufacturing companies have demonstrated gains in rate and quality of output, reduced turnover, and improvements in overall work efficiency (summarized in Peters [15] and Lawler [2]). There are also cases where work teams in these establishments did not survive. This appears to be most evident in companies having a more traditional management approach.

#### *8.2.2.4 Gain Sharing*

Gain sharing acknowledges worker participation in efforts to improve company economic performance through increasing the sales value of production relative to labor costs [3]. In one such plan, a ratio of the two factors is set, based upon the past year's experience, that, if exceeded, will result in cost savings to be shared by the employees and management in accordance with some agreed-upon formula. Another plan sets a production/performance standard that, if met in fewer than the expected work hours, yields savings that will be distributed. The participative structure in each instance uses a formal suggestion system that invites workers to submit in writing their ideas for improving work efficiency. Department production and screening committees made up of worker and management representatives review these inputs and select those to be implemented.

Company experiences with gain sharing and other incentive plans as reported in the literature show a roughly 20% increase in productivity but at the same time much variability in these results [16]. In some cases the plans yielded a 75% increase in output, and in others a 5% decrease. Success seems to be a function of many factors such as whether the market can absorb the increased production, whether product costs are controllable by employees, whether there is top management commitment and supervisory support for the plan, and whether management is open in sharing financial results and giving other evidence of trustworthiness.

#### *8.2.2.5 Levels of Participation*

Worker participation can also be viewed along a number of different dimensions. Liker et al. [17], for example, offer models reflecting variations in two dimensions. One is the locus of decision making, whether at the management level with consultations sought from affected individuals or groups or delegated downward with little management involvement. The second dimension is the manner of employee input into such processes—whether each person in an affected group has direct involvement or is represented by a subgroup. Quality circle and work teams as described previously would appear to fit the model in which all workers are involved and have authority to make and carry out decisions. In contrast, joint labor-management committees would be categorized as representative in makeup with authority limited to making recommendations, not actual decisions. By itself, the formal suggestion system inherent in gain sharing would offer opportunity for direct input but no decision-making power, this being assumed by other committees or retained by management.

As noted by Liker et al. [17], Lawler [2], and others, the success of worker participation efforts in solving workplace problems and enhancing productivity, worker motivation, and satisfaction is not dependent on any one form of involvement but on what is best suited to the issues to be addressed and the situational factors that are present. Also, certain forms may evolve into others as conditions change, which may be important to sustain or further the positive effects seen in such practices.

## 8.3 WORKER PARTICIPATION APPROACHES IN WORKPLACE HAZARD CONTROL

Evidence indicating the effectiveness of worker involvement in efforts to reduce work-related risks of injury and disease is the subject of this section. Such participation has taken various forms akin to those previously mentioned. Reports documenting the importance of these approaches in cause-and-effect terms as well as defining factors of major consequence to successful outcomes are not numerous. Indeed, field studies in this area do not allow for easy isolation or manipulation of these variables or for comparisons with adequate control or nontreatment conditions. Due caution is thus advised in either interpreting or generalizing results. In this section, worker involvement in general injury and disease control problems is first described, followed by a discussion of efforts directed to controlling ergonomic hazards. The literature reviewed in these cases is admittedly selective and to some extent summary in nature. Its purpose is to illustrate worker participation approaches as applied to these kinds of concerns, highlighting certain aspects of their implementation, and resultant findings.

### 8.3.1 Joint Labor-Management Safety and Health Committees

The most common institutionalized form of worker participation in workplace safety and health matters is though membership on joint labor-management committees set up for that purpose [14]. Collective bargaining agreements between unions and management, especially after the passage of the Occupational Safety and Health Act of 1970, contained provisions for the establishment of these committees that were attempts to resolve these issues on a local level. The Bureau of National Affairs [18] reported that in 1970, 31% of industrial contracts covering 1000 or more workers had such provisions; this proportion rose to 39% in 1975 and to 45% in 1983. Boden et al. [19], in a survey of manufacturing companies having 500 or more employees in one state (Massachusetts) in the early 1980s, found that 67% of the unionized establishments had a joint labor-management committee addressing safety and health issues and that 49% of nonunion workplaces had similar groups with employee-management representations. A 1993 national poll by the National Safety Council [20] found that 66% of the respondent companies had joint committees. The survey acknowledged sampling and other limitations that led the authors to feel that this figure might be higher than the national average.

The more cogent question, however, is whether the existence of these committees has had a positive impact on worker safety and health. The literature suggests mixed findings. For example, Cooke and Gautschi [21] used data from the state of Maine for compensable injuries and OSHA citations in 113 manufacturing firms during the period 1970–1976. Controlling for the size of the production workforce, business cycle effects, and OSHA citation experience, they found that the presence of joint labor-management safety and health committees was associated with a small and statistically insignificant decrease in lost-time injuries over the period in question. Similarly, Boden et al. [19] found virtually no effect in a study of whether the existence of a joint safety and health committee was correlated with either the number of OSHA complaints or serious hazards as measured by citations for 127 Massachusetts manufacturing firms. More detailed study of a subsample of companies with these committees, however, showed these outcome measures to covary in inverse fashion with the number of the powers of the committee to act, its opportunities to access and review different types of data (hazard, injury, and medical reports), and perceptions of a strong management commitment to worker health and safety. The authors concluded that maintaining a joint health and safety committee as a formality yields little results on company safety and health experience and that its impact is a function of activity level and a company environment truly supportive of its efforts.

Reinforcing the point just made, California in 1984 [18] reported the benefits of organizing joint labor-management committees to conduct self-inspections of safety and health conditions at major construction sites in the state as part of a voluntary compliance program. For work at three sites that employed 200–2600 workers, the injury and illness incidence rate dropped far below those averaged for the construction industry as a whole or the individual employer's rate at other similar projects. At one site, the decrease was from 7.4 cases per 100 full-time workers per year at program start-up to 4.2 cases afterward. Project managers attributed the improved safety performance to increased awareness of hazards by employers and employees, better communications between the parties, and a belief by the workers that they could influence safety on the job.

Evidence that the effectiveness of joint health and safety committees in reducing injury and enforcement rates and in enhancing problem solving in workplaces is dependent on added considerations is also seen in the survey results from two Canadian provinces (Ontario and Quebec [22]). In analyzing performance on such indicators between the sample sites surveyed relative to all other workplaces in their peer group, it was found that a *committee capacity* factor was an important determinant in attaining a positive outcome. Elements of committee capacity, based on factor analyzes, were found to consist of opportunities to conduct a broad scope of activities, use of institutionalized procedures in decision making, policies supporting improved resources for member training in safety and health, and sharing of information. While showing the apparent benefits of legislation mandating the establishment of joint labor-management health and safety committees in Canadian businesses, the authors acknowledge that the committee capacity factor accounts for only a limited amount of variance and that workforce experience and stability may be even more important factors in the final analysis.

Joint labor-management committees by themselves do not appear to be a major determinant in studies contrasting organizational program practices in companies that have exemplary safety and health records with those of more poorly performing cohorts. While perhaps facilitating worker participation, other direct means for promoting worker inputs into the program seem to be more influential than a formal committee. For reasons stated by Boden et al. [19], committees can vary greatly in their activities and roles that can affect workplace safety and health. Most studies comparing program factors in companies with good versus poor safety performance lack details as to whether there are functional differences between the committees found in the contrasting samples and details of their relationship to other participative efforts that may be of consequence. A commonly expressed view about safety and health committees is that without their establishment workers would have little means for involvement in any safety and health activities [20].

Joint labor-management health and safety committees have also been formed nationally to support continuing education of their respective members and to sponsor research work to address pressing health and safety problems of mutual benefit.

### 8.3.2 Work Teams for Hazard Control

Case studies and other reports in the popular and technical literature illustrate how work teams, safety circles, or equivalent groups, each of small size and composed of worker members engaged in similar jobs and from the same area, have made positive contributions to hazard control efforts [12,23–25]. Typical is a report by Edwards [12], who studied the impact of a quality circle (QC) technique on safety issues in a large surface mine. Setup elements included (1) forming a screening committee of department heads and a QC-trained facilitator to set ground rules for the plan, (2) composing QCs of five to eight persons from worker volunteers in four selected departments, and (3) giving QC members plus mine safety committee persons 8 h of training on subjects such as brainstorming, data collection, and group dynamics. Subsequent 1 h weekly meetings were held during which the QCs focused on problems that would be expected at most mines, that is, tool shortages, poor communications, unavailability of parts or supplies, lack of support equipment, inadequate housekeeping, etc. The circle members chose a problem they wanted to solve, collected data for delineating its nature, and then offered possible solutions taking into account cost-effectiveness considerations. A number of recommendations were implemented that had a significant effect on both productivity and safety. With regard to the latter, QC members experienced 58% of the accident occurrences on their shift in the 6 months before the program and 31% in the following 6 month period. For departments with circles, the accident frequencies decreased by 18% in before/after comparisons over the same 6 month periods.

Some difficulties in organizing or maintaining work team efforts directed to hazard control have also been noted. For example, a county engineering department reported marked improvement in the safety performance of work crews in one section when a quality circle approach was adopted to elicit worker inputs into ways of making their operations safer and feedback and incentives were used to reinforce the program efforts [24]. Injury frequency dropped by 52% and their associated cost by 92% after the plan was instituted for work crews who previously had the worst safety record of the various sections in the department. However, expanding this program to another division within the engineering department proved problematic for a number of reasons. The job routines of these workers did not require a natural team effort, and workers enjoyed their independence in fulfilling their specific responsibilities. As a consequence, the team problem-solving effort was viewed more as a "gimmick" of management. The program was nevertheless implemented wherein team members began blaming each other and management for failure to achieve any positive results. As a remedy, and at the suggestion of the workers with their supervisors, the teams were redrawn to take account of mutual needs for working relationships and compatibility among the partners. This worker input into the program helped reduce the earlier resistance. An 18% drop in injuries was noted after the revamped teams were formed though costs remained unchanged.

Peters [15], in reviewing research on organizational and behavioral factors associated with mine safety, mentions a study assessing the benefits of a self-regulated work team as introduced in a Pennsylvania coal mine on an experimental basis. The miners received training to make them each capable of performing any job in their section and to teach them about mine safety laws and violations. Periodic meetings and feedback were used to motivate worker interest in safety, and the autonomous nature of the group made each miner responsible for maintaining safe working conditions. Supervisors had responsibility and authority for the safety of their work crews with lesser concerns for production. This group experienced fewer violations and shutdowns than others in the same mine. The work crews also put into place more safe work practices and were more proactive toward safety than they were before the intervention. Despite these positive findings, however, an effort to expand the program to other mine sections was voted down by the union. One reason for the rejection was the perception that the special treatment given to the experimental group created an elitist attitude among its members that was resented by the miners in the other sections. This effect was unintended, and efforts to overcome the negative fallout either were not taken or were not sufficient to correct the situation. Peters [15] notes that the intervention effort in the mine disappeared 4 years after it was first initiated.

### 8.3.3 Direct Worker Inputs in Hazard Control

Reports of direct worker inputs having been formally solicited into hazard control programs, as contrasted with using a team or committee approach, are not that common. One case study of this type, conducted by Lin and Cohen [26], is important in showing both the merits of worker involvement for this purpose and some of its limitations. The site for the work was a 500-bed hospital with 1800 full- and part-time employees where a worker hazard detection program was put into place on a trial basis. Employees were first surveyed to determine their current level of awareness of workplace hazards and the means to control them. This was followed by a campaign to motivate employee reports of hazards by locating forms at convenient places, requiring a prompt follow-up response by safety staff to all such submissions, and highlighting actions taken in newsletters and posters.

Comparisons were made of the hazard reporting rates of employees, the number of recorded staff injuries or illnesses of staff members, and the content of the hazardous conditions reported by the employees as related to their recorded injuries and illnesses during 12 month periods before and after the start of this worker-based reporting system. Results showed that the frequency of hazard reporting increased during the intervention period and the frequency of actual injuries and illness declined during the last 6 months of the trial, by the end of which most of the hazard control recommendations had been implemented. This finding suggested an increased safety consciousness among the workers and a consequent reduction in the number of job mishaps. In analyzing the content of injury and illness records with the hazard recognition reports, there were instances of the hazard reports far exceeding the recorded cases of related injuries, which in turn became a basis for prioritizing control needs. Indeed, in several instances during the trial period, accident risk factors identified in worker reports were not acted upon soon enough to prevent an injury that occurred soon afterward.

On the other hand, there were also instances where some hazards resulting in a high percentage of injury cases went undetected by the workers. Needle puncture wounds and physical exertion back injury from patient lifting were particularly notable. Because these mishaps are inherent in job routines and procedural in nature, their risks appeared less obvious to the workers than those posed by fixed physical features in their work environment. This omission indicated the need for employee training in appreciating functional kinds of hazards as a means of improving their overall hazard recognition skills. The latter was one of the basic recommendations agreed to by management, who, being satisfied with the overall findings of the trial, decided to adopt this worker participation effort as a permanent hospital program.

### 8.3.4 Worker Participation in Ergonomic Problem Solving

Ergonomics addresses the interaction of job demands and worker capabilities, the aim being to design the work requirements and/or workplace conditions in ways that will optimize productivity and at the same time preserve the health and safety of the workforce. Although the subject is much broader in scope [27], the rising incidence of musculoskeletal disorders of the upper extremities and the unabated numbers of costly low back problems in U.S. industry have focused ergonomic concerns on these types of problems in the main. Much is already known about occupational risk factors for these kinds of disorders, the major ones being forceful exertions, awkward body postures, local contact stresses, and repetitive motions [28]. Some efforts at controlling these hazards through the redesign of tools, improvement of Workstation layouts, and the use of less fatiguing work organization methods have been reported, and guidelines for carrying out these actions have also been publicized [29–31]. Worker involvements in such activities and aspects of their participation are described in the following.

The automobile and auto parts industry has been the primary site for participatory ergonomics programs in the United States as well as in other countries. Indeed, the tradition of assembly line work with numerous workers engaged in short-cycle tasks requiring repetitive turning and twisting actions with tools and/or frequent lifting or other forms of manual materials handling make it a natural candidate for ergonomic study and problem solving. Reports in the popular literature cite a number of cases where worker participation has been instrumental in achieving successful outcomes. LaBar [32], for example, describes how the introduction of quality circles in a U.S. tire manufacturing plant following a takeover by a Japanese corporation turned around sagging production levels and an increasing injury incidence rate. The quality circles, referred to as Employee Involvement Groups (EIGs), were set up in different departments and run in accordance with Japanese practices with a steering committee overseeing their activities. While addressing a variety of safety, production, and quality control topics, a sampling of improvements made or recommended by these groups indicated a focus on ergonomic problems and solutions. One improvement noted was replacing 18 stitches per tire procedure with one requiring just two stitches, thus reducing problems of repetitive motions believed responsible for the excess cases of carpal tunnel syndrome and tendinitis found for workers engaged in this task. Another was installing hydraulic systems to lift and turn 115 lb tires for inspection as opposed to having workers lift them, and using similar powered systems to lift heavy sheets of rubber. The apparent benefits were reductions in the incidence and severity of back injury. Overall, these and other types of hazard control measures in the plant caused a fivefold reduction in the incidence rate of worker injury over a 4 year period after the introduction of the EIGs. Inquiries of senior level management and union persons who remained with the company after the takeover credit these and other positive changes to listening to workers' suggestions and getting workers more involved in company activities. Quality circle concepts were instrumental in accomplishing this purpose.

LaBar's [33,34] descriptions of ergonomics efforts in two other automobile assembly plants emphasize the need to train the workforce at all levels with respect to recognizing relevant risk factors and early symptoms, the importance of engineering controls, and the role of employees in identifying problem areas and developing solutions. Regarding the latter, mention is made of over 200 suggestions received in 1 year from employees at one plant for ergonomic improvements, many of which were implemented and some carried over to a new plant facility as well. However, the reports are not clear in defining whether there are recognized formal groups where workers interact with others in providing this input or whether the suggestions are made strictly on an individual basis. References to teams, committees, and task forces acknowledge persons from the medical, safety, and engineering departments who appear to spearhead the hazard control program with workers advised to report problems to them. Nevertheless, successes are noted. One plant [34] reported a 50% drop in the number of ergonomics-related injuries 1 year after the training program, and the second [33] a 27% reduction.

Unlike the reports cited previously, which offer popularized accounts of worker participation efforts in ergonomic activities within the auto industry, Liker et al. [35] provide a detailed critical analysis of such experiences in two auto plants, one engaged in stamping auto parts and the other in machining and assembling auto chassis. The programs, as described, grew out of collaborations between the nation's largest automobile manufacturers and the auto workers union to study ergonomic issues in their work operations. For this purpose, it was agreed to engage outside parties to offer needed training and consultations. It is noted that university faculty and staff with specialties in this area played a large role in facilitating the development of programs within the two plants.

The study was undertaken to determine if a participatory ergonomics approach could yield benefits in reducing work-related injury, given downsizing and the need for the workforce to quickly adapt to new and different production technologies. At the time of the study, both plants were under threat of closing as a cost-saving measure and apparently were kept open only by management and labor efforts to come up with innovative plans that would keep them competitive. The two plants were each subdivided into two major areas with separate ergonomic groups to address their respective problems, propose solutions, and put them into place. An advisory committee was also established at each plant to provide direction for the overall effort and monitor progress. Three stages of ergonomic program development are described at each plant, referred to as laying the groundwork (stage 1), program development (stage 2), and maintenance (stage 3). The authors describe how differences in leadership style, makeup and motivation of the advisory committee and members of the ergonomics group, training in and use of job analytical methods, and experience in group decision making affected the processes in each of these stages and the resultant outcomes of the program. For example, leaders who were most trained in ergonomics but poor at facilitating group processes did little to engage the rest of their group members and thus lost their contributions. Others committed to ergonomics and participative management practices were most effective as judged by the satisfaction ratings of members attending meetings and observer ratings of ergonomic project reports and accomplishments at each meeting. Having connections to secure or lobby for outside resources was considered an added leader asset in that implementation of some of the approved changes required support from plant departments other than those involved in this program.

In another example, managers and engineer members of ergonomics groups who used their formal authority to assert their views in meetings were found to stifle the inputs of production level members who took a more backseat role. Attendance at regular meetings ultimately dropped off despite efforts to break this pattern of domination. While the few who remained active made recommendations that improved operations, their outputs paled in comparison to the number of workstation improvements made by other groups whose efforts took account of the ideas and views of all group members.

In still another example, the ergonomics group that achieved the most active involvement of its members showed more deliberateness in undertaking job analyzes and in reaching a consensus on a problem-solving strategy than those groups in which the level of participation was less apparent. Though the former group's effort took more time, it yielded more in-depth changes per workstation and a greater number actually implemented than the latter groups' efforts. Further mention of the Liker et al. [35] report will be made in a later section dealing with key factors in worker participation efforts to effect ergonomic improvements.

Aside from experiences in the automobile manufacturing industry, descriptions of ergonomic problem-solving activities in warehousing, textile manufacture, and shipping and mail delivery operations have appeared in which worker involvement has been emphasized [25,34]. Of these cases, only the warehousing example is described here because it offers the most detail and has other features deserving mention. Embodying a companywide program for gaining worker input into efforts aimed at enhancing product quality, operational efficiency, and workplace safety, a team formed of seven storekeepers who receive, stock, and then move raw materials from the warehouse to the production assembly line noted two problems posing potential hazards. One was that employees engaged in materials movement work were subject to undue numbers of injuries. Using a problem-solving process that included analyzing accident and medical reports, it was found that back injuries from lifting constituted the major hazard. Team brainstorming sessions plus the use of consultants in materials handling identified major vendor contributions to the problem. Specifically, it was found that vendors routinely exceeded both package weight and size specifications in their deliveries. Some cartons weighed twice the load limit, and others were so large that they had to be broken down to fit the tote boxes of the materials handling systems that were in use. These factors not only increased the risk of overexertion injuries but also required extra labor. Steps recommended by the storekeeper team to remedy this problem consisted of debiting vendors for any deliveries received that did not meet the packaging limits and tagging cartons in violation to alert workers to take added precautions in handling them. Both of these recommendations were accepted by management with estimates that back injuries could be cut by 50%, which, along with the net gain from the debit charged back to vendors for packaging violations, would result in substantial cost savings for this operation.

A second potential hazard noted by forklift operators in this warehouse was that their route of travel posed a risk of pedestrian accidents, especially to other workers who were engaged in product testing and other operations in the same area. During peak times many of these workers were present and stood in the aisles to do their jobs. Adding to the problem was the many blind alleys and intersections where approaching vehicles could not be seen by pedestrians until they were almost directly in front of them. Although there was not a single accident to cite, the forklift truck operators felt strongly that this was a problem that had to be addressed. They proceeded to log near-miss incidents, which occurred at a rate of at least one per day. They set a goal of reducing near-misses by 75% and through team brainstorming sessions drew up a list of solutions that were agreed to by consensus. Relocating product test stations, installing mirrors to aid viewing around corners, and redesigning pedestrian walkways were among the remedies offered. After implementing these and other solutions, near-miss observations were repeated and found that the goal set forth had been achieved. Through the reaction of one team member, the report acknowledges the team-building experience that took place during this problem-solving effort. Indications of the growth of interactive skills and increasing trust based upon ratings by team members taken over the course of team meetings are mentioned though no data are actually presented.

Mention was made earlier that this discussion of participatory approaches was taken from a NIOSH report [1] that described three case studies using this form of intervention in addressing ergonomic hazards in meat packing. The observations from these three studies viewed the processes of team building and follow-through activities in each case against a set of pointers or guidelines that are presented in Section 8.4. The results are summarized in Section 8.5.

In sum, the aforementioned reports of employee involvement in solving workplace health and safety problems in general and ergonomic hazards in particular show the merits of such an approach. At the same time, conclusions and generalizations from these results require tempering. For example, because popular as well as scientific periodicals are more prone to publish work showing positive results, cases where worker participation efforts may have failed to produce successful outcomes go unreported. Also, most cases have not included controls for other influences that could be affecting results apart from worker participation per se. Increased management attention to worker groups and related affairs, irrespective of any efforts to solicit their inputs into work conditions, can produce positive effects on job performance (see the Hawthorne studies described in Schermerhorn et al. [4]). However, these and other criticisms notwithstanding, the cases speak for themselves in demonstrating worker contributions to positive hazard control accomplishments in the workplace.

## 8.4 CRITICAL FACTORS AFFECTING RESULTS

In viewing the literature on worker participation as a whole, certain elements appear common to many of the documented reports on successful application of this approach to workplace issues or problems. The more prominent of these elements, reflecting organizational factors as well as methodology, are elaborated in the following. While there have not yet been any systematic efforts to study and assess the significance of these elements in facilitating both the process and outcomes of worker participation, some supportive evidence for their importance is noted based upon the cases reviewed earlier as well as other references to be cited. Most of the commentary focuses on a work team approach to worker participation with special attention to ergonomics-type problems.

### 8.4.1 Commitment and Responsiveness of Top Management and Supervisory Staff

Before even beginning discussion of a worker participation program, top management's commitment to the program is necessary, as is the support of supervisory personnel, union officials, and other worker leaders. Expressions of commitment can take various forms. The presence of such officials serving on committees that set the overall goals for the program and monitor progress is one mode of expression. Another is a policy that formally delegates authority downward, allowing more worker input into decisions on working conditions via their participation on teams or other working groups set up for that purpose. Still another is a positive response to recommendations from such groups and the supply of resources to implement acceptable solutions. Liker et al. [35], in analyzing the ergonomic program experiences at two auto plants, note that committees serving steering or oversight functions for lower level groups should not overreach their roles. They mention how one such committee undertook some job analyzes and dictated suggestions for change that proved unfeasible. A top-down approach nullifies the whole concept of worker participation and was perceived in that way by the workers. It was later rectified. The support of middle-level supervisors for worker participation efforts can be problematic if they see their usual responsibilities being diluted. Many quality circle efforts stated in many U.S. plants, though showing some initial benefits, did not last, the suspicion being that the resistance of middle managers was one of the factors that led to the program demise. Supervisors who remain supportive see their role as that of mentoring workers on ways to improve their job performance and helping worker groups refine their suggestions and presentations to top management committees to secure approval.

### 8.4.2 Management and Worker Training

Organizational changes enabling front-line workers to have more input into decisions on workplace conditions and the manner of implementation necessitates additional training of both management and the affected workers that goes beyond the usual subjects of instruction. For workers, one major need is to improve their communication skills and their ability to interact with others in group projects. As Lawler [2] notes, quality circles and work teams, in particular, require numerous meetings during which positive interactions among the worker members and other parties can be critical to effective group action. Worker training in empowerment techniques now being offered in union-sponsored safety and health courses stresses these and other objectives in efforts to promote change for reducing the risks of injury and disease [36].

Managers at various levels may also need training in the listening and feedback skills necessary to work with groups of workers who are assuming responsibilities for decision making. Cascio [3] notes that both groups need to learn the basic interpersonal skills necessary to build respect for each other. On the technical side, and where emergent problems are at issue, special training for workers, managers, and supervisory staff may be warranted. Ergonomic hazards fall into this category, and most of the reports reviewed previously mention some form of additional instruction given to both the workers and managers to facilitate efforts in defining ergonomic risk factors and ways to control them. Resources for covering assorted training needs must be a consideration in a worker participation program, including provisions for outside consultants if necessary.

Aside from the subject of training, increasing attention is being paid to the manner of instruction in the area of occupational safety and health [36,37]. Adult learning techniques stressing active forms of instruction through case study and demonstrations and targeting issues directly related to the trainees' experiences appear to have the most merit. Special needs of some who because of language problems or other deficiencies have trouble in comprehending material are also being met by the use of interpreters or visual aids.

### 8.4.3 Composition of Participants

As already noted, there is no single form of worker participation that meets all needs. The approach depends on the nature of the problem to be addressed, whether local to a group or having wide-range implications, the skills and abilities of those to be involved, and the penchant of the organization for joint labor-management or participative approaches in problem-solving ventures. By their very nature, ergonomic problems, though perhaps specific to a given job or operation, typically require a response that cuts across a number of organizational units. Indeed, hazard identification through job analyzes, records of injuries, or symptom surveys and the development of control measures and their implementation can necessitate inputs from safety/hygiene, human resource, engineering, maintenance, and medical staff persons plus ergonomics specialists. These people plus workers and management representatives are considered essential players in a meaningful program effort. In listing possible members of an ergonomics team, Vink et al. [38] also include members from purchasing units as the issues raised can have implications for procurement actions, for example, added or revised specifications on new equipment orders.

Drawing front-line workers or their representatives for any work team approach to ergonomic problem solving from the real or suspected problem areas or operations is the natural choice. For reasons already stated, the intimate knowledge of the job scene and insights into problems possessed by these individuals can be tapped for decision making and can facilitate implementation. Emphasizing the importance of this kind of input, some recommend that workers themselves prioritize all proposed solutions in making final decisions or before a final review by experts [38]. Supervisors and specialist members of a work team must be careful not to dominate discussions or to allow their stature or expert knowledge to intimidate the workers, as either will limit their contribution to the group process. Consultants brought in to advise on a problem also present this risk. Rather than dictating solutions to those who know the job through everyday experience, consultants who make an effort to work with the group in formulating procedures for defining and solving problems are far more likely to produce successful outcomes. Benefits from these kinds of services can be experiences that build in-house resources for tackling future concerns.

While there is no "correct" size for a work team, the range appears to be 7–15 members for obtaining optimal results. Larger groups present difficulties in creating effective group interactions and cohesiveness, both considered critical to effective decision making [2]. A need for larger representation may be met by setting up parallel smaller groups and establishing a second-level steering or coordinating group to monitor the overall effort as necessary.

### 8.4.4 Information Sharing

Effective worker participation in problem solving requires access to information. In terms of addressing hazard control issues, records of accidents, injury data, and cost figures for proposed control measures need to be made available to the teams who are expected to come up with realistic recommendations for solving such problems in their work areas or operation. Knowledge of other departmental functions and business matters in general may also be essential if the problem at hand and its solution have broader implications. As already noted, ergonomic issues readily transcend the areas of immediate impact, which gives greater importance to communication and cooperation between the various organizational units and parties involved. Even more important is that management motivations for establishing or maintaining work teams be made clear to the participants and that the value of their activities be appropriately recognized and rewarded. Misinformation or misperceptions here can be damaging. The perception that management opts for suggestions from work teams that cut costs or improve productivity without demonstrating equal regard for those that benefit worker welfare can extinguish the program. Cascio [3] notes that in order for workers to be convinced that working harder and smarter will not cost them their jobs, they must be assured of job security.

### 8.4.5 Activities and Motivation

OSHA inspections, citations for violations, and work-related injury or illness statistics can prompt organizations to take actions for hazard control. Teams or groups formed for that purpose follow a common set of steps, typically holding discussions to define the problem, gathering and analyzing data to sort out key elements, and developing and agreeing on recommendations for control actions and plans for implementation. According to the reports of Liker et al. [35] and Lewis et al. [25], actions taken by groups that reflect deliberate discussions of ideas, more orderly forms of data collection, and the use of analytical techniques have the best chance of furnishing effective solutions to problems. But these points aside, what can drive the activity level of a work team and motivate its members to be responsive to its tasks or objectives? The psychology literature indicates that setting a goal and providing frequent feedback marking progress toward attainment of that goal are potent ways for bringing about behavioral actions toward the prescribed ends. Applying these ideas, a wealth of studies exist in the occupational safety and health literature that show the merits of goal setting and feedback in enhancing safety performance among worker groups who are at risk [37,39,40]. Similarly, several of the worker participation cases described in this chapter make mention of goal setting by the work team and evaluations to determine if and when the goal was met. Other factors are more subtle but nevertheless important. The commitment of the workers themselves and the team leader in the belief that their efforts are going to make a difference can be a driver. Liker et al. [35] note how the success of worker groups in the ergonomic study at the two auto plants was shaped by leaders who were totally committed to the process of group problem solving. Management's recognition and rewards for accomplishments of the work teams in solving problems can serve to reinforce these actions and further the team's efforts to tackle other issues. The literature notes too that worker participation programs are perceived positively by those members who participate directly; those not involved do not share that view.

### 8.4.6 Evaluation

References to feedback and goal attainment presume that some measurable indicators of team performance are being applied. The ergonomic cases in the auto plants reviewed previously used observer and participant ratings of team meetings in terms of satisfaction with their accomplishments, number of work situations studied for problems, and recommendations made and/or actually implemented. These represent process-type measures. Continuation of the program also represents this type of measure although not expressly mentioned in these cases. Outcome indicators such as changes in frequency and severity data of work-related injury and illness before and after the formation of work teams for addressing ergonomic hazards have also been used but have limitations. For one thing, unless applied to large databases, that is, employers with large labor forces, musculoskeletal disorders from ergonomic hazards remain statistically rare events and lack sufficient variability for meaningful evaluation. For another, the use of these measures can necessitate an extended time frame to determine whether the intervention has had any beneficial effects. Other influential factors apart from work team efforts may occur in this time period that can confound observations of this type. The cases cited in the general occupational safety and health literature have used surrogate indicators for assessing interventions such as near-misses for accident potential, extent or adherence to safe work practices, and/or the use of personal protective devices as evidence of reduced exposure and risk for more chronic disorders [37]. In this regard, data on the actual reduction of risk factors or levels of exposure to them could also serve to indicate the before and after benefits of ergonomic interventions stemming from work team efforts. In addition, surveys indicating fewer complaints or less fatigue or discomfort among workers following changes instituted by the work team could be taken as a positive sign of ergonomic job improvement. Of course, without baseline data or control groups to rule out intervening influences, there will be questions as to whether any of the aforementioned changes are truly due to the work team's actions. It is to be stressed that judgments of the efficacy of worker participation in team approaches to ergonomic hazard control or other endeavors will require data collection on measures that are valid reflections of this type of intervention.

Table 8.1 offers a series of pointers or guidelines for framing worker participation and general team-building programs that summarize the major thoughts of this section. Table 8.2 captures the main observations of the NIOSH case studies [1] using a work team approach to ergonomic problem solving at three meatpacking plants as viewed against these guidelines. Both similarities and differences are noted among these findings with

TABLE 8.1 Important Pointers in Framing Worker Participation and Team-Building Approaches to Workplace Problem Solving as Suggested by the Literature

| Issue | Pointer |
|---|---|
| Management commitment | 1. Top management's commitment and support of worker participation approaches to company problem-solving needs are critical, as is the cooperation of lower level supervisors and union officials or recognized worker leaders |
| | 2. Policy declarations on the importance of participative approaches in addressing workplace issues require follow-up management actions to prove credibility. Those having merit are worker memberships on existing or newly formed groups at various levels within the organization, including those that have authority to make decisions in local areas of operation; providing timely responses to worker generated proposals for problem solving and resources to implement |
| | 3. Efforts will be needed to redefine the roles of midlevel supervisors as mentors to workers, to work with them in promoting ideas for work improvement and ways in which they can be implemented |
| Training | 1. Workers and management staff plus others who may be formed into a work team, task group, or committee will require added training to ensure effective joint actions. Workers will need training in communication skills and abilities to interact in group problem-solving tasks; managers, in listening and feedback skills |
| | 2. Both workers and managers plus other participant members of a work team or task group should be given the necessary technical training to appreciate the targeted problems at issue. Resources for this and other add-on training should include provisions for outside consultants or experts as may be necessary |
| | 3. Training practices should stress active forms of instruction focused on issues relevant to the trainees' experience. Special needs of those having language difficulties or other impediments to comprehension should be addressed |
| Composition | 1. No single form of worker participation can effectively fit all needs. Approaches depend upon the problem(s) to be addressed, whether limited to one group, area, or operation or having broader ramifications, the abilities of the workforce involved, and the climate of the organization in terms of using participative approaches in problem solving |
| | 2. Teams formed to address workplace problems that cut across different units in an organization should include representatives from such groups in addition to impacted workers, management persons, and technical consultants as needed. Groups of 7–15 persons can afford ample interactions and cohesiveness in actions |
| | 3. Precautions should be taken to prevent supervisors/managers, specialists, and consultant persons on a team from intimidating front-line worker members of a team or dominating discussion |
| Information sharing | 1. Effective worker participation and team efforts to solve problems demand access to information germane to the issues in question |
| | 2. As the team participants may represent different operations and be at different staff levels, the success of group efforts can hinge on sharing information |
| | 3. Management must be up-front and honest in communicating their support for participative decision making and in acknowledging possible consequences of actions that may be proposed. Worker concerns for job security are certain to raise questions |

TABLE 8.1 (continued) Important Pointers in Framing Worker Participation and Team-Building Approaches to Workplace Problem Solving as Suggested by the Literature

| Issue | Pointer |
|---|---|
| Activities and motivation | 1. Team-building activities invariably include meetings to clarify aspects of the problem, doing data gathering and analyzes to isolate causal or contributing factors, and developing remedial suggestions and planned efforts at implementation. Procedures reflecting orderly, systematic ways for dealing with each of these elements offer the best chances for success |
| | 2. Goal setting and frequent feedback to mark progress in a group's problem-solving efforts loom as key ways for motivating performance |
| | 3. Team leader commitments to the objectives of the group can facilitate accomplishments |
| | 4. Management's recognition and rewards for team success in problem-solving work can reinforce and sustain the continued interest of team members |
| Evaluation | 1. Team performance efforts need to be evaluated, and suitable process and/or outcome measures should be used for that purpose |
| | 2. Surrogate indicators may offer alternatives to more basic measures in cases where the latter data do not satisfy conditions for meaningful evaluations |

*Source:* Gjessing, C.C. et al., *Participatory Ergonomic Interventions in Meatpacking Plants*, National Institute for Occupational Safety and Health, Cincinnati, OH, 1994.

regard to the guideline considerations. Most adhere to the pointers; some add important qualifiers or other considerations. An extensive analysis of these results and the lessons learned from the exercise is found in the NIOSH report [1].

## 8.5 EMERGENT ISSUES AND NEEDS

### 8.5.1 Political and Economic Factors of Consequence

Both political and economic factors continue to give increasing importance to worker input into decisions affecting company business matters and operations. OSHA reform legislation, the adoption of total quality management (TQM) concepts, and the downsizing and restructuring of businesses are particularly relevant to the topic of this report, and brief comments stressing the connection are noted in the following.

#### *8.5.1.1 OSHA Reform Legislation*

Discussions of OSHA reforms in Congress have considered a provision requiring companies with 11 or more workers to create joint management and employee safety committees [41]. The rationale is that forming such a group would enhance both the employers' and employees' commitment to addressing workplace hazards. By-products of this experience are also noted, such as greater workforce morale, increased worker responsibility for their own safety, and improved trust and cooperation between management and employees. A National Safety Council survey [20] found that responses from companies without such committees agreed with these views. At the same time, both these respondents and others who have existing worker-management safety committees indicated that the formation of such committees was not the only way to increase worker participation in safety and health matters. Other forms were surveys, group meetings, and the solicitation of individual suggestions. Perhaps the issue is not so much the form of worker involvement as the provision

TABLE 8.2 Observations in Meatpacking Case Reports: Pointers in Worker Participation/Team Approaches to Ergonomic Problem Solving as Suggested by the Literature

| Condition | Observations |
|---|---|
| Management commitment | *Case 1*: Formalized policy on ergonomics hazard control efforts involving worker participation. Plantwide committee formed to deal with such problems comprising department heads, worker representatives, and others instrumental in accomplishing goals. Made resources available to implement team-proposed solutions in a minimal time period |
| | *Case 2*: Instituted program in 1986. Issued formal policy on worker participation in ergonomics problem solving. Designated an ergonomics program coordinator to oversee multiplant efforts who sat on the top decision-making group of the corporation. Ergonomics committees formed in each plant with representatives from management, worker groups, and others in position to effect proposed changes |
| | *Case 3*: Offered resources to support team-building activities including overtime pay for workers to attend meetings. Ranking managers/directors sat on ergonomics teams with workers |
| Training | *Case 1*: Provisions made for training team members in both team building and ergonomics problem solving, the latter including opportunities for practicing methods and techniques. General awareness training on ergonomic problems given to all plant employees. Company safety officer capable of handling efforts with university investigator assistance |
| | *Case 2*: Specialty training on ergonomic issues given to team members. Awareness training on ergonomic hazards given to all employees, even office staff, as part of overall corporate policy. Corporate ergonomics coordinator assumed responsibility for all such training |
| | *Case 3*: Formal training limited in time and focused completely on ergonomic issues. No in-house expertise; handled exclusively by outside university consultant |
| Composition | *Case 1*: Team memberships ensured inputs from production workers engaged in the problem jobs, supervisory and engineering personnel, plus maintenance persons from the same department, or a combination who could facilitate data gathering, development, and implementation of proposals. Teams were 7–9 members in size and apparently small enough to be effective, considering overall results reported. Second-level, plantwide ergonomics committee representatives included the purchasing head, which is a recommended practice, and other members who provided close team support (e.g., nurse member supplied injury/medical data in defining problem jobs) |
| | *Case 2*: With two exceptions, departmental teams were formed similar to Case 1 as was a plantwide ergonomics committee at the intervention site. One difference was the presence of the corporate ergonomics coordinator, who served in an advisory capacity at both the team and plant committee level. The presence of the ergonomics coordinator at this site and other plants in the corporation suggested close oversight of all company ergonomics activities and possible limits on individual team/plant autonomy |
| | *Case 3*: Teams as formed did include production workers assigned to the problem jobs plus supervisory staff and maintenance people from the areas of concern. Also members were top plant officials whose presence could have limited openness of discussion and inputs from production workers, although one top official was intentionally absent from many meetings so as to not to exert disproportional influence on the team. The teams experienced some turnover in production worker members and had to cope with language/literacy limitations of some participants. Reasonable efforts were made to deal with some of these problems |

TABLE 8.2 (continued) Observations in Meatpacking Case Reports: Pointers in Worker Participation/ Team Approaches to Ergonomic Problem Solving as Suggested by the Literature

| Condition | Observations |
|---|---|
| Information sharing | *Case 1*: Individual teams received company information on CTD prevalence, workmen's compensation claims and costs, sick absence, and employee turnover to assist in defining problem jobs, though the means of access and/or mode of data presentation were not described. A more direct way for workers to track injuries was recommended. Opportunities to collect other data reflecting risk factors, interviewing workers as to complaints, were freely granted. Varied efforts made to publicize and keep all plant employees informed of team's activities, progress, and accomplishments |
| | *Case 2*: It is intimated that teams shared data similar to that noted in Case 1 for the jobs that were preselected by management and the corporate ergonomics coordinator for study at the plant intervention site. Also the teams had access to ergonomics risk factor information and could collect other information that went into the decisions to focus on these jobs. Monthly and quarterly reports on the team's progress were circulated to other plants in the corporation |
| | *Case 3*: Team members were provided injury statistics and workers' compensation data at the start of the project, but the teams did not review these records as the project progressed. Team activities were publicized in a quarterly newsletter distributed to all employees |
| Activities and motivation | *Case 1*: Teams attempted to follow an orderly approach in defining and rank-ordering jobs through the use of injury/medical record data and risk factor evidence, then brainstorming and prioritizing ideas for improvement along with means for implementation. These experiences should build team member skills and lay a strong foundation for future efforts. Proposed solutions took account of ease of implementation, feasibility, and cost and opted primarily for engineering changes, a preferred approach. Those actually implemented proved to have positive effects but did not meet the expectations of some teams and the workforce as a whole. This resulted in feelings of dissatisfaction with the overall program. More realistic goal setting would seem indicated |
| | *Case 2*: Procedures used customized forms, checklists for data gathering on risk factors, and decisions on solutions developed by the company. These gave order to team activities. Teams focused efforts on preselected problem jobs that were recognized as posing difficult problem-solving elements based on earlier attempts. Easier job targets could have provided the teams with some early success and positive motivation; the teams expressed disappointment that proposed changes would take some time to implement |
| | *Case 3*: Two jobs for study were preselected by management and the outside investigator-consultant. A team was formed for each job. Team activities almost solely directed to brainstorming preselected jobs for solutions, which were then prioritized as to feasibility and cost factors. Approach jumps to solution without allowing for much team understanding of the problem. Although some improvements were made to the jobs, certain elements were intractable, making it difficult for the teams to sense success |
| Evaluation | *Case 1*: Data collection addressed both team-building and performance issues in ways that showed changes over time, including first indications of positive results of team-generated ergonomic improvements following implementation. Both subjective survey methods and traditional objective measures were included in the evaluation, with efforts made to tap not just team responses but the workforce as a whole and to analyze the results in terms of those whose jobs were affected and those not affected |

(*continued*)

TABLE 8.2 (continued) Observations in Meatpacking Case Reports: Pointers in Worker Participation/Team Approaches to Ergonomic Problem Solving as Suggested by the Literature

| Condition | Observations |
|---|---|
| | *Case 2*: Data collection included self-report surveys of team members on how well meetings were run, productivity, representations, quality of leadership, and other team-building issues. Data also collected in symptom surveys to corroborate problems and risk factors and set a baseline for determining benefits of improvements along with the more traditional injury/medical data points. None of the proposed solutions could be implemented and/or evaluated within the study time frame |
| | *Case 3*: Surveys of teams concentrated on aspects of member interaction and team effectiveness, as well as responses to the objectives of the program as a morale builder, some given at the beginning and end of the study period. Data analyzed by different representative groups to show differences in views between management/supervisory staff and production worker team members. Besides symptom surveys, a plan was included to collect measurements of hand/wrist motions and forces before and after some proposed job improvements so as to offer quantitative indications of the potential benefits of certain job changes in more immediate ways |

*Source:* Gjessing, C.C. et al., *Participatory Ergonomic Interventions in Meatpacking Plants*, National Institute for Occupational Safety and Health, Cincinnati, OH, 1994.

of appropriate and effective mechanisms to ensure worker input. OSHA's current guidelines [42] for meatpacking plants in establishing a program to deal with ergonomic hazards cites the need for employee involvement including membership in safety and health committees who could process information to target problem areas, analyze risk factors, and make recommendations for corrective action. Regardless of the outcome of the legislative process, the push for worker involvement in company safety and health program practices committees is apparent.

#### *8.5.1.2 Total Quality Management Movement*

Adding impetus to worker participation approaches in industrial management practices is the growing acceptance of TQM principles first introduced by Deming and others [43–45]. Empowering workers to solve problems, help improve processes, and foster ongoing teamwork to ensure quality efforts at each stage of producing a product or providing a service is a key element in the TQM plan. Others are provisions for education, retraining, and self-improvement of the workforce, leadership roles that support or enable workers to do a better job, and continual striving to improve company operations and productivity. Auditing of performance at all stages is implicit to attaining the goal of a total quality effort. Safety and health objectives can be readily folded into the TQM program wherein cases of work-related injury and illness are treated as defects in the quality of the work process. Signs of unsafe conditions, poor work practices, and risky worker behaviors are targets for joint worker-management actions aimed at their elimination. Millar [44] and others, in extolling the virtues of TQM with respect to occupational safety and health, report that companies who have adopted this style of management show both a reduction in work injuries and number of lost workdays and an increase in productivity.

*8.5.1.3 Downsizing and Restructuring of Businesses*

The need to remain competitive in global markets and to maintain profitability has caused many U.S. businesses to reduce their workforces and restructure their operations. As a streamlining, cost-saving move, layers of middle management or supervision have been removed in many cases, giving work units at lower levels more autonomy in directing operations including those concerned with workplace safety and health. Greater worker involvement is seen as a key to success in making this change. LaBar [46, p. 30] paraphrases the statements of one executive of a major U.S. corporation:

> We used to have supervisors watching people, and if something wasn't being done right, the supervisor would walk over and correct it. With fewer management people around, self-directed worker groups must assume responsibility for everything—productivity, quality, safety.

Added training for workers is considered crucial to getting workers involved in safety as well as other issues. It is recognized too that garnering worker involvement in these efforts can be complicated if layoffs are also occurring in their ranks, which can produce morale problems. Labor-management cooperation on ways to resolve this conflict will be necessary.

### 8.5.2 Other Needs

The political and economic factors just described make apparent the trend for workers to have greater input into defining and solving workplace problems, and the literature takes account of the merits of such an approach and factors of consequence. What remains is to expand the knowledge base of applications, given that forms of worker participation, problems at issue, and situational circumstances may all vary. Recognizing that ergonomic issues cut across so many different work situations, it is important that any efforts made to solve such problems through worker participation approaches be reported. The examples documented so far offer only selective views of the processes involved in the efforts, and thoughts about generalizability and robustness await further confirmation.

Moreover, the successes being reported are largely based on first or early observations and relate to problems that are most amenable to solution. Even here, and as shown in the NIOSH meatpacking cases [1], worker expectations about having their recommendations approved, implemented, and producing results often can exceed realistic time lines. This suggests the need for techniques to overcome frustrations and sustain worker participation and interest where circumstances slow the adoption of recommended actions for ergonomic improvement or where the problems under study are complicated and do not lend themselves to any immediate solutions. Perhaps redirecting efforts to other, more tractable kinds of workplace issues may be one tactic to maintain the motivation of those involved, given delays or difficulties in the ergonomics area. Longer term study of worker participation efforts directed to ergonomic problem solving or other hazard control matters would help clarify the dynamics of this situation.

As already mentioned, employee participation in company efforts to solve problems raises the motivation and productivity levels of those involved. This finding together with the fact that most ergonomic improvements in work operations also result in productivity gains ought to be a big selling point for promoting this type of intervention. How best to use this approach in overcoming the usual business concerns of added cost and "bottom line" impacts on competitive market position in companies having suspected ergonomic problems needs to be considered.

## REFERENCES


1. C. C. Gjessing, T. F. Schoenborn, and A. Cohen, *Participatory Ergonomic Interventions in Meatpacking Plants*, DHHS (NIOSH) Publ. No. 94-124, National Institute for Occupational Safety and Health, Cincinnati, OH, 1994.
2. E. E. Lawler III, *High Involvement Management*, Jossey-Bass, San Francisco, CA, 1991.
3. W. F. Cascio, *Applied Psychology in Personnel Management*, Prentice-Hall, Englewood Cliffs, NJ, 1991, Chapter 6.
4. J. R. Schermerhorn, Jr., J. G. Hunt, and R. N. Osborn, *Managing Organizational Behavior*, Wiley, New York, 1985, Module A, pp. 5–8.
5. G. LaBar, Is your safety committee legal? *Occup. Hazards 11*:35–38 (1993).
6. U.S. Dept. of Labor and U.S. Dept. of Commerce, *Commission on the Future of Worker-Management Relations, Fact Finding Report*, U.S. Dept. of Labor and U.S. Dept. of Commerce, Washington, DC, May 1994.
7. N. Krigsman and R. M. O'Brien, *Organization Behavior Management and Statistical Process Control*, Haworth, Binghamton, NY, 1987, pp. 67–82.
8. M. L. Marks, The question of quality circles, *Psychol. Today 3*:36–45 (1986).
9. K. I. Miller and P. R. Monge, Participation, satisfaction and productivity: A meta-analysis, *Acad. Manage. J. 29*(4):727–753 (1986).
10. R. W. Griffin, Consequences of quality circles in an industrial setting: A longitudinal assessment, *Acad. Manage. J. 31*(2):338–358 (1988).
11. H. H. Cohen, Employee involvement: Implications for improved safety management, *Prof. Saf. 6*:30–35 (1983).
12. S. Edwards, Quality circles are safety circles, *Natl. Safety News*, June 1983, pp. 31–35.
13. National Safety Council, *Safety Circles*, Occup. Safety and Health Data Sheet 738 Rev. 93, Natl. Safety Council, Itasca, IL, 1993.
14. Office of Technology Assessment, *Preventing Illness and Injury in the Workplace*, OTA-H-256, U.S. Congress, Washington, DC, 1985, pp. 315–323.
15. R. H. Peters, *Review of Recent Research on Organizational and Behavioral Factors Associated with Mine Safety*, Bur. of Mines Inf. Circ. 9232, U.S. Dept. of Interior, Washington, DC, 1989.
16. R. A. Guzzo, R. D. Jette, and R. A. Katzell, The effects of psychologically based intervention programs on worker productivity, *Personnel Psychol. 38*:275–281 (1985).
17. J. K. Liker, M. Nagamachi, and Y. R. Lifshitz, A comparative analysis of participatory ergonomics programs in the U.S. and Japan manufacturing plants, *Int. J. Ind. Ergon. 3*:185–199 (1989).
18. Bureau of National Affairs, Lower injury rates, costs seen as a result of voluntary compliance, California/OSHA states, Washington, DC, January 5, 1984.
19. L. I. Boden, J. A. Hall, C. Levenstein, and L. Punnett, The impact of health and safety committees, *J. Occup. Med. 26*(11):829–834 (1984).
20. T. W. Planek and K. P. Kolosh, Survey shows support for safety and health committees, *Saf. Health 1*:76–79 (1993). (Also summarized in *Survey of Employee Participation in Safety and Health*, An executive summary available from the National Safety Council, 1121 Spring Lake Drive, Itasca, IL.)
21. W. Cooke and F. Gautschi, OSHA, plant safety programs and injury reduction, *Ind. Relations 20*(3):245–257 (1981).
22. C. Touhy and M. Simard, *The Impact of Joint Health and Safety Committees in Ontario and Quebec*, Spec. Rep. to the Canadian Association of Administrators of Labour Law, Toronto, Canada, August 1992.

23. K. J. Saarela, An intervention program utilizing small groups: A comparative study, *J. Safety Res. 21*:149–156 (1990).
24. E. B. Lanier, Jr., Reducing injuries and costs through team safety, *Prof. Saf. 7*:21–25 (1992).
25. H. B. Lewis, A. S. Imada, and M. M. Robertson, Xerox leadership through quality: Merging human factors and safety through employee participation, *Proc. 32nd Annual Meeting*, Anaheim, CA, 1988, Vol. 2, pp. 756–759.
26. L. Lin and H. H. Cohen, *Development and Evaluation of an Employee Hazard Reporting and Management Information System in a Hospital*, Safety Sciences Contract Rep. No. 210-81-3102, Division of Safety Research, National Institute for Occupational Safety and Health, Morgantown, WV, 1983.
27. A. Cohen and F. Dukes-Dobos, Applied ergonomics, in *Biological Responses—Patty's Industrial Hygiene & Toxicology*, 2nd edn., Vol. 3B, Wiley Interscience, New York, 1985, pp. 375–430.
28. W. Keyserling, T. Armstrong, and L. Punnett, Ergonomic job analysis: A structured approach for identifying risk factors associated with overexertion injuries and disorders, *Appl. Occup. Environ. Hyg. 6*:353–363 (1991).
29. S. Ulin, T. J. Armstrong, and G. G. Herrin, Preferred tool shapes for various horizontal and vertical work locations, *Appl. Occup. Environ. Hyg. J. 7*:327–337 (1992).
30. T. R. Waters, V. Putz-Anderson, A. Garg, and L. J. Fine, Revised NIOSH equation for the design and evaluation of manual lifting tasks, *Ergonomics 36*:749–776 (1993).
31. E. Grandjean, *Ergonomics in Computerized Offices*, Taylor & Francis, London, England, 1987.
32. G. LaBar, Employee involvement yields improved safety record, *Occup. Hazards 5*:101–104 (1989).
33. G. LaBar, Ergonomics, the Mazda way, *Occup. Hazards 4*:43–46 (1990).
34. G. LaBar, Succeeding with ergonomics, *Occup. Hazards 4*:29–33 (1992).
35. J. K. Liker, B. S. Joseph, and S. S. Ulin, Participatory ergonomics in two U.S. automotive plants, in *Participatory Ergonomics*, K. Noro and A. Imada, Eds., Taylor & Francis, London, England, 1991, Chapter 6.
36. N. B. Wallerstein and M. Weinger, Health and safety education for worker empowerment, *Am. J. Ind. Med. 22*(5):619–635 (1992).
37. A. Cohen and M. J. Colligan, *Assessing Occupational Safety and Health Training: A Literature Review*, Div. Training and Manpower Development, Natl. Inst. for Occupational Safety and Health, Cincinnati, OH, 1993.
38. P. Vink, E. Lourijsen, E. Wortel, and J. Dul, Experiences in participatory ergonomics: Results of a round-table session during the 11th IEA Congress, Paris, July 1991, *Ergonomics 35*(2):123–127 (1992).
39. J. S. Chhokar and J. A. Wallin, Improving safety through applied behavior analysis, *J. Safety Res. 13*:141–151 (1984).
40. B. Sulzer-Azaroff, T. C. Harris, and K. B. McCann, Beyond training: Occupational performance management techniques, in *Occupational Safety and Health Training—State of the Art Reviews*, M. J. Colligan, Ed., Hanley & Belfus, Philadelphia, PA, 1994, pp. 321–339.
41. M. P. Weinstock, OSHA reform: The push for worker involvement, *Occup. Hazards 12*:37–39 (1991).
42. Occupational Safety and Health Administration, *Ergonomic Program Guidelines for Meatpacking Plants*, OSHA Rep. 3123, U.S. Department of Labor, Washington, DC, 1993.
43. J. Roughton, Integrating a total quality management system into safety and health programs, *Prof. Saf. 6*:32–37 (1993).
44. J. D. Millar, Valuing, empowering employees vital to quality health and safety management, *Occup. Health Saf. 9*:100–101 (1993).
45. S. M. Mottzko, Variation, system improvement and safety management, *Prof. Saf. 8*:17–20 (1989).
46. G. LaBar, Safety management in tight times, *Occup. Hazards 6*:27–30 (1993).

# II

## Application of Ergonomic Principles

CHAPTER 9

# Job Analysis

Katharyn A. Grant

CONTENTS

## 9.1 INTRODUCTION

Ergonomic job analysis is the methodology used by engineers and safety professionals to describe work activities for the purpose of comparing existing task demands to human capabilities. Modern job analysis is founded in the work of early industrial engineers such as Frederick Taylor and Frank and Lillian Gilbreth [1]. At its inception, job analysis was primarily viewed as a tool for improving efficiency and productivity. However, in recent years job analysis techniques have been applied increasingly to identify work conditions and job demands associated with the onset of fatigue, overexertion, injuries, and chronic musculoskeletal disorders.* There is increasing evidence that if job analysis techniques are used to identify these conditions, effective control methods can be introduced to eliminate or reduce the risk of injury to the worker [2].

## 9.2 BACKGROUND AND SIGNIFICANCE TO OCCUPATIONAL ERGONOMICS

Renewed interest in job analysis techniques for identifying ergonomic hazards has its roots in several sources. First, there is research to indicate that the benefits of well-designed jobs include improved efficiency, safety, and satisfaction for employees. Many industries faced with rising workers' compensation costs and disability insurance premiums for musculoskeletal disorders have found job analysis and ergonomic redesign effective for reducing losses associated with musculoskeletal injuries [3–5]. Further, as companies face pressures to enhance quality and remain competitive in a global marketplace, there is hope that ergonomic improvements will ultimately translate into greater productivity and profitability.

Second, the growing interest in ergonomics parallels the increasing impact of society's expectations about occupational health. Although ergonomics was not mentioned by name in the Occupational Safety and Health Act of 1970 (OSH Act), the OSH Act states that it is the general duty of all employers to provide their employees with a workplace free of recognized hazards, whether or not they are regulated by specific federal standards [6]. In recent years, the OSHA has taken an active role in investigating ergonomic hazards under authority of the General Duty clause. Furthermore, recognition of the significance of ergonomic problems in the workplace has led OSHA to call for an "ergonomics standard" to define requirements for a comprehensive ergonomic safety and health management program for general industry [7]. Under the proposed model, employers will be required to implement a system for identifying risk factors associated with musculoskeletal injuries and determining if these factors have been eliminated or reduced to the extent feasible [8]. If this proposal is enacted, job analysis will become increasingly important as a component of this program.

* Fatigue is usually transient; if given sufficient recovery time, muscles can overcome fatigue without permanent damage. Overexertion injuries can be caused by a single event and include most sprains and strains and some back injuries. Chronic musculoskeletal disorders are usually caused by repeated trauma to the musculoskeletal system.

## 9.3 FUNDAMENTALS OF JOB ANALYSIS

The steps involved in performing an ergonomic job analysis depend, in part, on the purpose and scope of the study. Some situations will demand a comprehensive survey of all jobs in a facility; in other circumstances, the evaluation may be limited to a specific group of workers in response to a specific complaint. Job analysis procedures are also used to evaluate the effectiveness of various control measures. In general, however, the fundamental elements of job analysis are as follows [9]:

Identifying potential hazards

Preparing for field study

Conducting the field study

Interpreting the results

### 9.3.1 Identifying Potential Hazards

The process of identifying hazards arising from poor work design begins with two activities:

1. Reviewing records to identify jobs associated with a high rate of accidents and injuries
2. Becoming familiar with the processes and job activities that are performed in each work area

#### *9.3.1.1 Review Injury Data*

The most obvious indication that a job poses excessive work demands is a high rate of injury or absenteeism among workers who perform the job. Excessive turnover can also indicate that a job poses difficulties for a large percentage of the workforce. Information about injuries, absenteeism, and turnover can come from two sources: (1) existing plant records (e.g., accident reports, dispensary logs, OSHA 200 reports, and worker compensation records) and (2) surveys of current employees (see Section 9.4 for information on medical surveillance for ergonomic programs).

Based on surveillance information, jobs should be selected for analysis based on the number of injuries or complaints associated with the job, the severity of those injuries and complaints, and the number of workers affected by current job conditions [10]. If significant hazards are identified during the analysis and control methods are introduced, continued surveillance can later provide a measure of the effectiveness of these interventions.

#### *9.3.1.2 Review Processes and Job Activities*

Before the investigator can begin the process of collecting data needed for the analysis at the work site, background information about the job processes and work activities is needed. Basic information should be obtained by reviewing process and job descriptions, interviewing supervisors and employees, and conducting a "walk-through" survey of the work site. In some cases a checklist may be used to identify potential risk factors that may be present at the work site and require further investigation. The information collected should answer questions such as

1. How many workers are employed in each job?
2. What are the characteristics of the workforce (e.g., gender, age, education level)?
3. What are the primary tasks involved in each job? Do workers perform the same tasks over and over throughout the work shift, or do they perform a large number of different tasks?
4. Is there an established work rate? If so, how is the work rate determined (e.g., line paced, time standards, etc.)?
5. Is there an opportunity for workers to rotate to other jobs?
6. How much do workers employed in this job earn? What type of pay system is used (e.g., hourly wage, piece rate)?
7. How many hours do employees work per week? Is work organized into shifts? How much overtime is worked per week?

Although these questions do not address the presence of specific hazards, the information gained from responses is often useful in identifying resources needed for the evaluation, developing an analysis strategy, and interpreting data that are gathered later in the evaluation.

### 9.3.2 Preparing to Conduct a Job Analysis

#### *9.3.2.1 Gather Necessary Equipment*

Most information needed to identify and evaluate ergonomic hazards is collected through observation and direct measurement. Equipment needs depend on the nature of the suspected hazards and the desired level of sophistication in the measurement; however, in most cases, the following items are useful for gathering data for job analysis [11]:

1. Cameras and film for recording workers' postures and motions during job activities
2. Tape measures and rulers for measuring workstation dimensions and reach distances
3. Force gauges and spring scales for measuring the force of exertions (e.g., the force needed to push or pull a hand cart) and the weight of tools or objects
4. Stopwatches for measuring the duration of work activities, breaks, etc.

*Videotape* can be invaluable as a tool for documenting biomechanical stressors in the workplace. Guidelines for recording work activities on videotape are provided in Table 9.1. After work activities are recorded on videotape, one or more analysts can review tapes

TABLE 9.1 Guidelines for Recording Work Activities on Videotape

- If the video camera has the ability to record the time and data on the videotape, use these features to document when each job was observed and filmed. Recording the time on videotape can be especially helpful if a detailed motion study will be performed at a later date (time should be recorded in seconds). Make sure the time and date are set properly before videotaping begins.
- If the video camera cannot record time directly on the film, it may be useful to position a clock or a stopwatch in the field of view.
- At the beginning of each recording session, announce the name and location of the job being filmed so that it is recorded on the film's audio track. Restrict subsequent commentary to facts about the job or workstation.
- For best accuracy, try to remain unobtrusive; that is, disturb the work process as little as possible while filming. Workers should not alter their work methods because of the videotaping process.
- If the job is repetitive or cyclic in nature, film at least 10–15 cycles of the primary job task. If several workers perform the same job, film at least two or three workers performing the job to capture differences in work method.
- If necessary, film the worker from several angles or positions to capture all relevant postures and the activity of both hands. Initially, the worker's whole body posture should be recorded (as well as the work surface or chair on which the worker is standing or sitting). Later, close-up shots of the hands should also be recorded if the work is manually intensive or extremely repetitive.
- If possible, film jobs in the order in which they appear in the process. For example, if several jobs on an assembly line are being evaluated, begin by recording the first job on the line, followed by the second, third, etc.
- Avoid making jerky or fast movements with the camera while recording. Mounting the camera on a tripod may be useful for filming work activities at a fixed workstation where the worker does not move around much.

using slow-motion or real-time playback to accurately measure task durations or detect subtle or rapid movements [12]. Analysts can also record job activities on video before and after changes are made to determine if changes were effective in reducing exposure to hazards. Finally, videotape can be useful as a visual aid for ergonomic training and demonstration purposes.

More sophisticated methods for analyzing motions and evaluating responses to work demands are available, although most are rarely used outside of the research laboratory (Table 9.2). The primary disadvantage of most of these measurement systems is that the equipment tends to be expensive and not well suited to industrial environments. In addition, many of these methods require extensive training and calibration, and data reduction and analysis can be time consuming [13]. However, under the right conditions, some of these methods have been used effectively in industrial settings. For example, Armstrong et al. [14] successfully applied electromyography (EMG) to examine forearm muscle forces during poultry-cutting tasks. Habes [15], Christensen [16], Aaras and Westgaard [17], Gomer et al. [18], and Milerad et al. [19] also used EMG to examine muscle load and fatigue among assembly line workers, postal workers, and dentists. Marras et al. [20] employed a triaxial electrogoniometer to document the three-dimensional (3-D) trunk motion characteristics associated with over 400 industrial lifting jobs in 48 industries. Garg et al. [21] used portable devices for measuring heart rate and oxygen consumption to examine the physiological responses of grocery warehouse workers to their work tasks.

TABLE 9.2 Direct Measurement Systems for Characterizing Ergonomic Hazards

| Equipment | Use | Applications | Limitations |
|---|---|---|---|
| Accelerometers | Describe acceleration of body segments | Tasks involving high-speed motions | Six or more accelerometers required at each joint—can interfere with motion |
| Electromyographs (EMG) | Assess muscle activation patterns, muscle force or effort, and fatigue | Tasks involving static muscle exertions | Applicable only to static work; requires calibration; signal subject to noise |
| Finger force sensors | Assess finger forces, evaluate contact stresses | Manual tasks requiring force exertions with the hands and fingers | Sensors must be attached to tool or equipment surface or to fingertips (may affect tactile sensitivity) |
| Electrogoniometers | Describe posture and motion (velocity and acceleration) | Tasks requiring highly dynamic movements | Goniometer may affect motion pattern. Data must be telemetrized or logged to permit free movement |
| 2-D and 3-D video motion analysis systems | Describe posture and motion (velocity and acceleration) | Lifting tasks, static tasks, and tasks with highly dynamic movements | Require good lighting; 3-D requires multiple cameras and adequate floor space |
| Pendulum potentiometers | Describe trunk and shoulder posture (continuous) | Static tasks | Inertial effects can induce errors; provide no data about trunk loading |
| Heart rate monitors | Describe metabolic load and physical stress | Hot tasks, highly static or dynamic tasks, and tasks with mental demands | Signal may be subject to noise |
| Oxygen uptake monitors | Describe metabolic load/ physical stress | Tasks requiring whole body exertion | Equipment often bulky; require calibration |
| Wrist activity monitors | Describe work–rest patterns, movement frequency and intensity | Repetitive tasks, tasks with high metabolic demands and shift work | Additional work needed to link data to specific tasks; movement intensity data are limited |

#### *9.3.2.2 Identify Strategy*

The primary objective of the field study should be to collect sufficient information to allow the analyst to completely describe the job as it is currently being performed (i.e., what the worker is doing and how it is being performed in time). Generally, this requires the analyst to observe the job during a "typical" work period under "normal" operating conditions. The number of workers employed in a job, the number of tasks, the type of rotation system employed, and the hours of work all have an impact on the sampling strategy employed in gathering data. If workers perform a well-established set of tasks that tend to be repeated in a fixed sequence, the time needed for data collection is usually minimized. However, if jobs do not involve repetitive cycles or if irregular elements that occur only once or twice a day (e.g., machine setup) are a concern, then it may be necessary to observe workers at randomly selected times over an extended period (i.e., several days) to gather information about all of the various activities they perform. If necessary, work-sampling procedures can be used to estimate the percentage of time workers spend performing various activities [1].

## 9.3.3 Conducting the Field Study

### *9.3.3.1 Observe Work Processes*

In almost all cases, the bulk of job-related information will be collected through direct observation of the work tasks at the work site. For best accuracy, the work process should be disturbed as little as possible during the observation period. Because workers who perform the same job may use different methods due to differences in training, stature, or strength, several workers who perform the same job should be observed. The analyst should also attempt to observe the same task at different times during the work shift to determine if fatigue affects workers' performance or if the workload changes during the workday.

Formalized procedures for organizing and recording observational data have been developed and are described in Table 9.3. Because many of these systems are applicable to only one or two of the risk factors that may be present at a work site, they should be implemented only after a preliminary evaluation indicates that specific problems (e.g., awkward

TABLE 9.3 Observational Systems for Characterizing Ergonomic Hazards

| System | Hazards Described | Applications |
|---|---|---|
| Posturegram [22] | Static postures (head, trunk, upper and lower limbs) | Tasks involving static postures |
| Ovako working posture analysis system (OWAS) [23] | Static postures (back, upper and lower limbs) | Tasks involving static postures (e.g., steel production, construction) |
| Posture targeting [24] | Static postures (head, trunk, upper and lower limbs) | |
| Postural stability diagram (PSD) [25] | Static exertions with hands and feet | Tasks requiring static exertions (e.g., push–pull forces) |
| AET [26] | Static postures, heavy muscular work, strenuous or repetitive exertions | "All purpose" system—tasks presenting both physical and mental loads |
| Upper extremity posture analysis system [14] | Awkward upper limb postures, manual force exertion | Highly repetitive manual activities (e.g., poultry processing) |
| ARBAN [27] | Postures (neck, shoulders, trunk, legs), muscle stress (static and dynamic), vibration and shock | Repetitive work; can be used to compare alternative work methods |
| Classification system for the trunk and shoulders [28] | Awkward postures (trunk, shoulders) | Repetitive nonseated jobs (e.g., automotive assembly) |
| VIRA [29] | Awkward postures (neck, shoulders) | Seated, repetitive arm work |
| Posture recording [30] | Awkward/static sitting postures | Seated work |
| Physical work stress index (PWSI) [31] | Posture (whole body, hand), acceleration and/or vibration, thermal load | Jobs with significant physiological demands (e.g., lifting and digging) |
| Hand exertion classification system [32] | Repetitive or forceful manual exertions; awkward shoulder, arm, wrist, and hand postures; contact stresses; vibration | Repetitive manual activities (e.g., manufacturing and warehouse jobs) |
| Rapid upper limb assessment (RULA) [33] | Awkward posture (neck, trunk, arm, leg, wrist), muscle use (static and dynamic), muscle force | Repetitive tasks (e.g., manufacturing, grocery VDT, and microscope work) |

upper extremity postures) are present. For most of these systems, there is also an important trade-off between the time required to analyze and record the data and the level of detail produced by the analysis. For example, although the Ovako Working Posture Analysis System (OWAS) [23] is easily learned and applied, the procedure provides only a rough description of posture. Systems developed by Priel [22], Corlett et al. [24], and Armstrong et al. [14] provide a more detailed description of posture but require more time for training and data analysis.

#### *9.3.3.2 Interview Supervisors and Employees*

Informal discussions with workers and supervisors can also provide useful information for the analysis. Workers who perform a job on a daily basis are often the best source of information about the specific elements that may pose a hazard. Supervisors can provide a "big picture" view of operations as well as insight into the feasibility of proposed changes. Interviews also give workers the opportunity to participate and to provide input to the job analysis process. Soliciting workers' input may increase their willingness to accept changes to their job if redesign is necessary.

#### *9.3.3.3 Take Measurements*

Observing the job also provides the analyst the opportunity to make measurements at the job site. A sketch or drawing of the workstation is useful for identifying the location of fixtures, equipment items, etc. Sketches can be labeled with dimensions indicative of work surface heights, reach distances, walking distances, and so on [11]. If workers handle tools or objects, the size, shape, and weight of these items should be recorded.

### 9.3.4 Interpreting the Results

Once data collection is completed, the first step in analyzing the results is to break each job into a series of tasks and subtasks. The goal of this analysis is to link excessive job demands to a specific aspect of the job or work environment. Once jobs are broken into tasks, the work methods, the workstation, and the tools and equipment required to perform each task can be examined to determine if one or more of these elements contributes to the biomechanical stress of the task [34]. Specifically, each task should be examined in terms of the following dimensions [10,35]:

*Forces required to perform the task*: Force requirements have a direct impact on the muscular effort that must be expended by the worker to perform a task. As muscular effort increases in response to heavy task loading, circulation to the muscle decreases, causing more rapid muscle fatigue. In addition, high contact forces that create pressure over one area of the body can inhibit nerve function and blood flow. If possible, the force required to perform a task should be measured at the work site and compared to the known capacities of the workforce (see the following). Contact points between the body and work surfaces or tools should also be identified and described.

*Postures assumed during the task*: Postures determine which muscles are used in an activity and how forces are translated from the muscles to the part or tool being handled.

In general, muscle strength about a joint is maximized near the midpoint of the range of motion. Therefore, performing a task in a neutral posture generally requires less effort than performing the same task in a more extreme posture. Furthermore, postures that are sustained through active muscle contraction consume energy and produce waste metabolic products that can cause fatigue and pain. The analyst should note situations where the tool design, workstation layout, or equipment design cause excessive reaches or sustained awkward shoulder, elbow, wrist, or trunk postures.

*Frequency of muscle activation (repetition rate)*: Frequent repetition of the same work activities can exacerbate the effects of awkward work postures and forceful exertions. Tendons and muscles can often recover from the effects of stretching or forceful exertion if sufficient time is allotted between exertions. However, if tasks are repeated frequently, fatigue and muscle strain can accumulate, producing permanent tissue damage. There is evidence that even low-level exertions, if repeated at a very high rate, can cause muscle fatigue and injury if sufficient recovery time is denied. If a job requires workers to repeat a simple task over and over throughout the workday, repetition can be quantified by counting the number of similar motion patterns that occur within a specific time period or estimated subjectively using qualitative descriptions [34].

*Duration of work and recovery periods*: Work duration can have a substantial effect on the likelihood of both localized and general fatigue. Even at low levels of effort, biomechanical or physiological strain will accumulate over time; therefore, the provision of frequent rest breaks is encouraged to give tissues the opportunity to recover from exertions during work. Work duration and the number of breaks allotted to employees during the work shift can usually be determined through observation or interview.

*Exposure to vibration and cold*: Exposure to whole body and segmental vibration has been associated with increased rates of back pain and upper extremity musculoskeletal disorders, although the pathogenic mechanism is not yet well understood. Exposure is frequently associated with activities such as the operation of motor vehicles and the use of powered hand tools [11]. Cold temperatures can reduce the dexterity and sensibility of the hand, causing workers to apply more grip force to tool handles and objects than necessary. There is also evidence that cold tends to exacerbate the effects of segmental vibration. Because of these relationships, vibrating or cold objects and surfaces should be identified and the frequency and duration of exposure should be recorded.

In some cases, it will be obvious that task demands exceed the capabilities of the workers performing the job. For example, most analysts will recognize a job as being too repetitive if the worker has difficulty keeping up with the required pace or if the body is in constant motion. However, in other instances it may not be as clear what force or postural demands are appropriate for the population of workers performing a task. For example, a workforce of young males may be able to tolerate job demands that would be inappropriate for an older or predominantly female population. Currently, it is not known how much force or repetition can be endured safely by a majority of workers. However, it may be possible to estimate the percentage of workers for whom a job may be difficult. Information that can be used for this purpose is discussed in the following paragraphs.

#### *9.3.4.1 Strength Data*

Strength data for various populations and muscle groups have been published in a number of sources [35–37]. If the force requirements of a task are known, it may be possible to compare these requirements with existing strength data to estimate the percentage of the population for which the job may be difficult. Successfully applying strength data requires knowledge of the subject population upon which the data are based, the posture in which the measurements were made, whether the measurement was static or dynamic, and how long the effort was sustained [12]. Recently, computerized *biomechanical models* have been developed to predict the percentage of males and females capable of exerting static forces in certain postures. The advantage of these models is that most recognize that a worker's capacity for force exertion is rarely dependent on the strength of a single muscle group. Rather, the capacity for exerting force is dependent on the moment created at each joint by the external load and the muscle strength at that joint. The models compute the moment created at each joint by an exertion and compare the moments against static strength data to estimate the percent of the population capable of performing a specific exertion for each joint and muscle function. Currently, both 2-D and 3-D models are available for these analyses [38]. (Additional information about biomechanical models is contained in Chapter 4 of this book.)

#### *9.3.4.2 Anthropometry*

Anthropometric data on body size and range of joint motion can be used to assess the appropriateness of workplace, equipment, and product designs relative to workforce capacities for reach, grasp, and clearance [11]. Compilations of anthropometric data for various populations are available from numerous sources [39,40]. Anthropometric data have also been used to generate computerized man-models that can be incorporated into existing CAD programs for direct application in the design and evaluation of new and existing workstations [41]. (For additional information on anthropometry, see Chapter 1.)

#### *9.3.4.3 Physiological Data*

In activities such as repetitive lifting and load carrying, large muscle groups perform submaximal, dynamic contractions. During these activities, a worker's endurance is primarily limited by the capacity of the oxygen transport and utilization systems (maximum aerobic power) [42]. Data compiled on the maximum aerobic capacities of working populations and the energy demands of common industrial tasks can be found in a number of sources [12,42,43]. In general, maximum aerobic capacity declines with age, increases with physical fitness level, and is 13%–30% lower for women compared to men [12,44]. Several researchers suggest that the maximum energy expenditure rate for an 8 h workday should not exceed 33% of maximum aerobic power [12]. Limits of 5.2 kcal/min for average healthy young males or 3.5 kcal/min for populations containing women and older workers have been proposed based on this recommendation [45–47].

Because table values provide only a rough approximation of the metabolic costs of a given job, models to predict metabolic energy expenditure for simple tasks based on a combination of personal and task variables have also been developed [48–50]. It has been demonstrated that the energy expenditure rate of complex jobs can be predicted if the energy expenditure rates of the simple tasks that make up the job and the time duration of the job are known. Comparison of measured and model-predicted rates for 48 tasks indicated that models can account for up to 90.8% of the variation in measured metabolic rates [50].

#### 9.3.4.4 Psychophysical Data

Difficulties in applying strength data to dynamic tasks involving more than one muscle group, and the recognition that motivational factors play an important role in determining an individual's capacity for physical work, have led to the use of *psychophysics* to develop guidelines for the evaluation and modification of repetitive work tasks [51–53]. Psychophysical limits are generally based on data derived from laboratory simulations of a specific task in which the participants are allowed to adjust their workload to a level subjectively defined as the maximum acceptable. Limiting workload in this manner should allow workers to perform work tasks without overexertion or excessive fatigue [54]. Although there are few data to indicate how psychophysically derived limits relate to the risk of injury during work, many researchers believe that use of these limits may be the most accurate method of determining whether a given task is acceptable [55]. Psychophysical limits for various lifting, pushing, and pulling tasks and other manual operations have been developed and are widely available for application [56–58].

#### 9.3.4.5 Integrated Models

More sophisticated evaluation techniques that integrate biomechanical, physiological, and psychophysical considerations to assess the appropriateness of job tasks are also available. Probably the best known method for evaluating the demands of materials handling tasks is the *NIOSH lifting equation*. The NIOSH lifting equation was first published in 1981 to assist safety and health practitioners in evaluating sagittal plane lifting tasks [43]. The equation has recently been revised to reflect new research findings and provide methods for evaluating asymmetrical lifting tasks, lifts of objects with less than optimal hand-container couplings, and jobs with a larger range of work durations and lifting frequencies [47]. Using criteria from the fields of biomechanics, psychophysics, and work physiology, the equation defines a recommended weight limit (RWL) based on specific task parameters (e.g., the location of the load relative to the body and the floor, the distance the load is moved, the frequency of the lift). The RWL represents a load that nearly all healthy workers can lift over a substantial period of time without placing an excessive load on the back, causing excess fatigue, or otherwise increasing the risk of low back pain. The actual weight of lift can be compared to the RWL for a given task to derive an estimate of the risk presented by the task and to determine whether measures to reduce the risk of injury to workers are needed. (Additional information about the NIOSH lifting equation can be found in Chapter 31 of this handbook.)

### 9.3.5 Developing Solutions

Ultimately, the results of a job analysis should suggest ways to eliminate, reduce, or control ergonomic hazards through modification of equipment, workstations, or work methods that contribute to excessive work demands. In all cases, the best ergonomic solutions are those in which safe work is a natural result of the job design and are independent of specific worker capabilities or work techniques. However, in situations where design changes are infeasible, it may be possible to limit exposure to ergonomic hazards by reducing production rates, providing additional rest breaks, periodically rotating workers to less stressful jobs, and increasing the number of employees assigned to specific work tasks. In some situations, control of ergonomic hazards may require a combination of engineering or administrative tactics. When possible, the proposed solution should be tested among a small group of workers to allow adjustments to be made before widespread changes are implemented. Finally, follow-up job analyses should be performed to ensure that the solution effectively reduces the hazard without imposing new demands on the worker. (Additional information about ergonomic control strategies is contained in Chapter 21.)

### 9.3.6 Documentation

The importance of documenting the results of an ergonomic job analysis cannot be underestimated. Benefits of careful documentation include the following [58]:

Documentation permits the analyst to track the progress of the analysis, that is, to determine what has been done and what remains to be done.

Careful documentation provides a compendium of facts for use in justifying requests for resources and intervention.

Adequate documentation allows the analyst to keep track of successful solutions that may work again in later projects and to track mistakes that should be avoided.

Good documentation forms the foundation for evaluating the overall success of an ergonomics project. Data obtained before and after the introduction of control measures can be compared to determine whether the interventions were effective in reducing worker exposures to hazards, the occurrence of complaints or injuries, and the cost of medical insurance and workers' compensation.

The final report of a job analysis should describe why the study was done, the tools and equipment used in the study, the problems identified, and the solutions recommended to correct these problems. An outline for this type of report is shown in Table 9.4.

### 9.3.7 Job Analysis Example

The following example illustrates how the methods and information discussed in the previous sections were used to identify potential hazards associated with jobs in one department of a manufacturing facility that produces printed circuit boards.

#### *9.3.7.1 Background*

The production of printed circuit boards is a multistep process. One step in the process is a cleaning procedure to remove impurities from the surface of each panel before it is printed and etched. This procedure is performed by a computer-operated "preclean" machine.

TABLE 9.4 Job Analysis Report Format

| |
|---|
| Introduction |
| Purpose of report |
| Definition of problem |
| Area layout (including dimensions and photos) |
| Worker population description |
| Injury statistics |
| Discomfort survey results |
| Interview findings |
| Procedure |
| Tools and equipment used |
| Work sampling plan |
| Analysis and discussion |
| Problems identified |
| Conclusions |
| Recommendations |
| Recommended solutions to problems identified (prioritized) |
| Associated costs |
| Associated benefits |
| References |
| Appendix |
| Examples of questionnaires/survey forms used |
| Analysis charts |
| Computer output (if computer software was used in the analysis) |
| Other supporting data and analyses |

Panels are automatically loaded into the machine from a cart and are transported through the machine on a conveyor belt. An operator monitors the operation of the machine and performs required maintenance activities (e.g., adding water or chemicals to the machine).

#### *9.3.7.2 Problems Identified*

As part of a plant-wide screening survey to identify early indications of musculoskeletal injuries, a body part discomfort survey (Figure 9.1) was administered to 12 machine operators in the preclean area. Despite the job's seemingly low physical demands, the survey indicated that 8 of the 12 workers experienced arm pain, 6 of 12 experienced back pain, and 9 of 12 reported moderate to severe foot pain.

A site visit was conducted to gather background information from operators working in the area. Interviews with employees and supervisors revealed that the machine operators' job is nonstereotypical; that is, operators perform a variety of tasks at different times throughout the workday. Therefore, the job was observed again at three separate times during a single work shift for periods of 10–15 min each. In addition, operators demonstrated a task (adding pumice to the machine) that is performed only once or twice during the work shift. Measurements were made of machine dimensions, reach distances, and work heights. In addition, the weights of objects handled by the operators (hoses, pumice bags) were obtained. These measurements were compared to recommended workstation

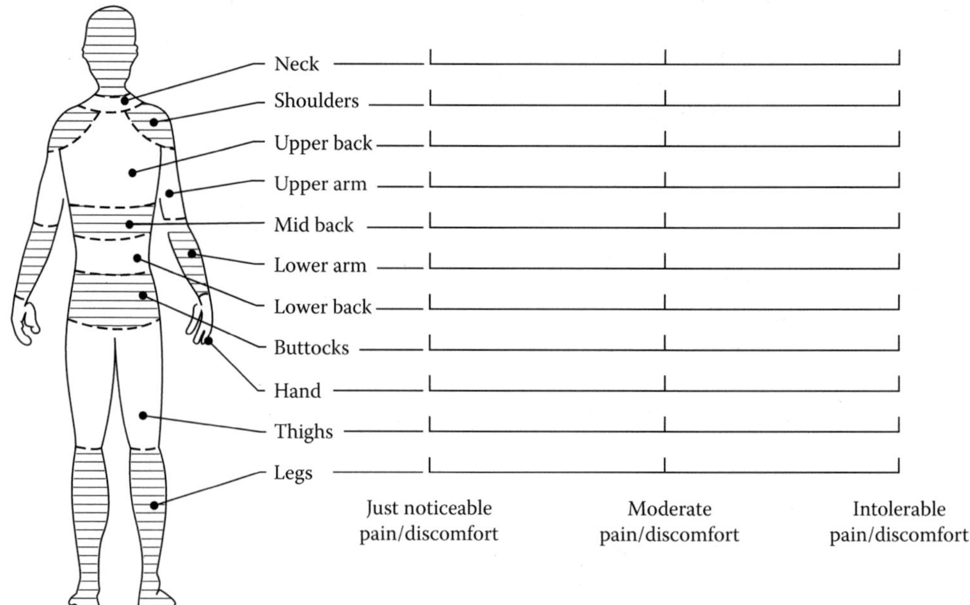


FIGURE 9.1 Body part discomfort survey used to collect data in the preclean area.

dimensions based on anthropometric information and used to compute RWLs using the 1991 NIOSH lifting equation.

From discussions with the operators and information obtained at the work site, several ergonomic problems were identified. These are summarized in Table 9.5. The principal hazards included awkward lifts and excessive reaches. Based on an evaluation of the frequency with which each task was repeated, the workers' average yearly exposures to these awkward motion patterns were estimated. Although tasks such as adding pumice and water to the preclean machine are performed only a few times each day, comparisons with NIOSH RWLs indicated that workers who performed these tasks were at risk for back injury.

#### *9.3.7.3 Recommended Solutions*

Methods of reducing or eliminating the number of awkward lifts and reaches were proposed to mitigate the hazards posed by this job. Recommendations included changing the way supplies (i.e., pumice and water) are provided to the operator and providing additional equipment (e.g., hoses and stools) in the preclean area (Table 9.6). As indicated, the recommended changes were anticipated to eliminate many of the workers' exposures to biomechanical hazards for musculoskeletal injury. Follow-up is necessary to determine if interventions are actually effective in reducing exposures to anticipated levels.

## 9.4 CRITICAL REVIEW OF CURRENT STATUS

### 9.4.1 Use of Checklists

Increases in the demand for ergonomic expertise in industrial settings have led to the development of various checklists for use in conducting workplace ergonomic audits. Generally, checklist evaluations should not be substituted for a full-scale ergonomic evaluation of the

TABLE 9.5 Problem Summary—Preclean Machine Operator

| | |
|---|---|
| Preclean operator task(s) | NIOSH RWL (back of pallet): 16.7 lb lift index: 2.10 |
| 1. Test chemicals in the preclean machine | |
| 2. Add pumice or water to the machine | NIOSH RWL (front of pallet): 21.7 lb lift index: 1.61 |
| 3. Perform machine maintenance or housekeeping as needed | |
| Body part discomfort data | Ergonomic Costs |
| 12 workers surveyed | 35,000 lb lifted per year |
| 6 of 12 neck pain | 3,900 extended reaches per year |
| 7 of 12 shoulder pain | 3,900 trunk flexions per year |
| 4 of 12 wrist pain | Machine lid |
| 8 of 12 back pain | Object weight: 22 lb |
| 9 of 12 foot pain | Horizontal distance: 24 in. |
| 1991 NIOSH lifting guide data | Final distance: 10 in. |
| Pumice bag (lift from pallet) | Origin of lift: 44 in. |
| Object weight: 35 lb | Distance traveled: 16 in. |
| Horizontal distance (back of pallet): 32 in. | Frequency: 2 lifts/8 h |
| Horizontal distance (front of pallet): 18 in. | Asymmetric angle: 30° |
| | Coupling: poor |
| Final distance: 11 in. | NIOSH RWL: 14.4 lb |
| Origin of lift: 19 in. | Lift index: 1.53 |
| Distance traveled: 17 in. | |
| Frequency: 2 lifts/8 h | |
| Asymmetric angle: 0° | |
| Coupling: poor | |
| Problems identified | |

1. To add pumice to the preclean machine, the operator must lift pumice bags from a pallet to the top of the machine. The weight of the pumice bags exceeds the NIOSH RWL; therefore, the operator may be at risk for back injury. Additionally, bags often leak, creating dust in the area
2. Spraying water inside the machine requires the operator to remove plastic covers from the top of the machine. Each lid weighs 22 lb, which exceeds the NIOSH RWL. Removing the lids from the machine also requires an extended reach of 24 in. with the trunk flexed. According to anthropometric tables in Eastman Kodak [40], most people cannot reach more than 18 in. in front of the body without bending or leaning
3. Only one water hose is provided in the preclean area. Adding water to the preclean machine or washing the floor requires the operator to uncoil the hose and drag it up to 30 ft, placing stress on the operator's back and shoulders. An extended reach of 24 in. is required to access the machine with the hose
4. The operator is occasionally required to activate controls and valves located on a panel above the preclean machine. An extended reach of 24 in. is required to access the panel. A large machine located next to the preclean machine hampers the operators' ability to obtain test samples
5. Operators are required to stand on concrete surface or metal gratings throughout their work shift. Prolonged standing may cause foot/leg discomfort

workplace; most checklists are designed to help investigators document their efforts in a systematic fashion and target areas for further analysis (see Section 9.3.1.2). They can also be used as a rapid screening tool to allow investigators with limited training to identify common ergonomic hazards in a short period of time.

The limitations of checklists are that they usually do not provide enough information to identify specific work attributes responsible for excessive stress nor do they

TABLE 9.6 Proposed Solutions—Preclean Machine Operator

| |
|---|
| Ergonomic savings |
| 35,000 lb lifted per year |
| 3,900 extended reaches per year |
| 3,900 trunk flexions per year |
| Problem solutions |
| 1. Instead of buying pumice in 35 lb bags, pumice should be purchased in drums, which can be stored next to the preclean machine. Providing pumice in drums instead of bags will prevent operators from having to lift the bags and should permit better dust control (leaky bags will be eliminated). A hoist would be used to lift the drums when pumice is needed. An in-line scale should be provided to allow operators to measure the amount of pumice required to maintain concentrations at proper levels |
| 2. The design of the machine covers should be modified to improve access to the inside of the machine. Lids should be hinged or should slide over or under one another so that lifts are not required |
| 3. Additional hoses with longer nozzles are needed. A longer nozzle would reduce the reach needed to spray water into the machine, and an additional hose (stored on a spring-loaded coil) would allow the operators to access the machine more easily (i.e., with less stress on the back and shoulders) from all directions |
| 4. The control panel over the preclean machine should be moved so that the operator can access the panel with a reach of less than 18 in. The machine next to the preclean machine should be relocated so that the operator can more easily access the valves needed for testing |
| 5. Provide a chair or stool in the preclean area to allow operators to sit occasionally. Providing antifatigue mats around the machine may also alleviate foot and leg discomfort |

provide insights as to how jobs can be redesigned to reduce biomechanical stress [59]. Furthermore, checklists do not substitute for an understanding of the risk factors for musculoskeletal disease. The danger in relying on checklists is that hazards may be overlooked if they are not specifically described by the checklist. For example, ergonomic hazards in an office environment are likely to be different from those in a manufacturing facility.

To ensure that the questions are appropriate to the workplace of interest, checklists should be customized and evaluated in the walk-through survey before the actual evaluation is conducted. Examples of various ergonomic checklists are found in the Appendix of this handbook.

### 9.4.2 Worker Assessment versus Expert Assessment

Traditionally, the job analysis task has been left to experts with professional training in ergonomics or industrial engineering. Because many industrial facilities lack in-house ergonomic expertise, some companies have filled this void with consultants. A drawback to this approach is that consultants may lack an understanding of the intricacies that can impact the effectiveness of various control methods; also, their involvement usually ends before the intervention process is complete. In these situations, companies are frequently left with little or no mechanism to ensure that changes will be implemented or that interventions will be effective.

It is increasingly recognized that the workers who perform a particular job possess a large knowledge base that could be helpful in recognizing ergonomic problems and implementing effective changes. As a result, several companies have established "ergonomics committees" composed of representatives of management, labor, engineering, and safety to identify and resolve ergonomic problems in the workplace. A participatory approach to the control of ergonomic hazards offers at least two advantages. First, ergonomic committees frequently possess the variety of experience needed to solve problems with multiple causation. Second, allowing employees to participate in decision-making processes creates ownership and improves commitment to solutions [60].

According to Joseph, participative programs can be an effective mechanism for implementing ergonomic changes in the workplace [61]. However, the success of a participatory ergonomics program will depend on the commitment of management to the principles of prevention through ergonomic job design. Furthermore, training is necessary to ensure that employees are informed of ergonomic hazards to which they may be exposed and are able to participate actively in their own protection. Supervisors and managers should also be trained to recognize ergonomic hazards and to understand their medical consequences. Finally, engineers and maintenance personnel should be trained to prevent and correct ergonomic hazards through design and maintenance procedures.

## 9.5 FUTURE CONCERNS

Job analysis is only one element in an overall program to control and prevent musculoskeletal disorders in the workplace. To be useful, information derived from job analysis activities must lead to appropriate intervention strategies for eliminating ergonomic stresses or reducing stresses to acceptable levels. It is hoped that future research will provide additional evidence to demonstrate that a program that includes a systematic method of identifying ergonomic hazards can be effective in reducing the incidence and cost of musculoskeletal injury. Furthermore, it is anticipated that future developments will result in new data-gathering techniques and the development of more accurate models to predict human capabilities in relationship to job demands.

## REFERENCES

1. B. W. Niebel, *Motion and Time Study*, 8th edn., Irwin, Homewood, IL, 1989.
2. T. J. Armstrong, R. G. Radwin, D. J. Hansen, and K. W. Kennedy, Repetitive trauma disorders: Job evaluation and design, *Hum. Factors 28*(3):325–336 (1986).
3. D. T. Ridyard, A successful applied ergonomics program for preventing occupational back injuries, in *Advances in Industrial Ergonomics and Safety*, Vol. 2, B. Das, Ed., Taylor & Francis, Philadelphia, PA, 1990, pp. 125–132.
4. F. McKenzie, J. Storment, P. Van Hook, and T. Armstrong, A program for control of repetitive trauma disorders associated with hand tool operations in a telecommunications manufacturing facility, *Am. Ind. Hyg. Assoc. J. 46*:674–678 (1985).
5. G. Lutz and T. Hansford, Cumulative trauma disorder controls: The ergonomics program at Ethicon Inc., *J. Hand Surg. 12A*(2Pt2):863–866 (1987).
6. Occupational Safety and Health Act, 1970, PL 91–596.
7. Code of Federal Regulation, 29 CFR 1910, Ergonomic Safety and Health Management Program, 1991.
8. U.S. Department of Labor, *Ergonomics Program Management Guidelines for Meatpacking Plants*, OSHA 3123, Occupational Safety and Health Administration, Washington, DC, 1990.
9. A. D. Hosey, General principles in evaluating the occupational environment, in *The Industrial Environment—Its Evaluation and Control*, G. D. Clayton, Ed., National Institute for Occupational Safety and Health, Cincinnati, OH, 1973, p. 95.

10. S. H. Rodgers, Functional job analysis technique, in *Occupational Medicine: State of the Art Reviews,* J. S. Moore and A. Garg, Eds., Hanley & Belfus, Philadelphia, PA, 1992, p. 680.
11. W. M. Keyserling, T. J. Armstrong, and L. Punnett, Ergonomic job analysis: A structured approach for identifying risk factors associated with overexertion injuries and disorders, *Appl. Occup. Environ. Hyg.* 6:353–363 (1991).
12. Eastman Kodak Co., *Ergonomic Design for People at Work,* Vol. 2, Van Nostrand Reinhold, New York, 1986.
13. L. Punnett and W. M. Keyserling, Exposure to ergonomic stressors in the garment industry: Application and critique of job-site work analysis methods, *Ergonomics 30*(7):1099–1116 (1987).
14. T. J. Armstrong, J. A. Foulke, B. S. Joseph, and S. A. Goldstein, Investigation of cumulative trauma disorders in a poultry processing plant, *Am. Ind. Hyg. Assoc. J. 43*:103–116 (1982).
15. D. J. Habes, Use of EMG in a kinesiological study in industry, *Appl. Ergon. 15*(4):297–301 (1984).
16. H. Christensen, Muscle activity and fatigue in the shoulder muscles of assembly-plant employees, *Scand. J. Work Environ. Health 12*:582–587 (1986).
17. A. Aaras and R. H. Westgaard, Further studies of postural load and musculoskeletal injuries of workers at an electro-mechanical assembly plant, *Appl. Ergon. 18*(3):211–219 (1987).
18. F. E. Gomer, L. D. Silverstein, W. K. Berg, and D. L. Lassiter, Changes in electromyographic activity associated with occupational stress and poor performance in the workplace, *Hum. Factors 29*(2):131–143 (1987).
19. E. Milerad, M. O. Ericson, R. Nisell, and A. Kilbom, An electromyographic study of dental work, *Ergonomics 54*(7):953–962 (1991).
20. W. S. Marras, S. A. Lavender, S. E. Leurgans, S. L. Rajulu, W. G. Allread, F. A. Fathallah, and S. A. Ferguson, Industrial quantification of occupationally related low back disorder risk factors, *Proceedings of the Human Factors Society 36th Annual Meeting,* Human Factors Society, Santa Monica, CA, 1992.
21. A. Garg, G. Hagglund, and K. Mericle, A physiological evaluation of time standards for warehouse operations as set by traditional work measurement techniques, *IIE Trans. 19*:235–245 (1986).
22. V. Z. Priel, A numerical definition of posture, *Hum. Factors 16*:576–584 (1974).
23. O. Karhu, P. Kansi, and I. Kuorinka, Correcting working postures in industry: A practical method for analysis, *Appl. Ergon. 8*:199–210 (1977).
24. E. N. Corlett, S. J. Madeley, and I. Manenica, Posture targeting: A technique for recording working posture, *Ergonomics 22*(3):357–366 (1979).
25. D. W. Grieve, The postural stability diagram (PSD): Personal constraints on the static exertion of force, *Ergonomics 22*(10):1155–1164 (1979).
26. W. Rohmert and K. Landau, *Das Arbeitswissenschaftliche Erhebungsverfahren zur Tatigkeitsanalyse* (*AET*), Verlag Hans Huber, Bern, Switzerland, 1979.
27. P. Holzmann, ABRAN—A new method for analysis of ergonomic effort, *Appl. Ergon. 13*(2):82–86 (1982).
28. W. M. Keyserling, Postural analysis of the trunk and shoulders in simulated real time, *Ergonomics 29*(4):569–583 (1986).
29. A. Kilbom and J. Persson, Work technique and its consequences for musculoskeletal disorders, *Ergonomics 30*(2):273–279 (1987).
30. H. J. C. Gil and E. Tunes, Posture recording: A model for sitting posture, *Appl. Ergon. 20*(1):53–56 (1989).
31. J. G. Chen, J. B. Peacock, and R. E. Schlegel, An observational technique for physical work stress analysis, *Int. J. Ind. Ergon. 3*(3):167–176 (1989).
32. D. Stetson, W. M. Keyserling, B. A. Silverstein, T. J. Armstrong, and J. A. Leonard, Observational analysis of the hand and wrist: A pilot study, *Appl. Occup. Environ. Hyg. 6*:927–937 (1991).

33. L. McAtamney and E. N. Corlett, RULA: A survey method for investigation of work-related upper limb disorders, *Appl. Ergon. 24*(2):91–99 (1993).
34. D. J. Habes and V. Putz-Anderson, The NIOSH program for evaluating biomechanical hazards in the workplace, *J. Saf. Res. 16*:49–60 (1985).
35. S. S. Ulin and T. J. Armstrong, A strategy for evaluating occupational risk factors of musculoskeletal disorders, *J. Occup. Rehab. 2*(1):35–50 (1992).
36. E. Kamon and A. Goldfuss, In-plant evaluation of muscle strength of workers, *Am. Ind. Hyg. Assoc. J. 39*:801–807 (1978).
37. V. Mathiowetz, N. Kashman, G. Volland, K. Weber, M. Dowe, and S. Rogers, Grip and pinch strength: Normative data for adults, *Arch. Phys. Med. Rehab. 66*:69–72 (1985).
38. D. B. Chaffin, Biomechanical modeling for simulation of 3D static human exertions, in *Computer Applications in Ergonomics, Occupational Safety and Health,* M. Mattila and W. Karwowski, Eds., Elsevier, Amsterdam, the Netherlands, 1992, pp. 1–11.
39. NASA, *Anthropometric Source Book,* Vols. 1–3 (Ref. Pub. 1024), NASA Scientific and Technical Information Office, Yellow Springs, OH, 1978.
40. Eastman Kodak Co., *Ergonomic Design for People at Work,* Vol. 1, Van Nostrand Reinhold, New York, 1983.
41. S. S. Ulin, T. J. Armstrong, and R. G. Radwin, Use of computer aided drafting for analysis and control of posture in manual work, *Appl. Ergon. 21*(2):143–151 (1990).
42. P. O. Astrand and K. Rodahl, *Textbook of Work Physiology: Physiological Bases of Exercise,* 3rd edn., McGraw-Hill, New York, 1986.
43. NIOSH, *Work Practices Guide for Manual Lifting,* NIOSH Technical Report No. 81–122, U.S. Dept. of Health and Human Services, National Institute for Occupational Safety and Health, Cincinnati, OH, 1981.
44. J. V. G. A. Durnin and R. Passmore, *Energy, Work and Leisure,* Heinemann, London, U.K., 1967.
45. B. Bink, The physical working capacity in relation to working time and age, *Ergonomics 5*(1): 25–28 (1962).
46. F. H. Bonjer, Actual energy expenditure in relation to the physical working capacity, *Ergonomics 5*(1):29–31 (1962).
47. T. R. Waters, V. Putz-Anderson, A. Garg, and L. Fine, Revised NIOSH equation for the design and evaluation of manual lifting tasks, *Ergonomics 36*(7):749–776 (1993).
48. B. Givoni and R. F. Goldman, Predicting metabolic energy cost, *J. Appl. Physiol. 30*:429–433 (1971).
49. W. H. VanderWalt and C. H. Wyndham, An equation for prediction of energy expenditure of walking and running, *J. Appl. Physiol. 34*:559–563 (1973).
50. A. Garg, D. B. Chaffin, and G. D. Herrin, Prediction of metabolic rates for manual materials handling jobs, *Am. Ind. Hyg. Assoc. J. 39*:661–674 (1978).
51. S. H. Snook and C. H. Irvine, Psychophysical studies of physiological fatigue criteria, *Hum. Factors 11*(3):291–300 (1969).
52. F. Gamberale, Perception of effort in manual materials handling, *Scand. J. Work Environ. Health 16*(Suppl. 1):59–66 (1990).
53. V. Putz-Anderson and T. L. Galinsky, Psychophysically determined work durations for limiting shoulder girdle fatigue from elevated manual work, *Int. J. Ind. Ergon. 11*(1):19–28 (1993).
54. S. H. Snook, The design of manual handling tasks, *Ergonomics 27*(12):963–985 (1978).
55. D. B. Chaffin and G. B. J. Andersson, *Occupational Biomechanics,* Wiley, New York, 1991.
56. S. H. Snook and V. M. Ciriello, The design of manual handling tasks: Revised tables of maximum acceptable weights and forces, *Ergonomics 21*:1197–1213 (1991).
57. S. H. Snook, D. R. Vaillancourt, V. M. Ciriello, and B. S. Webster, Psychophysical studies of repetitive wrist flexion and extension, *Ergonomics 38*(7):1488–1507 (1995).

58. UAW-Ford National Joint Committee on Health and Safety, *The UAW-Ford Job Improvement Guide,* Detroit, MI, 1988, p. 6-3.
59. W. M. Keyserling, D. S. Stetson, B. A. Silverstein, and M. L. Brouwer, A checklist for evaluating risk factors associated with upper extremity cumulative trauma disorders, *Ergonomics 36*(7):807–831 (1993).
60. A. S. Imada, Macroergonomic approaches for improving safety and health in flexible, self-organizing systems, in *The Ergonomics of Manual Work,* W. S. Marras, W. Karwowski, J. L. Smith, and L. Pacholski, Eds., Taylor & Francis, Philadelphia, PA, 1993, pp. 477–480.
61. B. S. Joseph, A participative ergonomic control program in a U.S. automotive plant: Evaluation and implications, PhD dissertation, University of Michigan, Ann Arbor, MI, 1985.

CHAPTER 10

# Workstation Evaluation and Design

David R. Clark

CONTENTS

## 10.1 INTRODUCTION

This chapter presents methods and recommendations for the evaluation of existing or proposed workstations and their design or redesign to minimize the risk of potential ergonomics problems. The reader is cautioned that the comprehensive evaluation and design of workstations often requires the application of wide varieties of knowledge available on ergonomics. Further, the purpose of this chapter is not to repeat that which is covered in much more detail elsewhere within this book, but to stress the importance of taking a systematic approach to the synergistic application of that knowledge. Although traditional examples of workstation design criteria are given, this chapter avoids the common practice of providing detailed checklist-oriented guidelines [1].

## 10.2 BACKGROUND AND SIGNIFICANCE TO OCCUPATIONAL ERGONOMICS

There can be no more fundamental aspect of occupational ergonomics than the concern with the design of the "local" space in which workers must spend considerable time and effort in doing their jobs. In this respect, it is a very personal process in that the workers feel a sense of ownership of the workstation; and the morale, health and safety, and productivity of workers can be significantly affected both positively and negatively by the success or failure of design efforts.

Ergonomics, or the "laws of work," by definition and tradition, has focused primarily on the physical aspects of work, such as force and energy requirements. That these are inexorably intertwined with the specific design of workstations hardly need be mentioned. The spatial and temporal arrangements of the tasks being done in a workstation define the parameters necessary for biomechanical analysis and design of work–rest cycles, among other things. In some sense, this part of workstation design and evaluation can be thought of as "applied anthropometry."

But one must look beyond traditional ergonomics to traditional human factors engineering. While the former focuses on the physical aspects of work, the latter examines the psychophysical interface between the worker and the task. Here, one is concerned with the input–output loop of stimulus and response involving information processing and action initiation, that is, the senses and motor control. Workers usually have to sense something about the dynamics of the process around them, make decisions, and take appropriate actions. The design of the workstation to facilitate this is critical [2,3].

Therefore, with respect to workstations, the traditional boundary between ergonomics and human factors engineering is blurred. It is interesting to note that the U.S.-based Human Factors Society changed its name to the Human Factors and Ergonomics Society in 1993 to reflect the fact that, to many practitioners, the terms are synonymous and are used interchangeably.

## 10.3 MODELS FOR WORKSTATION EVALUATION AND DESIGN

### 10.3.1 Human–Machine Interface Model

To properly appreciate the importance of the design of the workstation, a "system" must be defined that includes not only the "hardware" but also the worker and the interface between the two. Figure 10.1 illustrates the components of a model of the human–machine interface system [3,6].

#### *10.3.1.1 Human*

Humans can play a multitude of roles in systems, which can be arranged dimensionally as shown in Figure 10.2. In the dimension of *control*, or the degree to which humans determine which way the process will operate, their role can vary from one extreme, that of simply being a monitor or supervisor, to the other extreme, that of a controller, meaning that no action takes place in the process without intimate human involvement. In many systems, the control role will be somewhat less extreme, most commonly as an initiator, humans being able to choose the course and timing of the system's next action based on its current state and the knowledge and experience of the worker. Traditional human factors engineering is most likely to be applied in the analysis of the dimension of control.

A second dimension of the human role in systems is that of *physical work*, or the degree to which the worker is required to exert force or expend energy for the process to continue properly. At one extreme, the human may not be required to provide any significant force or energy, such as in the case of activating "zero-force" and/or "zero-displacement" controls.

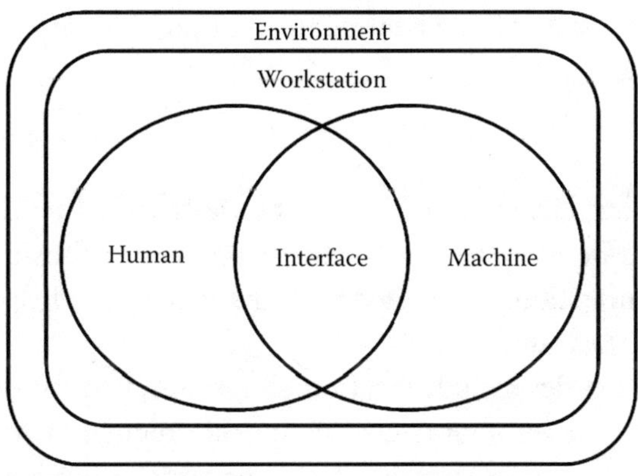


FIGURE 10.1 The human–machine interface model of the workstation system.

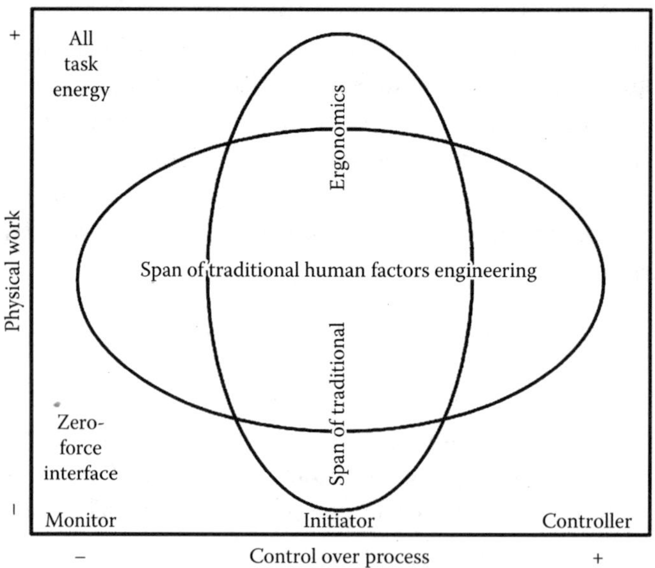


FIGURE 10.2 The human role in the workstation system.

It should be noted here that studies have shown that mental processes such as information processing and decision making do not significantly increase energy expenditure. At the other extreme of the physical work dimension would be found those cases in which the worker provides all the force and energy necessary for the task, such as in hammering a nail. More likely is the scenario of a combination of subtasks involving both extremes, mixed with those in which the worker provides some of the force or energy necessary to get the process moving, such as through the use of mechanical aids, servomechanisms, or simply letting gravity take over. It is in this dimension that traditional ergonomics is applied.

#### *10.3.1.2 Machine*

The concept of the machine, as used here, is quite broad and includes not only the hardware of the process itself, including tools, but also that of the product, including raw materials, finished work, and by-products. The associated information that is necessary for the system to operate is also part of the machine. In addition, the sources of energy other than the human are included.

#### *10.3.1.3 Interface*

The interface between human and machine can be viewed as either the physical characteristics of the connection between the two or the process of establishing, maintaining, and terminating such a connection. A detailed task analysis is very helpful in identifying the important elements of the interface.

The physical characteristics include location, size, shape, texture, color, force, and movement. The connection process characteristics include regularity and foreseeability (i.e., operation vs. service vs. supervisory tasks), frequency (including repetitiveness), duration, and competing tasks (i.e., primary vs. secondary tasks).

#### *10.3.1.4 Environment*

Although often difficult or impossible to control due to both technical and nontechnical issues, the environment in which the process is operating can play an important role in the success of the system. Environmental issues include those of engineering concern such as noise and vibration, temperature and humidity, and light and other radiation. But in addition there are sometimes subtle factors at work, such as workplace esthetics, privacy, communication, sense of community; management goals, strategies, policy, work rules, and performance measures; and worker goals, motivations, and morale.

### 10.3.2 Input–Process–Output Model

Another modeling approach is to consider the "what" and "how" of the tasks to be accomplished. The basic components of such a model are (1) the inputs to the process, (2) the process itself, and (3) the outputs of the process. The inputs and outputs will often simply be outputs and inputs of other processes. Figure 10.3 shows the components of this model.

#### *10.3.2.1 Input*

System inputs consist of (1) information, (2) raw material, and (3) energy. Raw material is to be transformed. Information is either to be transformed or provides the dynamic "instructions" as to how transformations are to take place. The nonhuman energy powers these transformations, either totally or in concert with human efforts.

#### *10.3.2.2 Process*

Processing components consist of (1) workers, (2) methods, (3) tools, and (4) work-in-process. Workers control transformation, albeit at different levels as discussed in Section 10.3.2.1. Methods are the static "instructions" as to how transformations are to take place. Tools are the "hardware" used in the transformation and can include machines, jigs, fixtures, measurement devices, and manual and powered hand tools. Work-in-process is any altered state of process inputs prior to becoming process outputs. This is important in that if the

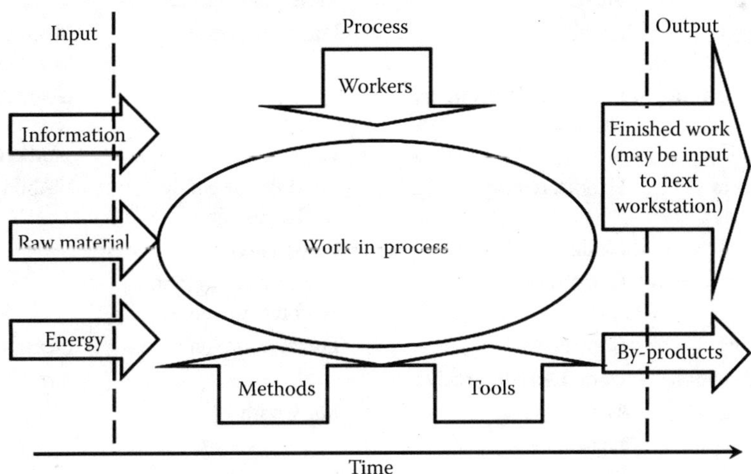


FIGURE 10.3 The input–output–process model of the workstation system.

process is interrupted, the state of work-in-process can be critical to the methods of resuming the process (e.g., retaining vs. discarding partially completed work).

#### *10.3.2.3 Output*

System outputs consist of (1) finished work and (2) by-products. The distinction between these two outputs is often in the eye of the beholder. A narrow view of a flow-through "super-system" of many connected systems might view all outputs not required as subsequent inputs as by-products. A broader view must take into account that the overall enterprise is responsible for the proper disposal of by-products, and thus, they are simply inputs to auxiliary systems charged with that function and subject to their own process definitions.

### 10.3.3 Object-Oriented Process Model

One could also consider the ergonomics attributes of the requirements of the work as found in the objects and processes associated with the workstation. Some of such attributes are shown in Table 10.1. Note that for each category—information, objects, and energy—a "flow" exists. For example, the flow of energy through the workstation entails the potential sequence: enable, control, apply, disable, and dissipate.

TABLE 10.1 Ergonomics Attributes of Workstation Requirements

| | **Workstation Requirements** | | **Ergonomics Attributes** | |
|---|---|---|---|---|
| **"Do..."** | **"With ..." or "to What ..."** | **"When ..." or "How ..."** | **Physical** | **Sensory** |
| Input | Information | Interruption-driven (push) | Access, connection | Signal-to-noise ratio |
| | | Process-driven (pull) | | Parsimony |
| | Objects | Locate | Location, reach | Amplitude, direction |
| | | Select | Separability | Texture, frequency |
| | | Grasp, control | Size, orientation, handhold, and hand clearance | |
| | | Move | Mass, path, and velocity | |
| | Energy | Enable | Location, action | Identification, assessment |
| Process | Information | Detect, understand, and decide | | Distraction, confusion, capacity, and error-proofing |
| | Objects | Handle, regrasp | Location, assembly order, and fit | Visibility |
| | | Modify | Force, energy | |
| | | Combine | Cycle time, duration, and repetitiveness | |
| | Energy | Control, apply | Location, action | Identification |
| Output | Information | Communicate, archive | | |
| | Objects | Move | Mass, path | |
| | | Locate | | |
| | | Release | Destination | |
| | Energy | Disable, dissipate | Location, action | Identification |

Thus, for each requirement of the work, the designer must consider all the appropriate ergonomics factors, both individually and in combination with others, to reduce the possibility of undue strain.

### 10.3.4 Ergonomics Stressors Model

Another way of assessing the human's role in the system is to look at the sources of stress, that is, the causes of strain. These can be divided into three categories: physical, psychophysical, and psychological. Not surprisingly, many of the factors already noted fall into these categories [7].

#### *10.3.4.1 Physical Stressors*

Physical sources of stress include force, movement, and repetition.

10.3.4.1.1 Force Force can be exerted statically or dynamically. In static force exertion, where the force generation is often localized, isolated muscle groups may be working very hard while others are not working at all. In addition, when muscles maintain contraction under static exertion, the blood vessels within the muscles are constricted, leading to reduced blood flow, which in turn reduces the amount of oxygen being delivered to the muscle and reduces the ability of the circulatory system to remove the by-products of static muscle work, notably lactic acid. This leads to generally lower forces and shorter fatigue times than would be present for dynamic work.

In the case of dynamic work, the often rhythmic body movements, such as in walking, cause elements of paired muscle groups to activate in an alternating fashion, reducing the amount of time any one muscle group is activated in comparison with static work. The associated pattern of blood vessel constriction and relaxation, in concert with the one-way "check valves" in the venous system, actually assists the heart in pumping blood back to the heart and lungs [4,5].

10.3.4.1.2 Movement Moving a mass as a result of exerting a force is the definition of work. In the case of static and many rhythmic dynamic force exertions, no net external work is performed because the distance moved is zero, although the internal work can be considerable. Also of concern is the necessity for the worker to effect whole body movement, such as is the case when reach capability is exceeded. The long-term concern would be energy expenditure, especially over the course of a workday. In tasks involving significant worker movement, it can often be shown that most of the energy expended goes to just moving the worker around.

10.3.4.1.3 Repetition Whenever the worker must repeat the same force exertions or movements, especially within short time frames, the repetitiousness of the events can become of primary concern. In fact, even very small forces or movements can be significantly stressful if they are repeated often enough. The types of strains that the body manifests in these cases can be broadly described as *cumulative trauma disorders* or *repetitive motion disease* or some other similar term and include such notable specific strains as carpal tunnel syndrome, tendinitis, and epicondylitis [5].

### *10.3.4.2 Psychophysical Stressors*

Psychophysical sources of stress include components of the "input–output" or "stimulus–response" loop: displays and other stimuli, information processing and decision making, and controls.

10.3.4.2.1 Displays and Other Stimuli Stimuli can be thought of as useful plants and weeds. Useful stimuli, like plants, must be sufficient for attention getting (here I am!), discriminability (this is what I am), and appropriateness (you need me), while not being crowded out by "weedy," distracting, or masking stimuli. Parsimony, or the keep it simple yet sufficient (KISS) principle, applies here but is often complicated by the lack of adequate control over the surrounding environment.

10.3.4.2.2 Information Processing and Decision Making Any useful stimulus received by the worker must be processed, even if only minimally, and a decision reached as to the appropriate response. The factors that affect this process include the sufficiency of the stimulus from the worker's perspective and the availability of a corresponding, objective decision rule. Worker education, training, and experience, as well as the use of visual controls in the workstation can greatly affect this part of the process.

10.3.4.2.3 Controls One of the possible responses of a worker in the stimulus–response loop is to operate a control. A control is defined as a discrete mechanism provided in the workstation designed to cause the system to operate in a specific way. As in stimuli, controls must have sufficiency, including attention getting (can I find it?), discriminability (do I know what it is for and how it operates?), and appropriateness (will it do what I need?) [4].

### *10.3.4.3 Psychological Stressors*

Psychological sources of stress include nonoccupational versus occupational stressors, motivation and reward systems, and the environment.

10.3.4.3.1 Nonoccupational versus Occupational Stressors In a perfect world, workers would arrive at their workstations without the "baggage" of what is happening to them outside of work. This is obviously not the case, and while these outside stresses cannot be controlled, they can be recognized, appreciated, and, often, accommodated. In a robust workstation design, reasonable variance in worker performance, whatever its cause, should not lead to catastrophic results. If it could and the worker was aware of the possibility, it would result in even more stress on the worker.

10.3.4.3.2 Motivation and Reward Systems Most workers want to do a good job. However, work and workstations are often designed in such a way that the worker has no awareness of the quality of the work being done (due to the lack of a standard of performance, a measurement system, or feedback) and/or no control over the quality of that work (due to the inability to remedy or stop output of poor quality work).

TABLE 10.2 Types of Product and Process Characteristics

| Type | Fundamental | Nominal | Discretionary |
|---|---|---|---|
| Definition | Cannot be changed without an overall change in design | Can be changed with localized change in design | Can be changed without any significant change in design |
| Product example | Snap-fit vs. screw-fastened cover | Self-drilling vs. self-tapping screws | Hex- vs. Torx-head screw |
| Process example | Batch build vs. one-piece flow | Manual vs. automatic part unloading | Dedicated staffing vs. job rotation |

10.3.4.3.3 Environment Outside the immediate workstation environment, over which control may be substantial, there is most likely a larger, more uncontrollable environment that contains a "meta" version of the same types of stressors as those discussed in the previous paragraphs. The extent to which they can affect individual workers affects the overall success of workstations and, therefore, the efforts that must go into their analysis and design.

### 10.3.5 Product–Process Model

Another view regarding workstations is that they are just a reflection of the characteristics of the product design and the basic processes that are necessary to produce that product (or service). It is important to isolate those characteristics that are truly fundamental and cannot be changed without undue effort from those that can be changed to accommodate more ergonomically designed tasks. Table 10.2 defines and gives examples of such characteristics.

### 10.3.6 Life Cycle Task Model

Comprehensive workstation evaluation and design should address the ergonomics of associated tasks throughout the life cycle of the workstation. The tasks of the worker include not only those that occur regularly and consistently, but also those that do not but are nonetheless foreseeable, such as dealing with breakdowns in the process due to bad materials, tools, controls, and information. The span of time over which such tasks can be considered can be from conception to disposal, that is, the entire workstation life cycle. A list of life cycle tasks is shown in Table 10.3. The tasks that the workstation analyst or designer should focus on are those shown in boldface—operation, service, quality assurance, and repair.

## 10.4 CRITERIA FOR WORKSTATION EVALUATION AND DESIGN

### 10.4.1 Criteria Development

Inevitably, the practitioner desires fairly specific criteria for the evaluation and design of a workstation. This is always the most problematic part of these discussions, because the wide range of applications and conditions makes for almost limitless combinations. Much information must be gathered before a proper analysis can be performed, which, when coupled with general engineering methods, will often lead to the selection of fairly specific

TABLE 10.3 Tasks Associated with the Life Cycle of a Workstation

| Timing | Regular Tasks | Irregular Tasks<br>Product-Oriented | Process-Oriented |
|---|---|---|---|
| More predictable or schedulable | Operation **Service** including job setting and scheduled maintenance | Conception | Conception |
| | | Design | Design |
| | | Prototyping | Construction |
| | | Validation | Validation |
| | | Pilot building | Distribution |
| | **Quality assurance** | Validation | Installation |
| | | | **Repair** |
| Less or not predictable or schedulable | | | Shutdown |
| | | | Mothballing |
| | | | Demolition |
| | | | Transportation from site |
| | | | Recycling |
| | | | Disposal |

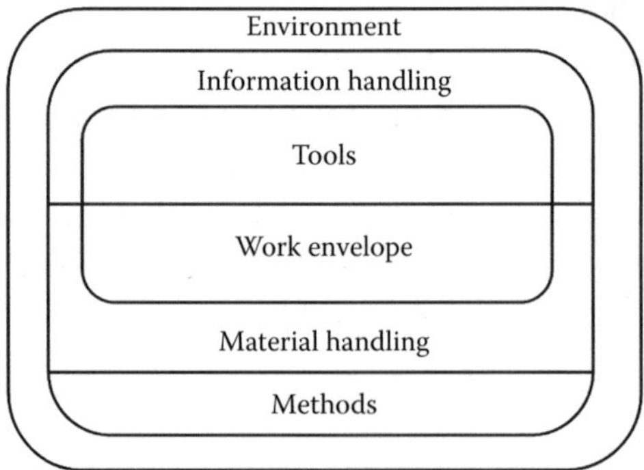


FIGURE 10.4 Applied workstation evaluation and design model.

design criteria. To organize the following discussion of information gathering, the model shown in Figure 10.4 will be used, in a somewhat outside-in manner.

### *10.4.1.1 Environment*

The presumption with regard to the environment is that it is not subject to redesign. If this is not the case, then the analysis/design process must move up hierarchically to include other workstations affected by change. Understanding the environment can affect workstation design decisions because the environment can act on (1) the motivation and (2) the capability of workers to perform their tasks. Some of the questions that should be answered to gain this understanding are as follows:

Does the worker feel a sense of isolation or sense of community, with respect to both the workforce and the overall process of which the workstation is but a part?

Is the worker afforded privacy as well as lines of communication when needed?

Is there workstation access to and from the general environment so as to promote the timely flow of needed people, information, equipment, and material?

Are the worker's senses (e.g., vision, hearing) overwhelmed so as to prevent or degrade the proper transfer of necessary information to and from the worker not only about that worker's workstation but also about adjacent workstations and the environment itself?

#### *10.4.1.2 Methods*

A well-designed workstation includes prescribed methods, where possible, for how the work is to be performed. Questions to be answered are as follows:

What general skills, specific skills, and training are or will be necessary for the worker prior to deployment?

Will periodic training or skills assessment/enhancement will be necessary after deployment?

What information (e.g., visual controls) will be readily accessible to the worker within the workstation regarding the methods to be used?

How much latitude will the worker have in following the method? This is a double-edged sword, because a totally inflexible method may render the worker incapable of reacting to irregular or unforeseen events, whereas too much flexibility may result in no method at all.

Does the design of the product and the process provide sufficient controls as to method? For example, do two parts go together in a single, unambiguous fashion?

What is the worker's span of control over his/her work?

How much variety is provided? What is the frequency of fundamental motions (i.e., bending, twisting, gripping, etc.)?

#### *10.4.1.3 Information Handling*

Information is any stimulus in the environment that provides the worker with necessary data about how to perform his/her task successfully. This includes static information about products and processes as well as dynamic information about their current state. Questions to be answered include the following:

What information must be obtained by the worker? Where is this information obtained? Does the information exhibit persistence, or is it of momentary quality? What is the predictability of the location, frequency, and form of this information? What is the proper response if information is not available when expected or needed?

What information must be provided by the worker? To whom or what does this information flow? What is the predictability of the location, frequency, and form of this information flow? What is the proper response if information flow is held up?

What decision-making processes must the worker perform to convert incoming information and existing knowledge into outgoing information and action? What is the duration of these processes?

What is the likelihood that environmental "noise" can affect any of what was mentioned earlier?

#### *10.4.1.4 Materials Handling*

Materials include any objects that flow into, are consumed within, or flow out of the workstation. Although most materials are incorporated into the product, there are others that must be considered, such as consumable tools (e.g., sandpaper) and fuels (e.g., batteries for a test gauge). Questions to be answered are as follows:

What materials must be obtained by the worker? What is the location, frequency, and characteristic of these materials? How predictable are they? What is the proper response if material is not available when expected or needed?

What materials are produced by the worker? To whom or what do these materials flow? What is the predictability of the location, frequency, and characteristics of this materials flow? What is the proper response if materials flow is held up?

What materials handling processes must the worker perform to convert incoming materials into outgoing materials? What is the duration of these processes?

What is the proper response if any material qualities adversely affect the workstation tasks? What mechanical aids are necessary?

#### *10.4.1.5 Tools*

Tools are defined broadly to include both portable and fixed tools, equipment, and machinery. They are further characterized as being more or less assigned to a workstation (or a related group of workstations) and are usually found within or in the vicinity of the workstation. Questions include the following:

What tools, equipment, or machines must be obtained by the worker? What are their location, frequency, and characteristics? How predictable are these? What is the proper response if the tools are not available or suitably operational when expected or needed?

What are the prescribed methods for using tools, equipment, or machines? What is their duration?

What is the proper response if any material qualities adversely affect the workstation tasks?

What are the energy sources? What controls are provided? Are controls unambiguous?

#### *10.4.1.6 Work Envelope*

It is not accidental that the work envelope is the last of the workstation model components to be discussed. It is often the most visible aspect of workstation design, sometimes at the expense of the other components. Further, it is impossible to lay out a work envelope if complete information is not available about all of the tasks and the information, materials, and tools to be used. The questions to be asked at this point are actually quite simple:

Is there enough room for the worker to comfortably fit and move into, within, and out of the workstation, before, during, and after the task?

Can the worker "reach" without undue effort all of the necessary materials, information, tools, and controls in the workstation? (*Reach* here means both physically and with the senses, e.g., seeing and hearing.)

In spite of the simplicity of these questions, the answers require a basic understanding of the variability of the size of workers (i.e., anthropometry) and the use of anthropometric data. There are three basic choices when applying anthropometric data in the design of a workplace, each of which has its applications: design for extremes, design for the average, and provide for adjustability [12].

10.4.1.6.1 Design for Extremes The extremes are either the largest or smallest individual who may be expected to use a workplace.

*10.4.1.6.1.1 Design to Fit* Dimensions for the largest individual, the 95th or 99th percentile, are used to determine clearance dimensions such as the minimum height for overhead conveyors or the top of a doorway that will permit the majority of the population to walk beneath without hitting their heads. The large individual also determines the necessary width for aisles, the size of access openings for machine repair, and the necessary clearance between a chair seat and the bottom of a table or workbench.

*10.4.1.6.1.2 Design to Reach* Dimensions for the smallest person, the 5th or 1st percentile, are used to determine reach dimensions, such as the maximum height of shelves or controls, the location of parts bins in the workplace, or the height of nonadjustable chair seats that will permit the small person's feet to touch the floor.

10.4.1.6.2 Design for the Average Designing for average body dimensions is usually a mistake. An average dimension may lead to discomfort for a large proportion of the workforce. The smaller person in the workforce will not be able to reach a part or control placed at the reach distance of the average person, and the larger person in the workforce may not fit in the workplace or may have to assume unusual postures to perform the task. A conveyor line set to the reach height of the average person will have the smaller individual reaching up to work, causing arm and shoulder fatigue, while the larger operators will be bending down, causing neck and back fatigue.

10.4.1.6.3 Provide for Adjustability The previous two design criteria assume that the workplace components (conveyors, workbenches, parts bins, etc.) are fixed and cannot be adjusted. The alternative to these approaches is to provide adjustability in the workplace [8,9]. The advantage of this approach is that a large proportion of the workforce can be accommodated. The disadvantages include higher design and manufacturing costs, time of adjustment, and greater likelihood of the adjustable component breaking.

Where adjustment is a practical approach, the proportion of the population to be accommodated must be determined. Typically the range of adjustment will include the middle 90% or 95% of the population. For example, a standing workplace for light assembly work should be at approximately elbow height. Obtaining the required percentile dimensions from a typical anthropometric table, accommodating the 5th percentile female through the 95th percentile male, might require the following range of adjustment:

Female 5th percentile = 96.8 cm

Male 95th percentile = 118.1 cm

Range of adjustment = 21.3 cm

Only the smallest 5% of the female workforce and the largest 5% of the male workforce will not be able to adjust this workplace properly.

To accommodate a different percentage of the population might require using the appropriate Z value from a normal probability table and the dimension's mean and standard deviation from a typical anthropometric table. For example, accommodating 0.5th percentile female to the 99.5th percentile male:

For 0.5th–99.5th percentile: $Z = \pm 2.576$

Female: 50th percentile = 102.6 cm, s = 3.6 cm

0.5th percentile = 102.6 − (2.576)(3.6) = 93.5 cm

Male: 50th percentile = 110.5 cm, s = 4.6 cm

99.5th percentile = 110.5 + (2.576)(4.6) = 122.2 cm

Range of adjustment = 28.7 cm

As can be seen, the range of adjustment would have to be increased by 7.4 cm to accommodate this additional part of the workforce. Often this is not practical or may be too costly; therefore, even in adjustable workplaces, the proportion of those to be accommodated must be reduced.

10.4.1.6.4 Alternatives to Adjustment Accommodating individuals in fixed height workstations can often be accomplished through appropriate placement of workers with respect to the tasks. The assembly of large items, such as car bodies or engines, usually involves component installations at a variety of work heights. By matching operators to tasks with inherently more appropriate work heights rather than by random assignment, a greater percentage of operators can be accommodated.

Where fixed work heights must be used, yet all work is performed at approximately the same height, such as work with presses or welders, designing the work height to fit the large individual may be the practical alternative. To fit smaller individuals, adjustable platforms can be moved into the workplace when needed.

Job rotation can also be used to minimize an individual's exposure to uncomfortable postures caused by improper work heights.

### 10.4.2 Examples of Specific Design Criteria

#### *10.4.2.1 Sitting Workstations*

10.4.2.1.1 Advantages Energy expenditure is approximately 20% less for sitting than for standing. A sitting workstation provides a high degree of body stability and reduces fatigue when work periods exceed 2 h. Workers can use feet for control actions [7].

10.4.2.1.2 Indications and Contraindications A seated workplace can be used under the following conditions:

1. All items needed to perform the task can be easily placed within the seated workspace.
2. Frequent reaches above the work surface are not required.
3. Tasks requiring fine manipulative hand movements are predominant.
4. The handling of parts or application of force does not exceed 45 N.
5. There is no interference with leg clearance.

10.4.2.1.3 Criteria The commonsense criterion for physical access is the ability to reach, which in turn defines the *reach envelope*, which consists of that spatial volume that can be reached by the worker while remaining seated (i.e., static) in a "normal" chair without undue stretching, bending, or twisting.

*10.4.2.1.3.1 Work Surface Height* In any workplace, work height is often a trade-off between (1) a close view of the work and the need for precise movements and (2) the freedom to perform gross movement and the ability to generate force. The normal work height for seated tasks is about 5 cm below elbow height. This clearance provides a reasonably close view of the work while permitting movement not interfered with by obstructions on the work surface. This clearance can then be increased for more movement and force capability, and decreased, even to where work height is above elbow height, for precise physical and visual tasks. In this last case, it may be necessary to provide elbow support for prolonged tasks.

It is important to carefully consider what constitutes the work surface. Often, it is not the physical desk or benchtop but the location of the hands while working, such as locations on machines, fixtures, or parts. In the case of computer keyboards, the home row of the keyboard is the work surface. It is recommended that the home row be at elbow height to facilitate the relatively precise movements required.

To select which height dimension within the workstation to set first, one should identify which dimension is the least flexible from such possible dimensions as the heights of points of operation, the desk or benchtop, the elbow, or the seating. The subsequent determination of the initial height of that dimension will then drive the process of finding most other height dimensions. For instance, using the work surface height as the initial dimension will most likely require accommodating workers through adjustable seat and foot rest dimensions.

*10.4.2.1.3.2 Reach Dimensions* Normally, workers will have to reach away from the normal point of operation to complete their task, perhaps to obtain material or tools, or to operate controls. Reaches should be kept within certain regions depending upon the frequency of the reach, as illustrated later in Figure 10.7. The majority of work should be performed within 15 cm of the work surface. Although both should be kept to a minimum, reaches below the level of the work surface are preferable to those above the work surface; reaches above shoulder level and behind the body should be kept to a minimum. Repetitive work should be performed with the shoulders and upper arms relaxed and the elbows bent at approximately 90°. Padded forearm rests should be provided for precision work.

*10.4.2.1.3.3 Clearances* Clearance below the work surface for legs and toes is necessary. Sufficient thigh clearance (25–30 cm) should be provided between the seat pan and the underside of the work surface. Sufficient clearance also permits the worker to vary posture to reduce fatigue.

*10.4.2.1.3.4 Seat Dimensions* The seat area should be large enough to permit movement to relieve pressure points [8,9,11]. The seat pan and back rest should be padded and covered with a permeable material that will allow absorption of sweat. The back support should adjust both horizontally and vertically. Figure 10.5 gives typical dimensions for a seated workplace.

#### *10.4.2.2 Standing Workstations*

10.4.2.2.1 Advantages When operators are required to stand for long periods of time, fatigue can become an important factor in their job performance. To minimize the effects of fatigue, it is important that the task be designed to eliminate excessive reaches, stooping, bending, or twisting motions, and unnatural head positions because of the visual requirements of the task.

10.4.2.2.2 Indications and Contraindications A standing workplace is needed under the following work conditions:

1. Mobility or walking is required.
2. Handling of objects weighing more than 45 N is frequently required.
3. Extended reaches are often required.
4. Large downward forces must be exerted.
5. The workplace has insufficient leg clearance for a seated operator.

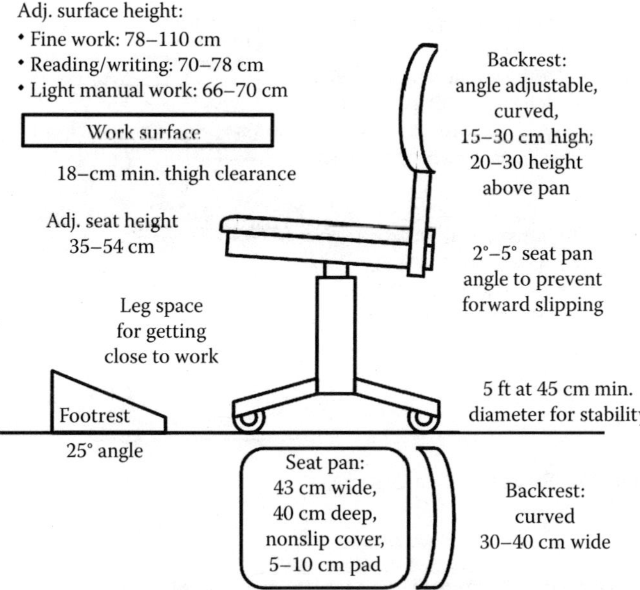


FIGURE 10.5 Seated workplace dimensions.

10.4.2.2.3 Criteria The standing reach envelope is a bit larger and extends the sitting reach envelope to include that which can be reached by more effective bending forward and twisting than is available while seated, and by a nominal repositioning of the feet that does not result in a significant movement away from "home" position of the workstation.

*10.4.2.2.3.1 Work Surface Height* Normal work height for light assembly tasks is just below elbow height. For more precise work, the work surface should be raised above elbow height and padded rests provided for the forearms. For heavy work where large downward or upward forces are required, the work surface should be 10–20 cm below elbow height. Remember, work height is where work is being done and is not necessarily the height of the work surface [4,5,7,8].

*10.4.2.2.3.2 Reach Distances* Allow for the shortest arms when reaching up or out. Keep frequent work within forearm reach and below mid-chest height. Avoid reaches behind the body.

*10.4.2.2.3.3 Clearances* Allow at least 200 cm of overhead clearance. Access into and out of the workplace should allow for the largest individual. Clothing and personal protective equipment require additional clearance. Allow toe space and knee clearance at workbenches and conveyors. Provide sufficient room for the operator to move about and change postures while working.

*10.4.2.2.3.4 Task Characteristics* Avoid the use of foot pedals in standing workplaces. Avoid placing displays or other visuals above eye height. In general, the primary visual task should be within 10° downward of the direct line of sight and within 15° right and left. Avoid or minimize work that requires fine manipulations or precision adjustment.

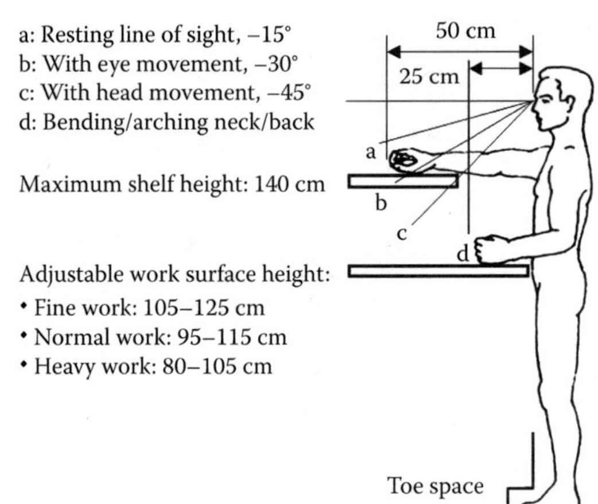


FIGURE 10.6 Dimensions for standing work.

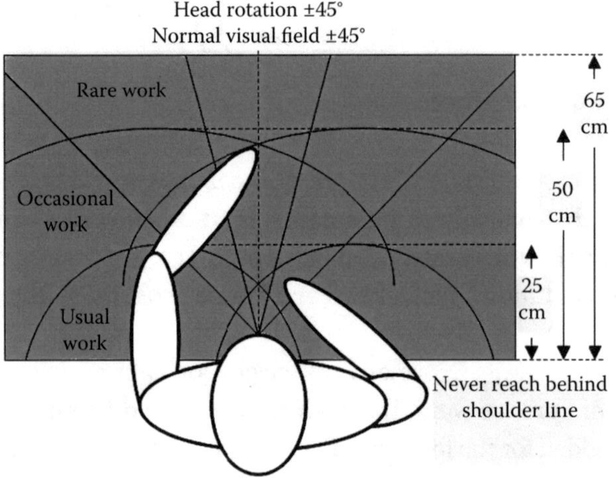


FIGURE 10.7 Reach dimensions for standing or seated work.

*10.4.2.2.3.5 Other Considerations* Providing floor padding or shoe inserts may improve worker comfort when continuous standing is required. Provide a stool or seat that the operator may use during breaks or temporary work stoppages. Figures 10.6 and 10.7 give typical dimensions for a standing workplace.

### *10.4.2.3 Sit/Stand Workstations*

10.4.2.3.1 Advantages A sit/stand workplace is often preferred from both a physiological and an orthopedic point of view and is recommended when the following work conditions are present:

1. Multiple tasks are performed, with some best performed while seated and others while standing.
2. Reach requirements occasionally extend more than 40 cm forward and/or more than 15 cm above the work surface.

10.4.2.3.2 Indications and Contraindications The minimum height for the work surface during standing tasks is 100 cm. However, this height may present a hazard from excessive chair height during seated work. A 90 cm work surface with a platform that folds down or slides onto the work surface for standing work can eliminate this problem. A work surface that adjusts between 90 and 105 cm would provide the best range of heights for both sitting and standing.

## 10.4.3 Examples of Workstation Evaluation and Design

This section presents two examples of applying the principles covered in this chapter. The first example is an evaluation of an industrial assembly workstation, and the second is the design of an office workstation. The steps in the evaluation/design process include the following: (1) definition of tasks, (2) application of workstation models and ergonomic principles, (3) layout, (4) input by worker(s) and modification, (5) prototyping, (6) refinement, and (7) utilizing feedback. The reader should note that all models and criteria presented in this chapter are not necessarily applicable or useful for every evaluation or design exercise.

### *10.4.3.1 Industrial Example of Evaluation*

This example is of an existing workstation that is part of a transfer line where radios are installed in automotive instrument panels. The nominal work height is governed by the transfer line and is 80 cm. Lateral space in the workstation is limited to 1.5 m. The overall cycle time for installation is 60 s. A layout of the existing workstation is shown at the top of Figure 10.8.

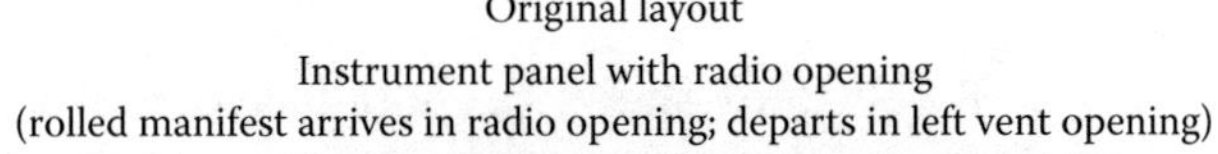


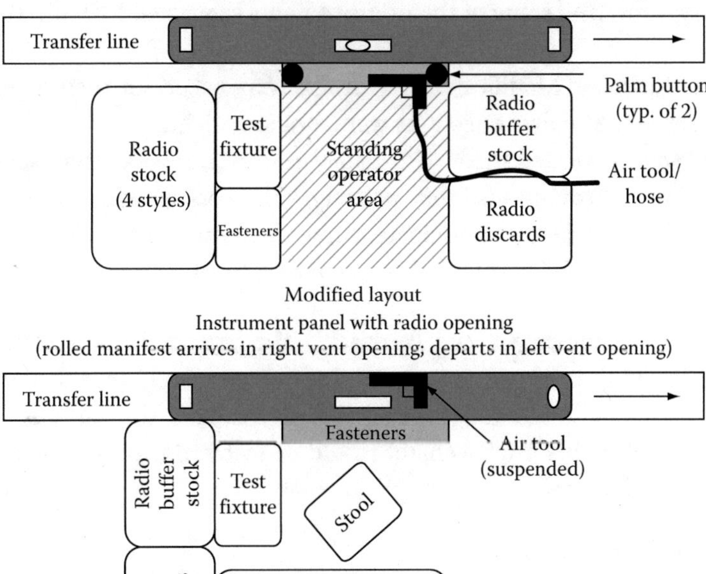


FIGURE 10.8 Old and proposed new layouts for industrial workstation example.

10.4.3.1.1 Task Definition The existing sequence of tasks is as follows:

1. Read manifest and determine radio option for next installation.
2. Get radio from stock.
3. Place in test fixture.
4. Connect test cable (contains power and signal).
5. Perform specified test sequence (power on, select station, volume increase/decrease, power off).
6. Remove radio from fixture (if necessary, discard failures and replace from pretested buffer stock).
7. Place radio in IP opening from bottom.
8. Secure radio from top with two fasteners using pistol-grip air tool.
9. Index transfer line using palm buttons.

10.4.3.1.2 Application of Workstation Models and Ergonomic Principles Table 10.4 presents a summary of possible analyses based on the models and principles shown in Tables 10.1 through 10.3.

10.4.3.1.3 Layout Figure 10.8 (bottom) shows proposed layout of a new workstation arrangement incorporating some of the improvements suggested in Table 10.4.

10.4.3.1.4 Worker Input and Modification Workers were asked for their assessment of the proposed workstation. The first questions were, not surprisingly, "what is wrong with the way it is?" and "who requested this?" Ultimately, however, several good suggestions were received, including improving visibility between this workstation and the two adjacent workstations to allow workers to anticipate things that might affect their job. As a result, modifications were proposed to materials storage rack heights.

10.4.3.1.5 Prototyping, Refinement, and Feedback The project is now set to proceed to appropriate prototyping to prove out spatial and temporal relationships and equipment selection. At least some small refinements will be realized during this process. After implementation, continuous improvement would be fostered through the use of feedback from the workers and others.

#### *10.4.3.2 Office Example of Design*

This example is of a proposed new workstation that will process requests for data entry to and retrieval from a computer database. Approximately 9 $m^2$ of floor space is available, but there are presently no serious constraints on the equipment that can be used [7,8,9,12].

TABLE 10.4 Summary of Analyses of Industrial Workstation Example

| No. | "Do ..." | "With ..." or "to What ..." | "When ..." or "How ..." |
|---|---|---|---|
| A1 | IO Wait | IP | From pervious palm buttons |
| A2 | IO Get | Manifest | From IP |
| A3 | II Read | Option | From manifest |
| A4 | IO Select | Radio | From workstation stock |
| A5 | PO Put | Radio | To test 70 fixture |
| A6 | PI Control | Test protocol | At loaded test fixture |
| A7 | II Get | Test result | During test protocol |
| A8 | PO Get | Radio | From test fixture |
| A9 | CO Put | Defective radio | From test result negative |
| A10 | IO Get | Pretested radio | From workstation stock |
| A11 | PO Put | Radio | Into IP |
| A12 | IO Get | Fasteners | From stock |
| A13 | PO Get, use, put | Air tool use | Radio in IP, fasteners in hand |
| A14 | II Sense | Installation complete | End of air tool use |
| A15 | OI Press | Palm buttons | End of installation complete |
| A16 | OO Wait | IP w/radio | From palm buttons |

(*continued*)

TABLE 10.4 (continued) Summary of Analyses of Industrial Workstation Example

| Product and Process Characteristics | | | | |
|---|---|---|---|---|
| **Fundamental** | **Nominal** | **Discretionary** | **Ergonomics Attributes** | **Possible Remediation/Improvements** |
| Frequency | Concurrent tasks | | Frequency, work–rest | Job rotation, task reallocation |
| Timing | Location | | Visual task, reach | Locate to leading end of IP to reduce waiting time/reach |
| Location | Coding | | Visual task, discrimination | KISS coding, if necessary |
| Number of types | | Quantity, location, and identification | Reach, force, posture, and discrimination | Multiple, single-part presentation in optimal location; visual controls |
| Fixture design | | Location | Reach, force, and posture | Optimal location out of way of conveyor |
| Procedure | | | Control design | Eliminate through 100% pretest |
| Criteria | | Sensing | Display design and noise | Environmental noise control |
| Fixture design | | Location | Reach, force, and posture | Optimal location out of way of conveyor |
| Not usable | | Location | Reach, force, and posture | Gravity discard |
| Number of types | | Quantity, location, and identification | Reach, force, and posture | Multiple, single-part presentation in optimal location |
| Location | | Orientation, holding | Reach, force, and posture | Rotating fixture on conveyor; design to hold radio in place |
| Type | | Quantity, location | Reach, force, and posture | Optimal location; auto feed to tool |
| | Energy source | Type, location | Reach, force, posture, noise, and vibration | Counterbalanced, torque-limited in-line with vibration dampening |
| | | | Decision making | Objective decision criteria |
| Signal | Type of control | Location | Reach, force, and posture | One-hand no/low force control with light curtain |
| Frequency | Concurrent tasks | | Frequency, work–rest | Job rotation, task reallocation |

10.4.3.2.1 Task Definition The existing sequence to tasks is as follows:

1. Batch enter data from hand-completed forms (approximately 100 letter-size forms per day, each of about 80 total characters in 10 fields; average input time about 2 min/form; total time about 200 min.
2. File forms after entry.
3. Receive individual customer requests (in-person, by mail, or by phone) for retrieval or change of existing data (in-person and phone requests handled while you wait; mail requests printed and mailed by next day; expectation is 10 in-person, 25 mail, and 25 phone requests per day; average service time about 4 min each; total time about 240 min).

10.4.3.2.2 Application of Workstation Models and Ergonomic Principles Table 10.5 presents a summary of possible analyses based on the models and principles shown in Tables 10.1 through 10.3.

10.4.3.2.3 Layout (New) Figure 10.9 shows the proposed layout of a workstation arrangement incorporating some of the improvements suggested in Table 10.5.

10.4.3.2.4 Worker Input and Modification In this case, no workers were directly available, so reliance will be placed on subsequent prototyping, refinement, and feedback processes.

10.4.3.2.5 Prototyping, Refinement, and Feedback Proceed as in first example, with greater emphasis on feedback because there was no worker input during the initial stages of the design process.

## 10.5 CRITICAL REVIEW OF CURRENT STATUS

### 10.5.1 Independent versus Nonindependent Design Criteria

As should be evident from a review of this book, ergonomics is a collection of a fairly large number of areas of concern. Taken separately, most of these areas have been systematically researched, and meaningful guidelines have been developed. However, the design of something as potentially open-ended as a workstation involves the simultaneous application of these multiple guidelines. As I see it, the primary problem in the effective design of workstations is the independent (albeit understandable) application of these nonindependent guidelines. No one would be foolish enough to suggest that one could apply anthropometry without considering biomechanics, or make decisions on task allocation between workers and machines independent of any knowledge of the effects of repetition on cumulative trauma disorders.

Unfortunately for the practitioner, the number of combinations and permutations of ergonomics stressors that can be present in the real world of work design is such that research has not had time to examine them all. When attempts are made to reduce complex

TABLE 10.5 Summary of Analyses of Office Workstation Example

| No. | "Do …" | "With …" or "to What …" | "When …" or "How …" |
|---|---|---|---|
| B1 | IO Get | Batch | 1/day at start of day |
| B2 | IO Get | Form | ~100/batch as need |
| B3 | II Sequence read | Character | ~80/form |
| B4 | OI Select, press | Key stroke | 1/character |
| B5 | II Read | Character | 1/key stroke |
| B6 | PI Select, press | Correct | Bad character |
| B7 | OO Put | Form | Successful |
| B8 | OO Put | File | Forms |
| B9 | II Communicate | Request | 10 in-person/day |
| B10 | II Communicate | Request | 25 phone/day |
| B11 | II Get | Request | 25 mail/day |
| B12 | PI Code | Request | As needed |
| B13 | OI Select, press | Key stroke | As needed per request |
| B14 | II Read | Character | 1/key stroke |
| B15 | PI Select, press | Correct | Bad character |
| B16 | OI Communicate | Verbal response | As needed per in-person and phone request |
| B17 | PI Select, press | Print response | As needed per mail request |
| B18 | OI Get, put | Printed response | As needed |

| Product and Process Characteristics | | | | |
|---|---|---|---|---|
| **Fundamental** | **Nominal** | **Discretionary** | **Ergonomics Attributes** | **Possible Remediation/Improvements** |
| Batching >1 day | Multiple batches per day | Time of day, location | Location and workload | Delivery to workstation |
| Form design | | Location | Location, repetition, visual task, and discrimination | Adjustable multiple-document holder |
| Field design | | Location | Location, visual task, and discrimination | Discrete information design (check off boxes, etc.) |
| Discrete entries | Coding | Location | Location, posture, and repetition | Equipment selection (including adjustability) |
| | Context checking | Location | Location, posture, and visual task | Adjustable video display terminal (VDT) with antiglare screen and/or appropriate task lighting |
| | Auto correct | | Repetition | Discrete information input (menu selection, check off boxes, etc.) |
| Form design | | Location | Location and repetition | Multiple-document holder |
| Filing > 1 day | Multiple filing per day | Time of day, location | Location, posture, and repetition | Active filing locations at reasonable distances and at heights minimizing reach |
| Option | Timing | Location, request, and drop-off | Location, repetition, energy expenditure, and auditory task | Workstation integrated into request counter, including height to eliminate sit/stand/sit |
| Option | Timing | Location, voice mail | Location, repetition, distraction, and auditory task | Telephone headset; noise control applied to environment (equipment selection, separation) |
| Option | Receipt timing | Processing timing | Location, repetition, and visual task | Delivery to workstation |
| Timing | Visual controls | | Identification, matching, and coding | Discrete information design |
| Discrete entries | Coding | Location | Location, posture, and repetition | Equipment selection (including adjustability) |
| | Context checking | Location | Location, posture, and visual task | Adjustable VDT with antiglare screen and/or appropriate task lighting |
| | Auto correct | | Repetition | Discrete information input (menu selection, check off boxes, etc.) |
| Option | Timing | Location, revisit, recall, or mail response | Location, repetition, energy expenditure, and auditory task | Workstation integrated into request counter, including height to eliminate sit/stand/sit |
| Form design, timing | | Location | Repetition | Discrete information input (menu selection, check off boxes, etc.), batching output |
| Option | Timing | Location | Locations, postures, and repetition | Printer at workstation; mail pick up at workstation |

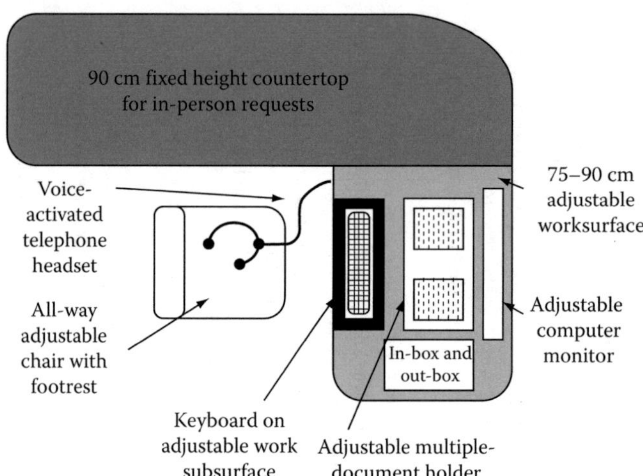


FIGURE 10.9 Proposed layout for office workstation example.

multidimensional phenomena to simplistic guidelines, much can be lost in their general applicability. For instance, the National Institute for Occupational Safety and Health (NIOSH) lifting guidelines sought to reduce the admittedly complex analysis of manual lifting tasks to a function of four (1981 version) or six (1991 version) easily measured variables. But the fine print in the "instruction manual" cautions about applying the guidelines in situations that are quite common, including highly dynamic lifting. It can be said without fear of contradiction that the use of these guidelines has often strayed beyond their limits.

What the practitioner must do, therefore, in the evaluation and design of workstations is to be thorough in the identification of ergonomics issues and in the analysis of each such issue. Common sense must then be applied [10].

## 10.6 FUTURE CONCERNS

### 10.6.1 Two-Sided Standards versus One-Sided Guidelines

In the application of guidelines that have been developed for engineers and workstation designers, a pattern has been observed that should give pause. It is natural for the users of guidelines to want definitive, objective criteria to divide the domain of designs into two all-inclusive, mutually exclusive regions, "OK" and "not OK" (i.e., two-sided standards). Real-world phenomena do not often succumb to such categorization, at least for the layperson using a relatively easy-to-measure dimension. The reality is that the distribution of workstations within the two regions along a given dimension generally overlaps as illustrated in Figure 10.10. As can readily be seen, at any value of the dimension, there are finite possibilities that a workstation may be in either region. Thus, when the question is asked, "Is the workstation OK at this value?" the answer is, "It depends." This can be very disconcerting to one trying to evaluate a workstation [13,14].

The response of those who prepare guidelines has generally been to pick a value at which the odds of a making a wrong "OK" decision are held to a reasonable level, accepting an

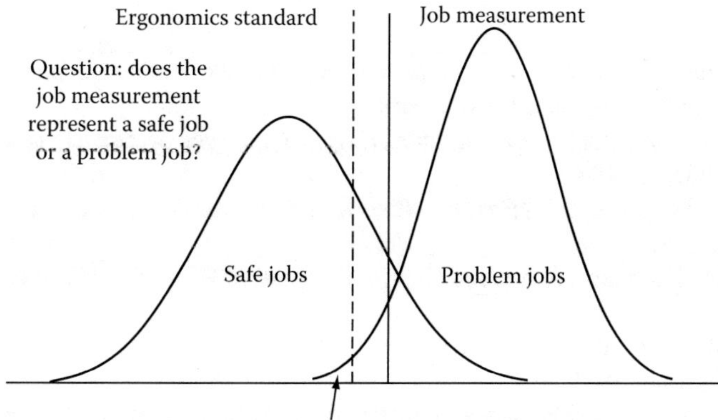


FIGURE 10.10 One-sided versus two-sided ergonomics standards.

increased chance of a wrong "not OK" decision (i.e., controlling α risk at the expense of β risk). As an example, a guideline might state that the clamp forces should not exceed 0.3 N or that energy expenditure should not exceed 3.5 kcal/min. Thus, associated decision rules become the following: If forces are less than 0.3 N, or energy expenditure is less than 3.5 kcal/min, the task is OK; else it depends.

What it depends on are values of other dimensions associated with the task; this is the very topic of Section 10.4.3. For instance, 1 N of force is acceptable in a preferred posture using a power grip; or 5 kcal/min is acceptable for a younger, male worker, This, then, is the one-sided guideline; that is, the decision is clear on only one side of the question.

This is the way it should work. However, it often results that practitioners start using the one-sided "OK-depends" guideline as a two-sided "OK-not OK" standard. This occurs for two reasons: (1) a lack of sufficient knowledge or training of the users on the basis for and the interpretation of the guidelines being used or (2) a lack of clear organizational policy as to the proper use of guidelines or the failure to follow such a policy. Thus, I see an unfortunate scenario occurring in many organizations that practice ergonomics, at least in part, in this fashion:

The inevitable: Using one-sided standards as two-sided

The only good news: Declaring "unsafe" jobs as unsafe

The bad news: Declaring some "safe" jobs "unsafe," causing resources to be invested to solve problems that do not exist while other "unsafe" jobs go uncorrected

The worse news: Institutionalizing the misuse of standards to the point where it becomes impossible to correct the misuse

The worst news: Giving ergonomics an undeserved reputation for being too easy on workers and compromising productivity, almost guaranteeing its ultimate failure to be an effective tool, both for worker health and safety and product and process design [10]

## REFERENCES

1. D. C. Alexander and B. M. Pulat, *Industrial Ergonomics: A Practitioner's Guide*, Industrial Engineering and Management Press, Atlanta, GA, 1985.
2. R. W. Bailey, *Human Performance Engineering: A Guide for System Designers*, Prentice-Hall, Englewood Cliffs, NJ, 1982.
3. J. H. Burgess, *Designing for Humans: The Human Factor in Engineering*, Petrocelli Books, Princeton, NJ, 1986.
4. Eastman Kodak Company, *Ergonomic Design for People at Work*, Vol. I, Van Nostrand Reinhold, New York, 1983.
5. Eastman Kodak Company, *Ergonomic Design for People at Work*, Vol. II, Van Nostrand Reinhold, New York, 1986.
6. A. Freivalds, *Niebel's Methods, Standards, and Work Design*, 12th edn., McGraw-Hill, New York, 2009.
7. E. Grandjean, *Fitting the Task to the Man*, 4th edn., Taylor & Francis, London, 1988.
8. Human Factors Society, *American National Standard for Human Factors Engineering of Computer Workstations, ANSI/HFS 100–2007*, Santa Monica, CA, 2007.
9. S. A. Konz and S. Johnson, *Work Design: Industrial Ergonomics*, 5th edn., Holcomb Hathaway Publishing, Scottsdale, AZ, 2000.
10. National Safety Council, *Making the Job Easier: An Ergonomics Idea Book*, Chicago, IL, 1988.
11. D. J. Oborne, *Ergonomics at Work*, 2nd edn., Wiley, New York, 1991.
12. S. Pheasant, *Bodyspace: Anthropometry, Ergonomics and the Design of Work*, 3rd edn., Taylor & Francis, London, 2006.
13. M. S. Sanders and E. J. McCormick, *Human Factors in Engineering and Design*, 7th edn., McGraw-Hill, New York, 1993.
14. W. E. Woodson, *Human Factors Design Handbook*, McGraw-Hill, New York, 1981.

CHAPTER 11

# Tool Evaluation and Design

Andris Freivalds

CONTENTS

## 11.1 INTRODUCTION

Tools are as old as the human race itself. The hands and feet could be considered tools given to the human by nature. However, tools as we know them were developed as extensions of the hands and feet to amplify the range, strength, and effectiveness of these limbs. Thus, the early human, by picking up a stone, could make the fist heavier and harder to produce a more effective blow. Similarly, by using a stick, a longer and stronger arm was created.

The exact time when humans began to use and make tools is not known. Leaky [1], during his excavations in Africa, uncovered evidence that more than a million years ago the prehistoric human was already a toolmaker using stones for chipping and bones for leatherwork. Similarly, Napier [2,3] indicated that with changing tasks, such as converting from the power grip to the precision grip, there was a similar change in the anatomy of the hand as well as the development of tools. An important milestone occurred when stone tools were provided with handles some 35,000 years ago. The addition of the handle increased the range and speed of action and increased the kinetic energy for striking tasks [4]. A still later change in tool development occurred with the change in tasks from food gathering to food production. New tools were required and accordingly developed. Surprisingly, many of these tools, with minor improvements and refinements, are still in use today. The reasons for such stagnation could be twofold: either the tool reached an optimal form very quickly with no room for improvement, or there was no impetus for further improvement. The latter is the resigned view that since a tool has been used by so many people for so many years, no further improvement is possible. The former view is obviously not true, because Lehmann [5] noted the existence of over 12,000 different styles of shovels in Germany in the 1930s, all essentially used for the same task. Indeed, the last great change in tool development occurred with the start of the Industrial Revolution when tasks changed from food production to the manufacture of goods.

The parallel development of tools with changing technology has given rise to another problem. The current technology explosion has proceeded too quickly to permit the gradual development of tools appropriate for the new industrial tasks. The instant demands for new and specialized tools to match the needs of technology have, in many cases, bypassed the testing needed to fit these tools to their human users. This has resulted in a variety of hand tool-generated work stressors, trauma, and chronic problems, reducing productivity, disabling individuals, and increasing medical costs for industry.

## 11.2 BACKGROUND AND SIGNIFICANCE TO OCCUPATIONAL ERGONOMICS

Poor design and excessive use of hand tools are associated with increased incidence of both acute and subacute cumulative trauma of the hand, wrist, and forearm [6,7]. Acute trauma includes burns, cuts, lacerations, abrasions, fractures, strains, sprains, dislocations, and even amputations caused by the upper extremity being caught, cut, or burned by the tool.

Cumulative trauma disorders (CTDs) are injuries to the musculoskeletal system that develop gradually as a result of repeated microtrauma. Because of the slow onset and relatively mild nature of the trauma, the condition is often ignored until the symptoms become chronic and more severe injury occurs. These problems are a collection of a variety of problems including repetitive motion disorders, carpal tunnel syndrome, tendonitis, ganglionitis, tenosynovitis, and bursitis, with these terms sometimes being used interchangeably. There are four major work-related factors that seem to lead to the development of CTD: (1) use of excessive force during normal motions, (2) awkward or extreme joint motions, (3) high amounts of repetition of the same movement, and (4) the lack of sufficient rest for the traumatized joint to recover. The most common symptoms associated with CTD are pain, restriction of joint movement, and soft tissue swelling. In the early stages, there may be few visible signs, but if the nerves are affected, sensory responses and motor control may be impaired. If left untreated, CTD can result in permanent disability.

Tenosynovitis is a disease of the tendon sheaths due to overuse or unaccustomed use of improperly designed tools. It is often experienced by trainees. Improperly designed tools causing ulnar deviation coupled with supination of the wrist may increase the occurrence of tenosynovitis. Repetitive motions and impact shocks may further aggravate the condition.

Carpal tunnel syndrome is a disorder of the hand caused by injury of the median nerve inside the wrist. The median nerve may be injured from extreme flexion and extension of the wrist, which causes inflammation and thickening of the tendon sheaths, which in turn compress the median nerve. Symptoms include impaired or lost nervous function in the first three and a half digits manifesting as numbness, tingling, pain, and loss of dexterity. Again, proper tool design is very important to avoid these extreme wrist positions [6].

Extreme radial deviations of the wrist are conducive to pressures between the head of the radius and the adjoining part of the humerus resulting in tennis elbow. Similarly, simultaneous dorsiflexion of the wrist with concurrent full pronation is equally stressful on the elbow. Tichauer and Gage [8] used the tennis elbow as a good example of the basic principle that work strain may affect a site distant from the site of work stress.

Trigger finger results from a work situation in which the distal phalanx of the index finger must be bent and flexed against resistance before more proximal phalanges are flexed. Excessive isometric forces impress a groove on the tendon, or the tendon enlarges due to inflammation and then snaps back into position when it moves within the sheath, usually with an audible click [8].

White finger results from excessive vibration from power tools inducing the constriction of arterioles within the digits. The resulting lack of blood flow appears as a blanching of the skin, with a corresponding loss of motor control. Other such CTDs are discussed by Meagher [9] and Stetson et al. [10].

Short-term fatigue and discomfort have also been considered risk factors and have been shown to be related to handle and work orientation in hammering [11,12] and to tool shape and work height in work with screwdrivers [13,14]. Also, it has been shown that poor design of the grip of a tool leads to exertion of higher grip forces [15,16] and to extreme wrist deviations [8] and therefore to more fatigue.

The cost of CTDs in U.S. industry, although not all due to improper tool design, is quite high. Data from the National Safety Council [17] suggest that 15%–20% of workers in key industries (meatpacking, poultry processing, auto assembly, and garment manufacturing) are at potential risk for CTD and that in 1991 some 223,600 cases or 61% of all occupational injuries were associated with repetitive actions. The worst industry was manufacturing, while the worst occupational title was butchering with 222 CTD claims per 100,000 workers [18]. With such high rates and average costs of $30,000 per case, National Institute for Occupational Safety and Health (NIOSH), in its *Year 2000 Objectives*, has targeted the reduction of CTD incidence from 82 to 60 cases per 100,000 overall workers and from 285 to 150 in certain manufacturing industries [19].

The proper selection, evaluation, and use of hand tools are the major ergonomic concerns. The following review discusses the basic principles involved in tool design, provides a general evaluation process for tools, and discusses the attributes desirable for specific tools.

## 11.3 TOOL EVALUATION AND DESIGN

### 11.3.1 Principles of Tool Design

#### *11.3.1.1 General Principles*

An efficient tool has to fulfill some basic requirements [4]:

1. It must perform effectively the function for which it is intended. Thus, an axe should convert a maximum amount of its kinetic energy into useful chopping work, separate wood fibers cleanly, and be easily withdrawn.
2. It must be properly proportional to the body dimensions of the operator to maximize the efficiency of human involvement.
3. It must be designed to match the strength and work capacity of the operator. Thus, allowances have to be made for the gender, age, training, and physical fitness of the operator.
4. It should not cause undue fatigue, that is, it should not demand unusual postures or practices that will require more energy expenditure than necessary.
5. It must provide sensory feedback in the form of pressure, some shock, texture, temperature, etc., to the user.
6. Its capital and maintenance costs should be reasonable.

#### *11.3.1.2 Anatomical Concerns*

11.3.1.2.1 Anatomy of the Hand and Types of Grip To better understand the design principles of hand tools, it is necessary to have a brief description of the anatomy and functioning of the human hand and some of the diseases that can result from its misuse. The human hand is a complex structure of bones, arteries, nerves, ligaments, and tendons (Figure 11.1). The fingers are controlled by the extensor carpi and flexor carpi muscles in the forearm. The muscles are connected to the fingers by tendons that pass through a channel in the wrist formed by the bones of the back of the hand on one side and the transverse carpal ligament on the other. Through this channel, called the carpal tunnel, pass also various arteries and nerves. The bones of the wrist connect to two long bones in the forearm, the ulna and the radius. The radius connects to the thumb side of the wrist, and the ulna connects to the little finger side of the wrist. The orientation of the wrist joint allows movement in only two planes, each at 90° to the other (Figure 11.2). The first gives rise to palmar flexion and dorsiflexion (or extension). The second movement plane gives ulnar and radial deviation.

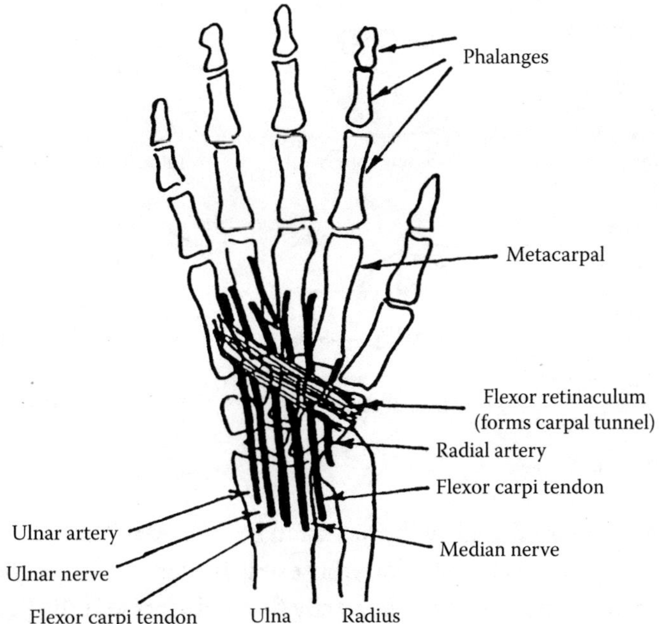


FIGURE 11.1 Anatomy of the hand.

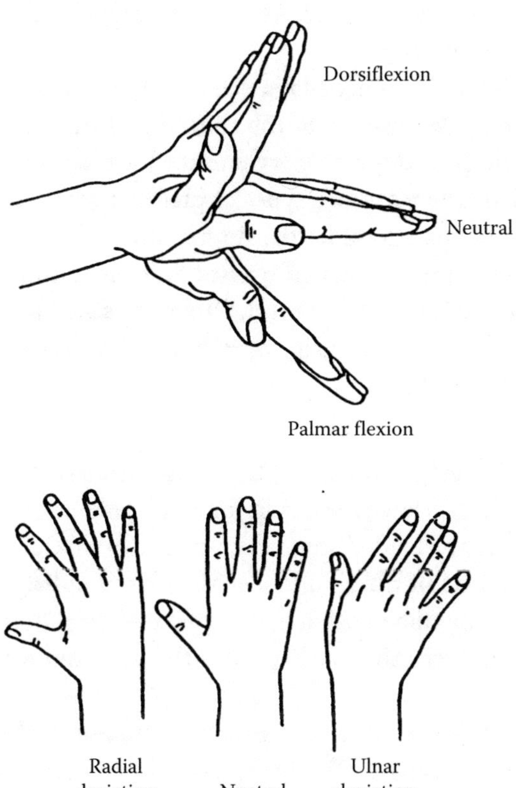


FIGURE 11.2 Types of wrist movement.

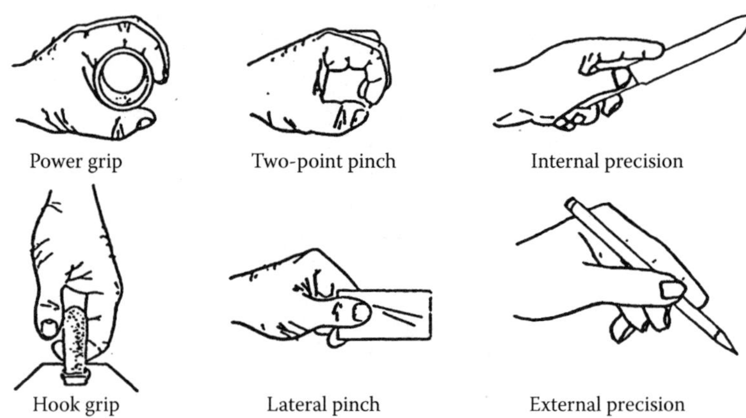


FIGURE 11.3 Types of grip.

The manual dexterity produced by the hand can be defined in terms of a power grip and a precision grip. In a power grip, the tool, whose axis is more or less perpendicular to the forearm, is held in a clamp formed by the partly flexed fingers and the palm, with opposing pressure being applied by the thumb (Figure 11.3). There are three subcategories of the power grip differentiated by the line of action of force: (1) force parallel to the forearm, as in sawing; (2) force at an angle to the forearm, as in hammering; and (3) torque about the forearm, as when using a screwdriver. As the name implies, the power grip is used for power or for holding heavy objects.

In a precision grip, the tool is pinched between the flexor aspects of the finger and the opposing thumb. The relative position of the thumb and fingers determines how much force is to be applied and provides a sensory surface for receiving feedback necessary to give the precision needed. There are two types of precision grip: (1) internal, in which the shaft of the tool (e.g., knife) passes under the thumb and is thus internal to the hand; and (2) external, in which the shaft (e.g., pencil) passes over the thumb and is thus external to the hand. The precision grip is used for control. Other grips are just variations of the power or precision grip and include the hook grip, for holding a box or handle, a two-point pinch, and a lateral pinch (Figure 11.3).

11.3.1.2.2 Static Muscle Loading When tools are used in situations in which the arms must be elevated or when they have to be held for extended periods, muscles of the shoulders, arms, and hands may be loaded statically, resulting in fatigue, reduced work capacity, and soreness. Abduction of the shoulder with corresponding elevation of the elbow will occur if work has to be done with a pistol grip tool on a horizontal workplace. An in-line or straight tool reduces the need to raise the arm and also allows for a neutral wrist posture [20].

Prolonged work with arms extended can produce soreness in the forearm for assembly tasks done with force. By rearranging the workplace so as to keep the elbows at 90°, most of the problem can be eliminated (Figure 11.4). Similarly, continuous holding of an activation switch can result in fatigue of the fingers and reduced flexibility.

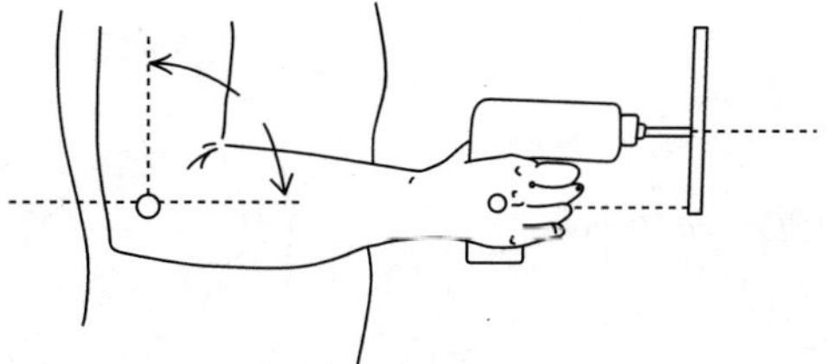

FIGURE 11.4 Optimum working posture with elbow bent at 90°.

11.3.1.2.3 Awkward Wrist Position As the wrist is moved from its neutral position, there is loss of grip strength. Starting from a neutral wrist position, pronation decreases grip strength by 12%, flexion/extension by 25%, and radial/ulnar deviation by 15% [21]. The percent of maximum grip strength available can be quantified by

$$\text{Grip} = 95.7 + 4.3\,\text{PS} + 3.8\,\text{FE} - 25.2\,\text{FE}^2 - 16.8\,\text{RU}^2, \tag{11.1}$$

where

PS = 1 if the wrist is fully pronated or supinated and 0 if in a neutral position
FE = 1 if the wrist is fully flexed or extended and 0 if in a neutral position
RU = 1 if the wrist is fully in radial or ulnar deviate and 0 if in a neutral position

Furthermore, awkward hand positions may result in soreness of the wrist, loss of grip, and, if sustained for extended periods of time, the occurrence of carpal tunnel syndrome. To reduce this problem, the workplace or tools should be redesigned to allow for a straight wrist—lowering work surface and edges of containers, tilting jigs toward the user (Figure 11.5), using a pistol grip on knives (Figure 11.6) [22], and using a pistol handle on

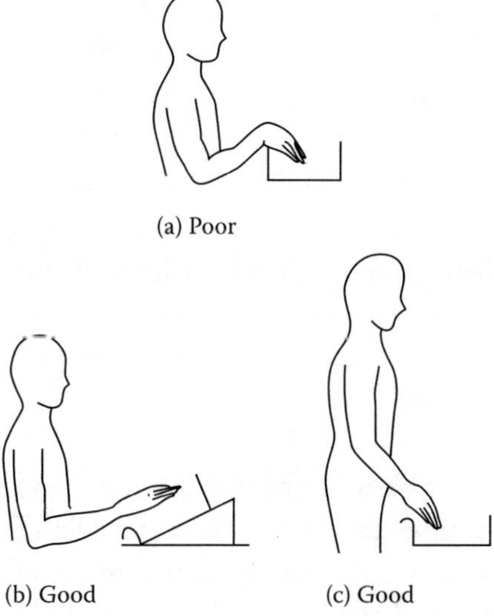


FIGURE 11.5 Improper orientation (a) and proper orientation (b,c) of jigs and containers.

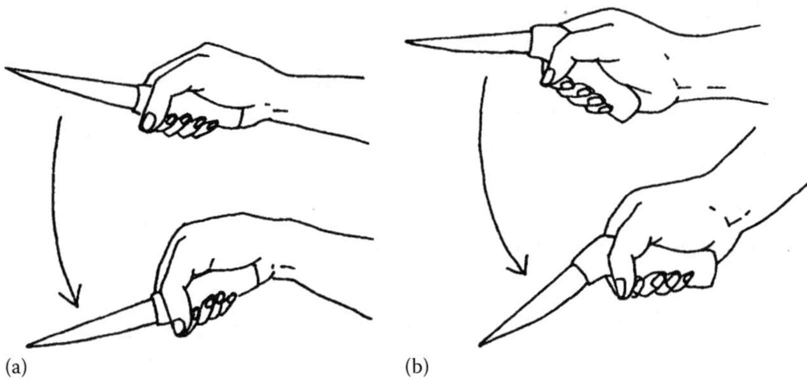


FIGURE 11.6 Example (b) of pistol grip types for knives. (Adapted from Armstrong, T.J. et al., *Am. Ind. Hyg. Assoc. J.*, 43, 103, 1982.)

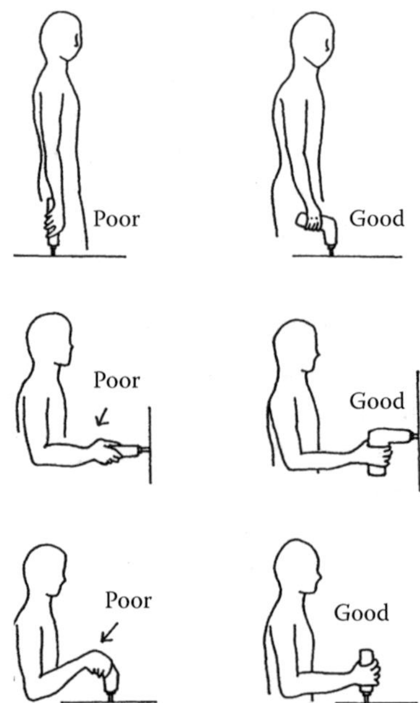


FIGURE 11.7 Proper orientation of power tools in the workplace.

powered tools for vertical surfaces and in-line handles for horizontal surfaces (Figure 11.7). Similarly, the tool handle should reflect the axis of grasp and should be oriented so that the eventual tool axis is in line with the index finger, for example, Tichauer's [23] "bent" plier handles (Figure 11.8).

11.3.1.2.4 Tissue Compression Often, in the operation of hand tools, considerable force is applied by the hand. Such actions can concentrate considerable compressive force on the palm of the hand or the fingers, resulting in ischemia, obstruction of blood flow to the tissues, and eventual numbness and tingling of the fingers. Handles should be designed to have large contact surfaces to distribute the force over a larger area (Figure 11.9) or to direct

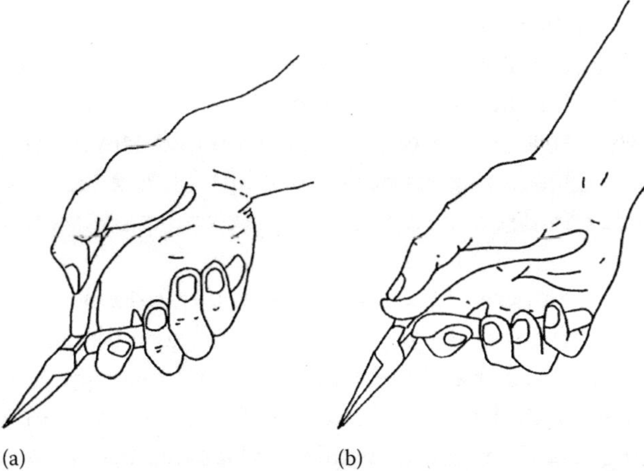


FIGURE 11.8 (a) Traditional and (b) redesigned pliers. (Adapted from Tichauer, E.R., *J. Occup. Med.*, 8, 63, 1966.)

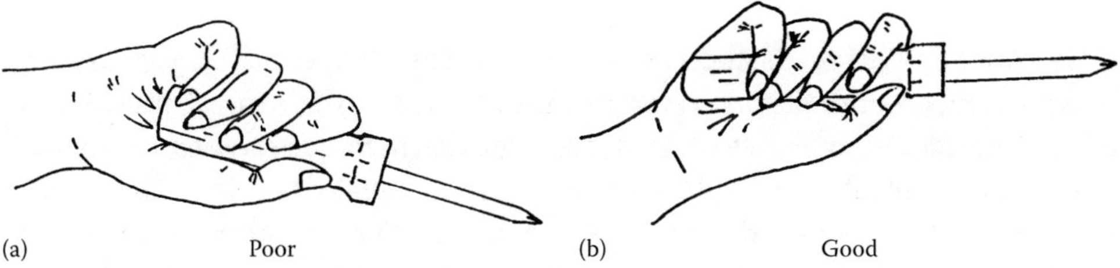


FIGURE 11.9 Avoiding tissue compression in tool design: (a) example of a poor tool handle design and (b) example of a good tool handle design.

it to less sensitive areas such as the tissue between the thumb and index finger. Similarly, finger grooves or recesses in tool handles should be avoided. Because hands vary considerably in size, the grooves will accommodate only a fraction of the population.

11.3.1.2.5 Gender Female grip strength typically ranges from 50% to 67% of male strength [24–26], that is, the average male can be expected to exert approximately 500 N, whereas the average female can be expected to exert approximately 250 N. An interesting survey by Ducharme [27] examined how tools and equipment that were physically inadequate for female workers hampered their performance. The worst offenders were crimpers, wire strippers, and soldering irons. Females have a twofold disadvantage—an average lower strength and an average smaller grip span. Ducharme concluded that women could be integrated more quickly and safely into the work if tools were designed to accommodate smaller dimensions.

On the other hand, Pheasant and O'Neill [28] challenged Ducharme's assertions on the basis of their findings that optimal performance for both males and females occurred at similar conditions. Males had sufficient strength to overcome the deficiencies in tool design, which posed much greater problems for females.

11.3.1.2.6 Handedness Alternating hands permits reduction of local muscle fatigue. However, in many situations this is not possible as the tool use is one handed. Furthermore, if the tool is designated for the user's preferred hand—which for 90% of the population is the right hand—then 10% are left out [29]. Laveson and Meyer [30] gave several good examples of right-handed tools that cannot be used by a left-handed person; these included a power drill with side handle on the left side only, a circular saw, and a serrated knife leveled on one side only.

A few studies have compared task performance using dominant and nondominant hands. Shock [31] indicated that the nonpreferred hand grip strength is 80% of the preferred hand grip strength. Miller and Freivalds [32] found that right-handed males show a 12% strength decrement in the left hand, whereas right-handed females show a 7% strength decrement. Surprisingly, both left-handed males and females had nearly equal strengths in both hands. He concluded that left-handed subjects were forced to adapt to a right-handed world. Using time study ratings, Konz and Warraich [33] found decrements ranging from 9% for an electric drill to 48% for manual scissors for ratings using the nonpreferred hand as opposed to the preferred hand.

11.3.1.2.7 Posture A series of studies were performed by Mital and colleagues to examine various tool and operator factors on torque capability [34–37]. In general, unless the posture is extreme, for example, standing versus lying down, torque exertion capability was not affected substantially. The height at which torque was applied had no influence on peak torque exertion capability. On the other hand, torque exertion capability decreased linearly with increasing reach distance. Another interesting requirement for proper tool use is the volume or space envelope generated during operation of the tool. Comprehensive data on a variety of tools were collected by Baker et al. [38].

11.3.1.2.8 Repetitive Finger Action If the index finger is used excessively for operating triggers, symptoms of trigger finger develop. Thus, trigger forces should be kept low, preferably below 10 N [20], to reduce the load on the index finger. Two- or three-finger-operated controls are preferable (Figure 11.10); finger strip controls or a power grip bar

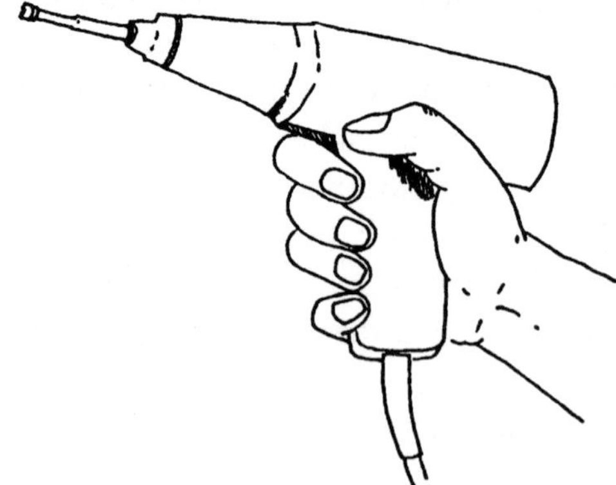

FIGURE 11.10 Three-finger trigger for power tools.

are even better. For a two-handled tool, a spring-loaded return saves the fingers from having to return the tool to its starting position [20]. In addition, the high number of repetitions must be reduced. Although critical levels of repetitions are not known, attempts have been made to identify the maximal number of exertions per hour or shift that can be tolerated. Most of these have concerned wrist movements but could reasonably apply to the fingers as well. Obolenskaja and Goljanitzki [39] observed high rates of tenosynovitis in tea packers with 7,000–12,000 hand movements per day. Luopajarvi et al. [40] found high rates of muscle–tendon disorders in assembly line packers with over 25,000 movements per day. NIOSH [41] found similar problems in workers exceeding 10,000 motions per day.

#### *11.3.1.3 Single Handles*

11.3.1.3.1 Theory of Cylindrical Grip One theory of gripping forces has been described by Pheasant and O'Neill [28] and by Grieve and Pheasant [42]. The hand gripping a cylindrical handle forms a closed system of forces in which portions of the digits and palm are used, in opposition to each other, to exert compressive forces on the handle (Figure 11.11). The strength of the grip (*G*) may be defined as the sum of all components of forces exerted normal to the surface of the handle.

When exerting a turning action on the handle, the maximum torque, as given at the moment of hand slippage, is given by

$$T = SD, \tag{11.2}$$

where

*T* is the torque (N-m)
*S* is the total shear/frictional force (N)
*D* is the handle diameter (m)

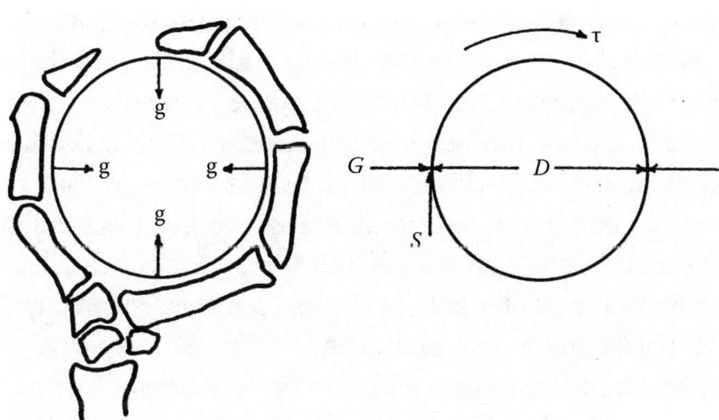


FIGURE 11.11 Theory of the cylindrical grip. (Adapted from Pheasant, S.T. and O'Neill, D., *Appl. Ergon.*, 6, 205, 1975.)

and where $S$ can be defined by

$$S = \mu G, \tag{11.3}$$

where
$\mu$ is the coefficient of friction
$G$ is the grip (N)

Thus, torque is directly dependent upon handle diameter because handle diameter determines the leverage of shear forces. This was confirmed experimentally by Pheasant and O'Neill [28]. For thrusting motions in the direction of the long axis of the handle, the diameter is not involved, and determination of maximum force is more complicated.

For handles larger than the grip span diameter, the gripped area no longer increases in proportion to the diameter. An analysis of such conditions was performed by Replogle [43], who concluded that for handles up to twice the grip span diameter, the relative ungripped area of the handle increases, reducing the effective gripped area. Torque then can be expressed by

$$T = \frac{3d^2(4-d)}{(d+2)^2}, \tag{11.4}$$

where $d$ is the ratio of handle diameter to grip span diameter. For larger handle diameters, the expression for torque becomes much more complicated [43].

11.3.1.3.2 Handle Diameter Power grips around a cylindrical object should surround more than half the circumference of the cylinder, but with the fingers and thumb barely touching [44]. For a power grip on a screwdriver, Rubarth [45] recommended a diameter of 40 mm. Basing their recommendations on empirical judgments of stair rails, Hall and Bernett [46] suggested 32 mm. Based on minimum electromyography (EMG) activity, Ayoub and LoPresti [47] found a 51 mm handle diameter to be best. However, based on the maximum number of work cycles completed before fatigue and the ratio of grip force to EMG activity, they suggested a 38 mm diameter. Pheasant and O'Neill [28] found that muscle strength deteriorates when handles greater than 50 mm in diameter are used. Rigby [48], for heavy loads and full encirclement of the hand, recommended 38 mm. For handles on boxes, Drury [49] found diameters of 31–38 mm to be best in terms of least reduction in grip strength. Using various handles of noncircular cross section, Cochran and Riley [50,51] found the greatest thrust forces in handles of 41.4 mm equivalent circular diameter (based on their 130 mm circumference) for both males and females. For manipulation, however, the smallest handles of 22 mm were found to be the best [52]. Replogle [43], in validating his grip model, found maximum torques with handle diameters of 50 mm. Eastman Kodak [20], based on company experience, recommends 30–40 mm with an optimum of 40 mm for power grips and 8–16 mm with an optimum of 12 mm for precision grips. Thus, one can summarize that handle diameters should be in the range of 31–50 mm, with the upper end of the range best for maximum torque and the lower end best for dexterity and speed.

11.3.1.3.3 Handle Length The length of the handle has been studied to a lesser extent. For cutout handles, there should be enough space to admit all four fingers. Hand breadth across the metacarpals ranges from 71 mm for a 5th percentile female to 97 mm for a 95th percentile male [53]. Thus, 100 mm may be a reasonable minimum, but 125 mm may be more comfortable [24]. Eastman Kodak [20] recommended 120 mm. If the grip is enclosed or gloves are used, even larger openings are recommended. For an external precision grip, the tool shaft must be long enough to be supported at the base of the first finger or thumb. A minimum value of 100 mm is suggested [24]. For an internal precision grip, the tool should extend past the palm, but not so far as to hit the wrist (Figure 11.9) [24]. It is interesting to note that screwdriver torque was experimentally found to be proportional to the handle grip length [54].

11.3.1.3.4 Handle Shape As early as 1928, Rubarth [45] investigated handle shape and concluded that for a power grip, one should design for maximum surface contact so as to minimize unit pressure of the hand. Thus, a tool with a circular cross section was found to give the greatest torque. Pheasant and O'Neill [28] concluded that the precise shape of the handle was irrelevant and recommended simple knurled cylinders. Evaluation of handle shape on grip fatigue in manual lifting (which is a different action than for tool use) did not indicate any significant differences in shapes [55]. Maximum pull force, though, was obtained with a triangular cross section, apex down. For thrusting forces, the circular cross section was found to be worst and a triangular best [50]. However, for a rolling type of manipulation, the triangular shape was slowest [52]. A more comprehensive study indicated that no one shape may be perfect and that optimal shape may be more dependent on the type of task and motions involved than was initially thought [51]. A rectangular shape of width/height ratios from 1:1.25 to 1:1.5 appeared to be a good compromise. A further advantage of a rectangular cross section is that the tool does not roll when placed on a table [24]. It should also be noted that handles should not have the shape of a true cylinder except for a hook grip. For screwdriver-type tools, the handle end is rounded to prevent undue pressure at the palm, and for hammer-type tools, the handle may have some flattening curvature to indicate the end of the handle.

In a departure from circular, cylindrically shaped handles, Bullinger and Solf [56] proposed a more radical design using a hexagonal cross section, shaped as two truncated cones jointed at the largest ends. Such a shape fits the contours of the palm and thumb best in both precision and power grips and yielded highest torques in comparison with more conventional handles. A similar dual truncated conical shape was also developed for a file handle. In this case, the heavily rounded square cross section was found to be markedly superior to more conventional shapes.

A final note on shape is that T-handles yield much better performance than straight screwdriver handles. Pheasant and O'Neill [28] reported as much as a 50% increase in torque. Optimum handle diameter was found to be 25 mm, and optimum angle was 60°, that is, a slanted T [57]. The slant allows the wrist to remain straight and thus generate larger forces.

11.3.1.3.5 Grip Surface, Texture, and Materials For centuries, wood was the material of choice for tool handles. Wood was readily available and easily worked. It has good resistance to shock, poor thermal and electrical conductivity, and good frictional qualities even when wet. Because wooden handles can break and stain with grease and oil, there has been a shift to plastic and even metal. However, metal should be covered with rubber or leather to reduce shock and electrical conductivity and increase friction [58]. Such compressible materials also dampen vibration and allow a better distribution of pressure, reducing the feeling of fatigue and hand tenderness [59]. The grip material, however, should not be so soft that sharp objects, such as metal chips, can become embedded in the grip and make it difficult to use. Grip surface area should be maximized to ensure a pressure distribution over as large an area as possible. Excessive localized pressure sometimes causes pain that forces workers to interrupt their work. Pressure/pain thresholds of around 500 kPa for females and 700 kPa for males have been found, with the thenar and so pisiform areas being most sensitive [60]. During maximal power grips, these values are greatly exceeded.

The frictional characteristics of the tool surface vary with the pressure exerted by the hand, the smoothness and porosity of the surface, and the type of contamination [61,62]. Sweat increases the coefficient of friction, whereas oil and fat reduce it. Adhesive tape and suede provide good friction when moisture is present [61]. The type of surface pattern as defined by the ratio of ridge area to groove area shows some interesting characteristics. When the hand is clean or sweaty, the maximum frictions are obtained with high ratios (i.e., maximizing the hand-surface contact area), whereas when the hand is contaminated, maximum frictions are obtained with low ratios (i.e., maximizing the capacity to channel away contaminants) [62].

11.3.1.3.6 Angulation of Handle As discussed previously, deviations of the wrist from the neutral position under repetitive load can lead to a variety of CTDs as well as decreased performance. Therefore, angulation of tool handles such as those of power tools may be necessary to maintain a straight wrist. The handle should reflect the axis of grasp, that is, about 78° from the horizontal, and should be oriented so that the eventual tool axis is in line with the index finger [58]. This principle has been applied to various tools such as pliers and soldering irons, as mentioned previously.

An interesting extension of this concept has been promoted as Bennett's handle [63]. Bennett developed this concept based on the angle formed by the index finger and the life line under the thumb. This angle of 19°, used for his handles, is claimed to maintain a straight wrist, generate increased strength and control, and reduce stress, shock, and fatigue [64]. Bennett's claims initially were supported by anecdotal evidence of improved performance [63]. Since then, Konz and his colleagues [65–67] have conducted a variety of tests to evaluate the effectiveness of Bennett's handle on a hammer in comparison with a standard hammer. In the second study, a variety of angled handles were evaluated, and subjects rated a 10° bend as being most preferred. In the third study, performance in driving nails was evaluated using various bent hammers. No performance difference was found, but the 10° bend was again rated significantly higher. In the final study, using a semantic differential questionnaire, Konz [67] concluded that although no significant performance effects were found, subjects preferred a slight (5°–10°) bend rather than the 19° of Bennett's handle. An independent study by Knowlton and Gilbert [68] used cinematography to evaluate curved and conventional claw hammers. Bilateral grip strength was measured before and after the task of nail driving. The curved hammer produced a smaller strength decrement and caused less ulnar deviation than the conventional hammer. Thus, a bent handle does give some benefits.

#### *11.3.1.4 Grip Span for Two-Handled Tools*

Grip strength and the resulting stress on finger flexor tendons vary with the size of the object being grasped. A maximum grip strength is achieved at about 45–50 mm on a dynamometer with parallel sides [25] and at about 75–80 mm on a dynamometer with handles angled inward [26]. At distances different from the optimum, percent grip strength decreases (Figure 11.12) as defined by

$$\text{Grip} = 100 - 0.11S - 10.2S^2 \tag{11.5}$$

where $S$ is the given grip span minus optimum grip span in centimeters.

Because of the large variation in individual strength capacities, and to accommodate 95% of the population, maximal grip requirements should be limited to less than 90 N [69].

#### *11.3.1.5 Weight*

The weight of the hand tool will determine how long it can be held or used and how precisely it can be manipulated. For tools held in one hand with the elbow at 90° for extended periods of time, Greenberg and Chaffin [69] recommended a load of not more than 2.3 kg. A similar value was suggested by Eastman Kodak [20]. For precision operations, tool

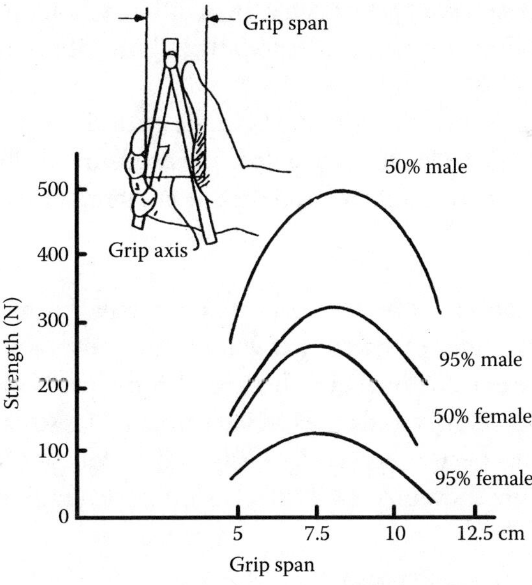


FIGURE 11.12 Grip strength as a function of grip span. (Reproduced with permission from Chaffin, D.B. and Andersson, G., *Occupational Biomechanics*, Wiley, New York, 1984.)

weights greater than 0.4 kg are not recommended unless a counterbalanced system is used. Heavy tools, used to absorb impact or vibration, should be mounted on a truck to reduce effort for the operator [20]. In addition, the tool should be well balanced, with its center of gravity as close as possible to the center of gravity of the hand (unless the purpose of the tool is to transfer force as in a hammer). Thus, the hand or arm muscles do not need to oppose any torque developed by an unbalanced tool.

#### *11.3.1.6 Gloves*

Gloves are often used with hand tools for safety and comfort. Safety gloves are seldom bulky, but gloves worn in subfreezing climates can be very heavy and interfere with grasping ability. The wearing of woolen or leather gloves may add 5 mm to the hand thickness and 8 mm to the hand breadth at the thumb, whereas heavy mittens add 25 and 40 mm, respectively [70]. More important, gloves reduce grip strength by 10%–20% [71–73] and manual dexterity performance times by 12%–64% [74]. Neoprene gloves slowed performance times by 12.5% over barehanded performance, terry cloth by 36%, leather by 45%, and polyvinyl chloride (PVC) by 64%. In some cases, by protecting the hand, gloves could improve operational speed [75,76]. On the other hand, gloves consistently reduced torque production [77]. Thus, there is a trade-off to be considered between increasing injury and reduced performance without gloves and reduced performance with gloves. Perhaps the tool should be redesigned even more so as to compensate for the glove effects.

#### *11.3.1.7 Vibration*

Vibration is a separate and very complex problem with powered hand tools. Vibration can induce white finger syndrome, the primary symptom of which is a reduction in blood flow to the fingers and hand due to vasoconstriction of the blood vessels, which lead to a loss of sensory feedback and decreased performance. In addition, vibration may contribute to the development of carpal tunnel syndrome, especially in jobs with a combination of forceful and repetitive exertions [78].

It is generally recommended that vibrations in the critical range of 40–130 Hz [79] or a slightly larger range of 2–200 Hz [80] be avoided. The exposure to vibration can be reduced through a reduction in the driving force and the use of vibration-damping materials [81].

#### *11.3.1.8 Rhythm*

The operation of hand tools involves repetition of a particular pattern of motion. A skilled operator requires a basic motor pattern that will be most economical in terms of energy expenditure and is thus one attribute of skill. Once this pattern is established, it is continued with very consistent velocity and acceleration through kinesthetic and aural feedback. Optimum rhythms have been observed by Drillis [4] as follows: filing, 78 strokes/min; chiseling, 60 strokes/min; shoveling, 14–17 strokes/min; and cranking, 35 rev/min.

#### *11.3.1.9 Other Aspects of Tool Design*

Tools should not have protruding sharp edges or corners. Two-handled tools should have stop limits to limit closure of the tools and prevent pinching of the fingers. Locking tools should not engage until the tool closes to the point where the fingers cannot be inserted. Tool surfaces should have matte surfaces to reduce glare [58].

### 11.3.2 Attributes of Common Industrial Hand Tools

#### *11.3.2.1 Shovels*

Shovels are used to lift, move, and toss loose dirt or sand or other material. The blade is fastened to the shaft through a socket, which, if stamped from a flat sheet, is generally rolled over to form a crimp known as a frog. The shaft may either taper to an end or have a handle. The handle traditionally has been of a T form but more lately is of a D form. Long shafts are generally 1.2–1.7 m long, whereas the short D handle is about 0.7 m long. However, for an unconstrained posture and task, long-handled shovels are 18% more efficient than short-handled shovels [82]. The angle of the shaft with respect to the horizontal, called the lift, provides the tool with added leverage. Based on a trade-off between smaller low-back compressive forces and lower energy cost, Freivalds [82] found the optimum lift angle to be approximately 32°. Shovel weight should be as small as possible, especially if one considers that this is unproductive weight that has to be lifted along with the load moved. Weights should be below 1.5 kg [83,84]. Blade size depends very much on the density of the material being handled; the less dense the material, the larger the blade size. For a given material (foundry sand), Freivalds and Kim [84] found that a blade size/weight ratio of 0.0676 $m^2$/kg is optimum; that is, for a 1.5 kg shovel, the optimum blade size is 0.1 $m^2$ with a load of 4.4 kg.

Optimum shoveling rate is in the range of 18–21 scoops/min [5,83,85]. The optimum shovel load ranges from 5 to 11 kg, depending on the decision criterion to be used [86,87]. For high rates of shoveling (18–20 scoops/min), the lower end of the load range (5–7 kg) may be more appropriate (which follows the principle of reducing static loading), whereas for lower rates (6–8 scoops/min), the higher end of the load range (8–11 kg) may be acceptable (which follows the principles of increasing efficiency with larger loads) [88]. Shoveling throw height is a trade-off between increased efficiency for higher heights and the cost of an increase in energy expenditure. Because shoveling performance stays reasonably constant up to a height of 1.3 m [89], an acceptable throw height may be as much as 1–1.3 m. The same conflicting criteria apply to throw distance. However, with shoveling performance remaining fairly constant up to a distance of 1.2 m [89], that would be an acceptable distance.

#### *11.3.2.2 Hammers*

Hammers are striking tools designed to transmit a force to an object by direct contact and thereby change its shape or drive it forward. The tool's efficiency in doing this may be defined as the ratio of the energy used in striking to the energy available in the stroke. This efficiency is maximized by placing the tool mass center as close as possible to the center of action, that is, increasing the mass of the tool head relative to that of the handle. Another aim is to transform as much as possible of the kinetic energy of the hammer into deforming an object's shape. Thus, the mass of the hammer should be small relative to the mass of the forging and anvil. On the other hand, in driving a nail, the intent is to transform the kinetic energy of the hammer into the kinetic energy of the nail. Then the mass of the hammer should be great in relation to the mass of the nail. The overall mechanical efficiency for hammering a 6 in. nail into a wooden block can be as high as 57% [4]. However, there is a limit to the weight that can be placed in the head of the hammer. Increasing the head weight decreases angular velocity and ultimately the total kinetic energy in addition to increasing physiological energy costs [90,91].

#### *11.3.2.3 Saws*

The action of heavy sawing requires a power grip with repetitive flexion and extension at the elbow, whereas the action of light sawing involves a precision grip with manipulation of the wrist. For the former, pistol grips are used, whereas for the latter a cylindrical screwdriver type of handle provides the best precision grip. Gläser [92] found that for forestry work, a two-handed action provided more force and better performance but at a higher energy cost. The most efficient was the kneeling posture because less torso support was needed and less energy was expended. Typically, western saws cut as they are pushed through the wood, whereas Japanese saws cut as they are pulled through the wood [93]. Although sawing times were not significantly different, energy expenditure is significantly lower for Japanese saws [93].

#### *11.3.2.4 Pliers*

Pliers and related tools—wire strippers, pincers, and nippers—are tools with a head in the form of jaws that can have a variety of configurations, that is, joints may be simple or complex. Although sometimes the handles are straight, more typically they are curved outward to conform roughly to the shape of the grasp. The grasp, depending on use, can be of the precision or power type. In their simple form, pliers are a very common tool and, if used casually for short periods of time, will give reasonable performance with little fatigue. However, the relationship of the handles to the head forces the wrist into ulnar deviation (Figure 11.8a), a posture that cannot be held repeatedly or for prolonged periods of time without fatigue or the occurrence of CTD. A further problem is that such a deviation reduces the range of wrist rotation by 50%, thus reducing productivity [94]. By bending the handles of the pliers (Figure 11.8b) instead of the wrist, Tichauer [94] was able to reduce the stress on the operator's wrists and reduce overall injury rates by a factor of 6 at Western Electric Co. A similar design (with similar success) was implemented for diagonal cutting pliers at Eli Lilly and Co. [95].

Other factors to be considered in the design of pliers were detailed by Lindstrom [96]: a working grip width of 90 mm for men and 80 mm for women, and a handle length of 110 mm for men and 100 mm for women. For repeated or continuous operation, the required working strength should not exceed 33%–50% of the individual's maximum strength. To minimize the applied pressure to the soft tissue of the palm (less than 200–400 kPa), the handles should be enlarged and flattened. Thus, indentation of the handles for the fingers is undesirable. Encasing the basic metal handles in a rubber or plastic sheath provides insulation and improves the tactile feel.

#### *11.3.2.5 Screwdrivers*

The handles of screwdrivers (and similar tools—files, chisels, etc.) can be used with either a precision grip for stabilization or a power grip for torque. The handle must also be capable of being approached equally effectively from all directions. Other crucial factors are the size, shape, and texture of the handle. In terms of size, applied torque increases with an increase in the diameter of the handle [28]. Differences in the precise shape of handles appear not to be significant as long as the hand does not slip around the handle [51]. Thus, knurled cylinders allow for significantly greater torque production than smooth cylinders. Further details can be found in Section 11.3.1.3.

#### *11.3.2.6 Knives*

Although a very old tool, the knife has recently appeared in the literature as a possible cause of the increase in CTDs suffered by food processors [22,97]. For poultry processing, Armstrong et al. [22] suggested a pistol-type grip (Figure 11.6) to allow the operator to hold the blade and the forearm horizontal so as to eliminate ulnar deviation and wrist flexion. A circular or elliptical handle with a large circumference (99 mm) and a strap was recommended to allow the hand to relax between exertions without losing a grip on the knife. Similarly, for the fish canning industry, Karlqvist [97] fitted some knives with a pistol grip handle and others with larger diameter handles for better balance and movement. Cochran and Riley [15] explained the injuries due to long hours of static loading on the forearm flexors resulting in fatigue. In addition, body fluids cause slippery handles, with the operator's hand slipping from the handle over to the blade. They recommended tangs (barriers on the handle perpendicular to the blade) of a minimum length of 1.5 cm to prevent injury.

### 11.3.3 Attributes of Common Industrial Power Tools

#### *11.3.3.1 Power Drills*

In a power drill or other power tool, the major function of the operator is to hold, stabilize, and monitor the tool against a workpiece, while the tool performs the main effort of the job. Although the operator may at times need to shift or orient the tool, the main function of the operator is to effectively grasp and hold the tool. A drill comprises a head, body, and handle, with all three, ideally, being in line. The line of action is from the line of the extended index finger so that in the ideal drill the head is off-center with respect to the central axis of the body. Handle configuration is important, with the choices being pistol grip, in-line, or right angle. As a rule of thumb, in-line and right angle grips are best for tightening downward on a horizontal surface, while pistol grips are best for tightening on a vertical surface (Figure 11.7), with the aim being to obtain a standing posture with a straight back, upper arms hanging down, and a straight wrist (Figure 11.4). For the pistol grip, this results in the handle being at an angle of approximately 78° to the horizontal [58].

Another factor of importance is the center of gravity. If it is too far forward in the body of the tool, a turning moment is created, which must be overcome by the muscle of the hand and forearm, creating muscular effort additional to that required for holding, positioning, and pushing the drill into the workpiece [69]. Greenberg and Chaffin recommend placing the primary handle directly under the center of gravity so that the body juts out behind the handle as well as in front. For heavy drills, a secondary supportive handle may be needed, either to the side or preferably below the tool, such that the supporting arm can be tucked in against the body rather than being abducted.

#### *11.3.3.2 Nutrunners*

Nutrunners, especially common in the automobile industry, are used to tighten nuts, screws, and other fasteners. They come in a variety of handle configurations, torque outputs, shutoff mechanisms, speeds, weights, and spindle diameters and are commercially available from a variety of sources. Torque levels range from 0.1 to 5000 N-m and, for pneumatic tools, are generally lumped into approximately 22 power levels (M1.6–M45) depending on motor size and the gearing required to drive the tool. The torque is transferred from the motor to the spindle through a variety of mechanisms in such a way that the power (often air) can be quickly shut off once the nut or other fastener is tight. The simplest and cheapest mechanism is a direct drive, which is under the operator's control but, because of the long time needed to release the trigger once the nut is tightened, transfers a very large reaction torque to the operator's arm. Mechanical friction clutches will allow the spindle to slip, reducing some of this reaction torque. A better mechanism for reducing the reaction torque is the airflow shutoff, which automatically senses when to cut off the air supply as the nut is tightened. A still faster mechanism is an automatic mechanical clutch shutoff [98]. The most recent mechanisms include the hydraulic pulse system where the rotational energy from the motor is transferred over a pulse unit containing an oil cushion (filtering off the high-frequency pulses as well as noise) and a similar electrical pulse system, both of which, to a large extent, reduce the reaction torque [99].

Variation of torque delivered to the nut depends on a variety of conditions such as the properties of the tool, the operator of the tool, and properties of the joint [i.e., the combination of the fastener and material being fastened (ranging from soft, with the materials having elastic properties such as body panels, to hard, when two stiff surfaces, such as pulleys on a crankshaft, are brought together)], and stability of the air supply. The torque experienced by the user (the reaction torque) depends on these factors plus the torque shutoff system [100]. In addition, other potential problems include noise from the pneumatic mechanism, which reaches levels as high as 95 dB(A), vibration levels exceeding 132 dB(V), and dust or oil fumes from the exhaust air [101].

Reaction torque, the excess torque produced by the nutrunner and transferred to the operator's hand and/or arm once the fastener is tightened and before the power source can be decreased, is believed to contribute to the development of CTDs [100]. A study by Freivalds and Eklund [98] examined relevant power tool parameters for their effect on operator stress levels as measured by EMG, wrist angle, and subjective ratings of perceived discomfort. In general, using electric tools at lower than normal rpm levels or under powering pneumatic tools resulted in larger reaction torques and more stressful ratings. Pulse-type tools produced the lowest reaction torques and were rated least stressful. It was hypothesized that the short pulses "chop up" or allow the inertia of the tool to resist the reaction torque. Also, subjective ratings correlated significantly with peak reaction torques, indicating that the operators were good judges of stress levels and that subjective ratings could be valid measures of operator stress.

Reaction torque bars should be provided if the torque exceeds 6 N-m for in-line tools used in a downward action, 12 N-m for pistol grip tools used in a horizontal mode, and 50 N-m for right-angled tools used in a downward or upward motion [102,103].

### 11.3.4 Tool Evaluation Checklist

The previous information can be summarized in the form on an evaluative checklist (Table 11.1). If the tool does not conform to the recommendations and desired features, it should be replaced or redesigned.

## 11.4 CRITICAL REVIEW OF THE CURRENT STATUS

Currently, the most important issue regarding tool design is the reduction in the potential for the development of CTDs. Until about 10 years ago, tool use was little changed from the days of the Industrial Revolution. The operator used tools manually in the manufacture of goods. The operations required considerable force, which was somewhat leveraged by the appropriate tool. Because of the manual nature of the tasks and the forces involved, the operations were fairly slow. With the advent of automation, the excessive force levels were eliminated and many task elements relegated to the human could be eliminated. The operator performs a smaller part of the original task, which now can be speeded up because the machine does most of the work. Unfortunately, the elements still left to the human operator become more limited in scope and thus more repetitious in nature. This incomplete and unergonomic automation has led to an upsurge in CTD cases, especially if the repetition is combined with excessive wrist deviations and forceful exertions. The threshold for injury based on frequency of repetition is not known. A couple of studies have indicated repetition rates ranging from 10,000 [41] to 25,000 [40] per shift. This is a critical number that undoubtedly depends also on the posture and force exerted and needs to be fine-tuned. In fact, the exact trade-offs between the three factors are not known but are very important to quantifying the exposure and threshold levels for CTD.

Tied in with frequency is the trade-off with productivity. Any reduction in frequency of tool use will have a direct result in decreasing productivity. One alternative to maintaining constant productivity is to rotate operators for a critically repetitive task. Then again, it is necessary to know threshold levels of frequency and to know how much rest, at what intervals, must be provided to the tool operator to ensure recovery from the trauma induced from repetitive tool use. Also, it is important to know whether performing a greater variety of movements than those in the injurious task will allow the body to recover or only delay recovery. These are all issues that have not been fully addressed.

Another issue not fully resolved is the trade-off between manual and power tool use. Most researchers [24,102,103], on the basis of information on the force capacity and greater fatigability of humans compared to machines, have advocated the use of power tools.

TABLE 11.1 Tool Evaluation Checklist

| | Yes | No |
|---|---|---|
| Basic Principles | | |
| 1. Does the tool perform the desired function effectively? | ☐ | ☐ |
| 2. Does the tool match the size and strength of the operator? | ☐ | ☐ |
| 3. Can the tool be used without undue fatigue? | ☐ | ☐ |
| 4. Does the tool provide sensory feedback? | ☐ | ☐ |
| 5. Are the tool capital and maintenance costs reasonable? | ☐ | ☐ |
| Anatomical Concerns | | |
| 1. If force is required, can the tool be grasped in a power grip (i.e., handshake)? | ☐ | ☐ |
| 2. Can the tool be used without shoulder abduction? | ☐ | ☐ |
| 3. Can the tool be used with a 90° elbow angle (i.e., forearms horizontal)? | ☐ | ☐ |
| 4. Can the tool be used with the wrist straight? | ☐ | ☐ |
| 5. Does the tool handle have large contact surfaces to distribute forces? | ☐ | ☐ |
| 6. Can the tool be used comfortably by a 5th percentile female operator? | ☐ | ☐ |
| 7. Can the tool be used in either hand? | ☐ | ☐ |
| Handles and Grips | | |
| 1. For power uses, is the tool grip 30–50 mm in diameter? | ☐ | ☐ |
| 1a. Can the handle be grasped with the thumb and fingers slightly overlapped? | ☐ | ☐ |
| 2. For precision tasks, is the tool grip 8–16 mm in diameter? | ☐ | ☐ |
| 3. Is the grip cross section circular? | ☐ | ☐ |
| 4. Is the grip length at least 100 mm (125 mm if gloves are worn)? | ☐ | ☐ |
| 5. Is the grip surface finely textured and slightly compressible? | ☐ | ☐ |
| 6. Is the handle nonconductive and stain-free? | ☐ | ☐ |
| 7. For power uses, does the tool have a pistol grip angled at 78°? | ☐ | ☐ |
| 8. Can a two-handled tool be operated with less than 90 N grip force? | ☐ | ☐ |
| 9. Is the span of the tool handles between 70 and 80 mm? | ☐ | ☐ |
| Power Tool Considerations | | |
| 1. Are trigger activation forces less than 10 N? | ☐ | ☐ |
| 2. For repetitive use, is a finger strip trigger present? | ☐ | ☐ |
| 3. Are less than 10,000 triggering actions required per shift? | ☐ | ☐ |
| 4. Is a reaction bar provided for torques exceeding | | |
| 6 N-m for in-line tools? | ☐ | ☐ |
| 12 N-m for pistol grip tools? | ☐ | ☐ |
| 50 N-m for right-angled tools? | ☐ | ☐ |
| 5. Does the tool create less than 85 dBA for a full day of noise exposure? | ☐ | ☐ |
| 6. Does the tool vibrate? | ☐ | ☐ |
| 6a. Are the vibrations outside the 2–200 Hz range? | ☐ | ☐ |
| Miscellaneous and General Considerations | | |
| 1. For general use, is the weight of the tool less than 2.3 kg? | ☐ | ☐ |
| 2. For precision tasks, is the weight of the tool less than 0.4 kg? | ☐ | ☐ |

TABLE 11.1 (continued) Tool Evaluation Checklist

| | Yes | No |
|---|---|---|
| 3. For extended use, is the tool suspended? | ☐ | ☐ |
| 4. Is the tool balanced (i.e., center of gravity on the grip axis)? | ☐ | ☐ |
| 5. Can the tool be used without gloves? | ☐ | ☐ |
| 6. Does the tool have stops to limit closure and prevent pinching? | ☐ | ☐ |
| 7. Does the tool have smooth and rounded edges? | ☐ | ☐ |

Unfortunately, power tools, whether powered electrically or pneumatically, produce some vibration. Vibration damping typically requires either an increase in the inertial mass, at the cost of increasing the weight of the tool and increasing the fatigue of the user, or vibration-absorbing systems that introduce a bit of "slop" in the hand/handle interface that absorbs the vibrations but at the cost of reducing control of the tool. Power tools also have a tendency to produce reaction torques, which can be reduced by using pulse-type tools [98] but at the cost of increasing vibration, or by using reaction bars, again at the cost of limiting the control or maneuverability of the tool. These are issues that need to be clarified further.

## 11.5 FUTURE CONCERNS

Recent research indicates that power grip capabilities can be increased through a better understanding of the pressure distribution of the hand during tool use [60,104] or by improving the frictional characteristics of the tool handle surface [62]. Perhaps the development of new polymers for application to tool handles can improve the efficiency of tool use. Also, new ways of measuring the hand/handle interface, such as the "data glove" of Yun [104], can provide more accurate information on this topic.

Most current work addresses the power grip for tools. However, most power requirements are being fulfilled by machines, leaving the human operator to perform more precise tasks that currently cannot be easily replicated by the machine. Unfortunately, there is very little information on precision or pinch grips and the precision aspects of tools. Questions on grip design and force exertion capabilities for precision grips and on occupational injury risk during work with high demands on precision need to be studied further.

Epidemiological considerations are also important in substantiating proper ergonomic designs. Unfortunately, at present, there are few good studies that support good ergonomic tool design or clearly indicate the deficiencies in such designs. More morbidity data for both hand and powered tools are needed.

A final but very important consideration is the adaptation of tools for a more diverse population. For example, with the aging of the worker population and the passage of the Americans with Disabilities Act, it is imperative that tools also be usable by individuals with a wide range of capabilities. This is both a challenge and an opportunity for ergonomists and tool designers to put their skills to effective use.

## REFERENCES


1. L. S. B. Leaky, Finding the world's earliest man, *Natl. Geogr. 118*:420–435 (1960).
2. J. Napier, The evolution of the hand, *Sci. Am. 207*:56–62 (1962).
3. J. Napier, Early man and his environment, *Discovery 24*:12–18 (1963).
4. R. J. Drillis, Folk norms and biomechanics, *Hum. Factors 5*:427–441 (1963).
5. G. Lehmann, *Praktische Arbeitsphysiologie*, Thieme Verlag, Stuttgart, Germany, 1953, pp. 182–197.
6. T. J. Armstrong, *An Ergonomic Guide to Carpal Tunnel Syndrome*, American Industrial Hygiene Association, Akron, OH, 1983.
7. F. Aghazadeh and A. Mital, Injuries due to handtools, *Appl. Ergon. 18*:273–278 (1987).
8. E. R. Tichauer and H. Gage, Ergonomic principles basic to hand tool design, *Am. Ind. Hyg. Assoc. J. 38*:622–634 (1977).
9. S. W. Meagher, Hand tools: Cumulative trauma disorders caused by improper use of design elements, in *Trends in Ergonomics/Human Factors*, W. Karwowski, Ed., Elsevier, Amsterdam, the Netherlands, 1986, pp. 581–587.
10. D. S. Stetson, T. J. Armstrong, L. J. Fine, B. A. Silverstein, and K. Tannen. A survey of chronic upper extremity disorders in an automobile upholstery plant, in *Trends in Ergonomics/Human Factors*, W. Karwowski, Ed., Elsevier, Amsterdam, the Netherlands, 1986, pp. 623–630.
11. R. W. Schoenmarklin and W. S. Marras, Effects of handle angle and work orientation on hammering, Part I. Wrist motion and hammering performance, *Hum. Factors 30*:397–411 (1989).
12. R. W. Schoenmarklin and W. S. Marras, Effects of handle angle and work orientation on hammering, Part II. Muscle fatigue and subjective ratings of body discomfort, *Hum. Factors 30*:413–420 (1989).
13. S. S. Ulin and T. J. Armstrong, Effect of tool shape and work location on perceived exertion for work on horizontal surfaces, *Proc. Int. Ergon. Assoc.*, 1991, pp. 1125–1127.
14. S. S. Ulin, C. M. Ways, T. J. Armstrong, and S. H. Snook, Perceived exertion and discomfort versus work height with a pistol-shaped screwdriver, *Am. Ind. Hyg. Assoc. J. 51*:588–594 (1990).
15. D. J. Cochran and M. W. Riley, The effects of handle shape and size on exerted forces, *Hum. Factors 28*:253–265 (1986).
16. Å. Kilbom and J. Ekholm, *Handgreppsstyrka*, MUSIC study, Stockholm, Sweden, 1991.
17. National Safety Council, *Accident Facts*, Chicago, IL, 1993.
18. V. Putz-Anderson, *Cumulative Trauma Disorders*, Taylor & Francis, London, U.K., 1988.
19. NIOSH, *Occupational Safety and Health, Year 2000 Objectives*, National Institute for Occupational Safety and Health, Centers for Disease Control, Atlanta, GA, 1989.
20. Eastman Kodak Co., *Ergonomic Design for People at Work*, Lifetime Learning, Belmont, CA, 1983, pp. 140–159.
21. R. Terrell and J. Purswell, The influence of forearm and wrist orientation on static grip strength as a design criterion for hand tools, *Proc. Hum. Factors Soc. 20th Annual Meeting*, Santa Monica, CA, 1976, pp. 28–32.
22. T. J. Armstrong, J. A. Foulke, B. S. Joseph, and S. A. Goldstein, Investigation of cumulative trauma disorders in a poultry processing plant, *Am. Ind. Hyg. Assoc. J. 43*:103–116 (1982).
23. E. R. Tichauer, Some aspects of stress on forearm and hand in industry, *J. Occup. Med. 8*:63–71 (1966).
24. S. Konz, *Work Design*, Publishing Horizons, Worthington, OH, 1990, pp. 237–258.
25. S. T. Pheasant and S. J. Scriven, Sex differences in strength—Some implications for the design of handtools, in *Proceedings of the Ergonomics Society*, K. Coombes, Ed., Taylor & Francis, London, U.K., 1983, pp. 9–13.

26. D. B. Chaffin and G. Andersson, *Occupational Biomechanics*, Wiley, New York, 1984, pp. 355–368.
27. R. E. Ducharme, Problem tools for women, *Ind. Eng.* 46–50 (September 1975).
28. S. T. Pheasant and D. O'Neill, Performance in gripping and turning—A study in hand/handle effectiveness, *Appl. Ergon.* *6*:205–208 (1975).
29. S. Konz, Design of handtools, *Proc. Hum. Factors Soc. 18th Annual Meeting*, Santa Monica, CA, 1974, pp. 292–300.
30. J. K. Laveson and R. P. Meyer, Left out "lefties" in design, *Proc. Hum. Factors Soc. 20th Annual Meeting*, 1976, pp. 122–125.
31. N. Shock, The physiology of aging, *Sci. Am. 206*:100–110 (1962).
32. G. Miller and A. Freivalds, Gender and handedness in grip strength, *Proc. Hum. Factors Soc. 31st Annual Meeting*, Santa Monica, CA, 1987, pp. 906–909.
33. S. Konz and M. Warraich, Performance differences between the preferred and non-preferred hand when using various tools, in *Ergonomics International '85*, I. D. Brown, R. Goldsmith, K. Coombes, and M. A. Sinclair, Eds., Taylor & Francis, London, U.K., 1985, pp. 451–453.
34. A. Mital, Effects of tool and operator factors on volitional torque exertion capabilities of individuals, in *Ergonomics International '85*, I. D. Brown, R. Goldsmith, K. Coombes, and M. A. Sinclair, Eds., Taylor & Francis, London, U.K., 1985, pp. 262–264.
35. A. Mital, N. Sanghavi, and T. Huston, A study of factors defining the operator-hand tool system at the work place, *Int. J. Prod. Res. 23*:297–314 (1985).
36. A. Mital, Effects of body posture and common hand tools on peak torque exertion capabilities, *Appl. Ergon. 17*:87–96 (1986).
37. A. Mital and N. Sanghavi, Comparison of maximum volitional torque exertion capabilities of males and females using common hand tools, *Hum. Factors 27*:283–294 (1986).
38. P. T. Baker, J. M. McKendry, and G. Grant, Volumetric requirements for hand tool usage, *Hum. Factors 2*:156–162 (1960).
39. A. J. Obolenskaja and I. Goljanitzki, Die seröse Tendovainitis in der Klinik und im Experiment, *Deut. Z. Chirurg. 201*:388–399 (1927).
40. T. Luopajarvi, I. Kuorinka, M. Virolainen, and M. Holmberg, Prevalence of tenosynovitis and other injuries of the upper extremities in repetitive work, *Scand. J. Work Environ. Health 5*(Suppl. 3):48–55 (1979).
41. NIOSH, *Health Hazard Evaluation—Eagle Convex Glass Co.*, HETA-89-137-2005, National Institute of Occupational Safety and Health, Cincinnati, OH, 1989.
42. D. Grieve and S. Pheasant, Biomechanics, in *The Body at Work*, W. T. Singleton, Ed., Cambridge University Press, Cambridge, U.K., 1982, pp. 142–150.
43. J. O. Replogle, Hand torque strength with cylindrical handles, *Proc. Hum. Factors Soc. 27th Annual Meeting*, Santa Monica, CA, 1983, pp. 412–416.
44. B. Jonsson, T. Lewin, P. Tomsic, G. Garde, and P. Forssblad, *Handen som Arbetsredskap*, Arbetarskyddsstyrelsen, Stockholm, Sweden, 1977.
45. B. Rubarth, Untersuchung zur Festgestaltung von Handheften für Schraubenzieher und ähnliche Werkzeuge, *Ind. Psychotech. 5*:129–142 (1928).
46. N. B. Hall and E. M. Bernett, Empirical assessment of handrail diameters, *J. Appl. Psychol. 40*:381–382 (1956).
47. M. Ayoub and P. LoPresti, The determination of an optimum size cylindrical handle by use of electromyography, *Ergonomics 14*:509–518 (1971).
48. L. V. Rigby, Why do people drop things? *Qual. Prog. 6*:16–19 (1973).
49. C. G. Drury, Handles for manual materials handling, *Appl. Ergon. 11*:35–42 (1980).
50. D. J. Cochran and M. W. Riley, An evaluation of handle shapes and sizes, *Proc. Hum. Factors Soc. 26th Annual Meeting*, Santa Monica, CA, 1982, pp. 408–412.

51. D. J. Cochran and M. W. Riley, An evaluation of knife handle guarding, *Hum. Factors 28*: 295–301 (1986).
52. D. J. Cochran and M. W. Riley, An examination of the speed of manipulation of various sizes and shapes of handles, *Proc. Hum. Factors Soc. 27th Annual Meeting*, Santa Monica, CA, 1983, pp. 432–436.
53. J. Garrett, The adult human hand: Some anthropometric and biomechanical considerations, *Hum. Factors 13*:117–131 (1971).
54. R. Magill and S. Konz, An evaluation of seven industrial screwdrivers, in *Trends in Ergonomics/ Human Factors*, W. Karwowski, Ed., Elsevier, Amsterdam, the Netherlands, 1986, pp. 597–604.
55. W. L. Scheller, The effect of handle shape on grip fatigue in manual lifting, *Proc. Hum. Factors Soc. 27th Annual Meeting*, Santa Monica, CA, 1983, pp. 417–421.
56. H. J. Bullinger and J. J. Solf, *Ergonomische Arbeitsmittel-gestaltung, II. Handgeführte Werkzeuge-Fallstudien*, Bundesanstalt für Arbeitsschutz und Unfallforschung, Dortmund, Germany, 1979.
57. C. Saran, Biomechanical evaluation of T-handles for a pronation supination task, *J. Occup. Med. 15*:712–716 (1973).
58. T. M. Fraser, *Ergonomic Principles in the Design of Hand Tools*, International Labor Office, Geneva, Switzerland, 1980.
59. G. L. Fellows and A. Freivalds, Ergonomics evaluation of a foam rubber grip for tool handles, *Appl. Ergon. 22*:225–230 (1991).
60. C. Fransson-Hall and Å. Kilbom, Sensitivity of the hand to surface pressure, *Appl. Ergon. 24*:181–189 (1993).
61. B. Buchholz, L. J. Frederick, and T. J. Armstrong, An investigation of human palmar skin friction and the effects of materials, pinch force and moisture, *Ergonomics 31*:317–325 (1988).
62. O. Bobjer, S. E. Johansson, and S. Piguet, Friction between hand and handle. Effects of oil and lard on textured and non-textured surfaces; perception of discomfort, *Appl. Ergon. 24*:190–202 (1993).
63. J. T. Emanual, S. J. Mills, and J. F. Bennett, In search of a better handle, *Proc. Symp. Hum. Factors Ind. Design Consumer Products*, Tufts University, Meford, MA, 1980, pp. 34–40.
64. Bennett Ergonomic Labs, *Why Bennett's Biocurve*, Minneapolis, MN, 1983.
65. R. Krohn and S. Konz, Bent hammer handles, *Proc. Hum. Factors Soc. 26th Annual Meeting*, Santa Monica, CA, 1982, pp. 413–417.
66. S. Konz and B. Streets, Bent hammer handles performance and preferences, *Proc. Hum. Factors Soc. 28th Annual Meeting*, Santa Monica, CA, 1984, pp. 438–440.
67. S. Konz, Bent hammer handles, *Hum. Factors, 27*:317–323 (1986).
68. R. G. Knowlton and J. C. Gilbert, Ulnar deviation and short term strength reductions as affected by a curve handled ripping hammer and a conventional claw hammer, *Ergonomics 26*:173–179 (1983).
69. L. Greenberg and D. B. Chaffin, *Workers and Their Tools*, Pendell Press, Midland, MI, 1976.
70. A. Damon, H. W. Stoudt, and R. A. McFarland, *The Human Body in Equipment Design*, Harvard University Press, Cambridge, MA, 1966.
71. H. Hertzberg, Engineering anthropometry, in *Human Engineering Guide to Equipment Design*, H. Van Cott and R. Kincaid, Eds., U.S. Government Printing Office, Washington, DC, 1973, pp. 467–584.
72. D. J. Cochran, T. J. Albin, M. W. Riley, and R. R. Bishu, Analysis of grasp force degradation with commercially available gloves, *Proc. Hum. Factors Soc. 30th Annual Meeting*, Santa Monica, CA, 1986, pp. 852–855.
73. L. R. Sudhakar, R. W. Schoenmarklin, S. A. Lavender, and W. S. Marras, The effects of gloves on grip strength and muscle activity, *Proc. Hum. Factors Soc. 32nd Annual Meeting*, Santa Monica, CA, 1988, pp. 647–650.
74. B. Weidman, *Effect of Safety Gloves on Simulated Work Tasks*, AD 738981, National Technical Information Service, Springfield, VA, 1970.

75. J. V. Bradley, Effects of gloves on control operation time, *Hum. Factors 11*:13–20 (1969).
76. J. V. Bradley, Glove characteristics influencing control manipulability, *Hum. Factors 11*:13–20 (1969).
77. A. D. Swain, G. G. Shelton, and L. V. Rigby, Maximum torque for small knobs operated with and without gloves, *Ergonomics 13*:201–208 (1970).
78. B. A. Silverstein, L. J. Fine, and T. J. Armstrong, Occupational factors and carpal tunnel syndrome, *Am. J. Ind. Med. 11*:343–358 (1987).
79. D. E. Wasserman and D. W. Badger, *Vibration and the Worker's Health and Safety*, NIOSH TR-77, U.S. Government Printing Office, Washington, DC, 1973.
80. R. Lundstrom and R. S. Johansson, Acute impairment of the sensitivity of skin mechanoreceptive units caused by vibration exposure of the hand, *Ergonomics 29*:687–698 (1986).
81. E. R. Andersson, Design and testing of a vibration attenuating handle, *Int. J. Ind. Ergon. 6*:119–125 (1990).
82. A. Freivalds, The ergonomics of shoveling and shovel design—An experimental study, *Ergonomics 29*:19–30 (1986).
83. E. A. Müller and K. Karrasch, Die grösste Dauerleistung beim Schaufeln, *Int. Z. Angew. Physiol. Einschl. Arbeitsphysiol. 16*:318–324 (1956).
84. A. Freivalds and Y. J. Kim, Blade size and weight effects in shovel design, *Appl. Ergon. 21*:39–42 (1990).
85. C. H. Wyndham, J. F. Morrison, C. G. Williams, R. Heyns, E. Margo, A. M. Brown, and J. Astrup, The relationship between energy expenditure and performance index in the task of shovelling sand, *Ergonomics 9*:371–378 (1966).
86. G. Kommerell, Die Schaufelarbeit in gebückter Haltung, *Arbeitsphysiologie 1*:278–295 (1929).
87. G. Dressel, K. Karrasch, and H. Sptizer, Arbetisphysiologische Untersuchungen beim Schaufeln, Steinetragen un Schubkarreschieben, *Zentralbl. Arbeistwiss. Soz. Betriebsprax. 3*:33–48 (1985).
88. A. Freivalds, The ergonomics of tools, *Int. Rev. Ergon. 1*:43–75 (1987).
89. A. G. Stevenson and R. L. Brown, An investigation on the motion study of digging and the energy expenditure involved, with the object of increasing efficiency of output and economising energy, *J. Roy. Army Med. Corps 40*:39–45, 99–111, 340–349, 423–434 (1923).
90. C. J. Widule, V. Foley, and F. Demo, Dynamics of the axe swing, *Ergonomics 21*:925–930 (1978).
91. D. L. Corrigan, V. Foley, and C. J. Widule, Axe use efficiency—A work theory explanation of an historical trend, *Ergonomics 24*:103–109 (1981).
92. H. Gläser, Beiträge zur Form der Waldsäge und zur Technik des Sägens, PhD Dissertation, Eberswalde, Germany, 1933.
93. A. S. Bleed, P. Bleed, D. J. Cochran, and M. W. Riley, A performance comparison of Japanese and American hand saws, *Proc. Hum. Factors Soc. 26th Annual Meeting*, Santa Monica, CA, 1982, pp. 403–407.
94. E. R. Tichauer, Biomechanics sustains occupational safety and health, *Ind. Eng. 8*:46–56 (1976).
95. T. A. Yoder, R. L. Lucas, and C. D. Botzum. The marriage of human factors and safety in industry, *Hum. Factors 15*:197–205 (1973).
96. F. E. Lindstrom, *Modern Pliers*, Bahco Verktyg, Enköping, Sweden, 1973.
97. L. Karlqvist, Cutting operations at canning bench—A case study of handtool design, *Proc. 1984 Int. Conf. Occup. Ergon.*, Human Factors Association of Canada, Rexdale, ON, 1984, pp. 452–456.
98. A. Freivalds and J. Eklund, Reaction torques and operator stress while using powered nutrunners, *Appl. Ergon. 24*:158–164 (1993).
99. Atlas Copco, *Industrial Power Tools* (Catalog), Atlas Copco Tools AB, Box 81510, Stockholm, Sweden, 1990.
100. R. G. Radwin, E. VanBergeijk, and T. J. Armstrong, Muscle response to pneumatic hand tool torque reaction forces, *Ergonomics 32*:655–673 (1989).

101. M. Vuori, M. Rauko, and S. Herranen, *Evaluating and Choosing Pneumatic Screwdrivers and Nutrunners*, The Finnish Work Environment Fund, Helsinki, Finland, 1989.
102. A. Mital and Å. Kilbom, Design, selection and use of hand tools to alleviate trauma of the upper extremities: Part I. Guidelines for the practitioner, *Int. J. Ind. Ergon. 10*:1–5 (1992).
103. A. Mital and Å. Kilbom, Design, selection and use of hand tools to alleviate trauma of the upper extremities: Part II. The scientific basis for the guide, *Int. J. Ind. Ergon. 10*:7–21 (1992).
104. M. H. Yun, A hand posture measurement system for evaluating manual tool tasks, *Proc. Hum. Factors Ergon. Soc. 37th Annual Meeting*, Santa Monica, CA, 1993, pp. 754–758.

CHAPTER 12

# Manual Materials Handling

Thomas R. Waters

CONTENTS

## 12.1 INTRODUCTION

The intent of this chapter is to (1) provide basic information to identify and quantify risk factors for work-related musculoskeletal disorders (MSDs) resulting from manual material handling (MMH) and (2) discuss typical methods used to reduce or eliminate these factors so that MSDs can be prevented. The chapter was designed to meet the needs of those individuals who are responsible for providing technical occupational ergonomics support but lack formal training in ergonomics. The chapter includes sections describing:

Who is at risk of MMH-related MSDs?

What risk factors are associated with MMH-related MSDs?

What methods are available for identifying and analyzing MSD risk factors?

What are the best approaches for preventing MMH-related MSDs?

## 12.2 HEALTH RISK OF MMH

### 12.2.1 Adverse Health Effects

Despite the spread of mechanization in industry, MSDs attributed to MMH are still a major cause of lost work time, increased costs, and human suffering in the American workforce. These MSDs include a variety of injuries or disorders of the wrist, arm, shoulder, neck, and back as well as the lower extremities.

Problems with the low back are especially problematic, and low back disorders resulting in low back pain (LBP) continue to be one of the leading occupational health and safety issues facing preventive medicine. Despite efforts at control, including programs directed at both workers and jobs, work-related back injuries still account for a significant proportion of human suffering and economic cost to this nation.

According to the Bureau of Labor Statistics (BLS) and the National Safety Council, of all injuries and illnesses recorded in 2007 and 2011 [1,2], the following statistics illustrate the significant health problems associated with MMH:

1. Sprains and strains were the most frequently occurring type.
2. Overexertion injuries accounted for 264,930 cases resulting in days away from work.
3. The body parts most often affected by work injuries were the trunk and back, accounting for approximately 33% of all injuries.
4. The median number of days away from work from all occupations was 7 days, but certain occupations had more average days away from work. For example, transportation and warehousing reported an average of 14 days away from work per case, twice the national average.
5. Injuries involving the shoulder took workers a median of 18 days to recuperate, on average, with over half the shoulder injuries attributed to overexertion.

6. Laborers and freight, stock, and material movers experienced the highest number of days-away-from-work injuries and illnesses, with 79,000 cases. This was followed by heavy and tractor-trailer truck drivers (57,050 cases), nursing aides, orderlies, and attendants (44,930 cases), construction laborers (34,180 cases), and light or delivery service truck drivers (32,930 cases).
7. Nursing aides, orderlies, and attendants have an MSD rate of 252 cases per 10,000 workers, a rate more than seven times the national average for all occupations.

These high-risk occupations all have one thing in common: workers in these industries are often involved in MMH on a daily basis.

### 12.2.2 Who Is at Risk?

Although the percentage of the American workforce who are routinely engaged in jobs that require them to perform MMH task is not precisely known, results from two recent surveys of American workers, one conducted in 2002 and another in 2006, found that 46% of respondents reported that they were required to perform heavy manual lifting in their work [3,4]. The survey was designed to be representative of virtually all the nonagricultural, nonmining, and nongovernmental businesses covered under the Occupational Safety and Health Act of 1970. Studies of exposures to various physical hazards and agents at work, such as frequent bending and/or unaided lifting, hand/wrist manipulations, arm movement, noise, and vibration typically report that lifting activities and arm transport motions are the most frequently identified exposure hazards.

Industries with the greatest number of workers exposed to the hazards of MMH typically are the same as those with the highest number of reported back and shoulder disorder cases. These include the health services, special trade contracting, general building contracting, food and kindred products, and trucking and warehousing industries. In general, these industries employ higher percentages of men than women, but the health services industry, which employs approximately four times as many female workers as males, also reports high numbers of workers who perform physically demanding MMH tasks, such as when they move or handle patients, lift and move heavy equipment, or push or pull heavy beds. All major industry sectors have identified MMH and work-related MSDs as a significant problem that needs to be addressed.

### 12.2.3 Summary

It is clear that a tremendous number of workers are routinely exposed to the physical hazards of MMH, and many of them will develop one or more serious work-related MSDs during their working lifetime. Nevertheless, it should be recognized that these hazards can be identified and controlled so that work-related MSDs can be prevented.

## 12.3 RISK FACTORS

There are a variety of MMH activities that increase a worker's risk of developing a work-related MSD, including jobs that involve a significant amount of manual lifting, pushing, pulling, or carrying. In addition to these frequent patterns of exposure, there are a variety

TABLE 12.1 Risk Factors Associated with MMH MSDs

| Personal Factors | Environmental Factors | Job-Related Factors | Psychosocial Factors |
|---|---|---|---|
| Gender | Humidity | Location of load relative to the worker | Occupational stress |
| Anthropometry (height and weight) | Light | Distance object is moved | Job satisfaction |
| Physical fitness and training | Noise | Frequency and duration of handling activity | Monotonous work |
| Lumbar mobility | Vibration | Bending and twisting | Social support at work |
| Strength | Foot traction | Weight of object or force required to move object | High perceived work demands |
| Medical history | | Stability of the load | Job control |
| Years of employment | | Postural requirements | |
| Smoking | | | |
| Anatomical abnormality | | | |

of personal and environmental factors that may increase a worker's risk of developing an MSD. Table 12.1 contains a list of the more noteworthy work-related risk factors for developing MSDs.

*Personal risk factors* are conditions or characteristics of the worker that may affect the probability that an overexertion injury may occur. Personal risk factors include attributes such as age, level of physical conditioning, strength, and medical history.

*Environmental risk factors* are conditions or characteristics of the external surroundings that may affect the probability that an overexertion injury may occur. Environmental risks include factors such as temperature, lighting, noise, vibration, and friction at the floor.

*Job-related risk factors* are conditions or characteristics of the MMH job that may affect the probability that an overexertion injury may occur. Job-related risk factors include attributes such as the weight of the load being moved, the location of the load relative to the worker when it is being moved, the size and shape of the object moved, and the frequency of handling. These factors may be the most important for prevention because they directly affect the magnitude of physical hazard to the worker.

*Psychosocial or organizational factors* are factors related to the social or organizational environment, such as occupational stress, job satisfaction, monotonous work, social support at work, high perceived work demands, etc. The mechanism for how these factors might increase the risk of MSDs is not fully understood, but it is believed that these factors lead to both physiological and biomechanical responses that increase the risk of MSDs.

Work-related MSDs are typically attributed to a direct trauma, a single exertion ("over-exertion"), or multiple exertions ("repetitive trauma"). It is not always possible, however, to determine the specific cause of the injury, and the pathophysiology of many types of MSDs is poorly understood. Moreover, it is important to note that personal characteristics such as age, physical conditioning, and concomitant diseases can modify the way the body responds to stressful exertions. Therefore, an injury could occur at different loading levels for different workers, and even for an individual worker, load may be tolerable 1 day and excessive on another day due to fluctuations in muscular strength and aerobic fitness [5].

A 2001 report by National Research Councils' Institute of Medicine contains a detailed review of the current literature on work-related risk factors for LBP [6]. The authors concluded that "there is a clear relationship between back disorders and physical load imposed by manual material handling, frequent bending and twisting, physically heavy work, and whole-body vibration." Although a large volume of research literature exists in which multiple factors contributing to LBP have been studied, knowledge about how job-related, environmental, psychosocial, and personal risk factors affect risk of MSDs is not complete.

A considerable volume of literature exists suggesting that there is a significant relationship between psychosocial factors, such as job control and satisfaction, and risk of MSD [6–8]. In a review conducted by Ferguson and Marras, the authors suggested that the role psychosocial factors play in affecting the risk of injury may be as important as the actual physical demands of the job [9]. The data indicate that as the perceived demands of the job increase and the workers' control over those demands and social support decrease, the rate of injury increases. It is probable that these psychosocial factors lead to an increase in the physical effects on the worker such as increased muscle tension, lower endurance, or modified body mechanics. More research is needed to develop a comprehensive control strategy that considers these psychosocial factors.

In summary, there is ample evidence that exposure to MMH can result in MSDs, especially to the low back, and that these injuries are attributable to numerous risk factors that must be considered in the design of a safe workplace. The interpretation of the research linking MSDs, and manual handling is problematic, because of the high prevalence of certain injuries such as LBP in the general population and their frequent association with nonoccupational factors. In addition, the relationship is further obscured by the wide range of disorders, the nonspecific nature of the condition, and the general lack of objective data relating different risk factors to overexertion injury. What is generally recognized, however, is that work-related MSDs are a function of a complex set of variables that include aspects of job design, work environment, psychosocial, and personal factors. How all of these factors are combined to precisely determine the overall risk associated with specific MMH jobs remains to be resolved.

## 12.4 HAZARD IDENTIFICATION/JOB ANALYSIS

A variety of analytical tools are available for the ergonomic evaluation of MMH tasks, especially manual lifting tasks. These tools range in complexity from simple checklists that are designed to provide a general indication of the physical stress associated with a particular MMH job to complicated computer models that provide detailed information about specific risk factors. Although not exhaustive, Table 12.2 summarizes the advantages and disadvantages of a variety of ergonomic assessment tools. As described in the following, these tools provide objective information about the physical demands of MMH tasks so that risks can be reliably assessed and an effective prevention strategy can be developed.

In general, these tools are based on scientific studies that provide a relationship between sources of physical stress and risk of musculoskeletal injury, particularly when those stressors exceed the physical capacity of the workers [10]. Any assessment of physical stress or human capacity is complicated by the influence of a variety of factors including work performance, motivation, expectation, and fatigue tolerance.

TABLE 12.2 Ergonomic Assessment Tools

| Assessment Tool | Advantages | Disadvantages |
|---|---|---|
| Checklists | • Simple to use<br>• Best suited for use as preliminary assessment tools<br>• Applicable to a wide range of MMH jobs<br>• Provide detailed estimates of mechanical forces on musculoskeletal components | • Do not provide detailed information about the specific risk factors<br>• Do not quantify the extent of the risk factors |
| Biomechanical models | • Can identify specific body structures exposed to high physical stress<br>• Can identify tasks that exceed specific tissue tolerance limits (i.e., tasks that create excessive joint loading) | • Not applicable for estimating effects of repetitive activities<br>• Difficult to verify accuracy of estimates<br>• Rely on a number of simplifying assumptions |
| Psychophysical tables | • Provide population estimates of worker capacities that integrate biomechanical and physiological stressors<br>• Applicable to a wide range of MMH activities | • Reflect more about what a worker will "accept" than about what is "safe"<br>• May over- or underestimate demands for infrequent or highly repetitive activities |
| Physiological models | • Provide detailed estimates of physiological demands for repetitive work as a function of duration<br>• Applicable to a wide range of MMH activities | • Not applicable for estimating effects of infrequent activities<br>• Lack of strong link between physiological fatigue and risk of injury |
| Integrated assessment models | • Simple to use<br>• Based on the most appropriate criterion for the specified task | • Require a significant number of assumptions<br>• Limited range of application |
| Videotape assessment | • Economical method of measuring postural kinematics<br>• Can be used to analyze a large number of samples | • Labor-intensive analysis generally required<br>• Limited to 2D analysis for field measurements |
| Exposure monitors | • Provide direct measures of posture and kinematics during MMH activities<br>• Applicable to a wide range of MMH activities | • Require the worker to wear a device on the body<br>• Lack of data linking monitor output and risk of injury |

According to the Occupational Safety and Health Administration (OSHA) [11], the objectives of the worksite analysis are to identify existing physical hazards and conditions, operations that create hazards, and areas where hazards may develop. Worksite analysis not only includes a systematic method to identify those work tasks or operations requiring hazard control but may also include a comprehensive review and analysis of injury and illness records to find any evidence of musculoskeletal injuries or to identify any trends in injury patterns relating to particular departments or jobs. This information is necessary to identify those jobs that are believed to possess a significant risk of injury and require a detailed quantitative analysis of physical hazards.

NIOSH Hazard Evaluation Checklist For Lifting, Carrying, Pushing, or Pulling

| Risk Factors | | Yes | No |
|---|---|---|---|
| 1. General | | | |
| 1.1 | Does the load handled exceed 50 lbs? | [ ] | [ ] |
| 1.2 | Is the object difficult to bring close to the body because of its size, bulk, or shape? | [ ] | [ ] |
| 1.3 | Is the load hard to handle because it lacks handles or cutouts for handles, or does it have slippery surfaces or sharp edges? | [ ] | [ ] |
| 1.4 | Is the footing unsafe? For example, are the floors slippery, inclined, or uneven? | [ ] | [ ] |
| 1.5 | Does the task require fast movement, such as throwing, swinging, or rapid walking? | [ ] | [ ] |
| 1.6 | Does the task require stressful body postures, such as stooping to the floor, twisting, reaching overhead, or excessive lateral bending? | [ ] | [ ] |
| 1.7 | Is most of the load handled by only one hand, arm, or shoulder? | [ ] | [ ] |
| 1.8 | Does the task require working in environmental hazards, such as extreme temperatures, noise, vibration, lighting, or airborne contaminants? | [ ] | [ ] |
| 1.9 | Does the task require working in a confined area? | [ ] | [ ] |
| 2. Specific | | | |
| 2.1 | Does lifting frequency exceed 5 lifts per minute? | [ ] | [ ] |
| 2.2 | Does the vertical lifting distance exceed 3 ft? | [ ] | [ ] |
| 2.3 | Do carries last longer than 1 min? | [ ] | [ ] |
| 2.4 | Do tasks which require large sustained pushing or pulling forces exceed 30 s duration? | [ ] | [ ] |
| 2.5 | Do extended reach static holding tasks exceed 1 min? | [ ] | [ ] |

Comment: "Yes" responses are indicative of conditions that pose a risk of developing low back pain. The larger the percentage of "yes" responses, the greater the possible risk.

FIGURE 12.1 MMH checklist.

## 12.4.1 Checklists

A checklist may be the first choice for a rapid ergonomic assessment of a particular workplace. Checklists are designed to provide a general evaluation of the extent of hazard associated with an MMH task or job. A checklist usually consists of a series of questions that are indicative of physical hazards. For example, questions about frequent bending, heavy lifting, awkward or constrained postures, poor couplings at the hands or feet, and hazardous environmental conditions. Some checklists use a yes/no format, whereas others use a numerical rating format. Checklists are easy to use, but they lack specificity and they are imprecise. An example of a manual handling checklist is presented in Figure 12.1. For more information on checklists, see Grandjean [12], Eastman Kodak [13], Alexander and Pulat [14], or the National Occupational Health and Safety Commission [15].

## 12.4.2 Biomechanical Models

Biomechanical assessment involves the systematic application of engineering concepts to the functioning human body to predict the distribution of internal musculoskeletal forces resulting from the interaction with externally applied forces of the task. Biomechanical modeling provides a method for predicting the pattern and magnitude of these internal forces during manual handling. These predicted forces can then be compared to predetermined tissue tolerance limits to assess the biomechanical stress associated with specific loading conditions. When a worker is performing a manual lifting task, for example, the internal reaction forces that are needed to provide equilibrium between the body segments and the external forces are supplied by muscle contractions, tendons, and ligaments at the body joints (i.e., the human body acts as a lever system). Specifically, the external forces at the hands and the body segments create rotational moments or torques at various body joints, especially the low back. The skeletal muscles exert forces that result in moments about the joints so as to counteract the moments due to external load and body segment weights. Since the moment arms of the muscles (and ligaments) are much smaller than the moment arms of the external forces and body segment weights, small external forces can produce large muscle, tendon, ligament, and joint reaction forces. On the other hand, the muscles can produce large motions with small degrees of shortening. The concept of muscles loading skeletal structures is extremely important in the biomechanics of MSDs, because handling light loads in certain postures can create large mechanical loads on the muscles, ligaments, and joint surfaces.

The complexity of mathematical formulation and ease of use of biomechanical models varies, depending on which factors are included. Some important factors to consider when using a biomechanical model are

1. The mechanical nature of the model (static or dynamic)
2. Dimensionality of the model (2D or 3D)
3. Accuracy of the representation (e.g., single or multiple muscles, intra-abdominal pressure, muscular cocontraction, and active and passive elements)
4. Complexity of the input needed to use the model (e.g., musculoskeletal geometry, mechanical parameters, and physiological measures of muscle function)

Some of the more complex biomechanical models of the low back, for example, integrate 3D mechanical data and electromyographic (EMG) data from the back muscles as input to estimate internal loading on the spine [16].

Although biomechanical models are typically used to provide guidance for the design of infrequent stressful activities with high levels of exertion, recent studies have provided some guidance for the use of biomechanical models in assessing repetitive lifting tasks. In studies of spinal compression tolerance, for example, Brinckmann et al. [17] showed that repeated compression loading of the spinal motion segments causes them to fail at lower forces than those required in a single loading cycle.

For more information on the use of biomechanical models to assess MMH tasks, refer to Chaffin et al. [18], National Research Council [19], or Garg [20].

### 12.4.3 Psychophysical Tables

Psychophysics is a branch of psychology that examines the relationship between the perception of human sensations and physical stimuli. Stevens [21] and Snook and Ciriello [22] suggest that a worker's actual level of physical stress can be assessed by her or his subjective judgment or perception of the physical stress. In other words, the worker's perception of workload is used to assess the combined effects of physiological and biomechanical stress created by various MMH factors.

Typical psychophysical studies of MMH capacity consist of measurement of acceptable levels of workload for various task conditions and for various workers. The results are usually provided in tables of acceptable weights of lift or carry or acceptable forces of pushing and pulling [22]. These psychophysical measures are contained in databases that provide acceptable workloads for various segments of the population. A portion of a psychophysical database for maximum acceptable sustained push force is shown in Table 12.3. According to the database, the push force acceptable to 50% of the female industrial subjects for a 15.2 m push (50 ft) at a height of 89 cm (35 in.) and a rate of 1 push every 5 min was 14 kg (31 lb).

Most psychophysical research involving MMH activities has emphasized lifting tasks, although the use of psychophysical techniques is not restricted to lifting. Psychophysics also is applicable to lowering, pushing, pulling, holding, and carrying activities. The use of psychophysical data to assess the physical demands of manual handling is most appropriate for repetitive activities that are performed more often than once a minute. For more information on psychophysical databases, see Snook and Ciriello [22] or Ayoub and Mital [23].

Other psychophysical assessment methods have been developed to assess various MMH activities. For example, self-report measures such as rating of perceived exertion (RPE) [24] and body part discomfort (BPD) [25] have been used to assess a variety of lifting jobs. These assessment measures provide useful information about the worker's perception of the physical demands of the job. Moreover, RPE and BPD compare favorably with measures of physical demand.

Databases containing whole body and segmental strength measures have also been developed for the design of MMH tasks. These include isometric, isokinetic, and isoinertial strength databases for whole body activities such as lifting and various databases for the arms, legs, and back. For more information on strength measurement, detailed summaries of these databases are available in Chaffin et al. [18] and Ayoub and Mital [23].

TABLE 12.3 Psychophysical Table for Maximum Acceptable Push Force

| | | Maximum Acceptable Forces of Sustained Push (kg) for 15.2 m Push | | | | | | |
|---|---|---|---|---|---|---|---|---|
| | | One Push Every | | | | | | |
| **Height from Floor to Hands (cm)** | **Percent of Industrial Females** | **6 s** | **12 s** | **1 min** | **2 min** | **5 min** | **30 min** | **8 h** |
| 89 | 90 | 5 | 6 | 6 | 7 | 7 | 8 | 10 |
| | 75 | 7 | 8 | 9 | 10 | 11 | 11 | 14 |
| | 50 | 9 | 11 | 13 | 13 | 14 | 15 | 19 |
| | 25 | 12 | 14 | 16 | 16 | 18 | 19 | 24 |
| | 10 | 14 | 17 | 19 | 19 | 21 | 23 | 28 |

*Source:* Adapted from Snook, S.H. and Ciriello, V.M., *Ergonomics*, 34, 1197, 1991.

It should be noted that psychophysics relies on self-report from subjects, and consequently the perceived "acceptable" limit may differ from the "safe" limit. Also, the psychophysical approach is not valid for some task conditions, for which other assessment measures may be more appropriate [10].

### 12.4.4 Physiological Models

One goal in designing an MMH task is to avoid the accumulation of physical fatigue that may contribute to an MSD. This fatigue can affect specific muscles or groups of muscles, or it can affect the whole body by reducing the aerobic capacities available to sustain work. The physiological factors that affect the suitability of a manual handling task at the local muscle effort level include (1) the duration of force exertion and (2) the frequency of repetition per minute. Local muscular fatigue will develop if a heavy effort is sustained for a long period. With heavy loads, the muscles will need a substantially longer recovery period to return to their previous state. Small changes in workplace layout or handling heights, however, can often solve a local muscle fatigue problem through a reduction in holding duration. In addition, local muscle fatigue associated with maintaining awkward postures or with constant bending can reduce the capacity of the muscles needed for lifting and therefore can increase the potential for an MSD injury to occur.

Although fatigue in one or more muscles at the local level may limit the acceptable workloads for MMH tasks that are performed for short, intensive periods during a work shift, it is the energy expenditure demands of the muscles, as measured by oxygen consumption, that have the most profound effect on what a worker is able to do. Energy expenditure demands depend on the extent of muscular exertion, frequency of activity, and duration of continuous work. A worker's limit for physiological fatigue is often affected by a combination of discomfort in local muscle groups and more centralized (systemic) fatigue associated with oxygen demand and cardiovascular strain [26].

To assess the cardiovascular demands of MMH tasks, physiological parameters such as heart rate (HR), oxygen ($O_2$) consumption, and ventilatory rate may be used. EMG assessments and blood lactate provide a relative measure of the instantaneous level of physiological status and muscular fatigue. As described in the following, these measures can be used to help prevent MSD by predicting the limits of fatigue for repetitive handling tasks.

Physiological models provide a method for estimating the cardiovascular demands associated with a specific MMH activity. One such model developed by Garg [27] allows the analyst to estimate the energy expenditure demands associated with a complex MMH job. First, the job is separated into distinct elements or subtasks such as standing and bending, walking, carrying, vertical lifting or lowering, and horizontal arm movement, for which individual energy expenditure values can be predicted by the model. Next, the amount of time engaged in each of the subtasks is determined, and a subtotal for the energy expenditure requirements for each subtask is computed by taking the product of the incremental expenditures and the amount of time spent performing that activity. Finally, the total energy expenditure of the job, which determines the overall physiological demands of the job, is estimated by taking the sum of the subtotals of the subtasks.

Direct measures of oxygen consumption may provide the best estimate of physiological demand. For this type of measurement, it is assumed that the consumption of 1 L of oxygen per minute is approximately equal to an energy expenditure of 5 kcal/min. As mentioned earlier, HR is also useful in predicting physiological demand but is less reliable than direct oxygen consumption measures due to individual differences in the relationship between HR and energy expenditure for different people. To simplify measurement, portable monitors are available for acquiring HR and oxygen consumption data during MMH tasks to accurately determine physiological demands. For information on assessing physiological demands, see Astrand and Rodahl [28] or Eastman Kodak [13].

### 12.4.5 Integrated Assessment Models

An integrated assessment model involves a unique approach that considers all three of the primary stress measures—biomechanics, physiology, and physchophysics. The integrated approach provides a method of estimating the relative magnitude of physical demand for a specific MMH task, which is based on the most appropriate stress measure for that unique set of task factors. The result of the assessment is typically represented as a weight or force limit or as an index of relative severity. An integrated model considers the combined effects of the various task factors and uses the most appropriate stress measure to estimate the magnitude of hazard associated with each task factor. For example, the model will predominantly rely on the biomechanical assessment for an infrequent heavy lift but will rely more heavily on the physiological assessment for a highly repetitive lifting task.

Examples of integrated assessment models include Ayoub's job severity index (JSI) [29] and the National Institute for Occupational Safety and Health (NIOSH) recommended weight limit (RWL) equation [10]. Details on the revised NIOSH lifting equation are presented in Chapter 33 of this book.

### 12.4.6 Videotape Assessment

Most ergonomic assessments include the use of videotape analysis, where a video camera is used to record the work activity for later analysis. Videotape recordings make it easier to stop or freeze the action so that various measures of body posture or workplace layout may be evaluated. Videotape analysis may consist of general observation by an analyst, which results in subjective estimates of physical hazards, or more detailed video assessment may be used to objectively quantify the extent of the hazard.

Complex computerized video analysis systems are available that are capable of automatically capturing individual frames from videotape recordings of workers performing MMH activities. These video frames can then be analyzed to provide more detailed assessments of spatial or dynamic biomechanical hazards that may not be apparent from the observational approach. These systems may include automatic digitizing capabilities for marking body joint locations and/or automatic scoring for assessing joint or body segment positions. Automated analysis systems are uniquely suited for certain types of assessments in which large amounts of videotape must be analyzed to determine the extent of the physical hazards.

Operationally, videotape analysis is generally easy to use, and the output data are presented in a form that is easy to understand and apply. Video analysis is limited, however, in its capability to analyze activities that occur outside of the camera's focal plane (the plane parallel to the face of the camera lens). For example, when a body segment or group of segments move outside the camera's focal plane, the joint angles and positions measured from the digitized frame are distorted. The amount of distortion depends on the degree of displacement of the joint or segment from the focal plane. In an attempt to compensate for this problem, some systems have incorporated distortion correction algorithms. Correcting distortion problems is problematic, however, and requires a high degree of technical training.

Guidelines for videotape job analysis that have been developed by NIOSH researchers are provided elsewhere in this book. For more information on videotape analysis and motion analysis, see Eastman Kodak [13] or Chaffin et al. [18].

### 12.4.7 Exposure Monitors

Monitoring devices have been developed to measure various aspects of physical activity, including position, velocity, and acceleration of movement. Some monitors can even measure 3D joint angles in real time. These systems consist of mechanical sensors that are attached to various parts of the worker's body such as the wrist, back, or knees. The mechanical sensors convert angular displacement (rotation) into voltage changes that can be displayed in real time and/or saved to a computer for later analysis. The position measures acquired from the sensors can then be differentiated to obtain rotational velocities and acceleration components. These movement characteristics can be used to estimate the extent of risk associated with a particular task and help to identify potential ergonomic solutions. Examples of positional monitoring equipment include potentiometer-based wrist motion monitors and back motion monitors, such as the lumbar motion monitor (LMM) developed at the Ohio State University (see Figure 12.2) [30,31]. Other equipment

FIGURE 12.2 LMM.

for measuring postures includes strain gauge-based strip goniometers such as those developed by Penny and Giles (Santa Monica, CA). In addition to development of the LMM, Marras and coworkers conducted large-scale epidemiological studies from which they developed a comprehensive database of MMH risk factors that are associated with LBP. Based on this database, they derived a logistics regression model with five factors that can be used to determine the probability that a specific job has similar task characteristics to jobs that were associated with historically high rates of LBP [32,33]. Theoretically, the higher the probability of high-risk group membership, the more likely it would be that the worker would experience LBP.

Another device that has been used to assess the extent of exposure to repetitive movement is an accelerometer-based motion-recording system or activity monitor. An activity monitor consists of one or more accelerometers mounted within a small aluminum container attached to a Velcro strap that is attached to a worker's wrist, leg, or trunk. The accelerometers are sensitive to the movements of the body and are capable of counting and recording rapid movements inherent in a specific task or activity. These measures are important because "highly dynamic" movements that occur over an extended period of time, such as an 8 h work shift, are believed to increase workers' risk of musculoskeletal injury. The data acquired from the activity monitor are typically plotted as a series of temporal histogram plots showing the extent of dynamic movement as a function of time. When the total dynamic activity observed is high, the height and density of the sequential histogram bars will be high and the potential for MSDs will be increased [34].

It is important to note that the output from exposure monitors alone cannot provide all the information that is needed to assess the overall extent of physical demand associated with an MMH task. The weight of the load handled and its position, velocity, and acceleration relative to the body during the task are also important in assessing the demands of the task. This approach may be best suited for repetitive or high-speed MMH tasks where the internal forces on the body may be affected more by extreme postures or rapid movements rather than the weight or position of the external load.

## 12.5 PREVENTION AND CONTROL

In the past, attempts to prevent MSDs associated with MMH focused on adopting arbitrary weight limits for lifting loads, hiring strong workers, or using training procedures that emphasized "correct" (but not necessarily safe) lifting techniques. None of these approaches, however, has proven to be effective in significantly reducing overexertion injuries [35,36]. More recently, industry has recognized the risks associated with MMH and, to reduce costs and increase productivity, companies have begun to implement ergonomic programs or practices aimed at preventing these injuries.

Recently, experts from the NIOSH, California Occupational Safety and Health Administration (Cal/OSHA), CNA Insurance, and the Material Handling Industry of America (MHIA) collaborated on the development of a comprehensive guideline for MMH [37]. The guideline provides information on risk factors associated with MMH that will help you to recognize high-risk MMH work tasks and choose effective options for reducing their physical demands. The document illustrates a wide range of ergonomic solutions for MMH tasks ranging from simple approaches such as eliminating lifting from the floor and using simple transport devices like carts or dollies to use of lift-assist devices like scissors lift tables or load levelers and more sophisticated equipment like powered stackers, hoists, cranes, or vacuum-assist devices. The information provided in the document can help guide your choice of equipment by analyzing and redesigning workstations and workflow. The document covers the following topics:

*Improving Manual Material Handling in Your Workplace.* This section lists the benefits of improving your work tasks. It also contains information on risk factors, types of ergonomic improvements, and effective training and sets out a four-step proactive action plan. The plan helps you identify problems, set priorities, make changes, and follow-up. Specific topics include what is MMH, why improve your workplace, what to look for, types of ergonomic improvements, training, and a proactive action plan.

*Improvement Options.* This section provides ways to improve lifting, lowering, filling, emptying, or carrying tasks by changing work practices and/or the use of equipment. Guidelines are provided for using safer work practices and ideas for using equipment instead of manually handling individual containers. Specific topics in this section include easier ways to manually lift, lower, fill, or empty containers, easier ways to manually carry containers, and alternatives to manual handling of individual containers.

*Resources.* This section contains additional information on administrative improvements, work assessment tools, and comprehensive analysis methods. It also includes an improvement evaluation tool and a list of professional and trade organizations related to material handling. Specific topics in this section include administrative improvements, assessment "tools," analysis methods, improvement evaluation "tools," and a listing of organizations with helpful information.

Ergonomic programs typically rely on guidelines provided by the federal government. One of the first guidelines of this type was a management guideline for implementing an ergonomics program for "red meat" packing plants [Standard Industrial Classification (SIC) code 2011] that was published by OSHA [11]. The guideline was designed to cover both cumulative trauma disorders (CTDs) of the upper extremities and low back disorders. The "red meat" packing guideline may serve as the model for development of a broader ergonomics guideline or standard that would apply to many industries.

The OSHA "red meat" packing guideline describes the basic requirements of an effective ergonomics program. According to OSHA, an effective ergonomics program requires

1. A *commitment by the employer* to provide visible involvement of top management, so that all employees, from management to line workers, fully understand that management has a serious commitment to the program
2. A *written program* that establishes clear goals and objectives that are communicated and understood by all members of the organization

3. A commitment to provide for and *encourage employee involvement* in the identification and control of ergonomic problems
4. A regular program of *review and evaluation* to evaluate the implementation of the program and to monitor the progress of the program accomplishments

According to OSHA, hazard prevention and control is an essential element of an ergonomics program.

More recently, OSHA has published a number of occupation-specific ergonomic guidelines, including guidelines for nursing homes [38], poultry processing [39], retail grocery stores [40], and shipyard work [41].

### 12.5.1 Hazard Prevention and Control

Once hazards have been identified through systematic worksite analysis, design measures must be implemented to prevent or control these hazards. According to the OSHA guideline [11], ergonomic hazards are prevented primarily by effective design of the workstation, tools, and job. To be effective, an employer's program should use appropriate engineering and administrative controls to correct or control ergonomic hazards. These controls should include design or modification of workstations, tools, or work methods to eliminate excessive exertion and awkward postures and modification of work schedules and rates to reduce the duration, frequency, and severity of exposures to ergonomic stressors.

Attempts at ergonomic control of MMH problems have included both worker-directed and workplace-directed programs. Worker-directed approaches primarily deal with attempts to maintain a match between the worker's capacity and the demands of the job. This is usually attempted through worker screening, increases in the physical capacity of the worker, or protection of the worker from the physical hazard in some way. Consequently, these approaches are developed on a worker-to-worker and job-to-job basis. More importantly, this approach requires detailed information about the capacity of each worker as well as the demands of the job. Moreover, worker-directed approaches fail to directly reduce the extent of the physical hazard associated with the task or job.

In comparison, workplace-directed approaches rely on changes in the design of the work to eliminate or minimize the effects of the MMH problem. Consequently, these approaches are more generalized and less dependent on having detailed knowledge of the capacity of the specific worker performing the task or job. Workplace-directed approaches include elimination or MMH from the process through automation, reduction of the amount of physical exertion required to perform the MMH activity by using mechanical aids to assist the worker, or modification of the layout of the workplace to reduce the physical hazard.

### 12.5.2 Workplace-Directed Approaches

#### *12.5.2.1 Automation*

Workplace automation should be a first approach when the work requires high physical demands, is highly repetitive, or is performed in a hazardous environment. Automation may consist of one or more machines or machine systems such as conveyors, automated handling lines, automated storage and retrieval systems, and robots. This approach is best suited for the design of new work processes or activities or for the design of highly stressful tasks. Since automation may require large capital expenditures, this approach could be prohibitively costly for small companies with few workers.

#### *12.5.2.2 Mechanical Aids*

In cases where the physical demands are high and automation is not practical, mechanical aids should be used to reduce the extent of physical demands of a task or job. Mechanical handling aids include machines or simple devices such as hand trucks, cranes, hoists, lift tables, powered mobile equipment and lift trucks, overhead handling and lifting equipment, and vacuum lift devices that provide a mechanical advantage during the MMH task. The NIOSH ergonomic guideline for MMH contains extensive information about mechanical aids that can be used to reduce risk of MSDs for workers who must perform MMH tasks [37]. Figure 12.3a through f shows a range of mechanical aids that can be used to reduce the risk of MSDs from MMH tasks.

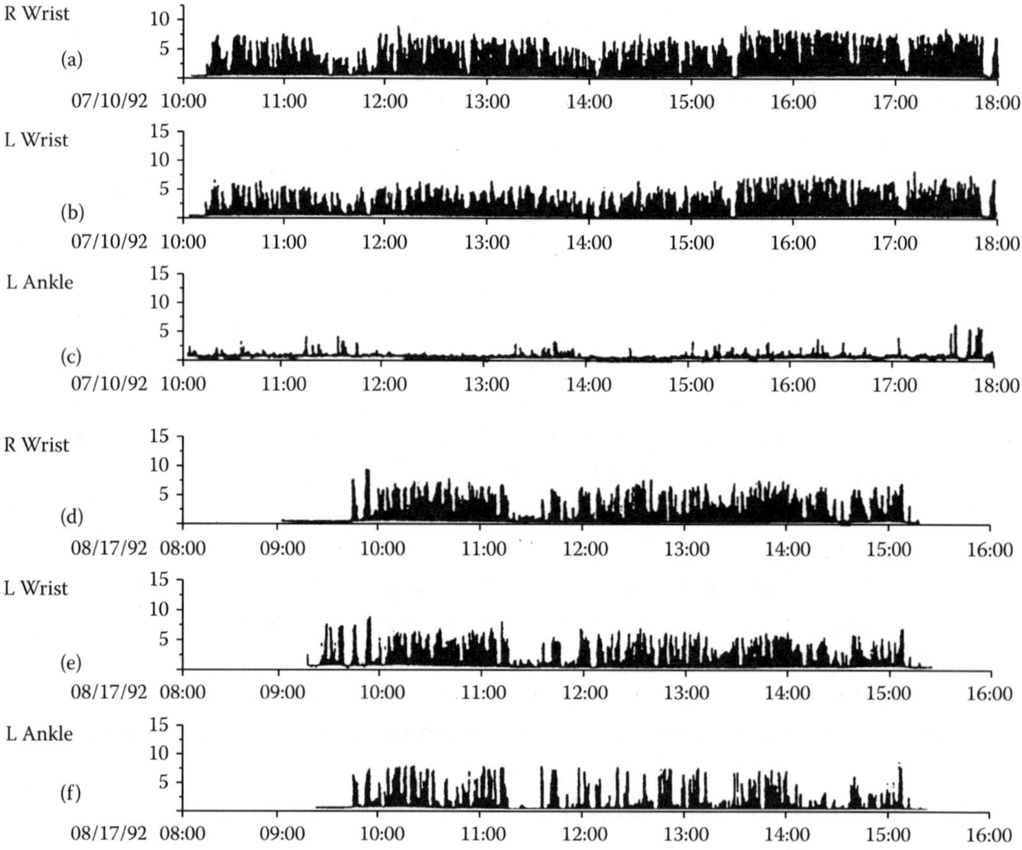


FIGURE 12.3 Temporal histogram plots from activity monitor show the extent of dynamic movement as a function of time on 07/10/92: (a) right wrist, (b) left wirst, and (c) left ankle and on 08/17/92: (d) right wrist, (e) left wrist, and (f) left ankle.

#### *12.5.2.3 Job Modification through Ergonomic Design*

The use of job modification through ergonomic design has been a basic tenet of occupational safety and health practice for achieving hazard control in preference to other methods such as worker selection and testing, training in safe work practices, or the use of personal protective equipment (PPE), which are less reliable and often less effective. Tasks and tools are modified by using ergonomic principles to reduce the effects of biomechanical stress. For MMH jobs, ergonomic design or redesign may be accomplished by modifying the job layout and procedures to reduce bending, twisting, horizontal extensions, heavy lifting, forceful exertions, and repetitive motions.

The ergonomic approach is largely based on the assumption that work activities involving less weight, less repetition, less time in awkward postures, and less applied force are less likely to cause injuries and disorders. The ergonomic approach is desirable because it seeks to eliminate potential sources of problems within the job itself. Ergonomics also seeks to make safe work practices a natural result of the tool and worksite design.

There are at least four advantages to adopting an ergonomic design/redesign strategy; these include the following:

1. An ergonomic approach does not depend on specific worker capabilities such as strength and conditioning.
2. Human biological factors and their variations are accounted for in ergonomic approaches because there is consideration of design data accommodating large segments of the population.
3. Compared to worker-directed approaches, which may be temporary, ergonomic intervention is relatively permanent because the workplace hazard is eliminated.
4. To the extent that sources of biomechanical stress at the worksite are eliminated or significantly reduced, the difficult issues involving potential worker discrimination, lifestyle modifications, or attempts at changing behavioral patterns of workers (training and education issues) will be of lesser practical significance.

Although this approach is preferred over worker-directed approaches, it should be noted that no single job modification will always be sufficient; numerous adjustments may be required to properly redesign a job.

### 12.5.3 Worker-Directed Approaches

#### *12.5.3.1 Training and Education*

According to OSHA [11], the purpose of training and education is to "ensure that employees are sufficiently informed about the ergonomic hazards to which they may be exposed and thus are able to participate actively in their own protection." Training should include general instruction on the types of injuries that may occur, what risk factors may contribute to these injuries, how to recognize and report symptoms, and how to prevent these injuries. Training should also include job-specific instruction on the proper use of essential MMH equipment and techniques for the specific task or job. The training program should include the following individuals:

All exposed workers

Supervisors

Managers

Engineers and maintenance personnel

Health care providers

The term *training* has also been used to describe two distinctly different approaches to injury prevention and control: (1) instructional training in "safe" material handling and (2) fitness training (e.g., conditioning, strengthening, or work hardening). The basic premise of instructional training in safe materials handling is that people can safely handle greater loads when they perform the task correctly than if they perform the task incorrectly. Fitness training, on the other hand, is based on the premise that people can safely handle greater loads when their strength or aerobic capacity is increased.

12.5.3.1.1 Instructional Training Although the use of instructional training is fundamentally sound and should be encouraged, there are some potential problems in its application. First, care must be taken to ensure that appropriate, scientifically based work practices are being taught. This is not always the case. In some training courses, for example, workers are commonly taught to lift with a straight back and bent legs (i.e., squat posture) rather than bending over to pick up the load (i.e., stoop posture). There is evidence, however, that this recommendation does not always provide the "safest" lifting style [36]. From a biomechanical perspective, for example, the horizontal distance of the load from the worker may be more important in determining spinal loading forces than the amount of forward bending. For this reason, NIOSH does not recommend a single, correct lifting style for all manual lifting tasks, but rather suggests that a free-style lift, in which workers choose whatever style they prefer, may be appropriate in most instances. Second, the instructional approach relies on the worker's ability to comply with a set of recommended practices that may be forgotten or changed from time to time.

Regardless of the potential problems associated with MMH training, all workers who perform MMH activities should receive basic instructional training in the recognition of hazardous tasks and should have a thorough knowledge of what to do when a hazardous task is identified. Furthermore, the instructional training should provide information to workers on how they can become involved in the process of preventing and controlling injuries on the job.

12.5.3.1.2 Fitness Training Unlike instructional training, the basic premise of fitness training is that a worker's risk of injury will decrease if his strength or fitness increases. Although this seems like a reasonable assumption, it is not easy to determine the relationship between an individual's strength and his risk of MSD. Certainly a worker's capacity to perform heavy work might be increased, but there is controversy about the relationship between worker's strength and risk of injury [42]. For example, stronger workers may not actually be able to tolerate stressful lifting demands any better than weak workers. This issue is discussed in more detail in Section 12.5.3.2. Finally, it is not known how the soft tissues of the body respond to increased loads associated with stronger muscles.

Although both types of training programs have been used to prevent MMH injuries, the effectiveness of training in preventing or controlling injuries is unclear at the present time. Therefore, training programs should be used primarily as a supplement to workplace-directed approaches. Additional information on training programs is provided by Ayoub and Mital [23] and Eastman Kodak [13].

#### *12.5.3.2 Employee Screening*

Some ergonomic experts advocate the use of screening methodologies, which rely on the assessment of one or more physical characteristics of the worker, to select specific workers for certain MMH jobs. In general, screening approaches are designed to (1) identify workers with a high risk of MSDs or (2) screen workers according to some preselected set of strength or endurance criteria in an attempt to match the capacity of the worker to the demands of the job.

12.5.3.2.1 Risk Assessment Screening Attempts to identify workers with a high risk for MSDs have included such activities as spinal radiographs (x-rays), psychological testing, genetic testing, and medical examinations. These approaches are conceptually designed to provide an objective basis for excluding certain individuals from stressful MMH jobs. None of these methodologies, however, has been shown to be reliable in predicting an individual's risk of developing a work-related MSD [35]. It is now widely accepted that the medical risks from radiation associated with radiography far outweigh any potential benefit derived from routine spinal x-ray screening [43]. Although no psychological tests have been found to quantify a worker's risk of overexertion injury, psychological testing may provide an indication of how a worker might respond to a severe injury. Similarly, other than identifying workers with previous low back injury, medical examinations also have failed to reliably identify workers who may have an above-average risk of overexertion injury.

12.5.3.2.2 Physical Capacity Screening Another type of screening approach that has been used to select workers for MMH tasks includes individual testing of physical characteristics such as strength, aerobic capacity, or functional capability. The underlying basis for using tests such as these to screen workers for MMH jobs is the belief that the risk of injury is dependent on the relationship between the capacity of the worker and the demands of the job. When the physical demands of the job exceed the capacity of the worker, then the worker is at risk of developing a musculoskeletal injury (MSD). Thus, it has been proposed that workers should be matched to jobs according to the demands of the job.

A number of studies have been conducted to develop databases of maximum strength capacities (i.e., population averages) that could be used to rate an individual's physical capacity. These studies, however, disagree as to which of the three principal testing methods—isometric (static), isokinetic (constant velocity), or isoinertial (constant acceleration)—is most useful for determining strength capacity guidelines.

Maximum isometric lifting strength (MILS) has been studied and reported extensively in the research literature, has a well-established testing procedure [44,45], and has been reported in some field tests to predict risk of injury [46]. Extensive measures of MILS have been made for various work postures and activities. In one study, for example, the isometric strength of 1239 workers in rubber, aluminum, and electronic component industries was measured [45]. In another study, the standardized isometric strengths (i.e., arm lift, torso lift, and leg lift) of 2178 aircraft manufacturing workers were measured [42], and the employees were followed for more than 4 years to document back pain complaints. The investigators concluded that worker height, weight, age, and gender are poor predictors of standardized isometric strength, a finding that is in agreement with most other studies. They also concluded, however, that standardized isometric strength is a poor predictor of reported back pain (a finding that does not agree with some other studies; cf. [46]). More data are needed to clarify this area of controversy. If isometric strength is a poor predictor of risk of injury, then care should be taken in using the results of isometric strength tests to place workers in stressful MMH jobs.

Some researchers argue that traditional isometric lifting strength measurements, by which thousands of workers have been tested, are limited in assessing what workers can do under dynamic task conditions. These researchers suggest that dynamic strength testing is more appropriate than static strength testing for determining strength capacity [47], because dynamic tests replicate actual job requirements better than static tests. Other researchers claim that isokinetic lifting strength measurements probably have no greater inferential power to predict risk of injury or job performance than any other form of testing [48]. On the other hand, Kroemer [49] suggested that the isoinertial strength testing method is the most appropriate for MMH because it most closely matches actual lifting conditions. Isoinertial methods, however, have not been generally validated regarding their ability to predict the risk of injury [50].

In terms of measurement issues, guidelines have been proposed for assessing dynamic muscle performance. Marras et al. [51] published an ergonomics guide for assessing dynamic measures of low back performance. This guide provides information on elements of dynamic performance, techniques to assess dynamic performance, and relationships between testing techniques and internal forces. Little information is available, however, for other MMH tasks.

#### *12.5.3.3 Personal Protective Equipment*

PPE is defined as devices or items used by workers to protect themselves from recognized hazards such as heat, cold, vibration, and other physical hazards. PPE includes gloves, shoes, arm guards, and protective clothing. Musculoskeletal stressors should be considered when selecting PPE, because the use of protective devices could contribute to extreme postures or excessive force. For example, the use of gloves may cause workers to unnecessarily use greater grip forces than needed to perform the job, which may increase their risk of MSD.

There is little published evidence that certain devices that have been advertised as PPE, such as braces, splints, back belts, and other similar devices, provide any realistic protection from injury for healthy workers performing MMH activities. For this reason, OSHA does not consider them to be PPE [11]. In fact, there is some evidence that prolonged use of back belts by healthy workers may actually increase the risk of low back injury above what it would have been if the workers had not used the back belt [52]. In addition, NIOSH conducted a review of studies aimed at examining the potential reduction of risk of MSDs provided by industrial back belts that was summarized in a 1994 report [53]. The review team concluded that "the effectiveness of using back belts to lessen the risk of back injury among healthy uninjured workers remains unproven." They also indicated that "back belts do not mitigate the hazards to workers posed by repeated lifting, pushing, pulling, twisting, or bending." The report did not recommend the use of back belts to prevent injuries among uninjured workers and they did not consider back belts to be PPE.

Another approach that has been used to protect workers from the risk factors associated with MMH includes a variety of devices designed to force workers to use "good body mechanics." For example, a back inclinometer alarm, which sounds an alarm when the worker's back exceeds a certain flexion angle, was developed to prevent excessive back flexion. These devices, however, have not been shown to reduce the incidence or severity of MMH injuries and cannot replace careful ergonomic job design.

### 12.5.4 Summary

In summary, although there is controversy about what the most appropriate approach should be for preventing MMH-related MSDs, it is clear that solutions do exist. Ideally, workplace-directed approaches should be the first choice, followed by worker-directed approaches when necessary.

## 12.6 CRITICAL REVIEW OF CURRENT STATUS

Few would argue that certain manual MMH activities such as excessive lifting, pushing, pulling, and carrying continue to represent a serious hazard for development of work-related MSDs for many workers. Historically, however, these disorders were simply considered one of the costs of doing business. This is no longer considered an acceptable viewpoint, especially when the staggering costs in human suffering as well as the financial costs are considered. Perhaps such costs have increased to the point where injuries can no longer be tolerated and action will be taken to eliminate hazards in the workplace. Information about how to determine the economic costs versus the benefits of specific ergonomic interventions for MMH problems can be found in another chapter in this book. Regardless of the cost, however, we must come to the realization that these work-related disorders can and must be prevented.

In this chapter, we have attempted to describe the MMH problem and provide an overview of analytical tools useful in preventing work-related MSDs. The analytical tools are needed to identify physical hazards that may result in tissue injury and are effective in reducing the potential for risk of work-related MSDs.

A discussion of medical management of musculoskeletal injuries is presented elsewhere in this book. Special attention, however, must be given to the medical management of LBP, which is the leading health disorder associated with MMH.

## 12.7 FUTURE CONCERNS

Although much is known about the assessment of the physical hazards of MMH, difficult questions remain that require further research. A number of research gaps have been identified by various national expert groups [6,8]. A recent report by Marras et al. [54] succinctly summarized the most important research gaps.

For example,

What level of accuracy is attainable in measuring a worker's maximum safe capacity? This is an important question, because worker selection approaches, which are generally believed to be less effective in reducing MSDs than ergonomic approaches, rely on the assumption that it is possible to accurately determine a worker's safe capacity. Also, how much does a worker's capacity vary from year to year, month to month, day to day, and hour to hour?

If a worker's maximum safe capacity cannot be accurately determined, what are appropriate "margins of safety" for specific MMH jobs? What criteria should be used for determining the acceptable margin of safety? What are the implications of designing jobs for large segments of the population?

How do performance standard or incentive programs affect workers' risk of injury? If performance standards must be used, what criteria should be used to ensure that they are safe?

What is the accuracy of existing assessment tools for the design and evaluation of MMH tasks (e.g., biomechanical, psychophysical, physiological, and integrated models)? Also, what are the limitations of the assessment tools?

What are the relative contributions of the various risk factors and how do the risk factors interact to modify the risk of MSDs resulting from exposure to MMH jobs? For example, how physical and psychosocial factors relate to the development of LBP?

There are also numerous questions remaining concerning methods for preventing MMH injuries. For example,

What criteria should be used to decide when and how much automation is needed in a job?

What is the effectiveness of equipment designed to reduce the stress of MMH problems?

Little has been done to encourage technology transfer between industries. What can be done to encourage the sharing of solutions to similar MMH problems between industries? Databases are needed to provide access to practical, economical solutions to MMH problems for industries in a competitive environment.

The workforce is aging, and more women are employed than ever before. How will workforce changes affect our approach to prevention?

How will be the Americans with Disabilities Act affect future research efforts to develop hazard control solutions?

These questions illustrate the complexity of many of the issues that safety and health experts and industrial leaders must resolve to reduce work-related injuries. As demonstrated in the past, no effort will be successful without the full cooperation of management, organized labor, government, and the workers themselves. The solution must be a team effort with a total commitment to identification and elimination of hazardous materials handling tasks.

Due to space limitations, it has not been possible to discuss all of the important issues related to MMH. We have not, for example, discussed issues such as slip, trip, or fall hazards or certain MMH activities such as one-handed lifting. Information on these issues, however, is provided in Refs. [13,18,23].

## REFERENCES

1. Bureau of Labor Statistics, *Nonfatal Occupational Injuries and Illnesses Requiring Days Away from Work, 2009,* Department of Labor, Bureau of Labor Statistics, retrieved from http://www.bls.gov/news.release/osh2.nr0.htm, 2009.
2. National Safety Council, *Injury Facts® 2011 Edition: A Complete Reference for Injury and Death Statistics*, 1121 Spring Lake Drive Itasca, IL 60143-3201.
3. T. R. Waters, R. B. Dick, J. Davis-Barkley, and E. F. Krieg, A cross-sectional study of risk factors for musculoskeletal symptoms in the workplace using data from the general social survey (GSS), *J. Occup. Environ. Med.* 49:172–184 (2007).
4. T. R. Waters, R. B. Dick, and E. F. Krieg, Trends in work-related musculoskeletal disorders: A comparison of risk factors for symptoms using quality of work life data from the 2002 and 2006 general social survey, *J Occup Environ Med*, 53(9):1013–1024 (2011).
5. M. H. Pope, J. W. Frymoyer, and G. Andersson, *Occupational Low Back Pain,* Praeger, New York, 1984.
6. National Research Council, Institute of Medicine, *Musculoskeletal Disorders and the Workplace: Low Back and Upper Extremities*, National Research Council and Institute of Medicine—Panel on Musculoskeletal Disorders and the Workplace. Commission on Behavioral and Social Sciences and Education, National Academy Press, Washington, DC, 2001.
7. NIOSH, *Musculoskeletal Disorders and Workplace Factor: A Critical Review of Epidemiologic Evidence for Work-Related Musculoskeletal Disorders of the Neck, Upper-Extremity, and Low Back*, B. Bernard (Ed.), U.S. Department of Health and Human Services, Public Health Service, Centers for Disease Control and Prevention, National Institute for Occupational Safety and Health, DHHS (NIOSH) Publication No. 97-141, Cincinnati, OH, 45226, 1997.
8. NIOSH, *National Occupational Research Agenda for Musculoskeletal Disorders Research Topics for the Next Decade: A Report by the NORA Musculoskeletal Disorders Team*, U.S. Department of Health and Human Services, Public Health Service, Centers for Disease Control and Prevention, National Institute for Occupational Safety and Health, DHHS (NIOSH) Publication No. 2001-117, Cincinnati, OH, 45226, 2001.
9. S. A. Ferguson and W. S. Marras, A literature review of low back disorder surveillance measures and risk factors, *Clin. Biomech.* 2:211–226 (1997).
10. T. R. Waters, V. Putz-Anderson, A. Garg, and L. J. Fine, Revised NIOSH equation for the design and evaluation of manual lifting tasks, *Ergonomics* 36(7):749–776 (1993).

11. OSHA, *Ergonomics Program Management Guidelines for Meatpacking Plants,* OSHA Document No. 3123, Occupational Safety and Health Administration, Department of Labor, Washington, DC, 1990.
12. E. Grandjean, *Fitting the Task to the Man,* Taylor & Francis, London, 1982.
13. Eastman Kodak, *Kodak's Ergonomic Design for People at Work,* John Wiley & Sons, Incorporated, Hoboken, NJ, 2003.
14. D. C. Alexander and B. M. Pulat, *Industrial Ergonomics: A Practitioner's Guide,* Industrial Engineering and Management Press, Institute of Industrial Engineers, Norcross, GA, 1985.
15. National Occupational Health and Safety Commission, *Safe Manual Handling: Discussion Paper and Draft Code of Practice,* Australian Government Publishing Service, Canberra, 1986.
16. W. S. Marras and C. M. Sommerich, A three dimensional model of loads on the lumbar spine. Part I. Model structure, *Hum. Factors* 33:123–137 (1991).
17. P. Brinckmann, M. Biggemann, and D. Hilweg, Fatigue fracture of human lumbar vertebrae, *Clin. Biomech.* 1(Suppl. 1):1–23 (1988).
18. D. B. Chaffin, G. B. J. Andersson, and B. Martin, *Occupational Biomechanics,* 4th ed., Wiley, New York, 2006.
19. National Research Council, *Ergonomic Models of Anthropometry, Human Biomechanics, and Operator-Equipment Interfaces, Proceedings of a Workshop,* K. H. E. Kroemer, S. T. Snook, S. K. Meadows, and S. Deutsch, Eds., National Research Council, Washington, DC, 1986.
20. A. Garg, *The Biomechanical Basis for Manual Lifting Guidelines,* National Technical Information Service, Rep. No. 91-222-711, Alexandria, Virginia, 1991.
21. S. S. Stevens, The psychophysics of sensory function, *Am. Sci.* 48:226–253 (1960).
22. S. H. Snook and V. M. Ciriello, The design of manual handling tasks: Revised tables of maximum acceptable weights and forces, *Ergonomics* 34:1197–1213 (1991).
23. M. M. Ayoub and A. Mital, *Manual Materials Handling,* Taylor & Francis, London, 1989.
24. G. Borg, Psychophysical scaling with applications in physical work and the perception of exertion, *Scand. J. Work Environ. Health* 16(Suppl. 1):55–58 (1990).
25. E. N. Corlett and R. P. Bishop, A technique for assessing postural discomfort, *Ergonomics* 19:175–182 (1976).
26. S. H. Rodgers, J. W. Yates, and A. Garg, *The Physiological Basis for Manual Lifting Guidelines,* National Technical Information Service, Rep. No. 91-227-330, Alexandria, Virginia, 1991.
27. A. Garg, *A Metabolic Rate Prediction Model for Manual Materials Handling Jobs,* PhD Dissertation, University of Michigan, Ann Arbor, MI, 1976.
28. P. O. Astrand and K. Rodahl, *Textbook of Work Physiology,* 3rd edn., McGraw-Hill, New York, 1986.
29. M. M. Ayoub, N. J. Bethea, S. Deivanayagam, S. S. Asfour, G. M. Bakken, and D. Liles, *Determination and Modeling of Lifting Capacity, Final Report,* Department of Health and Human Services, National Institute of Occupational Safety and Health, Grant No. 5-RO1-OH-00545002, 1978.
30. W. S. Marras and F. A. Fathallah, Accuracy of a three dimensional lumbar motion monitor for recording dynamic trunk motion characteristics, *Int. J. Ind. Ergon.* 9(1):75–87 (1992).
31. W. S. Marras and R. W. Schoenmarklin, Wrist motions in industry, *Ergonomics* 36(4):341–351 (1993).
32. W. S. Marras, S. A. Lavender, S. E. Leurgans, S. L. Rajulu, W. G. Allread, F. A. Fathallah, and S. A. Ferguson, The role of dynamic three-dimensional trunk motion in occupationally-related low back disorders. The effects of workplace factors, trunk position, and trunk motion characteristics on risk of injury. *Spine* 18:617–628 (1993).
33. W. S. Marras, S. A. Lavender, S. E. Leurgans, F. A. Fathallah, S. A. Ferguson, W. G. Allread, and S. L. Rajulu, Biomechanical risk factors for occupationally related low back disorders, *Ergonomics* 38:377–410 (1995).

34. K. A. Grant, T. L. Galinsky, and P. W. Johnson, Use of the actigraph for objective quantification of hand/wrist activity in repetitive work, *Proceedings of 37th Annual Human Factors Meeting*, Seattle, WA, 1993.
35. S. H. Snook, R. A. Capanelli, and J. W. Hart, A study of three preventive approaches to low back injury, *J. Occup. Med.* 20(7):478–481 (1978).
36. A. Garg, What basis exists for training workers in "correct" lifting technique? in *The Ergonomics of Manual Work*, W. S. Marras, W. Karwowski, J. L. Smith, and L. Pacholski, Eds., Taylor & Francis, London, 1993.
37. NIOSH, *Ergonomics Guidelines for Manual Material Handling*, DHHS (NIOSH) Publication No. 2007-131. National Institute for Occupational Safety and Health, Cincinnati, OH, 2007.
38. OSHA, *Guidelines for Nursing Homes, Ergonomics for the Prevention of Musculoskeletal Disorders*, U.S. Department of Labor, Occupational Safety and Health Administration, OSHA Document No. 3182-3R. Occupational Safety and Health Administration, Department of Labor, Washington, DC, 2003 (revised March 2009).
39. OSHA, *Guidelines for Retail Grocery Stores, Ergonomics for the Prevention of Musculoskeletal Disorders*, U.S. Department of Labor, Occupational Safety and Health Administration, OSHA Document No. OSHA 3192-06N. Occupational Safety and Health Administration, Department of Labor, Washington, DC, 2004.
40. OSHA, *Guidelines for Poultry Processing, Ergonomics for the Prevention of Musculoskeletal Disorders*, U.S. Department of Labor, Occupational Safety and Health Administration, OSHA Document No. 3213-09N. Occupational Safety and Health Administration, Department of Labor, Washington, DC, 2004.
41. OSHA, *Guidelines for Shipyards, Ergonomics for the Prevention of Musculoskeletal Disorders*, U.S. Department of Labor, OSHA Document No. 3341-03N. Occupational Safety and Health Administration, Department of Labor, Washington, DC, 2008.
42. M. C. Batti'e, S. J. Bigos, L. D. Fisher, T. H. Hansson, M. E. Jones, and M. D. Wortley, Isometric lifting strength as a predictor of industrial back pain reports, *Spine* 14(8):851–856 (1989).
43. NIOSH, *Work Practices Guide for Manual Lifting*, NIOSH Tech. Rep. No. 81–122. U.S. Department Health and Human Services, National Institute of Occupational Safety and Health, Cincinnati, OH, 1981.
44. D. B. Chaffin, Ergonomics guide for the assessment of human static strength, *Am. Ind. Hyg. Assoc. J.* 36:505–511 (1975).
45. W. M. Keyserling, G. D. Herein, and D. B. Chaffin, An analysis of selected work muscle strength, *Proceedings of 22nd Annual Meeting of the Human Factors Society*, Detroit, MI, 1978.
46. W. M. Keyserling, G. D. Herein, and D. B. Chaffin, Isometric strength testing as a means of controlling medical incidents on strenuous jobs, *J. Occup. Med.* 22(5):332–336 (1980).
47. K. H. E. Kroemer, Testing individual capability to lift material: Repeatability of a dynamic test compared with static testing, *J. Safety Res.* 3(6):4–7 (1985).
48. J. M. Rothstein, R. L. Lamb, and T. P. Mayhew, Clinical uses of isokinetic measurements, *Phys. Ther.* 67(12):1840–1844 (1987).
49. K. H. E. Kroemer, An isoinertial technique to assess individual lifting capacity, *Hum. Factors* 25(5):493–506 (1983).
50. K. H. E. Kroemer, Matching individuals to the job can reduce manual labor injuries, *Occup. Safety Health News Dig.* 3(6):4–7 (1987).
51. W. S. Marras, J. D. McGlothlin, D. R. McIntyre, M. Nordin, and K. H. E. Kroemer, *Dynamic Measures of Low Back Performance: An Ergonomics Guide*, American Industrial Hygiene Association, Fairfax, VA, 1993.

52. C. R. Redell, J. J. Congleton, R. D. Huchingson, and J. F. Montgomery, An evaluation of a weightlifting belt and back injury prevention training class for airline baggage handlers, *Appl. Ergon.* 23(5):319–329 (1992).
53. NIOSH, *Workplace Use of Back Belts Review and Recommendations*, NIOSH Publication No. 1994-122, DHHS. National Institute for Occupational Safety and Health, Cincinnati, OH, 1994.
54. W. S. Marras, R. G. Cutlip, S. E. Burt, and T. R. Waters, National occupational research agenda (NORA) future directions in occupational musculoskeletal disorder health research, *Appl. Ergon.* 40:15–22 (2009).

CHAPTER 13

# Manual Materials Assist Devices

Jeffrey C. Woldstad and Roderick J. Reasor (deceased)

CONTENTS



## 13.1 INTRODUCTION

Many manual materials handling tasks pose a significant musculoskeletal risk to workers as they are currently performed. The analytical methods discussed in earlier chapters of this book provide the means to identify these potentially harmful tasks and to prioritize them for intervention efforts. Unfortunately, ergonomic efforts to reduce the risk associated with manual materials handling tasks usually require fundamental changes to the work area and the component work tasks. These changes are often expensive, and, if not done carefully, they can create new engineering and ergonomic problems.

This chapter addresses one of the most common and least expensive solutions proposed to address harmful manual materials handling problems, that being the use of a manual materials assist device. For the purposes of this chapter, manual materials assist devices are defined as devices that can be used to assist the worker in performing some component of a manual materials handling task. This includes work components related to lifting, lowering, carrying, and holding a load. We restrict our discussion to devices intended to assist workers in performing these work activities as opposed to automation intended to replace human workers.

The intent of this chapter is to review both the ergonomic and manufacturing issues associated with the selection of a manual materials assist device. To accomplish this, we first review some of the many different types of devices that can be classified as manual materials assist devices using the aforementioned definition. Then, we discuss the ergonomic issues that should be considered when selecting and installing a manual materials assist device.

## 13.2 BACKGROUND AND SIGNIFICANCE TO OCCUPATIONAL ERGONOMICS

The use of manual materials assist devices is recommended by NIOSH [1] and others [2–5] as an engineering solution to potentially harmful manual materials handling tasks. Practitioners new to occupational ergonomics often mistakenly assume that installing such a device immediately solves the ergonomics and biomechanics problems associated with a work task. Although the installation of these devices is often successful, just as often devices are disliked by workers, underutilized, and eventually discarded in favor of the old manual methods. In our experience, it is rare to go into a large U.S. industrial facility and not see one or more manual materials assist devices lying unused in a corner of the plant.

To ensure that manual materials assist devices are used to their fullest potential, it is necessary for the ergonomist to go beyond just suggesting that such a device be employed. Ergonomists must become actively involved in the design, selection, and installation of these systems. This chapter is an initial attempt at providing some basis for this process.

## 13.3 TYPES OF MANUAL MATERIALS ASSIST DEVICES

A variety of equipment is available to assist operators with the manual movement of materials in the workplace. This equipment is made by many companies in the United States and throughout the world. Although our discussion will not be able to adequately reflect this diversity, three general categories of equipment will be briefly reviewed. These are industrial manipulators, lift tables, and cart and trolley systems.

### 13.3.1 Industrial Manipulators

Industrial manipulators are often employed to assist operators in lifting, supporting, and moving heavy, awkward, or fragile loads [6]. The manipulator grasps the load and then uses pneumatic, hydraulic, or vacuum force to overcome the effects of gravity. An operator is then able to move the load to a desired location while maintaining control of the load throughout its movement. Specialized manipulators can permit the operator to roll the load over, tilt the load, and pour or dump the load.

There are a large variety of industrial manipulators available on the market (see Figures 13.1 through 13.4 for some examples). These devices are typically capable of handling loads up to 1700 lb within a work area of 8–20 ft. Of the many different designs, three basic types of manipulators are often identified: triaxial manipulators, balancing hoists, and articulated jib cranes [7]. Each of these is briefly discussed in the following.

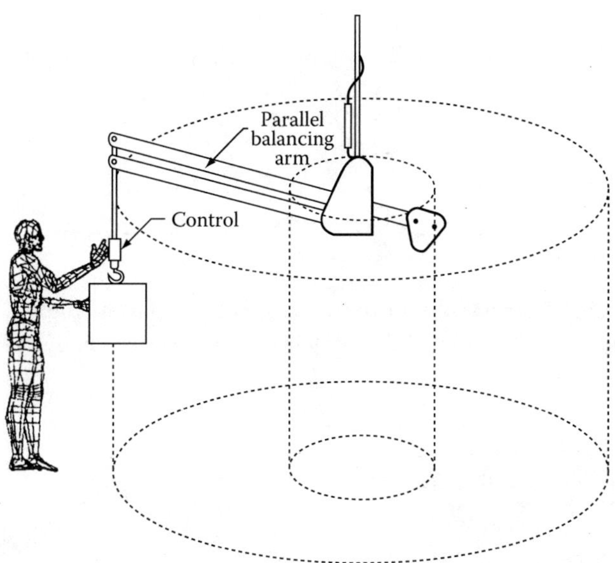


FIGURE 13.1 Triaxial manipulator. (From Castleberry, G.A., Industrial manipulators, in *Material Handling Handbook*, 2nd edn., Kulwiec, R.A., Ed., Wiley, New York, 1985.)

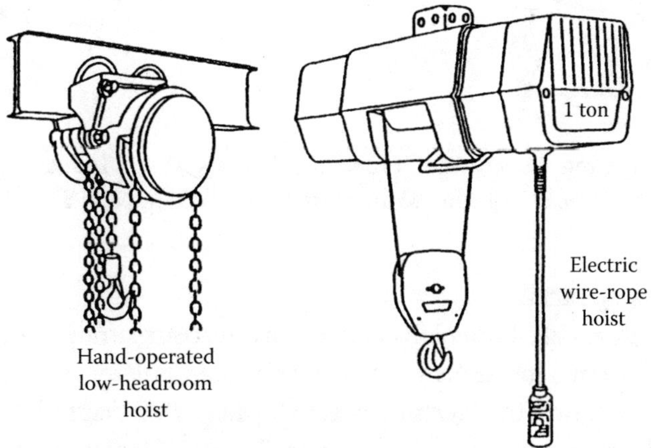


FIGURE 13.2 Trolley-suspended hoists. (From Konz, S.A., *Facility Design: Manufacturing Engineering*, 2nd edn., Publishing Horizons, Scottsdale, AZ, 1994.)

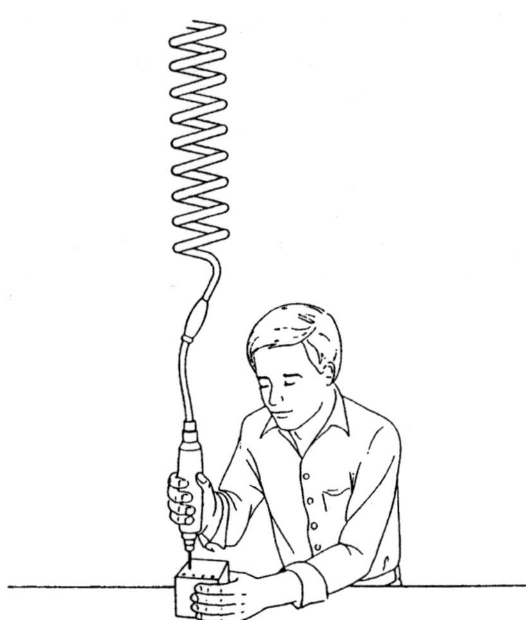

FIGURE 13.3 Balancing hoist—tool and fixture suspension. (From Konz, S.A., *Facility Design: Manufacturing Engineering*, 2nd edn., Publishing Horizons, Scottsdale, AZ, 1994.)

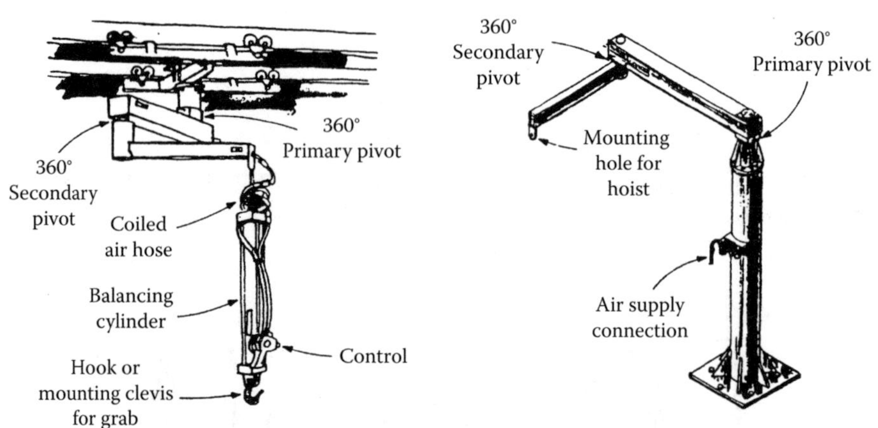


FIGURE 13.4 Articulating jib cranes. (From Castleberry, G.A., Industrial manipulators, in *Material Handling Handbook*, 2nd edn., Kulwiec, R.A., Ed., Wiley, New York, 1985.)

#### *13.3.1.1 Triaxial Manipulators*

A triaxial manipulator is used when objects vary in size, configuration, or weight and when the pick-up and put-down locations vary. A triaxial manipulator has structural members arranged in a parallelogram that permit loads to be moved vertically, horizontally, or diagonally. They can be fixed overhead as shown in Figure 13.1 or mounted to an overhead trolley system. Besides supporting the load, often these devices are effective in extending an operator's reach capability and permit more precise load positioning.

There are several factors that must be considered when selecting a triaxial manipulator for a given application [9]. These include the following:

*Headroom requirements.* Typically, 12 ft of headroom is needed unless a low-profile manipulator is used.

*Reach and lifting capacity.* Lifting capacity will vary with the design of the device and should be closely tailored to the application. Reach is a function of lifting capacity and is defined as the total horizontal travel.

*Reach-in capability.* The manipulator may need the ability to be extended beyond the normal work envelope for special applications.

*Side-loading capability.* The ability to insert loads from the side can be limited; if required, the capability of a given unit must be determined.

##### *13.3.1.2 Balancing Hoists*

Balancing hoists float a load (overcome the effects of gravity) much like a triaxial manipulator, but they have no reach-in capability. Their primary function is load positioning by an operator. They are typically mounted on an overhead trolley running on a jib or bridge crane and can handle loads of up to 1400 lb. Figure 13.2 illustrates a common trolley-suspended hand hoist. Figure 13.3 shows a simple tool and fixture suspension (where the tool remains on the hook all the time).

##### *13.3.1.3 Articulated Jib Cranes*

A jib crane has a rotating boom attached to a vertical column. The vertical column can be mounted from an overhead trolley, on a floor pedestal, or from a wall bracket. Jib cranes typically use hoists for lifting and lowering a load. An articulated jib crane has an additional pivoting arm that provides additional reach capabilities as shown in Figure 13.4. Articulating jib cranes have a working area of up to 40 ft in diameter. They are especially useful when overhead space is limited, when loads are heavy or awkward, or when smooth, lateral movement along a conveyor is essential [9].

##### *13.3.1.4 End-of-Arm Tooling*

Industrial manipulators are used to lift a wide variety of loads. These include sacks of flour, TV tubes, computers, carpeting, heavy steel coils, furniture, and more. Thus, the end-of-arm tooling is almost always custom designed for the application. Common end-of-arm tooling includes vacuum cups, clamps, magnets, C-hooks, scissor grabs, core grips, and grabs. Figure 13.5 illustrates the variety of end-of-arm tooling commonly used.

### 13.3.2 Lift Tables

Lift tables provide a versatile means to lift, lower, or hold material. They are often an effective means of reducing the horizontal and vertical distances associated with lifting tasks (see Chapter 4, Section 4.2 and Chapter 3, Section 3.6). Lift tables use a mechanical, hydraulic, or pneumatic mechanism to move a platform up or down (see Figure 13.6). The lifting

FIGURE 13.5 End-of-arm tooling for industrial manipulators. (From Torok, D.B., *Mater. Handl. Eng.*, 48(1), 1991.)

FIGURE 13.6 Lift table. (From Quinn, D.J., Lift tables, in *Material Handling Handbook*, 2nd edn., Kulwiec, R.A., Ed., Wiley, New York, 1985.)

mechanism most commonly operates in a scissor-like manner, but it may act directly on the platform. Lift tables are available in a wide variety of sizes (table tops from 3 to 240 sq ft) and capacities (200–100,000 lb). Table tops can be simple flat surfaces, or they can be designed to tilt, rotate, or contain some special-purpose superstructure that supports the handling of tools, coils, and special equipment (such as wheels, rollers, or ball transfers).

Lift tables have been used in a variety of applications. These fall into five major areas: stacking and unstacking, lift and tilt, work positioning, upenders and downenders, and dock lifts. Each of these application areas is briefly discussed here:

*Stacking and unstacking.* Product is brought to the work height and then stacked or unstacked by hand. Stacking is easier if the top surface of the table or stacked pile is slightly below the work height. Conversely, unstacking is easier if the stack height is slightly above the work height.

*Lift and tilt.* Primarily used for dumping or feeding small, individual materials in bulk or mass.

*Work positioning.* Work is positioned at the ideal height for a given operator.

*Upenders and downenders.* Change the vertical or horizontal axis of a product or process.

*Dock lifts.* Equalize the height difference between a loading/unloading dock and a shipping/receiving container (e.g., truck or railcar).

### 13.3.3 Industrial Hand Trucks

Industrial hand trucks are widely used to assist workers in transporting or carrying the load from one location to another. They represent a versatile method of performing intermittent movements over a variable path. Figure 13.7 shows examples of several types of industrial trucks.

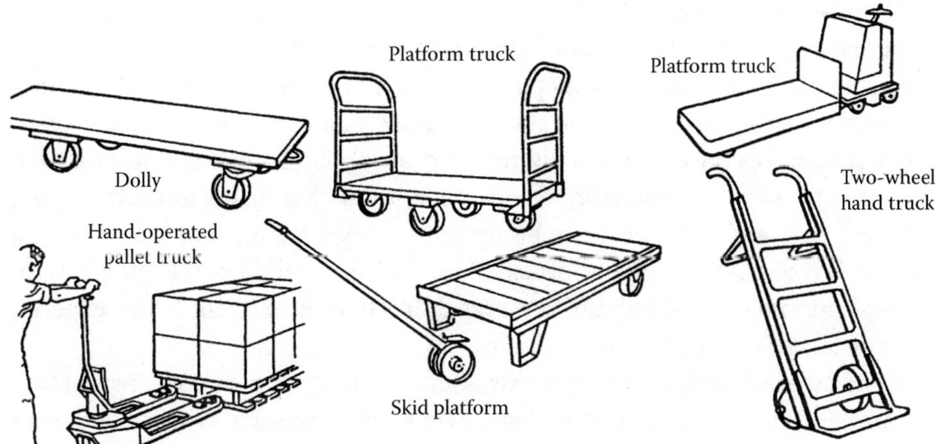


FIGURE 13.7 Industrial hand trucks. (Adapted from Kulwiec, R.A., *Advanced Material Handling*, Material Handling Institute, Charlotte, NC, 1983.)

Hand trucks are also known as floor trucks. Hand trucks are used in warehousing, manufacturing, shipping, and distribution. They are most commonly used when the travel distance is relatively short, the volume and frequency of moves are low, physical limitations (narrow aisles, doors, etc.) prohibit the use of other types of equipment, or low initial cost and maintenance requirements are constraining considerations. Hand trucks can be classified as two-wheel hand trucks, multiple-wheel floor trucks, and hand lift trucks.

A two-wheel hand truck is the most common and most widely used type of hand truck. It basically operates as a lever with wheels and is used to move materials that are too heavy or too awkward to be carried. Typical capacity is up to 1000 lb. Examples include barrel trucks, bag trucks, beverage trucks, appliance trucks, and a wide variety of special-purpose trucks (cylinder trucks, cotton trucks, grain trucks, etc.).

Multiple-wheel floor trucks typically consist of a rectangular, load-carrying platform with four or six wheels. They are used to transport up to 4000 lb. These platforms may have attached superstructures ranging from four corner stakes to completely enclosed containers. The platforms can be interconnected and used in towline operations. Examples include dollies, platform trucks, semi-live skid platforms, tilt- and nontilt-type trucks, and wagon-type trucks.

Hand lift trucks contain a wheeled load carrier and a lifting device. The load carrier (a platform or forks) can be rolled under a unit load (pallet or skid). The lifting device can be hydraulically or mechanically operated and is typically used to raise the unit load only 2–3 in. above the floor for transport. Hand lift trucks are commonly used to handle 2500–5000 lb.

## 13.4 ERGONOMICS ISSUES IN DESIGN, SELECTION, AND INSTALLATION

This section describes some of the ergonomic issues relevant to the design, selection, and installation of manual materials assist devices. The discussion is directed primarily at different types of industrial manipulators and hand trucks. Many of the issues relevant to the installation of lift tables have been covered in previous chapters of this text (see Chapter 2).

### 13.4.1 Horizontal Pushing and Pulling

Most manual materials assist devices function to support the load vertically during lift, carry, position, and lower tasks. By supporting the load in this manner, they eliminate the static force required to counteract the effects of gravity on the load. However, to manipulate these devices, operators are usually required to supply the horizontal push and pull forces needed to move the load from one position to another. Unfortunately, the magnitude of these push and pull forces can become sizable depending upon the configuration of the assist device and the work performed. Peak push and pull forces ranging up to 500 N and 80% of the subject's maximum force capability have been recorded in experiments with manual materials assist devices [12,13].

Similar to other whole body strength exertions, the ability of operators to generate horizontal push and pull forces is highly dependent on factors such as individual strength capability and body posture. In attempting to establish limits for push and pull activities, several methods have been used, including strength evaluations [14,15], intra-abdominal pressure [16–18], biomechanical models [19,20], and psychophysical methods [21,22].

For those engineers interested in establishing push and pull force limits, the article by Snook and Ciriello [22] provides a relatively complete table of recommended maximum push and pull force limits based on a number of psychophysical evaluations.

### 13.4.2 Floor Surface Coefficient of Friction

The definition of coefficient of friction (COF) and some requirements for various tasks have been addressed earlier in this text (see Chapter 5). For pushing and pulling activities, it is clear that the COF should exceed the ratio of the horizontal push or pull force and the operator's body weight [23]. As an example, if we have an operator who weighs 170 lb and the operator is required to produce 100 lb of horizontal push force, then a minimum COF of 100/170 = 0.58 would be reasonable.

It might seem that this value would be an absolute minimum requirement if 100 lb of force is to be produced. However, high push and pull forces can be generated with relatively low COF values if the handle used for pushing and pulling is high enough [24]. With high handle heights, operators are able to generate a force vector with both an upward component and a horizontal component. The vertical force component will serve to increase the force normal to the floor surface and, as a result, the frictional force. Using this method, operators are able to generate much higher initial push or pull forces than anticipated; unfortunately, the same procedure does not work very well in stopping the device.

In an investigation of the required COF for push and pull tasks, Redfern and Andres [25] found that increased cart resistance and faster pushing speeds produced a statistically significant increase in the required COF. However, they were unable to find any statistically significant effect of handle height or exertion direction (push or pull). In addition, they note that the required COF was extremely variable between the different subjects tested. Clearly, more experimental evaluations are required in this area before concrete COF guidelines can be established for pushing and pulling tasks.

In summary, push and pull tasks are clearly tasks that demand a high COF between the shoe and the floor surface. When installing manual materials assist devices, ergonomists should evaluate the slipperiness of the floor surface and consider the installation of one of the many high-traction flooring surfaces available on the market. Such precautions are especially important if high push and pull forces will be required, if pushing or pulling will be done along an inclined surface, or if floor contaminants are present in the work area (see Chapter 5).

### 13.4.3 System Mass and Inertia

Several recent studies have shown that the force generated in using a manual materials assist device and the operator's perception of task difficulty are substantially increased by the inertial mass of the load being moved [12,13]. This is a particular concern with manual materials assist devices because of the large loads that are sometimes transported with the systems and the mass of the devices themselves.

The inertial mass of a device can be thought of as the characteristic of the device that resists motion or acceleration when a force is applied. Technically, mass is the quantitative measure of inertia; however, throughout this chapter, we use the terms mass, inertia, and inertial mass interchangeably. The larger the mass of the device, the more force that will be required to produce a constant acceleration. This relationship is quantified in Newton's second law of motion:

$$F = m \times a, \tag{13.1}$$

where

$F$ is the applied force
$m$ is the inertial mass of the system
$a$ is the acceleration [26]

In practical terms, this means that to keep push and pull forces within reasonable bounds, operators must accelerate and decelerate an assist device with a large inertial mass much more slowly than a similar device with a small inertial mass. In many cases, operators simply refuse to control their motions to the extent required, and high acceleration and deceleration forces result. In still other cases, time standards make a reduction in movement time impossible.

In an investigation of these issues, Woldstad and Chaffin [13] recently conducted an experiment that measured the peak hand forces associated with the movement of a small overhead balancing hoist. Four inertial loads were considered, ranging from 81.7 to 217.9 kg. The experimental task had seven inexperienced operators push the load from a standstill and position it over a specified target. Peak acceleration and deceleration forces for this experiment are shown for the 2.74 m condition in Figure 13.8. Note that the forces for both acceleration and deceleration were much higher for the larger inertial loads. Movement times, shown in Figure 13.9, indicated that subjects did slow down for the higher inertia trials but not to the extent required to maintain moderate acceleration and deceleration forces. In fact, a careful review of the average forces demonstrated in Figure 13.8 will show

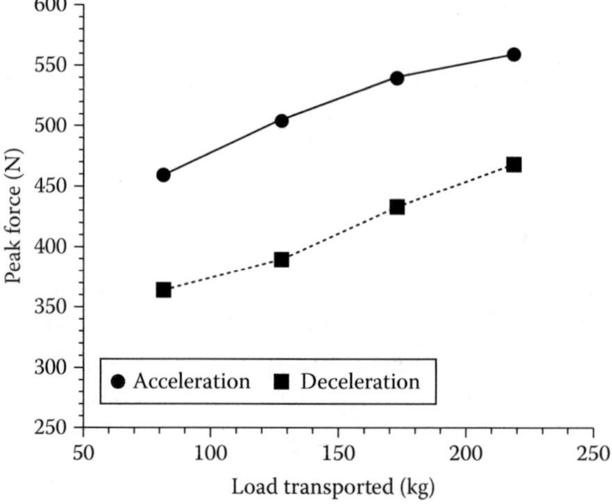


FIGURE 13.8 Peak acceleration and deceleration forces (in newtons) for hoist movements using different loads. (From Woldstad, J.C. and Chaffin, D.B., *IIE Trans.*, in press.)

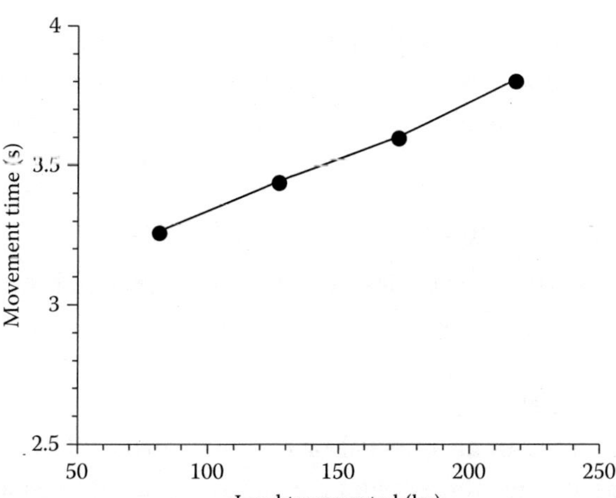


FIGURE 13.9 Movement times (s) for hoist movements using various loads. (From Woldstad, J.C. and Chaffin, D.B., *IIE Trans.*, in press.)

that at the highest inertial loads, the deceleration forces approached maximum recommended levels for push and pull activities [22].

A similar set of experiments were recently conducted by Resnick [12]. The first two of these experiments used a triaxial manipulator assist device and measured peak hand force during acceleration and deceleration for symmetric and asymmetric positioning tasks. In addition to hand force, the experiment measured the subjects' perceived level of exertion using the Borg CR-10 scale. The results of manipulating the inertial mass of the system were mixed for these experiments. For sagittally symmetric manipulations, increasing the load manipulated (loads considered were 0, 22.5, 45, and 67.5 kg) significantly increased both the peak push force and the peak pull force but did not affect the CR-10 rating. For asymmetric manipulations, changes to the load had no significant effect on any of the three measures. The third experiment reported by Resnick considered three different types of assist devices: a triaxial manipulator, a hoist with a fixed pivot, and a hoist on an overhead rail. A 10 kg and a 30 kg load were used in this experiment. The results showed significant increases in both push and pull forces for the heavier load. This was especially true for the hoist with the fixed pivot, where average push forces increased by approximately 220% and average pull forces increased by approximately 200%.

Although the results reported earlier are from only a small set of investigations, they coincide well with the experience of many operators and engineers. Operators often complain of fatigue when moving systems with large inertial masses, especially if the load must be moved rapidly or over a large distance. Similarly, performance times are often much longer for large mass systems. To avoid these problems, every attempt should be made to reduce the inertial mass of manual materials assist devices. This includes avoiding devices that use counterweights to compensate for the load and selecting a device that is specifically designed for the size of load to be handled. In addition, both operators and industrial engineers should recognize that large inertial loads have the potential for creating large acceleration and deceleration forces. Operators need to be trained to accelerate and decelerate large loads slowly and smoothly, even though this may require longer movement and cycle times. At the same time, industrial engineers may need to consider the increase in time required to safely manipulate large loads when rating performance and developing time standards.

### 13.4.4 System Friction or Resistance to Movement

Frictional forces result when objects come into contact with each other. They are the forces that resist a sliding of one object over another. In manual materials assist devices, there are frictional forces that resist motion at any point where two components come into contact. This includes joints and axles where rotations occur, and also areas where wheels in the device come into contact with an internal or external surface. Frictional forces are often separated into a static frictional force component and a dynamic frictional force component. For manual materials assist devices, the static frictional force will resist any attempt to put the assist device in motion from a stopped or resting position, and the dynamic frictional force will resist the motion of the device as it is moved. For a number of reasons, the static frictional component is usually much larger than the dynamic component (see Chapter 5).

To evaluate the effects of frictional forces on manual materials assist devices, the movement of these devices must be considered in separate phases or components. Consider that the typical manual movement of an assist device consists of an acceleration phase during which the operator accelerates the device to walking speed, a transport phase during which the operator provides only the amount of force necessary to maintain a walking velocity, a deceleration phase during which the operator provides the required force to stop the assist device at the intended position or target, and a final adjustment phase during which the operator provides any final positioning of the load. As we argued in the previous section, increasing the inertial mass of the assist device will increase the force used for all of these different movement components, except perhaps the middle transport component. The effect of frictional forces, however, will be dramatically different for different phases of the movement.

For the acceleration phase of the movement, the static frictional force will resist initial movement of the device, and the dynamic frictional force will resist the attempt by the operator to accelerate the load. If the operator has a desired acceleration profile, then frictional forces will increase the force used during acceleration. For the transport phase of the movement, the dynamic frictional forces will determine the force that must be used by the operator. Because we usually assume that the operator is not accelerating or decelerating the load during the phase, the only force that is required to maintain the desired velocity is the force needed to oppose the frictional forces.

Although the dynamic frictional forces will increase the effort required of the operator to accelerate and transport the manual materials assist device, they will decrease the force required to decelerate the system. For the deceleration phase of the movement, the frictional forces act in the same direction as the forces the operator must supply. The higher the frictional force, the lower the deceleration force. The final positioning phase of the movement is usually composed of several small accelerations and decelerations of the load. Similar to the accelerations and decelerations described, increased resistance or frictional force will increase the force associated with the accelerations and decrease the force associated with deceleration. As a general rule, increased frictional forces will increase the effort associated with the final positioning component. This is of particular concern if the work task requires precise positioning of the load.

Woldstad and Chaffin [13] and Resnick [12] (as described previously in Section 13.4.3) have tested several of these assumptions in the laboratory by measuring the force profiles created for subjects using different manual materials assist devices while manipulating the frictional forces associated with the system. In the first experiment [13], a small overhead balancing hoist with a rigid connection between the overhead trolley and the load was used. The results of this experiment showed that increasing the system friction (in this case the friction of the rail system) from 24.2 to 48.4 N had no significant effect on the force used for accelerating the load but did significantly reduce the force used to decelerate the load. The time taken for these movements under the higher friction condition was not affected for short distance movements (0.9 m) but was significantly longer for longer distance movements (2.7 m).

Resnick [12] used a triaxial manipulator assist device and controlled system friction by using disk brakes at the joints of the articulated arm. The frictional resistance levels used were 0 and 25 N. For sagittally symmetric movements, the results showed significantly higher forces for both acceleration and deceleration and significantly higher subjective exertion ratings (using the Borg CR-10) with the high-friction condition. For sagittally asymmetric movements, the results showed significantly higher acceleration forces and subjective exertion ratings (using the Borg CR-10) but no significant difference in deceleration force. A curious result of these experiments was that, in contrast to the results reported by Woldstad and Chaffin [13], the increase in frictional forces increased or did not affect the deceleration. However, in evaluating the force profiles created by subjects in this experiment, Resnick did not separate the deceleration component of the movement from the final positioning component. Further, he notes that peak deceleration forces occurred during push components of the positioning task as opposed to during the decelerations.

In summary, it is clear that larger frictional forces within manual materials assist devices will increase the push and pull forces generated by workers for the greater part of the work cycle. The exception to this will be during deceleration of the load. Although frictional forces in general should be minimized to decrease the hand forces required, there may be situations where the deceleration of the load is of such concern that system friction is desirable, as when extremely large inertial loads are being transported. In these situations, operators can accelerate the load to a level where they have difficulty controlling and stopping the system within the work area. Increased frictional forces in these situations will decrease the operator's ability to accelerate the system and increase their ability to decelerate it. However, in this same situation, the addition of some type of brake to the device would likely achieve the same level of control while not increasing the effort associated with acceleration.

Although high levels of system friction are clearly not desirable for a majority of applications, the effects of low levels of friction are not yet clear. Evidence from investigations using purely inertial loads suggests that the addition of a small amount of system friction may improve operators' ability to control manual materials assist devices [27]. It is hoped that future investigations will provide us with a better understanding of these issues.

### 13.4.5 Physical Layout and Linkage

The large variety of manual materials assist device designs presents a variety of problems in the physical placement of these devices within the workplace. For many devices, the system resistance and inertial mass are often different for different directions of movement. As an example, consider the simple overhead cable hoist depicted in Figure 13.10. For this device, large movements parallel to the dual overhead rails require that the entire cross-bridge assembly be moved. Because the load is connected to the cross-bridge assembly via a chain or cable as opposed to a rigid connection, the cross-bridge assembly will lag behind the operator as shown in the figure. When the operator stops, the overhead assembly will pass over the top of the operator and produce a delayed jerk as it reaches the end of its travel envelope (see Figure 13.10). This jerk can be substantial if the mass of the cross-bridge assembly is large. For this same device, the dynamics of the movement are much different if the movement is perpendicular to the dual rails. In this situation, only the lift mechanism must be moved and not the entire cross-bridge assembly.

Similar problems exist with many triaxial manipulators. Typical manipulations of these devices often combine rotational and linear translations of components within the device. Device components often have different inertial mass and frictional resistance characteristics in different movement directions, resulting in a dynamic response of the device that is highly dependent on the movement trajectory. In addition, many linkage systems prevent straight-line paths between different points within the work envelope. To help alleviate many of these trajectory problems, some designers have increased the degrees of freedom or number of moving joints in the system. Although this provides more flexibility in the

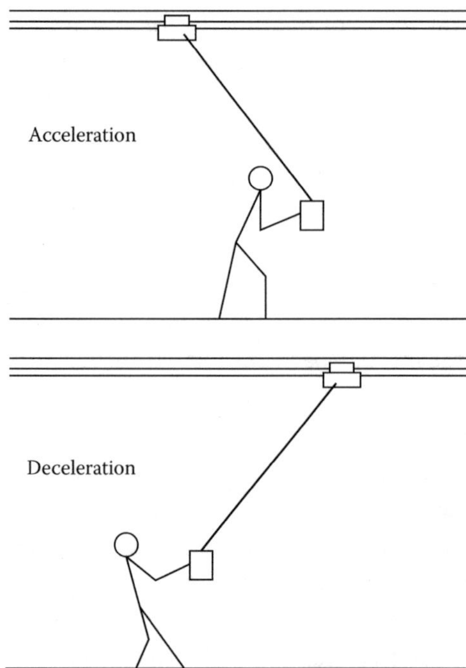


FIGURE 13.10 Position of the overhead carriage of a balancing hoist during acceleration and deceleration.

number of possible trajectories of the device, it often creates a problem similar to that described earlier for overhead hoist systems, where components of the device continue to move after the load has stopped moving. In addition, operators are often not able to predict the response of the device for different force inputs when it has many degrees of freedom, thus creating a safety problem.

There have been no quantitative investigations of the linkage and arrangement problems described earlier. In many cases, such research will be difficult due to the variety of devices on the market. However, on the basis of a number of years of working with these devices, the following practical recommendations are proposed:

1. Use the device with the smallest number of moving components that will still do the job. Every attempt should be made to use the simplest possible design if human operators are to control the device.
2. Attempt to provide as rigid a link as possible between major mass components of the device and the operator's hands. Large moving mass components controlled through multiple-link arm assemblies, cables, or chains should be avoided.
3. When installing a device, identify the trajectories within the work area that will be associated with the work tasks that will use the device. Orient the device in a manner that allows linear movement and minimizes the inertial mass along these trajectories.
4. Carefully consider work height and control design for these devices. More detail on work height considerations can be found in Chapter 2, and more on control design in Chapter 3.
5. Do not be constrained by previous work methods when redesigning a work area to incorporate a manual materials assist device. These devices present new problems, but they also present new opportunities. In most cases, assist devices are able to manipulate much larger loads than human workers. Designing work activities to take advantage of these differences often can result in substantial improvements in productivity.

## 13.5 CRITICAL REVIEW OF CURRENT STATUS

This chapter has attempted to provide a broad overview of some of the issues that should be considered in the design, installation, and use of manual materials assist devices. Experimental research in this area has only just begun. In reviewing the literature, we were able to locate only two sets of experimental investigations that involved the actual human manipulation of manual materials assist devices, and both of these were conducted in the same laboratory. Clearly, more evaluations are necessary before many of the issues discussed in this chapter can be resolved. However, while this work is proceeding, our present lack of information should not inhibit ergonomists from taking an active role in the engineering of these devices. The use of manual materials assist devices has increased dramatically in the last 10 years, and this trend is likely to continue into the foreseeable future. Attention to the effects of these devices on their human operators is critical to successfully integrating this technology into the workplaces of the future.

## 13.6 FUTURE CONCERNS

A great many areas need to be investigated with respect to manual materials assist devices. We believe the following to be the most critical:

1. A better understanding of the biomechanics of pushing activities is needed. We have learned a great deal about lifting tasks in the last decade, but very little research has been directed at understanding pushing and pulling activities. Most manual materials assist devices replace lifting work activities with pushing and pulling activities.

2. More information is needed on how people physically control complex machinery. We have only a very limited understanding of how people use their musculoskeletal system to control and direct complicated mechanical systems. Manual materials assist devices are currently being developed with redundant degrees of freedom, braking systems, and a wide variety of levels of automation. At present, we are unable to predict whether such changes will improve or degrade operator performance.

3. Finally, the practical experiences of ergonomists and engineers working with manual materials assist devices need to be more effectively communicated. Because of the lack of established theory in this area, anecdotal and case study reports are extremely valuable. Unfortunately, few if any reports of this type are currently available.

## REFERENCES


1. National Institute of Occupational Safety and Health, *A Work Practice Guide for Manual Lifting*, Tech. Rep. No. 81-122, U.S. Dept. of Health and Human Services, Washington, DC, 1981.
2. D. B. Chaffin and G. Andersson, *Occupational Biomechanics*, 2nd edn., Wiley, New York, 1991.
3. Eastman Kodak Company, *Ergonomic Design for People at Work*, Vols. 1 and 2, Van Nostrand Reinhold, New York, 1986.
4. S. A. Konz, *Work Design: Industrial Ergonomics*, 3rd edn., Publishing Horizons, Worthington, OH, 1990.
5. K. Kroemer, H. Kroemer, and K. Kroemer-Elbert, *Ergonomics: How to Design for Ease and Efficiency*, Prentice-Hall, Englewood Cliffs, NJ, 1994.
6. D. B. Torok, Industrial manipulators lend a hand to your material handling applications, *Mater. Handl. Eng. 48*(1): (1991).
7. G. Schwind, Manipulators: Power and reach to the worker, *Mater. Handl. Eng. 48*(1): (1993).
8. S. A. Konz, *Facility Design: Manufacturing Engineering*, 2nd edn., Publishing Horizons, Scottsdale, AZ, 1994.
9. G. A. Castleberry, Industrial manipulators, in *Material Handling Handbook*, 2nd edn., R. A. Kulwiec, Ed., Wiley, New York, 1985.
10. D. J. Quinn, Lift tables, in *Material Handling Handbook*, 2nd edn., R. A. Kulwiec, Ed., Wiley, New York, 1985.
11. R. A. Kulwiec, *Advanced Material Handling*, Material Handling Institute, Charlotte, NC, 1983.
12. M. L. Resnick, *Biomechanics, Kinematics, Psychophysics and Motor Control in the Application of Material Handling Devices (MHDs)*, PhD dissertation, University of Michigan, Ann Arbor, MI, 1993.
13. J. C. Woldstad and D. B. Chaffin, Dynamic push and pull forces while using a manual materials handling assist device, *IIE Trans.*, in press.
14. M. M. Ayoub and J. W. McDaniel, Effects of operator stance on pushing and pulling tasks, *AIIE Trans. 6*(3): 185–195 (1974).

15. K. H. E. Kroemer, Horizontal push and pull forces: Applied when standing in working positions on various surfaces, *Appl. Ergon.* 5(2): 94–102 (1974).
16. P. R. Davis and D. A. Stubbs, Safe levels of manual forces for young males, 1, *Appl. Ergon.* 8(3): 141–150 (1977).
17. P. R. Davis and D. A. Stubbs, Safe levels of manual forces for young males, 2, *Appl. Ergon.* 8(3): 219–228 (1977).
18. P. R. Davis and D. A. Stubbs, Safe levels of manual forces for young males, 3, *Appl. Ergon.* 8(4): 33–37 (1978).
19. D. B. Chaffin, G. D. Herring, W. M. Keyserling, and A. Garg, A method for evaluating the biomechanical stresses resulting from manual materials handling jobs, *AIHA J. 38*: 662–675 (1977).
20. J. B. Martin and D. B. Chaffin, Biomechanical computerized simulation of human strength in sagittal-plane activities, *AIIE Trans. 4*(1): 19–28 (1972).
21. S. H. Snook, The design of manual handling tasks, *Ergonomics 21*: 963–985 (1978).
22. S. H. Snook and V. M. Ciriello, The design of manual handling tasks: Revised tables of maximum acceptable weights and forces, *Ergonomics 34*(9): 1197–1213 (1991).
23. W. F. Fox, *Body Weight and Coefficient of Friction as Determiners of Pushing Capability*, Human Eng. Spec. Stud. Ser. No. 17, Lockheed Co., Marrietta, GA, 1967.
24. D. W. Greive, Slipping due to manual exertion, *Ergonomics 26*(1): 61–72 (1983).
25. M. S. Redfern and R. O. Andres, The analysis of dynamic pushing and pulling: Required coefficients of friction, *Proc. Int. Conf. Occup. Ergon.*, Toronto, Ontario, 1984.
26. N. Özkaya and M. Nordin, *Fundamentals of Biomechanics: Equilibrium, Motion, and Deformation*, Van Nostrand Reinhold, New York, 1991.
27. R. M. Pagulayan, *Determining the Human Ability to Judge Inertia during a Dynamic Pushing Task*, MS thesis, Virginia Polytechnic Institute and State University, Blacksburg, VA, 1994.

CHAPTER 14

# Matching the Physical Qualifications of Workers to Jobs

Rick Wickstrom

CONTENTS



## 14.1 INTRODUCTION

Safety and productivity are impacted by the extent that work stresses the capacities or limits of the body's biomechanical, physiological, or psychological systems. When physical stress is insufficient, physical fitness tends to decline, and productivity may suffer. On the other hand, too much physical stress may cause overexertion injuries or cumulative trauma disorders to develop over time. Epidemiologic investigations of lifting strength ratio [1–4], job severity index [5–7] have revealed significant correlations between injury and physical capacities of workers.

A common problem with many methods that assess the physical qualifications of workers for jobs is that selection of equipment and methods is rarely based on a clear rationale about the class of task or skill represented. This chapter discusses the need for job-match taxonomy improvements to better compare the physical abilities of workers with the physical demands of jobs, provides examples of objective and relevant methods for assessing workers, and introduces a new model for fitness-for-duty screening to prevent injuries and motivate higher risk workers to participate in appropriate wellness interventions.

## 14.2 SIGNIFICANCE OF TAXONOMY TO ERGONOMICS

A common task-descriptive language (or taxonomy) is helpful to conceptualize and classify the factors or qualifications associated with the kinds of tasks that individuals can perform. Table 14.1 contains a summary of important scientific/theoretical and applied/practical benefits of having a common task-descriptive language, as reported by Fleishman and Quaintance [8].

## 14.3 CRITICAL REVIEW OF CURRENT STATUS

The most widely used taxonomy and database for comparing the physical abilities of workers to the physical demands of jobs is the Dictionary of Occupational Titles (DOT) system that was developed by the U.S. Department of Labor Employment and Training Administration [9,10]. Unfortunately, the DOT database has not been updated since 1991 because the U.S. Department of Labor (DOL) implemented a different occupational information network (O*NET) database to meet its user needs for workforce development, economic development, career development, academic and policy research. The Social Security Administration (SSA) has continued to rely on the aging DOT system because O*NET was not designed to meet SSA's data needs related to disability benefits adjudication. Occupational health and vocational rehabilitation professionals prefer the DOT taxonomy because it supports a more individualized approach to evaluate the impact of disabling health conditions on job placement and vocational rehabilitation.

In response to increasing concerns that DOT system no longer accurately reflects the national existence and incidence of work since its last update in 1991, the Commissioner of SSA established the Occupational Information Development Advisory Panel (OIDAP) on December 9, 2008 [11]. The OIDAP was charged to provide independent advice and

TABLE 14.1 Benefits of a Common Task-Descriptive Language

*Scientific/Theoretical Benefits*

1. Conducting literature reviews
2. Establishing better bases for conducting and reporting research studies to facilitate their comparison
3. Standardizing laboratory methods for studying human performance
4. Generalizing research to new tasks
5. Exposing gaps in knowledge
6. Assisting in theory development

*Applied/Practical Benefits*

1. Job definition and job analysis
2. Human–machine system design
3. Personnel selection, placement, and human resource planning
4. Training
5. Performance measurement and enhancement
6. Development of retrieval systems and databases

recommendations on plans and activities to replace the DOT with a new occupational information system that is tailored specifically for SSA's disability programs and adjudicative needs. In the OIDAP preliminary September 2009 report, the panel recommended creation of a new occupational information system that would build on the best existing features of DOT, O*NET, and other systems to meet the need for individualized assessment and legal defensibility in the SSA adjudication process [12]. With respect to physical demands factors, the OIDAP 2009 Report stated that "We know that while the physical demand work traits of the DOT represent some of its best features, there is room to refine and expand the traits, include additional discrete elements, and make their measurement more realistic." The OIDAP and the 2010 review of O*NET by the National Academy of Sciences were in agreement that O*NET in its current form is not suitable for disability adjudication because its data are not specific enough to support a forensic decision-making process that requires a more individualized assessment of person–job match [13,14].

While efforts are underway to develop a new occupational information system, the DOT system remains the most comprehensive set of occupational characteristics currently available [15]. This chapter describes some of the problems identified by the OIDAP and stakeholders with existing DOT physical demand factors, as well as suggestions for improvement to promote a better taxonomy to match the physical qualifications of workers with the physical demands of jobs. A complete list of DOT physical demand factors is summarized in Table 14.2. This table also demonstrates distribution across each rating level for the strength factor, physical demand factors 2–20, and the five physical aptitude factors for the 12,741 job titles reported in the 1991 *Dictionary of Occupational Titles* [10].

The DOT strength factor has been widely used as a general classifier for the overall physical demands of work as sedentary, light, medium, heavy, and very heavy based on the amount of standing, walking, sitting, lifting, carrying, pushing, and pulling required [9,10]:

- *Sedentary*: Exerting up to 10 lb of force occasionally or a negligible amount of force frequently to lift, carry, push, pull, or otherwise move objects, including the human body. Sedentary work involves sitting most of the time but may involve walking or standing for brief periods of time. A job is sedentary if walking and standing are required only occasionally and all other sedentary criteria are met.
- *Light*: Exerting up to 20 lb of force occasionally, up to 10 lb of force frequently, or a negligible amount of force constantly to move objects. Physical demand requirements are in excess of those for sedentary work. Even though the weight lifted may be only a negligible amount, a job should be rated as light work (1) when it requires walking or standing to a significant degree, (2) when it requires sitting most of the time but entails pushing or pulling of arm or leg controls, or (3) when it requires working at a production rate pace entailing the constant pushing or pulling of materials, even though the weight of those materials is negligible. Note: The constant stress and strain of maintaining a production rate pace, especially in the industrial setting, is physically demanding of a worker even though the amount of force exerted is negligible.

- *Medium*: Exerting 20–50 lb of force occasionally or 10–25 lb of force frequently or greater than negligible up to 10 lb of force constantly to move objects. Physical demand requirements are in excess of those for light work.
- *Heavy*: Exerting 50–100 lb of force occasionally or 25–50 lb of force frequently or 10–20 lb of force constantly to move objects. Physical demand requirements are in excess of those for medium work.
- *Very heavy*: Exerting in excess of 100 lb of force occasionally or in excess of 50 lb of force frequently or in excess of 20 lb of force constantly to move objects.

Unfortunately there is no validity reference to justify the definitions for strength levels in the Department of Labor's *Handbook for Analyzing Jobs* [9]. One criticism with the existing

TABLE 14.2 Physical Demand Factor Distribution

| Strength/Levels | Very Heavy | Heavy | Medium | Light | Sedentary |
|---|---|---|---|---|---|
| 1. Strength | 0.7% | 9.1% | 29.6% | 49.6% | 11.0% |

| Frequency/Levels | Constant | Frequent | Occasional | None |
|---|---|---|---|---|
| 2. Climbing | 0.0% | 3.1% | 10.2% | 86.7% |
| 3. Balancing | 0.1% | 1.8% | 5.3% | 92.8% |
| 4. Stooping | 0.1% | 11.5% | 23.5% | 64.9% |
| 5. Kneeling | 0.0% | 3.7% | 10.7% | 85.6% |
| 6. Crouching | 0.0% | 6.1% | 15.6% | 78.3% |
| 7. Crawling | 0.0% | 0.5% | 2.5% | 97.0% |
| 8. Reaching | 10.6% | 81.0% | 7.5% | 0.9% |
| 9. Handling | 11.1% | 81.1% | 7.0% | 0.8% |
| 10. Fingering | 4.6% | 50.5% | 28.9% | 4.6% |
| 11. Feeling | 0.4% | 4.8% | 10.7% | 84.7% |
| 12. Talking | 0.9% | 25.5% | 9.5% | 64.1% |
| 13. Hearing | 1.0% | 26.4% | 12.5% | 60.1% |
| 14. Tasting/smelling | 0.0% | 0.3% | 0.6% | 99.2% |
| 15. Near acuity | 5.9% | 66.9% | 13.5% | 13.7% |
| 16. Far acuity | 0.4% | 6.2% | 5.3% | 88.1% |
| 17. Depth perception | 1.2% | 30.7% | 14.7% | 53.4% |
| 18. Accommodation | 1.7% | 16.6% | 20.1% | 61.7% |
| 19. Color vision | 0.7% | 10.3% | 26.9% | 62.1% |
| 20. Field of vision | 0.4% | 3.8% | 3.2% | 92.6% |

| Physical Aptitude/Levels | Extremely High | High | Medium | Low | Negligible |
|---|---|---|---|---|---|
| K, motor coordination | 0.0% | 3.2% | 48.1% | 48.1% | 0.6% |
| F, finger dexterity | 0.4% | 4.6% | 36.0% | 58.2% | 0.8% |
| M, manual dexterity | 0.1% | 4.6% | 67.8% | 26.9% | 0.6% |
| E, eye/hand/foot coordination | 0.1% | 0.6% | 4.9% | 21.0% | 73.4% |
| C, color discrimination | 0.2% | 1.5% | 8.3% | 30.0% | 60.0% |

*Source:* U.S. Department of Labor, Dictionary of Occupational Titles, 4th edn., U.S. Govt. Printing Office, Washington, DC, 1991.

TABLE 14.3 Decline in Maximum Acceptable Lift as Repetition Increases

| Task<br>Repetition Rate | 1 Lift Every 8 h (kg) | 1 Lift Every 2 min (kg) | % Decline |
|---|---|---|---|
| Floor-to-knuckle lift, 50% males | 40 | 40 | 75 |
| Floor-to-knuckle lift, 50% females | 22 | 15 | 68 |
| Shoulder to overhead, 50% males | 29 | 23 | 79 |
| Shoulder to overhead, 50% females | 14 | 11 | 79 |

strength definition is that expressing the upper limits for frequent forces as 50% of the upper limit for occasional forces and for constant forces at 20% of occasional forces is inconsistent with manual materials handling research. Frequent and constant weights or forces acceptable to workers have been shown to be a much higher percentage of occasional weights/forces in normative tables for materials handling tasks from research sponsored by Liberty Mutual Insurance Company, recommendations by the National Institute of Occupational Safety and Health (NIOSH), and threshold limit values (TLVs®) recommended by the American Conference of Governmental Industrial Hygienists (ACGIH) [16–19]. Table 14.3 illustrates an average ratio of 75% for frequent to occasional weights when comparing data examples from the 1991 Liberty Mutual norm tables for one lift every 8 h to one lift every 2 min [17]. Application of the NIOSH lifting equation to compare similar conditions would result in an 81% reduction in capacity based on its frequency multiplier table [18].

Another criticism for DOT strength definition is that the terms occasional, frequent, and constant are not operationally defined to specify associated repetition ranges for materials handling tasks. The Physical Demands Subcommittee of OIDAP report [12] stated that "most agree that some sort of classification system of the extent of repetition as well as duration should be included.

Repetition:

- Low repetition: 1–12 times per hour
- Moderate: 13–30 times per hour
- High: 31 to 60 times per hour."

Specifying low repetition frequency as 1–12 times per hour is consistent with the ACGIH TLV definition for infrequent lifting that stipulates circumstances of ≤12 lifts per hour for activities ≥2 hours per day. The moderate repetition of 13–30 times per hour is consistent with the ACGIH TLV definition for moderately frequent lifting that stipulates circumstances of >12 and ≤30 lifts per hour for activities lasting ≥2 hours per day.

Yet another criticism by DOT users is that the range for levels of occasional force for heavy and very heavy strength levels is too high based on consideration of materials handling loads that represent a risk for injury according to ergonomic risk assessment models [18–20]. The practical usefulness of the very heavy strength level is also questionable because so few jobs even exist that require a very heavy level of strength.

In a recent normative study of functional capacities for workers employed at different strength levels [21], the data for workers performing heavy and very heavy physical demands had to be combined into one category because only a small number of subjects in the sample were employed at these levels. In this study, no female subjects in the normative sample were even working at the heavy or very heavy physical demands categories. Ergonomic risk assessment tools such as the ACHIG TLVs and the Washington's hazard zone checklist identify upper limits for infrequent lifting of 70 lb due to safety risk [19,20]. For a lower lift strength task, this is reasonably consistent with epidemiologic studies that demonstrate an increased odds ratio for lower back injury at two times the recommended weight limit (RWL) using the NIOSH lifting equation [22]. These ergonomic risk assessment models would support recommendations by DOT users to lower the upper limit of the "heavy" strength for occasional lifting from 100 to 70 lb.

Finally, many DOT users and the OIDAP commission have advocated for revision of strength level definitions to limit this factor only to materials handling characteristics and categorize postural demands such as standing and sitting as separate physical demand factors. Examples of revised "strength" scale definitions proposed by the Occupational Health Special Interest Group of the American Physical Therapy Association are:

*Very light*: 1–10 lb maximum and a negligible amount of weight frequently

*Light*: 11–25 lb maximum, up to 10 lb frequently, or a negligible amount of weight constantly

*Medium*: 26–40 lb maximum, 11–25 lb frequently, or up to 10 lb of weight constantly

*Heavy:* 41–70 lb maximum, 26–40 lb frequently, or up to 15–25 lb of weight constantly

*Exceptional*: >70 lb maximum, >40 lb frequently, or >25 lb of weight constantly

When placing a worker in a specific job, one must know the zone for lifting in addition to frequency and weight of the load handled. A worker may be unable to lift 10 lb overhead due to a shoulder injury yet still be well matched for a job that requires lifting up to 50 lb at waist level or below. A worker with a back or shoulder injury can often be accommodated by being placed in a job where heavier items are lifted between waist and chest level. The *Applications Manual for the Revised NIOSH Lifting Equation* [18,22] may be used to characterize and evaluate lifting tasks based on the maximum and average force exerted, the location of the hands and angle of asymmetry at the origin and destination of the lift, the frequency and duration of lifting, and coupling to the object. This revised NIOSH lifting equation may be used to justify ergonomic redesign or to validate the design of a job-specific work fitness test.

Commercially available systems for functional capacity evaluation (FCE) use different methods to determine the end point for tests of worker abilities during manual materials handling tasks. The Liberty Mutual normative studies used a psychophysical method whereby the opinion of the subject being evaluated determines the testing end point by self-monitoring feelings of exertion or fatigue and adjusting the weight accordingly. With kinesiophysical methods, the evaluator relies on functional movement criteria to determine the end point of testing, with the examiner using observational skills to more objectively determine when the subject is struggling with the load as the end points. The kinesiophysical method has some distinct advantages compared to psychophysical testing for subjects that have a poor perception of their own ability, have psychological factors such as fear avoidance behavior, or are suspected of not providing accurate reports or not demonstrating a full effort. Using this method, an examiner's systematic observational rating of effort was shown to be the single best indicator of sincere effort by Jay et al. [23]. Another investigation by Reneman [24] showed that effort level can be determined with validity on materials handling tasks by visual observation.

Variations in box dimensions, coupling, lifting zones, and repetitions between systems account for different results when comparing the results of tasks administered by different FCE systems. For example, Ijmker et al. [25] explored the concurrent validity of test results of upper lifting tasks of the Ergo-Kit FCE and the Isernhagen work systems (IWS) FCE and concluded that upper lifting tasks of the Ergo-Kit FCE and the IWS FCE do not meet the criteria for concurrent validity and can, therefore, not be used interchangeably. In this investigation, it was not a surprising conclusion that the mean maximum lifting capacity was substantially higher for the Ergo-Kit method because the two kinesiophysical methods were quite different in the zone of lift (Ergo-Kit upper lift to chest level versus Isernhagen lift to overhead level) and in the number of repetitions required (Ergo-Kit upper lift strength test is a one-repetition task versus Isernhagen waist-to-overhead lift (WOL) task that requires achievement of five repetitions at each load progression). The Isernhagen WOL task requires multiple repetitions in a more physically challenging lifting zone. Differences between FCE methods are further illustrated in the concurrent validity study by Rustenburg et al. [26] that demonstrated significantly higher mean maximum lifting capacity for the Ergo-Kit lower and upper lifting tasks compared to the ERGOS work simulator. This conclusion was also not surprising, given the differences in these FCE systems in end point methodology (ERGOS is a psychophysical method, whereas Ergo-Kit is a kinesiophysical method) and repetitions required (ERGOS work simulator requires three repetitions, whereas Ergo-Kit requires only one repetition). In this investigation, the zones of lifting were more similar, but a higher frequency of repetitions and psychophysical method used with ERGOS work simulator method was more likely to result in fatigue and a lower acceptable weight for the subjects, compared to the Ergo-Kit method. These studies suggest that use of multiple repetitions to determine an end point may represent a frequent test rather than an occasional test of materials handling capacities. This speaks to the need for a common taxonomy to provide a better framework of referencing tests of workers' manual materials handling abilities.

Physical demand factors 2–20 in the DOT represent work tolerances that are classified according to a frequency scale [9,10] based on percentage of an 8 h work shift:

Constantly (C): Activity or condition exists for 2/3 or more of the time.

Frequently (F): Activity or condition exists from 1/3 to 2/3 of the time.

Occasionally (O): Activity or condition exists up to 1/3 of the time.

Not present (N): Activity or condition does not exist.

The OIDAP Physical Demands Subcommittee reported that "most agree that some sort of classification system of the extent of repetition as well as duration should be included…It may be that the number of repetitions would vary depending on whether one is classifying upper extremity vs. trunk repetition… The length of time a physical demand is performed and the length of a workday should be captured in the data gathering process." DOT users have recommended further delineation of the frequency scale to specify repetition ranges and to add one or two new levels of frequency that correspond to "rarely" (corresponding to 1%–5% of an 8 h shift) and "exceptional" (corresponding to >8 h of exposure during a work shift). The "exceptional" category would apply to standing tolerance of workers such as nurses in a hospital setting that commonly work 12 h shifts or to sitting tolerance of over-the-road truck drivers that drive for 11 or more hours during a given shift. This is important because jobs that require constant driving in a constrained sitting posture for extended periods of time are much harder for workers with health conditions that limit sitting tolerance that jobs that involve constant sitting in an office setting where the worker has more flexibility to change work postures. DOT users have advocated for separation of postural factors such as standing and sitting from the existing DOT strength factor definition to better assess the impact of these work tolerance factors.

The DOT taxonomy contains five physical worker–job match factors under aptitudes that are rated using an aptitude or skill-level scale:

*Extremely high*: The top 10% of the population

*High*: The highest third exclusive of the top 10% of the population

*Medium:* The middle third of the population

*Low*: The lowest third exclusive of the bottom 10% of the population

*Very low*: The lowest 10% of the population

Some adjustment in these aptitude definitions may be warranted to better distinguish between people and jobs that have varying abilities. The practical usefulness of the "extremely high" aptitude levels for the physical aptitude factors is low for disability adjudication purposes, given that only a small percentage of jobs existed in 1991 (0.0%–0.4%) that required this level of aptitude or skill for the five physical aptitudes. For example, adding a "none" category may help better distinguish jobs that have no requirement for a given aptitude factor. *The Revised Handbook for Analyzing Jobs* includes many examples of functional anchors that provide examples of task performance that correspond to specific aptitude levels.

The Occupational Health Special Interest Group of the American Physical Therapy Association recommended that a number of physical demand factors evaluated by health care professionals would be better rated by the degree of aptitude or skill required than by a frequency scale that is more oriented toward tolerance during the work shift. Examples of specific factors recommended for rating as an aptitude include ambulation agility, ambulation stamina, climbing, near vision acuity, far vision acuity, hearing acuity, and keyboarding speed.

The world of work has changed considerably since the last update of the DOT database in 1991. Changes to manufacturing, distribution, and service processes have further lowered the physical demands of many jobs. It is anticipated that ongoing efforts of the OIDAP will result in an improved occupational information system that results in further refinement of rating scales, as well as addition or modification of specific factors to improve person-side measurements based on well-defined functional levels. Commercial FCE systems must evolve with improvements in worker–job match taxonomies and evidence-based practice research.

## 14.4 EVOLUTION OF THE WORKABILITY EJOBMATCH TAXONOMY

In this section, operational definitions and related examples of worker assessment methods are described for the physical ability taxonomy developed for the WorkAbility eJobMatch System to illustrate a new direction for future worker–job match taxonomy.

### 14.4.1 Materials Handling Abilities

WorkAbility eJobMatch includes the following materials handling ability factors:

1. *High lift strength (>52 in.)* is the maximum load that is raised or lowered using one or both hands while reaching from shoulder level to overhead. This relates to Physical demand factor 1 (strength) as defined by the U.S. Department of Labor [9]. A dynamic high lift strength task is commonly administered during many FCEs. For most healthy subjects, shoulder strength is often the limiting factor. The *WorkAbility high lift strength test* is depicted in Figure 14.1. This method involves raising the bottom of the tote pan to a vertical mark that corresponds to 12 in. above shoulder level. Use of inexpensive painter's tape to mark the desired vertical end point for the bottom of the tote

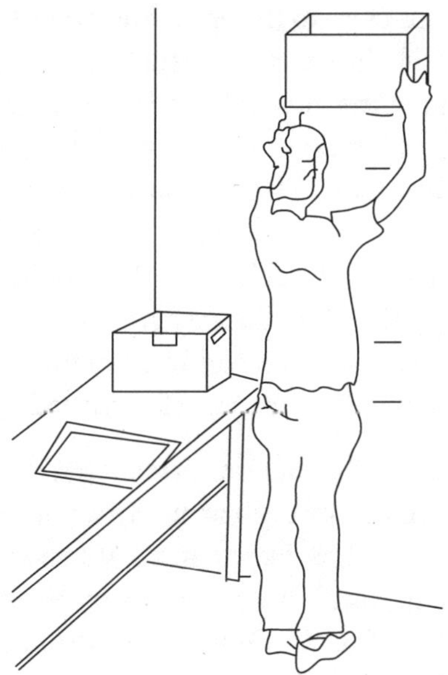

FIGURE 14.1 WorkAbility high lift strength test.

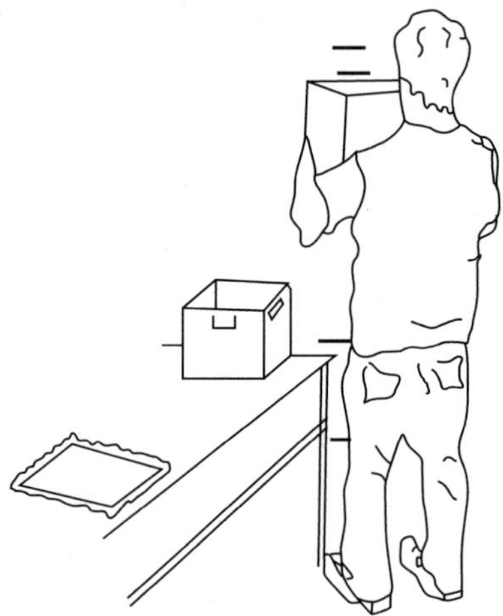

FIGURE 14.2 WorkAbility chest lift strength test.

pan eliminates the need for fixed or adjustable height shelving that cannot be readily transported within a clinic or to other locations. Setting the destination limit at 12 in. above shoulder level is consistent with upper reach limit recommended for lifting in the ACGIH publication of TLVs, which also references the High lift zone as 52–72 in. [19].

2. *Chest lift strength (30–52 in.)* is the maximum load that is raised or lowered using one or both hands while reaching from waist to below shoulder level. This relates to physical demand factor 1 (strength) as defined by the U.S. Department of Labor [9]. A dynamic lift strength task at this mid-range zone is commonly administered during many FCEs and often called a knuckle-to-shoulder lift. For most healthy subjects, elbow and grip strength is the limiting factor, because the subject can usually protect the shoulders by positioning the upper arms close to the body. The *WorkAbility chest lift strength test* is depicted in Figure 14.2 [13]. This method involves lifting a tote pan to a vertical mark that positions the bottom of the tote and hands at a vertical height of 52 in. The 52 in. height was selected because it marks the upper range of a reference task used by Snook [16,17] to communicate the maximum acceptable loads for workers performing two-handed symmetrical lifting. The 52 in. height is related to the upper vertical limit for the chest zone in the ACGIH TLV publication [19]. The WorkAbility chest lift strength is quite similar to the Ergo-Kit upper lift strength test [27,28], except for protocol modification to foster complete mobility and addition of examiner ratings of effort that are similar to the research on the EPIC lift capacity method by Jay et al. [23]

3. *Carry strength (28 ft or less)* is the maximum load that is transported by walking, usually by holding the load in one or both arms. This relates to physical demand factor 1 (strength) as defined by the U.S. Department of Labor [9]. The limiting factor during carry is usually arm strength. A dynamic carry strength task for short distances is common to many FCEs. The *WorkAbility two-arm carry strength test* is depicted in Figure 14.3. This method involves carrying a tote pan with progressive weights for a distance of 28 ft. This distance was selected because it was one of the reference

FIGURE 14.3 WorkAbility two-arm carry strength test.

distances used for carry tasks in the normative studies by Liberty Mutual [16–17] to investigate the maximum acceptable loads for workers performing two-handed symmetrical carry tasks. Two-handed carrying tasks that occur over long distances are not routinely assessed in most FCE systems, because it is usually possible to reduce these physical demands by using a cart or to use another device such as a back pack to transport the load. Keeping the two-handed carry distance short for a job task also limits the risk of tripping because carrying an object in both arms tends to obstruct the subject's view of the ground and impose greater agility challenges. If the subject has a unilateral problem affecting only one arm, then a comparative assessment may be warranted to evaluate right versus left carrying strength.

4. *Knee lift strength (15–30 in.)* is the maximum load that is raised or lowered using one or both hands while reaching between knee and waist level. This zone of lifting is tested less often in other FCEs; however, this zone is important to because it usually represents the strongest lifting zone because the subject can use the strongest muscle groups in more optimal positioning of joints to the greatest biomechanical advantage. The *WorkAbility knee lift strength test* is depicted in Figure 14.4. This method involves lowering and raising a tote pan from table level to and from a vertical mark that puts the hands at 18 in. This knee lift strength definition is reasonably comparable to the ACGIH zone from 12 to 30 in. that is referenced in the TLV.

5. *Low Lift strength (<15 in.)* is the maximum load that is raised or lowered using one or both hands while reaching below knee level. A test of dynamic low lift strength is commonly included in most FCEs. The *WorkAbility low lift strength test* is depicted in Figure 14.5. This method requires the subject to lower a tote pan from table level to a destination mark that puts the bottom of the tote pan and hands at a vertical height of 8 in. The vertical height of 8 in. was chosen because this corresponds to a vertical

FIGURE 14.4 Ergo-Totes Knee Lift Strength Test.

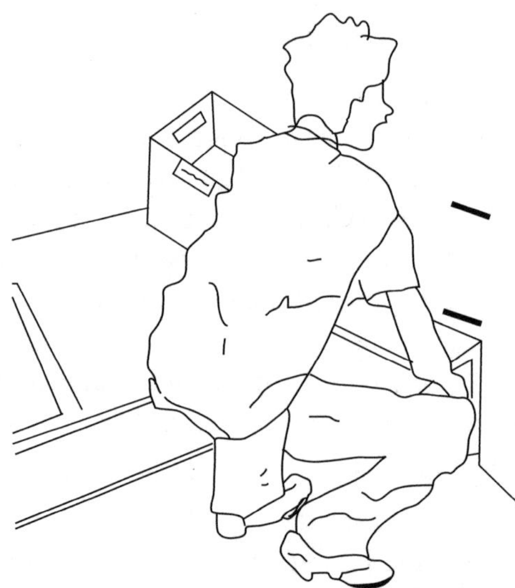

FIGURE 14.5 Ergo-Totes Low Lift Strength Test.

reach height to grasp items on a pallet. This low lift strength definition is comparable to the ACGIH ankle zone that is referenced in the TLV. For most healthy subjects, knee extensor strength is usually the limiting factor. The WorkAbility low strength test method is quite similar to the Ergo-Kit lower lift strength test [27,28], except for protocol modification to foster complete mobility and addition of examiner ratings of effort that are similar to the research on the EPIC lift capacity method by Jay et al. [23]

6. *Push strength* is the maximum force that is exerted to move an object away from the direction of the force. This relates to physical demand factor 1 (strength) as defined by the U.S. Department of Labor [9]. This may be measured in the job in terms of frequency (percent of the day), maximum and average force exerted during transporting (pounds), average and maximum distance of transporting (yards), and number

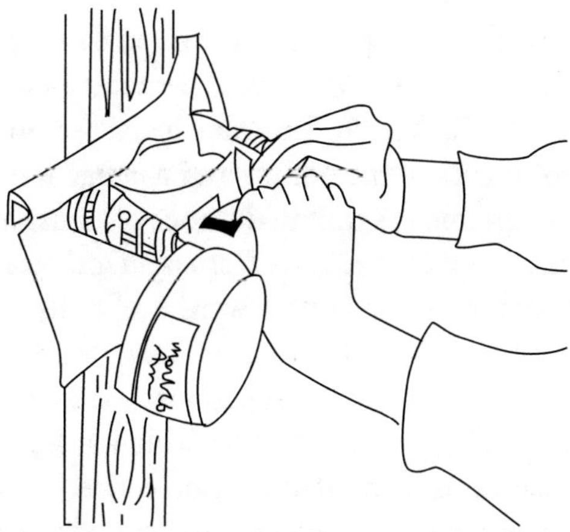

FIGURE 14.6 Dynamometer.

of repetitions. The *WorkAbility push strength test* posture is depicted in Figures 14.6 and 14.7 [13]. The subject pushes with both arms statically against a dynamometer that is stabilized against a stationary surface at chest level. A static push method is preferred by many professionals over a dynamic push against a weighted sled while walking because a static push can be readily measured for direct measurement of a job tasks with an inexpensive dynamometer that is completely mobile for use in multiple locations. Healthy subjects are most commonly limited on this test by shoulder and abdominal muscle strength. Unilateral difference in extremity loading can be assessed by having the subject perform the test with the right versus left leg forward, because the back leg assumes most of the weight bearing force.

FIGURE 14.7 WorkAbility push strength test.

7. *Pull strength* is the maximum force that is exerted to move an object in the same direction as the force. This relates to physical demand factor 1 (strength) as defined by the U.S. Department of Labor [9]. This may be measured on job tasks in terms of frequency (percent of the day), maximum and average force exerted during transporting (pounds), average and maximum distance of transporting (feet or yards), and number of repetitions. The *WorkAbility pull strength test* posture is depicted in Figure 14.8. The subject stands with one foot forward and pulls with both arms statically against a dynamometer that is affixed to a stationary surface such as the handle on the opposite side of a closed door. A static pull method tends to be preferred by many professionals over a dynamic pull because walking while pulling against a weighted sled presents both vision and trunk stabilization concerns. A static pull test can also be readily accomplished using an inexpensive dynamometer and strap that is completely mobile for use in multiple locations. This test also provides a gross functional measure of the individual's composite trunk, hip, and knee extensor strength.

8. *Frequent lift/carry (13–30 times per hour)* is the average of loads lifted or carried between 13 and 30 repetitions per hour. It is usually determined by averaging the heaviest 13 loads during a given hour. When frequent or continuous lifting is the critical job demand, lifting endurance may be measured with the WorkAbility chest lift frequent and knee lift frequent tests [13]. These tests are administered using a four-repetition protocol that is comparable to the PILE protocol [29,30], except for equipment modification to allow complete mobility and some minor to replicate the same postures and midrange zone as for the WorkAbility chest lift and knee lift strength tests.

FIGURE 14.8 WorkAbility pull strength test.

9. Constant *lift/carry (>30 times per hour)* is the average of loads lifted or carried more frequently than 30 repetitions during a given hour. It is usually determined on the job by sampling the heaviest 31 loads for each hour during a representative 8 h work shift. Constant lifting is often not practical to evaluate during a job-specific or comprehensive FCE. It is usually best to monitor an injured worker's psychophysical ratings and physiological parameters while the worker is participating in a progressive transitional work program at the jobsite or in a clinic-based work rehabilitation program.

### 14.4.2 Work Tolerances

WorkAbility eJobMatch includes the following work tolerances factors:

1. *Sit or stand work option* is performing work that can be done with a choice of either sitting or standing.
2. *Standing only* is remaining on one's feet in an upright position at a workstation or when moving about. This relates to physical demand factor 1 (strength) as defined by the U.S. Department of Labor [9]. This may be measured in terms of frequency (percent of the day) and maximum and average time spent standing for any given function.
3. *Sitting only* is remaining in a seated position. Includes driving done while sitting. This relates to physical demand factor 1 (strength) as defined by the U.S. Department of Labor [9]. This may be measured in terms of frequency (percent of the day) and maximum and average time spent sitting for any given function.
4. *Operating foot controls* is performing work activities that operation of controls with one or both feet. This relates to physical demand factor 1 (strength) as defined by the U.S. Department of Labor [9]. This may be measured in terms of frequency (percent of the day), type of foot movement or coordination required, and maximum and average duration and speed of foot use. It is important to note whether the dominant foot, other foot, both feet, and either foot may be used for the task. Foot use ability is usually estimated from performance on other tests.
5. *One-handed work option* is performing work that can be done with a choice of using either hand. One-handed work options may be evaluated by individual assessment of the right versus left arms during tests of finger dexterity, manual dexterity, lifting, and carrying. One of the greatest shortcomings of the existing DOT is its database limitations prevent assessment of opportunities for workers that have substantial physical limitations involving only one upper extremity.
6. *High reaching above shoulder* is extending either arm to reach from shoulder level to overhead. This relates to physical demand factor 8 (reaching) as defined by the U.S. Department of Labor [9] that is based on the percentage of time a worker is extending hand(s) and arm(s) in any direction. DOT users have advocated for more specific classification of reaching in vertical zones, because of confusion created by this operational definition. Reaching above shoulder level may be further delineated on a

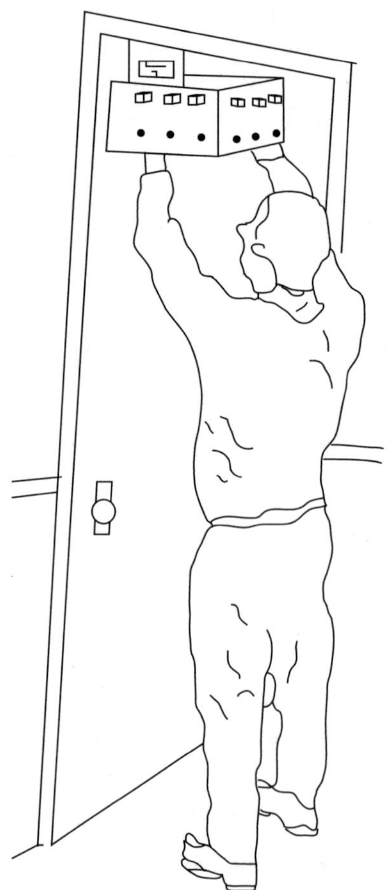

FIGURE 14.9 Total-Body Dexterity Tester Overhead Manipulation Test.

functional job analysis by maximum and average duration for reaching with the arms in the upper range. It is important to note whether the dominant arm, other arm, both arms, or either arm may be used for reaching overhead. Reaching above shoulder level may be evaluated by tolerance in performing a work sample such as the *Total-Body Dexterity Tester* [31,32] *Overhead Manipulation Test* that is shown in Figure 14.9.

7. *Head turning >45°* is rotating the head 45° or more either way with respect to the upper torso. This may be evaluated by measuring cervical rotation with a goniometer and observing the behavior of subjects when distracted and performing a work sample such as the *Total-Body Dexterity Tester* [31,32] as depicted in Figures 14.9 through 14.12.

8. *Forward bending/stooping* is bending downward more than 20° at the waist or hips while standing. This relates to physical demand factor 4 (stooping) as defined by the U.S. Department of Labor [9]. This may be measured in terms of frequency (percent of the day) and maximum and average duration of stooping. When stooping is required occasionally, this can be measured with the *Total-Body Dexterity Tester* during the *Forward Manipulation Test* [31,32] as shown in Figures 14.11 and 14.12. This ability can also be inferred from other physical ability tests such as the *Ergo-Totes*

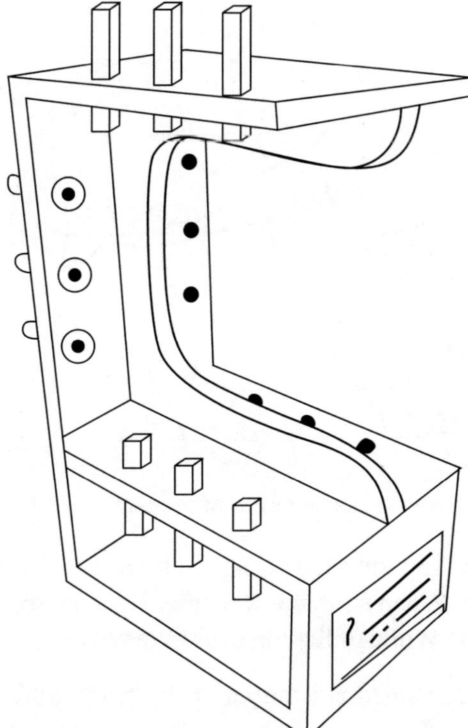

FIGURE 14.10 Total-Body Dexterity Tester.

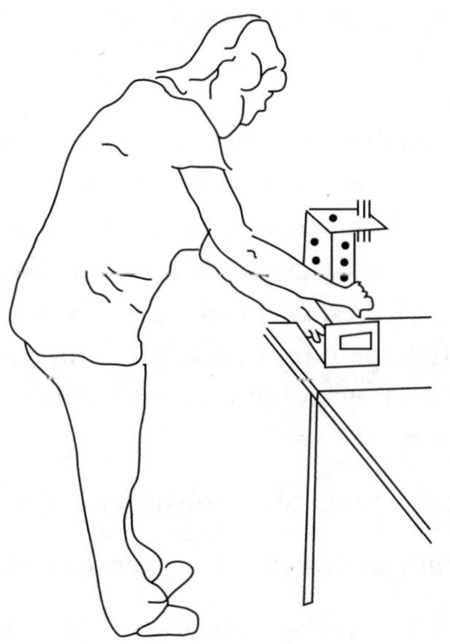

FIGURE 14.11 Total-Body Dexterity Tester Forward Manipulation Test.

FIGURE 14.12 Total-Body Dexterity Tester Lower Manipulation Test.

*Low Lift Strength Test* (Figure 14.5) or *Ergo-Totes Knee Lift Strength Test* (Figure 14.4) as previously shown. When stooping is required frequently, fitness for stooping may be measured with the WorkAbility knee lift frequent test.

9. *Low work (e.g., kneel/crouch)* is bending at the trunk and knees to work with the hands below knee level in a kneeling, squatting, or seated posture. This relates to physical demand factor 5 (kneeling) and physical demand factor 6 (crouching) as defined by the U.S. Department of Labor [9]. This may be measured in terms of frequency (percent of the day) and maximum and average duration of working at a low level. In many instances, it is up to the worker to decide what method of working to employ at a lower level, e.g., whether to kneel, crouch, or stoop. When low work is required, the *Total-Body Dexterity Tester Lower Manipulation Test* [31,32] may be administered in the subject's preferred work posture (Figure 14.12).

### 14.4.3 Physical Aptitudes

1. *Ambulation agility* is the ability to quickly move about for short periods of time while walking, jogging, or running. This relates to physical demand factor 3 (balancing), which is defined by the U.S. Department of Labor [9] using a frequency scale based on the percentage of time a worker is maintaining equilibrium when standing in place or moving on varying surfaces. Ambulation agility would be better conceptualized as a skill-based rating that spans the continuum from walking very slowly with use of an assistive device to sprinting and cutting on an athletic field, based on functionally described rating examples:

    a. None: Ambulation not required, all job duties can be done from a wheelchair

    b. Very low: Walking only short distances at a very slow pace (e.g., <0.8 m/s or <1.8 mph)

    c. Low: Walking at a below average rate of speed (0.8 to <1.3 m/s or 1.8 to <3.0 mph)

    d. Medium: Normal walking speed (1.35 to 1.9 m/s or 3.0 to <4.25 mph)

e. High: Jogging (1.9 to <2.45 m/s or 4.25 to <5.5 mph)

f. Extra high: Running or sprinting with or without changes in direction such as during competitive sports such as football (=>2.45 m/s or =>5.5 mph)

Ambulation demands can be further detailed in a job analysis by frequency (percent of the day), maximum and average distance ambulated, and maximum and average speed of ambulation. Examples of worker functional capacity tests for ambulation agility include timed performance of five cycles of stepping back and forth during the *Two Square Agility Test* as shown in Figure 14.13 [32,33] or timed walk tests [34,35].

2. *Ambulation stamina* is the ability to move about for long periods of time while walking, jogging, or running. This measure related to aerobic capacity and becomes particularly important when a person has a significant musculoskeletal disorder affecting the lower extremities or cardiopulmonary system that interferes with sustained walking, jogging, or running. There are many tests of ambulation endurance, including a step test [36–39], self-paced walking test [40], 12 min run test [41], or treadmill tests. Although a treadmill test is the most accurate method [42–44], this mode of testing is less common for field applications due to expense and lack of portability. Standardized tests of cardiorespiratory fitness may also be used to screen workers for physiologically demanding jobs [45,46]. Submaximal step tests are often selected for this purpose because they have been shown to be inexpensive, safe, and simple to administer and tend to be most similar to lifting demands. To avoid excessive cardiorespiratory fatigue, NIOSH recommends energy expenditure limits of 50%, 40%, and 33% of maximum aerobic capacity for tasks lasting up to 1, 1–2, and 2–8 h, respectively [18]. The *WorkAbility submaximal step test* is illustrated in Figure 14.14 with an 8 in. step and in Figure 14.15 with a 2.5 in. accommodated step height. This completely mobile protocol has the advantage of accommodation of the step height from 8 in. to low step of 2.5 in. for individuals that have substantial weakness of effecting leg.

FIGURE 14.13 Two Square Agility Test [one of five cycles shown].

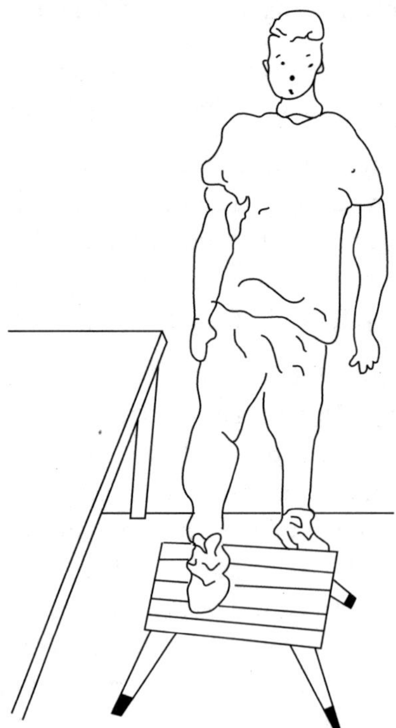

FIGURE 14.14 WorkAbility Submaximal Step Test 8 in.

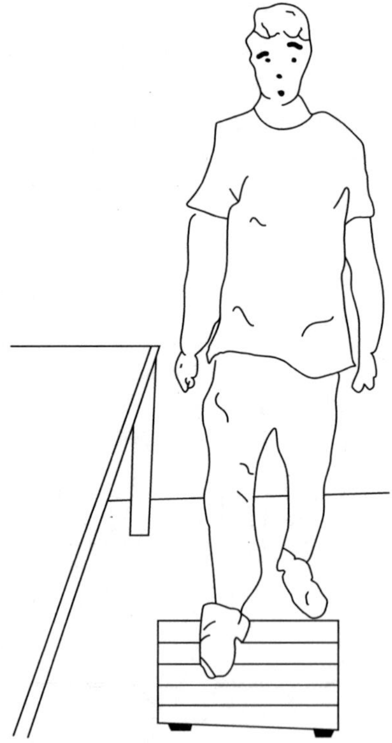

FIGURE 14.15 WorkAbility Submaximal Step Test 2.5 in. Accomodation.

3. *Climbing* is the ability to ascend or descend ramps, stairs, ladders, scaffolding, poles, and other structures. This relates to physical demand factor 2 (climbing) as defined by the U.S. Department of Labor [9] using a frequency scale that is based on the percentage of time a worker is ascending or descending ladders, stairs, scaffolding, ramps, poles, and the like, using feet and legs or hands and arms. Climbing demands would be better conceptualized by using a skill-based rating that spans the continuum from climbing up or down from a low step or ramp to complex climbing up the side of a mountain with poor hand and foot holds, based on functionally described rating examples:

   a. None: Climbing not required, all job duties can be done on level ground

   b. Very low: Climbing limited to a low step or Americans with Disabilities Act (ADA)-approved ramp, with or without handrail assist

   c. Low: Climbing up or down multiple steps that are up to 8 in. each without handrail assistance

   d. Medium: Climbing steps higher than 8 in. or short ladders with good hand and foot holds for up to 6 ft above ground level (e.g., climbing into a semitruck cab)

   e. High: Climbing up or down ladders or scaffolding higher than 6 ft above ground level (OSHA fall protection is typically required)

   f. Extra high: Exceptional climbing that involves poor hand or foot holds (e.g., climbing a telephone pole with spikes)

   This factor may be measured in more detail on a functional job analysis in terms of frequency (percent of the day); height, steepness, and type of structure climbed; and maximum and average duration of climbing. This is typically evaluated on the person by administering a representative work sample. For example, climbing up and down from a low 2.5 in. platform or ramp might be consistent with very low agility, climbing up a single 8 in. step without a handrail is consistent with low agility. Climbing a higher step or a less than 6 ft up a ladder is consistent with medium agility. Grip strength [47,48] and tests of cardiorespiratory fitness such the *WorkAbility submaximal step test* illustrated in Figures 14.14 and 14.15 are important when evaluating higher aptitude levels for climbing.

4. *Keyboarding speed* is the ability to move the fingers rapidly and accurately during keyboarding tasks. Keyboarding is a learned skill that is more closely related to the DOT aptitude factor M (motor coordination) than to DOT aptitude factor F (finger dexterity). This is evaluated during the WorkAbility Systems' FCE by administration of a 2 min test of typing speed. Given the importance of keyboarding to productivity and successful employment in many jobs, it is anticipated that other FCE systems will eventually include tests of keyboarding speed as a common component.

5. *Finger dexterity* is the ability to manipulate with the fingers of one or both hands rapidly and skillfully manipulate small objects. This definition is similar to aptitude

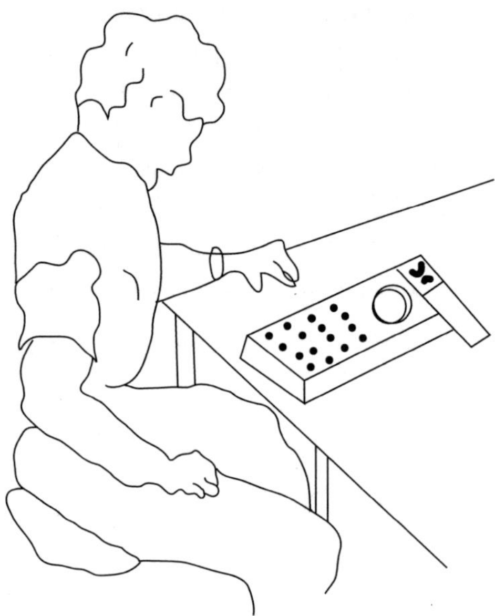

FIGURE 14.16 WorkAbility grooved pegboard test.

factor F (finger dexterity) for the DOT [9] and also relates to physical demand factor 10 (Fingering) and physical demand factor 11 (Feeling) that is defined by the U.S. Department of Labor [9] using a frequency scale based on the percentage of time a worker is feeling, picking, pinching, or otherwise working primarily with the fingers rather than with the whole hand as in handling. The *grooved pegboard test* is illustrated Figures 14.16 and 14.17 as one example of a finger dexterity test that can differentiate right versus left hand performance [49,50]. The grooved pegboard placing task is routinely included in many neuropsychological test batteries because it is sensitive to brain functions such as TBI, stroke, or neurotoxic exposures that impact precise manipulation with the fingers. The grooved pegboard return task is a task that has been shown to provide different insight as to motor coordination that may be important for functions such as keyboarding speed [44]. It may be relevant to express the worker's timed performance on finger dexterity tasks as a ratio to a predetermined

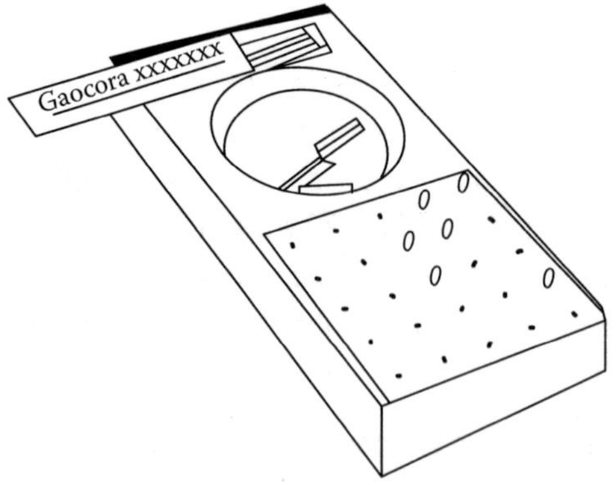


FIGURE 14.17 Grooved pegboard.

industrial time standard such as Modular Arrangement of Predetermined Time Standards (MODAPTS) as a real-world depiction of worker's productively on tasks that require finger dexterity [51,52].

6. *Manual dexterity* is the ability to move one or both hands rapidly and skillfully to perform gross grasping, placing, and turning motions. It includes hand writing. This definition is similar to aptitude factor M (manual dexterity) for the DOT [9] and also relates to physical demand factor 9 (handling) that is defined by the U.S. Department of Labor [9] using a frequency scale based on the percentage of time a worker is seizing, holding, grasping, turning, or otherwise working with the hands. Manual dexterity tasks may be further delineated in a functional job analysis in terms of frequency (percent of the day), type of movement or coordination required, speed, and maximum and average duration and speed of hand/wrist use. The evaluator should note whether the dominant hand, other hand, both hands, or either hand may be used for handling. The *PAT total-body dexterity forward manipulation test* [31,32] is illustrated in Figure 14.11 as one example of a manual dexterity test. This can be performed sitting or on an elevated tray table if stooping tolerance is a limiting factor to performance. One advantage of manual dexterity tests is that the workers' performance can be expressed as a ratio to the industrial time standard such as MODAPTS to evaluate whether the worker can work productively on production tasks that require manual dexterity [52].
7. *Near vision acuity* is clarity of vision at 20 in. or less. Aptitude ratings are based on Snellen equivalent (font size) results for both eyes [53]: (1) very low: <20/200 (26 pt), (2) low: 20/200 (26 pt) to <20/70 (10 pt), (3) medium: 20/70 (10 pt) to <20/40 (6 pt), (4) high: 20/40 (6 pt) to <20/25 (4 pt), (5) extra high: 20/25 (4 pt) or better. Levels of aptitude for near vision may be evaluated in Snellen equivalents using the hand-held vision card or professional vision tester (PVT) to measure acuity both eyes "near," acuity right eye "near," and acuity left eye "near" tests [54].
8. *Far vision* acuity is clarity of vision at 20 ft or more. Aptitude ratings are based on Snellen equivalent results [53]: (1) very low: <20/200, (2) low: binocular vision of 20/200 to <20/70 or monocular vision of 20/200 to <20/60, (3) medium: binocular vision of 20/70 to <20/40 or monocular vision of 20/60 to <20/30 in only one eye (restricted to daytime driving only), (4) high: binocular vision of 20/40 to <20/20 or monocular vision of 20/30 or better, and (5) extra high: binocular vision of 20/20 or better. "Binocular" means visual acuity of 20/200 or better in both eyes and "Monocular" means visual acuity of 20/200 or better in only one eye, on a Snellen chart with or without corrective lenses. Federal Motor Carrier Safety Regulations [55] require truck drivers to have far vision acuity of at least 20/40 (Snellen) in each eye without corrective lenses or visual acuity separately corrected to 20/40 (Snellen) or better with corrective lenses and far vision acuity of at least 20/40 (Snellen) using both eyes with or without corrective lenses. The Department of Transportation also requires truck drivers to have a field of vision of at least 70° in the horizontal meridian in each eye. Levels of peripheral vision may be evaluated with the PVT peripheral

test—right and peripheral test—left [36]. The lights flash at 85°, 70°, and 55° temporally and approximately 35° nasally, so a possible total of 120° for each eye can be attained (highest temporal reading plus highest nasal reading). Federal Motor Carrier Safety Regulations [34] require truck drivers to be able to recognize the colors of traffic signals and devices showing standard red, green, and amber. Levels of color vision may be evaluated with the PVT color discrimination test [36].

9. *Hearing sensitivity* is the ability to detect or tell the differences between sounds that vary in pitch and loudness. This relates to physical demand factor 13 (hearing) as defined by the U.S. Department of Labor [9]. Aptitude ratings are based on pure tone average for the frequencies 500–4000 Hz for the best ear with or without a hearing aid at 500, 1000, and 2000 Hz as follows: (1) very low: 80 dB HL or more, (2) low: 60–79 dB HL, (3) medium: 40–59 dB HL, (4) high: 20–39 dB HL, and (5) extra high: 1–19 dB HL. This may also be measured in the job in terms of frequency (percent of the day); however, aptitude ratings are usually reported only of worker evaluations. Federal Motor Carrier Regulations [55] require that truck drivers be able to first perceive a forced whispered voice in the better ear at not less than 5 ft with or without the use of a hearing aid or, if tested by use of an audiometric device, to not have an average hearing loss greater than 40 dB at 500, 1000, and 2000 Hz frequencies with or without a hearing aid when the audiometric device is calibrated to American National Standard (formerly ASA Standard) Z24.5—1951 [56]. A more elaborate and expensive setup is required to meet the rigid specifications by the Occupational Safety and Health Administration (OSHA) [57] for initial and periodic audiograms on workers exposed to 90 dB (A) for 8 h exposure. OSHA hearing exams must be done in a suitable quiet environment, the testing apparatus (audiometer, headphones, test room) must be calibrated, a wider range of frequencies must be evaluated (500, 1000, 2000, 3000, 4000, and 6000 Hz), the procedure must be standardized, and the person responsible must be properly trained in test administration, interpretation of results, and provision of ear protection.

## 14.5 FUTURE CONCERNS

### 14.5.1 Impact of Civil Rights Legislation

Methods of assessing a worker's fitness for duty have been influenced by civil rights legislation and court decisions related to employment practices. Title VII of the Civil Rights Act of 1964 prohibits employers from discriminating against applicants and workers on the basis of race, color, religion, sex, or national origin. Section 503 of the Rehabilitation Act of 1973 requires employers with government contracts of $2500 or more to take affirmative action to employ and advance qualified handicapped individuals. Title I of the Americans with Disabilities Act (ADA) of 1990 [58] as amended in 2008 [59] couples many of the principles of the Rehabilitation Act of 1973 with the remedies and procedures set forth in the Civil Rights Act of 1964 for other types of discrimination. ADA prohibits covered entities with 15 or more employees from discriminating against a qualified individual with a disability, where that individual, with or without reasonable accommodation, can perform the essential job functions. The new ADA Amendments Act of 2008 expressly overturns several landmark Supreme Court decisions that narrowly interpret the definition of "disability," imposing more responsibility on employers to engage in a more interactive process with anyone alleging a disability or need for reasonable accommodation.

The ADA includes a number of provisions that impact methods for assessing physical qualifications:

1. *Before making a job offer*, medical examinations and inquiries are not permitted. However, an employer may give a physical agility test to determine physical qualifications, provided the test is given to all similarly situated applicants or employees.
2. *After making a conditional job offer and before the individual starts to work*, unrestricted examinations and medical inquiries are permitted; however, the employer may not reject a person with a disability unless the reason is job related and consistent with business necessity. Post-offer examinations must be given to all entering employees within the same job category.
3. *After employment*, any examination or medical inquiry required of employees must be job related and consistent with business necessity, except voluntary examinations by employee health programs and examinations that are mandated by federal laws.

The Equal Employment Opportunity Commission's Uniform Guidelines on Employee Selection Procedures [60] state that selection procedures having adverse impact constitute discrimination unless validated. Records concerning impact are to be maintained by sex, race, and ethnic groups, and a selection rate for any race, sex, or ethnic group that has less than four-fifths of the rate for the group with the highest rate will generally be regarded as evidence of adverse impact but must be statistically significant.

Employers may use selection procedures that are not fully supported if (1) substantial evidence of validity exists and (2) a study is in progress to provide additional evidence of validity within a reasonable time. Procedures for conducting acceptable types of validity studies are detailed in other publications.

## 14.6 CONCLUSION

The WorkAbility eJobMatch taxonomy presented in this chapter may be used to identify the critical physical demands of jobs and tasks. Developing a "functional" profile of job demands is a critical first step in the job modification and selection process. Objective, performance-based tests of an individual's ability to perform the critical demands of the job help to bridge the gap in our understanding of the relationship between medical impairment and job demands. To avoid legal pitfalls, it is important that our selection of equipment and methods for functional screening be based on a clear rationale about the class of task or skill represented. The OIDAP initiative to develop a new occupational information system to replace the DOT represents an unparalleled opportunity to implement a better taxonomy to match the measured abilities of workers to the demands of jobs base on DOT user feedback. Performance-based functional capacity testing of workers is still a relatively new area of research, and there have been only a limited number of studies to suggest that any one method of functional testing is more accurate than another. It is important that selection or development of objective and job-relevant tests be guided by NIOSH's criteria for evaluating employee screening programs—safety, reliability, validity, practicality, and utility [61].

## REFERENCES


1. D. B. Chaffin, Human strength capability and low back pain, *J. Occup. Med.* 16:248–254, 1974.
2. D. B. Chaffin, G. B. Herrin, and W. M. Kesserling, *Pre-Employment Strength Testing in Selecting Workers for Materials Handling Jobs,* NIOSH Tech. Rep. 77-163, 1977.
3. D. B. Chaffin and K. S. Park, A longitudinal study of low-back pain as associated with occupational weight lifting factors, *Am. Ind. Hyg. Assoc. J.* 34:513–535, 1973.
4. W. M. Keyserling, G. D. Herrin, and D. B. Chaffin, Isometric strength testing as a means of controlling medical incidents on strenuous jobs, *J. Occup. Med.* 22(5):332–336, 1980.
5. M. M. Ayoub, N. J. Bethea, S. Deivanayagam, S. S. Asfour, G. M. Bakken, and D. Liles, *Determination and Modeling of Lifting Capacity*, DHHS (NIOSH) Grant No. 5-R01-0H-00545-02, 1978.
6. D. H. Liles, S. Deivanayagam, M. M. Ayoub, and P. Mahajan, A job severity index for the evaluation and control of lifting injury, *Hum. Factors* 26(6):683–693, 1984.
7. M. M. Ayoub, R. Dryden, J. McDaniel, R. Knipfer, and D. Dixon, Predicting lifting capacity, *Am. Ind. Hyg. Assoc. J.* 40(12):1075–1084, 1979.
8. E. A. Fleishman and M. K. Quaintance, *Taxonomies of Human Performance,* Academic Press, Orlando, FL, 1984.
9. U.S. Department of Labor, *The Revised Handbook for Analyzing Jobs,* U.S. Govt. Printing Office, Washington, DC, 1991.
10. U.S. Department of Labor, *Dictionary of Occupational Titles,* 4th edn., U.S. Govt. Printing Office, Washington, DC, 1991.
11. Social Security Administration, *Charter: Occupational Information Development Advisory Panel (OIDAP).* Approved on December 9, 2008. Congressional filing date: January 15, 2009. Retrieved from http://www.ssa.gov/oidap/Documents/FINAL–OIDAP%20CHARTER%20COSS%20&%20DATED%20OIDAPCharter.pdf
12. Occupational Information Advisory Panel, *Content Model and Classification Recommendations for the Social Security Administration—Report to the Commissioner of Social Security,* Accessed on December 14, 2011 from http://www.ssa.gov/oidap/panel_documents.htm
13. Occupational Information Advisory Panel, *Findings Report: A Review of the National Academy of Sciences Report A Database for a Changing Economy: Review of the Occupational Information Network (O*NET).* Report date 6/28/2010. Accessed on December 14, 2011 from http://www.ssa.gov/oidap/panel_documents.htm
14. N. T. Tippins and M. L. Hilton. *A Database for a Changing Economy: Review of the Occupational Information Network (O*NET),* The National Academies Press, Washington, DC. Accessed on December 14, 2011 from http://www.nap.edu/catalog.php?record_id=12814, 2010.
15. A. R. Miller, D. J. Treiman, P. S. Cain, and P. A. Roos (Eds.) *Work, Jobs, and Occupations: A Critical Review of the Dictionary of Occupational Titles,* The National Academies Press, Washington, DC, 1980.
16. S. H. Snook, The design of manual material handling tasks, *Ergonomics,* 21:963–985, 1978.
17. S. H. Snook and V. M. Ciriello, The design of manual material handling tasks: Revised tables of maximum acceptable weights and forces, *Ergonomics,* 34(9):1197–1213, 1991.

18. T. R. Waters, V. Putz-Anderson, and A. Garg, Applications manual for the revised NIOSH lifting equation, DHHS (NIOSH) Publ. No. 94-110, 1993.
19. American Conference of Governmental Industrial Hygienists (ACGIH), *Threshold Limit Values (TLVs) for Lifting*, TLVs and BEIs Book, ACGIH, Cincinnati, 2007.
20. Washington State, *Hazard Zone Jobs Checklist*. Accessed on December 14, 2011 at http://www.lni.wa.gov/wisha/ergo/evaltools/hazardzonechecklist.pdf
21. R. Soer, C. P. van der Schans, J. H. Geertzen, J. W. Groothoff, S. Brouwer, P. U. Dijkstra, and M. F. Feneman, Normative values for a functional capacity evaluation, *Arch. Phys. Med. Rehabil.* 90:1785–1794, 2009.
22. T. R. Waters, S. L. Baron, L. A. Piacitelli, V. P. Anderson, T. Skov, M. Haring-Sweeney, D. K. Wall, and L. J. Fine, Evaluation of the revised NIOSH lifting equation: A cross-sectional epidemiologic study, *Spine* 24(4):386–394, 1999.
23. M. A. Jay, J. M. Lamb, R. L. Watson, I. A. Young, F. J. Fearon, J. M. Alday, and A.G. Tindall, Sensitivity and specificity of the indicators of sincere effort of the EPIC lift capacity test on a previously injured population, *Spine* 25(11):1405–1412, 2000.
24. M. F. Reneman, A. S. Fokkens, P. U. Dijkstra, J. H. Geertzen, and J. W. Groothoff, Testing lifting capacity: Validity of determining effort level by means of observation, *Spine* 30(2):E40–E66, 2005.
25. S. Ijmker, E. H. Gerrits, and M. F. Reneman, Upper lifting performance of healthy young adults in functional capacities evaluations: A comparison of two protocols, *J. Occup. Rehabil.* 13(4):297–305, 2003.
26. G. Rustenburg, P. P. F. M. Kuijer, and M. H. W. Frings-Dresen, The concurrent validity of the ERGOS™ work simulator and the Ergo-Kit® with respect to maximum lifting capacity, *J. Occup. Rehabil.* 14(2):107–118, 2004.
27. V. Gouttebarge, H. Wind, P. P. Kuijer, J. K. Sluiter, M. H. Frings-Dresen, Intra- and interrater reliability of the Ergo-Kit functional capacity evaluation method in adults without musculoskeletal complaints, *Arch. Phys. Med. Rehabil.* 86(12):2354–2360, 2006.
28. V. Gouttebarge, H. Wind, P. P. Kuijer, J. K. Sluiter, M. H. Frings-Drensen, Reliability and agreement of 5 Ergo-Kit functional capacity evaluation lifting tests in subjects with low back pain, *Arch. Phys. Med. Rehabil.* 87(10):1365–1370, 2006.
29. T. G. Mayer, D. Barnes, N. D. Kishino, G. Nichols, R. J. Gatchel, H. Mayer, and V. Mooney, Progressive isoinertial lifting evaluation I. A standardized protocol and normative database, *Spine* 13(9):993–997, 1988.
30. T. G. Mayer, R. Gatchel, D. Barnes, G. Nicols, N. D. Kishino, K. Coval, B. Piel, D. Hoshino, and R. J. Gatchel, Progressive isoinertial lifting evaluation. II. A comparison with isokinetic lifting in a disabled chronic low-back pain industrial population, *Spine* 15(1):998–1002, 1990.
31. WorkAbility Systems, Inc., *Instruction Manual*, Total-Body Dexterity Tester, 2011.
32. R. Wickstrom, Test-retest reliability of WorkAbility fitness tests, *Proceedings of the APTA Combined Sections Meeting*, Boston, MA, 2007.
33. WorkAbility Systems, Inc., *Instruction Manual*, PAT Agility Mat, 2011.
34. R. Rossier and D. T. Wade, Validity and reliability comparison of 4 mobility measures in patients presenting with neurologic impairment, *Arch. Phys. Med. Rehabil.* 82(1):9–13, 2001.
35. T. Steffen and M. Seney, Test-retest reliability and minimal detectable change on balance and ambulation tests, the 36-Item short-form health survey, and the unified Parkinson disease rating scale in people with parkinsonism. *Phys. Ther.* 88(6):733–746, 2008.
36. R. J. Shephard, The current status of the Canadian home fitness test, *Br. J. Sports Med.* 14(2–3): 114–125, 1980.
37. M. Jette, J. Campbell, J. Mongeon, and R. Routhier, The Canadian home fitness test as a predictor of aerobic capacity, *CMAJ*. 114:680–682, 1976.
38. Canadian Society for Exercise Physiology, *Canadian Standardized Test of Fitness (CSTF) Operations Manual,* 3rd edn., Ottawa: Fitness and Amateur Sport, Canada, 1986.

39. S. F. Siconolfi, C. E. Garber, T. M. Lasater, and R. A. Carleton, A simple valid step test for estimating maximal oxygen uptake in epidemiologic studies, *Am. J. Epidemiol.* 121(3):382–390, 1985.
40. E. J. Bassey, P. H. Fentem, I. C. MacDonald, and P. M. Scriven, Self-paced walking as a method for exercise testing in elderly and young men, *Clin. Sci. Mol. Med.* 51:609–612, 1976.
41. K. H. Cooper, *New Aerobics,* Bantam Books, New York, 1970.
42. M. L. Pollock and J. H. Wilmore, *Exercise in Health and Disease: Evaluation and Prescription for Prevention and Rehabilitation*, 2nd edn., W. B. Saunders Co., Philadelphia, PA, 1990.
43. N. L. Jones, *Clinical Exercise Testing,* 3rd edn., W. B. Saunders, Philadelphia, PA, 1988.
44. American College of Sports Medicine, *ACSM's Guidelines for Exercise Testing and Prescription*, 7th edn., Lippincott Williams & Wilkins, Philadelphia, PA, 2006.
45. G. E. Caple, Energy expenditure modeling in the return-to-work decision process, *Appl. Ind. Hyg.* 3:348–352, 1988.
46. C. K. Anderson and M. J. Catterall, The impact of physical agility testing on incidence rate, severity rate, and productivity, in *Trends in Ergonomics/Human Factors,* Vol. 4, S. S. Asfour, ed., North-Holland/Elsevier, Amsterdam, the Netherlands, 1987, pp. 577–584.
47. V. Mathiowetz, N. Kashman, G. Volland, K. Weber, M. Dowe, and S. Rogers, Grip and pinch strength: Normative data for adults, *Arch. Phys. Med. Rehabil.* 66:69–74, 1985.
48. Lafayette Instrument JAMAR Hydrolic Hand Dynamometer, Model No. J00105, Lafayette Instrument Co., Lafayette, IN, 2004.
49. P. J. Bryden and E. A. Roy, A new method of administering the Grooved Pegboard Test: Performance as a function of handedness and sex, *Brain Cogn.* 58:258–268, 2005.
50. R. M. Ruff and S. B. Parker, Gender- and age-specific changes in motor speed and eye–hand coordination in adults: Normative values for the finger tapping and grooved pegboard tests. *Percept. Mot. Skills* 76(3 Pt 2):1219–1230, 1993.
51. R. J. Wickstrom and M. Shinnick, *MODAPTS Study of Grooved Pegboard Placing and Return Tasks.*
52. P. Carey and G. C. Heyde, *Heyde's MODAPTS: A Language of Work,* P. Carey, ed., Heyde Dynamics, Brisbane, 2001.
53. G. W. Good, *Occupational Vision Manual: Part II: Performance and Productivity in the Work Place*, American Optometric Association. Accessed on December 14, 2011 from www.aoa.org/x5361.xml, St. Louis, 2007.
54. Stereo Optical Co., *Reference and Instruction Manual, "Optec" Vision Tester www.stereooptical.com,* Chicago, IL.
55. U.S. Department of Transportation, *Federal Motor Carrier Safety Regulations, Part 391*, 1990.
56. American Speech Language Hearing Association, *Hearing Assessment.* Accessed from http://www.asha.org/public/hearing/testing/assess.htm, September 29, 2010.
57. *Code of Federal Regulations,* Title 29, Sec. 1910.95 (General Industry), 2008.
58. Equal Employment Opportunity Commission, *A Technical Assistance Manual on the Employment Provisions (Title 1) of the Americans with Disabilities Act,* 1990.
59. Americans with Disabilities Act of 1990 as amended in 2008. Accessed on December 14, 2011 from http://www.ada.gov/pubs/adastatute08.pdf
60. Equal Employment Opportunity Commission, *Uniform Guidelines on Employee Selection Procedures,* Sec. 60-3, 43 FR 38295 (Aug. 25):126–153, 1978.
61. NIOSH, *Work Practices Guide for Manual Lifting,* NIOSH Tech. Rep. No. 81-122, U.S. Dept. of Health and Human Services, Natl. Inst. Occupational Safety and Health, Cincinnati, OH, 1981.

CHAPTER 15

# Office Ergonomics

Mary O'Reilly

CONTENTS

## 15.1 INTRODUCTION

Since the first edition of this book not only have computers become ubiquitous, but also the very structure and definition of "office" has evolved. Desktop computers still basically have a keyboard, a screen, and a mouse; but each of these devices has morphed, often for the better, over the past 25 years. Keyboards and mice come in a variety of shapes and configurations and may be cordless. The screen has evolved into the flat panel that is more easily adjustable and much less prone to glare problems. Laptops are almost as ubiquitous as desktop computers, and handheld devices are commonplace. The concept of office as a place where everyone has a desk, a chair, and an assigned work area has often been replaced with a laptop that can be used anywhere, at home, in the car, or almost anywhere else.

Despite all these changes, musculoskeletal disorders (MSDs), also referred to as cumulative trauma disorders (CTDs) or repetitive strain injuries (RSIs), are unfortunately still with us. The main areas of the body involved with computer-associated injuries include the upper extremity, the shoulder and neck, as well as the low back. In addition to MSDs, computer users often encounter eyestrain.

The variety of office equipment including chairs, mice, keyboards, and monitors has also expanded, and the Internet and networking groups have made access to this information available to almost everyone. Some of the products touted as ergonomic may not be well designed. Even well-designed products do not fit every situation. The interaction between computer and user is complex. Changing one variable affects all the other relationships. For example, changing the height of the seat pan on the chair affects how the upper extremity interacts with the keyboard and mouse, and how the head and eyes are positioned to view the display screen.

## 15.2 METHODS

### 15.2.1 Risk Factors

Ergonomic assessments are done in a variety of ways. Dr. Thomas J. Armstrong and his associates (Silverstein et al., 1987; Armstrong, 1988) at the University of Michigan identified occupational risk factors associated with the development of MSDs in their studies of the meat-processing plants of the Midwestern United States (Armstrong et al., 1982). These occupational risk factors include repetition, force, sustained or awkward postures, contact stress, vibration, and cold. The factors relevant to office work are the first four.

Personal risk factors have also been recognized as being associated with carpal tunnel syndrome (Cannon et al., 1981; Silverstein et al., 1986; Silverstein et al., 1987; Kroemer, 1992). Personal risk factors are independent of the workplace and include genetics, previous injury, and lifestyle choices. Genetic risk factors include an individual's likelihood of contracting a disease such as diabetes, thyroid deficiency, or rheumatoid arthritis. In a study conducted at the Mayo Clinic, individuals with diabetes were up to 15% more likely to develop carpel tunnel syndrome than individuals without diabetes (Posch and Marcotte, 1976). Hormonal changes during pregnancy have also been associated with increased development of carpal tunnel syndrome (Gould and Wissinger, 1978).

In a controversial study, Nathan et al. (1992) reported that body mass index was one of the best predictors of who would develop carpel tunnel syndrome 5 years into the future. The subjects were divided into five groups from underweight to grossly obese. Although some people (factory workers) in each group developed carpel tunnel syndrome, people in the obese and grossly obese groups had about a 40% chance of developing carpel tunnel syndrome compared to about a 15% chance in the underweight group. The same study looked at exercise patterns. The lowest chance of developing carpel tunnel syndrome (about 10%–15%) was in the group that exercised every day, while the highest chance (about 40%) of developing carpel tunnel syndrome was in the group that never exercised (Nathan et al., 2001). Others have confirmed these observations (Becker et al., 2002). These are intriguing results for two reasons. First, they put CDTs in the same class as heart attacks, stroke, high blood pressure, and diabetes, in that healthy lifestyle choices help to reduce the risk of contracting them. Second, people have some choice over the lifestyle choices they make and, therefore, can be proactive.

Although the idea that stress can result in illness is not new (Selye, 1956), the implication that work organization, often referred to as occupational psychosocial factors, can be associated with the development of MSDs, particularly in the office setting was strengthened in the late 1980s and throughout the 1990s (Caryon-Sainfort, 1992; Caryon, 1993; Sauter and Swanson, 1996; Smith and Carayon, 1996). Examples of psychosocial stressors include monotonous work, high-perceived workload, time pressure, low control, and unsupportive social support (Bongers and deWinter, 1992; Bongers et al., 1993).

### 15.2.2 Risk-Based Protocols

In the 1990s, several groups both in the United States and England developed risk-based scoring schemes that rated the probability an individual would develop an MSD. Moore and Garg (1995) proposed the strain index to quantify the risk of distal upper extremity disorder while performing a variety of tasks at a pork-processing plant in the United States. The strain index methodology rates six task variables (intensity of exertion, duration of exertion per cycle, efforts per minute, wrist posture, speed of exertion, and duration of task per day) on a scale of less than or equal to 1 to greater than or equal to 8. The rating value is multiplied by an assigned multiplier to obtain a final score for each of the six variables. If the sum of the final scores is greater than 7, the job is probably hazardous.

The rapid upper limb assessment (RULA) methodology was developed in England specifically for computer users (McAtamney and Corlett, 1993). This work was based on three groups of factors (external load factors, factors that varied between individuals, and internal factors specific to each individual) described by McPhee (1987). The external factors include numbers of movements, static muscle work, force, work posture determined by the equipment and furniture, and time worked without a break. The factors that vary from individual to individual include adopted work postures, unnecessary use of static muscle work, speed and accuracy of movements, and the frequency and duration of work breaks. The internal factors include age, experience, workplace environmental factors, and psychosocial variable. The RULA methodology focuses on the external factors and gives a score to six body parts (upper arms, lower arms, wrist, neck, trunk, and legs) based on observed postures. Scores for static muscle exertion and load are also assigned. The integration of these scores produces a final score between 1 and 7. Scores of 1 and 2 are acceptable. Scores of 2 or 3 indicate a possible need to investigate further. Scores of 4 through 7 indicate remediation is required soon or immediately.

The rapid entire body assessment (REBA) methodology (Hignett, 1998; Hignett and McAtamney, 2000) incorporates similar scoring schemes to evaluate the entire body. In some ways, it is similar to the NIOSH lifting formula (Waters et al., 1993) but focuses on the entire body including upper and lower extremities rather than primarily on the back. The U.S. OSHA standard that was enacted in November 2000 and rescinded in April 2001 also follows a similar methodology. These approaches measure exposure to a variety of musculoskeletal stressors, or risk factors, and identify a level above which intervention should occur. This is a semiquantitative risk-based approach and is different from the original work which identified the risk factors but did not quantify exposure levels or identify at which level risk of injury is significantly increased.

### 15.2.3 Standards

Other methodologies include design specifications for offices and management system standards. The ANSI/HFES 100 (1988) outlines human factors engineering principles and practices to design and configure computer workstations. The document offers quantitative specifications for a variety of computer workstations.

The ANSI/AIHA Z10 (2005) Occupational Health and Safety Management System Standard is a voluntary consensus standard that provides critical management systems requirements and guidelines for improvement of occupational health and safety and includes consideration of human factors. The Z10 standard follows the structure of ILO standards such as ILO 9000 and ILO 14000 as well as the OHSAS 18000. In the 1990s, the ANSI Z-365 committee convened to develop a standard on the control of work-related CTDs. This standard has not been published. The American Industrial Hygiene Association, however, has developed an Ergonomics Program Guidance Document (2008) that incorporates the information developed by the ANSI Z-365 committee and integrates it into a format compatible with the Z-10 standard.

### 15.2.4 Other Sources of Information

The OSHA requires employees to keep records of employee illnesses and injuries. These records can provide a way to identify the incidence of MSDs as well as what jobs have MSDs associated with them. More detailed and often more accurate information can be obtained from medical insurance and worker compensation records. These records are not, however, publicly available because of privacy concerns. The information obtained from this type of record provides retrospective data only.

Symptom surveys are used to get proactive data. Most symptom surveys consist of rating the amount of pain or discomfort in specific parts of the body such as hand, wrist, shoulder, and neck. A rating scale typically between either 1 and 5 or 1 and 10 is used to convey the intensity and/or the duration of pain and discomfort. The value of symptom surveys is that they can identify employees with incipient musculoskeletal injuries so that effective intervention can prevent the development of more serious injury.

## 15.3 EVALUATION OF THE WORKPLACE

Evaluation of existing workplaces requires identification of risk factors associated with musculoskeletal injuries and the implementation of a remediation plan to reduce or eliminate them. In addition, an ergonomics program that includes employee training, medical surveillance (when necessary), follow-up, and purchasing guidelines for new equipment should be established to ensure maximum benefit. In situations where large numbers of workstations are evaluated, it is important to prioritize the work. The risk-based methods discussed earlier may be useful to accomplish this.

One of the most effective ways to evaluate a workstation is to watch the person working at his or her computer for a period of time (so they become comfortable with being observed) and note the risk factors. The most common risk factors include force, repetition, posture (awkward and/or sustained), mechanical or contact stress, eyestrain, and psychosocial stress.

### 15.3.1 Force

Although evaluation of industrial workers has indicated that excessive force is associated with an increased incidence of MSDs (Armstrong and Chaffin, 1979; Keyersling et al., 1991), muscular exertion is usually not associated with computer use. Computer users, however, may use more force than is necessary when keying (Armstrong et al., 1991a). When this happens, the extra force is transferred to the tendons and soft tissue of the hand and forearm and may be significant especially when coupled with the number of repetitions throughout the day (Armstrong et al., 1991b; Feuerstein et al., 1997; Amell and Kumar, 2000; Gerard et al., 2002). Even when force remains constant, joint torque can be reduced when the fingers are flexed (Harding et al., 1993; Jindrich et al., 2004) and/or when the direction of key travel is away from the user (Balakrishnan et al., 2006). Definition of an acceptable level of force for the digits is, however, elusive (Nussbaum and Johnson, 2002).

The use of excessive force when keying can be the result of habit, internal mood state of the computer user, or deadlines and other job stress, and is related to musculoskeletal symptoms (Szeto et al., 2005). For example, monitoring of performance without fully involving the employee in the monitoring process can lead to increased stress, which is then translated to increasingly forceful exertions while keying.

While the use of excessive force may be behavioral, it also appears that different keyboards require different amounts of force to depress the keys (Armstrong et al., 1991a). A keyboard with a light touch reduces the amount of force required to activate the keys, but this translates into less stress on the upper extremity only if the computer user makes the behavioral modifications that result in the application of less forceful exertions to depress the keys.

Other examples of forceful exertions in office work may be the use of manual staplers and the physical arrangement of the workstation. When an office worker must lift a heavy manual from an overhead shelf or retrieve a large file from under the work surface, forceful exertions are necessary.

### 15.3.2 Repetition

Repetition is associated with an increased incidence of MSDs (Luopajarvi et al., 1979; Cannon et al., 1981; Armstrong, 1988). Keying is the most obvious example of repetitive movements associated with the use of computers. When keying at a rate of about 70 words/min, an individual performs approximately 2100 exertions per hour with each finger, assuming an average of five letters per word and an equal number of key strokes for each of the 10 fingers. One of the earliest recommended limits for human tendons is 1500–2000 repetitions per hour (Hammer, 1934).

In a study of 29 data entry personnel everyone keying at a rate greater than 11,000 keystrokes per hour reported pain, whereas only 60% of individuals keying a rate of less than 11,000 keystrokes per hour reported pain (Thatcher and Brophy, 1999).

With increasing keying speed, both the finger force and the electromyographic activity of finger flexors and extensors increase, but the ratio of the force produced by the flexors and extensors changes. The relative contribution of force from the finger extensor muscles decreases in individuals reporting no increase in pain during the increased keying speeds (Sillanpaa et al., 2003).

The amount of total repetition also depends on the length of time spent using the computer. Musculoskeletal symptoms typically increase with increased continuous time of use (Matias et al., 1998; Demure et al., 2000), but all do not agree (Sillanpaa et al., 2003).

### 15.3.3 Posture

Certain postures are associated with an increased incidence of musculoskeletal pain and/or injury. The postures do not cause MSDs but, if the design of a workstation requires that a person using that workstation conform to an awkward shape for extended period of time, the probability increases that an injury will occur. In addition, any posture rigidly held for an extended period of time increases the probability of musculoskeletal injury. Our bodies were made for movement. Sitting and standing for extended periods of time increases the likelihood of adverse effects.

Insufficient recovery time is one of the contributing factors to muscle fatigue (Sjogaard et al., 1986, 1988; Guyton, 1992). In sustained static exertions, the activity is continuous. The muscle is acting isometrically producing tension but not changing in length. Some examples of activities that require sustained static exertion are prolonged sitting, holding the hands above the keyboard, prolonged gripping of the computer mouse or phone handset, and maintaining the shoulders in a set position while keying.

The extensor muscles of the forearm and fingers are activated when the hands are held over the keyboard. These muscles must continue to hold the hands in position when the flexor muscles of the fingers contract to strike the keys. The extensor muscles of the fingers are performing eccentric contractions (as opposed to isometric or isokinetic contraction) during keying and this may account for the pain and/or discomfort reported by some computer users in the backs of their hands and wrists. Eccentric contraction is the lengthening of a muscle against its own tension, for example, when the pointer finger flexes to hit the "H" key while the hand is held steady over the keyboard. The higher the keyboard, the harder the extensor muscles have to work to position the hands during keying.

*Note*: Another example of eccentric contraction is walking down a mountain. Hikers know that muscles, particularly the quadriceps on the fronts of the thighs, hurt more from walking down the mountain than from walking up the mountain. This is because the quadriceps is lengthening under tension. The same phenomenon occurs walking downstairs.

Awkward postures have been associated with both muscle-tendon disorders and nerve compression disorder (Marklin, 2003). In an office setting, awkward postures are most likely to occur in the neck, shoulder, wrist/hand, and low back. In order to identify awkward postures, it is important to know the neutral position for the neck, shoulder, wrist/hand, and low back. There is a range of movement around the neutral position that is also considered neutral, but when the body is forced to assume and maintain a position that deviates beyond this range of motion in the neutral zone, the probability of developing an MSD increases. Both RULA and REBA define a neutral zone and identify postures outside that zone as problematic.

#### *15.3.3.1 Neck Concerns*

An area of concern among computer users is neck pain or discomfort. Because it is essential that the computer user be able to see the information on the display screen, the neck often places and holds the head in awkward positions. Neck flexion brings the chin closer to the sternum or chest. Neck extension makes the chin jut out and raises the gaze upward. The neck can also turn the head to either side. The neutral position for the neck is with the crown of the head drawn upward and the chin tucked in but not bent toward the chest. This is the position that requires the least muscle activity to maintain.

Studies show that neck flexion between 0° and 15° produces a minimal level of extensor muscle exertion (Chaffin, 1973; Chaffin et al., 1999), but that 30° of flexion produces up to 50% of the muscle force generated in the extensor muscles during full neck flexion (Kumar and Scaife, 1979). Several models predict that neck flexion of 15° or less would use less than 5% of the maximum exertion strength of the neck extensor muscles, but 30° of flexion would require 13% of the maximum exertion strength of the neck extensor muscles (Snyder et al., 1975; Moroney et al., 1988; Vasavada et al., 1998). One of the main reasons for neck flexion during computer use is a work surface that is too low.

Neck extension, on the other hand, increases muscle exertion within the first few degree deviation from neutral. The exertion of the extensor muscles increases as the angle of flexion increases. One of the two main reasons for neck extension while viewing the display screen is to position the head so that the gaze can be directed through lens located in the lower area of glasses frames (Ferguson and Duncan, 1974; Life and Pheasant, 1984; Sauter et al., 1991; Marklin, 2003). The other main reason for neck extension during computer use is that the display screen is too high.

Monitors that are placed in a position that requires neck extension, that is, looking up, have been associated with reports of increased discomfort and pain in the neck and shoulders (Grandjean et al., 1983; Ankrum and Nemeth, 1995). In both these cases, the chin of the computer user juts out when the neck is extended even slightly. Placing the top of the monitor at or below the line of sight resolves these issues and also provides the computer user with more options for slight postural changes (Ankrum et al., 1995).

Maintaining the head twisted either to the right or the left also requires increased muscle activity and has been associated with neck pain. The display screen should be placed directly in front of the computer user. In addition, any copy material should also be placed so that the computer user does not have to twist to view the material to be copied. Document holders should be placed in the same line of sight as the display screen.

#### *15.3.3.2 Shoulder Concerns*

The neutral position for the shoulder is for the arm to be hanging loosely at the side of the torso. Flexion of the arm at the shoulder is used to position the hands for keying when the keyboard is too high and too far forward. If the keyboard is too high, the shoulders are often scrunched (abduction of the deltoids and contraction of the trapezius muscles) in an effort to raise the hands to the appropriate level. Scrunched shoulders can result in pain and/or discomfort in both the neck and the shoulder. When keyboard placement is inappropriate, the pain and/or discomfort typically occurs in both shoulders and across the neck.

When the mouse is placed on the desk next to the keyboard, the mousing arm has to be both flexed and abducted. The extensor muscles of the mousing forearm and hand operate under eccentric contractions. This position is often held for long periods of constant mouse use and results in shoulder pain and/or discomfort in the mousing arm. The awkward static posture associated with reaching for the mouse can be associated with pain and/or discomfort in the entire mousing upper extremity.

Cradling the telephone receiver with the shoulder is another posture associated with shoulder and neck pain. People who are on the telephone or any other communication network should have a cordless headset that allows them to communicate in a neutral position.

#### *15.3.3.3 Hand/Wrist Concerns*

Extending or flexing the wrists while keying as well as bending the wrist sideways away from the thumb (ulnar deviation) while trying to reach the keys or numbers on the number pad has been associated with an increased incidence of MSDs. Extreme flexion or extension of the wrist increases the pressure in the carpal tunnel (Gelberman et al., 1981; Armstrong et al., 1991b; Rempel et al., 1992; Keir et al., 2007). Under extreme conditions, the pressure in the carpal tunnel is increased 20–30 times.

One common cause of flexion or extension of the wrists when keying is improper adjustment of the keyboard height. If the keyboard is too high, computer users will often extend their wrists to compensate. A statistically significant increase in muscle activity was reported in a tapping task when the work surface was raised 2 in. (Zennaro et al., 2004).

Extrinsic extensor forearm muscles are used to position the hands over the keyboard. Keir and Wells (2002) have developed a model to evaluate the contribution of static loading to the development of upper extremity pain. Their research suggests that a neutral wrist position will minimize the risk of developing a musculoskeletal injury while keying. Others suggest sloping the keyboard away from the user while keying to minimize wrist extension (Hedge and Powers, 1995; Simoneau and Marklin, 2001).

A type of movement unique to the forearm-hand unit is supination and pronation. In the supine position, the hand is rotated so that the thumb is lateral, that is, away from the torso with the palm facing forward or upward. In the pronated position, the thumb is medial next to the torso with the palm facing backward or downward. The neutral position is somewhere in the middle between extreme pronation and extreme supination. Rotation of the forearm is performed by one supinator muscle and two pronator muscles located in the forearm. When the hands are positioned over the keys of a traditional keyboard, they are in extreme pronation.

#### *15.3.3.4 Low Back Concerns*

The amount of time spent in a seated position has dramatically increased with the advent of modern transportation, entertainment, and computer use. Both awkward postures and sustained immobility may contribute to low back problems associated with sitting.

In the office, computers are typically used in the seated position. Although sitting is generally preferred because it uses less energy than standing and the lower limb does not have to bear the entire weight of the body, it also introduces specific physiological problems such as flattening of the lower back and decreased muscle tone in the abdominal muscles. When the lumbar curve is flattened, there is an increase in pressure in the intervertebral discs in the lumbar region (Nachemson and Morris, 1964; Okushima, 1970; Nachemson and Elfstrom, 1970; Tzivian et al., 1971; Andersson and Ortengren, 1974; Chaffin and Andersson, 1984). Slumped sitting and unsupported forward sitting increase the intervertebral disc pressure even further.

The intervertebral discs themselves are cartilaginous and have no blood vessels or nerves. Pain associated with injured discs is due to the disc compressing adjacent nerves and not to pain originating in the disc itself. Without blood vessels, the cells that make up the discs depend on gross body movement to facilitate diffusion of oxygen into the tissue and metabolic wastes out of the disc tissue. When the same position is maintained for prolonged periods of time, the rate of diffusion decreases and the discs are oxygen deprived and bathed in their own metabolic waste products. Neither of these conditions promotes healthy cells.

Sustained posture also requires protracted isometric contraction of the interspinous muscles and other supporting musculature. The increased endomuscular pressure is associated with edema and decreased blood flow through the contracting muscles. The decreased blood flow results in hypoxia and the buildup of waste metabolites which may, over time, lead to muscle and/or tendon soreness (Grieco and Molteni, 2003). Even when seating is optimal, prolonged sitting may be associated with discomfort and, eventually, musculoskeletal injury (Graf et al., 1991, 1993, 1995; Grieco and Molteni, 1999, 2003). Short periods of sitting cross-legged, slumping, slouching, and forward-leaning are all better than prolonged sitting in one posture (Kramer, 1973).

### 15.3.4 Contact Stress

Localized mechanical stress, or contact stress, is caused by the continuous pressure of a sharp or hard surface against the soft tissue of an upper or lower limb. For example, injury of the ulnar nerve in a data entry clerk has been described as the result of the desk edge pressing against the soft tissues of the upper extremity (Todaro, 1990). When the back of the thigh presses against the hard edge of the front of a chair seat, contact stress occurs in the lower limb, which can result in decreased blood flow and nerve compression.

Even if the chair has a rounded or waterfall edge, contact stress can still be experienced if the feet of the chair user cannot reach the floor or if the seat pan is too long. Compression of the buttocks and backs of the thighs is often associated with swelling of the lower extremities (Noddeland and Winkel, 1988; Stranden, 2000; Chester et al., 2002). Because leg discomfort and swelling is related to immobility as well as contact stress, anything that encourages even slight adjustment in position can help to minimize leg discomfort (Pottier et al., 1969; Winkel and Jorgensen, 1986; Sauter et al., 1991). Another approach to relieving leg discomfort is to properly adjust the height of the seat pan (Burandt and Grandjean, 1963). Chairs that encourage continuous passive movement reduce lower leg swelling associated with discomfort (van Deurson et al., 2000). The use of a stability ball for seated office work, however, is not recommended in the workplace (Gregory et al., 2006).

A footrest may be helpful in situations where the chair cannot be adjusted. The problem with footrests, however, is that they often limit leg and foot movement by restricting where they can be placed (ANSI/HFES 100, 2007).

### 15.3.5 Eyestrain

Eyestrain is the most frequent physical symptom suffered by computer users (Smith et al., 1981; Dain et al., 1988). The ocular symptoms reported by computer users are usually neither permanent nor disabling, but an accurate estimate of their incidence is not available. Many optometrists believe that a significant percentage of the patients they evaluate have sought medical help for symptoms that are associated with computer use (Sheedy, 1992).

Using a computer is primarily a visual task, and the user will typically assume whatever posture is necessary to read the display screen. Not only is placement of the screen critical, but evaluation of the properties of the screen itself is also essential (Arndt, 1983; Helander et al., 1984; Laubli and Grandjean, 1984; Anshel, 1998; Dainoff, 2003). The visual risk factors associated with computer use can be divided into three categories, namely, the surrounding environment, the inherent properties of the human visual system, and the properties of the display screen itself (O'Reilly et al., 2007).

#### *15.3.5.1 Surrounding Environment*

Surroundings include air quality, temperature range, lighting quality and quantity, noise, and other indoor environmental quality considerations. All of these are topics beyond the scope of this chapter. However, there are a few points that are particularly relevant to computer use.

Although moderate levels of lighting in the range of 200–500 lux are recommended (Illuminating Engineering Society, 1981; ANSI/HFES, 1988, 2007), the appropriate quantity of light is dependent on the age of the worker, the type of printed material being used, and personal preferences. Higher light levels have been associated with increased complaints (Nemecek and Grandjean, 1971); others prefer lower levels of about 100 lux (van der Heiden et al., 1984). In general, older workers require higher levels of lighting.

Flat-screen displays have minimized concerns about glare by including built-in filters and by being quite easy to reposition. If glare is present, however, it is important to recognize the two types. Direct glare results from light shining directly into the eyes. Indirect glare results from light bouncing off shiny or reflective surfaces, such as display screens or glossy paper. Shielding or changing the spatial relationship between the light source and the glare is the easiest way to reduce or eliminate direct glare. Indirect glare can be reduced or eliminated by changing the angle of the screen or document or shielding them from the light source.

The ratio between the brightness of the screen and the surrounding objects (clothing, desk, paper, walls, and furnishings) can be a source of discomfort during computer use (Luckiesh and Guth, 1949; Guth, 1958, 1981). The Illuminating Engineering Society (Illuminating Engineering Society, 1981) suggests a ratio of 3:1 for screen luminance to immediate surroundings and a ratio of 10:1 for screen luminance to more general surroundings (Anshel, 1998).

Artificial light sources have an inherent flicker rate. Although humans cannot cognitively recognize flicker when it is a higher rate than between 50 and 60 Hz, eyestrain and headache occurred significantly less often when high-frequency ballast lighting with a flicker rate of 3200 Hz was used instead of normal fluorescent lighting with a flicker rate of about 100 Hz (Wilkins et al., 1989).

#### *15.3.5.2 Human Vision*

Although even a cursory review of the human vision system is way beyond the limits of this chapter, the concepts of accommodation and convergence are extremely important to understand the interaction between sight and computer use (Gur et al., 1994).

Accommodation is the process through which the eye focuses incoming light on the retina. Involuntary muscles control the shape of the lens. The muscles contract and allow the lens to assume a more spherical shape to view close objects (Guyton, 1992). As people age, the lens becomes rigid and cannot change shape despite the contraction of the involuntary muscles. This is why people with normal vision begin to require glasses to perform close work as they age (typically between 40 and 45 years of age). When the corrective lens is placed in the lower portion of the glasses frame, the person wearing the glasses has to tilt his or her head backward to look out of the corrective lens. Computer glasses avoid this problem by having the corrective lens fill the entire glasses frame. Computer glasses cannot be used for distance viewing, but can be used to view the display screen without tilting the head backward.

Convergence is the ability to use the voluntary muscles of the eye to focus both eyes simultaneously on a close object. Each eyeball has six small voluntary muscles that usually work in tandem to move the eyeballs in a coordinated fashion. When viewing close objects, such as reading material or the display screen, the voluntary eye muscles must sustain force on the eyeballs so that the gaze converges on the nearby object.

Constantly focusing on nearby objects can fatigue both the involuntary and the voluntary eye muscles. Looking back and forth between two nearby objects that are different distances from the eyes requires refocusing and can also result in muscle fatigue. This often happens when a document is much closer or farther from the eyes than the display screen.

Dry eyes are a frequent complaint of computer users. Blinking is one way to keep eyes moist. Several studies have shown that humans blink at significantly lower rates when viewing display screens (Acosta et al., 1990; Yaginuma et al., 1990; Patel et al., 1991; Tsubota and Nakamori, 1993; Nakaiski and Yamada, 1999; Doughty, 2001; Schlote et al., 2003). Monitor placement also affects the percentage of eyeball exposed to ambient air. When the monitor is lower, the eyelid can cover more of the eyeball and thus reduce the surface area from which evaporation can occur (Anshel, 1998).

Some visual complaints from computer users may also be associated, at least in part, with underlying psychological stress (Mocci et al., 2001).

### 15.3.6 Psychosocial Stress

Psychosocial stress is a significant factor in the development of MSDs, including low back pain, as well as in the length of time away from work once an MSD has been recognized (Hult, 1954; Lawrence, 1955; Maeda et al., 1980; Smith et al., 1981; McPhee, 1982; Oxenburgh et al., 1985; Green and Briggs, 1989; Waersted et al., 1991; Moon and Sauter, 1996). Stress also influences behavior such as the amount of force used during keying, reporting of medical problems, the tendency to maintain a rigid posture, and the amount of sick time taken (Scalet, 1987).

Stress in computer users is multifactorial, and may be as much related to the total job and organizational structure as to computer use itself. Job characteristics that have been linked to psychosocial stress include monitoring and other forms of performance pressure, rigid work procedures, relentless schedules, lack of social support, job insecurity, job dissatisfaction, and monotony (Grandjean et al., 1984; Pickett and Lees, 1991; Demure et al., 2000). For example, the relative risk of pain in both the neck and the shoulder was three times greater in a group of 210 secretaries who reported poor working conditions compared with a group of 210 secretaries doing similar jobs but who reported good working conditions (Linton and Kamwendo, 1989).

### 15.3.7 Multiple Risk Factors

When two or more risk factors are present simultaneously, they often act synergistically. This is true for awkward and sustained postures, as well as for other risk factors, for example, repetition and force. Because using a computer requires long periods of minimal movement, it is very difficult to remove the sustained postures. That makes it even more important to provide a workplace that allows the computer user to work in a neutral posture.

## 15.4 DESIGN

When an office is designed from scratch, several initial steps need to be taken: identify the user population, specify performance criteria, and understand underlying health, safety, and comfort concerns. Identification of the user population is important in the global context because groups of individuals have significantly different physical characteristics, as do males and females. Adjustability is the underlying key concept. Unfortunately, most ergonomic evaluations are done in offices that are already established.

Although people typically have assigned workstations in offices, the user of the workstation often changes over time. The height of working surfaces must be adjustable to accommodate individuals that may vary significantly in height and size. The surface height in workstations assigned to an individual may be set at the appropriate height. That height, however, will need to be adjusted for the next individual assigned to that workstation unless he or she is the same size. Many offices have surfaces that can be adjusted in height by maintenance personnel to fit the assigned user, but cannot be adjusted on a daily basis. File cabinets that slide under work surfaces may preclude lowering the surface to a height appropriate for smaller individual users.

Some computer workstations are used by many individuals during the course of a day, or may be shared by two or three individuals, for example, during shift work. These workstations must be fully adjustable so that each individual user can be easily accommodated.

For a discussion on chairs please, see Section 15.5.

Once the user population has been identified and the work flow designed, the general design principles should be applied (Sanders and McCormick, 1993). This is where the entire range of human variability is considered. Most anthropometric variables follow the familiar bell-shaped curve distribution. Design can be for the average or the extreme. The average can include both male and female dimensions, or it can be based on male or female dimension only. In designing for the average, the designer chooses where to place the cutoff for the tails. For example, if the designer wants to include 90% of the population, the cutoff will exclude 5% of the smallest and 5% of the largest dimensions.

Functionality needs to be considered when deciding which extreme to use. If the function is something that must be easily reached, for example, a brake pedal or stop button, or in the office the keyboard, the smallest person performing the function needs to be considered. If the function is something that should include everyone, for example, seat pan width, the largest person should be considered. The term used to denote the first approach is reach; the second approach, clearance.

Even when the design of a parameter allows for adjustability, for example, seat pan height, it is usually not possible to include individuals at the very extreme ends of the distribution curve. This is why it is often useful to have different sizes available for the individuals at the extreme. Goldilocks and the three bears is a deep-rooted example of this type of design.

There are advantages and disadvantages to each approach. For example, how high should the desks be to allow everyone to place their knees underneath? The clearance principle would suggest that the underside of the work surface should be high enough so that the tallest person can fit his knees under it. If this principle is followed, however, the height of the desktop (inextricably related to the height of the desk underside) will be too high for the smallest user. If the desk surface height is designed for the average user, both the tallest and the smallest workers will be unable to use it, and most of the workers will be uncomfortable using it. The use of adjustable keyboard trays helps to address this problem. Adjustability, however, cannot solve all the concerns, especially for those individuals at either end of bell shaped distribution of size parameters.

## 15.5 EQUIPMENT

### 15.5.1 Chair

The single most important piece of office furniture is arguably an easily adjustable chair. When people use computers for the majority of their time at work, it is essential that the chair be adjustable in height, provide lumbar support, and have an adjustable seat pan and backrest, as well as rounded edges to reduce contact stress. Armrests are desirable only if they are adequately adjustable. Chair adjustability is also important to avoid leg discomfort which is associated with static posture, compression of the back of the thighs, and seat pans that are too high for the user (Burandt and Grandjean, 1963).

Although many different chair designs are available, many do not meet even minimal standards (Sanders and McCormick, 1993; Corlett, 1995). Helander et al. (1995) identify 10 separate features as essential in evaluating the functionality of an office chair, namely, seat height, seat depth, lumbar support height, lumbar support depth, armrest height, seat pan tilt, seat pan tilt tension, backrest height, backrest angle, and backrest tension.

A well-designed office chair minimizes leg discomfort associated with compression of the back of the thigh and postural immobility (Winkel and Jorgensen, 1986; Pottier et al., 1969; Sauter et al., 1991) by providing rounded edges on the seat front as well as adequate cushioning over a firm support. Some research indicates that the seat height contributes to leg discomfort (Burandt and Grandjean, 1963). For example, if the seat height is too high, it will compress the back of the thighs.

A chair designed to tilt, as opposed to a fixed position one, encourages continual small muscle changes to adjust posture. Some chairs flex or hinge to follow the user through different positions; others are larger than necessary to allow the user to shift position often (Callaghan and McGill, 2001; deLooze et al., 2003). Chairs that encourage continuous passive motion have been shown to reduce lower extremity swelling (van Deursen et al., 2000). The use of the stability ball in an office setting, however, was not recommended (Gregory et al., 2006).

It is important to provide lumbar support in the sitting position. This support restores at least some of the lumbar curve lost in the sitting position. Many chairs have a lumber support built into the seat back. For this support to be effective, it must be at the appropriate level, that is, at the level of the L5 disc. This is approximately at, or slightly below, belt level. A rolled towel or contoured pillow can be used in chairs without a lumbar support as an interim corrective measure. The thickness of the roll should be less than 2 in. (5 cm). Research indicates that back rolls that are too thick have resulted in increased muscle activity in the lower back (Zacharkow, 1988). Chairs that provide adequate pelvic support also reduce muscle fatigue (Cram and Vinitzky, 1995).

One way to decrease the compressive force on the lumbar discs in the sitting position is to open the angle between the thighs and the torso. This can be done by tilting the seat pan forward or reclining the backrest. Reclining the backrest at least 10° rotates the pelvis to help restore the lumbar curve and also transfers some of the torso weight to the backrest (Keegan, 1953; Schoberth, 1962; Bendix, 1987; Lengsfeld et al., 2000; ANSI/HFES 100, 2006, 2007). Harrison et al. (1999) report that seating with lumbar support and a backrest angle of 110°–130° provides the lowest compressive force on the lumbar discs as well as the lowest muscle activity in the spinal musculature.

Another way to transfer some of the body weight from the spine is to use armrests (Nachemson and Elfstrom, 1970; Andersson and Ortengren, 1974; Chaffin and Andersson, 1984; Congleton et al., 1985, 1988; Chaffin et al., 1999; Makhsous et al., 2003). These must be used at the proper height, which is approximately elbow height when the arms are hanging freely at the side of the torso. Chairs with armrests that are not adjustable may cause more problems than they solve. When armrests are too low, the person using the chair tends to slouch or bend to make the body fit the chair. If the armrests are too high, they may force the chair user to maintain abducted arms, which can contribute to neck and shoulder pain. The armrests also need to adjust low enough to clear the work surface so that the user can get positioned to use the keyboard in a neutral position.

All the best features in a chair will not contribute to worker health, comfort, and productivity if the chair does not fit the individual and is not adjusted properly. It is important to emphasize that not only should the chair be adjusted to each individual, but that each individual may need to make small adjustments periodically during the workday. The best designed chair works well only if workers understand the need to, as well as how to, adjust their chairs. In addition to being well designed, chairs should also be aesthetically appealing (Legg et al., 2002; Helander and Zhang, 1997; Helander, 2003).

### 15.5.2 Keyboard

Traditional flat keyboards require close to full pronation of the forearm typically with ulnar deviation of both wrists for the fingers to reach the lateral keys. If the keyboard is too high, wrist extension may also contribute to awkward wrist posture. Ergonomically designed keyboards address these issues in a variety of ways. Some do this by splitting the keyboard in the middle; others are designed so that the keyboard is raised in the middle like a tent; still others may offer both configurations (Simoneau and Marklin, 2001).

Because ulnar deviation of the right wrist to position the fingers over the number pad is one of the most awkward postures commonly seen, many keyboard designers have a detachable number pad. If the computer user needs to do a lot of work with numbers, the number pad can be repositioned so as to minimize ulnar deviation and other awkward postures of the upper extremity.

### 15.5.3 Mouse

Mouse use has increased with various computer applications including e-mail, CAD design, and web surfing. Many office workers use the mouse for half or more of the time they spend using a computer each day. The length of time using the mouse each day seems to have less impact than the overall amount of time spent using a computer (Blatter and Bongers, 2001; Jensen et al., 2002).

Several different sizes and types of mice are available to use, but different mice are more useful for some applications than others. For example, CAD design requires precise placement control and some input devices do not provide that.

As discussed in the section on Evaluation of the Workplace reaching to hold the mouse positioned on a desk surface requires static awkward positioning of the upper extremity and the reporting of increased musculoskeletal symptoms (Jensen et al., 1998; Cook et al., 2000). The awkward postures include abduction of the arm, activation of the forearm/hand extensor muscles, full pronation of the forearm, and, quite often, ulnar deviation. These positions can also contribute to musculoskeletal symptoms in the shoulder and neck. When the right hand is used to operate the mouse, adverse musculoskeletal effects can be exacerbated (Cook and Kothiyal, 1998) because the number pad to the right of the alphabet section of the keyboard requires the mouse be placed even further to the right requiring increasingly awkward posture (further reaching with greater ulnar deviation).

Positioning the mouse lower and toward the midline of the user can help to reduce postural stress. Some keyboards do this by attaching a mouse pad that can be positioned along the edge facing the user. Use of a touch pad or roller device that can be positioned in the center between the computer user and the keyboard also achieves this goal. Forearm support while mousing is another way to reduce musculoskeletal stress (Conlon et al., 2008).

Rotation of the typical two button mouse so that it is oriented vertical rather than flat with respect to the desk is another way to reduce full pronation and ulnar deviation (Lee et al., 2007).

### 15.5.4 Monitor

Most humans prefer to read from paper than from display screen for many reasons (Dillon, 1992). Resolution, measured in dots per inch (dpi), is one factor that affects readability. Paper provides a resolution of about 225 dpi. Display screens typically provide resolutions between 60 and 120 dpi. High-resolution monitors (greater than 90 dpi) may promote better performance (Ziefle, 1998), but they also produce smaller images that may be perceived as harder to read (Gould et al., 1987; Harpster et al., 1989; Miyao et al., 1989).

Flicker is dependent on the refresh rate of the display screen and is measured in cycles per second or hertz (Hz). Humans can consciously detect flicker at rates below 50–60 Hz (Wilson and Corlett, 1995; Anshel, 1998). Flicker in the 10 Hz frequency may precipitate seizures in epileptics (Howarth, 1995). As discussed in the section on Human Vision, headaches and blurred vision have been associated with flicker rates that are not consciously perceived, but nonetheless have a physiological effect.

Computer users need to know what adjustments they can make on their display screens to adjust focus, brightness, and contrast. Focus is affected by internal and external adjustments. If the image seems blurred, screen focus should be checked first. It may be caused by too high settings of brightness or contrast, or it might be the result of internal problems. Font size and refresh rate can be adjusted through internal computer settings (HFES, 2006). Special software programs are available that can be used by individuals with vision impairments.

Flat-panel display units are replacing the cathode ray tube monitors that were in use in the 1990s. The newer monitors control glare very well and are easier to position for both height and slant.

### 15.5.5 Desk

Typical office desks are 29–30 in. high. Although this works well for many tasks, such as reading, writing on paper, and talking on the phone, it is not an appropriate height for keying or mousing for most people. Unless the workstation is to be used exclusively for computer work, the traditional surface height works well for most people. The exceptions are very tall or very short people, the two ends of the bell-shaped curve for height distribution.

Many types of modular furniture have adjustable desk surface heights, but the adjustment requires tools and effort and is, therefore, not an option for small or daily changes. It does, however, allow for a change in height of the work surface when the occupant of the workstation is reassigned.

Another option for changing the work surface height is an easily adjustable worktable. The adjustment can be done either with a crank or a hydrolic mechanism. The entire work surface with all its contents, keyboard, mouse, monitor, phone, paper, and whatever else is on the desk, can be adjusted up or down during the course of the workday. Another advantage of the adjustable height table is that it allows the computer user to easily change from working in a seated position to working in a standing position. The adjustable height worktables are expensive ($1500–3000 in 2010 American dollars).

### 15.5.6 Keyboard Tray

There are a variety of keyboard and mouse trays available. The first step is to make sure that the mechanism that attaches the keyboard tray to the underside of the desk is easily adjustable for both height and slant, and sturdy enough to hold the position set by the computer user. Then users have to become comfortable with their ability to make small adjustments to the keyboard tray throughout the work day.

## 15.6 IMPLEMENTATION OF OFFICE ERGONOMICS PROGRAMS

### 15.6.1 Cost

Good ergonomics is good economics (Hendricks, 1996). Ergonomics is not just about making people comfortable and healthy, but it is also about giving people the tools to do their jobs efficiently and effectively. Several reports have documented this (Webster and Snook, 1994; Oxenburgh, 1997; Alexander and Albin, 2003; Getty, 2003).

It is difficult to measure productivity in the office with a single tool because the product that many office workers produce is often intangible, such as customer satisfaction or accurate record keeping (medical, legal, and other types of records). In addition, office work is not a homogeneous type of work but encompasses a variety of tasks and goals. Data entry is very different from billing or journalism despite the use of computers for each application.

Although the costs of designing and implementing effective ergonomics interventions are apparent, the cost savings often are less obvious particularly in an office setting where productivity is more difficult to measure compared with manufacturing operation, for example, where widgets are countable (Sommerich, 2003).

Other indications such as absenteeism and turnover rates (including severance, hiring, and training costs) can be used to calculate the cost benefits from ergonomics implementation. Consider a person who makes $52,000 a year and is absent 5 days a year due to musculoskeletal pain. Assume for the sake of simplicity that wages are, therefore, equal to $1000 per week. If an intervention costs $1000, including easily adjustable chair and work surfaces, and the employee is no longer absent 5 days a year because the source of the work-related musculoskeletal pain has been removed, the cost of the intervention is recouped in 1 year based on wages alone. Even if the number of days absent is reduced from 5 to 2, the cost of the intervention is recouped in 3 years.

Each workplace will be different, but this type of analysis needs to be done and presented as part of ergonomics intervention plans. In some situations, the cost of absenteeism may be the driver, in others the costs of hiring and retraining individuals may be the most expensive facet of poor workplace ergonomics. What is clear is that poor design always has a cost because it prevents people from doing the best job possible.

Another consideration is the direct cost of injuries. This often requires cooperation among many departments that typically operate independently within the corporation or organization.

### 15.6.2 Purchasing

Because office equipment and furniture is constantly turning over and new items are bought throughout the year, it is essential that people in purchasing understand ergonomics principles and embrace the importance of incorporating them in new purchases. Least initial cost does not always equate to best value over the long run. Every organization has a collection of purchases that have either not functioned as expected or broken down once they are placed in a workstation. The cost of these unusable items needs to be included in any evaluation of an ergonomics program.

A purchasing guide for computer chairs is included to help minimize the losses associated with buying inadequate chairs.

### 15.6.3 Education and Training

The best designed chair will not benefit its user if he or she does not know how to use it (Amick et al., 2003). Ergonomically designed chairs often have several adjustments. Employees have to be taught how to use the adjustments, even the height adjustment. After considering workplace ergonomics, one CEO decided that all the chairs in his organization would be put at the lowest height at the start of each shift for a week. When his employees started their shift, they would be forced to adjust their chairs. The goal was to help the employees make a habit of adjusting chair height and, by association, other chair parameters.

Another value of training is that employees recognize the early signs of MSDs and can take the appropriate steps to prevent full-blown injury. Even small adjustments in, for example, mouse height and placement can result in significant posture improvements. Employees need to know the importance of working in a neutral posture as well as the importance of making small postural adjustments while working.

## 15.7 SUMMARY AND CONCLUSION

Although one size and one solution do not fit every situation, there is ample empirical and theoretical support for the application of ergonomic principles to design and/or redesign productive, efficient, and sustainable computer workstations in a variety of settings. The physiological and the cognitive capabilities and limitations of computer users have been documented. These principles will apply even as the specifics of the design change.

The variety of equipment available is sufficient to create healthy workstations for almost all individuals. Good ideas eventually become mainstream ideas. The challenge is to facilitate the widespread acceptance and implementation of ergonomic principles to reduce, as quickly as possible, the number of individuals who needlessly suffer with MSDs associated with computer use.

### 15.7.1 Chair Buying Guidelines

The chair may be the single most important piece of office equipment that you buy. Although individual preferences are certainly part of choosing a chair, the following are essential features that should be evaluated before a new chair is purchased:

1. *Lumbar support.* When people sit down the lumbar curve flattens out, increasing the compressive force on the intervertebral discs. The lumbar support should be positioned at about the belt level and should feel comfortable as well as support the lumbar curve in the seated position.

2. *Seat pan height adjustability.* The rule of thumb is that the seat pan height should be approximately the same distance from the floor as the creases at the back of your knees. This distance will change with the height of the heel on your shoes, and you should be able to adjust the chair height accordingly.

3. *The seat pan length.* The length of the seat pan should correspond with the length from the back of the knees to the buttocks of the person sitting in the chair. Differences in seat pan length can be achieved by making the chair in different sizes, or by designing the chair so that the seat back can be adjusted forward to shorten the seat pan length or backward to lengthen the seat pan length.

4. *Armrests.* Armrests should be adjustable both in and out and up and down. The armrest should support the elbow and forearm when the arm is in the neutral position. The neutral position for the arm is at the side of the trunk. If the armrests do not fit the person using the chair, it is better not to have them.

Other features that are essential in an ergonomically designed chair include the following:

1. A five-point base
2. A waterfall edge to the front of the seat pan
3. Supportive cushioning
4. Breathable upholstery material

Many chairs come with a 10 year warranty that can offset a higher initial price. Of course, the chair should be of overall good quality and the controls should be intuitive and easy to use.

Following are the additional features to consider:

- *Seat pan tilt.* By changing the tilt of the seat pan, the angle between the thighs and the back opens up and decreases the compressive force on the intervertebral discs.
- *Seat back angle.* Another way to open up the angle between the thighs and the back is to adjust the seat back so that the user can lean back in the chair. This posture is usually not comfortable for extensive keying but can be used while reading material or talking on the phone. The change in posture decreases the accumulated physiological and musculoskeletal stress.

## PROBLEMS OF OFFICE ERGONOMICS

Each answer has at least one correct answer, but some have more than one correct answer. Please identify all correct answers.

The types of force exerted on the lumbar discs include

a. Compressive
b. Shear
c. Centrifugal
d. Acceleration
e. Torsion

Section III
The intervertebral discs do not contain

a. Living cells
b. Blood vessels
c. Gelatinous material
d. Nerves
e. Extracellular material

Section III
Some differences between risk factor identification and risk-based methodologies are as follows:

a. No difference
b. Risk-based methodologies identify an acceptable level of risk
c. Risk factor identification methods identify an acceptable level of risk
d. Risk is equal to the hazard and the exposure to the hazard

Section II
The relationship between research methodologies and application methodologies could be characterized as

a. Research results need statistical validity
b. Application results may be individual or small group
c. Research results inform intervention
d. Intervention results inform theoretical models

Section II
Some diseases that are often present in people who have carpal tunnel syndrome include

a. Diabetes
b. Thyroid disease
c. Myasthenia gravis
d. Rheumatoid arthritis
e. Anemia

Section III
The most common types of MSD are

a. Muscle/tendon disorders
b. Vascular disorders
c. Nerve entrapment disorders
d. Nerve degenerative disorders
e. Metabolic muscle disorders

Section III
The structures that go through the carpal tunnel include

a. Ulnar nerve

b. Median nerve

c. Tendons of the wrist/hand extensor muscles

d. Tendons of the wrist/hand flexor muscles

e. Tendons of the thumb

Section III
If you design a chair whose seat pan adjusts between 14 and 19 in. for a population of 500,000 office workers, how many people will be smaller or larger than the design specifications?

a. 50,000

b. 25,000

c. 10,000

d. 5,000

e. 2,500

Section IV
Which of the following design principle was the primary principle used in the design of fast-food restaurant seating?

a. Design for the extreme

b. Adjustable design

c. Design for the average man

d. Design for the average woman

e. Design for the average (50% male/50% female)

Section V
The typical computer mouse was designed according to which of the following principles?

a. Design for the extreme

b. Adjustable design

c. Design for the average man

d. Design for the average woman

e. Design for the average (50% male/50% female)

Section V
If a company spends $500 on a chair for an employee who makes $50,000, what increase in productivity must be achieved to obtain the entire return on investment (ROI) in a year?

a. 10%

b. 5%

c. 2.5%

d. 1%

e. 0.5%

Section VI

## REFERENCES

Acosta, M.C., J. Gallar, and C. Belmonte. 1990. The influence of eye solutions on blinking and ocular comfort at rest and during work at video display terminals. *Experimental Eye Research* 68(6): 663–669.

Alexander, D.C. and T.J. Albin. 2003. Economic justification of the ergonomic process. In *Occupational Ergonomics: Design and Management of Work Systems*. W. Karwowski and W.S. Marras (eds.). New York: CRC Press, pp. 7-1–7-11.

Amell, T.K. and S. Kumar. 2000. Cumulative trauma disorders and keyboarding work. *International Journal of Industrial Ergonomics* 25(1): 69–78.

Amick, B.C., M.M. Robertson, K. DeRange, L. Bazzini, A. Moore, T. Rooney, and R. Harrist. 2003. Effect of office ergonomics intervention on reducing musculoskeletal symptoms. *Spine* 28(24): 2706–2711.

Andersson, G.B. and R. Ortengren. 1974. Lumbar disc pressure and myoelectric back muscle activity during sitting. *Scandinavian Journal of Rehabilitation Medicine* 3: 104–121.

Ankrum, D.R., E.E. Hansen, and K.J. Nemeth. 1995. The vertical horopter and viewing distance at computer workstations. In *Symbiosis of Human and Artifact*. Y. Anzai, K. Agawa, and H. Mori (eds.). New York: Elsevier Science.

Ankrum, D.R. and K.J. Nemeth. 1995. Posture, comfort and monitor placement. *Ergonomics in Design* (April): 7–9.

Anshel, J. 1998. *Visual Ergonomics in the Workplace*. London: Taylor & Francis Group.

ANSI/AIHA Z-10. 2005. *Occupational Health and Safety Management System Standard*. Fairfax, VA: American Industrial Hygiene Association.

ANSI/HFES 100. 1988. *Human Factors Engineering of Computer Workstations*. Santa Monica, CA: Human Factors and Ergonomics Society.

ANSI/HFES 100. 2007. *Human Factors Engineering of Computer Workstations*. Santa Monica, CA: Human Factors and Ergonomics Society.

Armstrong, T.J. 1988. Ergonomics and cumulative trauma disorders of the hand and wrist. In *The Worker and Performer in a Hand Rehabilitation Setting*. Philadelphia, PA: W.B. Saunders, pp. 1175–1191.

Armstrong, T.J. and D. Chaffin. 1979. Carpal tunnel syndrome and selected personal attributes. *Journal of Occupational Medicine* 21: 481–486.

Armstrong, T.J., J. Foulke, B.S. Joseph, and S. Goldstein. 1982. An investigation of cumulative trauma disorders in a poultry processing plant. *American Industrial Hygiene Association Journal* 43: 103–116.

Armstrong, T.J., J. Foulke, B. Martin, and D. Rempel. 1991a. An investigation of finger forces in alpha-numeric keyboard work. In *Proceedings from the 11th Congress of the International Ergonomics Association*, Vol. 3. New York: Taylor & Francis Group, pp. 75–76.

Armstrong, T.J., R.A. Werner, J.P. Waring, and J.A. Foulke. 1991b. Intra-carpal canal pressure in selected hand tasks. In *Proceedings of the 11th Congress of the International Ergonomics Association,* Vol. 3. New York: Taylor & Francis Group.

Arndt, R. 1983. Working posture and musculoskeletal problems of video display terminal operators: Review and reappraisal. *American Industrial Hygiene Association Journal* 44(6): 437–446.

Balakrishnan, A.D., D.L. Jindrich, and J.T. Dennerlein. 2006. Keyswitch orientation can reduce finger joint torques during tapping on a computer keyswitch. *Journal of the Human Factors and Ergonomics Society* 48: 121–129.

Becker, J., D.B. Nora, I. Gomes, F.F. Stringari, R. Seitensus, J.S. Panasso, and J.A.C. Ehlers. 2002. An evaluation of gender, obesity, age and diabetes mellitus as risk factors for carpal tunnel syndrome. *Clinical Neurophysiology* 113(9): 1429–1434.

Bendix, T. 1987. *Adjustment of the Seated Workplace: With Special Reference to Heights and Inclinations of Seat and Table*. Copenhagen, Denmark: Lægeforeningens Forlag, pp. 1–23.

Blatter, B.M. and P.M. Bongers. 2001. Duration of computer use and mouse use in relation to musculoskeletal disorders of neck or upper limb. *International Journal of Industrial Ergonomics* 30: 295–306.

Bongers, P.M. and C.R. deWinter. 1992. Psychosocial factor and musculoskeletal disease: A report of the literature. (Report 92.028), Leiden, the Netherlands: TNO Institute of Preventive Health Care.

Bongers, P.M., C.R. deWinter, M.J. Kompier, and V.H. Hildebrandt. 1993. Psychosocial factors at work and musculoskeletal disease. *Scandinavian Journal of Work, Environment and Health* 19(5): 297–312.

Burandt, U. and E. Grandjean. 1963. Sitting habits of office employees. *Ergonomics* 6: 217–228.

Callaghan, J.P. and S.M. McGill. 2001. Low back joint loading and kinematics during standing and unsupported sitting. *Ergonomics* 44: 280–294.

Cannon, L.J., E.J. Bernacki, and S.D. Walter. 1981. Personal and occupational factors associated with carpal tunnel syndrome. *Journal of Occupational Medicine* 23(4): 255–258.

Caryon, P. 1993. Job design and job stress in office workers. *Ergonomics* 36(5): 463–477.

Caryon-Sainfort, P. 1992. The use of computers in footsies: Impact of task characteristics and worker stress. *International Journal of Human-Computer Interaction* 4(3): 245–261.

Chaffin, D.B. 1973. Localized muscle fatigue: Definition and measurement. *Journal of Occupational Medicine* 15(4): 346–354.

Chaffin, D.B. and G.B.J. Andersson. 1984. *Occupational Biomechanics*. New York: John Wiley & Sons, pp. 289–323.

Chaffin, D.B., G.B.J. Andersson and B.J. Martin. 1999. *Occupational Biomechanics* (3rd edn.). New York: Wiley & Sons.

Chester, M.R., M.J. Rys, and S.A. Konz. 2002. Leg swelling, comfort and fatigue when sitting, standing, and sit/standing. *International Journal of Industrial Ergonomics* 29: 289–296.

Congleton, J.J., M.M. Ayoub, and J.L. Smith. 1985. The design and evaluation of the neutral posture chair for surgeons. *Human Factors* 27: 589–600.

Congleton, J.J., M.M. Ayoub, and J.L. Smith. 1988. The determination of pressures and patterns for the male human buttocks and thigh in sitting utilizing conductive foam. *International Journal of Industrial Ergonomics* 2(3): 193–202.

Conlon, C.F., N. Krause, and D.M. Rempel. 2008. A randomized controlled trial evaluating an alternative mouse and forearm support on upper body discomfort and musculoskeletal disorders among engineers. *Occupational and Environmental Medicine* 65: 311–318.

Cook, C.J., R. Burgess-Limerick, and S. Chang. 2000. The prevalence of neck and upper extremity musculoskeletal symptoms in computer mouse users. *International Journal of Industrial Ergonomics* 26(3): 347–356.

Cook, C.J. and K. Kothiyal. 1998. Influence of mouse position on muscular activity in the neck, shoulder and arm in computer users. *Applied Ergonomics* 29(6): 439–443.

Corlett, E.N. 1995. The evaluation of industrial seating. In *The Evaluation of Human Work*. J.R. Wilsona and E.N. Corlett (eds.). London: Taylor & Francis Group, pp. 621–636.

Cram, J.R. and I. Vinitzky. 1995. Effects of chair design on back muscle fatigue. *Journal of Occupational Rehabilitation* 5: 101–113.

Dain, S.J., A.K. McCarthy, and T. Chan-Ling. 1988. Symptoms in VDU operators. *American Journal of Optometry and Physiological Optics* 65: 162–167.

Dainoff, M. 2003. Ergonomics of seating and chairs. In *Occupational Ergonomics: Design and Management of Work Systems*. W. Karwowski and W.S. Marras (eds.). Boca Raton, FL: CRC Press, pp. 20-10–20-18.

Demure, B., R. Luippold, C. Bigelow, D. Ali, K.A. Mundt, and B. Liese. 2000. Video display terminal workstation improvement program: I. Baseline associations between musculoskeletal discomfort and ergonomic features of workstations. *Journal of Occupational and Environmental Medicine* 42(8): 783–791.

deLooze, P.M., F.M. LottieKuiftevers, and J. van dieen. 2003. Sitting comfort and discomfort and the relationships with objective measure. *Ergonomics* 46: 985–997.

Dillon, A. 1992. Reading from paper versus screens: A critical review of the empirical literature. *Ergonomics* 35: 1297–1326.

Doughty, M.J. 2001. Consideration of three types of spontaneous eyeblink activity in formal humans: During reading and video display terminal use, in primary gaze, and while in conversation. *Optometry and Vision Science* 78(10): 712–725.

Ergonomics Program Guidance Document Aligned with ANSI/AIHA Z10. 2008. Rostykus, W. (ed.). Fairfax, VA: American Industrial Hygiene Association.

Ferguson, D. and J. Duncan. 1974. Keyboard design and operating posture. *Ergonomics* 17(6): 731–744.

Feuerstein, M., T. Armstrong, P. Hickey, and A. Lincoln. 1997. Computer keyboard force and upper extremity symptoms. *Journal of Occupational Medicine* 39: 1144–1153.

Gelberman, R.P., A. Hergenmoeder, A. Hargans, G. Lundborg, and W. Akeson. 1981. The carpal tunnel syndrome—A study of carpal canal pressures. *Journal of Bone and Joint Surgery-American Volume*, 63A: 380–383.

Gerard, M.J., T.J. Armstrong, B.J. Martin, and D.A. Rempel. 2002. The effects of work pace on within-participant and between-participant keying force, electromyography, and fatigue. *Journal of the Human Factors and Ergonomics Society* 44: 51–61.

Getty, R.L. 2003. Quality and ergonomics: Application of ergonomics to continuous improvement is integral to the goals of business. In *Occupational Ergonomics: Design and Management of Work Systems*. W. Karwowski and W.S. Marras (eds.). New York: CRC Press, pp. 4-1–4-14.

Gould, J.D., L. Alfaro, R. Finn, B. Houpt, and A. Minuto. 1987. Reading from CRT displays can be as fast as reading from paper. *Human Factors* 29: 497–517.

Gould, J.S. and A. Wissinger. 1978. Carpal tunnel syndrome in pregnancy. *Southern Medical Journal* 71(2): 144–145.

Graf, M., U. Guffenbuhl, and H. Kruger. 1991. Movement dynamics of sitting behaviour during different activities. In *Proceedings of the 11th Congress of the International Ergonomics Association*. Y. Queinnec and F. Daniellou (eds.). London: Taylor & Francis Group, pp. 15–17.

Graf, M., U. Guggenbuhl, and H. Krueger. 1993. Investigations of the effects of seat shape and slope on posture, comfort and back muscle activity. *International Journal of Industrial Ergonomics* 12: 91–103.

Graf, M., U. Guggenbuhl, and H. Krueger. 1995. An assessment of seated activity and postures at five workplaces. *International Journal of Industrial Ergonomics* 15: 81–90.

Grandjean, E., W. Hunting, and E. Nishiyama. 1984. Preferred VDT workstation settings, body posture and physical impairment. *Applied Ergonomics* 15: 99–104.

Grandjean, E., W. Hunting, and M. Pidermann. 1983. VDT workstation design: Preferred setting and their effects. *Human Factors* 25: 161–175.

Green, R.A. and C.A. Briggs. 1989. Anthropometric dimensions and overuse injury among Australian keyboard operators. *Journal of Occupational Medicine* 31: 747–750.

Gregory, D.E., N.M. Dunk, and J.P. Callahan. 2006. Stability ball versus office chair: Comparison of muscle activation and lumbar spine posture during prolonged sitting. *Human Factors* 48: 142–153.

Grieco, A. and G. Molteni. 1999. Seating and posture in VDU work. In *Occupational Ergonomics Handbook*. Karwowski, W. and W. Marras (eds.). Boca Raton, FL: CRC Press, pp. 1779–1791.

Grieco, A. and G. Molteni. 2003. Seating and posture in VDT work. In *Occupational Ergonomics: Design and Management of Work Systems*. W. Karwowski and W.S. Marras (eds.). Boca Raton, FL: CRC Press, pp. 21-1–21-12.

Gur, S., S. Ron, and A. Heicklen-Klein. 1994. Objective evaluation of visual fatigue in VDU workers. *Occupational Medicine* 44(4): 201–204.

Guth, S.K. 1958. Light and comfort. *Industrial medicine and Surgery* 27: 570–574.

Guth, S.K. 1981. The science of seeing; a search for criteria. *American Journal of Optometry and Physiological Optics* 58: 870–875.

Guyton, A.C. 1992. Contraction of skeletal muscle. In *Human Physiology and Mechanisms of Disease* (5th edn.). London: W.B. Saunders Company, pp. 55–66.

Hammer, A. 1934. Tenosynovitis. *Medical Record* 140: 353–355.

Harding, D.C., K.D. Brandt, and B.M. Hillberry. 1993. Finger joint force minimization in pianists using optimization techniques. *Journal of Biomechanics* 26(12): 1403–1412.

Harpster, J.L., A. Freivalds, S.L. Shulman, and H.W. Leibourtz. 1989. Visual performance on CRT screens and hard copy displays. *Human Factors* 31: 247–257.

Harrison, D.D., S.O. Harrison, A.C. Croft, D.E. Harrison, and S.J. Troyanovich. 1999. Sitting biomechanics part I: Review of the literature. *Journal of Manipulative and Physiological Therapeutics* 22: 594–609.

Hedge, A. and J.R. Powers. 1995. Wrist postures while keyboarding: Effects of a negative slope keyboard system and full motion forearm supports. *Ergonomics* 38: 508–517.

Helander, M.G. 2003. Forget about ergonomics in chair design? Focus on aesthetics and comfort! *Ergonomics* 46: 1306–1319.

Helander, M.G., P.A. Billingsley, and J.M. Schurik. 1984. An evaluation of human factors on VDTs in the workplace. *Human Factors Reviews* 55: 129.

Helander, M.G., L. Zhang, and D. Michel. 1995. Ergonomics of ergonomic chairs: A study of adjustability features. *Ergonomics* 38: 2007–2029.

Helander, M.G. and L. Zhang. 1997. Field studies of comfort and discomfort in sitting. *Ergonomics* 40: 895–915.

Hendricks, H.W. 1996. *Good Ergonomics is Good Economics*. 1996 HFES Presidential Address.

Hignett, S. 1998. Ergonomics. In *Rehabilitation of Movement: Theoretical Basis of Clinical Practice*. J. Pitt-Brooke, H. Reid, J. Lockwood, and K. Kerr (eds.). London: W.B. Saunders Company, pp. 480–486.

Hignett, S. and L. McAtamney. 2000. Rapid Entire Body Assessment (REBA). *Applied Ergonomics* 31: 201–205.

Howarth, P.A. 1995. Assessment of the visual environment. In *Evaluation of Human Work*. J.R. Wilson and E.N. Corlett (eds.). London: Taylor and Francis, pp. 445–482.

Hult, L. 1954. Cervical, dorsal and lumbar spinal syndromes. *Acta Orthopaedica Scandinavica Supplementum* 17: 1–102.

Illuminating Engineering Society. 1981. *IES Lighting Handbook*. New York: Illuminating Engineering Society.

Jensen, C., V. Borg, L. Finsen, K. Hansen, B. Juul-Kristensen, and H. Christensen. 1998. Job demands, muscle activity and musculoskeletal symptoms in relation to work with the computer mouse. *Scandinavian Journal of Work, Environment & Health* 24: 417–443.

Jensen, C., L. Finsen, K. Sagaard, and H. Christensen. 2002. Musculoskeletal symptoms and duration of computer and mouse use. *International Journal of Industrial Ergonomics* 30: 265–275.

Jindrich, D., A. Balakrishnan, and J. Dennerlein. 2004. Effects of keyswitch design and finger posture on finger joint kinematics and dynamics while tapping on computer keyswitches. *Clinical Biomechanics* 19: 600–608.

Keegan, J.J. 1953. Alterations of the lumbar curve related to posture and seating. *Journal of Bone and Joint Surgery* 35: 567–589.

Keir, P.J., J.M. Bach, M. Hudes, and D.M. Rempel. 2007. Guidelines for wrist posture based on carpal tunnel pressure thresholds. *Journal of Human Factors and Ergonomics Society* 49(1): 88–99.

Keir, P.J. and R.P. Wells. 2002. The effects of typing posture on wrist extensor muscle loading. *Human Factors* 44(3): 392–403.

Keyersling, W.M., T.J. Armstrong, and L. Punnett. 1991. Ergonomic job analysis: A structured approach for identifying risk factors associated with overexertion injuries and disorders. *Applied Occupational and Environmental Hygiene* 6: 353–363.

Kramer, J. 1973. *Biomechanische Veranderungen im Lumbalen Bewegungssegment*. Sturrgart, Germany: Hippokrates.

Kroemer, K.H.E. 1992. Avoiding cumulative trauma disorders in shops and offices. *American Industrial Hygiene Association Journal* 53: 596–604.

Kumar, S. and W.G.S. Scaife. 1979. A precision task, posture and strain. *Journal of Safety Research* 11(1): 28–36.

Laubli, T. and E. Grandjean. 1984. The magic of control groups in VDT field studies. In *Ergonomics and Health in Modern Offices*. London, U.K.: Taylor & Francis Group, pp. 105–112.

Lawrence, J.S. 1955. Rheumatism in coal miners. Part III. Occupational factors. *British Journal of Industrial Medicine* 12: 249–261.

Lee, D.L., J. Fleisher, H.E. McLoone, K. Kotani, and J.T. Dennerlein. 2007. Alternative computer mouse design and testing to reduce finger extensor muscle activity during mouse use. *Human Factors* 49(4): 573–584.

Legg, S.J., H.W. Mackie, and W. Milicich. 2002. Evaluation of a prototype multi-posture office chair. *Ergonomics* 45: 153–163.

Lengsfeld, M., A. Frank, D.L. van Deursen, and P. Griss. 2000. Lumbar spine curvature during office chair sitting. *Medical Engineering & Physics* 22: 665–669.

Life, M.A. and S.T. Pheasant. 1984. An integrated approach to the study of posture in keyboard operation. *Applied Ergonomics* 15: 83–90.

Linton, S.J. and K. Kamwendo. 1989. Risk factors in the psychosocial work environment for neck and shoulder pain in secretaries. *Journal of Occupational Medicine* 31(7): 609–613.

Luckiesh, M. and S.K. Guth. 1949. Brightness in visual field at borderline between comfort and discomfort. *Illumination Engineering* 44: 650–670.

Luopajarvi, T., I. Kuorinka, M. Virolainen, and M. Holmberg. 1979. Prevalence of tenosynovitis and other injuries of the upper extremities in repetitive work. *Scandinavian Journal of Work and Environmental Health* 5: 48–55.

Maeda, K., W. Hunting, and E. Grandjean. 1980. Localized fatigue in accounting machine operators. *Journal of Occupational Medicine* 22: 810–816.

Makhsous, M., F. Lin, R.W. Hendrix, M. Hepler, and L.Q. Zhang. 2003. Sitting with adjustable ischial and back supports: Biomechanical changes. *Spine* 28(11): 1113–1121.

Marklin, R.W. 2003. Biomechanical aspects of CTDs. In *Occupational Ergonomics: Engineering and Administrative Controls*. Boca Raton, FL: CRC Press, pp. 3-1–3-38.

Matias, A.C., G. Salvendyand, and T. Kiczek. 1998. Predictive models of carpal tunnel syndrome causation among VDT operators. *Ergonomics* 41(2): 213–226.

McAtamney, L. and J.N. Corlett. 1993. RULA: A survey method for the investigation of work-related upper limb disorders. *Applied Ergonomics* 24(2): 91–99.

McPhee, B. 1982. Deficiencies in the ergonomic design of keyboard work and upperlimb and neck disorders in operators. *Journal of Human Ergology* 11: 31–36.

McPhee, B.J. 1987. Work-related musculoskeletal disorders of the neck and upper extremities in workers engaged in light, highly repetitive work. In *Proceedings of the International Symposium on Work-related Musculoskeletal Disorder.* U. Osterholz, W. Karmaus, B. Hullman, and B. Ritz (eds.). Bonn, Germany, pp. 244–258.

Miyao, M., S. Hacisalhzade, J. Allen, and L. Stark. 1989. Effects of VDT resolution on visual fatigue and readability: An eye movement approach. *Ergonomics* 32: 603–614.

Mocci, F., A. Serra, and G.A. Corrias. 2001. Psychological factors and visual fatigue in working with video display terminals. *Occupational and Environmental Medicine* 58: 267–271.

Moon, S.D. and S.L. Sauter. 1996. *Beyond Biomechanics: Psychosocial Aspects of Musculoskeletal Disorders in Office Work.* London, U.K.: Taylor & Francis Group.

Moore, S.J. and A. Garg. 1995. The Strain Index: A proposed method to analyze jobs for risk of distal upper extremity disorders. *American Industrial Hygiene Association Journal* 56: 443–458.

Moroney, S.P., A.B. Schultz, and J.A.A. Miller. 1988. Analysis and measurement of neck loads. *Journal of Orthopaedic Research* 6: 713–720.

Nachemson, A. and G. Elfstrom 1970. Intravital dynamic pressure measurements in lumbar discs. *Scandinavian Journal of Rehabilitation Medicine–Supplement* 1: 1–40.

Nachemson, A. and J.M. Morris. 1964. In vivo measurements of intradiscal pressure. *Journal of Bone and Joint Surgery* 46: 1077.

Nakaiski, H. and Y. Yamada. 1999. Abnormal tear dynamics and symptoms of eyestrain in operators of visual display terminals. *Occupational and Environmental Medicine* 56: 6–9.

Nathan, P.A., R.C. Kenniston, L.D. Myers, and K.D. Meadows. 1992. Obesity as a risk factor for slowing of sensory conduction of the median nerve in industry: A cross-sectional and longitudinal study involving 429 workers. *Journal of Occupational Medicine* 34(4): 379–383.

Nathan, P.A., A. Wilcox, P.S. Emerick, K.D. Meadows, and A.L. McCormack. 2001. Effect of an aerobic exercise program on median nerve conduction and symptoms associated with carpal tunnel syndrome. *Journal of Occupational and Environmental Medicine* 43(10): 840–843.

Nemecek, J. and E. Grandjean. 1971. Das Frossraumburo in Arbeitsphysiologischer Sicht. *Ubdystruekkt Irgabusatuib* 40: 233–243.

Noddeland, H. and J. Winkel. 1988. Effects of leg activity and ambient barometric pressure on foot swelling and lower limb skin temperature during 8 hours of sitting. *European Journal of Applied Physiology* 57: 409–415.

Nussbaum, M.A. and H. Johnson. 2002. Determination and evaluation of acceptable force limits in single-digit tasks. *Human Factors* 44(4): 545–546.

Okushima, H. 1970. Study on hydrodynamic pressure of lumbar intervertebral discs. *Nippon Geka Hokan–Archiv Für Japanische Chirurgie* 39: 45.

O'Reilly, M., B. Finder, and M. Werrell. 2007. *Ergonomics Guide to Computer Workstations.* Fairfax, VA: American Industrial Hygiene Association.

Oxenburgh, M.S. 1997. Cost-benefit analysis of ergonomics programs. *American Industrial Hygiene Journal* 58(2): 150–156.

Oxenburgh, M.S., S.A. Rowe, and D.B. Douglas. 1985. Repetition strain injury in keyboard operators. *Journal of Occupational Health and Safety* 1: 106–112.

Patel, S., R. Henderson, L. Bradley, B. Galloway, and L. Hunter. 1991. Effect of visual display units on blink rate and tear stability. *Optometry and Visual Science* 68(11): 888–892.

Pickett, C.W.L. and R.E.M. Lees. 1991. A cross-sectional study of health complaints among 79 data entry operators using video display terminals. *Journal of the Society of Occupational Medicine* 41: 113–116.

Posch, J. and D. Marcotte. 1976. Carpal tunnel syndrome: An analysis of 1201 cases. *Orthopedic Review* 5: 25–35.

Pottier, M., A. Dubreuil, and H. Monod. 1969. The effects of sitting posture on the volume of the foot. *Ergonomics* 12: 753–758.

Rempel, D., T. Bloom, R. Tal, A. Hargens, and L. Fordon. 1992. A method of measuring intracarpal pressure and elementary hand maneuvers. In *Arbete Och Halsa* 1992: 17, *International Scientific Conference of Prevention of Work-Related Musculoskeletal Disorder (PREMUS)*. Stockholm, Sweden, May 12–14, 1992. Published by Arbets Miljo Institute (National Institute of Occupational Health), Stockholm, [English] pp. 249–251.

Sanders, M.S. and E.J. McCormick. 1993. Applied anthropometry, work-space and design and seating. In *Human Factors in Engineering and Design* (7th edn.). New York: McGraw-Hill, pp. 415–454.

Sauter, S.L., L.M. Schiefer, and S.J. Knutson. 1991. Work, posture, workstation design and musculoskeletal discomfort in a VDT data entry task. *Human Factors* 33: 151–167.

Sauter, S.L. and N.G. Swanson. 1996. An ecological model of musculoskeletal disorders in office work. In *Beyond Biomechanics: Psychosocial Aspects of Musculoskeletal Disorders in Office Word.* S.D. Moon and S.L. Sauter (eds.). Bristol, PA: Taylor & Francis Group, pp. 3–21.

Scalet, E.A. 1987. *VDT Health and Safety*. Lawrence, KS: Ergosyst.

Schlote, T., G. Kadner, and N. Freudenthaler. 2003. Marked reduction and distinct patterns of eye clinking in patients with moderately dry eyes during video display terminal use. *Graefe's Archive for Clinical and Experimental Ophthalmology* 242(4): 306–312.

Schoberth, H. 1962. *Stizhaltung, Sitzchaden, Sitzmobel*. Berlin, Germany: Springer.

Selye, 1956. *The Stress of Life*. New York: McGraw-Hill.

Sheedy, J.E. 1992. Vision problems at video display terminals: A survey of optometrists. *Journal of the American Optometric Association* 63: 687–692.

Sillanpaa, J.S., M. Huikko, P. Nyberg, P. Kivi, P. Laippala, and J. Uitti. 2003. Effect of work with visual display units on musculoskeletal disorders in the office environment. *Occupational Medicine* 54: 443–451.

Silverstein, B.A., L.J. Fine, and T.J. Armstrong. 1986. Carpal tunnel syndrome: Causes and preventive strategy. *Seminars in Occupational Medicine* 1(5): 213–221.

Silverstein, B.A., L.J. Fine, and T.J. Armstrong. 1987. Occupational factors and carpal tunnel syndrome. *American Journal of Industrial Medicine* 11: 343–358.

Simoneau, G.G. and R.W. Marklin. 2001. Effect of computer keyboard slope and height on wrist extension angle. *Human Factors* 43(2): 287–298.

Sjogaard, G., B. Kiens, K. Jorgensen, and B. Saltin. 1986. Intramuscular pressure, EMG, and blood flow during low-level prolonged static contraction in man. *Acta Physiologica Scandinavica* 128: 475–484.

Sjogaard, G., G. Savard, and C. Joel. 1988. Muscle blood flow during isometric activity and its relation to muscle fatigue. *European Journal of Applied Physiology and Occupational Physiology* 57: 327–335.

Smith, M. and P. Carayon. 1996. Work organization, stress, and cumulative trauma disorders. In *Beyond Biomechanics: Psychosocial Aspects of Musculoskeletal Disorders in Office Work*. S.D. Moon and S.L. Sauter (eds.). Bristol, PA: Taylor & Francis Group, pp. 23–42.

Smith, M.J., B.G.F. Cohen, and L.W. Stammerjohn. 1981. An investigation of health complaints and job stress in video display operations. *Human Factors* 23: 387–400.

Snyder, R.G., D.B. Chaffin, and D.R. Foust. 1975. Bioengineering study of basic physical measurements related to susceptibility to cervical hyperextension-hyperflexion, Technical Report UM-HSRI-BI-75-6, Highway Safety Research Institute, University of Michigan, Ann Arbor, MI.

Sommerich, C.M. 2003. Economic analysis for ergonomics programs. In *Occupational Ergonomics: Design and Management of Work Systems*. W. Karwowski and W.S. Marras (eds.). New York: CRC Press, pp. 6-1–6–20.

Stevens, J.C., C.M. Beard, W.M. O'Fallon, and L.T. Kurland. 1992. Conditions associated with carpal tunnel. *Mayo Clinic Proceedings* 67(6): 541–548.

Stranden, E. 2000. Dynamic leg volume changes when sitting in a locked and free-floating tilt office chair. *Ergonomics* 43: 421–433.

Szeto, G.P.Y., L.M. Straker, and P.B. O'Sullivan. 2005. The effects of typing speed and force on motor control in symptomatic and asymptomatic office workers. *International Journal of Industrial Ergonomics* 35(9): 779–795.

Thatcher, S.S. and M.O. Brophy. 1999. Risk factors and reported pain in VDT users. *CybErg*. Perth, Australia.

Tsubota, K. and K. Nakamori. 1993. Dry eyes and video display terminals. *New England Journal of Medicine* 328: 584.

Tzivian, L.L., V.H. Rayhinstein, V.F. Motov, and F.F. Oveychik. 1971. Results of clinical study of pressure within the intervertebral lumbar discs. *Otrop Traumatol Protex* 6: 31.

van der Heiden, G., U. Brauninger, and E. Grandjean. 1984. Ergonomic studies on computer aided design. In *Ergonomics and Health in Modern Offices*. London, U.K.: Taylor & Francis Group, pp. 119–128.

van Deursen, D.L., L.L.J.M. van Deursen, C.J. Snijders, and R.H.M. Goossens. 2000. Effect of continuous rotary seat pan movements on physiological oedema of the lower extremities during prolonged sitting. *International Journal of Industrial Ergonomics* 29: 521–526.

Vasavada, A.N., S. Li, and S.L. Delp. 1998. Influence of muscle morphometry and moment arms on the moment-generating capacity of human neck muscles. *Spine* 23(4): 412–422.

Waersted, M., R.A. Bjorlund, and R.H. Westgaard. 1991. Shoulder muscle tension induced by two VDU-based tasks of different complexity. *Ergonomics* 34: 137–150.

Waters, T.R., V. Putz-Anderson, A. Garg, and L.J. Fine. 1993. Revised NIOSH equation for the design and evaluation of manual lifting tasks. *Ergonomics* 36(7): 749–776.

Webster, B.S. and S.H. Snook. 1994. The cost of compensable upper extremity cumulative trauma disorders. *Journal of Occupational Medicine* 36: 713–717.

Wilkins, A.J., I. Nimmo-Smith, A.I. Slater, and I. Bedocs. 1989. Fluorescent Lighting, headaches and eyestrain. *Lighting Research and Technology* 211: 11–18.

Wilson, J.R. and E.N. Corlett. 1995. *Evaluation of Human Work: A Practical Ergonomics Methodology* (2nd edition). London: Taylor and Francis.

Winkel, J. and K. Jorgensen. 1986. Swelling of the foot, its vascular volume and systemic hemoconcentration during long-term constrained sitting. *European Journal of Applied Physiology* 55: 162–166.

Yaginuma, Y., H. Yamada, and H. Nagai. 1990. Study of the relationship between lacrimation and blink in VDT work. *Ergonomics* 33(6): 799–809.

Zacharkow, D. 1988. *Posture: Sitting, Standing, Chair Design and Exercise*. Springfield, IL: Charles C. Thomas, pp. 338–342.

Zennaro, D., T. Laubli, D. Krebs, H. Krueger, and A. Klipstein. 2004. Trapezius muscle motor unit activity in symptomatic participants during finger tapping using properly and improperly adjusted desks. *Human Factors* 46(2): 252–266.

Ziefle, M. 1998. Effects of display resolution on visual performance. *Human Factors* 40: 554–567.

CHAPTER 16

# Shift Work and Long Work Hours*

Claire C. Caruso

CONTENTS



* The findings and conclusions in this chapter are those of the author and do not necessarily represent the views of the National Institute for Occupational Safety and Health.

## 16.1 INTRODUCTION

In today's society, services around the clock are needed from many types of workers and at times long work hours are not avoidable. As a result, shift work and long work hours are common in many workplaces. Research evidence is building that these types of work schedules are associated with several health and safety risks for workers. The risks also extend to their families, employers, and society when they make fatigue-related errors or show declines in performance. This chapter will discuss the wide range of risks that are linked to these schedules, theories about why these risks occur, and strategies to reduce risks.

### 16.1.1 Characteristics of Shift Work and Long Work Hour Schedules

Shift work involves work outside the daytime hours of approximately 7:00 AM to 6:00 PM. According to the U.S. Bureau of Labor Statistics, almost 15% of full-time workers (or 15 million Americans) work on evening, night, rotating, split or employer-arranged irregular shifts [1]. Both societal needs and economic factors drive the use of shift work. Society needs critical services around-the-clock for police and fire protection, health care, transportation, communications, public utilities, and military service. Other industries require workers around-the-clock because their production processes and operations need to be continuous to optimize capital investment in machinery, or the manufacturing process cannot be interrupted for production reasons. As a result, certain types of work are likely to always require shift workers.

Shift work schedules can be characterized by whether they are fixed or rotating schedules, the time of the shift, and if rotating shifts, the direction and speed of rotation. Fixed shifts have consistent work times. Rotating shifts change schedule, for example, from day shift to evening shift. Evening shifts are shifts with most of the work hours between 2:00 PM and midnight, and night shifts have most of the hours worked between 9:00 PM and 8:00 AM [1]. Some industries use split shifts, with two work periods separated by a longer break of a few hours to accommodate predicted peaks in activity. For example, a restaurant worker may work for the lunchtime hours, have a few hours off, and then return to work the dinnertime hours. Rotating shifts can change every few days (a fast rotation), weekly, or every 2 weeks or more (a slow rotation). The direction of the rotation can be forward or backward. Forward rotations change from day shift to evening shift to night shift. This pattern is recommended over a backward rotation because the shift change allows longer time off between shifts and has progressively later start times which allow the workers to get up later [2]. Backward rotations move from night shift to evening shift to day shift. This is a more difficult schedule because at the shift change the time off between shifts is short and the work times progressively become earlier which require the workers to waken earlier. A short number of hours off between shifts (e.g., rotating from evening shift to day shift) is called a "quick change" and is difficult because of the reduced time available to commute, prepare for sleep, sleep, prepare for work, and commute back to work. Quick changes do not allow for 7 or more hours of sleep that are recommended for adults [3].

Long work hours include extended work shifts and schedules with more than 40 h per week. Extended work shifts may involve compressed work weeks, with fewer days of work and more days off and the weekly work hours remaining about 40 per week (e.g., 12 h day shifts with 3–4 work days per week). The point at which the number of hours becomes too long may not be a fixed number, but rather may be influenced by the demands of the job, the work environment, characteristics of the worker, and other work schedule characteristics such as the time of the shift and within-shift rest breaks [4,5].

### 16.1.2 Theory for Health and Safety Risks

Researchers theorize that shift work and long work hours increase health and safety risks by disturbing sleep and circadian rhythms and increasing strains on family and social relationships [4,6]. In addition, long work hours increase exposures to hazards at work and reduce recovery times. The immediate effects of demanding work hours can include stress, fatigue, negative mood, discomfort, and physiological dysfunction. Fatigue and stress may increase poor health behaviors such as smoking, low physical activity, and obesity. These in turn could lead to illnesses and injuries for the worker. The risks can also extend to families, for example, through conflicting work—family demands. Risks can extend to employers through reduced productivity and increases in worker errors which negatively impact goods and services. Mistakes by fatigued workers can have broad-reaching negative impacts to the community such as medical errors, motor vehicle crashes, and industrial disasters. Research indicates that the influence of shift work and long work hours on health and safety may involve a complex interaction of work schedule characteristics, as well as work demands, worker characteristics, compensation, commute time, occupational exposures, and the nature of worker control over their work schedule and work tasks.

## 16.2 METHODOLOGICAL CONSIDERATIONS FOR STUDIES OF WORK HOURS

Field studies examining the influence of work schedules on health and safety have several challenges in controlling for many factors that may influence outcomes. Knutsson [7] points out the methodological issues for studies of shift work, which are also applicable to long work hours: adequate comparison groups; need for complete descriptions of the workers, the job, and the work schedule; outcomes measures; and accounting for other influential factors.

The first challenge in studying work schedules is reliably measuring their various features to determine which patterns most impact health. Caruso et al. [4] list several characteristics of work schedules to consider when assessing studies. The time of the shift is a critical factor: for example, risks for accidents and incidents tend to be the lowest for day shift, intermediate for evening shift, and highest for night shift [5]. Start times of 6:00 AM, or earlier, truncate sleep and are associated with more fatigue and sleepiness [8]. Backward rotations show higher risk than forward rotations [9–11]. Another important feature is the number of consecutive work days to rest days. Long series of work days without a day off increase risk. Work shifts having rest breaks only every 4 or 6h are also associated with higher risks than those with breaks every 2h [5]. Other factors to consider are the length of the shifts, total hours per week (or total hours over longer periods), distribution of days off, schedule predictability, and degree of worker control over their schedule. Research shows that schedules which combine difficult work scheduling factors (e.g., 12h night shifts) tend to show increased risks [5,12]. All of these details are critical to account for when studying work schedules to better identify the key work schedule characteristics influencing risk and requiring modification.

Another research challenge is the control group. A desirable control group would consist of day-shift workers in jobs with similar occupational exposures and demands as the shift work or long work hour group. This is often difficult to achieve because the work that needs to be done sometimes differs at different times of the day. Also, the best comparison group would have no previous history of shift work or long work hours, because this history may be associated with continued adverse effects on health even after moving to a regular daytime schedule [13,14]. In addition, other work-related factors can vary as a function of time of day such as supervisor-to-employee ratio, pressures to perform, number of coworkers and other people at the worksite, activities at the worksite, types of food available, and work environment (lighting, noise, environmental temperature and humidity, and chemical exposure). These other factors may influence health and safety outcomes.

The "survivor effect" is another challenge for studies of work schedules: workers tend to self-select their work schedule and job, which influences findings. People who feel they could not tolerate a nonstandard work schedule would tend to avoid jobs requiring shift work or long work hours, leaving workers who are more likely to tolerate these scheduling demands. As a result, any significant finding is particularly noteworthy.

A wide range of health outcomes have been examined in studies. Outcomes have included simple symptom surveys, laboratory tests for evidence of beginning pathological processes, whether or not the individual had sought medical care, diagnoses, absenteeism, workers' compensation claims, disability, and death. Outcomes can have limitations: for example, absenteeism may be influenced by not taking time off from work when sick for a variety of reasons, as well as taking sick time for reasons other than sickness. Some studies have used standardized health indicators, which improves the ability to examine the complex interactions between work scheduling patterns and risk factors across studies.

Clear and complete description of the job demands helps enlighten the influence of work scheduling patterns on outcomes. Jobs vary in physical, cognitive, emotional, and social demands as well as the ability of the worker to control the timing, speed, and type of work performed. For example, physical factors such as heavy lifting, pushing and pulling, awkward postures, or repeated forceful exertions play a major role in development of work-related musculoskeletal disorders and are critical to account for in studies of work hours and these outcomes. Similarly, other types of job demands may be critical factors for developing other types of health and safety problems.

Seniority and age impact shift and job assignment. With seniority, workers tend to move to day shift, so day shift is associated with increased years of employment and increased age [15]. As a result, effect estimates will be reduced when comparing health in younger shift workers (who tend to be healthier), with older day workers (who tend to have more health problems). This will lessen the likelihood of finding differences in health across shifts. Other factors that appear to influence the relationship between work schedules and health outcomes include gender, compensation (as pay increased, negative outcome decreased), macro-economic conditions (recession or depression), commute time, and occupational exposures (chemicals, temperature extremes, noise, and job demands) [4,12].

In summary, field studies examining the influence of shift work and long work hours on health and safety have many challenges regarding controlling for a variety of other factors that influence risk. However, other related research topics contribute evidence, such as laboratory and field studies examining sleep deprivation, circadian rhythms, jet lag, and sleep disorders. The evidence from all these areas suggests shift work and long work hours are associated with several health and safety risks.

## 16.3 HEALTH AND SAFETY RISKS ASSOCIATED WITH SHIFT WORK AND LONG WORK HOURS

### 16.3.1 Sleep Findings

Shift work and long work hours can lead to difficulties with sleep because of the need to sleep at irregular times and at times out of phase with normal circadian rhythms. In addition, the time available for sleep may be reduced because of short periods between work shifts or demands from work and family.

To examine sleep in shift workers, Drake et al. [16] surveyed a representative sample drawn from the working population in the Midwest. They found 32% of night-shift workers and 26% of rotating-shift workers reported long-term insomnia and excessive sleepiness. Ingre and Åkerstedt [17] studied 169 monozygotic twin pairs who were discordant on exposure to night shift work. History of night shift significantly increased risk for sleep disturbances into retirement years. A meta-analysis by Pilcher et al. [18] of 36 studies found that permanent day-shift workers slept an average of 7.0 h per day, permanent and rotating evening-shift workers slept 7.6–8.1 h, permanent night-shift workers slept 6.6 h, and rotating night-shift workers slept the least, 5.85 h. As mentioned earlier, Healthy People 2020 recommends adults sleep 7 h or more each day [3].

In a review of studies examining extended work shifts, Knauth [19] reported 13 studies showed shifts longer than 8 h had negative effects on sleep, 8 studies showed mixed results, and 4 studies showed positive effects. Long weekly work hours and overtime have been associated with shorter sleep duration or sleep disturbances in several studies [20–24]. Basner et al. [25] conducted a study of a national representative sample of Americans and reported 1 additional hour of work a day was associated with 1.5 h less sleep.

### 16.3.2 Declines in Performance

Sleep is a biological need for life and health, similar to the need for food and water [26]. Most adults need 7–9 h of good-quality sleep each day [27]. Good-quality sleep means the major sleep episode does not have frequent arousals and is long enough for the individual. Laboratory studies provide evidence for a wide range of cognitive declines associated with sleep deprivation, which are discussed as follows [28].

Without adequate sleep, people feel sleepy, are less alert, and can experience microsleeps (short episodes of sleep lasting a few seconds). During microsleeps, the person may appear to be awake (eyes open), but his or her brain will not process information and lapses in attention occur. A sleep-deprived person cannot control the onset of microsleeps and is often not aware that they are occurring. Microsleeps can be identified on electroencephalograms and can be dangerous if they occur at critical times [29].

Sleep deprivation impairs performance on tasks that require intense or prolonged attention [30]. Performance is unstable with increased errors of omission (failing to respond to a stimulus) and commission (responding when a stimulus is not present). These errors occur more frequently over time. When the sleep-deprived person can pace activities without time pressure, cognitive slowing will be evident and response time slows, but accuracy can be preserved. Under time pressure, however, cognitive slowing will produce errors. Also, performance may begin well, but deteriorates as time-on-task increases and the effort to remain behaviorally effective increases.

Sleep deprivation negatively affects information processing and learning [28]. Both short-term recall and working memory performance decline. Tasks that require information to be accurately held in memory for brief periods will be impaired.

Communication skills decrease as sleep deprivation increases [30,31]. A worker may fumble trying to find the right word, say the wrong word when meaning to use another, or may misinterpret what another person said or wrote.

Sleep deprivation leads to impaired cognitive flexibility, negatively affects innovation, and decreases insight [28]. In situations during which "thinking outside of the box" is needed, sleep-deprived workers will tend to use routine solutions. They may fixate on an ineffective solution and may not be able to recognize better alternatives, even if these are clearly available.

Sleep deprivation reduces a person's ability to assess risks, and risk-taking behaviors increase [32]. A sleep-deprived person is less able to learn from the negative consequences of risky decisions. Activities judged to be nonessential are increasingly neglected because situational awareness is lost. For example, a worker may misjudge how much time has passed.

Sleep deprivation can result in impaired mood including irritability, anxiety, and depression. The sleep-deprived worker may behave in a rude or hostile manner because he or she views the situation in a more negative light. Sleep deprivation may lead to inappropriate behavior such as giggling or laughing in serious situations.

Several studies show evidence of similarities between fatigue-related impairment in mental and physiological functioning and impairments because of alcohol intoxication [33–35]. In these studies, participants' performances were tested and compared after staying awake for extended periods of time and after drinking alcohol to a certain blood alcohol concentration (BAC). Participants staying awake for 17 h showed performance deficits similar to a BAC of 0.05%, and after 24 h awake, the deficits were similar to having a BAC of 0.10%. The United States defines legal intoxication for purposes of driving as having a BAC of 0.08% or greater. However, driving impairments are seen at lower levels, and some countries use a 0.05% or lower cutoff [36].

A common misconception is that a person can overcome the effects of sleep deprivation through motivation, professionalism, training, or experience. Recent research provides evidence to the contrary [33,37]. A study of medical residents did not provide evidence that they adapted to sleep deprivation better as their training years progressed. In fact, researchers found that sleep-deprived participants did not reliably recognize the extent of their declines in performance.

Performance declines vary for individuals when they are sleep deprived [28]. Testing the same person over time showed consistent patterns in his or her response to sleep loss. However, some people showed a greater decline in performance than others, which may reflect an individual trait or sensitivity.

### 16.3.3 Increased Risk for Worker Injury and Errors

Declines in neurocognitive performance could lead to higher rates of fatigue-related injury and worker errors. There is evidence that certain work scheduling patterns are associated with higher risks. When compared with the day shift, risks for accidents and errors were 15% higher for evening shifts, and 28% higher for night shifts [5]. When compared with 8 h shifts, 10 h shifts increased the risk by 13%, and 12 h shifts increased this risk by 28% [5]. As the number of successive night shifts worked increased, risk increased (by 17% for the third night shift, and 36% for the fourth). Dembe et al. [38] reported a possible "dose response"—as weekly work hours increased, injury rates to workers increased. Only a few studies have examined mandatory overtime but the evidence available raises concern. Studies of nurses linked mandatory overtime with higher risk for needlestick and other work-related injury, work-related illness, and missing more than 2 days of work because of these [39,40].

### 16.3.4 Poor Health Behaviors and Metabolic Syndrome Linked to Shift Work and Long Work Hours

Research links shift work and long work hours to poor health behaviors. Bushnell et al. [41] found that of nine work schedule patterns, the patterns predicted to be the most difficult were associated with the highest rates of smoking, obesity, and low physical activity. A systematic review by Frost et al. [42] concluded that smoking was generally more frequent in shift workers. Workers may be drawn to smoking as a countermeasure for the fatigue, sleepiness, and stress linked to demanding work schedules. A review of obesity and shift work by Antunes et al. [43] concludes there is considerable epidemiological evidence for a link between shift work and obesity. A few studies have examined long work hours and obesity, and some report a significant positive relationship [44,45]. The mechanism could be through poor-quality sleep and short sleep duration promoting an imbalance in appetite hormones that increase feelings of hunger and metabolic changes leading to obesity, insulin resistance, and reduced lipid tolerance [46]. Bushnell et al. [41] found there are few studies to date that have examined exercise and shift work or long work hours, and these have reported mixed findings.

### 16.3.5 Diseases Associated with Shift Work

Along with sleepiness, mood disturbances and gastrointestinal complaints are common initial symptoms that occur after disturbances to sleep. This set of complaints is frequently reported by people experiencing quick changes to their sleep/activity cycle due to traveling across several time zones [47]. Shift workers commonly complain of gastrointestinal symptoms including abdominal pain, gas, diarrhea, constipation, nausea, vomiting, change in appetite (either eating more or less than usual), indigestion, and heartburn [48,49]. A variety of stomach and digestive disorders have been reported in shift workers. Schernhammer et al. [50] reported an increased risk of colon cancer in nurses working 3 or more nights per month for 15 or more years. Possible mechanisms for gastrointestinal disturbances include disturbances to circadian rhythms involved in eating and digesting food, sleep deprivation leading to stress response, changes in immune function, or the types of foods that are available during these shifts [48,49]. Shift workers commonly report psychological complaints including bad mood, depression, irritability, anxiety, personality changes, and difficulty with personal relationships [51].

Shift workers are at somewhat higher risk for cardiovascular disorders. A review by Puttonen et al. [52] reported the epidemiologic data support a possible link with cardiovascular disease, including myocardial infarction, chest pain, and high blood pressure. Brown et al. [53] recently reported an increased risk for ischemic strokes. The mechanism could be through metabolic changes and smoking. Frost et al. [42] concluded there is some evidence that workers are at higher risk for metabolic disturbances and increases in smoking after starting shift work.

In 2007, the International Agency for Research on Cancer of the World Health Organization announced sufficient evidence supports that shift work with circadian rhythm disruption is a probable carcinogen [54]. The data from human studies suggest the link is possible, while the evidence from animal studies is sufficient. The most studied cancer in shift workers is breast cancer. A meta-analysis by Megdal et al. [55] of 13 studies reported night work increased cancer risk by 48%. Two of three studies found an increase in prostate cancer [56–58]. Costa et al. [59] discuss mechanisms that could possibly lead to cancer: exposure to light at night reduces melatonin levels, repeated shifting of the times of sleep and activity, and depression of immune surveillance.

Shift work is associated with adverse reproductive outcomes for women. A review by Frazier and Grainger [60] reported that shift work was associated with a modest increase in spontaneous abortion, preterm birth, and reduced ability for women to become pregnant. Shechter et al. [61] reported increased medical office visits for menstrual cycle symptoms and cycles that were shorter, longer, or irregular.

Persons with chronic diseases may have more difficulty managing their symptoms and disease progression when working shift work. Some medications have changes in effectiveness across the circadian cycle which can lead to more difficulty in determining the appropriate dosage and time for administration when working shift schedules. Shift work can also interfere with treatment regimens that involve maintaining regular sleep times, avoiding sleep deprivation, and controlling amounts and times of meals and exercise. Sood [62] warned that the following conditions could get worse: heart disease, high blood pressure, stomach and intestinal disorders, sleep disorders, insulin-dependent diabetes, seizure disorders, asthma requiring medication, psychiatric diseases, and alcohol/drug abuse.

Some people have many symptoms, do not get relief from medical therapies, and appear unable to adjust to shift work. Harrington [63] estimated 20% of workers who tried shift work were unable to adjust. The reason is not fully understood, but it may be due to an inherited tendency to be more sensitive to sleep loss and disruption of circadian rhythms.

### 16.3.6 Diseases Associated with Long Work Hours

The number of studies examining the influence of long work hours on health has been growing recently but is substantially less than the number of studies examining shift work. Several studies report long work hours are associated with increased fatigue, poor mood, poor recovery from work, and reduced perceived health [20,24,64–68]. Higher risk for depression is reported [69]. Suwazono et al. [70] reported better scores for depression and fatigue when work hours were less than 12 per day and 58 per week. Long work hours are associated with increased risk for heart disease, high blood pressure, and diabetes although not all studies consistently report a significant increase [20,22,65,71–73]. A systematic review identified eight studies with controls for physical demands that link long work hours to adverse musculoskeletal disorders [74]. This may be due to longer exposure to physical demands and less recovery time between work periods. Possible adverse reproductive outcomes include increased time to pregnancy [75]. A meta-analysis of six studies suggested a possible weak relationship with preterm birth [76].

## 16.4 IMPACTS TO FAMILY, EMPLOYER, AND COMMUNITY

Fatigue-related impairments can lead to reduced performance on the job. Rogers et al. [77] reported that the odds for a nurse making an error at work increased by three times when work shifts lasted 12.5 h or longer, compared with 8.5 h shifts. Interns working long hours made more serious medical errors in intensive care units [78]. Circadian, a shift work consulting firm, estimates fatigue-related accidents, lost productivity, and health care expenses costs employers $116.5 billion per year [79]. Similarly, Rosekind et al. [80] estimated annual fatigue-related productivity losses for four employers were $1967 for each of their employees, a total of $54 million.

Worker fatigue is a problem not only in the workplace, but also in the community. Experts who examined drowsy driving crashes listed shift work and long work hours in the top five factors that increase risk for a motor vehicle crash [81]. As a consequence, drowsy workers could also endanger others on the road. Worker fatigue has also been linked to several well known industrial disasters including the 2005 BP Texas City explosion, the Buffalo jet crash, and the Exxon Valdez oil spill [5,82,83]. In addition, society may be negatively impacted by a shortage of workers in critical jobs because too few people are willing to work demanding hours, as has been reported in the nursing workforce [84].

## 16.5 OVERVIEW OF SLEEP, CIRCADIAN RHYTHMS, AND FATIGUE PHYSIOLOGY

Sleep and circadian rhythm physiology provides insights into why workers make errors and develop chronic illnesses when they work demanding hours and do not get adequate sleep. Although science does not have the full answer for the function of sleep, evidence is mounting that critical processes occur during sleep that are connected with maintaining health and life. For example, a laboratory study reported rodents who were kept awake for a few weeks died as a result of sleep deprivation [26]. Sleep deprivation took just a little longer than time until death due to starvation.

Insufficient sleep is associated with reduced immune functioning and impaired mental functioning [85]. Spiegel et al. [86] linked sleep loss with insulin resistance, changes in appetite hormones that increase hunger and appetite, increases in sympathetic activity, and increases in evening cortisol. In the study, young healthy men restricted their sleep to 4h a night for 6 nights. At the end of this time, their test results showed clinically diagnosable impaired glucose tolerance. These findings suggest that not only do sleep deprived people feel sleepy, but other changes occur in their bodies that could speed the development of a broad range of chronic health problems such as obesity, diabetes, and cardiovascular disease. Sleep researchers warn that sleep appears to be as important for living as food and water, and recommend that people get the amount of sleep they need.

### 16.5.1 Regulation of Sleep: Simple Two-Process Model

Basic scientific information about sleep regulation provides insights into why workers have difficulty with shift work and long work hours and also guides coping strategies. Human physiology promotes sleep at consistent times at night and wakefulness and activity during the daylight hours. As a consequence, working during the night and sleeping during the daylight hours or sleeping at irregular times goes against human physiology and is difficult.

The sleep/wake cycle is driven by two processes: (1) homeostatic sleep drive and (2) circadian rhythms [87]. These two processes interact to determine the time of sleep and time of wakefulness, as well as the stability of waking neurocognitive functions (e.g., how well the person feels when awake).

The homeostatic sleep drive refers to a buildup of pressure to sleep in the brain the longer a person is awake, which gets stronger the longer a person is awake. Pressure to sleep decreases during sleep and reach a low after a full night of good-quality sleep. On awakening, the homeostatic process begins to build again the longer the person is awake. If a person remains awake for long periods, sleep pressure builds even further, putting higher pressure on the person to fall asleep.

The second process driving the sleep/wake cycle is circadian rhythms, which are internally-driven cyclic rises and falls of biochemical, physiological, and behavioral processes of living beings that occur across the 24h day. There are numerous circadian rhythms in the human body; in fact, it is unusual to find a physiological phenomenon that does not display a circadian rise and fall across the 24h day. A pair of circadian pacemakers in the hypothalamus of the brain orchestrates the timing of the circadian rhythms so they work together. Circadian rhythms cause a predictable rise and fall in wakefulness across the 24h day. A peak in wakefulness occurs in the evening a few hours before usual bedtime for a person on a normal, daytime activity schedule. This can be experienced as enhanced alertness in the early evening, even after being awake for long periods. The evening peak in alertness, called the "forbidden zone" or "wake maintenance zone," makes it difficult to go to bed earlier [88].

Circadian rhythms promote sleepiness before usual bedtime, initiate sleep, and begin promoting wakefulness before usual waking time in the morning. Circadian rhythms also promote a shorter period of sleepiness in the middle of the afternoon, a time when people tend to want to take a nap. The pacemakers have an internally-driven 24 h rhythm which tends to run longer than 24 h, but is entrained (reset) every day by external timing cues to keep the cycle at 24 h. Researcher theorize this is why when changing sleep times, it is easier to get up later than it is to get up earlier—because the circadian pacemakers naturally run longer than 24 h. The strongest external cue is the light/dark cycle of the sun [89]. Exercise and melatonin also appear to exert an influence on the timing of the pacemakers, but they have less influence than light. Normally, light from the sun enters the eye (through closed eyelids while asleep) and signals to the pacemakers the time for awakening and activity and time for sleep. In addition to sleep and wakefulness, pacemakers orchestrate the timing for numerous circadian rhythms that regulate physiological and behavioral processes in our bodies such as: hormones, body temperature rhythm, and times of physiological processes connected with eating and digesting food.

Circadian rhythms have an important purpose: they prepare the body for expected changes in the environment (e.g., the time for activity, time for sleep, and times to eat). External cues have a strong influence on the timing of the circadian pacemakers. The strongest cue is the sun's light/dark cycle; however, artificial light also influences the pacemakers [90,91].

The circadian system needs time to adjust to new sleep times and cannot shift to a new schedule quickly, such as when switching shifts [90]. So, although workers can try to force themselves to be awake and work at night, they will be working against the sun's light/dark cycle and their circadian pacemakers that are strongly promoting sleep at night and wakefulness during daylight hours. Researchers are testing manipulating timing of light exposure to promote adjustment of the circadian rhythms to working at night [92].

Night work and shift work often result in de-synchronization or dissociation of the homeostatic and circadian processes so they no longer work together, negatively impacting sleep. As a result, sleep during the daytime tends to be lighter, shorter, more fragmented, and less recuperative than sleeping at night [93]. Workers on night shift commonly build up pressure to sleep during their work shift so they tend to fall asleep quickly upon their return home. However, because circadian rhythms promote wakefulness during daylight hours, they tend to awaken around noon and have difficulty going back to sleep. Because of this waking pressure, small environmental disturbances such as outside noise, or minor discomfort, can awaken them before they have slept sufficiently. During night shift, workers often have to fight the strong pressure to sleep due to circadian rhythms that promote sleepiness. This is especially true between 2:00 and 6:00 AM [94].

Individual factors influence a person's ability to sleep at different times, and how the person performs when he or she does not get enough sleep. Some people have more difficulty than others. One factor researchers are studying is the "morningness/eveningness" trait. People with "morningness" or "lark" trait are naturally early-to-bed and early-to-rise and tend to have more difficulties working night and evening shifts. Persons with the "night owl" or "eveningness" trait are naturally late-to-bed and late-to-rise and tend to have fewer difficulties working night and evening shifts. Some people will show the "intermediate" trait, between "larks" and "owls."

The morningness/eveningness trait may be inherited, or part of the functioning of the person's circadian pacemakers [95]. Teenagers and young adults tend to be "evening" types, and researchers theorize this may be due to brain development at those ages [96]. Morningness increases with age, so older adults tend to be "morning" types [97]. This trait may have consequences for shift preferences and coping capabilities. However, according to Sack et al. [98], sufficient research evidence is not currently available to support the validity and reliability of a morningness/eveningness questionnaire score to predict a person's ability to adapt to night shift.

There is individual variability in the length of sleep each person needs [95]. Most adults need 7–9 h of good-quality sleep per night. Sleep researchers report the amount of sleep needed by each person may be an inherited trait like eye color and hair color and recommend that each person gets the amount they need each day.

Research links some factors to more vulnerability to sleep loss. People with an inherited trait show more neurocognitive performance declines with sleep loss [28]. Another factor is age. Younger people (children, teens, and young adults) tend to be able to sleep during the daytime and have more flexibility in the times they can sleep. They also tend to have a longer sleep length need. Aging is associated with less tolerance of shift work, which may be due to age-related changes in sleep that may make it more difficult for older people to initiate and maintain sleep at different times of the day [99]. As a consequence, some workers may experience more difficulties as they grow older, even those who adjusted adequately to shift work at younger ages.

### 16.5.2 Drowsiness and Fatigue

Long periods awake and poor-quality sleep build up "sleep debt"; extra sleep is needed to "pay down" this debt. For example, a person needing 8 h of sleep, but only getting 6 h, would accumulate a "sleep debt" of 2 h that day. If the person has a habit of restricting their sleep, they could experience ongoing fatigue, sleepiness, and other negative effects during their wake time hours. To keep the sleep debt down, a person should use good sleep behaviors to prepare for sleep and sleep long enough to feel rested shortly after awakening.

Research suggests building up a sleep debt and paying if off later is risky. Although the level of sleepiness may return to normal after a day or two of recovery sleep, neurocognitive performance will still be reduced compared to baseline [100,101]. Thus, getting adequate sleep on a regular basis, rather than episodically, is important. This is especially important for workers working 12 h shifts, who do not sleep much between the 12 h shifts [19], then try to catch-up on sleep after the block of shifts ends.

Sleep deprivation increases pressure for the brain to fall asleep. The transition from wakefulness to sleep is abrupt and rapid and occurs in an on-off-type switch in the brain stem [102]. The nature of this on-off-type switch transition to sleep can lead to dangerous situations if sleep occurs during critical times such as when driving or operating equipment.

The warning signs that the brain is close to falling asleep include: difficulty focusing; frequent blinking or heavy eyelids; daydreaming or wandering and disconnected thoughts; yawning repeatedly or rubbing eyes; difficulty keeping head up; and feeling restless and irritable [103]. When driving, the following symptoms indicate sleep may be imminent: trouble remembering the last few miles driven, missing exits or traffic signs, drifting from a lane, tailgating, or hitting a shoulder rumble strip. If these symptoms occur, the person should stop as soon as possible and take action quickly to protect self and others from the dangers of drowsiness while driving or during work.

Wesensten et al. [94] discussed three key factors that promote fatigue-related errors: time of day, time awake, and time-on-task. The time of day or circadian factor predicts that the strongest decrease in wakefulness occurs during the early morning hours (2:00–6:00 AM). Other vulnerable times are in the middle of the afternoon (2:00–5:00 PM) and from midnight to 2:00 AM. The time-awake factor predicts the longer a person has been awake (or the poorer the quality and shorter the amount of sleep the person had recently), the greater their risk for fatigue. Lastly, the time-on-task factor predicts the longer the time spent on a task, particularly one that is monotonous, the more likely a person will feel fatigue. If two or three factors occur at the same time, risk for fatigue increases.

## 16.6 STRATEGIES TO REDUCE RISKS

Strategies to reduce risks from shift work and long work hours focus on correcting disruptions to sleep and circadian rhythms and strains on family and social life. Workers, managers, and employers share the responsibility of implementing strategies to reduce risks. Strategies for managers include instituting policies and systems in the workplace to reduce fatigue. Strategies for workers include adopting good sleep practices and behaviors and allowing sufficient time to get enough sleep.

### 16.6.1 Improving the Workplace

There is no single best work schedule pattern: hundreds of shift scheduling patterns exist. However, researchers offer several recommendations for improving work schedules [2,104,105]. A common recommendation is to give workers input on the design of their schedule since this promotes schedules that better meet the worker's needs. Work scheduling patterns that are associated with higher risks for fatigue-related accidents and errors can be avoided. Very early morning start times (6:00 AM or earlier) are not recommended because these tend to truncate sleep. Examine work start and end times and consider whether the evening and nighttime hours are required for the products and services produced. Avoid quick changes with less than 10–11 h off between shifts because this does not give enough time for a worker to commute to work, get ready for work, and get the 7–9 h of sleep most people need every day. Night shifts are associated with higher risks, so researchers commonly recommend using fixed nights with caution and limiting the consecutive number of night shifts to three for shift rotations [2,5,105]. Since longer shifts are also associated with higher risks, Folkard and Lombardi [5] recommend using shorter more frequent work shifts rather than longer, less frequent shifts. Avoid a long series of work days followed by a long series of days off (a mini-vacation without use of vacation days) because long series of work days tend to build up fatigue. Researchers recommend using a forward rotation for rotating shifts (day shift to evening shift to night shift) over a backward rotation (evening shift to day shift to night shift) [2,106]. They recommended avoiding a weekly rotation, and suggested using rotations that change shifts several times per week, or every 2 weeks or more [2,106,107]. Researchers recommend maintaining a regular and predictable schedule because it helps the worker prepare for work, sleep, and other responsibilities. The inability to plan makes unscheduled mandatory work periods more difficult. Keep overtime to a minimum. Schedule 1 or 2 full weekends off each month to help workers maintain contact with family and friends [105].

Extended shifts require special consideration for the worker's needs due to the long number of consecutive hours worked. Knauth [19] recommended the following conditions be met for using extended shifts:

- The nature of the work and workload are suitable and do not include heavy physical demands, dangerous work, fast-paced demands, or high stress.
- Work periods include adequate rest breaks.
- The shift system, work tasks, and other features of the job are designed to minimize fatigue.
- There are adequate arrangements for covering absentees so that overtime will not be added.
- Toxic exposure is limited.
- Complete recovery after work is possible.
- Assessments are systematically carried out to identify adverse effects early.
- Demographic factors that are linked to more difficulties with long shifts are not present, such as an aging workforce.

The combination of difficult scheduling patterns or long shifts with heavy work demands is associated with higher risks. For example, a review by Caruso et al. [12] reports a pattern of adverse findings when 12 h shifts combine with more than 40 h a week, when 12 h shifts combine with very early start times, and when high workloads or hot work environments combine with 12 h shifts. Managers and workers can carefully consider the combined scheduling and job demands and modify schedules and workload to reduce risks.

Managers can work toward moving their organizations to make use of naps during work shifts. This includes policies to allow naps during the work shift, good nap environments, systems to schedule naps and awaken napping workers, and adequate staffing to cover the work.

Frequent rest breaks during the work shift help reduce risks. By pooling findings from several studies, Folkard and Lombardi [5] estimated that rest breaks every 2 h reduced risks more than 4 or 6 h rest breaks. Managers can improve their workplace approach for taking rest breaks.

Managers can schedule employees to work with others rather than working alone. They can also consider the timing of demanding tasks and try to schedule these at times when there is less chance for fatigue [105]. Long periods working on monotonous tasks increase risk for fatigue [94], so it is useful to break up time on these tasks with other activities. In addition, managers can assess if staffing levels are appropriate for workload needs.

Managers can also consider the timing of services, environmental conditions, and social programs [105]. Since workers on shift work and long work hours may be at higher risk for certain diseases, managers can encourage regular health examinations to identify and treat any problem early. It would be helpful to schedule occupational health and counseling services so they are convenient for workers on evening and night shifts. The environmental conditions during nonstandard shifts should be considered and maintained so they are comfortable during the evening and night-shift hours. Workplaces often organize social programs during the holidays and at other times. These events should be planned to include shift workers and scheduled at times that are convenient for them.

There is widespread lack of appreciation and knowledge about the risks connected with insufficient sleep and demanding work hours, and as a consequence, workers and workplaces often lack systems to protect against these risks [108]. Training programs for workers and managers are recommended by the National Occupational Research Agenda (NORA) Long Work Hours Team, the American Petroleum Institute, and other organizations [4,109]. Content for training programs can include these nine topics:

1. Basic information about sleep, circadian rhythm, and fatigue physiology
2. Good sleep practices and behaviors to get adequate sleep
3. Sleep disorders and other sources of sleepiness and the importance of identifying and treating them
4. Safety risks linked to insufficient sleep
5. Risks for development and exacerbation of acute and chronic diseases from insufficient sleep and shift work and long work hours
6. Strategies to reduce fatigue-related injuries and errors
7. Signs of fatigue and workplace policies and strategies when workers are fatigued at work
8. Workplace interventions to reduce safety and health risks associated with shift work and long work hours
9. Training for family members about the special needs of workers on shift work and long work hours so they can provide better understanding and support

Workplaces can conduct systematic assessments to examine the influence of work schedules on factors on- and off-the-job including: performance, alertness, sleep, unintentional injury, worker errors, near misses, illnesses, and off-the-job responsibilities. An anonymous, no-blame reporting system can collect incident and near miss reports. Include information about the time of the incident, shift details, number of consecutive shifts worked, number of hours the person has been awake, number of hours of sleep in the previous 24 h, and normal or unusual circumstances such as overtime because of a weather emergency [109]. Analyses of several incident reports may reveal factors that can be modified to reduce risks. Workplace fatigue-related policies can guide workers and managers on things such as the appropriate actions when a worker appears to be too fatigued to work safely, and the level of work hours that may be too high for safety reasons.

### 16.6.2 Improving Sleep Habits and Routines

The key strategy for reducing risks is getting enough good-quality sleep every day. Good sleep habits and routines, such as making the sleep environment more conducive to sleep and preparing oneself for sleep, improve sleep quality. Information on sleep hygiene, sleep disorders, and strategies for shift workers is available from several sources [98,110–113]. Figure 16.1 summarizes frequently recommended suggestions for improving sleep.

- Get the amount of sleep needed each day; most people need 7–8 h of sleep.
- Sleep at about the same times every day to help improve sleep.
- Use the bedroom only for sleeping and intimacy to foster associating it with relaxation and sleep. Do not use the television or work in the bedroom. Sleep in a very dark, quiet, well-ventilated space with a comfortable, cool temperature. Have a comfortable bed and pillows.
- For night and evening workers who need to sleep during daylight hours, create a completely dark bedroom by covering the windows with opaque coverings and block light from coming under the doorways. Light bulbs emitting dim red light or very dim night lights can be used by the path to the restroom. An alternative is to use an eye mask during the entire sleep episode. Prevent noise by using a white noise machine (like a fan) or ear plugs. Take measures to prevent awakening because of phone calls or the door bell.
- Relax before going to bed. Keep the light levels lower 2 h before bedtime and avoid household chores or stimulating activities for at least 2 h before bedtime. Avoid bright computer screens during this time period (or wear dark sunglasses to reduce this light exposure). Follow a relaxing bedtime routine within an hour or more of sleep to signal the brain it is time to sleep. Brushing the teeth and getting ready for bed facilitate relaxation and sleep. Some suggestions to help relax are a warm bath, listening to soothing music, meditating, breathing exercises, and other relaxation techniques.
- Avoid before bedtime.
    - i. Heavy or spicy meals (2–3 h before bed).
    - ii. Liquid intake (to avoid getting up to go to the bathroom).
    - iii. Alcohol (causes early awakening and sleep disturbances).
    - iv. Caffeine (avoid for several hours according to person's own sensitivity).
    - v. Nicotine (avoid for several hours).
    - vi. Going to bed on an empty stomach.
- Face the clock away from the bed to avoid focusing on the time if awaking before wake-up time.
- If unable to fall asleep within 30 min of going to bed or awaking early and being unable to get back to sleep, get out of bed and try a relaxing activity such as reading until becoming drowsy, but keep the light level low because bright light will promote wakefulness.
- Exercise regularly to improve sleep. However, end exercise at least 5 h before bedtime.

FIGURE 16.1 Suggestions for getting a good night's sleep. (From Patlak, M., *Your Guide to Healthy Sleep*, U.S. Department of Health and Human Services, National Institutes of Health, National Heart, Lung, and Blood Institute, NIH Publication No. 06-5271, 2005.)

Because sunlight and artificial light have an alerting effect and a strong influence on the timing of sleep, workers who need to sleep during daylight hours should take additional measures to prevent light exposure to their eyes during their sleep period and prevent disturbances from noise.

### 16.6.3 Correcting Other Causes of Sleepiness

Another strategy for reducing risks from fatigue is to identify and manage other factors that increase daytime sleepiness, including sleep disorders and a wide range of other chronic disorders that disturb sleep. Sleep-related problems are common, afflicting 50–70 million Americans, but are often undiagnosed and, therefore, not treated [108]. Insomnia increases with age. According to the National Sleep Foundation poll, "Sleep in America," only 33% of women ages 18–24 had a sleep problem a few nights a week, but 48% of women ages 55–64 did [114]. Sleepiness is a common side effect of several medications, such as antidepressants, cold tablets, and antihistamines. Sleep deprivation from shift work or long work hours can exacerbate underlying sleep disorders [115]. Because workers with fatigue who work these demanding schedules will be particularly vulnerable to fatigue-related risks, they should consider seeking an evaluation from a sleep disorders specialist if symptoms are present.

### 16.6.4 Using Naps

The Standards Practice Committee for sleep medicine recommends planned naps before or during night shifts, because research reflects a high degree of clinical certainty for their effectiveness [111]. Takahashi's [116] findings support the positive effects of naps during work shifts. Research indicates that a brief period of sleep (15–20 min), not just a quiet period, has an alerting effect [117]. If time is available, longer naps of 1.5 h or more can reduce sleep debt and provide further gains. However, after naps of greater than 45 min, there will be a period of grogginess, so it is important to allow time for this to pass before resuming dangerous or critical tasks [118]. When workers are sleepy after the work shift, a short nap may help restore alertness before the drive home.

### 16.6.5 Using Caffeine

Caffeine is a commonly used stimulant that takes approximately 30 min to reach peak plasma levels [119]. The equivalent of two cups of coffee can increase alertness for a few hours [120]. However, heavy daily use of caffeine may reduce its effectiveness. Instead, workers can use it strategically when the stimulant effect is needed.

### 16.6.6 Using a Combination of a Nap and Caffeine

If experiencing drowsiness while driving, experts suggest stopping driving, consuming caffeine, and then quickly taking a short nap [113]. On awakening, the person will gain the benefits from both. The combination of a 2.5 h nap before a night shift and caffeine at the beginning of the shift had positive effects on alertness [98,121].

### 16.6.7 Using Light

Researchers are testing the effectiveness of light treatments to improve adaptation to shift work. Light increases alertness and is a strong signal for circadian rhythms [91,110]. Many questions remain about the optimum color of light, intensities, and length of time for exposure to light to use to improve adjustment of the circadian rhythms [122]. Sunlight is most effective, but common levels of artificial room-level light (100 lux or more) also increase alertness [110]. For night workers, some researchers suggest intermittently going into well-lit areas during the first several hours of the shift and reducing light exposure a few hours before bedtime to help prepare the body for sleep when home [98,111].

### 16.6.8 Diet and Exercise

Many research questions remain about the effects of diet when working shift schedules and long work hours. A few studies suggest that simple sugars may increase sleepiness, so it is best to avoid sugar-rich beverages and foods when prone to sleepiness and there is a need to stay alert [123–125]. There is not yet enough evidence to clearly determine the best time to eat when working night shift and to give detailed information about what to eat to reduce metabolic disturbances and optimize alertness and performance [126]. Circadian rhythms involved in eating and digesting food are timed for eating during the day and fasting during the night. Researchers theorize that shift workers experience gastrointestinal symptoms, reduced glucose and lipid tolerance, and other gastrointestinal problems because of the mismatch in these circadian rhythms and the timing of food intake, sleep, and activity. Given the evidence currently available, Lowden et al. [126] offer the following dietary guidelines for shift workers:

- Avoid eating or reduce food intake between midnight and 6:00 AM. Use the normal daytime and nighttime pattern of meal timing as much as possible and divide intake into three meals per 24 h period.
- Eat more frequently when there are higher energy needs.
- During the shift, eat high-quality foods such as vegetables, salads, vegetable soups, fruits, wholegrain sandwiches, yogurt, cheese, eggs, nuts, and green tea. Avoid sugar-rich products, and low-fiber carbohydrate foods.
- Eat with colleagues in a pleasant environment away from the work site.
- For general dietary suggestions, avoid alcohol and foods with added sugar. Eat complete meals with animal products and starchy vegetables, or protein-rich plant foods (beans plus cereals, rice, pasta, plus roots) and non-starchy vegetables and fruits, or vegetarian meals consisting of protein-rich plant foods, fruits, and vegetables.
- For night workers, eat breakfast before day sleep to avoid waking due to hunger.
- Avoid large meals 1–2 h before the main sleep episode.

Although the evidence about the influence of exercise on shift work and long work hours leaves many questions, researchers offer some guidelines. A review by Caldwell et al. [110]

concluded that short periods of moderate exercise can increase alertness for brief periods (30 min), while longer periods of intense exercise may increase sleepiness. Sleep researchers commonly recommend regular exercise because it promotes better sleep. However, they recommend ending exercise at least 2–5 h before the main sleep episode [110,112].

### 16.6.9 Avoiding Ineffective Strategies

Researchers cautioned that several common strategies people use when feeling sleepy while driving actually do not help, such as putting down the window, turning the air conditioner on full blast, and playing the radio loudly [127]. Resorting to these strategies indicates that someone is struggling to stay awake and is too sleepy to continue driving safely [113].

### 16.6.10 Cautioning about Substances to Promote Sleep or Alertness

Many people can be helped by good sleep habits and routines, so it is best to use these behavioral strategies fully. Stimulants and sleep medications can help people who have difficulty with sleep or daytime sleepiness, but these substances can have side effects, can become addictive, and are generally just for short-term use [110,128]. In addition, they do not completely remedy the effects of inadequate sleep, and questions remain about their long-term use [111]. Given that caution, it is advisable for a worker to see a sleep disorders specialist if experiencing frequent sleep disturbances or excessive daytime sleepiness.

### 16.6.11 Gain Cooperation and Help from Family and Friends

Several studies link shift work and long work hours to strains on personal relationships and work/family conflict [51,129,130]. For families with children, Presser [131] reported divorce and separation are six times higher for men on night shift and three times higher for women on night shift. Mood disturbances from poor sleep and disturbances to circadian rhythms could be the source of some of these strains. Also, evening and night shifts and weekend work make it difficult to participate in normal social and family activities and give workers less quality time to spend with family and friends. Adopting good sleep practices and improving sleep would help reduce mood disturbances that strain relationships. Workers also can educate their family and friends about their special needs to gain their understanding and willingness to be flexible in juggling activities and demands [105]. Another strategy is to adopt systems to maintain communications such as writing notes and calling during the work shifts and scheduling time together.

## 16.7 FUTURE CONCERNS

There are many unanswered questions about risks linked to shift work and long work hours, the best work schedule patterns to use, and strategies to reduce risks. Because society requires many types of workers to provide vital services around the clock, it is not possible to eliminate shift work. At times, long work hours cannot be avoided. Future work settings and society would benefit from many types of research on this topic: (1) research to better explain the physiology underlying the health risks of circadian disturbances and sleep deprivation as well as strategies to reduce these risks; (2) studies to identify individual factors that account for the wide range of tolerances to sleep deprivation and circadian rhythm disturbances; (3) epidemiology studies to assess the risks using better methods to handle underestimates from factors such as drop outs and the healthy worker effect; and (4) intervention studies using laboratory experiments as well as real world settings [4,93,132].

1. Research needs to provide a clear and complete description of the work schedules, the workers, and the work environment to better compare findings across studies, build evidence, and move the science forward.

2. Research needs to measure and control for factors that may influence the relationship between shift work and long work hours and outcomes.

3. Research should explore a wider range of possible outcomes: a variety of symptoms and functional deficits associated with having less time for sleep, work recovery, and other life necessities as well as longer exposures to occupational hazards and demands; the development of chronic diseases as well as acute injuries and illnesses; risks for the employer such as reduced productivity and quality of goods and services, increased absenteeism, higher health care and workers' compensation costs, and less successful recruitment and retention; short and long term impacts on marital partners, family life, and children; and impacts on the community.

4. Research should examine shift work and long work hours in occupations involving public safety (e.g., health care, law enforcement, transportation, utilities) since the consequences of errors due to fatigue can be especially great.

5. Research should examine workers in vulnerable groups: women who are pregnant and couples who wish to become pregnant; older workers; workers with chronic illnesses including symptom management and disease progression; workers exposed to occupational hazards; and socio-economically disadvantaged workers.

6. More regular, systematic tracking and assessment of work schedules in the United States is needed.

7. Research should develop and test interventions to reduce risks including but not limited to: various work scheduling patterns; manipulating light exposure, pharmacology agents, diet regimes, and other work organization strategies such as supervisor and coworker support; workplace or corporate interventions including fatigue-related policies, fatigue risk management systems, and education campaigns for workers and managers; models to predict risks; and studies of the impact of broader public policy measures (e.g., impact of the state laws limiting mandatory overtime for nurses).

## REFERENCES

1. T. M. McMenamin, A time to work: Recent trends in shift work and flexible schedules, *Mon Labor Rev. 130*(12): 3–15 (2007).
2. P. Knauth and S. Hornberger, Preventive and compensatory measures for shift workers, *Occup Med (Lond). 53*: 109–116 (2003).
3. Healthy People 2020. Sleep health. www.healthypeople.gov/2020/topicsobjectives2020/objectiveslist.aspx?topicid=38 (accessed December 3, 2010).

4. C. C. Caruso, T. Bushnell, D. Eggerth et al., Long working hours, safety, and health: Toward a national research agenda, *Am J Ind Med. 49*: 930–942 (2006).
5. S. Folkard and D. A. Lombardi, Modeling the impact of the components of long work hours on injuries and 'accidents,' *Am J Ind Med. 49*: 953–963 (2006).
6. J. Barton, E. Spelten, P. Totterdell et al., The standard shiftwork index: A battery of questionnaires for assessing shiftwork-related problems, *Work Stress. 9*: 4–30 (1995).
7. A. Knutsson, Methodological aspects of shiftwork research, *Chronobiol Int. 21*: 1037–1047 (2004).
8. R. R. Rosa, M. Harma, K. Pulli et al., Rescheduling a three shift system at a steel rolling mill: Effects of a one hour delay of shift starting times on sleep and alertness in younger and older workers, *Occup Environ Med. 53*: 677–685 (1996).
9. C. A. Czeisler, M. C. Moore-Ede, and R. H. Coleman, Rotating shift work schedules that disrupt sleep are improved by applying circadian principles, *Science. 217*: 460–463 (1982).
10. K. Orth-Gomer, Intervention on coronary risk factors by adapting a shift work schedule to biologic rhythmicity, *Psychosom Med. 45*: 407–415 (1983).
11. C. J. Poole, G. R. Evans, A. Spurgeon, and K. W. Bridges, Effects of a change in shift work on health, *Occup Med. 42*: 193–199 (1992).
12. C. C. Caruso, E. M. Hitchcock, R. B. Dick et al., *Overtime and Extended Work Shifts: Recent Findings on Illnesses, Injuries, and Health Behaviors*, Department of Health and Human Services, Public Health Service, Centers for Disease Control and Prevention, National Institute for Occupational Safety and Health, Cincinnati, OH, DHHS NIOSH Publication No. 2004-143, 2004.
13. D. Angersbach, P. Knauth, H. Loskant et al., A retrospective cohort study comparing complaints and diseases in day and shift workers, *Int Arch Occup Environ Health. 45*: 127–140 (1980).
14. L. Nylén, M. Voss, and B. Floderus, Mortality among women and men relative to unemployment, part-time work, overtime work, and extra work: A study based on data from the Swedish twin registry, *Occup Environ Med. 58*: 52–57 (2001).
15. T. M. Beers, Flexible schedules and shift work: Replacing the '9-to-5' workday? *Mon Labor Rev. 123*: 33–40 (2000).
16. C. L. Drake, T. Roehrs, G. Richardson et al., Shift work sleep disorder: Prevalence and consequences beyond that of symptomatic day workers, *Sleep. 27*: 1453–1462 (2004).
17. M. Ingre and T. Åkerstedt, Effect of accumulated night work during the working lifetime, on subjective health and sleep in monozygotic twins, *J Sleep Res. 13*: 45–48 (2004).
18. J. J. Pilcher, B. J. Lambert, and A. I. Huffcutt, Differential effects of permanent and rotating shifts on self- report sleep length: A meta-analytic review, *Sleep. 23*: 155–163 (2000).
19. P. Knauth, Extended work periods, *Ind Health. 45*: 125–136 (2007).
20. T. Hayashi, Y. Kobayashi, K. Yamaoka, and E. Yano, Effect of overtime work on 24-h ambulatory blood pressure, *J Occup Environ Med. 38*: 1007–1011 (1996).
21. T. Kageyama, N. Nishikido, T. Kobayashi et al., Long commuting time, extensive overtime, and sympathodominant state assessed in terms of short-term heart rate variability among male white-collar workers in the Tokyo megalopolis, *Ind Health. 36*: 209–217 (1998).
22. N. Nakanishi, K. Nakamura, S. Ichikawa et al., Lifestyle and the development of hypertension: A 3- year follow-up study of middle-aged Japanese male office workers, *Occup Med. 49*: 109–114 (1999).
23. J. Park, Y. Kim, H. K. Chung, and N. Hisanaga, Long working hours and subjective fatigue symptoms, *Ind Health. 39*: 250–254 (2001).
24. T. Sasaki, K. Iwasaki, T. Oka et al., Effect of working hours on cardiovascular-autonomic nervous functions in engineers in an electronics manufacturing company, *Ind Health. 37*: 55–61 (1999).
25. M. Basner, K. Fomberstein, F. M. Razavi et al., American time use survey: Sleep time and its relationship to waking activities, *Sleep. 30*: 1081–1091 (2007).

26. C. A. Everson, Comparative research approaches to discovering the biomedical implications of sleep loss and sleep recovery, in *Basics of Sleep Guide*, C.J. Amlaner, D. Phil, and P.M. Fuller, Eds., Sleep Research Society, Westchester, IL, 2009.
27. National Sleep Foundation, How much sleep do we really need? www.sleepfoundation.org/how-much-sleep-do-we-really-need (accessed June 7, 2010).
28. N. Goel, H. Rao, J. S. Durmer, and D. F. Dinges, Neurocognitive consequences of sleep deprivation, *Semin Neurol. 29*: 320–339 (2009).
29. H. J. Moller, L. Kayumov, E. L. Bulmash et al., Simulator performance, microsleep episodes, and subjective sleepiness: Normative data using convergent methodologies to assess driver drowsiness, *J Psychosom Res. 613*: 335–342 (2006).
30. J. S. Durmer and D. F. Dinges, Neurocognitive consequences of sleep deprivation, *Semin Neurol. 25*: 117–129 (2005).
31. J. J. Pilcher, L. E. McClelland, D. D. Moore et al., Language performance under sustained work and sleep deprivation conditions, *Aviat Space Environ Med. 785*: Suppl B25–B38 (2007).
32. V. Venkatraman, Y. M. L. Chuah, S. A. Huettel, and M. W. L. Chee, Sleep deprivation elevates expectation of gains and attenuates response to losses following risky decisions, *Sleep. 305*: 603–609 (2007).
33. J. T. Arnedt, J. Owens, M. Crouch et al., Neurobehavioral performance of residents after heavy night call vs. after alcohol ingestion, *JAMA. 2949*: 1025–1033 (2005).
34. D. Dawson and K. Reid, Fatigue, alcohol and performance impairment, *Nature. 388*(6639): 235 (1997).
35. A. M. Williamson and A. M. Feyer, Moderate sleep deprivation produces impairments in cognitive and motor performance equivalent to legally prescribed levels of alcohol intoxication, *Occup Environ Med. 57*: 649–655 (2000).
36. National Highway Traffic Safety Administration, On DWI laws in other countries. www.nhtsa.gov/people/injury/research/pub/DWIothercountries/dwiothercountries.html (accessed July 7, 2010), 2000.
37. H. P. Van Dongen, G. Maislin, J. M. Mullington, and D. F. Dinges, The cumulative cost of additional wakefulness: Dose-response effects on neurobehavioral functions and sleep physiology from chronic sleep restriction and total sleep deprivation, *Sleep. 262*: 117–126 (2003).
38. A. E. Dembe, J. B. Erickson, R. G. Delbos, and S. M. Banks, The impact of overtime and long work hours on occupational injuries and illnesses: New evidence from the United States, *Occup Environ Med. 62*: 588–597 (2005).
39. B. de Castro, K. Fujishiro, T. Rue et al., Associations between work schedule characteristics and occupational injury and illness, *Int Nurs Rev. 57*(2): 188–194 (2010).
40. R. R. M. Gershon, J. M. Pearson, M. F. Sherman et al., The prevalence and risk factors for percutaneous injuries in registered nurses in the home health care sector, *Am J Infect Control. 37*: 525–533 (2009).
41. P. T. Bushnell, A. Colombi, C. C. Caruso, and S. Tak, Work schedules and health behavior outcomes at a large manufacturer, *Ind Health. 48*: 395–405 (2010).
42. P. Frost, H. A. Kolstad, and J. P. Bonde, Shift work and the risk of ischemic heart disease—A systematic review of the epidemiologic evidence, *Scand J Work Environ Health. 35*: 163–179 (2009).
43. L. C. Antunes, R. Levandovski, G. Dantas et al., Obesity and shift work: Chronobiological aspects, *Nutr Res Rev. 23*: 155–168 (2010).
44. K. Nakamura, S. Shimai, S. Kikuchi et al., Increases in body mass index and waist circumference as outcomes of working overtime, *Occup Med* (*Lond*). *48*: 169–173 (1998).
45. M. Shields, Long working hours and health, *Health Rep. 11*: 33–48 (1999).
46. K. Spiegel, E. Tasali, R. Leproult, and E. Van Cauter, Effects of poor and short sleep on glucose metabolism and obesity risk, *Nat Rev Endocrinol. 5*: 253–261 (2009).
47. R. L. Sack, Jet lag, *N Engl J Med. 362*: 440–447 (2010).

48. C. C. Caruso, S. L. Lusk, and B. W. Gillespie, Relationship of work schedules to gastrointestinal diagnoses, symptoms, and medication use in auto factory workers, *Am J Ind Med. 46*: 586–598 (2004).
49. A. Knutsson and H. Boggild, Gastrointestinal disorders among shift workers, *Scand J Work Environ Health. 36*: 85–95 (2010).
50. E. S. Schernhammer, F. Laden, F. E. Speizer et al., Night-shift work and risk of colorectal cancer in the nurses' health study, *J Natl Cancer Inst. 95*: 825–828 (2003).
51. S. M. Rohr, S. G. Von Essen, and L. A. Farr, Overview of the medical consequences of shift work, *Clin Occup Environ Med. 3*: 351–361 (2003).
52. S. Puttonen, M. Harma, and C. Hublin, Shift work and cardiovascular disease—Pathways from circadian stress to morbidity, *Scand J Work Environ Health. 36*(2): 96–108 (2010).
53. D. L. Brown, D. Feskanich, B. N. Sanchez et al., Rotating night shift work and the risk of ischemic stroke, *Am J Epidemiol. 169*(11): 1370–1377 (2010).
54. K. Straif, R. Baan, Y. Grosse et al., Carcinogenicity of shift-work, painting, and fire-fighting, *Lancet Oncol. 8*: 1065–1066 (2007).
55. S. P. Megdal, C. H. Kroenke, F. Laden et al., Night work and breast cancer risk: A systematic review and meta-analysis, *Eur J Cancer. 41*: 2023–2032 (2005).
56. M. Conlon, N. Lightfoot, and N. Kreiger, Rotating shift work and risk of prostate cancer, *Epidemiology. 18*: 182–183 (2007).
57. T. Kubo, K. Ozasa, K. Mikami et al., Prospective cohort study of the risk of prostate cancer among rotating-shift workers: Findings from the Japan Collaborative Cohort Study, *Am J Epidemiol. 164*: 549–555 (2006).
58. J. Schwartzbaum, A. Ahlbom, and M. Feychting, Cohort study of cancer risk among male and female shift workers, *Scand J Work Environ Health. 33*: 336–343 (2007).
59. G. Costa, E. Haus, and R. Stevens, Shift work and cancer—Considerations on rationale, mechanisms, and epidemiology, *Scand J Work Environ Health. 36*: 163–179 (2010).
60. L. M. Frazier and D. A. Grainger, Shift work and adverse reproductive outcomes among men and women, *Clin Occup Environ Med. 3*: 279–292 (2003).
61. A. Shechter, F. O. James, and D. B. Boivin, Circadian rhythms and shift working women, *Sleep Med Clin. 3*(1): 13–24 (2008).
62. A. Sood, Medical screening and surveillance of shift workers for health problems, *Clin Occup Environ Med. 3*: 339–349 (2003).
63. J. M. Harrington, Health effects of shift work and extended hours of work, *Occup Environ Med. 58*: 68–72 (2001).
64. S. L. Ettner and J. G. Grzywacz, Workers' perceptions of how jobs affect health: A social ecological perspective, *J Occup Health Psychol. 6*: 101–113 (2001).
65. K. Iwasaki, T. Sasaki, T. Oka, and N. Hisanaga, Effect of working hours on biological functions related to cardiovascular system among salesmen in a machinery manufacturing company, *Ind Health. 36*: 361–367 (1998).
66. S. P. Proctor, R. F. White, T. G. Robins et al., Effect of overtime work on cognitive function in automotive workers, *Scand J Work Environ Health. 22*: 124–132 (1996).
67. O.-L. Siu and I. Donald, Psychosocial factors at work and workers' health in Hong Kong: An exploratory study, *Bull Hong Kong Psychol Soc. 34/35*: 30–56 (1995).
68. M. van der Hulst, M. van Veldhoven, and D. Beckers, Overtime and need for recovery in relation to job demands and job control, *J Occup Health. 48*: 11–19 (2006).
69. E. Kleppa, B. Sanne, and G. S. Tell, Working overtime is associated with anxiety and depression: The Hordaland Health Study, *J Occup Environ Med. 50*: 658–666 (2008).
70. Y. Suwazono, S. Nagashima, Y. Okubo et al., Estimation of the number of working hours critical for the development of mental and physical fatigue symptoms in Japanese male workers—Application of benchmark dose method, *Am J Ind Med. 50*(3): 173–182 (2007).

71. N. Kawakami, S. Araki, N. Takatsuka et al., Overtime, psychosocial working conditions, and occurrence of non-insulin dependent diabetes mellitus in Japanese men, *J Epidemiol Community Health. 53*: 359–363 (1999).
72. Y. Liu, H. Tanaka, and The Fukuoka Heart Study Group, Overtime work, insufficient sleep, and risk of non-fatal acute myocardial infarction in Japanese men, *Occup Environ Med. 59*: 447–451 (2002).
73. S. Sokejima and S. Kagamimori, Working hours as a risk factor for acute myocardial infarction in Japan: Case-control study, *BMJ. 317*: 775–780 (1998).
74. C. C. Caruso and T. R. Waters, A review of work schedule issues and musculoskeletal disorders with an emphasis on the healthcare sector, *Ind Health. 46*: 523–534 (2008).
75. P. Tuntiseranee, J. Olsen, A. Geater, and O. Kor-anantakul, Are long working hours and shift-work risk factors for subfecundity? A study among couples from southern Thailand, *Occup Environ Med. 55*: 99–105 (1998).
76. E. L. Mozurkewich, B. Luke, M. Avni, and F. M. Wolf, Working conditions and adverse pregnancy outcome: A meta-analysis, *Obstet Gynecol. 95*: 623–635 (2000).
77. A. E. Rogers, W. Hwang, L. D. Scott et al., The working hours of hospital staff nurses and patient safety, *Health Aff. 23*: 202–212 (2004).
78. C. P. Landrigan, J. M. Rothschild, J. W. Cronin et al., Effect of reducing interns' work hours on serious medical errors in intensive care units, *N Engl J Med.* 351: 1838–1848 (2004).
79. W. G. Sirois, The secret cost of fatigue. www.circadian.com/download/15_the_secret_cost_of_fatigue.pdf (accessed March 2008), 2007.
80. M. R. Rosekind, K. B. Gregory, M. M. Mallis et al., The cost of poor sleep: Workplace productivity loss and associated costs, *J Occup Environ Med. 521*: 91–98 (2010).
81. National Sleep Foundation, Who is at risk? http://drowsydriving.org/about/ whos-at-risk/ (accessed July 7, 2010).
82. Baker Panel, *The Report of the BP U.S. Refineries Independent Safety Review Panel.* www.bp.com/liveassets/bp_internet/globalbp/globalbp_uk_english/SP/STAGING/local_assets/assets/pdfs/Baker_panel_report.pdf (accessed December 3, 2010), 2007.
83. M. M. Mitler, M. A. Carskadon, C. A. Czeisler et al., Catastrophes, sleep, and public policy: Consensus report, *Sleep. 11*: 100–109 (1988).
84. Peter D. Hart Research Associates, *The Nurse Shortage: Perspectives from Current Direct Care Nurses and Former Direct Care Nurses.* www.aft.org/pubs-reports/healthcare/Hart_Report.pdf (accessed on June 7, 2006), 2001.
85. P. C. Zee and F. W. Turek, Sleep and health—Everywhere and in both directions, *Arch Intern Med. 166*: 1686–1688 (2006).
86. K. Spiegel, R. Leproult, and E. Van Cauter, Impact of sleep debt on metabolic and endocrine function, *Lancet. 354*: 1435–1439 (1999).
87. F. A. J. L. Scheer and S. A. Shea, Fundamentals of the circadian system, in *Basics of Sleep Guide*, 2nd edn., C.J. Amlaner and P.M. Fuller, Eds., Sleep Research Society, Westchester, IL, 2009.
88. P. Lavie, Ultrashort sleep-waking schedule. III. 'Gates' and 'forbidden zones' for sleep, *Electroencephalogr Clin Neurophysiol. 63*: 414–425 (1986).
89. P. C. Zee and B. S. Lu, Insomnia and circadian rhythm sleep disorders, *Psychiatr Ann. 38*: 583–589 (2008).
90. J. Arendt, Shift work: Coping with the biological clock, *Occup Med* (*Lond*). *60*: 10–20 (2010).
91. C. Cajochen, Alerting effects of light, *Sleep Med Rev. 116*: 453–464 (2007).
92. V. L. Revell and C. I. Eastman, How to trick mother nature into letting you fly around or stay up all night, *J Biol Rhythms. 20*: 353–365 (2005).
93. T. Kantermann, M. Juda, C. Vetter, T. Roenneberg, Shift-work research: Where do we stand, where should we go? *Sleep Biol Rhythms. 8*: 95–105 (2010).
94. N. J. Wesensten, G. Belenky, D. R. Thome et al., Modafinil vs. caffeine: Effects on fatigue during sleep deprivation, *Aviat Space Environ Med. 756*: 520–525 (2004).

95. H. P. Landolt and D. J. Dijk, Genetic basis of sleep in healthy humans, in *Principles and Practice of Sleep Medicine*, 5th edn., M.H. Kryger, T. Roth, and W.C. Dement, Eds., Elsevier, St. Louis, MO, 2011, p. 175.
96. S. J. Crowley, C. Acebo, and M. A. Carskadon, Sleep, circadian rhythms, and delayed phase in adolescence, *Sleep Med. 8*: 602–612 (2007)
97. J. F. Duffy, D. J. Dijk, E. B. Klerman et al., Later endogenous circadian temperature nadir relative to an earlier wake time in older people, *Am J Physiol.* 275(5 Pt 2): R1478–R1487 (1998).
98. R. L. Sack, D. Auckley, R. R. Auger et al., Circadian rhythm sleep disorders: Part 1, basic principles, shift work and jet lag disorders, *Sleep. 30*: 1460–1483 (2007).
99. J. F. Duffy, Shift work and aging roles of sleep and circadian rhythms, *Clin Occup Environ Med. 3*: 311–332 (2003).
100. J. Axelsson, G. Kecklund, T. Åkerstedt et al., Sleepiness and performance in response to repeated sleep restriction and subsequent recovery during semi-laboratory conditions, *Chronobiol Int. 25*(2): 297–308 (2008).
101. G. Belenky, N. J. Wesensten, D. R. Thorne et al., Patterns of performance degradation and restoration during sleep restriction and subsequent recovery: A sleep dose-response study, *J Sleep Res. 12*(1): 1–12 (2003).
102. J. R. Schwartz and T. Roth, Neurophysiology of sleep and wakefulness: Basic science and clinical implications, *Curr Neuropharmacol. 64*: 367–378 (2008).
103. National Sleep Foundation, Detection and prevention. http://drowsydriving.org/about/detection-and-prevention/ (accessed July 7, 2010).
104. S. Pallesen, B. Bjorvatn, N. Mageroy et al., Measures to counteract the negative effects of night work, *Scand J Work Environ Health. 36*: 109–120 (2010).
105. R. R. Rosa and M. J. Colligan, *Plain Language About Shiftwork*, Department of Health and Human Services, Public Health Service, Centers for Disease Control and Prevention, National Institute for Occupational Safety and Health, Division of Biomedical and Behavioral Science, Cincinnati, OH, DHHS NIOSH Publication No. 97-145, 1997.
106. M. Sallinen and G. Kecklund, Shift work, sleep, and sleepiness—Differences between shift schedules and systems, *Scand J Work Environ Health. 36*(2): 121–133 (2010).
107. T. H. Monk, What can the chronobiologist do to help the shift worker? *J Biol Rhythms. 15*: 86–94 (2000).
108. H. R. Colten and B. M. Altevogt, *Sleep Disorders and Sleep Deprivation an Unmet Public Health Problem*, National Academies, Washington, DC, 2006.
109. American Petroleum Institute, *Fatigue Risk Management Systems for Personnel in the Refining and Petrochemical Industries*, American Petroleum Institute, Washington, DC, 2010.
110. J. A. Caldwell, M. M. Mallis, J. L. Caldwell et al., Fatigue countermeasures in aviation, *Aviat Space Environ Med. 801*: 29–59 (2009).
111. T. I. Morgenthaler, T. Lee-Chiong, C. Alessi et al., Practice parameters for the clinical evaluation and treatment of circadian rhythm sleep disorders, *Sleep. 30*: 1445–1459 (2007).
112. M. Patlak, *Your Guide to Healthy Sleep*, U.S. Department of Health and Human Services, National Institutes of Health, National Heart, Lung, and Blood Institute, NIH Publication No. 06-5271, 2005.
113. National Sleep Foundation, www.sleepfoundation.org (accessed July 7, 2010), n.d.d.
114. National Sleep Foundation, Summary of findings of 2007 sleep in America poll. www.sleepfoundation.org/sites/default/files/Summary_Of_Findings%20-%20FINAL.pdf (accessed July 7, 2010), n.d.e.
115. A. Sharifian, M. Firoozeh, G. Pouryaghoub et al., Restless legs syndrome in shift workers: A cross sectional study on male assembly workers, *J Circadian Rhythms. 7*: 12 (2009).
116. M. Takahashi, The role of prescribed napping in sleep medicine, *Sleep Med Rev. 73*: 227–235 (2003).

117. M. Takahashi, H. Fukuda, and H. Arito, Brief naps during post-lunch rest: Effects on alertness, performance, and autonomic balance, *Eur J Appl Physiol Occup Physiol. 782*: 93–98 (1998).
118. A. T. Wertz, J. M. Ronda, C. A. Czeisler, and K. P. Jr. Wright, Effects of sleep inertia on cognition, *JAMA. 195*: 163–164 (2006).
119. T. Roehrs and T. Roth, Caffeine: Sleep and daytime sleepiness, *Sleep Med Rev. 122*: 153–162 (2008).
120. National Sleep Foundation, Getting enough sleep can save your life on the roads. http://drowsydriving.org/2010/05/getting-enoughsleep-can-save-your-life-on-the-roads/ (accessed July 10, 2010), n.d.f.
121. P. K. Schweitzer, A. C. Randazzo, K. Stone et al., Laboratory and field studies of naps and caffeine as practical countermeasures for sleep-wake problems associated with night work, *Sleep. 291*: 39–50 (2006).
122. M. Terman, Blue in the face, *Sleep Med. 103*: 277–278 (2009).
123. C. Anderson and J. A. Home, A high sugar content, low caffeine drink does not alleviate sleepiness but may worsen it, *Hum Psychopharmacol. 215*: 299–303 (2006).
124. D. Burdakov, L. T. Jensen, H. Alexopoulos et al., Tandem-pore K+ channels mediate inhibition of orexin neurons by glucose, *Neuron. 505*: 711–722 (2006).
125. A. Lowden, U. Holmback, T. Åkerstedt et al., Performance and sleepiness during a 24 h wake in constant conditions are affected by diet, *Biol Psychol. 653*: 251–263 (2004).
126. A. Lowden, C. Moreno, U. Holmbäck et al., Eating and shift work—Effects on habits, metabolism and performance, *Scand J Work Environ Health. 362*: 150–162 (2010).
127. W. Vanlaar, H. Simpson, D. Mayhew, and R. Robertson, Fatigued and drowsy driving: A survey of attitudes, opinions and behaviors, *J Safety Res. 393*: 303–309 (2008).
128. J. A. Caldwell and J. L. Caldwell, Fatigue in military aviation: an overview of U.S. military-approved pharmacological countermeasures, *Aviat Space Environ Med. 76*(7 Suppl): C39–C51 (2005).
129. D. S. Carlson and P. L. Perrewe, The role of social support in the stressor-strain relationship: An examination of work-family conflict, *J Manag. 25*: 513–540 (1999).
130. J. H. Greenhaus, A. G. Bedeian, and K. W. Mossholder, Work experiences, job performance, and feelings of personal and family well-being, *J Vocat Behav. 31*: 200–215 (1987).
131. H. B. Presser, Nonstandard work schedules and marital instability, *J Marriage Fam. 62*: 93–110 (2000).
132. H. Bøggild, Settling the question—The next review on shift work and heart disease in 2019, *Scand J Work Environ Health. 35*(3): 157–161 (2009).

CHAPTER 17

# Design and Evaluation of a Musculoskeletal and Work History Questionnaire

Grace Kawas Lemasters and Margaret R. Atterbury

CONTENTS



## 17.1 INTRODUCTION

The cornerstone of many studies investigating occupationally related disorders is the development of a quality survey instrument. Accuracy of the conclusions, however, depends on how carefully a questionnaire has been constructed and evaluated. Before developing a survey instrument, the investigators must become thoroughly acquainted with the topic under investigation, knowledgeable about the disease process, and familiar with the targeted population. In ergonomic field studies, a questionnaire is a useful tool for assessing the presence of symptoms of work-related musculoskeletal disorders (WMDs), for characterizing job factors associated with the development of these disorders, and for assessing symptoms and associated outcomes before and after an ergonomic intervention.

## 17.2 BACKGROUND AND SIGNIFICANCE TO OCCUPATIONAL ERGONOMICS

The design and complexity of a musculoskeletal questionnaire depends on its intended use, method of administration, and time availability. The significance of this chapter to occupational ergonomics is that it describes the process of designing and evaluating a symptom and work history questionnaire associated with musculoskeletal disorders. The questionnaire described here was used to estimate the prevalence of WMDs among over 500 currently employed union carpenters and as part of an ergonomic intervention study among journeymen carpenters. In addition, this instrument is being used in a longitudinal study of apprentice carpenters. Apprentices are being followed prospectively to determine the incidence of WMDs.

This instrument was pretested, pilot tested, and evaluated for its reliability and validity. In Sections 17.3 through 17.10, we describe the steps in its development and results of the evaluation. The final questionnaire is presented in Appendix 17.A.

## 17.3 REVIEW OF OTHER INSTRUMENTS

The first step in questionnaire construction begins before the first question is written. In addition to a literature review of relevant musculoskeletal disorders, a review of available instruments is critical. First we reviewed prior studies of musculoskeletal disorders [1–4]. Next, several questionnaires used in NIOSH Health Hazard Evaluation (HHE) studies of WMDs were reviewed. These included the *Los Angeles Times*, *Newsday*, Shoprite, Perdue, and Morrell questionnaires [5–9]. These instruments were administered to office workers, newspaper writers and editors, grocery clerks, and meatpackers. Therefore, they needed to be tailored to our study population of carpenters. Additionally, changes were made to adapt those that were self-administered to one that was to be administered over the telephone by trained interviewers who were unfamiliar with the fields of medicine and ergonomics.

Several of the questionnaires reviewed began with the question “In the past year, have you had pain, aching, stiffness, burning, numbness, or tingling in the area shown in this diagram?” and the respondent was referred to a body diagram. Because our interview was to be administered via telephone, a multicolored body part diagram was mailed to the carpenter’s homes. However, realizing that the respondent might not have the diagram in hand at the time of interview, the questionnaire was designed to function independently of the body figure. In fact, only half the respondents actually had the diagram present when contacted. Therefore, each body part was specified by name, with some degree of misclassification inevitable. For example, the respondent might be confused about where the shoulder begins and the neck ends. Since we were concerned about misclassification, the upper and lower back were not differentiated but were referred to as the “back.”

Another difference in our questionnaire was that questions about the quality of symptoms were asked separately; that is, the respondent could answer just “burning” or “pain” or "numbness or tingling" or any combination of these symptoms. This separation of symptoms was done to determine the prevalence of specific types of disorders. For instance, symptoms of nerve-related disorders were likely to include numbness and tingling.

In asking about frequency and duration of symptoms, the format used was similar to the one found in the NIOSH HHE instruments, but the responses were simplified by reducing the number of categories from seven to four. This simplification was necessary because the questions and possible responses were asked over the telephone, and it was thought that subjects might not remember such a long list. Additionally, it was thought that subjects might recall symptoms over the past year as occurring daily, weekly, or monthly rather than every 2 or 3 weeks or once every 2 months.

Another modification was made to expand the section on injuries. Injuries were hypothesized to be a major cause of musculoskeletal symptomatology among carpenters, in contrast to what might be expected among office workers, news reporters, editors, meatpackers, or those in manufacturing. In the questionnaires reviewed, only one injury question was included for the purpose of excluding that respondent's symptoms as a WMD. Since many carpenters have on-the-job injuries, additional questions were included to further characterize types and causes of injuries.

We composed an entirely original work history section. Characterizing job tasks and musculoskeletal stressors in carpentry and construction was complex because of the wide variety of tasks performed by carpenters in contrast to manufacturing where workers repeat the same task over and over. Input from retired and active union carpenters was invaluable in designing this section; in particular, they assisted in constructing a comprehensive list of specialties.

In summary, the review of these questionnaires helped by building upon the successes and avoiding the mistakes of others. Thus, the instrument review, though time-consuming, was infinitely more useful than having to start from the beginning.

## 17.4 KNOWING THE TARGET POPULATION

In the initial phase of questionnaire construction an in-depth understanding of the study group is needed. It is important to know the general knowledge base of the respondents. Discussions with individuals in the occupation under investigation help in understanding their health concerns, work activities, knowledge base, and literacy level. Input solicited from several carpenters and visits to construction sites were crucial to gaining an understanding of the nature of the work tasks and possible health problems associated with these tasks.

Site visits to construction areas helped identify the body areas at greatest risk of damage. A body parts diagram and symptom check adapted from previous studies [10,11] was given to a group of carpenters to complete at the end of their work shift. Reviewing the results of this initial walk-through was critical in designing the larger in-depth instrument. It not only helped define the body regions with the most discomfort for carpenters, but also familiarized the study team with the variety of activities performed by carpenters. Specifically, from these data we identified that the hip area and the upper extremities may be important target areas of damage. It was learned, for example, that carpenters wear tool belts weighing from 6 to 15 lb and that much of their work requires the use of power tools with arms extended overhead. Interviews with carpenters and visits to work sites facilitated the process of gaining knowledge about the study population, their work, and their potential clinical problems.

## 17.5 DEFINING WMDs

The term *work-related musculoskeletal disorders* (WMDs) is used in reference to conditions that are also called cumulative trauma disorders (CTDs), repetitive strain injuries (RSIs), or overuse syndromes. All these terms refer to conditions that involve muscles, tendons, and/or nerves and usually manifested as pain, discomfort, and/or tingling in a body area. Many studies of these conditions have investigated symptoms in the neck and upper extremities [3,12,13]. Some WMDs such as carpal tunnel syndrome (CTS) and de Quervain's tenosynovitis are relatively specific clinical diagnoses. Others have less specific signs and symptoms (e.g., myofascial pain, tension neck syndrome, lumbar strain) and may encompass a broader anatomical area and/or lack specific criteria for diagnosis. The term WMD is preferred because of the multifactorial nature of these problems [13]. Most of these disorders have been associated with occupational factors (e.g., force, repetition, awkward or static postures, vibration) and/or nonoccupational factors such as sports activities, hobbies, underlying disease, and personal factors. These additional risk factors also need to be addressed in the questionnaire.

We defined an individual as having a WMD if all the following criteria were satisfied:

1. Onset of symptoms after starting work as a carpenter
2. Severity reported as mild, moderate, severe, or the worse discomfort ever
3. Symptoms occurring at least once per week (frequency) or lasting at least 1 week (duration)
4. No history of injury to that body part

NIOSH used a similar case definition in their HHE studies of WMDs [5–8].

Defining WMDs by questionnaire is useful for surveillance purposes because a range of severity of symptoms can be assessed and localized to a particular body area in a specified population. Although numerous studies have been carried out using symptoms and standardized physical exams to identify WMDs [1,3,5,14], these have not been validated as diagnostic for specific clinical disorders.

## 17.6 QUESTION CONSTRUCTION

There are two general approaches to question development. The first is unstructured and open-ended. The unstructured question allows more flexibility; the question content, sequence, and wording are under the control of the interviewer, and the answers of the respondents often direct the interview. The second approach, the standardized closed-ended instrument, has structured questions with a fixed format. The latter approach was selected for the carpenter study in order to (1) increase efficiency in terms of time, (2) eliminate difficult coding problems and a need for interpretation by judges, and (3) increase the ease of administration by a nonmedical interviewer group.

The advantage of asking standardized questions is that it provides greater uniformity of measurement responses and the investigator can obtain the responses that the question was intended to address. For example, when asking questions concerning an injury, we wanted to know about specific types of injuries and therefore provided a list: "a fall, an object hitting or falling on you [as is common in the carpentry trade], an auto accident, sports accident, and other sudden injury" (Part II, neck, Q12). This question was followed by one asking for information on the type of injury: "a broken bone, whiplash, ruptured disc, burn or other." These fixed, generally closed-ended responses reduced the need for coding and provided information that helped in determining if an individual symptom was related to a WMD or an injury at the workplace or elsewhere. In contrast, an open-ended question such as "What type of injury did you have?" requires the review and coding of every response.

Several types of standardized questions were used in this instrument. A combination of strict-choice and item checklist was used to gather prior medical history (Part II, Q8). The respondent was asked to answer yes or no (strict choice) to each disease condition provided in a checklist of medical conditions. The disease conditions that were listed are risk factors that may be associated with WMDs. The checklist is probably the most direct approach for providing a memory jog for prior medical conditions possibly related to a WMD.

The medical symptom questionnaire portion used a rating scale to determine the degree of pain that the individual was experiencing. The rating scale defined pain ranging from mild, moderate, and severe to the worst discomfort ever experienced. Originally it was felt that only pain at the moderate or greater level should be considered in our case definition for a WMD. As our experience with the carpenters grew, it became apparent that this group was relatively stoic, accustomed to a high degree of physical labor and to considerable pain as a part of their job. Hence, the decision was made in the analysis stage to include those reporting mild pain in the case definition.

In reviewing other instruments we occasionally found a "double-barreled" question, one that asks for a single response to a combination of questions. An example of a double-barreled question is as follows: "In the past year, did you have severe pain lasting longer than a week?" That question should be divided and asked as follows: "How long did the symptom last?" (Part II, neck, Q4) and "How severe is the pain?" (Q5).

One common problem in question construction concerns the response set. A response set is a tendency of a subject to answer in a certain way, such as "never" or "always," irrespective of the content of the question. We were particularly concerned about a potential response set problem in the work history section (Part IV, Work History, Q1). The first question asked, "From when you started as a carpenter at 19— have you done any of the following jobs?" The goal was to learn about each specialty (flooring, welding/burning, form work, drywall, scaffolding, framing, piledriving, finish, ceiling, millwright, fixtures, supervision, and office work) that the carpenters had ever performed. The follow-up question asked, "How often have you done this job in the past year?" with the categories ranging from "never" to "all the time." If both questions were asked sequentially for each specialty, the concern was that the participant would grow tired of answering the "how often" question and would say "no" to the subsequent specialties in order to end the interview. Therefore, we decided to first ask a strict-choice response of "yes" or "no" to each specialty area for part A. Then we returned to each of those having an affirmative response and asked the second question (B). Though this may not have completely avoided a response set, it helped to minimize the problem.

In summary, there are some general tenets of question construction that should be followed:

1. Questions should be asked in an unbiased manner with no leading.
2. Every question should be clear, unambiguous, and understandable for both the respondent and the interviewer. Avoid using unfamiliar medical items and ambiguous words such as always, may, large, many, important, some, most, much, near, several, generally, occasionally, and often.
3. *Every* word in a question is important; *each* should be examined for possible elimination. Questions should not exceed 20 words in length.
4. Ask easy questions first and save the difficult and sensitive ones for last.

In formulating questions, the investigator needs to think backward from the analysis and forward from the hypothesis. Before writing the first question, it is of utmost importance to know specifically what information is needed in order to answer the overall research question(s).

## 17.7 QUESTIONNAIRE COMPONENTS

In order to assess prevalence or incidence as well as characterize other issues pertaining to WMDs, the questionnaire needed to contain the following sections: (1) demographics and medical history, (2) musculoskeletal symptoms and injuries, and (3) a work history.

### 17.7.1 Demographics and Medical History

The demographics and medical history section included factors such as age, sex, height, weight, and hand dominance, which may play a role in the development of musculoskeletal disorders. The question concerning income was saved until last. Sports activities and hobbies were probed as possible risk factors for musculoskeletal disorders. The medical history (Part II, Q8) establishes the presence of self-reported physician-diagnosed medical conditions related to the development of WMDs. These may have been either other risk factors (e.g., diabetes mellitus related to the development of CTS) [15] or conditions that may have caused symptoms similar to those of WMDs (e.g., cervical spine disease) [16]. Additionally, prior diagnoses of CTS, arthritis, tendinitis, or other conditions that could be work related were specified. These items specifically referred to "physician-diagnosed" conditions in order to eliminate inaccurate self-diagnosis that may occur for such conditions as rheumatoid arthritis (RA).

Diabetes mellitus, systemic lupus erythematosus ("lupus"), RA, and hypothyroidism are systemic illnesses that are associated with the development of CTS [15,17]. Gout, lupus, and RA are inflammatory disorders that also cause musculoskeletal disease, especially arthritis and/or tendinitis. A herniated or "ruptured" cervical disc with nerve root compression may cause symptoms (paresthesias and weakness in the hand) similar to CTS, though usually neck and arm pain are present. Cervical nerve root compression also has been associated with coexistent CTS [15,18]. CTS is the most common nerve entrapment syndrome [16,17] that has been associated with repetitive motion and force at work [1,19]. Thoracic outlet syndrome (TOS) which is much less common, may result in symptoms of numbness and tingling in the hand, usually in the ulnar distribution. TOS is often difficult to diagnose and is usually a result of anatomical conditions [20] but may be aggravated by work postures [21]. Raynaud's phenomenon, which is paresthesias and white or blue discoloration of the fingers on exposure to cold temperatures, may be an isolated syndrome, part of a systemic illness or a result of vibration exposure at work [22].

A smoking history was included in the questionnaire because of its association with increased risk of low back pain and intervertebral disc disease [23]. Likewise, pregnancy [15] and the use of oral contraceptives [24] were included because these factors have been associated with CTS. Finally, questions on alcoholism and diabetes mellitus were included because these are common causes of diffuse peripheral neuropathies that may mimic nerve-related WMDs such as CTS.

### 17.7.2 Musculoskeletal Symptoms

We were most interested in assessing the presence of persistent musculoskeletal symptoms, so the interview began with a screening question (see Part II, Q12): "Within the past 12 months have you experienced any recurring symptoms such as pain, aching, numbness, or any other symptom in your…"; specific body areas were then listed. An affirmative response generated further questions characterizing musculoskeletal symptoms in the identified area. The same questions were asked for each body area in order to establish onset, quality, frequency, duration, and intensity of symptoms, and a history of injury to that body part. These questions addressed musculoskeletal symptom characteristics needed in establishing whether or not a respondent met the case definition for a WMD.

Questions were added to assess a respondent's opinion as to whether or not a musculoskeletal problem was temporally associated with and/or caused by carpentry work. Severity of symptoms and resulting disability were ascertained by questions on whether medical attention was sought, number of workdays lost, and job task modification, and on whether or not the symptoms awoke the respondent from sleep.

### 17.7.3 Injuries

Injuries were hypothesized to be a major cause of musculoskeletal symptomatology among carpenters. Consequently, the questionnaire had a group of questions characterizing injuries to individual body areas. These questions were asked only if the patients had had symptoms in the past 12 months, and subsequently assessed only injuries with recent recurring symptoms or with symptoms in that area due to another cause (e.g., a coexistent WMD).

Injury was defined as a sudden event such as a fall or an object hitting a worker. Our intent was to separate symptoms due to sudden injuries at work from those due to WMDs. We did not want to exclude strains (e.g., low back) that may have been a result of cumulative job-related musculoskeletal stresses but were manifested following a single event. Therefore, we chose not to include "strain" as a sudden injury category but had "other" with an open-ended option.

### 17.7.4 Work History

When designing the work history, it is crucial that investigators have a predefined plan by which to model exposures. Will surrogate measures of exposure such as duration, latency, and job tasks be used, or will intensity measures also be made? Because of the variety of tasks performed by carpenters and the job movement from one specialty to another, it was decided that duration of employment in the carpentry trade was to be the primary measure of exposure. Also of interest was knowing if one specialty (e.g., finish, ceiling, form) was more ergonomically hazardous than other specialties. For example, it was hypothesized that those who did ceiling work might report more shoulder disorders. Therefore, as mentioned earlier, the carpenters were queried about how much time they had spent in each specialty in the last month, in last year, and in their entire career. Because this study also included an intervention phase, other general questions were asked. These had to do with what tools they wanted to change and how they might institute changes.

Workplace psychosocial stress also has received attention as a risk factor for musculoskeletal disorders [25]. In a study by Leino [26], musculoskeletal disorders as measured by rheumatic-type symptoms were associated with occupational stress in both men and women and for all occupational class groups. These findings are in support of the demand–control model proposed by Karasek [27]. This model is based on the combined joint effects of task demands and individual task control as predictors of stress-related illness. This theory hypothesizes that a worker is at greatest risk for a stress-related illness when the demands of the job are high and the worker's control over the tasks is low. Hence, as part of the work histories, questions were included that focused on job demands and job control; these questions were drawn from another instrument shown to have good reliability [28]. The first set of psychosocial questions asked about the level of influence the individual had over the "amount of work, availability of materials, policies and procedures, pace of your work, quality of the work, and hours or schedule" (Part IV, Work History, Q7). Demands of the job were addressed with questions on how fast and hard they had to work, how physically exhausted they felt, and the extent of their work load.

In summary, the work history section served several purposes. Exposure models were developed that addressed duration of time in the trade and the development of a WMD and identified the relationship between a WMD and work in particular specialties. In addition, questions were formulated that assisted in devising possible interventions in either tool design or application. Finally, the work history included psychosocial elements related to job function that might also be risk factors for a WMD.

## 17.8 PRETEST AND PILOT TEST

Pretesting an instrument is a means of evaluating its effectiveness before using it in the final study. The pretest is usually conducted on a small group in an interactive manner. Its goal is to determine if the questionnaire is clear, is understandable, and contains important elements as viewed from the subject's perspective. A pretest was performed on this instrument with the help of carpenters at a local union meeting. In teams of two investigators and one subject, we administered the questionnaire to five volunteers. The pretest identified questions that were inappropriate, delineated those that were misleading, and identified questions that should be added.

With a "final" draft in hand, the instrument was then pilot tested. A pilot study is a small-scale test using the exact procedures planned for the larger study. For the pilot study, carpenters belonging to a union in another city were recruited. They were sent a letter and a body part diagram and were interviewed by telephone using the same interview team planned for the larger study. Thus, the pilot study not only helped to refine the instrument but also assisted in interviewer training.

Though much was gleaned from the pilot study, one major finding is worth discussing. Originally it was planned to ask carpenters about musculoskeletal symptoms by systematically proceeding through body regions one by one. During the pilot study, it was learned that once the carpenters became aware of how many additional questions there were when an affirmative response was given for one body area, (approximately 15), they became more and more reluctant to report symptoms. There was a decreasing frequency of affirmative responses progressing from the neck to the ankles. To prevent this from occurring in the larger study, a screening question (Part II, Q12) that asked about recurring musculoskeletal symptoms for all eight body areas was added. Then for each affirmative response, the extra unit of questions were asked (for an example, see neck questions 1–15).

Without the pilot study and the addition of the screening questions, the conclusions of the study could have been biased simply due to the way the instrument was formatted. The pilot study also revealed that on average the questionnaire could be administered in 20 min.

## 17.9 RELIABILITY AND VALIDITY OF THE INSTRUMENT

The reliability of a questionnaire is the extent to which results obtained by a procedure, in this case telephone interviews, can be replicated [29]. A questionnaire may lack reliability for several reasons: The factor under investigation lacks stability, there is divergence among interviewers in the way the instrument is administered, or question construct is faulty.

Instrument reliability was assessed using the test-retest methodology. Portions of the questionnaire were administered to approximately 9% (*N*-49) of a subgroup randomly selected from the original population. The reinterview (retest) occurred between 1 and 3 weeks after completion of the original interview. A detailed report of this reliability assessment is given in Ref. [30]. The areas addressed in the reliability assessment included recall of reported symptomatology, prior medical conditions, prior injuries, and work histories. The observed agreements ranged from 75.5% to 95.7%. Not surprisingly, prior medical history (94.7%), injury to body region (88.4%), and job specialty performed (87.9%) showed the highest agreement with kappa (κ) estimates ranging from 0.70 to 0.77. A kappa estimate of 1.0 indicates perfect reproducibility, and one of 0.0 indicates no agreement.

The observed agreement for the musculoskeletal symptoms was 75.5% (κ = 0.46). Thirty-seven of the 49 reported exactly the same symptoms as they did in the original interview; they agreed that they did ($N = 26$) or did not ($N = 11$) have a musculoskeletal symptom at a particular body region. Eight changed from yes to having a symptom to no on the retest, and four did the reverse. Certainly, some of this change is due to instability of the response. In contrast to items like prior medical diagnoses and prior injuries that are established past events, current symptomatology is in a dynamic state with changes expected over time. Levine et al. [31] showed a higher agreement of report of symptoms on a test-retest evaluation with a Pearson correlation coefficient of 0.91. In their study, however, the retest was done the day after the original interview, maximizing the stability of a response by minimizing the elapsed time between interviews. Overall, our questionnaire was found to have acceptable reliability.

The validity of the instrument also was evaluated. Validity is the extent to which an instrument measures what it purports to measure. There are three basic types of validity: content, construct, and criterion. All three forms of validity assessment were undertaken.

The least powerful, although certainly not less important, is content validity. Content validity implies that the instrument incorporates the domain of the phenomena under study, that is, WMDs. Since our instrument was adapted from those developed by other experts in the field of WMD, content validity was, in part, gained by association. A panel of experts also was convened to further evaluate the content validity of this instrument.

Construct validity is defined as the extent to which the measurement corresponds to theoretical constructs. One test of construct validity is to determine the instrument's predictive ability. In this study, duration of employment as a carpenter was associated with a significant increase in the relative risk of having a WMD. As an example of construct validity, the prevalence study demonstrated that, compared to the baseline group having less than 10 years in the trade, those with 20 years in the trade as a carpenter showed a significant increase in shoulder disorders, relative risk = 2.8 (95% confidence interval 1.5, 5.3) [32]. Another example of construct validity is found in a study that compared median sensory distal amplitude and latency using an ergonomic checklist to define the exposed and unexposed groups [33]. Those who were identified on the checklist as gripping greater than 6 lb were shown to have significantly smaller amplitudes and longer latencies. Thereby, the checklist was able to define a theoretical construct of an ergonomic hazard.

The third form of validity measure is criterion. Criterion validity is the degree to which the measures correlate with an external criterion commonly referred to as a gold standard. Finding a criterion for evaluation of symptoms of musculoskeletal disorders is problematic because there is no gold standard in diagnosing these conditions. Clinical diagnoses require the presence of a combination of symptoms, physical signs, often results of electrodiagnostic or other laboratory testing, and sometimes response to treatment. Lack of money and time limit the use of a complete clinical evaluation when evaluating a cohort of workers. When determining the prevalence of WMDs among a large group, WMD cases may be defined by symptom criteria alone or with physical findings.

Symptoms are by definition subjective. Since symptoms are thought to be the earliest clinical manifestation of a musculoskeletal disorder, establishing the presence and severity of symptoms is critical to evaluating the prevalence of WMDs. A common feature among WMDs is that they manifest as discomfort in various anatomical locations and result in a broad range of disabilities. Measures of symptoms and disability using a self-administered questionnaire have been found to be reproducible and consistent in assessment of CTS [31].

Physical examination outcomes have limited use as a gold standard for several reasons. First, most physical examination maneuvers have not been tested for sensitivity and specificity for specific disorders when the examiners are blinded to symptoms. Second, most examination techniques used in the evaluation of musculoskeletal disorders require a subjective rating of discomfort on the part of the examinee. Finally, there may be intra- and interexaminer variation in examination techniques.

Nerve conduction outcomes do provide an objective measure of median nerve function and are often referred to as a gold standard in the diagnosis [20] of CTS. Nerve conduction studies (NCSs) can quantitate median nerve function, measuring the latency, or the time required for a nerve impulse to travel a set distance, and the amplitude of sensory and motor nerves. These electrodiagnostic tests are used to establish whether or not a median nerve abnormality, or mononeuropathy, at the wrist (i.e., CTS) exists. Even NCSs have limited use as a gold standard in diagnosing CTS. There is some variation in electrodiagnostic criteria for diagnosing a median mononeuropathy. Further, these tests were developed to be part of the entire clinical evaluation, as an extension of the physical examination, to further assess nerve function. Electromyographers use patients' clinical history, physical exam findings, and NCS results in making diagnoses. And finally with respect to CTS, this diagnosis requires the presence of particular symptoms as well as physical findings and/or NCS findings of a median mononeuropathy.

Thus far there have been relatively few published studies that have examined the validity of a broad musculoskeletal symptom questionnaire [34,35]. A few studies have addressed the validity of a variety of surveillance case definitions for CTS by comparing symptoms and/or physical findings [36,37] with nerve conduction results.

In our study, 100 carpenters who were either asymptomatic or met the questionnaire WMD case definition for hand/wrist, shoulder, or knee underwent a standardized musculoskeletal examination performed by two physician examiners. This examination included Phalen's test and Tinel's tests, which are commonly used in the diagnosis of CTS. All 25 hand/wrist cases and 29 asymptomatic carpenters underwent nerve conduction studies performed by a single electromyographer blinded to case status.

Thus far we have analyzed the data with respect to hand/wrist symptoms, physical exam signs of CTS, that is, Phalen's and Tinel's tests, and electrodiagnostic findings. With respect to positive physical findings of Phalen's or Tinel's tests among hand/wrist cases, we found fair agreement over that which might occur by chance. Forty-eight percent of cases had a positive response to one of these tests. Overall agreement, or percent of cases with positive findings and noncases with negative findings, was 70% [38].

Hand/wrist cases were found to have mean NCS values representing a median mononeuropathy at the wrist (including prolonged mean median sensory and motor distal latencies and diminished median sensory and motor amplitudes) compared to noncases. Approximately 80% of the hand/wrist cases met the electrodiagnostic criteria for a median mononeuropathy at the wrist [39]. There was greater agreement between hand/wrist case status and a median mononeuropathy than between hand/wrist case status and physical findings.

In summary, given the aforementioned limitations, there were more physical examination and NCS findings of CTS among the questionnaire-defined hand/wrist cases than among noncases. NCS results in particular indicate that the questionnaire case definition, based on symptoms, was useful in estimating the prevalence of CTS among carpenters.

## 17.10 ANALYSIS STRATEGY

The data analysis plan should be developed prior to questionnaire administration. Without an a priori plan, the investigator may find that an important risk factor needed for the analysis has been overlooked. There is usually minimal recourse for recovery of missing information after interviewing has been completed. The symptom survey described in this chapter was conducted using direct computer entry during the telephone interview. Hence, the data cleanup step was shortened by eliminating the need to prepare the data for keypunch, have it keypunched, usually with 100% verification (double entry), and then check for errors associated with the keypunch process.

At the end of our questionnaire the interviewers were asked to evaluate the respondent's recall abilities and level of cooperation. This evaluation is, of course, the interviewer's subjective interpretation. The interviewers were asked the following questions: (1) "How well did the respondent seem to remember his/her health information?" "How well did the respondent seem to remember his/her work history?" and "How cooperative was the respondent?" Respondents who were scored as "not very well" or "not at all cooperative" on all three questions were excluded. This evaluation resulted in the exclusion of only eight individuals but was important to ensure higher quality data.

We were then ready to begin data analysis and to proceed logically from the simple to the complex. The first analytic step was the generation of simple descriptive statistics—means, medians, proportions, standard deviation, and ranges. These statistics serve not only to provide a quick assessment of data quality (i.e., ensuring that the minimum and maximum observed values for each variable are within acceptable ranges), but also to provide descriptive characteristics of the study population.

The next phase of analysis was to compare exposure (the independent variable) to the outcome WMDs (the dependent variable). Based solely on questionnaire data, two "surrogate" measures of exposure to ergonomic risk, duration of time in the trade and primary carpenter specialty, were used. The number of years in the trade was divided into three categories, less than 10, greater than 10 but less than 20, and greater than 20. These categories were chosen because they divided the study sample into three approximately equal groups. Those meeting the case definition were compared with those who did not. Mantel Haenzel and relative risk estimates were calculated comparing the two longer duration categories to the baseline of less than 10 years. An "unexposed" noncarpentry group would have been preferred as a baseline comparison group, however. Using the "low" tenure group probably underestimates the true risk. The final analysis step was to use a multivariate approach. For this study, a multivariate logistic regression analysis was conducted using duration of time in the trade as a continuous variable while adjusting for other risk factors, for example, chronic medical conditions and age. In conclusion, the analytic procedures were more complex that was presented in this brief summary.

## 17.11 SUMMARY

This chapter discusses some of the principles in questionnaire development and the methods used to construct questions. The NCS findings indicated that the questionnaire case definition based on symptoms was useful in estimating the prevalence of CTS among carpenters. Though a questionnaire is not a substitute for clinical evaluation in the diagnosis and treatment of individuals, it was shown to be effective in estimating the prevalence of musculoskeletal disorders, at least in this population.

## 17.A APPENDIX

Musculoskeletal Symptom and Work History Questionnaire

**Version 1 © GCOHC, 1992**

PART I

DATE: (of completing questionnaire) [ | ] Month [ | ] Day 19[ | ] Year [ ] Card 1–80

Interviewer identification: [ | | ] (interviewer enter your three initials)

Carpenter identification: [ | | ]

START TIME (use military time) ____ ____:____ ____

“Hello, Mr./Ms. ____________, this is __________ calling for the Greater Cincinnati Occupational Health Center. Did you recently receive a letter from us about our health study on carpenters?

(If no) May I read you the letter? (READ LETTER)

(If yes) Then you know we are trying to find out some things about the strains and pains that carpenters get because of their work.”

(THEN READ FOLLOWING STATEMENT)

“Also, this interview is completely voluntary. If you need to stop at any time just let me know and we can call you back to finish when it is convenient for you. Would this be a convenient time for you?”

> Have you been retired for more than a year? (If no continue)
>
> (If yes) Well today we are talking only to people who are currently working as a carpenter but thank you for your time.

"Do you happen to have the diagram of the body we sent you?"

1. Yes
2. No

1. What is your date of birth? (DK/REFUSED = 99)

   | | | | | | |
   |---|---|---|---|---|---|
   | | | | | | |

   (Month) (Day) (Year)

2. What is your height in stocking feet? (DK/REFUSED = 9,99)

   [ ] ft [ ][ ] in.

3. What is your weight without shoes? (DK/REFUSED = 999)

   [ ][ ][ ] lb

4. Are you
   1. Right-handed?
   2. Left-handed?
   3. Both-handed?
   4. Refused
5. Which hand do you use most at work?
   1. Right
   2. Left
   3. Both equally
   4. Refused
6. Are you 1. Male or 2. Female?

   (Circle appropriate response)

7. What is your ethnic background? (read list)

   1. _____Black, not of Hispanic origin
   2. _____Hispanic
   3. _____White, not of Hispanic origin
   4. _____Other
   5. _____DK/REFUSED

PART II

**"Now we would like to get some information on your medical background."**

8. "Have you ever been told by a physician that you had any of the following?"

| | | | | | |
|---|---|---|---|---|---|
| a. | Diabetes | 1 Yes | 2 No | | |
| b. | Gout | 1 Yes | 2 No | | |
| c. | Hypothyroidism or underactive thyroid | 1 Yes | 2 No | | |
| d. | Lupus | 1 Yes | 2 No | | |
| e. | Ruptured disc in the neck | 1 Yes | 2 No | | |
| f. | Ruptured disc in the back | 1 Yes | 2 No | | |
| g. | Rheumatoid arthritis | 1 Yes | 2 No | | |
| h. | Carpal tunnel syndrome | 1 Yes | 2 No | | |
| | Side: 1 left | 2 Right | 3 Both | | |
| i. | Thoracic outlet syndrome | 1 Yes | 2 No | | |
| | Side: 1 left | 2 Right | 3 Both | | |
| j. | Raynauds syndrome | 1 Yes | 2 No | | |

9. "Have you ever been told by a PHYSICIAN that you had any there type of arthritis, tendinitis, or other joint problem?"

   1. Yes
   2. No (GO TO Q 10)

For each yes circle body location and write the year condition began.

| IF YES (Read Each) Did This Problem Occur in Your | | | | | | Did This Occur in the (Left, Right, Middle) Part of Your Body? | | | | | | What Year Was the Condition Diagnosed by a Doctor? |
|---|---|---|---|---|---|---|---|---|---|---|---|---|
| 1 | Neck? | 1 | Yes | 2 | No | 1 | Left | 2 | Right | 3 | Middle | 19__ |
| 2 | Shoulder? | 1 | Yes | 2 | No | 1 | Left | 2 | Right | 3 | Both | 19__ |
| 3 | Elbow/forearm? | 1 | Yes | 2 | No | 1 | Left | 2 | Right | 3 | Both | 19__ |
| 4 | Hand/wrist? | 1 | Yes | 2 | No | 1 | Left | 2 | Right | 3 | Both | 19__ |
| 5 | Back? | 1 | Yes | 2 | No | 1 | Left | 2 | Right | 3 | Middle | 19__ |
| 6 | Hip/thigh? | 1 | Yes | 2 | No | 1 | Left | 2 | Right | 3 | Both | 19__ |
| 7 | Knee? | 1 | Yes | 2 | No | 1 | Left | 2 | Right | 3 | Both | 19__ |
| 8 | Ankle? | 1 | Yes | 2 | No | 1 | Left | 2 | Right | 3 | Both | 19__ |

10a. "Have you ever smoked cigarettes regularly (at least 1 pack per week)?"

1. Never smoked
2. Ex smoker
3. Current smoker

10b. "On average, how many alcoholic beverages do you have a week?"

_____ _____ _____

**IF FEMALE**

11. Are you currently

| | | | | |
|---|---|---|---|---|
| a. Pregnant | 1 | Yes | 2 | No |
| b. Using birth control pills | 1 | Yes | 2 | No |

12. "Within the past 12 months have you experienced any recurring symptoms such as pain, aching, numbness, or any other symptom in your"

    1. Neck?
    2. Shoulder?
    3. Elbow/forearm?
    4. Hand/wrist?
    5. Back?
    6. Hip/thigh?
    7. Knee?
    8. Ankle?

CIRCLE ALL THAT APPLY AND ASK THE SERIES OF QUESTIONS FOR EACH BODY PART CIRCLED.

IF NO SYMPTOMS SKIP TO WORK HISTORY SECTION.

***Now I would like to ask you a few questions about your NECK.***

1. "In the PAST YEAR have you experienced any of the following in your NECK?" (CIRCLE ALL THAT APPLY)

    1. Pain
    2. Aching
    3. Burning

4. Numbness or tingling
5. None of the above (IF NONE GO TO Q 12)

2. "Please tell me the month and year you first had this <u>NECK</u> problem(s)."

   __ __/19__ __

   Month Year

3. "How often have you had this <u>NECK</u> problem in the <u>PAST YEAR?</u>"
   1. Daily
   2. About once a week
   3. About once a month
   4. Generally less than once a month
4. "In the <u>PAST YEAR</u> when you had this <u>NECK</u> problem, on average how long did it usually last?"
   1. Less than a week
   2. Between a week and 1 month
   3. Longer than a month but not constant
   4. Constant
5. "Would you call this discomfort mild, moderate, severe, or just about the worst discomfort you have ever suffered in your life?"
   1. Mild discomfort
   2. Moderate discomfort
   3. Severe discomfort
   4. Worst discomfort ever
6. "Did you become aware of this problem while employed as a carpenter?"
   1. Yes
   2. No
7. "Do you feel this problem was caused by your work as a carpenter?"
   1. Yes
   2. No

8. "Have you seen a doctor, nurse, or other health care provider for the problem you've had with your NECK?"

    1. Yes. If yes, what was the diagnosis?______________________________

    2. No ______________________________________[ ][ ][ ]code

9. "In the past year how many days of work did you miss because of this problem?"
   _____ _____ _____ days

10. "Did you have to change your job task(s) because of this problem?"

    1. Yes

    2. No

11. "Does this problem you have with your NECK wake you from sleep at night?"

    1. Yes

    2. No

**INJURIES**

12. "Did you ever have an injury to your NECK as a result of" (RECORD MOST SEVERE)

    1. A fall?

    2. An object hitting you or falling on you?

    3. An auto accident?

    4. A sports activity?

    5. Other sudden injury? (PLEASE SPECIFY)

    ______________________________________________

    ______________________________________________

    ______________________________________________[ ][ ][ ]code

    6. None of the above? (IF NONE GO TO NEXT BODY PART WITH A YES ON Q 12 Page 3)

13. "When did this injury occur?"

    __ __/19__ __

    Month Year

    (IF MORE THAN ONE INJURY TAKE DATE OF THE MOST SEVERE)

14. "What type of injury was this?" (CIRCLE MOST SEVERE)

    1. Broken bone

    2. Whiplash

3. Ruptured disc

4. Burn

5. Other ____________________________ [ ][ ][ ]code

15. "Do you think your current NECK problem is a result of this injury?"

1. Yes

2. No

***Now I would like to ask you a few questions about your SHOULDER.***

1. "In the PAST YEAR have you experienced any of the following in your SHOULDER?" (CIRCLE ALL THAT APPLY)

1. Pain

2. Aching

3. Burning

4. Numbness or tingling

5. None of the above (IF NONE GO TO Q 13)

2. "In which SHOULDER does (did) this occur?"

1. Left

2. Right

3. Both

3. "Please tell me the month and year you first had this SHOULDER problem(s)."

__ __/19__ __

Month Year

4. "How often have you had this SHOULDER problem in the PAST YEAR?"

1. Daily

2. About once a week

3. About once a month

4. Generally less than once a month

5. "In the PAST YEAR when you had this SHOULDER problem, on average how long did it usually last?"

1. Less than a week

2. Between a week and 1 month

3. Longer than a month but not constant
4. Constant

6. "Would you call this discomfort mild, moderate, severe, or just about the worst discomfort you have ever suffered in your life?"
   1. Mild discomfort
   2. Moderate discomfort
   3. Severe discomfort
   4. Worst discomfort ever
7. "Did you become aware of this problem while employed as a carpenter?"
   1. Yes
   2. No
8. "Do you feel this problem was caused by your work as a carpenter?"
   1. Yes
   2. No
9. "Have you seen a doctor, nurse, or other health care provider for the problem you've had with your <u>SHOULDER</u>?"
   1. Yes. If yes, what was the diagnosis? ______________________________

   ____________________________________[ ][ ][ ]code

   2. No
10. "In the past year how many days of work did you miss because of this problem?"
    _____ _____ _____ days
11. "Did you have to change your job task(s) because of this problem?"
    1. Yes
    2. No
12. "Does this problem you have with your <u>SHOULDER</u> wake you from sleep at night?"
    1. Yes
    2. No

**INJURIES**

13. "Did you ever have an injury to this SHOULDER as a result of" (RECORD MOST SEVERE)

    1. A fall?
    2. An object hitting you or falling on you?
    3. An auto accident?
    4. A sports activity?
    5. Other sudden injury? (PLEASE SPECIFY)

    ______________________________________________

    ______________________________________________

    ______________________________________________[ ][ ][ ]code

    6. None of the above? (IF NONE GO TO NEXT BODY PART SECTION WITH A YES ON Q 12 Page 3)

14. "When did this injury occur?"

    __ __/19__ __

    Month Year

    (IF MORE THAN ONE INJURY TAKE DATE OF THE MOST SEVERE)

15. "What type of injury was this?" (CIRCLE MOST SEVERE)

    1. Broken bone
    2. Dislocation
    3. Severe bruise
    4. Cut
    5. Burn
    6. Other__________________________________ [ ][ ][ ]code

16. "In which SHOULDER did this injury occur?"

    1. Right
    2. Left

17. "Do you think your current SHOULDER problem is a result of this injury?"

    1. Yes
    2. No

***Now I would like to ask you a few questions about your ELBOW/FOREARM.***

1. "In the PAST YEAR have you experienced any of the following in your ELBOW/FOREARM?"
   (CIRCLE ALL THAT APPLY)
   1. Pain
   2. Aching
   3. Burning
   4. Numbness or tingling
   5. None of the above (IF NONE GO TO Q 13)
2. "In which ELBOW/FOREARM does (did) this occur?"
   1. Left
   2. Right
   3. Both
3. "Please tell me the month and year you first had this ELBOW/FOREARM problem(s)."

   __ __/19__ __

   Month Year

4. "How often have you had this ELBOW/FOREARM problem in the PAST YEAR?"
   1. Daily
   2. About once a week
   3. About once a month
   4. Generally less than once per month
5. "In the PAST YEAR when you had this ELBOW/FOREARM problem, on average how long did it usually last?"
   1. Less than a week
   2. Between a week and 1 month
   3. Longer than a month but not constant
   4. Constant

6. "Would you call this discomfort mild, moderate, severe, or just about the worst discomfort you have ever suffered in your life?"

    1. Mild discomfort
    2. Moderate discomfort
    3. Severe discomfort
    4. Worst discomfort ever

7. "Did you become aware of this problem while employed as a carpenter?"

    1. Yes
    2. No

8. "Do you feel this problem was caused by your work as a carpenter?"

    1. Yes
    2. No

9. "Have you seen a doctor, nurse, or other health care provider for the problem you've had with your <u>ELBOW/FOREARM</u>?"

    1. Yes. If yes, what was the diagnosis?______________________________
    ________________________________________________[ ][ ][ ]code
    2. No

10. "How many days of work did you miss because of this problem?"
    _____ _____ _____ days

11. "Did you have to change your job task(s| because of this problem?"

    1. Yes
    2. No

12. "Does this problem you have with your <u>ELBOW/FOREARM</u> wake you from sleep at night?"

    1. Yes
    2. No

**INJURIES**

13. "Did you ever have an injury to your <u>ELBOW/FOREARM</u> as a result of" (RECORD MOST SEVERE)

    1. A fall?
    2. An object hitting you or falling on you?

3. An auto accident?

4. A sports activity?

5. Other sudden injury? (PLEASE SPECIFY)

________________________________________________

________________________________________________

________________________________________________[ ][ ][ ]code

6. None of the above? (IF NONE GO TO NEXT BODY PART SECTION WITH A YES ON Q 12 Page 3)

14. "When did this injury occur?"

__ __/19__ __

Month Year

(IF MORE THAN ONE INJURY TAKE DATE OF THE MOST SEVERE)

15. "What type of injury was this?" (CIRCLE MOST SEVERE)

1. Broken bone
2. Dislocation
3. Severe bruise
4. Cut
5. Burn
6. Other______________________________________ [ ][ ][ ]code

16. "To which <u>ELBOW/FOREARM</u> did this injury occur?"

1. Right
2. Left
3. Both

17. "Do you think your current <u>ELBOW/FOREARM</u> problem is a result of this injury?"

1. Yes
2. No

***Now I would like to ask you a few questions about your <u>HANDS and WRISTS.</u>***

1. "In the <u>PAST YEAR</u> have you experienced any of the following in your <u>HAND/WRIST?"</u> (CIRCLE ALL THAT APPLY)

   1. Pain
   2. Aching

3. Burning
4. Numbness or tingling in the cold
5. Numbness or tingling any other times
6. Fingers turning white or blue
7. None of the above (IF NONE GO TO Q 13)

2. "In which HAND/WRIST does (did) this occur?"
   1. Left
   2. Right
   3. Both

3. "Please tell me the month and year you first had this <u>HAND/WRIST</u> problem(s)."

   __ __/19__ __

   Month Year

4. "How often have you had this <u>HAND/WRIST</u> problem in the <u>PAST YEAR?</u>"
   1. Daily
   2. About once a week
   3. About once a month
   4. Generally less than once a month

5. "In the <u>PAST YEAR</u> when you had this <u>HAND/WRIST</u> problem, on average how long did it usually last?"
   1. Less than a week
   2. Between a week and 1 month
   3. Longer than a month but not constant
   4. Constant

6. "Would you call this discomfort mild, moderate, severe, or just about the worst discomfort you have ever suffered in your life?"
   1. Mild discomfort
   2. Moderate discomfort
   3. Severe discomfort
   4. Worst discomfort ever

7. "Did you become aware of this problem while employed as a carpenter?"

    1. Yes
    2. No

8. "Do you feel this problem was caused by your work as a carpenter?"

    1. Yes
    2. No

9. "Have you seen a doctor, nurse, or other health care provider for the problem you've had with your <u>HAND/WRIST</u>?"

    1. Yes. If yes, what was the diagnosis?___________________________
    _________________________________________[ ][ ][ ]code
    2. No

10. "How many days of work did you miss because of this problem?"
    _____ _____ _____ days

11. "Did you have to change your job task(s) because of this problem?"

    1. Yes
    2. No

12. "Does this problem you have with your <u>HAND/WRIST</u> wake you from sleep at night?"

    1. Yes
    2. No

**INJURIES**

13. "Did you ever have an injury to your <u>HAND/WRIST</u> as a result of" (RECORD MOST SEVERE)

    1. A fall?
    2. An object hitting you or falling on you?
    3. An auto accident?
    4. A sports activity?
    5. Other sudden injury? (PLEASE SPECIFY)
    ________________________________________________
    ________________________________________________
    ________________________________________________[ ][ ][ ]code
    6. None of the above? (IF NONE GO TO NEXT BODY PART WITH A YES ON Q 12 Page 3)

14. "When did this injury occur?"

   __ __/19__ __

   Month Year

   (IF MORE THAN ONE INJURY TAKE DATE OF THE MOST SEVERE)

15. "What type of injury was this?" (CIRCLE MOST SEVERE)

   1. Broken bone
   2. Dislocation
   3. Severe bruise
   4. Cut
   5. Burn
   6. Other________________________________ [ ][ ][ ]code

16. "In which <u>HAND/WRIST</u> did this injury occur?"

   1. Right
   2. Left
   3. Both

17. "Do you think your current <u>HAND/WRIST</u> problem is a result of this injury?"

   1. Yes
   2. No

***"Now I would like to ask you a few questions about your <u>BACK</u>."***

1. "In the <u>PAST YEAR</u> have you experienced any of the following in your <u>BACK?"</u> (CIRCLE ALL THAT APPLY)

   1. Pain
   2. Aching
   3. Burning
   4. None of the above (IF NONE GO TO Q 13)

2. a. "In which part of your BACK does (did) this occur?"

   1. Left?
   2. Right?

3. Middle?
4. Both sides?
5. Whole back?

b. "Did this occur in the"

1. Upper back (above the waist)?
2. Lower back?
3. Whole back?

3. "Please tell me the month and year you first had this BACK problem(s)."

__ __/19__ __

Month Year

4. "How often have you had this BACK problem in the PAST YEAR?"

1. Daily
2. About once a week
3. About once a month
4. Generally less than once a month

5. "In the PAST YEAR when you had this BACK problem, on average how long did it usually last?"

1. Less than a week
2. Between a week and 1 month
3. Longer than a month but not constant
4. Constant

6. "Would you call this discomfort mild, moderate, severe, or just about the worst discomfort you have ever suffered in your life?"

1. Mild discomfort
2. Moderate discomfort
3. Severe discomfort
4. Worst discomfort ever

7. "Did you become aware of this problem while employed as a carpenter?"

    1. Yes
    2. No

8. "Do you feel this problem was caused by your work as a carpenter?"

    1. Yes
    2. No

9. "Have you seen a doctor, nurse, or other health care provider for the problem you've had with your BACK?"

    1. Yes. If yes, what was the diagnosis?________________________
    ______________________________________[ ][ ][ ]code
    2. No

10. "How many days of work did you miss because of this problem?"
    _____ _____ _____ days

11. "Did you have to change your job task(s) because of this problem?"

    1. Yes
    2. No

12. "Does this problem you have with your BACK wake you from sleep at night?"

    1. Yes
    2. No

**INJURIES**

13. "Did you ever have an injury to your BACK as a result of" (RECORD MOST SEVERE)

    1. A fall?
    2. An object hitting you or falling on you?
    3. An auto accident?
    4. A sports activity?
    5. Other sudden injury? (PLEASE SPECIFY)
    ________________________________________________
    ________________________________________________
    ________________________________________________[ ][ ][ ]code
    6. None of the above? (IF NONE GO TO NEXT BODY PART WITH A YES ON 0.12 Page 3)

14. "When did this injury occur?"

    __ __/19__ __

    Month Year

    (IF MORE THAN ONE INJURY TAKE DATE OF THE MOST SEVERE)

15. "What type of injury was this?" (CIRCLE MOST SEVERE)

    1. Broken bone
    2. Ruptured/slipped disc
    3. Severe bruise
    4. Cut
    5. Burn
    6. Other_______________________________ [ ][ ][ ]code

16. "Do you think your current <u>BACK</u> problem is a result of this injury?"

    1. Yes
    2. No

***Now I would like to ask you a few questions about your <u>HIP/THIGH.</u>***

1. "In the <u>PAST YEAR</u> have you experienced any of the following in your <u>HIP/THIGH?"</u> (CIRCLE ALL THAT APPLY)

    1. Pain
    2. Aching
    3. Burning
    4. Numbness or tingling
    5. None of the above (IF NONE GO TO Q 13)

2. "In which HIP/THIGH does (did) this occur?"

    1. Left
    2. Right
    3. Both

3. "Please tell me the month and year you first had this <u>HIP/THIGH</u> problem(s)."

    __ __/19__ __

    Month Year

4. "How often have you had this HIP/THIGH problem in the PAST YEAR?"

    1. Daily
    2. About once a week
    3. About once a month
    4. Generally less than once a month

5. "In the PAST YEAR when you had this HIP/THIGH problem, on average how long did it usually last?"

    1. Less than a week
    2. Between a week and 1 month
    3. Longer than a month but not constant
    4. Constant

6. "Would you call this discomfort mild, moderate, severe, or just about the worst discomfort you have ever suffered in your life?"

    1. Mild discomfort
    2. Moderate discomfort
    3. Severe discomfort
    4. Worst discomfort ever

7. "Did you become aware of this problem while employed as a carpenter?"

    1. Yes
    2. No

8. "Do you feel this problem was caused by your work as a carpenter?"

    1. Yes
    2. No

9. "Have you seen a doctor, nurse, or other health care provider for the problem you've had with your HIP/THIGH?"

    1. Yes. If yes, what was the diagnosis? ______________________________
    ______________________________________________[ ][ ][ ]code
    2. No

10. "How many days of work did you miss because of this problem?"
    _____ _____ _____ days

11. "Did you have to change your job task(s) because of this problem?"

    1. Yes
    2. No

12. "Does this problem you have with your HIP/THIGH wake you from sleep at night?"

    1. Yes
    2. No

**INJURIES**

13. "Did you ever have an injury to your HIP/THIGH as a result of" (RECORD MOST SEVERE)

    1. A fall?
    2. An object hitting you or falling on you?
    3. An auto accident?
    4. A sports activity?
    5. Other sudden injury? (PLEASE SPECIFY)

    ______________________________________________

    ______________________________________________

    ______________________________________________ [ ][ ][ ]code

    6. None of the above? (IF NONE GO TO NEXT BODY PART WITH A YES ON Q 12 Page 3)

14. "When did this injury occur?"

    __ __/19__ __

    Month Year

    (IF MORE THAN ONE INJURY TAKE DATE OF THE MOST SEVERE)

15. "What type of injury was this?" (CIRCLE MOST SEVERE)

    1. Broken bone
    2. Dislocation
    3. Severe bruise
    4. Cut
    5. Burn
    6. Other______________________________________ [ ][ ][ ]code

16. "Do you think your current HIP/THIGH problem is a result of this injury?"
    1. Yes
    2. No

***"Now I would like to ask you a few questions about your KNEE."***

1. "In the PAST YEAR have you experienced any of the following in your KNEE?" (CIRCLE ALL THAT APPLY)
    1. Pain
    2. Aching
    3. Swelling
    4. Locking
    5. Giving out or giving way of your knee
    6. None of the above (IF NONE GO TO Q 13)
2. "In which KNEE does (did) this occur?"
    1. Left
    2. Right
    3. Both
3. "Please tell me the month and year you first had this KNEE problem(s)."

    __ __/19__ __

    Month Year
4. "How often have you had this KNEE problem in the PAST YEAR?"
    1. Daily
    2. About once a week
    3. About once a month
    4. Generally less than once a month
5. "In the PAST YEAR when you had this KNEE problem, on average how long did it usually last?"
    1. Less than a week
    2. Between a week and 1 month
    3. Longer than a month but not constant
    4. Constant

6. "Would you call this discomfort mild, moderate, severe, or just about the worst discomfort you have ever suffered in your life?"

    1. Mild discomfort
    2. Moderate discomfort
    3. Severe discomfort
    4. Worst discomfort ever

7. "Did you become aware of this problem while employed as a carpenter?"

    1. Yes
    2. No

8. "Do you feel this problem was caused by your work as a carpenter?"

    1. Yes
    2. No

9. "Have you seen a doctor, nurse, or other health care provider for the problem you've had with your KNEE?"

    1. Yes. If yes, what was the diagnosis?______________________________
       ________________________________________________[ ][ ][ ]code
    2. No

10. "How many days of work did you miss because of this problem?"
    _____ _____ _____ days

11. "Did you have to change your job task(s) because of this problem?"

    1. Yes
    2. No

12. "Does this problem you have with your <u>KNEE</u> wake you from sleep at night?"

    1. Yes
    2. No

**INJURIES**

13. "Did you ever have an injury to your <u>KNEE</u> as a result of" (RECORD MOST SEVERE)

    1. A fall?
    2. An object hitting you or falling on you?
    3. An auto accident?

4. A sports activity?
5. Other sudden injury? (PLEASE SPECIFY)

__________________________________________________

__________________________________________________

__________________________________________________[ ][ ][ ]code

6. None of the above? (IF NONE GO TO NEXT BODY PART WITH A YES ON Q 12 Page 3)

14. "When did this injury occur?"

__ __/19__ __

Month Year

(IF MORE THAN ONE INJURY TAKE DATE OF THE MOST SEVERE)

15. "What type of injury was this?" (CIRCLE MOST SEVERE)
    1. Broken bone
    2. Severe bruise
    3. Torn ligament
    4. Torn meniscus
    5. Dislocated kneecap
    6. Other?______________________________________ [ ][ ][ ]code
16. "In which KNEE did this injury occur?"
    1. Right
    2. Left
    3. Both
17. "Do you think your current KNEE problems are a result of this injury?"
    1. Yes
    2. No

***"Now I would like to ask you a few questions about your ANKLE."***

1. "In the PAST YEAR have you experienced any of the following in your ANKLE?" (CIRCLE ALL THAT APPLY)
    1. Pain
    2. Aching
    3. Swelling
    4. None of the above (IF NONE GO TO Q 13)

2. "In which ANKLE does (did) this occur?"
    1. Left
    2. Light
    3. Both
3. "Please tell me the month and year you first had this ANKLE problem(s)."

    __ __/19__ __

    Month Year

4. "How often have you had this ANKLE problem in the PAST YEAR?"
    1. Daily
    2. About once a week
    3. About once a month
    4. Generally less than once a month
5. "In the PAST YEAR when you had this ANKLE problem, on average how long did it usually last?"
    1. Less than a week
    2. Between a week and 1 month
    3. Longer than a month but not constant
    4. Constant
6. "Would you call this discomfort mild, moderate, severe, or just about the worst discomfort you have ever suffered in your life?"
    1. Mild discomfort
    2. Moderate discomfort
    3. Severe discomfort
    4. Worst discomfort ever
7. "Did you become aware of this problem while employed as a carpenter?"
    1. Yes
    2. No
8. "Do you feel this problem was caused by your work as a carpenter?"
    1. Yes
    2. No

9. "Have you seen a doctor, nurse, or other health care provider for the problem you've had with your ANKLE?"

    1. Yes. If yes, what was the diagnosis?______________________________
    ____________________________________________[ ][ ][ ]code

    2. No

10. "How many days of work did you miss because of this problem?"
    _____ _____ _____ days

11. "Did you have to change your job task(s) because of this problem?"

    1. Yes

    2. No

12. "Does this problem you have with your ANKLE wake you from sleep at night?"

    1. Yes

    2. No

**INJURIES**

13. "Did you ever have an injury to your ANKLE as a result of" (RECORD MOST SEVERE)

    1. A fall?

    2. An object hitting you or falling on you?

    3. An auto accident?

    4. A sports activity?

    5. Other sudden injury? (PLEASE SPECIFY)

    ______________________________________________
    ______________________________________________
    ______________________________________________[ ][ ][ ]code

    6. None of the above? (IF NONE GO TO PART III)

14. "When did this injury occur?"

    __ __/19__ __

    Month Year

    (IF MORE THAN ONE INJURY TAKE DATE OF THE MOST SEVERE)

15. "What type of injury was this?" (CIRCLE MOST SEVERE)

    1. Broken bone

    2. Severe bruise

3. Torn ligament or sprain
4. Dislocated ankle
5. Other________________________________ [ ][ ][ ]code

16. "In which <u>ANKLE</u> did this injury occur?"
    1. Right
    2. Left
    3. Both

17. "Do you think your current <u>ANKLE</u> problems are a result of this injury?"
    1. Yes
    2. No

**PART III**

1. "Of the problems you have just described, which ONE do you consider to be the most serious or troublesome?"
    1. Neck
    2. Back
    3. Shoulder
    4. Elbow/forearm
    5. Hand/wrist
    6. Hip/thigh
    7. Knee
    8. Ankle

2. "Have you ever filed a worker's compensation claim for problems in any of the following body parts?"
    1. Neck (CIRCLE ALL THAT APPLY)
    2. Back
    3. Shoulder
    4. Elbow/forearm
    5. Hand/wrist
    6. Hip/thigh

7. Knee
8. Ankle
9. None of the above

**PART IV**

***Now we would like to ask you some questions about your carpentry work.***

1. "When were you first employed as a carpenter?" _____ _____ MO 19_____ _____
2. "Over the past 5 years, on average, how many weeks per year were you employed as a carpenter?" _____ _____ weeks

   (IF LAID OFF FOR A LONG TIME ASK: WHEN YOU WERE WORKING AS A CARPENTER ON AVERAGE HOW MANY WEEKS WERE YOU EMPLOYED?)
3. a. "Since you began working as a carpenter did you leave the trade or have you been unemployed for 6 months or longer?"
      1. Yes
      2. No (IF NO GO TO Q 4)

   b. (If yes) "What is the total amount of time you were not working as a carpenter?" _____ _____ _____ years/months
4. a. "Are you currently working in the carpentry industry?"
      1. Yes
      2. No (IF NO GO TO PART D)

   b. "What is your current primary job activity?" (CIRCLE APPROPRIATE NUMBER)
      1. Flooring
      2. Welding/burning
      3. Form work
      4. Drywall
      5. Scaffolding
      6. Framing
      7. Piledriving
      8. Finish
      9. Ceiling
      10. Millwright

11. Fixtures
12. Supervision with no actual carpentry work
13. Office work
14. Something else (specify)

____________________________________________________

____________________________________________________[ ][ ][ ]

c. "What is the most physically demanding task on your current job?" (try for one answer)

1. Lifting
2. Carrying
3. Climbing
4. Other (please specify)

____________________________________________________

____________________________________________________

5. DK/Refused

d. "When was the last time you worked as a carpenter?"

__ __/19__ __

Month Year

**WORK HISTORY:**

1. "We would like to get some information about the jobs you have done over the years.

FROM WHEN YOU STARTED AS A CARPENTER IN 19_____ _____, have you done any of the following jobs?"

(*Read whole job list then ask weeks for each yes*)

| | | | How Often Have You Done This Job in the PAST YEAR? | | | | |
|---|---|---|---|---|---|---|---|
| | | | (1) | (2) | (3) | (4) | (5) |
| | | | Never | Seldom | Sometimes | Most of the Time | All the Time |
| Flooring | 1 Yes | 2 No | _____ | _____ | _____ | _____ | _____ |
| Welding/burning | 1 Yes | 2 No | _____ | _____ | _____ | _____ | _____ |
| Form work | 1 Yes | 2 No | _____ | _____ | _____ | _____ | _____ |
| Drywall | 1 Yes | 2 No | _____ | _____ | _____ | _____ | _____ |
| Scaffolding | 1 Yes | 2 No | _____ | _____ | _____ | _____ | _____ |
| Framing | 1 Yes | 2 No | _____ | _____ | _____ | _____ | _____ |

| | | | | | | | |
|---|---|---|---|---|---|---|---|
| Piledriving | 1 Yes | 2 No | _____ | _____ | _____ | _____ | _____ |
| Finish | 1 Yes | 2 No | _____ | _____ | _____ | _____ | _____ |
| Ceiling | 1 Yes | 2 No | _____ | _____ | _____ | _____ | _____ |
| Millwright | 1 Yes | 2 No | _____ | _____ | _____ | _____ | _____ |
| Fixtures | 1 Yes | 2 No | _____ | _____ | _____ | _____ | _____ |
| Supervision with no actual carpentry work | 1 Yes | 2 No | _____ | _____ | _____ | _____ | _____ |
| Office work other (specify) | 1 Yes | 2 No | _____ | _____ | _____ | _____ | _____ |
| ___________ | 1 Yes | 2 No | _____ | _____ | _____ | _____ | _____ |
| ___________ | 1 Yes | 2 No | _____ | _____ | _____ | _____ | _____ |
| ___________ | 1 Yes | 2 No | _____ | _____ | _____ | _____ | _____ |

2. "In JUST the past 1 month which one of the just mentioned jobs have you done the most?"

    1. Flooring
    2. Welding/burning
    3. Form work
    4. Drywall
    5. Scaffolding
    6. Framing
    7. Piledriving
    8. Finish
    9. Ceiling
    10. Millwright
    11. Fixtures
    12. Supervision with no actual carpentry work
    13. Office work
    14. Other(specify)

    ____________________________________________________________

    ____________________________________________________________[ ][ ][ ]

3. "In your entire career as a carpenter what specialty/job have you done the most?" (NAME TOP 3 IN RANK ORDER)

    1. Flooring
    2. Welding/burning

3. Form work
4. Drywall
5. Scaffolding
6. Framing
7. Piledriving
8. Finish
9. Ceiling
10. Millwright
11. Fixtures
12. Supervision with no actual carpentry work
13. Office work
14. Other (specify)

_____________________________________________

_____________________________________________[ ][ ][ ]

TOP THREE

1. _____
2. _____
3. _____

4. "On average, during a typical year, how many weeks did you work overtime, that is more than 5 days or more than 40 hours?"

   _____weeks

5. "If you could change any of the tools or equipment to make your job easier on your body, which tools would you change?"

   1. _____________
   2. _____________
   3. _____________
   4. _____________
   5. _____________

6. "How would you change these tools?" (PROBE AND CLARIFY)

   _____________________________________________

   _____________________________________________

   _____________________________________________

   _____________________________________________

7. "The next series of questions asks HOW MUCH INFLUENCE you felt you have had in your work situation over the past 12 months. By influence, we mean the degree to which YOU CONTROL what is done by others and have freedom to determine what you do yourself.

   Please answer the following questions with a number on a scale from 1 to 5, with 1 being little to no control, and 5 being very much control."

   Scale: None or very little 1

   A little .................... 2

   A moderate amount 3

   Much .......................4

   Very much .................5

   "OVER THE PAST 12 MONTHS," (CIRCLE CORRECT NUMBER)

   a. How much influence do you have over the AMOUNT OF WORK you do?
   1 2 3 4 5

   b. How much influence do you have over the AVAILABILITY OF MATERIALS and EQUIPMENT you need to do your work? 1 2 3 4 5

   c. How much do you influence the POLICIES AND PROCEDURES in your work group? 1 2 3 4 5

   d. How much influence do you have over THE PACE OF YOUR WORK, that is, how FAST or SLOW you work? 1 2 3 4 5

   e. How much influence do you have over THE QUALITY OF THE WORK that you do? 1 2 3 4 5

   f. How much influence do you have over THE HOURS OR SCHEDULE that you work? 1 2 3 4 5

8. "The next series of questions asks HOW OFTEN certain things happen at your job, over the past 12 months.

   Please answer the following questions with a number on a scale from 1 to 5, with 1 being rarely, and 5 being very often."

   Scale: Rarely .................... 1

   Occasionally .............. 2

   Sometimes .................3

   Often ................. 4

   Very often ................. 5

"OVER THE PAST 12 MONTHS,"

a. How often does your job require you to work VERY FAST?
1 2 3 4 5

b. How often does your job require you to work VERY HARD?
1 2 3 4 5

c. How often are you PHYSICALLY exhausted at the end of the work day?
1 2 3 4 5

d. How often does your job leave you little time to get things done at work?
1 2 3 4 5

e. How often is there a great deal to be done at work?
1 2 3 4 5

9. "All in all, HOW SATISFIED are you with being a carpenter?"
   1. Very satisfied
   2. Somewhat satisfied
   3. Not too satisfied
   4. Not at all satisfied

10. "If you could change any aspect of your job, what would you change?"

    ______________________________________________________

    ______________________________________________________

    ______________________________________________________

    ______________________________________________________

11. "Do you do any of the following sports activities or hobbies 3 or more hours per week?"

    Racket sports .............. 1. Yes 2. No

    Musical instrument ....... 1. Yes 2. No

    Ball sport .................... 1. Yes 2. No

    Fish ............................ 1. Yes 2. No

    Knit/sew ...................... 1. Yes 2. No

    Hunt/shoot .................. 1. Yes 2. No

12. "Is your family income about"
    1. Less than $15,000 per year?
    2. Between $15,000 and $25,000?
    3. Between $25,000 and $50,000?
    4. Over $50,000?

**This concludes our interview. Do you have any additional questions or comments? We will be sending you a summary of our findings. Thank you very much for your participation.**

VERIFY NAME/ADDRESS/PHONE AND RECORD CORRECT INFORMATION

STOP TIME _____ _____:_____ _____

INTERVIEWER:

How well did the respondent seem to remember his/her health information?

1. Very well
2. Fairly well, some problems
3. Not very well

How well did the respondent seem to remember his/her work history?

1. Very well
2. Fairly well, some problems
3. Not very well

How cooperative was the respondent?

1. Very cooperative
2. Fairly cooperative
3. Not at all cooperative

## REFERENCES


1. B. Silverstein, L. Fine, and T. Armstrong, Occupational factors and carpal tunnel syndrome, *Am. J. Ind. Med. 11*:343–358 (1987).
2. I. Kuorkina, B. Jonsson, A. Kilbom, H. Vinterberg, F. Biering-Sorinsen, G. Andersson, and K. Jorgensen, Standardised Nordic questionnaires for the analysis of musculoskeletal symptoms, *Appl. Ergon. 18*(3):233–237 (1987).
3. L. Punnett, J. Robins, D. Wegman, and W. M. Keyserling, Soft tissue disorders in the upper limbs of female garment workers, *Scand. J. Work Environ. Health 11*:417–425 (1985).
4. E. Holmström, *Musculoskeletal Disorders in Construction Workers*, Dept. of Physical Therapy, Lund University, Lund, Sweden, 1992.
5. NIOSH, *Health Hazard Evaluation Report*, Los Angeles Times, Los Angeles, CA. U.S. Dept of Health and Human Services, Public Health Service, Centers for Disease Control, Natl. Inst. Occupational Safety and Health, Cincinnati, OH, NIOSH Rep. No. HETA 90-132277, 1993.
6. NIOSH, *Health Hazard Evaluation Report*, Newsday, Inc., Melville, New York. U.S. Dept. of Health and Human Services, Public Health Service, Center for Disease Control, Natl. Inst. Occupational Safety and Health, Cincinnati, OH, NIOSH Rep. No. HETA 89-250-2046, 1990.
7. NIOSH, *Health Hazard Evaluation Report*, Shoprite Supermarkets, New Jersey–New York. U.S. Dept. of Health and Human Services, Public Health Service, Centers for Disease Control, Natl. Inst. Occupational Safety and Health, Cincinnati, OH, NIOSH Rep. No. HETA 88-344-2092, 1990.

8. NIOSH, *Health Hazard Evaluation Report*, Perdue Farms, Inc. Lewiston, North Carolina and Robersonville, NC. U.S. Dept. of Health and Human Services, Public Health Service, Center for Disease Control, Natl. Inst. Occupational Safety and Health, Cincinnati, OH, NIOSH, Rep. No. HETA 89–307–2009, 1990.
9. NIOSH, *Health Hazard Evaluation Report*, John Morrell & Co., Sioux Falls, SD. U.S. Dept. of Health and Human Services, Public Health Service, Centers for Disease Control, Natl. Inst. Occupational Safety and Health, Cincinnati, OH, NIOSH Rep. No. HETA 88–180–1958, 1990.
10. E. Corlett and R. Bishop, A technique for assessing postural discomfort, *Ergonomics 19*:175–182 (1976).
11. S. Sauter and L. Schliefer, Work posture, work station design and musculoskeletal discomfort in a VDT data entry task, *Hum. Factors 33*(2):151–167 (1991).
12. M. Hagberg and D. H. Wegman, Prevalence rates and odds ratios of shoulder-neck diseases in different occupational groups, *Br. J. Ind. Med. 44*:602–610 (1987).
13. T. J. Armstrong, B. Buckle, L. Fine, M. Hagberg, B. Jonsson, A. Kilborn, I. Kuorinka, B. Silverstein, G. Sjogaard, and E. Viikara-Juntur, A conceptual model for work related neck and upper limb disorders, *Scand. J. Work Environ. Health 19*:73–84 (1993).
14. S. R. Stock, Workplace ergonomic factors and the development of musculoskeletal disorders of the neck and upper limbs: A meta-analysis, *Am. J. Ind. Med. 19*:87–107 (1991).
15. B. Dorwart, Carpal tunnel syndrome: A review, *Semin. Arthritis Rheum. 14*(2):134–140 (1984).
16. M. Bleeker, Medical surveillance for carpal tunnel syndrome in workers, *J. Hand Surg. 12A*(2 Pt 2):845–848 (1987).
17. R. T. Katz, Nerve entrapments: An update, *Orthopedics 12*(8):1097–1197 (1989).
18. A. R. M. Upton and A. J. McComas, The double crush in nerve entrapment syndromes, *Lancet* 18:359–360 (August 1973).
19. B. Silverstein, L. Fine, and T. Armstrong, Carpal tunnel syndrome: Causes and a preventive strategy, *Semin. Occup. Med. 1*(3):213–221 (1986).
20. D. Dawson, Entrapment neuropathies of the upper extremities, *N. Engl. J. Med. 329*(27):2013–2018 (1993).
21. P. Waris, Occupational cervicobrachial syndromes, *Scand. J. Work Environ. Health 6*(Suppl. 3):3–14 (1980).
22. D. Rempel, R. Harrison, and S. Barnhart, Work related cumulative trauma disorders of the upper extremity, *J. Am. Med. Assoc. 267*(6):838–842 (1992).
23. J. L. Kelsey, A. L. Golden, and D. J. Mundt, Low back pain/prolapsed lumbar intervertebral disc, *Rheum. Dis. Clin. N. Am. 16*(3):699–716 (1990).
24. M. Sabour and H. Fadel, The carpal tunnel syndrome: A new complication ascribed to the pill, *Am. J. Obstet. Gynecol. 107*:1265–1267 (1970).
25. K. O. Anderson, L. A. Bradley, L. D. Young, L. K. McDaniel, and C. M. Wise, Rheumatoid arthritis: Review of psychological factors related to etiology, effects and treatment, *Psychol. Bull. 98*:358–387 (1986).
26. P. Leino, Symptoms of stress predict musculoskeletal disorders, *J. Epidemiol. Community Health 43*:293–300 (1989).
27. R. Karasek, Control in the workplace and its health-related aspects, in *Job Control and Worker Health*, S. Sauter, J. Hurrell, and C. Cooper, Eds., Wiley, New York, 1989, p. 130.
28. J. Hurrell and M. McLaney, Exposure to job stress—A new psychometric instrument, *Scand. J. Work Environ. Health 4*(1):27–28 (1988).
29. J. M. Last, *A Dictionary of Epidemiology*, 2nd edn., Oxford University Press, New York, 1988.
30. A. Booth-Jones, G. Lemasters, P. Succop, A. Bhattacharya, M. Atterbury, H. Applegate, and R. Stinson, A test-retest reliability study of work-related musculoskeletal disorder questionnaire, *Am. Ind. Hyg. Conf. Expo.*, Anaheim, CA, 1994.

31. D. W. Levine, B. P. Simmons, M. J. Koris, L. H. Daltroy, G. G. Hohl, A. H. Fossel, and J. N. Katz, A self-administered questionnaire for the assessment of severity of symptoms and functional status in carpal tunnel syndrome, *J. Bone Joint Surg. 75-A*(11): 1585–1592 (1993).
32. G. Lemasters, M. Atterbury, H. Applegate, A. Bhattacharya, R. Stinson, Y. Li, H. Pierson, A. Booth-Jones, and C. Forrester, Prevalence of work related musculoskeletal disease in carpenters, *Am. Ind. Hyg. Conf. Expo.*, Anaheim, CA, 1994.
33. D. S. Stetson, B. A. Silverstein, W. M. Keyserling, R. A. Wolfe, and J. W. Albers, Median sensory distal amplitude and latency: Comparisons between nonexposed and managerial professional employees and industrial workers, *Am. J. Ind. Med. 34*:175–189 (1993).
34. E. Holmström and M. Ulrich, Low back pain—Correspondence between questionnaire interview and clinical examination, *Scand. J. Rehab. Med. 23*:119–125 (1991).
35. K. Ohlsson, R. Attewell, B. Johnsson, A. Ahlm, and S. Skerfving, An assessment of neck and upper back extremity disorders by questionnaire and clinical examination, *Ergonomics 37*(5):891–897 (1994).
36. J. N. Katz, M. Larson, A. Fossel, and M. Liang, Validation of a surveillance case definition of carpal tunnel syndrome, *Am. J. Public Health 81*(2):189–193 (1991).
37. A. Franzblau, R. Werner, J. Valle, and E. Johnston, Workplace surveillance for carpal tunnel syndrome: A comparison of methods, *J. Occup. Rehab. 3*(1):1–4 (1993).
38. M. Atterbury, G. Lemasters, J. Limke, Y. Li, H. Applegate, R. Stinson, H. Pierson, and C. Forrester, Physical examination and nerve conduction results in carpenters with hand wrist symptoms, *Am. Ind. Hyg. Conf. Exp.*, Anaheim, CA, 1994.
39. M. Atterbury, J. Limke, G. Lemasters, Y. Li, H. Applegate, R. Stinson, H. Pierson, and C. Forrester, Nerve conduction studies in carpenters with hand/wrist symptoms, *Am. Occup. Health Conf.*, Chicago, IL, 1994.

CHAPTER 18

# Fall Prevention in Industry Using Slip Resistance Testing

Mark S. Redfern and Timothy P. Rhoades

CONTENTS



## 18.1 INTRODUCTION

Occupational slips and falls are a major ergonomic problem across most industries. Occupational falls are estimated to cause about 13% of all work-related deaths in the United States and between 10% and 20% of all accidental injuries [1]. Only overexertion and musculoskeletal injuries and being struck by an object rank higher. Falls have been a particularly serious problem in the construction industry [2–4]. In a survey of worker injuries involving disability, the National Safety Council [5] found that the construction industry had the highest percentage at 22%. Other industries, such as service, transportation, agriculture, and mining, are also highly susceptible, with 16%–19% of injuries involving disabilities due to falls.

Causes of falls in the workplace are multidimensional and complex, involving environmental and human factors. Environmental factors include characteristics of work surfaces, elevation, and changes in elevation. The dominant environmental factor in falls is the slip resistance characteristic of the floor [6]. Changes in frictional characteristics of the floor surface can cause a loss of traction resulting in a fall.

Injury can also result from a slip without a fall. Often a slip occurs but stability is recovered; yet, injury can still occur from striking an object or from a muscular strain. Anderson and Langerhof [7] found that 37% of nonfalling accidents were reported to involve a foot slip in a particular manufacturing environment. Low back injuries can also be initiated by a foot slip [8,9]. Injuries of this type are usually not reported to be related to a slip; thus, the number of low back injuries caused by slips is underestimated.

## 18.2 BACKGROUND AND SIGNIFICANCE TO OCCUPATIONAL ERGONOMICS

Due to the high frequency of slips as a cause of injury in the workplace, slip prevention is critical if musculoskeletal injuries are to be reduced. Slip prevention must address both human movement characteristics and environmental factors. This chapter focuses on methods used to prevent slips in the workplace. The topic is addressed from an ergonomic perspective that evaluates the relationship between the worker and the work environment. Engineers have primarily sought solutions to reduce slip hazards by their design of the workplace, but environmental designs need to be developed with the human movement requirements in mind. The biomechanics of slips and falls are presented to describe how slips occur. This information then forms the basis for a discussion of preventive ergonomic measures.

## 18.3 BIOMECHANICS OF WALKING AND SLIPS

Slips that occur during walking and carrying loads on a level surface are related to forces generated at the feet and frictional properties of the shoe/floor interface. A slip occurs when the shear forces at the shoe/floor interface exceed the frictional capabilities of that interface. Since the shear forces are highest at the heel contact and push-off phases of walking, these are the points where slips most often occur. Heel contact is the critical phase where slips can result in falls. During heel contact, body weight is being transferred to the lead foot. Should a slip occur at heel contact, the new base of support is not able to accept the body weight. The result is a fall.

The ratio of shear to normal foot force components has been used to understand the biomechanics of slips. This value, described by Strandberg and Lanshammer [10] as the instantaneous "friction used," may also be thought of as the "required" coefficient of friction (RCOF) for those phases of gait where heel movement is not expected [11]. These RCOF data can be used to assess frictional requirements of the shoe/floor interface. The RCOF during a normal step is biphasic with maximum resistance to forward movement just after heel contact and maximum resistance to rearward movement just prior to toe off (Figure 18.1). Heel movement will occur if the "friction use" values at heel contact exceed the coefficient of friction (COF) of the shoe/floor interface. The peak of these values during the heel contact phase has been used to predict slip potentials for various gait activities [1,12,13]. When proper allowances are made for normal heel strike and toe-off displacement, the remaining "friction use" data represent the RCOF for the movement analyzed. It is believed that the peak RCOF value represents the maximum frictional requirement (in terms of shoe/floor COF) during the task.

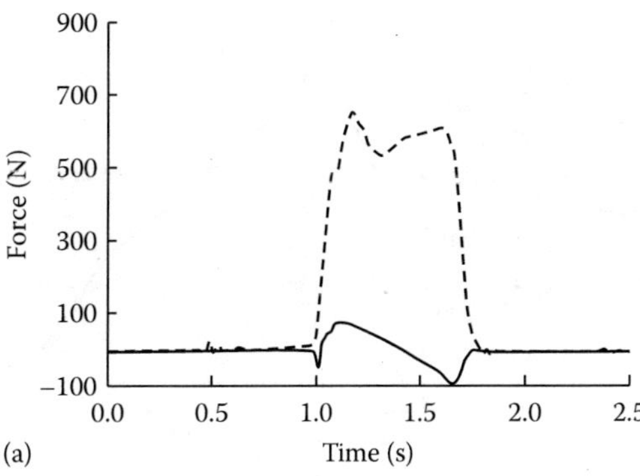


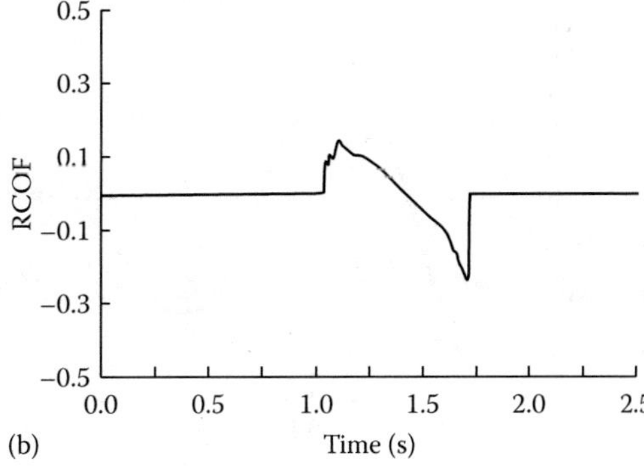


FIGURE 18.1 Foot force reactions taken by a force plate during walking at a natural cadence. (a) Vertical (dashed) and anteroposterior (solid) forces during a step; (b) the estimated required coefficient of friction (RCOF) for the same step.

Since heel contact is most often the critical point in the gait cycle for slips that result in falls, the dynamics of heel contact are important biomechanical factors to consider in understanding the mechanisms of a fall. Even during normal walking, heel contact is a dynamic event. Small slips of the heel commonly occur at heel contact but do not result in falls or loss of stability. These "microslips," as Perkins [14] calls them, are usually less than 2 cm [15,16]. Slips lead to falls when the length of the slip is greater than about 10 cm, termed a "slide" by Leamon and Lee [17]. Slips with recovery occur for heel movements of about 2–10 cm.

Heel dynamics of individuals who were carrying loads were investigated by Redfern et al. [16]. This study was undertaken to determine precisely how the foot moves during the heel contact phase. Motion of the heel during heel contact was measured while individuals were carrying boxes of varying weights at differing speeds. Table 18.1 shows the results of this study. Microslips occurred, and their length was affected by cadence or walking

TABLE 18.1 Characteristics of Heel Contact Dynamics during Gait

| | **Results (by Cadence)** | | | | **Strandberg and Lanshammar [10]** |
|---|---|---|---|---|---|
| **Steps/Min** | **70** | **90** | **110** | **Avg.** | **90,100,110,120** |
| Microslip (cm) | 1.13 | 1.34 | 1.56 | 1.34 | 1.2 (mini) |
| | (0.34) | (0.39) | (0.55) | (0.47) | (0.4) |
| Angle at heel strike (deg.) | 22.9 | 24.2 | 25.3 | 24.2 | 22.0 (grips) |
| | (3.5) | (3.3) | (4.2) | (3.8) | (5.3) |
| Velocity *(Y)* at heel strike (cm/s) | 13.9 | 19.0 | 24.1 | 19.1 | 28.2 (grips) |
| | (9.9) | (15.1) | (20.8) | (16.4) | |
| Deceleration *(Y)* before heel/ strike (cm/s$^2$) | 1577.3 | 2931.3 | 3831.2 | 2780.5 | |
| | (502.0) | (426.4) | (517.6) | (1056.1) | |
| Velocity *(Y)* 100 ms before heel strike (cm/s) | 171.7 | 312.2 | 407.3 | 297.1 | |
| | (53.1) | (40.6) | (55.9) | (110.3) | |
| Time of foot flat (s) | 0.122 | 0.115 | 0.102 | 0.113 | |
| | (0.027) | (0.045) | (0.044) | (0.040) | |

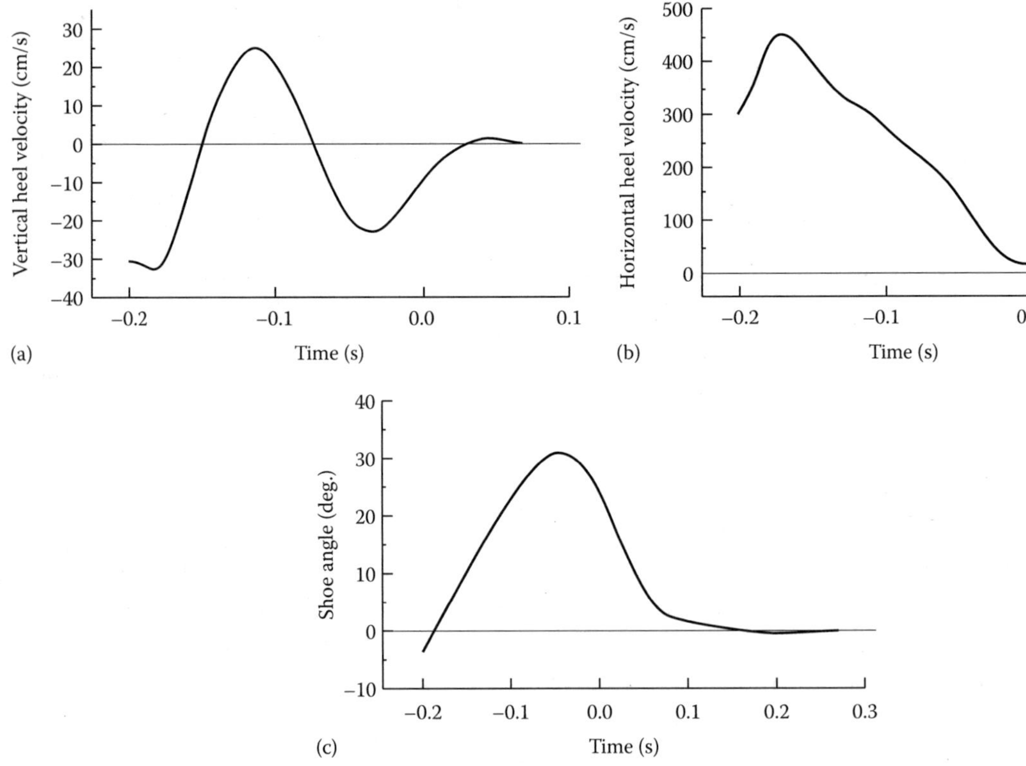


FIGURE 18.2 Dynamics of the foot during heel contact. Heel contact is at $t = 0$ s. (a) Vertical heel velocity; (b) horizontal anteroposterior heel velocity; (c) shoe sole angle to the floor.

speed. The heel rapidly decelerates just prior to heel contact (Figure 18.2). The horizontal velocity decreases from a preheel contact maximum of about 450 cm/s to between 14 and 24 cm/s at heel contact. The exact heel contact velocity was dependent on cadence (see Table 18.1). During a microslip condition, the heel comes to a complete stop about 100 ms after impact. Shoe angle to the floor also changes rapidly. A maximum of almost 25° is established just prior to heel contact, and the transition to foot flat occurs about 100 ms after contact. The shoe angle at heel contact is about 20° but is changing rapidly at that time. These results are similar to those found by Strandberg and Lanshammar [10] during different gait experiments in terms of microslips, velocities, and shoe angles and indicate the highly dynamic nature of walking and carrying. This dynamic quality of walking and carrying has direct implications for the relevance of slip resistance testing, which will be discussed in the following section.

During an actual slip and fall while walking, the foot rarely comes to a stop but instead continues to move after heel contact [15,18]. Under these conditions, a foot slip at heel contact can be considered dynamic and without a static component. Once a slip is initiated and continues beyond about 10 cm, the leading foot accelerates and moves out in front of the body. This causes an increase in the ratio between horizontal and vertical forces that lessens the chance that a slip can be stopped. Since the lead foot is the base of support for the anticipated step, a slip causes a loss of that base of support. Substantial compensatory actions (movements of the arms and other leg) are used to attempt to reestablish a base of support, often with limited success. In this sliding situation the center of mass of the body is in front of the trailing foot during the period when the slip of the leading foot occurs, normally resulting in a loss of any base of support and a fall.

## 18.4 SLIP RESISTANCE TESTING

### 18.4.1 Slip Resistance Measurement

Slip and fall prevention has focused on measurement of the "tractive" or slip resistance properties of flooring and shoes. Most slip resistance tests attempt to evaluate the coefficient of friction (COF) of the shoe/floor interface under various relevant conditions such as dry, wet, oily, etc. These measurements are then used to rate the slip potential of an industrial environment.

The two basic measures used for rating slip resistance of the shoe/floor interface are the static and dynamic coefficients of friction. A static coefficient of friction (SCOF) is most often used and can be defined as the shear force required to initiate sliding of the shoe material over the floor divided by the vertical force on that same material. Numerous devices have been developed to measure the SCOF of floor surfaces using various shoe materials. Some of these devices have been tested for usability and reliability [19,20]. Two basic types of SCOF measurement devices are the drag sled tester and the articulated strut tester (Figure 18.3). Probably the most widely used field device is the sled-type tester. In this test, a weight (usually 10 lb) is applied to a sample of shoe sole material that is placed on the floor. A force gauge is then used to apply a force in the horizontal direction until the sled begins to move. The horizontal force is recorded and divided by the vertical force to calculate the SCOF. In contrast, articulated strut systems have a weight mounted at the top of an articulated arm located above the shoe sole material. The articulated arm moves

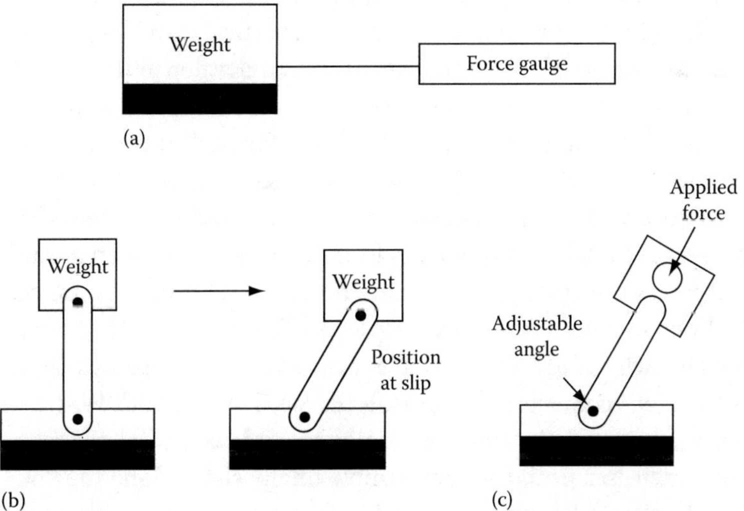


FIGURE 18.3 Diagrams of three types of static COF testing devices. (a) Drag tester; (b) articulated strut; (c) inclined strut.

laterally, altering the ratio of normal to shear forces on the shoe sole material. The angle of the articulated arm when the sole material slips is noted and directly related to the SCOF.

The choice of a SCOF tester is difficult for the safety practitioner because SCOF measurement values can vary substantially across devices. In other words, different SCOF testers will give different SCOF values under the same conditions. As a result, there is considerable controversy regarding which device or devices provide a valid measure. As testers are expensive to purchase and no one tester has been established as best for field use, many practitioners choose to conduct inexpensive, manually controlled drag-sled SCOF tests using homemade test fixtures. This measurement method has several shortcomings, but the practitioners' reluctance to invest in equipment that may not stand the test of time is understandable.

Among the most widely known static testers are the horizontal pull slipmeter (HPS), the James machine, and the NBS-Brungraber tester. The HPS measures SCOF in a manner like the generic drag-sled test described earlier, except that the HPS uses a motor to apply horizontal force in a controlled manner and can also provide a dynamic COF measurement in addition to the SCOF measure. The James machine, an articulated strut design developed in the early 1940s, is the most established SCOF tester. The James machine is particularly important because it is used to test floor finishes used nationwide in both industrial and commercial settings. The SCOF acceptance criteria of 0.5 that seem to be gaining consensus is most closely tied to the James machine. Unfortunately, the James machine is too large to permit field testing. To accommodate field testing, the NBS-Brungraber tester (also called the Mark I) was developed. This device successfully incorporates the articulated strict design concept of the James machine into a portable device and has had increasing acceptance as a field measurement device.

A major limitation of the HPS, James machine, and NBS-Brungraber tester is that they each suffer from an adhesion-related problem that can often result in readings that are too high under wet conditions. In particular, with each of these testers there is a period of time prior to application of a horizontal force when only a vertical load is applied to the shoe sole sample. During this period of time adhesive forces can develop at the shoe/ floor interface under wet conditions. Thus, when the horizontal force is applied to move the shoe sole material it must overcome adhesive forces that are not found during heel strike, and a falsely high SCOF reading is possible. To address this problem one could restrict testing to dry conditions, as is done with the James machine, or resort to dynamic coefficient of friction (DCOF) measures. Yet a third possibility is to measure SCOF in such a way that horizontal and vertical forces are applied at the same time. This approach is the design principle behind two inclined strut testers recently developed. These devices, the Brungraber Mark II and the Ergodyne, each feature an adjustable inclined strut that defines the vector of travel for the shoe sample as it contacts the floor material (see Figure 18.3c). With a vertical vector, no slip resistance is required. As the vector moves toward horizontal, the ratio of horizontal to vertical forces increases until the shoe sample finally slides along the floor surface. The SCOF is defined by the shallowest angle at which the sole material will not slip.

Although new, the inclined strut tester may overcome the problem of adhesion found with other SCOF testers. There is debate, however, about whether SCOF is the appropriate measure of slip resistance under all conditions [15]. This debate stems from studies of gait and actual slips such as those discussed in the previous section. On the basis of these studies, many researchers believe dynamic COF measures better reflect the actual frictional properties of the shoe/floor interface for slips during walking or load carrying.

A dynamic COF (DCOF) is simply the shear force required to sustain movement of the shoe material divided by the vertical force. Some DCOF devices that have been used in evaluating the shoe/floor interface are the HPS [21], the SATRA frictional tester [18,22], the PSRT [23], the finish slip tester [24], and the Tortus [19]. The HPS and the Tortus are portable devices that can be used in the field. The SATRA, Gronqvist, and PSRT testers are laboratory-based machines. In general, dynamic measurements can vary greatly from static COFs, with static measures being higher than dynamic measures. Some tests, however, have shown dynamic COF actually higher than the static measures for certain shoe/floor interfaces under specific contaminant conditions.

Although no thorough comparisons among dynamic devices have been made, there do appear to be differences. These differences are due to a number of factors or device characteristics. Variables such as the angle of the shoe sole material to the floor, sliding velocity, and vertical force can all have an effect on the resulting COF. Redfern et al. [25] investigated the effects of velocity and vertical force on DCOF recordings using a sled-type device [26]. It was found that for this type of device, pulling velocity has a significant influence on the resulting DCOF. Vertical force was statistically significant but accounted for only small changes in DCOF. Although this work provided important information about the effects of vertical force and velocity, the sled device used was limited in its ability to incorporate other important parameters. Other parameters, such as angle of the shoe, contact time before testing, and the contact application point of the shoe have been found to affect test results [22,24] and also to be relevant to the biomechanics of slips and falls [10].

Laboratory devices that incorporate some of these parameters have recently emerged [22,24,27]. These devices move a shoe over a floor surface at a velocity typical of heel contact. Figure 18.4 shows a diagram of the device used by Redfern and Bidanda [23], called the programmable slip resistance tester (PSRT). This device allows slip resistance tests to be performed under varying conditions that better reflect the interactions of the shoe and floor during human gait. The effects of shoe materials, floors, shoe angle, heel velocity, vertical force, and contaminants on DCOF recordings using the PSRT were investigated. Test parameters that were found to be significant were the shoe velocity and vertical force as they interacted with the contaminants and shoe materials used. In general, higher test velocities produced lower DCOF measures; however, this was not always true. Thus, it is important in using the PSRT to choose the proper velocities, shoe angles, and vertical forces based upon knowledge of the basic biomechanics of tasks to be investigated.

Once the DCOF test parameters have been chosen, the effects of floors, shoes, and contaminants can be evaluated. Redfern and Bidanda [23] showed the effects of shoe material, floor type, and contaminant on DCOF measures (Figure 18.5). Floor effects are mixed and depend upon the shoe. A harder shoe sole [in this case polyvinyl chloride (PVC)] performed best on the rougher concrete surface under contaminated conditions. The softer rubber-soled shoes appeared to be better on smoother surfaces for either dry or

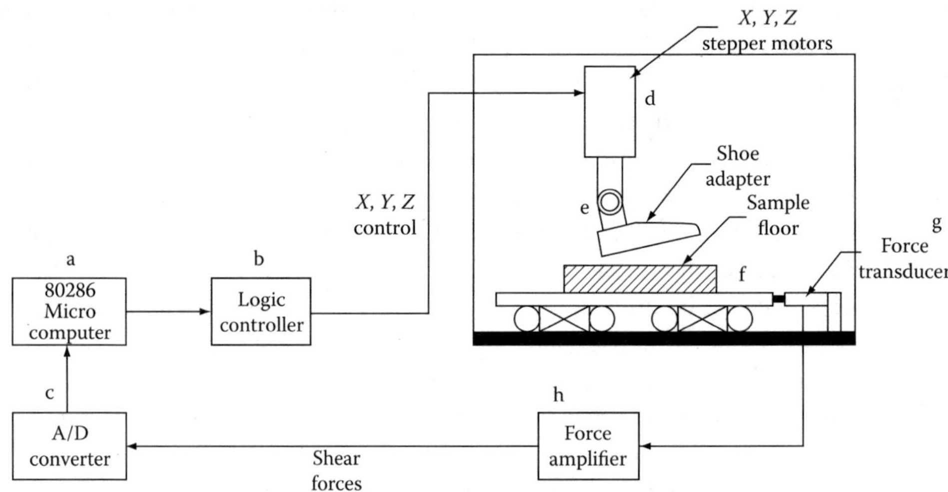


FIGURE 18.4 Diagram of the programmable slip resistance tester. The components of the system are (a) microcomputer; (b) logic controller; (c) analog-to-digital converter; (d) *X, Y, Z* stepper motors; (e) shoe adapter; (f) floor base; (g) force transducer; and (h) force amplifier. (Taken from Bidanda, B. and Redfern, M., *Comput. Ind. Eng. J.*, In Press.)

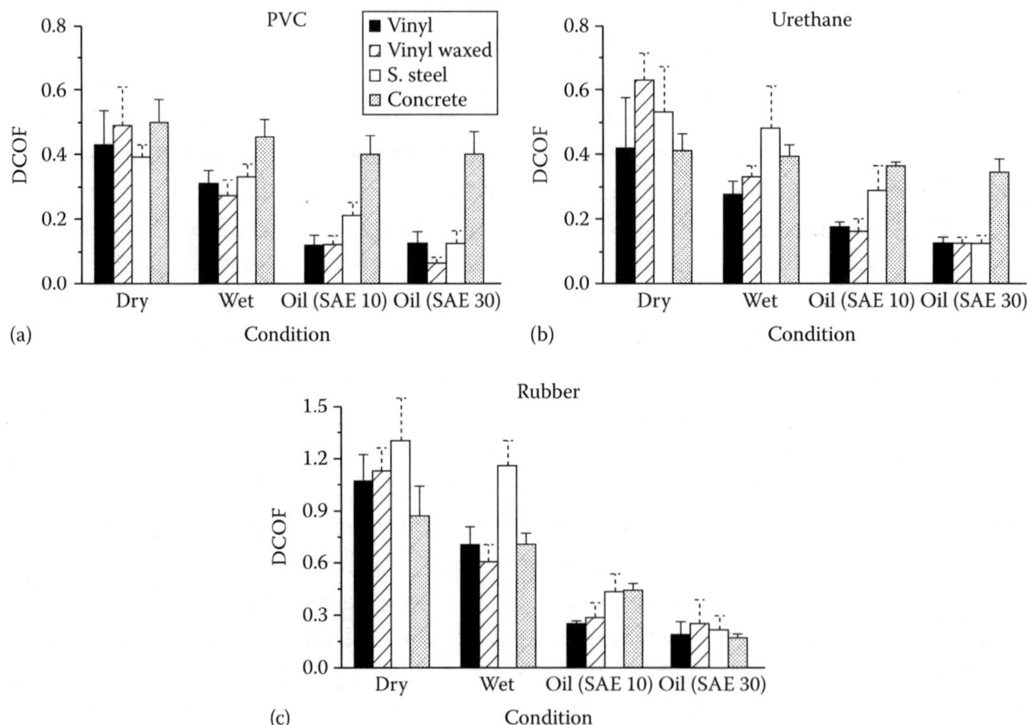


FIGURE 18.5 DCOF measurements using the PSRT under containment conditions for three shoe sole materials (a) PVC (b) Urethane (c) Rubber on four types of floors. (Taken from Redfern, M. S. and Bidanda, B., *Ergonomics*, 37(3), 511–524, 1994.)

wet conditions. Interactions of shoes, floors, and contaminants thus form complex results that can be compared using these test methods.

### 18.4.2 Slip Resistance Guidelines

Attempts have been made to establish safety classifications of shoe/floor conditions for both SCOFs and DCOFs. The first limits were developed for the SCOF in the 1940s using the James machine. A SCOF criterion of 0.5 as measured by the James machine was established at that time by underwriter's laboratory (UL) using a leather shoe material on floor polishes under dry conditions. Since that time, a SCOF level of 0.5 has become increasingly accepted as "safe" for a variety of testers and conditions, with values less than 0.5 being increasingly more hazardous. Two scales for SCOF safety levels are from the American Society for Testing and Materials (ASTM) [28,29] and Rosen [30] (Table 18.2). In 1975, the ASTM suggested gradual limits for SCOF, with 0.5 being the minimum standard and SCOF of less than 0.3 hazardous [28]. Rosen [30] proposed a SCOF rating level with slightly different gradations, but with 0.5 as the key value to define safe walkways. Note that these SCOF classification schemes and others have been based upon consensus processes rather than scientific evidence [31]. As previously discussed, a major problem with a poorly defined 0.5 criterion is that COF measurements can vary substantially between devices, particularly under wet, oily, or other contaminant conditions [21].

Dynamic COF safety limits have also been explored. In the 1940s, Sigler developed a pendulum-type DCOF tester, now known as the Sigler pendulum tester, and suggested a DCOF minimum of 0.4 using this machine [32,33]. This guideline of 0.4 again seems to be promoted more from consensus and experience than from scientific investigation. The British Standard Institute (BSI) later developed a DCOF rating scale with a minimum DCOF level of 0.4 for slip prevention [34,35] (see Table 18.2).

More recently, experiments that investigated the required COFs during gait and other tasks were performed [13,14] and related to slip resistance criteria. Perkins [14] showed that the peak RCOF at heel contact is about 0.28 for normal gait. Thus, the minimum required COF to prevent any slipping is suggested to be 0.28. While this level is regarded by some as a limit for static friction, it is probably better regarded as a limit for dynamic friction [10].

TABLE 18.2 Proposed Coefficient of Friction Recommendations for Safe Walkways

| ASTM [28] | | Rosen [30] | | BSI [34] | | Gronqvist et al. [24] | |
|---|---|---|---|---|---|---|---|
| **SCOF** | **Description** | **SCOF** | **Description** | **DCOF** | **Description** | **DCOF** | **Description** |
| 1.00 | Very good | >0.60 | Very safe | >0.75 | Very good | >0.30 | Very slip resistant |
| 0.80 | Good | 0.50–0.59 | Relatively safe | 0.40–0.75 | Good | 0.20–0.29 | Slip resistant |
| 0.50 | Standard for nonhazardous walkway | 0.40–0.49 | Dangerous | 0.40 | Minimum | 0.15–0.19 | Unsure |
| 0.40 | Poor | 0.35–0.39 | Very dangerous | 0.40–0.20 | Poor | 0.05–0.14 | Slippery |
| 0.30 | Hazardous | <0.34 | Unusually dangerous | <0.20 | Hazardous | <0.05 | Very slippery |

Gronqvist et al. [24] developed a DCOF classification based upon the studies of Strandberg and Lanshammar [10] that agrees with Perkin's [14] observations (see Table 18.2). This classification was focused on DCOF measures made with the tester that Gronqvist et al. [24] developed. They propose a minimum DCOF level of 0.20 to be "slip resistant." This guideline is much lower than traditional SCOF criteria and lower than previous DCOF limits suggested by BSI [34] and Sigler [2]. Redfern and Bidanda [23] stated in their study of slip resistance using a similar device that the Gronqvist [24] guidelines may be low. Redfern and Bidanda [23] found that DCOF levels for dry floors (believed to be safe) were all greater than 0.3. For oily floors, however, DCOF values were between 0.2 and 0.3 and are probably hazardous based on experience in industry. One possible explanation for this discrepancy is the fact that the Gronqvist scale [24] was based on normal gait. However, in industrial environments there are a wide variety of tasks and floor surface conditions that may increase frictional requirements to prevent slips. These tasks include walking on ramps, turning, and pushing/pulling. For example, frictional requirements of the surface on ramps can be more than double those of level surfaces [12]. Also, recent data suggest that turning may also increase frictional requirements by as much as 50% [36]. Thus, slip resistance guidelines suggesting frictional requirements need to consider biomechanical requirements of the tasks.

### 18.4.3 Slip Resistance Application

Despite the potential problems, slip resistance testing can be an effective method for evaluating both the current workplace and possible redesign of it to reduce slip and fall potentials. The approach incorporates basic ergonomic principles of evaluation and design. Jobs are documented and evaluated in terms of worker tasks and the environment. Second, any constraints in the work area are noted. These constraints may be structural requirements of flooring, sanitation and maintenance conditions, and durability of the materials. Once this information has been collected, slip resistance tests can be used to establish a relative slip potential of the environment as it currently stands and any possible alternative designs.

One example of an industrial application of slip resistance testing is given by the evaluation and redesign of a balcony area within a manufacturing facility. There was a history of serious accidents due to slips and falls from the maintenance balconies in this area. It was determined that the balconies needed to be redesigned. First, the environment and worker tasks were recorded. The current flooring was a diamond plate steel floor. Contaminants were found to consist of water and hydraulic fluids. Worker tasks included routine maintenance of equipment and occasional emergency repair due to mechanical failure of the manufacturing equipment. During these tasks, it was not uncommon for hydraulic fluid to be present, particularly for the emergency repairs. The required worker tasks pertinent to slips on the balconies involved walking, leaning on the rails, pulling, and twisting. The amount of time spent on the balconies was relatively short. Structural constraints of the balconies imposed a need for solid steel floors. Based on this information, a number of candidate floor coverings were identified. These floors were then evaluated by using slip resistance tests in the laboratory under conditions found at the work site. These slip resistance measurements indicated that two floors had significantly higher slip resistance than other candidate floors and that there was no difference in slip resistance between the two.

One floor was chosen on the basis of a significant difference in cost and availability. Since the installation of this new flooring on the balconies, fall injuries have been dramatically reduced and the same design has been used in other plants.

Another application of the effective use of slip resistance tests was in the footwear selection and even the design of shoe sole materials for a given environment. Although the development of a specific shoe design is usually not warranted, this may be a viable option for extreme environments where flooring cannot be changed. In one industry, a number of slips and falls off equipment were noted. The constraints were that the surface of the equipment where the workers walked could not be changed. A decision was made to focus on the shoe materials and attempt to develop a shoe that could reduce slips in these areas. A shoe manufacturer was contacted, and a design procedure was implemented. There were other constraints on the shoe materials that were conveyed to the shoe manufacturer, and a number of shoe sole materials were chosen as candidates based upon these criteria. Shoe soles were designed with different materials and different tread designs to be tested by using dynamic and static slip resistance tests. The contaminant conditions of greatest concern were wet and wet with detergent. Environmental temperature was also a concern and was included in the tests. All shoe soles were tested on the given floor with contaminants. The results of the slip resistance tests were compared across materials and treads and indicated that certain materials performed better under wet conditions and others under wet with detergent conditions. One tread pattern was generally superior, although the increase in slip resistance was small. Temperature had a large effect on some materials and less on others. After comparing all shoe soles and treads over all conditions, one sole configuration was chosen. The shoe was designed and worn in field tests with very positive results to date.

This example of designing specific shoes for industry is not common, but it has great potential for reducing slips and falls. In some ways it is analogous to the way in which tool manufacturers design hand tools to prevent hand and wrist disorders. Shoe manufacturers with the capability for flexible manufacturing could be used to develop specific shoes for difficult slip and fall areas that have specific contaminant and task requirements.

*********

A warning with respect to the use of slip resistance tests in the design of shoes and floors for industry: Slip resistance is only one ergonomic factor to consider. High slip resistance work surfaces are not always appropriate, because very high slip resistance can lead to other problems such as trips or stumbles. Other ergonomic shoe/flooring considerations are fatigue and chronic musculoskeletal disorders of the lower extremities. Standing and walking on an extremely high friction surface can cause lower extremity and even whole body fatigue. Also, knee and hip pain can result from twisting and turning on surfaces with high frictional properties. In the balcony design example, workers were on these surfaces infrequently and falls were the overwhelming concern. The same floors would not be appropriate in an area where workers stand for long periods of time. Thus, the appropriate flooring depends on the ergonomic factors of the workplace.

Another concern in floor designs to reduce falls is that many falls occur at a point of transition in flooring characteristics. Transitions from high to low slip resistance can result in a slip. When going from low slip resistance to high, a trip can occur. Gait patterns appropriate for one frictional surface may not be appropriate for another. Gait appropriate for low to moderate slip resistant areas can lead to midswing trips on a high-friction surface, particularly if the floor surface is raised or rough. Transitions from a high-friction surface to a slippery surface can lead to heel slips if a person's gait is not changed. In designing the flooring system, transitions should be minimized, particularly in high-traffic areas, and clearly marked when they are unavoidable.

## 18.5 CRITICAL REVIEW OF CURRENT STATUS

Occupational slips and falls are an often overlooked important ergonomic problem. Ergonomic assessment of the workplace should include evaluation and interventions to reduce falls. Current slip measurement methods can be used to evaluate slip potentials through COF testing. Guidelines have been proposed and used that give levels of COF values to determine slip and fall potentials. However, there is not full agreement on what level of slip resistance is "safe." There is also an ongoing debate in the ergonomics community on which COF measure, static or dynamic, should be used. A further disagreement exists regarding which device should be used within SCOF and DCOF testing procedures. Although these controversies exist, slip testing can and is used effectively for evaluating workplace slip potential and determining ergonomic interventions. Comparisons of alternative solutions using slip evaluations can quantify potential improvements to the workplace.

## 18.6 FUTURE CONCERNS

More research is required to understand the relationship between slip resistance measurements and actual slip and fall accidents. These studies should involve biomechanical research under various slippery conditions correlated to results with slip resistance measurements. This will lead to an understanding of the effectiveness of these measures in evaluating foot slip potential. This could also lead to improved COF measurement systems or possibly some agreement on which current measurement systems are most appropriate for various conditions and work environments.

## 18.7 CONCLUSIONS

Occupational falls are a continuing ergonomic problem that cuts across all industrial sectors. The causes of these falls are complex and involve environmental and human movement factors. Slip prevention needs to address both factors and be performed within the scope of other ergonomic concerns (i.e., fatigue, stability during manual tasks, etc.) to be effective. Foot forces generated during various tasks can vary greatly and must be considered in environmental design of work surfaces. The primary environmental concern for slip prevention is the frictional property of the shoe/floor/contaminant interfaces. Dynamic and static COF measures are currently used to evaluate slip resistance of the shoe/floor interface. There is no consensus in the ergonomics community on which measure is better;

however, there is an increasing belief that dynamic testing may be more relevant to tasks that require walking and carrying. Criteria to establish safe levels of SCOF and DCOF are also not agreed upon throughout the field. Despite these shortcomings, slip resistance tests are useful in determining the relative slip resistance of the shoe/floor interface under various conditions. Slip resistance tests are particularly useful in evaluating current conditions and comparing alternative solutions. These tests should be used as one component of an ergonomic evaluation and design of the shoe/floor system in the workplace. Other factors such as worker task requirements, fatigue from long-term standing, inadequate lighting, and tripping hazards also need to be incorporated into ergonomic evaluations of workplaces.

## REFERENCES


1. F. L. Buczek, P. R. Cavanagh, B. T. Iulakowski, and P. Pradhan, Slip resistance needs of the mobility disabled during level and grade walking, in *Slips, Stumbles and Falls: Pedestrian Footwear and Surfaces,* B. E. Gray, Ed., ASTM STP 1103, ASTM, Philadelphia, PA, 1990, pp. 39–54.
2. R. G. Snyder, *Occupational Falls,* UM-HSRI-77-51, National Institute for Occupational Safety and Health, Cincinnati, OH, 1977.
3. P. Britt, Construction safety: Wave goodby to work-site falls, *Saf. Health 148*(3):54–57 (1993).
4. Construction Safety Association of Ontario (CSAO), *Human Factors Engineering Report on Mounting and Dismounting Construction Equipment,* September 1980.
5. National Safety Council, *Accident Facts,* 1991 edition, NSF, Chicago, IL, 1991.
6. *Federal Register 55*:13341–13360 (1990).
7. R. Andersson and E. Lagerlof, Accident data in new tech Swedish information system on occupational injuries, *Ergonomics 26*(1):33–42 (1983).
8. D. P. Manning and H. S. Shannon, Slipping accidents causing low-back pain in a gearbox factory, *Spine 6*(1):70–72 (1981).
9. D. G. Troup, J. W. Martin, and D. C. E. F. Lloyd, Back pain in industry: A prospective survey, *Spine 6*(1):61–69 (1981).
10. L. Strandberg and H. Lanshammar, The dynamics of slipping accidents, *J. Occup. Accidents 3*:153–162 (1981).
11. T. P. Rhoades and J. M. Miller, Measurement and comparison of "required" versus "available" slip resistance, *Proceedings of the 21st Annual Meeting of the Human Factors Association of Canada,* Edmonton, Alberta, Canada, September 14–16 (1988).
12. E. J. McVay and M. S. Redfern, Rampway safety: Foot forces as a function of rampway angle, *Am. Ind. Hyg. Assoc. J* 55(7):626–634 (1994).
13. M. S. Redfern and R. O. Andres, The analysis of dynamic pushing and pulling: Required coefficients of friction, *Proc. Int. Conf. Occup. Ergon.,* 1984, pp. 569–571.
14. P. J. Perkins, Measurement of slip between the shoe and ground during walking, in *Walkway Surfaces: Measurement of Slip Resistance,* C. Anderson and J. Senne, Eds., ASTM STP 649, ASTM, Philadelphia, PA, 1978, pp. 71–87.
15. L. Strandberg, On accident analysis and slip-resistance measurement, *Ergonomics 26*(1):11–32 (1983).
16. M. S. Redfern, M. A. Holbein, D. Gottesman, and D. B. Chaffin, Kinematics of heelstrike during walking and load carrying: Implications for slip testing, *Ergonomics.* In Press.
17. T. B. Leamon and K. W. Lee, Microslip length and the perception of slipping, *Proceedings of 23rd International Congress of Occupational Health,* Montreal, QC, Canada, 1990, p. 17.
18. P. J. Perkins and M. P. Wilson, Slip resistance testing of shoes—New developments, *Ergonomics 26*(1):73–82 (1983).

19. R. O. Andres and D. B. Chaffin, Ergonomic analysis of slip-resistance measurement devices, *Ergonomics 28*(7):1065–1079 (1985).
20. B. T. Kulakowski, F. L. Buczek, P. R. Cavanagh, and P. Pradham, Evaluation of performance of three slip resistance testers, *J. Testing Eval. 17*(4):234–240 (1989).
21. W. English, Improved tribology on walking surfaces, in *Slips, Stumbles and Falls: Pedestrian Footwear and Surfaces,* B. E. Gray, Ed., ASTM STP 1103, ASTM, Philadelphia, PA, 1990, pp. 73–81.
22. M. P. Wilson, Development of SATRA slip test and tread pattern design guidelines, in *Slips, Stumbles and Falls: Pedestrian Footwear and Surfaces,* B. E. Gray, Ed., ASTM STP 1103, ASTM, Philadelphia, PA, 1990, pp. 113–123.
23. M. S. Redfern and B. Bidanda, Slip resistance of the shoe-floor interface under biomechanically relevant conditions, *Ergonomics 37*(3):511–524 (1994).
24. R. Gronqvist, J. Roine, E. Jarvinen, and E. Korhonen, An apparatus and a method for determining the slip resistance of shoes and floors by simulation of human foot motions, *Ergonomics 32*(8):979–995 (1989).
25. M. S. Redfern, A. Marcotte, and D. B. Chaffin, The effects of velocity and applied vertical force on the dynamic coefficient of friction: A dynamic slip resistance study, *Ergonomics.* In Press.
26. M. S. Redfern, A. Marcotte, and D. B. Chaffin, A dynamic coefficient of friction measurement device for shoe/floor interface testing, *J. Safety Res. 21*:61–65 (1990).
27. B. Bidanda, and M. Redfern, Development of a microcomputer based slip resistance tester, *Comput. Ind. Eng. J.,* In Press.
28. American Society for Testing and Materials (ASTM), *Standard Method of Test for Static Coefficient of Friction of Polish-Coated Floor Surfaces as Measured by the James Machine,* ASTM Designation D 2047–75, ASTM, Philadelphia, PA, 1975.
29. American Society for Testing and Materials, (ASTM), *Standard Test Method for Static Coefficient of Friction of Shoe Sole and Heel Materials as Measured by the James Machine,* reprinted from the *Annual Book of ASTM Standards,* ASTM, Philadelphia, PA, 1977.
30. S. I. Rosen, *The Slip and Fall Handbook,* Hanrow Ltd., Columbia, MD, 1983.
31. J. M. Miller, Slippery work surfaces: Towards a performance definition and quantitative coefficient of friction criteria, *J. Safety Res. 14*(4):145–158 (1983).
32. P. A. Sigler, *Relative Slipperiness of Floor and Deck Surfaces,* U.S. Bureau of Standards-Building Materials and Structures, Report BMS 100, 1943.
33. P. A. Sigler, M. N. Geib, and T. H. Boone, Measurement of slipperiness of walkway surfaces, *J. Res. Natl. Bur. Stand. 40*:339–346 (1948).
34. British Standards Institution (BSI), *British Standard Code of Practice for Stairs,* BS5395, British Standards Institution, London, U.K., 1977.
35. P. E. Ballance, J. Morgan, and D. Senior, Operational experience with a portable friction testing device in university buildings, *Ergonomics 28*(7):1043–1054 (1985).
36. M. S. Redfern and E. McDonald, *The Effects of Turning on Frictional Forces at the Feet,* Technical Report, University of Pittsburgh, Pittsburgh, PA, 1995.

# III

# Medical Surveillance for Ergonomics Programs

CHAPTER 19

# Record-Based ("Passive") Surveillance for Cumulative Trauma Disorders

Shiro Tanaka

CONTENTS



## 19.1 INTRODUCTION

Surveillance means different things to different people. A formal definition of surveillance for cumulative trauma disorders (CTDs) may be stated as "the ongoing and systematic collection, analysis and interpretation of exposure and health data necessary for the planning, implementation, and evaluation of programs for prevention and control of CTDs" (from Ref. [1], with slight modification).

Surveillance is a noun derived from the verb "survey," which means "to take a general or comprehensive view of a situation" and "to view in detail in order to ascertain a condition, value, etc." Further, the word *surveillance* implies the attitude of preparedness to make appropriate responses to the surveyed situation. Therefore, in a very broad sense, surveillance is a basic survival function of any living organism. Even a small creature performs surveillance of its surroundings and responds by flight from an attacker or a dash toward food. In our current context, business organizations, both small companies and large corporations, must maintain surveillance of production, customer's needs, market trends, and employee health status to survive and maintain a healthy existence.

## 19.2 TYPES OF SURVEILLANCE FOR CTDs AND DEFINITIONS

Since there are many types of surveillance, it is worthwhile to clarify the position of this chapter in the overall surveillance scheme. In the fields of public health and epidemiology, the term *active surveillance* has been traditionally used to indicate the activities of data collection (case finding) by reaching out to, and searching through, patients' records at doctors' offices and hospitals. In contrast, *passive surveillance* has been used to mean the data collection activity of waiting for and receiving the disease reports at the public health office, usually at the Department of Health of a state or city [2].

In recent years, however, a new meaning of active and passive surveillance has been added within the discipline of CTD epidemiology [3]. In this instance, the term active surveillance is used to denote the activities of generating the data, finding cases by administering (musculoskeletal) health questionnaires, and/or conducting physical examinations among workers. In contrast, the term passive surveillance is used in CTD epidemiology to denote reviewing and analyzing preexisting records such as OSHA 200 logs and workers' compensation (WC) claims, which are usually kept at the employer's office. As a result, the terms may refer to entirely different surveillance activities.

In view of the increasing public health importance of CTDs and increasing interactions between the disciplines of public health and ergonomics, it is highly desirable to avoid confusion by clarifying these terms and definitions. Therefore, I have devised more descriptive terms:

*Record-based surveillance*: the use of available records such as OSHA 200 logs

*Data-generating surveillance*: activities such as questionnaire administration and/or health examinations

To place our topic in perspective, various types of surveillance programs that can be performed for various purposes are listed in Table 19.1. In this chapter, most of the discussions will focus on record-based surveillance [B.1(a), Table 19.1] for an in-plant ergonomic

TABLE 19.1 Types and Definitions of Surveillance for CTDs

1. Exposure variables (measured in ergonomics surveys). Repetition of task, force, posture, duration of work/rest periods, vibration, workstation design, etc. (discussed elsewhere in this book)
2. Outcome variables. Health/morbidity status of workers
   - (a) Source of surveillance data
     - (i) Record-based surveillance (passive surveillance[a])—the analysis of existing records and/or data
     - (ii) Data-generating surveillance (active surveillance[a])—the generation of data by the use of questionnaires, medical examinations, etc. (discussed in Chapter 23)
   - (b) Scope of surveillance
     - (i) In-plant surveillance—private industrial effort
     - (ii) Public health surveillance—governmental effort

[a] In this chapter, the author has avoided using the terms "active" and "passive" surveillance because they may be confusing or misleading to readers coming from different disciplines.

program [B.2(a)]. At the same time, readers are reminded that the entire ergonomic program in the plant, from reviews of records to the assessment of effectiveness of ergonomic intervention (feedback), can also be included in a broad sense of surveillance, as shown in Figure 19.1.

Depicted in the upper part of the flowchart are the surveillance activities for CTDs in a narrower sense. These begin with record-based surveillance, with a systematic review of OSHA 200 logs, WC claims, sickness and absenteeism records, and other health-care records. This analysis will assist identification of high CTD risk jobs.

The results of such an analysis should be compared with the results of various other assessment activities, which are presented in the middle of the flowchart. Such activities

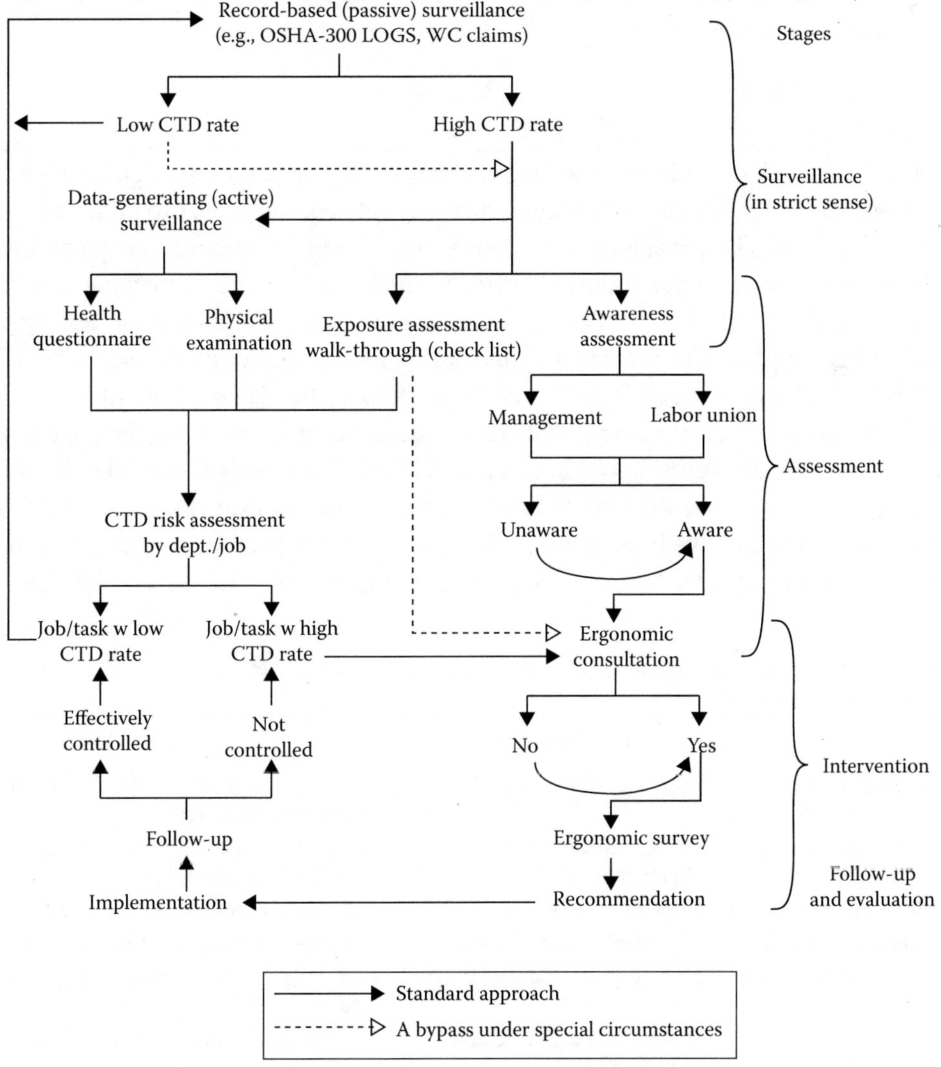


FIGURE 19.1 A flowchart of comprehensive CTD surveillance in industry. (Solid arrow: standard approach. Dashed arrow: a bypass under special circumstances.)

would include examination of the awareness of ergonomic issues among management and labor, assessment of the type and degree of exposure (e.g., ergonomic walk-through), and data-generating surveillance such as musculoskeletal health questionnaires and physical examinations. These results usually corroborate one another, leading to the identification of the problem areas and jobs in the plant as illustrated in Case Report 2.

Shown in the bottom third of the flowchart is the intervention stage, in which various ergonomics intervention measures are implemented to abate the problem. When a follow-up assessment is performed after a sufficient time period, it can be expected that the data from assessment activities will show a decreasing trend of CTDs (provided that the cause of the problem was properly identified and the intervention was correctly targeted and implemented).

In a very broad sense, all of the stages depicted in Figure 19.1 may be included in a comprehensive surveillance scheme.

## 19.3 BACKGROUND AND SIGNIFICANCE TO OCCUPATIONAL ERGONOMICS

Surveillance activities should provide the backbone of any ergonomics program. However, for people who are not familiar with ergonomics, it is not necessarily easy to understand and conceptualize the CTD surveillance functions in a plantwide ergonomics program. People who have been well trained in industrial hygiene, epidemiology, or occupational medicine may find that their learned principles and techniques are not quite suitable for dealing with CTDs in the workplace. The reasons for this may be illustrated by comparing surveillance methods for lead exposure and its health effects with those for CTDs (Table 19.2).

Lead has been known for many centuries to be toxic to humans. In the past several decades, much has been elucidated about the biochemical and toxicological details of lead poisoning. As a result, we now know how much exposure to lead can cause what toxic effects, and standards have been developed to control the exposure by setting the maximum limit for atmospheric lead as well as the blood lead level. Also, several biochemical

TABLE 19.2 Comparison of Exposure/Effect Assessments for Lead Poisoning and Cumulative Trauma Disorders

| | **Lead Poisoning** | **CTDs** |
|---|---|---|
| Exposure agent | Lead, lead compounds | Physical stress (force, repetition, posture, duration/lack of rest) |
| Assessment methodology | Established and specific (lead in air or blood) | Being developed but not established (e.g., ergonomic checklist) |
| Quantification | Precise | Difficult (except for repetition and duration) |
| Nonoccupational exposure | Identifiable and distinguishable | Identifiable but not easily distinguishable |
| Outcome/effect indicators | Nerve damage; kidney damage | Local fatigue, pain, discomfort (e.g., health questionnaire) |
| Diagnosis | If suspected, usually simple and definitive | Fairly simple for presumptive diagnosis; tests such as EMG and NCV are costly[a] |
| Technology for prevention | Known and available | Still much unknown or under development |

[a] EMG = electromyography; NCV = nerve conduction velocity.

indicators of lead exposure and absorption have been identified and used as exposure-monitoring tools. Today, a very effective surveillance for occupational lead exposure and absorption can be conducted by using these refined methodologies to monitor lead levels in atmospheric and biological samples [4].

In contrast, the state of our current monitoring capability in medical surveillance for CTDs does not have the advantage of sensitive and accurate measures. This situation is analogous to that of several decades back in history when we had to wait until the manifestation of frank symptoms of lead poisoning such as lead colic and radial nerve paralysis or signs such as "lead lines" on the gum or in skeletal radiographs. Now we have sensitive tools to monitor exposure to lead. In contrast, this is not the case for CTDs. For example, despite the fact that carpal tunnel syndrome (CTS) is probably the most studied CTD and criteria for its surveillance have been developed [5], medical evaluation of CTS is still largely subjective. Although the measurement of nerve conduction velocity (NCV) is available as an objective method to test dysfunction of the median nerve [6], its high cost (due to the need for skilled technicians and professional interpretations) is rather prohibitive for the routine use of NCV measurement as an industrial screening tool.

Furthermore, the technology for the ergonomic assessment of exposure factors is still in its developmental stage, although some noteworthy advances have been made in recent years [7]. There have been a few documented success stories in which epidemiological and ergonomic investigations have led to some definitive intervention strategies [8]. These cases typically involved specific and obvious physical stresses such as the use of the knee kicker by carpet layers to stretch carpet [9,10]. However, to date, most epidemiological attempts to simply establish certain occupations or jobs as the cause of certain CTDs (e.g., CTS) have been unsuccessful [11]. Such failures are not surprising when one realizes that the causative agents are not the occupations or jobs per se, but rather the physical stresses demanded by the job and how the job or task is performed. In other words, almost any manual job can lead to CTS or tendinitis depending on how the task is performed, while even meat-packing jobs that have been known for their severe musculoskeletal stresses [12] could be performed, theoretically, without incurring a CTD if appropriate ergonomic practices were put into effect.

Therefore, we have to come to the realization that these difficulties are inherent in the medical and ergonomic surveillance of CTDs. Nonetheless, we must use currently available methodologies to perform the needed quantitative assessment and hope for technological advancements in coming years. It is not difficult for an epidemiological study to identify occupations or industries with an elevated risk of CTDs. However, epidemiological techniques alone cannot determine what exposure factors in the work are really causing the problem and how these can be controlled. Thus, cooperative and coordinated efforts of both ergonomists and epidemiologists are much needed to achieve this goal [13].

## 19.4 RECORD-BASED SURVEILLANCE

In-plant record-based surveillance ("passive" surveillance) for CTDs involves reviewing and analyzing existing records or data systems that are normally kept by the employer [14]. Typically, available records would include OSHA 200 logs and WC claims, which will be the main subjects of discussion in this chapter. Record keeping of OSHA 200 logs is required by law for the purpose of surveillance, and the details are described in official publications [15]. In contrast, the use of WC claims for this purpose is for convenience, because the WC system was not created for surveillance purposes. Some companies keep so-called sickness and accident (S&A) records from which data for CTDs can be extracted. Medical records may be kept at the company's health unit or at the health-care provider's office. Also, the health insurance records of employees have been used for CTD surveillance [16]. More detailed medical information is usually available from these health-care record systems than from OSHA 200 logs or WC claims. However, routine extraction of necessary data from health insurance records is not always easy or simple unless the data are computerized and suitably coded. The insurance carriers are usually reluctant to release the data owing to the confidentiality issue and the proprietary nature of information.

For the purpose of basic CTD surveillance at the place of employment, periodic (preferably monthly) review of OSHA 200 logs and WC claims is recommended and should be effective as long as the information is recorded honestly and without bias or interference. Equally important are the keen sense and ability of the person in charge of surveillance to recognize workers' physical complaints or remarks, to detect abnormal trends, and to respond appropriately (see Section 19.7.1, Case Report 1).

The quality and usefulness of a record system for the purpose of CTD surveillance will depend on various factors. To calculate incidence (i.e., number of new cases) rates, the record should include, at a minimum, date of occurrence, personal identifier, department, job title, part of the body affected, and preferably the diagnosis. Conditions reportable in column 7-f of OSHA 200 logs are defined as "disorders associated with repeated trauma (DART)" with examples such as CTS, synovitis, tenosynovitis, bursitis, Raynaud's phenomenon, and noise-induced hearing loss [15]. Also, the denominator data, such as the number of workers by department and hours worked, should be available for calculation of incidence rates (IRs). The record should be easily accessible to and retrievable by authorized personnel for the purpose of conducting surveillance. Computerized data processing would make the analysis fast and simple.

An IR per 100 full-time employees (FTEs) of illnesses may be computed by using the formulas [15]

$$\text{IR} = \frac{\text{Number of illness} \times 200{,}000^{*}}{\text{Employee hours worked}}$$

In this calculation, only new cases are counted for a given time period. If one worker experienced the same CTD more than once during the reporting period, the occurrences are counted separately as long as there was a period of complete recovery in between. The severity or duration of the disorder is not considered in IR.

* 200,000 h worked per year = 100 FTEs.

The IR should be computed for the entire company or plant, for each department, and for each section if within a large department. It may sometimes happen that only a small number of workers are performing highly repetitive manual work within a large company or department. In such a situation, the IR for the company or department may not be high, and the CTD hazard for the small group of exposed workers can be overlooked (a dilution effect). Calculation by job/task groups, combined with an ergonomic evaluation, should be able to point out the problem areas (see Section 19.7.2, Case Report 2).

The calculation of the total hours worked can be obtained from computerized payroll data in most companies. If the number of hours worked is not computerized, it must be hand calculated using a calculator. For the purpose of obtaining a rough estimate, the average or usual number of workers may be used instead of hours worked, as long as the number of employees remained fairly stable during the year. (e.g., if a lathe operator quit after working 6 months and was replaced by another lathe operator who worked the remaining 6 months of the year, they can be counted as 1 person-year.) However, it should be kept in mind that this rough estimation tends to underestimate the IR if each worker did not work a full 2000 h a year due to part-time work, vacation, illness, layoff, etc.

For official reporting purposes, annual IRs are required by OSHA. However, for the purpose of maintaining an effective in-plant ergonomics surveillance program, it is not recommended to wait for 12 months if one wants to detect an upsurge of CTDs at the earliest possible stage. Therefore, in addition to being in compliance with the OSHA requirement, examination of the current data and trend should be performed at least on a quarterly, and preferably on a monthly, basis.

In addition to the IR, which considers the number of new cases in a set time period, the prevalence (rate) is sometimes used. The prevalence measures the frequency of all current cases of a disease, both new and old (continuing from a previous period), at a given point in time (point prevalence) or for a prescribed period of time (period prevalence). Thus,

$$\text{Point prevalence} = \frac{\text{Number of new and old cases at a given point in time}}{\text{Number of workers at the same point in time}}$$

$$\text{Period prevalence} = \frac{\text{Number of new and old cases during a given time period}}{\text{Number of workers at the mid-interval in the same time period}}$$

Period prevalence is of limited usefulness [17], as it does not distinguish new cases from old ones nor does it count repeated episodes of the same disease of the same person occurring in the time period.

For the purpose of estimating the overall seriousness of CTDs by departments or disorder and for setting the intervention priority, the severity index (SI) may be useful. The SI may be calculated for all CTD cases or by diagnosis, or by part of the body affected, using the following formula:

$$\text{SI} = \frac{\text{Total number of workdays lost due to the defined disorder(s)}}{\text{Total number of workers or hours worked in a time period}}$$

However, it must be kept in mind that the SI may be influenced by such factors as the type (conservative or invasive) of medical management [18], sickness benefit, and opportunity for transfer to less stressful jobs. It may also be skewed by unusually long illnesses experienced by a small number of employees.

## 19.5 ADVANTAGES AND DISADVANTAGES OF RECORD-BASED SURVEILLANCE

Major advantages of record-based surveillance using OSHA 200 logs or WC claims are its low cost and easy accessibility. Since the employer is required to collect and maintain the record, the reviewer's main task is to tabulate and analyze the data. This effort is small compared to the data-generating surveillance, which involves administration of a health questionnaire and/or physical examinations, and subsequent data analysis [3].

A major shortcoming of record-based surveillance is said to be its underreporting. Fine and others [3] estimated that record-based surveillance detected only one-seventh of potential CTD cases that were uncovered by data-generating surveillance. There are several reasons for this underestimation. First, the mere presence of pain or other symptoms is neither reportable in OSHA 200 logs nor sufficient for filing a WC claim. Second, some employers may be reluctant to list all reportable cases in the OSHA 200 logs. There were well-publicized cases of deliberate underreporting by some meat-packing companies in the mid-1980s. After OSHA started imposing large fines for such neglect or concealment, the number of reported cases started to increase. Third, employees may hesitate to report their symptoms or illnesses to their supervisor, particularly at a time when jobs are scarce.

In contrast, musculoskeletal questionnaires or physical examinations of data-generating surveillance (described in more detail in Chapter 23) typically elicit reports of symptoms and signs in various body parts in a confidential manner. Pains and discomfort above certain levels and lasting more than a certain number of days are counted as positive cases. Therefore, it is not surprising that data-generating surveillance (particularly for research purposes) can detect many times more cases than record-based surveillance.

WC claims are filed under specific rules and regulations that vary from state to state. Therefore, record-based surveillance using these databases may be subject to an equal or higher degree of underestimation compared to the review of OSHA 200 logs. On the other hand, WC claims usually contain a wealth of information related to each CTD claim, including the body part affected and diagnosis [19]. Under special circumstances such as for research, health insurance records can be obtained and analyzed for detection of potential CTDs [16]. However, widespread or routine use of health insurance records as a surveillance tool may not always be feasible or practical, particularly for small companies.

The primary objective of conducting an in-plant CTD surveillance is early detection and intervention. For this purpose, the concept of Sentinel Health Event—Occupational (SHE-O), which was first proposed by Rutstein and others [20], can be applied. For investigation of various occupational diseases, including CTDs, a single case could trigger more focused examinations of the health status of coworkers and the work area where the index case was detected [21]. Even if record-based surveillance may detect only a small portion of CTD cases, it can be said that one case detected by record-based surveillance may lead to several times as many unreported cases.

## 19.6 CRITICAL REVIEW OF CURRENT STATUS AND FUTURE CONCERNS

As presented earlier, record-based surveillance for in-plant monitoring of CTDs can be conducted fairly rapidly and easily once the system is in place and managed by a knowledgeable and responsible individual. With the support of responsive management, this person can play an important role for timely intervention and control of CTD problems in the plant. At the plant level, DART conditions reportable in column 7-f of OSHA 200 logs seem to be specific enough to include the part of the body in the data.

Nationwide surveillance for CTDs is very important for making effective policy decisions for prevention but difficult to conduct for various reasons. As discussed in the beginning of this chapter, record-based CTD surveillance is plagued by the ill-defined nature of the disorders and the difficulty of exposure assessment. The current OSHA-required record-keeping system (which is the basis of the Bureau of Labor Statistics' (BLS) annual report) has been criticized for its tendency to underestimate the incidence of CTDs [3]. Also, when the data were compiled by BLS to prepare the annual report, all of the DART conditions were lumped together. This process reduced its usefulness as a nationwide surveillance tool, although DART accounted for 56% of the total cases of occupational illnesses reported by private industry in 1990.

Since 1992, BLS has been using a redesigned occupational injury and illness surveillance program to collect more detailed information such as the demographics of the affected workers and the circumstances of the incident for lost workday cases. This new method has generated information that is more useful toward prevention of CTDs, which was not available under the old reporting system [22].

Also, at the time of this writing, OSHA's effort to propose an ergonomic protection standard for general industries is stalled in a political process. Detailed methods of surveillance for exposure to musculoskeletal stress and health effects are described in the proposed draft [23]. Whether or not such a standard is eventually promulgated, the surveillance methods described in the OSHA draft are very useful for implementation of an in-plant ergonomics program on a voluntary basis. I am also confident that the basic aspects of record-based CTD surveillance described in this chapter would be applicable regardless of the type of standard or guidelines that might be finalized by OSHA.

## 19.7 CASE REPORTS

Two cases have been selected to illustrate some of the points made in this chapter.

### 19.7.1 Case Report 1

This case is based on a paper by Luopajärvi et al. [24] reporting on a food (unspecified) production factory where approximately 200 female workers performed packing tasks on assembly lines in the 1970s. As shown in Table 19.3, the number of cases of occupational hand disease in 1972 was only 1 with 42 lost workdays. In 1973, this jumped to 12 cases with

TABLE 19.3 Surveillance Record of a Food Packing Company (Case Report 1)

| Year | Number of Cases | Lost Workdays | Hindsight Remarks |
|---|---|---|---|
| 1972 | 1 | 42 | } Something happened here, but no intervention? |
| 1973 | 12 | 1117 | |
| 1974 | 16 | 1446 | } Many of these cases could have been prevented if problem had been identified and intervention begun in 1973 |
| 1975 | 46 | 3670 | |
| 1976 | 51 | 5288 | |
| 1977 | 20 | 840 | |
| 1978 | 5 | 201 | |
| 1979 | 1 | 24 | |
| 1980 | 0 | 0 | |

*Source:* Data from BLS: Survey of Occupational Injuries and Illnesses, 1993, BLS Form-9300 NO 4, U.S. Department of Labor, Washington, DC, by permission.

1117 lost days, and over the following years the numbers continued to increase. In 1976, the situation had become so serious that a project was started to deal with the problem with seven working groups directed by a multiprofessional leadership. The epidemic was eventually brought under control by 1979. However, during the 6-year span 1973–1978, a total of 150 cases and 12,562 lost workdays due to hand disorders were recorded.

Although the intent of the paper was to report on the effectiveness of various intervention measures to contain the epidemic, it seems to present a very interesting case study for surveillance. First, there was at least a basic surveillance system in this plant to record occupational illnesses and lost workdays. Through 1972, the packaging was probably done by a slow, old-fashioned method. Although it was not described in the paper, something happened during the year 1973. A likely scenario might be that a new company policy was implemented to increase the production rate by way of increased quota, possibly accompanied by the introduction of partial automation. Workers had to adjust the speed of manual work to that of the machine, but obviously it was far beyond their physical capacity.

The sudden increase of morbidity must have caught the eye of plant management in 1973. Strangely, however, nothing was done to deal with the surge of new cases until 1976. By hindsight, if the intervention effort had been initiated in 1973, they could possibly have prevented up to 130 cases and 11,000 lost workdays. This case amply illustrates that the failure of early detection of the problem by surveillance and resultant lack of timely intervention allowed the problem to continue and even increase in size for several years.

The moral of this case would be that an increased rate of CTDs must be examined to determine the reason for the increase and dealt with promptly to prevent an epidemic, which can be very costly.

## 19.7.2 Case Report 2

This case is based on my own experience at the National Institute for Occupational Safety and Health (NIOSH). In a joint surveillance project by NIOSH and Ohio Bureau of Workers' Compensation (BWC), WC claims were analyzed for "inflammation or irritation of joints, tendons, or muscles" resulting from "overexertion occurring over a protracted time period" for various parts of the body [19]. As a result, we were able to identify companies with a high IR of such cases for the hand/wrist including CTS. (Before 1985, Ohio BWC did not include a separate code for CTS.) Subsequent telephone contacts with seven of these companies all confirmed the existence of, and the management's concern with, work-related CTS among their employees. One company, a manufacturer of garden, automotive, and industrial hoses, cooperated with NIOSH in an ergonomic and epidemiological investigation of their CTD problem [25].

The investigation consisted of the review of WC claims and OSHA 200 logs, administration of a questionnaire, and an ergonomic walk-through assessment and analysis of videotapes. Table 19.4 shows a summary of crude IR of CTS (and related disorders) by department.

It can be noted in Table 19.4 that positive responses by questionnaire (data-generating surveillance) were far more frequent than those detected by WC claims or OSHA 200 logs (record-based surveillance). This is consistent with the previous discussion on the varying degree of detection by different surveillance methods. However, regardless of surveillance methodology, the interdepartmental comparisons showed a consistently higher rate for a specific department (Department A). A later ergonomic survey revealed that the CTDs in this department were indeed caused by very forceful and repetitive manual work, which required frequent bending and twisting of the wrist, elbow, and shoulder.

It is interesting to note that this company was initially selected because of a very high overall IR of WC claims for CTDs. However, it was later found that the high rate was spurious and was due to the erroneously small number of employees (denominator) listed in an industrial directory for the state. The company would not have been selected if the number of employees had been listed correctly. Nevertheless, upon further inquiry and site visit, it was revealed that Department A had a very high IR of CTDs compared to other departments.

TABLE 19.4 Comparison of Crude Incidence Rates of Carpal Tunnel Syndrome (CTS) and Related Disorders by Various Surveillance Methods at a Hose Manufacturing Plant (Case Report 2)[a]

| | | | From OSHA 200 Logs | | From Questionnaire | | |
|---|---|---|---|---|---|---|---|
| **Department** | **Number of Workers** | **WC Claims** | **CTS or Like** | **Sprains/ Strains** | **Neck/ Arm (%)** | **Hand Wrist (%)** | **Nocturnal Hand Pain (%)** |
| A | 67 | 2.4 | 1.5 | 13.4 | 31 | 31 | 32 |
| B–J | 388 | 0.2 | 0.1 | 0.9 | 22 | 17 | 6 |
| Office | 156 | 0 | 0 | 0 | 10 | 3 | 2 |
| Total | 611 | 0.4 | 0.2 | 2.0 | 20 | 15 | 8 |

*Source:* Condensed from tables in Occupational Safety and Health Administration: OSHA Draft Proposed Ergonomic Protection Standard: Summaries, Explanations, Regulatory Text, Appendices A and B, *Occupational Safety and Health Reporter,* 24(42) Special Supplement, pp. S1–S248, March 20, 1995, The Bureau of National Affairs, Inc., Washington, DC.

[a] Rates are per 100 employees per year or percent as indicated.

This case illustrates that for the purpose of in-plant surveillance for CTDs, WC claims and OSHA 200 logs can be used effectively to detect high-risk departments or jobs. At the same time, however, it was learned that for the purpose of public health surveillance, an overall low or moderate rate of CTDs determined by a simple calculation may be misleading, because the problem areas may be identified only after a detailed in-plant investigation.

## REFERENCES


1. D. N. Klaucke, J. W. Buehler, S. B. Thacker, R. G. Parrish, F. L. Trowbridge, R. L. Berkelman, and the Surveillance Coordination Group, Guidelines for evaluating surveillance systems, *Morbidity and Mortality Weekly Report* (*MMWR*) *37*(S-5):1–17 (1988).
2. W. A. Orenstein and R. H. Bernier, Surveillance for the control of vaccine-preventable disease, in *Public Health Surveillance,* H. Halperin, E. L. Baker, and R. R. Monson, Eds., Van Nostrand Reinhold, New York, 1992, pp. 80–82.
3. L. J. Fine, B. A. Silverstein, T. J. Armstrong, C. A. Anderson, and D. S. Sugano, Detection of cumulative trauma disorders of the upper extremity in the workplace, *J. Occup. Med. 28*: 674–678 (1986).
4. P. J. Seligman, W. E. Halperin, R. J. Mullan, and T. M. Frazier, Occupational lead poisoning in Ohio: Surveillance using workers' compensation data, *Am. J. Public Health 76*:1299–1302 (1986).
5. Centers for Disease Control (CDC), Occupational disease surveillance: Carpal tunnel syndrome, *MMWR 38*:485–488 (1989).
6. J. N. Katz, M. G. Larson, A. Sabra, C. Krarup, C. R. Stirrat, R. Sethi, H. M. Eaton, A. H. Fossel, and M. H. Liang, The carpal tunnel syndrome: Diagnostic utility of the history and physical examination findings, *Ann. Intern. Med. 112*:321–327 (1990).
7. W. S. Marras and R. W. Schoenmarklin, Wrist motions in industry, *Ergonomics 36*:341–351 (1993).
8. S. Tanaka, S. T. Lee, W. E. Halperin, M. J. Thun, and A. B. Smith, Reducing knee morbidity among carpetlayers, *Am. J. Public Health 79*:334–335 (1989).
9. M. J. Thun, S. Tanaka, A. B. Smith, W. E. Halperin, S. T. Lee, M. E. Luggen, and E. V. Hess, Morbidity from repetitive knee trauma in carpet and floor layers, *Br. J. Ind. Med. 44*:611–620 (1987).
10. A. Bhattacharya, M. Mueller, and V. Putz-Anderson, Traumatogenic factors affecting the knees of carpet installers, *Appl. Ergon. 16*:243–250 (1985).
11. S. R. Stock, Workplace ergonomic factors and the development of musculoskeletal disorders of the neck and upper limbs—A meta analysis, *Am. J. Ind. Med. 19*:87–107 (1991).
12. V. R. Masear, J. M. Hayes, and A. G. Hyde, An industrial cause of carpal tunnel syndrome, *J. Hand Surg. 11A*:222–227 (1986).
13. S. Tanaka and J. D. McGlothlin, A conceptual quantitative model for prevention of work-related carpal tunnel syndrome (CTS), *Int. J. Ind. Ergon. 11*:181–193 (1993).
14. M. Burke, *Applied Ergonomics Handbook,* Lewis Publishers, Boca Raton, FL, 1992.
15. Bureau of Labor Statistics (BLS), *Recordkeeping Guidelines for Occupational Injuries and Illnesses,* U.S. Department of Labor, Washington, DC, 1986.
16. R. M. Park, N. A. Nelson, M. A. Silverstein, and F. E. Mirer, Use of medical insurance claims for surveillance of occupational disease—An analysis of cumulative trauma in the auto industry, *J. Occup. Med. 34*:731–737 (1992).
17. B. MacMahon and T. F. Pugh, *Epidemiology: Principles and Methods,* Little, Brown, Boston, MA, 1970.
18. N. M. Hadler, Arm pain in the workplace—A small area analysis, *J. Occup. Med. 34*:113–119 (1992).

19. S. Tanaka, P. Seligman, W. Halperin, M. Thun, C. L. Timbrook, and J. J. Wasil, Use of workers' compensation claims data for surveillance of cumulative trauma disorders, *J. Occup. Med. 30*:488–492 (1988).
20. D. D. Rutstein, R. J. Mullan, T. M. Frazier, W. E. Halperin, J. M. Melius, and J. P. Sestito, Sentinel health events (occupational): A basis for physician recognition and public health surveillance, *Am. J. Public Health 73*:1054–1062 (1983).
21. R. J. Mullan and L. I. Murthy, Occupational sentinel health events—An up-dated list for physician recognition and public health surveillance, *Am. J. Ind. Med. 19*:775–799 (1991).
22. T. Luopajärvi, I. Kuorinka, and R. Kukkonen, The effects of ergonomic measures on the health of the neck and upper extremities of assembly-line packers—A four year follow-up study, *Proc. 8th Congr. Int. Ergon. Assoc.* Tokyo, Japan, 1982, pp. 160–161.
23. S. Tanaka and D. Habes, *Health Hazard Evaluation*, Report No. 87–428–2063, Anchor Swan Division, Harvard Industries, Inc., Bucyrus, OH; National Institute for Occupational Safety and Health (NIOSH), Cincinnati, OH, 1990.
24. Bureau of Labor Statistics, Survey of Occupational Injuries and Illnesses, BLS Form-9300 NO 4, U.S. Department of Labor, Washington, DC, 1993.
25. Occupational Safety and Health Administration: OSHA Draft Proposed Ergonomic Protection Standard: Summaries, Explanations, Regulatory Text, Appendices A and B. *Occupational Safety and Health Reporter,* 24(42), Special Supplement, pp. S1–S248; The Bureau of National Affairs, Inc., Washington, DC, March 20, 1995.

CHAPTER 20

# Active Surveillance of Work-Related Musculoskeletal Disorders

## *An Essential Component in Ergonomic Programs*

Norka Saldaña

CONTENTS

## 20.1 INTRODUCTION

Identification of work-related musculoskeletal problems in industry at an early stage allows for early control, a safer environment, and a healthier workforce. OSHA defined ergonomics surveillance as "the ongoing, systematic collection, assessment and interpretation of health incidence and exposure data in the process of describing and monitoring the circumstances which may be related to ergonomic hazards or the presence thereof" [1]. Thus, ergonomic surveillance techniques are used (1) to detect, monitor, and control patterns of health and disease among employees; (2) to identify work-related risk factors and related ergonomic hazards that may cause, precipitate, or aggravate a condition; and (3) to evaluate the effectiveness of the ergonomic interventions in reducing the incidence or prevalence of musculoskeletal disorders.

Ergonomic surveillance attempts to anticipate work-related health problems before they occur. When ergonomic hazards cannot be eliminated completely, then ergonomic surveillance serves as a control mechanism aimed at identifying a disorder in its early stages before it becomes a more serious problem. To accomplish this, ergonomic surveillance should address the following questions:

1. What types of health problems are the employees experiencing?
2. What is causing these problems?
3. How can these health problems be eliminated?

Work-related musculoskeletal disorders (WMDs) accounted for 60% of all occupational illnesses among U.S. workers in 1991 [2]. Consequently, ergonomic surveillance concentrates mostly on the identification of WMDs referred to as WMD surveillance. The underlying assumption for WMD surveillance is that if musculoskeletal discomfort symptoms persistently recur from job exposure and there is no intervention and no provision for sufficient recovery time, then a musculoskeletal disorder may develop.

This chapter concentrates on WMD surveillance. There are two basic methods for collecting health data that can help determine past or present occurrences of musculoskeletal problems. These two methods are (1) passive surveillance and (2) active surveillance. The words *passive* and *active* relate to the time of occurrence of the events being studied relative to the researcher's place in time. Passive surveillance looks at past cases, whereas active surveillance anticipates new cases. Passive surveillance is discussed briefly here (for more information, refer to Chapter 22). Then, two approaches for active surveillance are presented and discussed in terms of the effectiveness of each approach for the surveillance of WMDs.

## 20.2 BACKGROUND AND SIGNIFICANCE TO OCCUPATIONAL ERGONOMICS

Surveillance is from the French for keeping a watch over. It was first used by French police to describe the way in which criminals or potential criminals were watched [3]. In the medical arena, the term is applied to the observation of infectious diseases.

Thus surveillance involves the planned follow-up of individuals at risk of developing a disease or individuals with a disease in an early stage who need close observation as the condition develops. In general, the objective of medical surveillance is to prevent diseases from claiming new victims.

Although medical surveillance programs are commonly established in a variety of industrial settings, the quality and effectiveness of such programs vary widely [4]. These programs incorporate medical examinations as well as laboratory tests and work-site exposure data. The purpose of occupational surveillance in large part is to detect adverse health conditions and to determine whether employees are at risk from continued exposure [5]. Ergonomic surveillance of upper extremity WMDs is most commonly used to identify the jobs with high rates of disorders so that an effective control program may be developed. However, identification of these disorders is very difficult due to many complex etiologic factors, long latency, effects of aging, and lack of standardized diagnostic criteria [6].

### 20.2.1 Definition of Work-Related Musculoskeletal Disorders

WMDs are conditions that affect the soft tissues. Other commonly used terms include ergonomic disorders (EDs), cumulative trauma disorders (CTDs), and repetitive strain injuries (RSIs). WMDs are a class of musculoskeletal disorders that include damage to the tendons, tendon sheaths, and synovial lubrication of tendon sheaths, and to the related bones, muscles, and nerves of the hands, wrists, elbows, shoulders, neck, and back. These disorders develop gradually over periods of weeks, months, or even years due to repeated exertions and movements of the body. These musculoskeletal disorders belong to a collection of health problems that are more prevalent among the working class than among the general population. The more frequently occurring WMDs in the workplace include carpal tunnel syndrome, epicondylitis (tennis elbow), tendinitis, tenosynovitis, synovitis, stenosing tenosynovitis of the finger, DeQuervain's disease, and low back pain. Refer to Putz-Anderson [7] for more information on WMDs.

Early symptoms of musculoskeletal disorders are referred to as unpleasant sensations or discomfort associated with fatigue, perceived exertion, and poor posture [8]. Aches, numbness, and burning are some symptoms associated with discomfort. At the beginning these discomfort symptoms may be transient and occur mostly at night. As the disorder develops, they become more persistent and painful. Musculoskeletal pain arises from injury, irritation, or inflammation and may be considered a condition affecting the soft tissues [9]. Thus, discomfort symptoms may be predictive of musculoskeletal problems.

### 20.2.2 Significance of Work-Related Musculoskeletal Disorder Surveillance

WMD constitutes a major source of employee disability and lost wages. The National Institute for Occupational Safety and Health (NIOSH) has identified WMDs as one of the 10 leading occupational health problems of workers [10]. Officers of NIOSH have stated that over 5 million workers (some 4% of the American workforce) suffer WMDs each year, with predictions that the figure will reach 50% of the workforce by the year 2000. Compensation and disability claims for some of the most severe cases of WMDs that required extensive treatment and surgery ranged from $30,000 to $60,000 [11]. Carpal tunnel syndrome strikes 23,000 workers per year, costing about $3500 in benefits and rehabilitation per person and about $30,000 per person if the injury requires surgery [12]. Thus, active surveillance of WMDs should constitute an essential component in an ergonomic program used to control WMDs and reduce human suffering, lost workdays and wages, and compensation claims. As a result, the active surveillance of WMDs can have a significant impact on the overall safety and health of employees and on the overall safety and health costs to employers.

## 20.3 PASSIVE SURVEILLANCE OF WORK-RELATED MUSCULOSKELETAL DISORDERS

Passive surveillance is characterized by the collection of past data from available records. Thus passive surveillance involves looking first at the effects and then at the causes. Among the available records used for passive surveillance of WMDs are OSHA 200 logs, worker's compensation and insurance records, plant medical records, safety and accident reports, and payroll records. Records are examined to identify past cases of WMDs. This information can be used to determine which departments or jobs pose a risk to the workers. Thus, the emphasis of a retrospective approach is on the health effects of poor ergonomic design. The health information collected from a records review may provide an indication of jobs that may need immediate attention or modification.

There are various techniques for analyzing the data on medical and safety records to identify priority jobs. Three popular statistics used to monitor WMD experience are frequency of cases, incidence rate, and severity rate.

The frequency of cases is the number of cases reported during a specific time period. This statistic is useful only when it specifies the group (e.g., a department, a specific job, a particular operation, or the whole plant) in which the count is made and the type of cases (e.g., carpal tunnel syndrome, all shoulder problems, all ergonomics-related problems) counted. When the numbers of employees per group differ substantially, the frequency count is not enough to provide an indication of the priority jobs. In these cases, the frequency must be related to the number of people in each group before groups are compared. The computation of the incidence rate makes this adjustment and gives an indication of the number of new cases among a specific number of workers or for a given number of worker hours. Finally, the severity rate can be determined based on such factors as the number of days lost, the medical costs, and additional operational costs.

To determine which rates are high, baseline rates may be established. Frequently used baseline rates are the plantwide incidence and severity rates for work-related musculoskeletal injuries and illnesses or those for nonproduction personnel. Using these baselines, the analyst may determine the number of cases that could have been avoided in a specific group in relation to the baseline group.

The major benefits of reviewing available records for passive surveillance include (1) low cost, (2) the possibility of using a systematic approach to obtain and code the data, (3) the possibility of performing historical analyses, and (4) the possibility of providing a basis for comparing the health effects due to the lack of an adequate ergonomic design among departments, operations, and areas of the plant. Moreover, it provides a means for measuring and evaluating the impact of ergonomic interventions.

Reviews of available records at a worksite have been used for ergonomic surveillance to identify jobs with a high incidence of musculoskeletal problems. For example, workers' compensation claims data were found to be useful in locating high-risk operations causing CTD [13]. Plant medical records have also been used in retrospective studies to demonstrate associations between musculoskeletal disorders and occupational risk factors [14] and to demonstrate the effects of ergonomic intervention on the development of WMDs [15].

One major disadvantage of the passive surveillance approach is that records are not complete in the sense that minor injuries and discomfort are often ignored or underreported in health records. The nature of WMDs is such that initial symptoms can be somewhat vague and discomfort symptoms may develop gradually over a period of weeks, months, or years. Thus, recording a condition in its early stage depends on employees taking the first step to report symptoms. As a result, the analysis of the injury and illness incidence rates does not adequately reflect the relative risks of the operation [16–18]. To overcome this problem, an alternative model of injury analysis was developed to predict subsequent major injury using work history and medical records [19].

Another disadvantage of analyzing health records retrospectively for passive surveillance is that the rates of recordable cases are not updated in a timely manner. Thus, it is not always possible to determine the present health status of the jobs. To determine whether the health status of a job was "under control," an injury-monitoring test was developed based on updating the incidence and severity rates in a timely manner [20].

Furthermore, the possibility of doing a retrospective study for passive surveillance depends on the availability and reliability of records on both the risk factor and the outcome factor. Striking differences in the plantwide incidence of WMDs were found depending on which source was examined (OSHA logs, workers' compensation claims, medical absences, and plant medical cases) [22]. This result may also be related to the lack of a standardized method for reporting WMDs or inconsistent diagnostic criteria. For example, a case of carpal tunnel syndrome may be reported as an industrial injury rather than as an illness [23].

Although passive surveillance is relatively low-cost, collecting information from records is very time-consuming and the data are subject to misclassification at the time they are recorded as well as when they are retrieved from the record. Furthermore, records used for passive surveillance rarely have exposure information such as the tools used, the employee's department and job, the length of time on the same job, overtime hours worked, previous jobs, or additional jobs. Thus, it is necessary to pursue this exposure information by other means.

In general, the main disadvantage of passive surveillance is that even when records are complete, consistent, and available for review, they only reflect past occurrences of health problems and may not give an accurate picture of the present situation.

## 20.4 ACTIVE SURVEILLANCE OF WORK-RELATED MUSCULOSKELETAL DISORDERS

### 20.4.1 Traditional Pencil-and-Paper Surveys

Active surveillance involves looking first at causes and then at effects by actively seeking and collecting relevant information for a specific purpose and in a standardized manner. In an ergonomic program, the prevention of WMDs is a main goal. Active surveillance may offer the means to identify musculoskeletal problems early and thus prevent more serious disorders. Active surveillance involves the design of a data collection system and the collection of the necessary information from the workforce. One system commonly used in active surveillance for WMDs uses discomfort symptoms questionnaires or pencil-and-paper surveys, also referred to as "shop floor surveillance" [21], that are completed by all employees in a particular problem area or department or in the entire plant. This traditional survey method allows for active investigation of health problems by obtaining information directly from the employees about their current discomfort symptoms.

Questionnaires used for active surveillance may include the following information:

1. Location of discomfort
2. Onset of discomfort
3. Severity of discomfort
4. Frequency of discomfort episodes
5. Duration of discomfort episodes
6. Employee perception of what aggravates discomfort
7. Job-related information (e.g., shift, job title, time on the job)

OSHA [22] has included in the *Ergonomics Program Management Guidelines for Meatpacking Plants* an example of a symptoms survey for an ergonomics program developed by Silverstein [23].

Some questions used for WMD surveillance use anatomical discomfort charts depicting the anterior view, the posterior view, or both views of the human body to identify the location of discomfort. The location of discomfort is indicated by shading the affected areas. This information may be useful for identifying the elements of the job that may be causing discomfort in specific body areas.

Also, some questionnaires used for WMD surveillance may request an indication of the severity of discomfort in the affected areas. This information may be useful to (1) identify employees who need referral for medical supervision or, where needed, timely initiation of treatment, (2) identify employees particularly vulnerable to a certain health hazard, (3) evaluate the effectiveness of medical interventions and ergonomic interventions on the severity of discomfort, and (4) identify trends in discomfort severity with respect to time. The severity of discomfort is usually assessed in the surveys by means of rating scales.

Pain intensity levels have been assessed with a variety of rating scales [24]. Rating scales are particularly useful because they can be used quickly and easily with minimal instruction. There are two popular types of scales: visual analog scales and numerical or verbal scales. Visual analog scales are straight lines with ends defined by the extreme limits of the sensation or response to be measured. Verbal or numerical scales are straight lines with descriptive terms placed at intervals along the line. Numerous studies have shown that pain or discomfort can be mapped and rated using various scales [25–34].

Several investigators have adapted a method developed by Corlett and Bishop [35] that combines a rating scale and a pictorial presentation of body diagrams to assess the relationship between discomfort and postural loading at the joints [28,30,36], to assess the relationship between discomfort and working postures [8,37,38], and to assess the incidence and severity of musculoskeletal disorders in the workforce [39,40].

### 20.4.2 Real-Time Computerized Surveys

#### *20.4.2.1 Computerized Data Collection Systems*

Automated health data collection from patients is not a new idea. The first reported computerized questionnaire was developed in 1966 [41] to elicit information on allergy symptoms from patients. This initial computerized interviewing system consisted of a small general-purpose digital computer with a keyboard with four special function keys. Questions were displayed to patients on a small cathode ray oscilloscope. The patient would respond to questions by pressing one of the four special keys. This first computerized questionnaire established three important points (1) that computers could obtain information from patients, (2) that computers could do this in a manner that was well received by patients, and (3) that patients could interact with a computer "interviewer" without any prior training or knowledge of computing.

Other investigators have developed and evaluated computerized questionnaires for health data collection [42–46]. Among the computerized questionnaires for health data collection reviewed, there were two that differed from the others in that they made use of a graphic display terminal and an electronic light pen to select a rectangular response area on the screen. These are the automated medical history (AMH) [46] and a computer-based interview system for patients with back pain [47]. These two systems incorporated advanced computer technologies such as graphics and a penlike input device to facilitate the interaction between the computer and the user.

Questions in general are routinely used to collect data that are usually graded by computers. Traditional computerized questionnaires have required patients to input information directly to the computer, thus reducing the time spent coding and entering the data for grading. However, these computerized questionnaires required patients to interact with the computer only once, to report health information. Examples where patients routinely enter health data directly into a computer have not been reported so far. The availability of personal computers now makes this possible.

#### *20.4.2.2 Discomfort Assessment System*

The Discomfort Assessment System (DAS) is a computerized system developed for the collection of perceived musculoskeletal discomfort [48,49]. This system allows for real-time active surveillance of musculoskeletal discomfort data. It consists of a computerized questionnaire operated directly by the worker. The system has a graphical interface that makes extensive use of diagrams, windows, buttons, and menus to display information on the screen. There is no keyboard in this system.

The interface simply consists of a series of screens and a light pen. There are screens that request the patient identification number, job title, shift number, and length of time in the job, screens that request the location of discomfort, and screens that request the severity of discomfort in each area affected. When items on the screen (buttons, menu alternatives, body areas, etc.) are selected with the light pen they become highlighted.

DAS displays the anterior and posterior views of the human body with various body areas delineated to facilitate the recognition and selection of affected areas. The posterior view of the human body was adapted from the Nordic Questionnaire [50]. When a user selects a body area by pointing with the light pen, the area becomes highlighted.

The severity of discomfort is indicated by pointing at a desired level on numerical a scale with verbal anchors (a modification of Borg's scale [25]) that looks like a thermometer. The scale is vertical, and the numbers range from 0 to 10 with the words "nothing at all" [26] and "worst imaginable" [34] used as anchors at 0 and 10, respectively. When a level on the scale is selected, the "mercury" fills up to the selected level. Figure 20.1 shows the screen that requests the severity of discomfort displaying some information entered during an interaction with the system.

#### *20.4.2.3 Case Study*

A musculoskeletal discomfort survey in Michigan demonstrated the potential capability of DAS for collecting discomfort symptoms information directly from the workforce [51]. In one study, musculoskeletal discomfort data were collected in a 5 day study of 11 rural mail carriers using DAS. Participants were asked to submit five reports, one at the beginning of the day and each of the remaining four at the end of a specific task during their workday. It was observed that different tasks required different muscle groups. The purpose of requesting a discomfort report after completing specific subtasks was to investigate the location of discomforts resulting from these and to determine the patterns of discomfort that occurred with time.

The data revealed that more than 50% of the participants were experiencing musculoskeletal discomfort in one or more anatomical areas (neck, shoulders, elbows, hands, back, and legs). This result was consistent with the high incidence rate of musculoskeletal disorders among rural mail carriers. The discomfort data revealed differences in the frequency of discomfort reports between body areas depending on the task preceding the report; this indicated that reported discomfort was task-related and not random. This result showed that data from a discomfort survey may be useful for identifying job hazards that may be stressing a particular body area in a significant portion of the employees. Finally, discomfort data collected from this case study were used to compare the patterns of discomfort

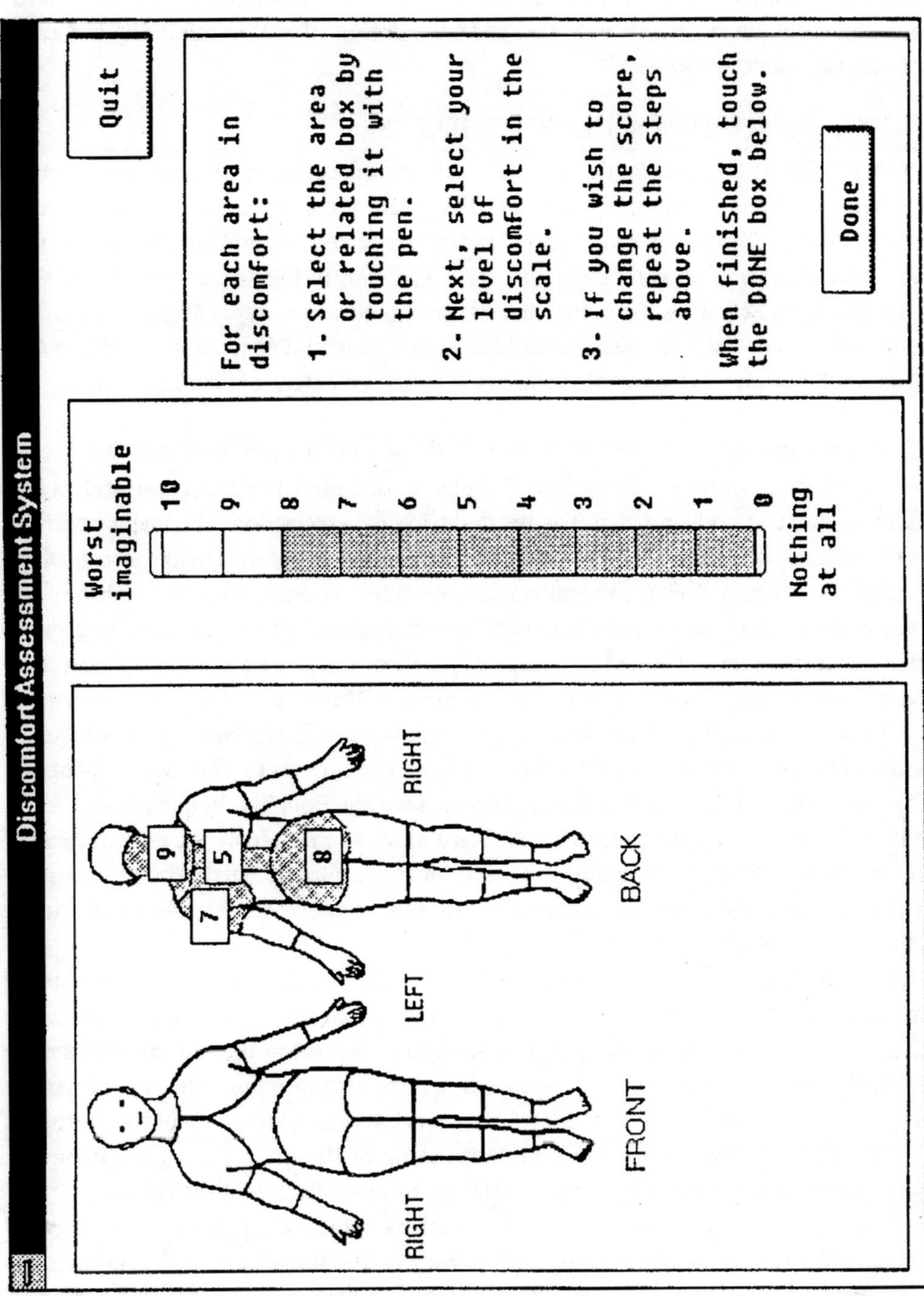


FIGURE 20.1 Example of a discomfort severity screen captured during an interaction with DAS.

severity reported by a healthy individual, an individual with carpal tunnel syndrome, and an individual who had undergone surgery for carpal tunnel syndrome. The results of the data analysis revealed significant patterns in discomfort severity among these three individuals. Thus active surveillance may also provide a means of evaluating individual cases and monitoring their progress.

## 20.5 CRITICAL REVIEW OF CURRENT STATUS

Complete elimination of ergonomic hazards in the workplace may not be a feasible alternative in some jobs. Thus ergonomic programs will operate as control mechanisms to identify musculoskeletal problems early enough to prevent more serious problems from developing. The effectiveness of an early detection system in identifying and controlling WMDs depends to a large extent on the completeness and timeliness of reporting and recording a person's symptoms data. At present, the best technique available to control WMDs uses active surveillance [21]. Yet active surveillance is in an early stage of development.

### 20.5.1 Advantages and Disadvantages of Traditional Pencil-and-Paper Surveys

Active surveillance allows the investigator to plan and control the collection of data and therefore be aware of problems that may result during data collection. The major benefit of active surveillance of the workforce is that it may be possible to obtain a better picture of the current health situation. This is so because active surveillance reaches out to employees and elicits information from them. Traditional survey data collection methods are advantageous because they can be administered to a large population and can be used to assess a large variety of health-related factors. Furthermore, reporting biases related to the definitions of different musculoskeletal disorders and requirements of specific records may be reduced.

Discomfort symptoms information obtained from the surveys can be used to identify a job or a department that needs further ergonomic investigation. The frequency and severity of discomfort symptoms can also be used as a baseline to evaluate the effectiveness of ergonomic interventions in reducing the discomfort symptoms. Furthermore, measurement of discomfort from time to time yields discomfort series (time series) that can be investigated for predictability [52].

The success of an active surveillance program also depends on the time taken to respond to the results of the survey. If feedback or interventions are delayed, then some conditions may become severe and employees may lose interest in reporting their discomfort symptoms. Furthermore, surveys rely on employees' willingness to report their health conditions to coworkers or supervisors. The level of awareness of employees about the objectives of the surveillance program and their understanding of the impact of their discomfort symptoms reports are extremely important for the success of active surveillance.

Traditional survey data collection methods such as pencil-and-paper surveys or questionnaires have several limitations associated with active surveillance. The administration of a discomfort symptoms survey may be costly because it requires trained personnel for the collection, interpretation, and coding of the data. On the other hand, if personnel involved in active surveillance are not properly trained, the effectiveness of the program will be reduced. Furthermore, surveys also rely on the employees' recall or recognition of their health conditions. However, surveys are conducted sporadically and, because in the early stages of WMDs many of the symptoms are transient and occur mostly at night, the employee may not recall or recognize the significance of symptoms during the survey.

Nevertheless, a study [53] showed that a pencil-and-paper questionnaire used in addition to physical examination data was as sensitive as an excellent plant medical department in detecting upper extremity musculoskeletal disorders. Other investigators [23,54,55] have also used pencil-and-paper questionnaires to assess employees' musculoskeletal discomforts.

Furthermore, when the purpose of the surveillance program is to detect early cases and some "false positive" are allowed, then pencil-and-paper questionnaires administered periodically to a random sample of employees to collect information about potential confounders as well as symptoms and exposure histories would provide sufficient information for an early warning system [56]. However, health surveillance information gathered at prescheduled intervals by means of questionnaires may not present a realistic picture of the number of cases because workers are exposed to job stresses that may change on a day-to-day basis, thus incorporating reporting biases [57].

### 20.5.2 Advantages and Disadvantages of Real-Time Computerized Surveys

The availability of personal computers and the rapid advances in computer technologies make the use of computers possible now for developing real-time computerized surveys like DAS. Automated surveys are an alternative to traditional methods involving pencil-and-paper surveys. Computerized surveys have the same advantages as the traditional surveys for active surveillance of WMDs. However, real-time computerized surveys like DAS may gather day-to-day information, reducing reporting biases and presenting a more realistic picture of musculoskeletal problems. Real-time computerized surveys may overcome some of the limitations of traditional pencil-and-paper surveys. For example, they gather information in a timely manner, thus allowing for the immediate investigation of the patterns of health and disease among groups of employees or within one individual.

Furthermore, real-time computerized surveys eliminate the need for trained personnel to administer the questionnaire and interpret and code the data. Thus, in that sense they may be less costly. Because the data are automatically coded, they are readily available for grading, which accelerates the intervention process.

Nevertheless, such surveys also have several shortcomings. First, they require computer hardware. This full implementation of a real-time computerized survey system in a large facility may be costly. Second, similar to traditional pencil-and-paper surveys, such a survey relies on a person's self-assessment of discomfort and willingness to report discomfort symptoms, in this case to a computer instead of to a coworker or supervisor. Third, computerized surveys need further testing in the field for validation.

Recent advances in technology have brought forth a competitive market of computer hardware and software. Thus, the implementation of a real-time computerized survey system for active surveillance of WMDs may be economically feasible for many companies. Furthermore, results obtained in 1966 [41] demonstrated that people were capable of interacting with a computer system without any prior training or knowledge of computers. With the proliferation of computers over the intervening years, that conclusion is even more valid today.

## 20.6 FUTURE CONCERNS

The effectiveness of an early detection system to identify and control WMDs depends to a large extent on the completeness and timeliness of reporting and recording symptoms data. Real-time computerized systems that take advantage of new computer technologies should be developed for the collection of data in active surveillance programs. A real-time active surveillance system of WMDs should allow an employee to report discomfort symptoms at any time. That is to say, employees should have access to a computer terminal to input a report any time they experience discomfort. Furthermore, the discomfort reports database should automatically update itself after each report, thus providing real-time analysis of the data.

Because real-time computerized surveys depend on the employee's self-assessment of discomfort and willingness to report, employees should receive education on WMD prevention, with emphasis on early recognition of symptoms as well as on the importance of an early report to aid in control of the problem.

Furthermore, new methodologies should be developed for the analysis of the data collected that will indicate whether the patterns of discomfort symptoms are "in control" and provide warnings when the health status is "out of control" [48].

## REFERENCES

1. U.S. Department of Labor, Occupational Safety and Health Administration, *Ergonomic Safety and Health Management*, Washington, DC, Proposed Rule 57 FR 34192-34200, 1992.
2. U.S. Department of Labor, Bureau of Labor Statistics, *BLS Report on the Survey of Occupational Injuries and Illnesses in 1991*, Washington, DC, USDL Pub. No. 92-731, 1991.
3. R. E. Yodaiken, Surveillance, monitoring, and regulatory concerns, *J. Occup. Med. 28*(8): 569–571 (1986).
4. D. K. Parkinson and M. J. Grennan, Establishment of medical surveillance in industry: Problems and procedures, *J. Occup. Med. 28*(8):772–777 (1986).
5. B. W. Mintz, Medical surveillance of employees under the Occupational Safety and Health Administration, *J. Occup. Med. 28*(10):913–919 (1986).
6. J. D. Millar, Summary of proposed national strategies for the prevention of leading work-related diseases and injuries, Part 1, *Am. J. Ind. Med. 13*:223–240 (1988).
7. V. Putz-Anderson, *Cumulative Trauma Disorders: A Manual for Musculoskeletal Diseases of the Upper Limbs*, Taylor & Francis, New York, 1988.
8. I. Kuorinka, Subjective discomfort in a simulated repetitive task, *Ergonomics 26*:1089–1101 (1983).
9. R. Cailliet, *Soft Tissue Pain and Disability*, F. A. Davis, Philadelphia, PA, 1988.
10. U.S. Department of Health and Human Services, *Centers for Disease Control*, National Institute for Occupational Safety and Health, Atlanta, GA, DHHS (NIOSH) Pub. No. 89-129, 1989.
11. R. Hiltz, Fighting work-related injuries, *Natl. Underwriter 89*(13):15 (1985).
12. The Bureau of National Affairs, Inc., *Occupational Safety and Health Reporter*, Washington, DC, BNA Pub. No. 5-31-89, 1989, p. 2165.
13. S. Tanaka, P. Seligman, W. Halperin, M. Thun, C. L. Timbrook, and J. J. Wasil, Use of worker's compensation claims data for surveillance of cumulative trauma disorders, *J. Occup. Med. 30*(6):448–492 (1988).
14. R. H. Westgaard and A. Aaras, Postural muscle strain as a causal factor in the development of musculo-skeletal illnesses, *Appl. Ergon. 15*:162–174 (1984).
15. R. H. Westgaard and A. Aaras, The effect of improved workplace design on the development of musculo-skeletal illnesses, *Appl. Ergon. 16*:91–97 (1985).

16. B. A. Silverstein, L. J. Fine, and T. J. Armstrong, Occupational factors and carpal tunnel syndrome, *Am. J. Med. 11*:343–358 (1987).
17. T. J. Armstrong, L. J. Fine, S. A. Goldstein, Y. R. Lifshitz, and B. A. Silverstein, Ergonomic considerations in hand and wrist tendinitis, *J. Hand Surg. 12A*(5; Part 2):830–837 (1987).
18. W. Margolis and J. F. Kraus, The prevalence of carpal tunnel syndrome symptoms in female supermarket checkers, *J. Occup. Med. 29*(12):953–956 (1987).
19. M. K. Chung, S. H. Wu, and G. D. Herrin, The use of a mixed Weibull model in occupational injury analysis, *J. Occup. Accid. 7*:239–250 (1986).
20. A. H. Boyd and G. D. Herrin, Monitoring industrial injuries: A case study, *J. Occup. Med. 30*(1):43–48 (1988).
21. B. A. Silverstein, Shop floor surveillance to identify and control work-related musculoskeletal disorders, *Presented at the Occupational Disorders of the Upper Extremities Course*, March 29–30, 1990, Ann Arbor, MI.
22. U.S. Department of Labor, Occupational Safety and Health Administration, *Ergonomic Program Management Guidelines for Meatpacking Plants*, Washington, DC, OSHA 3123, 1991 (reprinted).
23. B. A. Silverstein, The prevalence of upper extremity cumulative trauma disorders in industry, Doctoral Dissertation, Univ. Michigan, Ann Arbor, MI, 1985.
24. C. R. Chapman, K. L. Casey, R. Dubner, L. M. Foley, R. H. Gracely, and A. E. Reading, Pain measurement: An overview, *Pain 22*:1–31 (1985).
25. G. Borg, A category scale with ratio properties for intermodal and interindividual comparisons, in *Psychophysical Judgment and the Process of Perception*, H. G. Geisler and P. Petzold, Eds., VEB Deutscher Verlag der Wissenschaften, Berlin, Germany, 1982, pp. 25–34.
26. F. Gaston-Johansson, Pain assessment: Differences in quality and intensity of the words pain, ache and hurt, *Pain 20*:69–76 (1984).
27. K. Harms-Ringdahl, H. Brodin, L. Eklund, and G. Borg, Discomfort and pain from loaded passive joint structures, *Scand. J. Rehab. Med. 15*:205–211 (1983).
28. K. Harms-Rindgahl, On assessment of shoulder exercise and load-elicited pain in the cervical spine, *Scand. J. Rehab. Med. Suppl. 14*:1–40 (1986).
29. K. Harms-Ringdahl, A. M. Carlsson, J. Ekholm, A. Raustorp, T. Svensson, and H. Toresson, Pain assessment with different intensity scales in response to loading of joint structures, *Pain 27*:401–411 (1986).
30. K. Harms-Ringdahl and J. Ekholm, Intensity and character of pain and muscular activity levels elicited by maintained extreme flexion position of the lower cervical-upper-thoracic spine, *Scand. J. Rehab. Med. 18*:117–126 (1986).
31. R. Melzack, The McGill pain questionnaire: Major properties and scoring methods, *Pain 1*:277–299 (1975).
32. D. D. Price, P. A. McGrath, A. Raffi, and B. Buckingham, The validation of visual analogue scales as ratio scale measures for chronic and experimental pain, *Pain 17*:45–56 (1983).
33. J. Scott and E. C. Huskisson, Graphic representation of pain, *Pain 2*:175–184 (1976).
34. R. A. Seymour, J. M. Simpson, J. Charlton, and M. E. Phillips, An evaluation of length and end-phrase of visual analogue scales in dental pain, *Pain 21*:177–185 (1985).
35. E. N. Corlett and R. P. Bishop, A technique for assessing postural discomfort, *Ergonomics 19*:175–182 (1976).
36. M. Boussenna, E. N. Corlett, and S. T. Pheasant, The relation between discomfort and postural loading at the joints, *Ergonomics 25*:315–322 (1982).
37. V. Bhatnager, C. G. Drury, and S. G. Schiro, Posture, postural discomfort, and performance, *Hum. Factors 27*:189–199 (1985).
38. S. F. Wiker, Effects of relative hand location upon human movement time and fatigue, Doctoral Dissertation, Univ. Michigan, Ann Arbor, MI, 1986.

39. P. W. Buckle, D. A. Stubbs, and D. Baty, Musculo-skeletal disorders (and discomfort) and associated work factors, in *The Ergonomics of Working Postures*, N. Corlett, J. Wilson, and I. Manenica, Eds., Taylor & Francis, London, U.K., 1984, pp. 19–30.
40. B. A. Silverstein and L. J. Fine, *Evaluation of Upper Extremity and Low Back Cumulative Trauma Disorders: A Screening Manual*, Tech. Rep., Univ. Michigan School of Public Health, Environmental and Industrial Health Department, Ann Arbor, MI, 1984.
41. W. V. Slack, G. P. Hicks, C. E. Reed, and L. J. Van Cura, A computer based medical history system, *N. Engl. J. Med. 274*(4):194–198 (1966).
42. W. I. Card, M. Nicholson, G. P. Crean, G. Watkinson, C. R. Evans, J. W. Wilson, and D. Russell, A comparison of doctor and computer interrogation of patients, *Int. J. Biomed. Comput. 5*:175–187 (1974).
43. J. Diaz, O. M. Z. Miranda, A. Faundes, and J. A. Pinotti, Preliminary experiment with computerized anamnesis in gynecology and reproductive health, *Int. J. Gyn. Obstet. 24*:285–290 (1986).
44. W. E. Hammond and W. W. Stead, The evolution of computerized medical information system, *Proc. Tenth Annual Symp. Comput. App. Med. Care 10*:147–156 (1986), Computer Society of the IEEE, Washington, DC.
45. R. W. Lucas, W. I. Card, R. P. Knill-Jones, G. Watkinson, and G. P. Crean, Computer interrogation of patients, *Br. Med. J. 2*:623–625 (1976).
46. J. G. Mayne, W. Weksel, and P. N. Sholtz, Toward automating the medical history, *Mayo Clinic Proc. 43*:1–25 (1968).
47. A. M. C. Thomas, J. T. C. Fairbank, P. B. Pynsent, and D. J. Baker, A computer-based interview system for patients with back pain, *Spine 14*:844–846 (1989).
48. N. Saldaña, Design and evaluation of a computer system operated by the workforce for the collection of perceived musculoskeletal discomfort: A tool for surveillance, Doctoral Dissertation, Univ. Michigan, Ann Arbor, MI, 1991.
49. N. Saldaña, DAS: A graphical computer tool for the collection of musculoskeletal discomfort information from the workforce, *Comput. Ind. Eng. 23*:215–218 (1992).
50. I. Kuorinka, B. Jonsson, A. Kilbom, H. Vinterberg, F. Biering-Sorensen, G. Andersson, and K. Jorgensen, Standardised Nordic questionnaires for the analysis of musculoskeletal symptoms, *Appl. Ergon. 18*(3):233–237 (1987).
51. N. Saldaña, G. D. Herrin, T. Armstrong, and A. Franzblau, A computerized method for assessment of musculoskeletal discomfort in the workforce: A tool for surveillance, *Ergonomics 37*:1097–1112 (1994).
52. G. Affleck, H. Tennen, S. Urrows, and P. Higgins, Individual differences in the day-to-day experience of chronic pain: A prospective daily study of rheumatoid arthritis patients, *Health Psychol. 10*:419–426 (1991).
53. L. J. Fine, B. A. Silverstein, T. J. Armstrong, C. A. Anderson, and D. S. Sugano, Detection of cumulative trauma disorders of upper extremities in the workplace, *J. Occup. Med. 28*(2):674–678 (1986).
54. B. A. Silverstein, T. J. Armstrong, A. Longmate, and D. Woody, Can in-plant exercise control musculoskeletal symptoms? *J. Occup. Med. 30*(12):922–927 (1988).
55. S. L. Sauter and L. M. Schleifer, Work posture, workstation design, and musculoskeletal discomfort in a VDT data entry task, *Hum. Factors 33*(2):151–167 (1991).
56. B. A. Silverstein, Patterns of cumulative trauma disorders in industry, *Presented at the Engineering Summer Conferences: Occupational Ergonomics Course*, June 15–19, 1987, Ann Arbor, MI.
57. J. D. McGlothlin, An ergonomic program to control work-related cumulative trauma disorders of the upper extremities, Doctoral Dissertation, Univ. Michigan, Ann Arbor, MI, 1988, p. 119.

# IV

# Ergonomic Case Studies

CHAPTER 21

# Development and Implementation of an Ergonomics Process in the Automotive Industry

## *Reactive and Proactive Processes*

Bradley S. Joseph and Glenn Jimmerson

CONTENTS

## 21.1 INTRODUCTION

Ergonomics examines the interaction between the worker and the work environment including such factors as machinery, the workstation, and climate. If the match between worker and work environment is poor, the worker's ability to perform the job will be severely compromised. Over a short period of time, this poor match may lead to fatigue and worker discomfort. If conditions persist, physical injury and disability may result.

## 21.2 BACKGROUND AND SIGNIFICANCE TO OCCUPATIONAL ERGONOMICS

Large industrial firms are struggling to incorporate occupational ergonomics into their operating practices. Pressure from outside (pending legislation from the federal government) and inside (aging workforce, increasing costs of medical care, pressure to become more efficient through productivity gains) are forcing manufacturing companies to use ergonomics information. To accomplish this task, manufacturing must develop a model to begin incorporating ergonomics into general operating procedures. Kvalseth's [1] research indicated that several factors were important in implementing an ergonomics program. They include: management's perception for the need of ergonomics, management knowledge of the benefits or ergonomics, and the degree of cooperation between the workers and management. These factors show how important it is to "walk" a plant before attempting to implement the program. This chapter helps develop a plan that can be used for a general ergonomics process. It has two components—reactive and proactive processes.

To begin to develop this model, it is important to develop a clear and concise working definition of ergonomics. Ergonomics literally means the study of work. Since this definition can encompass a wide range of issues, manufacturing companies need to narrow the scope of occupational ergonomics. In the case of Ford Motor Company, occupational ergonomics means fitting jobs to people.

It is important to note that even though the process described in this chapter was used in a large manufacturing company, the concept and lesson can be used in most manufacturing industries large or small.

## 21.3 DESCRIPTION OF REACTIVE AND PROACTIVE ERGONOMICS PROCESSES AT FORD MOTOR COMPANY

### 21.3.1 History of Ergonomics at Ford Motor Company

Practical ergonomics has always been a consideration at Ford Motor Company. In the past, ergonomics analysis concentrated on the interaction between vehicle design and customer satisfaction. Designers incorporate ergonomics principles when determining reach distances to critical controls, display and display layout, seat dimension, and interior volume (leg room, shoulder room, etc.). Then in the late 1970s and early 1980s, it became apparent that these same principles should be applied when designing a vehicle for ease of manufacture and assembly. Industrial engineering began to focus more on human capability rather than on just the time required to complete a task, and ergonomics became one of several issues. Today, ergonomic principles are being considered for serviceability issues.

Several years ago the need emerged for a more focused effort in ergonomics of manufacturing. Special programs were undertaken in several divisions of Ford Motor Company. For example, in 1982 body and assembly operations (B&AO) began a long-term agreement with The University of Michigan Center for Ergonomics. Several goals were outlined, including the following:

- Establish a university organization to facilitate and develop ergonomics in the manufacturing and assembly process.
- Develop an education program to disseminate ergonomics principles to engineers, managers, supervisors, and suppliers.
- Coordinate research efforts to improve engineering specifications and guidelines for use in the design and evaluation of existing and future B&AO.
- Provide consultation services as requested by B&AO personnel.

This activity ran continually for 6 years (1982–1988) and involved a number of faculty members, research assistants, and graduate students. Even though the direct benefit of this relationship was difficult to measure, the relationship had a positive impact for both Ford Motor Company and the university. Following is a list of outcomes from the project:

- Eighty technical reports and scientific papers.
- Three videotapes—ranging from an overview of low back pain and upper extremity cumulative trauma disorders to in-plant studies.
- A 250-page ergonomics technical manual.
- A guide to developing and implementing in-plant ergonomics programs. (The guide and manual are incorporated into the UAW/Ford Ergonomics Process.)
- Fifteen different training courses attended by over 500 Ford personnel from 25 B&AO plants in the last year.

It is particularly important to note that even though over 500 persons within the company were trained, ergonomics did not become an integral part of the operating culture. Even though there were instances when engineers applied ergonomics principles, in most cases, they did not apply them as a general practice. There are many explanations for this, but the simplest and most correct is time—the engineers had many things to do to complete their jobs and less and less time to do them. Adding ergonomics evaluation was something they did not want or recognize as a priority issue.

During the course of this university/Ford relationship, it became apparent that if ergonomics was to become truly part of the operating culture at Ford, the company would have to attack it from within. Several pilot projects with The University of Michigan, inside and outside body and assembly division, were begun to study better ways to implement a program that would be owned by Ford Motor Company. Most of these projects were jointly sponsored with the United Auto Workers Union. These projects include the following:

1983—The Milan and Saline Plastics Plant. A joint UAW–Ford in-plant ergonomics program was initiated in cooperation with The University of Michigan [1].

1984—The Van Dyke Chassis Plant. Research efforts concentrated on organizational issues related to a joint labor-management ergonomics team.

These and other programs at Ford demonstrated the benefits resulting from the use of ergonomics principles on the plant floor. They have yielded real job improvements as well as valuable scientific information. In 1987, UAW–Ford contract negotiations resulted in an agreement to expend significant effort in ergonomics. In essence, the collective bargaining agreement states that "the Company believes that applying ergonomics principles to the work environment, if done properly, can lead to reduced injuries, improved quality, and greater productivity." To achieve these goals, The Union and the Company intended to increase awareness of ergonomics principles and practices by jointly developing and implementing and/or sponsoring education and training programs. More important, the agreement states that "each assembly and manufacturing unit of 125 or more employees may establish a Local Ergonomics Committee" [1].

At that point, the company began developing a formal action plan in ergonomics. The program involves a two-prong approach—reactive and proactive. The first prong was the development of the joint UAW–Ford Ergonomics Program. It is a program that reacts to existing concerns by reviewing problems jobs and redesigning existing equipment at the plant. The backbone of the program is the local ergonomics committee (LEC).

### 21.3.2 Development of the Reactive Program

Late in 1987, the National Joint Committee on Health and Safety initiated discussions on the best way to implement a reactive ergonomics process at Ford. Members of this committee identified three major requirements for success:

1. Ergonomics would be available to all levels of the plant.
2. Ergonomics would coexist with existing programs.
3. Ergonomics would be developed with a participative approach.

These discussions yielded a process that was jointly developed by the UAW and Ford with the cooperation of the University of Michigan and with the services of Maritz Communications Company. This process is discussed in detail in the book *The UAW–Ford Ergonomics Process,* which consists of two volumes—*The UAW–Ford Implementation Guide* [2] and *The UAW–Ford Job Improvement Guide* [3]. The process has three steps: process start-up, job improvement cycle, and long-term development (Figure 21.1). The *Implementation Guide* is an instruction manual that outlines the methodology to implement an ergonomics process in an industrial facility. It assumes that cooperation exists between labor and

Process start-up → Job improvement cycle → Long-term development

FIGURE 21.1 The UAW–Ford ergonomics process.

management and that a joint labor-management team will have the authority to identify, evaluate, develop, and implement ergonomics changes. The *Job Improvement Guide* is a manual to show LECs the detailed methodology to evaluate ergonomics issues and develop and implement solutions. Following is a discussion of the process.

#### *21.3.2.1 Process Start-Up*

Process start-up is the first stage of the Ergonomics Process. It has five substeps (see Figure 21.2). The first three are securing leadership commitment, selecting LEC members, and training. Securing leadership commitment is to be accomplished locally at each plant. The National Joint Committee on Health and Safety supports this task effort through the development of an *Ergonomics Process Leadership Commitment Videotape* and training for health and safety engineers and union representatives [4]. The video was distributed during an Ergonomics Process orientation seminar at the December 1988 UAW–Ford health and safety conference.

Members of the LEC received at least 4 days of professional training. In April and May 1989, 73 LECs (over 600 participants) went through a 4 day jointly developed introductory training course. The course had two goals:

1. To teach LEC members how to use the guide and execute the steps of this cycle
2. To teach LEC members an introductory level of technical ergonomics information to enable them to recognize ergonomics problems and develop solutions

#### *21.3.2.2 Job Improvement Cycle*

Figure 21.3 shows the substeps of the job improvement cycle. It is a six-step approach to practical ergonomics that includes methods for identifying priority jobs to fix, evaluating job Stressors, developing and implementing job improvements, and documenting and following up on individual projects. Since this cycle is so important, a *Job Improvement Guide* [3] and videotape were developed that detailed these steps [5].

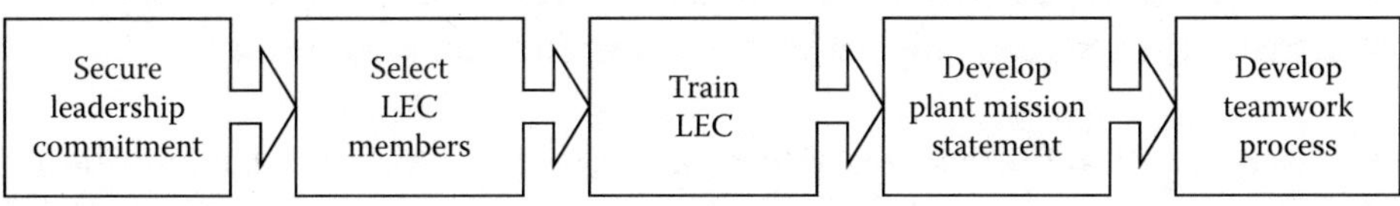


FIGURE 21.2 Process start-up.

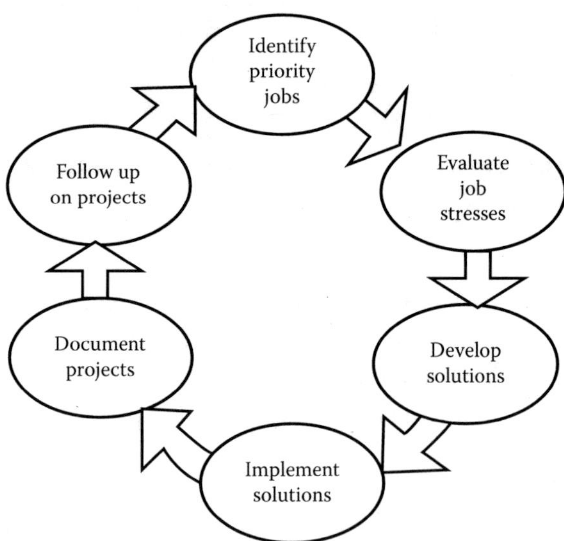


FIGURE 21.3 The job improvement cycle.

### *21.3.2.3 Long-Term Development*

No successful plant ergonomics program can survive without a long-term strategy and vision. Much thought has already been given to the long-term development of the UAW–Ford Ergonomics Process. Three areas of critical need have been identified:

1. *A need for specialized training*: As the LEC becomes more and more involved in the process, members will require advanced and specialized training. More and more expertise will be required from plant and division engineers. Consequently, a core of engineers will have to be trained in the UAW–Ford Ergonomics Process and in advanced basic ergonomics principles.
2. *A need to communicate ergonomics*: A main reason for bringing together the ergonomics committee is the multidisciplinary nature of the science. Successful implementation of ergonomics principles is both reactive (changes to existing jobs) and proactive (ergonomically designing new jobs). Successful activities in identifying and correcting poorly designed jobs must be communicated to the engineers and others who are responsible for developing new processes in the plant, so the same "mistakes" will not be repeated.
3. *A need to review constantly and improve the process*: As time goes on, many Ford sites will recognize the benefits of ergonomics. They will want to increase their plant's level of activity. The LEC can only perform so many functions at one time.

The current ergonomics efforts are reactive in nature; that is, the LECs will mainly be reviewing existing problems and redesigning existing equipment at the plant. The information generated by a LEC is isolated within its own plant. A more effective and cost-efficient approach is for the company to be proactive—for ergonomics principles to be applied during product research, design, and plant layout in order to prevent ergonomic Stressors. In this way, problems will be designed out of the product and process before reaching the operation phase. These efforts will help establish Ford Motor Company as the continued leader in ergonomics applications and research. However, it is essential to the success of these efforts that all sectors of product development, including research, design, and engineering, become an integral part of the ergonomics process.

### 21.3.3 Development of the Proactive Program

The proactive ergonomics process begins with an analysis of the way in which new equipment is engineered and installed in a facility. Figure 21.4 shows a summary of a typical process required for the installation and maintenance of industrial equipment. The process consists of a series of specialized functions performed by separate organizations or units. In general, the flow of information is from the corporation to the divisions to the plants. Units at the divisional level typically perform process study and design functions or phases; units at the plant perform implementation, operation, and maintenance phases. Often, these units are separated geographically and organizationally (e.g., division and plant), making it difficult to coordinate efforts.

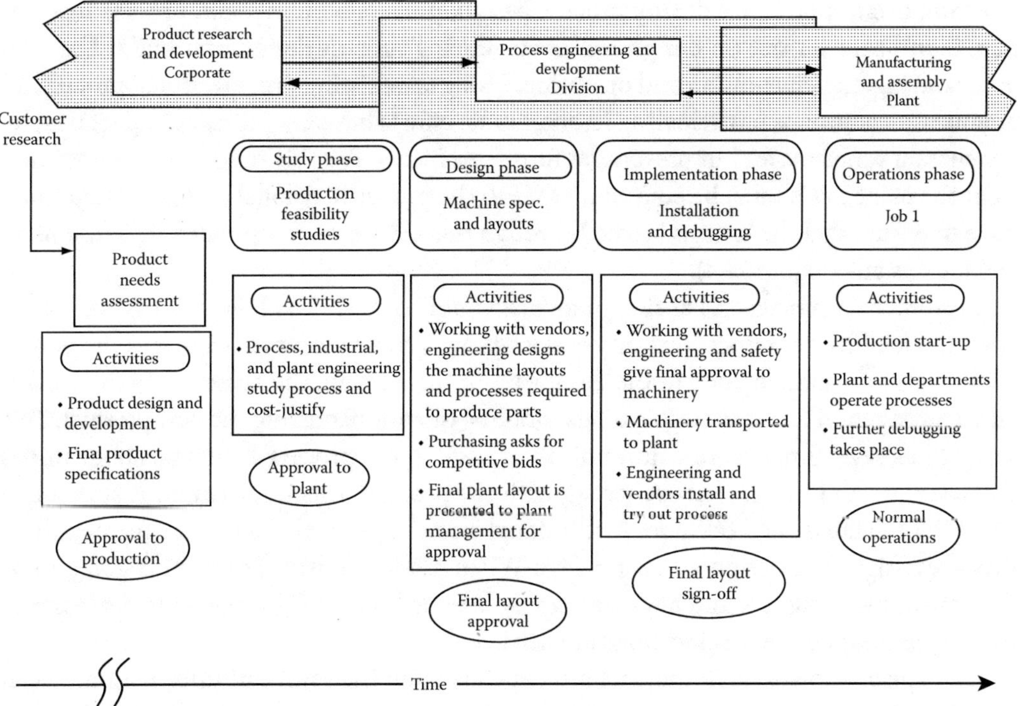


FIGURE 21.4 Product organizational and design process.

The phases differentiate periods of time in which groups of units have to complete a task before going to the next phase. First, new processes are studied (study phase) and designed (design phase) at division engineering. Little plant input is solicited during this phase. Next, division and plant engineering together install and debug machinery (implementation phase). This procedure involves a complex series of actions whereby process machinery plans are sent to selected vendors, the plans are interpreted, and the machinery is built to specifications, delivered, and installed in the plant using resources from the plant, vendor, and division. Unless the plant is willing to bear costly delays and excessive expenditures, few changes can be made on machinery, for ergonomic or other reasons, between the time the vendor builds the machinery and the time he delivers it. Also, the vendor has to sign off on machine specifications and is under contract to build it to standards agreed upon by the plant and division. Any changes to the design must be negotiated before the vendor can implement them. Consequently, the plant may wait until the machines are delivered and operating under those specifications (known as final sign-off) before changes are made. After debugging, normal operation and maintenance proceeds (operation phase). Often there is a need for process improvements or other redesign to update equipment (redesign phase). Depending on the cost, the plant usually controls these activities. However, due to limitations on cash and human resources, lost production from shutting down the machines, and other plant priorities, this activity is often limited and takes a considerable amount of time to complete [4].

There are several points in time when the ergonomics information can be applied. The most important points are during process design, process building, and process installation. Significant impact on design can occur during process design and build. Less can occur during implementation and operation. Using the existing organizational process, all organizational groups participating in ergonomic efforts have a significant impact in solving current problems and in preventing future problems. For this to happen, information from the plant LECs must be communicated to the proper divisional and corporate functions to ensure that the lessons learned through the LECs are incorporated into new plant and job designs (Figure 21.4).

This process is referred to as design for ergonomics (DFE). The DFE process was developed to look like a product warranty system in which the customer who uses the equipment feeds information back to the designers. In this system, the customer is the worker on the production line responsible for assembling or manufacturing the new product. This worker has several means of communicating back to the process designers about things that are right and things that are wrong with the process. This information is stored in a data bin or database and becomes institutional knowledge for the engineers the next time they are assigned to design a new process. When a new industrial process is going to be developed, the designers and engineers apply lessons learned from previous processes to the new process before specifications are finalized.

The type of knowledge in the data bin depends upon the source of information. Recall that the LECs are primarily responsible for identifying ergonomics issues and developing and implementing solutions in the workplace. The outcome from this process is a data file indicating where ergonomics issues occurred. These issues are recorded in the

Concern Log, and the details are recorded in a problem-solving record. At Ford this record is referred to as the 8-D form or eight-discipline problem-solving form.

*21.3.3.1 Implementing the Design for Ergonomics Process: Process Start-Up*

Implementing the DFE process is similar to implementing the reactive process. First leadership commitment must be secured. This usually occurs through a meeting with the project or program manager. During this meeting, the manager must be convinced of the necessity for ergonomics and the need for the engineers on the project to participate.

Second, the DFE team is selected. This team usually resembles a simultaneous engineering team consisting of facility, process, manufacturing, and industrial engineers; equipment vendors; product engineers; hourly representatives; and employee relations personnel. It is easy to see the reason for the engineering and vendors to participate. However, it is often difficult to explain the employee relations role in the process. Employee relations' primary role is to tell the group about the human side of the production process. It represents the voice of the workers (the voice of the customer) and may include injury/illness data, employee complaints, and absenteeism data.

Third, the DFE team must be trained. Recently, a new DFE training program was developed. During the training, teams are brought together about 2 years before the job is to be completed. They are trained in basic methods of ergonomics analysis and methods to integrate ergonomics into their timing plans. Finally, the team forms a mission statement.

*21.3.3.2 Implementing the Design for Ergonomics Process: Design Improvement Cycle*

The design improvement cycle is similar to the job improvement cycle except that the process is evaluated proactively during the study, design, and implementation phases. To accomplish this, the team must have a process to identify and evaluate and develop solutions to its problems. Figure 21.5 shows a detailed flowchart of the design improvement cycle.

The chart shows a large cycle beginning at the top left-hand corner and continuing counterclockwise. The design improvement cycle begins with product development. As product design is released, engineers begin to develop a process to build the product. These processes are generally developed in two parts—a functional flow and a functional allocation. The function flow is an operation-by-operation overview of the steps necessary to manufacture and assemble a product. A function allocation is a summary of the tasks necessary to complete each operation. Engineers are required to evaluate each operation to determine if the process is cost-effective and feasible.

These data are valuable to the proactive ergonomics process. Figure 21.6 shows a summary sheet from a function flow for a front-end chassis component of a small truck. Figure 21.7 shows a function allocation for one operation within the process. Reviewing these documents can help engineers make reliable decisions regarding potential ergonomics issues. For example, according to the function allocation (Figure 21.7), operators are required to lift two 17 lb parts to the machine 70 times an hour. These parts are going to be lifted from and to an overhead monorail.

These data can be used to compare or "benchmark" the process with other processes already being carried out within the plant. For example, spindles are a necessary

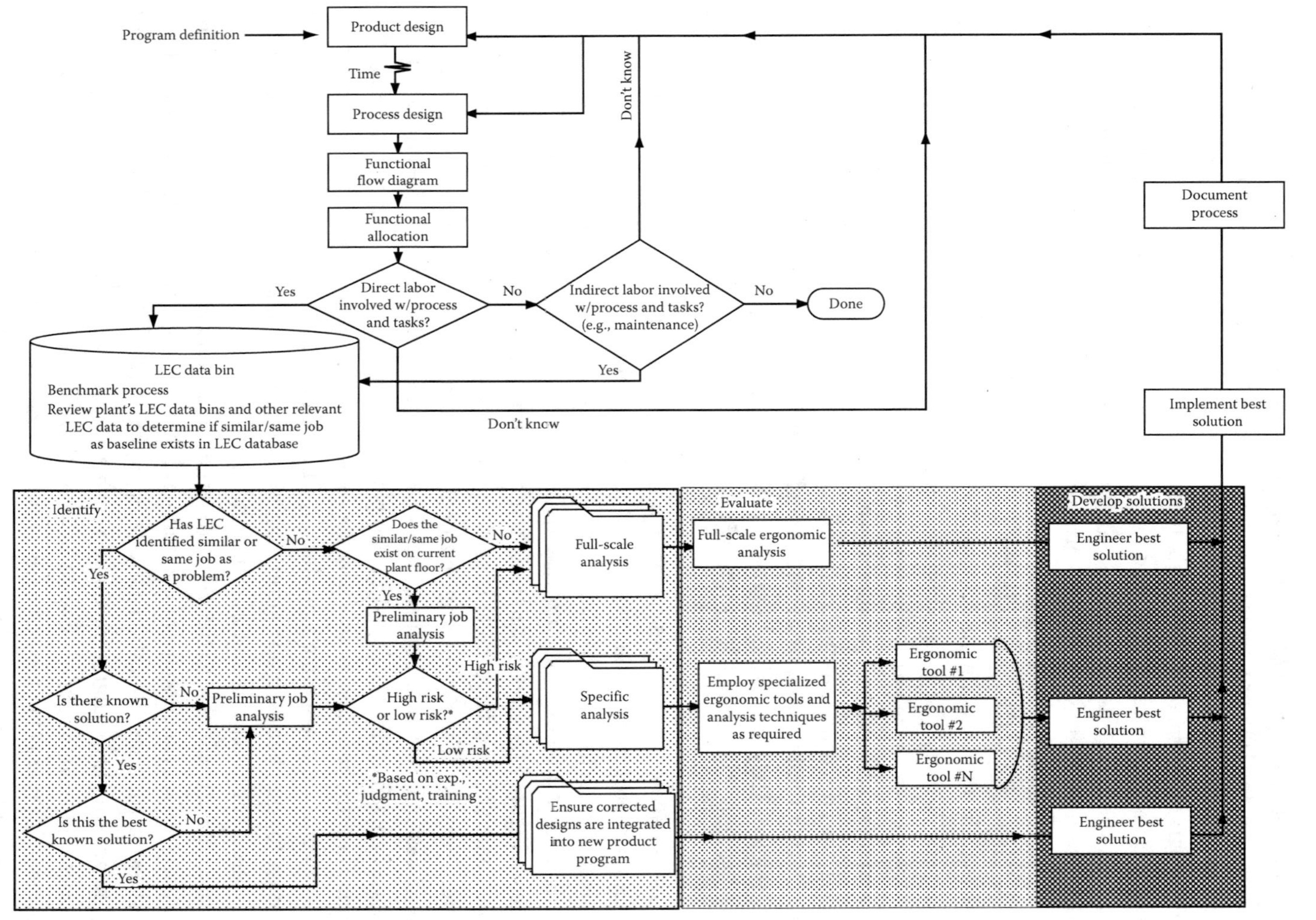


FIGURE 21.5 The design improvement cycle for process design and build.

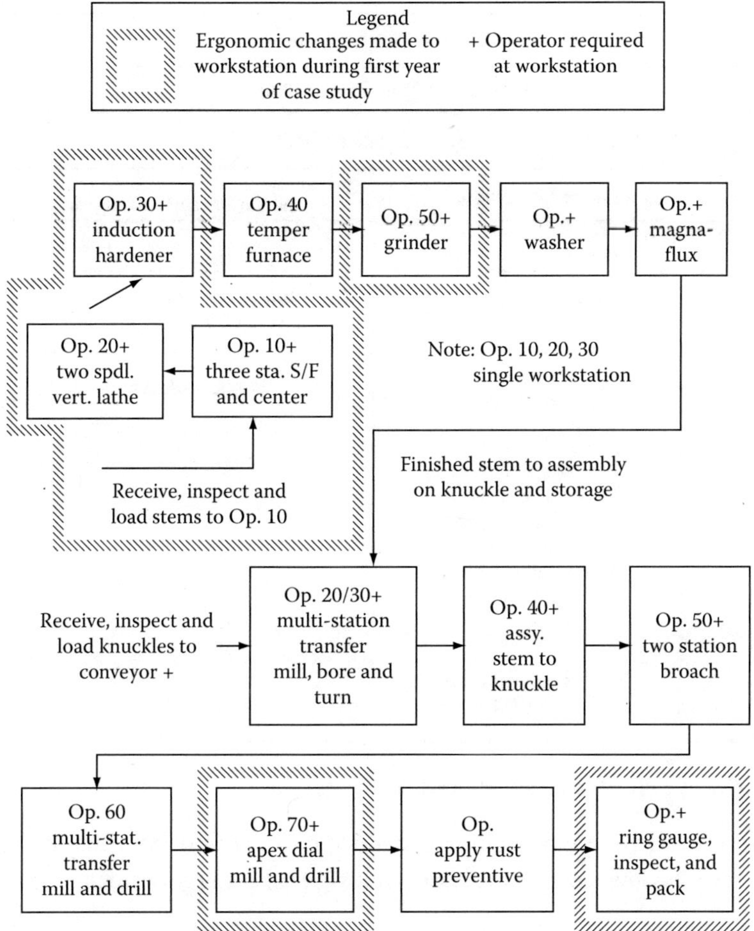


FIGURE 21.6 Functional flow of light truck spindle department.

component on almost every light truck vehicle. The LEC can review operations that are in production and determine if these are ergonomics issues. These data can be collected by the LEC through the 8-D process and given to the engineers. These data are referred to as the LEC data bin. If a similar allocation has an issue and the LEC has identified the problem, then the engineers should be informed of the issue before final specifications are made for the new process design. According to the flowchart and depending on whether a feasible solution is available, the engineer may be required to perform a complete or partial analysis or implement the solution already suggested by the LEC.

A complete analysis involves looking at four key activities of the process for potential risk—manual handling, assembly and/or disassembly, inspection, and machine operation. Various ergonomic analysis tools can be used to aid in the analysis. Solutions already developed by the LEC and proven to work can be specified directly to the vendor for the new process. It is important to note that, if possible, at least three separate ergonomics evaluations should occur. Recall that there are three phases during which an engineer can impact upon the design of a new process. Also recall that the first two phases (process design and

<table>
<tr><td colspan="3">PC/HR = (3916(2)/123) 1.09 = 70</td></tr>
<tr><td colspan="3">Projected weight of part = 17 LBS.</td></tr>
<tr><td colspan="3">Process estimate sheet</td></tr>
<tr><td colspan="2">Program or ERC number</td><td>Part name<br>Spindle and spindle assy. -frt.</td></tr>
<tr><td colspan="2">For models<br>Light truck</td><td>Material<br>Iron</td></tr>
<tr><td>Oper. no.</td><td>Operation description</td><td>Tool-machine-equipment description-tool or B.T. no.</td></tr>
<tr><td>50</td><td>Broach brake surfaces<br><br>Operator removes assy. from OHM, loads into RH fixture, cycles. LH fixture is unloaded and reloaded while RH ram is cycling<br><br>Reload part onto OHM</td><td>15 Ton 90-dual ram</td></tr>
</table>

FIGURE 21.7 Cover page of function allocation/process sheet.

process build) are where the most impact can occur. It is important that at each phase the responsible group review the specifications and sign off for ergonomics.

#### *21.3.3.3 Case Study: Chassis Components Plant, Spindle Line, Light Truck*

21.3.3.3.1 Plant Description and Layout The chassis components plant (Study Plant) was a captive automotive supplier for a major automotive concern in the United States. Located in Michigan, it occupies over 1.75 million sq. ft; 1.5 million sq. ft for floor space is devoted to the manufacturing processes necessary to make chassis components. At the time of the case study, the manufacturing area was divided into two separate production areas: the rear drive manufacturing and assembly area (areas A: employing approximately 520 hourly workers) and the front and rear suspension components area (area B: employing approximately 603 hourly workers). In addition, there was a plant central maintenance area exists (area C: employing about 200 hourly workers).

Area A manufactured and assembled most of the parts for rear drive axles. Castings were brought into the plant from outside suppliers and machined to specification. Gear, tubes, brackets, bearings, and castings are subassembled then transferred to the final assembly lines. Axles are assembled at the rate of 4000 per day. These axles varied in size and made for vehicles ranging from midsize cars to medium duty trucks. The finished parts are stored in racks and shipped to the auto/truck assembly plants.

Area B manufactured the components necessary for the suspensions of front and rear drive cars and light duty trucks. A stamping area forms suspension arms that house spring and shock assemblies. Some subassembly is required for the parts before they are packaged and shipped. A machining area machines spindles and knuckles from raw castings. These components are part of the front and rear suspensions and are responsible for steering, ride, handling, and holding the wheels to the car.

Area C existed to service, fabricate, and build equipment to operate machinery in both production areas. It consisted mostly of skilled trades, maintenance, cleaning, and fabrication. Its duties included building platforms and special equipment to assist in machine operations and tool changes, submitting proposals for the redesign and modernization of existing machinery to improve efficiency, and maintaining equipment to ensure high quality and lowest possible downtime.

This case study shows three workstations that were altered during the build and implementation phase of the DFE process in area B of the production facility. Area management anticipated that this approach would have two main effects. First, correction of problems before a process was at full production would increase the effectiveness of the program. However, for this method to be most efficient, problems on the jobs would have to be anticipated before they were in full production. Such anticipation would involve job simulation and still risk not fully experiencing all potential problems. Second, making changes to existing jobs is a drain on area resources. As the process was new, the division responsible for installing the equipment was also responsible for any changes to the equipment until final sign-off at the plant. Therefore, the use of divisional resources in conjunction with plant personnel to make the changes would minimize the drain on area B resources.

Area B combined the resources in one large DFE group because combining the resources would eliminate one communication link that might create unnecessary overhead.

21.3.3.3.2 Operations 10, 20, 30 of Stem Line Operations 10, 20, 30 are all in one workstation, and the redesign reflects changes made to all three operations. In general, there was no risk of low back injury for any job because the parts weighed well under 5 lb and all handling was done at or above waist height. However, workplace characteristics existed that could lead to stressful postures of the upper extremity, and modifications were made to correct them (see Figure 21.8, * marking), including installation of an adjustable platform and automatic doors on the lathe, redesign of load chutes, and repositioning of palm buttons. In addition, the front edge of the lathe was rounded to reduce the risk of striking the elbows when loading and unloading parts. These changes reduced the amount of upper extremity stress to the operator.

21.3.3.3.3 Operation 50 of Stem Line On operation 50 of the stem line, the major area of ergonomics Stressors was again the upper extremities. In particular, the gauging portion of the job was quite stressful to the wrist and hands of the operator because the task required rotation of the stem in the pot gauge. The spindle group reduced the magnitude of the force and virtually eliminated the stressful postures by designing a tool to turn the stem in the gauge.

Also, to facilitate using the tool, the vendor redesigned the pot gauge (see Figure 21.9, * marking). Finally, to improve the general workplace characteristics of the job, an amplifier was lowered to aid the operator in quality control.

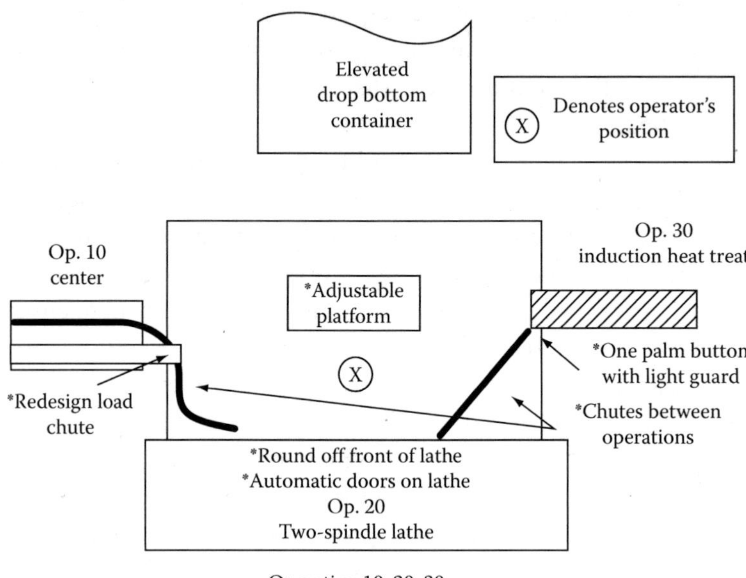


FIGURE 21.8 Stem operations 10, 20, 30.

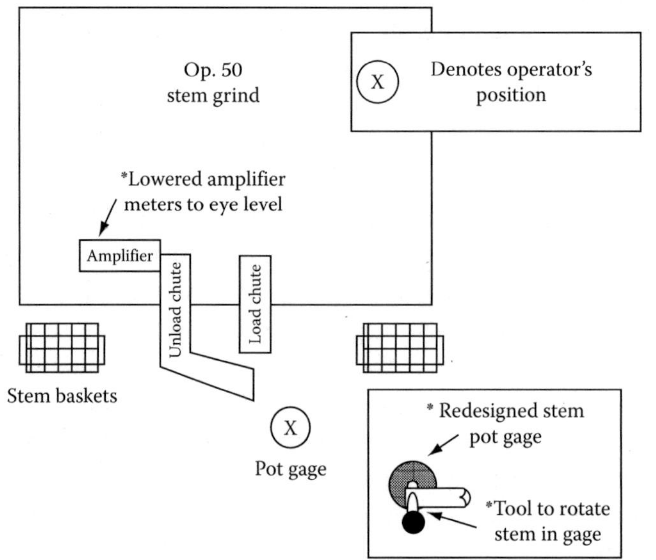


FIGURE 21.9 Stem operation 50—grinder.

21.3.3.3.4 Operation 70 of Spindle Line After operation 40, where the knuckle and stem were assembled, the spindle group made changes for two additional workstations. On operation 70, the operator was required to get left and right spindles, corresponding to the right and left sides of the automobile, from a pallet on a conveyor and load them into the machine. After the machine cycled, the finished spindles were unloaded and moved to the washer conveyor (Figure 21.10). Because of the weight of the spindle and its location on the conveyor, there was a potential for low back stress. In particular, because of the

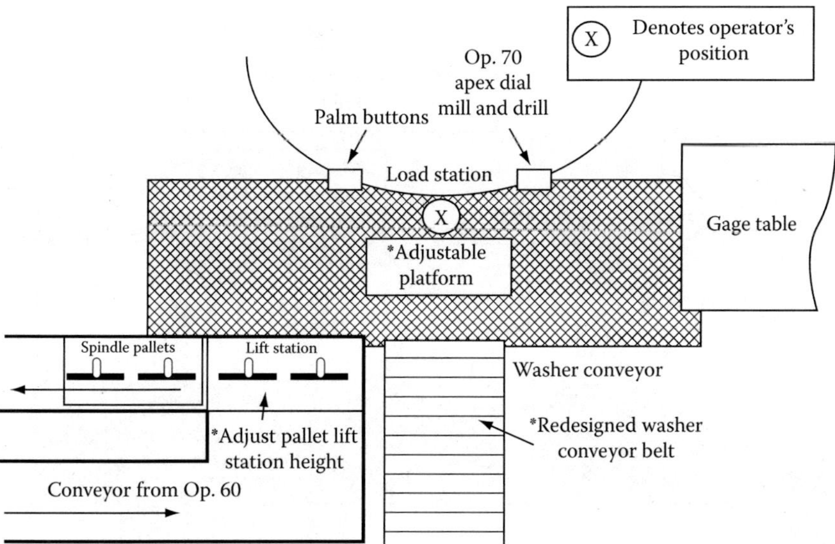


FIGURE 21.10 Spindle operation 70—apex mill and drill.

height at which the spindles were delivered, a source of back stress existed when spindles were being removed from the conveyor and moved to operation 70. To reduce potential stress, a lift table was provided on the conveyor to bring the spindles up to an acceptable height. The spindle group decided on the correct height of this table.

Another area of possible back strain existed when the spindle was removed from the machine and placed on the washer conveyor. Due to the shape of the parts, spindles had to be positioned upright to ensure proper spray dispersion in the washer. Fixtures were provided on the conveyor to aid in holding the spindle upright through the washing. However, these fixtures required the operator to precisely position the spindle on the conveyor. Because of the degree of precision required, the operator was forced to hold the spindle for an extended period of time, leading to back and shoulder stress. The group decided that removing the fixtures and placing the spindles on their side would reduce the strain. To do this required adjusting the aim of the spray nozzles in the washer. This suggestion was completed after a quality control and time study determined its feasibility. Finally, an adjustable platform aided the operator in moving the parts into and out of the machine and onto the washer conveyor. It should be noted that the changes outlined not only affected the low back but also reduced stress to the upper extremities. Note the reductions in stressful upper extremity postures.

21.3.3.3.5 Wash, Inspect, and Pack The last operation on the spindle line where ergonomic changes were made was the packing area. Prior to formation of the ergonomics group, division engineering determined that reaching to the back of the baskets could be excessively stressful to the low back, and rotary tables were installed. However, the height of the tables in relation to the washer conveyor and platform still presented a stressful lifting condition. In response, the spindle group determined that lowering the rotary tables

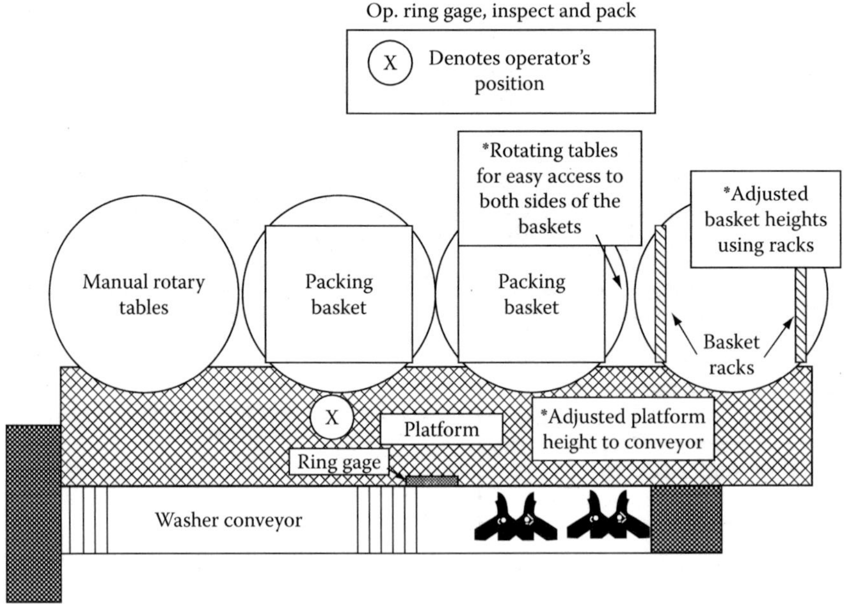


FIGURE 21.11 Spindle operation—ring gauge, inspect, and pack.

and platform while raising the baskets would dramatically reduce the stress on the back. In addition, these same changes, along with repositioning the spindles on the washer conveyor (see changes to Apex workstation), reduced the amount of stressful postures affecting the upper extremities (see Figure 21.11, * marking).

## 21.4 CRITICAL REVIEW OF CURRENT STATUS OF REACTIVE AND PROACTIVE PROCESSES

### 21.4.1 Reactive Process

Once the program was under way, a process of constant improvement was adopted. After two years, several weaknesses appeared in the process. A summary of key areas of improvement follows.

It is almost impossible to directly measure the success of this process throughout the entire company. Jobs and people are constantly shifting, making it difficult to study the data from a critical baseline. However, it should be noted that in past studies of the joint labor-management ergonomics process, results have been very favorable. Three variables were measured in the study plant: training, group process, and ergonomic outcomes. Overall, these variables led to the following conclusions:

> This single-site project demonstrates that a participative program is an effective way to implement ergonomics changes in the workplace. All of the implemented projects involved both worker and management participation. The top-down approach, however, was unsuccessful at implementing change; none of the changes using this approach were implemented. The affected worker was a valuable data resource and

> his assistance was essential to enhance change efforts. Obviously, a major problem with the participative approach is that the time spent in meetings is time away from the principal work of the organization. The participative approach can also be more time consuming than the traditional top-down approach. However, in the long run, these negative implications are often more than offset by job changes that produce a healthier and more satisfied workforce [6,7].

Currently, the UAW–Ford process is being evaluated as to how well the LECs are performing. This is done by means of an annual audit and a self-report form. The annual audit is completed by an internal team that reviews each plant's progress and makes suggestions for improvement. At the beginning of each calendar year, the LEC is required to submit an action plan to their division coordinator for review. This information, along with the audit, should help direct the LEC along a positive path.

In addition to the internal audit, at least two external groups have reviewed the process. The first reviewed the one plant's data for a 1 year period and found that on a case-by-case basis (a case being a job), injury rates dropped dramatically and the financial pay-back, based on injury data, was realized in one and a half to two years. This type of study was very encouraging and will be continued. The second group reviewed the LEC's ability to analyze jobs. The results indicated that, in general, the LECs had a tendency to miss key areas of risk on the job. The cause of this may be twofold—either the LEC's were not adequately trained or the training encouraged them to review only risks whose existence was suggested by the medical data. For example, if the medical data indicated shoulder issues, then they would concentrate their evaluation on the shoulder. In any case, the results of the study indicated that a more disciplined approach to the job analysis was required. Currently, the researcher is applying the new methodology in a pilot program, and, if successful, it will be implemented.

These issues resulted in a review of the entire documentation process. The original method of documentation did not allow for the information gathered from the LECs to be used by any other group. Therefore, a team was developed to explore better methods of documentation. Two major forms were developed to better document changes. The first is called a concern log. This document records each instance where an ergonomics concern is identified. It does not record much detail; rather it is a listing of the concerns. The second document is referred to as an 8-D form. The "8-D" stands for eight disciplines of problem solving. This system uses eight unique but connected steps to identify and solve problems. Completion of the eight steps ensures complete root cause analysis. Together these documents resulted in a better and more thorough system to communicate ergonomics information within the company. Manufacturing and assembly facilities with similar problems can now share the solutions. In addition, the system resulted in better institutional knowledge for the engineers. When they design and develop new processes, they have a baseline to review to help define the things that went right and those that went wrong (from an ergonomics perspective) with the previous process. This knowledge is essential if ergonomics issues are going to be fixed through the design of new equipment.

### 21.4.2 Proactive Process

Currently, the proactive process is being employed in several pilot programs within the company. It is essential to capture the program at the correct time to have the most impact. Once this is done, it is also important to convince senior program management that the process is important and should be prioritized as highly as other key program objectives. Most important, the analysis tools for ergonomics should be user-friendly and accessible. As stated before, times is the biggest issue in the success of this process. Engineers seem to be willing to try ergonomic analysis if it can be done simply and conveniently. Therefore, the tools should be placed in a simple-to-use toolbox. One method being studied at Ford is the use of computers to do ergonomic analyses. Today, most engineering functions are tied to large computer systems that aid engineers in performing their job functions. These functions include purchasing, computer-aided design, quality, etc. A study has begun at Ford to evaluate the effectiveness of putting the ergonomics analysis tools into the same system. One possible approach is the development of an ergonomics worksheet that will be accessible to engineers during their workplace specification. This worksheet will have most of the state-of-the-art tools available at the push of a button. It is expected that with this system, less time will be required and engineers will be more likely to consider ergonomics design attributes during the early phases of process design.

### 21.4.3 Conclusions and Future Concerns

How effective is the process? At this time, because of various issues noted here, this question cannot be answered. Current trends are encouraging. However, a better data acquisition system is necessary to properly measure performance [8]. Recently, a new health data analysis system was launched. This system is tied to the person, the job, and medical and workers' compensation data. Over the next several years, data will be collected that will enable the analysis of morbidity, cost, quality, and productivity data simultaneously, allowing for a more complete analysis of the ergonomics process.

There are several areas of concern. Currently, the reactive part of the program is driven by passive surveillance—reviewing existing data sources to pinpoint areas needing ergonomic intervention. Although this approach has been successful in changing many jobs, it will never be completely successful in primary prevention—preventing the injury before it happens. This can only be accomplished by incorporating active surveillance—the real-time collection of employee complaints and injuries to enable the estimation of the prevalence of various types of injuries—into the process. Active surveillance can be carried out in several ways, ranging from simply using a simplified questionnaire that is taken to the job site to using noninvasive diagnostic examinations involving highly trained medical professionals. In either case, an estimate of risk and employee complaints should be used to estimate exposure.

As one might suspect, even though active surveillance systems are more accurate than passive surveillance, they can be costly to administer. Also, they are time-consuming and disruptive to normal plant operations. In addition, to fully utilize the proactive process, the data generated from the reactive process must be available to persons involved in proactive ergonomics. This institutional knowledge of "things gone right" and "things gone wrong" should help engineers and designers understand where improvements are needed in new program designs. To aid in this process, databases will have to be developed and made accessible to the designers. This will require standardizing data input and output so that the data are easy to interpret and use. Consequently, the reactive and proactive processes are fully dependent on one another, having one without the other will cause the ergonomics process to slow and possibly to fail.

Therefore, the challenge to the manufacturing ergonomics process is to develop an integrated approach that finds and fixes existing issues and prevents future ones from being designed into new equipment. Only then will we get full control of the problem.

## ACKNOWLEDGMENTS

The United Auto Workers Union's full name is United Automobile, Aerospace and Agricultural Implement Workers of America. The UAW–Ford National Joint Committee on Health and Safety is jointly chaired by members of the company and the union to oversee certain activities in health and safety for UAW Master Agreement Plants. We thank the UAW–Ford National Joint Committee on Health and Safety for their support and guidance in developing this chapter.

## REFERENCES

1. T. O. Kvalseth, Factors influencing the implementation of Ergonomics: An empirical study based on psychophysical scaling technique, *Ergonomics 23*(8):821 (1980).
2. UAW–Ford National Joint Committee on Health and Safety, *The UAW–Ford Ergonomics Process—Implementation Guide,* University of Michigan, Ann Arbor, MI, 1988.
3. UAW–Ford National Joint Committee on Health and Safety, *The UAW–Ford Ergonomics Process—Job Improvement Guide,* University of Michigan, Ann Arbor, MI, 1989.
4. UAW–Ford National Joint Committee Health and Safety, *The UAW–Ford Ergonomics Process Leadership Commitment Videotape*, 1988.
5. UAW–Ford National Joint Committee on Health and Safety, The UAW–Ford ergonomics process, *Segments* :1–5, 1989.
6. J. K. Liker, B. S. Joseph, and S. S. Ulin, Participatory Ergonomics in two U.S. automotive plants, in *Participatory Ergonomics,* K. Noro and A. Imada, Eds., Taylor & Francis, London, U.K., 1991.
7. B. S. Joseph, A participative ergonomics control program in a U.S. automotive plant: Evaluation and implications, Dissertation, The University of Michigan; Ann Arbor, MI, 1986.
8. B. S. Joseph, *In-Plant Ergonomics Programs: Development, Implementation, and Evaluation*, Final Report, Motor Vehicles Manufacturers Association, Detroit, MI, 1988.

CHAPTER 22

# Ergonomic Control Measures in the Health Care Industry

Arthur R. Longmate

CONTENTS

## 22.1 INTRODUCTION

Whether a company is large or small, potential ergonomics-related problems can be found in every workplace where people are an integral component. Johnson & Johnson (J&J) is a highly decentralized corporation composed of many large and small individual companies. The key to effective ergonomics programs at J&J is *proactivity*—identification and resolution of potential problems before the occurrence of serious medical cases. Effective programs to control ergonomics-related disorders require the education of, cooperation of, and input from each person involved in the process. Success in the prevention of serious medical cases depends on the degree of success in the following categories:

- Early documentation of ergonomic risk factors and reporting of medical symptoms before the medical cases become chronic and severe
- Initial sound ergonomic design of processes and equipment using ergonomic job design guidelines
- Prompt intervention steps taken to alleviate identified stresses
- Provision of comprehensive medical case management to prevent medical cases from becoming more severe
- Ergonomics program components that have been found to be successful at J&J are the focus of this chapter, illustrated by several ergonomics engineering case examples.

## 22.2 SETTING REALISTIC ERGONOMICS PROGRAM GOALS

Due to the wide range of personal and work-related ergonomic risk factors, it is virtually impossible to prevent all ergonomics-related medical disorders. Thus, it is critical to set attainable program goals and objectives so that realistic program expectations are maintained. A common mistake is to adopt a "zero defect" mentality with the goal of eliminating all ergonomics-related injuries. An important initial pill to swallow is to accept the fact that some level of ergonomics complaints and medical symptoms will always be present in any business environment—particularly those requiring extensive repetitive motion, manual materials handling, and/or use of computers. A frequent basis for OSHA fines is the failure to report ergonomics-related medical cases when it is readily apparent that ergonomic risk factors are present in the workplace and/or when employee interviews reveal that discomfort and medical symptoms are being experienced. Inaccurate or misleading recording practices are often the result of initially unrealistic program goals. Formulation and acceptance of realistic program expectations, up front, will avoid many problems later on in the process.

It is not uncommon for a well-meaning upper management representative to ask the question, "What will it take to make these problems go away?" The simple answer to this question is the following. If the repetitive manual work, the related ergonomic risk factors, and the people performing this work go away, so will the related medical cases and discomfort symptoms! Since this does not generally happen, the setting of unrealistic expectations has often led to the unfortunate end result of creative record keeping to create the desired impression that the problems are under control. To prevent this occurrence, performance-oriented goals and objectives should be created rather than those geared to specific incidence rates or to complete elimination of problems. If all elements of a comprehensive ergonomics control program are in place, then improvements in the numbers will be realized in time.

One realistic goal is to attempt to eliminate all *serious* medical cases. This requires the establishment of a sensitive surveillance system, an objective medical case severity rating system, and periodic case reevaluation to allow case tracking. Another realistic goal should be continuous improvement and not the expectation of complete problem eradication. If a manufacturing facility claims that ergonomic injuries or illnesses are not being experienced at any level, one or more of the following factors may be the underlying cause:

- Employees (management, professional, and hourly) may not have received sufficient training to recognize ergonomics-related disorders.
- Occupational medicine professionals may not have had sufficient training to recognize the work relatedness of ergonomics-related disorders.
- The threshold of company ergonomic injury/illness recording criteria may be set too high to detect low-severity cases.
- The company may be spending excessive resources to reduce ergonomic stresses to an unreasonably low and financially unjustifiable level.

A key premise to achieving an effective ergonomics control program is to accept the fact that ergonomic medical cases will occur at some baseline level in all environments where ergonomic stresses are present. A key objective for an ergonomics program should be to minimize the incidence and severity of these cases, not necessarily to completely eliminate them. Performance-oriented goals are preferable to the setting of specific incidence rate targets, which may typically lead to creative bookkeeping schemes.

## 22.3 ERGONOMICS TEAM STRUCTURE

A typical first step in the establishment of an ergonomics program is the creation of an ergonomics task force composed of team members from the various aspects of the business, such as hourly employees, management, engineering, human resources, medical, and product development representatives. In a simple manufacturing environment, a single ergonomics task force may be effective in identifying and resolving existing problems throughout the entire business. In many of today's complex business environments, it may be necessary to create a network of specialized groups to address problems in various diverse segments of the business. For example, office-related problems may be of little interest to associates working in the manufacturing environment; however, both environments probably do have serious ergonomics issues that need to be addressed.

At one J&J corporate headquarters location, five ergonomics committees have evolved to address issues in different segments of the business. This structure allows a clear focusing of efforts on the problems in a specific area and does not waste the time of a larger group of individuals whose primary interest or responsibility lies in other areas. A brief description of the jurisdiction of the various committees follows.

### 22.3.1 Ergonomics Management Advisory Team

A key component of an ergonomics program is to secure the interest, involvement, and commitment of upper management. The ergonomics/safety management advisory team is composed primarily of department directors who represent each functional area of the business and are champions for each of the focused ergonomics teams. In addition, key health professionals such as the ergonomics engineer and medical, safety, industrial hygiene, and human resources personnel are available resources to the committee. The key objectives of this committee are to secure upper management awareness of current ergonomics issues and to ensure that adequate resources are provided to resolve them. The management advisory team also provides a forum to share the efforts being undertaken by each of the other teams.

### 22.3.2 Manufacturing/Engineering Team

The manufacturing/engineering team, informally known as the safe workplace action team (SWAT), is made up of hourly associates from each functional manufacturing team. The same group of technical support personnel assists the team as technical support is needed. Each SWAT member receives comprehensive training in the recognition of ergonomics problems and functions as the day-to-day eyes, ears, and mouth of the ergonomics program, identifying and reporting problems as they occur. The group also functions as a body of local experts who can provide valuable, practical input in the design of new equipment and processes. This group is also responsible for a variety of other health-related functions, such as

- Conducting periodic surveys of all department associates to determine current ergonomic problems and concerns
- Conducting monthly departmental safety meetings
- Investigating all illnesses and injuries and completing related accident reports
- Reviewing all new equipment and/or processes to identify potential problems prior to implementation
- Assisting with the evaluation of the effectiveness of all new ergonomic materials such as chairs, equipment, and tools

Equipment and process engineers from each product development team provide SWAT members with technical support to ensure that all ergonomics equipment issues are promptly resolved and to ensure that ergonomics equipment design guidelines and specifications have been provided to all equipment suppliers.

### 22.3.3 Product Development Laboratory Team

This organization has many different types of development and testing laboratory environments. Problems experienced with the specialized test equipment found in these laboratories are unlike the typical problems found in other functional areas. As a result, a separate team was formed, with members from each laboratory area, to focus on issues specific to these areas.

### 22.3.4 Office/Computer Ergonomics Team

Office issues are sufficiently unique to warrant a separate, focused ergonomics team. The most frequent source of critical office ergonomics issues is the ever-growing and almost constant user interface with computers. This team adopted the name office safe workplace action team (OSWAT) to piggyback off the success of the manufacturing team. The main objective of this team is once again, to be the eyes, ears, and mouth of the office ergonomics program to identify and resolve issues in a preventive mode.

### 22.3.5 Surgical Training Facility Ergonomics Team

A key business objective in the minimally invasive surgery marketplace is the training of surgeons and sales representatives in new endoscopic surgical procedures. To accomplish this strategic objective, a separate training facility was constructed. The ergonomic issues involved with this surgical training facility are once again sufficiently unique to warrant a separate ergonomics team.

### 22.3.6 How Many Teams Are Too Many?

The existence of five ergonomics teams at one primary location is probably quite unusual. In today's competitive and hectic business environment, there is no room for wasted effort. If teams can be effectively combined and remain focused on the specific issues, all the better. The downside of a multiteam structure is that the health professionals who support the focused teams sometimes get stretched somewhat thin attending to the activities of each team. On the other hand, if a group of individuals has a unique set of problems, a separate team is probably justified. Another advantage of multiple teams is that the number of issues facing an individual team will usually be smaller and more manageable than the number that a larger, all-encompassing team must deal with. The smaller focused teams may need to meet only infrequently, such as on a quarterly basis, if the number of issues is small and if all outstanding issues have been resolved. The team meeting may follow a walk-through inspection to determine if any new problems have surfaced. A potential problem for larger teams is to carry too long a list of projects. A smaller team with only a few key issues has a higher probability of success than one whose list is lengthy and frustrating. In either case, issues must be prioritized and addressed within the constraints of resource limitations.

## 22.4 OTHER J&J ERGONOMICS PROGRAM COMPONENTS

J&J initiated its first ergonomics program efforts approximately 12 years ago at Ethicon, Inc. and J&J Products, Inc. (now J&J Consumer Products, Inc.) [1,2]. The goals of these programs have been to establish necessary ergonomics program components. A brief survey of these components follows.

### 22.4.1 Surveillance Techniques

Surveillance techniques are needed to identify existing and potential problems. These techniques consist primarily of passive surveillance methods such as simple review and summarization of available records. In addition, active surveillance techniques such as questionnaires, surveys, and comprehensive medical examinations are often utilized. The most common surveillance techniques used at J&J are as follows:

1. Thorough analysis of medical visit data and OSHA logs.
2. Ergonomic review and risk factor documentation of new and existing equipment and workstations.
3. Informal employee surveys—a proactive approach to identifying problems in the early stages before medical symptoms have developed. Why wait for complaints, when you can simply ask the employees which jobs (a) cause discomfort during or after performing the work, (b) are associated with the highest turnover rates, or (c) experience problems meeting expected production goals?
4. Postural discomfort questionnaires [3] to evaluate the effect of specific program interventions before and after implementation.

### 22.4.2 Reporting Systems

Reporting systems are needed to document the incidence and severity of problems and promptly disseminate critical information to parties responsible for problem intervention. The ergonomic teams in each area are responsible for the completion of accident/injury reports that identify causal factors and the formulation of action plans to eliminate the problems.

- Medical case incidence and severity data must be maintained in a user-friendly information system to ensure that data are provided to interested parties in the most effective format. It must be easy to customize the system to meet any future information need. Available systems may be purchased or a system can be set up using dBASE III or a similar database software system.
- Ergonomic complaints/medical cases are reported to responsible managers and engineers within 48 h to facilitate formulation of effective action plans.
- Monthly injury/illness summary reports, by department, are distributed to each manager and ergonomics team member along with a cover sheet indicating the current monthly trend compared to recent historical data. A summary, with observed trends and currently active intervention projects, is also provided with each monthly report.
- Quarterly and year-end summary reports are also provided to indicate trends, successes, and possible failures.

### 22.4.3 Intervention Techniques

Intervention techniques can generally be classified into the following categories:

*Comprehensive medical case management*:

- Conservative medical treatment protocol (surgery is considered only as a last resort)
- Physical therapy (most effective when conducted on site)
- Effective medical job restriction protocol
- Methods modifications to reduce risk factors
- Case-specific ergonomics engineering interventions (equipment/process modifications)
- Work method retraining [methods analysis/modification, possible use of biofeedback monitoring (EMG) to make person aware of least stressful work methods]
- Administrative procedures (e.g., job enlargement, job rotation, imposed limitations on overtime)

*Complaint/problem follow-up*:

- Team problem resolution (brainstorming) techniques
- Problem resolution tracking—action items with assigned responsibility and target completion dates

*Short-term and long-term action plans*:

- Simple administrative/method changes made up front
- Long-term job reengineering changes made later

### 22.4.4 Prevention Techniques

A prevention technique is generally one of the following.

*Ergonomics training*: Each and every employee, from the hourly worker to the company president, must receive some level of ergonomics training.

Ergonomics committee members must receive the most extensive training in all phases of ergonomic job assessment, design criteria, prevention, and intervention techniques. The training may be conducted using internal resources or through the use of comprehensive ergonomics training programs offered by academic institutions or private consultants. After initial training, regular follow-up sessions should be conducted to gain critical experience through brainstorming and the sharing of experiences.

Upper management representatives should receive basic training in ergonomics principles and periodic briefings on the extent of problems, goals, and objectives of the ergonomics program, and the support required to achieve the objectives. Selected management champions should be identified and invited to attend comprehensive training.

Equipment, process and product development engineers, and designers should receive extensive training in ergonomic design guidelines to prevent future problems. An effective training technique is to get technical support personnel closely involved with medical case follow-ups to allow them to directly interface with persons experiencing the problems. Requiring engineers to work for extended periods actually performing the work tasks they have designed provides an excellent learning experience.

Managers, supervisors, and hourly employees should receive basic ergonomics training to help them understand the types and symptoms of problems that might be experienced and aid in early detection and reporting. The goals of the ergonomics program should be stressed along with specific actions that the company is taking to prevent and manage these problems to minimize the effect on their lives and on the business. Since implementation of ergonomic intervention is a team effort, the team approach should be emphasized along with what is expected of the employees and their managers to address ergonomic issues.

*Ergonomic job/equipment design guidelines*: It is critical to establish written job and equipment design guidelines to function as a constant reference and reminder during all phases of the design process. These guidelines are a critical component of the engineer and designer training referenced earlier.

*Ergonomic equipment checklists and reviews*: In addition to comprehensive ergonomics guidelines, it is helpful to have simple ergonomics checklists to act as a simple reminder at critical times during the equipment design and development and implementation processes. The complexity and length of the checklists should be minimized to simply encourage the thought process and not to focus on the completion of the checklist itself.

*Exercise programs*: To be discussed in detail in the following section.

While all of the previous components are critical to a comprehensive ergonomics program, the remainder of this chapter focuses on two critical elements of J&J's ergonomics program strategy: on-the-job ergonomics exercise programs and effective ergonomics engineering intervention efforts.

## 22.5 ON-THE-JOB EXERCISE PROGRAMS

A key component of J&J's proactive ergonomics program philosophy is on-the-job exercise. The commitment to on-the-job exercise began over 10 years ago at Ethicon, Inc., one of the corporation's largest domestic operating companies. It should be noted that ergonomic exercises should be considered only after other critical program components (e.g., effective surveillance networks, training programs, problem documentation, and implementation of interventions to alleviate stress) are established. The main objective of the exercise programs is the prevention of ergonomics-related injuries and illnesses. When used as a proactive component of an overall comprehensive ergonomics program, and developed and administered by qualified health professionals, exercise programs can be a valuable tool. On the other hand, when used as primary control strategy in situations where problems are severe and well established, dramatic results may not be achieved [4].

Important benefits of exercise that have been documented are the following:

1. Increased blood flow to active muscle/tendon systems: Many ergonomics-related disorders, such as tendon and muscle strains and inflammation, are partially caused by the inability of the body to transport sufficient nutrients to the site of active tendon/muscle groups. When muscles are contracted or tensed, blood flow to the area is restricted. In the case of static postural situations or highly repetitive tasks, sufficient recovery time is not provided to allow this critical replenishment process to occur to the necessary extent. Each exercise period is, in effect, a break from the repetitious or static postural work regimen. Every break in the work routine allows the muscles to relax and blood flow to be reestablished [3,5]. Whether exercise sessions or simple rest periods are more effective in reducing discomfort and other factors such as mood state is not clear; however, exercise may be more effective in reducing decreases in productivity that occur over the course of the workday. Research focused on selected repetitive tasks has indicated that more widely distributed rest or exercise minibreaks can lead to reduced postural discomfort and increased productivity [6,7].
2. Increased range of motion (flexibility): Periodic muscle/tendon stretching exercises, throughout the worker's entire range of motion, may diminish the risk of developing various types of motion specialization disorders such as trigger finger. It is well known within the sports medicine community that stretching and the resultant increase in range of motion and flexibility are critical factors in the reduction of injury [5].
3. Increased strength: This has been a recognized and well-documented benefit of exercise regimens designed for these purposes. When an individual's maximum strength increases, the relative level of job stress is reduced because the person is working at a lower percentage of maximum strength. This relative reduction in stress lowers the person's risk of injury.
4. Exercise of antagonistic muscle groups: Specific exercises are selected for the program that exercise antagonistic muscles, which act in the opposite direction than those primarily being used while the work is being performed. If work content requires gripping/squeezing motions, it makes little sense to take an exercise break that requires additional gripping and squeezing. A more sensible alternative would be to select an exercise that requires extension of the fingers to exercise the antagonistic muscle groups.
5. Reduction in the level of overall mental and physical tension/stress: While not fully understood, physical exercises are successful in relaxing the tensions that build from performing physical and mental tasks over extended periods.

6. Increase in the level of attention: A person's level of attention becomes degraded by the performance of repetitive tasks over extended periods. Exercise breaks are often referred to as "energy breaks" because they have the desirable effect of getting the blood flowing and bringing renewed feelings of energy to the individual. This may explain why people have the perception that the exercises are "good" and generally claim that they feel better even while reductions in discomfort are not consistently reported [4].

7. Other benefits, such as reductions in visual strain and improvement in intellectual facilities and sensorimotor skills, have also been documented when exercise breaks are taken during work hours [8,9].

The immediate benefits from identifying and eliminating ergonomic risk factors in the workplace are easy to comprehend and have been clearly documented. In comparison, the benefits of exercise, intangible terms, are often more difficult to document. Although many important effects of exercise have been documented, the critical scientific correlation between on-the-job exercise and reduced medical case incidence rates has been elusive.

Although preliminary studies have indicated that the productivity lost during exercising can be made up because employees are "energized" and are more highly motivated, it may be overly optimistic to assume that this can always be expected. In today's competitive business environment, the first question often asked by management is how much the implementation of a program will cost and how it will translate into financial savings and/or a reduction in future injuries and illnesses. Since an exercise program performed on company time can potentially result in a loss of 3%–4% (approximately 15 min) of productive working time, it is natural for managers to ask what they will get in return. At one J&J facility employing approximately 2000 people, management representatives suggested that 70 new people might have to be hired to make up for the 3.5% conversion of available work time to exercise time. Based upon the assumption that each new employee would cost the company approximately $60,000 (including benefits), the additional annual cost would take $4.2 million for this facility. Whereas the argument could be made that this "lost time" could be made up, it is clear that the potential exists for exercise programs to demand substantial investment to make up for lost production time. It is clear that further research is necessary to explore and establish cost/benefit relationships.

## 22.6 ENGINEERING CONTROL CASE EXAMPLES

### 22.6.1 Trocar Assembly

The following case study illustrates the benefits of many aspects of a comprehensive ergonomics program. During development and production of an endoscopic trocar device, the aggressive project implementation schedule dictated that assembly processes would be primarily manual with the exception of several ultrasonic welding operations. This is a common decision path in the minimally invasive surgery industry due to the requirement of speed to get the product introduced into the marketplace and the critical requirement for frequent quality inspections.

Trocar assembly steps were the following:

1. Get trocar casing bottom half and put into assembly nest.
2. Get knife assembly and compression spring.
3. Slide compression spring over knife stem.
4. Get anti-backup spring.
5. Insert tip of knife shaft through the hole in anti-backup spring while overcoming the pressure of the compression spring by pushing the hands toward each other.
6. Install knife/spring subassembly into trocar casing half to engage slot in knife onto boss in trocar casing while maintaining compression of spring.
7. Get buckling spring and install into trocar assembly.
8. Get two latches and install into trocar assembly.
9. Get trocar casing top half and install onto assembly.
10. Remove completed assembly from nest and place in tray.

During the initial process review, several ergonomic risk factors were noted with this process, particularly the following:

- *Repetitiveness*: Associates were expected to complete approximately 550 assemblies per day.
- *Force*: Substantial pinch force was required to hold the anti-backup spring between the thumb and index finger during compression of the main spring while inserting the knife shaft through the hole in the anti-backup spring.
- *Static exertion*: The pinching of the anti-backup spring was statically maintained from the time the main spring was compressed until the knife shaft was inserted through a hole in the anti-backup spring and installed into the trocar casing.
- *Contact with sharp edges*: The edges of the anti-backup spring were quite sharp and caused local discomfort to the fingers during the spring compression process.

The provision of some type of mechanical assembly assist fixture was considered. However, a feasible design had not been finalized by the time the manufacturing process was launched.

Since all associates had been trained in the recognition of ergonomic disorders and a sensitive surveillance network was in place to detect medical cases in the early stages, the decision was made to proceed in a manual assembly mode and to monitor the process carefully. Efforts also continued to identify an effective mechanical assist fixture.

An additional confounding factor had occurred during the engineering pilot process. For the instrument to function properly, it became necessary to provide a more forceful main compression spring. This incremental increase in the force requirement appeared to be the straw that broke the camel's back.

Over the next 3 months, 14 associates visited the medical department to report upper extremity symptoms, with one interesting pattern—all were left-hand-specific. All cases were relatively minor (most received severity ratings of 1 or 2 on a scale from 1 to 5, with 1 representing the least serious symptoms and 5 the most severe symptoms). As the cases began to appear, comments from assemblers verified that the problems centered on pinching and statically holding the anti-backup spring in the left hand during the somewhat tedious manual assembly process. It was also felt that if a mechanical fixture could be developed to compress and hold the main spring throughout the remainder of the assembly process, then the related physical stress would be greatly reduced. The project to develop a mechanical assist fixture received even greater attention when it was discovered that the tedious process of squeezing the anti-backup spring was causing damage to the spring and was resulting in occasional instrument failure.

Development of the spring compression fixture was placed at the highest priority, and several prototypes were developed and evaluated. Some of the critical design criteria that were determined through interaction with the instrument assemblers were the following:

1. The main spring should be mechanically compressed and locked into position to allow the rest of the assembly to be completed without the necessity of maintaining this static force.
2. Because different assemblers demonstrated many slightly different methods, the angle of the fixture was made adjustable to more effectively accommodate these individual differences.
3. As some of the assemblers were left-handed, the fixture was designed to be adaptable for convenient use by either left- or right-handed individuals.

The fixtures were designed and built over an approximately 6 week time frame at a cost of approximately $800 per fixture. The relative simplicity of the fixtures is the primary reason for the quick turnaround on custom-designed equipment. It is typical for more complicated fixturing to require several months to design and build.

After implementation of the new assembly fixtures, the stressful static holding of the anti-backup spring was eliminated. The job steps involved with the new trocar assembly fixture are the following:

1. Get knife assembly and position under fixture.
2. Push in knife assembly retaining knob.
3. Get compression spring and slide over knife stem and into position in front of spring compressor bar.
4. Pull lever to right to compress main spring until latch is engaged.

5. Get anti-backup spring and slide over end of knife stem.
6. Get trocar casing bottom half and position under assembly.
7. Gently squeeze anti-backup spring and position into casing while engaging slot in knife onto boss in trocar casing.
8. Pull out knife assembly retaining knob.
9. Pull assembly down to disengage from spring compressor bar and place on assembly table.
10. Get buckling spring and install into trocar assembly.
11. Get two latches and install into trocar assembly.
12. Get trocar casing top half and install onto assembly.
13. Remove completed assembly from nest and place in tray.

It should be noted that three extra job elements were added through the use of the assembly fixture. However, the stressful job elements associated with compressing the main spring were eliminated. Since the degree of assembly difficulty was reduced significantly, no productivity was lost in conjunction with implementation of the fixtures even though a few simple elements were added.

The fixtures were very positively received by the majority of operators as soon as they were introduced. This highly positive response was better than expected in light of the fact that people had been performing the old method for several months. Usually, a longer adjustment period is required to allow people to acclimate to a new method. Involving the assemblers in the fixture design process clearly played an important role in the feeling of ownership and acceptance of the fixtures.

Within 3 months of implementation of these fixtures, virtually all CTD symptoms and other chronic hand overuse disorders had disappeared. The fixtures also made it possible for all people on the line to perform the job and cleared the way for the elimination of several medical restrictions that had specifically related to this job.

### 22.6.2 Low-Force Activation Buttons

On several ultrasonic welding operations, an elevated incidence of thumb tendinitis was observed. Risk factor assessment and interviews with associates indicated that the 5 lb of force required to fully depress the welder activation buttons in a very dynamic environment might be a primary cause of these problems. The standard activation buttons were positioned in the front plate of the welder base as depicted in Figure 22.1. The job elements at the welding station were as follows:

1. Get assembled stapler from tray.
2. Put the stapler into the welder nest. (Instruments are automatically clamped into position when required.)

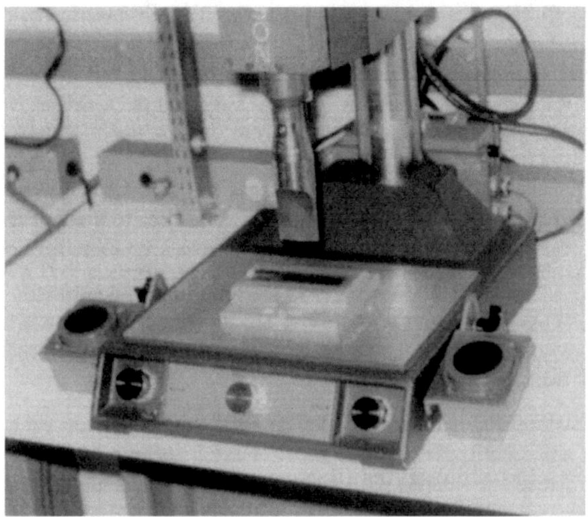

FIGURE 22.1 Standard welder activation buttons (in front) and presence sensing, capacitive activation buttons (at sides).

3. With both hands, fully depress and hold the cycle activation buttons until the vibratory horn contacts the instrument.
4. Get the instrument from the nest and put into a finished instrument tray.

Due to the location of the buttons and the pace of the operation, most welder operators use the thumbs to activate the buttons. To overcome the 5 lb of force quickly, substantially more than 5 lb was actually applied.

Several options were explored to address these issues.

Presence-sensing, capacitive-coupled activation buttons were purchased and installed into auxiliary electrical boxes attached to each side of the welder base (also see Figure 22.1). This initially appeared to be the perfect solution, because no force on the activation pad was required. Several problems arose during initial evaluations.

The 3 in. of space occupied on each side by the auxiliary electrical enclosures added 3 in. of reach distance each time the associate reached for parts. This was a very undesirable ergonomic consideration in the use of this type of activation control.

Substantial time was required for associates to learn and remember that forceful pressing on these controls was not required. Since the majority of machine activation controls throughout the rest of the facility were push buttons, new operators coming into the area had to be constantly reminded that the controls did not require the application of force. In fact, some associates so expected movement of the "button" that they imparted more force to the stationary sensing pad than would have been required by the original push button control. Also, since the stationary pads were not able to absorb force from quickly moving fingers, more actual force may have actually been imparted by the fingers under dynamic conditions.

The presence-sensing controls also had an inherent problem in that the welder cycle would unexpectedly interrupt for unknown reasons. This required a simple reactivation of the controls after a short pause but was perceived to be a major inconvenience by many operators.

The cost of the presence-sensing controls was also a negative factor compared to the much simpler and less costly push button controls. A project to convert all controls in all facilities to the presence-sensing variety would have been extremely burdensome.

The second option investigated was to evaluate the new photoelectric light beam sensor activation controls. Although these devices are marketed with the claim of eliminating ergonomic stresses by eliminating the pushing of buttons, once again this did not prove to be entirely true, for the following reasons:

1. When working in a dynamic environment, an operator cannot be expected to quickly extend one finger and calmly place it into the U-shaped opening of the control as is often depicted in the advertisements. Generally, the entire open hand makes dynamic contact, and the fingers generally bump into the relatively sharp edges on the inside edges of the finger cutout area.
2. When the photoelectric control was oriented upward, in several instances the machine cycle was observed to be accidentally initiated when parts of an operator's clothing were inadvertently dragged over the controls. Safety concern resulted in the necessity to turn the controls 90° to face outward. This modification had three undesirable results:
    a. A 6 in. reach was added to the side for components because parts trays had to be located beyond the controls with enough clearance for hands to have access to controls.
    b. It was necessary to rotate the hands and forearms 90° from the palm-down posture used to place components to the palm-inward posture used to activate the controls.
    c. More complex, two-directional motion (out and back) was needed to move the hands to the controls. It was felt that these motions require both more time and more muscle activity.

Efforts were undertaken to identify other, more traditional button activation controls with reduced activation force and larger button contact surface. A survey of many available push button controls determined that most required approximately the same amount of force—in the 5 lb range. Also, most of these controls were permanently sealed to prevent the removal of internal springs.

One push button switch was identified that could be easily disassembled and the main compression spring removed to reduce the required activation force. An additional spring, built into the contact set, provided sufficient force to return the button to its original position after pressure on it was released. The force required to activate the modified button was approximately 1 lb. This represents a highly significant 80% reduction in required force.

When an application called for two contact sets, the required activation force was approximately 2 lb, which still represented in a highly significant 60% reduction in required finger-pressing force. A standard low-force push button switch requiring 2 lb of finger or thumb actuation force was eventually developed by the Allen Bradley Company. This switch was preferred because it did not require in-house modification to reduce the actuation force. Dukane Corporation, manufacturer of the ultrasonic welders, agreed to equip future welders with the low-force Allen Bradley buttons as a standard feature.

### 22.6.3 Knob Twist Fixtures

In departments assembling intestinal staplers, CTD incidence rates were running at 2–3 times the overall plant average. Review of the OSHA log revealed that about 67% of these illnesses were related to various types of elbow tendinitis. Risk factor assessment and interviews with associates indicated that the problem was more than likely related to the repetitive twisting of the adjustment knob at the end of the instrument. Some of the operations required only slight knob adjustment (less than one complete knob rotation), whereas others required extensive twisting, (up to 10 twists, occasionally in both directions). The majority of the twisting required negligible torque but was highly repetitive—occasionally over 10,000 90°–110° forearm rotations per 8 h work shift. On one operation, the operator had to forcefully twist the knob loose to break weld adhesions after the casing halves were ultrasonically welded together. This operation required substantial torque—up to 40 in.-lb.

When the probable cause of the illnesses was identified, a crude powered knob twist device was built that used an air motor mounted inside a square aluminum box that sat on top of the work surface. This device was effective in turning the knob but created additional problems for the shoulders and arms because it was necessary for the operator to reach approximately 18 in. above the workstation each time an instrument was placed into the fixture. The device also required the operator to press and hold a button to turn on the motor, which resulted in discomfort to the hands and fingers. Also, it did not have any torque-limiting capability, which resulted in occasional damage to the instrument.

Input was solicited from employees, engineers, and management in the area to determine the most critical features for the next-generation fixture. This was an extremely critical process to maximize the changes that employees would use the fixture, that it would satisfy process constraints, and that production requirements would be met. Based on this input, a more sophisticated fixture was designed and built that had the following features necessary for effective process control:

1. Orientation of the fixture such that the instrument could be inserted into the nest with a lateral move of about 2–3 in. and without instrument reorientation. This was accomplished by using drive belts from the motor to the instrument chuck rather than driving the chuck directly from the motor shaft.

2. Layout of the fixture to consume minimum prime working space. The fixture was designed to be extremely narrow and took up only about 8 in. to the side of the assembler.

3. Selectable number of turns to the nearest quarter turn. The machine would automatically stop when the specified number of turns for a particular application was reached.
4. Selectable autoreverse feature in case assembler had to alternate direction of rotation.
5. Adjustable speed control.
6. An adjustable clutch to prevent instrument overtorquing.

A 300-instrument fixture validation study was conducted to ensure that the fixtures did not affect the functionality of the instruments. Employees were somewhat slow in warming to the new fixture, which took up over 8 in. of their workspace, but after about 2 weeks most people said they didn't know how they ever survived without it.

Within 6 months of implementation of these devices, the incidence of new cases of elbow trauma was essentially eliminated (reduced by approximately 95%). Utilization of the new fixtures also allowed employees with preexisting conditions to work on many jobs from which they were previously restricted due to the requirement of repetitive twisting.

In addition to eliminating the repetitive motions responsible for the occurrence of elbow tendinitis, an additional benefit was noted, during the follow-up evaluation, when the new fixture was used on a calibration torquing operation. The old method required manually twisting the instrument several times and then mounting it in a fixture. A manual torque wrench was then used to prestress the instrument prior to final calibration. Using the new fixture, the twisting was performed automatically and the torquing was also coincidentally achieved through the built-in torque clutch. Once again, a validation study was conducted to ensure that the torque achieved by the fixture was within process limits. Provision of these fixtures at this operation resulted in much lower physical stress for the operator plus an approximately fourfold increase in productivity. Productivity was increased on most other jobs by 10%–15% through the use of these fixtures.

### 22.6.4 Instrument Test Fire Fixtures

Hand firing refers to power gripping a handle or trigger instrument configuration and squeezing to provide the power to form surgical staples or ligating clips. For some surgical instrument end users, hand firing of instruments requires a high percentage of their available strength (e.g., some instruments require approximately 50th percentile maximum female grip strength). For the end user, the redeeming factor is that the instrument firing task is performed only once every few minutes over a short period in the operating room and certainly less than 100 times over the course of the day.

During the manufacture of endoscopic surgical instruments, it is critical to ensure that instruments function properly during these critical medical procedures. As a result, each instrument must typically be hand fired several times during the manufacturing process to ensure its functionality in all ranges of future potential use. On production lines where hundreds and perhaps thousands of these instruments are assembled and tested daily, thousands of hand firings may be required.

Surgical instrument assemblers were experiencing a high incidence of upper extremity disorders (tendinitis and other overuse syndromes) that seemed to be related to the highly repetitive and forceful hand firing. Incidence was higher on lines assembling instruments that required higher firing force (e.g., up to 50 lb of grip force). Attempts were initiated to provide alternative powered fixtures to perform this repetitive manual function.

One of the problems was that it was difficult to know how much force a given person was applying to the instrument. The verbal instructions given during training were to squeeze until the trigger of the instrument just bottomed out with the handle. It was found that many of the more highly motivated employees were experiencing problems on a more frequent basis. Upon questioning and observation, it appeared that these individuals were unconsciously squeezing the instrument much harder than the minimum requirement because they were trying to do the best job possible. Training efforts were undertaken to demonstrate that excessive squeezing was a non-value-added function and was indeed increasing the risk of their experiencing a cumulative trauma disorder. To augment these training efforts, training fixtures were also provided that read out how much force an individual was actually applying. These fixtures were modified to have a light indicate when "sufficient" force limit was reached.

An additional argument was offered that powered fixtures would provide a more effective SPC (statistical process control) quality evaluation tool than the hand firing method. Process parameters such as the magnitude of force applied to the trigger and trigger closure speed could be tightly controlled and manipulated in a powered mechanical fixture. Control of these parameters during manual test firing was much more difficult due to the wide range of capabilities within the employee population. Even with training, much greater variability existed with the hand method, particularly over the course of the workday as fatigue began to take its toll. Subsequent tests, using prototype mechanical fixtures, verified the opportunity for the manufacturing process to be controlled to much tighter tolerances.

Efforts to provide powered test fire fixtures were temporarily delayed by the argument that instruments should be test fired by hand because that is the method used in the operating room. In cases where this argument remained an obstacle, a compromise was reached to the effect that at least one of the test firings would be performed manually and the test would be performed on the powered test fixture. For product lines where this argument remained unresolved, efforts were undertaken to provide training aids as described earlier that trained personnel to limit their exertions.

On instruments designed for multiple firings, such as skin staplers, hernia staplers, or ligation clip appliers, quality procedures required that several instruments out of each batch be completely fired out to comprehensively test all instrument functions. Personnel responsible for this function were known to experience an extremely high incidence of upper extremity disorders; in fact, nearly all persons performing this function had experienced some type of medical symptom. Powered test fire fixtures were provided for these operations to greatly reduce the repetition of test firing. In most cases, about 95% of the manual firing could be eliminated because only one or two of the firings were now performed by hand to test the actual "feel" of the instrument. Elimination of the repetitive firedown requirement virtually eliminated all incidence of CTDs.

In the interim, and in situations where firing fixtures were delayed, a job rotation matrix was constructed to guide supervisors in the required frequency of job rotation based upon the actual force requirement and the frequency of firing. Very often, the inconvenience of frequent job rotation provided added incentive to remove obstacles that were delaying implementation of powered test fire fixtures.

An additional benefit to the implementation of the powered test fire fixtures was the elimination of many injuries involving staples being accidentally fired into the fingers and hand during the manual test fire procedure. Limiting manual test firing also greatly limited the occurrence of this unpleasant event.

## 22.7 CRITICAL REVIEW OF CURRENT STATUS

Although many ergonomics problems can be eliminated through engineering redesign of processes and equipment, many others must be addressed through administrative and other control methods due to financial and feasibility constraints. The discussed ergonomic program components will improve the chances of proactively preventing problems, detecting existing problems early, providing a responsive management support environment, and, most importantly being able to implement solutions within a supportive, involved, and educated workforce. These components are

- Employee empowerment to identify and solve problems (focused ergonomics committees)
- Realistic, performance-oriented program objectives
- Upper management support and involvement
- Comprehensive medical case management
- Effective medical case reporting and tracking system
- Ergonomic job design guidelines
- Focus on prevention of problems (critical review of new processes and equipment—don't wait for problems to come to you)
- Tailored ergonomics training programs
- On-the-job exercise programs

The key to improving the probability of achieving a successful program and maintaining long-term management commitment lies in setting realistic, attainable, performance-oriented goals. Improvements will come, and goals will be accomplished, but only after a lot of hard work has been put into creating a strong foundation upon which the framework of a comprehensive ergonomics program can be built.

Setting goals strictly aimed at reducing the incidence of ergonomic medical cases, in the absence of performance-oriented objectives to initiate and maintain critical program components, often leads to the development of creative bookkeeping schemes that lead to force-fitting incidence rates to meet objectives. The unfortunate side effect of trying to force incidence numbers down is that the legitimacy of each medical case is often questioned. This phenomenon usually leads to the development of mistrust and negative perceptions between employer and employees. The basic management philosophy must be to accept and believe what employees are saying and to work directly with them to identify and resolve problems. If employees are continually scrutinized in an attempt to disprove work relatedness and company responsibility, a company philosophy of basic mistrust will be very quickly understood.

While J&J's ergonomics philosophy includes on-the-job exercise, for which many benefits have been documented in the literature, very limited scientific evidence has linked exercise with documented reductions in the incidence of ergonomics-related medical disorders. Although participants say that they enjoy the exercise programs and that they "feel better," significant reductions in postural discomfort and in the severity of medical symptoms have not been scientifically documented. Although exercise clearly has favorable preventative effects on the body, it has not been shown to be a successful therapeutic intervention when medical problems are well entrenched. Additional research is needed to better document the benefits of on-the-job exercise.

## 22.8 FUTURE CONCERNS

One of the most significant barriers confronting ergonomics practitioners, in selling the need for ergonomic changes, is the shortage of objective, practical guidelines governing the science. If more precise guidelines were available regarding the number of repetitions or the force or posture deviation levels at which problems will begin to precipitate, the ergonomist would be in a better position to know when to react with strong action plans in the best interests of the company and its employees. With current knowledge, there is always some level of guesswork and risk associated with any proposed action plan. An overly conservative approach may place the company under unfair economic constraints in an already highly competitive marketplace. If the ergonomist is too lenient in allowing stressful work practices to persist, CTDs and other ergonomics-related medical cases and employee complaints will begin to appear. There are always a multitude of possible solutions and very few precise answers when trying to determine to what extent actions must be taken to manage ergonomic risk. As discussed, it is critical to have an educated workforce whose members can recognize problems in the early stages and to create a sensitive surveillance network that facilitates early detection.

It is clear that many of the ergonomic "epidemics" that have occurred in various geographical locations around the world have resulted from lack of education of the workforce, which has allowed emotions to get out of control and myths to run rampant. The more the working population is educated on the recognition of the risk factors and medical symptoms relating to ergonomic problems, the more efficiently these problems can be understood, prevented, identified, and resolved. It is clear that the earlier the potential problems are identified, the better the chances are of resolving them with minimal effort and cost to the company along with minimal suffering to the people performing the work.

## REFERENCES

1. A. Longmate and T. Hayes, Making a difference at Johnson & Johnson: Some ergonomic intervention case studies, in *Industrial Ergonomics: Case Studies*, B. Pulat and D. Alexander, Eds., Industrial Engineering and Management Press, Norcross, GA, 1991, p. 181.
2. A. Longmate and C. Welker, Components of an industrial ergonomics program: The Johnson & Johnson experience, in *Industrial Ergonomics: A Practitioners Guide*, D. Alexander and B. Pulat, Eds., Industrial Engineering and Management Press, Atlanta, GA, 1985, p. 129.
3. T. Hansford, H. Blood, B. Kent, and G. Lutz, Blood flow changes at the wrist in manual workers after preventive interventions, *J. Hand Surg. 11A*:4 (1986).
4. B. Silverstein, T. Armstrong, A. Longmate, and D. Woody, Can in-plant exercise control musculoskeletal symptoms, *J. Occup. Med. 30*:12 (1988).
5. G. Lutz and T. Hansford, Cumulative trauma disorder controls: The ergonomics program at Ethicon Inc., *J. Hand Surg. 12A*:863 (1987).
6. N. Swanson and S. Sauter, The effects of exercise on the health and performance of data entry operators, in *Work with Display Units '92*, H. Luczak, A. Cakir, and G. Cakir, Eds., Elsevier, New York, 1993, p. 288.
7. K. Lee and L. Humphrey, Comparison of the relative effectiveness of ergonomic design and physical exercise, in *Ergonomics International '85*, I. D. Brown, Ed., Taylor & Francis, London, U.K., 1985, p. 556.
8. W. Laporte, The influence of a gymnastic pause upon recovery following post office work, *Ergonomics 9*:6 (1966).
9. D. Thompson, Effect of exercise breaks on musculoskeletal strain among data entry operators: A case study, in *Promoting Health and Productivity in the Computerized Office: Models of Successful Ergonomic Interventions*, S. Sauter and M. Dainoff, Eds., Taylor & Francis, London, U.K., 1986, p. 118.

CHAPTER 23

# Health Care Ergonomics*

Thomas R. Waters

## CONTENTS



* The findings and conclusions in this chapter are those of the author(s) and do not necessarily represent the views of the National Institute for Occupational Safety and Health. This work was done by a U.S. Govt. employee as part of his/her work and is not subject to copyright.

## 23.1 INTRODUCTION

Workers in the health care industry who perform physically demanding tasks as part of their jobs, such as nurses, nurses' aides, physical therapists, and health care technicians, are at high risk for the development of work-related musculoskeletal disorders (MSDs). These workers are exposed to major risk factors for MSDs, such as lifting and moving heavy patients and equipment, pushing and pulling heavy equipment, working in extreme postures, and standing for long periods of time without adequate rest periods. When the demands of the job (physical demands, work environment, and workplace culture) exceed the capacity of the worker, the risk of an MSD is increased, and is highest when exposures are intense and prolonged and when there are several risk factors present at the same time [1–3].

Although health care workers have widely reported that injuries to the shoulder and neck have prevented them from doing their work and caused them to leave their job, by far, the most common occupational injuries in the health care industry involve back disorders [4]. Nurses working in the private sector had 11,800 MSDs reported in 2001, with the majority (nearly 9,000) of the reported injuries to the back, and more than a third (36%) of the injuries resulting in lost time from work were reported to be back injuries. Another study revealed that 12% of nurses planning to leave the profession indicated back injuries were either a main or contributing factor [5].

Work-related MSDs are very important in the health care industry. Employees that experience pain and fatigue are less productive, less attentive, more prone to make consistent mistakes, more susceptible to injury, and may be more likely to affect the health and safety of others. Nurses suffering from disabling back injuries and fear of getting injured have contributed to nurses leaving the profession, thus increasing the nursing shortage [5]. Workplaces with high incidences of MSDs report increases in lost/modified workdays, higher staff turnover, increased costs, and adverse patient outcomes [6]. The extent of the problem is likely worse than reported in official injury reporting records due to likely widespread underreporting of injuries. In a study by Cato et al. [7], it was reported that 78% of nurses with back pain in the previous 6 months did not report it to management. In another study of nurses, Owen [8] reported that 67% of nurses who reported low-back pain related to work did not report the incident in writing. Others have consistently demonstrated that injury reports on the Occupational Safety and Health Administration (OSHA) log are underreported.

In the past decade, tremendous strides have been made in identifying high-risk tasks and in developing and implementing solutions for reducing the risk. Programs have been developed and many states have passed legislation mandating implementation of these safety programs. Ergonomic guidelines have been developed to provide useful information to the industry that will help them efficiently implement the programs that have been shown to be effective in reducing risks and costs associated with work-related MSDs. Most importantly, easy-to-use equipment that reduces the amount of physical demand required to perform these tasks has been developed for many of the high-risk tasks and is widely available. Also, studies have been conducted demonstrating that implementation of these programs and use of this equipment is cost effective, often paying for itself in less than 3 years time. Lastly, there has been recent evidence that implementation of a safe patient-handling (SPH) program can also increase the quality of care provided to the patient [9].

It is clear that a large number of workers are routinely exposed to work-related risk factors for MSDs, often at very high levels of exposure. It is also clear that many of them develop one or more serious work-related MSDs as a consequence of their work during their working lifetime. Nevertheless, it should be recognized that these hazards can be identified and controlled so that work-related MSDs can be prevented.

## 23.2 HIGH-RISK TASKS AND RISK FACTORS

By far, the most common risk factor for MSDs is performance of patient-handling tasks. Patient handling occurs in numerous acute care health care environments including, critical care, orthopedic, surgical, rehabilitation, and radiology units of hospitals, as well as long-term health care facilities, nursing homes, and in home care environments. Transferring partially or fully dependent patients from beds to chairs or wheelchairs, sliding patients up in bed, transferring patients from stretcher to stretcher, or anybody positioning activity requiring extreme muscular exertions, or work in awkward or extremely bent or twisted postures have been shown to be very high risk for the development of work-related MSDs [10].

Another significant risk factor for MSDs is pushing and pulling heavy equipment, such as occupied beds, unoccupied surgical beds, diagnostic equipment, and floor-based patient lifting devices fully loaded with a patient, food carts, and carts for dispensing medicine. Often these tasks require pushing and pulling of objects weighing several hundred pounds, sometimes over long distances, across uneven or carpeted surfaces. These tasks have been identified as high risk for work-related MSDs [11–13].

An additional risk factor for health care workers is working in awkward or extreme body postures. This may occur during specific nursing tasks, such as when inserting IVs or catheters, or during treatment activities, such as applying embolism stockings, or setting up treatment equipment at the bedside. Bed height is a critical factor associated with the extent of exposure to awkward postures. If the bed or table height is not adjusted properly, the worker will have to bend or reach in order to perform the specific task. For example, in an operating room, a health care worker may have to spend many hours standing in a static, flexed posture with the back and arms in a specific posture while retracting tissues or holding assistive devices during the performance of the surgical procedure.

A variety of other risk factors are also present in the health care industry, including long work hours, mandatory overtime, shift work, exposure to work stress, and other work organizational risk factors that have been shown to increase the incidence and severity of work-related MSDs [1,2,14,15].

Research evidence related to high-risk tasks in specific health care settings is limited, but there is evidence that the highest risk settings include operating rooms, medical/surgical wards, psychiatry, rehabilitation/spinal cord injury, critical care units, trauma/emergency units, long-term care, and home care [16,17].

## 23.3 OTHER FACTORS AFFECTING RISK OF MSDs

### 23.3.1 Overreliance on "Body Mechanics" as a Prevention Strategy

From about 1945 to the present, the health care industry began recommending reliance on "body mechanics" as a way to protect the caregiver from risk of injury due to patient-handling activities. Although there was little or no empirical evidence that this approach would actually reduce the risk of work-related MSDs, it was widely adopted because other solutions were not available. Recently, researchers have shown that there is no safe way to manually lift and transfer a fully dependent patient weighing as little as 110 lb, even with two caregivers performing the task [10]. Unfortunately, many schools of nursing and physical therapy continue to teach these outdated patient-handling methods.

### 23.3.2 Majority of Exposed Population Is Female

The direct patient care segment of the health care industry employs the highest percentage of female employees of any industry. Because females generally have lower strength and lifting capacity than males, on average, they must work at a higher percentage of their maximum physical capabilities than males when performing the same physically demanding tasks. Thus, the risk to the worker likely would be greater for females than for males when performing most patient care tasks. Therefore, tasks must be designed such that the strength demands are sufficiently low (i.e., designed for female workers), so that nearly all workers will be able to safely perform the task.

### 23.3.3 Shortage of Nursing Staff

It has been reported that there is a shortage of nursing staff at many health care facilities [18]. There is little doubt that work-related MSDs contribute to the critical nursing shortage in the health care industry that also leads to more overtime for working nurses. According to a 2006 report by the Health Resources and Services Administration (HRSA), by the year 2020, the supply of RNs is projected to fall 29% below predicted requirements [18]. Factors that contribute to the shortage include professional burnout, unappealing work climates, lack of job satisfaction, quality of care issues, and lack of managerial support [18]. It is likely that excessive physical workloads associated with direct patient care, such as manual patient handling, increasing patient weights, and use of heavy equipment contribute to the shortage of nurses. Thus, a reduction in the physical demands of the work tasks by using assistive technology will help keep nurses in the profession, thereby helping to maintain nurse staffing levels.

### 23.3.4 Aging Workforce

As in most industries in the United States, the health care industry workforce is getting older [19], and this trend is likely to continue as the retirement age to qualify for full Social Security benefits increases. The American Nurses Association (ANA) has reported the average age of a registered nurse in the United States to be 46.8 years of age. From an ergonomics perspective, increasing worker age is a concern because on average, physical strength and muscular endurance generally decreases with age, especially above 50 years of age. Since the risk of MSD is a function of the ratio of strength demands-to-worker capacity, any reduction in physical strength would increase MSD risk for a specific task. This is because when the workers strength is diminished, the workers will have to expend a higher percentage of their maximum capacity. Although older workers often have developed more knowledge and better skills, these improvements cannot compensate for decreased physical capacity, especially when the physical demands are high. Therefore, an effective ergonomic approach would be to provide assistive technology to reduce job demands so that aging workers can continue to safely perform their job. This approach has the added benefit of broadening the pool of workers, which will alleviate the nursing shortage discussed previously [20].

### 23.3.5 Obesity Epidemic

Another changing demographic characteristic of concern to the health care industry is the rapid increase in body weight of the U.S. population over the past three decades, including patients and caregivers [21]. The current epidemic of obesity in the United States is playing a major role in the increased risk of MSDs for health care workers. The percentage of U.S. citizens who are classified as obese has increased from 15% in 1980 to 35% in 2006 [22]. The percentage of citizens who are reported to be either overweight or obese (BMI $\geq$ 25) was 68.0% in 2007–2008. The percentage of U.S. citizens who are classified as extremely obese (BMI of 40 or higher) has increased from 1.4% in 1980 to 6.2% in 2006 [22]. Patients are becoming heavier, and it is not uncommon to see patients requiring hospitalization who weigh over 400 lb. As the weight of the patient increases, the risk of injury to the patient and the health care worker who must transfer, move, and treat the patient also increases. Also, many tasks that may have been considered acceptable to perform manually in the past are no longer safe for the caregiver, such as lifting an arm or leg for treatment or reaching across a patient to perform a task.

### 23.3.6 Earlier Hospital Discharge Times

In addition to the risk factors listed earlier, shorter hospital stay times may play a role in increasing the risk of MSDs for health care workers. Due to spiraling health care costs, patients are not staying in the hospital following surgery and other treatments as long as they did in the past. In 1980, for example, the average length of hospital stay was 7.5 days compared with only 4.9 days in 2001 [23]. This reduction in hospital stay time has resulted in two significant effects. First, there is a concentration of acute patient needs associated with patient transfers and movement while in the acute care environment. Second, there is now a need for a higher level of patient transfer assistance in the home care environment at an earlier stage of recovery than was previously required. Unfortunately, the home care environment is often lacking in the availability of assistive patient-handling technology. Both of these factors have increased the risk of work-related MSDs for health care workers.

## 23.4 PREVENTION AND CONTROL

In the past, attempts to prevent work-related MSDs due to heavy physical exertion associated with lifting, pushing, pulling, and carrying focused on adopting arbitrary weight or force limits for handling loads, hiring strong workers, or using training procedures that emphasized "correct" (but not necessarily safe) lifting techniques. None of the approaches noted earlier, however, has proven to be effective in significantly reducing risk of overexertion injuries [24]. As previously noted, since the 1940s, the health care industry has relied almost exclusively on teaching "good body mechanics" for controlling the risks of patient-handling injuries. This approach, however, has been shown to be ineffective in reducing risk of work-related MSDs. More recently, industry has recognized the risks associated with performance of physically demanding tasks, and to reduce costs and increase productivity, companies have begun to implement ergonomic programs or practices aimed at preventing these injuries. The core element of these programs is reliance on use of state-of-the-art equipment designed to assist the worker in carrying out the prescribed task. Two excellent resource books that provide comprehensive and detailed descriptions of effective evidence-based SPH programs are available. The first is a Practical Guide for Safe Patient Handling and Movement for Health Care Professionals [16] and the second is The Illustrated Guide to Safe Patient Handling and Movement [25]. For guidelines on solving nonpatient-handling manual material–handling problems, the reader is referred to the National Institute for Occupational Safety and Health (NIOSH) Guidelines for Manual Material Handling [26].

### 23.4.1 SPHM Algorithms

The patient safety center of inquiry, a research arm of the Veterans Health Administration (VHA), in collaboration with a number of partners, has developed a series of algorithms for assisting in assessing the safest and most effective approach to various patient-handling tasks. These algorithms are simple decision logic that allows caregivers to quickly determine when assistive technology should be used and how many caregivers are needed to perform a specific patient-handling task. The algorithms cover patient-handling tasks such as bed to chair, lateral transfers between beds, movement on a bed, lift from the floor, etc. The algorithms can be found at the Veterans Health Administration (VA) patient-handling website [27].

### 23.4.2 Guidelines for Bariatric Patient Handling

The VHA, in collaboration with partners, has developed a bariatric toolkit for assessing patient-handling tasks for bariatric patients that can also be found at the VA website [27]. A bariatric patient is defined as an individual who exceeds standard capacity equipment (300 lb) with a Body Mass Index (BMI) of 50 or higher. The BMI for a person is calculated by multiplying their weight (kg) times their height squared ($m^2$). The bariatric guidelines, which can be found at the VHA website, provide patient-handling algorithms for (1) transfer to bed to chair, chair to toilet, or chair to chair; (2) bariatric lateral transfer to and from bed to stretcher or trolley; (3) bariatric reposition in bed: side to side, up in bed; (4) bariatric reposition in chair: wheelchair, chair, or dependency chair; (5) patient-handling tasks requiring sustained holding of limb to access body parts; (6) bariatric transporting (stretcher, wheelchair, and walker); and (7) bariatric toileting tasks.

### 23.4.3 Ergonomic Recommendations for Nursing Home and Home Health Care

Researchers at NIOSH have developed recommendations for prevention of work-related MSDs in nursing homes [28,29]. Patient handling in nursing homes typically involves work with older, increasingly frail patients, who may have significant impairments, complicating even simple transfer tasks. The NIOSH recommendations demonstrate the effectiveness of implementing SPH programs in nursing homes, show how use of technology in the context of an overall program significantly reduce risk of MSDs, and the economic benefits of purchasing equipment and implementing an SPH program. In fact, the cost of obtaining patient-handling equipment and implementing an SPH program can be recovered in as few as 3 years of implementation.

OSHA recently published an ergonomics guideline that provided an overview of the risks of work-related MSDs in nursing homes, provided information about the most effective approaches for mitigating or reducing those risks, and discussed training needs [6]. One of the most important statements in the OSHA nursing home guideline was that "manual lifting of residents be minimized in all cases and eliminated when feasible." OSHA also recommended that employers implement an effective ergonomics process that provides management support, involves employees, identifies problems, implements solutions, addresses reports of injuries, provides training, and evaluates ergonomics efforts. The document also provided a section titled "Identifying Problems and Implementing Solutions for Activities Other Than Resident Lifting and Repositioning," which included sections dealing with storage and transfer of food, supplies, and medications; mobile medical equipment; working with liquids in housekeeping; working with liquids in kitchens, hand tools, and linen carts; handling bags; reaching into sinks; loading or unloading laundry; and cleaning rooms (wet method and electrical).

### 23.4.4 Ergonomic Guidelines for Home Health Care Settings

NIOSH researchers have also developed a series of guidelines for application of ergonomics in the home health care environment [30–33]. These guidelines describe the high-risk environment associated with patient handling in the individual's home. Home health care workers are at additional risk of MSDs due to lack of available assistance, space restrictions, lack of assistive technology, and lack of control over the workplace environment. These guidelines provide a range of ideas and solutions for reducing risk to home health care workers due to patient handling.

### 23.4.5 Weight Limit Recommendations for Patient Lifting

Recently, based on the Revised NIOSH Lifting Equation, NIOSH researchers assessed typical patient-handling tasks and suggested that no caregiver should manually lift more than 35 lb of person's body weight for a vertical lifting task [34]. The author recommended that when the weight to be lifted exceeded this limit, then assistive devices should be used. This recommendation has been adopted by the VHA and incorporated into their current patient-handling algorithms.

### 23.4.6 NAON Ergonomics Guidelines

The National Association for Orthopaedic Nursing also recently developed ergonomics guidelines for prevention of work-related MSDs in the orthopedic setting [35–40]. The guidelines included recommendations for (1) turning a patient in bed from side to side [36];

(2) performing a vertical transfer of a postoperative total hip replacement patient from a bed to chair, chair to toilet, chair to chair, or car to chair [37]; (3) performing a vertical transfer of a patient with an extremity cast/splint [38]; (4) assisting an orthopedic patient with ambulation [39]; and (5) lifting a leg or an arm with a cast [40].

### 23.4.7 AORN Ergonomics Guidelines

A task force organized by the VA and the Association for periOperative Registered Nurses (AORN) recently developed ergonomic guidelines for prevention of work-related MSDs in the operating room setting [13,41–46]. These guidelines were developed by a task force including representatives from AORN, the NIOSH, the Patient Safety Center of Inquiry at the James A. Haley Veterans Administration Medical Center (VMAC) in Tampa, and the ANA. The task force was formed to identify high-risk tasks performed in the perioperative area and to develop evidence-based solutions to minimize the risk of MSDs. The guidelines consist of seven clinical tools/algorithms to guide perioperative registered nurses and other perioperative team members incorporating current ergonomic safety concepts, scientific evidence, and available technology including SPH equipment and devices. The seven tools include lateral transfer from stretcher to operating room (OR) bed [41]; repositioning patients on OR beds [42]; lifting and holding legs, arms, and heads for prepping [43]; prolonged standing [44]; holding retractors for extended periods of time [45]; lifting and carrying supplies/equipment [46]; and pushing, pulling, and moving equipment on wheels [13].

### 23.4.8 APTA Guidelines

Physically demanding patient-handling and transferring tasks in the rehabilitation setting have been shown to present high risk for the development of work-related MSDs for therapists. Factors that have been shown to increase risk of work-related MSDs in the rehabilitation setting include patient transfers, patient repositioning, working in bent or twisted postures, joint mobilization, soft tissue work, PROM treatments, and job strain [47]. Traditional tasks have a practical goal, such as transferring a patient from a bed to wheelchair, while therapeutic tasks have more targeted goals of facilitating patient function and independence. Therapeutic patient-handling tasks likely present greater risk of WMSD to the caregiver than do typical patient-handling tasks due to the longer times of exposure to the high mechanical loads on the spinal tissues of the worker. The American Physical Therapy Association (APTA) recognized the importance of ergonomics in prevention of work-related MSDs during rehabilitation treatment. The APTA in collaboration with the VHA developed a white paper outlining the need for ergonomics in patient handling for the rehabilitation setting [48]. The APTA white paper contained six important recommendations, including the following:

1. Implement the *OSHA Ergonomics for Prevention of Musculoskeletal Disorders: Guidelines for Nursing Homes.*
2. Build and support a culture of safety in rehabilitation settings that protects staff as well as patients.

3. Improve communication channels between nurses and physical therapists to facilitate SPH and movement tasks.

4. Develop policies and procedures for the therapeutic use of patient-handling equipment, including selection of equipment that first provides safety for staff and patients and equipment with features, as appropriate, that allow for or promote active use of the assistive equipment by the patient for therapeutic benefit.

5. Develop competency-based assessments that demonstrate proficiency for use of all patient-handling equipment used on the respective patient care unit, including return demonstration (i.e., training through role playing and teaching other staff members).

6. Encourage research that supports the improvement of patient and staff safety while maximizing patient rehabilitation potential, including investigation of the cost-effectiveness of ergonomics interventions, impact of injury-risk reduction to physical therapists, and determination of the efficacy of patient-handling equipment when integrated into therapeutic activities.

### 23.4.9 Safe Patient-Handling Training for Schools of Nursing

In an attempt to provide information on SPH concepts to nurses at an early stage in their careers, NIOSH recently published curricular materials for a training program on SPH for use in schools of nursing [49]. The training materials include a narrated slide show with imbedded video and a booklet with the actual training materials that can be downloaded for free from the NIOSH website: www.cdc.gov/niosh/. The training materials have been shown to be effective in changing the knowledge, attitude, and beliefs regarding evidence-based SPH concepts and procedures [50].

### 23.4.10 State-of-the-Art Ergonomic Technology

Recently, the health care industry has recognized the risks associated with performance of physically demanding patient-handling tasks, and to reduce costs and increase productivity, companies have begun to implement ergonomic programs or practices aimed at preventing these injuries. The core element of these programs is reliance on use of state-of-the-art ergonomically designed equipment to assist the worker in carrying out the prescribed task. As an added incentive to adopt technology-based patient-handling practices, OSHA recently published an ergonomics guideline that provided an overview of the risks of work-related MSDs in nursing homes. The guideline provided information about the most effective approaches for mitigating or reducing those risks, and discussed training needs [6]. One of the most important statements in the OSHA nursing home guideline was that "manual lifting of residents be minimized in all cases and eliminated when feasible." This is best accomplished by implementing a technology-based SPH program.

A wide range of ergonomic patient-handling technology is available that can be applied to most types of patient-handling activities [51–54]. The most common patient-handling activities include (1) *lateral transfers* of patients between two lateral surfaces, such as bed to bed or between a bed and an examining table; (2) *vertical transfers* of patients, such as from a bed to a chair or between chairs; (3) *standing a patient from a sitting position*; (4) *ambulation*; (5) *transporting heavy equipment*, such as pushing or pulling occupied or unoccupied beds, heavy treatment equipment, OR beds, etc.; (6) *repositioning people in bed*, side to side or up and down; and (7) *working in extreme or awkward postures* for long periods of time, including standing. Each specific health care setting may have unique requirements for patient-handling technology, such as operating rooms, critical care environments, orthopedic, home care, and rehabilitation settings. Examples of common state-of-the-art patient-handling devices are shown in Figure 23.1a–f.

FIGURE 23.1 Examples of common state-of-the-art patient-handling devices.

TABLE 23.1 Categories of Ergonomic Patient-Handling Technology Available

| | |
|---|---|
| Air-assisted lateral transfer aids | Bed systems (frames/surfaces/assist devices/other bed safety devices) |
| Car lifts/vehicle extraction | Ceiling lifts |
| Dependency/Geri/specialty chairs/wheelchairs | Floor-based lifts |
| Friction reducing lateral transfer aids | Gait belts w/handles |
| Lifts (other) | Mechanical lateral transfer aids |
| Other mobility aids | Powered standing lifts |
| Repositioning devices | Sliding boards |
| Slings | Standing assist aids |
| Transport devices/powered beds | |
| *Bariatric devices (patient BMI > 40)* | |
| Bariatric ambulatory/mobility aids | Bariatric bathing equipment (other) |
| Bariatric beds/mattresses/transportation | Bariatric ceiling lifts |
| Bariatric commodes/shower chairs | Bariatric lateral transfer systems |
| Bariatric repositioning systems | Bariatric powered lifts |
| Bariatric standing assist aids | Bariatric transfer/dependency chairs and cushions |
| Bariatric wheelchairs | |

A wide range of assistive devices have been developed for each type of patient-handling activity. The VHA has developed a technology resource guide that lists a wide range of technology for use in SPH and movement. The technology resource guide classifies patient-handling technology into the broad categories listed in Table 23.1. Brief descriptions of the most important categories of technology are provided later. The VHA technology resource guide can be found at the VA website [27]. The OSHA Ergonomics Guidelines for Nursing Homes also provides an overview of technology available for SPH [6].

## 23.5 CRITICAL REVIEW OF CURRENT STATUS

### 23.5.1 Gaps in Technology

Although a wide range of equipment is available, gaps in technology remain. For example, ceiling lift devices for vertical transfer of patients in operating rooms where the patient may need to be moved laterally but also may need to be turned from supine to prone position are lacking. Also, more devices specifically designed for use in rehabilitation settings where the handling equipment must provide dual function are needed. That is, the equipment must be designed so that it can provide adequate assistance for patient transfers when needed, but also must be flexible enough so that it can quickly be adjusted for use during therapeutic procedures where the patient is required to use more and more of their own capabilities as they recover. Also, more research is needed to identify issues related to equipment usage. There are three areas of gaps in technology application. These include the following:

1. Equipment that is needed, but has not been developed.
2. Equipment that has been developed, but facilities have not obtained it.
3. Equipment that has been developed and facilities have obtained it, but it is not being used.

These issues need to be addressed in order to maximize the effectiveness of technology-based solutions in health care.

### 23.5.2 Implementation and Sustainability of SPH Programs

There are a number of research questions that need to be answered. These include the following:

- What elements are most important to make SPH programs effective?
- How can SPH programs be made more sustainable?
- What are the barriers to implementation of SPH programs and how can they be overcome?
- What is the impact of regulation on implementation and sustainability of SPH programs?
- How do SPH programs affect the quality of care for patients?
- Can quality-of-care issues be a driver for increased implementation of SPH programs?
- How does cost–benefit analysis impact implementation and sustainability of SPH programs?

## 23.6 FUTURE CONCERNS

Problems with physical therapy
Aging and need for higher levels of care for older individuals
Health care costs

In conclusion, there is strong evidence that (1) manual patient-handling presents high risk for the development of work-related MSDs; (2) reliance on body mechanics alone is not effective in reducing the risk of patient-handling injuries; (3) effective technological solutions are available to reduce or eliminate the risk of these health problems; and (4) implementation of an SPH program that relies on use of state-of-the-art technology is cost effective and, on average, will provide a positive return on investment within 3 years of implementation. Unfortunately, many health care settings are still relying on unsafe manual lifting techniques rather than investing in newer, safer patient-handling technology. Recent state legislation and possible U.S. federal legislation may force implementation of programs, but proactive ergonomics based on implementation of a SPH program will provide a safe work environment for both workers and for patients.

## REFERENCES

1. NIOSH, *Musculoskeletal Disorders (MSDs) and Workplace Factors*, National Institute for Occupational Safety and Health, DHHS (NIOSH) Publication No. 97-141, Cincinnati, OH, 1997.
2. National Research Council and Institute of Medicine, *Musculoskeletal Disorders and the Workplace: Low Back and Upper Extremities*, Panel on Musculoskeletal Disorders and the Workplace Commission on Behavioral and Social Sciences and Education, Washington, DC, 2001.

3. T. R. Waters, Introduction to ergonomics for healthcare workers, *Rehabilitation Nursing*. 35(5): 185–191, 2010.
4. BLS, *Nonfatal Occupational Injuries and Illnesses Requiring Days Away From Work*, Retrieved from Bureau of Labor Statistics website on February 2, 2010: http://www.bls.gov/iif/oshwc/osh/case/osnr0029.pdf, 2007.
5. D. A. Stubbs, P. W. Buckle, M. P. Hudson, and D. Baty, Backing out: Nurse wastage associated with back pain, *International Journal of Nursing Studies*. 23(4): 325–336, 1986.
6. OSHA, *Ergonomics for the Prevention of Musculoskeletal Disorders: Guidelines for Nursing Homes*, U.S. Department of Labor, Occupational Safety and Health Administration, Document No. OSHA 3182-3R, Washington, DC, 2009.
7. C. Cato, D. K. Olson, and M. Studer, Incidence, prevalence, and variables associated with low back pain in staff nurses, *AAOHN Journal*. 37(8): 321–327, 1989.
8. B. D. Owen, The magnitude of low back problems in nursing, *Western Journal of Nursing Research*. 11(2): 234–242, 1989.
9. A. Nelson, J. Collins, K. Siddharthan, M. Matz, and T. Waters, Link between safe patient handling and patient outcomes in long-term care, *Rehabilitation Nursing*. 33(1): 33–43, 2008.
10. W. S. Marras, K. G. Davis, B. C. Kirking, and P. K. Bertsche, A comprehensive analysis of low-back disorder risk and spinal loading during the transferring and repositioning of patients using different techniques, *Ergonomics*. 42(7): 904–926, 1999.
11. W. S. Marras, G. G. Knapik, and S. Ferguson, Lumbar spine forces during manoeuvring of ceiling-based and floor-based patient transfer devices, *Ergonomics*. 52(3): 384–397, 2009.
12. M. Rice, S. Wooley, and T. Waters, Comparison of required operating forces between floor-based and overhead-mounted patient lifting devices, *Ergonomics*. 52(1): 112–120, 2009.
13. T. Waters, J. Lloyd, E. Hernandez, and A. Nelson, AORN Ergonomic Tool 7: Pushing, pulling and moving equipment on wheels in the perioperative setting, *AORN Journal*. 94(3): 254–260, 2011.
14. T. Waters, J. Collins, T. Galinsky, and C. Caruso, NIOSH research efforts to prevent musculoskeletal disorders in the healthcare industry, *Orthopaedic Nursing*. 25(6): 380–389, 2006.
15. C. Caruso and T. Waters, Work schedule issues and musculoskeletal disorders with an emphasis on the healthcare sector, *Industrial Health*. 46: 523–534, 2008.
16. A. Nelson, *Safe Patient Handling and Movement: A Practical Guide for Health Care Professionals*, Springer Publishing Co. Inc., New York, 2006.
17. T. Waters, A. Nelson, and C. Proctor, Patient handling tasks with high risk for musculoskeletal disorders in critical care, *Critical Care Nursing Clinics of North America*. 19: 131–143, 2007.
18. CHWS, *The Impact of the Aging Population on the Health Workforce in the United States: Summary of Key Findings*, Center for Health Workforce Studies, School of Public Health, University at Albany, Rensselaer, NY, pp. 1–19, 2006.
19. J. Elwood, The aging nursing population, Article in *Springfield Business Journal*, Accessed online at Springfield Business Journal online on January 13, 2011, at: http://sbj.net/main.asp?Search=1&ArticleID=79035&SectionID=48&SubSectionID=108&S=1, 2007.
20. T. R. Waters and L. A. MacDonald, Ergonomic job design to accommodate and prevent musculoskeletal disabilities, *Assistive Technology*. 13: 88–93, 2001.
21. K. M. Flegal, M. D. Carroll, C. L. Ogden, and L. R. Curtin, Prevalence and trends in obesity among US adults, 1999–2008, *JAMA*. 303(3): 235–241, 2010.
22. National Center for Health Statistics, *Prevalence of Overweight, Obesity and Extreme Obesity among Adults: United States, Trends 1960–1962 through 2005–2006*, Retrieved from NCHS website on February 2, 2010, at: http://www.cdc.gov/nchs/data/hestat/overweight/overweight_adult.htm, 2008.
23. National Center for Health Statistics, *2005 National Hospital Discharge Survey*, Retrieved from NCHS website on February 2, 2010, at: http://www.cdc.gov/nchs/data/ad/ad385.pdf, 2007.

24. S. H. Snook, Approaches to the control of back pain in industry: Job design, placement, and education/training, *SPINE: State of the Art Reviews*. 21(1): 45–59, 1987.
25. A. Nelson, K. Motacki, and N. Menzel, *The Illustrated Guide to Safe Patient Handling and Movement*, Springer Publishing Co. Inc., New York, 2009.
26. NIOSH, *Ergonomic Guidelines for Manual Material Handling*, National Institute for Occupational Safety and Health, DHHS (NIOSH) Publication No. 2007-131, Cincinnati, OH, 2007.
27. Veterans Health Administration, *Safe Patient Handling and Movement Website*, Accessed on January 20, 2011, at: http://www.visn8.va.gov/patientsafetycenter/safePtHandling/default.asp, 2011.
28. J. W. Collins, L. Wolf, L. J. Bell, and B. Evanoff, An evaluation of a best practices musculoskeletal injury prevention program in nursing homes, *Injury Prevention*. 10: 206–211, 2004.
29. NIOSH, *Safe Lifting and Movement of Nursing Home Residents*, National Institute for Occupational Safety and Health, DHHS (NIOSH) Publication No. 2006-117, Cincinnati, OH, 2006.
30. K. S. Parsons, T. L. Galinsky, and T. R. Waters, Suggestions for preventing musculoskeletal disorders in home healthcare workers, Part 1: Lift and transfer assistance for partially weight-bearing home care patients, *Home Healthcare Nurse*. 24(3): 158–164, 2006.
31. K. S. Parsons, T. L. Galinsky, and T. R. Waters, Suggestions for preventing musculoskeletal disorders in home healthcare workers, Part 2: Lift and transfer assistance for non-weight-bearing home care patients, *Home Healthcare Nurse*. 24(4): 228–233, 2006.
32. NIOSH, *NIOSH Hazard Review Occupational Hazards in Home Healthcare*, National Institute for Occupational Safety and Health, DHHS (NIOSH) Publication No. 2006-117, Cincinnati, OH, 2010.
33. T. Galinsky, T. Waters, and B. Malit, Overexertion injuries in home health care workers and need for ergonomics, *Home Health Care Services Quarterly*. 20(3): 57–73, 2001.
34. T. R. Waters, When is it safe to manually lift a patient? *American Journal of Nursing*. 107(8): 53–59, 2007.
35. C. A. Sedlak, M. O. Doheny, A. Nelson, and T. R. Waters, Development of the National Association of Orthopaedic Nurses guidance statement on safe patient handling and movement in the orthopaedic setting, *Orthopaedic Nursing*, 28(2S): S2–S8, 2009.
36. C. M. Gonzalez, C. M. Howe, T. R. Waters, and A. Nelson, Recommendations for turning patients with orthopaedic impairments, *Orthopaedic Nursing*. 28(2S): S9–S12, 2009.
37. C. M. Gonzalez, C. M. Howe, T. R. Waters, A. Nelson, and N. Hughes, Recommendations for vertical transfer of a postoperative total hip replacement patient (bed to chair, chair to toilet, chair to chair, or car to chair), *Orthopaedic Nursing*. 28(2S): S13–S17, 2009.
38. M. Patterson, P. Mechan, N. Hughes, and A. Nelson, Safe vertical transfer of patient with extremity cast or splint, *Orthopaedic Nursing*. 28(2S): S18–S23, 2009.
39. S. M. Radiwiec, C. Howe, C. M. Gonzalez, T. R. Waters, and A. Nelson, Safe ambulation of an orthopaedic patient, *Orthopaedic Nursing*. 28(2S): S24–S27, 2009.
40. T. R. Waters, C. A. Sedlak, C. M. Howe, C. M. Gonzalez, M. O. Doheny, M. Patterson, and A. Nelson, Recommended weight limits for lifting and holding limbs in the orthopaedic setting, *Orthopaedic Nursing*. 28(2S): S28–S32, 2009.
41. T. R. Waters, A. Baptiste, M. Short, L. Plante-Mallon, and A. Nelson, AORN Ergonomic Tool 1: Lateral transfer of a patient from a stretcher to an OR bed, *AORN Journal*. 93(3): 334–339, 2011.
42. T. R. Waters, M. Short, J. Lloyd, A. Baptiste, L. Butler, C. Petersen, and A. Nelson, Ergonomic Tool 2: Positioning and repositioning the supine patient on the OR bed, *AORN Journal*. 93: 445–449, 2011.

43. T. R. Waters, P. Spera, C. Petersen, A. Nelson, E. Hernandez, and S. Applegarth, AORN Ergonomic Tool 3: Lifting and holding legs, arms, and heads for prepping in a perioperative setting, *AORN Journal*. 93: 589–592, 2011.
44. N. Hughes, M. Matz, and A. Nelson, AORN Tool for assessing prolonged standing, *AORN Journal*. 93: 767–774, 2011.
45. P. Spera, J. Lloyd, E. Hernandez, N. Hughes, D. Spratt, and A. Nelson, AORN guideline for retractor use, *AORN Journal*. 94(1): 54–58, 2011.
46. T. Waters, M. Short, A. Baptiste, L. Mallon, and A. Nelson, AORN Ergonomic Tool 6: Lifting and carrying supplies/equipment in a perioperative setting, *AORN Journal*. 94(2): 173–179, 2011.
47. T. Waters and K. Rockefeller, Safe patient handling and movement from a physical therapy perspective, *Rehabilitation Nursing*. 35(5): 216–222, 2010.
48. APTA, *Improving Patient and Health Care Provider Safety: Task Force Develops Recommendations on Patient Handling*, White Paper published by the American Physical Therapy Association, Accessed from APTA web site on October 10, 2008, at: http://www.apta.org/AM/Template.cfm?Section=Home&CONTENTID=18516&TEMPL..., 2004.
49. NIOSH, *Safe Patient Handling Training for Schools of Nursing: Curricular Materials*, National Institute for Occupational Safety and Health, DHHS (NIOSH) Publication No. 2009-127, Cincinnati, OH, 2009.
50. A. Nelson, T. Waters, N. Menzel, N. Hughes, P. Hagan, G. Powell-Cope, C. Sedlak, and V. Thompson, Effectiveness of an evidence-based curriculum module for schools of nursing targeting safe patient handling and movement, *International Journal of Nursing Education and Scholarship*. 4(1), Article 26, 2007.
51. A. Baptiste, *New and Emerging Technologies for Safe Patient Handling and Movement*, Campden Publishing, Ltd., London, U.K., 2005.
52. A. Baptiste, Technology solutions for high-risk tasks in critical care, *Critical Care Nursing Clinics of North America*. 19(2): 177–186, 2007.
53. A. Baptiste, E. Meittunen, and G. Bertschinger, Technology solutions for bariatric populations, *Journal of Association of Occupational Health Professionals in Healthcare*. XXIV(2): 18–22, 2004.
54. A. Baptiste, S. Boda, A. Nelson, J. Lloyd, and W. Lee, Friction-reducing devices for lateral patient transfers: A clinical evaluation, *American Association of Occupational Health Nurses*. 54(4): 173–180, 2006.

CHAPTER 24

# Injuries and Ergonomic Applications in Construction

Hongwei Hsiao and Ronald L. Stanevich

CONTENTS



## 24.1 INTRODUCTION

Construction work is very hazardous. Workers frequently perform their tasks at elevations, work with heavy construction machinery, face exposure to various types of hazardous energy such as electrical energy, or manually handle a wide variety of bulky, heavy materials. In addition, at construction sites, the work environment, the work to be done,

TABLE 24.1 Incidence Rate[a] per 100 Full-Time Workers in Major Industries

| Industry | 1986 | 1987 | 1988 | 1989 | 1990 |
|---|---|---|---|---|---|
| Construction | 15.2 | 14.7 | 14.6 | 14.3 | 14.2 |
| Manufacturing | 10.6 | 11.9 | 13.0 | 13.1 | 13.2 |
| Agriculture | 11.2 | 11.2 | 10.9 | 10.9 | 11.6 |
| Transportation | 8.2 | 8.4 | 8.9 | 9.2 | 9.6 |
| Mining | 7.4 | 8.5 | 8.8 | 8.5 | 8.3 |
| Trade[b] | 7.7 | 7.7 | 7.8 | 8.0 | 7.9 |
| Service | 5.3 | 5.5 | 5.4 | 5.5 | 6.0 |
| Finance[c] | 2.0 | 2.0 | 2.0 | 2.0 | 2.4 |

*Source:* From U.S. Department of Labor, *Occupational Injuries and Illnesses in the United States by Industry, 1987,* Bureau of Labor Statistics Bull. 2328, U.S. Department of Labor, Washington, DC, 1989; U.S. Department of Labor, *Occupational Injuries and Illnesses in the United States by Industry, 1989,* Bureau of Labor Statistics Bull. 2379, U.S. Department of Labor, Washington, DC, 1991; U.S. Department of Labor, *Occupational Injuries and Illnesses in the United States by Industry, 1990,* Bureau of Labor Statistics Bull. 2399, U.S. Department of Labor, Washington, DC, 1992.

[a] Includes fatalities, injuries, and illnesses.

[b] Wholesale and retail trade.

[c] Finance, insurance, and real estate.

and the composition of crews are subject to continuous change. These are some of the factors that make construction a high-risk industry.

During the period 1986–1990, the U.S. construction industry had an average annual employment of 5.2 million workers [1–3]. The occupational injury and illness rate per 100 full-time construction workers per year was estimated to be 15.2 cases in 1986 with a steady decline to 14.2 cases in 1990. The corresponding lost workday incidence rate per 100 workers during this period was 6.9 in 1986 and 6.7 in 1990. Although slightly in decline during 1986–1990, the construction industry incidence rates were still the highest among the major industries (Table 24.1) [1–3]. In addition, during the period 1980–1990, the U.S. construction industry had an overall fatality rate of 25.6 per 100,000 full-time workers. This rate was more than 3.5 times the occupational fatality rate for all other industries in the United States for the same period [4].

In 1987, 21 states reported 180,876 compensable cases of construction injury and illness into the Bureau of Labor Statistics' (BLS) Supplementary Data System (SDS) (Table 24.2) [5]. This represented a compensable incidence rate of 9.5 per 100 construction workers in these 21 SDS states. Of these injuries, 25.5% involved overexertion, 20.1% struck by, 15% falls from elevations, 7.8% struck against, 6.6% bodily reaction, and 6.6% falls from the same level. Many of these injuries could have been prevented through ergonomic design of the tools, work procedures, and work environments.

TABLE 24.2 Participating States in the SDS Program in 1987[a]

| State | Injury Frequency | Percent |
|---|---|---|
| Alaska | 1,033 | 0.6 |
| Arizona | 4,418 | 2.4 |
| California | 83,860 | 46.4 |
| Colorado | 5,996 | 3.3 |
| Hawaii | 2,414 | 1.3 |
| Iowa | 3,723 | 2.1 |
| Indiana | 8,610 | 4.8 |
| Kentucky | 6,752 | 3.7 |
| Maryland | 12,226 | 6.8 |
| Maine | 6,651 | 3.7 |
| Missouri | 9,303 | 5.1 |
| Mississippi | 2,701 | 1.5 |
| Nebraska | 1,672 | 0.9 |
| New Mexico | 889 | 0.5 |
| Oklahoma | 3,112 | 1.7 |
| Oregon | 4,804 | 2.7 |
| Tennessee | 4,888 | 2.7 |
| Virginia | 5,320 | 2.9 |
| Washington | 6,569 | 3.6 |
| Wisconsin | 5,285 | 2.9 |
| Wyoming | 650 | 0.4 |

*Source:* U.S. Department of Labor, *Supplementary Data System,* Bureau of Labor Statistics, U.S. Department of Labor, Washington, DC, 1987–1988.

[a] Twenty-five states/territories participated in the SDS program in 1987. The data from Louisiana, Michigan, and Ohio were excluded from the analysis because their waiting period (7 days) for compensable claims is different from those of other states (0–3 days). Data from the Virgin Islands were also excluded due to lack of occupation-specific population data.

## 24.2 BACKGROUND AND SIGNIFICANCE OF THE CHAPTER TO OCCUPATIONAL ERGONOMICS

In 1990, the U.S. Public Health Service (PHS) developed and published *Healthy People 2000: National Health Promotion and Disease Prevention Objectives.* The construction industry was identified as a "special population target" [6]. Ergonomics-related issues such as cumulative trauma, traumatic injuries, and low back disorders were some of the target topics listed as applicable to the construction industry.

The focus of most ergonomics research has been on highly repetitive manufacturing jobs, traditionally related to the cause of cumulative trauma disorders. Construction workers face a host of other ergonomic risk factors for both traumatic and musculoskeletal injuries. Their jobs are marked by long work cycles, changing work environments, and varied tasks. The demands of construction activities include repetitive and forceful exertions, awkward postures, prolonged static activities, and the stress of being struck by falling objects or falling from an elevated work surface. Despite the importance that has been placed on worksite construction safety and the high injury rate suffered by the workers in this industry, research into construction ergonomics in the United States has been relatively limited.

This chapter focuses on construction work-related nonfatal injuries. It addresses some of the injury sources, their ergonomic solutions, and the research needs for ergonomic applications in the construction industry. First, we identify the high-risk U.S. construction occupations with their injury characteristics. Then, we present a summary of general ergonomic principles for construction work and their application to two selected construction occupations. Finally, we recommend five areas for ergonomic research emphasis in the future.

## 24.3 OCCUPATIONAL INJURY CHARACTERISTICS OF THE CONSTRUCTION INDUSTRY

This section identifies the occupational groups in the U.S. construction industry that experience high rates of injury. Priority-target occupations and the prevalent injury characteristics of these occupations are discussed.

### 24.3.1 Data Resources

Injury data obtained through the U.S. Department of Labor, BLS SDS, and the occupational employment data obtained from the U.S. Department of Labor and the U.S. Department of Commerce construction census reports were used for this investigation [5,7,8].

The SDS program is a federal-state cooperative system that provides information on occupational injuries and illnesses reported to the participating states' workers' compensation programs. The SDS does contain a small percentage of data on occupational illness and fatalities. Because there were so few occupational illnesses and fatalities in these data, results and discussion of the SDS analysis in this chapter focus on work-related nonfatal injuries. The data used for this assessment are based on reports provided by the 21 participating states (Table 24.2). Both the 1987 and 1988 SDS data tapes were searched to identify the cases that occurred in 1987 [5]. Data on the state-specific employment levels for the major occupational groups in 1987 were obtained through the U.S. Departments of Labor [7] and Commerce [8] industry census reports. These data provided a means for estimating occupation-specific injury rates in the 21 participating states.

### 24.3.2 Analysis Methods

The following procedure was used in developing occupation-specific injury frequencies and injury rates and determining research priorities for the occupational groups.

#### *24.3.2.1 Determination of Occupation-Specific Injury Frequencies*

The occupation-specific injury frequencies of workers' compensation cases were established from the SDS database. The SDS database consists of 17 key variables and 13 optional

TABLE 24.3 Key Variables and Optional Variables in the SDS Program

| Key Variables | Optional Variables |
|---|---|
| State code | Time of accident |
| Reference year | Time workday began |
| Year of occurrence | Hour of shift |
| Month of occurrence | Associated object or substance |
| Day of occurrence | Age |
| Year of receipt | Duration of employment |
| Occupation | Wage code |
| Industry | Weekly wage |
| Ownership | Kind of insurance |
| Nature of injury or illness | Indemnity compensation code |
| Part of body affected | Medical payments code |
| Source of injury or illness | Medical payments |
| Type of accident or exposure | Case type |
| Sex | |
| Extent of disability | |
| Indemnity compensation | |
| Sample weight | |

*Source:* From U.S. Department of Labor, *Supplementary Data System Microdata Files User's Guide, 1985 Edition*, Bureau of Labor Statistics, U.S. Department of Labor, Washington, DC, 1989.

variables (Table 24.3). Key variables include state code, reference year, occupation, and industry. Optional variables include age of employee, duration of employment, and weekly wages. A frequency analysis was performed using the Statistical Analysis System (SAS) program to generate the frequencies of injuries by occupations for all sectors of the construction industry, for incidents occurring in 1987.

#### *24.3.2.2 Determination of Occupation-Specific Employment*

Specific occupational employment figures for the 1987 SDS states were derived through the equation

$$E_{\mathrm{O}} = E_{\mathrm{NO}} \times \left( \frac{E_{\mathrm{C}}}{E_{\mathrm{NC}}} \right)$$

where

- $E_{\mathrm{O}}$ is the estimated employment by occupation in the 21 participating SDS states for 1987
- $E_{\mathrm{NO}}$ is the national employment by occupation for 1987 according to [7]
- $E_{\mathrm{C}}$ is the construction industry employment in the 21 SDS states for 1987 (obtained by summing the 21 SDS states' construction employment, as listed in [8])
- $E_{\mathrm{NC}}$ is the national construction industry employment for 1987 (obtained from [8])

#### *24.3.2.3 Determination of Injury Rates*

Occupation-specific injury rates for 1987 were determined by dividing the number of reported workers' compensation cases by the corresponding established occupational employment and multiplying by 100 to obtain the number of cases per 100 workers. The overall construction industry injury rate (9.5) for 1987 was determined by dividing the total number of construction injuries (180,876 cases) reported in the 21 SDS states in 1987 by the construction employment (1,897,040 workers) in the same 21 states.

#### *24.3.2.4 Determination of Priority Occupations*

A frequency-weighted occupation-specific rating technique was developed to prioritize occupations as targets for injury prevention. The rankings were obtained through the equation

$$\text{FW\_RK} = \text{Rank of}\left[\text{ORDER} \times \left(\frac{\text{FREQ}_{\text{SO}}}{\text{FREQ}_{\text{AO}}}\right)\right],$$

where

FW_RK is the frequency-weighted rank of priority occupations (listed from high to low priority)

ORDER is the sequence rank of injury rates (The lowest injury rate was assigned ORDER = 1)

$\text{FREQ}_{\text{SO}}$ is the injury frequency of a specific occupation

$\text{FREQ}_{\text{AO}}$ is the injury frequency of all occupations in construction

### 24.3.3 Results

The priority ranking was applied to 30 occupations. Each of these 30 occupations was involved in more than 0.3% of the total number of compensation claims (180,876) filed in the 21 SDS states for the construction industry in 1987. The occupations with the highest frequency of injury were construction laborer (25.8%), carpenter (14.8%), electrician (5.9%), and plumber (5.6%). Occupations with the highest injury rate (number of cases per 100 workers) were miscellaneous materials-moving equipment operators (47.6), construction laborer (43.3), miscellaneous construction trades workers (32.2), and carpet installer (32.0) (Table 24.4).

Seven occupations were identified as priority research targets on the basis of the frequency-weighted occupation-specific rating technique. They were construction laborer, carpenter, roofer, drywall installer, plumber, electrician, and structural metal worker (Table 24.4). These seven occupations represented an estimated 30.7% of the U.S. construction industry workforce and 65% of the total compensable injuries reported for construction. The combined injury rate of the seven target occupations (20.2 cases per 100 workers) was more than twice that for all construction workers combined (9.5 cases per 100 workers).

A cross-analysis was performed with injuries typed by injury source, injured body part, and nature of injury for the seven occupations (Table 24.5). This was done to

TABLE 24.4 Potential High-Risk Construction Occupations and Ranking of Safety Research Priorities (1987 SDS, 21 States)

| OCC3 | Occupation | FREQ | EMP21 | RATE | ORDER | FW_RK |
|---|---|---|---|---|---|---|
| 869 | Construction laborers | 46,838 | 108,116 | 43.3 | 29 | 1 |
| 567 | Carpenters | 26,857 | 200,188 | 13.4 | 18 | 2 |
| 595 | Roofers | 7,731 | 31,987 | 24.2 | 25 | 3 |
| 573 | Drywall installers | 7,327 | 26,688 | 27.5 | 26 | 4 |
| 585 | Plumbers, pipefitters | 10,148 | 79,312 | 12.8 | 16 | 5 |
| 575 | Electricians | 10,738 | 102,984 | 10.4 | 11 | 6 |
| 597 | Structural metal workers | 4,187 | 21,457 | 19.5 | 23 | 7 |
| 599 | Misc. construction trade workers | 3,707 | 11,495 | 32.2 | 28 | 8 |
| 534 | Heating and air cond. mechanics | 4,470 | 36,942 | 12.1 | 14 | 9 |
| 579 | Painters and paperhangers | 5,462 | 55,496 | 9.8 | 9 | 10 |
| 596 | Sheet metal duct installers | 2,414 | 13,770 | 17.5 | 20 | 11 |
| 593 | Insulation workers | 2,672 | 20,238 | 13.2 | 17 | 12 |
| 566 | Carpet installers | 1,654 | 5,166 | 32.0 | 27 | 13 |
| 584 | Plasterers | 1,783 | 7,789 | 22.9 | 24 | 14 |
| 804 | Truck drivers, heavy | 3,654 | 36,578 | 10.0 | 10 | 15 |
| 516 | Heavy equipment mechanics | 1,529 | 8,342 | 18.3 | 22 | 16 |
| 783 | Welders and cutters | 1,759 | 10,102 | 17.4 | 19 | 17 |
| 653 | Sheet metal workers | 2,570 | 21,480 | 12.0 | 13 | 18 |
| 550 | Construction supervisors | 5,901 | 79,925 | 7.4 | 5 | 19 |
| 565 | Tile setters | 1,128 | 6,211 | 18.2 | 21 | 20 |
| 589 | Glaziers | 1,298 | 10,223 | 12.7 | 15 | 21 |
| 844 | Operating engineers | 2,488 | 33,013 | 7.5 | 7 | 22 |
| 859 | Misc. materials moving equip. operators | 573 | 1,203 | 47.6 | 30 | 23 |
| 563 | Brick masons | 2,849 | 38,202 | 7.5 | 6 | 24 |
| 588 | Concrete finishers | 2,411 | 39,402 | 6.2 | 4 | 25 |
| 643 | Boiler makers | 486 | 4,424 | 11.0 | 12 | 26 |
| 544 | Millwrights | 611 | 6,654 | 9.2 | 8 | 27 |
| 853 | Excavating machine operators | 814 | 18,683 | 4.4 | 3 | 28 |
| 019 | Managers, nec and manage. occ. | 2,013 | 141,550 | 1.4 | 1 | 29 |
| 855 | Grader, dozer, scraper operators | 745 | 22,252 | 3.3 | 2 | 30 |
| All occupations | | 180,876 | 1,897,040 | 9.5 | | |

OCC3 = three-digit occupation code
FREQ = frequency (reported SDS cases in the construction industry, 1987)
EMP21 = construction employment in the selected 21 states (1987 SDS)
RATE = FREQ/EMP21*100
ORDER = sequence rank of RATE (from least to greatest)
FW_RK = frequency-weighted rank of research priorities, that is, the rank of [ORDER*(FREQ of a specific occupation/FREQ of all occupations)]

TABLE 24.5 Major Injury Characteristics of Seven Selected Construction Occupations

| Occupation | Type | Source | Parts | Nature |
|---|---|---|---|---|
| Construction laborers | Overexertion [26.2%] | Tools (6.0%)<br>Containers (3.5%)<br>Metal (3.4%)<br>Wood (2.7%) | Back (15.8%) | Sprain/strain (22.4%) |
| | Strack by [24.8%] | Tools (6.3%)<br>Metal items (6.3%)<br>Wood items (2.8%) | Finger (4.7%)<br>Leg (3.2%)<br>Foot (2.5%)<br>Toe (2.3%) | Cut laceration (8.0%)<br>Fracture (6.5%)<br>Contusion (4.9%) |
| | Fall from elevation [12.0%] | Working surface (10.8%) | Back (1.6%)<br>Leg (1.6%)<br>Multiple (1.9%) | Fracture (3.7%)<br>Sprain/strain (3.3%) |
| Carpenter | Strack by [27.5%] | Tools (0.7%)<br>Metal items (6.5%)<br>Wood items (5.6%) | Finger (8.4%)<br>Hand (3.3%)<br>Leg (3.3%)<br>Eye (2.2%) | Cut laceration (14.3%) |
| | Overexertion [22.4%] | Structure (3.5%)<br>Tool (3.4%)<br>Metal (1.8%)<br>Wood (6.5%) | Back (13.3%) | Sprain/strain (18.8%) |
| | Fall from elevation [18.6%] | Working surface (16.7%) | Back (2.4%)<br>Leg (1.8%)<br>Ankle (1.8%)<br>Multiple (3.2%) | Fracture (7.1%)<br>Sprain/strain (4.0%) |
| Roofer | Fall from elevation [24.7%] | Working surface (22.4%) | Back (3%)<br>Leg (2.4%)<br>Ankle (2.4%)<br>Foot (2.3%)<br>Multiple (5.0%) | Fracture (11.1%)<br>Sprain/strain (5.5%) |
| | Overexertion [23.6%] | Lifting containers (8.6%)<br>Tools (3.0%) | Back (14.8%) | Sprain/strain (20.5%) |
| | Struck by [13.6%] | Tools (4.3%)<br>Metal item (3.8%) | Finger (2.9%)<br>Leg (2.6%) | Cut laceration (7.6%) |
| | Contact with extreme | Coal/oil product (9.3%) | Hand (3.2%) | Burn (heat) (10.1%) |
| | Temperature [10.4%] | | Arm (1.9%)<br>Upper ext. (1.1%)<br>Multiple (2.2%) | |
| Dry wall installer | Overexertion [28%] | Miscellaneous (14.4%)<br>Containers (4.2%) | Back (15.2%) | Sprain/strain (24.3%) |
| | Fall from elevation [24.7%] | Working surface (22.7%) | Multiple (3.4%)<br>Back (2.7%)<br>Leg (2.8%)<br>Ankle (2.7%)<br>Wrist (2.7%) | Fracture (8.3%)<br>Sprain/strain (6.8%) |

TABLE 24.5 (continued) Major Injury Characteristics of Seven Selected Construction Occupations

| Occupation | Type | Source | Parts | Nature |
|---|---|---|---|---|
| | Struck by [15.1%] | Tools (4.8%) | Finger (5.4%) | Cut laceration (8.3%) |
| | | Metal items (4.2%) | Hand (2.4%) | |
| Plumber | Overexertion [29.2%] | Metal items (8.8%) | Back (17.0%) | Sprain/strain (24.4%) |
| | | Tools (6.6%) | Shoulder (1.7%) | |
| | Struck by [18.6%] | Metal items (7.3%) | Finger (3.8%) | Cut laceration (5.9%) |
| | | Tools (4.7%) | Eye (2.3%) | |
| | Fall from elevation [14.6%] | Working surface (4.7%) | Back (1.7%) | Fracture (3.0%) |
| | | | Multiple (1.5%) | Sprain/strain (3.9%) |
| Electrician | Overexertion [25.7%] | Electric apparatus (4.4%) | Back (14.7%) | Sprain/strain (20.9%) |
| | | Metal item (4.3%) | Shoulder (1.6%) | |
| | | Containers (4.3%) | | |
| | Fall from elevation [14.6%] | Working surface (4.7%) | Multiple (2.5%) | Fracture (3.0%) |
| | | | Back (1.9%) | Sprain/strain (3.9%) |
| | | | Leg (1.9%) | |
| | | | Ankle (1.4%) | |
| Structural metal workers | Overexertion [28.4%] | Metal (17.8%) | Back (18.4%) | Sprain/strain (24.0%) |
| | | | Shoulder (1.8%) | |
| | Struck by [17.7%] | Metal (11.3%) | Finger (2.1%) | Cut laceration (4.7%) |
| | | | Foot (2.0%) | Fracture (4.5%) |
| | | | Eye (1.8%) | Contusion (3.5%) |
| | Fall from elevation [16.6%] | Working surface (15.3%) | Back (1.9%) | Fracture (3.7%) |
| | | | Leg (1.8%) | Sprain/strain (2.9%) |
| | | | Multiple (4.1%) | |

[] Indicates percentage of injury frequency of a specific injury type to overall injury frequency in a specific occupation.

() Indicates percentage of injury frequency of a specific category (source, parts, and nature) within a specific injury type to overall injury frequency in a specific occupation.

identify problem activities where there was a high rate of injury. In all seven occupations, the sum of overexertion, struck by, and falls from elevation accounted for more than 50% of the injuries. Roofers and drywall installers are most frequently injured as a result of fall-from-elevation and overexertion incidents. For the remaining five occupational groups, overexertion and struck-by incidents caused the greatest number of injuries. Overexertion and fall-from-elevation incidents most frequently result in sprain/strain and fracture injuries to the back; struck-by incidents typically result in cuts and lacerations to the fingers and legs and foreign objects in the eyes.

### 24.3.4 Discussion

Differences in state workers' compensation coverage and reporting requirements have been cited as limitations of the SDS program [9]. This study used data available from 21 states to determine the priority rankings. In spite of these limitations, the broad database and uniform coding format employed by the SDS, combined with employment data from the Department of Labor and the Department of Commerce on construction, allow a more comprehensive analysis of the occupational characteristics associated with work-related injuries in construction than presently exists in the literature. It may not be perfect, but it is the best available.

The censuses of construction industries were performed by the Bureau of the Census, U.S. Department of Commerce every 5 years (for all years ending in 2 and 7) and were available for 1982 and 1987. Occupational employment data in the construction industry were available for 1981, 1984, and 1987. These were separately published by the BLS, U.S. Department of Labor, in 1984, 1986, and 1989. The SDS database contains annual workers' compensation data for 1981–1989. The states participating in this program varied from year to year. Based on the available data sets, 1987 was selected as the best data year for this study.

These analyses provide an approach for determining priority occupations for injury prevention research and analysis of the occupational characteristics associated with work-related injuries in the construction industry.

## 24.4 ERGONOMIC PRINCIPLES AND APPLICATIONS FOR WORKER SAFETY

The analysis of the BLS workers' compensation data presented in Section 24.3 shows that construction workers are vulnerable to overexertion, fall, and struck-by injuries. Manual handling of construction materials, which are usually heavy and bulky, presents one of the problems. Inappropriate design or operation of hand tools, which are used in almost all construction operations, can result in injury. Heavy equipment, which is used in many construction operations, also presents several problems. Whole body vibration, seat design, operator field of vision, and access systems are some of the potential ergonomic issues. Additionally, many construction workers perform tasks above shoulder level or below knee height and are thereby exposed to situations hazardous to the musculoskeletal system.

Although ergonomics has been increasingly employed in occupational settings and in safety and health disciplines, little ergonomic research has been done in the United States on its application to construction work. Previous studies have concentrated on manufacturing and office workers. Nevertheless, some of the risk factors that have been studied in manufacturing settings and other industries are characteristic of many construction tasks. These factors include (1) forceful exertion, (2) repetitive motions, (3) awkward or static posture, (4) direct external stress over the skin and muscle tissue, (5) vibration, (6) extreme temperature, (7) psychosocial stresses, (8) noise, (9) task difficulty, and (10) illumination. Some of these factors may be more closely associated with overexertion problems in manufacturing settings than with the construction industry. However, in construction, they may result in overexertion, fall, and struck-by incidents or combinations thereof.

### 24.4.1 General Ergonomic Principles

In general, a three-tiered injury control program using ergonomic principles as outlined in [10] can be applied to the construction industry. These approaches are as follows:

1. Engineering controls—ergonomic design changes to tools, handles, equipment, work methods, or other aspects of the workplace
2. Administrative controls—changes in work practices or organizational and management policies
3. Use of personal protective equipment (PPE)

However, this approach may be more difficult at a construction site because the work environment, the work to be done, and the composition of crews are subject to continuous change. The ergonomic approaches to reducing construction injuries need to be more creative in this setting. Table 24.6 summarizes the general ergonomic principles in construction.

### 24.4.2 Ergonomic Applications for Selected Construction Occupations

Seven occupations have been identified as having priority for ergonomic research. The most hazardous activities for two of the seven occupations (roofer and drywall installer) and potential ergonomic countermeasures are discussed in this section. These two occupations are presented as examples of potential ergonomic applications in the construction industry.

#### *24.4.2.1 Roofers*

Roofers are primarily concerned with covering roofs and exterior walls of structures with slate, asphalt, aluminum, rubber, wood, and related materials. Most roofers work for special trades contractors (Standard Industrial Classification Code SIC 176) at various building construction sites. Roofers represent 1.7% of the construction workforce. Their incidence rate (24.2 cases per 100 workers) was more than 2.5 times that of the average construction worker. The leading exposures for roofers were falls from elevation (24.7%), overexertion (23.6%), being struck by an object (13.6%), and contact with extreme temperature (10.4%).

Figure 24.1 presents samples of some roofing activities. Studies of roofing-related injuries have shown that the most hazardous activities of roofers are associated with (1) transporting materials or equipment while ascending and descending ladders [42], (2) using unsecured/inadequate ladders and scaffolds [43–45], (3) working at the edge of the roof or slipping/losing balance while on the roofing surface [42], (4) working on roofs that contain holes or thin sheeting [46,47], and (5) manually handling heavy and/or bulky material or equipment [48–50]. These bulky materials include bundles of shingles (weighing approximately 80 lb), 5 gal buckets of hot tar, reels/rolls, and containers [51,52].

Ergonomic approaches to reducing injuries during roofing work have been proposed by some researchers and institutes. One general approach is to use mechanical lifting devices to transport materials and equipment between the ground and the roof. This can reduce the

TABLE 24.6 General Ergonomic Principles for Construction Safety

| Suggested Action | Benefit | References |
|---|---|---|
| 1. Engineering controls | | |
| a. Use mechanical aids, torque control devices, or lifting aids to transport construction materials and equipment | Reduce frequency and duration of heavy lifting | [11–13] |
| b. Automate certain construction tasks | Reduce the chance of overexertion due to repetitive and overreach work | [14] |
| c. Optimize the weight of construction tools and containers | Reduce biomechanical stress | [11,15–18] |
| d. Improve mechanical advantages of tool designs or work procedures (such as hoisting and pulleys) for handling heavy loads | Reduce required reaction force | [13] |
| e. Improve visibility during heavy construction equipment operation by modifying vehicle design | Improve work postures; reduce injury incidents due to poor visibility | [19–21] |
| f. Improve heavy construction vehicle access systems by cab step redesign or mounting and dismounting procedure design | Reduce fall injuries | [13,22,23] |
| g. Optimize size and shape of construction tool handles and strength requirement to operate the tool | Reduce biomechanical stresses due to awkward postures and forceful grip exertions | [24–26] |
| h. Design hand-held tools and containers for balance movement on the spine during lifting and carrying of construction materials | Reduce rotation on the forearm or lateral bending on the trunk | [27,28] |
| i. Select construction tools/equipment with minimum vibration; use isolation/damping for equipment that operates above the point of resonance; adjust tool speed to avoid resonance | Reduce vibration exposure | [29,30] |
| j. Select methods to minimize surface and edge finishing | Reduce repetitive motions and static loadings | [14] |
| k. Use full hand grip, not just fingertips, when lifting heavy objects | Avoid external mechanical stress | [13,31,32] |
| l. Alter position of tools or worker to avoid awkward working postures | Reduce pronounced strains on back, shoulders, and arms | [33] |
| m. Provide good couplings between workers and tools/environment, such as texturing handles to prevent hands slipping and using appropriate slip-resistant materials for shoe soles | Reduce overexertion, slipping, and being struck by objects | [32,34] |
| 2. Administrative controls | | |
| a. Use biomechanical principles such as reducing the distance between the worker and the load, reducing the distance the load must be moved, keeping all movement close to the body, and keeping movements between the knuckles and the shoulder | Reduce biomechanical stress | [11,18] |
| b. Remove all obstacles between the worker and the load by practicing good housekeeping procedures | Reduce slipping and tripping hazards and awkward working posture exposures | [35,36] |

TABLE 24.6 (continued) General Ergonomic Principles for Construction Safety

| Suggested Action | Benefit | References |
|---|---|---|
| c. Keep construction tools in good condition and in a safe place | Reduced hand forces and the chance of tool slipping and striking the worker | [37] |
| d. Seek help when handling large and/or bulky loads | Balance materials being handled and lessen the load distribution during lifting and carrying work | [13] |
| e. Provide frequent rest breaks to offset undue fatigue in jobs requiring heavy labor or high performance/production rates | Offset undue fatigue | [38] |
| f. Limit overtime work and periodically rotate workers to less stressful jobs | Job stress is spread over a longer time and over different body parts, resulting in less strain | [14] |
| g. Vary work tasks or broaden job responsibilities | Offset boredom and sustain worker motivation | [14] |
| h. Train workers in techniques to avoid awkward postures and repetitive motions | Reduce construction injuries that result from lack of knowledge and experience | [39] |
| 3. Use of personal protective equipment | | |
| a. Use safety shoes and protective eye wear | Reduce injuries due to being struck by objects | [40,41] |
| b. Use fall arrest devices | Reduce injuries due to falls | [40,41] |
| c. Use vibration-attenuating gloves | Reduce vibration exposures | [40,41,109–112] |

possibility of overexertion and the chance of losing a grip on manually handled materials, which could result in dropped materials. The Occupational Safety and Health Administration (OSHA) and many other safety agencies recommend that workers practice ladder safety to prevent falls and overexertion. The OSHA recommendations include (1) securing the ladder properly at the roof top to prevent horizontal movement, (2) resting the ladder on a solid base, (3) extending the ladder 3 ft above the point where a worker steps onto the roof, (4) placing the base of the ladder a distance of one-fourth of its working length from its vertical line of support, (5) employing the concept of three-point contact with the ladder when ascending and descending ladders, and (6) not carrying objects while ascending and descending ladders [53].

Other ergonomic and safety countermeasures for fall prevention in roofing work include, but are not limited to, (1) using guard rails [54–56], (2) using fall protection anchoring devices and protective scaffolds [57–59], (3) covering and securing openings in roofs and roof decks with plywood or other suitable materials [53], and (4) employing a safety line with a shock absorber [60]. Medical examination to detect defects of worker equilibrium may also have a positive impact in preventing roofer fall injuries [61].

Furthermore, roofers should practice good housekeeping [35,42] and practice the philosophy of "good couplings" to reduce struck-by, slipping, and tripping hazards. This involves such practices as removing all obstacles between the worker and the load, using steel toe safety shoes [34], selecting shoes with appropriate slip-resistant characteristics [34,62,63], and providing handles on the items being lifted. The most frequent roofer injury scenarios involve a worker being struck by objects such as rolls, barrels, buckets, cans, asphalt,

FIGURE 24.1 Some ergonomics-related at-risk activities of roofing work. (a) Transporting heavy materials; (b) working at the roof edge; (c) prolonged kneeing work; (d) unsafe overreach work.

shingles, or equipment that fell, slipped, or was dropped by another worker; alternatively, the worker may have lost his grip on the material he was handling and injured himself [42].

Tool selection is another important ergonomic concern during roofing work. The principles are to select the correct/best tools for the job, use them correctly, maintain them in good condition, and keep them in a safe place [37]. Tools such as knife blades or chisels should be sharpened regularly. Dull tools require greater hand forces, which can trigger the development of hand, wrist, and upper extremity cumulative trauma disorders and can also result in the tool slipping and striking the worker.

Use of appropriate tools can reduce the biomechanical stress created by some job tasks. Roofers who fasten single-ply roofing systems in place frequently do so in a bent-over posture. A new Dutch device for fastening single-ply roofing has been designed to allow for less bent-over posture during that task, thereby reducing stresses on the musculoskeletal system [64].

Training is also an important countermeasure. Workers should be trained in techniques to avoid awkward postures and repetitive motions and in appropriate procedures for new work methods or equipment [55]. Prior to being assigned work on a roof, new employees should be given safety and health training to control/reduce hazardous exposures [48].

Roofing workers appear to have a high risk of occupation-related burn injuries [65]. In California during 1979, burns and scalds in roofing work occurred 21 times more often than in the construction industry as a whole [50]. Transferring hot asphalt in a can or bucket and filling the melt kettle are hazardous work activities for roofers. Ergonomic design can reduce the hazard of burn injuries. Half-lids installed on cans and buckets can reduce splashing during carrying and pouring tasks. Better designs for placing solid asphalt into the melt kettles can further reduce the incidence of burns and scalds [66]. In addition, improved materials for PPE must be identified to effectively protect roofers using asphalt. Parsons and Pizatella [52] found that roofers who wore gloves were absent from work longer for recovery from asphalt burn injuries to the hands than workers whose hands were burned but were not protected by gloves. It was surmised that the protective gloves were of the cotton variety and actually caused the workers' hands to be exposed to the hot asphalt for a longer time, resulting in more severe burns.

#### *24.4.2.2 Drywall Installers*

Drywall installers are primarily concerned with installing, taping, and surfacing plasterboard and drywall in the interior of a structure. Typical drywall comes in sheets measuring 4 ft wide by 8, 10, or 12 ft long and 1/4, 1/2, 3/4, or 5/8 in. thick. The sheets weight 55–84 lb each. In 1987, the U.S. construction industry employed 70,510 drywall installers—1.41% of the construction workforce. Their compensable injury rate (27.5 cases per 100 workers) was almost three times the average rate of 9.5 for all construction occupations combined. The injury data revealed that overexertion (28%), fall from elevation (24.7%), and being struck by an object (15.1%) were the leading types of injury. Most drywall installers work for special trades contractors (SIC 174).

The hazardous exposures for drywall installers include manually transporting drywall; applying glue to studs; hanging drywall on ceilings; applying tape to joints and corners; applying skim coats of joint compound to joints and corners; and sanding wall joints, attachment indentions, and corners (see Figure 24.2). All of these activities involve overhead work, heavy lifting, repetitive motion, forceful cutting, or work at elevated heights. These activities may put considerable biomechanical stress on the worker or expose drywall workers to tripping/falling and struck-by hazards. During taping, compound application, and finishing activities, stilts are frequently used. The stilts are 2-ft-high structures that can be strapped to the workers' legs for quickly moving around while completing the upper part of drywall work (Figure 24.2). Although no studies have yet been reported, the use of stilts may place workers at increased risk for knee injuries [67] and falls.

Some countermeasures have been developed and used to reduce the potential hazards associated with drywall work. The biomechanical stress associated with transporting the dry-wall sheets to the wall can be reduced by the use of carts and dollies [68]. Plasterboard lifts can be used to raise the board to the desired height and to hold it in place for fastening the drywall [69]. The biomechanical stress also can be reduced by reducing the width of drywall sheets. Studies have shown that the reduction of drywall board width from just under 4 ft to slightly less than 3 ft did reduce stress on workers' backs and resulted in slower heart rates among workers [70]. This reduction in width makes it considerably easier to handle the sheets of drywall because of the reduction in both weight and size.

FIGURE 24.2 Some ergonomics-related at-risk activities of drywall installation. (a) Hanging drywall on ceiling; (b) applying tape to joints and corners; (c) finishing wall joints, nail indentions, and corners; (d) use of stilts while taping compound.

The increased material consumption, mainly studs and screws, can be compensated for by lower working costs and more rapid fitting of the 3-ft-wide sheets [70].

Screw guns have been developed with a swiveling handle to reduce wrist flexion in fastening drywall to metal studs in overhead and under-knee positions [38]. Using "micropauses" or short breaks after installation of a few screws has also been suggested to reduce stress and increase productivity [38]. Proper training and the use of carrier handles also reduced biomechanical stress among the drywall installers [71]. Finally, sanding high and low areas requires awkward body-part angles for applying proper sanding force. Use of an electric sander may reduce the force required but requires a static hold to support the sander [67].

## 24.5 CRITICAL REVIEW OF CURRENT STATUS OF CONSTRUCTION ERGONOMIC RESEARCH AND FUTURE CONCERNS

Ergonomic research in the United States specific to the seven selected construction occupations is limited. Review of construction ergonomics literature related to the seven occupations indicated that most studies presented retrospective analyses of injury frequencies collected from historical records [35,42,50,52,72–84,108] and subjective questionnaire/observation surveys of safety-related matters [46,48,64,68,72–74,84–89]. Occupation-specific injury rates and evaluation of worker tasks associated with resulting injuries were very limited [90]. This limitation might have resulted from the fact that technologies for construction hazard exposure assessment were themselves limited in number. Recently, a few ergonomic-assessment methods such as postural measurement [91], postural balance evaluation [92], biomechanical measurement [70,91], physiological measurement [70,93], electromyography [94–96], sensorimotor behavior test [39], movement performance [94,95], vibration measurement [97,98], and audiometric evaluation [92] were applied in some construction ergonomics studies. In addition, syntheses of overexertion problems, metabolic cost, and muscle fatigue during construction work were presented by some researchers [67,99–106]. Injury prevention approaches such as training [39,107], tool redesign [91,96,98], and work design [68,70] were proposed in some of the studies. A number of these activities were from Bygghalsan of Sweden [100] and the Construction Safety Association of Ontario [13]. It appears that more injury prevention intervention and ergonomic evaluation studies of construction occupations are needed. To effectively reduce U.S. construction worker injuries, the five following areas are recommended for future research.

1. Occupation-specific tasks associated with injuries must be identified and evaluated. Past studies have emphasized retrospective analyses of injury frequency data for the overall construction industry. This chapter has identified injury rates of construction occupations. Active surveillance programs must be initiated to investigate occupation-specific tasks and activities for the priority construction occupations so that effective interventions can be developed.

2. There is a need for more emphasis on in-depth study of injury prevention strategies for high-risk tasks in the seven occupations. Current studies on existing work techniques and injury prevention strategies of these occupations are limited. More in-depth laboratory studies as well as field evaluation to determine and overcome barriers to the use of existing injury prevention technologies are desired.

3. The methods for measuring construction worker exposure to biomechanical and other stressors need to be improved because construction tasks differ from tasks performed by other industrial workers. The environment, the work to be done, and the composition of crews change continually. Studies on measurement methodologies such as telemetry and other technologies would be very useful for researchers and safety professionals in the construction industry.

4. Emphasis should be placed on evaluating and redesigning tools, equipment, assistive devices, and work procedures. Change of workplace, assignment, and crew composition is the nature of construction work. However, tools, equipment, and work procedures remain fairly consistent. Therefore, redesign of equipment or assistive devices will be more effective and should be emphasized for reducing ergonomics-related injuries in construction.

5. Studies of fall prevention should not just look at passive protective devices; other creative solutions need to be investigated and continued development of new technologies to evaluate and design better fall prevention solutions is also needed.

## 24.6 CONCLUSION

This chapter addressed injury sources, ergonomic solutions, and research needs in the construction industry.

The study has identified construction laborers, carpenters, roofers, drywall installers, plumbers, electricians, and structural metal workers as being at an elevated risk of nonfatal injuries. They are likely to be injured from overexertion, fall from an elevation, or being struck by an object. Some ergonomic solutions have been proposed to reduce these injuries in these selected occupations. Clearly, there is a need to focus additional safety research and ergonomic programs on these occupations and their cause of injuries.

Continued research on the following subjects is recommended to further reduce construction injuries:

1. Identification of occupation-specific risk tasks and activities so that effective interventions can be developed
2. Systematic evaluation of existing injury prevention technologies to overcome the barriers to the use of these technologies
3. Improvement of biomechanical exposure assessment technologies to help identify risk factors
4. Improvement of construction equipment and assistive devices to reduce work-related stress
5. Implementation of new technologies to evaluate and develop fall prevention solutions that better protect construction workers

## REFERENCES


1. U.S. Department of Labor, *Occupational Injuries and Illnesses in the United States by Industry, 1987,* Bureau of Labor Statistics Bull. 2328, U.S. Department of Labor, Washington, DC, 1989.
2. U.S. Department of Labor, *Occupational Injuries and Illnesses in the United States by Industry, 1989,* Bureau of Labor Statistics Bull. 2379, U.S. Department of Labor, Washington, DC, 1991.
3. U.S. Department of Labor, *Occupational Injuries and Illnesses in the United States by Industry, 1990,* Bureau of Labor Statistics Bull. 2399, U.S. Department of Labor, Washington, DC, 1992.
4. S. M. Kisner and D. E. Fosbroke, Injury hazards in the construction industry, *J. Occup. Med. 36*(2):137–143 (1994).
5. U.S. Department of Labor, *Supplementary Data System,* Bureau of Labor Statistics, U.S. Department of Labor, Washington, DC, 1987–1988.
6. U.S. Department of Health and Human Services, *Healthy People 2000: National Health Promotion and Disease Prevention Objectives,* DHHS Pub. No. (PHS) 91–50212, U.S. Government Printing Office, Washington, DC, 1990.
7. U.S. Department of Labor, *Occupational Employment in Mining, Construction, Finance, and Services,* Bureau of Labor Statistics Bull. 2330, U.S. Department of Labor, Washington, DC, 1989.
8. U.S. Department of Commerce, *1987 Census of Construction Industries,* Geographic Area Series CC87-A-1 through CC87-A-9 and U.S. Summary CC87-A-10, Bureau of the Census, U.S. Department of Commerce, Washington, DC, 1990.

9. N. Root and D. McCaffrey, Providing more information on work injury and illness, *Mon. Labor Rev. 101*:16–21 (1978).
10. U.S. Department of Labor, *Ergonomics Program Management Guidelines for Meatpacking Plants,* OSHA 3123, U.S. Department of Labor, Washington, DC, 1990.
11. NIOSH, *Work Practices Guide for Manual Lifting,* Tech. Rep. DHHS Pub. No. 81–122, U.S. Government Printing Office, Washington, DC, 1981.
12. T. H. Hettinger, Occupational hazards associated with diseases of the skeletal system, *Ergonomics 28*(1):69–75 (1985).
13. Construction Safety Association of Ontario, *Stand, Lift, Carry,* rev. ed., Construction Safety Association of Ontario, Toronto, Ontario, Canada, 1993.
14. S. A. Knoz and A. Mital, Carpal tunnel syndrome, *Int. J. Ind. Ergon. 5*:175–180 (1990).
15. University of Michigan, *2D Static Strength Prediction Program,* Center for Ergonomics, University of Michigan, Ann Arbor, MI, 1990.
16. M. M. Ayoub, N. J. Bethea, S. Deivanayagum, S. S. Asfour, G. M. Bakken, D. Liles, A. Mital, and M. Sherif, *Determination and Modeling of Lifting Capacity*, Final Rep., HEW, Grant No. 5R01–OH–00545–02, Institute for Ergonomics Research, Texas Tech University, Lubbock, TX, 1978.
17. S. H. Snook, The design of manual handling tasks, *Ergonomics 21*(12):963–985 (1978).
18. T. R. Waters, V. P. Anderson, A. Garg, and L. J. Fine, Revised NIOSH equation for the design and evaluation of manual lifting tasks, *Ergonomics 36*(7):749–776 (1993).
19. F. Hella, M. Tisserand, J. F. Schouller, and M. Englert, A new method for checking the driving visibility on hydraulic excavators, *Int. J. Ind. Ergon. 8*(2):135–145 (1991).
20. J. Rouquie, A wide-angle rear-view periscope for dump trucks, *Protect. Civ. Secur. Ind. 184*:33–35 (1970) (French; English abstr.).
21. W. K. Miller, *Analysis of Haulage Truck Visibility Hazards at Metal and Nonmetal Surface Mines*, Mining Enforcement and Safety Administration, U.S. Department of the Interior, Inf. Rep. 1038, Washington, DC, 1976.
22. R. L. Stanevich, A study of earthmoving and highway construction machinery fatalities and injuries, *Ann. Am. Conf. Gov. lnd. Hyg. 14*:703–710 (1986).
23. Construction Safety Association of Ontario, *Mounting and Dismounting Procedures for Construction Equipment*, Construction Safety Association of Ontario, Toronto, Ontario, Canada, 1989.
24. D. B. Chaffin and G. B. J. Andersson, *Occupational Biomechanics,* Wiley, New York, 1984.
25. T. J. Armstrong, *An Ergonomics Guide to Carpal Tunnel Syndrome*, AIHA Ergonomic Guide Series, American Industrial Hygiene Association, Akron, OH, 1983.
26. D. B. Chaffin, Localized muscle fatigue—Definition and measurement, *J. Occup. Med. 15*(4):346–354 (1973).
27. A. Mital and M. Ilango, Load characteristics and manual carrying capabilities, *Proc. 27th Annual Meeting Hum. Factors Soc.,* Norfold, VA, 1983, pp. 274–278.
28. A. Mital and H. F. Fard, Psychophysical and physiological responses to lifting symmetrical and asymmetrical loads symmetrically and asymmetrically, *Ergonomics 29*:1263–1272 (1986).
29. M. J. Griffin, *Handbook of Human Vibration*, Academic, London, U.K., 1990.
30. D. E. Wasserman, The control aspects of occupational hand-arm vibration, *Appl. Ind. Hyg. 4*(8):22–26 (1989).
31. P. Tadano, A safety/prevention program for VDT operators: One company's approach, *J. Hand Ther. 3*(2):64–71 (1990).
32. C. G. Drury, Handles for manual material handling, *Appl. Ergon. 11*(1):35–42 (1980).
33. T. J. Armstrong, *Cumulative Trauma Disorders and Forceful Exertions, Occupational Ergonomics Summer Course Materials,* University of Michigan, Ann Arbor, MI, 1990.
34. Knollma, Safety shoes for roofers, *BAU-Berufsgen. Wuppertal Mitteilungsbl. 3*:147–149 (1988).
35. H. H. Cohen and D. M. J. Compton, Fall accident patterns—Characterization of most frequent work surface-related injuries, *Prof. Safety 27*(6):16–22 (1982).

36. Construction Safety Association of Ontario, *Construct. Safety 4*(3), Construction Safety Association of Ontario, Toronto, Ontario, Canada, 1993.
37. National Safety Council, *Accident Prevention Manual for Industrial Operations, Engineering, and Technology*, 9th edn., National Safety Council, Chicago, IL, 1988.
38. P. Andersson, *Manual Screw Tightening with and without Micro Pauses,* Bygghalsan Bull. 91–09–16, The Construction Industry's Organization for Working Environment, Occupational Safety and Health, Danderyd, Sweden, 1991, p. 39.
39. K. Prather, R. A. Crisera, and S. Fidell, *Behavior Analysis of Workers and Job Hazards in the High Risk Construction Occupation of Roofing*, NIOSH Contract 099–72–0121, Theodore Barry and Associates, Los Angeles, CA, 1975.
40. J. B. Moran and R. M. Ronk, Personal protective equipment, in *Handbook of Human Factors*, G. Salvendy, Ed., Wiley, New York, 1987, pp. 876–894.
41. Construction Safety Association of Ontario, *Fundamentals of Occupational Health in Construction*, Construction Safety Association of Ontario, Toronto, Ontario, Canada, 1993.
42. T. J. Parsons, T. J. Pizatella, and J. W. Collins, Safety analysis of high risk injury categories within the roofing industry, *Prof. Safety 31*(6):13–17 (1986).
43. State of California, *Work Injuries in Roofing and Sheet Metal Work, 1970*, Division of Labor and Statistics, San Francisco, CA, 1972.
44. Health and Safety Executive, HSE warns builders over standards for timer ladders, *Health and Safety Executive News Release,* E64, 1990, pp. 1–2.
45. J. Poirel, On roofs, *Cah. Com. Prevent. Batiment Trav. Public 17*(3):130 (1962).
46. R. A. Crisera, J. P. Martin, and K. L. Prather, *Supervisory Effects on Worker Safety in the Roofing Industry*, Division of Safety Research, NIOSH, Morgantown, WV, Contract No. 099–74–0035, 1977.
47. Sigeru, Repair of flat roofs, *Hochbau 78*(4):110–112 (1990).
48. NBS, *Roofing Accident Data Base*, Rep. No. NBS-GCR76–62, National Bureau of Standards, Washington, DC, 1975.
49. T. Barry, K. Prather, R. A. Crisera, and S. Fidell, *Behavioral Analysis of Workers and Job Hazards in the Roofing Industry*, NIOSH, Res. Rep., NIOSH Contract No. 099–72–0121, 1975.
50. K. O'Gara, *California Roofing and Sheet Metal Work: Analysis of Work Injuries and Illnesses*, Res. Bull. 6, Division of Labor Statistics and Research, San Francisco, CA, 1982.
51. U. Gustafsson and L. Wimnell, *An Ergonomic Study of Rooflayers*, Bygghalsan Bull. 1984–09–01, The Construction Industry's Organization for Working Environment, Occupational Safety and Health, Danderyd, Sweden, 1984, pp. 18–19.
52. T. J. Parsons and T. J. Pizatella, *Safety Analysis of High Risk Activities within the Roofing Industry*, Tech. Rep., NTIS PB-85163236, Government Printing Office, Washington, DC, 1985.
53. OSHA, *Construction Industry*, OSHA Safety and Health Standards, 29 CFR 1910.29 and 1926.500, Occupational Safety and Health Administration, Washington, DC, 1991.
54. *Health and Safety at Work*, Sloping roofs, *Health Safety Work 1*(11):32–33 (1979).
55. T. Niskanen and E. Seppanen, *Measures for Preventing Falls in the Construction Industry* (Engl. Abstr.), Tyoterveyslaitos, Julkaisumyynti, Topeliuksenkatu 41 a A, 00250, Helsinki, Finland, 1987.
56. Construction Safety Association of Ontario, *Safety Guidelines for Roofers*, Construction Safety Association of Ontario, Toronto, Ontario, Canada, 1993.
57. H. Moser, *Safety Aspects of Roofing and Work on Roofs*, Caisse Nationale Suisse d'Assurance en Cas d'Accidents, Sheet 22024, Luzern, Switzerland, 1982.
58. Federation of Industrial Mutual Accident Insurance Associations (FIMAIA), *Safety Rules for Work on Roofs and Corrugated Roofing*, Carl Heymanns Verlag KG (in German, Engl. Abstr.), 1980.
59. Construction Safety Association of Ontario, *Safety Tips—Sloped Roofs*, Construction Safety Association of Ontario, Toronto, Ontario, Canada, TP004, 1993.

60. P. O. Axelsson and L. Nilsson, *Hanging in There* (in Swedish). Rep. TRITA-AOG-0041. Royal Institute of Technology, Stockholm, Sweden, 1987. Cited by M. G. Helander, Safety hazards and motivation for safety work in the construction industry, *Int. J. Ind. Ergon. 8*:205–233 (1991).
61. D. Andreoni, in *Encyclopedia of Occupational Health and Safety*, 3rd edn., L. Parmeggiani, Ed., Vol. 2, International Labour Office Publications, Geneva, Switzerland, 1983, pp. 1960–1962.
62. S. M. Szymusiak and J. P. Ryan, Prevention of slip and fall injuries, Part I, *Prof. Safety 27*(6): 11–16 (1982).
63. S. M. Szymusiak and J. P. Ryan, Prevention of slip and fall injuries, Part II, *Prof. Safety 27*(7): 30–35 (1982).
64. P. Vink, Application problems of a biomechanical model in improving roof work, *Appl. Ergon. 23*(3):177–180 (1992).
65. W. Inancsi and T. L. Guidotti, Occupation-related burns: Five-year experience of an urban burn center, *J. Occup. Med. 29*(9):730–733 (1987).
66. NIOSH, *Health and Safety Guide for the Commercial Roofing Industry*, DHEW Pub. No. 78–194, U.S. Government Printing Office, Washington, DC, 1978.
67. S. Schneider and P. Susie, *Ergonomics and Construction: A Review of Potential Hazards in New Building Construction*, The Center to Protect Workers' Rights, Washington, DC, 1993.
68. M. Bjorklund, P. Helmerskog, M. Nordberg/Bohlin, U. Soderman, L. Holmqvist, B. Lindblad, J. Makynen, and S. Ahrman, Bygghalsan Bull. 91–09–16, The Construction Industry's Organization for Working Environment, Occupational Safety and Health, Danderyd, Sweden, 1991, pp. 35–36.
69. E. Rosenlund, B. Maenpaa, H. Nilsson, and I. Persson, *Ergonomic Equipment in Construction Work*, Bygghalsan Bull. 1987–05–01, The Construction Industry's Organization for Working Environment, Occupational Safety and Health, Danderyd, Sweden, 1987, pp. 38–39.
70. H. Isakson and J. Kling, *Fitting 90 mm and 1200 mm Plasterboard Sheets—An Ergonomic/Economic Comparison*, Bygghalsan Bull. 91–09–16, The Construction Industry's Organization for Working Environment, Occupational Safety and Health, Danderyd, Sweden, 1991, pp. 37–38.
71. M. Wahlin, *Work Methods for Handling Plasterboard and Chipboard in Building Training*, Bygghalsan Bull. 1981–07–01, The Construction Industry's Organization for Working Environment, Occupational Safety and Health, Danderyd, Sweden, 1981, pp. 20–21.
72. Bureau of Labor Statistics, U.S. Department of Labor, *Survey of Power Saw Accidents*, U.S. Government Printing Office, Washington, DC, 1980.
73. G. R. McCormack, *Work Injuries and Work-Injury Rates in the Highway and Street Construction Industry*, BLS Rep. No. 257, Bureau of Labor Statistics, U.S. Department of Labor, Washington, DC, 1961.
74. E. B. Holmstrom, J. Lindell, and U. Moritz, Low back and neck/shoulder pain in construction workers: Occupational workload and psychosocial risk factors. Part 1: Relationship to low back pain, *Spine 17*(6):663–671 (1992).
75. B. Salengro and F. Commandre, Musculoskeletal disorders at work in building constructions: Epicondylitis and low back pains, in *Musculoskeletal Disorders at Work*, P. Buckle, Ed., Taylor & Francis, London, U.K., 1987, pp. 254–258.
76. E. Broberg, *Ergonomic Injuries at Work*, National Board of Occupational Safety and Health, ISA Information System on Occupational Injuries, Rep. No. 1984:3E, Danderyd, Sweden, 1984.
77. J. R. Myers and R. B. Trent, Hand tool injuries at work: A surveillance perspective, *J. Safety Res. 19*(4):165–176 (1988).
78. S. Tanaka, A. B. Smith, W. Halperin, and R. Jensen, Carpet-layer's knee, *N. Engl. J. Med. 307*(20):1276–1277 (1982).
79. D. A. Stubbs and A. S. Nicholson, Manual handling and back injuries in the construction industry: An investigation, *J. Occup. Accidents 2*:179–190 (1979).

80. S. Kumar, Injury profile of the construction industry in Alberta in Canadian context: A case study of the impact of a non-monetary motivational factor in a multicenter Albertan company, *Int. J. Ind. Ergon.* 8(3):197–204 (1991).
81. T. Niskanen and J. Lauttalammi, Accidents in materials handling at building construction sites, *J. Occup. Accidents 11*(1):1–17 (1989).
82. T. Niskanen and O. Saarsalmi, Accident analysis in the construction of buildings, *J. Occup. Accidents* 5(2):89–98 (1983).
83. P. Barnard, *Injuries in the Erection Industry*, Construction Safety Association, Res. Publ. No. 14, Ontario, Appl. Res. Dept., Toronto, Ontario, Canada, 1970.
84. Bygghalsan, *Protect Your Knees*, Bygghalsan Broschyr 10(1), Stockholm, Sweden, 1983.
85. P. O. Axelsson, *Development of a Safety Device for Roof Work*, Tekniska Hogskolan, Arbets-olycksfalls-Gruppen, Stockholm, Sweden, 1983.
86. P. J. Leather, Attitudes towards safety performance on construction work: An investigation of public and private sector differences, *Work Stress* 2(2):155–167 (1988).
87. H. Bergenudd and B. O. Nilsson, Back pain in middle age; occupational workload and psychological factors, *Spine 13*(1):58–60 (1988).
88. H. Riihimaki, S. Tola, T. Videman, and K. Hanninen, Low-back pain and occupation. A cross-sectional questionnaire study of men in machine operating, dynamic physical work, and sedentary work, *Spine 14*(2):204–209 (1989).
89. R. J. Cleveland, Factors that influence safety shoe usage, *Prof. Safety 29*(8):26–29 (1984).
90. T. J. Parsons, *Establishment of a Construction Safety Research Plan*, Internal Rep., Division of Safety Research, NIOSH, Morgantown, WV, 1984.
91. R. G. Knowlton and J. C. Gilbert, Ulnar deviation and short-term strength reduction as affected by a curve-handled ripping hammer and a conventional claw hammer, *Ergonomics 26*(2):173–179 (1983).
92. K. H. Kilburn, R. H. Warshaw, and B. Hanscom, Are hearing loss and balance dysfunction linked in construction iron workers? *Br. J. Ind. Med. 49*(2):138–141 (1992).
93. I. Astrand, Degree of strain during building work as related to individual aerobic work capacity, *Ergonomics 10*(3):293–303, (1967).
94. E. Hammarskjold, K. Harms-Ringdahl, and J. Ekholm, Reproducibility of carpenters' work after cold exposure, *Int. J. Ind. Ergon.* 9(3):195–204 (1992).
95. E. Hammarskjold and K. Harms-Ringdahl, Effects of arm-shoulder fatigue on carpenters at work, *Eur. J. Appl. Physiol. Occup. Physiol. 64*(5):402–409 (1992).
96. R. Ortengren, T. Cederqvist, M. Lindgberg, and B. Magnusson, Workload in lower arm and shoulder when using manual and powered screwdrivers at different working heights, *Int. J. Ind. Ergon.* 8(3):225–235 (1991).
97. E. Hammarskjold, K. Harms-Ringdahl, J. Ekholm, and B. Samuelson, Effect of short-time vibration exposure on work movements with carpenters' hand tools, *Int. J. Ind. Ergon.* 8(2):125–134 (1991).
98. H. Wos, J. Lindberg, R. Jakus, and S. Norlander, Evaluation of impact loading in overhead work using a bolt pistol support, *Ergonomics 35*(9):1069–1079 (1992).
99. University of Iowa, *Work-Related Musculoskeletal Disorders: A Construction Bibliography*, The Center to Protect Workers' Rights, Washington, DC, 1993.
100. Bygghalsan, Bygghalsan Bull. 91–09–16, Bygghalsan, The Construction Industry's Organization for Working Environment, Occupational Safety and Health, Danderyd, Sweden, 1991.
101. T. D. Proctor and F. J. Rowland, Development of standards for industrial safety helmets—The state of the art, *J. Occup. Accidents* 8:181–191 (1986).
102. Health and Safety Commission, *Safety Helmets on Construction Sites: Recommendations of the Construction Industry Advisory Committee on Their Provision and Use*, Health and Safety Commission, London, U.K., 1979.

103. M. G. Helander, Safety hazards and motivation for safety work in the construction industry, *Int. J. Ind. Ergon.* *8*:205–233 (1991).
104. E. A. P. Koningsveld, Permissible loads for the Dutch construction industry, *Ergonomics* *28*(1):359–363 (1985).
105. A. D. F. Price, Calculating relaxation allowances for construction operatives. Part 1: Metabolic cost, *Appl. Ergon.* *21*(4):311–317 (1990).
106. A. D. F. Price, Calculating relaxation allowances for construction operatives. Part 2: Local muscle fatigue, *Appl. Ergon.* *21*(4):318–324 (1990).
107. B. Zimolong, Hazard perception and risk estimation in accident causation, in *Trends in Ergonomics/Human Factors II,* R. E. Eberts and C. G. Eberts, Eds., North-Holland/Elsevier, Amsterdam, the Netherlands, 1985, pp. 463–470.
108. U.S. Department of Labor, *Supplementary Data System Microdata Files User's Guide, 1985 Edition*, Bureau of Labor Statistics, U.S. Department of Labor, Washington, DC, 1989.
109. T. Miwa, Vibration-isolation systems for hand-held vibrating tools, in *Vibration Effects on the Hand and Arm in Industry*, A. J. Brammer and W. Taylor, Eds., Wiley, New York, 1981, pp. 303–310.
110. G. A. Hampel and W. J. Hanson, Hand vibration isolation: A study of various materials, *Appl. Occup. Environ. Hyg.* *5*(12):859–869 (1990).
111. E. R. Andersson, Design and testing of a vibration attenuating handle, *Int. J. Ind. Ergon.* *6*(2):119 (1990).
112. A. M. Steward and D. F. Goda, Vibration syndrome, *Br. J. Ind. Med.* *27*:19–27 (1970).

CHAPTER 25

# Ergonomic Hazards and Controls for Elevating Devices in Construction

Christopher S. Pan, Sharon S. Chiou, Hongwei Hsiao, and Paul Keane

CONTENTS

The construction industry sector has long been recognized for its high number and rates of nonfatal and fatal injuries. Data for 2007 from the survey of occupational injuries and illnesses, an occupational injury and illness database maintained by the U.S. Bureau of Labor Statistics (BLS), indicate that workers in the construction sector experienced the fourth highest number of injuries and illnesses with days away from work (135,350) and the highest rate of injuries and illnesses with days away from work (190.3 per 10,000 full-time workers) [1,2]. Falls accounted for 24% of the nonfatal injuries, while contact with objects and equipment and overexertion were associated with another 52% of the injuries in this sector. Preliminary fatality data for 2008 from the BLS, the Census of Fatal Occupational Injuries (CFOI), indicate that workers in the construction industry sector experienced the highest number of fatalities (969) and the fourth highest fatality rate (9.6 per 100,000 full-time equivalent workers) [3]. Nonetheless, the fatality rate in construction was still almost three times the overall fatality rate for all workers (3.6 per 100,000 full-time equivalent workers). Falls led to the largest number of fatalities within the construction industry sector (34%). Another 32% of the fatalities were due to contact with objects and equipment and highway incidents [4].

This chapter discusses the aforementioned injuries and their prevention for elevating devices at construction work sites. These devices—including aerial lifts, stilts, scaffolds, and mast work platforms—are widely used, and research further indicates increasing trends for their application in the construction industry. This chapter also identifies hazards and presents previous and current findings, focusing on four research studies conducted by National Institute for Occupational Safety and Health (NIOSH) for elevating devices. Because of the dynamic nature of construction tasks, safety professionals and ergonomists hesitate to evaluate and identify single risk factors associated with the use of elevating devices in the construction industry and prefer to consider the systematic risk involved with the use of this equipment [5]. This chapter promotes the recognition of these hazards and the use of the most advanced and current technologies for injury prevention and control, exposure simulation, and hazard evaluation—for example, computer simulation and force data collection through sensor technologies—to evaluate and control these hazards for this workforce.

## 25.1 AERIAL LIFTS

### 25.1.1 Injury Surveillance and Current State of Knowledge

Elevating devices are widely used for elevating workers, tools, and materials in various industries that require workers to work above ground level. The North American Industry Classification System does not have separate categories for elevating devices or aerial lifts but considers scaffold erection and dismantling, manufacturing, wholesaling, and rental as industrial classifications. Only scaffold erection and dismantling are considered specialty trade designations.

Because elevating devices are primarily and widely used in construction, many trade specialties (e.g., carpenters) and laborers in construction are involved with tasks related to the erection, dismantling, use, and deployment of elevating equipment. Because of these factors, a single industrial classification for elevating equipment does not exist, and research focused on fall prevention related to elevating equipment encompasses construction and related industries, rather than focusing on scaffolding or elevating equipment designations.

Aerial lifts constitute a common form of elevating equipment used in construction and related industries; this equipment is used for general construction tasks, masonry work, and related tasks. The applicable safety standard promulgated by the American National Standards Institute (ANSI) for safe use of this equipment is ANSI A92.2-1969. Because of the high degree of innovation and technical development in elevating devices at present, additional standards are under development.

The fall hazards associated with work on aerial lifts are well recognized within the scaffolding industry [6–9]. Surveillance reveals the increasing risk of severe injury and death associated with the adoption of this equipment in construction, telecommunications, and other industries. In addition, expert opinion within the user and manufacturing communities indicates great concern over the potential exposures involved with increasing vertical extension of newly developed equipment and the increased adoption of this equipment for nontraditional uses [10–13]. Industrial reports on aerial lifts indicate increasing use in various industries, including construction (7,583,000 employment), landscape services (705,900 employment), telecommunications (705,800 employment), warehousing and storage (595,700 employment), electric power transmission and generation (162,300 employment), and other industries [14].

There are two types of aerial lifts, mainly differing by design characteristics: boom lifts (articulating or telescopic), which support a work platform at the end of an extensible boom, or scissor lifts, which support a work platform at the top of an extensible, linked and folding support assembly (Figure 25.1). Aerial lifts can also be classified by their power sources,

Articulating boom lift telescopic boom lift

Scissor lift

FIGURE 25.1 Types of aerial lifts. (Reprinted with permission of Skyjack Inc., Guelph, Ontario, Canada.)

tire types, maximum elevated heights, wheelbases, and load capacities [6]. Tipovers of scissor lifts mostly occur while the lift is elevated over 15 ft, especially while the lift is moving [8]. Scissor lifts are available that can reach between 20 and 75 ft. At such heights, the stability of the lift and safety of the worker are of great concern. Scissor lifts that are designed to fit through doorways can have wheelbases as narrow as 17 in., decreasing their stability. The use of rollout platform extensions on the scissor lift platform also decreases stability since the extension extends beyond the wheelbase, affecting the center of gravity. Workers performing pulling or pushing activities while the platform is elevated can exert horizontal forces that the scissor lift is not designed to withstand [15]. This is a concern, especially when the platforms are extended horizontally beyond the base of the scissor lift [15].

NIOSH, in collaboration with the National Safety Council and the Center for Construction Research and Training, conducted a surveillance study of aerial platform falls/collapses/tipovers across all industry classifications. Three databases were used to analyze aerial lift fall incidents: BLS CFOI data (1992–2003), Occupational Safety and Health Administration (OSHA) incident investigation records (1990–2003), and NIOSH Fatality Assessment and Control Evaluation (FACE) reports (1985–2002). Pan et al.'s [9] review of these surveillance systems indicated that extensibility factors—the extended height of the lift or the vertical position of the worker as a result of extension of the lift—were significant contributing factors for fatal injury. These height factors accounted for 72% of the scissor lift cases in CFOI; 83% of cases investigated by OSHA/FACE involved falls/collapses/tipovers within the height categories of 10–19 and 20–29 ft [9]. In CFOI data, 45% of boom lift fatalities and 72% of scissor lift fatalities occurred in the construction industry; in OSHA/FACE data, 43% of boom lift fatalities and 74% of scissor lift fatalities occurred in construction (Figures 25.2 and 25.3). Based on the findings of this study [9] and the marketing lift sales information, NIOSH developed an aerial lift project focusing on a laboratory study of a commercially available, and the most popular, 19-ft electric scissor lift.

Another finding of interest from this study [9] relates to the human operator's stability/balance and related workstation design issues for aerial lifts. Workers typically position the

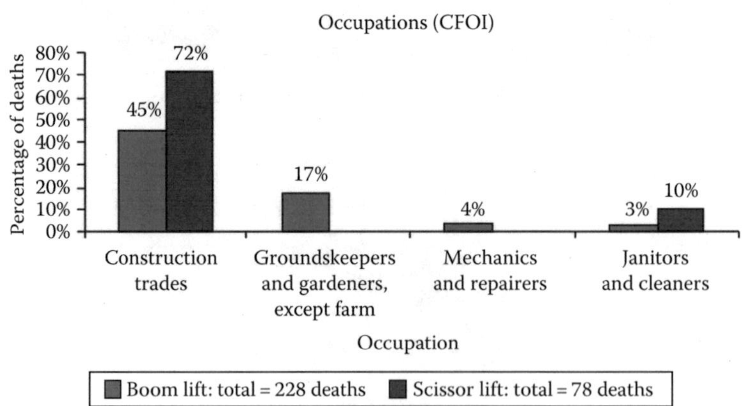


FIGURE 25.2 Occupations in aerial lift deaths, 1992–2003. (U.S. Bureau of Labor Statistics CFOI Research File.)

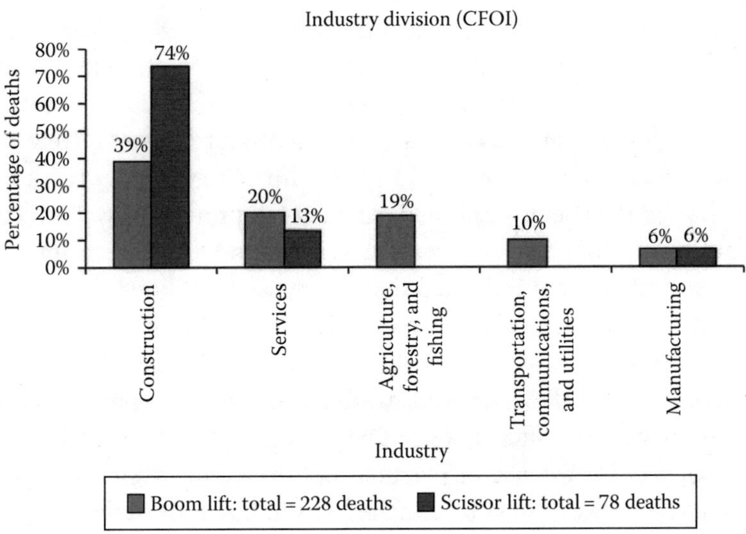


FIGURE 25.3 Industry divisions in aerial lift deaths (OSHA/FACE).

aerial lift platform directly adjacent to the work location to perform assigned work tasks at positions that are convenient and ergonomically adjacent to work, as opposed to positioning the platform beneath a work location and working overhead [16]. Work positions above shoulder height—especially for an increasingly aging workforce with musculoskeletal limitations—impose musculoskeletal demands that make it difficult to perform work while assuming this posture. Frequently, this workforce has additional limitations—for example, reduced vision requiring corrective bifocal lenses—which make overhead work more problematic when work tasks require changing visual focus. Common tasks for elevated workers frequently involve work practices that impose potentially destabilizing forces on the human body—for example, forces from excessive or awkward motions (jerking or forcing tools, excessive pressures on tools, over reaching, leaning over the edge or guardrail, and standing on the top rails or mid rails). These work practices and postural positions place the operator at increased risk of falls from the platform to a lower level and may additionally introduce overexertion hazards. Therefore, worker-preferred task locations/heights and work practices are extremely important to consider in the design of lifts. Also, since the platform/bucket constitutes a confined work space and operators need to perform tasks using awkward working postures, the use of fall protection systems (e.g., harness and lanyard) will generate significant issues in ergonomic design and health-impact considerations, especially during fall arrest situations [17–19]. When these awkward postural issues are combined with the application of significant task-related side forces in the horizontal direction, the potential for hazards associated with ejection and tripping incidents, as well as overexertion, significantly increases [5]. Improvement in control measures addressing operators' whole-body postural stability within the platform/bucket will be critical to prevent such fall/trip/ejection incidents [20].

The use of fall protection equipment on aerial lifts is generally recognized by aerial lift safety experts as one of the most effective safety control practices to reduce fall-risk exposure for operators [6]. However, results from this study [9] indicated that, for a significant percentage (82% for OSHA/FACE data) of fall-from-elevation incidents, safety controls did not protect workers because existing fall protection systems (e.g., guard rails, chains, gates/doors, belts, harnesses with or without lanyards) were not in use at the time of the incident. Only 4 out of 13 scissor lift injury/fatality cases from OSHA/FACE reports showed the use of additional personal fall protection systems (i.e., belts and harnesses). Findings from a field observation study found significant nonusage of existing fall protection systems and indicated that the majority of boom lift operators (16 out of 18) conducted tasks without using personal fall protection systems [21]. This finding supports data analysis of this study [9], which found 45% (18 of 39) of boom-lift-related fatalities recorded in OSHA/FACE reports involved nonuse of fall protection systems. This finding is also extremely significant, since OSHA regulations for working on boom lifts [1926.453(b)(2)(v)] require the use of personal fall protection systems, with the further requirement that the lanyard must be attached to the boom or basket [22]. However, no body of scientific knowledge establishes the efficacy of personal fall protection systems for use on scissor lifts, and the use of harnesses and lanyards is not required (OSHA-regulated scissor lifts are regulated as mobile scaffolds). Guardrails on the platforms are sufficient to meet the OSHA requirement [1926.451(g)(4)] for fall injury prevention for scissor lifts, and additional requirements for using personal fall protection systems currently are still under consideration by industry and standard committees (ANSI A92.6 and ANSI A10.29). The lack of universal standards regarding best practices for the use of fall protection standards represents a serious concern for the aerial lift industry. It should be noted that fall-arrest and fall-protection devices perform no function in the event of a tipover, and fall-protection devices do not constitute hazard-reduction devices in the event that the elevating platform tips over and propels the operator to the ground.

### 25.1.2 NIOSH Research Findings

NIOSH has conducted research into various conditions of exposure related to the use of aerial lifts. Constraints on this research chiefly are found in conducting research that puts human subjects at risk of fall injury. NIOSH researchers have responded to this risk by conducting a significant part of their research in the form of computer simulation and modeling of various hazardous exposures. Preliminary computer model development was conducted using input from the static test results of the center-of-gravity and horizontal-stability tests from the standard promulgated by ANSI A92.6 (2006) [23]. Additional human subject tests were conducted to collect side force information for the computer model [15]. The center-of-gravity (Figure 25.4) and horizontal-stability tests (Figure 25.5) generated data for developing a computer simulation model for the SkyJack model 3219 with manufacturer's lift part mass/geometric information using the SolidWorks™ software program. The computer model was generated based on data from automated dynamic analysis of mechanical systems software (*ADAMS*™ 2005, MSC Software Corporation, Santa Ana, CA). In order to calculate the center of gravity, four force plates (Bertec Corporation, Columbus, OH) were placed under the wheels of the scissor lift (Figure 25.4), and the lift was tilted using hand pump jacks and jack stands (Figure 25.4). Platform height

FIGURE 25.4 Center of gravity tests.

FIGURE 25.5 Horizontal stability tests.

was recorded using a cable-extension transducer (Model PT5A-250-N34-UP-500-C25, Celesco Transducer Products, Inc., Chatsworth, CA). A horizontal actuator (Series 247, MTS Systems Corporation, Eden Prairie, MN) was used to apply horizontal loads through a cable-and-sheave arrangement as shown in Figure 25.5. The sheave was hung from a 5-ton-capacity overhead crane. Load readings were taken via a load cell (Model 661.20e-02,

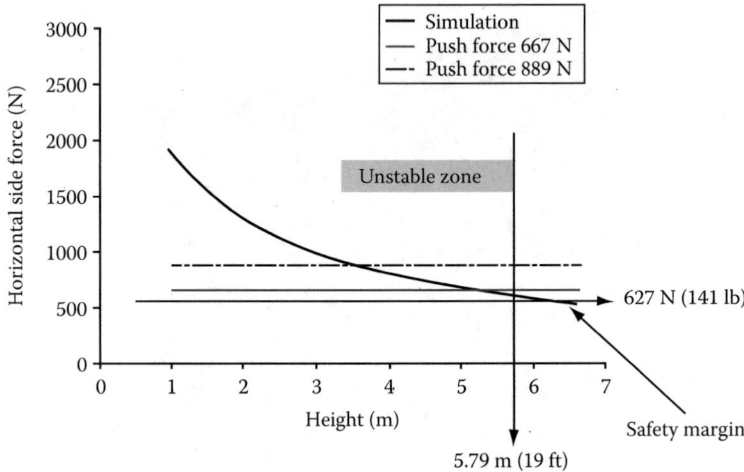


FIGURE 25.6 Stability prediction from the horizontal side forces and elevated heights or SkyJack Model 3219 scissor lift.

MTS Systems Corporation, Eden Prairie, MN) integrated with the hydraulic actuator. The modeling predictions agreed well with the experimental data with an error less than 1% for the whole range of the lift height variation in three orthogonal directions.

The NIOSH study team completed data collection for operations within the scissor lift platform. Results indicate that the scissor lift could lose static equilibrium when operated at an extended height above 5.49 m with the application of a horizontal force of 627 N, which is the maximum push force measured in the experiment simulating working conditions on the platform [15] (Figure 25.6). The study indicates that the scissor lift may tip over in the horizontal direction during normal operations with the excessive applied forces. If the applied forces are between 627 and 889 N, the scissor lift can be safely extended to a height between 5.49 and 3.49 m respectively [24].

### 25.1.3 Curb and Pothole Test Results

To quantify the fundamental dynamic characteristics of the lift, accelerations in three orthogonal directions on the platform and base were measured at the right-side base frame and main-platform frame, as shown in Figure 25.7. Two in-house packaged triaxial accelerometers (Model KXM52-1050, Kionix Inc., Ithaca, NY) were used for the measurement. A 20-cm-high granite slab of 956 kg was positioned tightly against a concrete curb (50 mm) at the back of the slab. As specified in the standards, a depression of 600 × 600 mm with a depth of 100 mm was constructed on the runway. In separate tests, each leading wheel was driven into the depression at its full speed for each given elevated height. The drive control was maintained at maximum until the leading wheel was driven into or over the depression. A data acquisition box was installed securely within the platform.

For the curb tests, the wheel of the lift did not climb up the curb, and tipover did not occur. A large phase difference between the accelerations on the base and platform suggests that there is some flexibility in the scissor lift structure. However, the speed on the base before the impact was practically identical to that on the platform. The extension of the platform

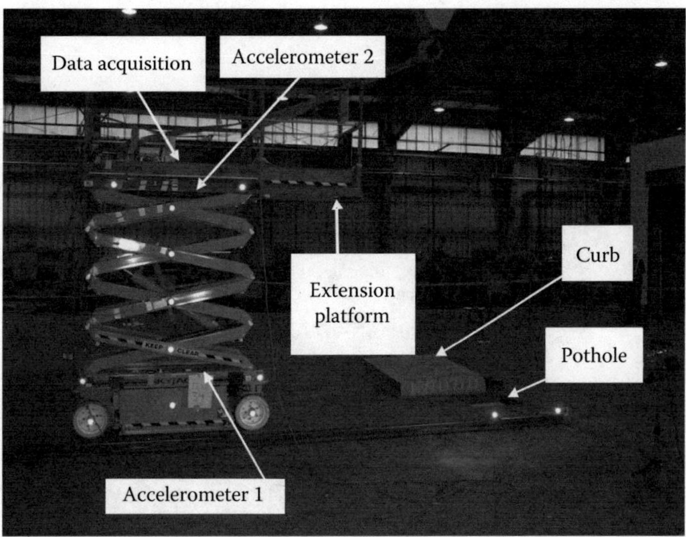


FIGURE 25.7 Two accelerometer locations, the curb test setup, and the pothole test setup.

reduced the bounce frequency ($F = 451$, $p < 0.001$). The impact angle affected the pitch and bounce resonance frequencies ($F \geq 13.5$, $p \leq 0.002$). The travel direction did not significantly influence the pitch and rolling resonance frequencies ($F \leq 1.6$, $p \geq 0.224$), but it marginally affected the bounce frequency ($F = 6.0$, $p = 0.026$). The maximum peak accelerations in the 90° impact test were larger than those in the 30° impact test ($F \geq 12.1$, $p \leq 0.001$). Because of the increased flexibility and reduced speed in the fully elevated test, the maximum peak accelerations in the full-height test were much less than those in the low-height test ($F \geq 450.3$, $p < 0.001$). The maximum rigid body roll angle was 2.3°, which was lower than the tipover angle (9.0°) calculated from the rationale for the theoretical criterion adopted in ISO 16368 [25,26]. The resonance frequencies of the lift at full height in the depression tests were also generally lower than those observed in the curb impact test ($p < 0.001$). This is because the lift base tilted when one wheel dropped into the depression such that the gaps between the wheels and ground were larger than those in the curb test. A very common scenario leading to a tipover event occurs as a general function of this situation: Lift stability is compromised as a result of a wheel entering a depression, causing the supporting plane of the lift base to move off a horizontal plane. Destabilization can occur as a result of this "tilted" plane, which translates into a greater degree of deflection for a platform at height.

The highest acceleration was found in the high-speed curb impact test. Therefore, the acceleration in a curb impact event is of major concern for the workers' stability within the platform. Appropriate fall protection systems should be used to prevent ejections. Another potential concern is that a tipover may occur while the scissor lift runs over a depression without a pothole guard [26].

### 25.1.4 Computer Simulation and Modeling Results

The computer simulation model (Figure 25.8) was developed based on experimental data obtained in standardized curb impact and pothole depression tests. These tests investigated

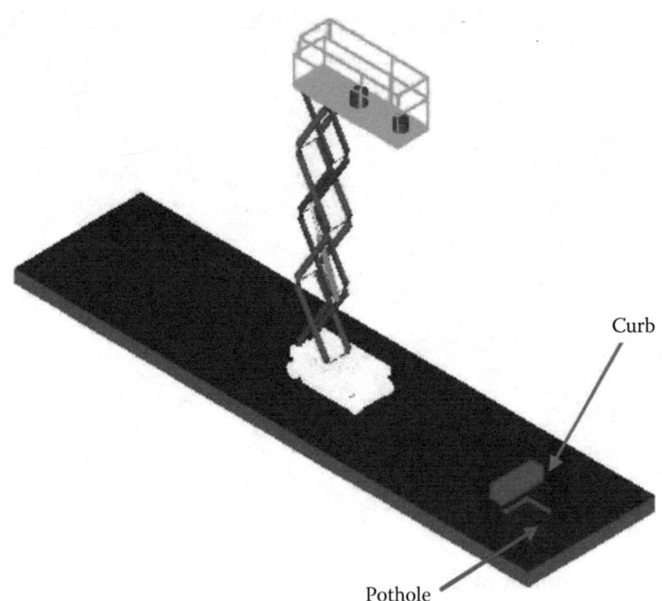


FIGURE 25.8 ADAMS model of the scissor lift with its full load and elevated at its full height (the curb block is de-actuated when the pothole test is simulated).

the effects of the lift structural flexibility on the tipover thresholds of road surface slope, curb impact speed, and pothole guardrail height. The mass distribution of the model was validated using lift center of gravity measured at four elevated heights. The connection stiffness and damping parameters of the model were estimated based on the experimental data obtained from an impact test and a depression test of the scissor lift. The model was also validated and refined using the time histories of the lift dynamic responses measured in these physical experiments [27]. The modeling results indicate that decreasing the stiffness of the scissor lift generally reduces both static and dynamic stabilities of the lift. This study showed that lift instability could be achieved by increasing the flexibility of the scissor-lift ground system, which includes severe wear (e.g., aging lifts) and decoupling of structural joints, damage to the joints resulting in decoupling of rigid frame members, and the use of the lift on deformable or uneven surfaces (e.g., soft ground). This information will be used to determine the future tasks associated with refining this multibody model simulation and may contribute to more finite element analyses and better evaluation techniques.

Simulated operator information was also incorporated into the completed scissor lift model using 2009 LifeMOD Biomechanics Human Modeler (LifeModeler Inc., San Clemente, CA), which is a plug-in to ADAMS. Using this joint human/lift model, various hazardous working conditions were simulated (i.e., pothole, curb, tilting/slope, gusting wind, and fall-arrest loading using a fall-protection harness/lanyard) (Figures 25.9 through 25.12). These hazardous working conditions cannot be tested on a human subject, even in well-controlled laboratory events; however, the simulation model of the lift/operator can appropriately assess the contributions of each individual hazard to the incidents or the combined effects from these hazards. In other words, outcomes of this study component can provide users with an effective tool for better incident investigations.

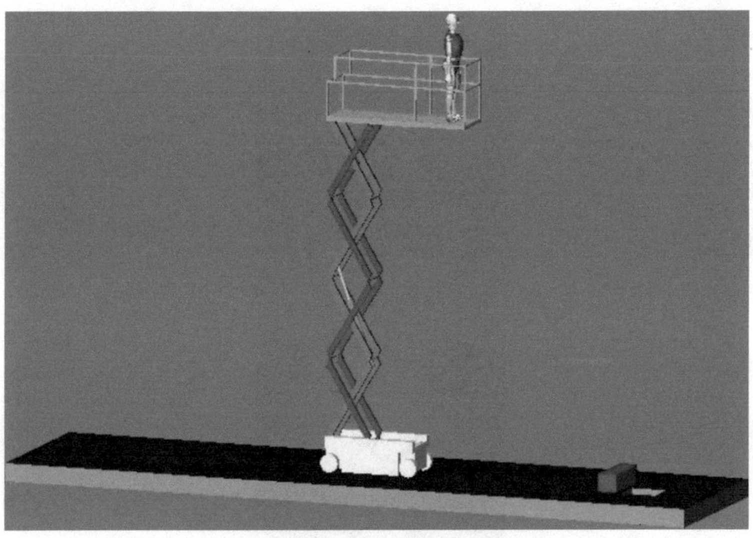

FIGURE 25.9 ADAMS/LifeMOD model of the scissor lift with an operator.

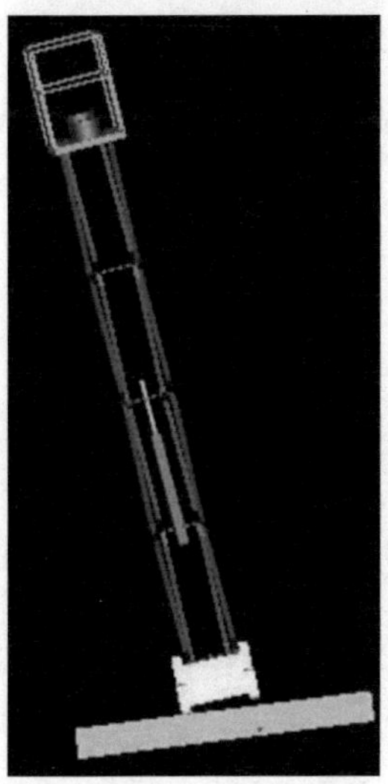

FIGURE 25.10 ADAMS simulation of tilting and slope conditions.

For example, a simulation was to determine the lateral tipover threshold for an applied wind load. For this simulation, the scissor lift was positioned on a level surface and no other external forces were applied. Based on ISO 16368 Section 5.2.3.3, the initial wind speed was assumed to be 12.5 m/s. The corresponding wind forces were calculated for the base, lifting mechanism, platform, guardrail assembly, and operator. Table 25.1 shows the wind forces calculated for various wind speeds.

It was assumed that an operator (113 kg) was on the main platform and another worker (113 kg) was on the extension platform. The wind forces were then applied horizontally to the center of each component. The scissor lift was raised to a maximum platform height of 5.8 m. At a wind speed of 12.5 m/s, the scissor lift did not tip over. The wind speed was then

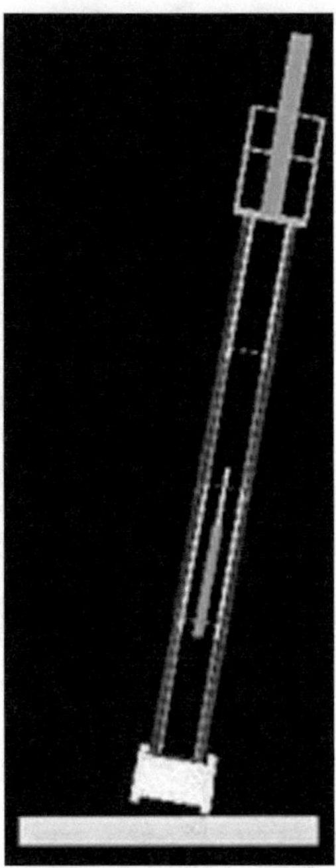

FIGURE 25.11 ADAMS simulation of gusting wind conditions.

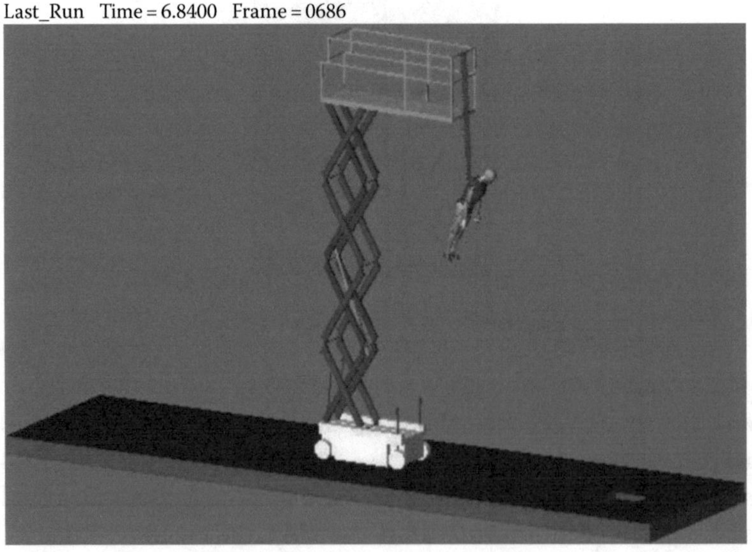


FIGURE 25.12 ADAMS/LifeMOD simulation of drop test in which a manikin is dropped from a fully elevated scissor lift to test lift stability. *Note*: Physical tests were done on different lift platform locations at NIOSH laboratories, not just the recommended anchorage points. Top rail is commonly used, but not recommended. Lift remained stable even under adverse (extreme) test conditions.

TABLE 25.1 Wind Forces for Components of the Scissor Lift

| | Wind Forces (N)—Adjusted for 1.1 Multiplication Factor | | | | | | | | | | |
|---|---|---|---|---|---|---|---|---|---|---|---|
| **Wind Speed (m/s)** | **Wind Pressure** | **Base** | **Lift Frame Member** | **Platform** | **Guard-rail 1** | **Guard-rail 2** | **Guard-rail 3** | **Guard-rail 4** | **Guard-rail 5** | **Guard-rail 6** | **Operator** |
| 12 | 92.16 | 109.72 | 19.87 | 22.24 | 2.05 | 1.46 | 3.51 | 3.51 | 8.20 | 8.65 | 70.96 |
| 12.5 | 100 | 119.05 | 21.56 | 24.14 | 2.23 | 1.58 | 3.81 | 3.81 | 8.90 | 9.39 | 77.00 |
| 14 | 125.44 | 149.34 | 27.04 | 30.28 | 2.79 | 1.98 | 4.78 | 4.78 | 11.17 | 11.78 | 96.59 |
| 16 | 163.84 | 195.05 | 35.32 | 39.54 | 3.65 | 2.59 | 6.24 | 6.24 | 14.58 | 15.38 | 126.16 |
| 18 | 207.36 | 246.86 | 44.71 | 50.05 | 4.61 | 3.28 | 7.90 | 7.90 | 18.46 | 19.47 | 159.67 |
| 19 | 231.04 | 275.05 | 49.81 | 55.76 | 5.14 | 3.65 | 8.80 | 8.80 | 20.57 | 21.70 | 177.90 |
| 20 | 256 | 304.77 | 55.19 | 61.79 | 5.70 | 4.05 | 9.75 | 9.75 | 22.79 | 24.04 | 197.12 |

increased until tipover occurred. Tipover occurred at a wind speed of 20 m/s (44.7 mph) (Figure 25.11). Also, the wind speed threshold increased to 23 m/s for one person (113 kg) on the main platform.

### 25.1.5 Drop Test Results (Dead Weight Drop) [28]

The object of this study was to measure structural and dynamic stability of aerial lift platforms under dynamic load conditions, through the application of weights dropped from the platform surface. A load cell (3000 lb S-type, Interface Inc., Scottsdale, AZ) was used to record the maximum arrest force, a string potentiometer (250 in., Model PT5D, Celesco Transducer Products, Inc., Chatsworth, CA) recorded positioning of the drop test fixture, an electromagnet (700 lb, Model SE-35352, Magnetic Products, Inc., Highland, MI) activated the drop test fixture, and data were logged onto a laptop computer equipped with a data acquisition card (Model DAQCard-6036E, National Instruments Corporation, Austin, TX) running the LabVIEW data acquisition application (National Instruments Corporation, Austin, TX).

Two fall conditions were tested: 6-foot falls and 11-foot falls. The latter represented a common misuse scenario, in which operators stand on the midrail. Fall arrest loads of 2400 lb were selected to exceed the ANSI Z359.1 standard, which calls for loads of 1800 lb. Various anchorage point locations were chosen in the platform. To determine the free-fall height for these tests, the lanyard attachment point to a fall arrest harness was assumed at approximately chest height for a 95th percentile worker, 53.7 in. (1.364 m) (MIL-STD-1472D, 1989). Total fall height was 139 in. (3.53 m) when anchored to the midrail and 122 in. (3.10 m) when anchored to the toprail. The scissor lift maintained structural and dynamic stability for all drop tests under various test conditions (Figure 25.13).

### 25.1.6 Drop Test Results (Manikin Drop) [29]

This study measured drop forces on an anthropomorphic manikin (ADAM™, Veridian, Dayton, OH) (Figure 25.14) under loads mediated by four safety harnesses. Acceleration forces were measured by embedded triaxial accelerometers (Kionix, Inc., Ithaca, NY), which were positioned at three locations—head, middle of spine, and torso. A load cell

FIGURE 25.13 Lift tilting angles setup during the drop tests (1.5° on the left and 3.5° on the right).

FIGURE 25.14 The advanced dynamic anthropomorphic manikin (ADAM) system.

(Model SSM-S, Series 1000, Interface, Inc., Scottsdale, AZ) was placed in line with the safety lanyard to measure the arrest force; lanyards were secured at an anchorage point at the bottom of the scissor lift platform. A high-speed camera (Phantom v4.3, Vision Research, Inc., Wayne, NJ) was used to record the drop test. An electromagnet (700 lb, Model SE-35352, Magnetic Products, Inc., Highland, MI) secured and released the drop test fixture to the hook of a 5-ton overhead crane, and data were collected on a laptop computer equipped with a data acquisition card (Model DAQCard-6036E, National Instruments Corporation, Austin, TX) running a LabVIEW data acquisition application (National Instruments Corporation, Austin, TX). The ADAM was dropped three times from each of the 2 heights, 6 and 11 ft (see dead weight drop tests in the previous section).

During the deceleration phase of a fall, the personal fall arrest system (PFAS) absorbs the kinetic impact energy, thereby reducing the impact force on the human body. The kinetic energy dissipated during the fall impact is an important parameter that characterizes the dynamic performance of the PFAS. The impact kinetic energy was either not considered or not correctly estimated in the literature. In the current study, we developed a systematic approach to evaluate the energy dissipated in the energy absorbing lanyard (EAL) and in the human body during the fall impact. The kinematics of the human body and EAL during the impact was derived using the data of the time histories of the arrest force, which was measured experimentally. We applied the proposed method to analyze the experimental data of a 6-ft drop test and an 11-ft drop test. The preliminary results indicate that the distribution of the kinetic energy in the EAL and the falling body depends on the intensity of the impact: The portion of the kinetic energy dissipated in the EAL for higher impact force is more than that for lower impact force [30].

### 25.1.7 Future Studies for Aerial Lifts

As stipulated in draft standard *A10.29* of the ANSI, workers may enter and exit an aerial platform at heights greater than 6 ft when the aerial platform surface is adjacent to the elevated surface. The standard further specifies that if the platform is adjacent to the elevated surface, there shall not be a vertical gap larger than 8 in. or a horizontal gap larger than 14 in. between the aerial lift platform and the adjacent surface. To date, there has been no published scientific justification on the manner in which the vertical and horizontal gaps were determined and how the distances between the lift platform and the adjacent surface may affect workers' postural stability and fall propensity. Additionally, there is a lack of quantitative data to demonstrate that potential risks may be associated with improper exiting and entering techniques, especially at these heights. The purpose of this ongoing study is to determine workers' postural stability under various conditions and methods of exiting and entering scissor lift platforms at elevations. In addition, partial results of this experiment—that is, maximum forces and moments resulting from exiting the scissor lift platform and on the positions of workers' centers of gravity—will provide information for the update and development of a computer simulation model for predicting the overall stability of aerial lifts. This would be considered in a long-term NIOSH aerial lift project.

## 25.2 STILTS

### 25.2.1 Background

Falls and overexertion are the leading causes of occupational traumatic injuries among construction workers in the United States [31]. In 2008, the construction industry reported an estimated 24,720 falls and 21,310 overexertion injuries involving days away from work, which accounted for approximately 21% and 18% of total injuries, respectively [32]. Among fall injuries, 63% of cases were falls to a lower level [32]. One specialty of the construction industry—drywall installers—is also traditionally plagued with high frequencies of overexertion and fall injuries. In a study of injury characteristics of drywall installers, falls were found to be the second leading cause of injury (overexertion was the leading cause) accounting for 32% of total injuries [33,34]. In a workers' compensation study of North Carolina residential construction workers, drywall installers were among the specialties with the highest rate for falls from a different level resulting in medical costs or lost work time [35]. The analyses of administrative data revealed that the most common mechanisms of injury for drywall installers involved being struck by objects, overexertion, and falls [36]. Similarly, in a large cohort study of union carpenters and drywall installers, the major causes of injury were being struck by or against something, manual materials handling, and falls [37]. This high frequency of overexertion and fall injuries of drywall installers is likely due to the main job demands for this workforce—constant handling of heavy and bulky drywall sheets and working at elevations to perform tasks close to ceilings or on the upper half of walls.

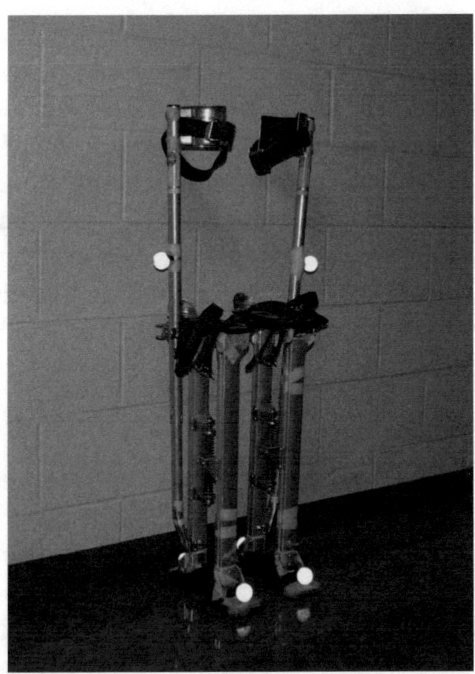

FIGURE 25.15 Construction stilts with four reflective markers attached to each side for motion study.

Falls from heights may occur from a variety of surfaces for carpenters and drywall installers [37]. One unique item of elevating equipment that is commonly used for drywall tasks is stilts (Figure 25.15). In a study of Washington State Workers' compensation data for the period of 1996–2001, stilts-related injuries resulted in a median of 73 lost workdays compared to 24 lost workdays for all claims combined [38]. Like scaffolds, stilts are an elevating tool that raises workers above the ground level to allow them to perform tasks on the ceiling or upper half of a wall. Stilts are composed of more than 50 small parts, providing mobility for workers to move from one location to another without the burden of erecting scaffolds or repositioning ladders. A pair of common commercially available construction stilts weighs about 7.26 kg (16 lb) and the height is typically adjustable from 0.6 to 1.02 m (24–40 in.). When wearing stilts, a shoe plate is strapped onto the bottom of the workers' shoes with a strut tube running up the side of the leg and strapped right below the knee. Two lower strut tubes connect the shoe plate with a nylon floor plate. The lower tubes are spring loaded to provide forward and rearward stilt-walking actions (Figure 25.15). The construction activities commonly performed by workers on stilts include drywall finishing, taping, and sanding. Stilts can also be used for interior tasks such as painting, plastering, insulation installation, acoustical ceiling installation, and any other light-duty building maintenance.

Earlier studies addressing safety and health issues of drywall installation have focused on asbestos, dust, or silica exposure from drywall sanding [39–42]. A limited number of studies have been conducted to examine the overexertion and fall injury mechanisms that may be associated with the use of stilts. The potential hazards associated with stilts in construction have been well recognized nationally and internationally. The state of California,

New York City, the provinces of Ontario, Canada, and Victoria, Australia, do not recommend, and/or have established legislation against, the use of stilts in construction as a preventive measure against occupational injuries. Several interested parties have made recommendations regarding the use of stilts; for example, one of the training guidelines of the International Union of Painters and Allied Trades recommend the maximum safe height for stilts be limited to 24 in. for painter apprentice training [43]. However, most of the guidelines and regulations were established based on experiences and perceptions of safe operating parameters. No quantitative and objective data demonstrated the injury mechanism associated with stilts until NIOSH embarked on a series of laboratory studies and computer simulations [44–47].

### 25.2.2 Musculoskeletal Injuries Associated with Stilts

Stilts add excessive weight to the mass moments of inertia to the lower limbs, requiring the stilt users to apply more efforts during walking. Consequently, workers may require excessive forces in their lower limbs when walking on stilts, which could contribute to cumulative trauma musculoskeletal disorders in the lower limbs for construction workers [48]. Increased musculoskeletal forces can increase the risk of injuries to tendon and muscle tissue [49–51], while increased joint loading converts to excessive contact pressure in articular cartilage. Numerous clinical studies indicated that excessive nontraumatic loading pressure on articular cartilage is the main cause of the initiation and development of osteoarthritis in old age [52–54]. The etiology of overexertion injuries can include the contribution of slip and fall injuries to tissue damage, leading to subsequent overexertion injury to soft body tissues. The mutual interaction of overexertion and slip-and-fall injury events in inducing further injury has been extensively reported [55–57].

#### *25.2.2.1 Lower Extremity Joint Forces*

Despite the need for further research articulated by Schneider and Susie [58], who hypothesized that the use of stilts may place workers at increased risk for knee injuries, the musculoskeletal loadings imposed on workers during stilts walking were not investigated until two stilt-walking models were developed by researchers at NIOSH [44,47]. This research was based on the emerging scientific techniques of multibody system dynamic modeling and simulation, which were increasingly found to be valuable techniques that could be used to evaluate joint loadings, fall scenarios, and gait-related issues. In 2002, using ADAMS and LifeMOD as a simulation platform, a human-stilts multibody system was first developed through a collaborative research project between NIOSH and Mechanical Dynamics Inc. Although computer simulations of human walking were well established [59–62], no published literature existed at that time on computer simulations of gait on stilts. After the stilt model was developed, it was further validated using kinematic and kinetic data of construction workers walking on stilts, collected using a motion-analysis system and force plates [44]. The validated model can be used to predict the mechanical stability, joint reaction forces, and trajectories of human center of mass for workers on stilts. Furthermore, it can be used to assess tripping hazards, struck by/against, sudden starts/stops, pivots while using stilts, and other traumatic injury scenarios.

The second stilt-walking model developed by NIOSH involved an inverse dynamic model to investigate musculoskeletal loadings in the lower limbs [47]. The model was established using musculoskeletal simulation software AnyBody (version 3.0, AnyBody Technology, Aalborg, Denmark) by modifying its existing three-dimensional gait model. Seventy muscles of the lower extremities, 35 muscles on each leg, were simulated to estimate the muscle forces. Eight muscle groups—soleus, gastrocnemius, gluteus maximus, vastii, rectus femoris, hamstring, posterior gluteus minimus/medius, and anterior gluteus minimus/medius—were compared for walking trials with or without stilts.

Results from the simulations revealed that the time-history patterns of the muscle forces for stilts walking were similar to those for no stilts. The use of stilts was found to induce force redistribution among the muscles. The forces in five out of eight muscle groups increased, whereas those in three muscle groups were decreased due to the stilts use. Injuries to the quadriceps group, which consists of the vastii and rectus femoris, are among the most common muscle injuries in sports [63] and also in some occupational activities. The loading levels increased substantially in the rectus femoris (180%) and slightly in vastus muscles (6%) due to the stilts use. The magnitudes of the muscle forces during stilts walking were lower than those that cause injuries [63]. However, the use of stilts may still have some effects on the muscle loading in the hamstring. The most common hamstring injuries occur after indirect trauma from excessive stretching or forceful contraction. Stilts may potentially cause an increase in loadings in five of eight major muscle groups in the lower extremities. The major increase was the force in the rectus femoris muscle, which was 1.8 times more than that of normal walking. The increase in muscle loadings during stilts walking may likely speed the muscle fatigue of the workers and increase the balance demand.

#### *25.2.2.2 Lower Extremity Postures and Ranges of Motion*

During normal walking, the hip flexes and extends, and the maximum extension is reached during the middle of the gait cycle with heel contact at 0% and the subsequent heel contact at 100%. The knee flexes in the loading response and the early part of the stance phase and fully extends before the initial foot contact of the next cycle. In comparison with normal gait, the hip and knee motion profiles are similar in curve shapes in general except for the smaller ranges of motion associated with gait on stilts (Figure 25.16). The ankle joint motions are quite different when walking on stilts. In particular, the angular changes of ankles during the weight acceptance and swing phases are largely reduced, with no plantar flexion observed (Figure 25.16). The ankles are dorsiflexed throughout the entire gait cycle to keep the feet on stilts and sustain the weight of stilts.

The major impact of stilts on joint motions is the restrictions of lower extremity joint movements during walking. Ankle, knee, and hip motions are restricted because of the 7.3 kg of extra weight of the stilts attached to the feet and the confinement of the straps around the feet and knees. The most restricted joint was the ankles, with 56% and 61% decrease in range of motion for 24 and 40 in. stilts, respectively. The reduction is 29%–41% for knees and 20%–30% for hips [45]. As the height of the stilts increases, the joint flexibility decreases, suggesting more muscle activities are needed to maintain

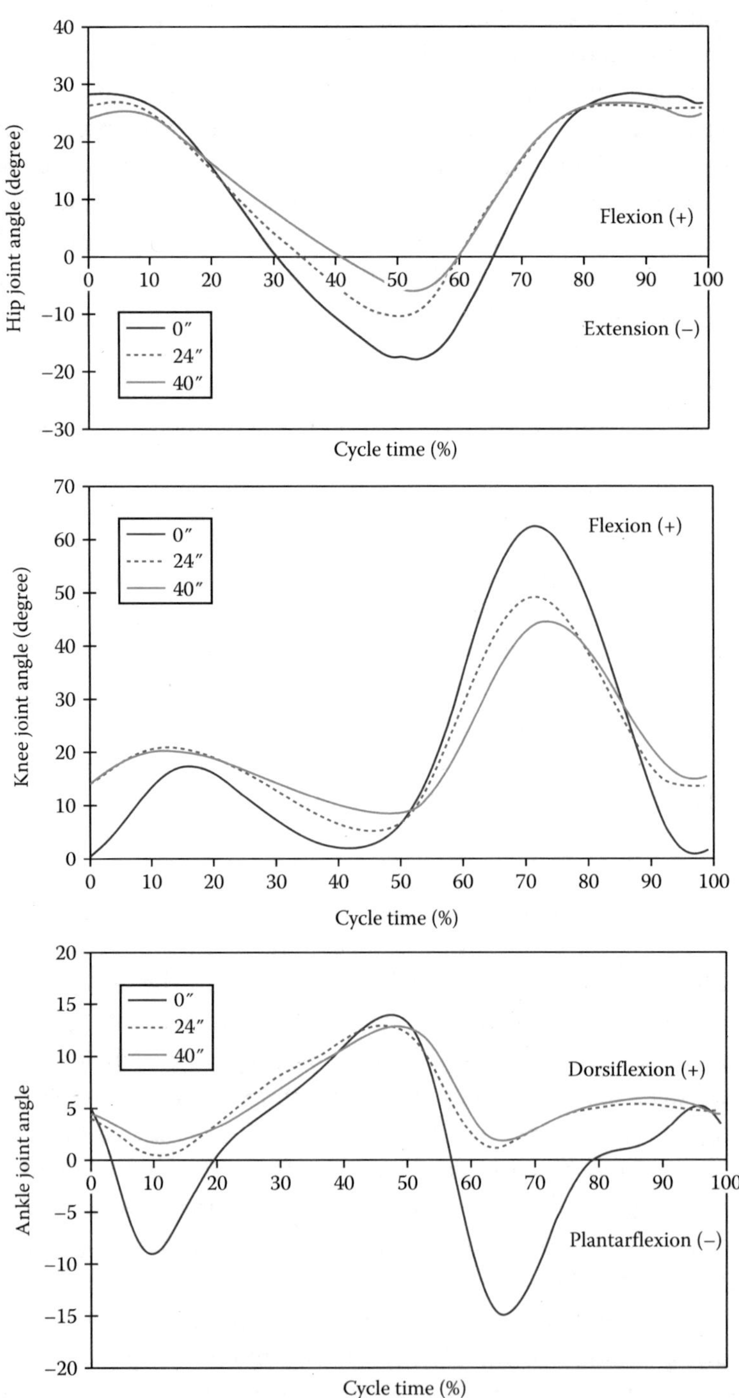


FIGURE 25.16 Hip, knee, and ankle joint angles for three stilt conditions averaged over all subjects. Joint angles were normalized to the gait cycle with the first foot contact at 0% and the second heel contact of the same foot at 100% of gait cycle.

stability on stilts and suggesting that this activity is more likely to result in accumulated muscle fatigue in the lower extremities.

### 25.2.3 Fall Injury Potential Associated with Stilts

When walking on stilts, construction workers are typically elevated 0.6–1.0 m above the ground. Due to the increased height, the postural stability of workers is likely to be influenced by visual or psychological perturbations at the construction work sites. A survey of carpenters and drywall installers indicated that workers perceived the greatest fall potential to be associated with the use of stilts, when compared to other common devices, such as scaffolds and ladders [64].

#### *25.2.3.1 Gait Characteristics*

In a study of recreational wood stilts, stilt walking was found to be generally faster than normal gait due to an increased stride length, in spite of a decrease in cadence [65]. Such recreational stilts weigh approximately 2 kg each. In contrast, the construction stilts are much heavier, with a weight of 3.64 kg each. Experienced workers were able to adapt to the added weight on their feet; however, they walked significantly slower on stilts [45]. Table 25.2 shows general gait parameters for different stilt heights in a gait study of construction workers. Gait on construction stilts is characterized by a decrease in speed and increases in stride length, step width, and double stance period (i.e., the percentage of the time when both feet are in contact with the floor during a gait cycle). Workers walk significantly slower on stilts, with the mean stride period increased from 1.24 s of no stilts to 1.59 and 1.69 s for 0.6 m (24 in.) and 1.02 m (40 in.) stilts, respectively. The mean step width for gait on stilts can be significantly increased to as much as 25 cm on stilts compared to 15 cm of normal gait. The stride length is significantly increased to 1.75 m on stilts compared to 1.55 m for normal gait; however, it is still within the normal gait limits (1.25–1.85 m) [66]. When negotiating a curved path, the walking speed on stilts is slower with the mean speed of 1.03 m/s compared to a straight path (1.25 m/s).

Figure 25.17 illustrates the position of the right heel marker in the mediolateral direction as a function of gait cycle. The position of heel for no-stilts conditions remained essentially the same mediolateral position—around 0.23 m. For gait on stilts, the stilt was at 0.23 m in the beginning of the gait cycle but swung 8 cm outward during the swing phase. The functional gait changes reflect the adaptations adopted by workers to maintain

TABLE 25.2 Means and Standard Errors for Gait Parameters

| Gait Cycle Parameters | Safety Shoes | Stilt Height 0.6 m | Stilt Height 1.02 m |
|---|---|---|---|
| Average speed (m/s) | 1.25 (0.16) | 1.08 (0.15) | 1.03 (0.16) |
| Stride period (s) | 1.24 (0.08) | 1.59 (0.15) | 1.69 (0.17) |
| Double stance period (%) | 20.0 (2.1) | 21.6 (3.5) | 22.6 (3.3) |
| Step width (m) | 0.15 (0.05) | 0.24 (0.08) | 0.25 (0.10) |

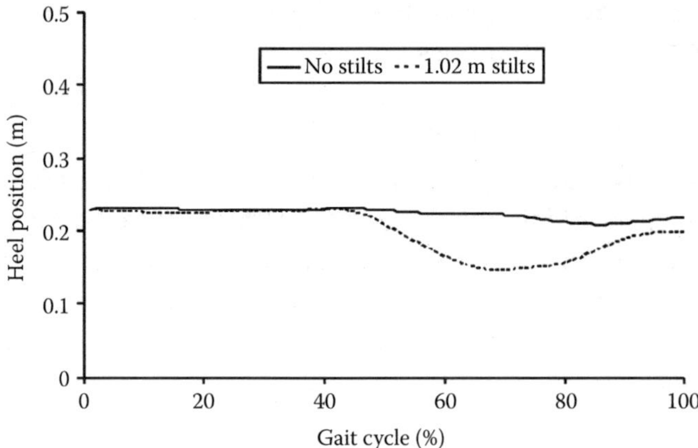


FIGURE 25.17 Mediolateral position of the left heel as a function of the gait cycle.

dynamic balance. The increase in balance demand is evident in the increases in step width and double stance period. The 67% increase in step width suggests a wider base of support was needed to improve stability on stilts. The lengthened double stance period indicates a longer period of time was required for the body to reestablish stability from one step to another. The trajectory of heel markers for gait on stilts suggests a "lurching style" walking pattern with the foot swung out laterally, providing a wider base of support.

#### *25.2.3.2 Postural Stability for Various Foot Placements on Stilts*

Walking on stilts may result in accumulated muscle fatigue in the lower extremities [47] and eventually produce postural instability while holding a prolonged standing position [58,64]. Other research findings have determined that postural stability and motor control mechanism were highly associated with various foot placements [67–69]. Independent research has attempted to identify and quantify issues related to standing and balance performance while conducting manual material handling, and various additional research efforts have attempted to determine and evaluate fall hazards associated with foot placement [67,70,71]. Workers' foot placements are known to influence whole-body postural stability and lower-extremity joint force [72–75]. As a result, alternative arrangements of foot placement are suggested as one of the main biomechanical implications from these studies, and the alternative foot placements are believed to increase postural stability and to reduce the incidence of fall-related injuries [67,74,76].

In a study of postural stability associated with stilts, a set of two by three stance conditions were investigated to determine the optimal foot placement for postural stability in a stationary standing position [43]. The six conditions include two levels of asymmetric foot placement direction (parallel versus forward) and three levels of foot placement width as scaled against shoulder width (SW) (0.5, 1, 1.5 SW) as shown in Figure 25.18. Findings from this study suggested stilt height, foot placement width, and foot placement direction all had significant effects on postural stability. Postural instabilities were increased with increasing stilt height and forward foot placement direction. Also, half-shoulder-width foot placement was the most hazardous condition in terms of fall-risk factors. A parallel foot

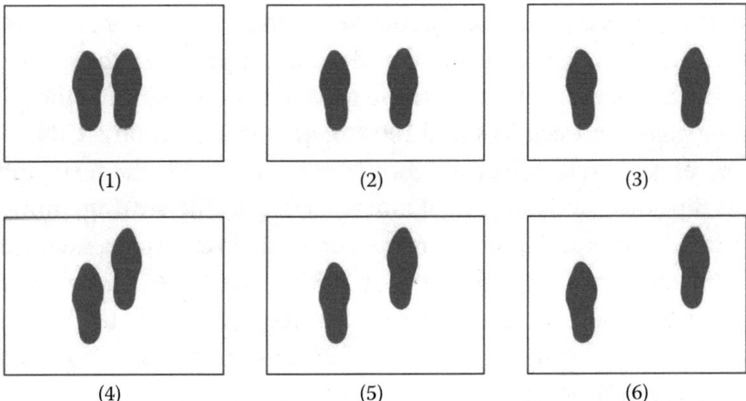


FIGURE 25.18 Six foot placements: (1), (2), and (3) are 2 ft parallel direction; (4), (5), and (6) are 1 ft forward direction. (1) Standing, with feet at half of the individual's shoulder width and directly beneath the body; (2) standing, with feet placed at participants' shoulder width and directly beneath the body; (3) standing, with feet placed at 1½ times shoulder width and directly beneath the body; (4) standing, with feet at half of the individual's shoulder width and with the left foot beneath the body and the right foot placed forward a distance of half the individual's foot length; (5) standing, with feet placed at participants' shoulder width and with the left foot beneath the body and the right foot placed forward a distance of half the individual's foot length; and (6) standing, with feet placed at 1½ times the participants' shoulder width and with the left foot beneath the body and the right foot placed forward a distance of half the individual's foot length.

placement directly beneath the body reduced postural instability more than a placement with the right foot placed forward to the left foot. These findings are especially important for those construction tasks involving confined work spaces, within which drywall carpenters or painters frequently perform tasks within shoulder-width range (e.g., in closets and bathrooms).

Construction workers wearing stilts had significantly higher postural sway and instability than those with no stilts. This instability increased substantially with increasing stilt height regardless of the foot placement (i.e., width and direction). This finding supports the views of the painter's union [43] that 24 in.—the lowest setting—is a safe height setting for occupational stilt use. Thus, it is suggested that users keep stilts at the lower height settings whenever possible and adopt proper standing posture with stilts, that is, keep feet parallel and directly beneath the body and place them one shoulder-width apart.

#### *25.2.3.3 Dynamic Stability and Fall Hazards Associated with Gait on Stilts*

Human gait requires an integration of a complex neuromuscular-skeletal system and the coordination of many muscles and joints. In most gait studies, joint moments are quantified to better understand how the work is distributed among joints; to determine the net muscular, ligament, and friction forces acting on the joints; and to obtain a clear picture of the balance mechanism during locomotion [77–79]. For gait on stilts, the general patterns of lower extremity joint moment are similar to those of normal gait with safety shoes; however, the peak joint moments are significantly greater on stilts. Findings from published and definitive studies [45,47] indicate that the peak joint moments for stilt walking are significantly greater in the weight acceptance and preswing phases of the gait cycle. There is no significant dependency on stilt condition in general when comparing the joint moments during the swing phase between 60% and 100% of gait cycle. In contrast, during the stance phase (0%–60% of gait cycle) when the foot is in contact with the floor, differences are observed in the hip, knee, and ankle joint moments due to stilt wearing. Immediately after initial foot contact, sudden and quick increases occur in hip extension, knee extension, and ankle dorsiflexion moments. Approximately 19%, 30%, and 110% increases are observed in peak joint moments at the hips, knees, and ankles, respectively. As the foot progressed to 50% of gait cycle in the preswing phase, 56% and 28% increases occur in peak hip and knee flexion moments, respectively [45,47].

The most pronounced differences in joint moments observed during stilt walking are the high magnitudes of ankle dorsiflexion moments and hip flexion moments. Ankle and hip muscles appear to play dominant roles in modulating the joint loadings to compensate for the weight of stilts and the increase in demand for balance. The high ankle dorsiflexion moments in the early gait cycle are needed to ensure a successful landing, while the high hip flexion moments in late stance are needed for propulsion of the body forward and upward. The increases in lower extremity joint moments could be interpreted as signs that gait on stilts is a particularly challenging task for the neuromuscular system [77]. Moreover, if any additional unexpected perturbations arise from environmental changes or job task demand—for example, slippery surfaces, excessive arm reach, and negotiating a sharp turn—the dynamic balance of the workers could be further challenged.

Walking on stilts, like normal gait, requires sufficient foot clearance of the swing limb as well as the stability of the body supported primarily by the stance limb. Previous gait studies have documented the minimum foot clearance of the midswing phase [80–83] as an indicator for examining the potential of tripping. When walking on stilts, workers' minimum foot clearances during midswing phase were found to be distinctly related to the stilt condition [45]. In general, the foot clearances for stilt conditions were consistently lower than those of no stilts during the entire gait cycle. An approximate 1-cm reduction was observed during midswing when the stilt swings directly beneath the body. The small foot clearance on stilts could come from insufficient knee flexion during swing phase, excessive knee flexion during stance phase, or the mechanism of the stilt floor plate. Due to the restricted joint motions and the weight of the stilts, workers may not pick up their feet as high as they would for a normal gait. Thus, workers are more likely to trip on any objects that might be on the ground at the job site when wearing stilts. The risks of tripping may be even greater since workers on 40-in. stilts may not visibly detect obstacles on the floor. Workers should inspect the work environment before putting on stilts, to ensure the floors are free of obstacles.

### 25.2.4 Future Direction for Stilts

Stilts appear to place greater demands on lower extremities to compensate for the limited joint motions caused from restriction by the binding straps as well as by the weight of the stilts abnormally loading the limbs. The computer simulations presented in this chapter suggested that the use of stilts may potentially cause increased musculoskeletal loadings in five of eight major muscle groups in the lower extremities, with the greatest increase suggested for the rectus femoris. Workers may need to limit the prolonged use of stilts, especially at high elevation, to alleviate the burdens on the joints and reduce muscle fatigue arising from increased muscle activities required to maintain balance. However, the theoretical analysis of muscle fatigue as well as energy expenditure is yet to be investigated by further clinical and experimental studies.

## 25.3 SCAFFOLDS

Scaffolds are defined by OSHA as any temporary platform, either suspended or supported, and its supporting structure, including its anchorage system, used for supporting materials or employees or both [84]. The definition is broad enough to include different types of scaffolding used in various industries; while it is common to think of scaffolds as construction-related devices, they are found in numerous industry groups and industrial applications. Across all industrial sectors, surveillance findings indicate that falls from scaffolds represent a serious occupational hazard; falls from scaffolds are one of the leading causes of work-related fall fatalities and injuries [85]. In 2006, approximately 14.2% (809 of 5703) of fatal occupational injuries were due to falls; of those, 10.9% (88) involved scaffolds or staging [86].

Three out of every four fatal and nonfatal injuries associated with scaffold use [87] occur in just one of the three major categories of scaffolds: supported scaffolds. Supported scaffolds are defined as platforms supported by outrigger beams, poles, frames, or similar rigid support systems [88]. Since the magnitude of injury events is much smaller within the other two categories of scaffolds—suspended and other categories—NIOSH summary research has focused on supported scaffolds, and the incidence of injury in supported scaffolds is what is reported herein. Emerging technology represents a significant component of the "other" category of scaffolds, and this emerging technology is chiefly represented in one category of scaffold, the aerial lift; research in aerial lifts chiefly represents original research and is reported separately in this document.

Given the high magnitude, incidence, and personal consequences of supported-scaffold, fall-related injuries, NIOSH has directed significant resources to determining the cause and characteristics of falls related to the use of scaffolds and to identifying risk factors and common scenarios related to the deployment and use of scaffolds at work sites. This research has encompassed numerous avenues to determine the state of the art: consultation with experts, extensive review of the state of knowledge published in peer-reviewed literature and user-oriented publications focused on industrial applications, as well as original NIOSH-directed research studies addressing under-researched areas directly related to scaffold safety. Areas in which original research has been conducted at NIOSH addressed issues related to the lack of published biomechanical results and findings related to common postures in scaffold disassembly, handling methods, body postures, and optimal hand placement. Basic research on the use of virtual reality techniques in simulating fall-hazard exposure for scaffolds also represents NIOSH contribution to the body of knowledge in this area.

NIOSH has reviewed various summary research findings and presented them before the ergonomics society, a leading research institute addressing the ergonomic component of occupational safety. The extensive findings constitute an overview of the current (2008) state of knowledge of scaffold safety; these findings have been published as a succinct review of the knowledge gaps, opportunities for directing research, and current knowledge of safety related to scaffolds. This summary, published as a chapter in *Contemporary Ergonomics 2008: Proceedings of the International Conference on Contemporary Ergonomics*, remains a valuable and still current summary of the research direction for much ongoing scaffold-safety research [89].

The strategies and recommendations from this research summary provides valuable information on (1) key fall-injury risk factors related to supported scaffold use, including OSHA's performance-based criteria to protect employees from scaffold-related hazards; (2) technologies for supported scaffold safety; (3) the state of knowledge about fall safety; and (4) remaining unanswered questions pertaining to supported scaffold safety and existing knowledge gaps. In addition to OSHA criteria, which can be accessed separately, two main summaries were presented and are reported here: current measures and technologies to control falls from scaffolds and knowledge gaps remaining to be answered.

The NIOSH review on the leading risk factors associated with falls from scaffolds found that the most common causes of falls could be organized into five major categories: (1) scaffold tipping or structure failure [90,91]; (2) planks breaking, slipping, and gapping [87,92]; (3) unguarded scaffolds [92,93]; (4) difficult access or transition onto or off a scaffold [93]; and (5) problems with erection and dismantling of scaffolds [90,93].

Scaffold tipping or structure failure was attributable to various factors, such as inadequate anchoring into the walls, improper scaffold assembly, improperly secured bracing, loading beyond designed capacity, and failure of scaffold components under stress [86,90,94]. Plank breaking, slipping, and gapping may result from a heavy load, physical damage to a plank, misinformation about the type of plank or its rating, inadequate overhang over the supporting bearers, unsecured planks (no cleat), sideways movement of planks, and missing planks [88,95,96]. Conditions of unguarded scaffolds include missing guardrail and inadequate cross bracing [92,93]. A variety of factors may influence safe access to the working level of the scaffold: the width of a run of integral ladders in scaffold end frames, the distance between two runs, and the difficulty transiting onto or off a scaffold platform to integral ladders due to the required plank overhang [92]. Problems with erection and dismantling of scaffold may result from environmental conditions, the weight of scaffold units, and the availability of handholds [95]. OSHA regulations addressing scaffold use were reviewed in this document; further information in this area is available from review of the original document or, because of the evolving nature of knowledge related to this risk exposure, preferably by reference to the current OSHA website (http://www.osha.gov/SLTC/scaffolding/index.html).

The current measures and control technologies were presented as follows. OSHA requires an anchorage to control scaffold tipping or failure when the scaffold reaches four times its width at the base [96]. Halperin and McCann (2004) [97] reported that properly erected scaffolds were correlated with supervision by a competent person.

Modern modular scaffold systems, which are simpler to erect, were reported to be safer because they reduced the possibilities for errors in construction [98]. The use of metal catwalks or platforms is also a solution to the problem. Plank locks are available in the market to prevent scaffold planks of solid sawn lumber from slipping [99]. Fabricated steel planks in 15- and 23-cm widths are also commercially available, thereby making it possible to arrange the planking to reflect the constraints of work situations involving irregular surface contact. Because of this fact, the maximum room between planks and the uprights of the scaffold frame can be lessened to below 7.5 cm, instead of the allowed 24 cm, to provide better slippage control [100].

Essentially all scaffold manufacturers currently provide guardrail components that can be used to ensure compliance with current OSHA requirements; some frame scaffold systems use a modular guardrail and midrail system that does not rely on cross bracing to provide fall protection. As to the control of difficult access or transition onto or off a scaffold, engineered scaffold planks that hook onto the horizontal scaffold member and do not extend beyond the end frame (as compared to the required 15-cm overhang to the supporting member for sawn lumbers) provide increased ease of access to the working surface from a ladder built into the end frame. Finally, lightweight scaffold components are available to ease the problem of falls during scaffold erection and dismantling [95]. Fabricated scaffold planks are also available that are lighter than wood planks and thus decrease the risk of imbalance-related falls during the erection and dismantling phase.

A current study based on the findings of OSHA fatality investigations may help improve our understanding of the extent to which scaffolds tip when they are not anchored or are improperly anchored. Industry literature generally suggests that fabricated planking or decking appears to offer certain advantages over sawn lumber planks; it may be useful to conduct a series of experiments to validate these putative safety differences and the life-cycle costs. In addition, if one considers exposure to scaffold falls as a function of time, then the initial access phase during which the worker accesses or steps onto the scaffold may be the most dangerous part of scaffold work. The challenges in negotiating overhanging planks or guardrail components bear serious consideration. Evaluations were also needed on whether certain scaffolding designs are safer to erect and safer to use, given current advances in scaffold technology and work practices. Moreover, what opportunities are there for scaffold erectors to use PFAS? How can technology be used to reduce fall exposures for this group? The development of a comprehensive set of "best scaffold safety practices" is suggested. Finally, the most recent detailed BLS survey on scaffold use and injuries dates to 1978 [86] and is now over three decades old; this survey may not be fully representative of the conditions under which workers perform tasks on scaffolding today. At the date of this publication an enhanced national survey on scaffold use and injuries remains long overdue.

In addition to the summary measures presented in the preceding study, various original analytic studies were conducted by NIOSH to address under-researched areas related to supported scaffold safety. NIOSH sponsored original research studies on pressing safety issues related to scaffolds: (1) biomechanical evaluation of scaffolding tasks, (2) a comparison of different postures for scaffold end frame disassembly, (3) optimal hand locations for safe scaffold end frame erection and dismantling, and (4) carrying strategies for scaffolds. In addition, various NIOSH research activities were undertaken to determine the value of virtual reality applications in relation to scaffold safety. This constituted a novel and emerging research area, with potential application to all areas of safety under elevated conditions.

A primary research component of NIOSH activity related to scaffold safety was that of a biomechanical assessment of erection and dismantling tasks [101]. This comprised a field study to identify tasks and activities that specifically increased the risk of overexertion injury associated with the erection and dismantling of frame scaffolds and to determine strategies that would prevent or reduce the worker's risk of injury. Twelve construction sites were visited involving a total of 29 workers. The investigation identified activities that increase the risk of overexertion injuries during task performance: lifting scaffold end frames, carrying end frames, handling scaffold planks, removing cross braces, and removing guardrails. This chapter focuses on end frame handling problems. Although the techniques used to handle end frames varied among the construction sites and subjects, six lifting and five carrying strategies were commonly used. Computer simulations of these work techniques show that considerable biomechanical stress occurs to most of the workers at their shoulders, elbows, and hips. To reduce overexertion injuries during erection and dismantling of frame scaffolds, design of an assistive device to lift scaffold end frames and modifications to the end frame fixtures are suggested. Future research areas for the prevention of injury during scaffolding work are also proposed. The relevance to industry is also discussed: The construction industry is characterized by high frequencies of musculoskeletal injuries. This chapter analyzes scaffolding, a common activity in construction; evaluates stresses associated with different methods used; and develops recommendations on changes in scaffold design, training, and accessory equipment necessary to reduce accidents and injuries.

A second study was conducted related to scaffold end frame erection and disassembly while adopting postural variations [102]. This study attempted to identify the most favorable scaffold end frame disassembly techniques and evaluate the associated slip potential by measuring whole-body isometric strength capability (Figure 25.19) and required coefficient of friction (RCOF) to reduce the incidence of injury. Forty-six male construction workers were used to study seven typical postures associated with scaffold end frame

FIGURE 25.19 Isometric strength-testing apparatus.

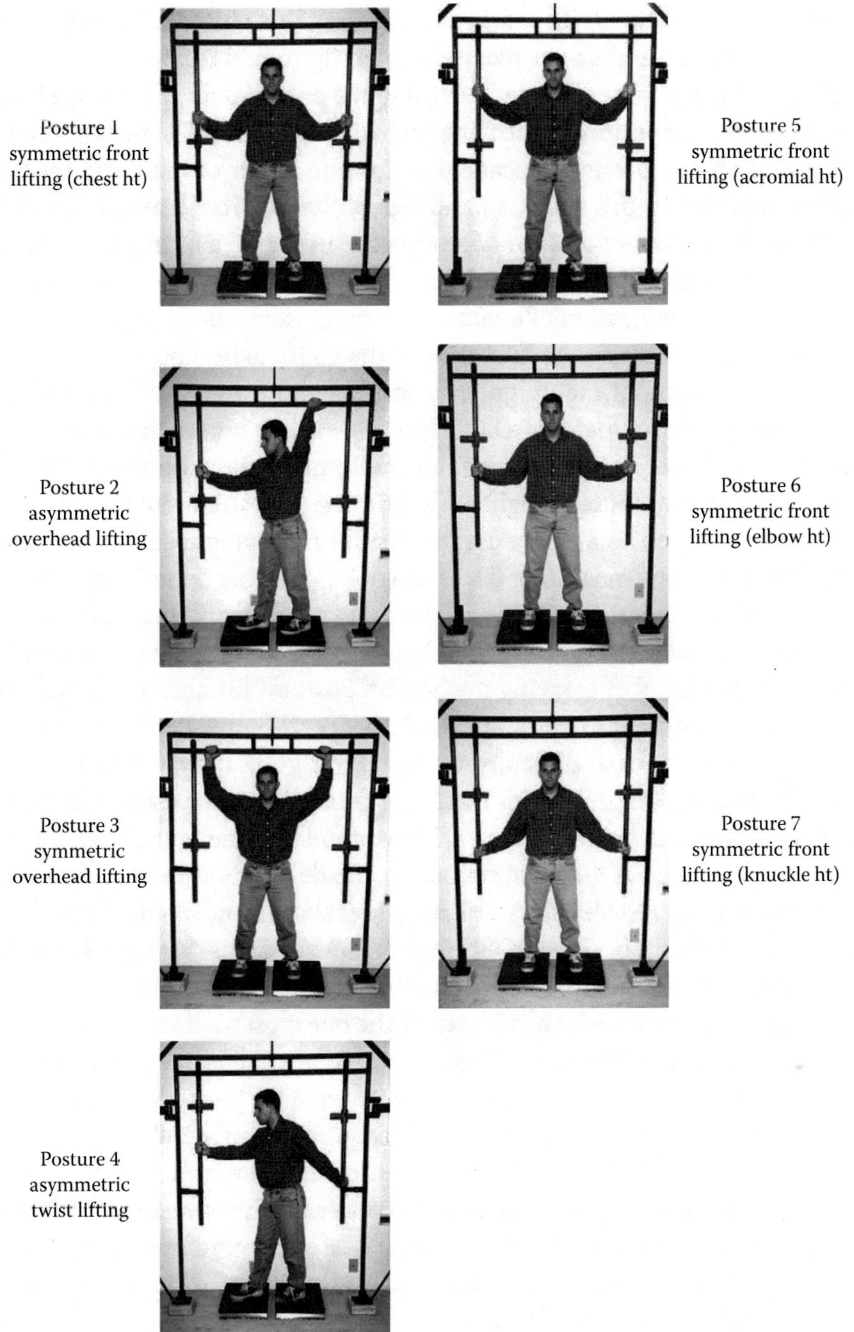


FIGURE 25.20 Scaffold disassembly postures.

disassembly (Figure 25.20). An analysis of variance showed that the isometric forces resulting from the seven postures were significantly different. Three of the disassembly postures resulted in considerable biomechanical stress to workers. The symmetric front-lift method with hand locations at knuckle height would be the most favorable posture; at least 93% of the male construction worker population could handle the end frame with minimum overexertion risk. The static RCOF value resulting from this posture during the disassembly phase was less than 0.2; thus, the likelihood of a slip should be low.

A third NIOSH study was conducted related to the potential for musculoskeletal injury from scaffold carrying, specifically focusing on hand location and isometric force generation [103]. Based on the location of scaffold end frame center of mass and the isometric strength data collected in this study, a hand separation of 46 cm between the elbow and chest heights would be an optimal hand location for an assistive lifting device for scaffold disassembly. This 46-cm separation would mitigate the likelihood of postural imbalance while allowing for the generation of a mean maximum isometric force of about twice the weight of a scaffold end frame. At least 95% of the construction population would have isometric forces in excess of the weight of a scaffold end frame. An alternative method without an assistive device would be a hand location slightly higher than the elbow height with a hand separation of 116.8 cm. This is a compromised situation that yields 2.4 times isometric strength of the scaffold weight with little risk of postural imbalance.

A final study examined commonly used methods of carrying scaffold end frames [95]. This study determined the most favorable strategy for carrying scaffold end frames while minimizing the risk of injuries from being struck by an object, falling, and overexertion. An experimentally validated method of isolating the effect of scaffolding-carrying methods established that the effects of carrying method on postural instability and task difficulty were significant for handling a commonly used 22-kg end frame. Response time, postural instability, and perceived task difficulty rating were significantly reduced when a 9-kg end frame was used as compared with the 22-kg frame. The conclusion drawn from this research was that the symmetric side-carrying method was the best option for handling 22-kg scaffold end frames. A 9-kg end frame (e.g., made of reinforced lightweight materials) had the potential to reduce injury risk among scaffold handlers during their scaffold erection and dismantling jobs. The NIOSH recommendation was that scaffold erectors may want to adopt the symmetric side-carrying method as the primary technique for handling the 22-kg scaffold end frame, which is currently the one most used in the industry.

NIOSH additionally performed significant research in establishing basic understanding of the validity and application of virtual reality techniques in modeling and simulating fall-safety exposures from scaffolding. NIOSH acquired a surround-screen virtual reality (SSVR) system in 1996 (Figure 25.21) to examine exposure conditions that could not be studied using human subjects, given safety concerns. This system employs projected computer-generated imagery to develop a visual sense of immersion on the part of the subject; participants wear goggles that integrate visual cues to give the impression that they are immersed within a simulated environment. Various published and unpublished methodological and applied research studies examined critical issues with the application of this simulation technology to elevated-use tasks involving scaffolds. A pioneering study by Hsiao et al. [104] investigated the effect of adding real planks, in virtual scaffolding models of elevation, on human performance in an SSVR system. Twenty-four construction workers and 24 inexperienced participants (controls) performed walking tasks on real and virtual planks at three virtual heights (0, 6, and 12 m) and two scaffolding platform width conditions (30 and 60 cm). Gait patterns, walking instability measurements, and cardiovascular

FIGURE 25.21 Simulated scene within SSVR.

reactivity were assessed. The results showed differences in human responses to real versus virtual planks in walking patterns, instability scores, and heart rate interbeat intervals; it appeared that adding real planks in the SSVR virtual scaffolding model enhanced the quality of SSVR as a human–environment interface research tool. In addition, significant differences were observed in performance between construction workers and the control group. The inexperienced participants were more unstable as compared to construction workers. Both groups increased their stride length with repetitions of the task, indicating a possibly confidence- or habit-related learning effect. The practical implications of this study are in the adoption of augmented virtual models of elevated construction environments for injury prevention research and the development of a program for balance-control training to reduce the risk of falls at elevation before workers enter a construction job.

A further study by Simeonov et al. [105] compared human perceptions of height, danger, and anxiety, as well as skin conductance and heart rate responses and postural instability effects, in real and virtual height environments. The 24 participants performed "lean-over-the-railing" and standing tasks on real and comparable virtual balconies, using the SSVR system. The results indicate that the virtual display of elevation provided realistic perceptual experience and induced some physiological responses and postural instability effects comparable to those found in a real environment. It appears that a simulation of elevated work environment in a SSVR system, although with reduced visual fidelity, is a valid tool for safety research. Potential applications of this study include the design of virtual environments for safe evaluation of human performance at elevation, identification of risk factors leading to fall incidents, and assessment of new fall prevention strategies.

This area of research remains ongoing within NIOSH, and future studies will assess exposure conditions under a variety of elevated conditions and for a variety of tasks within construction and other industrial conditions.

In summary, NIOSH research on supported scaffolds has consisted of original studies to address under-researched areas related to exposures from scaffold-related tasks, original research on the application of virtual reality to scaffold-related safety, as well as summary research on the most current and well-researched areas related to scaffold safety. NIOSH summary reviews on scaffolding safety, published in 2008, still constitute a valuable resource for determining the state of knowledge and the research gaps obtained from safety studies related to scaffolding. NIOSH research is part of an ongoing effort to reduce injury exposures from construction and other industrial tasks using this common elevating device.

## 25.4 MAST CLIMBING WORK PLATFORM

A NIOSH study initiated in April 2010 focuses on mast climbing work platforms (MCWP), an emerging form of scaffolding increasingly used throughout the construction industry. Originating in Europe in the 1970s, MCWPs have become more popular in North America in recent years. For this system, a platform is elevated and lowered along a mast by the use of an elevating unit. While the early units were powered by electric motors, gasoline engines have become standard for MCWPs used in North America. The mast can be used in a freestanding condition for heights up to 100 ft or in an anchored condition at greater heights. MCWPs have many advantages over other forms of scaffolding. One major advantage is its ability to elevate workers to working heights up to 1000 ft when anchored to a building. Also, the platforms can travel at speeds up to 40 ft/min, providing great advantage for the movement of materials. The platforms are able to support large loads (25,000 lb) and provide adequate work area. For wide facades, multiple platforms can be connected by bridges. This allows workers free access to the entire span of a wall, which reduces setup time. However, as MCWPs have increased in popularity, nonfatal and fatal incidents involving MCWPs have occurred.

In 2008, Ayub, Susi, and O'Shea individually presented data on these fatal and nonfatal injury incidents to the advisory committee on construction safety and health and identified the following potential hazards [106–108].

- One area of concern is the erection/dismantling of the mast. During the erection of the mast, sections can be added by the use of a crane located on the work platform or lifted in place by workers. Alternatively, several sections of mast can be assembled on the ground and positioned by a crane. If the worker is not properly restrained, manual placement of the sections of mast could result in a worker falling to the platform or to the ground. Also, the platform could rise above the highest section and fall to the ground. The top mast section is equipped with a limit switch and/or physical stop, but it is the only protection for the fully erected mast. A removable stop could be designed to prevent the platform from climbing over the mast during erection. Another potential hazard is when the mast is unsupported after the last anchor has been removed. Even when the unsupported mast height is less than the freestanding height, tipover could occur from overloading the platform with removed mast sections.
- Due to the low velocities of a MCWP, stability is mainly a static issue. However, studying the effects of wind and platform loading on stability would be of importance. For example, tipover could occur when a fully loaded platform, while being used in high winds, is subjected to a force at the guardrail. This could be especially dangerous

when dealing with an unsupported or freestanding mast. Dynamic analysis performed with software such as ADAMS could lead to a better understanding of MCWP stability. Overhead protection and winter weather enclosures need to be addressed by manufacturers during use and especially during erection and disassembly.

- The anchoring of the mast to the building is also a process that should be evaluated. Due to various wall materials, the proper fastener may vary from one job to the next. Also, some wall materials may not be suitable for anchoring the mast. Although the manufacturer may specify the fastener to be used for a certain condition, workers may substitute other fasteners. This could become more complicated when using MCWPs supplied by rental companies. Tie-in to the building structure should be reviewed and approved by the engineer of record for the building.
- In general, lifting of the platform is accomplished by two types of systems. Rack and pinion systems perform climbing/descending by a motor-driven pinion traveling along a rack that is connected to the mast. For this system, the rack and gears should be inspected for corrosion and wear. Also, the use of duplicate drives could provide safety from failure of a single mechanical component. Some North American manufactures use hydraulic cylinders to climb/descend. Fraco's system consists of a hydraulic cylinder that actuates a hook (Fraco is the largest mast scaffold manufacturer and NIOSH's research partner for this proposed study). The hook climbs a tower rung that is fixed to the mast. The elevating unit is also equipped with a safety hook that prevents the platform from lowering in case of main hook failure.
- In order to move the worker closer to the building, extension outriggers can be used. A few potential hazards could occur from their use. First, the worker is exposed to an unprotected edge of the platform when extending the outriggers and installing the planks. The use of a safety harness could protect the worker while installing the planks. Also, if the planks used on the outriggers are not fastened properly, the dynamic motion created by the workers could cause the planks to move off the outriggers. Lastly, the planks must be able to support the applied loads. Rather than using wood planks from the jobsite, engineered planks with load ratings could be used. These planks could also have an antislip surface for wet conditions.
- Some systems allow multiple platforms to operate on the same mast. Work can then be performed simultaneously at different levels of the building façade. This configuration could create several safety issues. Tools, materials, and debris could fall from the upper platform to the lower platform. Also, the platforms could collide during raising/lowering if proper safety controls are not implemented.

In conclusion, MCWPs are an efficient means of elevating workers and materials to extreme heights. In order to prevent incidents, regulation specific to MCWPs must be developed and implemented. Proper training is essential in providing workers with the knowledge to safely use MCWPs. Lastly, structural analysis of various components and dynamic simulations of different operating conditions could lead to safer design (Figure 25.22).

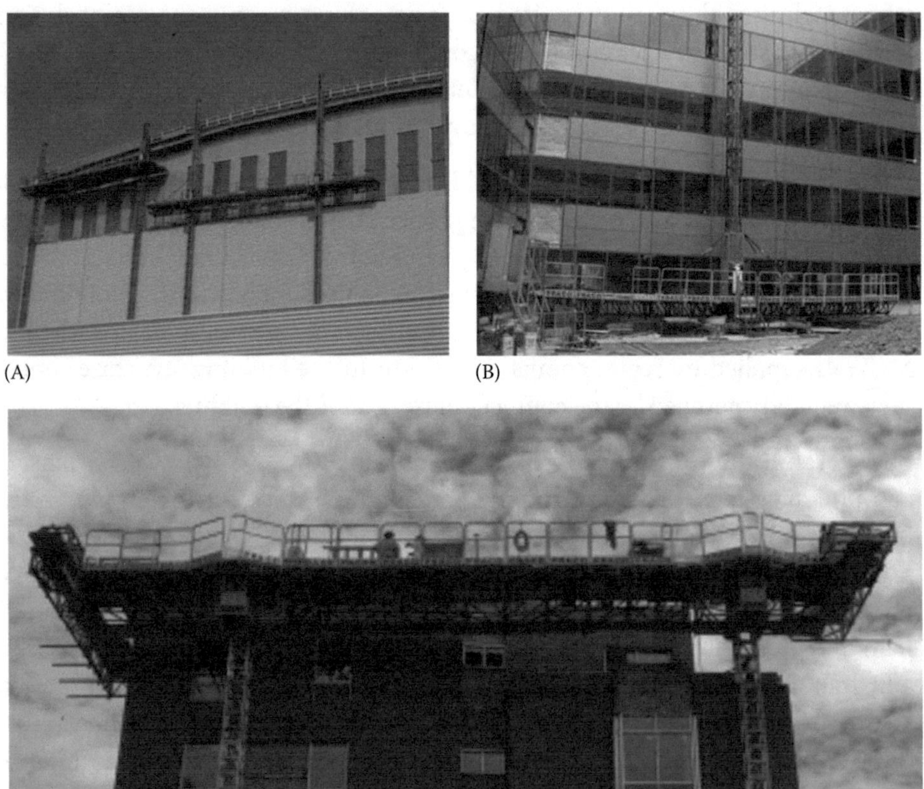

(A) (B)

(C)

(D)

(E)

FIGURE 25.22 Types of MCWPs (Reprinted with permission of Fraco Products Ltd., St-Mathias-sur-Richelieu, Quebec, Canada.) (A) and (B) are Fraco Model ACT8; (C–E) are Fraco Model FRSM-20 K.

## 25.5 FUTURE DIRECTIONS/RECOMMENDATIONS IN ELEVATED DEVICE RESEARCH

NIOSH is committed to supporting continuing research into elevating device safety, in a way that will support the need for advanced research into the most salient, anticipated conditions. While the continued use of frame scaffolding will surely continue for the foreseeable future, it is anticipated that increasing the use of mechanized scaffolds will transform the range of tasks associated with construction activities, as well as tasks in various industries requiring elevating equipment. The reasons for this are varied, but the combination of increased productivity factors, accessibility to mechanized scaffolds through rental agencies, increased familiarity with operational demands and requirements for skilled laborers at work sites, as well as other factors, supports a clear trend for the use of mechanized scaffolding, such as scissor lifts and mast scaffolds, in construction and related industries.

While NIOSH will continue to support pertinent research for frame scaffolding, an increasing awareness has emerged for issues related to mechanized scaffolding. Issues related to frame scaffolding include assembly, carrying, ingress and egress, disassembly, and task completion at heights. Certain issues remain under-researched for frame scaffolding, and NIOSH remains aware of the continued need for further research in these areas, but research into issues related to the emerging technology represented by mechanized scaffolding will call for increasing focus and attention.

The mechanized systems used to elevate workers in construction and other industries are more complex, subject to a wider range of mechanical issues and problems, and in general, involve more multifactorial exposures than frame scaffolds. This can be for multiple reasons: Mechanized elevating equipment can routinely reach greater heights than frame scaffolding and is subject to propulsion-related failures, hydraulic-system failures, anchorage issues, ingress–egress issues, stability-related failures, a range of fall-from-height issues, fall-arrest-induced destabilization and tipover, and other issues. A significant under-researched component is environmental conditions. The contribution of environmental conditions is largely unknown and can range from extreme and unanticipated weather conditions to catastrophic system failure. The contribution of misuse of equipment is similarly unknown, but it is known that fall-from-elevation injuries have occurred from equipment misuse in either a deliberate or unintended fashion. Other issues, such as location of self-arrest anchorage points, unsecured elevation of equipment, and task-destabilization-related issues (such as destabilization of scaffolds from drilling, cable pulling and similar tasks), call for continued and in-depth research.

Because of the complex and multifactorial nature of hazard exposures related to mechanized elevating equipment, it is unlikely that the research expertise necessary to conduct pertinent research on this wide range of issues will be found in private enterprises. The need for NIOSH to direct resources to this emerging issue is evident; however, NIOSH will not be able to address these issues singly. NIOSH will increasingly turn to stakeholders and interested parties in private enterprise, other research agencies with a national scope, consensus standard-development agencies and regulatory agencies, labor and manufacturing groups, and other interested parties to address the emerging hazards related to mechanized elevating equipment.

## ACKNOWLEDGMENTS AND DISCLAIMER

### Acknowledgments

We would like to acknowledge the contributions of SkyJack Inc., which provided the aerial lift researchers with the use of a new scissor-lift and other critical technical and design data. We would like to extend our appreciation for the contributions of the following companies: DBI/SALA, Elk River, Inc., MSA, SafeWaze, who generously provided new harnesses/lanyards for use by the aerial lift researchers. We acknowledge the actions of the International Safety Equipment Association in providing constructive comments at various stages of this aerial lift study. The authors also want to express our gratitude to Randall Wingfield and Gravitec Inc., who generously provided constructive comments on the drop test component of the aerial lift study. The authors further acknowledge the review and commentary from Fraco Inc. for the mast work platform study.

The authors would like to acknowledge Darlene Weaver for technical assistance in the preparation of this document, Suzanne Marsh for her assistance in performing surveillance data analysis and summary, and Cathy Rotunda for editorial review. The authors are grateful to John Powers for his valuable assistance with instrumentation and data collection for the aerial lift project. The authors also would like to acknowledge Matt Burkart and Michael McCann for reviewing this document.

### Disclaimer

The findings and conclusions presented herein are those of the authors and do not necessarily represent the views of NIOSH. Mention of any company names or products does not constitute the endorsement by NIOSH.

## REFERENCES

### Aerial Lifts

1. Bureau of Labor Statistics, U.S. Department of Labor, *Table 5, Number of Nonfatal Occupational Injuries and Illnesses Involving Days Away from Work by Selected Injury or Illness Characteristic and Major Industry Sector, 2007*, Washington, DC, 2009. Available at http://www.bls.gov/news.release/osh2.t05.htm (accessed January, 2011).
2. Bureau of Labor Statistics, U.S. Department of Labor, *Table 7, Incidence Rates for Nonfatal Occupational Injuries and Illnesses Involving Days Away from Work per 10,000 Full-Time Workers for Selected Characteristics and Major Industry Sector, 2007,* Washington, DC, 2009. Available at http://www.bls.gov/news.release/osh2.t07.htm (accessed January, 2011).
3. Bureau of Labor Statistics, U.S. Department of Labor, *Number and Rate of Fatal Occupational Injuries, by Industry Sector, 2008*, Washington, DC, 2009. Available at http://www.bls.gov/iif/oshwc/cfoi/cfch0007.pdf (accessed January, 2011).
4. Bureau of Labor Statistics, U.S. Department of Labor, *Table A-9, Fatal Occupational Injuries by Event or Exposure for All Fatalities and Major Private Industry Sector, All United States, 2008*, Washington, DC, 2009. Available at http://www.bls.gov/iif/oshwc/cfoi/cftb0240.pdf (accessed January, 2011).
5 C. S. Pan and S. Chiou, Slip and fall: Fall protection in construction safety, in *Industrial and Occupational Ergonomics: Users Encyclopedia,* A. Mital, M. Ayoub, S. Kumar, and M. J. Wang, Eds., International Journal of Industrial Engineers, Cincinnati, OH, 1999.
6. M. J. Burkart, M. McCann, and D. M. Paine, Aerial work platforms, in *Elevated Work Platforms and Scaffolding,* McGraw-Hill Companies Inc., New York, 2004.
7. F. G. Heath, Accident waiting to happen, *Lift and Access*, May–June 2006.
8. M. McCann, Death in construction related to personnel lifts, 1992–1999, *J. Saf. Res. 34*: 507–514 (2003).
9. C. S. Pan, A. Hoskin, M. Lin, D. Castillo, M. McCann, K. Fern, P. Keane, Aerial lift fall injuries: A surveillance and evaluation approach for targeting prevention activities, *J. Saf. Res. 38*: 617–625 (2007).

10. Skyjack delivers 100,000th scissor lift to Ahern equipment, *Lift and Access*, April 21, 2004.
11. Skyjack stacks the deck, *Lift and Access*, February 2, 2005.
12. Rental re-emergency, *Lift and Access*, February 10, 2006.
13. 100,000th JLG boom lift sold at private auction, *Lift and Access*, January 17, 2007.
14. Bureau of Labor Statistics, U.S. Department of Labor, *Occupational Injuries and Illnesses: Counts, Rates, and Characteristics, 2004*, Washington, DC, U.S. Department of Labor, Bulletin 2493 (2006).
15. C. S. Pan, S. Chiou, J. Powers, D. Cantis, M. Ronaghi, and B. Boehler, Effect of side forces on the stability of scissor lifts (abstract), presented at the *Fourth National Occupational Injury Research Symposium*, Pittsburgh, PA, October 2008.
16. J. N. Ellis, Aerial lifts and platforms, in *Introduction to Fall Protection*, American Society of Safety Engineers, Des Plaines, IL, pp. 115–118, 2001.
17. The Eastman Kodak Company, Work design, in *Ergonomic Design for People at Work*, John Wiley & Sons, Hoboken, NJ, pp. 449–452, 2004.
18. C. S. Pan, Aerial lifts and fall protection systems, presented at the *International Safety Equipment Association Fall Meeting*, Arlington, VA, November 12, 2009.
19. K. H. E. Kroemer and E. Grandjean, The design of workstations, in *Fitting the Task to the Human*, Taylor & Francis, Philadelphia, PA, p. 53, 1997.
20. C. S. Pan, A. Hoskin, M. Lin, D. Castillo, M. McCann, and K. Fern, Injuries due to aerial work platforms, in *Proceedings of XVIIth World Congress on Safety and Health at Work*, National Safety Council, Itasca, IL, 2005.
21. J. Weeks, Work practices in aerial lifts, presented at the *First Aerial Platform Safety Conference*, Dallas, TX, November 5, 2004.
22. Occupational Safety and Health Administration, Office of Statistical Studies and Analysis, *Selected Occupational Fatalities Related to Vehicle-Mounted Elevating and Rotating Work Platforms as Found in Reports of OSHA Fatality/Catastrophe Investigations*, Washington, DC, 1999.
23. American National Standards Institute, *ANSI A92.6 (2006): Self-Propelled Elevating Aerial Work Platforms*, New York, 2006.
24. M. Ronaghi, J. Z. Wu, C. S. Pan, J. Harris, D. Welcome, S. Chiou, B. Boehler, and R. Dong, Scissor lift safety: An initiative to model static stability, *Prof. Saf. 54*(4): 43–48 (2009).
25. International Organization for Standardization, ISO 16368, 2003(E): *Mobile Elevating Work Platforms—Design Calculations, Safety Requirements and Test Methods*, Geneva, Switzerland, 2003.
26. National Institute for Occupational Safety and Health, Scissor lift study design, by R. Dong, Morgantown, WV, internal document (2006).
27. J. Hartsell, The development of a dynamic scissor lift model, West Virginia University, Morgantown, WV, master's thesis, West Virginia University, 2010.
28. J. R. Harris, J. Powers, C. S. Pan, and B. Boehler, Fall arrest characteristics of a scissor lift, *J. Saf. Res. 41*(3): 213–220 (2010).
29. C. S. Pan, Health impacts during the arrested falls, presented at the *International Society of Safety Equipment Fall Meeting*, Arlington, VA, November 2009.
30. J. Z. Wu, J. R. Powers, J. R. Harris, and C. S. Pan, Estimation of the kinetic energy dissipation in fall-arrest system and human body during fall impact, *Ergonomics 54*(4): 367–379.

## Stilts

31. S. M. Kisner and D. E. Fosbroke, Injury hazards in the construction industry, *J. Occup. Med. 36*: 137–143 (1994).
32. Bureau of Labor Statistics, U.S. Department of Labor, *Occupational Injuries and Illnesses by Selected Characteristics*, Washington, DC, News Release USDL-09-1454, November 24, 2009. Available at http://www.bls.gov/news.release/osh2.htm (accessed January, 2011).

33. S. Chiou, C. S. Pan, and D. E. Fosbroke, Identification of risk factors associated with traumatic injuries among drywall installers, in *Advances in Occupational Ergonomics and Safety*, B. Das and W. Karwoski, Eds., IOS Press, Amsterdam, the Netherlands, pp. 377–380, 1997.
34. S. Chiou, C. S. Pan, and P. Keane, Traumatic injury among drywall installers, 1992–1995, *J. Occup. Environ. Med. 42*(11): 1101–1108 (2000).
35. J. M. Dement and H. Lipscomb, Workers' compensation experience of North Carolina residential construction workers, *Appl. Occup. Environ. Hyg. 14*: 97–106 (1999).
36. H. J. Lipscomb, J. M. Dement, J. S. Gaal, W. Cameron, and V. McDougall, Work-related injuries in drywall installation, *Appl. Occup. Environ. Hyg. 15*(10): 794–802 (2000).
37. H. J. Lipscomb, J. M. Dement, L. Li, J. Nolan, and D. Patterson, Work-related injuries in residential and drywall carpentry, *Appl. Occup. Environ. Hyg. 18*(6): 479–488 (2003).
38. C. Whitaker, Stilts injuries in construction, *Prof. Saf.* September: 39–44 (2006).
39. A. Fischbein, A. M. Langer, Y. Suzuki, and I. J. Selikoff, Carcinoma of the lung in a drywall taping worker report of a case, *Toxicol. Lett. 2*: 231–236 (1978).
40. A. Fischbein, A. N. Rohl, A. M. Langer, I. J. Selikoff, Drywall construction and asbestos exposure, *Am. Ind. Hyg. Assoc. J. 40*: 402–407 (1979).
41. J. A. Freed, A. Miller, R. E. Gordon, A. Fischbein, J. Kleinerman, A. M. Langer, Desquamative interstitial pneumonia associated with chrysotile asbestos fibres, *Br. J. Ind. Med. 48*: 332–337 (1991).
42. National Institute for Occupational Safety and Health, *Hazard Controls: Control of Drywall Sanding Exposure*, Washington, DC, DHHS (NIOSH) Publication No. 99-113 (1999).
43. International Union of Painters and Allied Trades, *Safety and Overview: Student's Guide*, Version 1, International Brotherhood of Painters and Allied Trades, Washington, DC, 1998.
44. C. S. Pan, K. M. Miller, S. S. Chiou, and J. Z. Wu, Evaluation of a computer-simulation model for human ambulation on stilts, *J. Mech. Med. Biol. 4*(3): 283–303 (2004).
45. S. Chiou, C. S. Pan, and A. Bhattacharya, Kinematics and kinetics of gait on stilts: Identification of risks associated with construction stilt use, *Ergonomics 51*(12): 1814–1829 (2008).
46. C. S. Pan, S. Chiou, T. Kau, A. Bhattacharya, and D. Ammons, Effect of foot placement on postural stability of construction workers on stilts, *Appl. Ergon. 40*: 781–789 (2009).
47. J. Z. Wu, S. S. Chiou, and C. S. Pan, Analysis of musculoskeletal loadings in lower limbs during stilts walking in occupational activity, *Ann. Biomed. Eng. 37*(6): 1177–1189 (2009).
48. M. Erdil and O. Dickerson, *Cumulative Trauma Disorders: Prevention, Evaluation, and Treatment*, Wiley, New York, 2001.
49. P. M. Clarkson and M. J. Hubal, Exercise-induced muscle damage in humans, *Am. J. Phys. Med. Rehabil. 81*(11): S52–S69 (2002).
50. D. J. Newham, D. A. Jones, and P. M. Clarkson, Repeated high-force eccentric exercise: Effects on muscle pain and damage, *J. Appl. Philo. 63*(4): 1381–1386 (1987).
51. J. K. Garrick and S. L. Lewis, Career hazards for the dancer, *Occup. Med. 16*(4): 609–618 (2001).
52. A. Bhattacharya, M. Mueller, and V. Putz-Anderson, Traumatogenic factors affecting the knees of carpet installers, *Appl. Ergon. 16*(4): 243–250 (1985).
53. P. G. J. Maquet, Mechanics and osteoarthritis of the patellofemoral joint, *Clin. Orthop. 144*: 70–73 (1979).
54. M. Thun, S. Tanaka, and A. B. Smith, Morbidity from repetitive knee trauma in carpet and floor layers, *Brit. J. Ind. Med. 44*: 611–620 (1987).
55. A. S. Hyde, G. M. Bakken, J. R. Abele, H. H. Cohen, and C. A. LaRue, *Falls and Related Injuries: Slips, Trips, Missteps, and Their Consequences*, Lawyers and Judges, Tucson, AZ, pp. 93–103, 2002.
56. L. Nashner, Fixed patterns of rapid postural responses among leg muscles during stance, *Exp. Brain Res. 30*: 13 (1977).
57. S. I. Rosen, *The Slip and Fall Handbook*, Hanrow Press, San Jose, CA, pp. 76–77, 2003.
58. S. Schneider and P. Susie, Ergonomics and construction: A review of potential hazards in new building construction, *Am. Ind. Hyg. Assoc. J. 55*(7): 635–649 (1994).

59. Y. Ashkenazy, J. M. Hausdorff, P. C. Ivanov, and S. H. Eugene, A stochastic model of human gait dynamics, *Phys. Stat. Mech. Appl. 316*: 662–670 (2002).
60. S. Chung and J. K. Hahn, Animation of human walking in virtual environments, in *Proceedings of the Computer Animation Conference*, Geneva, Switzerland, pp. 4–15, 1999.
61. R. L. Huston, Multi-body dynamics—Modeling and analysis methods, *Appl. Mech. Rev. 44*: 109–117 (1991).
62. A. Schneider, D. C. Ferencz, H. J. Chizeck, Quantitative measurement of stability in human gait through computer simulation and Floquet analysis, *Ann. Int. Conf. IEEE Eng. Med. Biol. Proc. 17*: 1489–1490 (1995).
63. W. Whiting and R. Zernicke, *Biomechanics of Musculoskeletal Injury,* Human Kinetics, Champaign, IL, 1998.
64. C. S. Pan, S. Chiou, H. Hsiao, J. T. Wassell, and P. R. Keane, Assessment of perceived traumatic injury hazards during drywall hanging, *Int. J. Ind. Ergon. 25*: 29–37 (1999).
65. P. Vaida, B. Anton-Kuchly, and P. Varene, Mechanics and energetic of stilt walking, *J. Appl. Physiol. 51*(2): 529–532 (1981).
66. M. W. Whittle, *Gait Analysis: An Introduction*, 3rd edn., Butterworth-Heinemann, Woburn, MA, 2002.
67. M. A. Holbein and D. B. Chaffin, Stability limits in extreme postures: Effects of load positioning, foot placement, and strength, *Hum. Factors 39*: 456–468 (1997).
68. D. A. Winter, F. Prince, J. S. Frank, C. Powell, and K. F. Zabjek, Unified theory regarding A/P and M/L balance in quiet stance, *J. Neurophysiol. 75*: 2334–2343 (1996).
69. D. A. Winter, A. E. Patla, F. Prince, M. Ishac, and K. Gielo-Perczak, Stiffness control of balance in quiet standing, *J. Neurophysiol. 80*: 1211–1221 (1998).
70. J. C. Gillette, N. E. Quick, and J. J. Abbas, Center of pressure measures to assess standing performance, *Biomed. Sci. Instrum. 38*: 239–244 (2002).
71. M. A. Holbein-Jenny, C. S. Mcdermott, and J. Demchak, Validity of functional stability limits as a measure of balance in adults aged 23–73 years, *Ergonomics 50*: 631–646 (2007).
72. L. Chiari, L. Rocchi, and A. Cappello, Stabilometric parameters are affected by anthropometry and foot placement, *Clin. Biomech. 17*: 666–677 (2002).
73. R. F. Escamilla, G. S. Fleisig, N. Zheng, J. E. Lander, S. W. Barrentine, J. R. Andrews, B. W. Bergemann, and C. T. Moorman III, Effects of techniques on knee biomechanics during the squat and leg press, *Med. Sci. Sports Exerc. 33*: 1552–1566 (2001).
74. J. C. Gillette and J. J. Abbas, Foot placement alters the mechanisms of postural control while standing and reaching, *IEEE Trans. Neural. Syst. Rehabil. Eng. 11*: 377–385 (2003).
75. R. Moraes and A. E. Patla, Determinants guiding alternate foot placement selection and the behavioral responses are similar when avoiding a real or virtual obstacle, *Exp. Brain Res.* 171(4): 497–510 (2005).
76. J. C. Gillette, C. A. Stevermer, N. E. Quick, and J. J. Abbas, Alternative foot placements for individuals with spinal cord injuries standing with the assistance of functional neuromuscular stimulation, *Gait Posture 27*: 280–285 (2008).
77. T. D. Royer and C. A. Wasilewski, Hip and knee frontal moments in persons with unilateral trans-tibial amputation, *Gait Posture 23*(3): 303–306 (2006).
78. L. Sharma, D. E. Hurwitz, E. J. Thonar, J. A. Sum, M. E. Lenz, and D. D. Dunlop, Knee adduction moment, serum hyaluronan level, and disease severity in medial tibiofemoral osteoarthritis, *Arthritis Rheum. 41*:1233–1240 (1998).
79. D. Winter, Degenerative gait in the elderly, in *The Biomechanics and Motor Control of Human Gait: Normal, Elderly and Pathological*, 2nd edn., Waterloo Biomechanics, Ontario, Canada, pp. 87–94, 1991.
80. R. K. Begg and W. A. Sparrow, Gait characteristics of young and older individuals negotiating a raised surface: Implications for the prevention of falls, *J. Gerontol. 55*(3): M147–M154 (2000).

81. R. Begg, R. Best, L. Dell'Ore, and S. Taylor, Minimum foot clearance during walking: Strategies for the minimization of trip-related falls, *Gait Posture 25*: 191–198 (2007).
82. B. J. MacFadyen and F. P. Prince, Avoidance and accommodation of surface height changes by healthy, *J. Gerontol. 57*(4): 166–174 (2002).
83. D. A. Winter, Foot trajectory in human gait: A precise and multi-factorial motor control task, *Phys. Ther. 71*(1): 45–55 (1992).


## Scaffolds


84. Occupational Safety and Health Administration, Department of Labor, Construction eTool: Scaffolding, 2010. Available at http://www.osha.gov/SLTC/etools/scaffolding/supported/index.html (accessed January, 2011).
85. M. A. Personick, New data highlight gravity of construction falls, in *Issues in Labor Statistics*, U.S. Government Printing Office, Washington, DC, 1996.
86. Bureau of Labor Statistics, U.S. Department of Labor, *1926 Subpart L App A, Scaffold Specifications*, Washington, DC.
87. Bureau of Labor Statistics, U.S. Department of Labor, *Work Injury Report: Survey of Scaffold Accidents Resulting in Injuries, 1978*, Washington, DC, 1983.
88. Occupational Safety and Health Administration, Department of Labor, Construction eTool: Scaffolding, Glossary, 2010. Available at http://www.osha.gov/SLTC/etools/scaffolding/glossary.html (accessed January, 2011).
89. P. D. Bust (Ed.), *Contemporary Ergonomics 2008: Proceedings of the International Conference on Contemporary Ergonomics*, Taylor & Francis, London, U.K., 2008.
90. Occupational Safety and Health Administration, Department of Labor, *Occupational Fatalities Related to Scaffolds as Found in Reports of OSHA Fatality/Catastrophe Investigations*, Washington, DC, 1979.
91. S. M. Whitaker, R. J. Graves, M. James, and P. McCann, Safety with access scaffolds: Development of a prototype decision aid based on accident analysis, *J. Saf. Res. 34*: 249–261 (2003).
92. D. B. Chaffin and T. J. Stobbe, *Ergonomic Considerations Related to Selected Fall Prevention Aspects of Scaffolds and Ladders as Presented in OSHA Standard 29 CFR 1910 Subpart D*, University of Michigan, Ann Arbor, MI, 1979.
93. G. Shepherd, Construction fatalities—The toll of gravitational energy, presented at the *ISF Symposium*, Orlando, FL, 2000.
94. A-1 Plank & Scaffold Manufacturing, Inc, A-1 Scaffold safety guidelines, Paramount, CA, 2007, http://www.a1scaffold.com/safety.htm (accessed January, 2011).
95. H. Hsiao, M. Hause, J. Powers, T. Kau, C. Hendricks, and P. Simeonov, Effect of scaffold-end-frame carrying strategies on worker stepping response, postural stability, and perceived task difficulty, *Hum. Factors 50*(1): 27–36 (2008).
96. National Association of Home Builders (NAHB), *Scaffold Safety Handbook*, NAHB, Washington, DC, 2004.
97. K. Halperin and M. McCann, An evaluation of scaffold safety at construction sites, *J. Saf. Res. 35*: 141–150 (2004).
98. U.S. Department of Energy, *Innovative Technology Summary Report: Excel Automatic Locking Scaffold*, Idaho Falls, ID, 1999, DOE/EM-0482.
99. Safway Steel Products Inc., *Safway Systems Scaffold Technical Manual*, 2007, http://www.safway.com/PDF/ORN203.pdf (accessed January, 2011).
100. Excel Modular Scaffold & Leasing Corporation, *Technical Manual—Savings through Safety, Quality, and Technology*, 2001, EMSLC-TSM-1001.
101. H. Hsiao and R. Stanevich, Biomechanical evaluation of scaffolding tasks, *Int. J. Ind. Ergon. 18*(5/6): 407–415 (1996).

102. R. Cutlip, H. Hsiao, R. Garcia, E. Becker, and B. Mayeux, A comparison of different postures for scaffold end-frame disassembly, *Appl. Ergon. 31*: 507–513 (2000).
103. R. Cutlip, H. Hsiao, R. Garcia, E. Becker, and B. Mayeux, Optimal hand locations for safe scaffold-end-frame disassembly, *Appl. Ergon. 33*: 349–355 (2002).
104. H. Hsiao, P. Simeonov, B. Dotson, D. Ammons, T.-Y. Kau, and S. Chiou, Human responses to augmented virtual scaffolding models, *Ergonomics 48*(10): 1223–1242 (2005).
105. P. I. Simeonov, H. W. Hsiao, B. W. Dotson, and D. Ammons, Height effects in real and virtual environments, *Hum. Factors 47*(2): 430–438 (2005).

## Mast Climbing Work Platform

106. Occupational Safety and Health Administration, U.S. Department of Labor, *Presentation for the Meeting of the Advisory Committee on Construction Safety and Health*, Washington, DC, 2008, Docket number: OSHA-2008-0013.
107. Occupational Safety and Health Administration, U.S. Department of Labor, *Presentation for the Meeting of the Advisory Committee on Construction Safety and Health*, Washington, DC, 2008, Docket number: OSHA-2008-0029.
108. Occupational Safety and Health Administration (OSHA), *Presentation for the Meeting of the Advisory Committee on Construction Safety and Health* (*ACCSH*), Washington, DC, 2008, Docket number: OSHA-2008-0029.

CHAPTER 26

# Ergonomics in the Agricultural Industry*

Thomas R. Waters, Kermit G. Davis, and Susan E. Kotowski

CONTENTS



* The findings and conclusions in this chapter are those of the author(s) and do not necessarily represent the views of the National Institute for Occupational Safety and Health.

## 26.1 INTRODUCTION

Agriculture represents one of the largest industrial sectors in the United States. In a recent (2002) census of agriculture, it was reported that more than 2.1 million farms were active in the United States, with total acreage surpassing 938 million acres (USDA, 2002). Although the majority (about 90%) of farms in the United States are small, family-owned enterprises, commercial farms account for more than 50% of farm production (USDA, 2002). However, these large, commercial farms represent less than one-third of the total acreage used for farming (USDA, 2002). In 2004, the average size of a farm was 443 acres (USDA, 2002).

A large portion of the background material for this chapter was adapted from a review by Davis and Kotowski (2007), who focused on ergonomic risks associated with musculoskeletal injury; herein, we discuss how those risks apply to agriculture. Agricultural production in the United States can be divided into two broad categories: animal production (40%) and crop production (60%) (USDOL, 2005). These two categories of farm production present different sets of risk factors for work-related injuries and illnesses. Risk factors in animal production include being struck or stepped on by an animal and performing heavy lifting (e.g., in milking, feeding, transferring, and caring for animals). On the other hand, risk factors for musculoskeletal disorders (MSDs) associated with crop production result from more repetitive activities that often involve extreme bending or stooping, working with the hands and arms in awkward positions, and heavy lifting of agricultural products and equipment. Regardless of the type of production, the majority of workers in this industry are manual laborers (USDOL, 2005).

A difficult problem with applying ergonomics to agriculture is that, in comparison with other industrial sectors, the specific exposures are unique; in addition, the tasks and exposures vary by season and type of agricultural production. This variation places farmers and farmworkers at increased risk of injuries and MSDs. Contributing to the uniqueness of risk exposures in agriculture are the special populations within the sector. The U.S. agricultural workforce is dominated by males, most likely because of the physical demands of the work. In addition, however, the industry includes three other at-risk populations of workers: females, children/adolescents, and migrants.

Females make up about 30% of the farm workforce and often play an active role in the daily work activities (Oliveira et al., 1993). They are involved in heavy physical work on the farm and perform many of the same tasks as males. Unfortunately, this role may put them at increased risk of MSDs because they generally have a lower strength capacity than males.

Furthermore, agriculture is one of the few industries in which children and adolescents are considered an integral component of the workforce, performing physically demanding jobs that are typically designed for adults. Children and adolescents represent a major segment of the agricultural workforce (Waters, 2002). Researchers estimate that this segment comprises between 1.2 and 3.4 million individuals under 20 years of age (Dacquel and Dahmann, 1993; Schulman et al., 1997; Myers and Hendricks, 2001). Among these young workers, the predominant age is between 16 and 19 years (53%) (Myers and Hendricks, 2001). Farm youth perform many physically demanding tasks, including lifting and moving materials and equipment, operating farm equipment, and doing jobs requiring moderate to high levels of strength and coordination (Bartels et al., 2000). Few studies, however, have evaluated the physical demands associated with jobs performed by children and adolescents, and even fewer studies have examined the magnitude and severity of risks that these jobs represent for young workers. Moreover, there are no surveillance systems in place to monitor and evaluate the magnitude of risk for this younger population. In addition, there is no scientific information about the potential long-term risk of chronic health problems (such as MSDs or osteoarthritis) that may occur in adulthood resulting from work during childhood.

One of the unique populations within the agricultural workforce is migrant workers, who account for a significant percentage (42%) of the crop workers in the United States. These workers are often overlooked with respect to health and safety issues because they travel between regions (often more than 75 miles apart) to meet seasonal farming demands (USDA, 2002). These workers perform many physically demanding jobs that usually involve heavy lifting, awkward postures, and long work hours in poor environments (e.g., heat, rain, and muddy fields). They have limited access to information about reducing exposures and treating health issues.

Another major issue with regard to understanding the magnitude of the problem of MSDs in agriculture is the lack of reporting of such incidents. Only 4%–10% of U.S. farms (both family and nonfamily) have more than 11 employees and therefore are subject to reporting requirements of the Occupational Safety and Health Administration (OSHA) for injuries and illnesses (Aherin et al., 1990; Purschwitz and Field, 1990; Baker et al., 1992; Waller, 1992; Zhou and Roseman, 1994; NSC, 1997). As a result, because the statistical analyses do not take into account the majority of the exposed population, their findings cannot be considered conclusive (Purschwitz and Field, 1990; Zhou and Roseman, 1994).

Musculoskeletal health outcomes for farmers and farmworkers fall under two broad categories: acute injuries occurring during a onetime event and longer-term MSDs resulting from cumulative exposure over time. Acute injuries result from single traumatic events such as slips and falls, kicks by animals, and lacerations; most MSDs likely develop from cumulative or repetitive exposures to a stressor. Although both types of injury have significant ramifications for the long-term health of farmers, there has been much less research on MSDs than on traumatic injuries and fatalities in farming.

## 26.2 ACUTE INJURIES

Acute farm-related injuries are often significantly underreported, and it is estimated that only 1 of every 31 farm residents seeks treatment for such an injury (Nordstrom et al., 1995). Only serious and disabling injuries are tracked, and these occur at significant rates in farming. An estimated 140,000 disabling injuries occurred in 1998, accounting for a rate of 40.58 per 1000 farmworkers (Gerberich et al., 2001). Researchers also have reported high rates for specific farm sectors. For instance, a rate of 166 injuries per 1000 workers per year was noted among diary farmers in New York (Pratt et al., 1992). One obvious factor related to the incidence of acute injuries is the amount of time spent working on farms. Researchers have reported an approximately 3% increase in injury rate for every hour worked (Kidd et al., 1996; Stueland et al., 1997; Carruth et al., 2002). Many of these acute injuries are major traumas to the body, including fractures, lacerations, and internal bleeding. Hwang et al. (2001) reported that more than 90% of farmers in the state of New York reported having a severe injury in the previous year. A breakdown of injuries showed rates for contusions were between 11% and 52%, and those for lacerations were between 12.5% and 34.0% (Swanson et al., 1987; Schulman et al., 1997; Stueland et al., 1997; Gerberich et al., 2001; Carruth et al., 2002; Munshi et al., 2002; Browning et al., 2003). The magnitude of the prevalence of injuries appears to depend upon the agricultural subsector (Table 26.1) with some of the highest rates for beef, hog, and sheep farmers, followed by cash grain, nursery, field crop, dairy, and poultry farmers, and (considered as a group) vegetable, fruit, and nut farmers (Meyers et al., 2000).

The top causes of these injuries included kicks, bites, and falls from farm animals (36%), falls from tractors and rollovers (24%), and falls from other farm machinery or attempts to unjam material from moving parts (28%) (Cogbill and Busch, 1985). In their review, McCurdy and Carroll (2000) summarized the sources of injuries as follows: machinery, 18%–35%; animals, 12%–33%; tractors, up to 40%; and falls, 16%–25%. Similar rankings have been reported by others (Cogbill et al., 1991; Brison and Pickett, 1992; Pickett et al., 1995; Browning et al., 1998; Hwang et al., 2001). Lacerations and punctures resulting from machinery account for about 26% of all injuries (Layde et al., 1995; Gerberich et al., 1998), and hand injuries make up the largest portion of these machine-related injuries (26%) (Layde et al., 1995). Tractor mounting and dismounting also were found to be major causes of farmers' injuries, accounting for about 40% of total injuries (Lee et al., 1996). Tractor rollover is a significant contributor to major injuries and a leading cause of machine-related deaths (McCurdy and Carroll, 2000). The type of injury appears to be highly dependent upon the type of machinery being utilized by the farmer, according to Hartling et al. (1997). For instance, they found that (1) balers tend to cause entanglements, (2) hay elevators or conveyors cause entanglements and can strike or pin the operator, (3) augers are associated with entanglements, (4) farm wagons result in fall hazards and can strike or pin a farmworker, and (5) combines and harvesters are associated with entanglements and falls. Of course, animal-related injuries are most prevalent on farms where livestock is raised or used for work. For example, farmers who are routinely in proximity with cattle and horses are four times more likely to be injured than farmers who are not

TABLE 26.1 Prevalence of Injuries in Agricultural Subsectors

| Farming Subsector | Injuries per 200,000 h Worked |
|---|---|
| Beef, hog, and sheep | 10.2 |
| Cash grain | 7.6 |
| Nursery | 7.3 |
| Field crop | 5.8 |
| Dairy | 4.4 |
| Poultry | 4.0 |
| Vegetable, fruit, and nut | 3.8 |

(Boyle et al., 1997). Other researchers have found beef cattle farmers suffer injuries twice as often as general farmers (Browning et al., 1998). However, tobacco farmers have been found to be more susceptible to falls (29.5%), lacerations or punctures (27.2%), and injuries related to machinery (10.2%) (Struttmann and Reed, 2002).

Another major cause of injuries is falls. Nordstrom et al. (1996) reported annual rates of approximately 7.5 falls per 1000 person-years. According to their study (1996), the rate of fall-related injuries is higher for males (2.4 times that for females) and with longer work hours (an increase of 2% for each extra hour worked). Reasons for an increased risk of falls include environmental factors such as mud and variability in terrain, as well as mounting and dismounting hazards presented by large vehicles.

## 26.3 MUSCULOSKELETAL DISORDERS

MSDs are defined as injuries to the musculoskeletal system (including nerves, tendons, muscles, and supporting structures such as intervertebral disks) resulting from cumulative exposure to physical stressors (NIOSH, 1997). MSDs are also referred to as overexertion/overuse injuries, cumulative trauma disorders, repetitive strain injuries, and sprains and strains. The impact of MSDs on the agricultural industry is significant. In 1995, sprains and strains accounted for approximately 56,000 lost-time injuries on U.S. farms (Myers, 2001) and were the leading cause of injuries on farms (Zhou and Roseman, 1994). Overexertion injuries account for more than 43% of all agricultural sprains and strains (Meyers et al., 1997). One research group found that more than 60% of farmers reported at least one MSD symptom (Rosecrance et al., 2006). As with acute injuries, certain subsectors are more prone to high prevalence rates of MSDs. For example, the rates are high (almost 50%) in horticultural and nursery specialties (Meyers et al., 1997). Also common are more long-term, debilitating musculoskeletal problems such as osteoarthritis and arthropathy (unspecified joint disorder) (Kirkhorn et al., 2003).

## 26.4 PRIMARY RISK FACTORS FOR MSDs

### 26.4.1 Low Back

Although the prevalence of low-back pain (LBP) is significant among farmers, generally the percentage reported in the literature varies between 37% and 50% (Walker-Bone and Palmer, 2002; Gomez et al., 2003; Rosecrance et al., 2006). However, Struttmann and Reed (2002) reported that tobacco farmers suffered LBP at significantly lower prevalence rates (about 9%). More severe pain such as chronic back pain that lasts more than 3 months was reported by 10% of the farmers investigated by Brackbill et al. (1994), and Myers (2001) found that lost-time low-back injuries accounted for about 14% of all such injuries. Estimates of the prevalence of LBP have generally been reported for farmers as a whole, but no doubt there is variation between subsectors of the agricultural industry in relation to the tasks they perform.

Two candidates for the most common causes of LBP on the farm are heavy lifting and performing tasks in awkward postures (Holmberg et al., 2003; Rosecrance et al., 2006). As can be seen in Figure 26.1A through C, working in stooped postures and lifting heavy loads occur in many types of farmwork.

(A) (B) (C)

FIGURE 26.1 Examples of agriculture jobs associated with a high risk for MSDs—in particular, LBP—include (A) working in a stooped posture, (B) carrying heavy objects overhead, and (C) lifting and carrying heavy objects.

Crop cultivation, which often requires heavy lifting in awkward postures, has been related to LBP, especially more severe cases such as sciatica (Manninen et al., 1995). Farmers who work in nurseries are often required to complete tasks with awkward postures and in poor lifting conditions and environments, for instance, when handling plant containers, pruning plants, weeding, labeling plants, and loading or unloading trucks (Meyers et al., 1997, 2000). One task associated with the most extreme postures is weeding, for which the worker stoops for extended periods; both poor posture and long duration are known risk factors for LBP, according to the National Institute for Occupational Safety and Health (NIOSH, 2001). Prolonged extreme trunk flexion (>60°) is commonly found in farm tasks such as pruning, weeding, labeling, and harvesting crops (Meyers et al., 1997, 2000; NIOSH, 2001). Fruit harvesters have also been found to work for significant durations in severe flexion, especially when trees are below 3.4 m in height (Janowitz et al., 2000; Meyers et al., 2000; Fulmer et al., 2002).

Lifting heavy objects such as fully loaded totes or bins is a common problem for vegetable and fruit farmworkers (Duraj et al., 2000; Meyers et al., 2000; NIOSH, 2001; Fulmer et al., 2002). Lifting more than 50 lb (22.7 kg) occurs routinely on farms and is considered a major risk to farmworkers (Chapman and Meyers, 2001). Carrying bags while picking fruit on ladders puts a significant static compressive load on the low back of workers, especially when they lean away from the ladder to reach for fruit (Meyers et al., 2000; Fulmer et al., 2002). Transferring containers of fruits, vegetables, and plants also has been identified as a risk factor for LBP (Duraj et al., 2000; Janowitz et al., 2000; Meyers et al., 2000; NIOSH, 2001; Chapman et al., 2004).

One additional, common risk factor for farmworkers is long-term exposure to whole-body vibration. Many farmers spend close to 500 h per year (equivalent to 12–15 h a day) driving machinery (Toren et al., 2002), mostly large farm vehicles such as tractors, combines, and hauling trucks. To make matters worse, this whole-body vibration is often accompanied by twisting when the driver turns to look back at the fields (Toren et al., 2002; Walker-Bone and Palmer, 2002; Holmberg et al., 2003). These two risk factors commonly occur together during silage chopping and plowing (Toren et al., 2002). Exposure to whole-body vibration has been associated with high rates of LBP in farmers: between 20% and 75%, with the average from the most recent studies being approximately 40% (Boshuizen et al., 1990). Certainly, the long-term impact of whole-body vibration, which often is due to lack of ergonomic controls, such as anti-vibrating devices on seats, is greater for older individuals (Gomez et al., 2003).

### 26.4.2 Upper Extremities (Hands, Wrists, Elbows, Arms, and Shoulders)

In the United States, upper-extremity injuries are another major risk on farms, accounting for about 22% of all lost-time injuries (Meyers et al., 2000). The prevalence of hand or wrist MSDs is roughly 25%–30% in many subsectors of the agriculture industry (Pratt et al., 1992; Struttmann and Reed, 2002; Gomez et al., 2003). A Swedish study found that dairy farms may have one of the highest prevalence rates of hand/wrist injuries, exceeding 50% (Stal, 2000). Shoulder injuries and shoulder pain due to farmwork, although not as prevalent as hand/wrist injuries, happen often, with a yearly prevalence of about 7%–15% (equal to or greater than the rate associated with other manual labor, 9.7%) (Struttmann and Reed, 2002; Walker-Bone and Palmer, 2002).

Nursery farmers are often exposed to risk factors for upper-extremity MSDs, such as work involving extreme wrist postures, highly repetitive gripping, and high-magnitude hand forces (specifically, pinch forces and contact stresses during pruning, plant propagation, and weeding) (Meyers et al., 1997; NIOSH, 2001). Harvesting presents similar risk factors involving wrist posture and hand repetitions (which can exceed 15 per minute), as well as static loads (Meyers et al., 2000; NIOSH, 2001). The combination of high hand force and repetitive hand exertions also occurs during pruning and harvesting at vineyards (Janowitz et al., 2000; Meyers et al., 2000; Wakula and Landau, 2000; Wakula et al., 2000). On many dairy farms where milking remains predominantly manual, intense hand and arm motions are still required (Stal, 2000). Even more-automated dairy operations require intensive handwork for attaching the milking devices to udders (Stal, 2000). Another common risk factor for hand/wrist disorders is the utilization of vibrating tools such as chain saws, weed eaters, and other mechanical hand equipment (Walker-Bone and Palmer, 2002).

Fruit and vegetable harvesters are exposed to stressors that increase the risk for shoulder problems. Using arms above the shoulder significantly increases the risk of such injuries. These elevated arm postures are common for tomato-vine trainers, who sometimes make up to 16,000 motions at or above shoulder height each week (Palmer, 1996). Fruit harvesters who pick from prune trees such as apples, pears, and oranges are exposed to shoulder stressors that include reaching up to pick, wearing a shoulder bag that puts a pressure point and load on the shoulder joint, carrying bins of fruits above the head, and carrying other heavy objects such as ladders (Duraj et al., 2000; Fulmer et al., 2002). Finally, many vineyard workers adopt poor postures (shoulders are typically flexed and abducted) during pruning (Meyers et al., 2000).

### 26.4.3 Lower Extremities (Hips, Legs, Knees, Ankles, and Feet)

Another type of MSDs in farmers is lower-extremity injuries, which account for approximately 28% of all lost-time injuries that occur on farms (Meyers et al., 2000). The most common of these injuries involve the knees and hips: 29% of farmers report having had knee injuries, and 15% report having had hip trouble (Gomez et al., 2003). Two subsectors that appear to be prone to high rates of lower-extremity injuries are dairy farmers (32%) (Pratt et al., 1992) and tobacco farmers (24%) (Struttmann and Reed, 2002). Another major debilitating cumulative disorder common among farmers is osteoarthrosis, particularly in the hip joint (Axmacher and Lindberg, 1993).

Multiple risk factors have been associated with lower-extremity MSDs. Whole-body vibration from driving farm machinery has been linked to osteoarthritis of the hip (Jacobsson et al., 1987; Thelin, 1990; Croft et al., 1992; Axmacher and Lindberg, 1993; Thelin et al., 1997; Walker-Bone and Palmer, 2002). The debilitating condition of osteoarthritis in the hips, knees, ankles, and feet has been linked to the following risk factors for farmers: increased age (>65 years), increased body weight (obese farmers, with a body mass index >30 $kg/cm^2$, are particularly vulnerable), highly repetitive and forceful work, and extreme bending and squatting (Kirkhorn et al., 2003). Furthermore, nonspecific knee and hip pain has been found to be associated with awkward and extreme work postures and heavy lifting on farms (Holmberg et al., 2003). One of the most unique risk factors for lower-extremity MSDs in farmers is walking in uneven, oftentimes wet fields. The harvest season commonly presents this stressor. Also, just as harvesting has been related to upper-extremity MSDs, it is associated with lower-extremity ones as well (Meyers et al., 2000; NIOSH, 2001). Carrying heavy bins, bags, or baskets of vegetables or fruits puts stress on the lower extremities, as does working in postures such as squatting to pick weeds or crops. Farmers sometimes spend long periods of time kneeling, causing contact stress and deep bending of the knee joints. Finally, one typically understudied risk factor for lower-extremity MSDs is prolonged standing on hard, uneven, or unstable surfaces such as muddy fields and concrete slabs.

### 26.4.4 Upper Back and Neck

Almost no research has been conducted on the prevalence of upper back and neck MSDs in the agricultural industry. One subsector of the agriculture industry that has significant neck injuries (20% of injuries) is dairy farming (Pratt et al., 1992). Although the literature on MSDs of the upper back and neck is minimal, one would expect that certain farming duties would contribute to such injuries.

Severe neck flexion and extension are the primary risk factors for upper back and neck MSDs in farming. Farming tasks that are performed low to the ground, such as harvesting small plants, weeding fields, and pruning below waist level, oftentimes induce severe neck flexion (Meyers et al., 2000). In addition, neck extension occurs when farmworkers do tasks above their heads, as in the aforementioned fruit harvesting and vineyard pruning (Meyers et al., 2000). Another risk factor may be prolonged neck twisting while driving tractors in the field or hauling loads.

## 26.5 ERGONOMIC INTERVENTIONS

Along with the vast potential for farmers to be exposed to risk factors for MSDs, there is an abundance of opportunities to develop and implement interventions to control these exposures. Caution must be exercised, however, in their implementation, because the effectiveness of an intervention will depend upon the specific situation in which it will be implemented and the person who will be using it (Davis and Kotowski, 2007). This section provides a summary of interventions that have been developed and used for controlling MSD risk factors.

Chapman et al. (2004) evaluated the use of mesh bags and standard containers for washing and transferring vegetables. They investigated these tools in terms of their efficiency for completing the tasks and reducing exposure to MSD risk factors. The mesh bags were lighter and provided less resistance in the water, because liquids could easily flow through the material. The bins had handles, specially designed to decrease lifting loads. One of the primary results was to show the importance of information dissemination with respect to adoption of and compliance with a potential intervention, especially when such information is linked to economic outcomes (e.g., profitability). Figure 26.2A shows how crops were processed before the mesh bags were recommended, and Figure 26.2B shows the use of a mesh bag.

In another study, by Janowitz et al. (2000), utilization of smaller, lighter bins during the harvesting of wine grapes was found to reduce biomechanical stresses on the body. This improvement was accomplished with the following features: (1) reducing (by 12%) the maximum amount of weight lifted, (2) designing the bin to be less wide so that it would be closer to the body and thus a smaller horizontal arm movement would be needed, (3) including handles on the bins, and (4) designing the bins to have smooth bottoms, which reduced the sliding forces (by 32%). On the basis of the revised NIOSH lifting equation, the MSD risk due to lifting was reduced by 23% with the intervention tubs, in comparison with the traditional tubs. The intervention had a nonsignificant impact on the productivity required to harvest the grapes. The original and redesigned tubs are shown in Figure 26.3.

(A)

(B)

FIGURE 26.2 Examples of washing vegetables (A) before and (B) after implementation of a mesh bag. Note that the posture in (A) is extremely bent compared to that in (B).

FIGURE 26.3 Original tub (left) and redesigned tub (right) for fruit harvesting. Note the smaller width, reduced volume, and incorporation of handles in the redesigned version.

The California-based research group of Duraj et al. (2000) has continued to investigate such interventions with respect to reducing risk for workers while maintaining productivity. They investigated the utility of a specially designed cart for transferring tubs between fields and trucks. The results of the study indicated that workers adopted better back and knee postures (less flexion) when using the cart. However, the cart failed to perform optimally when used in uneven fields or on muddy surfaces. More recent research by Duraj et al. (2001a,b) has focused on more-automated methods, involving a large machine that moved and dumped grapes into the collector bin. The automatic device was able to eliminate the need to manually dump the transfer bins at a rate of 15 containers per minute.

Simple educational programs have been relatively ineffective in improving safety and preventing MSDs on farms, but multifaceted approaches have shown some success (DeRoo et al., 2000). Multifaceted programs effectively reduced safety hazards, reduced worker-compensation claims, and increased knowledge among farmers.

Several interventions have been designed on the basis of engineering principles. One leading document, published by NIOSH (2001), is *Simple Solutions: Ergonomics for Farm Workers*. This document is free to farmers and introduces several potential interventions for a variety of farm tasks, including the following: an extended lifting tool for picking up plants in containers that reduces severe forward flexion of the trunk and eliminates the pinch grip; a weeding stand for nursery work with potted plants, which reduces prolonged stooping and neck flexion; standard picking bins with handles that reduce the weight lifted and provide a better grasping surface; a lightweight berry harvesting rake that reduces prolonged stooping postures; and a packing shed layout that adjusts the work area to the farmworker and increases productivity. These examples are only a few of the many interventions that could significantly reduce the risk and stressors that plague the agricultural

(A)

(B)

FIGURE 26.4 In these illustrations of a high-risk nursery task, potted plants are lifted (A) by the conventional method and (B) with the use of a new ergonomic tool.

(A)

(B)

FIGURE 26.5 In these illustrations of a high-risk farming task, ground crops are picked and weeded (A) by the conventional method and (B) with the use of a new ergonomic cart.

industry (Chapman and Meyers, 2001; Waters, 2002). Some examples of the interventions discussed in the NIOSH booklet are shown in Figures 26.4 and 26.5.

## 26.6 MSDs IN CHILDREN AND ADOLESCENT FARMWORKERS

As mentioned previously, children and adolescents make up a significant proportion of the farmworker population and have special issues relating to risk of MSDs. In a recent study by Allread et al. (2004), many of the tasks routinely performed by farm youth were identified as being risky, at some level, for developing LBP. Lifting of heavy objects such as water buckets, bales of hay, straw, or alfalfa, and feed bags as well as filling feed buckets were found to pose the greatest risk for low-back injuries. In addition to handling water and feed pails, several other tasks were found to have some level of risk for low-back injuries: lifting objects such as grass bales, milk crates, and hay flakes (partial hay bales); spreading hay; pushing and pulling wheelbarrows; shoveling manure or feed; connecting trailer hitches; raking stalls; and picking up rocks (Marlenga et al., 1978). Most of these tasks pose a risk to adults and therefore would be expected to pose an even greater one for youths. With each of these tasks, the key risk factors were high trunk movement (weight being lifted in combination with distance from the spine), awkward postures of the trunk, and rapidity of trunk motion (Allread et al., 2004).

Hartling et al. (2004) reviewed the literature regarding interventions for farm-related injuries in children. Overall, the results indicated that school-based and safety day camps that focus on reducing MSDs stressors are effective in disseminating such information to children.

Recently, Waters and Garg (2010) developed a two-dimensional biomechanical model for assessing the physical demands of manual handling tasks (e.g., pushing, pulling, and lifting) for youth. A series of screenshots from the computer program is shown in Figures 26.6 through 26.9. As can be seen from the figures, the model allows the analyst to input the youth worker's characteristics (age, gender, height, and weight), body posture (position of the limbs and trunk) during the work activity, and job characteristics (type of activity, direction of applied load, and magnitude of applied load). On the basis of these input data, the analyst can then estimate the load on the spine and determine what percentage of youths of that gender and age would be able to perform the activity. These data are useful in determining whether the youth should perform the work activity and how potential changes in the work design, such as in job layout or weights handled, might affect the physical demands of the job.

A series of studies supported by NIOSH have been conducted recently to examine the potential effectiveness of ergonomic interventions targeting youth farm tasks that have been identified as posing a high risk for MSDs. The first set of studies examined the use of

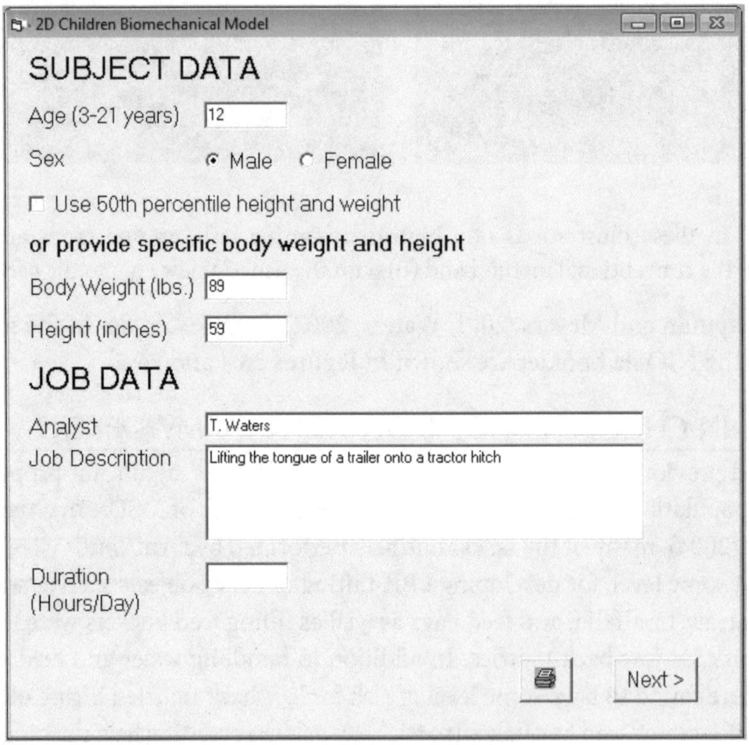

FIGURE 26.6 A sample input screen for anthropometry.

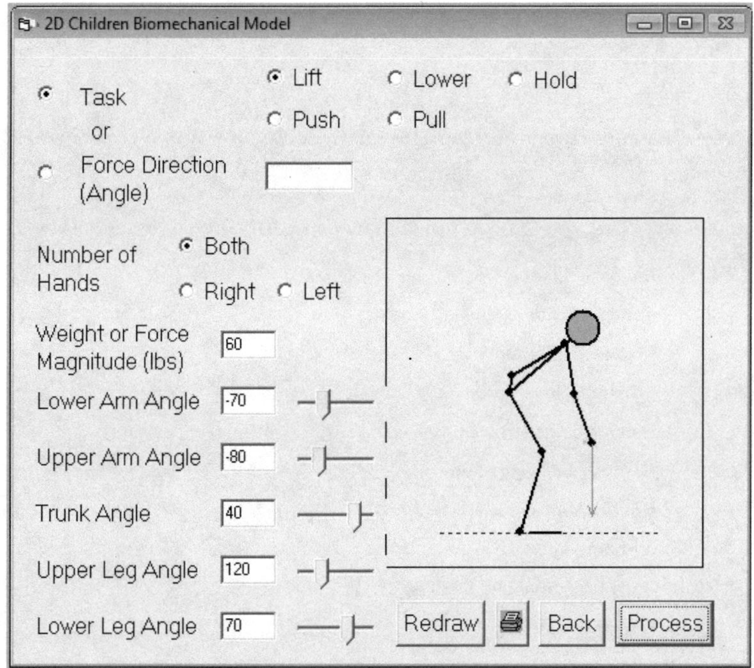


FIGURE 26.7 A sample input screen for task and posture data.

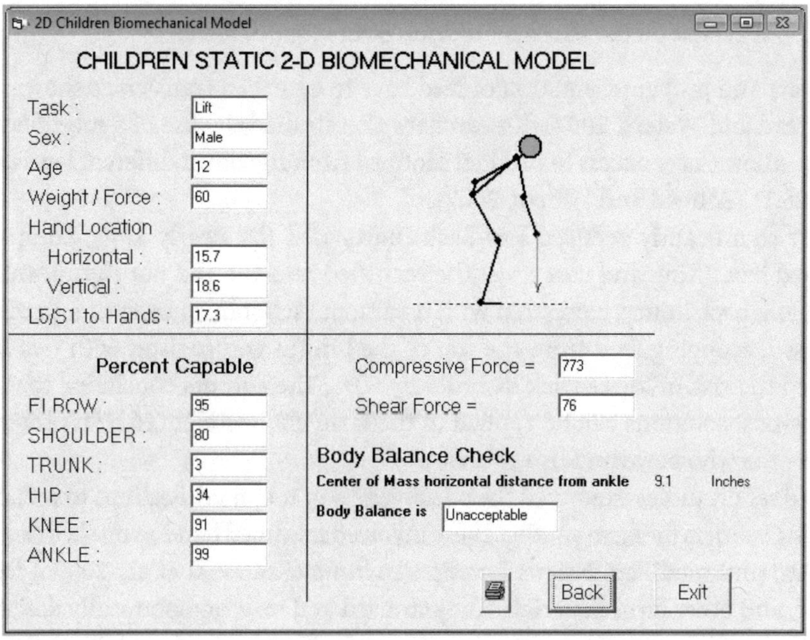

2D Children Biomechanical Model

CHILDREN STATIC 2-D BIOMECHANICAL MODEL

Task : Lift
Sex : Male
Age 12
Weight / Force : 60
Hand Location
Horizontal : 15.7
Vertical : 18.6
L5/S1 to Hands : 17.3

Percent Capable
ELBOW : 95
SHOULDER : 80
TRUNK : 3
HIP : 34
KNEE : 91
ANKLE : 99

Compressive Force = 773
Shear Force = 76

Body Balance Check
Center of Mass horizontal distance from ankle 9.1 Inches
Body Balance is Unacceptable

Back Exit

FIGURE 26.8 A sample output screen for a summary report.

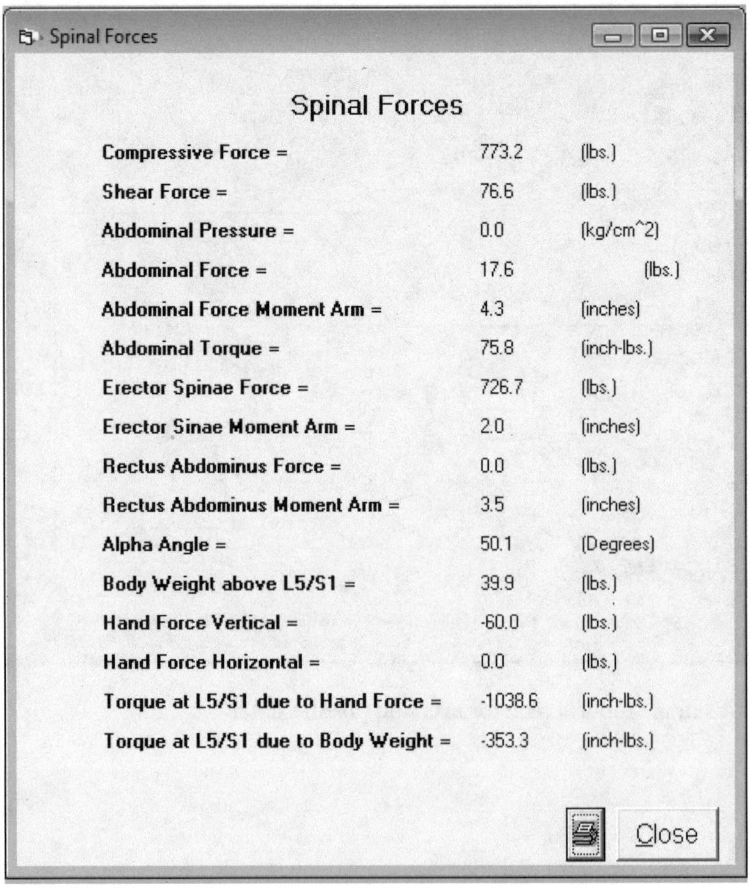

Spinal Forces

**Spinal Forces**

| | | |
|---|---|---|
| **Compressive Force =** | 773.2 | (lbs.) |
| **Shear Force =** | 76.6 | (lbs.) |
| **Abdominal Pressure =** | 0.0 | (kg/cm^2) |
| **Abdominal Force =** | 17.6 | (lbs.) |
| **Abdominal Force Moment Arm =** | 4.3 | (inches) |
| **Abdominal Torque =** | 75.8 | (inch-lbs.) |
| **Erector Spinae Force =** | 726.7 | (lbs.) |
| **Erector Sinae Moment Arm =** | 2.0 | (inches) |
| **Rectus Abdominus Force =** | 0.0 | (lbs.) |
| **Rectus Abdominus Moment Arm =** | 3.5 | (inches) |
| **Alpha Angle =** | 50.1 | (Degrees) |
| **Body Weight above L5/S1 =** | 39.9 | (lbs.) |
| **Hand Force Vertical =** | -60.0 | (lbs.) |
| **Hand Force Horizontal =** | 0.0 | (lbs.) |
| **Torque at L5/S1 due to Hand Force =** | -1038.6 | (inch-lbs.) |
| **Torque at L5/S1 due to Body Weight =** | -353.3 | (inch-lbs.) |

Close

FIGURE 26.9 A sample output screen for measurement of spinal forces.

a rugged cart and proper orientation of feed bags to ease feed transfer, as shown in Figure 26.10 (Allread and Waters, 2007). Researchers also studied the use of a modified feed bin, intended to allow easier access to product scooped from the bin at different levels, as shown in Figure 26.11 (Allread and Waters, 2007).

The cart significantly reduced low-back injury risk (by nearly 10%), compared with manual feed bag lifting and carrying. The modified feed bin did not significantly reduce the risk of low-back injury, compared with traditional scooping; however, regardless of the method used, scooping feed from the top of the bin (in comparison with two lower levels) reduced the risk of lower-back disorder by 50%. The authors concluded that relatively simple, low-cost solutions can be applied in the farm environment to help protect the low back for youths who perform farm chores.

A second set of studies examined the effectiveness of tool modifications to reduce the risk of low-back disorders for farm youths. These involved modified handles on shovels (Kotowski et al., 2009a) and modified designs for wheelbarrows (Kotowski et al., 2009b) for moving feed, waste, and other farm materials. The standard and new, ergonomically designed shovels and wheelbarrows that were tested are shown in Figures 26.12 and 26.13. The findings from the shovel study revealed that the alternative shovel designs did affect the way youths

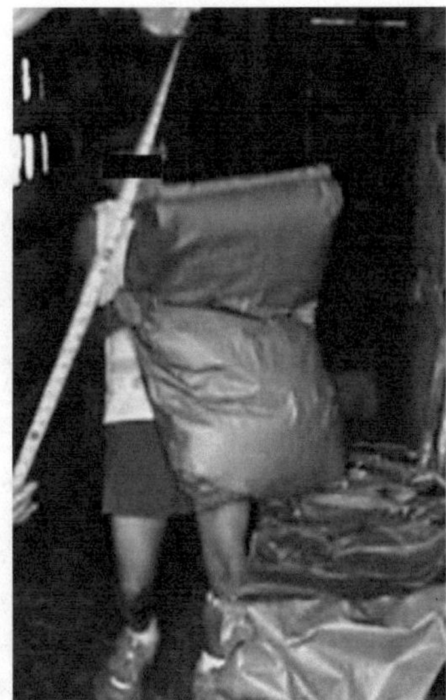

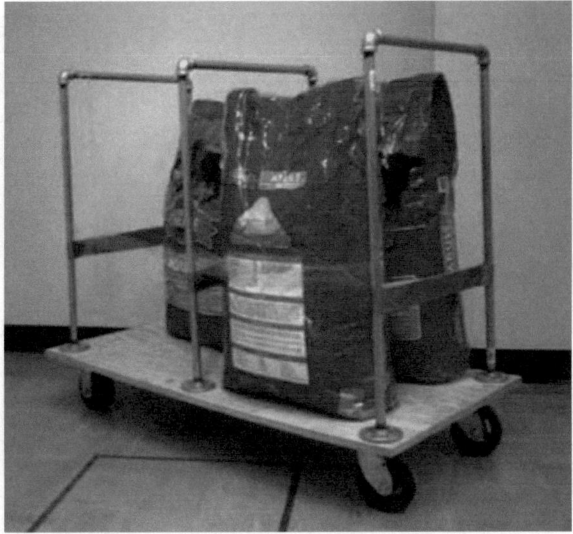

FIGURE 26.10 Carrying a feed bag, *left*, is an example of a high-risk task for farm youths. At right is an ergonomic cart for moving feed bags.

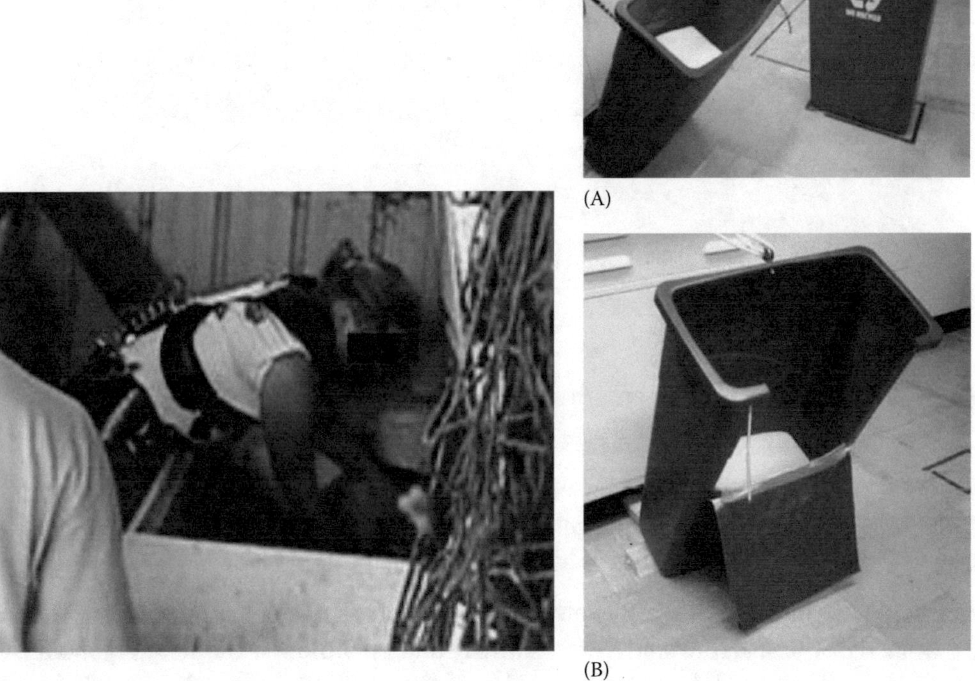

(A)

(B)

FIGURE 26.11 Scooping feed, *left*, is another high-risk farming task for youths. At right is an ergonomic bin for (A) storing and (B) scooping feed.

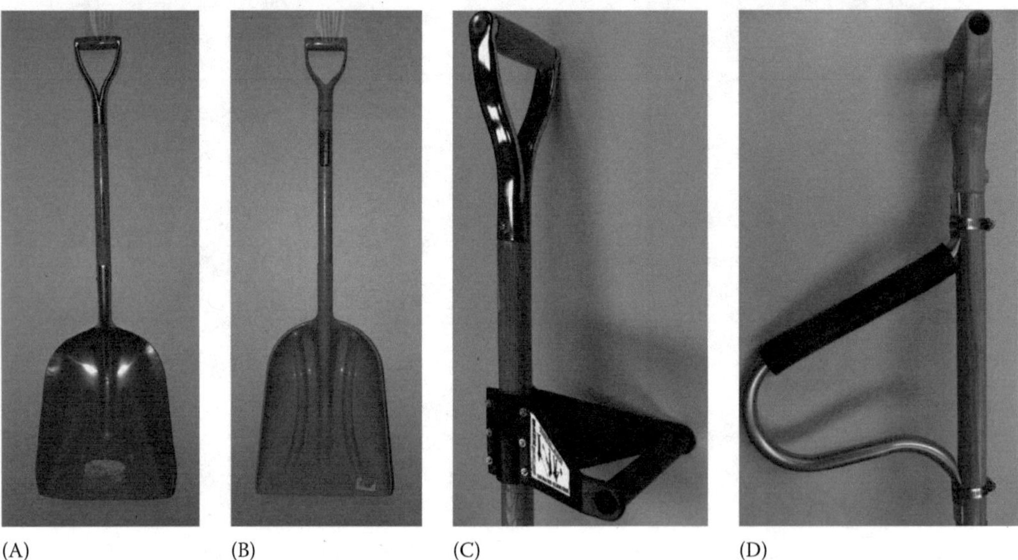

FIGURE 26.12 Various styles of ergonomic handles for shovels tested are shown (A: aluminium scoop shovel, B: plastic scoop shovel, C: triangle grip add-on handle, and D: D-grip add-on handle).

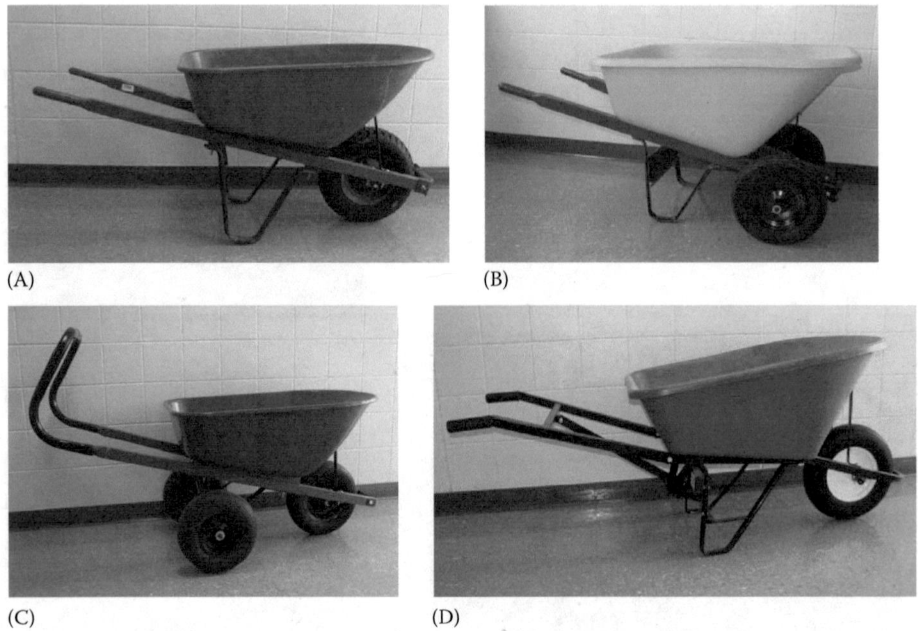

FIGURE 26.13 Various ergonomic designs for wheelbarrows tested are shown (A: regular, B: two-wheel, C: three-wheel, and D: adjustable handle).

performed tasks but did not significantly decrease their risk of low-back disorders. The findings from the wheelbarrow study were similar; that is, although the new, ergonomically designed wheelbarrows did affect the way the youths performed the tasks, overall they did little to reduce the risk of low-back disorders. The results of these studies suggested that perhaps training and more familiarity with the new devices might have improved the findings.

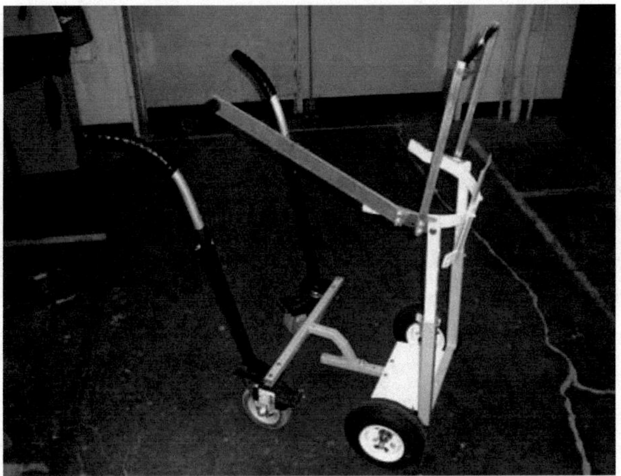

FIGURE 26.14 An ergonomic cart for moving large buckets.

FIGURE 26.15 An ergonomic handle for carrying and dumping buckets.

A third set of studies evaluated a cart for moving and dumping buckets and a new tool for lifting buckets, as shown in Figures 26.14 and 26.15 (Tang et al., 2006). Results from these studies showed that the ergonomic interventions effectively and significantly reduced risk of low-back disorders associated with these work tasks.

To examine potential indicators of long-term health effects associated with farmwork for youths, a study was conducted to compare bone mineral quality and quantity between youths who live and work on farms and youths who do not (Bhattacharya et al., 2007). Previous studies have shown that work-related factors such as repetitive kneeling, squatting, and heavy lifting are contributors to the development of osteoarthritis and other types of MSDs. The main hypothesis for this study was that bone mass, static biomechanical properties, and dynamic shock absorption properties do not differ between youths involved in physically demanding farm-related tasks compared with those who do not work on farms. The study showed that dynamic bone response measures provide preliminary evidence that farm-related physically demanding tasks lead to changes that may predispose teenagers to degenerative skeletal disorders later in life (Bhattacharya et al., 2007).

## 26.7 RESEARCH GAPS

### 26.7.1 Working in Stooped Postures

In 2004, a meeting was held in San Francisco to examine prevention of MSDs due to working in stooped postures. As a result, the authors of the proceedings (Fathallah et al., 2004) proposed a series of research topics and concluded the following:

> In order to improve our understanding of the relationship of stooped, squatting, and kneeling postures and MSDs and their prevention, we must increase and improve research focused on these risk factors. A necessary first step will require differentiation by the research community between stooped posture and stooped work. This may be achieved by determining at what exposure level assuming a flexed posture becomes "stooped work" and establishing a consensus definition of stooped work (e.g., work below knees >40% of time). Secondly, there is a need to develop practical and objective measures of exposure to stooped work (degree of bending, duration, frequency) and refine the epidemiological case definition of outcome (symptoms, physical findings, diagnoses) for a deeper focus on the effects of stooped work. To seriously begin to improve our understanding of the etiology and causal relationship between stooped and squatting postures and MSDs, we need to increase our understanding of the biomechanics of the spine and the lower extremities in these positions. There is a need for research studies designed to evaluate the effects of these postures on tissue responses under various conditions and loading patterns. Research is required to understand how the intervertebral disc, the meniscus of the knee, and other passive tissues respond to repetitive versus static loading. Such research may point the way to understanding the relationship of degrees of postural stress and disease and, similarly, how much postural relief is needed or useful in preventing disease. Epidemiology and biomechanics provide much of the basis for understanding the effects of working in stooped and squatting postures, and the knowledge gained from such study needs to be incorporated into the case definition of stooped work, and in job design criteria that take into account the tissue fatigue generated by static postures.
>
> Finally, there is need for an improved system of intervention research that both disseminates the evaluation of alternative strategic approaches in different workplaces and supports the development of workplace-specific adaptations of known approaches.

> Interventions that are not both acceptable to workers and employers and that fit the work system without serious detriment to productivity will not be widely adopted. Development of such interventions is neither an automatic nor a guaranteed result of publication of research results or successful demonstrations in other industries.

The specific topics identified at the conference on stooped work are as follows.

#### *26.7.1.1 Suggestions for Assessing High-Risk Jobs*

1. Evaluate the effectiveness of different methods of risk assessment, with attention to predictive ability and field utility.

#### *26.7.1.2 Suggestions for Surveillance Research*

1. Develop a national registry of musculoskeletal hazards and health outcomes.
2. Add supplements to existing surveillance systems for stooped, squatting, and kneeling postures.
3. Conduct surveys in high-risk industries (agriculture, construction, and mining).
4. Determine the number of workers exposed and what jobs they are doing.
5. Record the exposure in identified jobs: time in stooped and squatting postures, considered by "zones" of mild, moderate, and severe positions.
6. Identify the specific job or task elements associated with stooped postures and why they are necessitated.
7. Conduct cross-sectional and longitudinal studies to develop and validate a list of high-risk jobs and significant health outcomes.

#### *26.7.1.3 Suggestions for Intervention Research*

1. Develop new partnerships with agencies, academia, and industry to support intervention research focused on stooping, squatting, and kneeling work, including national and regional partnerships focused on industry- or task-specific applications.
2. Increase the number and range of intervention research projects under way.
3. Encourage higher-quality intervention evaluations using randomized trials, quasi-experimental designs, and blended evaluations.
4. Document and broaden the dissemination of successful, proven interventions.

#### *26.7.1.4 Suggestions for Etiological Research*

1. Conduct studies to improve understanding of specific biomechanical stresses and MSD development during stooping, squatting, and kneeling postures.
2. Develop and evaluate improved standards and methods for assessing exposure, health outcomes, and other etiological factors for stooped, squatting, and kneeling work.

3. Conduct population, clinical, and laboratory studies to evaluate the short-term impacts of different types of exposure to working in stooped, squatting, or kneeling postures, with regard to MSD development and early indicators of such disorders.

4. Conduct studies as described earlier, in number 3, except with a focus on long-term impacts.

### 26.7.2 Youth and Adolescents Working in Agriculture

In 2002, NIOSH and the Great Lakes Center for Agricultural Safety and Health at The Ohio State University cosponsored a conference that brought together national experts to discuss research needs regarding prevention of work-related MSDs for children and youths working in agriculture. These agricultural safety and health experts identified specific topic areas for which research is lacking, such as (1) identification of potentially high-risk jobs; (2) quantification of the level of risk for jobs performed by children and adolescents in agriculture; (3) developing, evaluating, and implementing surveillance systems for measuring and tracking the magnitude of health effects and risk for children and adolescents working in agriculture; and (4) developing and evaluating ergonomic interventions for reducing risk of MSDs for children and adolescents working in agriculture. Although the meeting focused on youths and adolescents working in agriculture, the research topics identified at the conference would likely apply to adult farmers as well. The suggestions for future research that were published in the meeting proceedings (NIOSH, 2004) are summarized as follows.

#### *26.7.2.1 Suggestions for Assessing High-Risk Jobs*

1. Develop an "enterprise classification" system and evaluate the risk of work-related MSDs according to this classification (e.g., determine risk by region, agriculture sector, or size of enterprise).

2. Determine the number of exposed youths and what jobs they are doing in each commodity area.

3. Identify the hazards or physical work factors in each job or task and determine the number of hours worked per year.

4. Evaluate the effectiveness of different methods of risk assessment, including self-assessment, professional judgment, and objective quantitative methods. Use health outcome or level of exposure as a measure of risk.

5. Evaluate risk in non-mechanized production (e.g., tool use in manual labor).

#### *26.7.2.2 Suggestions for Surveillance Research*

1. Develop a national registry of musculoskeletal hazards and health outcomes (e.g., with a National Health and Hazard Exam).

2. Supplement existing surveillance systems (e.g., NHIS, NHANES, BRFSS, California Department of Health, and prospective community-based surveys such as Keokuk and Iowa Safe Farm).

3. Conduct ad hoc population-based health and hazard surveys (e.g., by means of clinic- or school-based methods or face-to-face interviews).

4. Develop partnerships with individuals or agencies that interact regularly with children and adolescents working in agricultural settings.

5. Conduct cross-sectional and longitudinal studies to develop and validate a consensus list of high-risk jobs and significant health outcomes.

#### 26.7.2.3 *Suggestions for Intervention Research*

1. Develop private industry, academic industry, and state agency partnerships (e.g., establish a vocational agriculture awards program for interventions at the high school or college level).

2. Develop improved methods for disseminating information.

3. Conduct studies that address liability and cultural, ethical, and economic barriers to implementing interventions.

4. Encourage more high-quality evaluations of interventions (using randomized trials as well as quasi-experimental and blended evaluations).

5. Investigate the use of existing or modified models for increasing adoption of interventions based on similar successful models, such as the NIOSH hazard control hierarchy model for injury prevention or the tobacco model for increasing awareness of the benefits of smoking cessation.

#### 26.7.2.4 *Suggestions for Etiological Research*

1. Conduct studies to assess physical, cognitive, and developmental capabilities of children and adolescents.

2. Conduct studies to determine the magnitude of exposures and symptoms for children and adolescents in agriculture, including examination of multiple exposures (e.g., sports and second jobs).

3. Develop and evaluate improved methods for measuring exposure, health, and other outcomes.

4. Conduct population, clinical, and laboratory studies to evaluate the short-term impact of work-related MSD risk factors, such as effects of different types of exposures on MSD risk and early indicators such as biomarkers, bone density, stiffness, and pain.

5. Conduct population, clinical, and laboratory studies to evaluate the long-term impact of repeated exposure (e.g., studies to compare the health status of retired farmers and nonfarmers, and evaluations of the permanent effects of physical loading). Studies should include groups with maximal exposures.

## 26.8 SUMMARY

In the United States, farmers and farmworkers of all ages are constantly faced with hazards in their jobs, potentially some of the worst in private industry. There is significant research indicating farmers and farmworkers suffer many different injuries resulting from their work. There are also indications about the causes of these injuries. However, what percentage of farmers and farmworkers suffer from musculoskeletal conditions and whether they result from cumulative exposure to risk factors are unanswered questions. Researchers have only scratched the surface with respect to understanding the magnitude and causes of such musculoskeletal problems. In recent years, several major initiatives have been formulated to begin determining the magnitude of the injuries, the sources of the injuries, and whether certain populations are at more risk (e.g., females, farm youths, and migrant workers). These populations are especially vulnerable to MSDs but are also the most poorly understood in this respect. There have been numerous interventions for the stressful tasks involved in a variety of farming activities, but few have been rigorously investigated for effectiveness and implementation. In spite of the development of interventions, many stressors that plague farmers remain, because of a lack of knowledge and awareness and a lack of resources to implement these interventions. Although there have been great strides in ergonomics and safety in the agricultural industry over the past several decades, more research is needed to establish a safe and healthy work environment in the agricultural industry in the United States.

## REFERENCES


Aherin, R.A., D.J. Murphy, C. Westaby, *Changing Farm Worker Behavior: A Literature Review of Injury Control Strategies*, St. Joseph, MI: American Society of Agricultural Engineers, 1990.

Allread, W.G., T.R. Waters, Interventions to reduce low-back injury risk among youth who perform feed handling and scooping tasks on farms, *J Agri Saf Health* 13(4):375–393 (2007).

Allread, W.G., J.R. Wilkins, T.R. Waters, W.S. Marras, Physical demands and low-back injury risk among children and adolescents working on farms, *J Agri Saf Health* 10:257–274 (2004).

Axmacher, B., H. Lindberg, Coxarthrosis in farmers, *Clin Orthop* 287:82–86 (1993).

Baker, S.P., B. O'Neill, M.J. Ginsbury, L. Gupa, *The Injury Fact Book*, New York: Oxford University Press, 1992.

Bartels, S., B. Niederman, T.R. Waters, Job hazards for musculoskeletal disorders, *J Agri Saf Health* 6:191–201 (2000).

Bhattacharya, A., N.B. Watts, J. Gordon, R. Shukla, T.R. Waters, S. Bartels, R. Coleman, Bone quantity and quality of youths working on a farm: A pilot study, *J Agromed* 12(4):27–38 (2007).

Boshuizen, H.C., P.M. Bongers, C.T. Hulshof, Self-reported back pain in tractor drivers exposed to whole-body vibration, *Int Arch Occup Environ Health* 62:109–115 (1990).

Boyle, D., S.G. Gerberich, R.W. Gibson, G. Maldonado, R.A. Robinson, F. Martin, C. Reiner, H. Amandus, Injury from dairy cattle activities, *Epidemiology* 8:37–41 (1997).

Brackbill, R.M., L.L. Cameron, V. Behrens, Prevalence of chronic diseases and impairments among US farmers, 1986–1990, *Am J Epidemiol* 139:1055–1065 (1994).

Brison, R.J., C.W. Pickett, Non-fatal farm injuries on 117 eastern Ontario beef and dairy farms: A 1-year study, *Am J Ind Med* 21:623–636 (1992).

Browning, S.R., H. Truszczynska, D. Reed, R.H. McKnight, Agricultural injuries among older Kentucky farmers: The farm family health and hazard surveillance study, *Am J Ind Med* 33:341–353 (1998).

Browning, S.R., S.C. Westneat, C. Donnelly, D. Reed, Agricultural tasks and injuries among Kentucky farm children: Results of the farm family health and hazard surveillance project, *South Med J* 96:1203–1212 (2003).

Carruth, A.K., L. Skarke, B. Moffett, C. Prestholdt, Nonfatal injury experiences among women on family farms, *Clinic Nurs Res* 11:130–148 (2002).

Chapman, L., J. Meyers, Ergonomics and musculoskeletal injuries in agriculture: Recognizing and preventing the industry's most widespread health and safety problem, Baltimore, MD: *Agricultural Safety and Health Conference Proceedings*, 2001.

Chapman, L.J., A.C. Newenhouse, R.H. Meyer, A.D. Taveira, B.T. Karsh, J.J. Ehlers, T. Palermo, Evaluation of an intervention to reduce musculoskeletal hazards among fresh market vegetable growers, *Appl Ergon* 35:57–66 (2004).

Cogbill, T.H., H.M.J. Busch, The spectrum of agricultural trauma, *J Emerg Med* 3:205–210 (1985).

Cogbill, T.H., E.S. Steenlage, J. Landercasper, P.J. Strutt, Death and disability from agricultural injuries in Wisconsin: A 12-year experience with 739 patients, *J Trauma* 31:1632–1637 (1991).

Croft, P., D. Coggon, M. Cruddas, C. Cooper, Osteoarthritis of the hip: An occupational disease in farmers, *BMJ* 304:1269–1272 (1992).

Dacquel, L.T., D.C. Dahmann, *Residents of Farms and Rural Areas: 1991*, Washington, DC: U.S. Bureau of Census, Current Populations Report Series, Report No. P20, No. 472, 1993.

Davis, K., S. Kotowski, Understanding the ergonomic risk for musculoskeletal disorders in the United States, *Am J Ind Med* 50:501–511 (2007).

Deroo, L.A., H. Risto, R.H. Rautiainen, A systematic review of farm safety interventions, *Am J Prev Med* 18:51–62 (2000).

Duraj, V., J.A. Miles, J.M. Meyers, *Machine Handling of Winegrape Picking Containers*, St. Joseph, MI: American Society of Agricultural Engineers, 2001a.

Duraj, V., J.A. Miles, J.M. Meyers, *Continued Work on Machine Handling of Winegrape Picking Containers*, St. Joseph, MI: American Society of Agricultural Engineers, 2001b.

Duraj, V., J.A. Miles, J.M. Meyers, J.A. Faucett, I.L. Janowitz, M.E. Tarter, D.G. Tejeda, R.H. Smith, E.A. Weber, *Harvesting Aids for Reducing Ergonomics Risk Factors in Wine Grape Hand Harvesting*, St. Joseph, MI: American Society of Agricultural Engineers, 2000.

Fathallah, F.A., J.M. Meyers, I. Janowitz, *Conference Proceedings: Stooped and Squatting Postures in the Workplace*, Berkeley, CA: Center for Occupational and Environmental Health, University of California, 2004.

Fulmer, S., L. Punnett, D. Slingerland, A. Tucker, G. Earle-Richardson, Ergonomic exposures in apple harvesting: Preliminary observations, *Am J Ind Med* 42(Suppl 2):3–9 (2002).

Gerberich, S.G., R.W. Gibson, L.R. French, T.Y. Lee, W.P. Carr, L. Kochevar, C.M. Reiner, J. Shutske, Machinery-related injuries; regional rural injury study-I (RRIS-I), *Accid Anal Prev* 30:793–804 (1998).

Gerberich, S.G., R.W. Gibson, L.R. French, C.M. Renie, T.Y. Lee, W.P. Carr, J. Shutske, Injuries among children and youth in farm households: Regional rural injury study–I, *Inj Prev* 7:117–122 (2001).

Gomez, M.I., S. Hwang, A.D. Stark, J.J. May, E.M. Hallman, C.I. Pantea, An analysis of self-reported joint pain among New York farmers, *J Agric Saf Health* 9:143–157 (2003).

Hartling, L., R.J. Brison, E.T. Crumley, T.P. Klassen, W. Pickett, A systematic review of interventions to prevent childhood farm injuries, *Pediatrics* 114:483–496 (2004).

Hartling, L., W. Pickett, R.J. Brison, Non-tractor, agricultural machinery injuries in Ontario, *Can J Pub Health* 88:32–35 (1997).

Holmberg, S., A. Thelin, E. Stiernstrom, K. Svardsudd, The impact of physical work exposure on musculoskeletal symptoms among farmers and rural non-farmers, *Ann Agric Environ Med* 10:179–184 (2003).

Hwang, S.A., M.I. Gomez, A.D. Stark, T.L. St John, J.J. May, E.M. Hallman, Severe farm injuries among New York farmers, *Am J Ind Med* 40:32–41 (2001).

Jacobsson, B., N. Dalen, B. Tjornstrand, Coxarthrosis and labour, *Int Orthop* 11:311–313 (1987).

Janowitz, I., D.G. Tejeda, J.A. Miles, V. Duraj, F. Fathallah, J.M. Meyers, J. Faucett, Ergonomics interventions in the manual harvest of wine grapes, San Diego, CA: *Proceedings of the IEA 2000/HFES 2000 Congress*, 2000.

Kidd, P., T. Scharf, M. Veazie, Linking stress and injury in the farming environment: A secondary analysis of qualitative data, *Health Educ Q* 23:224–237 (1996).

Kirkhorn, S., R.T. Greenlee, J. Reeser, The epidemiology of agriculture-related osteoarthritis and its impact no occupational disability, *Wis Med J* 102:38–44 (2003).

Kotowski, S.E., K.G. Davis, T.R. Waters, Investigation of select ergonomic interventions for farm youth. Part 1: Shovels, *J Agromed* 14(1):33–43 (2009a).

Kotowski, S.E., K.G. Davis, T.R. Waters, Investigation of select ergonomic interventions for farm youth. Part 2: Wheelbarrows, *J Agromed* 14(1):44–57 (2009b).

Layde, P.M., D.L. Nordstrom, D. Stueland, L. Brand, K.A. Olson, Machine-related occupational injuries in farm residents, *Ann Epidemiol* 5:419–426 (1995).

Lee, T.Y., S.G. Gerberich, R.W. Gibson, W.P. Carr, J. Shutske, C.M. Reiner, A population-based study of tractor-related injuries: Regional rural injury study–I, *J Occup Environ Med* 38:782–793 (1996).

Manninen, P., H. Riihimak, M. Heliovaara, Incidence and risk factors of low-back pain in middle-aged farmers, *Occup Med* 45:141–146 (1995).

Marlenga, B., W. Pickett, R.L. Berg, Agricultural work activities reported for children and youth on 498 North American farms, *J Agric Saf Health* 7:241–252 (1978).

McCurdy, S.A., D.J. Carroll, Agricultural injury, *Am J Ind Med* 38:463–480 (2000).

Meyers, J.M., J.A. Miles, J. Faucett, I. Janowitz, D.G. Tejeda, V. Duraj, J. Kabashima, R. Smith, E. Weber, High risk tasks for musculoskeletal disorders in agricultural field work, San Diego, CA: *Proceedings of the IEA 2000/HFES 2000 Congress*, 2000.

Meyers, J.M., J.A. Miles, J. Faucett, I. Janowitz, D.G. Tejeda, J.N. Kabashima, Ergonomics in agriculture: Workplace priority setting in the nursery industry, *AIHA J* 58:121–126 (1997).

Munshi, K., D.L. Parker, H. Bannerman-Thompson, D. Merchant, Causes, nature, and outcomes of work-related injuries to adolescents working at farm and non-farm jobs in rural Minnesota, *Am J Ind Med* 42:142–149 (2002).

Myers, J.R., *Injuries Among Farm Workers in the United States, 1995*, Cincinnati, OH: National Institute for Occupational Safety and Health, Report No. 2001-153, 1–295, 2001.

Myers, J.R., K.J. Hendricks, *Injuries among Youth on Farms in the United States*, 1998, Cincinnati, OH: National Institute for Occupational Safety and Health, Report No. 2001-154, 1–179, 2001.

NIOSH (National Institute for Occupational Safety and Health), *Musculoskeletal Disorders and Workplace Factors: A Critical Review of Epidemiological Evidence for Work-Related Musculoskeletal Disorders of the Neck, Upper Extremity and Low Back*, Cincinnati, OH: DHHS (NIOSH) Publication No. 1997-141, 1997.

NIOSH (National Institute for Occupational Safety and Health), *Simple Solutions: Ergonomics for Farm Workers*, Cincinnati, OH: DHHS (NIOSH) Publication No. 2001-111, 1–53, 2001.

NIOSH (National Institute for Occupational Safety and Health), *Conference Proceedings: Prevention of Musculoskeletal Disorders for Children and Adolescents Working in Agriculture*, Cincinnati, OH: DHHS (NIOSH) Publication No. 2004-119, 2004.

Nordstrom, D.L., P.M. Layde, K.A. Olson, D. Stueland, L. Brand, M.A. Follen, Incidence of farm-work-related acute injury in a defined population, *Am J Ind Med* 28:551–564 (1995).

Nordstrom, D.L., P.M. Layde, K.A. Olson, D. Stueland, M.A. Follen, L. Brand, Fall-related occupational injuries on farms, *Am J Ind Med* 29:509–515 (1996).

NSC (National Safety Council), *Accident Facts*, Itasca, IL: NSC, 1997.

Oliveira, V.J., A.B.W. Effland, J.L. Runyan, S. Hamm, *Hired Farm Labor Use on Fruit, Vegetable and Horticultural Specialty Farms*, Agricultural Economic Report No. 676, Washington, DC: U.S. Department of Agriculture, 1993.

Palmer, K.T., Musculoskeletal problems in the tomato growing industry: 'Tomato trainer's shoulder'? *Occup Med* 46:428–431 (1996).

Pickett, W., R.J. Brison, H. Niezgoda, M.L. Chipman, Nonfatal farm injuries in Ontario: A population-based survey, *Accid Anal Prev* 27:425–433 (1995).

Pratt, D.S., L.H. Marvel, D. Darrow, L. Stallones, J.J. May, P. Jenkins, The dangers of dairy farming: The injury experience of 600 workers followed for 2 years, *Am J Ind Med* 21:637–650 (1992).

Purschwitz, M.A., W.E. Field, Scope and magnitude of injuries in the agricultural workplace, *Am J Ind Med* 18:179–192 (1990).

Rosecrance, J., G. Rodgers, L. Merlino, Low back pain and musculoskeletal symptoms among Kansas farmers, *Am J Ind Med* 49:547–556 (2006).

Schulman, M.D., C.T. Evensen, C.W. Runyan, L.R. Cohen, K.A. Dunn, Farm work is dangerous for teens: Agricultural hazards and injuries among North Carolina teens, *J Rural Health* 13:295–305 (1997).

Stal, M., Upper extremity disorders in professional female milkers, San Diego, CA: *IEA 2000/HFES 2000 Congress Proceedings*, 2000.

Struttmann, T.W., D.K. Reed, Injuries to tobacco farmers in Kentucky, *South Med J* 95:850–856 (2002).

Stueland, D.T., B.C. Lee, D.L. Nordstrom, P.M. Layde, L.M. Wittman, P.D. Gunderson, Case-control study of agricultural injuries to women in central Wisconsin, *Women Health* 25:91–103 (1997).

Swanson, J.A., M.I. Sachs, K.A. Dahlgren, S.J. Tinguely, Accidental farm injuries in children, *AJDC* 141:1276–1279 (1987).

Tang, S.C., F.A. Fathallah, T.R. Waters, Development and evaluation of ergonomic interventions for bucket handling on farms. *Proceedings of the Human Factors and Ergonomics Society 50th Annual Meeting*, Santa Monica, CA: Human Factors and Ergonomics Society, 2006.

Thelin, A., Hip joint arthrosis: An occupational disorder, *Am J Ind Med* 18:339–343 (1990).

Thelin, A., B. Jansson, B. Jacobsson, H. Strom, Coxarthrosis and farm work: A case-referent study, *Am J Ind Med* 23:497–501 (1997).

Toren, A., K. Oberg, B. Lembke, K. Enlund, A. Rask-Andersen, Tractor-driving hours and their relation to self-reported low-back and hip symptoms, *Appl Ergon* 33:139–146 (2002).

USDOL (United States Department of Labor), *Career Guide to Industries*, Washington, DC: Bureau of Labor Statistics, 2005.

USDA (United States Department of Agriculture), *Census of Agriculture*, Washington, DC: USDA, 2002.

Wakula, J., T. Bechmann, M. Hett, K. Landau, Stress-strain-analysis of grapevine pruning with powered and non-powered hand tools, San Diego, CA: *Proceedings of the IEA 2000/HFES 2000 Congress*, 2000.

Wakula, J., K. Landau, Ergonomic analysis of grapevine pruning and wine harvesting to define work and hand tools design requirements, San Diego, CA: *Proceedings of the IEA 2000/HFES 2000 Congress*, 2000.

Walker-Bone, K., K.T. Palmer, Musculoskeletal disorders in farmers and farm workers, *Occup Med* 52:441–450 (2002).

Waller, J.A., Injuries to farmers and farm families in a dairy state, *J Occup Med* 34:414–421 (1992).

Waters, T.R., Musculoskeletal disorders among children and adolescents working in agriculture, *J Agr Saf Health* 8:253–255 (2002).

Waters, T.R., A. Garg, Two-dimensional biomechanical model for estimating strength of youth and adolescents for manual material handling, *Appl Ergonomics* 41:1–7 (2010).

Zhou, C., J.M. Roseman, Agricultural injuries among a population based sample of farm operators in Alabama, *Am J Ind Med* 25:385–402 (1994).

CHAPTER 27

# Ergonomic Analysis and Abatement Recommendations to Reduce Musculoskeletal Stress in Warehousing Operations

## *Case Study*

Donald S. Bloswick and Emil Golias

CONTENTS

## 27.1 INTRODUCTION

Warehousing operations require considerable manual handling of loads. One of us was requested to assist Occupational Safety and Health Administration (OSHA) with hazard analysis and abatement recommendations relating to grocery warehouse operations in the eastern United States. Epidemiological data indicated a high musculoskeletal injury rate, which suggested the existence of manual materials handling hazards. A variety of abatement recommendations were made, ranging from those that required a relatively low cost change in work procedure to those that would require costly facility modification. These abatement recommendations relate to general warehousing operations as well as grocery storage. This chapter describes the actual abatement efforts made in two typical grocery warehouses, and discusses their impact on musculoskeletal injury rates.

## 27.2 BACKGROUND AND SIGNIFICANCE TO OCCUPATIONAL ERGONOMICS

The development of low back and other musculoskeletal problems is often associated with the magnitude and duration of spinal compressive loads, frequent repetition of manual materials handling activities, awkward postures, and torso rotation during the lift or load movement.

Since the purpose of a warehouse is to store material, the principal product of a warehouse is space. This often results in the requirement that loads be stacked high and be tightly packed together. These operational conditions combined with the fast work pace resulting from the piecework incentive pay programs result in an occupational setting in which the primary risk factors for back and musculoskeletal injuries exist: high loads, awkward postures, and high repetition.

## 27.3 CASE STUDY OF A TYPICAL GROCERY WAREHOUSING OPERATION

### 27.3.1 Epidemiological Information

The back injury history for one grocery warehouse is shown in Table 27.1. The tabulated data may include injuries to some "nonselectors." (A selector is a worker who selects requested items from the warehouse storage space.) If so, the effective rate for the selectors may actually be higher than noted in Table 27.1.

### 27.3.2 Task Description: Groceries and Perishables

Selectors are required to move cases and objects of varying sizes and weights from the pallet racks to pallet jacks. For one typical warehouse, the average frequency for this task was 1.9 cases/min with an average weight of 23.9 lb (10.8 kg) for groceries and 1.85 cases/min with an average weight of 28.3 lb (12.8 kg) for perishables.

A summary of the grocery and perishables workload for one typical grocery warehousing task is shown in Table 27.2. In the perishables area, some of the boxes of meat are very heavy, which increases the biomechanical hazard. A sample of 18 boxes of pork had an average weight of 77 lb (34.9 kg) with a box width of 34 in. (86.4 cm), and a sample of 10 boxes of beef had an average weight of 64.6 lb (29.3 kg) with a box width of 24 in. (61 cm).

TABLE 27.1 Musculoskeletal Injury Rate for a Typical Grocery Warehouse Selecting Operation

| Year | No. of Selector Lifting Injuries | No. of Selectors | Percent Selectors Injured |
|---|---|---|---|
| Grocery | | | |
| 1988 | 21 | 65 | 32 |
| 1987 | 26 | 68 | 38 |
| 1986 | 10 | 61 | 16 |
| 1985 | 26 | 55 | 47 |
| 1984 | 35 | 50 | 70 |
| Total | 118 | 299 | 39 |
| Perishables | | | |
| 1988 | 24 | 35 | 69 |
| 1987 | 28 | 38 | 74 |
| 1986 | 12 | 38 | 32 |
| 1985 | 17 | 35 | 49 |
| 1984 | 36 | 55 | 65 |
| Total | 117 | 201 | 58 |

TABLE 27.2 Grocery and Perishable Workload for a Typical Grocery Warehouse Selecting Operation

| | |
|---|---|
| Groceries | |
| Selector workload | 863 pcs per person per shift (7.5 h) = 1.9 cases/min |
| Average case size | 15.9 × 10.9 × 8.7 in. (40.4 × 27.7 × 22.1 cm) |
| Average case weight | 23.9 lb (10.8 kg) |
| Maximum case weight | Several cases in the range of 40–50 lb (18.1–22.7 kg) were observed |
| | Company time standards allow "80 lb (36.3 kg) and above." |
| Perishables (dairy, meat, fish, and produce) | |
| Perishables workload | 1027 lifts per person per shift (9.25 h) = 1.85 cases/min |
| Average case size | 16.7 × 11.9 × 7.8 in. (42.4 × 30.2 × 19.8 cm) |
| | Boxes of meat up to 34 in. (86.4 cm) wide were observed |
| Average case weight | 28.3 lb (12.8 kg) |
| Maximum case weight | 77 lb (34.9 kg), sample of 18 items on 2 pallets of pork |
| | 64.6 lb (29.3 kg), sample of 10 boxes of beef |

The workers were also required to wear additional clothing, which reduces the range of motion and increases the strain on the body. In some cases, the boxes were frozen together, which required considerable force to break the bond.

### 27.3.3 Task Analysis

#### *27.3.3.1 General Discussion of Acceptable Loads and Storage Locations*

When a load is lifted from the floor, additional stresses are exerted on the low back due to the body weight moment during stooping to pick the load up. Thus, heavy loads should not be stored on the floor but should be raised to about standing knuckle height [minimum 20 in. (50.8 cm)] to avoid the need to stoop and lift.

Mital [1,2] notes that the average maximum weights for male industrial workers for 8 and 12h shifts lifting medium-sized boxes at the rate of 1 lift/min are floor to knuckle 8h = 33.3lb (15.1kg), floor to knuckle 12h = 26.5lb (12kg), knuckle to shoulder 8h = 35.9lb (16.3kg), and knuckle to shoulder 12h = 28.3lb (12.8kg). Mital [3] also indicates that lifting capacity may decrease by up to 24% as the vertical lift height increases from 30in. (76.1cm) to 65in. (165.1cm).

The *Ergonomics Guide to Manual Lifting* [4] indicates that a 50lb (22.7kg) weight held 20in. (50.8cm) away from the body is acceptable for 75% of the male industrial population. Chaffin and Park [5] suggest that two-handed lifting in the sagittal plane of loads greater than about 35lb (15.9kg) is associated with increased incidence of low back pain if the load is close to the body. It should be noted that in many warehousing tasks, loads are not lifted in the sagittal plane nor can they generally be held close to the body. Herrin [6] notes that loads in excess of 50lb (22.7kg) cause an increase in medical incidents and a significant increase in days lost and days on restricted duty. Snook [7] notes that the average maximum weights for lifting medium-sized boxes at the rate of 1 lift/min to accommodate 75% of the male industrial population is approximately 49.2lb (22.3kg) for floor-to-knuckle height and 44.1lb (20kg) for knuckle-to-shoulder height. Snook also lists a maximum carry weight of 62.8lb (28.5kg) for an approximate 8ft (2.4m) carry. The limit for the lift *and* carry would, of course, be less than for the lift alone.

The International Labor Office [8] suggests a maximum of 55.1lb (25kg) for males 20–35 years old, 46.3lb (21kg) for males 35–50, and 35.3lb (16kg) for males over 50. Koningsveld [9] suggests that limits for frequent lifts by 21–44 year old males during the whole day be 39.7lb (18kg) for ideal conditions and 17.6lb (8kg) for aggravating conditions, and for males over 45 years, 33.1lb (15kg) for ideal conditions and 11lb (5kg) for aggravating conditions.

The Ergonomics Training Program by the National-American Wholesale Grocers' Association (NAWGA) [10] provides the following guidelines:

1. Try to eliminate items weighing more than 55lb (25kg).
2. Locate cherry picker so that heavier items can be lifted between knuckle and waist height and lighter items can be lifted between knuckle and shoulder height.
3. Limit weight on upper racks to 45lb (20.4kg).
4. Arrange picks so that light items are at the top of the pallet.
5. Arrange stock so that the heaviest picks are between knuckle and elbow height.

It would appear that research supports a lifting limit of between 30 and 50lb (13.6 and 22.7kg) independent of the load location. In light of the fact that the actual lift frequency is frequently higher than the 1 lift/min used in many of the studies noted earlier, and awkward postures are frequently required in handling material in warehouses, an approximate limit of 30lb (13.6kg) would be appropriate. It is suggested that consideration be given to the establishment of the guidelines shown in Table 27.3 for warehouse operations.

TABLE 27.3 General Lifting Recommendations for Warehouse Operations

| | |
|---|---|
| >50 lb (22.7 kg) | Not allowed |
| 30–50 lb (13.6–22.7 kg) | Implementation of control measures |
| 0–30 lb (0–13.6 kg) | Analysis and implementation of control measures if indicated by NIOSH *Work Practices Guide* or biomechanical analysis |

Note that the information cited earlier does not recognize the fatigue caused by highly repetitive lifting or the additional stress resulting from torso twisting and/or constrained working postures. These lifting guidelines do not relate to the net weight of the contents but to the gross weight of the box. In a sample of items weighed at one facility, the gross weight was approximately 40% higher than the net. In addition, these limits generally assume that the workforce is physically fit and do not recognize individual limitations due to age, physical disability, gender, or other personal factors.

#### *27.3.3.2 NIOSH Work Practices Guide*

Since this analysis was performed in 1989, an analysis based on the 1981 version of the NIOSH *Work Practices Guide for Manual Lifting* (WPG) [11] was performed for the warehousing operation. This version of the NIOSH WPG considers the horizontal distance that the load is held out from the body, the vertical location of the load at the beginning of the lift/lower, the vertical distance that the load is moved, and the frequency of lift/lower. The result of the NIOSH WPG is an action limit (AL) that indicates the load that can be lifted by most people without significant risk of injury and a maximum permissible limit (MPL) that is three times the AL and indicates the load that will be harmful for most people. The NIOSH WPG applies to populations who are physically fit and accustomed to physical labor.

To facilitate the general application of the NIOSH WPG, the grocery and perishable selection operations were synthesized into the tasks shown in Table 27.4. The result of the NIOSH WPG analysis is presented in Table 27.5. This analysis indicates that the movement of objects from the lowest and highest rack levels results in a situation where the load

TABLE 27.4 Synthetic Jobs for NIOSH *Work Practices Guide* Analysis

| |
|---|
| Lift from pallet racks to waist/thigh |
| Horizontal location out from body = 20 in. |
| Vertical location at beginning = 4, 36, 68 in. (10.2, 91.4, 172.7 cm) |
| Vertical location at end = 30 in. (76.2 cm) (objects moved to waist) |
| Frequency = 1.9 cases/min |
| Lift/lower to pallet jack from waist/thigh |
| Horizontal location out from body = 12 in. |
| Vertical location at beginning = 30 in. (76.2 cm) (objects moved from waist) |
| Vertical location at end = 4, 36, 68 in. (10.2, 91.4, 172.7 cm) |
| Frequency = 1.9 cases/min |

TABLE 27.5 Results of 1981 NIOSH *Work Practices Guide* Analysis[a]

| | AL | MPL | Load |
|---|---|---|---|
| Rack (4 in., 10.2 cm) to waist | 14.7 (6.7) | 44.1 (20.0) | 23.9 (10.8) |
| Rack (36 in., 91.4 cm) to waist | 24.4 (11.1) | 73.2 (33.2) | 23.9 (10.8) |
| Rack (68 in., 172.7 cm) to waist | 13.3 (6.0) | 39.9 (18.1) | 23.9 (10.8) |
| Waist to jack (4 in., 10.2 cm) | 37.2 (16.9) | 111.6 (50.6) | 23.9 (10.8) |
| Waist to jack (36 in., 91.4 cm) | 42.7 (19.4) | 128.1 (58.1) | 23.9 (10.8) |
| Waist to jack (68 in., 172.7 cm) | 33.3 (15.1) | 99.9 (45.3) | 23.9 (10.8) |
| Average | 20.9 (9.5) | 62.7 (28.4) | 23.9 (10.8) |

[a] Waist is 30 in. above the floor. Weights are in lb (kg).

exceeds the AL and may be hazardous. The following factors suggest that the actual hazard was higher than that noted earlier:

1. These lifts were often performed in a constrained posture with considerable twisting of the torso.
2. The analysis was performed with the estimated average load. This means that in 50% of the lifts the load was higher than used in the analysis, and these higher loads were often located on the lower levels where the hazard was highest.
3. The loads were carried to the pallet jack in addition to the lifting and lowering included in the task analysis.
4. The population may not be physically fit and accustomed to physical labor.

A revised version of the NIOSH WPG was published in 1991 that recognizes the effect of twisting the torso and the coupling between the hands and the load [12]. The application of the revised WPG results in a recommended weight limit (RWL) at both the beginning and end of the lift/lower, which is similar to the AL in the 1981 NIOSH WPG. The revised guide does not contain an MPL but does include a limit of three times the RWL. This indicates the load that will be harmful for most people.

The revised NIOSH WPG can also be applied to the same tasks noted in Table 27.4 with the assumption that the torso is rotated approximately 60° at the beginning and end of the lift and that the hands are gripping the bottom of the box. The result of the revised NIOSH WPG analysis is presented in Table 27.6.

This analysis indicates that the average load frequently exceeds the RWL and approaches 3 × RWL. This indicates that workers would be at risk during much of the manual materials handling activity.

#### *27.3.3.3 Biomechanical Analysis*

A biomechanical model analysis was run to establish the relationship between load, lift posture, and back stresses [13]. This was done to determine the biomechanical stress and establish acceptable storage locations for loads of different weights and involved the determination of the back compressive force for a 50th percentile male with a height of 70 in. (1.78 m) and a weight of 166 lb (75.3 kg). The results are noted in Table 27.7.

TABLE 27.6 Results of 1991 NIOSH *Work Practices Guide* Analysis Assuming That the Torso Is Rotated 60° and the Boxes Are Lifted from the Bottom, Waist Is 30 in. Above the Floor and Weights Are in lb (kg)

| | RWL | | 3 × RWL | | |
|---|---|---|---|---|---|
| | **Begin** | **End** | **Begin** | **End** | **Load** |
| Rack (4 in., 10.2 cm) to waist | 9.1 (4.1) | 9.6 (4.4) | 27.3 (12.3) | 28.8 (13.2) | 23.9 (10.8) |
| Rack (36 in., 91.4 cm) to waist | 12.8 (5.8) | 13.4 (6.1) | 38.4 (17.4) | 40.2 (18.3) | 23.9 (10.8) |
| Rack (68 in., 172.7 cm) to waist | 8.3 (3.8) | 11.7 (5.3) | 24.9 (11.4) | 35.1 (15.9) | 23.9 (10.8) |
| Waist to jack (4 in., 10.2 cm) | 19.8 (9.0) | 15.1 (6.8) | 59.4 (27.0) | 45.3 (20.4) | 23.9 (10.8) |
| Waist to jack (36 in., 91.4 cm) | 22.3 (10.1) | 21.2 (9.6) | 66.9 (30.3) | 63.6 (28.8) | 23.9 (10.8) |
| Waist to jack (68 in., 172.7 cm) | 19.4 (8.8) | 13.8 (6.3) | 58.2 (26.4) | 41.4 (18.9) | 23.9 (10.8) |
| Average | 15.3 (6.9) | 14.1 (6.4) | 45.9 (20.7) | 42.3 (14.2) | 23.9 (10.8) |

TABLE 27.7 Back Compressive Force as a Function of Load Weight and Location for an Average Male[a]

| | | Back Compressive Force (lb) | | |
|---|---|---|---|---|
| | | Horizontal Location (Inches from Center of Ankles) | | |
| **Vertical from Floor** | **Load (lbs)** | **12** | **20** | **28** |
| 68 | 10 | 184 | 226 | 327 |
| | 20 | 236 | 307 | 416 |
| | 30 | 287 | 383 | 499 |
| | 40 | 335 | 456 | 577 |
| | 50 | 383 | 524 | 649 |
| 52 | 10 | 151 | 215 | 300 |
| | 20 | 206 | 296 | 399 |
| | 30 | 258 | 372 | 491 |
| | 40 | 309 | 445 | 576 |
| | 50 | 359 | 514 | 655 |
| 36 | 10 | 139 | 399 | 485 |
| | 20 | 194 | 486 | 590 |
| | 30 | 248 | 569 | 689 |
| | 40 | 300 | 648 | **778** |
| | 50 | 351 | 723 | **867** |
| 20 | 10 | 433 | 568 | 642 |
| | 20 | 503 | 668 | **773** |
| | 30 | 572 | 766 | **902** |
| | 40 | 640 | **862** | **1029** |
| | 50 | 706 | **956** | **1154** |
| 4 | 10 | 591 | 649 | 679 |
| | 20 | 674 | 763 | **817** |
| | 30 | 757 | **878** | **956** |
| | 40 | **840** | **992** | **1094** |
| | 50 | **924** | **1106** | **1232** |

[a] Numbers in bold represent back compressive forces exceeding the NIOSH *Work Practices Guide* recommendations.

The back compressive forces noted in bold in Table 27.7 are above the NIOSH action level of 770 lb (350 kg). This level of back compressive force is thought to cause back problems in a significant portion of the workforce [11]. The additional factors mentioned in the earlier NIOSH WPG analysis would also suggest that the actual hazard is higher than that noted earlier.

### 27.3.4 Abatement Recommendations

It should be noted that not all recommended controls need necessarily be implemented for each employee in the operations evaluated. It is understood that often more than one operator uses these workstations and often operators have alternate jobs that may also contribute to their problems. The recommendations are not all-inclusive, and a more in-depth study may be required to determine safe operating parameters for particular employees. Other factors that should be included for any program to be effective include employee training and participation in planning, ergonomic abatement committees, supervisor training, hazard analysis, administrative controls, and medical monitoring.

The abatement recommendations relating to this typical grocery warehousing operation were divided into the following categories:

1. Abatement recommendations that require primarily changes in work methods or storage procedure. These abatement recommendations require relatively little capital expenditure but may affect time standards and/or storage capacity.
2. Abatement recommendations that require some capital expenditure for lift tables, hoists, and facility modification.
3. Abatement recommendations that require significant changes in storage facility layout and design.

The feasibility of the proposed recommendations is noted in light of state-of-the-art warehouse practices and examples.

*Abatement Recommendations That Require Primarily Changes in Work Methods or Storage Procedure*: (These abatement recommendations require relatively little capital expenditure for materials handling equipment (MHE) but may affect time standards and/or storage capacity.)

1. Store items weighing more than 30 lb (13.6 kg) between knuckle and shoulder height. The optimum storage location would be between 32.5 in. (82.6 cm), which is the approximate knuckle height of a 95th percentile male, and 51.8 in. (131.6 cm), which is the approximate shoulder height of a 5th percentile male. This can be accomplished by raising the height of the pallet rack “floor” and limiting the material on each pallet. Since this is a lower (as opposed to a lift), it may be appropriate to increase the upper height somewhat. The feasibility of this recommendation is also supported by the

Ergonomics Training Program of NAWGA, where it is noted that heavier items should be lifted between knuckle and waist height and lighter items between knuckle and shoulder height [10].

2. Store only light, nonbreakable items on the pallet racks above shoulder height. Be sure that hooks are provided at every pallet location to facilitate selection of items without excessive reaches.

3. Limit the pallet loads so that loads greater than 30 lb (13.6 kg) are stored, or moved to, below shoulder height. The placement of even lighter loads above shoulder height should also be minimized if possible. Also investigate the feasibility (and compliance with regulations) of maintaining the pallet on the pallet jack at a level that minimizes bending while putting the cases or objects on the pallet jack. OSHA regulations relating to transporting a pallet in the raised position must be clarified. The feasibility of this recommendation is supported by the Ergonomics Training Program of NAWGA, where it is noted that heaviest picks should be between knuckle and elbow height and picks should be arranged so that light items are at the top of the pallet [10].

4. Limit the item weight to 50 lb (22.7 kg). The feasibility of this recommendation is supported by the Ergonomics Training Program of NAWGA, where it is noted that items weighing more than 55 lb (24.9 kg) should be eliminated if possible [10].

5. Limit the use of the pallet racks with a 45 in. height to very slow-moving items. The lifting of items from these pallets requires a constrained posture that increases the musculoskeletal risk. The feasibility of this is clear. It is simply a change in administrative procedure.

6. Improve access to pallets by facilitating access to the pallet racks from three sides for items weighing more than 30 lb (13.6 kg) to decrease the horizontal distance. The pallet racks are generally 94 in. (2.39 m) wide. The pallets are generally 40 × 48 in. (1.02 × 1.22 m). One way to facilitate access to three sides of the pallet is to place only one pallet "sideways" with the 48 in. (1.22 m) side toward the aisle, instead of two pallets, in each bay for fast-moving and/or heavy items. If placed in this fashion, there would be approximately 23 in. (58 cm) on each side of the pallet to allow access of items on the sides and back of the pallet with a minimum of reaching or bending. This is just above the 22 in. (56 cm) found by Ayoub and Mital [14] to cause a 13% decline in carrying capabilities. Alternatively, the pallet could be placed in the "normal" way [with the 40 in. (1.02 m) side toward the aisle] with approximately 27 in. (68.6 cm) allowed on each side to access items. The feasibility of this is clear. It is simply a change in procedure. It might, however, decrease facility storage capacity. In addition, the feasibility of this recommendation is supported by the Ergonomics Training Program of NAWGA [10], where it is noted that far reaches may be avoided by placing the pallets with the long side on the aisle.

7. Improve access to pallets by facilitating access to the pallet racks from two sides for items in excess of 30 lb (11.8 kg). As noted earlier, the pallet racks are generally 94 in.

(2.39 m) wide and the pallets are generally 40 × 48 in. (1.02 × 1.22 m). One way to facilitate access to two sides of the pallet is to place the two pallets as far apart as possible within each bay. Although this is not optimum, it would allow 14 in. (35.6 cm) between the pallets to increase the accessibility of items on the sides and back of the pallet. The feasibility of this is clear. It is simply a change in procedure.

8. Specify that suppliers stack boxes on pallets in a "noninterlocking" fashion whenever possible. Although interlocking increases pallet stability, it increases the reaches if selectors reach to the rear of the pallet and increases the force required if they remove an interlocked pallet in the front. The feasibility of this is clear. It is simply a change in procedure. This may require a decrease in the number of items stored on a single pallet.
9. Require that pallets with heavy items be rotated to facilitate access. The feasibility of this recommendation is supported by the Ergonomics Training Program of NAWGA [10], where it is noted that far reaches may be avoided by rotating half-empty pallets of heaviest products 180°.
10. Sequence the pickup of items to ensure that moderate, sturdy objects are loaded on the bottom of the pallet jacks, heavier in the middle, and lighter on the top. This would minimize the musculoskeletal stress and maintain the stability of the load. The feasibility of this is clear. It is simply a change in procedure.
11. Reduce the importance of incentive-based pay systems, which encourage workers to work at excessive rates.

*Abatement Recommendations That Require Some Capital Expenditure for Lift Tables, Hoists, and Facility Modification*:

1. Redesign racks to allow the placement of two pallets in each bay with a minimum of 20.1 in. (51 cm) separation in the center. (This would allow the shoulder width of a 95th percentile male.) This would require a minimum of 100 in. (2.54 m) bay width in the pallet racks [40 in. (1.02 m) pallet + 20 in. (50.8 cm) gap + 40 in. (1.02 m) pallet = 100 in. (2.54 m)]. The feasibility of this is clear. It would, however, require changes in some pallet rack construction and would slightly decrease overall facility storage capacity.
2. Investigate the feasibility of using pullout or rolling platforms on which the pallets holding items weighing over 30 lb (13.6 kg) can be placed. This would allow the heaviest objects to be easily accessed without twisting and lateral bending. The feasibility of this approach is supported by the variety of devices available for this purpose. Care must be taken to ensure the pull forces are not stressful.
3. Place pallets in scissors lifts or lift tables to keep the lift start height at approximately knuckle height. The best abatement would be the combination of the lift with a turntable to allow the load to be lifted and rotated to obtain the optimum vertical location and keep the horizontal distance to a minimum. The feasibility of this

approach is supported by the variety of devices available for this purpose. It would, however, require widening of pallet racks and would slightly decrease overall facility storage capacity.

4. Provide small crane-type lift devices where required for heaviest loads. Optimally, this would be for loads of over 30 lb (13.6 kg) but would be needed for loads greater than 50 lb (22.7 kg). The feasibility of this approach is supported by the variety of devices available for this purpose. These devices can be attached to the pallet jack, attached to the pallet rack, or suspended on rails above the aisle and include powered and manual devices.

*Abatement Recommendations That Require Significant Changes in Storage Layout and Design*:

1. Install a conveyor between the pallet racks to transport the selected items to the palletizing/shipping area. The feasibility of this approach is documented by Johnson, *Material Handling Engineering Magazine*, and *Modern Materials Handling* [15–18]. An increase in efficiency is noted for this type of system is discussed in the *Material Handling Engineering Yearbook* [19].
2. Completely automate the selection and transportation of items from the pallet racks (or similar storage system) to the shipping area. This is obviously a very expensive alternative. Johnson notes that a "fully automated" warehouse may not be the best alternative for a grocery warehouse [15]. It has, however, been accomplished to some extent in at least one grocery warehouse [20].
3. One possible materials assist device is a robot palletizer [21]. This device would assist in palletizing items that have been loaded on a conveyor as discussed in some of the abatements noted earlier.

## 27.4 CRITICAL REVIEW OF CURRENT STATUS

In the facility reviewed earlier, the following controls were implemented [22]:

- Review and upgrade of lifting training for new hires and refresher training for employees who lift on a regular basis.
- Daily 10 min warm-up program for all selectors.
- Strength and endurance testing for all new hires.
- Consideration of ergonomic design factors when replacing existing material handling equipment.
- Consultation with an ergonomic expert when performing the preceding and other feasible, beneficial ergonomic controls.
- The company indicated that these controls were effective.

In another grocery warehouse, the following ergonomic abatements were implemented [23]:

- Train selectors and supervisors in back injury prevention.
- Annually review potential manual materials handling risks and lifting techniques.
- Move heavier items from top racks to lower racks.
- Use hooks to pull items closer to the body before lifting.
- Limit stack height on pallet jacks to the approximate shoulder height of an average-size person.
- Modify work procedures so that selectors need not reach past the centerline of the pallet.
- Require selectors to work at a steady pace through the shift instead of finishing an 8 h workload in fewer hours.
- Work with suppliers to reduce the weight of product containers.

After these abatements were implemented, the OSHA recordable back injuries decreased from 19 cases with 378 lost days in 1990 to 13 cases with 80 lost days in 1991.

## 27.5 FUTURE CONCERNS

It is important that ergonomic controls in warehousing operations focus on the design/redesign of the facility. Selection, training, and conditioning of applicants or workers may be part of an effective program, but primary consideration must be given to the reduction/elimination of ergonomic hazards. Of particular concern is the attractiveness of back-belts as an easy, inexpensive "solution." There is little support for the extensive use of back-belts as an effective abatement measure [24,25].

It is also important that warehouse operators realize that it is not uncommon for there to be an increase in the ergonomic incidence rate on the OSHA 200 log after the implementation of an ergonomics program. This indicates that workers are reporting early symptoms before they become serious and require costly medical care. There is also generally a corresponding decrease in the *cost* of ergonomics-related incidents because the early intervention is less expensive.

## REFERENCES

1. A. Mital, Comprehensive maximum acceptable weights of lift database for regular 8-hour work shifts, *Ergonomics 27*(11):1127–1138 (1984).
2. A. Mital, Maximum weights of lift acceptable to male and female industrial workers for extended work shifts, *Ergonomics 27*(11):1115–1126 (1984).
3. A. Mital, Task variables in manual material handling, *J. Safety Res. 12*(4):163–173 (1980).
4. American Industrial Hygiene Association, *Ergonomics Guide to Manual Lifting*, Ergonomics Guide Series, July–August 1970.
5. D. B. Chaffin and K. S. Park, A longitudinal study of low-back pain as associated with occupational weight lifting factors, *Am. Ind. Assoc. J. 34*(12):513–525 (December 1973).

6. G. D. Herrin, *A Taxonomy of Manual Materials Handling Hazards,* Report on International Symposium: Safety in Manual Materials Handling, SUNY Buffalo, July 18–20, 1976, pp. 6–15.
7. S. H. Snook, The design of manual handling tasks, *Ergonomics 21*(12):963–985 (1978).
8. International Labor Office, *Manual Lifting and Carrying,* Inf. Sheet No. 3, March 1962.
9. E. A. P. Koningsveld, Permissible loads for the Dutch construction industry, *Ergonomics 28*(1):359–363 (1985).
10. National American Wholesale Grocers' Association, *The Backbone of the Food Industry,* NAWGA, Falls Church, VA, 1989.
11. NIOSH, *NIOSH Work Practices Guide for Manual Lifting,* NIOSH Pub. 81-122, Printed by AIHA, 1983, 1987.
12. T. R. Waters, V. Putz-Anderson, A. Garg, and L. Fine, Revised NIOSH equation for the design and evaluation of manual lifting tasks, *Ergonomics 36*(7):749–776 (1993).
13. Two Dimensional Static Strength Model, Univ. Michigan, Ann Arbor, MI, 1989.
14. M. M. Ayoub and A. Mital, *Manual Materials Handling,* Taylor & Francis, London, U.K., 1989, p. 49.
15. J. P. Johnson, "Fully automated" isn't always the best solution to your warehouse operations, *Ind. Eng. 22*(2):30–33 (1990).
16. *Material Handling Engineering,* Twenty-six miles of palletflo live storage (Advertisement), *41*(10):14 (1986).
17. G. F. Schwind, Warehousing 87: Redefining the role, *Mater. Handl. Eng. 42*(3):57–65 (1987).
18. L. Beck, What's available for moving loads between levels? *Modern Mater. Handl. 41*(15):68–70 (1986).
19. Case studies: Distributor increases productivity with conveyor system, *Mater. Handl. Eng. Handb. Direct. 41*(13):122–123 (1989).
20. C. E. Witt, Automatic identification highlights distribution system, *Mater. Handl. Eng. 43*(10):49–52 (1988).
21. C. E. Witt, Electrolux: A lesson for American manufacturers, *Mater. Handl. Eng. 43*(1):55–61 (1988).
22. Settlement Agreement, Secretary of Labor v. First National Supermarkets, Inc., OSHRC Docket No. 89-2417, May 14, 1991.
23. Consent Agreement, Secretary of Labor v. Weis Markets, Inc., OSHRC Docket No. 89-3139, July 1991.
24. C. R. Reddell, J. J. Congleton, R. D. Huchingson, and J. F. Montgomery, An evaluation of a weightlifting belt and back injury prevention training class for airline baggage handlers, *Appl. Ergon. 23*(5):319–329 (1993).
25. S. Steers, To belt or not to belt? *Safety Health* 40–41 (November 1993).

# V

# Physical Agents in Workplace

CHAPTER 28

# Occupational Heat Stress

Thomas E. Bernard

CONTENTS



## 28.1 INTRODUCTION

Heat stress is the combined effect of climatic conditions (called the environment for simplicity), metabolic demands, and clothing. The obvious environmental factors are air temperature, humidity, radiant heat, and air movement. Metabolic demands are associated with the performance of external work, and the concern in heat stress is the rate of heat generated by muscle metabolism. Clothing material, construction, and usage affect the exchange of heat between the body and the environment.

Heat strain includes changes in body temperature, heart rate, and sweating. During exposures to heat stress, the temperature of most body compartments increases, and these changes facilitate the removal of heat from the body. Core temperature is an important measure of heat strain. Blood flow moves heat from the deep body tissues, including working muscles where it is generated, to the skin, where it is dissipated. Heart rate reflects the cardiovascular response to blood flow requirements and is a sensitive indicator of physiological strain due to both work demands and heat stress. Finally, sweating ensues to provide for evaporative cooling of the body.

In summary, as muscles work, they metabolize substrates to supply energy, which generate heat. The rate of blood flow through the muscles is in proportion to the metabolic demands. The heat generated in the muscles is distributed to the rest of the body by the blood, and body core temperature increases. The heat in the core compartment is moved to the skin by additional blood flow that is directly proportional to the metabolic rate and inversely proportional to the difference between the core and skin temperatures. There may be an additional heat load due to hot air and surfaces, and this gain is influenced by air temperature, humidity, and motion, by the temperature of the solid surroundings, and the clothing worn influences the rate. Heat is removed from the body primarily by evaporation of sweat, and the rate of evaporation is influenced by the ambient humidity and air motion as well as by the type and configuration of the clothing.

The evaluation of heat stress by the application of heat stress models and heat strain criteria are the primary emphasis of this chapter, which also provides a framework for viewing the principles of recognition and control. The goals of this chapter are to describe a basic model of heat stress, outline heat strain in response to the stress, describe ways that heat stress may be recognized in the workplace, discuss two conventional methods of heat stress assessment in industrial settings, highlight alternative assessment methods based on physiological response, and present an overall framework for the management of heat stress.

## 28.2 BACKGROUND AND SIGNIFICANCE TO OCCUPATIONAL ERGONOMICS

Briefly, ergonomics is the study of fitting a task to enhance performance and protect well-being. The first order of concern in occupational heat stress is well-being. Heat stress is recognized as a workplace hazard with heat stroke as a particular concern. In U.S. workplaces, there are about 28 heat-related fatalities per year [1], with crop workers among those at greatest risk.

Heat fatalities are not the only concern. Heat stress can cause other heat-related disorders as well as acute injuries. Ramsey et al. [2] demonstrated that the rate of unsafe acts increased with light and moderate metabolic rates above 23°C-wet bulb globe temperature (WBGT). At the Threshold Limit Value® (TLV®) for metabolic rate of 250 W (29°C-WBGT), the predicted increase in unsafe acts was 60% over the baseline. Fogleman et al. [3] reported a significant increase in acute injury (OR = 2.3) in an aluminum smelter as the ambient heat index increased above 32°C. From a subset of the published data, Bernard and Fogleman examined 174 acute outcomes related to heat stress [unpublished analysis]. Over 3 years, about 70% of the time the workplace exposures were below the TLV (none—reference level), another 20% of the time was between 0°C and 3°C-WBGT above the TLV (low), and the remaining 10% of

TABLE 28.1 Odds Ratios (95% Confidence Interval) for Heat-Related Disorders and Acute Injuries at Moderate and High Heat Stress levels

| Heat Stress Level (Reference Level is below TLV) | Heat-Related Disorders | Acute Injury | Acute Musculoskeletal Disorder |
|---|---|---|---|
| Low (0°C–3°C-WBGT above TLV) | 25 (3.3–202) | 1.4 (0.9–2.2) | 1.8 (1.1–2.9) |
| High (>3°C-WBGT above TLV) | 158 (21 to >1000) | 1.7 (1.0–2.9) | 2.4 (1.4–4.3) |

the time was greater than 3°C-WBGT above the TLV (high). The odds ratios and associated confidence intervals for heat-related disorders as well acute injuries and musculoskeletal disorders are provided in Table 28.1. The heat-related reports showed steep increases in risk from none (reference) to low to high heat stress levels. There were only two instances of heat-related injuries in the reference category, with a substantial increase in the rates in the low and high categories of heat stress. These were accompanied by significantly elevated odds ratios in the low and high categories. Although these results have substantial face validity, the odds ratios should be interpreted with caution due to the very low number of cases in the reference (none) category. There was a steady increase in the rate of acute injuries and musculoskeletal disorders with significantly elevated odds ratios for the high level for both and for the low exposure category for musculoskeletal disorders. It is worth noting that there was a link found between the risk of fatal gastrointestinal disorders and lifetime exposure to heat stress above the TLV [4].

By managing the stress or strain, performance is enhanced and the risk of disorders is controlled within acceptable limits. To fit a task requires either (1) a model for the stressors that reasonably predicts the resulting strain or risk of disorders or (2) a method of evaluating the strain. Heat stress becomes interesting in the application of occupational ergonomics when the factors of environment, work, and clothing come together to influence an individual's capacity for work and risk of suffering heat-related disorders. Occupational heat stress is well enough understood to use either stress models or strain evaluation for design of work.

## 28.3 OCCUPATIONAL HEAT STRESS

The management of occupational heat stress requires an understanding of the factors of heat stress and how it may affect a worker. The three job risk factors are environment, metabolic rate, and clothing, which combine to dictate the rate of heat gain and the amount of heat loss required to protect the worker.

Management of heat stress is directed to two fronts: general and job-specific controls. General controls include training, heat stress hygiene practices, surveillance, and an emergency response plan. The job-specific controls are directed toward containing or managing the risk of specific exposures.

### 28.3.1 General Model of Heat Stress

A description of heat exchange between the body and the environment is frequently used to qualitatively and quantitatively express the factors that affect the level of heat stress. The heat balance equation may be reported in different ways, but they have the same fundamental structure [5–9]:

$$S = (M - W) + C + R + K + (C - E)_{resp} - E \tag{28.1}$$

Each term in the equation represents a rate of heat flow for a different modality. By convention for *C*, *R*, and *K*, positive values mean that the flow is toward the body, and negative is away from the body. In other words, a positive value means that there is a tendency for body temperature to increase, and a negative value means that there is a tendency for it to decrease. *M* is always positive and a heat gain; *W* and *E* are always positive and a heat loss, hence the minus sign that precedes them. The sign convention for Equation 28.1 has varied over the years and among sources. Sometimes the heat balance equation is reported with a plus sign in front of *E* and *W* to agree with a more rigorous sign convention that the value (either positive or negative) represents the direction (i.e., gain or loss) and not the sign in front of the symbol.

*S* is the rate of heat storage in the body. As heat is stored, the temperature of the body increases. Over the course of a day, there is no net storage of heat, but over shorter periods of time there will be increases due to heat stress exposures. Under extreme conditions, the rate of storage may reach 500 W.

$M - W$ is the net heat gain due to metabolic rate (*M*) less the rate of external work accomplished (*W*). Because external work is only about 10% of the total metabolic rate [10], *W* is usually ignored. *M* is the internal source of heat during heat stress and is always positive and proportional to work rate. When no work is being performed, the internal heat generation is about 100 W, but this can easily increase to 500 W during heavy work.

*C* is the convective heat flow between the body surface and the surrounding air. It is directly proportional to the temperature gradient between the skin and air and to air speed. Clothing modifies the proportionality constant. *C* is positive when air temperature is greater than mean skin temperature (nominally 36°C in heat stress) and negative with lower air temperature. Whether the convective heat transfer is positive or negative, the value does not often exceed 50 W.

*R* is the radiant (infrared) heat flow between the body surface and the solid surroundings. It is proportional to the temperature gradient between the skin and the average surface temperature of the surroundings. Like convection, clothing modifies the proportionality constant. *R* is positive when the average temperature of the surroundings is greater than skin temperature. The value of radiant heat transfer does not often exceed 50 W.

*E* is the rate of heat loss from the evaporation of sweat. The skin is wetted by water from the sweat glands, and as the water evaporates the energy required for vaporization is taken from the skin. In this way, the body is cooled. The value of *E* in Equation 28.1 is always positive and with the minus sign represents a heat loss. Evaporative cooling is generally under physiological control and is adjusted to minimize storage. *E* can be limited by the environment and clothing as well as by physiological capability. Looking at the environment, the maximum rate of cooling is proportional to air speed (up to 2 m/s) and to the difference in water vapor pressure on the skin and in the air (humidity). Clothing modifies the proportionality constant. When the environmental and clothing conditions are favorable for evaporative cooling, the rate of cooling may be physiological control and can range from 0 to a physiological limit of 600 W.

*K* represents heat conduction from direct contact between the body and a solid surface (with the possibility of intervening clothing). Large contact surface areas are required before conduction becomes an important avenue of heat gain or loss in occupational evaluations of heat stress. Significant heat flow rates over smaller surface areas are more likely to affect local tissue by inducing extreme discomfort, pain, or a burn [9,11,12]. Conduction as a factor in heat stress will not be considered further in this chapter.

$(C + E)_{resp}$ represents the heat exchange in the respiratory track due to convection and evaporation. Although these paths are important under sedentary conditions, they represent a minor pathway of heat exchange for occupational heat stress. For the sake of discussion, this term will be ignored.

Equation 28.1 is reduced to the form

$$S = M + C + R - E \tag{28.2}$$

That is, the rate of heat storage is equal to the internal heat generation plus the gains or losses from convection and radiation less the loss due to sweat evaporation.

The heat balance expressed by either Equation 28.1 or 28.2 is a comprehensive means of describing the level of heat stress because the environment, work, and clothing factors can be accounted for. Recall that the major source of heat gain is the metabolic heat generated by performing the work and that the major avenue of heat loss is evaporation. Although radiation and convection are important, they have a much smaller contribution.

Ideally, heat stress is manageable if there is no storage of heat in the body ($S = 0$). In this case, the required level of evaporative cooling ($E_{req}$), which is under physiological control, is determined from the equation

$$E_{req} = M + C + R \tag{28.3}$$

Sometimes the value of $E_{req}$ is greater than the maximum evaporative cooling that can be sustained physiologically or by the combination of clothing and environment. In this case, there is a net gain in heat (positive value for storage rate), and an acceptable increase in core temperature is usually specified ($\Delta T_{re,max}$). The maximum acceptable amount of heat storage ($H_{max}$) then follows from the equation

$$H_{max}[\text{W}\bullet\text{h}] = 0.75 \times \text{body weight [kg]} \times \Delta T_{re,max}[^{\circ}\text{C}] \tag{28.4}$$

Then the safe exposure time ($t_{max}$) can be estimated from the storage rate and the maximum allowable increase in heat storage:

$$t_{max}\,[\text{h}] = H_{max}/S \tag{28.5}$$

For example, if $S = 200$ W and $H_{max} = 100$ W • h, then $t_{max}$ would be 0.5 h (30 min). This is the evaluation principle behind time-limited heat stress exposures based on rational heat stress models. The greater the rate of storage, the shorter is the safe exposure time for the same storage limit.

### 28.3.2 Description of Heat Strain

Physiological strain comprises the body's responses to heat stress. One response is an increase in body temperature as metabolic rate increases and heat is moved. Another is the cardiovascular response that is necessary to support both the metabolic demands and the demand to move heat from the core to the skin. The sweating response is necessary to support evaporative cooling.

Starting with the generation of internal heat from metabolism, there is an initial tendency to increase body core temperature at the onset of work, even in thermally neutral environments. A leveling off once the necessary adjustments are completed follows this. The amount of storage and the related core temperature depend primarily on the individual and the relative level of metabolic rate. The controlled level of storage is marked by the work-specific core temperature ($T_{\text{core-work}}$). A simple relationship that describes this increase is given by the following equation [13]:

$$T_{\text{core-work}}[^{\circ}\text{C}] = 36.5 + 3.0\, f V_{\text{O}_2,\text{max}} \quad (28.6)$$

where $f V_{\text{O}_2,\text{max}}$ is the fraction of the individual's maximum aerobic capacity that the work metabolic rate represents. For a given work demand, $T_{\text{core-work}}$ will increase with decreasing fitness levels of the workers, or for a given worker $T_{\text{core-work}}$ will increase with increasing demands.

Equation 28.6 is predictive of $T_{\text{core-work}}$ for a wide range of environments. If the environmental conditions begin to limit the ability of the physiological systems to remove the metabolic heat, then $T_{\text{core}}$ will increase out of proportion to the work demands [14,15]. Evaluation schemes for long-term exposures to heat stress attempt to limit the increase in core temperature to roughly that dictated by the work, that is, without an environmental contribution to heat storage [16].

So core temperature can be assessed as a measure of physiological strain in response to heat stress. The World Health Organization (WHO) suggested that long-term exposures to heat stress should be limited in such a way that core temperature does not exceed 38°C [17]. However, WHO recognized that there can be transient responses greater than 38°C that are safe for intermittent periods. For controlled and well-monitored exposures, 39°C is acceptable.

Blood flow from the core tissues to the periphery is responsible for moving the metabolic heat to the skin, where it can be dissipated to the environment. As a normal response to the metabolic demands of work, the cardiovascular system adjusts blood flow to the working muscles in proportion to the work. In addition, cardiac output must increase to meet the needs of heat transport from the deep body tissues to the skin, where the heat can be dissipated from the body. Under some circumstances, the gradient between the core and skin temperatures decreases. When this happens, more blood must be delivered to the skin to maintain the rate of heat transport. As an index of cardiovascular response to work and heat stress, heart rate is a valuable measure [18].

Several guideposts are available to judge whether heart rates are excessive. One is a daily average below 110 beats per minute (bpm) [17]. A second threshold on heart rate is a sustained rate (about 1 min) greater than 90% of the maximum heart rate [19]. (A popular rule of thumb for estimating maximum heart rate is 220—age.) A third guidepost is recovery heart rate [20,21]. The heart rate at 1 min of recovery ($HR_{rec,min}$) should be less than 110 bpm to avoid excessive cardiovascular strain [22].

Sweating is the remaining component of heat strain to be discussed in this chapter. Sweat rate is adjusted physiologically to a level necessary to support the required amount of evaporative cooling. The rate of sweating is limited either by the environment or by physiological capacity. Acclimatization to heat stress is a physiological adaptation to repeated heat stress exposures that increase the individual's capacity to secrete sweat and thus increases the rate of evaporative cooling [23]. It has the beneficial effects of lowering the levels of core temperature and heart rate for the same level of heat stress.

In terms of physiological strain, sweating is important as a potential cause of dehydration. Sweat losses on the order of 1 L/h or a total of 6 L in a day are limits within which most workers can replace the water lost to sweat. If the losses exceed these values or if there is not adequate water replacement, then dehydration may occur. Marked changes in physical work capacity are noted when dehydration exceeds 1.5% of body weight [24].

If the physiological strain, which is a normal response to heat stress, becomes excessive for an individual, then that individual has an increased risk for a heat-related disorder. Major heat-related disorders are described in the following list along with their symptoms and signs. The first three are a progression of events from a simple, subclinical dehydration to a serious medical emergency. The other three are common disorders associated with heat stress.

1. *Dehydration*: It is an excessive loss of water due to either inadequate replacement or an illness.

   a. *Symptoms*: no early symptoms (<1.5%); fatigue/weakness; dry mouth

   b. *Signs*: loss of work capacity; increased response time

   The signs and symptoms start when dehydration exceeds 1.5% of body weight, and they become more severe as the dehydration increases. Dehydration over 5% is incapacitating. The most harmful consequence of dehydration is the increased risk for heat exhaustion and heat stroke.

2. *Heat exhaustion*: It is marked by lowered work capacity and inability to continue. It is due to a reduced cardiovascular capacity associated with decreased systemic pressure associated with vasodilation and/or advanced dehydration.

   a. *Symptoms*: fatigue; weakness; blurred vision; dizziness and/or headache

   b. *Signs*: high pulse rate; profuse sweating; low blood pressure; *insecure* gait; pale face; collapse; normal or slightly elevated body temperature

3. *Heat stroke*: It is a medical emergency. It is the result of a breakdown in the thermoregulatory center that leads to very high core temperatures and subsequent effects on a broad range of organ systems. The signs and symptoms focus on failures of the central nervous system. The usual causes are subnormal tolerance, lack of acclimatization, excessive exposure to heat stress, or drug or alcohol abuse.
   a. *Symptoms*: chills; restlessness; irritability
   b. *Signs*: euphoria; disorientation; erratic *behavior*; shivering; collapse; unconsciousness; convulsions; body temperature ≥40°C
4. *Heat syncope*: It is faintness experienced when there is a significant drop in blood pressure, usually occurring when there is a sudden change in posture to an upright position or after prolonged maintenance of an upright posture. Under these circumstances, blood pools in the lower extremities and in the skin, where the blood vessels are dilated due to heat stress. This reduces blood flow to the head.
   a. *Symptoms*: blurred vision (gray-out); fainting (brief) (blackout)
   b. *Signs*: ear-fainting behavior; brief fainting; normal temperature
5. *Heat cramps*: These are muscle cramps associated with work under conditions of severe heat stress. They are sometimes attributed to electrolyte imbalances. The cramps will occur in fatigued muscle or the abdominal muscles, and the cramping may occur during work or after hours.
6. *Heat rash*: Prolonged, uninterrupted sweating along with inadequate cleaning of the skin may lead to an inflammation of the sweat gland ducts, thus reducing the amount of sweat secretion. The result is a skin rash marked by small red eruptions and itching sensations.

### 28.3.3 Recognition of Heat Stress in the Workplace

Recognition is the first step in the management of industrial hazards, and heat stress is not an exception. There are several perspectives from which someone may recognize that heat stress is an important factor in the performance of work. These perspectives include contributors in the workplace, worker behaviors, medical surveillance, and physiological sampling. Many times recognition is a qualitative judgment, and that is the way it is treated here. If there is reason to believe that sufficient heat stress is present to be a hazard, then evaluation is required.

The first perspective is recognition of the contributing factors to heat stress—hot environments, high work demands, or protective clothing. These are the very factors that are considered in the quantitative evaluation of heat stress. Generally, if the environment is commonly recognized as being hot, heat stress is likely. Sometimes the role of metabolic rate in heat stress is underestimated. If the metabolic rate is moderate or high, heat stress may be present in environments that are judged comfortable by less active people. Clothing is also an important factor. Most people recognize that vapor-barrier clothing may cause significant heat stress even in cool environments. To a lesser degree, other kinds of clothing will reduce the rate of evaporative cooling and thus increase the level of heat stress. Any time clothing is worn that is not constructed of lightweight cloth, the potential for heat stress should be considered. A clear sign that heat exposures are significant is when the clothing is soaked with sweat. Malchaire et al. [25] offered a structured approach to the qualitative exposure assessment.

Worker behaviors provide a second perspective on recognizing heat stress. These may be observed during a walk-around or through discussions with employees and supervisors. One behavior is to modify clothing materials or the way clothing is worn to enhance evaporative cooling, and another is to seek shade or other cooler locations. A second group of employee behaviors is directed toward reducing the metabolic rate by increasing the number of breaks, slowing the pace of work, not following safe work practices, and neglecting maintenance. Other features of employee behavior include increased irritability and absenteeism and decreased morale during periods of heat stress exposure. Sluggish decision making, increased number of errors, and decreased quality of work are further indicators of heat-induced behavioral changes.

Medical surveillance is a useful recognition tool. Assuming for the moment that heat-related disorders are not being clinically diagnosed, there are patterns of complaints that may be related to overexposure to heat stress. Looking through first aid or clinic logs or reviewing the types of complaints that employees present, the following may be seen: fatigue, faintness, nausea, headache, skin rashes, and muscle cramps. If body (e.g., oral) temperatures are taken routinely, they may be elevated, and, if asked, employees may report particularly concentrated urine (this is due to dehydration). There may also be a pattern of increased accidents during periods of heat stress exposure [2,3,26].

In the face of indicators mentioned in the preceding paragraphs, physiological sampling may provide the final and definitive clues that heat stress is at a level that justifies an evaluation. Remembering that physiological strain associated with heat stress includes increases in sweating, body temperature, and heart rate, these responses can be another recognition tool. Monitoring sweat loss can be troublesome, but an indicator of dehydration is the change in body weight between the beginning and end of a shift. If the weight loss exceeds 1.5% of total body weight measured at the beginning, then significant dehydration from sweating is likely [24]. To assess the potential for excessive body temperature, a reading of oral temperature or other acceptable surrogate for core temperature may be taken at the end of a work cycle. If the temperature corrected for differences from core temperature exceeds 38°C, then the heat stress exposures require evaluation and control [27]. At the same time work is interrupted to take a worker's temperature, recovery heart rate can be noted. This is accomplished by asking the worker to sit down at the end of the work cycle, starting the temperature measurement, and assessing the heart rate after about 1 min [21]. If the recovery heart rate ($HR_{rec,min}$) is less than 110 bpm, then the cardiovascular response is not excessive [22]. If it is greater than 110, and especially if it is greater than 120, then heat stress may affect the well-being of the individual [28].

### 28.3.4 Evaluation of Heat Stress

The evaluation of heat stress requires a characterization of the environment along with an estimate of the metabolic rate and some consideration of clothing. After outlining important environmental measures, two evaluation schemes will be presented. One scheme is empirical, and the other is rational. In both cases, the emphasis is on an assessment of heat stress as a predictor of physiological or heat strain.

#### *28.3.4.1 Environmental Measures*

The environmental contributions to heat stress are convection and radiation as well as limits on evaporative cooling. Environmental measures, therefore, need to encompass these contributors in some fashion. Common measures of the environment are described in this section [29].

Dry bulb temperature ($T_{db}$) is the measure of air temperature using a sensor freely suspended in the air. (A sensor with a significant profile like a bulb should be shielded from sources of radiant heat.) The primary value of $T_{db}$ is to help estimate the direction and magnitude of convective heat exchange. Alone, it is a very weak indicator of heat stress.

Psychrometric wet bulb temperature ($T_{pwb}$) is the value read from a temperature sensor that is covered by a wetted wick over which air is forced at a speed greater than 3 m/s. This temperature is a function of the air temperature decreased in proportion to the amount of evaporative cooling from the wick, which decreases with increasing water vapor pressure (absolute humidity) in the ambient air. In other words, as the humidity increases, $T_{pwb}$ approaches $T_{db}$. The combination of $T_{db}$ and $T_{pwb}$ is used to determine the level of humidity in the air. The water vapor pressure is a critical environmental parameter because it affects the maximum amount of evaporative cooling that can be supported by the environment.

Natural wet bulb temperature ($T_{nwb}$) is similar to $T_{pwb}$ except that the air motion over the wick is simply the ambient air motion; the air is not forced over the wick by a fan or other means. Because the wick is exposed to and the evaporation rate depends on the ambient conditions, $T_{nwb}$ is a good index of the ability of the environment to support evaporative cooling.

Globe temperature ($T_g$) is measured by a temperature sensor in the middle of a black-painted copper sphere. (The sphere was originally 6 in. in diameter, but new instruments use a smaller one.) It is used to estimate the average temperature of the solid surroundings for estimation of the radiant heat exchange. Because the globe is in intimate contact with the surrounding air, $T_g$ also responds to convective heat flow between the globe and the air. Empirically, $T_g$ is a good index of sensible heat exchange (combined effects of convection and radiation).

Air speed ($V_{air}$) is a measure of the air movement over the body. It affects the rate of convective heat flow as well as the rate of evaporative cooling. It is used to compute these avenues of heat exchange.

The WBGT index was originally designed for heat stress interventions during military training [30]. WBGT is affected by both sensible heat exchange and the ability of the environment to support evaporative cooling. It is the principal index for empirical methods of heat stress evaluation. Under most conditions,

$$\text{WBGT} = 0.7T_{\text{nwb}} + 0.3T_{\text{g}} \quad (28.7)$$

When the workplace is in direct sunlight,

$$\text{WBGT}_{\text{sunshine}} = 0.7T_{\text{nwb}} + 0.2T_{\text{g}} + 0.1T_{\text{db}} \quad (28.8)$$

### *28.3.4.2 Empirical Scheme for Heat Stress Evaluation*

A widely accepted method to assess the presence of heat stress in the workplace was first proposed by the ACGIH® in 1973 and the National Institute for Occupational Safety and Health (NIOSH) in 1986 [27]. The goal of this evaluation scheme was to limit body core temperature to 38°C using hourly time-weighted averages (TWAs) that could be sustained for long periods.

WBGT was selected as the index of the thermal environment because of its simplicity, ruggedness, and predictive value. Investigators then looked for a protective limit on WBGT at various levels of metabolic rate. The protective limit was a value that would limit the increase in core temperature to 38°C or less for about 95% of workers. The threshold then was a map of limiting WBGTs as a function of metabolic rate. Because acclimatization state affects the ability to physiologically control heat stress, separate exposure limits were developed for unacclimatized and acclimatized workers. The current ACGIH TLV calls the limit for unacclimatized workers the action limit because it marks the point at which heat stress controls should be implemented. The rationale for WBGT-based occupational exposure limits is described in the appendix.

The ACGIH TLV exposure limits are illustrated in Figure 28.1 [31]. In essence, the TLV line represents the points at which thermal equilibrium can be maintained at or below 38°C core temperature by most acclimatized people. Notice that as the rate of internal heat generation from energy metabolism increases, the environmental conditions must be cooler to maintain sufficient heat loss. The action limit is similar but for unacclimatized workers.

Clothing effects were not addressed directly by NIOSH other than to say that the thresholds apply to workers wearing customary, single-layer work clothes. A line of thinking initiated by Ramsey [32] and further developed for the Electric Power Research Institute (EPRI) by Bernard and Kenney has led to the use of clothing adjustment factors (CAFs) [31,33,34]. The factors in essence suggest what the effective increase in WBGT is when clothing other than ordinary work clothes is worn. Table 28.2 is a list of CAFs recommended by ACGIH and values I have estimated for other clothing configurations. In the evaluation of heat stress, the CAF is added to the prevailing WBGT in the workplace.

For multiple locations, multiple clothing ensembles, and multiple tasks, the effective WBGT and effective metabolic rate ($M$) are determined from a TWA. That is,

$$\text{WBGT}_{\text{effective}} = \frac{\{\Sigma[(\text{WBGT}_{\text{i}} + \text{CAF}_{i})t_{\text{i}}]\}}{T} \quad (28.9)$$

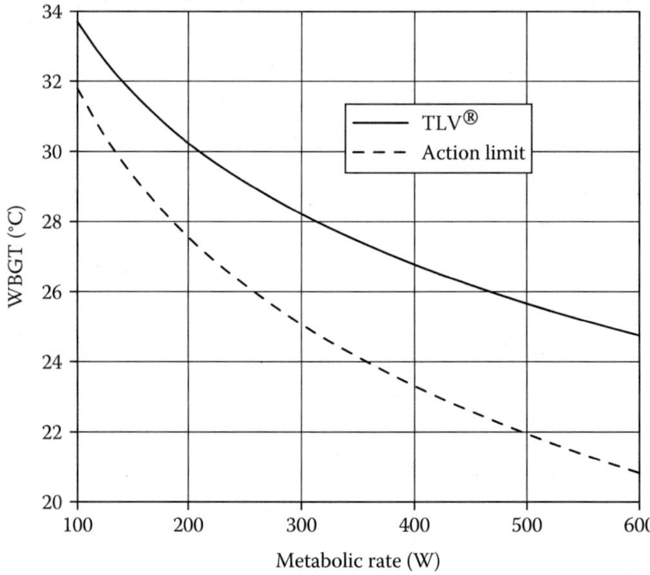


FIGURE 28.1 Thresholds for heat stress as a function of metabolic rate and WBGT to limit core temperature to 38°C for acclimatized and unacclimatized workers based on ordinary work clothes as described by NIOSH [27] and the ACGIH [31]. The TLV and the NIOSH REL (for acclimatized workers) are the same, as are the action limit and the NIOSH RAL (for unacclimatized workers).

TABLE 28.2 Clothing Adjustment Factors (CAFs) for Selected Clothing Ensembles[a]

| Ensemble | CAF from USF [°C-WBGT] | CAF from ACGIH [°C-WBGT] | Comments |
|---|---|---|---|
| Work clothes | 0 | 0 | Work clothes made from a woven fabric are the baseline ensemble |
| Cloth coveralls | 0 | 0 | Woven fabric |
| SMS Nonwoven coveralls as a single layer | −1 | 0.5 | SMS is a nonproprietary process to make nonwoven fabrics from polypropylene |
| Tyvek 1422A coveralls as a single layer | 2 | 1 | Tyvek is a proprietary fabric made from polyethylene. Tyvek 1422A has a somewhat higher CAF than 1424 and 1427, which were used for the TLV but not available |
| Double layer of woven clothing | | 3 | |
| NexGen® coveralls as a single layer | 2.5 | — | NexGen is a proprietary microporous fabric that is water-barrier, vapor-permeable. There is great variability in these types of fabrics and one CAF will not apply for another fabric |
| Vapor-barrier coveralls as a single layer | 10 | | No hood |
| Vapor-barrier coveralls with hood as a single layer | 11 | 11 | It was assumed that a hood would be worn with vapor-barrier coveralls |
| Hood | +1 | | Wearing a hood of any fabric with any clothing ensemble |

[a] The CAFs are added to the measured WBGT to obtain an effective WBGT.

TABLE 28.3 Illustration of the Use of the WBGT-Based Threshold and CAFs

| Task | Clothing | $M$ (W) | Measured WBGT (°C) | Effective WBGT (°C) | Threshold WBGT (°C) | Time (min) |
|---|---|---|---|---|---|---|
| 1. | Vapor-barrier with hood | 300 | 33 | 44 | | 10 |
| 2. | Coveralls | 400 | 33 | 33 | | 20 |
| 3. | Work clothes | 100 | 20 | 20 | | 10 |
| TWA | | 300 | 29.8 | 32.5 | 28.2 | 40 |
| Change task 3 time to 25 min: | | | | | | |
| TWA | | 245 | 27.1 | 29.1 | 29.2 | 55 |

and

$$M_{\text{effective}} = \frac{\{\Sigma M_i t_i\}}{T} \tag{28.10}$$

where $T = \Sigma t_i$.

Table 28.3 illustrates this simple process for determining whether a job is above the threshold. As can be seen, the overall demands are above the threshold. The time in the rest location (Task 3) can be increased to 25 min to bring the work cycle to the threshold.

The time interval encompassed for time weighting is usually 1–2 h. If the work is cyclic in nature and includes the rest place, a TWA of one work cycle up to a 2 h duration is the appropriate time base. If the work is very intermittent or irregular, the selection of the time base requires some discretion, but it should not be more than 2 h. For combinations of $\text{WBGT}_{\text{effective}}$ and $M_{\text{effective}}$ that are above the action limit line in Figure 28.1, unacclimatized workers may have difficulty coping with the heat stress. For this reason, heat stress can be considered a potential hazard above the action limit. When the exposure is above the TLV line, then some acclimated workers may not be able to fully thermoregulate. The further above the line the combination is, the greater the risk.

Although the evaluation is for time intervals between 1 and 2 h, the assumption is that the work is performed at these levels over the 8 h day. Shorter times at high levels of heat stress are also safe, but the WBGT-based method described here is not designed to evaluate those exposures. The U.S. Navy [35] has proposed WBGT-based methods to evaluate short-term exposures. The PHEL considers WBGT and metabolic rate in recommending a time limit for the exposure. It is reasonable to use a CAF for other clothing requirements. In these cases, heat stress is an important job factor if the work time exceeds the recommended exposure time. Rational methods of heat stress evaluation also provide safe exposure times.

#### *28.3.4.3 Rational Scheme for Heat Stress Evaluation*

Rational methods for heat stress evaluation use a different approach to the problem. The starting point is a model of heat exchange between the person and the environment similar to the one presented in Equations 28.1 and 28.2. For a given scheme, computations or estimations are described to determine values for the different avenues of heat exchange. Generally, if thermal balance can be achieved, then no storage occurs and the exposure is not limited. On the other hand, if $S > 0$, then the exposure is time-limited and the time limit can be determined from Equation 28.5. The time limit can be used to evaluate the level of heat stress with respect to the work time.

A recent rational scheme has been accepted by the ISO [7]. It uses two criteria to prescribe time limits. The first is a maximum amount of heat storage (refer to Equation 28.5), and the second is a limit on the total volume of sweat that is required to maintain thermal balance. The standard provides BASIC code to evaluate a single exposure in a normative appendix and Excel®-based methods are available [36,37].

The method requires the collection or estimation of a number of environmental and work parameters including dry bulb, psychrometric wet bulb, globe temperatures, air speed, metabolic rate, and clothing insulation and permeability (or evaporative resistance). If the psychrometric wet bulb is not known, it can be taken as 1°C lower than the natural wet bulb temperature. The ISO standard provides some information on clothing and there is an emerging database on insulation and vapor permeation for noncotton clothing that can be adapted for this method [38–40].

From these data, heat transfer coefficients can be estimated for convection, radiation, and maximum evaporative cooling. Then the values in Equation 28.3 can be determined to estimate the required evaporative cooling. Because the proportion of sweat that is evaporated decreases with increasing rates of sweating and evaporative cooling requirements, a level of sweating efficiency is estimated in order to estimate the total amount of sweating required to support the required evaporative cooling.

The next step in the ISO evaluation process is to select the criteria that will be used for the evaluation based on the acclimatization state. Knowing the requirements and the limits, it is possible to estimate the amount of evaporative cooling that is available and the amount of sweating it will require. Based on these predicted values, time limits for the exposure can be computed to avoid excessive heat storage or sweat loss (dehydration).

#### *28.3.4.4 Physiological Evaluation of Heat Stress*

An alternative method to evaluate the level of heat stress is to look at the heat strain associated with the work. This concept was introduced earlier as a means of recognition. It is also a valid means of evaluation when the effects are monitored instead of the exposure [27]. Physiological strain for heat stress exposures is well reflected in sweating, core temperature, and heart rate.

*Sweating*: Sweat losses are interesting in that they provide a way to anticipate dehydration. Over 1–2 h intervals, the average rate of sweating can be assessed as [13]

$$\text{Sweat rate [L/h]} = \frac{BW_{\text{initial}} - BW_{\text{final}} + \text{ingestion [kg]}}{\text{Time [h]}} \tag{28.11}$$

where

*BW* is the body weight in kilograms before and after the measurement interval

Ingestion is the weight of drink and food consumed during the same period

This equation assumes that there is no loss of body weight by excretion. The commonly accepted hourly sweat loss should be less than 1 L.

There is also a total sweat loss threshold of 6 L in the workday. Again, it is difficult to replace more than 6 L of water in 8 h. To monitor total sweat loss means that weight loss by excretion must also be assessed. As a practical measure, total sweat loss is not a routine measure. An estimate of dehydration over the workday, however, is available. This estimate is the percent change in body weight from the beginning to end of the work shift. As mentioned earlier, a loss of more than 1.5% of body weight (about 1–1.5 kg) is significant [24].

*Core temperature*: Core temperature is an abstraction, and there are various means to measure the "core" temperature. In the laboratory, the usual means are rectal, tympanic, and esophageal temperatures. Esophageal is very effective for assessing the physiological responses to heat stress and strain. It is less indicative, however, of the total thermal storage. Tympanic temperature was an early favorite for predicting thermoregulatory response but has yielded to esophageal. A personal preference is to use rectal temperature ($T_{re}$) as the best indicator of the three for total heat storage for industrial settings. A deep body temperature similar to rectal temperature can be measured with an ingestible sensor that transmits temperature data to a receiver. This method is coming into more common use, but still relatively expensive for routine monitoring.

The measures of core temperature are difficult to measure in the field for social and cost reasons. Therefore, a surrogate measure of core temperature is usually sought. These include oral and ear canal temperatures. Each method has its advantages and disadvantages. Oral temperature measurement is the most readily recognized method, with a long history of use for the evaluation of industrial heat strain. Modern devices have electronic means that allow them to respond more rapidly than mercury-in-glass thermometers. Disposable thermal strips are also available. Necessary precautions include (1) no eating or drinking for 15 min prior to measurement and (2) mouth closed during the measurement (no talking or mouth breathing). Generally, adding 0.5°C to the value is a good predictor of rectal temperature, although the average steady-state difference may reach 1°C. Any method should be calibrated prior to use. For the disposable thermal strips, a random sample from each lot should be calibrated.

An alternative measure is the temperature in the ear canal. The temperature sensor is placed in the ear canal near the tympanic membrane, and care is taken to insulate the ear canal from the outside environment. Insulating techniques have included cotton and hearing protectors. The ear canal temperature values are sensitive to the environmental conditions. For instance, in hot environments or with high radiant heat, the value may be greater than rectal temperature. Temperature in the ear canal is also more reactive to the level of heat stress than rectal temperature, which means that it will rise or fall faster.

Once a surrogate method of assessing core temperature is selected, then a sampling strategy can be developed. Basically, the TWA core temperature should be 38°C for the 8 h day, with a ceiling limit of 39°C.

*Heart rate*: Heart rate measures are relatively easy to obtain in a reliable and acceptable fashion. Recovery heart rate, peak values, and TWAs represent some of the methods that can be used to assess cardiovascular response to heat stress.

Historically, recovery heart rate was used to assess work demands and heat stress because palpation of the radial artery was a simple field method. Criteria have evolved from the initial work of Brouha in 1960 [20] through efforts by Fuller and Smith [21] to a single sample at 1 min of recovery [22,28,41]. A worker is stopped and asked to sit down and the recovery heart rate is noted at the end of 1 min. The original method called for a 30 s count from 30 s to 1 min, but recovery heart rate is best assessed by electronic meters near the end of the minute. If the value is less than 110 bpm, there is no excessive strain.

Data loggers in conjunction with heart rate monitors are now readily available. These enable investigators to record heart rate on a minute-by-minute basis for an entire work shift. Review of the heart rate data for a worker includes an evaluation of the average for the workday as well as an examination of peaks. The average over a workday should be less than about 110 bpm [17]. Peak heart rates (sustained for more than 1 min) should be less than 90% of the individual's maximum heart rate (an estimate for $HR_{max}$ is 195–0.67 [age: 25]) [42].

Because the average and peak heart rate evaluations can miss periods of significant cardiovascular strain, ergonomists will visually examine the patterns for heart rate creep or other indicators of failure to achieve complete control. Bernard and Kenney [42] proposed a quantitative method of assessing the heart rate pattern over the course of the day (or lesser periods). The analysis is based on moving-time averages (akin to TWAs on the same time base and continuously updated).

### 28.3.5 Management of Heat Stress

The management of heat stress is the natural and obligatory next step if the evaluation indicates that heat stress is a potential workplace hazard (e.g., above the ACGIH action limit). Management comes in two stages: general controls and job-specific controls. The following paragraphs outline when it is appropriate to implement general and specific controls and describe what the controls may be.

The controls are developed within the overall health and safety activity at the individual site. There are four important components of the heat stress management program:

*Policy statement*: The policy statement acknowledges the potential for heat stress exposure, the role of heat stress control in the overall health and safety program, and the availability of health and safety personnel to assist line management.

*Delineation of responsibilities*: The senior line managers *have* responsibility with some delegation of authority, first-line supervision has day-to-day responsibility to ensure compliance and provide a role model, and employees must follow guidelines and practice good heat stress hygiene.

*Workplace monitoring workplace*: Monitoring follows good industrial hygiene practice for monitoring hazards.

*Review and evaluation*: There should be a periodic review of the heat stress program to be assured that it is meeting its goals and to make adjustment as necessary.

The actual embodiment of the controls to manage heat stress is varied and adaptable in infinite combinations with the exercise of common sense and a little imagination. One valuable source is the current practices of workers.

#### *28.3.5.1 General Controls*

General controls are those actions required to reduce the risk of heat-related disorders among workers who may be exposed to heat stress in the workplace. General controls are the first stage of countermeasures, and they transcend the actual conditions in which the exposure may occur. Included in general controls are heat stress hygiene practices, training, and medical surveillance.

To make the decision about who might be exposed to heat stress and when, we must refer back to the evaluation step. Anytime workplace conditions combine to place the situation above the action limit in Figure 28.1, then heat stress is present at a level that should be considered a hazard (albeit a low-risk hazard at the threshold). If the calculation of time limits is used to evaluate the level of stress, then conditions that combine in the evaluation scheme to recommend an exposure time of less than 8 h are the threshold at which general controls are to be implemented. If only occasional samples are taken of the environment, they should be taken when the highest levels of heat stress are most likely.

*Heat stress hygiene practices*: These are the actions that an individual can take to lower the risk of heat-related disorders. They consist of the following:

*Self-determination*: The individual should interrupt a heat stress exposure once extreme discomfort or the initial symptoms of a heat-related disorder are detected.

*Fluid replacement*: Because thermal regulation depends on sweating and the *necessary* loss of water, the water must be replaced at frequent intervals to maintain acceptable hydration. Sometimes drinking is restricted or impossible (e.g., wearing a respirator), and then allowances for pre- and postexposure hydration must be considered.

*Lifestyle and diet*: Practicing a *generally* accepted healthy lifestyle (getting adequate sleep, limiting nonwork exposures to heat stress, exercising, not abusing alcohol or drugs, and eating a well-balanced diet) greatly reduces the risk of heat-related disorders.

*Health status*: Those with any chronic disease should inform their physician of occupational exposures to heat stress and follow the recommendations. Those with an acute illness should report the condition to a supervisor, and the heat stress exposures should be restricted or reduced.

*Acclimatization*: Because acclimatization requires at least 5 days, allowances must be made for those workers who are not acclimatized to the heat, and performance expectations should therefore be reduced.

It is clear that heat stress hygiene practices are the responsibility of individual workers but that management must minimize the barriers to the practice of heat stress hygiene.

*Training*: It is a fundamental health and safety practice for those who may be exposed to a hazard such as heat stress. Annual training that fits within the training practices of the site should be sufficient. The training should include discussions of heat stress and strain, heat-related disorders, heat stress hygiene practices, and controls used in the facility.

*Medical surveillance*: It encompasses physicals and monitoring of sentinel health events. Both preplacement and periodic physicals appropriate to evaluation of an individual's capacity to deal with heat stress are recommended. The physical should include comprehensive medical and work histories, comprehensive physical examination and tests, assessment of drug use, and ability to use personal protection. Good practice dictates that there be a written opinion as to the suitability of exposing the individual to heat stress.

Monitoring of sentinel health events includes monitoring individuals as well as the population. Relevant events are heat-related disorders, patterns of accidents, absenteeism, and chronic fatigue.

*Emergency response plan*: There are a number of factors that lead to poor outcomes in heat stroke cases. These include poor communications (often a language barrier but also training and supervision), new workers, working alone, and inadequate emergency response. The key elements to emergency response are recognition of signs of heat stroke followed by immediate and aggressive cooling. I recommend immersion in a cold water/ice bath as the most effective means of reducing body core temperature. The next most effective means are (1) packing ice near major superficial blood vessels (i.e., neck, axillaries, and groin areas) and/or (2) pouring/showering the person with cold water (the addition of fans may help). The plan should also include summoning EMS personnel and hospital transport.

#### 28.3.5.2 Specific Controls

Specific controls are those that apply to specific manifestations of heat stress. They follow the traditional hierarchy of engineering controls followed by administrative controls, with the use of personal protection as a last resort. Specific controls are appropriate for workers who are exposed to heat stress above the TLV in Figure 28.1 (or similar exposure limit), and only when they are exposed above the threshold.

*Engineering controls*: These change the conditions so that the level of heat stress is reduced, ideally below the exposure thresholds. Methods of engineering controls include the following.

*Reduce the metabolic rate*: A very effective means to reduce heat stress is to reduce the amount of internal heat generation. This can be done by spreading the workout, mechanization, and increasing staffing.

*Change clothing requirements*: A proper balancing of risks from heat stress and from agents requiring protective clothing may lead to the selection of different barrier fabrics with markedly different evaporative cooling capacities. For instance, some vapor-transmitting fabrics cause much less heat stress than vapor-barrier fabrics.

*Reduce temperature and humidity*: Reductions of air temperature and humidity are frequently achieved through *spot* or dilution ventilation. This is another method to significantly reduce the level of heat stress in the workplace. The ventilation systems can be temporary or permanent and may include mechanical cooling. Dramatic cost savings have been demonstrated with this method.

*Increase air motion*: Increasing air speed via fans is a time-honored method to enhance evaporative cooling, but it is of limited value once air speed exceeds 2 m/s. When air temperature is greater than 40°C, increasing air motion may actually increase heat stress.

*Control radiant heat*: When radiant heat is high, the effects can be reduced through combinations of *insulating* exterior surfaces and reducing surface emissivity. In addition, shields can be very effective.

*Administrative controls*: These manage the risk through work practices. They are relatively easy to implement, although they may not be the most cost-effective. Administrative controls include the following.

*Planned work time*: Limiting the heat stress exposure to a time period that would ensure that most workers are not overexposed is one way to limit the risk. The work time limit can be based on the WBGT or ISO methods of estimating safe exposure times discussed earlier for evaluation purposes.

*Self-determination*: *Giving* employees the opportunity to subjectively control the pace of work and the work time is frequently used as a means of controlling heat stress exposures. Self-pacing is a valuable means of reducing the physiological strain and improving efficiency. Subjective self-limitation, however, may not be reliable. Physiological monitoring to provide objective information on heart rate and body temperature will improve the reliability. Recommendations for personal monitoring include those physiological measures discussed earlier that can be performed and assessed in real time. The advantage of personal monitoring is that it allows the more heat-tolerant workers more exposure time. In this way, personal monitoring can improve productivity while controlling the risk of heat-related disorders.

*Recovery allowances*: It is *important* to provide adequate recovery from heat stress exposures. Including recovery times and locations in the analysis of overall evaluation of heat stress exposure provides insight as to whether or not the recovery allowance is adequate. If the exposure is above the threshold, the recovery is insufficient, and the exposure or recovery conditions or times must be adjusted.

*Scheduling work*: To the *extent* possible, scheduling work to times when the heat stress levels may be lower (e.g., night, during outages) is a useful way to control exposures.

*Personal Protection*: Personal protection for heat stress exposures means providing a microenvironment around the worker that allows a greater loss of heat. There are three types of personal cooling systems that have demonstrated effectiveness:

1. *Circulating air systems*: Venting air from supplied-air hoods or supplying breathing grade air directly under clothing enhances evaporative and convective cooling. Many times, the cooling is sufficient to virtually eliminate heat strain. The major disadvantage is that worker mobility is restricted with the airline.

2. *Liquid cooling systems*: This type of personal cooling is based on circulating cooling liquid (e.g., water) around some portion of the body within enclosed tubes or channels. The rate of cooling depends on the surface area of the body covered. The heat is taken up by a heat sink that is usually composed of ice but could theoretically be another material. The service time depends on the size of the heat sink or the ability to replenish the heat sink. The major disadvantage to these systems is the cost.

3. *Ice cooling garments*: Ice cooling garments cool the body by direct transfer of heat from the body to the heat sink by conduction. These are sometimes referred to as passive systems because there is no mechanical movement of air or liquid.

Under certain conditions of high radiant heat, reflective clothing over parts or all of the body can reduce the level of heat stress. The reflective clothing must be chosen carefully because it will reduce the rate of evaporative cooling.

In choosing personal protection it is crucial that the means be selected to best match the capabilities and requirements of the cooling system to the job mission. The decision should be confirmed by actual trials in the field including physiological measures to demonstrate a reduction in physiological strain.

## 28.4 CRITICAL REVIEW OF CURRENT STATUS

Two approaches to the evaluation of heat stress are described in this chapter. One approach is through the use of empirical and rational models such as WBGT-based thresholds and predicted heat strain (PHS). The other is monitoring physiological strain, which includes some combination of body temperature, heart rate, and sweat loss.

The heat stress models rely on the predictive ability of the model to indicate when excessive physiological strain is likely. Because individuals respond differently to heat stress, the criteria are selected to be protective of at least 95% of the healthy working population. They assume that if the predicted physiological strain is acceptable for most workers then the risk of a heat-related disorder is not significant. Many healthy workers are capable of safely working at levels of heat stress greater than the criteria levels, and these workers are more likely to self-select into the more stressful jobs. Experience makes workers and supervisors aware of this fact, which, in turn, makes it difficult for them to accept the criteria.

WBGT-based thresholds of ACGIH and NIOSH as well as the ISO are adequately protective based on the goal of establishing thermal equilibrium. From the progressive heat stress protocols performed at the University of Pittsburgh by Belding and Kamon, at the Penn State University by Kamon and Kenney, and at the University of South Florida (USF) by Bernard and Ashley, there was some evidence that the thresholds might be about 2°C–3°C-WBGT too low. Some unpublished work at USF suggests that the thresholds may be no more than 1°C-WBGT too low, and this puts the evaluation in the normal range of uncertainty for the measurements themselves. In light of the epidemiological data on heat disorders and accidents presented early in this chapter, there is little support for changing the current WBGT-based thresholds.

The PHS method was validated during its final development stages [8]. It is clear that the developers demonstrated that it was predictive of the mean response. This left open the issue of population protection. Some protection was afforded by a relatively low body core temperature threshold so that in most cases, the core temperature would be below 39°C at the predicted core temperature of 38°C criterion point. Another way of looking at the protective nature of PHS was to consider the safe exposure time [43]. In this case, PHS was protective about half the time. Ueno et al. [44] made changes to the PHS algorithm to account for clothing following ISO9920. A fortuitous outcome was the predicted safe exposure time was protective during most of the trial short-term exposures. While not an elegant solution, it works for the time being.

Clothing effects on heat stress are dictated by insulation, permeability, and ventilation [45], which in turn depend on the fabrics, construction, and use. The ACGIH recommended CAFs for the measured value of WBGT to account for some other clothing ensembles. These values are based on shifts in the upper limits of environmental conditions observed for establishing thermal equilibrium. CAFs are readily useful for those clothing ensembles that have been studied. There is some early evidence that CAFs can be estimated from the evaporative resistance of a clothing ensemble [38]. If this relationship holds and it can be demonstrated that the laboratory values for evaporative resistance are predicted by modifications to manikin values via ISO 9920, then there is an opportunity to predict CAF for a broader range of ensembles. This is the same process used in the modified version of PHS suggested by Ueno et al. [44].

Physiological evaluation examines the individual's response to heat stress. The advantage of monitoring physiological strain is that it removes the need to know well the environmental, work, and clothing factors as they contribute to heat stress. To the extent that a representative sample of the worker population is selected, the exposure to heat stress is sufficiently understood to guide decisions on the need for controls. It was the population response to heat stress in terms of core temperature and recovery heart rate that helped establish the WBGT thresholds, so it makes sense that these can be used to directly evaluate the work environment. By extension, physiological monitoring can be a tool to control exposures for individuals. The disadvantage is that the effectiveness of the controls cannot be estimated before they are first tested in actual or simulated work conditions.

## 28.5 FUTURE CONCERNS

There are three areas of heat stress exposure assessment that deserve special attention. These are (1) clothing models, (2) discriminate capability of physiological monitoring, and (3) epidemiological studies to related exposures to injury and illness outcomes.

The variety of clothing ensembles in the workplace will continue to grow. An efficient method to assess the thermal characteristics of the clothing that are most relevant to heat stress is required to meet the need. Recently, for instance, fabric porosity was demonstrated to play an important role but there is no clear way to assess its contributions to CAFs or evaporative resistance other than to run expensive progressive heat stress trials. A basic model that accounts for intrinsic insulation and evaporative resistance as well as convective pathways is needed to guide a research agenda.

Expanding the use of personal monitoring for evaluation of heat strain is appropriate and practical, and it may be the only way worker protection can be ensured during work with some common forms of protective clothing. Devices for monitoring core temperature and/or heart rate are available, and others can certainly be developed. With supporting data, these can be used to assess population responses to heat stress for the purposes of evaluation and also for improved self-assessment of individual exposures. The successful implementation of cost-effective personal monitoring can reduce the problems of conservative limits based on heat stress models and further reduce the risk of heat-related disorders. A critical step in the development of physiological measures is to distinguish between predictive validity based on a mean population response and the sensitivity and specificity of the method.

The case for heat stress as an occupational hazard is well established. What is less well understood is the relationship between exposure levels and health, safety, productivity, and quality outcomes. There is some data available to help inform a model that would guide an epidemiological study. A conceptual model and study are necessary to close the loop on heat stress assessment.

## 28.A APPENDIX: RATIONALE FOR WBGT-BASED EXPOSURE LIMITS

### 28.A.1 Initial Development of an Exposure Limit for Work Clothes

Current heat stress OELs based on WBGT have their genesis from the upper limit of the prescriptive zone (ULPZ) proposed by Lind [14]. In this seminal study, three male mine rescue team members, who were unacclimatized and wore a seminude ensemble (shorts, socks, and shoes), completed 1 h sessions in a range of fixed conditions at a metabolic rate of 350 W. Lind noted that the rectal temperature ($T_{re}$) was constant for environments below an effective temperature (*ET*) of 27.4°C. As the *ET* increased above 27.4°C, there was a steady increase in $T_{re}$ after 1 h. Lind referred to this threshold point as the ULPZ because the $T_{re}$ was not affected by changes in environmental heat below this point but over that point heat stress was overtaxing the thermoregulatory system. With two of the participants, Lind repeated the exposures at metabolic rates of 210 and 490 W. He reported ULPZ values of 30.2°C and 26.9°C-*ET*, respectively, for these participants. The change in ULPZ with work demands highlighted the interaction of two job risk factors: environment and work demands. The data also demonstrated that the $T_{re}$ below the ULPZ was higher with higher metabolic rate. Thus below the ULPZ, $T_{re}$ was driven by work demands alone; and above the ULPZ, it was driven by the environment.

In 1970, Lind [46] used unacclimatized seminude male soldiers to demonstrate the validity of the ULPZ at 350 W, which was 27.4°C-*ET* [14]. There were 4 exposure groups of about 25 participants in each. The groups were 21°C, 27°C, 28.9°C, and 31.1°C-*ET*. Each participant was asked to work for 3h. Everyone in the lowest group (21°C-*ET*) was able to complete the 3h session. At the environment just below the ULPZ (27°C compared to 27.4°C-*ET*), 24 of the 25 participants completed the 3h session. At the two conditions that were above the ULPZ (28.9°C and 31.1°C-*ET*), only 13 of 19 and 10 of 25 were able to complete the session. This offered a compelling argument in favor of the ULPZ for extended heat stress exposures of 27°C-*ET* at 350W for unacclimatized seminude personnel.

In a second study to validate the ULPZ, Lind et al. [47] studied six younger (average age of 27 year) and six older (47 year) miners, who were acclimatized and seminude. The exposure session was 8h at a metabolic rate of 350W. There were five environmental conditions at 20°C, 26°C, 28°C, 29.5°C, and 30.5°C-*ET*. For the first three conditions (20°C through 28°C-*ET*), there was no real difference in $T_{re}$ suggesting full thermoregulatory control. There was little difference due to age. For acclimatized seminude personnel at 350W, the ULPZ was 28°C-*ET*.

In summary, the initial study [14] posited values of the ULPZ at three levels of metabolic rate for unacclimatized workers. The intermediate metabolic rate of 350W was studied in more detail to validate the premise that the ULPZ could be used to set a threshold for sustained exposures to heat stress. For unacclimatized seminude workers, a value of 27°C–27.4°C-*ET* was well supported. When the worker was acclimatized, a value of 28°C-*ET* was supported with a little room to be slightly higher. Table 28.4 summarizes the results of these studies.

The next step was to translate the ULPZ into an OEL. NIOSH argued for WBGT as a simple and robust measure of the environment [27,48], and it has become the *de facto* method for environmental assessment for occupational exposures to heat stress in the United States. For this reason, the WBGT equivalent was provided in Table 28.4 and the remaining part of this appendix uses WBGT.

Because most of the data were from unacclimatized participants, an adjustment for acclimatization was necessary. Table 28.4 compares an unacclimatized population in one study of 3h sessions with a different population used in the 8h study. The difference was about 2°C-WBGT. As reported by Lind, Wyndom also saw a difference of 2.7°C-WBGT due to acclimatization in seminude workers (29.2°C–31.9°C-WBGT). These two studies were the primary basis for assigning an adjustment due to acclimatization. While not considered

TABLE 28.4 The Upper Limit of the Prescriptive Zone (ULPZ) at Three Levels of Metabolic Rate in Effective Temperature (ET) and Estimated WBGT for Seminude Clothing Condition

| | Unacclimatized | | Acclimatized | |
|---|---|---|---|---|
| **Metabolic Rate [W]** | **ET [°C]** | **WBGT [°C]** | **ET [°C]** | **WBGT [°C]** |
| 210 | 30.2[a] | 32.2 | | |
| 350 | 27.0[b] | 27.4 | 28.0[c] | 29.2 |
| 490 | 26.9[a] | 27.8 | | |

[a] Lind [51].
[b] Lind [46].
[c] Lind et al. [47].

originally, Kuhlemeier et al. [49] determined the ULPZ following Lind's method for several groups wearing work clothes, two of which were unacclimatized (non-heat-exposed workers in winter) and acclimatized (heat-exposed in summer). The difference was 1.8°C-WBGT at 345 W and 3.2°C at 490 W. Their data were supportive of the 2°C-WBGT effect due to acclimatization at a moderate metabolic rate. The second adjustment was to move from seminude to clothed workers. Following a progressive heat stress protocol, Belding and Kamon [50] found a 2°C-WBGT increase in critical WBGT for clothed compared to seminude participants. For practical purposes, unacclimatized seminude limits were the same as acclimatized work clothes limits.

In preparing the revised NIOSH criteria document for heat stress [27], Henschel and Dukes-Dobos looked first to the intermediate metabolic rate. They noted the 1°C-WBGT drop in ULPZ between Lind's original proposal [14] and the 3 h validity study [46]. They took the lower value as the starting point. To add a margin of safety at the higher metabolic rate, they decreased the ULPZ by 2°C-WBGT because the work-driven $T_{re}$ was near 38°C. At the lower metabolic rate, Henschel and Dukes-Dobos felt that no adjustment was necessary because there was an adequate margin of safety for $T_{re}$. The NIOSH recommended exposure limit [27] and the original 1973 ACGIH TLV were the same. They are described by the following equation [8]:

$$\text{TLV}[°\text{C} - \text{WBGT}] = 56.7 - 11.5 \log 10\, M[\text{W}] \quad (28.12)$$

Another feature of the occupational exposure limits based on WBGT was the TWA. Again, Lind [51] considered this for two nude acclimatized participants. He compared continuous work and a 50/50 work/rest cycle with an average metabolic rate of 305 W at three environmental conditions (equivalent to 25.7°C, 29.1°C, and 32.4°C-WBGT). The two participants exhibited good thermoregulatory control for the lower two conditions but a loss at the higher condition and a similar pattern of response for $T_{re}$ and heart rate between continuous and intermittent work when there was good control. The similar patterns supported the averaging of the work demands. The stability at 29.1°C-WBGT was consistent with the evidence that the ULPZ for acclimatized seminude at 350 W was 29.2°C-WBGT (Table 28.4).

### 28.A.2 Effects of Air Movement

Because air movement under the clothing reduces the effects of the clothing on heat stress, the degree of air motion was a concern. Belding and Kamon [50] examined air speed and critical WBGT using a single trial progressive heat stress exposure protocol in work clothes. Comparing the difference between low (0.83–1.0 m/s) and high (3.3 m/s) at three metabolic rates, they found differences of 2.4°C-WBGT at 220 W, −0.4 at 280 W, and 2.1 at 360 W. (As a note, there were little changes in critical WBGT over air speeds in the seminude trials.) The US OSHA review panel for a heat stress standard reviewed the first NIOSH criteria document [48]. That panel criticized several features of the NIOSH proposal, and one of the suggestions was two thresholds at different air speeds [52] because of the belief that WBGT did not adequately account for high and low air speeds. The ISO standard for

WBGT also included adjusted thresholds for different air speeds [53]. Kamon and Avellini [54] showed some differences from low to high air speed but concluded that air speed up to 4 m/s did not affect the threshold.

### 28.A.3 WBGT-Based Clothing Adjustment Factors

Because WBGT-based assessments are based on observed (empirical) relationships and not rational (biophysical) relationships, it is more difficult to account for clothing effects based on insulation and evaporative resistance. For this reason, offsets or adjustments for clothing in WBGT units have been sought. In essence, the adjustment factor represents an equivalent change of environmental WBGT. As mentioned earlier, CAFs were first proposed by Ramsey [32] and furthered by Kenney [55]; and first adopted by ACGIH in 1990 with changes over the following 15 years. O'Connor and Bernard [56] summarized the findings of a variety of studies performed at USF.

Prior to 2005, reports of CAFs were less reliable because experimental data were pooled from different laboratories at different times with limited number of observations; and often there was some professional judgment involved. Bernard et al. [34] reported CAFs for four clothing ensembles with a larger number of participants over a range of relative humidities. Of the trial ensembles, only the vapor-barrier ensemble exhibited different CAFs at the three humidity levels, and thus a high value was selected to represent vapor-barrier coveralls. These CAFs are provided in Table 28.2. Later, they demonstrated that CAFs were independent of metabolic rate [33]. Comparing Tyvek® 1424 and 1427 and vapor-barrier coveralls with and without an integral hood, Ashley and Bernard [57] found that an integral hood added about 1°C-WBGT to the ensemble CAF. In the same paper, they suggested that there were no differences between cotton, FR-treated cotton, and Nomex® coveralls. The ACGIH recommendations for CAFs are also given in Table 28.2. A CAF for SMS nonwoven coveralls is also provided in Table 28.2 based on two direct comparisons with Tyvek 1422A using the same participants and laboratory [56].

## ACKNOWLEDGMENTS

Some of the material presented was supported by CDC/NIOSH 1R01 OH03983. Other support came from Alcoa, DuPont, and USF. I would also like to thank the ACGIH for the opportunity to work on the Physical Agents Committee. It is important to note that I do not represent the views of any organization.

## REFERENCES

1. CDC, Heat-related deaths among crop workers—United States, 1992–2006, *Mortality and Morbidity Weekly Report* (*MMWR*) 57(24): 649–653 (2008).
2. J. Ramsey, C. Buford, M. Beshir, and R. Jensen, Effects of workplace thermal conditions on safe work behavior, *Journal of Safety Research* 14: 105–114 (1983).
3. M. Fogleman, L. Fakhrzadeh, and T. E. Bernard, The relationship between outdoor thermal conditions and acute injury in an aluminum smelter, *International Journal of Industrial Ergonomics* 35: 47–55 (2005).
4. C. K. Redmond, J. J. Emes, S. Mazumdar, P. C. Magee, and E. Kamon, Mortality of steelworkers employed in hot jobs, *Journal of Environmental Pathology and Toxicology* 2(5): 75–96 (1979).

5. H. S. Belding and T. F. Hatch, Index for evaluating heat stress in terms of resulting physiological strain, *Heating, Piping and Air Conditioning* 27(8): 129–135 (1955).
6. T. E. Bernard, Thermal stress, in *Fundamentals of Industrial Hygiene*, 5th edn., B. Plog, Ed., National Safety Council, Chicago, IL, 2002.
7. International Organization for Standardization, *ISO 7933: Ergonomics of the Thermal Environment—Analytical Determination and Interpretation of Heat Stress Using Calculation of the Predicted Heat Strain*, ISO, Geneva, Switzerland, 2004.
8. J. Malchaire, A. Piette, B. Kampmann, P. Mehnert, H. Gebhardt, G. Havenith, E. Den Hartog, I. Holmer, K. Parsons, G. Alfano, and B. Griefahn, Development and validation of the predicted heat strain model, *Annals of Occupational Hygiene* 45(2): 123–135 (2001).
9. K. C. Parsons, *Human Thermal Environments*, 2nd edn., Taylor & Francis, London, U.K., 2003.
10. J. Stegemann, *Exercise Physiology*, Year Book Medical Publishers, Chicago, IL, 1981.
11. International Organization for Standardization (ISO), *ISO 13732-1: Ergonomics of the Thermal Environment—Methods for the Assessment of Human Responses to Contact with Surfaces—Part 1: Hot Surfaces*, ISO, Geneva, Switzerland, 2006.
12. H. Siekmann, Recommended maximum temperatures for touchable surfaces, *Applied Ergonomics* 21(1): 69–73 (1990).
13. D. M. Kerslake, *The Stress of Hot Environments*, Cambridge University Press, London, U.K., 1972.
14. A. R. Lind, A physiological criterion for setting thermal environmental limits for everyday work, *Journal of Applied Physiology* 18: 51–56 (1963).
15. M. Nielsen, Die Regulation der Korpertemperatur bei Muskelarbeit, *Scandinavian Archives of Physiology* 79: 193–230 (1938).
16. F. N. Dukes-Dobos and A. Henschel, Development of permissible heat exposure limits for occupational work, *American Society of Heating, Refrigerating, and Air Conditioning Engineers Journal* 15: 57–62 (1973).
17. World Health Organization, *Health Factors Involved in Working Under Conditions of Heat Stress*, Technical Report Series 412, Geneva, Switzerland, 1969.
18. J. T. Krajewski, E. Kamon, and B. Avellini, Scheduling rest for consecutive light and heavy workloads under hot ambient conditions, *Ergonomics* 22(8): 975–987 (1979).
19. R. G. Holly and J. D. Shaffrath, Cardiorespiratory endurance, in *ACSM's Resource Manual for Guidelines for Exercise Testing and Prescription*, J. L. Roitman, Ed., Lippincott Williams & Wilkins, Baltimore, MD, 2001.
20. L. Brouha, *Physiology in Industry—Evaluation of Industrial Stresses by the Physiological Reactions of the Workers*, Pergamon Press, Oxford, U.K., 1960.
21. F. H. Fuller and J. P. E. Smith, Evaluation of heat stress in a hot workshop by physiological measurements, *American Industrial Hygiene Association Journal* 42(1): 32–37 (1981).
22. NIOSH, OSHA, USCG, and EPA, *Occupational Safety and Health Guidance Manual for Hazardous Waste Site Activities*, DHHS(NIOSH), Washington, DC, pp. 85–115, 1985.
23. A. R. Lind and D. E. Bass, Optimal exposure time for development of acclimatization to heat, *Federation Proceedings* 22: 704–708 (1963).
24. J. F. Greenleaf and M. H. Harrison, Water and electrolytes, in *Exercise, Nutrition and Health*, D. K. Layman, Ed., American Chemical Society, New York, 1985.
25. J. Malchaire, H. J. Gebhardt, and A. Piette, Strategy for evaluation and prevention of risk due to work in thermal environments, *The Annals of Occupational Hygiene* 43(5): 367–376 (1999).
26. H. S. Belding, T. F. Hatch, B. A. Hertig, and M. L. Riedesel, Recent developments in understanding of effects of exposure to heat, in *Proceedings of 13th International Congress on Occupational Health*, Craftsman, New York, 1961.
27. NIOSH, *Recommended Criteria for Exposures to Occupational Heat Stress—Revised*, DHHS (NIOSH), Washington, DC, pp. 86–113, 1986.
28. T. E. Bernard and W. L. Kenney, Heart rate recovery, in *American Industrial Hygiene Conference*, San Francisco, CA, 1988.

29. International Organization for Standardization (ISO), *ISO 7726: Ergonomics of the Thermal Environment—Instruments for Measuring Physical Quantities*, ISO, Geneva, Switzerland, 1998.
30. C. P. Yaglou and D. Minard, Control of heat casualties at military training centers, *A.M.A. Archives of Industrial Health* 16: 302–316 (1957).
31. ACGIH®, Heat stress and strain TLV®, in *Threshold Limit Values and Biological Exposure Indices for Chemical Substances and Physical Agents*, ACGIH®, Cincinnati, OH, 2011.
32. J. D. Ramsey, Abbreviated guidelines for heat stress exposure, *American Industrial Hygiene Association Journal* 39(6): 491–495 (1978).
33. T. E. Bernard, V. Caravello, S. W. Schwartz, and C. D. Ashley, WBGT clothing adjustment factors for four clothing ensembles and the effects of metabolic demands, *Journal of Occupational and Environmental Hygiene* 5(1): 1–5 (2008).
34. T. E. Bernard, C. L. Luecke, S. W. Schwartz, K. S. Kirkland, and C. D. Ashley, WBGT clothing adjustments for four clothing ensembles under three relative humidity levels, *Journal of Occupational and Environmental Hygiene* 2(5): 251–256 (2005).
35. U.S. Navy, Chapter 3: Prevention of heat and cold stress injuries, in *Manual of Naval Preventive Medicine* (*NAVMED P-5010-3* [*Rev. 2–2009*]), Washington, DC, 2009.
36. T. E. Bernard, Available from: http://personal.health.usf.edu/tbernard/thermal/index.html (accessed October 24, 2011).
37. J. Malchaire, Available from: http://www.deparisnet.be/chaleur/Chaleur.htm#programmes (accessed October 24, 2011).
38. V. Caravello, E. A. McCullough, C. D. Ashley, and T. E. Bernard, Apparent evaporative resistance at critical conditions for five clothing ensembles, *European Journal of Applied Physiology* 104(2): 361–367 (2008).
39. W. L. Kenney, D. J. Mikita, G. Havenith, S. M. Puhl, and P. Crosby, Simultaneous derivation of clothing-specific heat exchange coefficients, *Medicine and Science in Sports and Exercise* 25(2): 283–289 (1993).
40. International Organization for Standardization, *ISO 9920: Ergonomics of the Thermal Environment—Estimation of Thermal Insulation and Water Vapour Resistance of a Clothing Ensemble*, ISO, Geneva, Switzerland, 2007.
41. P. W. Logan and T. E. Bernard, Heat stress and strain in an aluminum smelter, *American Industrial Hygiene Association Journal* 60: 659–665 (1999).
42. T. E. Bernard and W. L. Kenney, Rationale for a personal monitor for heat strain, *American Industrial Hygiene Association Journal* 55(6): 505–514 (1994).
43. S. Ueno, R. Long, S. W. Schwartz, C. D. Ashley, S.-I. Sawada, and T. E. Bernard, Utility of predicted heat strain to limit short-term heat stress exposures, in *12th International Conference on Environmental Ergonomics*, Boston, MA, 2009.
44. S. Ueno, S.-I. Sawada, and T. E. Bernard, Modifications to predicted heat strain (PHS) (ISO7933), in *12th International Conference on Environmental Ergonomics*, Boston, MA, 2009.
45. T. Bernard, C. Ashley, J. Trentacosta, V. Kapur, and S. Tew, Critical heat stress evaluation of clothing ensembles with different levels of porosity, *Ergonomics* 53(8): 1048–1058 (2010).
46. A. R. Lind, Effect of individual variation on upper limit of prescriptive zone of climates, *Journal of Applied Physiology* 28(1): 57–62 (1970).
47. A. R. Lind, P. W. Humphreys, K. J. Collins, K. Foster, and K. F. Sweetland, Influence of age and daily duration of exposure on responses of men to work in heat, *Journal of Applied Physiology* 28(1): 50–56 (1970).
48. NIOSH, *Recommended Criteria for Exposures to Occupational Heat Stress*, DHHS (NIOSH), Washington, DC, pp. 72–10269, 1972.
49. K. V. Kuhlemeier, J. M. Miller, F. N. Dukes-Dobos, and R. Jensen, Determinants of the prescriptive zone of industrial workers, *Journal of Applied Physiology* 43(2): 347–351 (1977).
50. H. S. Belding and E. Kamon, Evaporative coefficients for prediction of safe limits in prolonged exposures to work under hot conditions, *Federation Proceedings* 32(5): 1598–1601 (1973).

51. A. R. Lind, Physiological effects of continuous or intermittent work in the heat, *Journal of Applied Physiology* 18: 57–60 (1963).
52. J. D. Ramsey, Heat stress standard: OSHA's advisory committee recommendations, *National Safety News* 68: 89–95 (1975).
53. International Organization for Standardization (ISO), *ISO7243: Hot Environments—Estimation of the Heat Stress on Working Man, Based on the WBGT-Index (Wet Bulb Globe Temperature)*, ISO, Geneva, Switzerland, 1989.
54. E. Kamon and B. Avellini, Wind speed limits to work under hot environments for clothed men, *Journal of Applied Physiology* 46(2): 340–345 (1979).
55. W. L. Kenney, WBGT adjustments for protective clothing, *American Industrial Hygiene Association Journal* 48: 576–577 (1987).
56. D. J. O'Connor and T. E. Bernard, Continuing the search for WBGT clothing adjustment factors, *Applied Occupational and Environmental Hygiene* 14(2): 119–125 (1999).
57. C. D. Ashley and T. E. Bernard, Effects of hoods and flame-retardant fabrics on WBGT clothing adjustment factors, *Journal of Occupational and Environmental Hygiene* 5(1): 59–62 (2008).

CHAPTER 29

# Occupational Human Vibration

Michael J. Griffin

CONTENTS

## 29.1 INTRODUCTION

People are exposed to vibration in many environments, yet understanding of the effects of oscillatory motion on the body is far from complete. This chapter summarizes human responses relevant to occupational exposures to vibration. More detail may be found elsewhere (e.g., Griffin, 1990).

Whole-body vibration occurs when the human body is supported on a surface that is vibrating, either when seated, standing, or lying. Whole-body vibration occurs in all forms of transport and may be associated with discomfort, interference with activities, and health risks.

Hand-transmitted vibration is the vibration that enters the body through the hands. It is caused by various processes in industry, agriculture, mining, and construction when vibrating tools are grasped or pushed by the hands or fingers. Occupational exposure to hand-transmitted vibration is mostly associated with the development of disorders of the fingers, hands, and arms.

## 29.2 BACKGROUND AND SIGNIFICANCE TO OCCUPATIONAL ERGONOMICS

All forms of transport and some industrial machines expose the body to whole-body vibration. Prior to the mechanization of transport, the body was exposed to motion from walking and running, riding on animals, and sailing in boats. The oscillations caused by traditional forms of transport were different from those occurring in modern transport, the tasks undertaken in transport have changed, and people, their postures, and their expectations have also changed. Furthermore, ergonomists can now design seats, controls, and displays to minimize the effects of vibration.

All powered hand tools generate hand-transmitted vibration. Prior to the mechanization of tools, hand-powered hammers exposed the fingers, hands, and arms to intermittent shocks, whereas tools powered by electricity, compressed air, hydraulic pressure, or an internal combustion engine produce shocks at a higher rate (i.e., vibration). Modern tools can produce vibration from the high-speed rotation of cutting or polishing surfaces (e.g., grinding tools), the impact of tool parts on material (e.g., percussive drills), or by the oscillation of an engine (e.g., chain saws). The hands of millions of workers are now exposed to types of vibration rarely experienced 100 years ago. The development of light and convenient hand-powered tools has also resulted in increased durations of tool use. The adverse health effects of the vibration from powered hand tools are therefore caused by the operation of man-made machines and were unknown before the twentieth century.

## 29.3 CHARACTERISTICS OF VIBRATION

### 29.3.1 Vibration Magnitude

The magnitude of an oscillation can be seen as the distance between the extremities reached by the motion (i.e., the peak-to-peak displacement), or the maximum deviation from some central position (i.e., the peak displacement). However, the magnitude of the vibration experienced by people is not usually expressed in terms of either the displacement or the peaks of the oscillation. The magnitude of vibration is expressed by an average measure of the acceleration of the oscillatory motion, usually the root-mean-square (r.m.s.) value expressed in meters per second per second (i.e., m $s^{-2}$ r.m.s.). For a sinusoidal motion, the r.m.s. value is the peak value divided by $\sqrt{2}$. For rotational vibration, the magnitude of vibration is expressed in radians per second per second (i.e., rad $s^{-2}$).

When observing vibration, it is sometimes possible to estimate the displacement caused by the motion. For a sinusoidal motion, the acceleration, $a$, can be calculated from the frequency, $f$, in hertz (Hz) and the displacement $d$:

$$a = (2\pi f)^2 d$$

So, for example, a sinusoidal motion with a frequency of 1 Hz and a peak-to-peak displacement of 0.1 m will have an acceleration of 3.95 m $s^{-2}$ peak-to-peak, 1.97 m $s^{-2}$ peak, and 1.40 m $s^{-2}$ r.m.s. The previous equation can be used to convert acceleration measurements to corresponding displacements throughout this chapter, although the conversion is only accurate when the motion has a sinusoidal waveform.

### 29.3.2 Vibration Frequency

The frequency of vibration is expressed in cycles per second using the SI unit hertz (Hz) (where 1 Hz = 1 cycle/s). Oscillations at frequencies less than about 0.5 Hz can cause motion sickness. The frequencies of greatest significance to whole-body vibration are usually at the lower end of the range from 0.5 to 100 Hz. Frequencies of hand-transmitted vibration as high as 1000 Hz have been suspected of causing detrimental effects on the fingers.

Mechanical systems have resonance frequencies at which they exhibit a maximum response to vibration. The human body is complex and has many modes of vibration, but it is mostly highly damped so that a significant response occurs at all frequencies and not

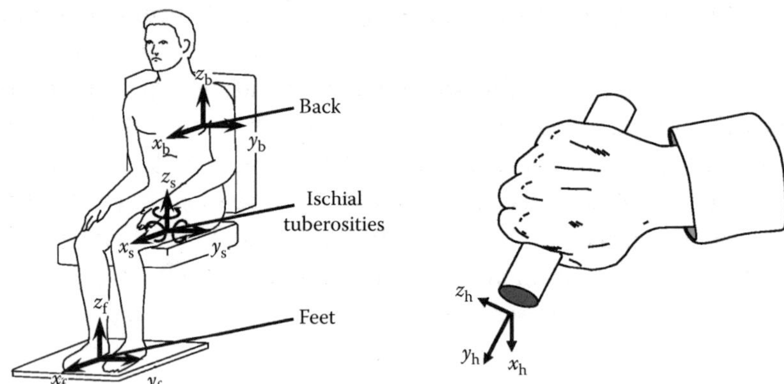


FIGURE 29.1 Axes of vibration used to measure exposures to whole-body vibration and hand-transmitted vibration.

only at resonance frequencies. Resonance frequencies of the body depend on the posture and orientation of the body and differ between individuals. Frequency weightings for both whole-body vibration and hand-transmitted vibration reflect, in a general way, the influence of various modes of vibration in the body, including resonance frequencies.

### 29.3.3 Vibration Direction

The responses of the body differ according to the direction of the motion. The three principal directions for seated and standing persons are fore-and-aft ($x$-axis), lateral ($y$-axis), and vertical ($z$-axis) (Figure 29.1). The vibration is usually measured at the interface between the body and the surface supporting the body (i.e., between the seat and the pelvis, at the ischial tuberosities, for a seated person; beneath the feet for a standing person). The vibration of a backrest and footrest can influence human responses to vibration, so the vibrations in the translational and rotational axes with an origin at the ischial tuberosities of a seated person, and between the back and a backrest, and the feet and a footrest are measured. To evaluate exposures to hand-transmitted vibration, the vibration is measured in three translational axes at both hands (Figure 29.1).

### 29.3.4 Vibration Duration

Many human responses to vibration depend on the total duration of exposure to vibration. Additionally, the duration of measurement may affect the measured magnitude of the vibration. The r.m.s. acceleration may not provide a good indication of vibration severity if the vibration is intermittent, contains shocks, or otherwise varies in magnitude from time to time. For these types of vibration, the VDV is preferred (see the following).

## 29.4 WHOLE-BODY VIBRATION

### 29.4.1 Effects on Comfort

The human body can detect magnitudes of vibration lower than those required to cause damage to structures. It is not practical to eliminate the perception of vibration in transport and it can be difficult to prevent the perception of vibration in fixed structures (e.g., buildings). It is possible to estimate the percentage of people who are likely to feel a specific vibration. For magnitudes of vibration greater than the threshold or perception, the extent of subjective reactions is suggested by a semantic scale of discomfort (see British Standard, 6841:1987; International Standard, 2631:1997).

The limit to prevent vibration discomfort varies between environments (e.g., between buildings and transport), between different types of transport (e.g., between cars, motorbikes, trucks), and between types of vehicles (e.g., sports cars and limousines). The design limit depends on external factors (e.g., cost and speed) and the comfort in alternative environments (e.g., competitive vehicles). There has been a gradual increase in the expectation of comfort as techniques have become available to make improvements.

#### *29.4.1.1 Effects of Vibration Magnitude*

Very approximately, the absolute threshold for the perception of vertical whole-body vibration in the frequency range of 1–100 Hz is 0.01 m $s^{-2}$ r.m.s.; a magnitude of 0.1 m $s^{-2}$ r.m.s. will be easily noticeable; magnitudes around 1 m $s^{-2}$ r.m.s. are usually considered uncomfortable; magnitudes of 10 m $s^{-2}$ r.m.s. are usually dangerous. The precise values depend on vibration frequency and exposure duration, and they are different for other axes of vibration (Morioka and Griffin, 2008).

To a useful approximation, a doubling of vibration magnitude corresponds to a doubling of discomfort. A halving of vibration magnitude can therefore produce a considerable improvement in comfort.

#### *29.4.1.2 Effects of Vibration Frequency and Direction*

The dynamic responses of the body and the relevant physiological and psychological processes dictate that human responses to vibration depend on vibration frequency and vibration direction (Morioka and Griffin, 2006a). Frequency weightings $W_b$ to $W_f$ are defined in British Standard 6841 (1987). International Standard 2631 (1997) defines similar weightings, except that for some applications weighting $W_k$ is used in preference to weighting $W_b$. Table 29.1 shows how the frequency weightings shown in Figure 29.2 should be applied to the 12 axes illustrated for whole-body vibration in Figure 29.1. The weightings $W_g$ and $W_f$ are not required to predict vibration discomfort: $W_g$ is used to evaluate the severity of vibration with respect to interference with activities and $W_f$ is used to predict motion sickness caused by vertical oscillation (see the following).

#### *29.4.1.3 Effects of Vibration Duration*

Discomfort tends to increase with increasing duration of exposure to a vibration. The precise rate of change may depend on many factors, but a simple fourth-power time dependency is used to approximate how discomfort varies with exposure durations from the shortest possible shock to a full day of vibration exposure (i.e., $(\text{acceleration})^4 \times \text{duration} = \text{constant}$). This time dependency appears to be more consistent with available information and expectations than the only current alternative, which is a second-power (sometimes called "energy") time dependence.

TABLE 29.1 Application of Frequency Weightings for Evaluating Whole-Body Vibration with Respect to Discomfort and Health in BS 6841 (1987) and ISO 2631-1 (1997)

| | | | Axis Multiplying Factor | | | |
|---|---|---|---|---|---|---|
| | | | Comfort | | Health | |
| **Input Position** | **Axis** | **Frequency Weighting** | **BS 6841** | **ISO 2631-1** | **BS 6841** | **ISO 2631-1** |
| **Seat** | $x_s$ | $W_d$ | 1.0 | 1.0 | 1.0 | 1.4 |
| | $y_s$ | $W_d$ | 1.0 | 1.0 | 1.0 | 1.4 |
| | $z_s$ | $W_b$ | 1.0 | 1.0 (or $W_k$) | 1.0 | |
| | | $W_k$ | — | 1.0 (or $W_b$) | | 1.0 |
| | $r_x$ (roll) | $W_e$ | 0.63 | 0.63 | | |
| | $r_y$ (pitch) | $W_e$ | 0.40 | 0.40 | | |
| | $r_z$ (yaw) | $W_e$ | 0.20 | 0.20 | | |
| **Seat back** | $x_b$ | $W_c$ | 0.80 | 0.80 | 0.8 | |
| | $y_b$ | $W_d$ | 0.50 | 0.50 | | |
| | $z_b$ | $W_d$ | 0.40 | 0.40 | | |
| **Feet** | $x_f$ | $W_b$ | 0.25 | 0.25 | | |
| | $y_f$ | $W_b$ | 0.25 | 0.25 | | |
| | $z_f$ | $W_b$ | 0.40 | 0.40 | | |

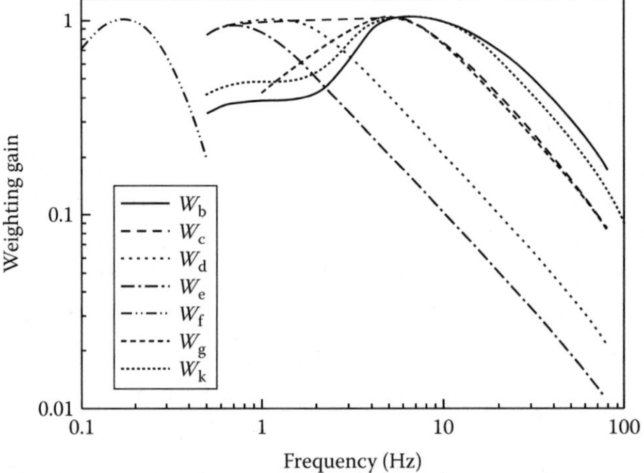


FIGURE 29.2 Acceleration frequency weightings for whole-body vibration and motion sickness as defined in BS 6841 (1987) and ISO 2631 (1997).

### *29.4.1.4 Vibration in Buildings*

Acceptable magnitudes for vibration in buildings are close to vibration perception thresholds. The effects of vibration in buildings are assumed to depend on the use of the building in addition to the frequency, direction, and duration of vibration. General guidance is given in International Standard 2631-2:2003. A procedure for evaluating and assessing the severity of vibration or shock in buildings using the "vibration dose value" is defined in British Standard 6472:2008.

### 29.4.2 Effects on Health

Acute injury to the body can be caused by short exposure to high magnitudes of whole-body vibration or shock. Chronic disease may arise from prolonged exposures to vibration.

Various studies have reported disorders among persons exposed to vibration from occupational, sport, and leisure activities. The studies do not all agree on either the type or the extent of disorders, and rarely have the findings been related to measurements of vibration exposure. However, it is widely believed that disorders of the back (back pain, displacement of inter-vertebral disks, degeneration of spinal vertebrae, osteoarthritis, etc.) may be associated with vibration exposure (e.g., Dupuis and Zerlett, 1986; Griffin, 1990; National Institute for Occupational Safety and Health, 1997; Bovenzi and Hulshof, 1999; Bovenzi, 2009).

For persons exposed to vibration, there are often other potential causes of some of the reported disorders of the back (e.g., poor sitting postures, heavy lifting, tobacco smoking). It is often difficult to decide whether a back disorder is solely, or primarily, caused by vibration.

Other disorders claimed to be due to occupational exposures to whole-body vibration include abdominal pain, digestive disorders, urinary frequency, prostatitis, hemorrhoids, balance and visual disorders, headaches, and sleeplessness.

#### *29.4.2.1 Evaluation of Whole-Body Vibration*

Epidemiological data alone are not sufficient to define how to evaluate whole-body vibration so as to predict the risks to health. A consideration of such data in combination with an understanding of biodynamic responses and subjective responses to vibration is used to provide current guidance. The manner in which the health effects of oscillatory motions depend on the frequency, the direction, and the duration of vibration is currently assumed to be similar to that for vibration discomfort. However, it is assumed that the "total" exposure, rather than the "average" exposure, is important and so a "dose" measure is used.

British Standard 6841 (1987) defines an "action level" based VDVs, and International Standard 2631 (1997) offers two alternative "health guidance caution zones," one based on r.m.s. acceleration and the other on VDVs.

The *vibration dose value* can be considered to be the magnitude of a 1 s duration of vibration that is equally severe to another duration of vibration. The VDV uses a "fourth power" time dependency to accumulate vibration severity over the exposure period from the shortest possible shock to a full day of vibration (see BS 6841, 1987; ISO 2631-1, 1997):

$$\text{Vibration dose value} = \left[\int_{t=0}^{t=T} a^4(t)\,dt\right]^{1/4}$$

If the exposure duration ($t$, seconds) and the frequency-weighted r.m.s. acceleration ($a_{rms}$, m s$^{-2}$ r.m.s.) are known for conditions in which the vibration characteristics are statistically stationary, it can be useful to calculate the "estimated vibration dose value," eVDV:

$$\text{Estimated vibration dose value} = 1.4\,a_{rms}t^{1/4}$$

The eVDV is not applicable to transients, shocks, or repeated shock motions in which the crest factor (peak value divided by the r.m.s. value) is high. The VDV can be used to evaluate the severity of transients, shocks, and repeated shock motions.

#### *29.4.2.2 Assessment according to BS 6841 (1987) and ISO 2631-1 (1997)*

There is no precise limit for preventing disorders caused by whole-body vibration, but standards and directives define useful methods of estimating vibration severity and deciding on the need for preventative measures. With severe vibration exposures, prior consideration of the fitness of the exposed persons and the design of adequate safety precautions may be required. The need for regular warnings and checks on the health of routinely exposed persons may also be considered.

British Standard 6841 (1987) offers the following guidance: "*High vibration dose values will cause severe discomfort, pain and injury. Vibration dose values also indicate, in a general way, the severity of the vibration exposures which caused them. However there is currently no consensus of opinion on the precise relation between vibration dose values and the risk of injury. It is known that vibration magnitudes and durations which produce vibration dose values in the region of 15 ms$^{-1.75}$ will usually cause severe discomfort. It is reasonable to assume that increased exposure to vibration will be accompanied by increased risk of injury.*" Figure 29.3 shows this action level for exposure durations from 1 s to 24 h.

Two different methods of evaluating vibration severity with respect to health effects are offered in International Standard 2631 (1997), and for both methods there are two

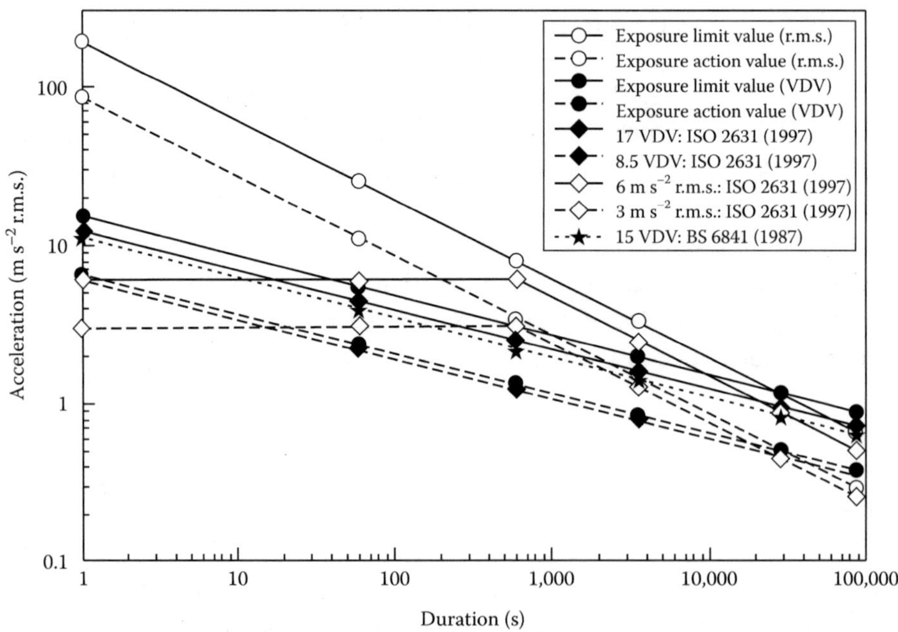


FIGURE 29.3 Comparison between the health guidance caution zones for whole-body vibration in ISO 2631-1 (1997) (3–6 m s$^{-2}$ r.m.s.; 8.5–17 m s$^{-1.75}$), 15 m s$^{-1.75}$ action level implied in BS 6841 (1987), the "exposure limit values" and "exposure action values" for whole-body vibration in the EU Physical Agents (Vibration) Directive.

boundaries. When using the VDV, it is suggested that for exposures less than a VDV of 8.5 m $s^{-1.75}$ *"health risks have not been objectively observed,"* between 8.5 and 17 m $s^{-1.75}$ *"caution with respect to health risks is indicated,"* and above 17 m $s^{-1.75}$ *"health risks are likely."* The two boundaries define a "VDV health guidance caution zone" (see Figure 29.3). The alternative method of evaluation uses a time dependency in which the acceptable vibration does not vary with duration between 1 and 10 min and then decreases in inverse proportion to the square root of duration from 10 min to 24 h. This method suggests an "r.m.s. health guidance caution zone," but the method is not fully defined in the text, it allows very high accelerations at short durations, it conflicts with the VDV method, and cannot be extended to exposure durations less than 1 min (Figure 29.3).

Unlike the use of r.m.s. acceleration, the use of the VDV does not allow extraordinarily high vibration magnitudes at short durations (i.e., a few or many minutes). Any exposure to continuous vibration, or intermittent vibration, or repeated shock may be compared with the action level by calculating the VDV. It would be unwise to exceed the guidance given by VDVs without consideration of the possible health effects of an exposure to vibration or shock.

#### *29.4.2.3 EU Machinery Safety Directive*

The Machinery Safety Directive of the European Community (89/392/EEC) states the following: *"machinery must be so designed and constructed that risks resulting from vibrations produced by the machinery are reduced to the lowest level, taking account of technical progress and the availability of means of reducing vibration, in particular at source"* (Council of the European Communities, 1989). Instruction handbooks for machinery causing whole-body vibration should specify the frequency-weighted acceleration if it exceeds a stated value [currently a frequency-weighted acceleration of 0.5 m $s^{-2}$ r.m.s. when measured according to ISO 2631 (1997)]. The vibration magnitude will depend on the test conditions and, although the test procedures must be specified, their selection leaves much control over the measured value to the manufacturer. Standardized test procedures are being prepared; the vibration magnitudes currently quoted by a manufacturer may not be representative of the vibration experienced when the machinery is operated in work.

#### *29.4.2.4 EU Physical Agents (Vibration) Directive*

The Parliament and Commission of the European Community have defined "minimum health and safety requirements" for the exposure of workers to the risks arising from vibration (The European Parliament and the Council of the European Union, 2002). For whole-body vibration, the Directive defines an 8 h equivalent "exposure action value" of 0.5 m $s^{-2}$ r.m.s. (or a VDV of 9.1 m $s^{-1.75}$) and an 8 h equivalent "exposure limit value" of 1.15 m $s^{-2}$ r.m.s. (or a VDV of 21 m $s^{-1.75}$). All member states of the European Union now have laws to enforce this Directive.

The Directive says that workers shall not be exposed to vibration in excess of the "exposure limit value." If the "exposure action values" are exceeded, the employer shall establish and implement a program of technical and/or organizational measures intended to reduce to a minimum exposure to mechanical vibration and the attendant risks. The Directive says workers exposed to vibration in excess of the "exposure action values" shall be entitled to appropriate health surveillance. Health surveillance is also required if there is any reason to suspect that workers may be injured by the vibration even if the "exposure action value" is not exceeded.

The probability of injury arising from occupational exposures to whole-body vibration at the "exposure action value" and the "exposure limit value" cannot be estimated because epidemiological studies have not yet produced dose–response relationships. However, it seems clear that the Directive does not define "safe exposures" to whole-body vibration: the r.m.s. values are associated with extraordinarily high magnitudes of vibration (and shock) when the exposures are short and these exposures may be assumed to be hazardous (see Figure 29.3; Griffin, 2004). The VDV procedure suggests more reasonable vibration magnitudes, especially for short duration exposures and shocks.

### 29.4.3 Effects on Performance

Vibration can interfere with the acquisition of information (e.g., by the eyes), the output of information (e.g., by hand or foot movements), or the complex central processes that relate input to output (e.g., learning, memory, decision making). Effects of oscillatory motion on human performance may impair safety.

Figure 29.4 illustrates the various component parts of human activity where vibration may affect performance (Griffin, 1990). The greatest effects, and the greatest understanding of the effects, occur with input processes (mainly vision) and output processes (mainly continuous manual control). In both cases there may be disturbance occurring entirely outside the body (e.g., vibration of a viewed display or vibration of a handheld control), disturbance at the input or output (e.g., movement of the eye or hand), and disturbance affecting the peripheral nervous system (i.e., afferent or efferent system). Central processes may also be affected by vibration, but understanding is currently too limited to make confident generalized statements.

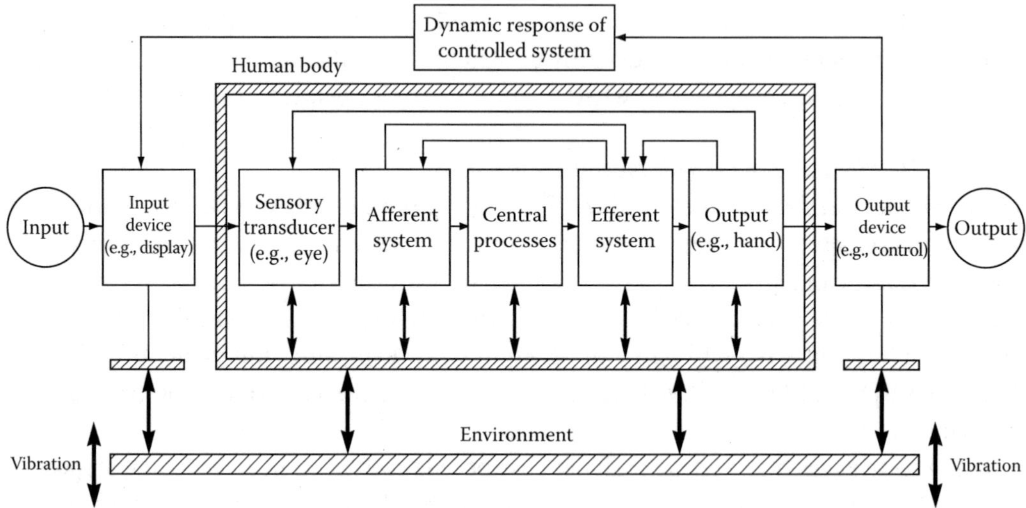


FIGURE 29.4 Information flow in a simple system and the areas where vibration may affect human activities. (From Griffin, M.J., *Handbook of Human Vibration*, Academic Press, London, U.K., 1990.)

The effects of vibration on vision and manual control are primarily caused by the movement of the affected part of the body (i.e., eye or hand). The effects may be reduced by reducing the transmission of vibration to the eye or to the hand or by making the task less susceptible to disturbance (e.g., increasing the size of a display or reducing the sensitivity of a control). Often, the effects of vibration on vision and manual control can be reduced by redesign of the task.

Reading from a book in a moving vehicle may be difficult if there is much vibration. This may arise because the book is vibrating, the eye is vibrating, or both the book and the eye are vibrating. There are many variables that affect visual performance in these conditions, with some variables being difficult to quantify. Published reports of experiments may appear to present a confused picture unless all the relevant variables are taken into account (Griffin and Lewis, 1978).

The most obvious effect of vibration on a control task is the direct mechanical jostling of the hand causing unwanted movement of the control. This is sometimes called "breakthrough" or "feedthrough." Because the errors produced by this means are directly caused by the vibration, they are statistically correlated with the vibration and are also called "vibration-correlated error." The inadvertent movement of the pencil caused by jostling while writing in a vehicle is a form of vibration-correlated error. Indeed, an attempt to draw a straight line will result in a graphical display on the paper of the vibration-induced movements of the hand.

Time series analysis techniques can be used to quantify how much tracking error in a control task is correlated with the vibration. In a simple tracking task in which the operator is required to follow movements of a target, some of the error will be correlated with the target movements. This is called "input-correlated error" and often mainly reflects the inability of an operator to follow the target without delays inherent in visual, cognitive, and motor activity. The part of the tracking error that is not correlated with either the vibration or the tracking task is called the "remnant." This includes operator-generated noise and any source of nonlinearity: Drawing a freehand straight line does not result in a perfectly straight line even in the absence of environmental vibration. The effects of vibration on vision can result in increased remnant with some tracking tasks. There are studies showing that vibration, usually at frequencies above about 20 Hz, can interfere with neuromuscular processes and also increase the remnant (McLeod and Griffin, 1989).

It seems probable that simple cognitive tasks (e.g., simple reaction time) are not affected by vibration other than by changes in arousal or motivation or by direct effects on input and output processes. This may also be true for some complex cognitive tasks. However, the scarcity of substantial experimental studies allows the possibility of real and significant cognitive effects of vibration (Sherwood and Griffin, 1992).

### 29.4.4 Motion Sickness

Motion sickness is not an illness but a normal response to motion that adversely affects many fit and healthy people. A variety of motions can cause sickness and thereby reduce the comfort, impede the activities, and degrade the well-being of both those directly affected and those associated with people who experience motion sickness.

Motion sickness is caused by motions at the low frequencies associated with normal postural control of the body. Various investigations have resulted in the formulation of a frequency weighting, $W_f$ (see Figure 29.2), and the definition of a "motion sickness dose value." The frequency weighting $W_f$ reflects greatest sensitivity to acceleration in the range 0.125–0.25 Hz, with a rapid reduction in sensitivity at greater frequencies. The motion sickness dose value (i.e., MSDV) predicts the probability of sickness from knowledge of the frequency and magnitude of vertical oscillation (see Lawther and Griffin, 1987; International Standard 2631-1, 1997):

$$\text{Motion sickness dose value } = a_{\text{rms}} t^{1/2}$$

where

$a_{\text{rms}}$ is the r.m.s. value of the frequency-weighted acceleration (i.e., m $s^{-2}$ r.m.s.)
$t$ is the exposure period (s)

The percentage of "unadapted" adults who are expected to vomit is given by $^1/_3$ MSDV. (These relationships have been derived from exposures in which up to 70% of persons vomited during exposures lasting between 20 min and 6 h.)

For vertical oscillation, the motion sickness dose value seems to provide useful predictions of sickness and may be useful where vertical oscillation is the prime cause of sickness, such as on ships, hovercraft and hydrofoil (Lawther and Griffin, 1988), and in some aircraft (Griffin, 2000; Turner et al., 2000). Vertical oscillation is not the principal cause of sickness in cars (Griffin and Newman, 2004), or in road coaches (Turner and Griffin, 1999), or in tilting trains (Donohew and Griffin, 2007). The motion sickness in surface transport seems to be mainly associated with horizontal acceleration arising from changes in speed and cornering and can be influenced by roll and pitch motions and vision. It is not yet possible to predict with confidence the probability of motion sickness in surface transport.

### 29.4.5 Control of Whole-Body Vibration

Whenever possible, vibration should be reduced at the source. In the case of vehicles, this may involve reducing the undulations of the terrain or reducing the speed of travel. Reduction at source is not always practicable or sufficient, so ergonomists may be asked to alleviate a problem by other means.

An understanding of the characteristics of the vibration environment and how vibration is transmitted to the body is needed. For example, the magnitude of vibration often varies with location: lower magnitudes will be experienced in some areas.

Seats and beds can be designed to attenuate vibration, but there is no equivalent method of reducing the transmission of the vibration of a floor to standing persons. Table 29.2 lists some preventive measures that may be considered.

Most seats exhibit a resonance at low frequencies, resulting in greater magnitudes of vertical vibration on the seat than on the floor at the resonance frequency. At much higher frequencies there is usually attenuation of vibration. The resonance frequencies of common seats are often in the region of 4 Hz, with the amplification at resonance influenced by the damping in the seat: Increases in damping tend to reduce the amplification at resonance

TABLE 29.2 Summary of Preventive Measures to Consider When Persons Are Exposed to Whole-Body Vibration

| Group | Action |
|---|---|
| Management | Seek technical advice |
| | Seek medical advice |
| | Warn exposed persons |
| | Train exposed persons |
| | Review exposure times |
| | Institute policy on removal from work |
| Machine manufacturers | Measure vibration |
| | Design to minimize whole-body vibration |
| | Optimize suspension design |
| | Optimize seating dynamics |
| | Practice ergonomic design to provide good posture, etc. |
| | Provide guidance on machine maintenance |
| | Provide guidance on seat maintenance |
| | Provide warning of dangerous vibration |
| Technical at workplace | Measure vibration exposure |
| | Provide appropriate machines |
| | Select seats with good attenuation |
| | Maintain machines |
| | Inform management |
| Medical | Perform pre-employment screening |
| | Perform routine medical checks |
| | Record all signs and reported symptoms |
| | Warn workers with predisposition |
| | Advise on consequences of exposure |
| | Inform management |
| Exposed persons | Use machine properly |
| | Avoid unnecessary vibration exposure |
| | Check that seat is properly adjusted |
| | Adopt good sitting posture |
| | Check condition of machine |
| | Inform supervisor of vibration problems |
| | Seek medical advice if symptoms appear |
| | Inform employer of relevant disorders |

*Source:* Adapted from Chapter 5 Griffin, M.J., *Handbook of Human Vibration*, Academic Press, London, U.K., 1990.

but increase the transmissibility at high frequencies. Large variations in transmissibility between seats result in significant differences in vibration exposure (Corbridge et al., 1989).

#### *29.4.5.1 Isolation Efficiency of Seats*

The "seat effective amplitude transmissibility" (SEAT) provides a simple numerical indicator of the isolation efficiency of a seat for a specific application (Griffin, 1990). A SEAT value greater than 100% indicates that, overall, the vibration on the seat is "worse" than the vibration on the floor. If a seat has a SEAT value less than 100%, it has provided some attenuation of the vibration. Seats should be designed to have the lowest possible SEAT value.

#### *29.4.5.2 Suspension Seats*

Suspension seats have a separate suspension mechanism located beneath the seat pan. These seats, used in some off-road vehicles, trucks, and coaches, have low resonance frequencies (often around 2 Hz) and so can attenuate vibration at frequencies greater than about 3 Hz. The transmissibilities of these seats are usually determined by the seat manufacturer, but their isolation efficiencies vary with operating conditions.

#### *29.4.5.3 Seat Testing Standards*

There are standards for the measurement of seat dynamic performance and the reporting of relevant values (e.g., International Standards, ISO 10326-1, 1992; ISO 7096, 2000; ISO 5007, 2003).

## 29.5 HAND-TRANSMITTED VIBRATION

Hand-transmitted vibration can cause discomfort, interfere with the performance of activities, and produce health disorders. All effects of hand-transmitted vibration are influenced by the biodynamic responses of the fingers, hands, and arms (Concettoni and Griffin, 2009). The discomfort or annoyance caused by hand-transmitted vibration may be estimated using frequency weightings determined from equivalent comfort contours (Morioka and Griffin, 2006b).

Prolonged regular exposure of the fingers or the hands to vibration or repeated shock can give rise to various signs and symptoms of disorder. The precise extent and interrelation of the signs and symptoms is not fully understood, but five types of disorders can be identified (see Table 29.3). The term "hand-arm vibration syndrome" (HAVS) is used to refer to some combination of these disorders. The onset and development of each disorder must be dependent on several variables, such as the vibration characteristics, the dynamic response of the fingers or hand, individual susceptibility to damage, and other aspects of the environment.

### 29.5.1 Vascular Disorders (Vibration-Induced White Finger)

The first published cases of vibration-induced white finger (VWF) are generally acknowledged to be those reported in Italy by Loriga (1911). Not many years later, cases were

TABLE 29.3 Five Types of Disorder Associated with Exposure to Hand-Transmitted Vibration

| Type | Disorder |
|---|---|
| A | Circulatory disorders |
| B | Bone and joint disorders |
| C | Neurological disorders |
| D | Muscle disorders |
| E | Other general disorders (e.g., central nervous system) |

documented at limestone quarries in Indiana (e.g., Hamilton, 1918). Among workers reported to be using pneumatic hammers in the Indiana quarries, there were attacks of numbness and blanching of the fingers that came on suddenly under the influence of cold and then disappeared. The prevalence of the symptoms was 86% among 50 men working on granite, 69% among 38 men working on marble, and 56% among 78 men working on limestone. It was reported that the condition appeared within about 1 year of first using the tools. A survey conducted 60 years later in the same quarry reported no change in the design of the air hammers used for stone-cutting and found an 80% prevalence of VWF among stonecutters (Taylor et al., 1984). VWF has been reported to occur in many other widely varied occupations in which the fingers are exposed to vibration (Taylor and Pelmear, 1975; Griffin, 1990).

##### 29.5.1.1 Signs and Symptoms

The condition now most commonly known as VWF is characterized by intermittent blanching of the fingers. The fingertips are first to blanch, but the affected area may extend to all of one or more fingers with continued vibration exposure. Attacks of blanching are precipitated by cold and therefore usually occur in cold conditions or when handling cold objects. The blanching lasts until the fingers are rewarmed and vasodilation allows the return of the blood circulation.

Many years of vibration exposure often occur before the first attack of blanching is noticed. Some persons have other signs and symptoms, such as numbness and tingling; cases of cyanosis and, rarely, gangrene have also been reported. It is not yet clear to what extent these signs are causes of, caused by, or unrelated to, the attacks of white finger.

##### 29.5.1.2 Diagnosis

An individual cannot be assumed to have VWF merely because there are attacks of blanching. It is necessary to exclude other known causes of similar symptoms (by medical examination) and also to exclude so-called primary Raynaud's disease (also called constitutional white finger). This exclusion cannot yet be achieved with complete confidence, but if there is no family history of the symptoms, if the symptoms did not occur before the first significant exposure to vibration, and if the symptoms and signs are confined to areas in contact with the vibration (e.g., the fingers, not the ears, etc.), they will often be assumed to indicate VWF.

The diagnosis of VWF is assisted by standardized tests of finger systolic blood pressure following finger cooling (ISO 14835-2, 2005b) and the measurement of finger rewarming times following cooling (ISO 14835-1, 2005b). Normal responses to these tests are also provided (Lindsell and Griffin, 2002).

The severity of the effects of hand-transmitted vibration is sometimes recorded by reference to the "stage" of disorder. The staging of VWF is based on the verbal statements of the affected person. In the Stockholm Workshop staging system of VWF, the staging is influenced by both the frequency of attacks of blanching and the areas of the digits affected by blanching (Table 29.4).

A "scoring system" is used to record the areas of the digits affected by blanching, or numbness, or tingling (Figure 29.5; Griffin, 1990). The scores correspond to areas of

TABLE 29.4 Stockholm Workshop Scale for the Classification of Vibration-Induced White Finger

| Stage | Grade | Description |
|---|---|---|
| 0 | — | No attacks |
| 1 | Mild | Occasional attacks affecting only the tips of one or more fingers |
| 2 | Moderate | Occasional attacks affecting distal and middle (rarely also proximal) phalanges of one or more fingers |
| 3 | Severe | Frequent attacks affecting all phalanges of most fingers |
| 4 | Very severe | As in stage 3, with trophic skin changes in the finger tips |

*Source:* Gemne, G. et al., *Scand. J. Work, Environ. Health,* 13, 275, 1987.
*Note:* If a person has stage 2 in two fingers of the left hand and stage 1 in a finger on the right hand the condition may be reported as 2L(2)/1R(1). There is no defined means of reporting the condition of digits when this varies between digits on the same hand. The scoring system is more helpful when the extent of blanching is to be recorded.

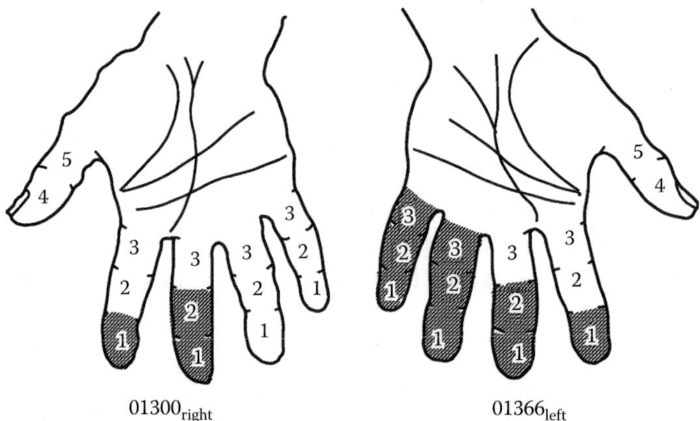


FIGURE 29.5 Method of scoring the areas of the digits affected by blanching. (From Griffin, M.J., *Handbook of Human Vibration*, Academic Press, London, U.K., 1990.) The blanching scores for the hands shown are $01300_{right}$, $01366_{left}$.

blanching on the digits commencing with the thumb. On the fingers a score of 1 is given for blanching on the distal phalanx, a score of 2 for blanching on the middle phalanx, and a score of 3 for blanching on the proximal phalanx. On the thumbs the scores are 4 for the distal phalanx and 5 for the proximal phalanx. The blanching score may be based on statements from the affected person or on the visual observations of a designated observer.

## 29.5.2 Neurological Disorders

Neurological effects of hand-transmitted vibration (e.g., numbness, tingling, and elevated sensory thresholds for touch, vibration, temperature, and pain) have become recognized as separate effects of hand-transmitted vibration that are not necessarily related to symptoms of VWF. A staging system for reporting the extent of vibration-induced neurological effects of vibration has been proposed (see Table 29.5). This stage is not directly related to the results of any specific objective test: it is the subjective impression of a physician that

TABLE 29.5 "Sensorineural Stages" of the Effects of Hand-Transmitted Vibration

| Stage | Symptoms |
|---|---|
| $0_{SN}$ | Exposed to vibration but no symptoms |
| $1_{SN}$ | Intermittent numbness with or without tingling |
| $2_{SN}$ | Intermittent or persistent numbness, reduced sensory perception |
| $3_{SN}$ | Intermittent or persistent numbness, reduced tactile discrimination and/or manipulative dexterity |

*Source:* Brammer, A.J. et al., *Scand. J. Work, Environ. Health,* 13, 279, 1987.

may be influenced by symptoms reported by the affected person and the results of any available clinical or scientific tests. The extent of neurological dysfunction is often assessed with vibrotactile thresholds and thermal thresholds and normal values are available for some tests (e.g., ISO 13091-1, 2001a; Seah and Griffin, 2008). A scheme for clarifying the measurement, evaluation, and reporting peripheral neurological disorders associated with hand-transmitted vibration has been proposed (Griffin, 2008).

### 29.5.3 Muscular Effects

Users of vibratory hand tools sometimes report difficulty with their grip, including reduced dexterity or reduced grip strength (Griffin, 1990). Many of the reports are derived from symptoms rather than signs and could be a reflection of neurological problems. Muscle activity may be of great importance to tool users because a secure grip may be essential to the performance of the job and safe control of the tool. The presence of vibration on a handle may encourage the adoption of a tighter grip than would otherwise be used, and a tight grip may increase the transmission of vibration to the hand. An effect of vibration that results in reduced grip may therefore help to protect an operator from further effects of vibration! Grip strength is often tested using the Jamar grip meter and dexterity is measured using the Purdue pegboard, for which there are normal values for persons of various ages (Haward and Griffin, 2002).

### 29.5.4 Articular Effects

Surveys of the users of handheld tools have found evidence of bone and joint problems, most often among those operating percussive tools, such as those used in metalworking jobs and mining and quarrying (Griffin, 1990). It is speculated that some characteristic of such tools, possibly the low-frequency shocks, is responsible. Some of the reported injuries relate to specific bones and suggest the existence of cysts, vacuoles, decalcification, or other osteolysis, degeneration, or deformity of the carpal, metacarpal, or phalangeal bones. Osteoarthrosis and olecranon spurs at the elbow and other problems at the wrist and shoulder have also been reported. There is not universal acceptance that vibration is the cause of these problems, and there is currently no dose–effect relation that predicts their occurrence. It seems that adherence to current guidance for the prevention of VWF may provide reasonable protection from articular problems.

### 29.5.5 Other Effects

The reported effects of hand-transmitted vibration on the body are not confined to the fingers, hands, and arms. Studies have found a high incidence of problems such as headaches and sleeplessness among tool users. Although these are real problems to those affected, these subjective effects are not accepted as real by all researchers. Some current research seeks to establish a physiological basis for such symptoms.

### 29.5.6 Causes of Hand-Transmitted Vibration

Exposures to vibration from tools vary greatly according to individual tool design and method of use, so it is often not possible to categorize individual tool types as "safe" or "dangerous." However, Table 29.6 lists tools and processes that are sometimes a cause for concern.

### 29.5.7 Control of Hand-Transmitted Vibration

The protection of workers exposed to hand-transmitted vibration involves actions from management, tool manufacturers, technicians, and medical doctors at the workplace and from the tool user. Table 29.7 summarizes some of the actions that may be appropriate. The optimum means of reducing vibration varies according to the cause of the vibration exposure. However, the first principles are similar: reduction at the source, reduced transmission, and reduced individual susceptibility.

Measurements of the vibration will determine whether any other tool or process could give a lower vibration severity. Reduction of exposure duration may include the provision of breaks from exposure during the day and, if possible, prolonged periods away from vibration exposure.

Gloves are sometimes recommended as a means of reducing the transmission of vibration to the hands. However, measurements using the frequency weightings in current standards show that commonly available gloves do *not* normally provide effective attenuation of vibration. Gloves and "cushioned" handles may reduce the transmission of high frequencies of vibration, but current standards imply that these are not usually the primary cause of disorders (Griffin, 1998). International Standard ISO 10819 (1996) defines a standard test for determining whether a glove can be considered an "anti-vibration glove." However, the underlying assumptions mean that a glove that passes the test might not provide useful attenuation whereas a glove that fails the test might provide useful attenuation.

Workers who are exposed to vibration magnitudes sufficient to cause injury should be warned of the possibility of vibration injuries, advised of the symptoms to look out for, and told to seek medical attention if the symptoms appear.

### 29.5.8 Standards and Regulations

International standards and government regulations provide guidance on the measurement, evaluation, and assessment of the severity of exposures to hand-transmitted vibration.

#### *29.5.8.1 Vibration Measurement*

Methods of measuring and evaluating hand-transmitted vibration on tools are defined in ISO 5349-1 (2001b) and ISO 5349-2 (2002). Guidance for the measurement of vibration

TABLE 29.6 Tools and Processes Associated with Vibration Injuries

| Type of Tool | Examples of Tool Type |
|---|---|
| Percussive metal-working tools | Riveting tools (ISO 8662-2) |
| | Caulking tools (ISO 8662-5) |
| | Chipping hammers (ISO 8662-2) |
| | Clinching and flanging tools (ISO 8662-10) |
| | Impact wrenches (ISO 8662-7) |
| | Impact screw drivers (ISO 8662-7) |
| | Nut runners (ISO 8662-7) |
| | Scaling hammers |
| | Needle guns (ISO 8662-14) |
| | Nibbling machines and shears (ISO 8662-10) |
| | Swaging |
| Grinders and other rotary tools | Pedestal grinders |
| | Handheld grinders (ISO 8662-4; ISO 8662-8) |
| | Handheld sanders (ISO 8662-4; ISO 8662-8) |
| | Handheld polishers (ISO 8662-4; ISO 8662-8) |
| | Flex-driven grinders/polishers |
| | Rotary burring tools |
| | Files |
| Percussive hammers and drills used in mining, demolition, road construction, and stone working | Hammer drill (ISO 8662-3) |
| | Rock drills (ISO 8662-3) |
| | Tampers and rammers (ISO 8662-9) |
| | Road breakers (ISO 8662-5) |
| | Stone working tools (ISO 8662-14) |
| Forest and garden machinery | Chain saws (ISO 7505) |
| | Anti-vibration chain saws (ISO 7505) |
| | Brush saws |
| | Mowers (ISO 5395) |
| | Hedge cutters and trimmers |
| | Barking machines |
| | Stump grinders |
| Other processes and tools | Nailing gun (ISO 8662-14) |
| | Stapling gun (ISO 8662-11) |
| | Pad saws (ISO 8662-12) |
| | Circular saws (ISO 8662-12) |
| | Scabblers |
| | Engraving pens |
| | Shoe-pounding-up machines |
| | Vibratory rollers |
| | Concrete vibro-thickeners |
| | Concrete leveling vibrotables |
| | Motorcycle handle bars |
| | Pedestrian-controlled machines |

TABLE 29.7 Some Preventive Measures to Consider When Persons Are Exposed to Hand-Transmitted Vibration

| Group | Action |
|---|---|
| Management | Seek technical advice |
| | Seek medical advice |
| | Warn exposed persons |
| | Train exposed persons |
| | Review exposure times |
| | Institute policy on removal from work |
| Tool manufacturers | Measure tool vibration |
| | Design tools to minimize vibration |
| | Use ergonomic design to reduce grip force, etc. |
| | Design to keep hands warm |
| | Provide guidance on tool maintenance |
| | Provide warning of dangerous vibration |
| Technical at workplace | Measure vibration exposure |
| | Provide appropriate tools |
| | Maintain tools |
| | Inform management |
| Medical | Perform pre-employment screening |
| | Perform routine medical checks |
| | Record all signs and reported symptoms |
| | Warn workers with predisposition |
| | Advise on consequences of exposure |
| | Inform management |
| Tool user | Use tool properly |
| | Avoid unnecessary vibration exposure |
| | Minimize grip and push forces |
| | Check condition of tool |
| | Inform supervisor of tool problems |
| | Keep warm |
| | Wear gloves when safe to do so |
| | Minimize smoking |
| | Seek medical advice if symptoms appear |
| | Inform employer of relevant disorders |

*Source:* Adapted from Griffin, M.J., *Handbook of Human Vibration*, Academic Press, London, U.K., 1990, Chapter 19.

on specific tools is given in appropriate parts of ISO 8662 or ISO 28927 using the principles specified in ISO 20643 (2005a).

#### *29.5.8.2 Vibration Evaluation*

All current standards use the same frequency weighting (called $W_h$) to evaluate hand-transmitted vibration over the frequency range 8–1000 Hz (Figure 29.6). This weighting is applied to measurements of vibration acceleration in each of the three translational axes

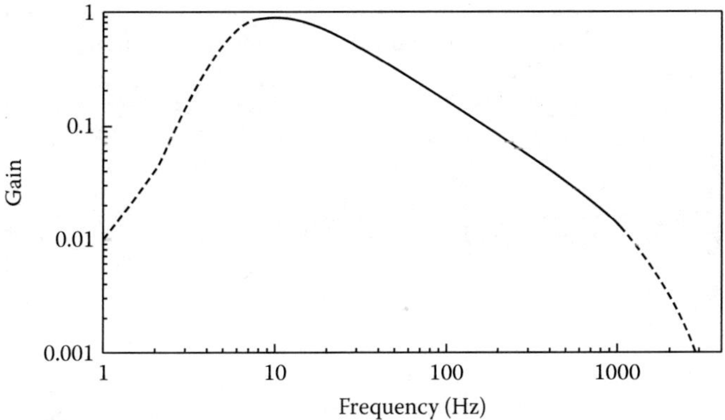


FIGURE 29.6 Frequency weighting $W_h$ for the evaluation of hand-transmitted vibration.

of vibration at the point of entry of vibration to the hand (see Figure 29.1). The standards indicate that the overall severity of hand-transmitted vibration should be calculated from the root-sums-of-squares of the frequency-weighted acceleration in the three axes.

To enable a daily exposure to be reported simply, the standards refer to an equivalent 8 h exposure:

$$a_{\mathrm{hw}(eq,8\mathrm{h})} = A(8) = a_{\mathrm{hw}}\left[\frac{t}{T_{(8)}}\right]^{1/2}$$

where

$t$ is the exposure duration to an r.m.s. frequency-weighted acceleration $a_{\mathrm{hw}}$
$T_{(8)}$ is 8 h (in the same units as $t$)

#### *29.5.8.3 Vibration Assessment according to ISO 5349 (2001)*

An annex to ISO 5349-1 (2001) offers a relation between the lifetime exposure to hand-transmitted vibration, $D_y$, (in years) and the 8 h energy-equivalent daily exposure $A(8)$ for conditions expected to cause 10% prevalence of finger blanching (Figure 29.7):

$$D_y = 31.8\,[A(8)]^{-1.06}.$$

In addition, the informative annex to ISO 5349 (2001) states the following: "*Studies suggest that symptoms of the hand-arm vibration syndrome are rare in persons exposed with an 8-h energy-equivalent vibration total value, A(8), at a surface in contact with the hand, of less than 2 m/s*$^2$ *and unreported for A(8) values less than 1 m/s*$^2$." The scientific evidence suggests very considerable doubts over the frequency weighting and time dependency in the standard, so it should not be expected that ISO 5349:1 will give precise predictions of risk (Griffin, 1997; Griffin et al., 2003).

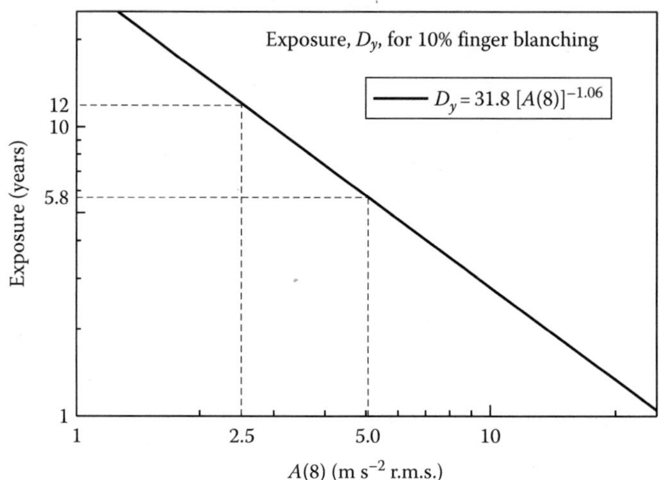


FIGURE 29.7 Relation between daily A(8) and years of exposure expected to result in 10% incidence of finger blanching according to ISO 5349-1:2001. A 10% probability of finger blanching is predicted after 12 years at the EU "exposure action value" and after 5.8 years at the EU "exposure limit value."

### 29.5.8.4 EU Machinery Safety Directive

The European Community Machinery Safety Directive (89/392/EEC) requires that instruction handbooks for handheld and hand-guided machinery specify the equivalent acceleration to which the hands or arms are subjected where this exceeds a stated value (currently a frequency-weighted acceleration of 2.5 m $s^{-2}$ r.m.s.). Many handheld vibratory tools exceed this value. Standard test conditions for the measurement of vibration on many tools (e.g., chipping and riveting hammers, rotary hammers and rock drills, grinding machines, pavement breakers, chain saws) have been defined in parts of ISO 8662 (1998) and ISO 28927.

### 29.5.8.5 EU Physical Agents (Vibration) Directive

The EU Physical Agents Directive defines an 8 h equivalent "exposure action value" of 2.5 m $s^{-2}$ r.m.s. and an 8 h equivalent "exposure limit value" of 5.0 m $s^{-2}$ r.m.s. for hand-transmitted vibration (Council of the European Communities, 1989). The Directive says workers shall not be exposed above the "exposure limit value." If the "exposure action values" are exceeded, the employer shall establish and implement a program of technical and/or organizational measures intended to reduce to a minimum exposure to mechanical vibration and the attendant risks. The Directive requires that workers exposed to mechanical vibration in excess of the exposure action values shall be entitled to appropriate health surveillance, although health surveillance is required wherever there is any reason to suspect that workers may be injured by the vibration, even if the action value is not exceeded.

According to ISO 5349-1 (2001), the onset of finger blanching would be expected in 10% of persons after 12 years at the EU "exposure action value" and after 5.8 years at the "exposure limit value." It is therefore clear that the exposure action value and the exposure limit value in the Directive do not define "safe exposures" to hand-transmitted vibration (Griffin, 2004).

## 29.6 CRITICAL REVIEW OF CURRENT STATUS

### 29.6.1 Whole-Body Vibration

The oscillatory motions to which the human body is exposed are complex. Vibration may affect comfort, the performance of many different activities, and the physiology and pathology of the body. Although understanding of the interactions between vibration and the body is far from complete, standards attempt to predict which types of vibration cause which effect. Some research effort is directed toward standardization rather than toward a basic understanding of the mechanisms involved. The complexities of the relevant phenomena may be too great to be incorporated within a reasonable standard.

Users of the relevant standards should be made aware of the uneven scientific foundations on which they are based. For example, there is no future in designing a hand control or a visual display for a vibration environment based solely on the information contained in current vibration standards. The design or evaluation process must always include an identification of the relevant variables. A review of the relevant standards may reveal that they do not consider these variables (e.g., control gain, control order, viewing distance, phase between motion of the body and motion of the control or display). Those considering the effects of whole-body vibration should therefore extend their reading beyond standards and into the scientific literature so as to try to understand the relevant phenomena.

### 29.6.2 Hand-Transmitted Vibration

The existence of a variety of disorders associated with hand-transmitted vibration is now recognized, but there may be confusion over what is known and what it is desirable to know. It is unlikely that current standards are close to accurately reflecting the true influence of the magnitude, frequency, direction, or duration of exposure to hand-transmitted vibration on the production of VWF. Current standards do not provide any guidance on how to predict the other disorders associated with occupational exposures to hand-transmitted vibration.

## 29.7 FUTURE CONCERNS

The complexity of the interactions between oscillatory motion (i.e., vibration and shock) and the functions of the body is great. Systematic multidisciplinary approaches are required to make parallel advances in relevant aspects of biodynamics, physiology, pathology, epidemiology, and psychology (i.e., subjective assessment and performance effects). Knowledge has not, in the past, been advanced by short-term attacks on the subject; there is no substitute for traditional science based on developing an understanding that leads to modeling and the testing of hypotheses. In this field there are many hypotheses to test en route to establishing the information needed to minimize the unwanted effects of occupational exposures to whole-body vibration and hand-transmitted vibration.

The output from a better understanding of the effects of whole-body vibration need not be new standards. The complex interactions between variables may be too great for a standard and not amenable to the compromise inherent in standardization. Ergonomic design guides may be a more appropriate means of conveying such information.

Current standards for hand-transmitted vibration provide guidance that cannot be ignored. However, those concerned with the design or evaluation of situations involving hand-transmitted vibration should anticipate that further understanding of the relevant pathology, physiology, and biodynamics may result in significant changes to the methods of assessing the safety of exposures to hand-transmitted vibration.

## REFERENCES


Bovenzi, M. 2009. Metrics of whole-body vibration and exposure-response relationships for low back pain in professional drivers: a prospective cohort study. *International Archives of Occupational and Environmental Health,* 82, 893–917.

Bovenzi, M. and Hulshof, C.T.J. 1999. An updated review of epidemiologic studies on the relationship between exposure to whole-body vibration and low back pain (1986–1997). *International Archives of Occupational and Environmental Health,* 72, (6), 351–365.

Brammer, A.J., Taylor, W., and Lundborg, G. 1987. Sensorineural stages of the hand-arm vibration syndrome. *Scandinavian Journal of Work, Environment and Health,* 13, (4), 279–283.

British Standards Institution. 1987. Measurement and evaluation of human exposure to whole-body mechanical vibration and repeated shock. British Standard, BS 6841.

British Standards Institution. 2008. Guide to the evaluation of human exposure to vibration in buildings. Part 1: Vibration sources other than blasting. British Standard, BS 6472-1.

Concettoni, E. and Griffin, M.J. 2009. The apparent mass and mechanical impedance of the hand and the transmission of vibration to the fingers, hand and arm. *Journal of Sound and Vibration,* 325, 664–678.

Corbridge, C., Griffin, M.J., and Harborough, P. 1989. Seat dynamics and passenger comfort, *Proceedings of the Institution of Mechanical Engineers,* 203, 57–64.

Council of the European Communities (Brussels). 1989. On the approximation of the laws of the member states relating to machinery. Council Directive (89/392/EEC). *Official Journal of the European Communities,* June, 9–32.

Donohew, B.E. and Griffin, M.J. 2007. Low frequency motions and motion sickness on a tilting train. *Proceedings of the Institution of Mechanical Engineers, Part F: Journal of Rail and Rapid Transit,* 211, 125–133.

Dupuis, H. and Zerlett, G. 1986. *The Effects of Whole-Body Vibration.* Springer-Verlag, Berlin, Heidelberg, New York, Tokyo. ISBN: 0-387-16584-3.

Gemne, G., Pyykko, I., Taylor, W., and Pelmear, P. 1987. The Stockholm Workshop scale for the classification of cold-induced Raynaud's phenomenon in the hand-arm vibration syndrome (revision of the Taylor-Pelmear scale). *Scandinavian Journal of Work, Environment and Health,* 13, (4), 275–278.

Griffin, M.J. 1990. *Handbook of Human Vibration.* Academic Press, London, U.K. ISBN: 0-12-303040-4.

Griffin, M.J. 1997. Measurement, evaluation and assessment of occupational exposures to hand-transmitted vibration. *Occupational and Environmental Medicine,* 54, (2), 73–89.

Griffin, M.J. 1998. Evaluating the effectiveness of gloves in reducing the hazards of hand-transmitted vibration. *Occupational and Environmental Medicine,* 55, (5), 340–348.

Griffin, M.J. 2004. Minimum health and safety requirements for workers exposed to hand-transmitted vibration and whole-body vibration in the European Union; a review. *Occupational and Environmental Medicine,* 61, 387–397.

Griffin, M.J. 2008. Measurement, evaluation, and assessment of peripheral neurological disorders caused by hand-transmitted vibration. *International Archives of Occupational and Environmental Health,* 81, (5), 559–573.

Griffin, M.J., Bovenzi, M., and Nelson, C.M. 2003. Dose response patterns for vibration-induced white finger. *Occupational and Environmental Medicine,* 60, 16–26.

Griffin, M.J. and Lewis, C.H. 1978. A review of the effects of vibration on visual acuity and continuous manual control, Part I: Visual acuity. *Journal of Sound and Vibration,* 56, (3), 383–413.

Griffin, M.J. and Newman, M.M. 2004. Visual field effects on motion sickness in cars. *Aviation, Space, and Environmental Medicine,* 75, (9), 739–748.

Hamilton, A. 1918. A study of spastic anemia in the hands of stonecutters, *U.S. Bureau Labor Statistics Bulletin* 236 (Part 19), 53–66.

Haward, B.M. and Griffin, M.J. 2002. Repeatability of grip strength and dexterity tests and the effects of age and gender. *International Archives of Occupational and Environmental Health,* 75, (1–2), 111–119.

International Organization for Standardization. 1992. Mechanical vibration—Laboratory method for evaluating vehicle seat vibration—Part 1: Basic requirements. International Standard, ISO 10326-1.

International Organization for Standardization. 1996. Mechanical vibration and shock—Hand-arm vibration—Method for the measurement and evaluation of the vibration transmissibility of gloves at the palm of the hand. International Standard, ISO 10819.

International Organization for Standardization. 1997. Mechanical vibration and shock—Evaluation of human exposure to whole-body vibration—Part 1: General requirements. International Standard, ISO 2631-1.

International Organization for Standardization. 1998. Hand-held portable tools measurement of vibration at the handle. Part 1: General, ISO 8662-1.

International Organization for Standardization. 2000. Earth-moving machinery—Laboratory evaluation of operator seat vibration. International Standard, ISO 7096.

International Organization for Standardization. 2001a. Mechanical vibration—Vibrotactile perception thresholds for the assessment of nerve dysfunction—Part 1: Methods of measurement at the fingertips. International Standard, ISO 13091-1.

International Organization for Standardization. 2001b. Mechanical vibration—Measurement and evaluation of human exposure to hand-transmitted vibration—Part 1: General requirements. International Standard, ISO 5349-1.

International Organization for Standardization. 2002. Mechanical vibration—Measurement and evaluation of human exposure to hand-transmitted vibration—Part 2: Practical guidance for measurement at the workplace. International Standard, ISO 5349-2.

International Organization for Standardization. 2003a. Agricultural wheeled tractors—Operator's seat—Laboratory measurement of transmitted vibration. International Standard, ISO 5007.

International Organization for Standardization. 2003b. Mechanical vibration and shock—Evaluation of human exposure to whole-body vibration—Part 2: Vibration in buildings (1 Hz to 80 Hz). International Standard, ISO 2631-2.

International Organization for Standardization. 2005a. Mechanical vibration—Hand-held and hand-guided machinery—Principles for evaluation of vibration emission. ISO 20643.

International Organization for Standardization. 2005b. Mechanical vibration and shock—Cold provocation tests for the assessment of peripheral vascular function—Part 1: Measurement an evaluation of finger skin temperature. International Standard, ISO 14835-1.

International Organization for Standardization. 2005c. Mechanical vibration and shock—Cold provocation tests for the assessment of peripheral vascular function—Part 2: Measurement an evaluation of finger systolic blood pressure. International Standard, ISO 14835-2.

Lawther, A. and Griffin, M.J. 1987. Prediction of the incidence of motion sickness from the magnitude, frequency, and duration of vertical oscillation. *The Journal of the Acoustical Society of America,* 82, (3), 957–966.

Lawther, A. and Griffin, M.J. 1988. Motion sickness and motion characteristics of vessels at sea. *Ergonomics*, 31, (10), 1373–1394.

Lindsell, C.J. and Griffin, M.J. 2002. Normative data for vascular and neurological tests of the hand-arm vibration syndrome. *International Archives of Occupational and Environmental Health*, 75, (1–2), 43–54.

Loriga, G. 1911. Il lavoro con i martelli pneumatici. The use of pneumatic hammers. *Boll, Ispett Lavoro*, 2, 35–60.

McLeod, R.W. and Griffin, M.J. 1989. A review of the effects of translational whole-body vibration on continuous manual control performance. *Journal of Sound and Vibration*, 133, (1), 55–115.

Morioka, M. and Griffin, M.J. 2006a. Magnitude-dependence of equivalent comfort contours for fore-and-aft, lateral and vertical whole-body vibration. *Journal of Sound and Vibration*, 298, (3), 755–772.

Morioka, M. and Griffin, M.J. 2006b. Magnitude-dependence of equivalent comfort contours for fore-and-aft, lateral and vertical hand-transmitted vibration. *Journal of Sound and Vibration*, 295, 633–648.

Morioka, M. and Griffin, M.J. 2008. Absolute thresholds for the perception of fore-and-aft, lateral, and vertical vibration at the hand, the seat, and the foot. *Journal of Sound and Vibration*, 314, 357–370.

National Institute of Occupational Safety and Health, (NIOSH), Bernard, B.P. (Ed.). 1997. Musculoskeletal disorders and workplace factors: A critical review of epidemiologic evidence for work-related disorders of the neck, upper extremities, and low back. U.S. Department of Health and Human Services, National Institute of Occupational Safety and Health, DHHS (NIOSH) Publication No. 97-141 Cincinnati, OH.

Seah, S.A. and Griffin, M.J. 2008. Normal values for thermotactile and vibrotactile thresholds in males and females. *International Archives of Occupational and Environmental Health*, 81, (5), 535–543.

Sherwood, N. and Griffin, M.J. 1992. Evidence of impaired learning during whole-body vibration. *Journal of Sound and Vibration*, 152, (2), 219–225.

Taylor, W. and Pelmear, P.L. (Eds.). 1975. *Vibration White Finger in Industry*. Academic Press, London, U.K., ISBN: 0 12 684550 6.

Taylor, W., Wasserman, D., Behrens, V., Reynolds, D., and Samueloff, S. 1984. Effect of the air hammer on the hands of stonecutters. The limestone quarries of Bedford, Indiana, revisited. *British Journal of Industrial Medicine*, 41, (3), 289–295.

The European Parliament and the Council of the European Union. 2002. On the minimum health and safety requirements regarding the exposure of workers to the risks arising from physical agents (vibration). Directive 2002/44/EC. *Official Journal of the European Communities*, July 6, 2002, L177/13–19.

Turner, M. and Griffin, M.J. 1999. Motion sickness in public road transport: The effect of driver, route and vehicle. *Ergonomics*, 42(12), 1646–1664.

Turner, M., Griffin, M.J., and Holland, I. 2000. Airsickness and aircraft motion during short-haul flights. *Aviation, Space, and Environmental Medicine*, 71, (12) 1181–1189.

CHAPTER 30

# Noise Exposure and Control

Sergey A. Grinshpun, Jay Kim, and William J. Murphy

CONTENTS

## 30.1 INTRODUCTION

For centuries, the ill effects of noise exposure in the workplace have been recognized as a principal cause of hearing loss. Evidence of hearing impairment in workers involved in, for example, blacksmithing and mining, has been documented in the Middle Ages (Berger, 2000) and cited in the medical literature of the nineteenth century (Fosboke, 1831). The problem of occupational hearing loss has been recognized as one of the most common work-related diseases in the United States. In 1998, the National Institute for Occupational Safety and Health (NIOSH) estimated approximately 5 million workers had daily time-weighted average (TWA) exposures in excess of 85 dBA (NIOSH, 1998). More recent estimates place this figure at 24 million workers based upon the National Health and Nutrition Examination Survey data (Tak et al., 2009). Using the National Occupational Exposure Health Survey data (NIOSH, 1988) as well as reports issued by the Booz-Allen Hamilton (1983) and the Bolt et al. (1976), Franks (1988) estimated that 95% of daily TWA exposures were less than 95 dBA. Although the TWA exposures represent an average over a day and are not necessarily indicative of instantaneous noise levels, such a statistic suggests that noise exposures can be reduced to a safe level.

In the 1970s, the U.S. Environmental Protection Agency (EPA) published the "levels" document (EPA, 1978), wherein they concluded that the safe level of exposure representing 1% risk of incurring a significant threshold shift was 70 dBA for 8 h a day for 40 years. The U.S. Occupational Safety and Health Administration (OSHA) concluded from a similar risk analysis that 20% of workers would incur a significant change in hearing at an exposure of 90 dBA for 8 h per day for 40 years. NIOSH conducted two separate analyses of the excess risk of material hearing impairment for an 8 h exposure per day for 40 years. The earlier analysis (NIOSH, 1972) suggested that 85 dBA represented a 15% risk. Later, this estimate was revised to 17% (Prince et al., 1997; NIOSH, 1998). NIOSH's earlier recommendation for a criterion for noise exposure was modified from 85 dBA with a 5 dB exchange rate to 85 dBA with a 3 dB exchange rate. Currently the OSHA regulation enforces a 90 dBA PEL and a 5 dB exchange rate. The exchange rate dictates the trading relationship between exposure time and level, for example, 90 dBA for 8 h, 95 dBA for 4 h, 100 dB for 2 h.

The problem of occupational hearing loss is multifaceted. Hearing loss is often identified through an employer's audiometric monitoring but can be caused by both occupational and recreational exposures. Noise-induced hearing loss (NIHL) may result from a single intense exposure or from prolonged exposure to lower levels of sound. Noise may be continuous and vary little with respect to time, or it may be intermittent and have large fluctuations in the instantaneous level. Exposure to intermittent and fluctuating noise may present a greater risk of producing hearing loss than the same equivalent levels of continuous noise (Dunn et al., 1991). Occupational hearing loss may also result from exposure to ototoxic chemicals, heavy metal contaminants, asphyxiants such as carbon monoxide, heat, and other physical or chemical agents (Morata, 2003; Sliwinska-Kowalska et al., 2007; Johnson and Morata, 2010).

Extensive research conducted during the last several decades has considerably enhanced our knowledge about noise-induced damage mechanisms of the inner ear. Efficient engineering and administrative control measures for reducing the noise exposure have been developed and widely implemented in industry. Various personal hearing protection devices (HPDs) are now available. Hearing conservation programs have been implemented widely in many countries, including a NIOSH-defined "hearing loss prevention program" in the United States (NIOSH, 1998), a "noise management program" in Australia, and others. Sophisticated sound monitoring devices such as sound level meters (SLMs) and sound dosimeters are at our disposal and are extensively used in occupational environments as critical tools for determining the workers' noise exposure.

The first part of this chapter introduces the fundamentals of sound propagation, description, and measurement. Subsequently, exposure- and health-related issues are discussed following by the determination of the allowable dose. The OSHA standard is described, including the elements of a hearing loss prevention program. Functional descriptions of the auditory system, measurement equipment, and noise control techniques are also provided.

## 30.2 FUNDAMENTALS

### 30.2.1 Generation and Propagation of Sound

Sound is a disturbance in an elastic medium that propagates through the medium at a speed that is a characteristic of the medium. The speed of sound in air at 20°C and 1 atm is approximately 343.1 m/s. Sound in air can be generated by vibration or impact of bodies, unsteady motion of air, or interaction of flow and structures. Sound pressure is measured as the variation relative to the static atmospheric pressure at sea level, $1.013 \times 10^5$ Pa (N/m$^2$). Sound pressures are generally much smaller than the static atmospheric pressure. For example, the sound pressure of 100 dB noise is only 0.002% of the atmospheric pressure.

Figure 30.1 illustrates how sound propagates in a one-dimensional (1D) duct of infinite length. The speaker attached to the left end of the duct induces a rarefaction or compression of air, which propagates along the duct with the speed of sound. Figure 30.1 can be considered as a snapshot of the pressure in the duct in a grayscale map (top) and by a transverse wave (bottom). Darker and lighter areas, respectively, represent compression and rarefaction. The pressure field is often depicted as a transverse wave because it is easier to show. Wavelength is the distance between the points of same pressure of adjacent cycles.

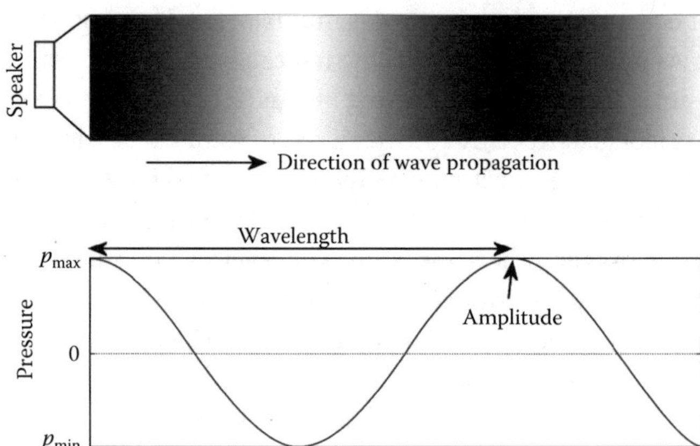


FIGURE 30.1 Sound propagation in a lossless 1D semi-infinite duct. Sound generated by the speaker at left travels to right. The degree of darkness (shown on the top figure) as well as the deviation from the background pressure (shown relative to $p = 0$ on the bottom figure) is proportional to the sound pressure. Wavelength is the distance between two adjacent points of the same magnitude of pressure.

While the wave shown in Figure 30.1 travels at the speed of sound, the sound pressure measured at any given point changes as a function of time. Figure 30.2 demonstrates how the pressure at point A in Figure 30.1 changes with the time. The frequency $f$, wavelength $\lambda$, and speed of sound $c$ are related as follows:

$$\lambda = \frac{c}{f} \tag{30.1}$$

The speed of sound in a gas is expressed as

$$c = \sqrt{\gamma RT} \tag{30.2}$$

where

$\gamma$ is the ratio of specific heats
$R$ is the ideal gas constant
$T$ is the absolute temperature (K)

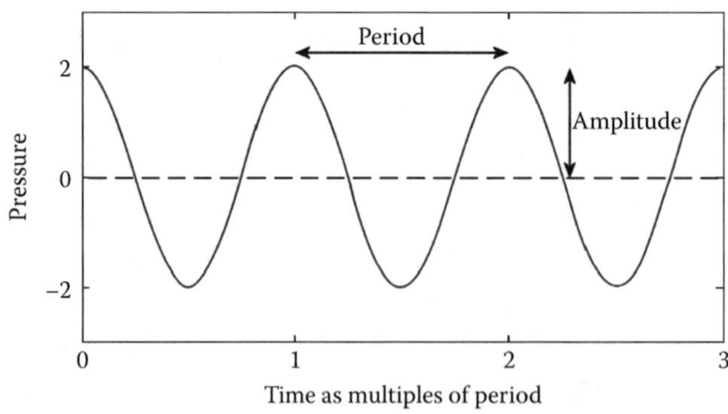


FIGURE 30.2 Variation of sound pressure as a function of time measured at a single point for the speaker operating in the duct at a single frequency.

Through substituting variables with numerical data as follows, $\gamma = 1.4$, $R = 287\,\text{N-m/(kg-K)}$, and $T = 293\,°\text{K}$, the speed of sound in air under the standard atmospheric conditions is calculated as $c = 343.1\,\text{m/s}$ (or 1130 ft/s).

To calculate the speed of sound in air at different temperatures, the following approximations can be adopted (in metric and English systems, respectively):

$$c = 331.5 + 0.58 \cdot (T, °\text{C})\ (\text{m/s}) \tag{30.2a}$$

$$c = 1054 + 1.07 \cdot (T, °\text{F})\ (\text{ft/s}) \tag{30.2b}$$

For example, according to this approximation, at 40°C (313 °K, or 104°F), $c$ = 354.7 m/s = 1165.3 ft/s.

In media of higher density, such as liquids and solids, the speed of sound is much higher, for example, approximately 1500 m/s in water, 4000 m/s in solid oak wood, and >5000 m/s in steel.

Using the speed of sound in air, the wavelength of sound wave of a specific frequency can be calculated.

Sample calculation 1:

Question: The frequency range of human hearing varies from one individual to another. The typically referred frequency range is $f$ = 20–20,000 Hz. What is the corresponding wavelength range under the standard atmospheric conditions?

Answer: According to Equation 30.1, the shortest wavelength corresponds to the highest frequency so that

$$\lambda = \frac{c}{f} = \frac{343.1\,\text{m/s}}{20{,}000\,\text{Hz}} \approx 0.0172\,\text{m} \approx 1.72\,\text{cm}$$

The longest one corresponding to the lowest frequency is

$$\lambda = \frac{c}{f} = \frac{343.2\,\text{m/s}}{20\,\text{Hz}} \approx 17.16\,\text{m}$$

In addition to the frequency, the amplitude of sound pressure (Figure 30.2) is a key characteristic representing the acoustic output. However, the maximum value of the oscillating quantity is generally not a good measure of its significance (except for a transient impulsive sound). An average value of sound pressure is zero, which itself is a useless quantitative characteristic of the sound source. An alternative—and useful—descriptor of the effective sound pressure is the root-mean-square (rms) value of the time-varying sound pressure. It is defined as a time-averaged magnitude of the pressure:

$$p_{\text{rms}} = \sqrt{\frac{1}{T}\int_0^T p^2(t)\,dt} \tag{30.3}$$

The sound power ($W$) of a source is the acoustic energy generated and propagated per second by the source. This quantity is a constant unless some attribute of the source is changed. The sound power is measured in watts (W).

The sound intensity ($I$) is defined as the time-averaged rate of energy transmission flowing through a surface of unit area:

$$I = \frac{1}{T}\int_{0}^{T} pu\,dt \tag{30.4}$$

where

$p$ is the instantaneous pressure
$u$ is the particle velocity normal to the unit surface

If the direction of the outward normal of a surface and the direction of the intensity flowing through that surface are the same, then the intensity is positive. If they are oppositely directed, then the intensity is negative. The sound intensity is measured in W/m$^2$.

For a point source operating in a free field (i.e., sound travels only in one direction without reflected waves), the intensity and rms pressure are related as follows:

$$I = \frac{p_{\text{rms}}^2}{\rho c} \tag{30.5}$$

where $\rho$ is the density of air. Assuming omnidirectional wave propagation, the sound intensity at a distance $r$ from the source relates to the sound power as follows:

$$I = \frac{W}{4\pi r^2} \tag{30.6}$$

In a typical noise environment, sources have directional properties and there are reflections. To consider these effects, Equation 30.6 is modified as follows:

$$I = \frac{WQ}{4\pi r^2} \tag{30.7}$$

where $Q$ is the directivity factor which is dependent on the source property as well as reflections from surfaces. If the acoustic wave from a point source located in free space propagates omnidirectionally, the radiation pattern is spatially uniform and $Q = 1$ (denoted as spherical radiation). If an ideal sound source is located on a floor in the center of a large room or at ground level outdoor, $Q = 2$ (hemispherical radiation). In this case the sound intensity at distance $r$ would be twice as great as that for spherical radiation because the surface area, over which the sound is spread, is reduced by a factor of 2. If the sound source is located at the intersection of the floor and a room wall, the radiation area is one-quarter of a sphere and $Q = 4$, reflecting that the spherical area is reduced by a factor of 4.

If the source is placed on the floor in a corner at an intersection of two walls, then $Q = 8$. In addition to the sound source location, directivity is influenced by the directional characteristics of the source itself.

### 30.2.2 Levels in Decibel: Sound Intensity, Power, and Pressure

In acoustics, most quantities are expressed in decibels (dB). The decibel is a logarithmic unit of measurement that expresses the magnitude of a physical quantity A (e.g., power or intensity) relative to a specified or implied reference level $A_0$:

$$L = 10 \cdot \log_{10} \frac{A}{A_0} \tag{30.8}$$

Since it expresses a ratio of two quantities with the same unit, it is a dimensionless characteristic.

The acoustic intensity can be expressed as a level above or below reference intensity:

$$L_I = 10 \cdot \log_{10} \frac{I}{I_0} \tag{30.9}$$

where

$L_I$ is the sound intensity level in dB
$I$ is the sound intensity in W/m$^2$
$I_0$ = 10–12 W/m$^2$ is the reference intensity

For example, the intensity of 1 W/m$^2$, based on Equation 30.9, corresponds to the intensity level of 120 dB, which represents relatively loud industrial noise. According to the definition, Equation 30.8, every increase in sound intensity by a factor of 10 results in an increase of 10 dB in the intensity level.

The sound power level is defined in a similar way as

$$L_W = 10 \cdot \log_{10} \frac{W}{W_0} \tag{30.10}$$

where

$W$ is the sound power in W
$W_0$ = 10 12 W is the reference power

The sound pressure level (SPL) is derived from the intensity of a plane wave (Equation 30.5) where the pressure is substituted for intensity and reference pressure is approximated as 20 μPa ($2 \times 10^{-5}$ N/m$^2$). The resulting equation for SPL is

$$L_p = 10 \cdot \log_{10} \left( \frac{p_{\text{rms}}}{p_0} \right)^2 = 20 \cdot \log_{10} \left( \frac{p_{\text{rms}}}{p_0} \right) \tag{30.11}$$

Note that the multiplier here is 20 and not 10 as with the sound intensity and power. According to Equation 30.11, doubling the sound pressure increases the SPL by 6 dB. The sound pressure increase by a factor of 10 results in a 20 dB increase of SPL.

The following examples are worth noticing: the sound pressure of $p = 1$ Pa corresponds to SPL = 94 dB; for $p = 1$ lb/ft², SPL = 128 dB; for $p = 1$ lb/in.², SPL = 171 dB; and for $p = 1$ atm, SPL = 194 dB.

The decibel scale is convenient for expressing sound intensity, power, and pressure over a very wide range. For example, exposures to noise ranging from 0 dB (below the threshold of a healthy human ear) to 140 dB (maximum impact noise allowable by OSHA) correspond to sound pressures covering seven decades: from 20 μPa to 200 Pa.

### 30.2.3 Decibel Multiplications

Decibel addition and subtraction is used when several independent and uncorrelated sound sources are considered. Examples of addition are (a) the estimation of the total SPL produced by several machines and (b) the estimation of the total SPL from a multifrequency source by adding frequency-band SPLs. Subtraction may be used to correct for the background noise that contaminates the field measurement. The equation for the total sound levels produced by $n$ random, uncorrelated independent sounds is

$$L_{\text{Total}} = 10 \cdot \log_{10}\left(10^{L_1/10} + 10^{L_2/10} + 10^{L_3/10} + \cdots + 10^{L_n/10}\right) \tag{30.12}$$

Sample calculation 2:

Question: Each of two machines produces SPL of 93 dB at the measuring location. What is SPL measured at this location if both machines are turned on?

Answer: According to Equation 30.10,

$$L_p = 10 \cdot \log_{10}(2 \times 10^{9.3}) \approx 96\,\text{dB}$$

that is, adding the second machine increased the pressure level only by 3 dB.

Sample calculation 3:

Question: Two different machines produce considerably different SPLs at the measuring location: 110 and 90 dB. What is SPL measured at this location if both machines are turned on?

Answer: According to Equation 30.12,

$$L_p = 10 \cdot \log_{10}(10^{11} + 10^{9}) \approx 110.04\,\text{dB} \approx 110\,\text{dB}$$

The example mentioned earlier shows that if two machines produce SPLs of significantly different level, their combined SPL is approximately equal to the higher SPL of the two. In other words, the lower one has little effect on the total (but increases in relative significance if the higher SPL source is quieted). Commonly one finds that changes in level of less than

TABLE 30.1 Combining Levels of Uncorrelated Sounds

| Numerical Difference between $SPL_1$ and $SPL_2$ (dB) | Amount to Be Added to the Higher SPL (dB) |
|---|---|
| 0.0–0.6 | 3.0 |
| 0.7–1.6 | 2.5 |
| 1.7–3.1 | 2.0 |
| 3.2–4.7 | 1.5 |
| 4.8–7.2 | 1.0 |
| 7.3–13.9 | 0.5 |
| >13.0 | 0.0 |

0.1 dB are insignificant. In addition to Equation 30.11, there is a "rule of thumb" for decibel addition, presented in Table 30.1.

### 30.2.4 Relationships between Sound Power and Sound Pressure Levels

In free-field environment, assuming propagation of a spherical wave of a specific frequency and no sound reflection, the following relationship exists between the sound power and pressure levels:

$$L_W = L_p + 10 \cdot \log_{10} r^2 + k - CF(T,P) \tag{30.13}$$

where

*r* is the distance from the source in meters or feet

*k* is a constant accounting for the unit system ($k = 11$ dB for metric units [with *r* expressed in meter], and $k = 0.5$ dB for English units [with *r* expressed in feet])

*CF* (*T*, *P*) is a correction factor, in decibels, accounting for atmospheric temperature and pressure

For most industrial noise situations involving steady-state sources, *CF* is negligible. Given that $L_W$ is constant for a given source, the SPLs measured in two locations from the source, SPL ($r_1$) and SPL ($r_2$), can be related as follows:

$$L_W = L_{p1} + 10 \cdot \log_{10} r_1^2 = L_{p2} + 10 \cdot \log_{10} r_2^2 \tag{30.14}$$

Equation 30.13 yields

$$L_{p2} = L_{p1} - 20 \cdot \log_{10}\left(\frac{r_2}{r_1}\right) \tag{30.15}$$

Equation 30.15 is known as the inverse-square law; it allows determining SPL at any distance ($r_2$) based on a single measurement performed at a specific distance ($r_1$). The equation produces an important "rule of thumb": each doubling of distance from the source results in a 6 dB loss of SPL.

Sample calculation 4:

Question: SPL = 100 dB is measured at 3 ft from the source. What is SPL measured at 12 ft from the source?

Answer: According to Equation 30.15,

$$L_{p2} = L_{p1} - 20 \cdot \log_{10}\left(\frac{12}{3}\right) \approx 100\,\text{dB} - 12\,\text{dB} = 88\,\text{dB}$$

Reflecting different conditions of the sound wave propagation, the directivity factor affects the relationship between sound power and pressure levels. Equation 30.13, derived from the omnidirectional propagation condition, can be modified to account for the directivity factor:

$$L_W = L_p + 10 \cdot \log_{10}\left(\frac{r^2}{Q}\right) + k - CF(T, P) \tag{30.16}$$

Sample calculation 5:

Question: A machine producing sound of a specific frequency is mounted on a steel floor in the center of a large room. SPL = 90 dB at 6 ft from the source. What would be SPL at 6 ft if the machine was moved into a reflecting corner?

Answer: According to Equation 30.16, first convert the SPL to sound power level using $r = 6$ ft, $Q = 2$, and $k = 0.5$:

$$L_W = L_p + 10 \cdot \log_{10}\left(\frac{6^2}{2}\right) + 0.5 \approx 103\,\text{dB}$$

For $Q = 8$, $r = 6$ ft, and $L_W = 103$ dB, solve for $L_p$ in Equation 30.16:

$$L_p = L_W - 10 \cdot \log_{10}\left(\frac{6^2}{8}\right) - 0.5 \approx 96\,\text{dB}$$

Note that the noise exposure at the same distance from the source increases once it is moved from the room center to a corner.

### 30.2.5 Bandwidth: Octave Band Principles

The acoustic energy is usually widely distributed over a range of frequencies. The frequency spectrum, or contents, can be expressed with frequency bands of fixed or proportional bandwidth. The spectrum with fixed band utilizes a constant frequency band of relatively narrow width for the entire frequency range. The most common proportional bands are the octave and one-third octave band. The octave band is such that the upper limit frequency ($f_2$) is twice of the lower limit frequency ($f_1$). The one-third octave band is such that the upper limit frequency ($f_2$) is $2^{1/3}$ times of the lower limit frequency ($f_1$). The center frequency ($f_c$) for a specific band is defined as the geometric mean of the lower and upper limit frequencies:

$$f_c = \sqrt{f_1 f_2} \tag{30.17}$$

Sample calculation 6:

Question: For the octave band centered at 1 kHz, find the lower and upper edge frequencies.

Answer: $f_2 = 2f_1$. According to Equation 30.17,

$$f_c = \sqrt{f_1 f_2} = \sqrt{2 f_1^2} = f_1\sqrt{2}, \quad \text{that is,} \quad f_1 = \frac{f_c}{\sqrt{2}} \approx 707\,\text{Hz}, \quad \text{and} \quad f_2 = \sqrt{2} f_c \approx 1414\,\text{Hz}.$$

## 30.3 MEASUREMENT AND INSTRUMENTATION

### 30.3.1 General

Measurement is an essential step in assessing the risk of an acoustic environment and developing exposure control measures. The sound measurements are conducted to generate information for predicting the impact of sound on humans as well as for mitigating this impact. A wide range of instruments are available today for acoustic measurement, including SLMs and sound dosimeters.

Performance of acoustic instrumentation is generally characterized by factors such as frequency response, dynamic range, crest factor capability, and response time (Lord et al., 1980). *Frequency response* refers to the range of frequencies that an instrument can accurately measure. *Dynamic range* refers to the range of amplitudes of the signal that can be measured by the instrument. For example, typical SLMs have a dynamic range of 60–80 dB. The *crest factor* is the ratio of the peak value to the rms value of the signal. *Crest factor capability* refers to the capacity of an instrument that can measure signals of high crest factor without overloading the instrument, which is important for measuring an impulsive sound. *Response time* refers to the time for an instrument to respond to a sudden change of the input signal.

### 30.3.2 Microphones

A microphone is an acoustic sensor that converts acoustic pressure variations into corresponding voltages. Dynamic, ceramic, electret, and condenser microphones are used; condenser and ceramic types are used most commonly (Lord et al., 1980). While high-precision SLMs (type 1, see the following text) use condenser microphones, conventional, general-purpose SLMs (type 2, see the following text) use ceramic ones.

Microphones are designed for optimum performance in specific sound fields (Wong and Embleton, 1994; Raichel, 2006). A free-field microphone is designed to be used whenever the sound comes mainly from one direction in a reflection-free field such as an open space or in an anechoic chamber. The free-field microphone should be pointed directly at the noise source for maximum accuracy. The diffuse field occurs when sound reflects multiple times in random directions, which is typical of a reverberant factory environment. A random-incidence microphone, also called omnidirectional microphone, provides better accuracy at the high frequencies in the diffuse field. When random-incidence microphones are used in a diffuse field, no corrections need to be made because the microphone response is designed to correct for the effects of sound arriving from all directions. However, if the source is known to exhibit primarily a single direction, correction curves are usually provided to allow the spectral measurements to be corrected for the angle of incidence. Microphones have to be calibrated along with the measurement instrumentation, for example, SLM should be calibrated by using an acoustic calibrator prior to and after a series of measurements.

### 30.3.3 Sound Level Meters

A typical SLM consists of a microphone, a preamplifier, a network of frequency filters, and a display. Modern SLMs used in the field are portable and battery operated. Figure 30.3 shows a commonly used handheld SLM. The current standard defines type 0, 1, and 2 SLMs, according to the American National Standard Institute (ANSI) and the International Electrotechnical Commission (IEC), respectively, referring to "laboratory standard," "precision," and "general purpose." Type 0 SLM is intended for use in the laboratory as a high-precision reference standard (error < 0.1 dB). Type 1 is intended for measurements in the field and laboratory; it is characterized by a relatively low error of ±1 dB or below. Type 2 is designed for a routine field use, particularly in applications where a high-frequency (>10 kHz) sound component is not dominant; its estimated error does not exceed ±2 dB. In addition, type S is defined as a "special purpose" SLM designed for operating in under special conditions (e.g., under water). Also, some documents refer to "type 3" SLMs, which are considerably less expensive and less accurate and used sometimes for rough survey work or for making preliminary measurements before more accurate surveys are initiated. Either type 1 or 2 is acceptable for enforcement of noise regulations. SLMs often come with A-, B-, or C-frequency weighting capabilities (explained in the following) and a setting for "fast" and "slow" response. A-weighting is the most commonly used as it reflects sensitivity of human ears for noises of 40 phon loudness. The "fast" setting responds more quickly to transient noises, but the setting makes it difficult to read the rapidly changing SPL. The "slow" setting reduces the response time; thus, the SPL of rapidly fluctuating sounds can be read more easily. Many modern SLMs come with capabilities to display one-third and

FIGURE 30.3 A handheld SLM. (Courtesy of Larson-Davis.)

full octave frequency spectra, a computer interface that enables downloading additional software and more sophisticated postprocessing.

### 30.3.4 Weighting Filters

Sound measuring instruments often employ frequency-selective weighting filters so that the data reflect how sound is actually perceived and how it can damage human hearing. The A-weighted filter is commonly used to emphasize frequencies around 3–6 kHz where the human ear is most sensitive, while attenuating very high and very low frequencies to which the ear is less sensitive. While A-weighting derives its shape from the 40 phon Fletcher-Munson equal-loudness contour (Fletcher and Munson, 1933), it is more useful to consider it as the transfer function of sound frequencies reaching the cochlea. The outer and middle ears provide amplification in the middle frequencies and enhance the amount of energy around 3–6 kHz. The B and C curves were intended for louder sounds (70 and 100 dB at 1000 Hz, respectively). The B-weighting is not extensively used in industrial hygiene monitoring. The C-weighting is utilized in applications that involve blast-type waveforms and often deployed for evaluating HPDs when considering the effects of predominantly low- or high-frequency environments. Table 30.2 shows the spectral corrections for estimating A- and C-weighted SPLs, dBA and dBC (Royster and Royster, 2002).

TABLE 30.2 A and C Frequency Weighting Curves

| **Octave Band Center Frequency (Hz)** | **A-Weighted Correction (dB)** | **C-Weighted Correction, (dB)** |
|---|---|---|
| 16 | −56.7 | −8.5 |
| 31.5 | −31.4 | −3.0 |
| 63 | −26.2 | −0.8 |
| 125 | −16.1 | −0.2 |
| 250 | −8.6 | 0.0 |
| 500 | −3.2 | 0.0 |
| 1,000 | 0.0 | 0.0 |
| 2,000 | 1.2 | −0.2 |
| 4,000 | 1.0 | −0.8 |
| 8,000 | −1.1 | −3.0 |
| 16,000 | −6.6 | −8.5 |
| 20.000 | −9.3 | −11.2 |

Question: The octave band survey revealed SPL values listed in the following table. What is the total SPL of the source (all bands)?

| Center frequency (Hz) | 31.5 | 63 | 125 | 250 | 500 | 1000 | 2000 | 4000 | 8000 |
|---|---|---|---|---|---|---|---|---|---|
| Octave-band SPLs (dB) | 85 | 86 | 90 | 94 | 98 | 101 | 102 | 104 | 105 |

Answer: First, the measured SPL are corrected as follows:

| Center frequency (Hz) | 31.5 | 63 | 125 | 250 | 500 | 1000 | 2000 | 4000 | 8000 |
|---|---|---|---|---|---|---|---|---|---|
| Octave-band SPLs (dB) | 85 | 86 | 90 | 94 | 98 | 101 | 102 | 104 | 105 |
| A-weighted correction (dB) | −39.4 | −26.2 | −16.1 | −8.6 | −3.2 | 0 | +1.2 | +1 | −1.1 |
| A-weighted octave-band SPL (dBA) | 45.6 | 59.8 | 73.9 | 85.4 | 94.8 | 101 | 103.2 | 105 | 103.9 |

### 30.3.5 Acoustical Calibrator

Acoustic calibrators are used to check that an SLM or microphone is in proper working condition. The acoustic calibrator produces a single tone sound of a precise reference sound level and fits over the microphone. One typical calibration frequency and SPL level are 1 kHz with 94 dB (rms pressure of 1 Pa). SPLs of 124 and 114 dB are also used for calibration, as well as a frequency of 250 Hz. When using a pistonphone to calibrate a microphone, an adjustment must be made for the barometric pressure.

Sample calculation 7:

Question: The octave band survey revealed SPL values listed in the table in the following. What is the total SPL of the source (all bands)?

Second, the total is calculated according to Equation 30.12:

$$L_{\text{Total}} = 10 \cdot \log_{10}\left(10^{45.6/10} + 10^{59.8/10} + 10^{73.9/10} + \cdots + 10^{103.9/10}\right) \approx 109.7\,\text{dBA}$$

### 30.3.6 Dosimeters

A noise dosimeter is a body-worn, battery-powered instrument designed to measure the percentage of the worker's noise exposure compared to the maximum allowable daily noise dose. A dosimeter essentially operates as an SLM but, in addition, utilizes software to express the measured exposure dose in percent. A typical dosimeter consists of a small microphone, a small amplifier and A-weighting filter, and a circuit that squares and integrates the signal in time (Equation 30.4) to obtain the cumulative exposure. An overload indicator is incorporated to indicate if the sound level exceeds the range of the meter. More sophisticated dosimeters are capable of tracking A- and C-weighted exposure levels and other characteristics.

### 30.3.7 Frequency Analyzers

Frequency analyzers are common tools to understand the spectral content of acoustic signals. A sound signal is measured with a microphone and preamplifier assembly and digitally sampled to represent the electrical signal. Different systems utilize a range of sampling rates and resolutions. These sampling rates and resolutions are sufficient to characterize the range of frequencies and levels that are typical of human hearing (Oppenheim et al., 1999; Bosi and Goldberg, 2003). Once the acoustic signal is digitally sampled, the available analyses are numerous. Fourier spectral analysis is a common means to understand the tonal content of a signal. Analyzers will provide a wide range of time window functions (e.g., Hanning, Blackman, flat-top, rectangular) to minimize the effects of sampling a limited portion of a long signal and to minimize signal processing errors. Octave band analysis can be done through the implementation of digital filters to condense the frequency information into a limited number of values typically used in room acoustics and industrial hygiene. Numbers such as equivalent level, A-weighted equivalent level, and other metrics can be computed from the same digitally sampled waveform.

Frequency analyzers are increasingly implemented in software applications that are used with data acquisition systems. Postprocessing of the digital waveforms allows analysis to be tailored to specific questions that may not be available with standard commercial instruments. Regardless of the analysis, frequency analyzers provide output to a visual display to represent the acoustic data. An acoustical consultant is often equipped with an analyzer or SLM that provides a real-time representation of the acoustic environment. Whether it is tethered to an AC electrical power outlet or runs on batteries, analyzers provide the frequency-related information and help identify critical acoustic components.

## 30.4 BASIC ANATOMY OF EAR AND CAUSES OF HEARING LOSS

### 30.4.1 Anatomy of Ear

Figure 30.4 schematically shows the human ear, comprising the outer, middle, and inner parts. The outer ear funnels the sound collected by the pinna through the ear canal to the ear drum (tympanic membrane). Vibration of the tympanic membrane induced by the incident sound is transmitted through three small bones (ossicles)—the malleus, incus, and stapes—to the oval window. The inner ear has two parts: the vestibular system including the semicircular canals, and the cochlea. The semicircular canals provide a sense of balance. The cochlea is the organ that senses the magnitude

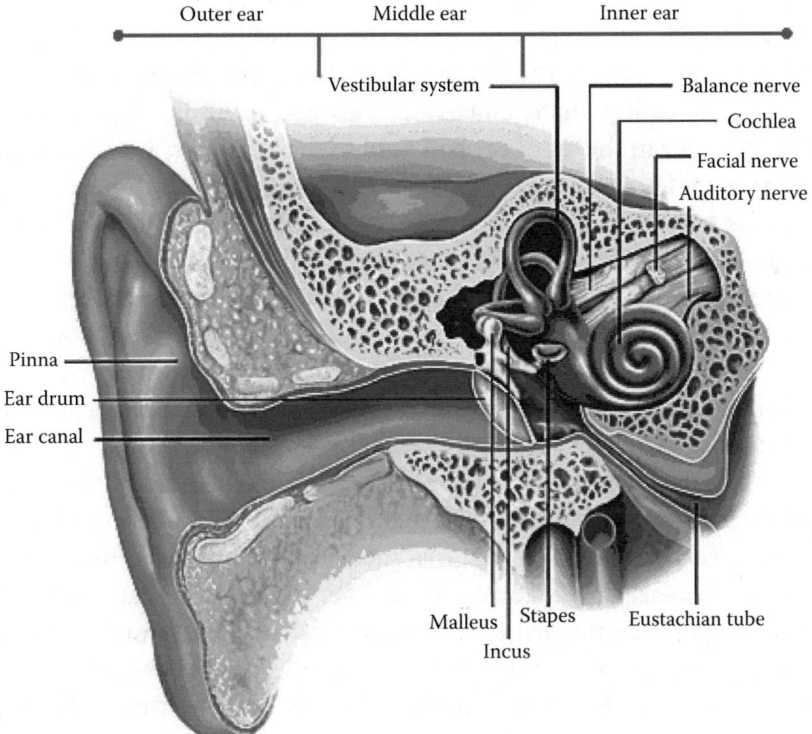


FIGURE 30.4 Anatomy of the ear. (Reproduced with permission from MED-EL Corporation.)

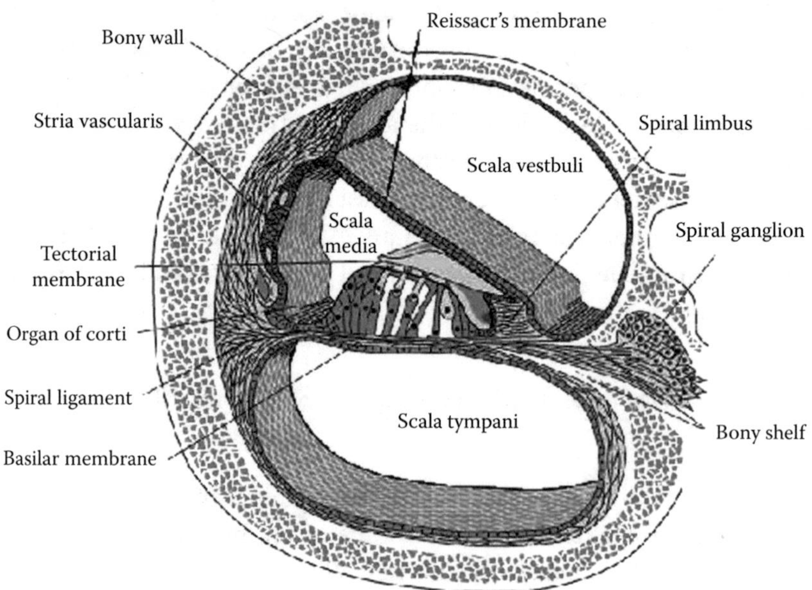


FIGURE 30.5 Structure of the cochlea. (Reprinted with permission from Richard, K., *Tissues and Organs: A Text-Atlas of Scanning Electron Microscopy*, W. H. Freeman & Co., New York, 1979.)

and frequency of sound. The cochlea is a tube-like structure wound about 2.7 turns in the shape of a snail shell and filled with fluid.

Figure 30.5 shows a cross section of the cochlea. The upper and lower scalae of the cochlea are connected by a small opening at the apex of the cochlea called the helicotrema (Durrant and Lovrinic, 1995). The ossicles maximize sound transmission to the inner ear by transforming air to fluid vibrations and minimize the impedance mismatch between the air in the outer ear and fluid in the inner ear. The disturbance of the fluid in the upper gallery induced by the stapes motion excites the motion of the basilar membrane while traveling along the cochlea. The basilar membrane has varying width and thickness, thus exhibiting the highest stiffness at the stapes side and the lowest stiffness at the apex of the cochlea. Due to this arrangement, the sound of a given frequency deflects the basilar membrane the most at a location uniquely corresponding to the frequency. A high-frequency sound deflects the basilar membrane close to the stapes and a low-frequency sound close to the apex of the cochlea. This frequency of sound is sensed accordingly.

The organ of Corti attached to the top of the basilar membrane contains four rows of hair cells spanning the entire length of the membrane. Cilia at the top of the hair cells are embedded in the bottom of the tectorial membrane. The displacement of the basilar membrane generates shearing force on the cilia that creates electrical potential that triggers electrical impulses in the neurons. The auditory nerve innervates the hair cells in the basilar membrane and transmits the afferent neural impulses toward the brain, which makes sound audible. Efferent neural impulses from the brain activate reflex responses of the middle ear and suppress hair cell responses in the cochlea.

TABLE 30.3 Definition of Hearing Losses

| Degree of Hearing Loss | Range of Hearing Loss (dB HL) |
|---|---|
| Normal | −10 to 15 |
| Slight | 16–25 |
| Mild | 26–40 |
| Moderate | 41–55 |
| Moderately severe | 56–70 |
| Severe | 71–90 |
| Profound | 91+ |

*Source:* Clark, J.G., *ASHA*, 23, 493, 1981.

### 30.4.2 Causes of Hearing Loss

There are three types of hearing loss: conductive hearing loss, sensorineural hearing loss, and mixed loss (Yost, 2006; Pickles, 2008). The first one is caused by damage in the transmission path in the outer and middle ear, which can often be repaired through surgery. The second one is due to damage to the inner ear or the neural pathway to the brain. Most occupational noise exposure results in sensorineural hearing loss, preceded by the metabolic failure of hair cells due to overloading of the hair cells for a prolonged period of time. Sensorineural hearing loss can also be caused by diseases, birth defects, ototoxic chemicals, ototoxic pharmaceuticals, and aging. Sensorineural hearing loss is a permanent loss that cannot be medically repaired. Mixed hearing loss is conductive hearing loss occurring in combination with sensorineural hearing loss.

Hearing loss is defined relative to the lowest SPL (dBA) that normal hearing persons can hear. Hearing threshold levels (HTLs) are defined for a highly screened population with no known abnormalities at each octave frequency 125–8000 Hz relative to the average hearing. For thresholds at 1000 Hz, 0 dB HTL = 0 dB SPL. Reference thresholds expressed as SPLs (decibel reference, 20 μPa) can be found in ANSI S3.6-1999. For example, an individual who has 30 dB hearing threshold at 1000 Hz is considered to have 30 dB hearing loss. Table 30.3 lists the ranges representing specific degrees of hearing loss defined by the American Speech-Language-Hearing Association (Clark, 1981).

## 30.5 ALLOWABLE EXPOSURE LIMITS

Existing regulations and guidelines concerning the noise exposure establish specific thresholds and relationships between the allowable sound level (an A-weighted $SPL_{exp}$) and the permitted exposure time ($T_{exp}$):

$$T_{exp} = \frac{8}{2^{\left((SPL_{exp} - SPL_0)/R_0\right)}} \tag{30.18}$$

Here $SPL_0$ is the criterion level representing a specific A-weighted threshold that corresponds to an 8 h permissible exposure, and $R_0$ is the exchange rate—the quantitative measure of the SPL change that allows for doubling the exposure time.

The OSHA guidelines (OSHA, 1983) specify $SPL_0$ = 90 dBA [defined as the permissible exposure limit (PEL)] and $R_0$ = 5 dBA so that

$$T_{exp} = \frac{8}{2^{\left(\left(SPL_{exp} - 90\right)/5\right)}} \tag{30.19}$$

Thus, OSHA guidelines allow for a maximum 8 h exposure of an unprotected worker to a 90 dBA noise. A louder noise leads to lower permitted exposure time intervals, for example, 4 h to SPL = 95 dBA, 2 h to SPL = 100 dBA, 1 h to SPL = 105 dBA, and 30 min to SPL = 110 dBA ($T_{exp}$ is reduced by a factor of 2 for every 5 dBA increase in SPL). The OSHA guidelines recognize a ceiling exposure level of 115 dBA, although there is evidence that this ceiling level is not being enforced (Suter, 2000). A much higher level of 140 dB of intermittent or impulsive noise is usually referred to as the upper limit of human exposure during any time duration (NIOSH, 1998).

The American Conference of Governmental Industrial Hygienists (ACGIH) adopted the threshold limit values (TLVs) for noise by utilizing the general form of Equation 30.18 but lowering $SPL_0$ to 85 dBA and changing $R_0$ from 5 to 3 dB. This resulted in the following relationship (ACGIH, 1994):

$$T_{exp} = \frac{8}{2^{\left(\left(SPL_{exp} - 85\right)/3\right)}} \tag{30.20}$$

Thus, one can calculate the allowable exposure times for a worker not wearing hearing protection for an 8 h work shift (85 dBA), 4 h shift (88 dBA), 30 min work (97 dBA), and so on. For an unprotected exposure of 139 dBA, the allowable exposure time is 0.11 s.

By comparing Equations 30.19 and 30.20, one would conclude that ACGIH allows a shorter exposure to the same noise level as compared to OSHA. Although the ACGIH recommendations do not directly affect the U.S. government regulations, they carry considerable weight in the scientific community and are adopted by some large companies (Suter, 2000). The ACGIH TLVs also match better the standards of exposure allowed outside the United States.

Although many agencies in the United States promulgating noise regulations primarily rely on OSHA guidelines, some variations exist. For instance, the U.S. Department of Transportation used the noise standard that calls for a maximum noise level of 90 dB at the driver's position (required to be measured with a 2 dB tolerance), that is, does not allow the driver exposure in excess of 90 dBA for any period of time (DOT, 1973). The U.S. Department of Defense's safety and health instructions are consistent with the ACGIH-recommended PEL and require an exchange rate at least as protective as 4 dB with a strong recommendation to use $R_0$ = 3 dB. Some variations take place in implementing the DoD instructions by the U.S. Army, Navy, and Air Force (Suter, 2000). Different countries use different standards and regulations for occupational exposure to noise. Most European counties have adopted PEL = 85 dBA and a 3 dB exchange rate.

The amount of actual exposure relative to the amount of allowable exposure is defined as dose. The dose of 100% and above represents exposures that are hazardous. For a specific SPL, the noise dose is calculated as the ratio of the actual exposure time ($T_{actual}$) to the permitted exposure time ($T_{exp}$ determined from Equation 30.18):

$$D = \frac{T}{T_{exp}} \times 100\% \quad (30.21)$$

If an individual is exposed to $n$ different noise levels ($SPL_1$, $SPL_2$,…, $SPL_n$) throughout the observation period (typically one day) during respective time intervals of $T_1$, $T_2$,…, $T_n$, the dose is calculated as follows:

$$D = \left( \frac{T_1}{T_{exp\,1}} + \frac{T_2}{T_{exp\,2}} + \cdots + \frac{T_n}{T_{exp\,n}} \right) \times 100\% \quad (30.22)$$

where $T_{exp\,1}$, $T_{exp\,2}$,…, $T_{exp\,n}$ are the permissible exposure times from Equation 30.18.

Sample calculation 8:

Question: The following worker's exposures were measured during the 8 h shift: $SPL_1 = 90$ dBA over the first 4 h, then 100 dBA over the next hour, 95 dBA over the next 2 h, and 110 dBA in the last hour. What is the noise dose assuming (a) the OSHA rules and (b) ACGIH TLV?

Answer:

(a) From Equation 30.20, the $T_{exp}$ values are 8, 2, 4, and 0.5 h, respectively. Thus, according to Equation. 30.22,

$$D_{OSHA} = \left( \frac{4}{8} + \frac{1}{2} + \frac{2}{4} + \frac{1}{0.5} \right) \times 100\% = 350\%$$

(b) From Equation 30.21, the $T_{exp}$ values are approximately 2.52, 0.25, 0.8, and 0.025 h, respectively. Thus, according to Equation 30.22,

$$D_{ACGIH} = \left( \frac{4}{2.52} + \frac{1}{0.25} + \frac{2}{0.8} + \frac{1}{0.025} \right) \times 100\% = 4808\%$$

As expected, the ACGIH-calculated dose is much greater than the one obtained based on the OSHA rules. Both are excessive (>100%).

A special cutoff level of 80 dBA is defined so that any noise exposure below this level is considered nonhazardous. This level corresponds to the risk of incurring a material hearing impairment of less than 3% for a 40 year occupational exposure. In assessing the dose, all SPL values below the referenced cutoff are replaced by zero.

The averaging of different noise exposure levels during an exposure period (an 8 h work shift) produces the TWA value defined as the constant sound level that would generate the same dosimeter reading over this exposure period as would the nonsteady noise. This sound level is also referred to as the dose-equivalent level ($L_{eq}$). Derived from the OSHA guidelines,

$$L_{eq} = 90 + \frac{5}{\log_{10} 2} \log_{10}\left(\frac{D}{100}\right) = 90 + 16.61 \times \log_{10}\left(\frac{D}{100}\right) \quad (30.23)$$

From the ACGIH TLVs,

$$L_{eq} = 85 + \frac{3}{\log_{10} 2} \log_{10}\left(\frac{D}{100}\right) = 85 + 9.67 \times \log_{10}\left(\frac{D}{100}\right) \quad (30.24)$$

The TWA values are calculated for nonsteady noise and compared to specific levels such the OSHA-defined action level (85 dBA) or PEL (90 dBA) in order to determine the course of action. Once TWA reaches the action level (dose = 50%), the hearing conservation ("hearing loss prevention") program must be established as explained in Section 30.6 in the following text. A higher TWA of 90 dBA (PEL, dose = 100%) requires establishing a hearing protection program. However, given the tolerance of measurement (±2 dB), the latter is enforced only if the measured value is equal to (or in excess of) 92 dBA (the corresponding dose is 132%).

## 30.6 ELEMENTS OF HEARING LOSS PREVENTION PROGRAM

### 30.6.1 Background

The Occupational Safety and Health Act and the Noise Control Act were passed by Congress and signed into law by President Nixon (U.S. Public Law 91–596, 1970; U.S. Public Law 92–574, 1972). In 1981, OSHA promulgated the Hearing Conservation Amendment to define the requirements of implementing a hearing conservation program within general industry (OSHA, 1981, 1983). In 1996, NIOSH published its recommendations for implementing a hearing conservation program (NIOSH, 1996). The NIOSH Practical Guide is available from the NIOSH web site (http://www.cdc.gov/niosh). The principal elements of a hearing conservation program are described in this section.

Since 2004, the U.S. Bureau of Labor Statistics (BLS) has been summarizing the number of recordable hearing loss injuries reported from the OSHA Form 300 for a sampling of industries in the United States. Hearing loss has consistently ranked as the second most prevalent nonfatal injury or illness with approximately 22,000 cases reported in 2007 (Hager, 2009). Although the specific cause of the cases reported is not provided within the BLS data, one may safely assume the primary cause to be noise exposure. Thus, the first element of a successful hearing conservation program is the protection against noise.

### 30.6.2 Monitoring Hearing Hazards

In addition to the classical risk of induced hearing loss, other less obvious issues that present safety hazards include the ability to communicate, to hear warning signals, or to identify auditory cues while operating machinery. Occupational exposures to solvents, chemicals, heavy metals, and even temperature extremes can contribute to worker hearing problems.

Noise exposure is the principal cause of hearing loss in most cases. Employers need to classify noise sources in the workplace through area noise surveys, personal dosimetry, or engineering surveys. Area surveys use an SLM to plot out the levels of noise within a workplace. A simple map of the factory and color-coded noise levels identifying different exposure levels (e.g., green for TWA < 85 dBA, yellow for TWA < 88 dBA, orange for TWA < 91 dBA, and red for TWA < 94 dBA) would inform workers of the risk of exposure. Areas with high exposures should be identified as mandatory use of hearing protection.

Personal dosimetry is important when the work force is mobile or the noise exposures during the workday vary. Mobile workers may be operating in several areas with a combination of high, moderate, or no exposure risk. For construction workers, the noise exposures vary considerably with the job task on a daily basis (Nietzel and Seixas, 2005). Workers should be monitored with personal noise dosimeters to determine their individual exposure. Two elements are critical to effective dosimetry: identification of job tasks and worker participation to record the exposures. Exposure levels are determined automatically in many dosimeters. Threshold levels within the dosimeter set the effective level above which the noise exposure is integrated into the dose. Typically, the threshold is set to the cutoff level (80 dBA) when estimating a noise dose. If workers spend significant time in quiet areas, the threshold can be set to zero to accurately assess the exposure.

Engineering surveys are typically more concerned with identifying particular features of a noise exposure that may be controlled. To accomplish these surveys, more sophisticated analysis of the spectral content may require octave band analyzers, sound level recorders, in situ analysis of sound power or sound intensity measurements, and vibration measurements with accelerometers to assess particular pieces of machinery that are targeted for engineering noise control.

### 30.6.3 Engineering and Administrative Controls

Engineering and administrative controls are essential to reducing exposures to hazardous levels of noise and increasing the effectiveness of a hearing loss prevention program. Within the NIOSH hierarchy of controls, the hazard should be removed or the worker should be removed from the hazard (NIOSH, 1996). Engineering noise controls can take several forms: reducing the noise at the source, interrupting the noise path, reducing the reverberation, and reducing the structure-borne vibration.

#### *30.6.3.1 Reducing Noise at the Source*

Engineering noise control identifies the noise sources and seeks a solution to reduce the contribution of the source to the exposure. In many cases, multiple sources exist that must be treated with a variety of techniques. For instance in coal mining, the continuous mining machine has a rotating cutter head at the face of the coal seam, fans that operate and ventilate the operator(s), and a conveyor bed and chain with flight bars to move the coal away from the seam. Each element requires unique noise controls to reduce their respective contributions to the exposure. The cutter head may be enclosed to reduce exposure; fans may be modified to move air with less turbulence; the conveyor bed may be treated with constrained layer damping to reduce impact noise; and the flight bars and chain may be coated to reduce the impact of coal with the metal bars (Camargo et al., 2008).

High-velocity fluid/air flows represent a common source of excessive noise, which can be treated by implementing different techniques from the elimination of leakage paths to the manipulation with the flow velocity. The latter is often effective given that, under common aerodynamic conditions, noise increases with the fifth power of flow speed. For the same flow rate, choosing a bigger duct cross-sectional area may thus considerably reduce the emitted sound energy. If a duct diffuser is utilized as a noise control technique, abrupt expansions should be avoided because rapid pressure changes may produce additional acoustic waves.

#### *30.6.3.2 Interrupting the Source*

Interruption of the path can be a highly effective means of controlling exposures. Noisy machinery may be housed in an enclosure designed to contain or absorb the acoustic energy from reaching the worker. For example, NIOSH's work with indoor firing ranges identified that adjacent rooms at one range had peak impulse SPL of about 130 dB. If walls are inadequately constructed, acoustic energy can be transmitted through direct and/or flanking pathways such as ventilation ducts, electrical chases, or inadequate isolation of the range (Kardous et al., 2003). High levels in adjacent spaces prevent full utilization of the spaces when weapons training occurs. At another firing range, the control rooms were isolated behind poured concrete walls with doors and windows that had sound transmission class (STC) ratings of 45 dB. During weapons training, conversations could be held in the control room. In the other spaces in the building, the activity on the range was audible but quieted due to the physical separation across a corridor and the control and preparation rooms (Kardous and Murphy, 2010).

#### *30.6.3.3 Reducing the Reverberant Exposure*

Reduction of reverberant exposure involves control of reflected sound, as opposed to control of direct sound (discussed earlier). Indirect sound is controlled by application of surface treatments. Absorptive treatments can significantly improve the usability of a noisy space. For instance, recent studies of the noise exposure of workers at animal shelters found maximum exposure levels of about 100 dBA (NIOSH, 2006a,b, 2007). Through the application of absorptive batting applied to the ceiling of the kennel area, the levels were reduced to about 3–4 dB. Before acoustical treatment, the workers experienced considerable difficulty communicating with one another while in the kennel area. After treatment, the reverberation times were reduced dramatically and speech intelligibility levels were improved.

#### *30.6.3.4 Structure-Borne Vibration*

Anytime a structure is vibrating, there is a potential of noise generation. In a work environment, tools may be poorly designed and have sympathetic vibrations produced by the equipment operation. For instance, table saws can be built with little more than a sheet metal enclosure to shield the worker from the rotating pulleys, belt, and blade. More expensive, cabinet-grade saws have enclosures that typically extend to the floor and are constructed from heavier stock to minimize vibrations of the saw blade. The unintended side effect is that cabinet-grade saws are often more quiet than portable contractor-grade table saws. The additional mass changes the resonance frequencies, the additional shrouding isolates the noise-producing elements, and the reduced vibrations result in a higher quality by reducing the tool marks on the workpiece.

#### *30.6.3.5 Quantification of Noise Reduction*

The sound reduction by a noise control treatment is characterized by absorption-reflection properties of its surfaces and transmission properties of the material. The absorption coefficient ($\alpha$) is used to describe the ability of a surface area to absorb sound energy. It is defined as the ratio of acoustic energy absorbed by a surface to the acoustic energy incident on the surface. The $\alpha$ values varies from 0 to 1 with $\alpha = 0$ corresponding to the case when all the incoming sound is reflected (no absorption) and $\alpha = 1$ representing the opposite case when the sound is fully absorbed (no reflection). Absorptivity of 0.5 means that the surface absorbs 50% of the sound energy. The transmission coefficient ($\tau$) of the barrier is defined as the ratio of the sound energy transmitted through its unit area to the sound energy incident on the barrier. The $\tau$ values varies from 0 to 1. Materials that can effectively absorb sound energy have $\tau$ close to 0 ($\tau = 0$ is designated for a perfect barrier), while $\tau = 1$ represents open space. Both absorption and transmission coefficients are dependent on frequency.

To express the ability of a given barrier to attenuate sound, a transmission loss (TL) occurring when the acoustic wave penetrates through the barrier is

$$\text{TL} = \text{SPL}_1 - \text{SPL}_2 = 10\log_{10}\frac{1}{\tau} \tag{30.25}$$

where $\text{SPL}_1$ and $\text{SPL}_2$ are the SPLs inside and outside the barrier, respectively. For example, if SPL on one side of a wall is 100 dB and on the other side it is 80 dB, the resulting TL is 20 dB. The material's transmission coefficient calculated for these conditions from Equation 30.25 is $\tau = 0.01$.

Adding barriers (such as walls and ceiling) around the noise source traps sound, thereby creating a reverberant sound field in the space containing the source. The size of the space and the reflectivity of its surfaces affect the reverberant sound field. If all the surfaces have the same absorption coefficient $\alpha$ and the total surface area is $S_t$, the total sound absorption area in the room is calculated as $A = \alpha \cdot S_t$. Wallace Clement Sabine, a Harvard physics professor, introduced the so-called room constant

$$R = \frac{\alpha \cdot S_t}{1 - \alpha} \tag{30.26}$$

as an indicator of the amount of sound absorption in the room. $R$ is expressed in Sabine square meters or Sabine square feet, depending on the unit chosen for the total surface area.

If the room is composed of wall surfaces of $n$ different acoustic properties, the average absorption coefficient ($\bar{\alpha}$) is calculated as follows:

$$\bar{\alpha} = \frac{S_1\alpha_1 + S_2\alpha_2 + \cdots + S_n\alpha_n}{S_1 + S_2 + \cdots + S_n} \tag{30.27}$$

and the room constant is

$$R = \frac{\bar{\alpha} \cdot S_t}{1 - \bar{\alpha}} \tag{30.28}$$

If a worker is exposed to a source located at a distance $r$ with directivity $Q$ in a room with a room constant $R$, the sound pressure relates to the sound power as follows (modified Equation 30.15):

$$L_p = L_W + 10 \cdot \log\left[\frac{Q}{4\pi r^2} + \frac{4}{R}\right] + 10.5 + CF(T,P) \tag{30.29}$$

The maximum noise reduction (in decibels) from adding absorption can be calculated as follows:

$$NR_{\max} = 10 \cdot \log_{10}\left(\frac{A_{\text{after treatment}}}{A_{\text{before treatment}}}\right) \tag{30.30}$$

where $A$ is the absorption area.

Sample calculation 9:

Question: A noise reduction treatment was applied to the room with an initial absorption of 270 Sabine square feet by introducing 1000 ft$^2$ of material with $\alpha = 0.6$. Determine the maximum noise reduction.

Answer: According to Equation 30.5, the maximum achievable noise reduction is

$$NR_{\max} = 10 \cdot \log_{10}\left(\frac{0.6 \times 1000}{270}\right) = 10 \cdot \log_{10} 2.22 = 3.47\,\text{dB}$$

The maximum noise reduction from a partial enclosure can be calculated through the percentage of the open area (area free to radiate):

$$NR_{\max} = -10 \cdot \log_{10}\left(\frac{\%\,\text{area free to radiate}}{100}\right) \tag{30.31}$$

Sample calculation 10:

Question: A partial enclosure is placed around the sound source so that 75% of the surface area is covered with absorptive material. Determine the maximum noise reduction.

Answer: The remaining area (free to radiate) accounts for 25%. Consequently, the maximum possible noise reduction can be calculated as

$$NR_{max} = -10 \cdot \log_{10} 0.25 \approx 6\,\text{dB}$$

#### *30.6.3.6 Administrative Controls*

Administrative controls can be defined as changes in the work schedule or operations that reduce noise exposure. Some equipment should be operated during different shifts to expose fewer workers. Shifting workers from noisier to quieter equipment is another way of reducing an individual's exposure. In the military, some weapons systems create highly intense exposures that cannot be avoided if the personnel are expected to become competent users. In such cases, the military has implemented administrative restrictions on the number of shots fired, kind and location of weapon, and the number of personnel around the weapons. One other practical administrative control is to provide a quiet area where workers may get relief from workplace noise. Break rooms or quiet rooms should have background noise levels limited to less than 70 dBA to allow for recovery from any noise insult. If the room is near the production line, then appropriate acoustic isolation is necessary to achieve lower background noise levels.

### 30.6.4 Audiometric Evaluation

Audiometric evaluation is critical for the execution of any successful hearing conservation program (NIOSH, 1996; Berger, 2000). Without an accurate assessment at the beginning (an end) of a worker's employment, the employer is effectively liable for any hearing loss the worker might have incurred prior to and after they have terminated employment. Thus, audiometric screening is recommended on the following five occasions:

1. Preemployment
2. Prior to an initial assignment to a hearing-hazardous area
3. Annually for as long as the employee is assigned to a noisy job (TWA > 85 dBA)
4. Whenever the employee is reassigned out of a noisy job
5. At the termination of employment

OSHA and NIOSH have different requirements and recommendations. OSHA currently requires that employees' hearing be tested at 0.5, 1, 2, 3, 4, and 6 kHz for occupational hearing screening. OSHA further requires that a standard threshold shift (STS) must be reported when the employee's average hearing at 2, 3, and 4 kHz shifts by more than 10 dB. NIOSH recommends that hearing also be tested at 8 kHz and that a significant threshold shift be considered to have occurred when hearing at any frequency has shifted by more than 15 dB and the shift is present at the same frequency in the same ear following an immediate retest. OSHA's occupational screening criteria use a higher maximum permissible ambient background noise limit (MPANL) than NIOSH. NIOSH recommends that the MPANL should not exceed the ANSI S3.1-1999 (2008) levels for audiometric test condition requiring ears be covered and test frequencies ranging from 500 to 8000 Hz. The background noise levels are critical to uncovering early threshold shifts that might signal the onset of occupational hearing loss. Whenever the test room is moved, the levels should be measured to assure compliance. If the testing is performed in a mobile unit, then every time the unit is set up in a new location, the levels should be measured and documented. NIOSH recommends daily checks of the MPANL. Conducted with a microphone installed in the test booth and linked to the audiometer, the testing can be suspended if the ambient room noise exceeds the prescribed limits.

Audiometric monitoring requires that a baseline audiogram be collected. The worker should not be exposed to noise levels at or above 85 dBA for a minimum of 12 h prior to the audiometric test. This test is critically important as all subsequent tests will be compared to this baseline. Annual testing is required for workers enrolled in the hearing loss prevention program. The audiograms should be conducted during the worker's normal work shift. Ideally the audiogram should be immediately examined to compare it to the baseline audiogram. Hearing testing software and computer-controlled audiometers are capable of flagging differences during the test to immediately retest suspicious thresholds. When monitoring identifies a shift, a retest should be conducted. Repositioning the headphones and reinstruction of the worker on how to take the audiometric test may affect the threshold that is measured. The important task is to identify a persistent threshold shift and recommend follow-up with an audiologist and/or occupational physician.

If a worker has a persistent threshold shift, a confirmation audiogram is required. NIOSH recommends that the confirmation testing take place within 30 days. Waiting another year or several months may result in further hearing loss. If a worker has an STS, then the employer is required to take appropriate action to protect the worker from additional hearing loss. Some of the actions include counseling on how hearing loss occurs, reinstructing the worker on the proper use of HPDs, or even reassignment of the worker to a quieter work area.

Finally, whenever a worker leaves employment, the employer should obtain an exit audiogram. The employee should be counseled regarding changes in the worker's hearing over the duration of employment.

## 30.7 PERSONAL HEARING PROTECTION DEVICES

HPDs have always been an essential element of a hearing loss prevention program. When engineering and administrative controls have not been practical solutions, HPDs have become the de facto fallback solution. Several aspects of hearing protector use must be considered when being incorporated into a hearing loss prevention program: determination of the necessary level of protection, selection of appropriate products for workers, training and education of workers to use HPDs, and verification of the use of HPDs by workers.

### 30.7.1 Hearing Protection for Hearing Loss Prevention

The OSHA mandates that "a variety of suitable hearing protectors" be made available to workers at no cost to the employees if they are exposed to an 8 h TWA level of 85 dBA (the action level). Once TWA reaches PEL = 90 dBA, the exposed workers are required to wear hearing protection. Responsible employers will do more than just provide protectors; they will include suitable training to educate the employees on protectors that are best fit to the head or ear canals. Recent research has demonstrated that the most effective tools for training are one-on-one sessions where the employee learns to competently fit the protector and understands the need to wear the protector whenever they are exposed to high levels of noise (Joseph et al., 2007; Witt, 2007; Murphy et al., 2010; Schulz, 2010).

### 30.7.2 Determination of Level of Protection

Currently, all HPDs sold in the United States are required to be labeled according to the U.S. EPA's regulation 40 CFR 211 subpart B (EPA, 1978). This regulation stipulates that the protectors shall be tested with a panel of 10 subjects where the experimenter fits the protectors on the subject's head or in the ear canals. This is referred to as the experimenter-fit protocol and was derived from the American National Standard ANSI S3.19-1974 for measuring the real ear attenuation of HPDs (ANSI S3.19-1974). While that standard has been in place since 1978, the U.S. EPA recently proposed a revision of the regulation to update the standard and provide specifications for additional tests for unique classes of HPDs. Both the current rating and the proposed rating are described in the following text.

In August 2009, EPA proposed significant revisions to the labeling regulation (EPA, 2009), including the utilization of a dual number rating system that informs the user of the range of attenuation that the product can provide (ANSI/ASA S12.68-2007, 2007). Products will be tested using an experiment-trained protocol whereby the experimenter will be allowed to teach the test subject how to properly fit the protector, but will not actually put the protector on the subject's head when the testing is carried out (ANSI/ASA S12.6-2008, 2008). Ratings will be required for two new classes of hearing protectors: active noise reduction (ANR) and level dependent (ANSI S12.42-2010). The proposed changes are expected to significantly affect hearing loss prevention programs.

#### *30.7.2.1 Noise Reduction Rating (1978)*

The noise reduction rating (NRR) provides the consumer with information regarding the potential performance capability of a hearing protector if it is worn in the manner in which it was tested. Shortly after the regulation was promulgated, the occupational safety and health community came to recognize that the NRR was the performance estimate rarely achieved by most users. As mentioned earlier, the use of an experimenter-fit protocol is the principal reason for the inflation of this estimate. When the product was tested in the laboratory, the evaluator ensured the best possible fit for every subject.

The NRR is a measure of attenuation that is designed to be used with C-weighted noise measurements. Industrial hygienists often collect noise measurements using the A-weighted scale, and therefore, the NRR must be converted from C- to A-weighting. In the EPA regulation, this conversion is accomplished by a spectral correction factor of 7 dB.

For example, suppose a noise exposure is 95 dBC and 91 dBA and that the worker is wearing a protector that has a 25 dB NRR. The estimated C-weighted exposure is 95 − 25 dB = 70 dBC while the estimated A-weighted exposure is 91 − (25 − 7 dB) = 73 dBA. As one can see, even a seemingly simple correction between C- and A-weighting yields two different answers.

The U.S. EPA requires that the mean and standard deviation of the octave band attenuations be provided for each product. These detailed attenuation values can be used to estimate exposure levels for specific noise exposures. Any hearing protector, when worn correctly, can provide 10–15 dB of protection and reduce the exposure to below 85 dBA. However, for high levels of noise exposure or for exposures that contain specific frequencies that must be reduced to protect the worker, the expanded methods must be applied. NIOSH has published several example calculations to describe the octave band method of estimating the protected exposure level (Kroes et al., 1975; Lempert, 1984; Franks et al., 2003). The last of these references is an online Compendium of Hearing Protection Devices that has been continually updated as manufacturers have developed more products.

#### *30.7.2.2 Noise Reduction Rating (2009)*

The proposed dual number on the primary label gives the user a sense of the range of protection that can be achieved if the protector is worn properly (see Figure 30.6). The lower value (19) is representative of what 80% of the test panel was able to achieve, and the greater value (27) is representative of what 20% of the test panel was able to achieve. Most trained users should be able to achieve between 19 and 27 dB of noise reduction. One difference between the NRR (1978) and the proposed rating is that the new rating is determined

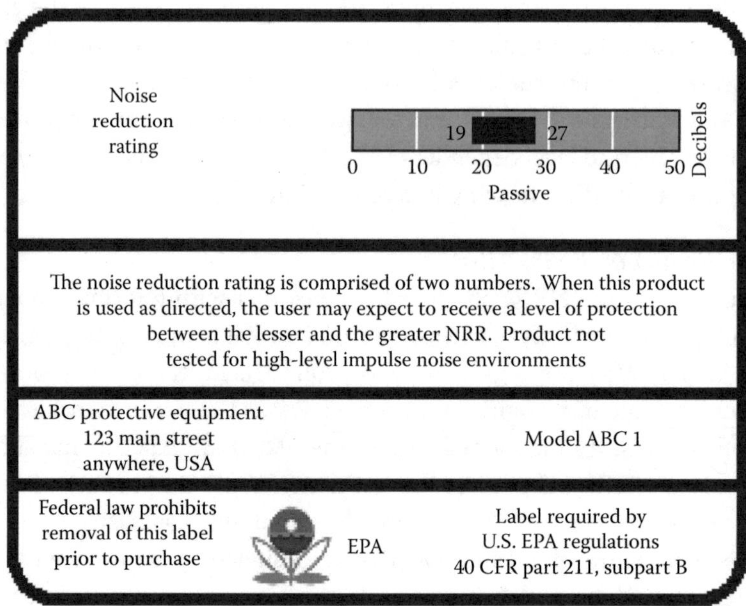


FIGURE 30.6 Proposed label for a typical passive HPD. Passive devices rely upon structural features to attenuate sound energy.

for A-weighted noise; in other words, it may be subtracted directly from an A-weighted noise level and not require adjustment for the C- to A-weighting scales. Again, assuming a 91 dBA noise exposure, the estimated exposure levels with the product worn will be between 72 dBA (=91–19) and 64 dBA (=91–27).

When more accurate assessments are required, the proposed regulation provides two methods for estimating the exposure. The first one employs a graph that illustrates how the attenuation of the protector changes as a function of the spectral balance (the difference between the C- and A-weighted measurements) of the noise. The second uses the octave band analysis. To provide the graphical information, the industrial hygienist should first measure the exposure with both the C- and the A-weighted scales and calculate the difference. Using the noise level from the previous example, the spectral balance is calculated as 95 – 91 = 4 dB. From the graph in Figure 30.7, the estimated noise reductions at the 80th and 20th percentiles are about 16 and 22 dB (as denoted with the red circles in the graph), respectively.

The estimated exposure level would be the A-weighted noise level minus the adjusted NRR for the 80th percentile: 91 – 16 = 75 dBA. The 20th percentile protection level would be 91 – 22 = 69 dBA.

This method provides the industrial hygienist a new means of understanding the effect of hearing protection interacting with the environmental noise exposure. First, the exposure is shown to be explicitly dependent upon the noise spectrum. As low-frequency content becomes more prevalent, the protector's effective protection is degraded from about 19–16 dB. In the worst-case noise, one dominated by energy around 125 Hz, the protection could be as low as 6 dB. Second, the range of protection can be used to determine whether

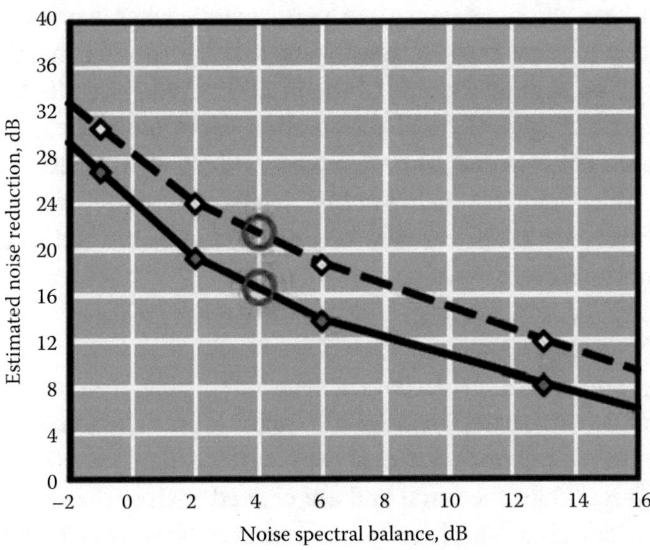


FIGURE 30.7 Estimated noise reduction as a function of spectral balance of a noise: the difference of C-weighted and A-weighted SPLs. Overall hearing protector performance is dependent upon both the attenuation of the protector and the noise in which it is worn. This particular protector has reduced performance when used in predominantly low-frequency noise.

the predicted protection is excessive. If a protector provides more attenuation than is needed, employees may tend not to fit the protector well or may remove the protector when communication becomes a problem. The upper value for the NRR or the graphical NRR provides an estimate of the maximum attenuation and therefore the potential for isolation and the appropriateness of protector selection.

### 30.7.3 Protector Types

#### *30.7.3.1 General*

Several styles of protection are available to meet just about every specific need in hearing protection. According to the NIOSH Compendium of Hearing Protection Devices, over 400 protectors are currently sold in the United States (Franks et al., 2003). Protectors come in three basic styles: earplugs, banded protectors, and earmuffs. In the EPA's proposed regulation, several different types of ratings cover specific features of the protectors. All protectors will be rated according to their basic performance due to the attenuation of the structural elements of the protector and without any electronics being activated. For those devices that incorporate noise cancellation, an NRR will be provided to demonstrate the additional protection that the protector provides when noise cancelling is operating. Another type of rating will be provided to describe the performance of devices in an impulsive noise environment. Some devices utilize electronic circuits to limit the exposure to high SPLs. Others are designed with physical acoustic features to vary the amount of attenuation with sound pressure. These protectors will be tested for a range of impulse levels (130–170 dB peak SPL) to estimate their performance.

#### *30.7.3.2 Earmuffs*

Earmuffs tend to have the most consistent performance when worn correctly. Workers must be careful that jewelry, safety glasses, hats, and hair do not interfere with the seal of the earmuff cushion against the side of the head. In studies conducted at NIOSH and other laboratories, muffs typically had the smallest variation in the overall attenuations measured for groups of naïve and trained subjects (Murphy et al., 2004, 2009). Earmuffs can be uncomfortable if one is working in high heat and humidity due to sweat and condensation under the muff. Muffs can also be perceived as bulky and be uncomfortable if the headband is not adjusted properly.

#### *30.7.3.3 Banded Protectors*

Banded protectors are something of a hybrid between earplugs and earmuffs. The earplugs or canal caps are attached to a flexible, springy band that provides force to seal the entrance of the ear canal with the cap or plug. Canal caps do not extend very far into the ear canal. They are slightly larger than the canal and are pressed against the entrance of the canal. Other banded protectors have earplugs that are meant to insert into the ear canal; these are also referred to as semi-insert devices. Banded protectors are popular with workers when the noise environment is changing or intermittent. Because they are banded, the user only needs to handle the band and not the actual plug; thus, transference of dirt on the hands and fingers to the ear canal is minimized. Banded protectors tend to be more comfortable than earmuffs when working in high heat and humidity. The attenuation of canal caps is often lower than that achievable by earplugs due to the smaller surface area. For some persons with irregularly shaped ear canal openings, the canal cap may not completely seal the entrance and may provide little if any protection.

#### *30.7.3.4 Earplugs*

Earplugs accounted for about 62% of sales of HPDs in the United States (Frost & Sullivan, 2009). Earplugs can be formable, premolded, or custom molded to ears. Formable earplugs are typically made from polyvinyl or polyurethane foams that need to be rolled into a tight cylinder and inserted into the ear canal. Premolded or no-roll-down earplugs have traditionally been made of a flexible polymer (rubber or vinyl) that has flanges to make contact with the ear canal walls. Recently, manufacturers have developed foam plugs that can be pushed into the ear canal by means of a solid stem surround by a flexible material. Custom earplugs are manufactured similar to hearing aids (with an impression of the user's ears and ear canals being taken). Another custom earplug manufacturer uses an inflatable bladder that is filled with a silicon gel that sets in the ear canal in a matter of minutes.

While earplugs are considered to be the most comfortable and least obtrusive forms of protection, they can also be the most ill-used product. Many users fail to properly roll or insert the plugs and achieve only a fraction of the intended attenuation. The workers require additional training to implement correct fitting and continuous use of the protectors.

#### *30.7.3.5 Specialty Products*

Specialty products such as sound restoration protectors, ANR protectors, and level-dependent hearing protectors are available. Sound restoration devices function similarly to a hearing aid. When the ambient noise levels are low, the electronic circuitry samples and amplifies the environmental sounds and reproduces them underneath the protector. In this manner a worker is able to hear low-level sounds without having to remove the protector.

Sound restoration devices work best in intermittent noise environments when the exposure is unpredictable and intense. Examples include use of pneumatic or powder-actuated tools on a construction site or the use of firearms at a firing range. The sound restoration circuitry quickly turns off the amplification when the ambient levels exceed 85 dB. Some devices use peak clipping circuits; others have compression circuits and still others may use voltage-controlled amplifiers to control the sound levels reproduced by the protector.

ANR protectors applied a sophisticated algorithm of sampling the ambient sound and reproducing it 180° out of phase underneath the protector. When the transmitted sound and the out-of-phase sound are combined underneath the protector, they cancel out one another and actively reduce the noise levels that reach the ear. Currently ANR devices are popular among airplane pilots as they can reduce strong tonal components produced by the engines. ANR algorithms can be tuned to remove specific components with high efficiency. The U.S. Air Force has developed ANR earplugs that are capable of providing a combined active and passive attenuation of 50 dB from 63 Hz to 8 kHz. ANR protectors tend to be more expensive than sound restoration and traditional protectors, but they yield attractive solutions for very specific noise environments.

Level-dependent protectors as a broad classification could encompass sound restoration devices. However, the acoustics of sound transmission through small orifices provides a completely passive means of providing attenuation at high levels without having to provide a power source. As sound passes through a small orifice, it encounters a viscous boundary layer near the walls of the orifice. As the sound pressure increases, the boundary layer extends further into the channel where the air is flowing back and forth through the orifice. The flow resistance increases with increasing sound pressure and the attenuation also increases (Allen and Berger, 1990; Franke et al., 1994; Parmentier et al., 2000; Berger and Hamery, 2008). The advantage of this style of protector is that it can provide almost as much attenuation as a solid protector of the same design but allows one to hear low-level sounds. The disadvantage is that the significant changes in attenuation do not occur until the noise reaches truly harmful levels such as about 140 dB or more. If one uses these protectors, the plug or muff must be fit tightly on the head or in the ear canal. Otherwise, leaks that bypass the orifice may permit excessive exposure. Such leaks could be formed due to the use of safety glasses that interrupt the seal of the earmuff cushion with the side of the head or a poorly fit earplug that has not sealed the ear canal. Most of these level-dependent products were designed for high-level impulsive noise. The protection they provide may just as easily be achieved by using a more traditional earmuff, banded protector, or earplug.

### 30.7.4 Training in Hearing Protector Use

Recent studies have evaluated the effect of training provided to hearing protection users and demonstrated that one-on-one training is the most effective means of teaching (Joseph et al., 2007; Murphy et al., 2010). Key to this training is the empowerment of the worker to understand what protector should be selected and be able to teach a coworker. Training does not require long sessions and examination. Rather, a simple video combined with individualized demonstration and evaluation by the instructor suffices to increase average attenuations by 7–15 dB (Joseph et al., 2007; Murphy et al., 2010). Even among experienced workers, training in proper rolling and insertion of formable earplugs can provide a 3–4 dB improvement (Murphy and Stephenson, unpublished).

Hearing conservationists have now developed applications that can provide a fit test of a user's hearing protectors. One system employs a dual microphone to sample the sound levels just outside the protector and the levels underneath the protector. The measured TL must be coupled with correction factors to account for the acoustics of the occluded environment (Voix and Laville, 2009). Another solution employs a psychophysical method called loudness balance to estimate the attenuation of each earplug separately (Schulz, 2010). Real ear attenuation at threshold using a headphone presentation measures the hearing threshold with and without earplugs to provide an estimate of the earplug attenuation (Edwards et al., 1978, 1983). While each system has advantages and disadvantages, the acceptance of fit testing has been increasing. Similar to fit-testing respirators for hazardous breathing environments, fit testing for hazardous noise environments is an important tool in hearing conservation. Visual examination of an earplug or earmuff alone is not sufficient to know whether protection is being provided to the worker. Everyone has a uniquely shaped head, pinna, and ear canal. Fit testing allows the employer to document that the employees have received counseling and training and demonstrated the ability to properly wear hearing protection. Finally, the hearing conservationist must recognize that the management should reinforce good practices at a workplace.

*Disclaimer: The findings and conclusions in this report are those of the authors and do not represent any official policy of the Centers for Disease Control and Prevention, The National Institute for Occupational Safety and Health, or the Environmental Protection Agency. Mention of company names and products does not constitute endorsement by CDC, NIOSH, or the EPA.*

## REFERENCES


Allen, C.H. and Berger, E.H. 1990. Development of a unique passive hearing protector with level dependent and flat attenuation characteristics. *Noise Cont. Eng. J.* 34(3): 97–105.

American Conference of Industrial Hygienists (ACGIH) 1994. *1994–1995 Threshold Limit Values for Chemical Substances and Physical Agents, and Biological Exposure Indices.* Cincinnati, OH: American Conference of Governmental Industrial Hygienists.

ANSI S3.19-1974 1974. *American National Standard Method for the Measurement of Real-Ear Protection of Hearing Protectors and Physical Attenuation of Earmuffs.* New York: American National Standards Institute.

ANSI S3.1-1999 (R2008) 2008. *American National Standard Maximum Permissible Ambient Noise Levels for Audiometric Test Rooms.* New York: American National Standards Institute.

ANSI/ASA S12.68-2007 2007. *American National Standard Methods of Estimating Effective A-Weighted Sound Pressure Levels When Hearing Protectors Are Worn.* New York: American National Standards Institute.

ANSI/ASA S12.6-2008 2008. *American National Standard for the Measuring Real-Ear Attenuation of Hearing Protectors.* New York: American National Standards Institute.

ANSI/ASA S12.42-2010 2010. *American National Standard of Hearing Protection Devices in Continuous or Impulsive Noise Using Microphone in Real Ear or Acoustic Test Fixture Procedures.* New York: American National Standards Institute.

Berger, E.H. 2000. Noise control and hearing conservation: Why do it? In *The Noise Manual,* E.H. Berger, L.H. Royster, J.D. Royster, D.P. Driscoll, and M. Layne (eds.). Fairfax, VA: AIHA Press, pp. 1–17.

Berger, E.H. and Hamery, P. 2008. Empirical evaluation using impulse noise of level-dependency for passive earplug designs. In *Acoustics 08,* Paris, France, July 2008. *J. Acoust. Soc. Am.* 123(5), pt2:3528.

Bolt, Beranek and Newman, Inc. 1976. *Economic Impact Analysis of Proposed Noise Control Regulation.* Report no. 3246, prepared for the U.S. Department of Labor, Occupational Safety and Health Administration (OSHA). Contract no. DOL-J-9-F-6-0019.

Booz, Allen & Hamilton, Inc. 1983. *Technical and Economic Analysis of Alternative Noise Standards.* Prepared for the Office of Regulatory Analysis, Occupational Safety and Health Administration. Washington, DC: U.S. Department of Labor.

Bosi, M. and Goldberg, R.E. 2003. *Introduction to Digital Audio Coding and Standards.* Springer, New York.

Brigham, E.O. 1998. *The Fast Fourier Transform and Its Application.* Upper Saddle River, NJ: Prentice Hall.

Camargo, H.E., Smith, A.K., Kovalchik, P.G., and Matetic, R.J. 2008. Noise source identification on a continuous mining machine. In *Proceedings of the 2008 National Conference on Noise Control Engineering,* C. Burroughs, T. Lim, J. Kim and G. Maling (eds.). Indianapolis, IN: Institute of Noise Control Engineering of the USA.

Clark, J.G. 1981. Uses and abuses of hearing loss classification. *ASHA* 23: 493–500.

Department of Transportation (DOT) 1973. Vehicle interior noise levels. Department of Transportation, Federal Highway Administration, Bureau of Motor Carrier Safety. *38 Fed Regist.,* 30880–30882.

Dunn, D.E., Davis, R.R., Merry, C.E., and Franks, J.R. 1991. Hearing loss in the chinchilla from impact and continuous noise exposure. *J. Acoust. Soc. Am.* 90(1): 1979–1985.

Durrant, J.D. and Lovrinic, J.H. 1995. *Bases of Hearing Science*, 3rd edn. Baltimore, MD: Lippincott Williams & Wilkins.

Edwards, R.G., Broderson, A.B., Green, W.W., and Lempert, B.L. 1983. A second study of the effectiveness of earplugs as worn in the workplace. *Noise Control Eng. J.* 20(1): 6–15.

Edwards, R.G., Hauser, W.P., Moiseev, N.A., Broderson, A.B., and Green, W.W. 1978. Effectiveness of earplugs as worn in the workplace, *Sound Vib.* 12(1): 12–30.

Fletcher, H. and Munson, W.A. 1933. Loudness, its definition, measurement, and calculations. *J. Acoust. Soc. Am.* 5: 82–108.

Fosboke, M.D.J. 1831. Practical observations on the pathology and treatment of deafness, No. II. *Lancet* 15(389): 645–648.

Franke, R., Parmentier, G., Buck, K., Kronenberger, G., and Beck, C. 1994. *Artificial Head for the Evaluation of the Effectiveness of Hearing Protectors with High Level Noises. Part 1: Development and Test in Impulsive Regime.* Research Report R-112/94. Saint-Louis Cedex, France: French-German Institute for Research of Saint-Louis (in French).

Franks, J.R. 1988. Number of workers exposed to occupational noise. *Seminars in Hearing* 9(4): 287–297.

Franks, J.R., Graydon, P.S., Jeng, C., and Murphy, W.J. 2003. *NIOSH Hearing Protector Device Compendium.* http://www2d.cdc.gov/hp-devices/hp_srchpg01.asp

Frost & Sullivan 2009. *U.S. Markets for Industrial Hearing Protection Products*, Pub ID: MC2275302, 101 pages.

Hager, L. 2009. BLS Occupational hearing loss report for 2007. *CAOHC Update* 27: 1.

Johnson, A.C. and Morata, T.C. 2010. *Occupational Exposure to Chemicals and Hearing Impairment. Nordic Expert Group for Criteria Documentation of Health Risks from Chemicals.* Report no. 2010; 44(4), Nordic Expert Group, Arbete och Halsa. 177pp.

Joseph, A., Punch, J., Stephenson, M.R., Paneth, N., Wolfe, E., and Murphy, W.J. 2007. The effects of training format on earplug performance. *Int. J. Audiol.* 46(10): 609–618.

Kardous, C.A. and Murphy, W.J. 2010. Noise control for indoor firing ranges. *Noise Control Eng. J.* 58(4): 345–356.

Kardous, C.A., Willson, R.D., and Hayden, C.S. 2003. Noise assessment and abatement strategies an indoor firing range. *Appl. Occup. Env. Hyg.* 18(8): 629–636.

Kroes, P., Fleming, R., and Lempert, B. 1975. *List of Personal Hearing Protectors and Attenuation Data*, Technical Report Publication No. 76-120, U.S. Department of Health Education and Welfare, Public Health Service, Centers for Disease Control and Prevention, National Institute for Occupational Safety and Health.

Lempert, B.L. 1984. Compendium of hearing protection devices. *Sound Vib.* 18(5): 26–39.

Lord, H.L., Gately, W.S., and Evensen, H.A. 1980. *Noise Control for Engineers.* New York: McGraw-Hill.

Morata, T.C. 2003. Chemical exposure as a risk factor for hearing loss. *J. Occup. Environ. Med.* 45(7): 676–682.

Murphy, W.J., Byrne, D.C., Gauger, D., Ahroon, W.A., Berger, E., Gerges, S.N.Y., McKinley R.L., Witt, B., and Krieg, E.F. 2009. Results of the National Institute for Occupational Safety and Health—U.S. Environmental Protection Agency Interlaboratory Comparison of American National Standards Institute S12.6–1997 Methods A and B. *J. Acoust. Soc. Am.* 125: 3262–3277.

Murphy, W.J., Franks, J.R., Berger, E.H., Behar, A., Casali, J.G., Dixon-Ernst, C., Krieg, E.F., Mozo, B.T., Ohlin, D.W., Royster, J.D., Royster, L.H., Simon, S.D., and Stephenson, C. 2004. Development of a new standard laboratory protocol for estimation of the field attenuation of hearing protection devices: Sample size necessary to provide acceptable reproducibility. *J. Acoust. Soc. Am.* 115: 311–323.

Murphy, W.J., Stephenson, M.R., Byrne, D.C., Witt, B., and Duran, J. 2010. Effect of training on hearing protector attenuation. *Noise Health* 13(51): 132–141.

National Institute for Occupational Safety and Health (NIOSH) 1972. *NIOSH Criteria for a Recommended Standard: Occupational Exposure to Noise.* Cincinnati, OH: U.S. Department of Health, Education, and Welfare, Health Services and Mental Health Administration, National Institute for Occupational Safety and Health. DHEW (NIOSH) Publication No. HSM 73-11001.

National Institute for Occupational Safety and Health (NIOSH) 1988. *National Occupational Exposure Survey (NOES), Field Guidelines.* Vol. 1, Cincinnati, OH: U.S. Department of Health and Human Services, Centers for Disease Control and Prevention, National Institute for Occupational Safety and Health. DHEW (NIOSH) Publication No. 88-106.

National Institute for Occupational Safety and Health (NIOSH) 1996. *Preventing Occupational Hearing Loss: A Practical Guide.* Cincinnati, OH: US Department of Health and Human Services, Centers for Disease Control and Prevention, National Institute for Occupational Safety and Health. DHHS (NIOSH) Publication No. 96-110.

National Institute for Occupational Safety and Health (NIOSH) 1998. *Criteria for a Recommended Standard: Occupational Noise Exposure; Revised Criteria.* Cincinnati, OH: U.S. Department of Health and Human Services, Centers for Disease Control, National Institute for Occupational Safety and Health. DHHS (NIOSH) Publication No. 98-126.

National Institute for Occupational Safety and Health (NIOSH) 2006a. *Health Hazard Evaluation and Technical Assistance Report: Kenton County Animal Shelter, Covington, Kentucky.* Cincinnati, OH: U.S. Department of Health and Human Services, Public Health Service, Centers for Disease Control and Prevention, National Institute for Occupational Safety and Health. NIOSH HETA No. HETA-2006-0212-3035.

National Institute for Occupational Safety and Health (NIOSH) 2006b. *Health Hazard Evaluation and Technical Assistance Report: Society for the Prevention of Cruelty to Animals, Cincinnati, Ohio.* Cincinnati, OH: U.S. Department of Health and Human Services, Public Health Service, Centers for Disease Control and Prevention, National Institute for Occupational Safety and Health. NIOSH HETA No. HETA-2006-0222-3037.

National Institute for Occupational Safety and Health (NIOSH) 2007. *Health Hazard Evaluation and Technical Assistance Report: Noise Exposures and Hearing Loss Assessments among Animal Shelter Workers.* Cincinnati, OH: U.S. Department of Health and Human Services, Public Health Service, Centers for Disease Control and Prevention, National Institute for Occupational Safety and Health, Louisiana Society for the Prevention of Cruelty to Animals, Algiers, Louisiana. NIOSH HETA No. HETA-2007-0068-3042.

Neitzel, R. and Seixas, N. 2005. The effectiveness of hearing protection among construction workers. *J. Occ. Env. Hyg.* 2: 227–238.

Occupational Safety and Health Administration (OSHA) 1981. *Occupational Noise Exposure: Hearing Conservation Amendment.* Washington, DC: U.S. Department of Labor, Occupational Safety and Health Administration, *46 Fed. Reg.* 4078–1179.

Occupational Safety and Health Administration (OSHA) 1983. *CPL 2-2.35A-29 CFR 1910.95(b)(1), Guidelines for Noise Enforcement; Appendix A.* Washington, DC: U.S. Department of Labor, Occupational Safety and Health Administration. OSHA Directive No. CPL 2-2.35A (December 19, 1983).

Oppenheim, A.V., Schaefer, R.W., and Buck, J.R. 1999. *Discrete-Time Signal Processing,* 2nd edn. Upper Saddle River, NJ: Prentice Hall.

Parmentier, G., Dancer, A., Buck, K., Kronenberger, G., and Beck C. 2000.Artificial head (ATF) for evaluation of hearing protectors. *Acta Acust.* 86: 847–852.

Pickles, J.O. 2008. *Introduction to the Physiology of Hearing,* 3rd edn. London, U.K.: Academic Press.

Prince, M.M., Stayner, L.T., Smith, R.J., and Gilbert, S.J. 1997. A re-examination of risk estimates from the NIOSH Occupational Noise and Hearing Survey (ONHS). *J. Acoust. Soc. Am.* 101(2): 950–963.

Raichel, D.R. 2006. *The Science and Application of Acoustics,* 2nd edn. New York: Springer.

Richard, K. 1979. *Tissues and Organs: A Text-Atlas of Scanning Electron Microscopy*. New York: W. H. Freeman & Co.

Royster, L.H. and Royster, J.D. 2002. *The Noise-Vibration Problem-Solution Workbook*. Fairfax, VA: AIHA Press.

Schulz, T. 2010. Individual fit-testing of earplugs: A review of uses. *Noise Health* 13(51): 152–162.

Sliwinska-Kowalska, M., Prasher, D., Rodrigues, C.A., Zamyslowska-Szmytke, E., Campo, P., Henderson, D., Lund, S.P., Johnson, A.C., Schaper, M., Odkvist, L., Starck, J., Toppila, E., Schneider, E., Moller, C., Fuente, A., and Gopal, K.V. 2007. Ototoxicity of organic solvents—From scientific evidence to health policy. *Int. J. Occup. Med. Environ. Health*. 20: 215–222.

Suter, A.H. 2000. Standards and Regulations. In *The Noise Manual*, E.H. Berger, L.H. Royster, J.D. Royster, D.P. Driscoll, and M. Layne (eds.). Fairfax, VA: AIHA Press, pp. 639–668.

Tak, S.W., Davis, R.R., and Calvert, G.M. 2009. Exposure to hazardous workplace noise and use of hearing protection devices among US workers—NHANES, 1999–2004. *Am. J. Ind. Med*. 53: 358–367.

U.S. Environmental Protection Agency 1978. CFR Title 40, subchapter G, 211, subpart B—*Hearing Protective Devices*. Washington, DC: U.S. EPA.

U.S. Environmental Protection Agency 2009. 40 CFR Part 211—Product noise labeling hearing protection devices; proposed rule. *Fed. Reg*. 74(149): 39150–39196.

U.S. Public Law No. 91–596, 84 Stat. 1590 1970. 91st Congress, S.2193 December 29, 1970, as amended through January 1, 2004 (1).

U.S. Public Law No. 92–574, 86 Stat. 1234 1972. Noise pollution and abatement act of 1972, codification amended at 42 U.S.C. 4901–4918 (1988) Section 8 [42 U.S.C. 4907] labeling.

Voix, J. and Laville, F. 2009. Prediction of the attenuation of a filtered custom earplug. *Appl. Acoust*. 70: 935–944.

Witt, B. 2007. Putting the personal back into PPE: Hearing protector effectiveness. *Occup. Health Saf*. 76(6): 90–94.

Wong, G.S.K. and Embleton, T.F.W. 1994. *AIP Handbook of Condenser Microphones: Theory, Calibration and Measurements*. Woodbury, MA: AIP Press.

Yost, W. 2006. *Fundamentals of Hearing: An Introduction*, 5th edn. Burlington, MA: Academic Press.

CHAPTER 31

# Nonionizing Radiation

John Cardarelli II

CONTENTS



## 31.1 NONIONIZING RADIATION

Nonionizing radiation (NIR) is a form of electromagnetic energy that is too weak to break chemical bonds in molecules. It takes many forms, including television and radio signals, radar, pager and cordless and cellular phone signals, microwaves, visible light, infrared and ultraviolet light, and lasers. Everyone is exposed to NIR from both naturally occurring and man-made sources. It can be beneficial or detrimental to those exposed. You cannot see it except for visible light (wavelength = 400–760 nm), taste it, or smell it; but you may be able to feel it by sensing heat or through electrostimulation. The phenomenon of hearing certain radiofrequencies is also a well-established biological effect with no known adverse health consequences. A quiet environment is needed for these radiofrequency-induced sounds (similar to other common sounds) to be heard. The presence of NIR is growing, fueling anxiety and speculation about its possible adverse health effects. Levels of exposure will continue to grow as technology advances and as society increasingly demands the conveniences it brings.

The electromagnetic spectrum includes ionizing and NIR (Figure 31.1). This chapter explains the characteristics of NIR radiation, with a focus on ways to measure it and its

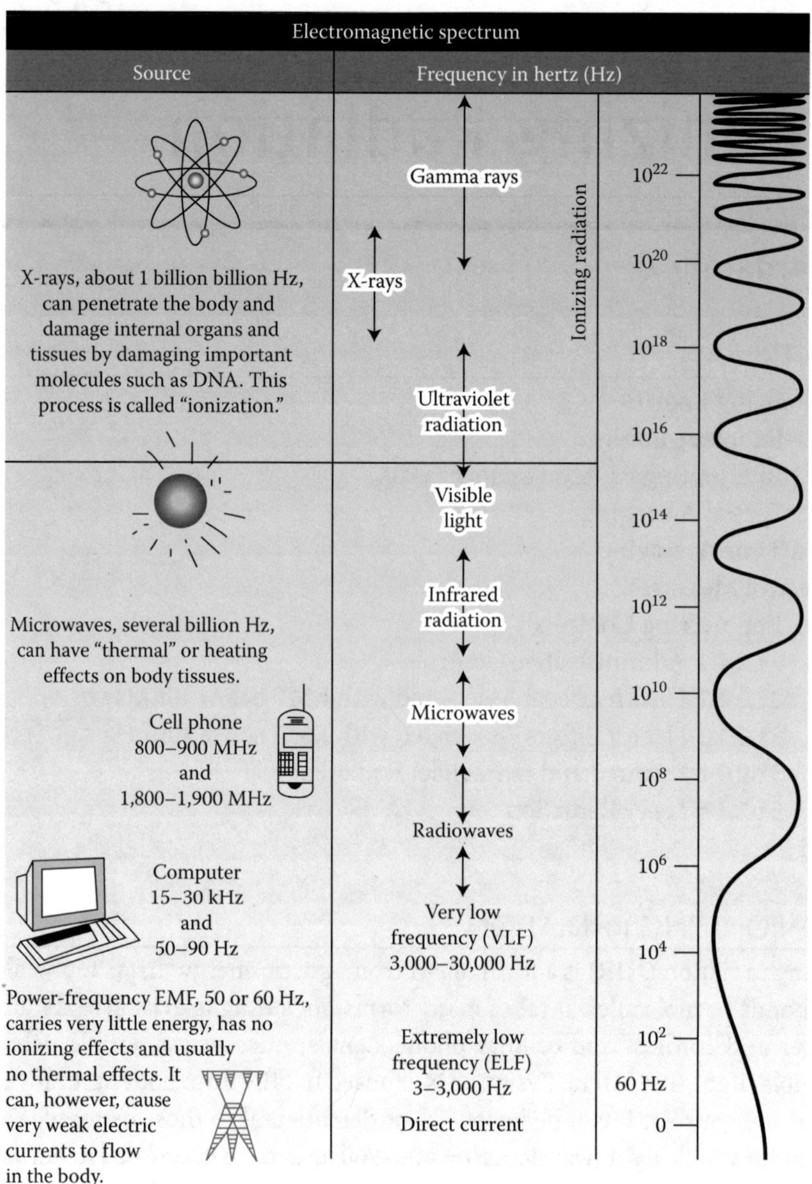


FIGURE 31.1 Electromagnetic spectrum. *Note*: ELF is defined as 3–300 Hz by NCRP Report 119. (From EMF in the Workplace. Department of Energy, National Institute for Occupational Safety and Health, and National Institute of Environmental Health Sciences, 1996.)

associated safety standards for occupational environments. The topics will progress from extremely low frequencies (ELF) to extremely high frequencies (laser energies), just below the ionization region of the electromagnetic spectrum. Common NIR units are shown in Table 31.1. Energy (Q) is the ability to do work and its unit is the joule (J). The rate at which energy is used, or work performed, is known as power (P) with units of joules per second (J/s). Energy represents the dose to a biologic system and power represents the dose rate. The watt (W) is the unit for power and 1 W = 1 J/s. All forms of NIR can be measured

TABLE 31.1 Common Nonionizing Radiation Units

| Quantity | Symbol | Unit |
|---|---|---|
| Frequency | f | Hertz; Hz (GHz, MHz, kHz) |
| Wavelength | Λ | Meter; m (mm, cm, km) |
| Energy | Q | Joule; J |
| Power | P | Watt; W or J/s |
| E-field strength | E | Volt/meter; V/m, $V^2/m^2$ |
| H-field strength | H | Ampere/meter; A/m, $A^2/m^2$ |
| Magnetic flux | B | Tesla; T or Gauss; G |
| Current density | J | Ampere/meter square; $A/m^2$ |
| Power density | W, S | Watt/meter square; $W/m^2$ |
| Specific absorption | SA | Joules/kilogram; J/kg |
| Specific absorption rate | SAR | Watts/kilogram; W/kg |

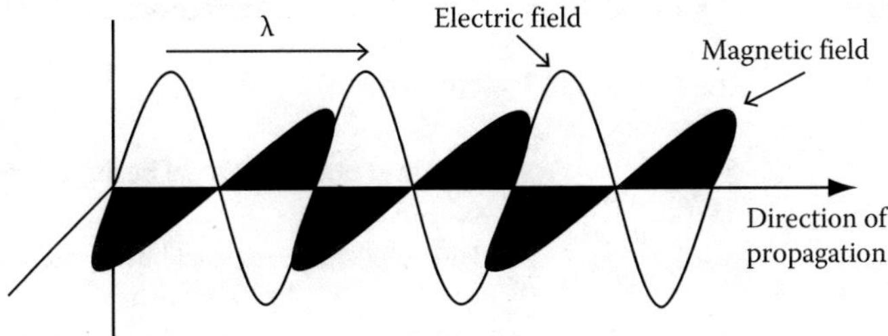


FIGURE 31.2 The propagation of energy by time-varying electric and magnetic fields. (http://www.fcc.gov/Bureaus/Engineering_Technology/Documents/bulletins/oet56/oet56e4.pdf)

in terms of energy or power and the way these parameters are expressed depends on the region of interest, or spectral regions, within the electromagnetic spectrum.

All NIR is presented in *electromagnetic fields* (EMFs), which propagate energy by time-varying electric and magnetic fields (Figure 31.2). Electromagnetic radiation is characterized by three related quantities: frequency, wavelength, and energy. Frequency is the number of complete cycles per second and is represented by the unit hertz (Hz) or (1/s; $s^{-1}$). By convention, frequency is the primary parameter used to describe the part of the electromagnetic spectrum from sub-extremely low frequency (SELF) to supra-extremely high frequency (SEHF) as defined in Table 31.2. Wavelength is the primary parameter used to describe the part of the electromagnetic spectrum from infrared (IR) to ultra violet (UV) frequencies. Frequency and wavelength are related by the following equation:

$$\lambda = \frac{c}{f} \tag{31.1}$$

where

λ is the wavelength in meters (m)

*c* is the velocity of light (about 300,000,000 m/s)

*f* is the frequency in cycles per second (Hertz, or Hz)

TABLE 31.2 Frequency Bands and Their Associated Biological Impacts

| Band | Frequency Range (Hz)[a] | Wavelength Range (m)[b] | Biological Impact |
|---|---|---|---|
| Sub-extremely low frequency (SELF) | 0–30 | $0–10^7$ | $0–10^5$ Hz<br>0–3000 m |
| Extremely low frequency (ELF) | 30–300 | $10^7–10^6$ | Electrostimulation (primary dosimetric parameter is internal current density) |
| Voice frequency (VF) | 300–3000 | $10^6–10^5$ | |
| Very-low frequency (VLF) | $3000–3 \times 10^4$ | $10^5–10^4$ | |
| Low frequency (LF) | $3 \times 10^4–3 \times 10^5$ | $10^4–10^3$ | $10^5–6 \times 10^9$ Hz |
| Medium frequency (MF) | $3 \times 10^5–3 \times 10^6$ | $10^3–10^2$ | 3000–0.05 m |
| High frequency (HF) | $3 \times 10^6–3 \times 10^7$ | $10^2–10$ | Specific absorption rates (heating effects) |
| Very-high frequency (VHF) | $3 \times 10^7–3 \times 10^8$ | 10–1 | |
| Ultra-high frequency (UHF) | $3 \times 10^8–3 \times 10^9$ | 1–0.1 | |
| Super-high frequency (SHF) | $3 \times 10^9–3 \times 10^{10}$ | $0.1–10^{-2}$ | Above $6 \times 10^9$ Hz, below 0.05 m: |
| Extremely high frequency (EHF) | $3 \times 10^{10}–3 \times 10^{11}$ | $10^{-2}–10^{-3}$ | Surface heating (radiant) |
| Supra-extremely high frequency (SEHF) | $3 \times 10^{11}–3 \times 10^{12}$ | $10^{-3}–10^{-4}$ | |
| Infrared radiation | IR-C | 0.3 μm–1 mm | Corneal burns, thermal skin burns |
| | IR-B | 0.14–0.3 μm | |
| | IR-A | 760–1400 nm | Retinal burns, cataracts of lens, thermal skin burns |
| Visible light | | 400–760 nm | Retinal burns, thermal skin burns |
| Ultraviolet radiation | UV-A | 400–320 nm | Cataract of lens, thermal skin burns |
| | UV-B | 320–280 nm | Corneal injuries, cataracts of lens, |
| | UV-C | 280–200 nm | photokeratitis, photoconjunctivitis, erythema |

[a] Frequency is used to describe these regions of the electromagnetic spectrum.
[b] Wavelength is used to describe these regions of the electromagnetic spectrum.

Hazards potentially associated with exposure to EMFs in various bands may result in (a) currents produced in the body by contact with energized sources or without such contact (electrostimulation), (b) increased core-body temperature, or (c) increased body surface temperature (Table 31.2). How efficient these fields interact with the body depends on several factors. For example, materials with high water content (muscles) absorb EMF energy at higher rates than dry materials. The absorption rate is higher when (a) the incident electric field is parallel to the body and (b) the incident magnetic field is perpendicular to a larger cross-sectional area. Sharp corners, edges and points concentrate electric fields. Depth of penetration of EMF energy decreases as conductivity or frequency increases, and as wavelengths decrease.

Electric fields (E) exist when electric charges exert forces on one another, whether in motion or not. E-fields describe the strength of forces on charges (volts per meter; V/m). They can be visualized as lines of force that emanate from a positively charged object to a negatively charged object. Since E-fields have a specific direction *and* magnitude, they are vector quantities, however only the magnitude component is reported as part of safety evaluations.

Magnetic fields ($H$), measured in ampere per meter (A/m), are associated with the strength of these additional forces on moving charges. An ampere is an SI unit for electric current. Magnetic fields are also vector quantities and contain a direction perpendicular to the electrical current as shown in Figure 31.2 and a magnitude proportional to amount of current present. Magnetic fields are related to another quantity called the magnetic flux density ($B$) by

$$B = \mu H \tag{31.2}$$

where
- $\mu$ is the permeability of the medium
- $B$ is the sum of the components of magnetic fields passing through a given area and is the quantity used for health and safety assessments

Its SI unit is the tesla (T) and the cgs unit is the gauss (G) (1 T = 10,000 G). A useful factor to convert between $B$ and $H$ is 1 G = 80 A/m.

The relationship between the E- and H-fields is described by the *power density,* which is related to the product of the E- and H-field components:

$$\mathbf{S}\left(\frac{W}{m^2}\right) = \mathrm{E}\left(\frac{V}{m}\right) \times \mathrm{H}\left(\frac{A}{m}\right) \tag{31.3}$$

The cross product of E and H vectors is called the Poynting vector and represents the magnitude (power density) and direction of the propagation of the EMF. Abbreviated as $S$, it can be calculated from E- or H-field measurements by the following equation:

$$S = \frac{\mathrm{E}^2}{377} \qquad S = 377\mathrm{H}^2 \tag{31.4}$$

where
- $S$ is the power density in watts per square meter (W/m$^2$ or VA/m$^2$)
- E is the electric field strength measurement (V/m)
- H is the magnetic field strength measurement (A/m)
- 377 is the constant = the impedance of free space (ohms, Ω or V/A)

Impedance describes the resistance experienced by the electromagnetic radiation travelling through space.

With increasing distance from its source, the quantitative nature of EMFs changes and requires specific measurement protocols to properly assess occupational exposures. When the distance from the source is small compared to the wavelength, the E- and H-fields are independent and should be considered separately. This is known as a near field, a distance from the source to about 1/6th the associated wavelength as shown in Figure 31.3. A power

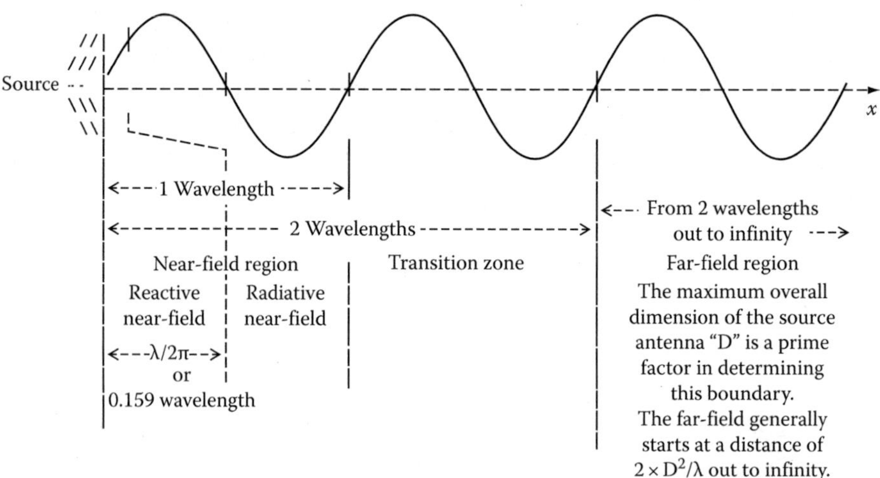


FIGURE 31.3 Antenna field regions for typical antennas. (http://www.osha.gov/SLTC/radiofrequencyradiation/electromagnetic_fieldmemo/electromagnetic.html#section_1)

density measurement for hazard assessments is inappropriate and the individual components of the E- and H-field strengths should be measured. At distances between 1/6th and less than about 2 wavelengths from the source, called the transition field, both the E- and H-fields must be used for safety assessments. Only at distances greater than about 2 wavelengths, known as the far-field, the use of power density for hazard assessments is appropriate. These three regions categorize EMF behaviors which vary even within each region and the boundaries are approximate rules-of-thumb. In this far-field region, the E- and H-field strengths decrease linearly with distance from the source and the power density decreases as the square of the distance from the source.

### 31.1.1 Exposure Limits

The transfer of energy from electric and magnetic fields in any material is described by the *specific absorption rate* (SAR). "Specific" refers to the normalization to mass of the material exposed, "absorption" refers to the absorption of the energy in a specific medium (tissue), and "rate" refers to the time rate of change of the energy absorption. SAR is the most reliable indicator or predictor of the potential for biological effects in test animals and a measure of what is happening inside the human body. It is expressed in units of watts per kilogram (W/kg) or milliwatts per gram (mW/g). Since SAR is difficult to evaluate or measure outside the laboratory, the measurable quantities of E- or H-field strengths and power density as well as induced and contact currents are used to define the radiofrequency (RF) environment. They have been correlated with SAR to determine the maximum permissible exposure (MPE) levels (Table 31.3). In the far field (greater than about two wavelengths from source), measuring field strengths or power density provides reliable exposure assessments. In the near field or in contact with RF sources and/or other metallic objects (where many occupational exposures occur), induced and contact current measurements provide the most reliable exposure evaluations (Figures 31.4 and 31.5). Measuring field strengths or power density is unreliable near or in contact with RF

TABLE 31.3 Maximum Permissible Exposure for the Occupational Environments[a]

| Frequency Range (MHz) | RMS E-Field[b] Strength (V/m) | RMS H-Field[b] Strength (A/m) | Power Density (S) E-Field, H-Field (W/m²) | Averaging Time $\|E\|^2$, $\|H\|^2$ or S (min) |
|---|---|---|---|---|
| 0.1–1.0 | 1842 | $16.3/f_M$ | $(9{,}000, 100{,}000/f_M^2)$[c] | 6 |
| 1.0–30 | $1842/f_M$ | $16.3/f_M$ | $(9{,}000/f_M^2, 100{,}000/f_M^2)$ | 6 |
| 30–100 | 61.4 | $16.3/f_M$ | $(10, 100{,}000/f_M^2)$ | 6 |
| 100–300 | 61.4 | 0.163 | 10 | 6 |
| 300–3,000 | | | $f_M/30$ | 6 |
| 3,000–30,000 | | | 100 | $19.63/f_G^{1.079}$ |
| 30,000–300,000 | | | 100 | $2.524/f_G^{0.476}$ |

**Induced and Contact Currents: Occupational Environment**

| | Maximum Current (mA)[d] | | | |
|---|---|---|---|---|
| | | | Contact | |
| | Through Both Feet | Through Each Foot | Grasp | Touch |
| 0.003–0.1 | $2.0 f_K$ | $1.0 f_K$ | $1.0 f_K$ | $0.5 f_K$ |
| 0.1–110 | 200 | 100 | 100 | 50 |

*Source:* Data from IEEE. *IEEE Standard for Safety Levels with Respect to Human Exposure to Radio Frequency Electronmagnetic Fields, 3 kHz to 300 GHz.* New York: Institute for Electrical and Electronics Engineers, Inc., 2005.

*Note:* $f_K$ = frequency in kHz; $f_M$ = frequency in MHz; $f_G$ = frequency in GHz.

[a] Occupational environment is also called a controlled environment and is an area where the occupancy and activity of those within are subject to control and accountability as established by an RF safety program for the purpose of protection from RF exposure hazards.

[b] For exposures that are uniform over the dimension of the body, such as certain far-field exposures, the exposure field strengths and power densities are compared with the MPEs in this Table. For nonuniform exposures, the mean values of the exposure fields, as obtained by spatially averaging the squares of the field strengths or averaging the power densities over an area equivalent to the vertical cross section of the human body, or a smaller area depending on the frequency, are compared with the MPEs in this table.

[c] These power density values are commonly used as a convenient comparison with MPEs at higher frequencies and are displayed on some instruments.

[d] The averaging time for determination of contact current limits is 6 min.

FIGURE 31.4 Obtaining RF contact current measurements.

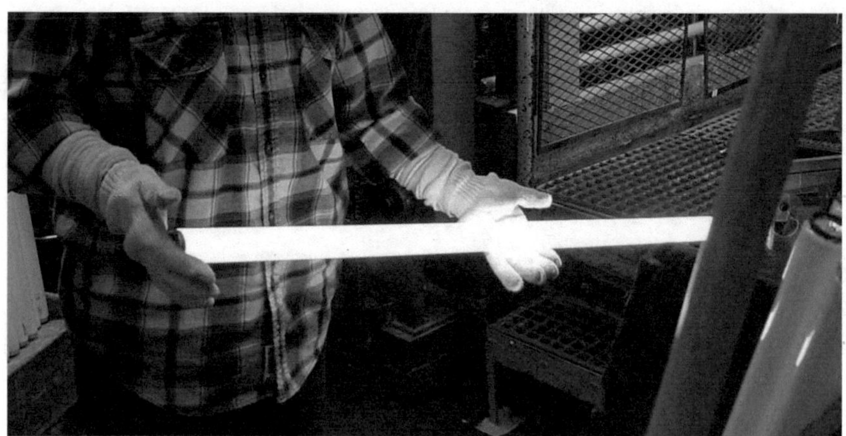

FIGURE 31.5 Personal protective equipment used in a lamp manufacturing process during a quality control check. RF energy activates the lamp. The gloves protect against sharp objects in case the bulb breaks and prevent unnecessary currents through the workers' limbs. Not shown is a rubber mat that insulates the body from acting as a electrical conduit to the ground.

sources or other metallic objects. The MPE values provided are those from the Institute of Electrical and Electronics Engineers, Inc., standard, which incorporate the latest scientific findings and recommendations for occupational exposures [1]. Guidelines for limiting RF exposure have also been developed by several other scientific organizations and government agencies, but the differences are minor and work is underway to harmonize the various exposure limits [2,3]. In the case of exposure of the whole body, a human adult (height = 175 cm) absorbs RF energy most efficiently when the wavelength is 40% of the long axis of the body and parallel to the incident E-field vector. This occurs at a frequency of about 70 MHz. The RF exposure limits, which are called Basic Restrictions, reflect this dependency on frequency and were derived from a SAR of 4 W/kg for those frequencies associated with heating affects (100 kHz to 3 GHz). In terms of human metabolic heat production, 4 W/kg represents a moderate activity level, such as with housecleaning. A safety factor of 10 was applied resulting in a RF exposure limit of 0.4 W/kg, virtually an indistinguishable heating affect from normal temperature variation, exercise, or exposure to the Sun. For localized exposures in an occupational environment where the field strength is more than 20 times the spatial average, the SAR should not exceed 10 W/kg. For extremities and the pinnae (the cartilaginous projection portion of the outer ear consisting of the helix, lobule, and anti-helix) the SAR should not exceed 20 W/kg. RF exposures below this level are intended to prevent adverse health effects. Exposures in excess of the limits are not necessarily harmful. However, without intended lifesaving or medical benefits, these situations are not recommended.

### 31.1.2 Interpreting RF Measurement Data

Occupational limits (sometimes referred to as a controlled environment) apply to persons exposed at work, provided they are fully aware of the potential for their exposure and can exercise control over it. One should understand there are three fundamental concepts when interpreting measurement data: (a) the difference between exposure and emission limits, (b) spatial averaging, and (c) time averaging.

*Emission limits* are the maximum power output authorized by government authorities for companies or individuals. However, these transmitting signals are often not emitted at the maximum power output. This is especially true for cell-phone base stations or towers, since the amount of power used is proportional to the number of calls handled. For this reason, it is important to note that the emission limit (maximum power output) may not be directly related to exposure potentials. Unlike emission limits, exposure guidelines apply to *exposure limits*, and they are relevant only to locations that are accessible by workers.

### 31.1.3 Spatial Averaging

The exposure limits are based on the concept that the exposures are applied to a whole-body averaged SAR. This means that spot measurements exceeding the stated exposure limits do not imply noncompliance or harmful exposure scenarios if the spatial average of RF fields over the body does not exceed limits. A spatial average measurement may consist of three or more measurements averaged together that span a length of an adult.

### 31.1.4 Time Averaging

Another feature of the exposure guidelines is that EMF exposures should be averaged over a 6 min period for workplaces (controlled environments). To properly apply field measurements to exposure limits, one must consider the length of time individuals are exposed. For example, during any given 6 min period, workers could be exposed to twice the applicable limit for 3 min as long as they are not exposed for the preceding or following 3 min. Similarly, a worker could be exposed at 3 times the limit for 2 min as long as no exposure occurs during the preceding or subsequent 4 min.

## 31.2 PROTECTIVE MEASURES

### 31.2.1 Engineering Controls

Protection of workers from unnecessary or excessive exposure to RF radiation is accomplished through engineering and administrative controls. Engineering controls are preferred since they eliminate or reduce the potential exposures at the source, but they require a sophisticated level of knowledge to install. Improperly installed controls may enhance worker exposures. Interlocks, shielding, bonding, grounding, and filtering are some of the more common controls employed. OSHA requires a lock-out/tag-out program for working with sources of hazardous energies, which may include installing many of the RF controls described earlier.

The effectiveness of shielding materials varies with the material, geometry, frequency, and where the field reduction is measured. Some are more effective for reducing electric fields while others are more suitable for reducing magnetic fields. One of the most recognizable types of shielding is that used on microwave ovens. The perforated screen is designed to allow penetration of visible light (wavelength about $0.7 \times 10^{-6}$–$0.4 \times 10^{-6}$ m, or 430 million to 750 million Hz), but prevents leakage of microwave radiation (wavelength about 12 cm, or 2450 MHz). Perforated or continuous shielding materials reduce exposures by reflection, absorption (attenuation), and internal reflection. The proper selection of material is complex and should be done by qualified individuals.

Techniques that may supplement the use of engineering controls include prudent placement of RF sources, resonant frequency shift, and personal protective equipment (shoes, clothing, and special suits). Consideration should be given to building-construction materials and layout when installing RF equipment to reduce or prevent unnecessary enhancement of reflected energy at the worker's location. If the operating frequencies are around 10–40 MHz, the whole-body SAR may be reduced by *resonant frequency shift*, separating the body from the ground plane by a small distance with electrically insulating materials. This measure reduces the worker's absorption characteristics by reducing the flow of current from the body to a grounded surface. Resonant frequency shift may be especially useful for dielectric-heaters (plastic-sealer) operators by having them stand on nonconductive platforms made of wood or rubber. For factory worksites, metal-reinforced concrete floors act as ground planes. Footwear that reduces the grounding effect also achieves the same effect as a resonant frequency shift. The level of RF exposure reduction is dependent on the RF frequency and the types of shoes and socks worn by the worker. Wool socks and rubber-soled shoes provide the greatest reduction for frequencies below 100 MHz (wavelengths above 3 m). RF protective suits may be helpful when work must be done in "hot" areas, such as continual-radar, onboard naval vessels and some communication and broadcast environments (Figure 31.6). Suit material is typically wool, polyester, or nylon impregnated with a highly conductive threaded metal. Some are more effective than others depending on frequency, orientation of the worker in the environment relative to the incident electric fields, and construction of openings for feet, hands, and head. Washing these suits may

FIGURE 31.6 Researchers conducting RF measurements near television broadcast antenna atop a high-rise building.

reduce their protective capabilities. Some experts recommend against use of RF-protective suits because they may be hazardous to individuals nearby the wearer, and may increase the hazard to the wearer by allowing closer proximity to open circuits that may act as secondary sources.

#### *31.2.1.1 Administrative Controls*

Administrative controls include increasing the distance between the source and workers (often used and easy to bypass), controlling the duration of exposure, restricting access, placing warning signs, providing training commensurate with the level of potential hazard, and real-time monitoring via dosimetry. Horizontal and vertical distance should be considered when determining the appropriate distance, which is often the distance that results in a radiation level equal to the limit (the *hazard distance*). There is no simple way to calculate the reduction of field strength with distance since the calculation depends on so many factors; however, some researchers measured magnetic field strengths which showed a reduction by $1/r^5$ for induction heaters [4]. Controlling the duration of exposure is achieved by applying the time-averaging technique discussed earlier. Finally, real-time monitoring devices (dosimeters) are especially useful in identifying potentially harmful exposures, allowing the recipient to take protective actions and reduce risk of injury. Dosimeters provide an audible and visual alarm when exposures exceed a predetermined level (usually 50% of the maximum permission exposure), and allow the wearer to quickly identify if changes occur during work activities.

#### *31.2.1.2 Health Effects Associated with EMF below 100 kHz*

Exposures to electric and magnetic fields emanating from the generation, transmission, and use of electricity have been studied extensively. Recommendations of various scientific organizations and regulatory agencies acknowledge controversy regarding the potential health effects of chronic low-level EMF exposures. However, there is no convincing evidence of a health risk [5,6]. One of the most comprehensive reviews of health effects associated with extremely low frequency (ELF) exposures was published by IARC [7] which found that

1. There is limited evidence in humans for the carcinogenicity of ELF magnetic fields in relation to childhood leukemia
2. There is inadequate evidence in humans for the carcinogenicity of ELF magnetic fields in relation to all other cancers
3. There is inadequate evidence in humans for the carcinogenicity of static electric or magnetic fields and ELF electric fields
4. There is inadequate evidence in experimental animals for the carcinogenicity of ELF magnetic fields
5. There are no available data for the carcinogenicity of static electric or magnetic fields and ELF electric fields in experimental animals

IARC concluded that ELF magnetic fields are possibly carcinogenic to humans, and that static electric and magnetic fields and ELF electric fields are not classifiable as to their carcinogenicity to humans.

#### *31.2.1.3 Health Effects Associated with EMF above 100 kHz*

More than 100 million Americans use wireless communication devices, with 50,000 new users daily [8]. If the use of wireless communication devices is ever associated with even the slightest increase in risk of adverse health effects, it could become a significant public health problem. At frequencies above 100 kHz, studies support the basic restrictions and MPEs recommendation described earlier. These recommendations were made on the basis of a comprehensive review of the scientific data to protect against established adverse health effects from RF exposures. An adverse health effect is defined as a harmful change in health that is supported by the consistent findings in the peer-reviewed literature, demonstrated by independent laboratories, and has consensus in the scientific community. The established adverse health effects associated with RF exposure above the basic restrictions and MPEs are (a) aversive or painful electrostimulation due to excessive RF internal electric fields, (b) RF shock or burns due to contact with excessively high RF voltages, (c) heating pain or tissue burns due to excessive localized RF exposures, and (d) behavioral disruption, heat exhaustion or heat stroke due to excessive whole body RF exposures [1]. It is worth noting that adverse effects do not include things like biological effects (sensations) without a harmful health effect, indirect effects caused by electromagnetic interference with electronic devices, or changes in subjective feelings of well-being that are a result of anxiety about RF effects or impact.

Debate continues on the level of protection necessary to prevent long-term health effects from RF exposures. WHO and many European countries promote a precautionary approach by discouraging the widespread use of mobile phones by children for non-essential calls because they may be more likely to develop adverse effect because their nervous system is still developing and they will face a lifetime of various hazardous exposures [9]. The Russian National Committee on Non-Ionizing Radiation Protection extends the WHO recommendations for children to pregnant women and to those suffering from specific diseases, and recommends that duration of cellular phone calls be limited to 3 min each with at least 15 min between calls. The United States does not necessarily endorse the precautionary approach because without clear, convincing epidemiologic evidence that a health hazard exists from RF exposures, this approach could negatively impact industry growth and development.

Cancer-related studies on animals provide no evidence of physiological, pathological, or disease-specific effects of long-term RF exposures. Likewise, epidemiological studies show no clear or consistent evidence to indicate a causal role of RF exposures in connection with human cancer or other disease endpoints at exposures below the basic restrictions and MPEs. However, it is scientifically impossible to prove absolute safety (the null hypothesis) of any physical agent. Many of original studies lacked adequate exposure assessment information and biological measures, and include confounding factors, such as multiple sources. More recent studies have benefited from improved dosimetry and modeling techniques, and better clinical testing protocols.

#### *31.2.1.4 Infrared and Ultraviolet Radiation*

Infrared radiation (IR) lies at frequencies higher than those of radar waves and microwaves (Table 31.3). Nearly half of the Sun's radiant energy is emitted as IR. IR is highly absorbed by water and the Earth's atmosphere and invisible to the eye. However, its warmth can be detected by the skin. All objects with temperatures above absolute zero emit IR. In industry, significant levels of IR are produced directly by lamps and indirectly by heat sources, such as heating and drying devices. The primary biological effect of IR is thermal due to absorption in the water within body tissues. For this reason, IR cannot penetrate the skin but leaves a sensation of heat, which often serves as an adequate warning sign to take protective action or risk skin burns. The lens of the eye is particularly vulnerable to IR because the lens has no heat sensors and a poor heat-dissipating mechanism. Cataracts may be produced by chronic IR exposure at levels far below those that cause skin burns. Occupations typically at risk of IR exposure include glass blowers, furnace workers, foundry workers, blacksmiths, solderers, oven operators, those who work near baking and drying heat lamps, and movie projectionists. Like RF radiation, IR exposure limits are frequency-based; however, they represent conditions under which it is believed that nearly all healthy workers may be repeatedly exposed without acute adverse effects. The limits for IR most recognized in the scientific community are published by ACGIH [10]. Control of IR hazard requires (a) shielding of the IR source and eye protection with appropriate IR filters, (b) maximizing the distance between workers and the IR source, and (c) reducing the time spent in areas with high levels of IR exposure.

Ultraviolet radiation (UVR) is produced by the sun and artificially by incandescent, fluorescent, and discharge types of light sources. It is characterized by three distinct energy bands known as UV-A (400–320 nm), UV-B (320–280 nm), and UV-C (280–200 nm). The first two bands are principal UV components in sunlight. Nearly all UV-A reaches the Earth's surface, but most UV-B is absorbed by the stratospheric ozone layer. UV-C is completely absorbed by the ozone layer and oxygen in the air, but it can be artificially produced. Industrial sources of UVR include welding arcs, plasma torches, electric arc furnaces (full spectrum of UVR), germicidal and black-light lamps (mostly UV-C), and certain type of lasers (full spectrum of UVR). Because wavelengths of UVR are so short, UVR presents a surface heating hazard.

The most common health effect from overexposure to UVR is sunburn (erythema). Chronic low-level UVR exposure from the Sun is also associated with various skin effects, including skin cancer (basal cell carcinoma, squamous cell carcinoma, and malignant melanoma), premature aging of the skin, solar elastosis (wrinkling), and solar keratoses (premalignant lesions). Basal cell carcinoma and malignant melanoma are more strongly associated with a history of multiple episodes of sunburn, whereas squamous cell carcinoma is associated with total exposure. UVR exposure has also been associated with suppressing the immune system and developing cortical cataracts (UV-B exposure). Photosensitizing agents, such as coal tar, plants (including figs, lemon and lime rinds, celery, and parsnips), and lemon and lime rinds (containing furocoumarins and psoralens), and pharmaceutical drugs (including chlorpromazine, chlorpropamide, and tolbutamide) can increase susceptibility to UVR. All these effects vary with individual susceptibilities (lighter skin is more susceptible than darker skin, and people on medicine for diabetes) and geographic location (UVR levels are highest near the equator, at higher altitudes, when the Sun is directly overhead, no cloud cover or ozone coverage, summer time, and in highly reflective environments). Acute high-level UVR exposures, especially from UV-B, result in eye injuries, which are often only recognized several hours after the exposure. Photokeratitis (inflammation of the cornea) and photoconjunctivitis (inflammation of the thin transparent mucous membrane lining the inner surface of the eyelids) are usually reversible within several days. Intense UVR exposure also has an indirect impact on health through its ability to cause photochemical reactions. Small amounts of oxygen and nitrogen can be converted into ozone and oxides of nitrogen, which are respiratory irritants. Halogenated hydrocarbon solvent vapors can decompose into toxic gases, such as perchloroethylene decomposing to hydrogen chloride and trichloroethylene decomposing to phosgene. Chronic low-level UVR exposures can be controlled by the use of protective clothing, eyewear, and sunscreen lotions, and by the reduction of duration of exposure. Controlling UVR from acute high-level photochemical exposures may require local exhaust ventilation and isolation of UVR sources from industrial processes that involve solvents. Only qualified personnel should determine the effectiveness of any particular form of personal protection.

#### *31.2.1.5 Laser Radiation*

*Laser* is an acronym for *l*ight *a*mplification by the *s*timulated *e*mission of *r*adiation. Uses in industry include heat treatment, glazing, alloying, cladding, cleaning, brazing, soldering, conduction welding, penetration welding, cutting, hole drilling, marking, trimming, and photolithography [11]. Health and safety decisions are based on the class of laser and the wavelength of the laser source. The hazard classification system places lasers into four categories depending on their potential to cause harm from direct beam exposures (Table 31.4). These exposures may result in at least four types of injury to the eyes and the skin, each requiring a special consideration for selecting the appropriate personal

TABLE 31.4 Laser Classification

| Class of Laser[a] | Hazard Potential |
|---|---|
| 1 | Pose no potential for injury. No safety measures required to either the eye or skin |
| 2, 2a | Visible beam posing no significant potential for injury. Blinking response limits exposure |
| 3, 3a, 3b | Modest potential for injury. Normal aversion response is not sufficient to limit eye exposure to a safe level. Skin hazards normally do not exist. May require safety precautions and personal protective equipment. Class 3b lasers require more safety precautions than Class 3a |
| 4 | Serious potential for injury of the eye and skin. Requires safety precautions and personal protective equipment. Diffuse reflection viewing hazard. Potential fire hazard. Most laser systems for cutting, heat treating, and welding are Class 4 |

[a] When Class 3 and 4 lasers are fully enclosed to prevent potentially hazardous laser radiation exposures, the system may be classified as a Class 1 system.

TABLE 31.5 Laser Injuries

| Type of Hazard | Laser Wavelength (nm) | Target Tissue | Comment |
|---|---|---|---|
| UV photochemical injury | 180–400<br>180–400<br>295–380 | Skin<br>Cornea<br>Lens | Eye protection is required whenever a bluish-white light is seen at the laser focal point |
| Blue-light photochemical injury | 400–550 | Retina | Retinal burn (has been termed "eclipse blindness") |
| Thermal injury | 400–1400 | Retina | Nd: YAG lasers pose the greatest risk because the beam image can be intensified in the order of 100,000 |
| | 1400 nm to 1 mm | Skin<br>Cornea<br>Conjunctiva | Most common injury from laser radiation exposure. Biggest concern with $CO_2$ lasers |
| Near-IR thermal injury | 800–3000 | Lens | Results from molten metal or large, heated surface during treatment. This hazard is only of concern for repeated, chronic exposures |

protective equipment (Table 31.5). However, nonbeam laser hazards constitute the greatest source of noncompliance with federal safety codes. Sources of nonbeam hazards include (a) improper electrical design or use of grounding, component or shielding, (b) lack of knowledge for production of laser-generated air contaminants (LGAC), (c) unwanted plasma radiation, (d) excessive noise levels, (e) inadequate ventilation controls, (f) fire hazards, (g) explosive hazards from high-pressure tubes, (h) exposure to toxic chemicals and laser dyes, and (i) fire hazards. Most of these hazards are associated with Class 3b and 4 lasers. In practice, it is always desirable to totally enclose the laser and beam path to prevent both direct-beam and nonbeam exposures.

Unlike most other workplace hazards, there is generally no need to perform workplace measurements for lasers because of highly confined beam dimensions, minimal likelihood of changing beam paths, and the difficulty and expense of using laser radiometers. However, measurements must be performed by manufacturers to ensure proper laser classification. Laser safety standards are published by government agencies, and by independent and industrial standards organizations. In the United States, the American National Standards Institute (ANSI) has developed the *Standard for the Safe Use of Lasers* (Z136.1) and publishes general safety requirements for users. Although this standards is not a law, it forms the basis for OSHA and many states' regulations. There are other laser safety standards and state-specific regulations, but they apply primarily to Class 3b and 4 installations and maintenance activities.

The ISO and International Electrotechnical Commission (IEC) have published standards similar to those in the United States. Two requirements in the ISO documents that affect manufacturers are that (a) all systems must be Class 1 during operation, and (b) manufacturers must specify which materials equipment is designed to process. A Class 1 laser rating can be achieved by installing appropriate engineering controls.

Controlling all aspects of potential laser exposures is complex and requires a qualified individual to assess direct and nonbeam hazards. Control measures include process isolation, local-exhaust and building ventilation, training and education, restricted access, proper housekeeping, preventive maintenance, and use of appropriate personal protective equipment.

## REFERENCES


1. IEEE. *IEEE Standard for Safety Levels with Respect to Human Exposure to Radio Frequency Electronmagnetic Fields, 3 kHz to 300 GHz.* New York: Institute for Electrical and Electronics Engineers, Inc., 2005.
2. NCRP. *Biological Effects and Exposure Criteria for Radio Frequency Electromagnetic Fields, Report 86.* Bethesda, MD: National Council on Radiation Protection and Measurements, 1986, pp. 1–382.
3. World Health Organization. *Framework for Developing EMF Standards.* International EMF project. World Health Organization. Geneva, Switzerland: WHO, 2003.
4. Conover DL, Murray WE, Lary JM, Johnson PH. Magnetic field measurements near RF induction heaters. *Bioelectromagnetics* 1986; 7: 83–90.
5. Institute of Electrical and Electronics Engineers (IEEE). Possible health hazards from exposure to power-frequency electric and magnetic fields—A COMAR technical information statement. *IEEE Engineering in Medicine and Biology Magazine* 2000; 19: 131–137.
6. AIHA. American Industrial Hygiene Association. Position statement on extremely low frequency (ELF) fields. Available at: http://www.aiha.org/Content/AccessInfo/gov/PSELF.htm (accessed on September 6, 2009).
7. International Agency for Research on Cancer. *IARC Monographs on the Evaluation of Carcinogenic Risks to Humans.* Vol. 80, Non-ionizing radiation, part 1: Static and extremely low-frequency (ELF) electric and magnetic fields. Lyon, France: IARC, 2002.
8. National Toxicology Program. Fact Sheet: Studies on radiofrequency radiation emitted by cellular phones. P.O. Box 12233, Maildrop A3-01, Research Triangle Park, NC 27709-2233, 2003.
9. Maisch D. Children and mobile phones … is there a health risk? *Journal of Australasian College of Nutritional & Environmental Medicine* 2003; 22: 3–8.
10. American Conference of Governmental Industrial Hygienists. *Threshold Limit Values for Chemical Substances and Physical Agents: Biological Exposure Indices.* Cincinnati, OH: ACGIH, 2004.
11. Ready JF (Ed.). *LIA Handbook of Laser Materials Processing.* Orlando, FL: Magnolia Publishing, Inc., 2001.

# VI

## Current Topics

CHAPTER 32

# Cumulative Trauma Disorders of the Upper Extremities*

Brian D. Lowe

CONTENTS



* This work is that of the author and does not necessarily reflect the views of the National Institute for Occupational Safety and Health. Mention of any company names or products does not constitute endorsement by the National Institute for Occupational Safety and Health.

## 32.1 INTRODUCTION

Upper extremity cumulative trauma disorders (CTDs) are regional musculoskeletal impairments that are associated with repetitive mechanical trauma occurring in the workplace [1]. CTDs of the upper extremities encompass a multitude of physical symptoms, pathology, and disability related to muscle tissue, ligaments, tendons, tendon sheaths, joints, and nerves. In various literature [2,3], CTDs have also been referred to as repeated trauma illness, repetitive motion injury, regional pain syndrome, repetitive strain injury (Australia), and occupational cervicobrachial disorder (Japan, Germany, and Scandinavia). They were once named according to the occupation or body part affected, for example, trigger finger, bricklayer's shoulder, pricer's palsy, mouse elbow, writer's cramp, stitcher's wrist, and carpenter's elbow [4]. These terms were descriptive but tended to minimize, even romanticize, the afflictions suffered by workers because of their jobs. The medical terms that are now more commonly used to describe CTDs (even among non-physicians) tend to add legitimacy to the symptoms that workers experience, and have likely increased workers' awareness regarding the work relatedness of overuse disorders.

Some authors (e.g., Ref. [5]) have advised against the use of inconsistent and imprecise terminology such as *cumulative trauma*, *repetitive strain injury*, and *overuse syndrome* in referring to musculoskeletal disorders (MSDs) of occupational origin. As such, the term work-related musculoskeletal disorders (WMSDs) may be more currently favored in the occupational health community; however, the term "cumulative trauma disorders" has been used interchangeably in this chapter. CTD is consistent with the terminology originally used in the first edition of this chapter, and the term distinguishes these conditions as a result of *chronic* overuse injury to the affected soft tissues. This contrasts with *acute* traumatic injuries, following the near instantaneous transfer of high energy, resulting in sprains, broken bones, cuts, lacerations, or amputations—which can also be viewed as "disorders" of the musculoskeletal system.

This chapter will address upper extremity CTDs (UECTDs) from an occupational ergonomics perspective, with emphasis on the approaches to identifying risk factors and characterizing a worker's exposure to them. The chapter will emphasize more recent work including job analysis and exposure assessment methods developed since the first edition of this text. The chapter will provide an update on MSD prevention programs (relevant to UECTDs) at Occupational Safety and Health Administration (OSHA) and

NIOSH (National Institute for Occupational Safety and Health), the U.S. federal agencies charged with protecting worker safety and health.

## 32.2 BACKGROUND AND SIGNIFICANCE TO OCCUPATIONAL ERGONOMICS

Ergonomics is a holistic approach to work layout that is rooted in achieving the best possible match of worker attributes and capabilities to the design and configuration of a work task. As such, application of ergonomic principles requires many considerations, including human anthropometry, biomechanics, work physiology, visual capabilities, and virtually every aspect of the physical environment [6]. Increasingly, there has been appreciation for the way that work is organized and prevention strategies that, in addition to physical stressors, also consider macro-ergonomic and psychosocial stressors and their influence on outcomes for WMSDs [7].

In industries where the work is physically demanding and many types of tools and powered equipment and machinery are in use, the classical approach of applying the many disciplines in the field of ergonomics is still necessary as a means to maximize worker health and efficiency. However, now that the health of workers is an issue in many workplaces that are quiet and environmentally controlled, the attention of ergonomic professionals (and indeed the overall perception of ergonomics by many people) has shifted from its historical multidisciplinary emphasis to the more limited area of WMSD prevention and control.

## 32.3 COMMON CTDs: THEIR SYMPTOMS AND DEVELOPMENT

UECTDs are classified as sprains, strains, inflammations, and irritations that affect tendon, muscle, nerve, vascular system, bursa, and bone/cartilage. This section presents an overview of the more common disorders to these various tissues. This is not intended to be a comprehensive description of these disorders. More detailed explanation of the diagnosis and clinical considerations of these conditions can be found in other references (e.g., Refs. [8,9]). A complete list of the MSDs that are considered to be related to work can be found in *The International Classification of Diseases*.

### 32.3.1 Tendon Disorders

The term *tendinitis* (or alternatively, *tendonitis*) generally refers to the inflammation of tendon, which may take on a variety of forms depending upon the location of the inflammation (see Figure 32.1). Insertional tendinitis occurs at the tendon-bone interface—a common example of which is observed as lateral epicondylitis ("tennis elbow") in which inflammation develops at the insertion of the extensor carpi radialis brevis muscle at the lateral epicondyle from repetitive forearm/elbow motions. *Peritendinitis* occurs in the more central region of the tendon, and *myotendinitis* is an inflammation at the muscle-tendon junction. *Tenosynovitis* refers to an inflammation in the sheath surrounding the tendon, which results in pain, swelling, and difficulty moving the affected joint. The inflammation results in a narrowing or stenosing of the tendon sheath, and tenosynovitis can lead to the condition of a trigger digit (trigger finger), in which a nodule forms on the tendon and inhibits normal gliding of the tendon within the sheath.

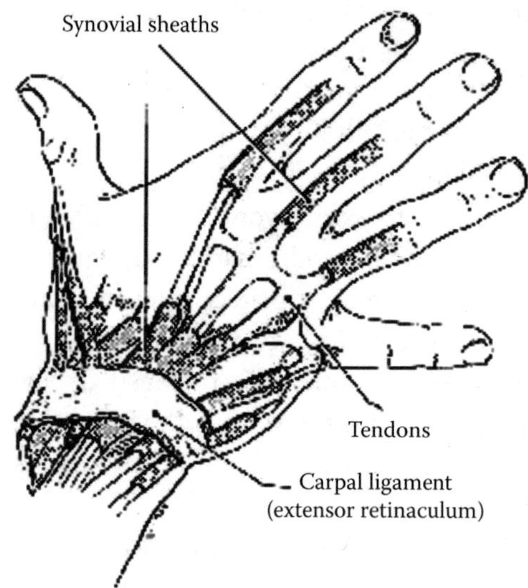


FIGURE 32.1 Dorsal (back) view of the hand showing the tendons and tendon sheaths. Also shown is the dorsal side of the carpal ligament.

De Quervain's syndrome describes a specific tenosynovitis of the tendon sheath of the thumb (abductor pollicis longus and extensor pollicis brevis). A modern day example of De Quervain's syndrome, illustrating how even contemporary technology can be implicated in the etiology of CTDs, is that coined in the lay media as "Blackberry Thumb" [10]. The seriousness of the musculoskeletal risk factors associated with the repetitive use of these hand-held mobile media devices was noted in the Consumer Education Alert released by The American Society of Hand Therapists [11].

### 32.3.2 Disorders of the Muscle

The term myalgia describes general tenderness and soreness of muscle tissue. Fibromyalgia is a condition characterized by chronic pain and a hypersensitivity to pressure on the affected tissue. Trigger points are localized points of spastic muscle, tender to the touch, that are surrounded by non-affected tissue. Pressure on these points elicits pain radiating along the extremity. Tension neck syndrome (TNS) is one of the more commonly reported myofascial syndromes, reported in a variety of occupations that involve prolonged static contraction of the trapezius (upper back) muscle for precision work involving visual task demands.

### 32.3.3 Nerve Disorders

Several nerve disorder syndromes are relevant to the discussion of UECTDs. A clinical syndrome refers to a conglomeration of concurrent symptoms that characterize a diagnosis. No upper limb nerve disorder seems to have received as much attention as carpal tunnel syndrome (CTS). CTS is a nerve entrapment syndrome in which the median nerve is compressed in the wrist cavity created by the carpal bones and the transverse carpal ligament (see Figure 32.2). Compression of the nerve in this tunnel results from an increase

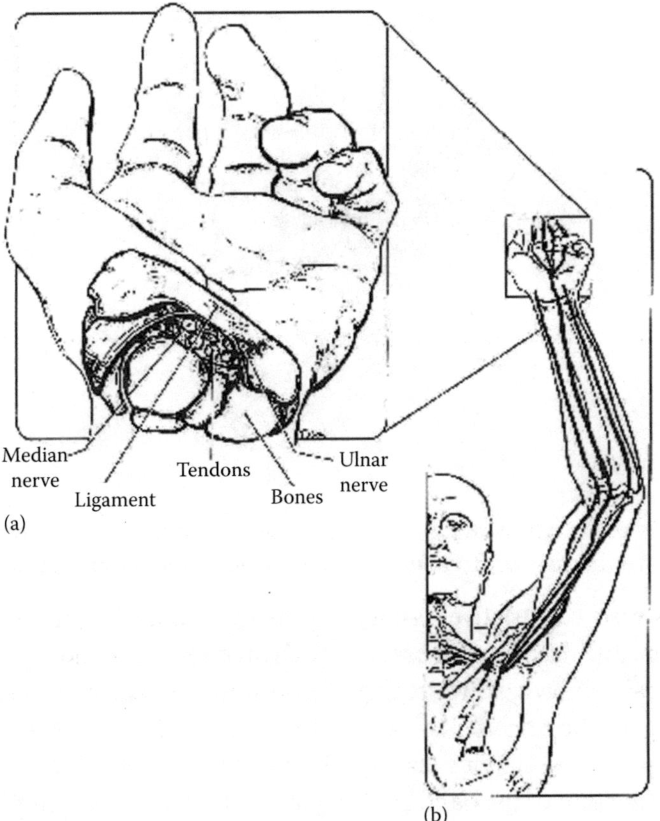


FIGURE 32.2 (a) Cross-sectional view of the carpal canal showing the tendons and nerves passing through it. (b) A full view of the three major nerves that originate in the neck and serve the arm and hand: the median, the ulnar, and the radial.

in the contents in this space and a number of other mechanical factors. Ischemia (restriction of blood supply to the tissue) and axonal demyelination (damage to the sheaths of the nerve) resulting from the compression contribute to the pathophysiology of CTS.

The ulnar and radial nerves are also subject to entrapment neuropathies. Ulnar nerve entrapment can occur at both the wrist (Guyon's canal syndrome) or the elbow (cubital tunnel syndrome). Radial tunnel syndrome is experienced as tenderness at the lateral aspect of the elbow and can mimic sensations of lateral epicondylitis. In the neck/shoulder region, thoracic outlet syndrome (TOS) is an umbrella term used to describe a number of entrapment neuropathies of the brachial plexus and/or the subclavian vessels as they pass through narrow passageways leading from the base of the neck to the armpit (See Figure 32.3). TOS is characterized by symptoms of numbness and tingling of the arm and hands and is commonly associated with occupational use involving overhead work and extensive elevation of the upper arm.

### 32.3.4 Vascular Disorders

A common vascular disorder of occupational origin is hand/arm vibration syndrome (HAVS). Use of vibrating powered hand tools exposes the user to a transfer of mechanical

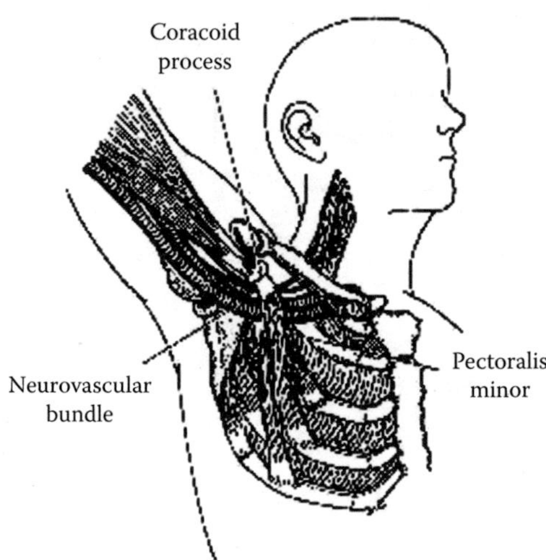


FIGURE 32.3 View of the nerves and blood vessels (neurovascular bundle) compressed between the neck and shoulder due to shoulder abduction (movement of the upper arm away from the body).

energy causing a vaso-constrictive response in the smooth muscle of the arterioles in the hands and fingers. This leads to a subsequent ischemic response and loss of blood flow to the digits. Classically known as "white finger" syndrome, the ischemia results in a blanching, or whitening, of the fingers. Raynaud's phenomenon describes bouts or "attacks" of white finger and painful sensations that are triggered by cold temperatures. Numbness and loss of precision function of the hand are symptoms of HAVS. Cold temperatures in combination with vibration exposure appear to magnify the risk for these disorders.

### 32.3.5 Bursa Disorders

Bursae are the fluid-filled sacs lined by synovial membrane that provide a cushion between bones and tendons and/or muscles around a joint. Bursae reduce friction and allow free joint movement. Bursitis refers to an inflammation of the synovial filled sacs and is commonly observed in the elbow (olecranon bursitis), shoulder (subacromial bursitis), and the knee (prepatellar bursitis). Other factors contributing to bursitis include blunt trauma, inflammatory disease, and infection.

### 32.3.6 Disorders of Bone and Cartilage

Osteoarthritis is a form of arthritis involving degenerative joint disease affecting articular cartilage and adjacent subchondral bone. It is believed to be precipitated or aggravated by joint overuse. Workplace factors such as pinch and power grip and impact loading have been identified as biomechanical risks for hand-wrist osteoarthritis [12].

Because many of the physical stresses attributed to UECTD development occur with regularity and in combination with high repetition in manual tasks in the workplace, they are known as work-related (or occupational) risk factors for CTDs. This term is not intended to diminish the multifactorial etiology, or cause, of CTDs but rather to emphasize the fact that work or occupation is a major factor in CTD development. The most common of the occupational upper extremity risk factors cited in the literature are the posture [13–16], the amount of muscular force associated with the activity [17,18], and the frequency or rate of repetition of the motions [19,20]. Working in cold environments [21,22] and using vibrating tools [23,24] are considered to increase the risk of CTDs.

The etiology and disease mechanisms associated with the development of CTDs are not fully understood, but it is believed that work activities which load the musculoskeletal system such as awkward posture, excessive force, and frequent repetition impose mechanical and physiological stress on the soft tissues of the upper extremity. For example, wrist deviations (flexion, extension, ulnar and radial deviation) shown in Figure 32.4 stretch the soft tissues across bones and ligaments, causing deformation of tendons and inflammation of tendon sheaths. Wrist deviations result in narrowing passageways that can place mechanical stress on tendons and entrap nerves. Nerve compression due to entrapment in a narrowed tunnel is thought to be one of the causes of CTS [25]. Under severe conditions, muscular force causes tendons and muscles to expand and swell, which can tear tissue and put pressure on nerves. Excessive repetitive movements can overcome the capacity of tendon sheaths to lubricate tendons and that of synovial membranes to lubricate joints. Motions and activities that comprise more than one or all of these occupational factors are more likely to result in the development of CTDs [26].

Regardless of the etiology and causation of UECTDs, there can be little doubt that affected individuals suffer in their physical, emotional, and social well-being. The financial burden of these conditions manifests at the establishment level through workers compensation costs and lost productivity. The financial burden on the individual has been shown to be even greater in terms of out of pocket medical expenses. Morse et al. [27] have shown that the majority of medical expenses for MSDs of the upper extremities are externalized from the workers' compensation system and that out of pocket costs to the afflicted worker averaged $500 per year. Lost productivity and quality of life in the home due to impaired function in activities of daily living

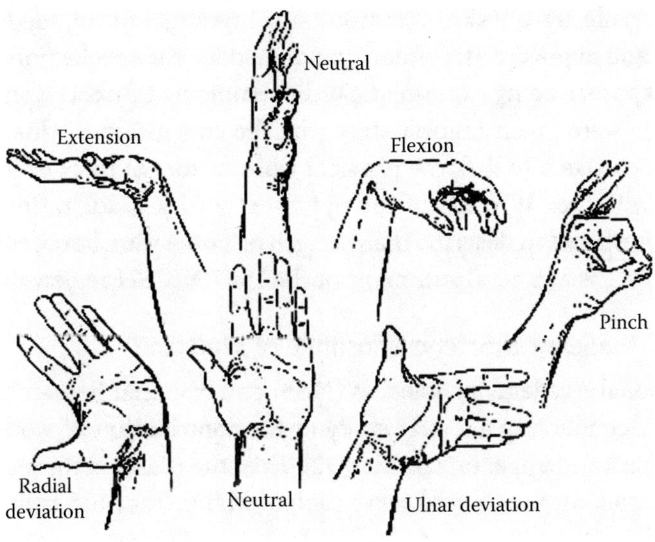


FIGURE 32.4 Deviated postures of the hand and wrist commonly associated with the development of UECTDs.

are difficult to calculate and are rarely included in these estimates. For these reasons UECTDs are of interest to the clinicians who treat the conditions, ergonomics specialists and engineers responsible for workspace layout and process design, and epidemiologists and public health practitioners in the surveillance and prevention of these disorders.

## 32.4 REVIEWS OF EVIDENCE FOR WORK RELATEDNESS

The term work-related musculoskeletal disorders (WMSDs) has emerged as the preferred terminology in the occupational health community. The "work relatedness" indicates that these conditions are precipitated by, or aggravated by, occupational activity. The World Health Organization [28] states that "...work-related diseases may be partially caused by adverse working conditions. They may be aggravated, accelerated, or exacerbated by workplace exposures, and they may impair working capacity." However, there are risk factors and physical exposures outside of the workplace that may contribute to the development of musculoskeletal symptoms and disorders, and similar MSDs exist in non-working populations. Thus, some have questioned the relative influence of workplace and non-workplace risk factors in the etiology of CTDs. This has motivated several significant epidemiologic efforts, published subsequent to the first edition of this text, to review the available body of evidence pertaining to the role of workplace exposures and non-occupational risk factors in the etiology of upper extremity WMSDs. Some of these major reviews are summarized here.

### 32.4.1 National Institute for Occupational Safety and Health: 1997 Review

In 1997, the NIOSH released a comprehensive review of the evidence of work relatedness of MSDs of the neck, upper extremity, and low back [29]. This publication, colloquially referred to as the "yellow book," reviewed over 600 published epidemiological studies to evaluate evidence for an association between physical work characteristics and WMSDs. Evidence was considered to be stronger for studies with participation rates greater than 70%, diagnoses made by physical examination, investigators blinded to case and/or exposure status, and exposure assessment conducted for the specific joint of interest and for the specific exposure being examined. Studies employing direct observation or physical measurements were given highest strength. The conclusion of this review was that sufficient evidence existed to link the physical work factors of force exertion, repetition, posture, and vibration to WMSDs of the upper extremity. Odds ratios and prevalence rate ratios were presented to describe the strength of association between risk factors and disorders. Table 32.1 is a general summary of the 1997 NIOSH review findings.

### 32.4.2 National Academy of Sciences/Institute of Medicine

In 1999, the National Academy of Sciences (NAS) and National Research Council (NRC) was charged with conducting a 2 year study of the contribution of workplace factors to MSDs of the low back and upper extremities [30]. This study came at the request of Congress to examine causation, diagnosis, and prevention of MSDs. The NAS committee identified 265 references related to work-related physical factors and upper extremity MSDs. Studies were reviewed based on criteria similar to that of the 1997 NIOSH review [29]. For UECTDs, 13 primary studies met the inclusion criteria and reported specific measures of exposure to

TABLE 32.1 Summary of NIOSH Findings in Its 1997 Review of Epidemiologic Studies of the WMSDs

| Body Region/Disorder (No. of Studies) | Posture | Force | Repetition | Vibration | Notes |
|---|---|---|---|---|---|
| Shoulder (40) | Evidence | Insufficient evidence | Evidence | Insufficient evidence | |
| Neck and neck/ shoulder (24) | Strong evidence | Evidence | Evidence | Insufficient evidence | |
| Elbow (22) | Insufficient evidence | Evidence | Insufficient evidence | | Strong evidence for combination of force w/posture or repetition |
| CTS (31) | Insufficient evidence | Evidence | Evidence | Evidence | Strong evidence for combination of force w/posture or repetition |
| Hand/wrist tendinitis (8) | Evidence | Evidence | Evidence | Evidence | Strong evidence for combination of force w/repetition |
| Hand-arm vibration syndrome (20) | | | | Strong evidence | |

*Source:* From Bernard, B., Musculoskeletal disorders and workplace factors: A critical review of epidemiologic evidence for work-related musculoskeletal disorders of the neck, upper extremity, and low back, DHHS(NIOSH) Pub. No. 97-141, 1997.

physical risk factors of manual materials handling, repetition, force, vibration, and combinations of force and repetition, and repetition and cold. The majority of the associations evaluated in these studies were significant positive associations, particularly for manual materials handling, vibration, and the interaction of repetition and force. The NAS review also reported 29 studies with less-specific exposure assessment measures. These studies confirmed the importance of these risk factors in contributing to CTDs. The NAS report indicated that vibration, force, and repetition have been most strongly and consistently associated with UECTDs, with relative risks ranging from 2.3 to 84.5.

### 32.4.3 Non-Occupational Risk Factors

Manual activities outside of the workplace and other individual biological factors have been associated with the occurrence of CTDs. These have been referred to as *non-occupational* risk factors or *personal* risk factors. Examples are physically intensive pastimes and hobbies that have similar postural and repetitive muscular exertion patterns to work tasks, such as racquet sports, knitting, sewing, or the playing of video games and musical instruments [15]. Additionally, some biological and medical conditions that may predispose one to CTDs (particularly CTS) are rheumatoid arthritis, gout, acromegaly, diabetes, wrist size and shape, and hormonal factors (menopause, use of oral contraceptives, pregnancy, and gynecological surgery) [25,31,32]. As a number of these conditions are unique to women, women's risk of CTS may be elevated. Although it is generally reported that the incidence of CTS is greater among women than men [32], studies that have compared the rates of CTS among men and women performing the same work tasks have found no difference in CTS prevalence [18].

A few recent studies have emphasized the contribution of non-occupational factors in the development of certain MSDs of the upper extremity. In particular, CTS has been scrutinized in the debate over the relative contribution of occupational and non-occupational factors in its etiology. There seems to be little disagreement that non-occupational factors contribute to CTS risk. The controversy lies in the relative contribution of occupational and non-occupational factors and the importance of this determination in medical compensation systems and policy in regard to workplace health. Recent reviews by Lozano-Calderón et al. [33] and Palmer et al. [34] illustrate divergent viewpoints on the relative influence of occupational exposure in the etiology of CTS. Palmer et al. [34] reviewed 38 articles meeting their inclusion criteria (12 of these published after the release of the 1997 NIOSH review) and concluded that a substantial body of evidence supports the influence of highly repetitive wrist-hand work, especially when combined with forceful grip, in the causation of CTS. Lozano-Calderón et al. [33] reviewed 107 publications that evaluated the association between either occupational factors or biological (personal) factors and CTS. Included papers were scored by the investigators based on nine criteria integral to the Bradford Hill [35] analysis of epidemiologic studies. The authors reported that 66% of studies examining repetitive hand use showed correlation with CTS and that vibration and stressful manual work showed correlation in 70% and 46% of studies, respectively. However, the percentage of studies showing correlation between biological factors (anthropometry, gender, age, genetics) and CTS were higher. These authors further assert that the strength of epidemiologic evidence for causative association for biological factors outweighs that for occupational factors.

The conflicting conclusions of these review studies, drawn from the same body of literature, illustrate the complexity of the etiology of a disease such as CTS and the futility of implicating any single factor in its causation. As noted by Punnett and Wegman [36], the fact that occupational factors cannot account for a large proportion of the MSD burden in the general population does not negate a causal relationship between occupational factors and preventable MSD risks in the workplace. Regardless of the relative contribution of occupational and non-occupational factors to CTS, or any other UECTDs from which a worker may suffer, a primary prevention approach can be adopted. A worker's biological makeup is unalterable, whereas the design of the work processes and workplace which he or she interacts is frequently well within the control of the employer to modify.

## 32.5 UPDATE ON SURVEILLANCE DATA

Throughout the 1980s and early 1990s, the number and percentage of total occupational illness cases documented by the U.S. Bureau of Labor Statistics (BLS) as attributable to repeated trauma were steadily increasing, and appear to have peaked in the mid-1990s at about the time the first edition of *Occupational Ergonomics: Theory and Application* was published. The increasing trend prior to the mid-1990s may have been partly a result of the increased awareness of these disorders that occurred followed by a leveling-off period in the mid to late 1990s. The percentage of total occupational injury and illness cases attributed to repeated trauma (roughly 63%) remained fairly constant between the mid-1990s (first edition) and 2001 (see Figure 32.5). BLS data for 1992–2007 suggest that the number of musculoskeletal injuries and illnesses involving days away from work has declined markedly

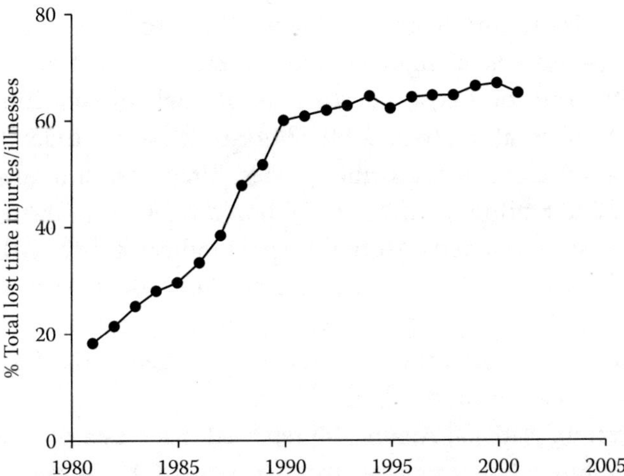


FIGURE 32.5 BLS data for repeated trauma illnesses (1981–2001) shown as a percentage of all lost time occupational injuries and illnesses.

from 782,000 to 335,000. However, significant changes in OSHA record keeping requirements occurring in 2001 have made it difficult to classify repeated trauma injury cases since 2001. Overall, MSDs still make up approximately 30% of the total non-fatal injuries and illnesses [37]. Specific to the upper extremity, the most recent BLS data [37] indicate that repetitive motion and overexertion cases that are not attributable to lifting reflect approximately 14% (162,000) of the 1.16 million illness and injury cases with lost days of work. CTS and tendinitis represent 3.6% and 1.1% of these 1.16 million cases, respectively.

It is commonly accepted that national U.S. injury and illness surveillance statistics underreport incidence of MSDs [38] and occupational illness overall [39]. As an example, BLS data indicate that CTS rates in the construction sector (per 10,000 workers) were 1.4, 1.7, 1.0, 1.3, and 0.4 for years 2003–2007. This is considerably lower than the 2.12% prevalence revealed for construction workers in the 1988 National Health Interview Survey (Occupational Health Supplement). Rosecrance et al. [40] found CTS rates in apprentice construction workers that were substantially higher—on the order of an 8.2% prevalence. It has been suggested that the changes in OSHA record keeping requirements in 2001 have resulted in lower reporting of MSDs. Friedman and Forst [41] have gone so far as to assert that 83% of the decline in occupational illness and injury observed since 2001 can be attributed to changes in the record keeping requirements.

Regardless of past trends, injury and illness rates suggest that workplace MSDs continue to be a public and occupational health problem. Their prevalence relative to all occupational injuries and illnesses is surpassed by their associated costs relative to other occupational injuries and illnesses. The total costs associated with WMSDs of the upper extremity are difficult to estimate and can include any combination of medical treatment and rehabilitation, lost time compensation, pensions for early retirement, or cost to rehire and to retrain new workers. In 1984, the American Academy of Orthopedic Surgeons conducted a study that included four data sources: (1) the National Health Interview Survey, (2) the National Hospital Discharge Survey, (3) the National Ambulatory Care

Survey, and (4) the Health and Nutrition Survey. The Academy estimated that annual direct costs for musculoskeletal injuries were in excess of $22 billion, indirect costs exceeded $5 billion, and total costs for all musculoskeletal conditions exceeded $65 billion [42]. Liberty Mutual, in their 2008 Workplace Safety Index, estimated direct workers' compensation costs in the United States. They cited a total workplace injury cost to industry of $48.6 billion, with $12.4 billion attributed to overexertion and $2.0 billion to repetitive motion. Liberty Mutual's report indicates that RMIs were associated with significant decline (down 35.3%) between 1998 and 2006. However, Rosenman et al. [43] found that only 25% of workers with work-related musculoskeletal conditions filed for workers' compensation. Thus, the true monetary burden of these disorders is underestimated by workers compensation claims.

In 1991, the Bureau of National Affairs [44] reported that a case of CTS could cost up to $30,000. In 2007, Washington state reported that the *average* CTS claim resulted in $22,000 in direct costs. Liberty Mutual Insurance Company indicated that one-third of workers' compensation (more than $10 billion in 1988) is paid out for RMIs in 1989, Liberty Mutual Insurance Company paid out an average of $5670 in workers' compensation for each case of CTD [44]. By 1991, that figure increased to $10,000 [45]. Between 1997 and 2005, the Washington State Workers' Compensation Fund accepted an average of 37,000 WMSD claims per year, with over $4 billion in direct costs resulting in more than 23.7 million lost work days [46]. The average compensable claims incidence rate of 92.9 per 10,000 full-time equivalents (FTEs) resulted in an average of 217 time-loss days per compensable claim. Of these compensable claims, 37% involved upper extremity MSDs.

It was noted in the first edition of this text that a fundamental reason for the increase in CTD incidence and cost through the 1980s and 1990s was the physical transformation of jobs and the workforce performing them. Technology advances such as the video display terminal in office environments, the laser scanner in grocery stores, and the bar code sorter in the post office have increased worker productivity, but have also increased the repetitive and stereotyped nature of jobs while depriving workers of the rest periods or microbreaks that were once inherent in most workstation designs. Many production jobs have been fragmented into repetitive, single-task activities that concentrate movements, forces, and mechanical stresses on small areas of the upper extremity. Such job designs are attractive to management and engineers because they simplify workplace layout, increase production output, facilitate work measurement, and minimize the amount of worker skill required. It is not surprising that some of the industries that have fragmented jobs the most have historically had the highest CTD rates, for example, meatpacking, poultry processing, and motor vehicle manufacturing.

## 32.6 JOB ANALYSIS AND EXPOSURE ASSESSMENT: METHODS FOR UECTDs

Since the first edition of this text, the discipline has seen a proliferation of checklist and observational-based methods for job analysis. These methods vary in their complexity and the resources required for their application. This section will focus on three approaches to job analysis and assessment of UECTD risk factors, classified as *risk factor checklists*, *systematic observational-based methods*, and *instrumentation-based measurement*. Checklist approaches afford the most rapid summary of the risk factors present, but with little sensitivity, precision, or ability to discriminate among exposure levels. Systematic observational-based approaches involve the application of a more rigorous structured assessment largely based on visual judgment of risk factor levels. They involve more analysis time than checklists, with the intent that the observed risk factor levels can be translated into an indication of risk for MSD outcomes. Direct-recording instrumentation-based measurement approaches are generally believed to be the most accurate methods of exposure assessment [47] but have not been widely adopted across the range of ergonomics practitioners.

### 32.6.1 Checklists

A number of ergonomic checklists have been developed in the last 15 years. Hamrick [48] described checklists as lists of risk factors or workplace conditions that help the user (or job analyst) ensure they have addressed areas of concern. Checklists generally serve as screening tools of jobs for the broad identification of risk factors and workplace conditions that are associated with increased risk for MSDs. They tend to require a lower level of user training and are well suited to situations where an analyst wishes to assess a large number of work processes to identify where ergonomic issues are present and to prioritize these work processes for more in-depth analysis. A common characteristic of checklists is that they rely on a dichotomous choice regarding the presence or absence of a physical risk factor, or whether or not the risk factor exceeds a certain threshold level. In general, checklist methods are applied in direct observation of the job, in contrast with more detailed event or time study-based methods that are more effectively applied during observation of a video recording of the job. The following are examples of checklist approaches appearing in the more recent literature on UECTD prevention.

Keyserling et al. [49] developed a two-page checklist addressing risk factors of repetitiveness, local contact stress, forceful exertions, awkward posture, and hand tool use. A qualitative rating scale was used to denote ergonomic exposures that are insignificant (denoted by a zero), moderate (denoted by a check mark), or substantial (denoted by a star). This checklist was applied to 335 jobs in an automotive facility as a screening tool, and the results were compared with those from an expert evaluation of a subset of 51 jobs. The checklist was designed to be highly sensitive to the presence of risk factors, and the comparison with experts' evaluations confirmed this intentional design bias. As an example, the checklist scores identified non-neutral wrist postures in 47 of the 51 jobs, whereas the expert evaluations identified this risk factor in 21 of the 51 jobs. All risk factors were more often shown to be present from the checklist results than by the expert evaluations. The authors suggest that a checklist should be designed for high sensitivity and a high rate of "false positives" in the screening phase. This way, few problematic jobs will be missed when jobs are screened and selected for more detailed ergonomic assessment.

PLIBEL [50] is a simple 17-item qualitative checklist that captures risk factors for all body regions. As with other checklists, many of the criteria are subjective—physical stresses pertaining to the upper extremity are particularly qualitative. (e.g., "Are tools and equipment unsuitably designed for the worker or the task?") The checklist questions are grouped according to relevant body regions so that specific risk factors for an injured body region can be identified. For the elbows, forearms, and hands, criteria include the presence of repetition of "similar work movement" and similar work movements beyond "comfortable reaching distance." Other criteria query the analyst about repetitive work with the forearm and hand performed with *twisting movements*, *forceful movements*, *uncomfortable hand positions*, *switches or keyboards*, and the presence of *awkward grasping of working materials or tools*.

The *Quick Exposure Checklist* (QEC) [51] combines assessments of risk factors made by the job analyst and the worker. Upper extremity stresses include the work height (categorized as below waist, chest height, above shoulder height); indication of a straight or bent wrist; whether arm movements are repeated infrequently, frequently, or very frequently; and the degree of repetitiveness of wrist hand motion patterns (three categories). Checklist criteria addressing the amount of force exertion required are directed to the worker and include the maximum weight handled (0–5, 6–10, and 11–20 kg) and the maximum force level exerted by one hand (<1 kg, 1–4 kg, >4 kg).

*Washington State Department of Labor and Industries* [52] developed an ergonomic assessment tool featuring "caution zone" and "hazard zone" checklists that identify upper extremity MSD risk factors. Jobs that exceed criteria for *caution zone* classification can be assessed against *hazard zone* criteria which represent greater severity. Criteria for the upper limb include working with the hands above the head more than two hours per day (caution), pinching an unsupported object weighing 2 or more pounds per hand, or pinching with 4 or more pounds of force more than 2 h per day (caution zone), and gripping unsupported objects weighing 10 lb or more per hand, or gripping with 10 lb or more force per hand, more than 2 h per day. Table 32.2 reproduces the checklist items for caution zone and hazard zone levels. Some of the mandatory language related to employer responsibilities for hazard zone jobs has been revised after the repeal of the Washington State Ergonomics Rule (see Section 32.7). The Washington state checklist is an example of one with more explicit quantitative criteria for the identification of UECTD risk factors.

The *ANSI A10.40* voluntary standard [53] is an industry-specific checklist developed for the construction industry. Because of this, there is an emphasis on the ergonomic aspects of the hand tools used. Checklist items address whether the tools are sharp and in good condition, whether the tools are very heavy or poorly balanced, or whether the tools vibrate excessively. Criteria address specific tools that must be used while in a difficult position (awkward posture), tools requiring bending of the wrists, and tools with poor handle design—defined as grips too big or too small, handles that are too short and dig into hands, handles with ridges that dig into hands, or slippery handles. Criteria also address the use of gloves and whether gloves make it hard to grip tools and whether there are other tools with a better design. A checklist item addresses repetitive motion through identification of tasks or jobs using the same motion dozens of times an hour for more than 1 h a day. The checklist addresses whether the number of repetitions can be reduced by job rotation or if rest breaks can be added. Awkward posture is addressed by identifying tasks or jobs that involve work above the shoulder or work at floor level or on the knees for more than 1 h per day and determining whether scaffolds, platforms, or other equipment cut down on the need

TABLE 32.2 Washington State Department of Labor and Industries Caution Zone Checklist

| | Caution | Hazard |
|---|---|---|
| Awkward posture | Working with the hand(s) above the head, or elbow(s) above the shoulder, for *more than 2h total per day* | *... more than 4h total per day*<br>Repeatedly raising the hand(s) above the head, or the elbow(s) above the shoulder(s) more than once per minute for more than 4h total per day |
| High hand force | Pinching an unsupported object or objects weighing 2 or more pounds per hand, or pinching with a force of 4 or more pounds per hand, more than 2h total per day | Pinching an unsupported object or objects weighing 2 or more pounds per hand, or pinching with a force of 4 or more pounds per hand:<br>• Plus highly repetitive motion for more than 3h total per day<br>• Plus wrist flexion (30°)/extension (45°) or ulnar deviation (30°) for more than 3h total per day<br>• For more than 4h total per day |
| | Gripping an unsupported object or objects weighing 10 or more pounds per hand, or gripping with a force of 10 or more pounds per hand, more than 2h total per day | Gripping an unsupported object or objects weighing 10 or more pounds per hand, or gripping with a force of 10 or more pounds per hand:<br>• Plus highly repetitive motion for 3h total per day<br>• Plus wrist flexion (30°)/extension (45°) or ulnar deviation (30°) for more than 3h total per day<br>• *For more than 4h total per day* |
| Highly repetitive motion | Repeating the same motion with the neck, shoulders, elbows, wrists or hands (excluding keying activities) with little or no variation every few seconds, for more than 2h total per day | Using the same motion with little or no variation every few seconds (excluding keying activities):<br>• Plus high forceful exertions with the hands and wrist plus deviated wrist posture for more than 2h total per day<br>• *For more than 6h total per day* |
| | Performing intensive keying more than 4h total per day | Performing intensive keying:<br>• Plus deviated wrist posture for more than 4h total per day<br>• *For more than 7h total per day* |
| Repeated impact | Using the hand (heel/base of palm) or knee as a hammer more than 10 times per h, more than 2h total per day | Using the hand as a hammer more than once per minute for more than 2h per day total |
| Hand-arm vibration | Using impact wrenches, carpet strippers, chain saws, percussive tools (jack hammers, scalars, riveting or chipping hammers) or other hand tools that typically have high vibration levels more than 30min total per day | |
| | Using grinders, sanders, jigsaws or other hand tools that typically have moderate vibration levels more than 2h per day | |

*Note:* The checklist items shown are those relevant to the upper extremity.

to work overhead. The checklist asks whether rotation or rest breaks can be used to reduce time in awkward postures.

A checklist with even greater emphasis on the design of the hand tools is the *California OSHA/NIOSH Hand Tool Checklist* [54]. This checklist was developed specifically for evaluating the non-powered hand tools used. The intent is to assist the worker in selecting the best tool for the particular job. The checklist includes quantitative and semi-quantitative criteria on the tool handle diameter, length, handle span, tool weight, handle surface contour, handle surface coating, and the degree to which the tool affords a neutral wrist posture of the user. A more detailed version of the checklist includes a proposed system for scoring tools on each of the checklist items to make comparisons among tools of the same type and function [55].

### 32.6.2 Systematic Observational-Based Methods

Checklists are generally inappropriate for quantifying the level of risk or likelihood of injury. The class of methods that exact more detail about risk factor levels and their translation to risk for MSDs of the upper extremities are referred to here as systematic observational-based methods. Since UECTD risk factors of awkward posture, repetitive motion, and the presence of hand force exertion (if not the force level) are visually observable, risk analysis approaches have been developed that are predicated on visual observation and estimation of these risk factors. Several structured approaches to observing and documenting these upper extremity risk factor levels have been developed and published since the first edition of this text. Analyses using these methods are frequently conducted from a video recording of the work so the analyst can view the work activity with slow motion, single frame advance, and pause capabilities.

This section will provide a brief summary of systematic observational-based methods for ergonomic assessment, emphasizing those specifically for UECTD risk factors and the whole-body methods that include the upper extremity. These methods have been summarized in other publications (e.g., Refs. [56,57]).

#### *32.6.2.1 Specific to the Upper Extremity*

Several observational-based measures have been developed specifically for the assessment of risk factors for the upper extremity, largely focusing on the hand and distal upper extremity. These are summarized in the following and in Table 32.3.

The *strain index* [58] is a method for evaluating physical risk factors and calculating a composite index that reflects the risk for distal UECTDs associated with a job. The strain index is suitable for jobs with relatively cyclic structure and is applicable to predict risk for CTDs of the elbow, forearm, wrist, and hand. Application of the strain index is based on observer judgment and rating of the intensity of exertion, duration of exertion, frequency of exertions (counted in efforts/min), hand/wrist posture, speed of work, and duration per day. Each of these variables is scaled with five rating levels and is assigned multipliers in proportion (inversely) to the rating. Exertion intensity scale levels are semantic (light, somewhat hard, hard, very hard, near maximal) but are anchored to quantitative percentages of maximum voluntary contraction (MVC) force. Conversely, the speed of work rating is a construct

TABLE 32.3 Summary of Four Observational-Based Job Analysis Methods Specifically for the Upper Extremities

| Instrument (Primary Reference) | Posture | Force | Repetitive Motion |
|---|---|---|---|
| Strain index (Moore and Garg [58]) | Wrist extension–5 categories (0°–10°, 11°–25°, 26°–40°, 41°–55°, >60°); wrist flexion–5 categories (0°–5°, 6°–15°, 16°–30°, 31°–50°, >50°); ulnar deviation–5 categories (0°–10°, 11°–15°, 16°–20°, 21°–25°, >25°) | Rating of intensity of exertion with 5 categories based on "% maximal strength" or rating of perceived exertion using the Borg CR-10 scale | Rating of "speed of work" based on how fast the worker is working relative to percentage of MTM-1 standard time |
| ACGIH TLV (ACGIH [79]) | Deferred to professional judgment | Rating of peak force exertion (visual analog scale from 0 to 10) | Visual analog scale rating of HAL with verbal descriptors of activity level, or, identifying hand force duty cycle and rest periods and referring to table in ACGIH documentation |
| CTD risk index (Seth et al. [68]) | Wrist angle used in equation to scale grip force capacity for types of grip/pinch<br>Elbow flexion/extension–3 categories (>10° flex, 10° flex–30° ext, >30° ext); shoulder flexion–4 categories (0°–20°, 20°–45°, 45°–90°, >90°); shoulder abduction–4 categories (0°–30°, 30°–60°, 60°–90°, >90°) | Percentage of endurance capacity calculated as a function of the time posture is held, rest period, total working time, and load (estimated as %MVC) | Assumes 10,000 hand motions can be performed in one day. The number of allowable hand motions are reduced by other multipliers reflecting forcefulness and posture |
| OCRA (Colombini [64]) | Wrist radial/ulnar deviation–2 categories; wrist flexion/extension–2 categories (0°–45°, >45°); forearm pronation/supination–2 categories; shoulder elevation–3 categories (0°–20°, 20°–60°, >60°) | The effort required to carry out a series of technical actions as expressed as a percentage of MVC | Assumes 30 technical actions per minutes are allowable. Allowable technical actions are reduced based on multipliers for posture and force |

derived from MTM (methods-time measurement) predetermined motion time systems and can involve significant subjective judgment on the part of the analyst. Moore and Garg [58] recommend that video recordings of the job be observed as the basis for the analysis.

The *American Council of Governmental Industrial Hygienists* (ACGIH) *Threshold Limit Value for Hand Activity Level* (HAL TLV) is a voluntary standard established for the determination of safe levels of upper limb activity in work performed for greater than 4 h per day [59]. Similar to the strain index, the HAL TLV is appropriate for "mono-task" work, characterized by a predictable cyclic pattern of work elements and jobs that do not have variable task exposure. The HAL TLV does not account for posture of the upper limb, which is deferred to the professional judgment of the job analyst. Application of the HAL TLV involves determination of two parameters: the peak hand force and the hand activity level (HAL).

Peak hand force can be derived by a number of methods. If continuous time measurements of force are obtained, the 90th percentile level for hand/finger force is used—the rationale being that the finger transmits force between the work object and the tendons and muscles in the hand, wrist, and forearm [60]. In the absence of direct measurement, the peak force can be estimated by the analyst, or estimated by the worker. Psychophysical approaches such as the Borg CR-10 scale [61], visual analog scale percentage of maximal exertion, and force matching have been applied to obtain worker estimates of hand force level. Using the force-matching approach, the worker is asked to estimate and reproduce the force exertion on a hand dynamometer in a similar configuration as the work situation dictates. The reliability of the estimates are improved when arm/wrist/hand postures and grip contact conditions more closely approximate those of the actual task [62]. It is recommended that individual worker estimates of peak force be normalized to the strength capability of the worker population for the job.

The second parameter, the HAL, reflects the repetitive nature of the work in the exertion of force. HAL can be derived quantitatively from the observed exertion frequency and exertion recovery time using a look up table (see Table 32.4). The HAL can also be rated by the job analyst using a 10-point Visual Analog Scale with verbal anchors describing the frequency of hand motions and/or repeated exertions of hand force [63]. This is a less quantitative but more rapid approach as the analyst makes a judgment about the motions and repeated forceful task elements of the hand(s). The HAL and peak force are plotted on a graph and can be interpreted in relation to the threshold limit value (TLV) and a more conservative action level (AL) indicating the need to initiate some form of intervention. In the example described in Section 32.4 and illustrated in Figure 32.7 (bottom panel), the solid line represents the TLV, and the dashed line, the action limit.

The *Occupational Repetitive Actions* (OCRA) method [64,65] is based on a count of "technical actions"—a term used to describe micro-motion elements in the task that would be considered in a methods-time measurement analysis. Colombini [64] cited this approach as an advantage of the method asserting that it is easy to define and recognize technical actions and that company technicians, who are experienced in production organization, can relate to this construct. Technical actions can be described in terms of their frequency of occurrence, and the OCRA frequency constant for allowable technical

TABLE 32.4 Hand Activity Levels for Combinations of Exertion Frequency and Duty Cycle

| **Frequency** | **Duty Cycle %** | | | | |
|---|---|---|---|---|---|
| **Exertions/s** | **0–20** | **20–40** | **40–60** | **60–80** | **80–100** |
| 0.125 | 1 | 1 | — | — | — |
| 0.25 | 2 | 2 | 3 | — | — |
| 0.5 | 3 | 4 | 5 | 5 | 6 |
| 1.0 | 4 | 5 | 5 | 6 | 7 |
| 2.0 | — | 5 | 6 | 7 | 8 |

*Source:* From American Conference of Governmental Industrial Hygienists (ACGIH), *Documentation of the TLVs and BEIs* (6th edn.), Cincinnati, OH: ACGIH Worldwide, 2005.

actions is 30 per min. The allowable number of technical actions is reduced by multipliers that account for force exertion, posture, recovery periods, and additional factors including exposure to vibration and/or cold, localized compression of soft tissues, requirements for accuracy, use of gloves, and hand impact. The formulation of the multipliers and how they reduce the allowable frequency of technical actions is similar to that of the original and revised NIOSH Lifting Equation [66,67]. The OCRA posture multiplier reduces the allowable technical actions based on theoretical assumptions about duration severity, which embodies both the duration of the awkward posture and its degree of deviation from neutral. Severity scores for the posture multiplier are proportional to the perceived discomfort elicited by the degree of postural deviation. Scaling of posture magnitude is based on 50% of the range of motion of the joints as shown in Table 32.3. The force multiplier decreases the frequency constant in proportion to the equivalent rating of perceived effort for the work cycle. Perceived effort is derived from the Borg CR-10 scale for perceived exertion.

A disadvantage of the OCRA method is that it can be time-consuming for complex tasks and multi-task jobs. Analyses almost always require observation of a video recording of the job and use of pause and single frame advance capabilities. A simplified OCRA checklist has subsequently been developed which reduces the complexity of the method.

The *CTD risk index* [68] has been less widely adopted, but is another detailed approach to assessing a job for increased risk for CTDs of the upper extremity. This method is based on established relationships between grip and pinch force capacity and wrist posture and discounting the allowable force exertion based on exertion duration and recovery periods. The method is somewhat similar to OCRA in that it involves counting the number of hand grip or pinch motions in a work cycle and converting this to a daily total based on the work cycle time and work duration. The daily count of grip motions is normalized to 10,000 per day, which is assumed to be the daily allowable limit. (The 10,000 daily hand motions compares favorably with 30 technical actions/minute, the frequency constant of the OCRA method.) The CTD risk index also adjusts the power and pinch grip force to the observed grip span based on equations expressing MVC capability as a function of grip span. The consideration given to posture in this method is complex and entails the use of equations for endurance capacity based on static hold time as a function of wrist posture, a recovery period between successive awkward postures, a point assignment for the severity of the postural magnitude (see Table 32.3), and the percentage MVC supported in relation to the 51 pound load constant of the NIOSH lifting guideline [66].

A simpler version of the CTD risk index appears in Niebel and Freivalds [69] which reduces the analysis complexity and time and is more appropriate for the evaluation of a larger number of jobs. This version calculates an index based on a hand motion frequency factor (relative to the allowable limit of 10,000 daily hand motions), a posture factor, a force factor, and miscellaneous factors such as use of gloves, presence of sharp edges on work contact surfaces, vibration exposure, and cold temperature.

#### *32.6.2.2 Whole-Body Methods Including the Upper Extremity*

In addition to the upper extremity–specific job analysis tools listed in the Section 32.6.2.1, other methods have been developed that assess ergonomic risk factors to all musculoskeletal regions.

These methods are classified here as *whole-body* methods. Ovako Working posture Analysis System (OWAS) and Rapid Upper Limb Assessment (RULA) were two of the earlier methods to be developed, and subsequent methods are similar in their approach. OWAS [70] emphasizes gross posture of the body and, in terms of the upper limbs, contains only a broad classification of shoulder posture based on the elevation of the elbows. The distal upper limb joints (elbows and wrists) are not considered in the OWAS method. The OWAS approach to the accumulation of points in proportion to the severity of individual body segment postures and the load handled has been adopted by other methods.

RULA is classified by many, including the original authors, as an upper limb assessment method [71]. It is grouped in this review with the whole-body methods because it includes assessment of lower extremity posture—unlike the methods discussed earlier that are specific to the upper extremity. RULA is largely based on an analysis of working posture. Severity points are accumulated for postures that deviate from neutral. There are four non-neutral posture categories for shoulder flexion/extension: >20° extension, 20° extension–45° flexion, 45°–90° flexion, >90° flexion, with one point added for abduction of the upper arm. There are two non-neutral categories for elbow flexion (0°–60°, >100°) and two non-neutral categories of wrist flexion/extension (beyond neutral to 15° flex/ext, >15° flex or ext) with one point added for a "deviated or twisted" wrist. Posture severity is classified according to the posture held for the longest duration or for which the highest loading occurs. Force exertion and repetition are established by classifying resistance or load handled according to levels of less than 2 kg, 2–10 kg, or greater than 10 kg and whether the load is static, intermittent, or repeated. The RULA grand score is derived from an accumulation of points for the severity of upper limb stresses in combination with the severity of neck, trunk, and leg stresses.

The Rapid Entire Body Assessment (REBA) [72] is a variation of RULA developed for application in the health care industry. REBA adds load coupling in a manner similar to the revised NIOSH Lifting Equation [66]. REBA is largely postural based, and the scoring is similar to that of RULA in the translation of the score to a risk or action level.

*Loading on the Upper Body Assessment* (LUBA) [73] differs from the previously described approaches in the rationale for assigning severity to postural deviation. LUBA posture severity scaling was based on a psychophysical study of perceived postural discomfort [74] and matching the ratio of severity scores to the ratio of discomfort levels associated with various postures. The LUBA method considers only posture, and its application considers only the psychophysical perception of postural discomfort as the determinant of acceptable working postures.

*PATH* [75,76] differs from the methods discussed earlier in that it is a work sampling-based approach that was developed more specifically for construction work, in which the postures, motions, and forces exerted are variable and non-cyclic in nature. Assessments using PATH consider postures (P), activities (A), tools (T), and handling (H) inherent in a number of construction trades. Because construction work is typically characterized by variable, non-cyclic patterns of exposure, jobs cannot be analyzed at the unit of a short duration fundamental work cycle. PATH relies on fixed-interval observations in real-time. PATH emphasizes trunk and leg posture in the framework of the OWAS method, but does include shoulder posture stresses when work is performed with elbows above shoulder height. Because the validity of work sampling increases when a large number of observations are included, the authors recommend that PATH observations be made in intervals less than 60 s.

#### *32.6.2.3 Validity Considerations*

Several studies have examined upper extremity observational-based job analysis methods for their predictive and/or external validity, which describe how well they predict the prevalence of upper extremity MSDs. Knox and Moore [77] assessed 28 jobs in a turkey processing plant with classifications of "hazardous" or "safe" based on a strain index score of 5 as a cutoff. Presence of MSD outcomes, or morbidity classification, was based on a physician review of OSHA 200 injury logs. Sensitivity, a measure of the percentage of jobs that are correctly predicted as being associated with morbidity, was reported as 0.91; specificity, a measure of the percentage of jobs correctly predicted as being associated with no morbidity, was reported as 0.83. An odds ratio of 50.0 was reported. Rucker and Moore [78] reported even stronger predictive validity of the strain index when applied in two manufacturing facilities.

In a larger cross-sectional study of 352 workers across three manufacturing facilities, Latko et al. [63] reported modest odds ratios for hand outcomes when considering only the ACGIH HAL rating as an exposure metric. Collapsing the continuous HAL scale into three categories (low, medium, high) resulted in odds ratios up to 3.33 (1.27–8.26, 95% CI) for the low vs. high category comparison in predicting tendinitis. The jobs evaluated in this study exhibited little variability in exposure level for hand posture or force exertion. Thus, the contribution of these stressors in predicting hand outcomes could not be determined. Franzblau et al. [79] conducted an even larger cross-sectional study with over 900 workers at 7 facilities to assess the predictive validity of the ACGIH TLV. TLV categories were positively associated with elbow/forearm tendonitis and CTS; however, sensitivity and specificity were shown to be modest for most outcomes, with the former being less than 0.6 for all outcomes. The authors noted a high prevalence of reported symptoms in jobs below the TLV action limit.

In a recent study of 567 workers, Spielholz et al. [80] evaluated the predictive validity of both the strain index and ACGIH TLV. The risk factors that were related to disorders of the dominant distal upper extremities were peak and most common hand force, a strain index greater than 7.0 vs. less than 3.0 (odds ratio = 2.33), and strain index greater than 7.0 vs. less than 7.0. For the nondominant hand, the HAL category less than 4.0 vs. HAL category greater than 4.0 (OR = 2.81) was the only significant relationship with health outcomes. The Spielholz et al. [80] study suggests that a strain index of 7 achieves roughly equivalent sensitivity and specificity in predicting MSD outcomes and was associated with significant odds ratios for distal upper extremity outcomes in the dominant hand. This finding is in line with the recommendations of Rucker and Moore [78] that the criterion value of 5.0 might be increased—perhaps to as high as 9.0, for manufacturing jobs. Spileholz et al. [80] reported some differences between the strain index and ACGIH TLV in terms of risk category classification (safe, action level, and hazardous zones). With a strain index of 7.0 as the hazard zone threshold, and 3.0 as the safe zone threshold, the strain index classified more jobs in the hazard zone than the ACGIH TLV, and that the TLV categorized more jobs as safe. This group [81] has suggested that the strain index may be more protective, particularly for jobs in the median exposure levels, and that when combining data from multiple studies, agreement in risk classification between the two methods was 75%. The studies of Bao et al. [81] and Armstrong et al. [60] suggest that the HAL TLV action limit could be lowered.

In addition to their external validity, that is, how well they are predictive of disease or morbidity, a consideration in the application of observational-based methods is their *internal validity.* Internal validity refers to how well the methods represent the exposure variable(s) they are intended to quantify. Methods for characterizing exposure based on visual observation and estimation of risk factors should be interpreted with an appreciation for the limitations in the ability of observers to make accurate estimates of the risk factor levels. This has been illustrated most clearly for the visual estimation of working posture. In general, the posture category boundaries of job analysis methods have not been selected based on empirical studies of what job analysts can reliably detect by visual observation.

Studies of the accuracy of posture estimation from video recording suggest that postural misclassification errors are inversely proportional to the size of the joint segments of interest [74,82]. When considering the width of angular posture categories (in degrees), a trade-off exists between the probability of error occurrence and the magnitude of the error when one is made [82,83]. van Wyk et al. [83] recently illustrated this concept empirically by deriving posture categories that optimize the probability of occurrence of misclassification errors and the magnitude of the errors as a function of the posture category width. The example shown in Figure 32.6 represents elbow flexion/extension and indicates an optimal category width of 25–30 degrees. Since elbow flexion/extension has approximately 120 degrees in the functional range of motion, van Wyk et al. [83] suggest that a four category scale in 30° increments is optimal in consideration of the accuracy of observer judgment. Table 32.5 shows the posture categories that optimize visual

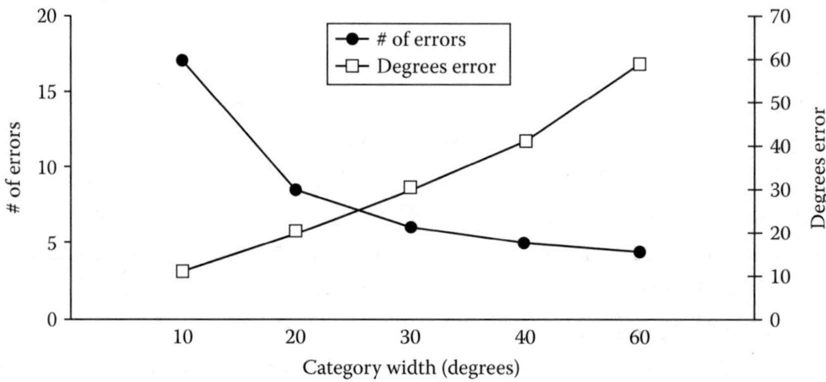


FIGURE 32.6 Optimized posture category size for elbow/flexion extension showing the trade-off between the probability of posture misclassification (number of errors made) and the misclassification error magnitude (in degrees). (Reproduced from van Wyk, P.M. et al., *Ergonomics,* 52, 921, 2009. With permission.)

TABLE 32.5 Optimal Posture Category Widths (in Degrees) and Number of Categories for Trunk, Shoulder, and Elbow Postures

| | Trunk Flexion | Trunk Lateral Bend | Shoulder Flexion | Shoulder Ab-/Adduction | Elbow Flexion |
|---|---|---|---|---|---|
| Category width | 30° | 15° | 30° | 30° | 30° |
| Optimal number of categories | 4 | 3 | 5 | 5 | 4 |

*Source:* From van Wyk, P.M. et al., *Ergonomics*, 52, 921, 2009.

judgment of posture based on the van Wyk et al. [83] study. While this represents only a single study and did not include wrist postures, these findings are in line with those of Bao et al. [81] who concluded that interrater reliability was superior with 30° posture category widths versus a smaller width, and that for most postures, 30° angle intervals appear to be appropriate. It should be noted that for radial/ulnar deviation of the wrist, an interval of 30° from neutral represents most, if not all, of the effective range of motion. Thus, wrist radial/ulnar deviation presents a dilemma in designing posture scales with categories that are small enough to capture differences in biomechanical risk, yet large enough that the angular intervals can be visually discriminated by observers conducting the posture analysis. Because of the relatively small size of the body segments, and smaller range of movement, visual discrimination of wrist radial/ulnar deviation is problematic for observational posture assessment [82,84].

#### *32.6.2.4 Example of Observational-Based Exposure Assessment*

An example of the application of the HAL TLV is as follows. A job on an automotive radiator assembly production line was video recorded, and the hand force profile for the dominant (right) hand was determined using two approaches. The first approach was with a wearable glove with thin profile force sensors attached to the palm surface to measure hand contact force with the dominant hand. Figure 32.7 (top panel) shows a force profile for a typical work cycle with a cycle time of 15 s. The second approach was using the Multimedia Video Task Analysis, or MVTA, system (see Section 32.6.4) and manually marking video frames at the transition points between hand force and no hand force exertion. Using MVTA, five periods of hand force exertion were identified for the right hand (see Figure 32.7, middle panel). These correspond closely to five distinct exertions of force evident from the direct-reading measurement (top panel). The total time duration of the force exertions sums to 7.6 s, resulting in a duty cycle of 50%. A 15 s work cycle with five exertions yields a force frequency of 0.33 exertions per second. Using Table 32.4, the HAL rating was obtained for this combination of exertion frequency and hand force duty cycle, and was determined to be greater than 3 and less than 4. Thus, a range of 3–4 was used. The normalized peak hand force in the grip of the radiator when lifting it from the pallet was estimated to be between 40% and 50% of MVC. By plotting the HAL and normalized peak force as shown in Figure 32.7 (bottom panel), it can be determined that the job falls above the action limit (dashed line) but below the TLV and can be characterized as a job in which some form of intervention should be considered.

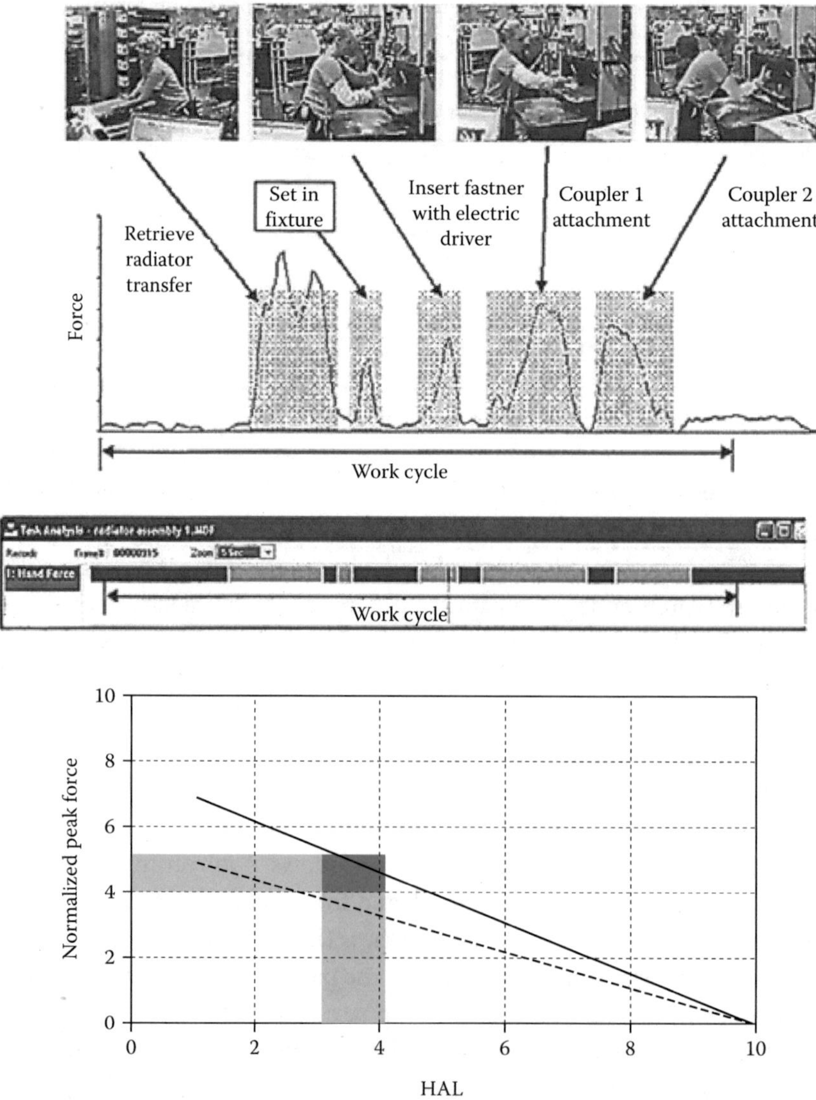


FIGURE 32.7 Example of HAL calculation (top and middle panels) and interpretation (bottom panel). The top panel shows a continuous recording of hand force and the middle panel an observational analysis of exertion duty cycle using the MVTA™ software.

## 32.6.3 Continuous-Recording Instrumentation-Based Methods

Though not the most widely adopted, it is generally believed that the most valid approach to quantifying physical risk factors for upper limb CTDs is by direct measurement with continuous-recording instrumentation [47,56]. UECTD risk factors have been quantified with a variety of continuous-recording measurement instrumentation, most notably those for measuring muscle electrical activity (electromyography), external muscle force (load cell/strain gauge transducers, thin profile pressure and force sensors), joint motion, or kinematics (electrogoniometry), limb segment orientation (inclinometry), and vibration (accelerometry). Motion capture systems, which are based on optical and magnetic technologies, have been used in studies conducted in more controlled environments, but are generally impractical in industrial settings.

The following is a brief overview of continuous-recording instrumentation applicable in the assessment of UECTD risk factors.

#### 32.6.3.1 Force Exertion

The surface electromyogram (SEMG) is a recording of the electrical activity in underlying skeletal muscle detected by electrodes attached to the skin surface over the belly of the muscle. SEMG recordings are typically subjected to data reduction and analysis of amplitude as a correlate of force exertion or amplitude and/or frequency spectra as indicators of muscle fatigue [85,86]. The amplitude of the detected electromyogram (EMG) signal is referenced (or normalized) to a known level of muscle contraction in a standardized exertion, and subsequent amplitudes can be expressed as percentages of the reference exertion force output. The reference exertion can be elicited as the maximum force the worker can exert (the *maximum voluntary contraction*, or MVC) or a submaximal exertion level that is controlled by standardizing a static load on the muscle. As an example, power grip force on a grip dynamometer can be calibrated (or normalized) to the SEMG amplitude of extrinsic muscles in the forearm that create forceful flexion of the fingers in the grip [87]. This SEMG to force relationship can be used to convert the SEMG measured in a gripping task to an equivalent dynamometer grip force level, provided that grip span and wrist posture are equivalent.

Normalized SEMG has also been interpreted with respect to the duration of intensity levels and their frequency of occurrence as a percentage of working time. This approach has been labeled exposure variation analysis [86,88] and has been applied to examine activity in non-cyclic work where force exertions vary and lack fundamental work cycles [89]. A simpler approach is to express the SEMG activity in terms of its amplitude probability distribution function (APDF), where the cumulative frequency or probability of occurrence is plotted against exertion level [90].

Some investigators have adopted or developed specialized systems for the continuous time measurement of external force exertion by the hand, which are particularly applicable to the grip force on tool handles. Two examples of such systems require either fabrication of special handles with embedded force transducers [91] or the application of thin flexible pressure sensors between the surface of the handle and the hand [92–94]. As an example of the former, Liberty Mutual has developed a Hand Tool Force Measurement System [91] based on an instrumented handle core with embedded strain gauges. The system has been used to measure multi-axis compression forces in the grip of the handle and moments at the cutting blade of a knife used in a variety of poultry processing jobs (see Figure 32.8a). As an example of the latter, Kong and Lowe [94] instrumented a wearable glove with thin profile force sensors on the phalangeal segments and metacarpal heads of the palm surface to measure finger segment contact force in the power grip of tool handles (see Figure 32.8b). A wearable force sensor system has the advantage of versatility, in that measurement of hand contact force is not limited to the particular tool that is instrumented. This contrasts with the embedded strain gauge approach where sophisticated fabrication of a single tool

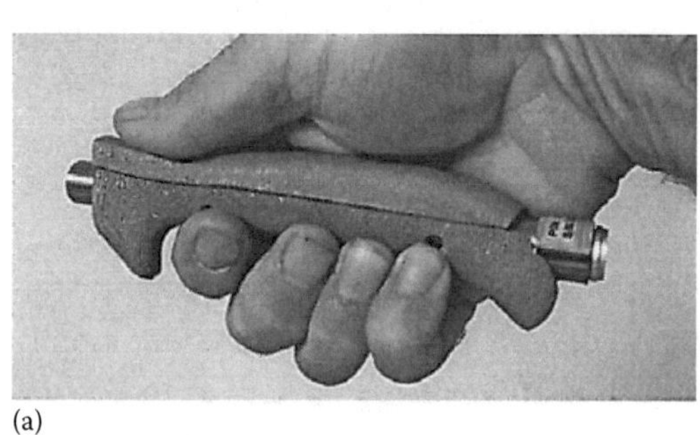
(a)

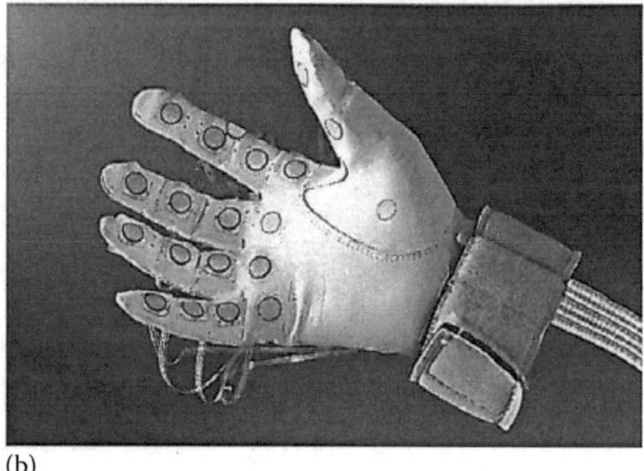
(b)

FIGURE 32.8 (a) Knife handle instrumented for grip forces in three axes and moments with force transducers embedded in the cylindrical core. (Photo courtesy of R. McGorry, Liberty Mutual Research Institute for Safety.) (b) Wearable (glove-attached) thin profile resistive sensors for measurement of contact force on the phalangeal segments and metacarpal heads.

handle can be costly and time-consuming. However, a disadvantage of the wearable thin profile sensor approach is that the measurement depends on the completeness of sensor coverage on the palm surface in contact with the object in the grip. The wearable glove approach is also intrusive, as the measurement device itself may alter the subject's interaction with the handle. Force transducers embedded in a tool handle accurately resolve the resultant forces on the handle unintrusively and are independent of the distribution of contact surface on the handle.

#### *32.6.3.2 Posture and Motion*

The most commonly reported instrumentation-based measurements of posture and motion of the hand, wrist, and forearm have been acquired using electrogoniometry. An electrogoniometer is a device that measures joint displacement using a transducer that spans the joint of interest. Wrist electrogoniometers commonly have an endblock attached on the metacarpal bones of the hand and on the forearm proximal to the wrist. The flexion/extension and radial/ulnar deviations of the wrist can be detected by these transducers (refer to Figure 32.4). Rotation of the forearm (supination/pronation) requires a torsional sensor in which twist between the device endblocks are calibrated to a degree of axial rotation.

An applied example of the use of electrogoniometry to quantify wrist and forearm motions is the study of Albers and Hudock [95]. This study evaluated risk factors associated with hand rebar tying with a traditional method using manual pliers and with a handheld battery-powered automated tying device. Electrogoniometry was used to quantify the reduction in wrist/forearm motions associated with the automated device relative to the hand-tying method. Ironworkers tied rebar intersection points at ground level in the grid of rebar that provided reinforcement to a poured concrete bridge surface (see Figure 32.9). The rebar segments intersected at 7 in. intervals, and 75% of the total intersection points required tying. Electrogoniometric recordings were made for approximately 30 min for each of three tying devices: the conventional manual method with pliers, an automated battery powered tying device, and the battery-powered device with an extension handle

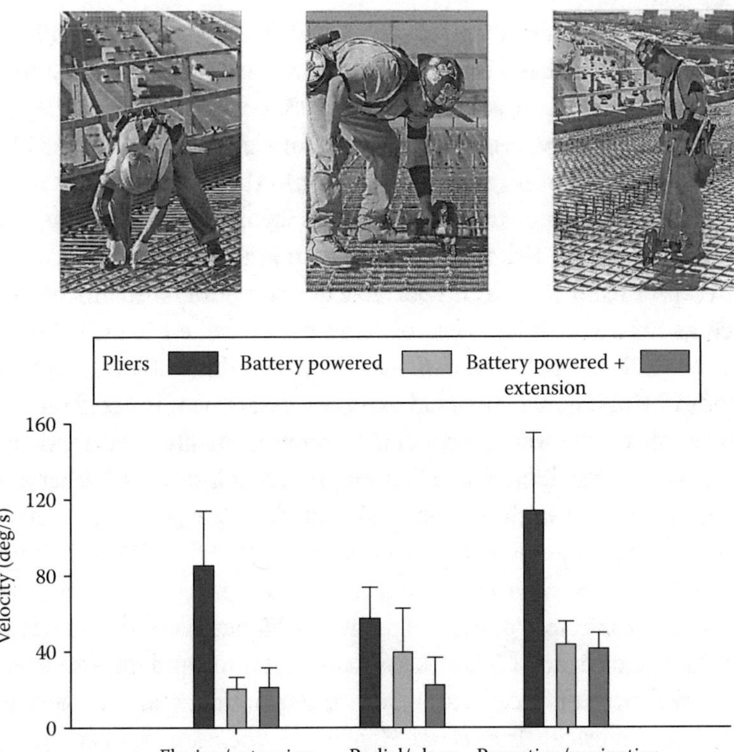


FIGURE 32.9 Example of electrogoniometry used to measure mean (+s.d.) wrist motion velocities for three methods of rebar tying: manual pliers tying, battery-powered tying, and battery-powered tying with the addition of an extension handle. The bar graph shows the substantial reduction in wrist motion velocities, particularly for flexion/extension and pronation/supination, with the battery-powered tying device. (From Albers, J.T. and Hudock, S.D., *Int. J. Occup. Saf. Ergon.*, 13, 279, 2007. With permission; photos by Earl Dotter.)

designed to reduce or eliminate trunk bending (see Figure 32.9). This was an ideal application for wrist electogoniometry since the traditional approach of hand tying rebar with pliers is associated with many repetitive high-velocity wrist and forearm motions that could not be reliably estimated by visual judgment. Albers and Hudock [95] reported that wrist motion velocities in the flexion/extension, radial/ulnar deviation, and forearm rotation (supination/ pronation) axes were reduced by 76%, 30%, and 63% respectively by adoption of the automatic rebar tying device. (Figure 32.9).

Inclinometry is a method for dynamically assessing the orientation of a limb segment with respect to gravity. One example is when the inclinometer is attached to the upper arm segment, the device can be calibrated to arm orientation and used to evaluate upper arm elevation [76].

#### *32.6.3.3 Direct-Reading Exposure Assessment Methods (DREAM)*

While direct-reading instrumentation is believed to be the most objective and quantitative method for assessing risk factor exposure, its use in occupational ergonomics appears to be highly concentrated in research applications. In a survey of Certified Professional

Ergonomist (CPE) practitioners, Dempsey et al. [96] found that 31% of CPEs reported having ever used EMG and, of these, only 2% rated EMG as "easy to use." Conversely, 52% of practitioners reported having used RULA, and 20% of these rated RULA as easy to use. A similarly low rating for *easy to use* was reported for electrogoniometry, which only 18.5% of CPEs reported as having ever used. A trend within these data appears to be that direct-reading measurement techniques are not viewed as favorably, or are not as widely adopted, by practitioners for assessing risk factor exposure in ergonomic analyses.

Advances in sensor technologies and portable direct-reading instruments for other physical agents such as noise, radiation, aerosols, and dust have led to an initiative in 2008 by NIOSH in the area of *Direct-Reading Methods* [97]. Under this initiative, the vision for direct-reading methods (DRMs) in occupational exposure assessment is based on a self-contained instrument, wearable by the worker that can "...provide on-site measurement of exposures in units (such as parts per million parts of air, or ppm) that indicate whether or not the exposures pose an occupational health or safety risk and if the prevention methods employed are actually providing the proper level of protection" [97]. This definition is clearly inspired by the monitoring of air-borne chemical agents and the sampling of hazards to which workers are exposed through respiratory or dermal pathways. Nonetheless, this initiative may benefit the area of MSD prevention and CTD risk assessment if novel and improved technologies for measuring the risk factors of force, posture, repetitive motion, and vibration result.

NIOSH and the American Industrial Hygiene Association sponsored a workshop in 2008 to address research needs related to DRMs for occupational exposures to hazards related to noise, radiation, aerosols, surface sampling and biomonitoring, gases and vapors, and ergonomics. The workshop titled "Direct-Reading Exposure Assessment Methods" (D.R.E.A.M.) served to solicit stakeholder input for the purpose of developing a research agenda for DRMs for exposure assessment. The ergonomics and vibration breakout session of the workshop focused on DRMs for the hazard of physical loading on the musculoskeletal system and the assessment of exposure to biomechanical risk factors for WMSDs. Discussion centered on the importance of immediate interpretation of exposure data, that is, to provide a real-time indication of exposure level. Real-time acquisition of data in the workplace (on-site) is necessary. However, the group was somewhat divided on the need for real-time *interpretation* of the data. Some participants felt that there was a need for DRMs to acquire the exposure data and interpret the measured exposure level in real-time. An example of such a scenario is when immediate feedback to the worker is desirable for changing work habits or technique, or as a form of biofeedback. Other participants believed that the real-time interpretation of exposure is not a necessary characteristic for a DRM and that post-processing of the data may be necessary for reconstruction and interpretation of exposure level. It was generally agreed that reducing data post-processing time is desirable. The discussion also emphasized the need to improve the usability, portability, and ruggedness of existing technologies or in the development of new technologies.

### 32.6.4 Computer Video–Based Task Analysis and Video Exposure Monitoring

Observational-based ergonomic assessments have traditionally been conducted by pencil and paper documentation of risk factors observed with the aid of analog video playback equipment. Some approaches to posture analysis have integrated computerized exposure documentation with observation from analog video tape recording [98]; however, prior to the digital video camcorder, these methods were less common. Advances in digital video recording technology have greatly simplified the integration of video playback and computer-assisted exposure assessment and time study. Digital video and computer-based time study and task analysis systems are now commonly used to aid the process of ergonomic job analysis. Commercial systems are available that provide a user interface to control the playback of digital video and facilitate manual event marking that allows an analyst to delineate exposure or task analysis events on a graphical time line. The software will calculate descriptive summary statistics for risk factor durations and category transitions. Summaries can include data such as the percentage of the work cycle with postures in non-neutral posture categories [99], or counts, durations, and frequency of hand force exertion [100].

One example of such a computer-based method is the MVTA™ [101,102]. MVTA has been used by several investigators to quantify upper limb risk factor levels in detailed observational-based exposure assessment. As an example, Bao et al. [81] described the continuous observation time-based posture analysis as one in which a posture is observed continuously and transitions between angular categories are documented on the timeline. Subsequent processing and analysis results in a distribution of posture among the categories. The advantages of a computerized system to control video playback and perform all of the timekeeping functions for this type of observation and recording of posture are obvious. As a result, computer-based systems such as MVTA are solidifying their place in a number of epidemiologic studies and ergonomic analyses of UECTD risk factors in which more complete exposure profiles are needed.

Video exposure monitoring (VEM) is related to computer video–based task analysis and time study. VEM can be broadly defined as the approach whereby a worker's specific activities can be visualized synchronously with quantitative data on exposure levels or exposure transitions [103–105]. In VEM, the video recording is made synchronously with continuous time exposure sampling, and the exposure level is overlayed graphically on the video image [103]. This was originally accomplished with analog video tape and a variable graphics array (VGA) card or adapter; however, a similar integrated system can now be accomplished with computer software and a digital video camera.

An example of a VEM system for visualizing the exertion of hand grip contact force is shown in Figure 32.10. This system was developed on the LabVIEW (National Instruments, Austin, TX) software platform and synchronized data acquisition from thin profile force sensors attached to the palmar surface of the hand with video capture from a consumer video camcorder [106]. Digital video frame grabbing was accomplished through an IEEE bus interface and a custom LabVIEW program which incorporates the Vision Development Module toolkit. In each iteration of the execution loop, a single video frame is grabbed with a sample from 20 force sensors. The frame grab images are appended in sequence to create a video file (.avi file format) in which each frame is synchronized with a force sample. A playback interface allows the viewer to scrub across the force time series trace while the corresponding video frame display is updated.

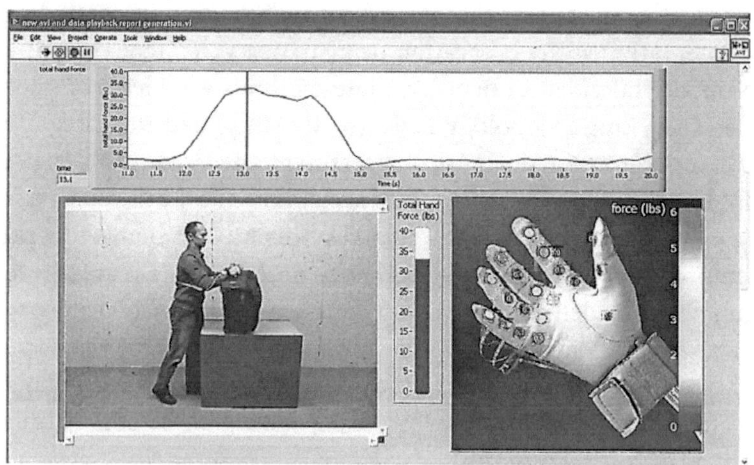


FIGURE 32.10 A VEM system that synchronizes video recording with direct measurement of hand contact force. The playback mode allows for scrolling the cursor across the total hand contact force time series trace (top) to continuously update the corresponding video frame (lower left) and the spatial force distribution representation on the hand (lower right).

## 32.7 REGULATORY ACTIVITY: OSHA AND NIOSH

### 32.7.1 Regulatory Activity and OSHA

The period of 1995–2000 resulted in a successful effort on the part of the OSHA in drafting and passing an Ergonomics Rule to address MSDs in general industry. In November 1999, OSHA issued a draft standard and the agency made significant changes from the original proposal, after listening to more than 700 witnesses during a nine-week public hearing and reviewing more than 8,000 public comments on the proposal. The final ergonomics program standard appeared in the November 14, 2000 edition of the Federal Register and took effect January 16, 2001. The rule had the following requirements:

*Management leadership and employee participation*: The employer was required to set up an MSD reporting and response system and an ergonomics program and provide supervisors with the responsibility and resources to run the program. The employer was also required to assure that policies encouraged, and did not discourage, employee participation in the program or the reporting of MSD signs, symptoms, and hazards. Employers were required to give employees the opportunity to participate in the development, implementation, and evaluation of the ergonomics program.

*Job hazard analysis and control*: If a job met an Action Trigger, the employer was required to conduct a job hazard analysis to determine whether MSD hazards existed in the job. If hazards were found, the employer was required to implement control measures directed at reducing hazards to the extent "feasible."

*Training*: The employer was required to provide training to employees in jobs that met the Action Trigger, their supervisors or team leaders and other employees involved in setting up and managing the ergonomics program. Training should include the hazards present in the problem job, and the signs and symptoms of the disorders that could result, while also encouraging early reporting of MSDs.

*MSD management*: Employees were to be provided, at no cost, with prompt access to a health care professional (HCP), evaluation and follow-up of an MSD incident, and any temporary work restrictions that the employer or the health care provider determined to be necessary.

*Work Restriction Protection*: Employers were required to provide work restriction protection (WRP) to employees who received temporary work restrictions. This meant maintaining 100% of earnings and full benefits for employees who receive limitations on the work activities in their current job or transfer to a temporary alternative duty job and 90% of earnings and full benefits to employees who were removed from work. WRP was to be good for 90 days, until the employee was able to safely return to the job, or until an HCP determined that the employee was too disabled to ever return to the job, whichever came first.

*Program evaluation*: The employer was required to evaluate their ergonomics program every three years to make sure it is effective.

*Record keeping*: Employers with 11 or more employees, including part-time employees, were required to keep written or electronic records of employee reports of MSDs, MSD signs and symptoms and MSD hazards, responses to such reports, job hazard analyses, hazard control measures, ergonomics program evaluations, and records of work restrictions and the HCP's written opinions.

The OSHA ergonomics rule was short-lived. In March of 2001, under a new administration, the ergonomics rule became the first and, to this date, the only federal regulation to be repealed under the Congressional Review Act, a 1996 bill that gives Congress the power to overturn federal regulations. This bill provided a mechanism for Congressional review and repeal of legislation, and even enables the retroactive repeal of existing legislation within a specific time frame, as was done in this case.

Following the 2001 repeal of the ergonomics rule, OSHA's emphasis shifted to a four-pronged approach which included the development of voluntary ergonomics *guidelines, enforcement efforts under the General Duty Clause, outreach and assistance efforts, and the formation of a National Advisory Committee on Ergonomics.* The emphasis on voluntary guidelines has resulted in the publication of industry-specific guidelines for nursing homes (2003), poultry processing (2004), retail grocery stores (2004), and shipyards (2008).

The repeal of the ergonomics rule in 2001 has not negated OSHA's ability to levy citations to employers with ergonomic hazards in their facilities. In the absence of an industry standard to address ergonomic hazards, OSHA had used, and can continue to use, Section 5(a) (1) of the Occupational Safety and Health Act as the authority to cite. Known as the General Duty Clause, Section 5(a) (1) states that "*…each employer shall furnish to each of his employees employment and a place of employment which are free from recognized hazards that are causing or are likely to cause death or serious physical harm to his employees.*" Opponents of the ergonomics rule had claimed that such a specific ergonomics regulation was unnecessary because the General Duty Clause was a mechanism by which OSHA could regulate for ergonomic hazards. However, since the repeal of the proposed rule in 2001, the number of citations for ergonomic-related hazards has seen a marked decrease. In 2009, OSHA's Directorate of Enforcement Programs indicated that the agency has issued 19 General Duty Clause citations for ergonomics since 2002. During the same time period, OSHA conducted 4500 ergonomic inspections and issued 640 hazard alert letters on ergonomics. In comparison, during the 10 year period between 1985 and the 1995 first edition of this text, OSHA had issued over 350 citations for either lifting or UECTD hazards.

*Outreach and assistance* efforts have placed emphasis on voluntary protection and a Voluntary Protection Program (VPP) based on incentivizing best practices to meet safety and health goals. However, concerns exist regarding the effectiveness of non-regulatory approaches to incentivizing occupational safety and health efforts, and such concerns may be well founded. Recently, the Government Accountability Office [107] identified OSHA VPP participants with high injury and illness rates—higher in fact than their industry averages. One employer had an injury and illness rate four times higher than the average rate for its industry. In response to this report, OSHA committed to conduct comprehensive evaluation of the VPP. VPP employers are supposed to have exemplary safety records. While participating industries remain subject to OSHA inspections following fatalities, serious injuries, or workers' complaints about safety or health hazards, they have been exempted from routine inspections. OSHA Alliance programs make up another component of outreach and assistance. These Alliances serve to help industry organizations build trusting, cooperative relationships with OSHA, network with others committed to workplace safety and health, leverage resources to maximize worker safety and health protection, and gain recognition as proactive leaders in safety and health. OSHA currently maintains Alliance Programs with 11 industry organizations (examples include the American Dental Association, American Society of Safety Engineers, Association of Occupational Health Professionals, Association of PeriOperative Registered Nurses, and the Brick Industry Association, among others). These Alliances have led to numerous industry-specific training tools and products.

The *National Advisory Committee on Ergonomics* was chartered in 2003–2004 and made several recommendations to OSHA for its MSD program. In addition to prioritizing the top 16 industries for which ergonomics guidelines should be developed, NACE advised OSHA to consider the following research gaps:

- More research is needed to examine the validity of techniques used to establish a diagnosis of MSDs.
- More research is needed to examine the role of psychosocial factors that contribute to or impact the development of MSDs.
- More studies are needed to develop additional animal models in which the effects of physical loading on living tissues can be studied in a controlled manner.

- More studies are needed to examine the validity and reliability of existing exposure assessment methods.
- More studies are needed to determine the economic impact to organizations of what are commonly described as ergonomic interventions.
- More studies are needed to address the multifactorial causes of MSDs, such as psychosocial, physical, occupational, and non-occupational factors, and their interactions.
- Additional studies are needed to describe the natural history of diseases or injuries, commonly known as MSDs.
- More studies are needed regarding factors in workers' compensation systems and other statutory payment mechanisms on findings of causation, diagnosis, duration of the disability, and other outcomes related to what are commonly known as MSDs.

Two states (California and Washington) have passed ergonomic standards enforceable at the state level. California's standard went into effect in July 1997 and requires that "…every employer subject to this section shall establish and implement a program designed to minimize repetitive motion injuries (RMIs). The program shall include a worksite evaluation, control of exposures which have caused RMIs and training of employees." California's rule applies to a job, process, or operation where an RMI has occurred to two or more employees. Washington state adopted an ergonomics standard in May 2000, with a phased-in enforcement that was scheduled to begin in July 2004. The Washington state standard required employers with "caution zone jobs" to find and fix ergonomic hazards instead of waiting for an injury to occur before taking action. On July 12, 2002, a county court rejected a business coalition's contention that the state exceeded its authority under state law, acted arbitrarily, and did not follow its rule-making requirements. During the process of appeal to the state supreme court, Washington state voters passed an initiative in the November 2003 election to repeal the state's ergonomics standard. Thus, California currently maintains the only ergonomics standard at the state level.

### 32.7.2 National Occupational Research Agenda

The National Occupational Research Agenda (NORA) was unveiled by the NIOSH in 1996 to provide a framework for research collaborations among universities, large and small businesses, professional societies, government agencies, and worker organizations. The NORA team charged with developing a research agenda for workplace MSDs published a report in 2001 [108] documenting the most important research priorities. These priorities included surveillance research, etiologic and medical research, intervention research, and efforts to improve the research process by strengthening communication between researchers and practitioners who apply research.

The agenda for improving surveillance research included such objectives as developing user-friendly, standardized workplace surveillance tools; increasing collaboration with federal, state, and non-governmental organizations to encourage comparability of data collection methods; and conducting an ongoing national hazard survey targeting physical workplace factors.

The research agenda for improving etiologic and medical research included the following:

- Refine instruments to detect and quantify the contribution of excessive force, awkward posture, movement, and vibration to the disease process.
- More clearly define stages of the MSD process, develop precise diagnostic tools, and provide guidelines for effective treatment and return to work.
- Clarify the interplay of the factors at different stages of causation, development, and treatment of MSD and measurement of risk factors.

Priorities for intervention research included evaluating the effects of the following on the development and prevention of MSDs:

- Alternative (product and/or tool) design criteria (force, spatial requirements of work)
- Optimization of mechanical work demands, such as force, movement, and posture, and temporal patterns of exposure
- Manual handling alternatives in posture, movement, force, productivity, and quality
- Ergonomic training and education
- Costs and benefits of ergonomic intervention
- Evaluate job assignment, selection, and choice on development of MSD
- Emerging technologies

Now in its second decade, NORA emphasizes meeting the occupational safety and health needs of the eight industry sectors, which are broken out according to the North American Industry Classification System (NAICS) code. Eight sector programs have been established to develop specific agendas that address the occupational safety and health needs of stakeholders in each industry sector. Strategic goals aimed at reducing the prevalence of MSDs for the low back and upper extremities are evident in all sectors. Upper limb CTDs have been identified as a high priority across all industry sectors.

## 32.8 FUTURE CONCERNS

If recent history is a good predictor, the regulatory landscape with respect to ergonomics and MSDs will continue to be shaped largely by political and economic factors. These factors are difficult to project beyond the short term, and it is difficult to predict the likelihood, much less the scope, of future regulatory activity affecting workplace prevention of UECTDs. However, there are clear trends in the U.S. workforce and labor market that can be anticipated to impact the way ergonomics professionals approach their discipline. The first, and most quantifiable, trend is the aging U.S. workforce. During the period 2006–2016, the number of workers aged 55–64 is expected to increase by 36.5%, and the number of workers over age 65 will increase by more than 80%. This will continue the current trend in which older workers make up a progressively larger percentage of the labor force. A second trend is the diversification of the workforce. Among those over 65 years of age, the percentage growth of women in the workforce greatly exceeds that of men. These changes in the demographics of the labor force underscore the need to consider the physical capabilities, such as strength, range of motion, dexterity, anthropometry, and metabolic work capacity of the specific worker population of interest.

No less significant of a future concern is the changing nature of work in U.S. industry. The way that work is organized affects exposures to physical and psychosocial risk factors for UECTDs. Automation and enhancements in process efficiency may result in a reduction of the highest biomechanical loads on the worker imposed by the gross handling of materials, but the positive effects of such automation may be negated by a resulting increased pace of work. It has been suggested that workplace exposures are shifting to less forceful but more frequent motions [109] performed in less-conventional environments. With increases in the service sector and warehousing distribution-related industries, fewer workers as a percentage of the labor force are employed in jobs organized around the traditional manufacturing assembly line. Physical exposures may be more difficult to assess in jobs with the characteristics observed in these sectors.

Evidence suggests that the changing nature of work has served to increase psychosocial stressors in the workplace. These stressors continue to be on the rise, driven by trends toward globalization, outsourcing, "right-sizing," longer work hours, and decreased job security for many workers. A growing body of evidence implicates psychosocial stresses in the etiology of UECTDs.

Trends toward non-traditional and flexible employment practices have raised concerns about the effect of such practices on worker safety and health. According to data from the BLS Current Population Survey, agency-supplied temporary workers and workers in other alternative employment arrangements (independent contractors, contractor-supplied labor, day laborers, and on-call workers) accounted for nearly 10% of the workforce in 2001 and represent a growing percentage of the labor force. The Current Employment Statistics Survey (CES) showed, for example, that the total number of jobs in the temporary help industry multiplied sixfold (to nearly 3 million) during the period 1982–1998, whereas total employment during this period grew by only 40%. It has been suggested that flexible employment practices are leading to a downward restructuring of the labor market with the temporary labor force becoming the group exposed to the most severe workplace hazards and health risks. This group of workers is the least trained to recognize and report ergonomic workplace hazards and CTD risk factors, the least protected by benefits and traditional employer obligations under labor law, and the most difficult for which to track physical exposure and prevalence of UECTDs. This may increase the difficulty of accurately representing the scope of UECTD problems in the workplace.

## ACKNOWLEDGMENT

The author would like to acknowledge the work of Daniel Habes, author of the Upper Extremity Cumulative Trauma Disorders chapter in the first edition of *Occupational Ergonomics: Theory and Application*. Some portions of the first edition chapter, written by Mr. Habes, have been preserved in this second edition. The assistance of John Sestito (NIOSH, Division of Surveillance, Hazard Evaluation, and Field Studies) is greatly appreciated, as is that of Steve Wurzelbacher (NIOSH), Jim Maddux (OSHA), Joanna Sznajder (OSHA), and Amit Bhattacharya who commented on a draft version of this chapter.

## REFERENCES

1. National Institute for Occupational Safety and Health (NIOSH), Proposed National Strategies for the Prevention of Leading Work-Related Diseases and Injuries, Part 1, 1986.
2. F.T. McDermott, Repetition strain injury: A review of current understanding, *Med J Aust, 144*(4), 196–200 (1986).
3. W.E. Stone, Occupational overuse syndrome in other countries, *J Occup Health Safety-Aust, 3*(4), 397–404 (1986).
4. G. Rosen, The worker's hand, *Ciba Symp, 4*(4), 1307–1322 (1942).
5. F. Gerr, R. Letz, and P.J. Landrigan, Upper-extremity musculoskeletal disorders of occupational origin, *Annu Rev Public Health, 12*, 543–566 (1991).
6. W. Singleton, *The Body at Work: Biological Ergonomics*, Cambridge, U.K.: Cambridge University Press, 1982.
7. L.A. MacDonald, R.A. Karasek, L. Punnett, and T. Scharf, Covariation between workplace physical and psychosocial stressors: Evidence and implications for occupational health research and prevention, *Ergonomics, 44*(7), 696–718 (2001).
8. K.G. Parker, and H.R. Imbus, *Cumulative Trauma Disorders*, Chelsea, MI: Lewis Publishers, 1992.
9. I. Kuorinka and L. Forcier, *Work Related Musculoskeletal Disorders (WMSDs): A Reference Book for Prevention*, London, U.K.: Taylor & Francis Group, 1995.
10. A. Joyce, For Some, Thumb Pain Is BlackBerry's Stain, *The Washington Post*, April 23, 2005.
11. American Society of Hand Therapists (ASHT), National Consumer Education Alert: *Heavy Use of Handheld Electronics Such as Blackberry, iPod Can Lead to Hand Ailments*: January 25, 2005.
12. C. Dillon, M. Petersen, and S. Tanaka, Self-reported hand and wrist arthritis and occupation: Data from the U.S. National Health Interview Survey-Occupational Health Supplement, *Am J Ind Med, 42*(4), 318–327 (2002).
13. W. Brain, A. Wright, and M. Wilkinson, Spontaneous compression of both median nerves in the carpal tunnel, *Lancet 1*, 277–282, 1947.
14. R.C. Tanzer, The carpal tunnel syndrome., *Clin Orthop, 15*, 171–179 (1959).
15. T.J. Armstrong and D.B. Chaffin, Carpal tunnel syndrome and selected personal attributes, *J Occup Med, 21*(7), 481–486 (1979).
16. E.R. Tichauer, Some aspects of stress on forearm and hand in industry, *J Occup Med, 8*(2), 63–71 (1966).
17. R. Muckart, Stenosing tendovaginitis of abductor pollicis longus and extensor pollicis brevis at the radial styloid (DeQuervain's disease), *Clin Orthop Relat Res, 33*, 201–208 (1964).
18. B.A. Silverstein, L.J. Fine, and T.J. Armstrong, Occupational factors and carpal tunnel syndrome, *Am J Ind Med, 11*(3), 343–358 (1987).
19. L. Hymovich and M. Lindholm, Hand, wrist, and forearm injuries. The result of repetitive motions, *J Occup Med, 8*(11), 573–577 (1966).
20. H. Ohara, S. Nakagiri, T. Itani, K. Wake, and H. Aoyama, Occupational health hazards resulting from elevated work rate situations, *J Hum Ergol (Tokyo), 5*(2), 173–182 (1976).
21. W.F. Fox, Human performance in the cold, *Hum Factors, 9*(3), 203–220 (1967).
22. J.M. Lockhart and H.O. Kiess, Auxiliary heating of the hands during cold exposure and manual performance, *Hum Factors, 13*(5), 457–465 (1971).
23. L.J. Cannon, E.J. Bernacki, and S.D. Walter, Personal and occupational factors associated with carpal tunnel syndrome, *J Occup Med, 23*(4), 255–258 (1981).

24. R.G. Radwin, T.J. Armstrong, and D.B. Chaffin, Power hand tool vibration effects on grip exertions, *Ergonomics, 30*, 833–855 (1987).
25. M.L. Bleecker, Medical surveillance for carpal tunnel syndrome in workers, *J Hand Surg Am, 12*(5 Pt 2), 845–848, (1987).
26. V. Putz-Anderson, *Cumulative Trauma Disorders: A Manual for Musculoskeletal Diseases of the Upper Limbs*, London; New York: Taylor & Francis Group, 1988.
27. T.F. Morse, C. Dillon, N. Warren, C. Levenstein, and A. Warren, The economic and social consequences of work-related musculoskeletal disorders: The Connecticut Upper-Extremity Surveillance Project (CUSP), *Int J Occup Environ Health, 4*(4), 209–216 (1998).
28. World Health Organization (WHO), Identification and control of work-related diseases, (WHO Technical Report Series No. 714), 1985.
29. B. Bernard, Musculoskeletal disorders and workplace factors: A critical review of epidemiologic evidence for work-related musculoskeletal disorders of the neck, upper extremity, and low back, DHHS(NIOSH) Pub. No. 97–141, 1997.
30. National Research Council and the Institute of Medicine, *Musculoskeletal Disorders and the Workplace: Low Back and Upper Extremities*, Panel on Musculoskeletal Disorders and the Workplace, Commission on Behavioral and Social Sciences and Education, Washington, DC: National Academy Press, 2001.
31. C.G. Barnes and H.L.F. Currey, Carpal tunnel syndrome in rheumatoid arthritis, a clinical and electrodiagnostic survey, *Ann Rheum Dis, 26*, 226–233 (1967).
32. M.S. Sabour and H.H. Fadel, The carpal tunnel syndrome–A new complication ascribed to the pill, *Am J Obstet Gynecol, 107*(3), 1265–1267 (1979).
33. S. Lozano-Calderon, S. Anthony, and D. Ring, The quality and strength of evidence for etiology: Example of carpal tunnel syndrome, *J Hand Surg Am, 33*(4), 525–538 (2008).
34. K.T. Palmer, E.C. Harris, and D. Coggon, Carpal tunnel syndrome and its relation to occupation: A systematic literature review, *Occup Med (Lond), 57*(1), 57–66 (2007).
35. B.A. Hill, The environment and disease: Association or causation? *Proc R Soc Med, 58*, 295–300 (1965).
36. L. Punnett and D.H. Wegman, Work-related musculoskeletal disorders: The epidemiologic evidence and the debate, *J Electromyogr Kinesiol, 14*(1), 13–23 (2004).
37. Bureau of Labor Statistics, *Annual Survey of Occupational Injuries and Illnesses*, Washington, DC: U.S. Department of Labor, 2007.
38. T. Morse, C. Dillon, E. Kenta-Bibi, J. Weber, U. Diva, N. Warren, et al., Trends in work-related musculoskeletal disorder reports by year, type, and industrial sector: A capture-recapture analysis, *Am J Ind Med, 48*(1), 40–49 (2005).
39. K.D. Rosenman, A. Kalush, M.J. Reilly, J.C. Gardiner, M. Reeves, and Z. Luo, How much work-related injury and illness is missed by the current national surveillance system? *J Occup Environ Med, 48*(4), 357–365 (2006).
40. J.C. Rosecrance, T.M. Cook, D.C. Anton, and L.A. Merlino, Carpal tunnel syndrome among apprentice construction workers, *Am J Ind Med, 42*(2), 107–116 (2002).
41. L.S. Friedman and L. Forst, The impact of OSHA recordkeeping regulation changes on occupational injury and illness trends in the US: A time-series analysis, *Occup Environ Med, 64*(7), 454–460 (2007).
42. T.L. Holbrook, K. Grazier, J.L. Kelsey, and R.N. Stauffer, *The frequency of occurrence, impact and cost of selected musculoskeletal conditions in the United States*, Chicago, IL: American Academy of Orthopadic Surgeons, 1984.
43. K.D. Rosenman, J.C. Gardiner, J. Wang, J. Biddle, A. Hogan, M.J. Reilly, et al., Why most workers with occupational repetitive trauma do not file for workers' compensation, *J Occup Environ Med, 42*(1), 25–34 (2000).
44. D.E. Elisburg, *Cumulative Trauma Disorders in the Workplace: Costs, Prevention, and Progress*, Washington, DC: Bureau of National Affairs (BNA), 1991.

45. G. Brogmus and R. Marko, The proportion of cumulative trauma disorders of the upper extremities in U.S. industry, *Paper Presented at the Human Factors Society 36th Annual Meeting*, Atlanta, GA, 1992.
46. Safety & Health Assessment & Research for Prevention (SHARP), *Work-related Musculoskeletal Disorders of the Neck, Back, and Upper Extremity in Washington State,* 1997–2005, Technical Report No. 40-11-2007, Olympia, WA: SHARP Program, 2007.
47. B. Juul-Kristensen, N. Fallentin, and C. Ekdahl, Criteria for classification of posture in repetitive work by observation methods: A review, *Int J Ind Ergon, 19*, 397–411 (1997).
48. C. Hamrick, *Overview of Ergonomic Assessment* (2nd edn.), Boca Raton, FL: CRC/Taylor & Francis Group, 2006.
49. W.M. Keyserling, D.S. Stetson, B.A. Silverstein, and M.L. Brouwer, A checklist for evaluating ergonomic risk factors associated with upper extremity cumulative trauma disorders, *Ergonomics, 36*(7), 807–831 (1993).
50. K. Kemmlert, A method assigned for the identification of ergonomic hazards—PLIBEL, *Appl Ergon, 26*(3), 199–211 (1995).
51. G. Li and P. Buckle, Current techniques for assessing physical exposure to work-related musculoskeletal risks, with emphasis on posture-based methods, *Ergonomics, 42*, 674–695 (1999).
52. Washington State Department of Labor and Industries, *Pocket Guide to Caution Zone Jobs*, 2003.
53. ANSI/ASSE, *Reduction of Musculoskeletal Problems in Construction and Demolition Operations, ANSI/ASSE A10.40-2007*, 2007.
54. NIOSH/Cal-OSHA, *Easy Ergonomics: A Guide to Selecting Non-Powered Hand Tools*, DHHS (NIOSH) Publication No. 2004-16, 2004.
55. A. Dababneh, B. Lowe, E. Krieg, Y.K. Kong, and T. Waters, Ergonomics: A checklist for the ergonomic evaluation of nonpowered hand tools, *J Occup Environ Hyg, 1*(12), D135–D145 (2004).
56. A. Kilbom, Assessment of physical exposure in relation to work-related musculoskeletal disorders–What information can be obtained from systematic observations? *Scand J Work Environ Health, 20 Spec No*, 30–45 (1994).
57. G. David, Ergonomic methods for assessing exposure to risk factors for work-related musculoskeletal disorders, *Soc Occupational Med, 55*, 190–199 (2005).
58. J.S. Moore and A. Garg, The strain index: A proposed method to analyze jobs for risk of distal upper extremity disorders, *Am Ind Hyg Assoc J, 56*(5), 443–458 (1995).
59. American Conference of Governmental Industrial Hygienists (ACGIH), *Documentation of the TLVs and BEIs* (6th edn.), Cincinnati, OH: ACGIH Worldwide, 2005.
60. T. Armstrong, M. Ebersole, A. Franzblau, S. Ulin, and R. Werner, The ACGIH TLV: A review of some recent studies, *Paper presented at the 2006 International Ergonomics Association Triennial Congress*, Maastricht, the Netherlands, 2006.
61. G. Borg, A category scale with ratio properties for intermodal and interindividual comparisons, In H.G. Geissler and P. Petzold (eds.), *Psychophysical Judgment and the Process of Perception*, Berlin, Germany: VEB Deutscher Verlag Der Wissenschaften, 1982.
62. S. Bao and B. Silverstein, Estimation of hand force in ergonomic job evaluations, *Ergonomics, 48*(3), 288–301 (2005).
63. W.A. Latko, T.J. Armstrong, A. Franzblau, S.S. Ulin, R.A. Werner, and J.W. Albers, Cross-sectional study of the relationship between repetitive work and the prevalence of upper limb musculoskeletal disorders, *Am J Ind Med, 36*(2), 248–259 (1999).
64. D. Colombini, An observational method for classifying exposure to repetitive movements of the upper limbs, *Ergonomics, 41*(9), 1261–1289 (1998).
65. E. Occhipinti, OCRA: A concise index for the assessment of exposure to repetitive movements of the upper limbs, *Ergonomics, 41*(9), 1290–1311 (1998).
66. T.R. Waters, V. Putz-Anderson, A. Garg, and L.J. Fine, Revised NIOSH equation for the design and evaluation of manual lifting tasks, *Ergonomics, 36*(7), 749–776 (1993).

67. NIOSH, *A Work Practices Guide to Manual Lifting,* Technical Report No. 81–122, Cincinnati, OH: U.S. Department of Health and Human Services (NIOSH), 1981.
68. V. Seth, R.L. Weston, and A. Freivalds, Development of a cumulative trauma disorder risk assessment model for the upper extremities, *Int J Ind Ergon, 23*, 281–291 (1999).
69. B.W. Niebel and A. Freivalds, *Methods, Standards, and Work Design* (11th edn.), Dubuque, IA: McGraw-Hill, 2003.
70. O. Karhu, P. Kansi, and I. Kuorinka, Correcting working postures in industry. A practical method for analysis, *Appl Ergon*, 8, 199–201 (1977).
71. L. McAtamney and E.N. Corlett, RULA: A survey method for the investigation of work-related upper limb disorders, *Appl Ergon, 24*(2), 91–99 (1993).
72. S. Hignett and L. McAtamney, Rapid entire body assessment (REBA), *Appl Ergon, 31*(2), 201–205 (2000).
73. D. Kee and W. Karwowski, *An Assessment Technique for Postural Loading on the Upper Body (LUBA)* (2nd edn.), Boca Raton, FL: CRC/Taylor & Francis Group, 2006.
74. A.M. Genaidy, A.A. Al-Shedi, and W. Karwowski, Postural stress analysis in industry, *Appl Ergon, 25*(2), 77–87 (1994).
75. A. Buchholz, V. Paquet, L. Punnett, D. Lee, and S. Moir, PATH: A work sampling-based approach to ergonomic job analysis for construction and other non-repetitive work, *Appl Ergon, 27*(3), 177–187 (1996).
76. V.L. Paquet, L. Punnett, and B. Buchholz, Validity of fixed-interval observations for postural assessment in construction work, *Appl Ergon, 32*(3), 215–224 (2001).
77. K. Knox and J.S. Moore, Predictive validity of the strain index in turkey processing, *J Occup Environ Med, 43*(5), 451–462 (2001).
78. N. Rucker and J.S. Moore, Predictive validity of the strain index in manufacturing facilities, *Appl Occup Environ Hyg, 17*(1), 63–73 (2002).
79. A. Franzblau, T.J. Armstrong, R.A. Werner, and S.S. Ulin, A cross-sectional assessment of the ACGIH TLV for hand activity level, *J Occup Rehabil, 15*(1), 57–67 (2005).
80. P. Spielholz, S. Bao, N. Howard, B. Silverstein, J. Fan, C. Smith, et al., Reliability and validity assessment of the hand activity level threshold limit value and strain index using expert ratings of mono-task jobs, *J Occup Environ Hyg, 5*(4), 250–257 (2008).
81. S. Bao, N. Howard, P. Spielholz, and B. Silverstein, Quantifying repetitive hand activity for epidemiological research on musculoskeletal disorders–Part II: Comparison of different methods of measuring force level and repetitiveness, *Ergonomics, 49*(4), 381–392 (2006).
82. B.D. Lowe, Accuracy and validity of observational estimates of shoulder and elbow posture, *Appl Ergon, 35*(2), 159–171 (2004).
83. P.M. van Wyk, P.L. Weir, D.M. Andrews, K.M. Fiedler, and J.P. Callaghan, Determining the optimal size for posture categories used in video-based posture assessment methods, *Ergonomics, 52*(8), 921–930 (2009).
84. S. Bao, N. Howard, P. Spielholz, B. Silverstein, and N. Polissar, Interrater reliability of posture observations, *Hum Factors, 51*(3), 292–309 (2009).
85. B. DeLuca, The use of surface electromyography in biomechanics, *J Appl Biomech, 13*, 135–163 (1997).
86. G.M. Hagg, A. Luttmann, and M. Jager, Methodologies for evaluating electromyographic field data in ergonomics, *J Electromyogr Kinesiol, 10*(5), 301–312 (2000).
87. M.J. Hoozemans and J.H. van Dieen, Prediction of handgrip forces using surface EMG of forearm muscles, *J Electromyogr Kinesiol, 15*(4), 358–366 (2005).
88. S.E. Mathiassen and J. Winkel, Quantifying variation in physical load using exposure-vs-time data, *Ergonomics, 34*, 1455–1468 (1991).
89. D. Anton, T.M. Cook, J.C. Rosecrance, and L.A. Merlino, Method for quantitatively assessing physical risk factors during variable noncyclic work, *Scand J Work Environ Health, 29*(5), 354–362 (2003).

90. B. Jonsson, Quantitative electromyographic evaluation of muscular load during work, *Scand J Rehabil Med, 6*, 69–74 (1978).
91. R.W. McGorry, A system for the measurement of grip forces and applied moments during hand tool use, *Appl Ergon, 32*(3), 271–279 (2001).
92. G.L. Fellows, and A. Freivalds, Ergonomics evaluation of a foam rubber grip for tool handles, *Appl Ergon, 22*(4), 225–230 (1991).
93. R.G. Radwin, S. Oh, T.R. Jensen, and J.G. Webster, External finger forces in submaximal five-finger static pinch prehension, *Ergonomics, 35*(3), 275–288 (1992).
94. Y.K. Kong and B.D. Lowe, Optimal cylindrical handle diameter for grip force tasks, *Int J Ind Ergon, 35*(6), 495–507 (2005).
95. J.T. Albers and S.D. Hudock, Biomechanical assessment of three rebar tying techniques, *Int J Occup Saf Ergon, 13*(3), 279–289 (2007).
96. P.G. Dempsey, R.W. McGorry, and W.S. Maynard, A survey of tools and methods used by certified professional ergonomists, *Appl Ergon, 36*(4), 489–503 (2005).
97. C. Coffey, *NIOSH Seeks Input on Direct Reading Exposure Assessment Methods (D.R.E.A.M.) Workshop*, Washington, DC, 2009 (September 4, 2009).
98. W.M. Keyserling, Postural analysis of the trunk and shoulders in simulated real time, *Ergonomics, 29*(4), 569–583 (1986).
99. A. Dartt, J. Rosecrance, F. Gerr, P. Chen, D. Anton, and L. Merlino, Reliability of assessing upper limb postures among workers performing manufacturing tasks, *Appl Ergon, 40*(3), 371–378 (2009).
100. S. Wurzelbacher, S. Burt, K. Crombie, J. Ramsey, L. Luo, S. Allee, et al., A comparison of assessment methods of hand activity and force for use in calculating the ACGIH TLV, *J Occup Environ Hygiene, 7*(7), 407–416 (2010).
101. T.Y. Yen and R.G. Radwin, A video-based system for acquiring biomechanical data synchronized with arbitrary events and activities, *IEEE Trans Biomed Eng, 42*(9), 944–948 (1995).
102. T.Y. Yen and R.G. Radwin, Multimedia video-based data acquisition and analysis applications for ergonomics research, *Paper Presented at the XIVth Triennial Congress of the International Ergonomics Association and 44th Annual Meeting of the Human Factors and Ergonomics Society*, San Diego, CA, 2000.
103. J.D. McGlothlin, M.G. Gressel, W.A. Heitbrink, and P.A. Jensen, Real-time exposure assessment and job analysis techniques to solve hazardous workplace exposures, in A. Bhattacharya and J.D. McGlothlin (eds.), *Occupational Ergonomics: Theory and Applications* (pp. xiv, 832 p), New York: Marcel Dekker, 1996.
104. J.D. McGlothlin, Occupational exposure visualization comes of age, *Ann Occup Hyg, 49*(3), 197–199 (2005).
105. G. Rosen, I.M. Andersson, P.T. Walsh, R.D. Clark, A. Saamanen, K. Heinonen, et al., A review of video exposure monitoring as an occupational hygiene tool, *Ann Occup Hyg, 49*(3), 201–217 (2005).
106. B.D. Lowe, Y. Kong, and J. Han, Development and application of a hand force measurement system, *Proceedings of the XVIth Triennial Congress of the International Ergonomics Association*, Maastricht, the Netherlands: Triennial Congress of the International Ergonomics Association, 2006.
107. Government Accountability Office (GAO), *OSHA's Voluntary Protection Programs: Improved Oversight and Controls Would Better Ensure Program Quality*, May 2009. GAO-09-395, 2009.
108. NIOSH, *National Occupational Research Agenda for Musculoskeletal Disorders: Research Topics for the Next Decade*, A Report by the NORA Musculoskeletal Disorders Team, DHHS (NIOSH) Publication No. 2001-117, 2001.

109. W.S. Marras, R.G. Cutlip, S.E. Burt, and T.R. Waters, National occupational research agenda (NORA) future directions in occupational musculoskeletal disorder health research, *Appl Ergon, 40,* 15–22 (2009).
110. M. DiNatale, Characteristics of and preference for alternative work arrangements, 1999, *Mon Labor Rev, 124*(3), 28–49 (2001).
111. J.P. Stephens, G.A. Vos, E.M. Stevens Jr., and J.S. Moore, Test-retest repeatability of the strain index, *Appl Ergon, 37*(3), 275–281 (2006).
112. E.M. Stevens, G.A. Vos, J.P. Stephens, and J.S. Moore, Inter-rater reliability of the strain index, *J Occup Environ Hyg, 1*(11), 745–751 (2004).

CHAPTER 33

# Revised NIOSH Lifting Equation

Thomas R. Waters

CONTENTS

## 33.1 INTRODUCTION

This chapter provides information about a revised equation for assessing the physical demands of certain two-handed manual lifting tasks that was developed by the National Institute for Occupational Safety and Health (NIOSH) and described earlier in an article by Waters et al. [1]. We discuss what factors need to be measured, how they should be measured, what procedures should be used, and how the results can be used to ergonomically design new jobs or make decisions about redesigning existing jobs that may be hazardous. We define all pertinent terms and present the mathematical formulas and procedures needed to properly apply the NIOSH lifting equation. Several example problems are also provided to demonstrate how the equations should be used. An expanded version of this chapter is contained in an NIOSH report [2].

Historically, NIOSH has recognized the problem of work-related back injuries resulting from manual lifting and, in response, published the *Work Practices Guide for Manual Lifting* (WPG) in 1981 [3]. The WPG contained a summary of the lifting-related literature up to 1981; analytical procedures and a lifting equation for calculating a recommended weight for specific two-handed, symmetrical lifting tasks; and an approach for controlling the hazards of low back injury from manual lifting. The approach to hazard control was coupled to the *action limit* (AL), a term that denoted the recommended weight derived from the lifting equation.

In 1985, NIOSH convened an ad hoc committee of experts who reviewed the current literature on lifting, with a special focus on the original NIOSH WPG.* The literature review

* The ad hoc 1991 NIOSH Lifting Committee members included M. M. Ayoub, Donald B. Chaffin, Colin G. Drury, Arun Garg, and Suzanne Rodgers. NIOSH representatives included Vern Putz-Anderson and Thomas R. Waters.

was summarized in a document containing updated information on the physiological, biomechanical, psychophysical, and epidemiological aspects of manual lifting [4]. Based on the results of the literature review, the ad hoc committee recommended criteria for defining the lifting capacity of healthy workers. The committee used the criteria to formulate the revised lifting equation.* Subsequently, NIOSH staff developed the documentation for the equation and played a prominent role in recommending methods for interpreting the results of the lifting equation. *The revised lifting equation reflects new findings and provides methods for evaluating asymmetrical lifting tasks and lifts of objects with less than optimal couplings between the object and the worker's hands. The revised lifting equation also provides guidelines for a more diverse range of lifting tasks than the earlier equation* [3].

The rationale and criterion for the development of the revised NIOSH lifting equation (RNLE) are provided in a journal article by Waters et al. [1]. We suggest that those users who wish to achieve a better understanding of the data and decisions that were made in formulating the RNLE consult that article. It provides an explanation of the selection of the biomechanical, physiological, and psychophysical criterion as well as a description of the derivation of the individual components of the RNLE. For those individuals, however, who are primarily concerned with the use and application of the RNLE, this chapter provides a more complete description of the method and its limitations.

Although the RNLE has not been fully validated, the recommended weight limits (RWLs) derived from the RNLE are consistent with, or lower than, those generally reported in the literature [1]. Moreover, the proper application of the RNLE is more likely to protect healthy workers for a wider variety of lifting tasks than methods that rely on only a single-task factor or single criterion.

Finally, it should be stressed that the NIOSH lifting equation is only one tool in a comprehensive effort to prevent work-related low back pain (LBP) and disability. Some examples of other approaches are described elsewhere [5]. Moreover, lifting is only one of the causes of work-related LBP and disability. Other causes that have been hypothesized or established as risk factors include whole body vibration, static postures, prolonged sitting, and direct trauma to the back. Psychosocial factors, appropriate medical treatment, and job demands may also be particularly important in influencing the transition of acute LBP to chronic disabling pain.

## 33.2 DEFINITION OF TERMS

This section provides the basic technical information needed to properly use the RNLE to evaluate a variety of two-handed manual lifting tasks. Definitions and data requirements for the RNLE are also provided.

### 33.2.1 Recommended Weight Limit

The RWL is the principal product of the RNLE. The RWL is defined for a specific set of task conditions as the weight of the load that nearly all healthy workers could perform

* For this remainder of this chapter, the revised 1991 NIOSH lifting equation will be identified simply as "the RNLE" [1,2]. The abbreviation WPG will continue to be used as the reference to the earlier NIOSH lifting equation, which was documented in the *Work Practices Guide for Manual Lifting* [3].

over a substantial period of time (e.g., up to 8 h) without an increased risk of developing lifting-related LBP. By "healthy workers" we mean workers who are free of adverse health conditions that would increase their risk of musculoskeletal injury.

The concept behind the RNLE is to start with a recommended weight that is considered safe for an "ideal" lift (i.e., load constant equal to 51 lb) and then reduce the weight as the task becomes more stressful (i.e., as the task-related factors become less favorable). The precise formulation of the RNLE for calculating the RWL is based on a multiplicative model that provides a weighting (multiplier) for each of six task variables:

1. Horizontal distance of the load from the worker (*H*)
2. Vertical height of the lift (*V*)
3. Vertical displacement during the lift (*D*)
4. Angle of asymmetry (*A*)
5. Frequency (*F*) and duration of lifting
6. Quality of the hand-to-object coupling (*C*)

The weightings are expressed as coefficients that serve to decrease the load constant, which represents the maximum recommended load weight to be lifted under ideal conditions. For example, as the horizontal distance between the load and the worker increases from 10 in., the RWL for that task would be reduced from the ideal starting weight.

The RWL is defined as

$$\text{RWL} = \text{LC} \times \text{HM} \times \text{VM} \times \text{DM} \times \text{AM} \times \text{FM} \times \text{CM}$$

where the term *task variables* refers to the measurable task-related measurements that are used as input data for the formula (i.e., *H, V, D, A, F,* and *C*), whereas the term *multipliers* refers to the reduction coefficients in the equation (i.e., HM, VM, DM, AM, FM, and CM).

### 33.2.2 Measurement Requirements

The following list briefly describes the measurements required to use the RNLE. Details for each of the variables are presented later in this chapter (see Section 33.4):

*H* is the horizontal location of hands from midpoint between the inner ankle bones. *H* should be measured at the origin and the destination of the lift (cm or in.).

*V* is the vertical location of the hands from the floor. *V* should be measured at the origin and destination of the lift (cm or in.).

*D* is the vertical travel distance between the origin and the destination of the lift (cm or in.).

*A* is the angle of asymmetry—angular displacement of the load from the worker's sagittal plane. *A* should be measured at the origin and destination of the lift (degrees).

*F* is the average frequency rate of lifting measured in lifts/min. Duration is defined to be ≤1 h, ≤2 h, or ≤8 h assuming appropriate recovery allowances (see Table 33.5).

*C* is the quality of hand-to-object coupling (quality of interface between the worker and the load being lifted). The quality of the coupling is categorized as good, fair, or poor, depending upon the type and location of the coupling, the physical characteristics of load, and the vertical height of the lift.

### 33.2.3 Lifting Index

The *lifting index* (LI) is a term that provides a relative estimate of the level of physical stress associated with a particular manual lifting task. The estimate of the level of physical stress is defined by the relationship between the weight of the load lifted and the RWL. The LI is defined by the equation

$$\text{LI} = \frac{\text{load weight}}{\text{recommended weight limit}} = \frac{L}{\text{RWL}}$$

where load weight (*L*) is the weight of the object lifted (lb or kg).

### 33.2.4 Miscellaneous Terms

*Lifting task*: The act of manually grasping an object of definable size and mass with two hands and vertically moving the object without mechanical assistance.

*Load weight* (*L*): Weight of the object to be lifted, in pounds or kilograms, including the container.

*Horizontal location* (*H*): Distance of the hands away from the midpoint between the ankles, in inches or centimeters (measure at the origin and destination of lift). See Figure 33.1.

*Vertical location* (*V*): Distance of the hands above the floor, in inches or centimeters (measure at the origin and destination of lift). See Figure 33.1.

*Vertical travel distance* (*D*): Absolute value of the difference between the vertical heights at the destination and origin of the lift, in inches or centimeters.

*Angle of asymmetry* (*A*): Angular measure of how far the *object* is displaced from the front (midsagittal plane) of the worker's body at the beginning or end of the lift, in degrees (measure at the origin and destination of lift). See Figure 33.2. The asymmetry angle is defined by the location of the load relative to the worker's midsagittal plane, as defined by the neutral body posture, rather than the position of the feet or the extent of body twist.

*Neutral body position*: Position of the body when the hands are directly in front of the body and there is minimal twisting at the legs, torso, or shoulders.

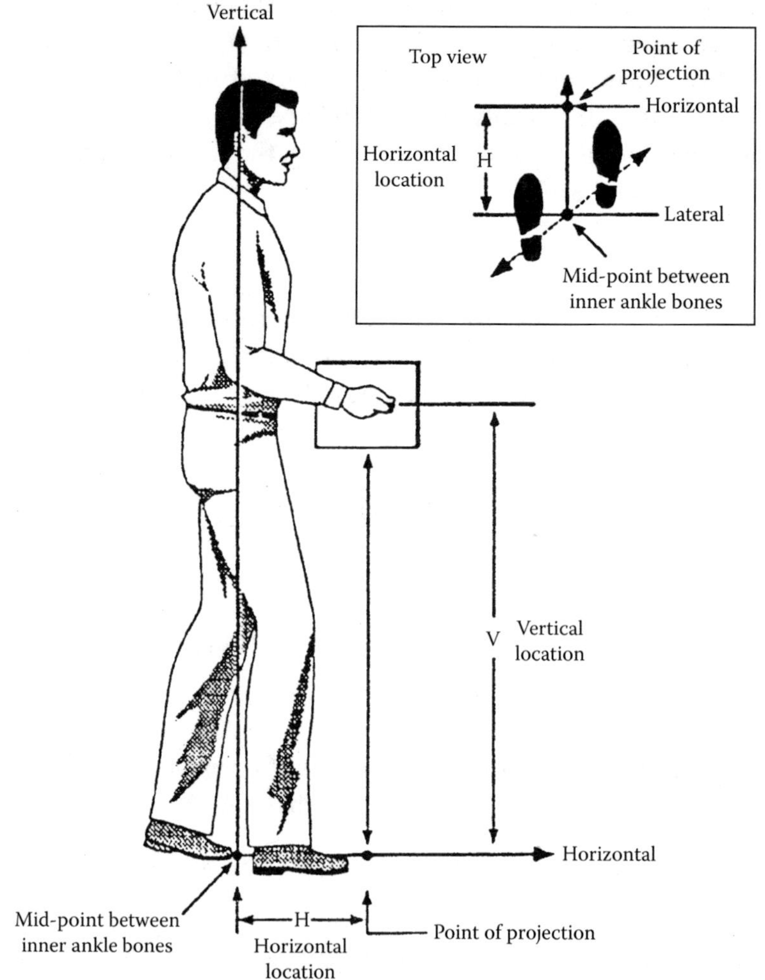


FIGURE 33.1 Graphic representation of hand location.

*Frequency of lifting* (*F*): Average number of lifts per minute over a 15 min period.

*Duration of lifting*: Three-tiered classification of lifting duration specified by the distribution of work time (WT) and recovery time (work pattern). Duration is classified as either short (1 h), moderate (1–2 h), or long (2–8 h), depending on the work pattern.

*Coupling classification*: Classification of the quality of the hand-to-object coupling (e.g., handle, cutout, or grip). Coupling quality is classified as good, fair, or poor.

*Significant control*: A condition requiring "precision placement" of the load at the destination of the lift. This is usually the case when (1) the worker has to regrasp the load near the destination of the lift, (2) the worker has to momentarily hold the object at the destination, or (3) the worker has to carefully position or guide the load at the destination.

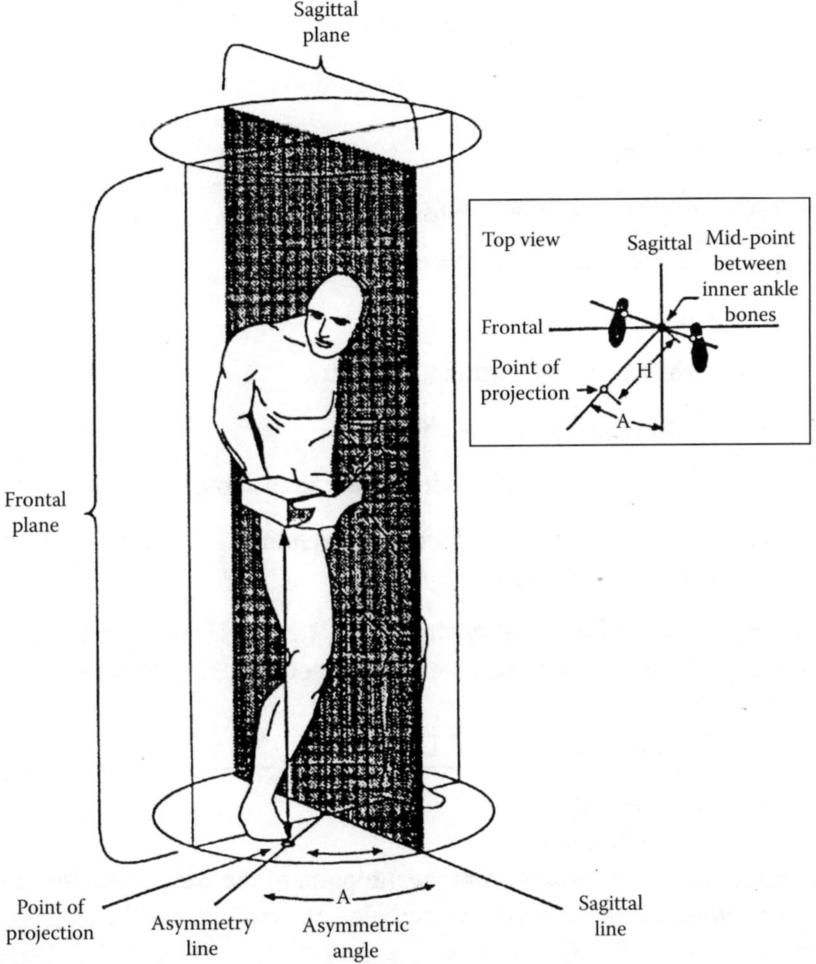


FIGURE 33.2 Graphic representation of angle of asymmetry (*A*).

## 33.3 LIMITATIONS OF THE EQUATION

The lifting equation is a tool for assessing the physical stress of two-handed manual lifting tasks. As with any tool, its application is limited to those conditions for which it was designed. Specifically, the lifting equation was designed to meet specific lifting-related criteria that encompass biomechanical, physiological, and psychophysical assumptions and data used to develop the equation. To the extent that a given lifting task accurately reflects these underlying conditions and criteria, this lifting equation may be appropriately applied.

The following list identifies a set of work conditions in which the application of the lifting equation could either under- or overestimate the extent of physical stress associated with a particular work-related activity. Each of the following task limitations also highlights research topics in need of further research to extend the application of the lifting equation to a greater range of real-world lifting tasks.

The RNLE does not apply if any of the following occur:

Lifting/lowering with one hand

Lifting/lowering for over 8 h

Lifting/lowering while seated or kneeling

Lifting/lowering in a restricted workspace

Lifting/lowering unstable objects

Lifting/lowering while carrying, pushing, or pulling

Lifting/lowering with wheelbarrows or shovels

Lifting/lowering with "high-speed" motion (faster than about 30 in./s)

Lifting/lowering with unreasonable foot/floor coupling (<0.4 coefficient of friction between the sole and the floor)

Lifting/lowering an unfavorable environment (temperature significantly outside 66°F–79°F (19°C–26°C) range; relative humidity outside 35%–50% range)

## 33.4 OBTAINING AND USING THE DATA

### 33.4.1 Horizontal Component

#### *33.4.1.1 Definition and Measurement*

Horizontal location (*H*) is measured from the midpoint of the line joining the inner ankle bones to a point projected on the floor directly below the midpoint of the hand grasps (i.e., load center), as defined by the large middle knuckle of the hand (Figure 33.1). Typically, the worker's feet are not aligned with the midsagittal plane, as shown in Figure 33.1, but may be rotated inward or outward. If this is the case, then the midsagittal plane is defined by the worker's neutral body posture as defined earlier. Also, if the worker leans over on one foot to perform the lift and uses the other leg/foot as a counterbalance, H is measured from directly below the foot that is supporting the body weight rather than from the midpoint between the two feet.* If significant control is required at the destination (i.e., precision placement), then *H* should be measured at both the origin and destination of the lift.

Horizontal distance (*H*) should be measured. In those situations, where the *H* value cannot be measured, then *H* may be approximated from the following equations:

*Metric* (all distances in cm):

$$H = 20 + W/2 \quad \text{for } V \geq 25 \text{ cm}$$

$$H = 25 + W/2 \quad \text{for } V < 25 \text{ cm}$$

* *Note:* The definition for measuring *H* when the worker leans over on one leg and uses the other leg as a counter balance is a recent change and was not published in the original Applications Manual (Waters et al. [1]).

*U.S. customary* (all distances inches):

$$H = 8 + W/2 \quad \text{for } V \geq 10\,\text{in.}$$

$$H = 10 + W/2 \quad \text{for } V < 10\,\text{in.}$$

where

$W$ is the width of the container in the sagittal plane
$V$ is the vertical location of the hands from the floor

#### 33.4.1.2 Horizontal Restrictions

If the horizontal distance is less than 10 in. (25 cm), then $H$ is set to 10 in. (25 cm). Although objects can be carried or held closer than 10 in. from the ankles, most objects that are closer than this cannot be lifted without encountering interference from the abdomen or hyperextending the shoulders. Although 25 in. (63 cm) was chosen as the maximum value for $H$, it is probably too great a distance for shorter workers, particularly when lifting asymmetrically. Furthermore, objects at a distance of more than 25 in. from the ankles normally cannot be lifted vertically without some loss of balance.

#### 33.4.1.3 Horizontal Multiplier

The horizontal multiplier (HM) is $10/H$ for $H$ measured in inches and $25/H$ for $H$ measured in centimeters. If $H$ is less than or equal to 10 in. (25 cm), the multiplier is 1.0. HM decreases with an increase in $H$ value. The multiplier for $H$ is reduced to 0.4 when $H$ is 25 in. (63 cm). If $H$ is greater than 25 in., then HM = 0. The HM value can be computed directly or determined from Table 33.1.

### 33.4.2 Vertical Component

#### 33.4.2.1 Definition and Measurement

Vertical location ($V$) is defined as the vertical height of the hands above the floor. $V$ is measured vertically from the floor to the midpoint between the hand grasps, as defined by the large middle knuckle. The coordinate system is illustrated in Figure 33.1.

#### 33.4.2.2 Vertical Restrictions

The vertical location ($V$) is limited by the floor surface and the upper limit of vertical reach for lifting (i.e., 70 in. or 175 cm). The vertical location should be measured at the origin and the destination of the lift to determine the travel distance ($D$).

#### 33.4.2.3 Vertical Multiplier

To determine the vertical multiplier (VM), the absolute value or deviation of $V$ from an optimum height of 30 in. (75 cm) is calculated. A height of 30 in. above floor level is considered "knuckle height" for a worker of average height (66 in. or 165 cm). The VM is $1 - (0.0075|V - 30|)$ for $V$ measured in inches $1 - (0.003|V - 75|)$, for $V$ measured in centimeters.

When $V$ is at 30 in. (75 cm), VM = 1.0. The value of VM decreases linearly with an increase or decrease in height from this position. At floor level, VM = 0.78, and at 70 in. (175 cm) height, VM = 0.7. If $V$ is greater than 70 in., then VM = 0. The VM value can be computed directly or determined from Table 33.2.

| | Metric | U.S. Customary |
|---|---|---|
| LC = Load constant | 23 kg | 51 lb |
| HM = Horizontal multiplier | $(25/H)$ | $(10/H)$ |
| VM = Vertical multiplier | $1 - (0.003\|V - 75\|)$ | $1 - (0.0075\|V - 30\|)$ |
| DM = Distance multiplier | $0.82 + (4.5/D)$ | $0.82 + (1.8/D)$ |
| AM = Asymmetric multiplier | $1 - (0.0032\ A)$ | $1 - (0.0032\ A)$ |
| FM = Frequency multiplier | From Table 33.5 | From Table 33.5 |
| CM = Coupling multiplier | From Table 33.7 | From Table 33.7 |

TABLE 33.1 Horizontal Multiplier

| *H* (in.) | HM | *H* (cm) | HM |
|---|---|---|---|
| ≤10 | 1.00 | ≤25 | 1.00 |
| 11 | 0.91 | 28 | 0.89 |
| 12 | 0.83 | 30 | 0.83 |
| 13 | 0.77 | 32 | 0.78 |
| 14 | 0.71 | 34 | 0.74 |
| 15 | 0.67 | 36 | 0.69 |
| 16 | 0.63 | 38 | 0.66 |
| 17 | 0.59 | 40 | 0.63 |
| 18 | 0.56 | 42 | 0.60 |
| 19 | 0.53 | 44 | 0.57 |
| 20 | 0.50 | 46 | 0.54 |
| 21 | 0.48 | 48 | 0.52 |
| 22 | 0.46 | 50 | 0.50 |
| 23 | 0.44 | 52 | 0.48 |
| 24 | 0.42 | 54 | 0.46 |
| 25 | 0.40 | 56 | 0.45 |
| >25 | 0.00 | 58 | 0.43 |
| | | 60 | 0.42 |
| | | 63 | 0.40 |
| | | >63 | 0.00 |

### 33.4.3 Distance Component

#### *33.4.3.1 Definition and Measurement*

The distance variable (*D*) is defined as the vertical travel distance of the hands between the origin and destination of the lift. For lifting, *D* can be computed by subtracting the vertical location (*V*) at the origin of the lift from the corresponding *V* at the destination of the lift (i.e., *D* is equal to *V* at the destination minus *V* at the origin). For a lowering task, *D* is equal to *V* at the origin minus *V* at the destination.

#### *33.4.3.2 Distance Restrictions*

*D* is assumed to be at least 10 in. (25 cm) and no greater than 70 in. (175 cm). If the vertical travel distance is less than 10 in. (25 cm), then *D* should be set to the minimum distance of 10 in. (25 cm).

TABLE 33.2 Vertical Multiplier

| *V* (in.) | VM | *V* (cm) | VM |
|---|---|---|---|
| 0 | 0.78 | 0 | 0.78 |
| 5 | 0.81 | 10 | 0.81 |
| 10 | 0.85 | 20 | 0.84 |
| 15 | 0.89 | 30 | 0.87 |
| 20 | 0.93 | 40 | 0.90 |
| 25 | 0.96 | 50 | 0.93 |
| 30 | 1.00 | 60 | 0.96 |
| 35 | 0.96 | 70 | 0.99 |
| 40 | 0.93 | 80 | 0.99 |
| 45 | 0.89 | 90 | 0.96 |
| 50 | 0.85 | 100 | 0.93 |
| 55 | 0.81 | 110 | 0.90 |
| 60 | 0.78 | 120 | 0.87 |
| 65 | 0.74 | 130 | 0.84 |
| 70 | 0.70 | 140 | 0.81 |
| >70 | 0.00 | 150 | 0.78 |
| | | 160 | 0.75 |
| | | 170 | 0.72 |
| | | 175 | 0.70 |
| | | >175 | 0.00 |

#### *33.4.3.3 Distance Multiplier*

The distance multiplier (DM) is $0.82 + 1.8/D$ for $D$ measured in inches $0.82 + 4.5/D$ for $D$ measured in centimeters. For $D$ less than 10 in. (25 cm), $D$ is assumed to be 10 in. (25 cm) and DM = 1.0. The distance multiplier, therefore, decreases gradually with an increase in travel distance. DM = 1.0 when $D$ is set at 10 in. (25 cm); DM = 0.85 when $D$ = 70 in. (175 cm). Thus, DM ranges from 1.0 to 0.85 as $D$ varies from 0 in. (0 cm) to 70 in. (175 cm). The DM value can be computed directly or determined from Table 33.3.

### 33.4.4 Asymmetry Component

#### *33.4.4.1 Definition and Measurement*

Asymmetry refers to a lift that begins or ends outside the midsagittal plane (see Figure 33.2). In general, asymmetric lifting should be avoided. If asymmetric lifting cannot be avoided, however, the RWLs are significantly less than those limits used for symmetrical lifting.*

An asymmetric lift may be required under the following task or workplace conditions:

1. The origin and destination of the lift are oriented at an angle to each other.
2. The lifting motion is across the body, such as occurs in swinging bags or boxes from one location to another.

---

* It may not always be clear whether asymmetry is an intrinsic element of the task or just a personal characteristic of the worker's lifting style. Regardless of the reason for the asymmetry, any observed asymmetric lifting should be considered an intrinsic element of the job design and should be considered in the assessment and subsequent redesign. Moreover, the design of the task should not rely on worker compliance but should rather discourage or eliminate the need for asymmetric lifting.

TABLE 33.3 Distance Multiplier

| *D* (in.) | DM | *D* (cm) | DM |
|---|---|---|---|
| ≤10 | 1.00 | ≤25 | 1.00 |
| 15 | 0.94 | 40 | 0.93 |
| 20 | 0.91 | 55 | 0.90 |
| 25 | 0.89 | 70 | 0.88 |
| 30 | 0.88 | 85 | 0.87 |
| 35 | 0.87 | 100 | 0.87 |
| 40 | 0.87 | 115 | 0.86 |
| 45 | 0.86 | 130 | 0.86 |
| 50 | 0.86 | 145 | 0.85 |
| 55 | 0.85 | 160 | 0.85 |
| 60 | 0.85 | 175 | 0.85 |
| 70 | 0.85 | >175 | 0.00 |
| >70 | 0.00 | | |

3. The lifting is done to maintain body balance in obstructed workplaces, on rough terrain, or on littered floors.

4. Productivity standards require reduced time per lift.

The asymmetric angle (*A*), which is depicted graphically in Figure 33.2, is operationally defined as the angle between the asymmetry line and the midsagittal line. The *asymmetry line* is defined as the line that joins the midpoint between the inner ankle bones and the point projected on the floor directly below the midpoint of the hand grasps, as defined by the large middle knuckle. The *sagittal line* is defined as the line passing through the midpoint between the inner ankle bones and lying in the midsagittal plane, as defined by the neutral body position (i.e., hands directly in front of the body, with no twisting at the legs, torso, or shoulders). *Note:* The asymmetry angle is not defined by foot position or the angle of torso twist, but by the location of the load relative to the worker's midsagittal plane.

In many cases of asymmetric lifting, the worker will pivot or use a step turn to complete the lift. Because this may vary significantly between workers and between lifts, we have assumed that no pivoting or stepping occurs. Although this assumption may overestimate the reduction in acceptable load weight, it will provide the greatest protection for the worker.

The asymmetry angle (*A*) must always be measured at the origin of the lift. If significant control is required at the destination, however, then angle *A* should be measured at both the origin and the destination of the lift.

#### *33.4.4.2 Asymmetry Restrictions*

The angle *A* is limited to the range 0°–135°. If $A > 135°$, then AM is set equal to zero, which results in an RWL of zero, or no load.

#### *33.4.4.3 Asymmetric Multiplier*

The asymmetric multiplier (AM) is 1–0.0032A AM has a maximum value of 1.0 when the load is lifted directly in front of the body and decreases linearly as the angle of asymmetry

TABLE 33.4 Asymmetric Multiplier

| *A*(°) | AM |
|---|---|
| 0 | 1.00 |
| 15 | 0.95 |
| 30 | 0.90 |
| 45 | 0.86 |
| 60 | 0.81 |
| 75 | 0.76 |
| 90 | 0.71 |
| 105 | 0.66 |
| 120 | 0.62 |
| 135 | 0.57 |
| >135 | 0.00 |

(*A*) increases. The range is from a value of 0.57 at 135° of asymmetry to a value of 1.0 at 0° of asymmetry (i.e., symmetric lift). If *A* is greater than 135°, then AM = 0, and the load is zero. The AM value can be computed directly or determined from Table 33.4.

## 33.4.5 Frequency Component

### *33.4.5.1 Definition and Measurement*

The frequency multiplier is defined by (1) the number of lifts per minute (frequency), (2) the amount of time engaged in the lifting activity (duration), and (3) the vertical height of the lift from the floor. Lifting frequency (*F*) refers to the average number of lifts made per minute as measured over a 15 min period. Because of the potential variation in work patterns, analysts may have difficulty obtaining an accurate or representative 15 min work sample for computing *F*. If significant variation exists in the frequency of lifting over the course of the day, analysts should employ standard work sampling techniques to obtain a representative work sample for determining the number of lifts per minute. For those jobs where the frequency varies from session to session, each session should be analyzed separately, but the overall work pattern must still be considered. For more information, most standard industrial engineering or ergonomics texts provide guidance for establishing a representative job sampling strategy (e.g., Eastman Kodak Company [6]).

### *33.4.5.2 Lifting Duration*

Lifting duration is classified into three categories based on the pattern of continuous WT and recovery time (i.e., light work) periods. A continuous WT period is defined as a period of uninterrupted work. Recovery time (RT) is defined as the duration of light work activity following a period of continuous lifting. Examples of light work include activities such as sitting at a desk or table, monitoring operations, and light assembly work. The three categories are short duration, moderate duration, and long duration.

*Short duration*: Short duration lifting tasks are those that have a work duration of 1 h or less followed by a recovery time equal to 1.0 times the WT [i.e., at least a 1.0 recovery time to work time ratio (RT/WT)].*

For example, to be classified as short duration, a 45 min lifting job must be followed by at least a 45 min recovery period prior to initiating a subsequent lifting session. If the required recovery time is not met for a job of 1 h or less and a subsequent lifting session is required, then the total lifting time must be combined to correctly determine the duration category. Moreover, if the recovery period does not meet the time requirement, it is disregarded for purposes of determining the appropriate duration category.

As another example, assume that a worker lifts continuously for 30 min, then performs a light work task for 10 min, and then lifts for an additional 45 min period. In this case, the recovery time between lifting sessions (10 min) is less than 1.0 times the initial 30 min WT (30 min). Thus, the two work times (30 and 45 min) must be added together to determine the duration. Since the total WT (75 min) exceeds 1 h, the job is classified as moderate duration. On the other hand, if the recovery period between lifting sessions were increased to 30 min, then the short-duration category would apply, which would result in a larger FM value.

A special procedure has been developed for determining the appropriate lifting frequency (*F*) for certain repetitive lifting tasks in which workers do not lift continuously during the 15 min sampling period. This occurs when the work pattern is such that the worker lifts repetitively for a short time and then performs light work for a short time before starting another cycle. For work patterns such as this, *F* may be determined as follows, as long as the actual lifting frequency does not exceed 15 lifts/min:

1. Compute the total number of lifts performed for the 15 min period (i.e., lift rate times WT).
2. Divide the total number of lifts by 15.
3. Use the resulting value as the frequency (*F*) to determine the frequency multiplier (FM) from Table 33.5.

For example, if the work pattern for a job consists of a series of cyclic sessions requiring 8 min of lifting followed by 7 min of light work, and the lifting rate during the work sessions is 10 lifts/min, then the frequency rate (*F*) that is used to determine the frequency multiplier for this job is equal to $(10 \times 8)/15$, or 5.33 lifts/min. If the worker lifted continuously for more than 15 min, however, then the actual lifting frequency (10 lifts/min) would be used.

When using this special procedure, the duration category is based on the magnitude of the recovery periods *between* work sessions, not *within* work sessions. In other words, if the work pattern is intermittent and the special procedure applies, then the intermittent recovery periods that occur during the 15 min sampling period are *not* considered as recovery periods for purposes of determining the duration category. For example, if the work pattern for a manual lifting job were composed of repetitive cycles consisting of 1 min of continuous lifting at a rate

* Note that the required recovery time to work time ratio (RT/WT) for the short duration category was changed from 1.2 to 1.0 since publication of the original Applications Manual for the Revised NIOSH Lifting Equation (Waters et al. [1]).

TABLE 33.5 Frequency Multiplier (FM) Table

| | Work Duration | | | | | |
|---|---|---|---|---|---|---|
| | ≤1 h | | >1 but ≤ 2 h | | >2 but ≤ 8 h | |
| **Frequency[a] Lifts/min (*F*)** | $V^b < 30$ | $V \geq 30$ | $V < 30$ | $V \geq 30$ | $V < 30$ | $V \geq 30$ |
| ≥0.2 | 1.00 | 1.00 | 0.95 | 0.95 | 0.85 | 0.85 |
| 0.5 | 0.97 | 0.97 | 0.92 | 0.92 | 0.81 | 0.81 |
| 1 | 0.94 | 0.94 | 0.88 | 0.88 | 0.75 | 0.75 |
| 2 | 0.91 | 0.91 | 0.84 | 0.84 | 0.65 | 0.65 |
| 3 | 0.88 | 0.88 | 0.79 | 0.79 | 0.55 | 0.55 |
| 4 | 0.84 | 0.84 | 0.72 | 0.72 | 0.45 | 0.45 |
| 5 | 0.80 | 0.80 | 0.60 | 0.60 | 0.35 | 0.35 |
| 6 | 0.75 | 0.75 | 0.50 | 0.50 | 0.27 | 0.27 |
| 7 | 0.70 | 0.70 | 0.42 | 0.42 | 0.22 | 0.22 |
| 8 | 0.60 | 0.60 | 0.35 | 0.35 | 0.18 | 0.18 |
| 9 | 0.52 | 0.52 | 0.30 | 0.30 | 0.00 | 0.15 |
| 10 | 0.45 | 0.45 | 0.26 | 0.26 | 0.00 | 0.13 |
| 11 | 0.41 | 0.41 | 0.00 | 0.23 | 0.00 | 0.00 |
| 12 | 0.37 | 0.37 | 0.00 | 0.21 | 0.00 | 0.00 |
| 13 | 0.00 | 0.34 | 0.00 | 0.00 | 0.00 | 0.00 |
| 14 | 0.00 | 0.31 | 0.00 | 0.00 | 0.00 | 0.00 |
| 15 | 0.00 | 0.28 | 0.00 | 0.00 | 0.00 | 0.00 |
| >15 | 0.00 | 0.00 | 0.00 | 0.00 | 0.00 | 0.00 |

[a] For lifting less frequently than once per 5 min, set $F = 0.2$ lift/min.
[b] $V$ is expressed in inches as measured from the floor.

of 10 lifts/min, followed by 2 min of recovery, then the correct procedure would be to adjust the frequency according to the special procedure [i.e., $F = (10 \text{ lifts/min} \times 5\text{ min})/15\text{ min} = 50/15 = 3.4$ lifts/min]. The 2 min recovery periods would not count toward the RT/WT ratio, however, and additional recovery periods would have to be provided as described previously.

*Moderate duration*: Moderate duration lifting tasks are those that have a duration of more than 1 h but not more than 2 h, followed by a recovery period of at least 0.3 times the WT [i.e., at least a 0.3 recovery time to work time ratio (RT/WT)].

For example, if a worker continuously lifts for 2 h, then a recovery period of at least 36 min would be required before initiating a subsequent lifting session. If the recovery time requirement is not met and a subsequent lifting session is required, then the total work time must be added together. If the total work time exceeds 2 h, then the job must be classified as a long duration lifting task.

*Long duration*: Long duration lifting tasks are defined as those that have a duration of 2–8 h, with standard industrial rest allowances (e.g., morning, lunch, and afternoon rest breaks). *Note*: No weight limits are provided for more than 8 h of work.

The difference in the required RT/WT ratio for the short (<1 h) duration category, which is 1.2, and the moderate (1–2 h) duration category, which is 0.3, is due to the difference in the magnitudes of the frequency multiplier values associated with each of the duration categories. Since the moderate category results in larger reductions in the RWL than the short category, there is less need for a recovery period between sessions than for the short duration category. In other words, the short duration category would result in higher weight limits than the moderate duration category, so larger recovery periods would be needed.

#### *33.4.5.3 Frequency Restrictions*

Lifting frequency (*F*) for repetitive lifting may range from 0.2 lift/min to a maximum frequency that is dependent on the vertical location (*V*) of the object and the duration of lifting (Table 33.5). Lifting above the maximum frequency results in RWL = 0.0 (except for the special care of discontinuous lifting discussed earlier, where the maximum frequency is 15 lifts/min).

#### *33.4.5.4 Frequency Multiplier*

The FM value depends upon the average number of lifts per minute (*F*), the vertical location (*V*) of the hands at the origin, and the duration of continuous lifting. For lifting tasks with a frequency less than 0.2 lifts/min, set the frequency equal to 0.2 lifts/min. Otherwise, FM is determined from Table 33.5.

### 33.4.6 Coupling Component

#### *33.4.6.1 Definition and Measurement*

The nature of the hand-to-object coupling or gripping method can affect not only the maximum force a worker can or must exert on the object, but also the vertical location of the hands during the lift. A "good" coupling will reduce the maximum grasp forces required and increase the acceptable weight for lifting, whereas a "poor" coupling will generally require higher maximum grasp forces and decrease the acceptable weight for lifting.

The effectiveness of the coupling is not static but may vary with the distance of the object from the ground, so that a good coupling could become a poor coupling during a single lift. The entire range of the lift should be considered when classifying hand-to-object couplings, with classification based on overall effectiveness. The analyst must classify the coupling as good, fair, or poor. The three categories are defined in Table 33.6. If there is any doubt about classifying a particular coupling design, the more stressful classification should be selected.

The decision tree shown in Figure 33.3 may be helpful in classifying the hand-to-object coupling.

#### *33.4.6.2 Coupling Multiplier*

Based on the coupling classification and vertical location of the lift, the coupling multiplier (CM) is determined from Table 33.7.

## 33.5 PROCEDURES

Prior to data collection, the analyst must decide (1) if the job should be analyzed as a single-task, multitask, sequential, or variable manual lifting job* and (2) if significant control is required at the destination of the lifts. This is necessary because the procedures differ according to the type of analysis required.

* Note, the sequential and variable methods have only recently been developed and were not published in the Applications Manual for the Revised NIOSH Lifting Equation (Waters et al. [1]).

TABLE 33.6 Hand-to-Container Coupling Classification

| Good | Fair | Poor |
|---|---|---|
| 1. For containers of optimal design, such as some boxes, crates, etc., a "good" hand-to-object coupling would be defined as handles or handhold cutouts of optimal design (see Notes 1–3). | 1. For containers of optimal design, a "fair" hand-to-object coupling would be defined as handles or hand-hold cutouts of less than optimal design (see Notes 1–4). | 1. Containers of less than optimal design or loose parts or irregular objects that are bulky, hard to handle, or have sharp edges (see Note 5). |
| 2. For loose parts or irregular objects, which are not usually containerized, such as castings, stock, and supply materials, a "good" hand-to-object coupling would be defined as a comfortable grip in which the hand can be easily wrapped around the object (see Note 6). | 2. For containers of optimal design with no handles or handhold cutouts or for loose parts or irregular objects, a "fair" hand-to-object coupling is defined as a grip in which the hand can be flexed about 90° (see Note 4). | 2. Lifting nonrigid bags (i.e., bags that sag in the middle). |

*Notes:*

1. An optimal handle design has 0.75–1.5 in. (1.9–3.8 cm) diameter, ≥4.5 in. (11.5 cm) length, 2 in. (5 cm) clearance, cylindrical shape, and a smooth, nonslip surface.
2. An optimal handhold cutout has the following approximate characteristics: ≥1.5 in. (3.8 cm) height, 4.5 in. (11.5 cm) length, semi-oval shape, ≥2 in. (5 cm) clearance, smooth nonslip surface, and ≥0.25 in. (0.60 cm) container thickness (e.g., double thickness cardboard).
3. An optimal container design has ≤16 in. (40 cm) frontal length, ≤12 in. (30 cm) height, and a smooth nonslip surface.
4. A worker should be capable of clamping the fingers at nearly 90° under the container, such as required when lifting a cardboard box from the floor.
5. A container is considered less than optimal if it has a frontal length > 16 in. (40 cm), height > 12 in. (30 cm), rough or slippery surfaces, sharp edges, asymmetric center of mass, unstable contents, or requires the use of gloves.
6. A worker should be able to comfortably wrap the hand around the object without causing excessive wrist deviations or awkward postures, and the grip should not require excessive force.

Definitions of the four types of manual lifting jobs are provided hereafter:

### 33.5.1 Single-Task Job

A manual lifting job is defined as a single-task job if the task variables do not differ from task to task or if only one task is of interest (e.g., single most stressful task). This may be the case if one of the tasks clearly has a dominant effect on strength demands, localized muscle fatigue, or whole body fatigue.

### 33.5.2 Multitask Job

A manual lifting job is defined as a multitask job if there are a distinct set of defined lifting tasks with a fixed set of task characteristics, but the task characteristics differ significantly between tasks. An example of a multitask job is a palletizing job in which each layer of the pallet can be treated as a unique set of lifts with fixed task characteristics. A multitask analysis is more difficult to perform than a single-task analysis because additional data and computations are required. The multitask approach, however, will provide more detailed information about specific strength and physiological demands.

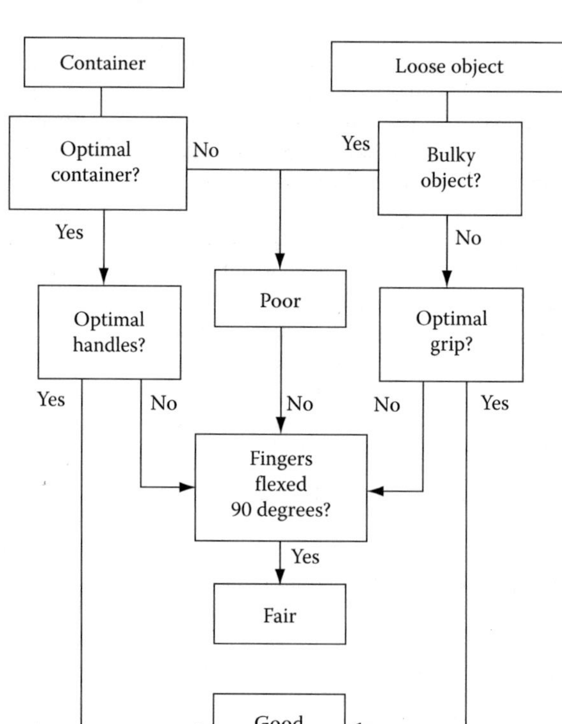


FIGURE 33.3 Decision tree for coupling quality.

TABLE 33.7 Coupling Multiplier

| | **Coupling Multiplier** | |
|---|---|---|
| **Coupling Type** | $V < 30$ **in. (75 cm)** | $V \geq 30$ **in. (75 cm)** |
| Good | 1.00 | 1.00 |
| Fair | 0.95 | 1.00 |
| Poor | 0.90 | 0.90 |

## 33.5.3 Sequential Lifting Job

Recently, NIOSH researchers developed the sequential lifting index (SLI), a new method for using the RNLE to assess jobs in which workers rotate between work stations [7]. When workers rotate between workstations and perform different lifting tasks at each work station, the SLI should provide a useful metric for determining which tasks and rotation slots are having the most effect on the physical demand for the job.

## 33.5.4 Variable Lifting Job

A team of researchers also recently developed the variable lifting index (VLI), a new method for using the RNLE to assess jobs in which the task characteristics frequently vary between lifts, such as might be seen in a warehouse or distribution center [8]. The VLI approach is similar to the CLI described previously, but for the VLI, the task characteristics can vary from lift to lift, whereas the CLI is restricted to a small set of defined task conditions.

For many lifting jobs, it may be acceptable to use any of the four approaches (i.e., either the single-task, multitask, sequential, or variable approach). The analyst should choose the method most appropriate for the job, keeping in mind that each type of analysis requires different amounts of data collection. The single-task analysis is the easiest to use, but when a job consists of more than one task and detailed information is needed to specify engineering modifications, then the more advanced procedures (multitask, sequential, or variable methods) provide a reasonable method of assessing the overall physical demands may be needed. The more advanced procedures are more complicated to use than the single-task procedure and require a greater understanding of assessment terminology and mathematical concepts. Therefore, the decision to use one of the approaches should be based on (1) the need for detailed information about all facets of the multitask lifting job, (2) the need for accuracy and completeness of data regarding assessment of the physiological demands of the task, and (3) the analyst's level of understanding of the assessment procedures.

The decision about whether control is required at the destination of the lift or not is important because the physical demands on the worker may be greater at the destination of the lift than at the origin, especially when significant control is required. When significant control is required at the destination, the physical stress is increased because the load will have to be accelerated upward to slow its descent. This acceleration may be as great as the acceleration at the origin of the lift and may create high loads on the spine. Therefore, if significant control is required, then the RWL and LI should be determined at both locations and the lower of the two values used to specify the overall level of physical demand.

To perform a lifting analysis using the RNLE, two steps are taken: (1) data are collected at the worksite as described in step 1 hereafter, and (2) the RWL and LI values (LI, CLI, SLI, or VLI) are computed using one of the four analysis procedures described in step 2 as follows.

*Step 1: Collect data*

The relevant task variables must be carefully measured and clearly recorded in a concise format. As mentioned previously, these variables include the horizontal location of the hands (*H*), vertical location of the hands, (*V*), vertical displacement (*D*), asymmetric angle (*A*), lifting frequency (*F*), and coupling quality (*C*). A job analysis worksheet, as shown in Figure 33.4 for single-task jobs or Figure 33.5 for multitask jobs, provides a simple form for recording the task variables and the data needed to calculate the RWL and LI values. A thorough job analysis is required to identify and catalog each independent lifting task in the worker's complete job. For multitask jobs, data must be collected for each task.

*Step 2: Calculate the LI, CLI, SLI, or VLI for the job*

#### *33.5.4.1 Lifting Index for Single-Task Jobs*

For a single-task procedure, step 2 consists of computing the RWL and the LI. This is accomplished as follows.

Calculate the RWL at the origin for each lift. For lifting tasks that require significant control at the destination, calculate the RWL at *both* the origin and the destination of the lift. The latter procedure is required if (1) the worker has to regrasp the load near the

**JOB ANALYSIS WORKSHEET**

| | | | |
|---|---|---|---|
| **DEPARTMENT** | ____________ | **JOB DESCRIPTION** | |
| **JOB TITLE** | ____________ | | ____________ |
| **ANALYST'S NAME** | ____________ | | ____________ |
| **DATE** | ____________ | | ____________ |

**STEP 1. Measure and record task variables**

| Object Weight (lb) | | Hand Location (in.) | | | | Vertical Distance (in.) | Asymmetric Angle (degrees) | | Frequency Rate lifts/min | Duration (h) | Object Coupling |
|---|---|---|---|---|---|---|---|---|---|---|---|
| | | Origin | | Destination | | | Origin | Destination | | | |
| *L* (AVG.) | *L* (Max.) | *H* | *V* | *H* | *V* | *D* | *A* | *A* | *F* | | *C* |
| | | | | | | | | | | | |

**STEP 2. Determine the multipliers and compute the RWL's**

**RWL** = **LC** × **HM** × **VM** × **DM** × **AM** × **FM** × **CM**

**ORIGIN** **RWL** = 51 × ☐ × ☐ × ☐ × ☐ × ☐ × ☐ = ☐ **lb**

**DESTINATION** **RWL** = 51 × ☐ × ☐ × ☐ × ☐ × ☐ × ☐ = ☐ **lb**

**STEP 3. Compute the LIFTING INDEX**

**ORIGIN** LIFTING INDEX = $\frac{\text{OBJECT WEIGHT (L)}}{\text{RWL}}$ = ______ = ☐

**DESTINATION** LIFTING INDEX = $\frac{\text{OBJECT WEIGHT (L)}}{\text{RWL}}$ = ______ = ☐

FIGURE 33.4 Single-task job analysis worksheet.

destination of the lift, (2) the worker has to momentarily hold the object at the destination, or (3) the worker has to position or guide the load at the destination. The purpose of calculating the RWL at both the origin and destination of the lift is to identify the most stressful location of the lift. Therefore, the lower of the two RWL values should be used to compute LI for the task, as this value would represent the limiting set of conditions.

The assessment is completed on the single-task worksheet by determining the LI for the task of interest. This is accomplished by comparing the actual weight of the load (*L*) lifted with the RWL value obtained from the lifting equation.

#### *33.5.4.2 Composite Lifting Index for Multitask Jobs*

For a multitask procedure, step 2 comprises three substeps:

1. Compute the frequency-independent recommended weight limit (FIRWL) and single-task recommended weight limit (STRWL) for each task.
2. Compute the frequency-independent lifting index (FILI) and single-task lifting index (STLI) for each task.
3. Compute the composite lifting index (CLI) for the overall job.

*Compute the frequency-independent recommended weight limits* (FIRWLs): Compute the FIRWL value for each task by using the respective task variables and setting the

**MULTI-TASK JOB ANALYSIS WORKSHEET**

| | | |
|---|---|---|
| DEPARTMENT | ________ | JOB DESCRIPTION |
| JOB TITLE | ________ | ________ |
| ANALYST'S NAME | ________ | ________ |
| DATE | ________ | ________ |

**STEP 1. Measure and Record Task Variable Data**

| Task No. | Object Weight (lb) | | Hand Location (in.) | | | | Vertical Distance (in.) | Asymmetry Angle (degrees) | | Frequency Rate lifts/min | Duration (h) | Coupling |
|---|---|---|---|---|---|---|---|---|---|---|---|---|
| | | | Origin | | Destination | | | Origin | Destination | | | |
| | $L$ (Avg.) | $L$ (Max.) | $H$ | $V$ | $H$ | $V$ | $D$ | $A$ | $A$ | $F$ | | $C$ |
| | | | | | | | | | | | | |
| | | | | | | | | | | | | |
| | | | | | | | | | | | | |
| | | | | | | | | | | | | |
| | | | | | | | | | | | | |

**STEP 2. Compute multipliers and FIRWL, STRWL, FILI, and STLI for Each Task**

| Task No. | LC × | HM × | VM × | DM × | AM × | CM | FIRWL × | FM | STRWL | FILI = L/FIRWL | STLI = L/STRWL | New Task No. | F |
|---|---|---|---|---|---|---|---|---|---|---|---|---|---|
| | 51 | | | | | | | | | | | | |
| | 51 | | | | | | | | | | | | |
| | 51 | | | | | | | | | | | | |
| | 51 | | | | | | | | | | | | |
| | 51 | | | | | | | | | | | | |

**STEP 3. Compute the Composite Lifting Index for the Job (After renumbering tasks)**

| | | | | | |
|---|---|---|---|---|---|
| CLI = | $STLI_1$ + | $\Delta FILI_2$ + | $\Delta FILI_3$ + | $\Delta FILI_4$ + | $\Delta FILI_5$ |
| | | $FILI_2(1\backslash FM_{1,2} - 1\backslash FM_1)$ | $FILI_2(1\backslash FM_{1,2,3} - 1\backslash FM_{1,2})$ | $FILI_2(1\backslash FM_{1,2,3,4} - 1\backslash FM_{1,2,3})$ | $FILI_2(1\backslash FM_{1,2,3,4,5} - 1\backslash FM_{1,2,3,4})$ |
| CLI = | | | | | |

FIGURE 33.5 Multitask job analysis worksheet.

frequency multiplier (FM) to a value of 1.0. The FIRWL for each task reflects the compressive force and muscle strength demands for a single performance of that task. If significant control is required at the destination for any individual task, the FIRWL must be computed at both the origin and the destination of the lift, as described earlier for a single-task analysis.

*Compute the single-task recommended weight limit* (STRWL): Compute the STRWL for each task by multiplying its FIRWL by the appropriate FM. The STRWL for a task reflects the overall demands of that task, assuming it was the only task being performed. *Note*: This value does not reflect the overall demands of the task when the other tasks are considered. Nevertheless, it is helpful in determining the extent of excessive physical stress for an individual task.

*Compute the frequency-independent lifting index* (FILI): The FILI is computed for each task by dividing the *maximum* load weight ($L$) for that task by the respective FIRWL. The maximum weight is used to compute the FILI because the maximum weight determines the maximum biomechanical loads to which the body will be exposed, regardless of the frequency of occurrence. Thus, the FILI can identify individual tasks with potential strength problems for infrequent lifts. If any of the FILI values exceeds a value of 1.0, then job design changes may be needed to decrease the strength demands.

*Compute the single-task lifting index* (STLI): The STLI is computed for each task by dividing the *average* load weight (*L*) for that task by the respective STRWL. The average weight is used to compute the STLI because the average weight provides a better representation of the metabolic demands, which are distributed across the tasks rather than being dependent on individual tasks. The STLI can be used to identify individual tasks with excessive physical demands (i.e., tasks that would result in fatigue). The STLI values do not indicate the relative stress of the individual tasks in the context of the whole job, but they can be used to prioritize the individual tasks according to the magnitude of their physical stress. Thus, if any of the STLI values exceeds a value of 1.0, then ergonomic changes may be needed to decrease the overall physical demands of the task. *Note*: It may be possible to have a job in which all of the individual tasks have an STLI less than 1.0 and yet is physically demanding due to the combined demands of the tasks. In cases where the FILI exceeds the STLI for any task, the maximum weights may represent a significant problem, and careful evaluation is necessary.

*Compute the composite lifting index* (CLI): The assessment is completed on the multitask worksheet by determining the CLI for the overall job. The CLI is computed as follows:

1. The tasks are renumbered in order of decreasing physical stress, from the task with the greatest STLI down to the task with the smallest STLI. The tasks are renumbered in this way so that the more difficult tasks are considered first.
2. The CLI for the job is then computed according to the formula

$$\text{CLI} = \text{STLI}_1 + \sum \Delta\text{LI}$$

where

$$\sum \Delta\text{LI} = \text{FILI}_2 \times \left(\frac{1}{\text{FM}_{1,2}} - \frac{1}{\text{FM}_1}\right) + \left[\text{FILI}_3 \times \left(\frac{1}{\text{FM}_{1,2,3}} - \frac{1}{\text{FM}_{1,2}}\right)\right]$$
$$+ \left[\text{FILI} \times \left(\frac{1}{\text{FM}_{1,2,3,4}} - \frac{1}{\text{FM}_{1,2,3}}\right)\right] + \cdots + \left[\text{FILI}_n \times \left(\frac{1}{\text{FM}_{1,2,3,4,\ldots,n}} - \frac{1}{\text{FM}_{1,2,3,\ldots,(n-1)}}\right)\right]$$

*Note*: (1) The numbers in the subscripts refer to the new task numbers and (2) the FM values are determined from Table 33.5, based on the sum of the frequencies for the tasks listed in the subscripts.

*An example*: The following example is provided to demonstrate this step of the multitask procedure. Assume that an analysis of a typical three-task job provided the results shown in Table 33.8.

To compute the CLI for this job, the tasks are renumbered in order of decreasing physical stress, beginning with the task with the greatest STLI. In this case, as shown in Table 33.8,

TABLE 33.8 Computations from Multitask Example

| Task No. | Load Weight (*L*) | Task Frequency (*F*) | FIRWL | FM | STRWL | FILI | STLI | New Task No. |
|---|---|---|---|---|---|---|---|---|
| 1 | 30 | 1 | 20 | 0.94 | 18.8 | 1.5 | 1.6 | 1 |
| 2 | 20 | 2 | 20 | 0.91 | 18.2 | 1.0 | 1.1 | 2 |
| 3 | 10 | 4 | 15 | 0.84 | 12.6 | 0.67 | 0.8 | 3 |

| **2H** | | **1H** | **1H** | **2H** | | **1H** | **2H** | |
|---|---|---|---|---|---|---|---|---|
| A | A | B | R | A | A | B | C | C |
| | MMH | | LUNCH | | MMH | | RECOV | |

**CASE 1**—Sequence AABC. Assuming each letter represents 1 h, the SLI for this job would be calculated as follows:

| Task | LC × HM × VM × DM × AM × CM | F Lifts/min | FIRWL | Load Wt | FM Actual Duration | LI | FM Total Duration | LImax | TF | Order by LImax |
|---|---|---|---|---|---|---|---|---|---|---|
| A | 23 × 0.83 × 0.78 × 0.88 × 0.80 × 0.90 | 8 | 9.43 | 10 | 0.35 | 3.03 | 0.18 | 5.89 | 0.46 | 1 |
| B | 23 × 1.00 × 0.93 × 0.93 × 0.88 × 0.90 | 4 | 15.76 | 15 | 0.84 | 1.13 | 0.45 | 2.11 | 0.23 | 2 |

**CASE 2**—Sequence ACBB. The SLI would be calculated as follows:

| Task | LC × HM × VM × DM × AM × CM | F Lifts/min | FIRWL | Load Wt | FM Actual Duration | LI | FM Total Duration | LImax | TF | Order by LImax |
|---|---|---|---|---|---|---|---|---|---|---|
| A | 23 × 0.83 × 0.78 × 0.88 × 0.80 × 0.90 | 8 | 9.43 | 10 | 0.60 | 1.77 | 0.35 | 3.03 | 0.23 | 1 |
| B | 23 × 1.00 × 0.93 × 0.93 × 0.88 × 0.90 | 4 | 15.76 | 15 | 0.72 | 1.32 | 0.72 | 1.32 | 0.46 | 2 |

**CASE 3**—Sequence AABB. The SLI would be calculated as follows:

| Task | LC × HM × VM × DM × AM × CM | F Lifts/min | FIRWL | Load Wt | FM Actual Duration | LI | FM Total Duration | LImax | TF | Order by LImax |
|---|---|---|---|---|---|---|---|---|---|---|
| A | 23 × 0.83 × 0.78 × 0.88 × 0.80 × 0.90 | 8 | 9.43 | 10 | 0.35 | 3.03 | 0.18 | 5.89 | 0.46 | 1 |
| B | 23 × 1.00 × 0.93 × 0.93 × 0.88 × 0.90 | 4 | 15.76 | 15 | 0.72 | 1.32 | 0.45 | 2.11 | 0.46 | 2 |

the task numbers do not change. Next, the CLI is computed according to the formula given earlier. The task with the greatest CLI is Task 1 (STLI = 1.6). The sum of the frequencies for Tasks 1 and 2 is 1 + 2, or 3, and the sum of the frequencies for Tasks 1, 2, and 3 is 1 + 2 + 4, or 7. Then, from Table 33.5, $FM_1 = 0.94$, $FM_{1,2} = 0.88$, and $FM_{1,2,3} = 0.70$. Finally,

$$CLI = 1.6 + 1.0(1/0.88 - 1/0.94) + 0.67(1/0.70) - 1/0.88 = 1.6 + 0.7 + 0.20 = 1.9$$

Note that the FM values were based on the sum of the frequencies for the subscripts, the vertical height, and the duration of lifting.

#### 33.5.4.3 Sequential Lifting Index

An example of a sequential lifting job would be one in which a worker lifts one product at the end of an assembly line for a fixed period of time (e.g., 1 h), and then changes products for a fixed period of time, and so on. Another example of a sequential lifting job would be one in which a worker performs a series of lifts at one workstation (e.g., a palletizing job) and then rotates to another workstation and performs a different set of lifting tasks (e.g., a different palletizing job). In such situations, the existing LI and CLI calculations cannot be used to estimate the physical demand of the job. For these types of jobs, each rotation position will be considered to be a job element.

The following steps are required to calculate the SLI for a job:

1. The work pattern or rotation pattern should be documented on a timeline of the shift, as shown in the following example. The pattern should show how the worker rotates among the different work stations or tasks. In the following example, A and B are manual lifting tasks, R represents a work break for lunch, and C represents a light duty task (i.e., a task without lifting).
2. When using this method, you must assume that no manual handling task (single or multiple) can last more than 4 h continuously without a recovery period (such as light duty or lunch). In cases where the total work time exceeds 8 h, you can calculate the SLI value for each 4 h period and then take the larger of the values to represent the SLI for the job as a whole.
3. Calculate the LI or CLI for each unique task category, as described by Waters et al. [2]. When a single task is performed for more than one continuous hour, the LI value for that task must be calculated using the duration category for that period, but when two identical tasks are separated by another task, then each category is calculated using its own duration. Considering the preceding example, if the order is ABAB, then the LI for each letter would be calculated using the short duration (1 h category) frequency multiplier (FM). If the sequence was AABB, however, then the FM for each would be calculated using the moderate duration (1–2 h category) FM. Similarly, for the sequence AAAB, the FM for each would be calculated using the long duration (8 h category).
4. For each task category, calculate the maximum lifting index (LImax), by taking the frequency multiplier as being relevant to task frequency related to the overall duration obtained by adding together the single-task duration times. For example, in the aforementioned sequences, the LImax for each task would be calculated using the long duration category since the total continuous work time exceeds the 2 h category. Once the LImax values are calculated for each task category, the tasks should be reordered from highest to lowest.

5. Calculate the time fraction (TF) for each task category by dividing the task duration in minutes by 240 min (e.g., for AABC, the TFA = 120/240 = 0.50 and TFB = 60/240 = 0.25).

6. Following the reordering of tasks, as already noted, determine the LI value for the new Task 1 (i.e., the task with the largest LImax) using its actual task duration. This is calculated by dividing the load weight of the task by the product of the FIRWL and the actual task duration FM value for that task. This value is then set equal to LI1. For example, assume that a job consists of two tasks and the task with the largest LImax has an LImax of 3.2, an FIRWL of 12.0 lb, a load weight of 18.0 lb, and an actual FM value of 0.6; the LI for this task would be 18.0/12.0 × 0.6 or 2.5. In this case, LI1 would be set equal to 2.5.

7. Calculate the SLI using the following formula:

$$\mathrm{SLI} = \mathrm{LI1} + (\mathrm{LImax} - \mathrm{LI1}) \times \mathrm{K}$$

$$\text{where } K = \frac{\sum \mathrm{LImax1} \times \mathrm{TF1} + \mathrm{LImax2} \times \mathrm{TF2} + \cdots + \mathrm{LImaxn} \times \mathrm{TFn}}{\mathrm{LImax1}}$$

Please note that

1. LImax1 is the highest LImax, and LI1 is its corresponding LI. It may be that LI1 is not the highest in absolute terms, but this does not bias the calculations that follow.

2. The time fraction (TF) is calculated on 240 min because in most actual working environments no continuous lifting task usually take place for periods longer than 4 h. Moreover, the use of 240 min as denominator in the calculation of TF enables a better evaluation of the SLI.

3. K is a weighting factor calculated by multiplying the LImax of each task by the corresponding TF, adding together the resulting values and then dividing them by the LImax1.

4. If the work pattern for an overall job is such that there are sufficient recovery periods to insure that the job always stays in the short duration category (i.e., sufficient recovery means that the duration of any task category is 1 h or less followed by a recovery period of equal duration), then the SLI is equal to the greatest of all LImax or CLImax. In this special case, the LImax1 becomes LI1, which cancels out the term (LImax1-LI) in the formula. Therefore, the SLI is equal to LI1. Since the LI1 is set to be the greatest LI among all tasks, the SLI for this job would be set to the greatest of the LI values for all the tasks. For example, if the job rotation is something like ACBC, from the previous example, where C is a sufficient recovery period for tasks A and B, then the LI for the job would be either the LI for Task A or Task B, whichever is greatest.

**Example**

Assume a job consists of sequential Tasks A and B, and a recovery or light duty Task C. The SLI is calculated for various sequences shown hereafter.

*Case 1*—Sequence AABC. Assuming each letter represents 1 h, the SLI for this job would be calculated as follows:

$$\text{SLI} = \text{LI1} + ((\text{LImax1} - \text{LI1}) \times K$$

$$\text{SLI} = \text{LI1} + (\text{LImax1} - \text{LI1}) \times \frac{\text{LImax1} \times \text{TF1} + \text{LImax2} \times \text{TF2}}{\text{LImax1}}$$

$$\text{SLI} = 3.03 + (5.89 - 3.03) \times \frac{5.89 \times 0.46 + 2.11 \times 0.23}{5.89}$$

$$\text{SLI} = 4.58$$

*Case 2*—Sequence ACBB. The SLI would be calculated as follows:

$$\text{SLI} = \text{LI1} + ((\text{LImax1} - \text{LI1}) \times K$$

$$\text{SLI} = \text{LI1} + (\text{LImax1} - \text{LI1}) \times \frac{\text{LImax1} \times \text{TF1} + \text{LImax2} \times \text{TF2}}{\text{LImax1}}$$

$$\text{SLI} = 1.77 + (5.89 - 1.77) \times \frac{5.89 \times 0.23 + 2.11 \times 0.46}{5.89}$$

$$\text{SLI} = 3.40$$

*Case 3*—Sequence AABB. The SLI would be calculated as follows:

$$\text{SLI} = \text{LI1} + ((\text{LImax1} - \text{LI1}) \times K$$

$$\text{SLI} = \text{LI1} + (\text{LImax1} - \text{LI1}) \times \frac{\text{LImax1} \times \text{TF1} + \text{LImax2} \times \text{TF2}}{\text{LImax1}}$$

$$\text{SLI} = 3.03 + (5.89 - 3.03) \times \frac{5.89 \times 0.46 + 2.11 \times 0.46}{5.89}$$

$$\text{SLI} = 4.82$$

Remember, that when a task is performed for more than 1 h, the FM value for the actual time period is used. That is, when AA was performed in the Case 1 mentioned earlier, the FMactual was 0.35, whereas in the Case 2 it was 0.6. Similarly, in Case 1, the FMactual for B was 0.91, whereas for Case 2 it was 0.84. Based on our assessment of the

SLI response across different sequence patterns, it appears to be sensitive to variations in the sequence of tasks across the work shift.

#### 33.5.4.4 *Variable Lifting Index*

The VLI, which is equivalent to the LI, CLI, or SLI for single-lifting, multi-lifting, or sequential lifting jobs, is computed using "probability data" collected at the worksite as input into the VLI equation. The input data for the VLI calculation will be obtained at the worksite through adjustable sampling methods of actual jobs, use of historical computerized production data obtained from the employer, when available, or some combination of the two sources of data. The sampling methods will be adjusted based upon the amount of variability observed in the task characteristics, such as the weight of load lifted, horizontal distance, asymmetry, etc. The greater the variability between lifts, the greater will be the requirement for data sampling.

The concept for the method is similar to the CLI method for multitask jobs. The difference is that rather than using individual task elements, all of the lifts will be distributed into a fixed number of FILI categories (one to nine categories), each with a variable frequency. These six FILI categories will then be weighted using the CLI equation. The frequency multiplier for each category is based on the average overall frequency for the six individual LI categories. The VLI should provide a reasonable estimate of the physical demand of the job that can be used to determine if the task is acceptable or not and how changes in the mix of tasks might affect the overall physical demand of the job. The steps are as follows:

1. Determine the range of FILI values for all of the sampled lifts.
2. Divide the range of FILI values into six categories, taking into account the variability of obtained results.
3. Determine the frequency of lifts in each of the six categories.
4. Apply the VLI using the CLI equation, but use the frequency data for each LI category to calculate the appropriate FM values for the calculation.

The VLI is computed as follows:

1. The task categories are renumbered in order of decreasing physical stress, beginning with the task category with the greatest single-task lifting index (STLI) down to the task category with the smallest STLI. The STLI is the defined as the LI value for each task, independent of the other tasks. The task categories are renumbered in this way so that the more difficult task categories are considered first.
2. The VLI for the job is then computed according to the following formula:

$$\text{VLI} = \text{STLI}_1 + \Sigma\,\Delta\text{LI}$$

where

$$\Sigma \Delta LI = \left( FILI_2 \times \left( \frac{1}{FM_{1,2}} - \frac{1}{FM_1} \right) \right)$$
$$+ \left( FILI_3 \times \left( \frac{1}{FM_{1,2,3}} - \frac{1}{FM_{1,2}} \right) \right)$$
$$+ \left( FILI_4 \times \left( \frac{1}{FM_{1,2,3,4}} - \frac{1}{FM_{1,2,3}} \right) \right)$$
$$\vdots$$
$$+ \left( FILI_n \times \left( \frac{1}{FM_{1,2,3,4,\ldots,n}} - \frac{1}{FM_{1,2,3,\ldots,(n-1)}} \right) \right)$$

Note that (1) the numbers in the subscripts refer to the new task category numbers, and (2) the FM values are determined from the frequency table published in the applications manual (Waters et al. [2]). The appropriate FM values are based on the sum of the frequencies for the task categories listed in the subscripts.

**Example**

A hypothetical example will demonstrate how the VLI equation might be applied. Assume that job sampling at a manufacturing plant revealed that the largest FILI sampled for any individual lift was 2.8. According to the VLI procedure, the range of FILI categories should be evaluated and a set of FILI categories should be chosen. For this example, six FILI categories were chosen and were defined as: 0–0.45, 0.46–0.61, 0.62–0.99, 1.0–1.51, 1.52–2.02, and 2.03–2.8. The choice of FILI categories is somewhat arbitrary, but we suggest choosing six categories. Also, assume that analysis of the sampled data revealed that the average FILI for individual lifts in each of these six categories and the percentage of tasks falling into the six cells are as shown in Table 33.9. As with the SLI approach, the VLI approach works best if the job is performed for a full 8 h shift [7]. If the overall frequency of lifting across an 8 h shift is 4/min, then the frequency of lifts for each category can be calculated (see Table 33.9).

TABLE 33.9 Hypothetical Data for VLI Example

| | LI Categories | | | | | |
|---|---|---|---|---|---|---|
| **Category Data** | **0–0.45** | **0.46–0.61** | **0.62–0.99** | **1.0–1.51** | **1.52–2.02** | **2.03–2.8** |
| Representative FILI within category | 0.33 | 0.53 | 0.79 | 1.20 | 1.66 | 2.8 |
| Renumbered | 6 | 5 | 4 | 3 | 2 | 1 |
| Percentage of tasks | 10% | 15% | 25% | 25% | 15% | 10% |
| Frequency (lifts/min) | 0.5 | 0.9 | 1.1 | 0.9 | 0.4 | 0.2 |

Based on the hypothetical data presented, the VLI for this job can be calculated, as follows:

$$\text{VLI} = \text{STLI}_1 + \sum \Delta\text{LI}$$

$$\text{STLI}_1 = 3.32$$

$$\Delta\text{FILI}_2 = 1.66((1/.80) - (1/.85)) = .116$$

$$\Delta\text{FILI}_3 = 1.2((1/.70) - (1/.80)) = .204$$

$$\Delta\text{FILI}_4 = .79((1/.60) - (1/.70)) = .205$$

$$\Delta\text{FILI}_5 = .53((1/.50) - (1/.60)) = .159$$

$$\Delta\text{FILI}_6 = .33((1/.46) - (1/.50)) = .069$$

$$\text{VLI} = \text{STLI}_1 + \Delta\text{FILI}_2 + \Delta\text{FILI}_3 + \Delta\text{FILI}_4 + \Delta\text{FILI}_5 + \Delta\text{FILI}_6$$

$$= 3.32 + .116 + .204 + .205 + .159 + .069 = 4.07$$

## 33.6 APPLYING THE EQUATIONS

### 33.6.1 Using RWL and LI, CLI, SLI, and VLI to Guide Ergonomic Design

The RWL and LI values (LI, CLI, SLI, or VLI) can be used to guide ergonomic design in several ways.

1. The individual multipliers can be used to identify specific job-related problems. The relative magnitude of each multiplier indicates the relative contribution of each task factor (e.g., horizontal, vertical, frequency).
2. The RWL can be used to guide the redesign of existing manual lifting jobs or to design new manual lifting jobs. For example, if the task variables are fixed, then the maximum weight of the load could be selected so as not to exceed the RWL; if the weight is fixed, then the task variables could be optimized so as not to exceed the RWL.
3. The LI values can be used to estimate the relative magnitude of physical stress for a task or job. The greater the LI value, the smaller the fraction of workers capable of safely sustaining the level of activity. Thus, two or more job designs could be compared.
4. The LI values can be used to prioritize ergonomic redesign. For example, a series of suspected hazardous jobs could be rank ordered according to LI values, and a control strategy could be developed according to the rank ordering (i.e., jobs with lifting indices above 1.0 or higher would benefit the most from redesign).

### 33.6.2 Rationale and Limitations for Lifting Index Values (LI, CLI, SLI, and VLI)

The NIOSH RWL equation and LI are based on the concept that the risk of lifting-related LBP increases as the demands of the lifting task increase. In other words, as the magnitude of the LI increases, (1) the level of the risk for a given worker would be increased and (2) a greater percentage of the workforce is likely to be at risk for developing lifting-related LBP. The shape of the risk function, however, is not known. Without additional data showing the relationship between LBP and LI, it is impossible to predict the magnitude of the risk for a given individual or the exact percent of the work population who would be at an elevated risk for LBP.

To gain a better understanding of the rationale for the development of the RWL and LI, consult [1], which provides a discussion of the criteria underlying the lifting equation and of the individual multipliers. It also identifies both the assumptions and uncertainties in the scientific studies that associate manual lifting and low back injuries.

### 33.6.3 Job-Related Intervention Strategy

The LI may be used to identify potentially hazardous lifting jobs or to compare the relative severity of two jobs for the purpose of evaluating and redesigning them. From the NIOSH perspective, it is likely that lifting tasks with LI > 1.0 pose an increased risk for lifting-related LBP for some fraction of the workforce [1]. Hence, to the extent possible, lifting jobs should be designed to achieve an LI of 1.0 or less.

Some experts believe, however, that worker selection criteria can be used to identify workers who can perform potentially stressful lifting tasks (i.e., lifting tasks that would exceed an LI of 1.0) without significantly increasing their risk of work-related injury above the baseline level [9,10]. Those who endorse the use of selection criteria believe that the criteria must be based on research studies, empirical observations, or theoretical considerations that include job-related strength testing and/or aerobic capacity testing. Even these experts agree, however, that many workers will be at a significant risk of a work-related injury when performing highly stressful lifting tasks (i.e., lifting tasks with LI > 3.0). Also, "informal" or "natural" selection of workers may occur in many jobs that require repetitive lifting tasks. According to some experts, this may result in a unique workforce that may be able to work above a LI of 1.0, at least in theory, without substantially increasing their risk of low back injuries above the baseline rate of injury.

### 33.6.4 Example Problems

Two simple example problems are provided to demonstrate the proper application of the lifting equation and procedures.* The procedures provide a method for determining the level of physical stress associated with a specific set of lifting conditions and assist in identifying the contribution of each job-related factor. The examples also provide guidance in developing an ergonomic redesign strategy. Specifically, for each example, a job description, job analysis, hazard assessment, redesign suggestion, illustration, and completed worksheet are provided.

* Additional example problems can be reviewed in Applications Manual for the RNLE [2].

To help clarify the discussion of the example problems, and to provide a useful reference for determining the multiplier values, the six multipliers used in the example problems are derived from the data contained in Tables 33.1 through 33.7.

A series of general design/redesign suggestions for each job-related risk factor are provided in Table 33.9. These suggestions can be used to develop a practical ergonomic design/redesign strategy.

**Example 1: Loading Supply Rolls**

*Job description*: With both hands directly in front of the body, a worker lifts the core of a 35 lb roll of paper from a cart and then shifts the roll in the hands and holds it by the sides to position it on a machine, as shown in Figure 33.6. Significant control of the roll is required at the destination of the lift. Also, the worker must crouch at the destination of the lift to support the roll in front of the body but does not have to twist.

*Job analysis*: The task variable data are measured and recorded on the job analysis worksheet (Figure 33.7). The vertical location of the hands is 27 in. at the origin and 10 in. at the destination. The horizontal location of the hands is 15 in. at the origin and 20 in. at the destination. The asymmetric angle is 0° at both the origin and the destination, and the frequency is 4 lifts/shift (i.e., less than 0.2 lift/min for less than 1 h; see Table 33.5).

Using Table 33.6, the coupling is classified as poor because the worker must reposition the hands at the destination of the lift and cannot flex the fingers to the desired 90° angle (e.g., hook grip). No asymmetric lifting is involved (i.e., $A = 0$), and significant control of the object is required at the destination of the lift. Thus, RWL should be computed at both the origin and the destination of the lift. The multipliers are computed from the lifting equation or determined from the multiplier tables (Tables 33.1 through 33.5 and 33.7). As shown in Figure 33.7, for this activity RWL = 28.0 at the origin and RWL = 18.1 lb at the destination.

*Hazard assessment*: The weight to be lifted (35 lb) is greater than the RWL at both the origin and destination of the lift (28.0 and 18.1 lb, respectively). At the origin, LI = 35 lb/28.0 lb = 1.3; and at the destination, LI = 35 lb/18.1 lb = 1.9. These values indicate that this job is only slightly stressful at the origin but moderately stressful at the destination of the lift.

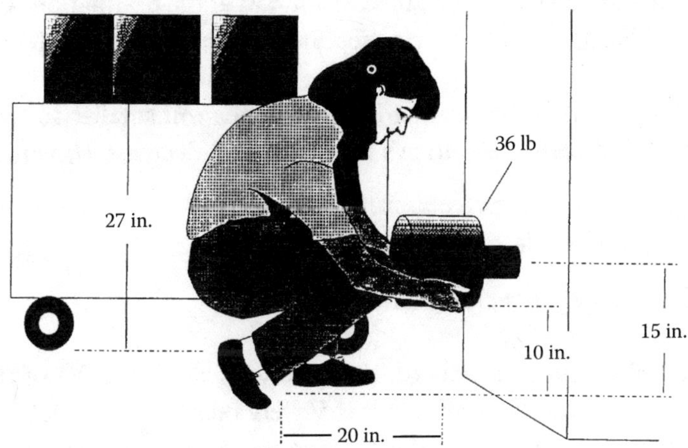


FIGURE 33.6 Loading supply rolls, Example 1.

**JOB ANALYSIS WORKSHEET**

| | | | |
|---|---|---|---|
| **DEPARTMENT** | Shipping | **JOB DESCRIPTION** | |
| **JOB TITLE** | Packager | Loading paper supply rolls | |
| **ANALYST'S NAME** | | | |
| **DATE** | | Example 2 | |

**STEP 1. Measure and record task variables**

| Object Weight (lb) | | Hand Location (in.) | | | | Vertical Distance (in.) | Asymmetric Angle (degrees) | | Frequency Rate lifts/min | Duration (h) | Object Coupling |
|---|---|---|---|---|---|---|---|---|---|---|---|
| | | Origin | | Destination | | | Origin | Destination | | | |
| *L* (AVG.) | *L* (Max.) | *H* | *V* | *H* | *V* | *D* | *A* | *A* | *F* | | *C* |
| 35 | 35 | 15 | 27 | 20 | 10 | 17 | 0 | 0 | <.2 | <1 | Poor |

**STEP 2. Determine the multipliers and compute the RWL's**

RWL = LC × HM × VM × DM × AM × FM × CM

**ORIGIN** RWL = 51 × .67 × .98 × .93 × 1.0 × 1.0 × .90 = 28.0 **lb**

**DESTINATION** RWL = 51 × .50 × .85 × .93 × 1.0 × 1.0 × .90 = 18.1 **lb**

**STEP 3. Compute the LIFTING INDEX**

**ORIGIN** LIFTING INDEX $= \dfrac{\text{OBJECT WEIGHT (L)}}{\text{RWL}} = \dfrac{35}{28.0} = 1.3$

**DESTINATION** LIFTING INDEX $= \dfrac{\text{OBJECT WEIGHT (L)}}{\text{RWL}} = \dfrac{35}{18.1} = 1.9$

FIGURE 33.7 Job analysis worksheet, Example 1.

*Redesign suggestions*: The first choice for reducing the risk of injury for workers performing this task would be to adapt the cart so that the paper rolls could be easily pushed into position on the machine without manually lifting them.

If the cart cannot be modified, then the results of the equation may be used to suggest task modifications. The worksheet displayed in Figure 33.7 indicates that the multipliers with the smallest magnitude (i.e., those providing the greatest penalties) are 0.50 for HM at the destination, 0.67 for HM at the origin, 0.85 for VM at the destination, and 0.90 for the CM value. Using Table 33.10, the following job modifications are suggested:

1. Bring the load closer to the worker by making the roll smaller so that the roll can be lifted from between the worker's legs. This will decrease *H*, which in turn will increase HM.
2. Raise the height of the destination to increase VM.
3. Improve the coupling to increase CM.

If the size of the roll cannot be reduced, then the vertical height (*V*) of the destination should be increased. Figure 33.8 shows that if *V* were increased to about 30 in., then VM would be increased from 0.85 to 1.0; the *H* value would be decreased from 20 to 15 in., which would increase HM from 0.50 to 0.67; DM would be increased from 0.93 to 1.0.

TABLE 33.10 General Design/Redesign Suggestions

| |
|---|
| If HM is less than 1.0, bring the load closer to the worker by removing any horizontal barriers or reducing the size of the object. Lifts near the floor should be avoided; if unavoidable, the object should fit easily between the legs. |
| If VM is less than 1.0, raise/lower the origin/destination of the lift. Avoid lifting near the floor or above the shoulders. |
| If DM is less than 1.0, reduce the vertical distance between the origin and the destination of the lift. |
| If AM is less than 1.0, move the origin and destination of the lift closer together to reduce the angle of twist, or move the origin and destination further apart to force the worker to turn the feet and step, rather than twist the body. |
| If FM is less than 1.0, reduce the lifting frequency rate, reduce the lifting duration, or provide longer recovery periods (i.e., light work period). |
| If CM is less than 1.0, improve the hand-to-object coupling by providing optimal containers with handles or handhold cutouts, or improve the handholds for irregular objects. |
| If the RWL at the destination is less than at the origin, eliminate the need for significant control of the object at the destination by redesigning the job or modifying the container/object characteristics. (See Section 33.5.2, p. 643.) |

**JOB ANALYSIS WORKSHEET**

| | | | |
|---|---|---|---|
| **DEPARTMENT** | Shipping | **JOB DESCRIPTION** | |
| **JOB TITLE** | Packager | Loading paper supply rolls | |
| **ANALYST'S NAME** | | | |
| **DATE** | | Modified Example 2 | |

**STEP 1. Measure and record task variables**

| Object Weight (lb) | | Hand Location (in.) | | | | Vertical Distance (in.) | Asymmetric Angle (degrees) | | Frequency Rate | Duration | Object Coupling |
|---|---|---|---|---|---|---|---|---|---|---|---|
| | | Origin | | Destination | | | Origin | Destination | lifts/min | (h) | |
| *L* (AVG.) | *L* (Max.) | *H* | *V* | *H* | *V* | *D* | *A* | *A* | *F* | | *C* |
| 35 | 35 | 15 | 27 | 15 | 30 | 3 | 0 | 0 | < .2 | < 1 | Poor |

**STEP 2. Determine the multipliers and compute the RWL's**

| | RWL = | LC | × | HM | × | VM | × | DM | × | AM | × | FM | × | CM | = | |
|---|---|---|---|---|---|---|---|---|---|---|---|---|---|---|---|---|
| **ORIGIN** | RWL = | 51 | × | .67 | × | .98 | × | 1.0 | × | 1.0 | × | 1.0 | × | .90 | = | 30.1 **lb** |
| **DESTINATION** | RWL = | 51 | × | .67 | × | 1.0 | × | 1.0 | × | 1.0 | × | 1.0 | × | .90 | = | 30.8 **lb** |

**STEP 3. Compute the LIFTING INDEX**

**ORIGIN** LIFTING INDEX $= \dfrac{\text{OBJECT WEIGHT (L)}}{\text{RWL}} = \dfrac{35}{30.1} = 1.2$

**DESTINATION** LIFTING INDEX $= \dfrac{\text{OBJECT WEIGHT (L)}}{\text{RWL}} = \dfrac{35}{30.8} = 1.1$

FIGURE 33.8 Modified job analysis worksheet, Example 1.

As shown in Figure 33.8, the final RWL would be increased from 18.1 to 30.8 lb, and LI at the destination would decrease from 1.9 to 1.1.

In some cases, redesign may not be feasible. In these cases, the use of a mechanical lift may be more suitable. As an interim control strategy, two or more workers may be assigned to lift the supply roll.

*Comments*: The horizontal distance (*H*) is a significant factor that may be difficult to reduce because the size of the paper rolls may be fixed. Moreover, redesign of the machine may not be practical. Therefore, elimination of the manual lifting component of the job may be more appropriate than job redesign.

**Example 2: Dishwashing Machine Unloading**

*Job description*: A worker manually lifts trays of clean dishes from a conveyor at the end of a dishwashing machine and loads them on a cart as shown in Figure 33.9. The trays are filled with assorted dishes (e.g., glasses, plates, bowls) and silverware. The job takes between 45 min and 1 h to complete, and the lifting frequency rate averages 5 lifts/min. Workers usually twist their body to one side to lift the trays (i.e., asymmetric lift) and then their body to the other side to lower the trays to the cart in one smooth continuous motion. The maximum amount of asymmetric twist varies between workers and within workers; however, there is usually equal twist to either side. During the lift the worker may take a step toward the cart. The trays have well-designed handhold cutouts and are made of lightweight materials.

*Job analysis*: The task variable data are measured and recorded on the job analysis worksheet (Figure 33.10). At the origin of the lift, the horizontal distance (*H*) is 20 in., the vertical distance (*V*) is 44 in., and the angle of asymmetry (*A*) is 30°. At the destination of the lift, $H = 20$ in., $V = 7$ in., and $A = 30°$. The trays normally weigh between 5 and 20 lb, but for this example, assume that all of the trays weigh 20 lb.

Using Table 33.6, the coupling is classified as "good." Significant control is required at the destination of the lift. Using Table 33.5, the FM is determined to be 0.80. As shown in Figure 33.10, RWL = 14.4 lb at the origin and 13.3 lb at the destination.

*Hazard assessment*: The weight to be lifted (20 lb) is greater than the RWL at either the origin or destination of the lift (14.4 and 13.3 lb, respectively). At the origin, LI = 20/14.4 = 1.4, and

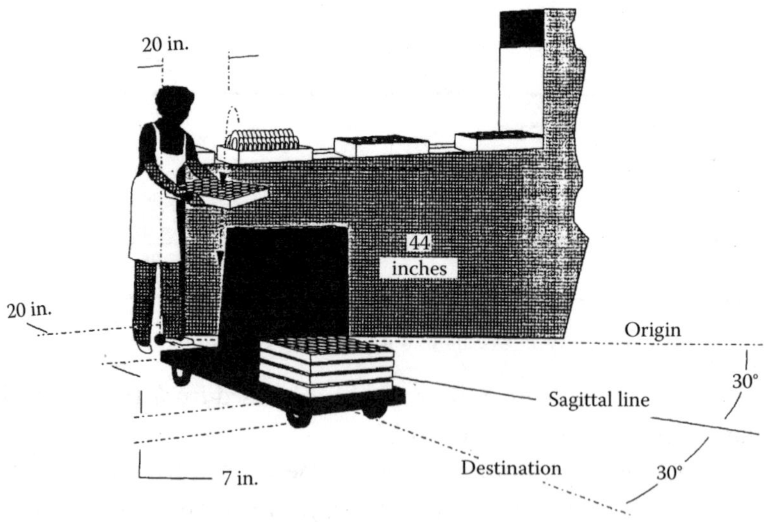


FIGURE 33.9 Dishwashing machine unloading, Example 2.

**JOB ANALYSIS WORKSHEET**

| | | | |
|---|---|---|---|
| DEPARTMENT | Food Service | JOB DESCRIPTION | Unloading a dish-washing machine Example 5 |
| JOB TITLE | Cafeteria Worker | | |
| ANALYST'S NAME | | | |
| DATE | | | |

**STEP 1. Measure and record task variables**

| Object Weight (lb) | | Hand Location (in.) | | | | Vertical Distance (in.) | Asymmetric Angle (degrees) | | Frequency Rate | Duration | Object Coupling |
|---|---|---|---|---|---|---|---|---|---|---|---|
| | | Origin | | Destination | | | Origin | Destination | lifts/min | (h) | |
| *L* (AVG.) | *L* (Max.) | *H* | *V* | *H* | *V* | *D* | *A* | *A* | *F* | | *C* |
| 20 | 20 | 20 | 44 | 20 | 7 | 37 | 30 | 30 | 5 | <1 | Good |

**STEP 2. Determine the multipliers and compute the RWL's**

RWL = LC × HM × VM × DM × AM × FM × CM

**ORIGIN** RWL = 51 × .50 × .90 × .87 × .90 × .80 × 1.0 = 14.4 lb

**DESTINATION** RWL = 51 × .50 × .83 × .87 × .90 × .80 × 1.0 = 13.3 lb

**STEP 3. Compute the LIFTING INDEX**

**ORIGIN** LIFTING INDEX $= \dfrac{\text{OBJECT WEIGHT (L)}}{\text{RWL}} = \dfrac{20}{14.4} = 1.4$

**DESTINATION** LIFTING INDEX $= \dfrac{\text{OBJECT WEIGHT (L)}}{\text{RWL}} = \dfrac{20}{13.3} = 1.5$

FIGURE 33.10 Job analysis worksheet, Example 2.

at the destination, LI = 1.5. These results indicate that this lifting task would be stressful for some workers.

*Redesign suggestions*: The worksheet shows that the smallest multipliers (i.e., the greatest penalties) are 0.50 for HM, 0.80 for FM, 0.83 for VM, and 0.90 for AM. Using Table 33.9, the following job modifications are suggested:

1. Bring the load closer to the worker to increase HM.
2. Reduce the lifting frequency rate to increase FM.
3. Raise the destination of the lift to increase VM.
4. Reduce the angle of twist to increase AM by either moving the origin and destination closer together or moving them further apart.

Since the horizontal distance (*H*) is dependent on the width of the tray in the sagittal plane, this variable can be reduced only by using smaller trays. Both DM and VM, however, can be increased by lowering the height of the origin and increasing the height of the destination. For example, if the height at both the origin and destination is 30 in., then VM and DM are 1.0, as shown in the modified worksheet (Figure 33.11). Moreover, if the cart is moved so that the twist is eliminated, AM can be increased from 0.90 to 1.00. As shown in

**JOB ANALYSIS WORKSHEET**

| | | | |
|---|---|---|---|
| **DEPARTMENT** | Food Service | **JOB DESCRIPTION** | |
| **JOB TITLE** | Cafeteria Worker | Unloading a dish-washing | |
| **ANALYST'S NAME** | | machine | |
| **DATE** | | Modified Example 5 | |

**STEP 1. Measure and record task variables**

| Object Weight (lb) | | Hand Location (in.) | | | | Vertical Distance (in.) | Asymmetric Angle (degrees) | | Frequency Rate lifts/min | Duration (h) | Object Coupling |
|---|---|---|---|---|---|---|---|---|---|---|---|
| | | Origin | | Destiantion | | | Origin | Destination | | | |
| *L* (AVG.) | *L* (Max.) | *H* | *V* | *H* | *V* | *D* | *A* | *A* | *F* | | *C* |
| 20 | 20 | 20 | 30 | 20 | 30 | 0 | 0 | 0 | 5 | < 1 | Good |

**STEP 2. Determine the multipliers and compute the RWL's**

| | | | | | | | | | |
|---|---|---|---|---|---|---|---|---|---|
| | RWL = | LC × | HM × | VM × | DM × | AM × | FM × | CM | |
| **ORIGIN** | RWL = | 51 × | .50 × | 1.0 × | 1.0 × | 1.0 × | .80 × | 1.0 = | 20.4 lb |
| **DESTINATION** | RWL = | 51 × | .50 × | 1.0 × | 1.0 × | 1.0 × | .80 × | 1.0 = | 20.4 lb |

**STEP 3. Compute the LIFTING INDEX**

**ORIGIN** LIFTING INDEX $= \frac{\text{OBJECT WEIGHT (L)}}{\text{RWL}} = \frac{20}{20.4} =$ 1.0

**DESTINATION** LIFTING INDEX $= \frac{\text{OBJECT WEIGHT (L)}}{\text{RWL}} = \frac{35}{20.4} =$ 1.0

FIGURE 33.11 Modified job analysis worksheet, Example 2.

Figure 33.11, with these redesign suggestions, RWL can be increased from 13.3 to 20.4 lb, and the LI values are reduced to 1.0.

*Comments*: This analysis was based on a 1 h work session. If a subsequent work session begins before the appropriate recovery period has elapsed (i.e., 1.2 h), then the 8 h category would be used to compute the FM value.

## REFERENCES


1. T. R. Waters, V. Putz-Anderson, A. Garg, and L. J. Fine, Revised NIOSH equation for the design and evaluation of manual lifting tasks, *Ergonomics 36*(7):749–776 (1993).
2. T. R. Waters, V. Putz-Anderson, and A. Garg, *Applications Manual for the Revised NIOSH Lifting Equation*, National Institute for Occupational Safety and Health Technical Report, DHHS(NIOSH) Pub. No. 94-110, 1994. Available from the National Technical Information Service, NTIS Doc. No. PB94-176930.
3. NIOSH, *Work Practices Guide for Manual Lifting*, NIOSH Technical Report No. 81-122, 1981, U.S. Department of Health and Human Services, National Institute for Occupational Safety and Health, Cincinnati, OH.
4. NIOSH, *Scientific Support Documentation for the Revised 1991 NIOSH Lifting Equation*, Technical Contract Reports, May 8, 1991, U.S. Department of Health and Human Services, National Institute for Occupational Safety and Health, Cincinnati, OH. Available from the National Technical Information Service, NTIS No. PB-91-226-274.
5. T. R. Waters, Manual material handling, Chapter 14, in *Occupational Ergonomics, Theory and Practice*. Eds. Bhattacharya and McGlothlin, Marcel Dekker, New York, 1996, pp. 329–349.

6. Eastman Kodak, *Kodak's Ergonomic Design for People at Work*, John Wiley & Sons, Incorporated, Hoboken, NJ, 2003.
7. T. R. Waters, M. L. Lu, and E. Occhipinti, New procedure for assessing sequential manual lifting jobs using the revised NIOSH lifting equation, *Ergonomics 50*(11):1761–1770 (2007).
8. T. R. Waters, E. Occhipinti, D. Columbini, E. Alvarez, and A. Hernandez, The variable lifting index (VLI): A new method for evaluating variable lifting tasks using the revised NIOSH lifting equation, *Proceedings of the 2009 IEA Conference*, Beijing, China, August 2009.
9. M. M. Ayoub and A. Mital, *Manual Materials Handling,* Taylor & Francis, London, U.K., 1989.
10. D. B. Chaffin, G. B. J. Andersson, and B. J. Martin, *Occupational Biomechanics,* 4th edn., Wiley, New York, 2006.

CHAPTER 34

# Americans with Disabilities Act*

## *Implications for the Use of Ergonomics in Rehabilitation*

Jerry A. Olsheski and Robert E. Breslin

CONTENTS



* Refer to Appendix H for updates on Americans with Disabilities Act.

## 34.1 INTRODUCTION

The Americans with Disabilities Act (PL 101–336), or ADA, is regarded by people with disabilities and those in the rehabilitation community to be the most significant piece of civil rights legislation enacted by Congress since the Civil Rights Act of 1964. It is designed to extend civil rights protection similar to that found in existing legislation related to race, sex, age, and ethnicity to individuals with disabilities [1]. Specifically, the ADA is intended to provide a federal mandate prohibiting discrimination based on disability in American society while establishing enforceable standards to address such discrimination and provide redress to those protected individuals who experience discrimination.

In contrast to the medical model, which equates disability with the functional limitations of the individual, ADA addresses the environmental side of disability. The act is specifically designed to remove attitudinal, architectural, and physical barriers that have the effect of excluding individuals with disabilities from places of employment and commerce. As more attention is given to the role that environmental factors play in creating and maintaining disability, the role of ergonomic interventions in disability prevention and management will assume more importance.

In this chapter, the passage of ADA is examined in terms of the impact that this legislation is having on the changing model of disability in American society, the vocational rehabilitation process, and the use of ergonomics in job accommodation.

## 34.2 LEGISLATIVE HISTORY OF THE ADA

Although most Americans became aware of the ADA relatively recently, particularly when it became an issue during the presidential election campaign of 1988, the genesis of the law dates back more than a decade to 1983 [2]. At this time the leadership of the National Council on the Handicapped began to explore methods for extending civil rights similar to those found in Title V of the Rehabilitation Act of 1973, which covered federal government agencies (Section 501), federal contractors (Section 503), and recipients of federal grants (Section 504), to the private sector and state and local governments. An initial draft of the ADA bill was introduced in 1988. This draft provided the 100th Congress with the opportunity to demonstrate the existence of broad bipartisan support for the underlying principles of the act.

After significant revision, the ADA was reintroduced to the 101st Congress in 1989. Throughout 1989 and 1990, a complicated series of negotiations took place among the House of Representatives, the Senate, the White House, and advocates for the disability and business communities. The result was a complex and multifaceted bill developed with extensive input from lobbyists and advocacy organizations representing those with affected interests. The ADA was passed in the Senate and the House of Representatives by overwhelming margins. The final version of the act was signed into law on July 26, 1990 by President George Bush.

### 34.2.1 Rationale for a New Civil Rights Law

It seems remarkable that a landmark piece of civil rights legislation was conceived and enacted in an era characterized by political conservatism and widespread calls for a diminished federal role in American social and economic life. One needs to look no further than the "Findings and Purposes" [Americans with Disabilities Act of 1991, PL 101–336, Section 2(a)] of the ADA, however, to understand why the vast majority of the legislators in Congress felt compelled to support the measure. McMahon and Shaw [3] have called this portion of the ADA "one of the single most eloquent and direct statements of societal culpability ever observed in print." The findings section of the law documents the prevalence of disability in U.S. society, estimating that 45 million Americans have some physical or mental impairment. These individuals have often been isolated, segregated, and discriminated against as the result of erroneous and stereotypical assumptions regarding their abilities. Congress sites data that document that "people with disabilities, as a group, occupy an inferior status in our society, and are severely disadvantaged socially, vocationally, economically, and educationally" [ADA, 1991, PL 101–336, Section 2(a) (7)]. This discrimination has cost the United States billions of dollars as a result of expenses associated with lost productivity and disability support programs.

## 34.3 ADA: A GUARANTEE OF EQUAL ACCESS FOR AMERICANS WITH DISABILITIES

The ADA contains five separate titles that guarantee equal access to employment, public accommodations, transportation, state and local government operations, and telecommunication services. Equal access to housing and air transportation were not included in the ADA because discrimination in these areas is prohibited by the Fair Housing Amendments Act of 1988 and the Air Carriers Access Act of 1986.

The following section provides a brief description of the scope of the coverage, dates of implementation, and specific protections afforded by the five titles.

### 34.3.1 Title I

Title I prohibits discrimination in *access to employment*. This portion of the law applies to private sector employers, labor unions, employment agencies, and all governmental bodies, with the exception of the federal government, who employ 15 or more employees in 20 or more calendar weeks during a calendar year. Employers who are exempt from ADA include corporations wholly owned by the federal government, Indian tribes, and tax-exempt private membership clubs. Provisions of the law contained in Title I were implemented in stages, with employers having 25 or more employees being subject to the provisions in July 1992. Employers having 15–24 employees were subject to the provisions of Title I in July 1994.

The most significant aspect of Title I is the extension of the "reasonable accommodation" provisions of the Rehabilitation Act of 1973 to the private sector. Under the ADA, an employer must consider whether an individual can perform the essential functions of the job with or without reasonable accommodation. Since the employment-related provisions of Title I will directly affect the provision of rehabilitation and ergonomic services in industry, these provisions are the primary concern of this chapter.

### 34.3.2 Title II

Title II prohibits discrimination by any public entity in providing *public services*, including transportation, to individuals with a disability. Public entities include agencies, special-purpose districts, departments, and other instrumentalities of state or local government and commuter authorities of the Rail Passenger Service Act. Title II became effective on January 26, 1992.

### 34.3.3 Title III

Title III prohibits discrimination on the basis of disability in places of *public accommodation*, including places of lodging, establishments for serving food and/or drink, places of exhibition or entertainment, places of public gathering, sales or rental establishments, service establishments, stations used for specified public transportation, places of public display, places of recreation, places of education, social service centers, and places of exercise [4]. Title III also covers all places of commerce. These provisions became effective July 26, 1993, although there are some exceptions for certain provisions that apply to existing business facilities based on size and financial status.

### 34.3.4 Title IV

Title IV contains the *telecommunications* provisions of the ADA. It requires that common carriers of wire or radio communications provide accommodations for individuals with hearing and speech impairments that will serve as the functional equivalent of telephone service for those individuals.

### 34.3.5 Title V

Title V contains *miscellaneous provisions* of the ADA, including those regarding the relationship of the Americans with Disabilities Act to other federal and state laws, those clarifying the effect of ADA on commercial insurance, and the specific prohibition of immunity for states under the 11th amendment to the Constitution of the United States. Title V also prohibits retaliation against an individual who takes action to exercise his or her rights under the ADA and outlines remedies and procedures as well as the plans for implementation of each of the first four titles. Title V outlines the technical assistance plan and makes some modifications to the definition of handicapped individual in the Rehabilitation Act of 1973 to exclude individuals currently using illegal drugs.

## 34.4 TITLE I EMPLOYMENT: IMPLICATIONS FOR AMERICAN INDUSTRY

As previously described, Title I of the ADA was designed to prohibit discrimination in employment practices against "qualified individuals with disabilities" by employers and other "covered entities [5]." The ADA regulates all aspects of the employment process, including application, testing, hiring and assignments, evaluation, disciplinary actions, training, promotion, and medical examinations, layoff and recall, termination, compensation, leave, and benefits.

### 34.4.1 Definition of Terms

To gain a more accurate understanding of Title I, it is necessary to define several key terms that operationalize the employment-related requirements and compliance process.

#### *34.4.1.1 Individual with a Disability*

The definition of disability in the ADA is a derivative of the definition found in the Rehabilitation Act of 1973. For the purposes of the ADA, an *individual with a disability* is any individual with a physical or mental impairment that substantially limits one or more major life activities. These major life activities include walking, speaking, breathing, performing manual tasks, seeing, hearing, learning, caring for oneself, and, perhaps most important from the perspective of both employers and individuals with disabilities, working.

Individuals may also fall under the protection of the act if they have a record of an impairment or if they can demonstrate that discrimination took place because they are regarded as having such an impairment. Finally, individuals may be protected by the act if they are discriminated against because of their relationship with an individual with a disability.

Several conditions that might be considered disabilities under certain circumstances or definitions are specifically excluded from protection under the ADA. These include transvestism, homosexuality and bisexuality, transsexualism, pedophilia, exhibitionism, voyeurism, gender identity disorders, compulsive gambling, pyromania, kleptomania, and current abuse of alcohol and drugs. Individuals with communicable diseases that can be transmitted through food handling may also be denied employment in jobs involving food handling if there is no reasonable accommodation that would eliminate that risk.

#### *34.4.1.2 Qualified Individual with a Disability*

In order to be considered a "qualified individual with a disability" under the ADA, the individual must be one who satisfies the requisite skill, experience, education, and other job-related requirements of the employment position that the individual holds or desires and who, with or without reasonable accommodation, can perform the essential functions of such position.

Thus, determining whether an individual with a disability is "qualified" under Title I of the ADA is a two-step process. The first step focuses on the qualities of the individual as a worker. That is, does the individual have the education, skills, work experience, licenses and certifications, and other prerequisites to perform the job successfully? This step should be present in any employment or hiring decision, regardless of the presence of an impairment.

The second step involves the evaluation of whether the individual can perform the "essential functions" of the job with or without reasonable accommodation. This step requires that the employer give attention to the work environment. Typically, a job analysis is conducted to quantify the functional demands and working conditions of a specific job. Job analysis data are used as the basis to identify the essential functions of the job. Second, the employer must evaluate whether the qualified individual with a disability can perform the essential functions of the job, as identified in the job analysis process, with or without reasonable accommodation. This second step, referred to as the job accommodation analysis process, is discussed in more detail later in the chapter.

#### *34.4.1.3 Essential Functions*

The ADA does not specifically define the term *essential functions* but rather provides guidelines to employers to aid in their identification. First and foremost, functions identified by an employer to be essential must actually be required of employees performing the job in question. They cannot be arbitrary, for example, requiring the candidate to type 80 words per minute in a job in which typing is rarely required of the worker.

The employer may also consider whether removing a particular function changes the job in a fundamental way. An employer is not required to accommodate an individual by removing a function if the job exists primarily to perform that function. The employer may also take into consideration the number of employees available to perform a particular function and the degree of specialization or special expertise required to perform the function.

#### *34.4.1.4 Reasonable Accommodation*

Perhaps the most significant of the employment-related provisions of the ADA is the legal requirement that employers who are "covered entities" consider reasonable accommodation as a component of employment decisions. *A reasonable accommodation* is defined by the EEOC (1992) as "any modification or adjustment to a job, an employment practice, or the work environment that makes it possible for an individual with a disability to enjoy an equal employment opportunity." Reasonable accommodation is required in order to guarantee equal opportunity in the application process, to enable a qualified individual with a disability to perform the essential functions of a job, and to allow an employee with a disability to enjoy equal "benefits and privileges" of employment. Reasonable accommodations in the application process are intended to guarantee all qualified individuals with disabilities an equal opportunity to be considered for employment. Examples of accommodations in the application process may include modifications that render the interview or application site accessible to an individual who is mobility impaired or the provision of assistance to an individual who is unable to complete the application process because of visual or manipulative impairments.

Reasonable accommodations to perform the essential functions of a job may include any type of modification of the work environment, including the work devices, the manner in which the job is typically performed, or the administrative policies governing the work and worker. Specific examples of the kinds of accommodations envisioned by Congress are listed in the act [Americans with Disabilities Act of 1991, PL 101–336,

Section 101(9) (a & b)] and the regulations developed by the EEOC (EEOC–29CFR, 1630, July 1992 (0) (1 & 2)]. Reasonable accommodations may include

- Making facilities readily accessible to and usable by an individual with a disability
- Restructuring a job by reallocating or redistributing marginal job functions
- Altering when or how an essential job function is performed
- Part-time or modified work schedules
- Obtaining or modifying equipment or devices
- Modifying examinations, training materials, or policies
- Providing qualified readers and interpreters
- Reassignment to a vacant position
- Permitting use of accrued paid leave or unpaid leave for necessary treatment
- Providing reserved parking for a person with a mobility impairment
- Allowing an employee to provide equipment or devices an employer is not required to provide

There are some limitations on the employer's obligation to provide reasonable accommodation. Only qualified individuals with a disability can request such an accommodation, and the employer is only required to accommodate the *known* limitations of the individual. The ADA does require, however, that covered entities post notices regarding the provisions of ADA including the reasonable accommodation requirement. In addition, the employer may request documentation of the functional limitations claimed by the individual requesting the accommodation. If it can be established that the qualified individual with a disability needs an accommodation, the employer is required to seek input from the individual regarding potential accommodation strategies and to consider his or her preferences for such an accommodation. The employer is not, however, required to select the accommodation chosen by the employee if one can be identified that is equally effective for the individual while better serving the needs of the employer.

#### *34.4.1.5 Undue Hardship*

An employer is not required to make an accommodation to a qualified individual with a disability if the employer can demonstrate that the provision of such an accommodation would impose an undue hardship on the business. The employer can use undue hardship as a defense in not making an accommodation if it places an unacceptable financial or other type of burden on the business or employer. This includes modifications that are unrealistic in terms of size, scope, or cost relative to the size and resources of the employer. Accommodations that would alter the fundamental nature or operation of the business or significantly disrupt the business may also be considered to pose an undue hardship.

## 34.5 TITLE I OF THE ADA: A SUMMARY OF THE PROHIBITION OF DISCRIMINATION IN EMPLOYMENT

Title I of the ADA prohibits a covered entity from discriminating on the basis of disability against a qualified individual with a disability, that is, an individual who can perform the essential functions of the employment position with or without reasonable accommodation. Employment discrimination is operationally defined in such a manner that Title I regulates the conduct of employers and their relationship with employees, job applicants, and other aspects of the employment environment.

The reasonable accommodation requirement of the ADA is especially significant in that it recognizes and addresses the environmental aspects of work disability. Although the act clearly addresses attitudinal barriers that unfairly stereotype individuals with disabilities as incapable workers, the reasonable accommodation provisions of the law place emphasis on those aspects of the vocational handicap that are external to the individual with an impairment. This shift, although inconsistent with the medical-legal perception of disability, is compatible with the orientations of the vocational rehabilitationist and the ergonomist.

## 34.6 CRITICAL REVIEW OF CURRENT STATUS

American society has primarily defined disability in medical terms and for the large part has ignored the role that environmental factors play in the creation and maintenance of disability. When disability is essentially defined as a medical condition, the focus of attention centers on the functional limitations of the individual [6,7]. The medical orientation to disability is personological or individualistic in nature [8]. This approach is based on the assumption that disability is a phenomenon dependent mainly on individual behaviors and individual decisions [9]. Consequently, this paradigm provides an incomplete picture of disability by belittling the political, social, ecological, and environmental variables that play an important role in mediating and determining disability.

The dominance of the medical model in the disability arena has influenced public policy, employment practices, the nature of vocational rehabilitation services, and claims management techniques. Public policy has tended to equate disability with impairment and has restricted disability to clinical criteria. Many employers have refused to reemploy workers with restrictions and have depended on the medical provider to restore the worker to "100%" capacity before the worker returns to work. Vocational rehabilitation in many instances has amounted to little more than an extension of medical interventions as evidenced by the increasing number of registered nurses involved in the "case management" of workers with disabilities and the rapidly escalating costs of medical services in the workers' compensation system. In response to runaway medical costs, insurers who underwrite workers' compensation coverage have developed the concept of *maximum medical improvement* as a short-term cost-containment technique. This concept indicates that the worker cannot benefit (i.e., make functional improvements) from additional medical services and the insurer will no longer pay for these types of interventions. Unfortunately, many workers who are evaluated as having achieved maximum medical improvement fail to return to work and instead enter long-term disability support systems [10]. Maximum medical improvement, without reemployment, actually signifies that medical interventions can only go so far in resolving disability. Likewise, the provision of additional medically oriented services under the guise of vocational rehabilitation has resulted in increased criticism of the effectiveness of vocational rehabilitation. A number of states have accordingly repeated mandatory rehabilitation provisions in their workers' compensation systems due to the perception legislators have that rehabilitation services amount to little more than additional medical costs with poor vocational outcomes [11].

### 34.6.1 ADA and the Changing Model of Disability in American Society

The passage of ADA marks a significant departure from the medical model of disability. Disability is defined more in environmental or sociopolitical terms as embodied in the various provisions of the ADA. The primary objective of this legislation is to alter the environmental side of disability. Thus, the emphasis or focus of change as mandated by ADA is on factors that are external to the individual. Whereas the medical model of disability aims at changing the person, the model of disability inherent in the provisions of ADA aims at changing the environment, and this is the essence of the accommodation process. The five titles of the ADA are concerned with removing *environmentally induced disability factors* in such areas as telecommunications, public services, public accommodations, and employment.

## 34.7 FUTURE NEEDS: AN ENVIRONMENTALLY ORIENTED MODEL OF VOCATIONAL REHABILITATION

From a theoretical perspective, the vocational rehabilitation process requires that equal attention be given to characteristics of the person and factors in the environment because disability can originate as much from environmental barriers as from the individual's impairment [12]. Although the environmental side of disability has long been recognized at the theoretical level, the influence of environmental factors in the rehabilitation process has largely been ignored at the (applied) service delivery level. Rehabilitation interventions that are aimed at altering the characteristics of the person are more widely used than interventions that focus on altering or restructuring the environment. The underutilization of environmentally focused interventions results from a rehabilitation model that continues to view disability as a pathological condition of the person. Pathological elements within the environment are overlooked. In this model, the goal of rehabilitation is to increase the functional capacities or in some fashion overcome the occupational liabilities of the person with a disability [6].

The most popular rehabilitation services for workers with disabilities are those that attempt to enhance physical and occupational capacities. The proliferation of work hardening and work capacity centers represents the medical orientation to disability in that such facility-based services focus almost entirely on the characteristics of the person and ignore the importance of environmental variables. Occupational disability is based solely on the capacities of the worker, and the factors in the work environment are not entered into the equation [13]. Thus, although every attempt is made to enhance the capacities of the worker, there is no corresponding attempt to enhance the capacity of the work environment in terms of job accommodation and retention.

### 34.7.1 Emerging Disability Management Model and the Role of Ergonomics

In response to the continued growth of the personal, social, and economic costs of disability, a revised model of disability and rehabilitation is beginning to emerge. This alternative model has been identified as *disability management* [13–16]. The disability management model recognizes that disability is a complex phenomenon that cannot be defined as a medical condition and resolved by clinical solutions alone [17]. Disability management represents a revitalization of the basic interactional nature of disability, which posits that work disability results from the interaction of the individual's characteristics with the physical, social, and organizational elements in the work environment. Recent research reveals the powerful influence of environmental and organizational variables in the occurrence and outcomes of work disabilities [18–20].

Disability management has been defined [14, p. 2] as

> A workplace prevention and remediation strategy that seeks to prevent disability from occurring or, lacking that, to intervene early following the onset of disability, using coordinated, cost-conscious, quality rehabilitation services that reflect an organizational commitment to continued employment of those experiencing functional work limitations.

A more detailed analysis of this definition highlights the differences between disability management and other types of traditional vocational rehabilitation programs. The differences in these two models are summarized in Table 34.1.

First, the disability management model contains a primary prevention component that seeks to minimize the possibility of the occurrence of disability. Second, disability management is considered a workplace- or employer-based approach in which labor and management actively participate in both the prevention and rehabilitation processes in the context of the *actual work environment*. Finally, disability management strategies require the organizational commitment to retaining and accommodating workers with permanent restrictions.

Unlike rehabilitation programs that are grounded in the medical model of disability, disability management goes well beyond the mere evaluation of the individual worker and analyzes the environmental or organizational structures within the worker's interactive system. It is within this interactive arena that the political nature of disability emerges as attempts to reemploy and accommodate workers with disabilities are negotiated.

Ergonomic interventions are of paramount importance in the disability management model. In general terms, *ergonomics* is defined as the scientific study of human work [21]. More specifically, ergonomics is the application of scientific information for the purposes

TABLE 34.1 Comparison of Traditional Vocational Rehabilitation and Disability Management Models

| Component | Traditional Vocational Rehabilitation | Disability Management |
|---|---|---|
| Concept of disability | Disability defined in medical terms, that is, functional impairment | Disability defined as interaction of person and environment |
| Focus of rehab | Alter capacity of the person | Balanced focus on worker capacities and capacities of work environment for accommodation and job retention |
| Typical rehab interventions | Physical therapy, work hardening, psychological and vocational testing | Worksite rehab, ergonomics, job analysis, organizational assessment |
| Objective of rehabilitation | Match person to job or no RTW | Match job to person via job accommodation |
| Roles of employer and union | Passive; rely on external provider | Active participation in rehab and RTW |
| Context of rehabilitation | External facility; simulated work | Actual work environment; real work |
| Role of PT/OT | Inclusion/exclusion function; that is, rule in or out of job | Negotiate job accommodation and retention |
| Source of job accommodation information | Simulated capacities of worker and third party job analysis data | Direct observation of worker and job |
| Role of case manager | External service coordinator; little influence in disability negotiation process | Empowered as internal negotiator; member of joint labor/management committee |
| Disability prevention | Externally generated treatment data are of little use in disability prevention because treatment team does not interact with actual work environment or influence safety policy | Use of clinical skills at worksite helps identify and resolve ergonomic risks and safety hazards to prevent future injuries |
| Role of physician | Typically establishes worker capacities based on type of impairment, limited RTW options | Provided with precise functional data that document the worker's ability to perform real work tasks; may release worker to RTW with restrictions |

RTW, return to work; PT, physical therapy; OT, occupational therapy.

of effectively matching jobs to workers. An effective match is one that maximizes working efficiency (productivity) without compromising the psychological or physical health of the worker. Ergonomic principles, when applied in the disability management model, are valuable tools in the primary prevention of disability and in negotiating the job retention of those with permanent residual limitations.

Ergonomics is a discipline that can help refocus the rehabilitation process so that individual and environmental factors are given equal attention. While medical and psychological interventions are valuable in evaluating the characteristics of the person, the primary value of ergonomic applications is in evaluating environmental factors. Thus, ergonomics and rehabilitation are complementary, for both disciplines strive to maximize the positive work adjustment of the individual [22].

## 34.8 ROLE OF ERGONOMICS IN WORKSITE REHABILITATION PROGRAMS AND JOB ACCOMMODATIONS

The disability management model serves as the philosophical and theoretical foundation for rehabilitation programs that occur in the workplace itself. Worksite rehabilitation programs focus on workers' capacities as they directly interact with the requirements of the job and other elements of the work environment. Transitional work return programs are an example of applying clinical, interdisciplinary rehabilitation skills at the worksite [23,24]. In transitional work programs, real work activities are used as therapeutic modalities to condition, readjust, and accommodate workers with functional limitations. Transitional work return programs have a unique set of features such as the following:

1. Transitional work programs have established time parameters so long-term or open-ended periods of light duty are avoided. Transitional work is not an occupational goal but an interim step in the physical, psychological, and vocational readjustment of the worker.
2. Workers who participate in transitional work are under the clinical supervision of a licensed physical or occupational therapist. These professionals design and monitor the task progression process with the objective of expeditiously returning the worker to full duty status.
3. Transitional work return programs involve the use of an interdisciplinary rehabilitation team including a vocational rehabilitation counselor, a therapist, and an ergonomist.
4. Since transitional work programs occur at the workplace, the professional rehabilitation team is complemented by the expanded rehabilitation team to include labor and management representatives, work supervisor, the worker, and coworkers.
5. Interaction among the professional and expanded rehabilitation team members facilitates the negotiation process that surrounds job retention and accommodation.
6. Transitional work is not a medically oriented intervention but rather a type of interactive rehabilitation service based on functional capacity and job analysis data. These sets of data form the basis of the job accommodation analysis as required by ADA.
7. The development of transitional work return programs requires a number of organizational interventions including staff development of management and labor representatives, formal policies and procedures that regulate the operational aspects of the program, joint labor/management committees to serve as the steering body for the program, ergonomic surveys of the work environment, and a program evaluation system.

Transitional work return programs provide labor and management with an excellent opportunity to comply with the requirements of ADA. Should the worker have any type of permanent functional limitation at the completion of the transitional work program, information is available to accurately identify specific areas of incompatibility between the worker's capacities and the essential functions of the job.

Certainly, ergonomics can be used as a tool for reducing or removing work–job incompatibilities. Collaboration among the ergonomist (environmental specialist), the therapist (functional capacity specialist), management and labor representatives (organizational specialists), and the worker lies at the core of the interactional model of disability. Information generated from this collaboration of specialists can be presented to joint labor/management committees to initiate the negotiation process and make decisions regarding the reasonableness of specific types of accommodations.

The primary role of the ergonomist in worksite rehabilitation and ADA compliance programs is the application of ergonomic skills in identifying, evaluating, and modifying environmental factors that influence the disability negotiation process. These environmental factors may include labor relations, safety practices, physical job design, job stress factors, coworker relationships, reemployment policies, adversarial claims management techniques, and other characteristics of the physical and psychosocial dimensions of the work environment. Table 34.2 presents examples of reasonable accommodations achieved through the use of ergonomic interventions in actual cases of workers who participated in transitional work return programs.

Specific job functions of the ergonomist in the disability prevention and management processes include the following:

1. Evaluating the work environment to identify risk factors that contribute to physical injuries or psychosocial stresses
2. Participating as a member of a professional, interdisciplinary (worksite) rehabilitation team; performing ergonomic job analysis for participants in transitional work return program when needed
3. As a member of the worksite rehabilitation team, advising physical/occupational therapist, vocational rehabilitation counselor, physician, and others regarding ergonomic issues
4. Serving as member of joint labor/management steering committee and advising members during job accommodation process; participating in ADA compliance policy development and making recommendations for specific job accommodations
5. Establishing and serving as member of ergonomic task force to identify and resolve risk factors; assisting in data collection process
6. Participating in staff development by providing educational programs for management, labor, employees, and rehabilitation professionals
7. Collaborating with internal engineering staff, supervisors, and workers in designing and implementing job modifications/accommodations

TABLE 34.2 Examples of Actual Ergonomic Hazard Prevention Methods and Engineering Controls

| Type of Industry | Problem | Control | Est. Cost |
|---|---|---|---|
| Manufacturer of electronic automobile parts | CTS from using a standard screw driver to "pry" distributor parts for adjustment | Covered screw driver handles with thermo plastic material that can be custom molded and remolded for each operator | \$25/handle |
| Manufacturer of electronic automobile parts | Hand/wrist tendinitis, CTS from cutting plastic tabs manually using diagonal cutters | Replaced manual cutters with an automatic trimming machine that uses a pneumatic rotary actuator and four hardened steel blades | \$1300 |
| Manufacturer of electronic automobile parts | CTS from removing automobile distributors from an assembly line, rotating them 90°, placing them in a fixture, and installing two screws using a suspended in-line powered driver | Replaced in-line screwdriver with a suspended pistol-grip model to install the screws while the distributor remains on the assembly line | \$1500 |
| Manufacturer of fiberglass compression-molded breaker boxes | CTS from manually filing to remove flashing | Covered all file handles with thermo plastic that can be custom molded for each operator | \$8/handle |
| Manufacturer of fiberglass compression-molded breaker boxes | CTS from manually filing to remove flashing | Purchased pneumatic belt sanders, bench-mounted and foot-controlled | \$400/unit |
| Manufacturer of fiberglass compression-molded breaker boxes | CTS from manually cutting fiberglass using straight knives | Replaced straight knives with 35° curved knives | \$8/knife |
| Manufacturer of fiberglass compression-molded breaker boxes | CTS from manually cutting fiberglass using straight-handled utility knives | Installed adjustable-height pneumatic "guillotines" | \$3800/unit |
| Manufacturer of large commercial trucks | Back injury from opening large sliding door with one hand | Replaced small door handle with a vertical two-hand "grab bar" | \$40 |
| Manufacturer of large commercial trucks | Left shoulder injuries from repeatedly lifting and pulling 40 lb lubrication and fuel hoses | Recommended overhead rail and balances system to suspend hoses and ease handling | \$6000 |
| Manufacturer of large commercial trucks | CTS and low back and shoulder pain from poor seating posture while installing dashboard wiring | Replaced metal stools with adjustable polyurethane workseat | \$260/stool |

*Source:* Courtesy of J. DeWees, Ergo Accommodations, Union, KY.
CTS, carpal tunnel syndrome.

## 34.9 CONCLUSION

Although many employers initially interpreted the employment-related provisions of ADA to be primarily concerned with preventing discrimination in the *hiring* of people with disabilities, recent EEOC data indicate that the majority of complaints filed thus far have originated from current or former workers with either preexisting or work-related disabilities [25]. ADA is accordingly having a significant impact on the reemployment practices of employers and unions who have traditionally refused to return a worker to the job unless he/she was able to perform all aspects of the pre-injury job. More emphasis must now be given to using reasonable accommodations not only in the personnel selection process but also as a method of job retention because many workers who sustain work-related injuries are protected by Title I of ADA.

In 1991, total workers' compensation costs in the United States were estimated at $60 billion per year, while medical costs for work-related injuries were increasing 50% faster than non-work-related injuries [26]. Employers and insurers may find it more prudent, from both the economic and legal standpoints, to invest more effort in seeking environmental solutions to disability. Research indicates that altering the characteristics of jobs rather than focusing only on the nature of the worker's impairment can substantially reduce work loss [27]. Many employers are beginning to realize the advantage of developing worksite rehabilitation programs as an alternative to the more costly approach of waiting for the worker to achieve "full recovery" in a nonwork environment. These employers are reporting significant reductions in lost work time and substantially lower compensation and rehabilitation costs [25,28,29].

Ergonomic interventions play a central role in the disability management model of prevention and rehabilitation. The science of ergonomics is helping to shift the context of industrial rehabilitation services from medically oriented environments to actual work environments. Work disability must be understood in the context of the employment situation, and the rehabilitation process must be extended as much as possible to the workplace [17].

Worksite rehabilitation programs allow for more accurate assessments of the worker's functional capacities as they interact with the demands of the job and characteristics of the work environment. A primary objective of delivering rehabilitation services in the actual work environment is to prevent or contain the phenomenon known in the disability research literature as "spread" [7,30]. The delivery of rehabilitation services in the context of the real work environment decreases the chances of inappropriately "spreading" or generalizing a specific functional limitation to all work tasks. Unfortunately, many workers have been deemed vocationally disabled on the basis of third party job analysis data and standardized functional capacity information that are characteristic of facility-based rehabilitation programs. The worksite clinician, however, has the advantage of observing the interaction between a worker's capacities and required work tasks. This type of clinical observation generates a worker–job compatibility profile and pinpoints very precisely what tasks the worker is capable or incapable of performing. This information forms the basis of the job accommodation analysis process and identifies the need for possible ergonomic interventions as a method of reducing or eliminating discrepancies between the worker's abilities and job requirements. Worksite rehabilitation programs mitigate the "all or nothing at all" approach that marks the reemployment practices of many employers and unions.

The use of ergonomic interventions in the rehabilitation process ensures that the environmental side of disability is addressed. As employers attempt to comply with the reasonable accommodation requirements of Title I, more emphasis will be placed on seeking environmental solutions to disability.

The ultimate goal of ADA is the total integration of people with disabilities into all aspects of the American society. The primary objective of vocational rehabilitation is the successful integration and retention of people with disabilities in the world of work. If either of these missions is to be fully realized, vocational rehabilitation interventions must reflect the true nature of disability and develop strategies that deal with organizational, environmental, and political variables that contribute to the high rate of unemployment among people with disabilities. The implementation of rehabilitation services at the actual workplace ensures the inclusion of environmental and organizational disability factors in the rehabilitation process. Conversely, industrial rehabilitation services that tend to segregate people with disabilities in facilities that rely on simulated work activities or synthetic work environments may ironically promote the discriminatory stereotype that people with disabilities do not belong in the workplace. Hopefully, the increased use of ergonomic interventions in rehabilitation will result in services that are more congruent with the ecology of the world of work and more effective in preventing environmentally induced disability.

## REFERENCES


1. P. Morrissey, The Americans with disabilities act, Report to the U.S. House of Representatives, Committee of Education and Labor, Washington, DC, 1990.
2. F. Bowe, Development of the ADA, in *The Americans with Disabilities Act: Access and Accommodation*, N. Hablutzel and B. McMahon, Eds., Paul M. Deutsch Press, Orlando, FL, 1992, pp. 3–10.
3. B. McMahon and L. Shaw, Considerations for the rehabilitation consultant, in *The Americans with Disabilities Act: Access and Accommodation*, N. Hablutzel and B. McMahon, Eds., Deutsch, Orlando, FL, 1992, pp. 199–211.
4. H. Gertsman, Environmental barriers: Questions and answers, in *The Americans with Disabilities Act: Access and Accommodations*, N. Hablutzel and B. McMahon, Eds., Deutsch, Orlando, FL, 1992, pp. 225–235.
5. EEOC, *A Technical Assistance Manual of the Employment Provisions (Title I) of the Americans with Disabilities Act*, U.S. Equal Employment Opportunity Commission, Washington, DC, 1992.
6. H. Hahn, A sociopolitical perspective, in *Economics and Equity in Employment of People with Disabilities*, R. Habeck, D. Galvin, W. Frey, D. Tate, and L. Chadderdon, Eds., University Center for International Rehabilitation, Michigan State University, East Lansing, MI, 1985, pp. 18–21.
7. C. DeLoach, When attitude begets impairment, in *The Americans with Disabilities Act: Access and Accommodations*, N. Hablutzel and B. McMahon, Eds., Deutsch, Orlando, FL, 1992, pp. 11–33.
8. D. K. Mitchell and S. Leclair, *Strategic Case Resolution, Enhancement, and Management Systems*, International Center for Industry, Labor, and Rehabilitation, Dublin, OH, 1988.

9. C. Castillo-Salgado, Assessing recent developments and opportunities in the promotion of health in the American workplace, *Soc. Sci. Med.* *19*(4):349–358 (1984).
10. E. Hester and P. Decelles, *The Worker Who Becomes Physically Disabled: A Handbook of Incidence and Outcomes*, Menninger Foundation, Topeka, KS, 1988.
11. M. Berkowitz, Should rehabilitation be mandatory in workers' compensation programs? *J. Disabil. Policy Stud.* *1*(1):63–80 (1990).
12. D. Hershenson, A theoretical model for rehabilitation counseling, *Rehabil. Counsel. Bull.* *33*(4):268–278 (1990).
13. D. K. Mitchell and S. Leclair, Building a working alliance with employers: The politics of work disability, in *Physical Medicine and Rehabilitation: Rehabilitation of the Injured Worker*, E. Johnson, Ed., Saunders, Philadelphia, PA, 1992, pp. 647–664.
14. S. Akabas, L. Gates, and D. Galvin, *Disability Management: A Complete System to Reduce Costs, Increase Productivity, Meet Employee Needs, and Ensure Legal Compliance*, Amacom, New York, 1992.
15. D. Galvin, Health promotion, disability management, and rehabilitation in the workplace, *Rehabil. Lit.* *47*(9):218–223 (1986).
16. R. Habeck, Implementing disability management: Programs in industry, *Paper Presented at the Meeting of the Center for Advancement of Industrial Rehabilitation and Evaluation*, Denver, CO, 1989.
17. R. Habeck, Managing disability in industry, *NARPPS J.* *6*(4):141–146 (1992).
18. R. Habeck, M. Leahy, H. Hunt, F. Chan, and E. Welch, Employer factors related to workers' compensation claims and disability management, *Rehabil. Counsel. Bull.* *34*(3):210–226 (1991).
19. E. Hester and P. Decelles, *The Effect of Employer Size on Disability Benefits and Cost-Containment Practices*, Menninger Foundation, Topeka, KS, 1990.
20. P. Rousmaniere, Too many cooks, no chefs, *Health Manage. Quart.* *11*:2 (1989).
21. S. Pheasant, *Ergonomics: Work and Health*, Aspen, Gaithersburg, MD, 1991.
22. S. Kumar, Rehabilitation: An ergonomic dimension, *Int. J. Ind. Ergon.* *9*(2):97–108 (1992).
23. D. Shrey and J. Olsheski, Disability management and industry-based work return transition programs, in *Physical Medicine and Rehabilitation: State of the Art Review*, C. Gordon and P. Kaplan, Eds., Belfus and Hanley, Philadelphia, PA, 1992, pp. 303–314.
24. D. Shrey and R. Breslin, Employer-based disability management strategies and work return transition programs, in *The Americans with Disabilities Act: Access and Accommodation*, N. Hablutzel and B. McMahon, Eds., Detusch, Orlando, FL, 1992, pp. 139–154.
25. R. Groepper, Structured return-to-work and the Americans with disabilities act, *Healthcost Monitor*, *2*:3 (1993).
26. R. Victor, Workers compensation at a crossroads, Fifth Annual John R. Commons Lecture, University of Wisconsin, Madison, WI, 1992.
27. E. Yelin, The myth of malingering: Why individuals withdraw from work in the presence of illness, *Milbank Quart.* *64*(4):622–647 (1986).
28. M. Weinstein, Many happy returns, *Risk and Insurance*, *6*(5):1–15, May 1995.
29. M. Padgett, E. Hollander, L. Warden, R. Coleman, and R. Schwartz, Evaluation of a model return-to-work program in Texas, *Work* *3*(3):21–41 (1993).
30. B. Wright, *Physical Disability: A Psychological Approach*, Harper and Row, New York, 1960.

CHAPTER 35

# Legal Aspects of Ergonomics

James J. Montgomery

CONTENTS



## 35.1 INTRODUCTION

On March 10, 1980, Michael Village had no idea that he was changing the course of workers' compensation law in the state of Ohio—all he knew was that his back ached. Village worked at an automobile assembly plant. He had worked there in various capacities since 1973, but for the last 5 days, he had been working as a battery securer: he chose batteries from a rack, each weighing between 20 and 40 lb, and installed them in automobiles. On March 10, 1980, he reported to the plant dispensary with a backache; on March 11, he was unable to get out of bed [1]. Village was a victim of a common ailment, in a common work setting.

Village filed a workers' compensation claim alleging that his injury occurred in the course and scope of his employment. Though the Industrial Commission allowed his claim, on the employer's appeal, the decision was reversed by the trial court, and the reversal was affirmed by the court of appeals. To be compensable, the lower courts ruled that "[T]he injury must be accidental in character as a result of [a] sudden mishap and not in the usual course of events." This rationale, imposed by the Supreme Court of Ohio as recently as 1978 [2], was the law in Ohio when Village developed his backache, and it was the law that the lower courts were bound to follow until it was changed by the Ohio State Supreme Court in *Village v. General Motors*. Reversing a long line of cases, the court [2] held that

> An injury which develops gradually over time as a result of the performance of the injured worker's job-related duties is compensable.

With the release of this decision, Ohio industry was suddenly exposed to an entirely new category of compensable injuries, significant expense, legal involvement, and an affirmative obligation to minimize or eliminate situations that create gradually developing injuries. Workers with such injuries were now able to obtain disability classifications and draw compensation for what used to be considered [2] predictable and expected noncompensable "normal wear and tear" injuries. After *Village v. General Motors*, Ohio industry had a tangible economic incentive to consider ergonomics.

## 35.2 BACKGROUND AND SIGNIFICANCE TO OCCUPATIONAL ERGONOMICS

Similar issues have been resolved, or are pending, in other states. Ergonomics, developing as a distinct discipline for decades, rose to prominence in the 1980s, in the field of occupational safety and health. With recent and evolving research on the identification and prevention of cumulative trauma disorders, industry is rapidly becoming sensitive to the need for ergonomic intervention to prevent injuries that occur over time in the normal course of work. This book is an example of that awareness: what constitutes "normal course of work" is being redefined and redesigned, from biomechanical, physiological, and psychophysical perspectives, on subjects as diverse as computer keyboards and materials handling systems.

Lawyers are now heavily involved. Since state courts began to recognize repetitive stress injuries as compensable, state workers' compensation systems have become flooded with claims. Numerous suits pend for repetitive stress injuries caused by the allegedly inadequate design and manufacture of keyboards, keypunches, alphanumeric machines, video display terminals, cash registers, supermarket workstations, stenographic machines, and computer "mouse" devices. Entire industries have been sued for noise-induced hearing loss—industries as diverse as pneumatic air tools, diesel engines, and tire-making equipment. Repetitive stress injury suits were unheard of in the 1970s; they literally did not exist. Now they constitute one of the fastest growing areas of the legal history. Practicing ergonomists and health and safety professionals must be aware of emerging legal precepts in the context of their disciplines. Ergonomists should expect and prepare for additional, significant involvement in administrative and litigation matters and should know and practice the standard of care to which they will be held.

## 35.3 WORKERS' COMPENSATION

Historically, state governments developed workers' compensation plans to meet the needs of injured workers who had difficulty recovering for injuries under the adversary system. In order to recover, injured workers had to prove some fault on the part of their employer; employers, in addition, could assert that the worker was negligent, or voluntarily assumed the risk of injury, to thwart these claims. Increasing recognition of the many uncompensated injuries led, in the first few decades of this century, to the widespread adoption of no-fault compensation schemes as the exclusive remedy for workers injured on the job. The workers did not have to establish fault; the employers did not have a defense. The workers received a certain and speedy recovery; the employers received a more limited liability [3].

For most of this century, the compensation systems required that the injury be traceable to a definite time, place, and cause [4], that is, a discernible "accident." This standard, by definition, excludes repetitive stress and cumulative trauma disorders, which, though scientifically recognized, were judicially and legislatively declared to be predictable, expectable, yet noncompensable elements of normal work. The only motivation, then, for ergonomic awareness was either a self-generated concern for the health of employees or the goal of increased production. If a worker simply "wore out," compensation was not available because he could not demonstrate a "precise moment of collapse and dysfunction" [5].

In the 1980s, decisions like that of *Village v. General Motors* became more commonplace. Ergonomic concerns appeared in reported decisions:

> We consider the implication of this rule for all of the employees, factory workers, supervisors, managers, secretaries, salespeople and others, working in Illinois in this technological age. In real life, the erosion of a bodily structure to the point of uselessness translates into arms that cannot lift, legs that cannot walk, knees that cannot bend, lungs that cannot breathe, and eyes chronically irritated or worse. But evidence of such work related injuries alone is not sufficient under the prior interpretations of "accidental injury." Instead, useless limbs, damaged organs and disabled bodies must be pushed to a precise moment of collapse and dysfunction. Then, and only then, according to these interpretations, may a court of this state find an employee eligible for compensation under the Act.
>
> The time has come to abandon an interpretation of "accidental" which fails to address documentable and medically recognizable risks faced by the individuals in connection with their employment. The risk of injury from repeated trauma and exposure endured by truck drivers, CRT operators, chemists and others must be recognized. The judicial interpretation of "accident" must be refined to reflect the purpose of the act and the reality of employees obligated to perform repetitive tasks [5].

Today, motivated by concern for worker health and safety and the economic implications of enlightened compensation systems, employers and ergonomists have an increased responsibility to eliminate or minimize the risk of cumulative trauma or repetitive stress injuries. A heightened awareness of the role of ergonomics is reflected in OSHA's proposed rule on ergonomic safety and health management [6], which is the federal response to the "significant increase in the reported cases of ergonomic disorders in the work place [which have] more than tripled since 1984" [6]. Purely altruistic motivation, while admirable, is simply not as effective as the prospect of punitive fines and awards, so the evolution in law described here should provide additional motivation. Ergonomists have a job to do. If they fail to perform that job effectively, workers will be at risk and employers will face significantly increased compensation costs, administrative regulation, fines, litigation, and potential punitive damage exposure.

## 35.4 INTENTIONAL TORTS

With ergonomic analysis and intervention, cumulative trauma disorders are quantifiable and predictable. For example, the ergonomic investigation in the John Morrell & Co meatpacking plant disclosed an upper extremity CTD incident rate of an astonishing 41.7 per 100 full-time workers per year and identified intermediate to high risk factors for 171 out of 185 analyzed jobs [7]. One of the conclusions advanced in the NIOSH report was that "ergonomic job analysis revealed the majority of jobs require tasks that are *known risk factors for developing upper extremity CTDs*" (emphasis added) [7]. The legal implications of such conclusions are staggering.

Consider Ohio as an example. As recently as the 1978 decision of *Bowman v. National Graphics* [2], the Ohio courts held that gradually developing disabilities may well be predictable and expected yet were not compensable under the workers' compensation system. From such decisions, industry could justifiably conclude that it was not responsible for normal wear and tear associated with the long-term performance of job activities. However, at the same time the *Bowman* decision was being written, a worker named Blankenship and seven of his coworkers were feeling wronged as a result of their claimed exposure to dimethyltin dichloride at the chemical plant at which they worked. These eight workers filed suit against their employer in 1979, claiming that they were entitled to file such a direct action, outside of the workers' compensation system, by virtue of the employer's allegedly intentional conduct in exposing the workers to hazardous chemicals [8]. The Supreme Court of Ohio held that the exclusive remedy provisions of the workers' compensation act did not preclude a civil suit against an employer for its intentional conduct. Although the *Blankenship* case did not involve ergonomic issues, it did engender a line of cases and legislation affirming the right to sue and defining the standard of proof to establish intent [9], thereby setting, if not springing, the trap for the unwary ergonomist who must evaluate and predict the occurrence of cumulative trauma disorders in an occupational setting.

An intentional tort claim has, as its major component, the allegation that the perpetrator—for our purposes, the employer—intended to harm the employee. With proof of intent comes the major reward of an intentional tort claim: punitive damages—damages awarded over and above the amount necessary to compensate injured parties for their loss and damages that are awarded to punish the offender and serve as an example to other potential offenders that such conduct will not be tolerated. Punitive damages are unpredictable, at best; the large awards generate publicity and provide impetus to otherwise latent claims of a similar nature.

Interpreted in the context of its plain and ordinary meaning, "intent" would not appear to present much of a problem for the ordinary employer–employee relationship. Short of an assault, no one ever actually "intends" to harm an employee. Unfortunately, the law is not accustomed to dealing with ordinary and simple understandings, and definitions of "intent" are very qualified, resulting in widely ranging interpretations.

Again, Ohio is representative of the national trend on these issues. To establish the necessary intent for an intentional tort claim against an employer, the employee must demonstrate

> (1) knowledge by the employer of the existence of a dangerous process, procedure, instrumentality or condition within its business operation; (2) knowledge by the employer that if the employee is subjected by his employment to such dangerous process, procedure, instrumentality or condition, then harm to the employee will be a substantial certainty; and (3) that the employer, under such circumstances, and with such knowledge did act to require the employee to continue to perform the dangerous task [10].

Mere knowledge, or appreciation of a risk short of substantial certainty, is not intent. However, if employers know that injuries to employees are certain or substantially certain to occur, and they still proceed, the law presumes intent. The qualifier, and perhaps sole comfort, in this equation, is contained in the following hierarchy:

> To establish an intentional tort of an employer, proof beyond that required to prove negligence, and beyond that to prove recklessness must be established. Where the employer acts despite his knowledge of some risk, his conduct may be negligence. As the probability increases that particular consequences may follow, then the employer's conduct may be characterized as recklessness. As the probability that the consequences will follow further increases, and the employer knows that injuries to employees are certain or substantially certain to result from the process, procedure or condition, and he still proceeds, he is treated by the law as if he had in fact desired to produce the result [10].

These quotes should be read carefully: they reflect court-imposed employer standards to determine if conduct is intentional and punitive damages are appropriate.

Consider the following situation. You are the ergonomist for a large insurance company that employs hundreds of clerical workers. Each clerical worker has a standard microcomputer used for data entry and text editing. You are aware of the variety of upper extremity disorders that can occur from the use of conventional keyboards in classic workstations, and you are trying to respond to the individual needs of the keyboard operators as inquiries and complaints come in. As a scientist and ergonomist, you know that, with a population as large as the one you are monitoring, there is a substantial certainty that at least one or some of the keyboard operators will experience a disabling upper extremity disorder as a result of repetitive stress or cumulative trauma. You continue to respond, as responsibly as you can, to the needs of the clerical population. Eventually a member of that population develops and reports a serious disabling condition stemming from the long-term use of the conventional keyboard.

The medical reports confirm that the disability is related to the long-term keyboard operation, just as you predicted. In addition to the claim for a permanent partial disability, the clerical worker sues your company, claiming an intentional tort and punitive damages.

The attorney for the clerical worker has obtained, through normal discovery, copies of your entire file, including those notations or memoranda that reflect your awareness that some portion of the clerical population is at risk for an upper extremity disorder. Using the standards mentioned before, the attorney contends that your actions were intentional because you knew that prolonged use of a conventional microcomputer keyboard had a potential for danger that was substantially certain to occur within the population you were monitoring. With this knowledge, you nonetheless permitted the work to continue.

You, of course, are astonished at these allegations, for you have been striving to do your job properly by being as responsive to the needs of the clerical population as you could be under the circumstances. You believe that you acted reasonably and do not believe that the statistical probabilities of your own science should be the predicate for your and your company's culpability. Unfortunately, the law is not that clear: you were aware of the problem and allowed the condition to continue even though someone was bound to be affected. Established case law supports the proposition that your actions were beyond reckless, simply because there was a substantial certainty that harm to an employee must result.

This is an uncomfortable example, for it is not a situation where a safety device was disabled or a chemical exposure tolerated, where circumstances would make it certain that injury would result. This is not a situation that was contemplated by the courts when they formulated the standards for determining intent. However, compensable injuries are the responsibility of the employer, and cumulative trauma or repetitive stress disorders are compensable. If you have a population where an injury is substantially certain to occur, then the potential for punitive exposure theoretically exists. Given a theoretical punitive exposure, there is also a substantial certainty that some lawyer is going to jump on that opportunity as soon as it arises. In order to combat such claims, ergonomists need to do their jobs in the professional and responsible manner in which they were trained. The court-imposed standards, admittedly subjective, will govern the analysis regarding responsibility and intent. At a minimum, the ergonomist has the affirmative obligation to act reasonably under the circumstances, for an act of negligence occurring after the employer or ergonomist realizes the substantial certainty of injury may be construed as intent.

The reasoning preserved in another case, *Cantrell v. GAF Corporation* [11], is instructive. This case involved the exposure of certain manufacturing plant workers to asbestos. The plaintiffs presented proof of the employer's awareness of hazards relating to asbestos exposure, including evidence of the employer's failure to heed the recommendations and warnings of a physician-hygienist who was specifically retained to investigate the problem. Instead, the employer terminated the physician-hygienist and failed to advise the employees of the risks associated with their asbestos exposure. The employer argued lack of specific intent to injure. The Sixth Circuit Court of Appeals placed great emphasis on the conduct of the employer:

> [the employer], with knowledge that injury and even death were substantially certain to result from asbestos exposure, continued to expose its employees to asbestos, and followed a policy of not explaining to its employees when medical screening revealed the existence of a disease [12].

With this focus on the employer's failure to warn and advise, and the employees' continuing exposure, the court held that "where the employer acts with knowledge that injury is a substantial certainty, the employer has gone beyond negligence, recklessness and wantonness and committed an intentional tort" [11]. The multimillion dollar punitive damage verdict against the employer was upheld.

The *Cantrell* court, when presented with the substantial certainty that a particular condition would result in injury over time, analyzed the *conduct* of the employer and its professional staff. Had the employer heeded the advice of its professionals and followed through with a program of education and prevention, there would have been no punitive verdict. Had the employer acted reasonably, the worst that would be anticipated would be the no-fault compensation claim for injuries sustained at work. If the employer and its professionals had acted in a responsible manner, the potential for punitive exposure would have been minimized.

## 35.5 STANDARD OF CARE FOR AN ERGONOMICS PROFESSIONAL

The ergonomist, as a professional, has an obligation to act, in the practice of his or her profession, with that degree of care, skill, and diligence commonly exercised by similar professionals under similar circumstances. If the professional does not act in accord with that standard, he or she is negligent, reckless, or worse. If that negligence, recklessness, or "intent" causes harm to a worker, lawyers appear. Simply stated, a cause of action is premised on (1) a duty, (2) which is breached, (3) causing injury.

It is relatively easy for a professional to avoid advancing from negligence to recklessness and beyond in the analysis and treatment of ergonomic problems: Do the job in a professional manner and be able to prove it. Consider, again, the example of the ergonomist charged with the responsibility of monitoring the clerical population. Ergonomics, since it analyzes the manner in which specific individuals work in specific environments, is customization in accord with generalized principles: The workstation for clerical worker No. 1 may not be suitable for worker No. 10, though there will be certain commonalities between them. In such a context, the professional must respond to the entire population with generalized precepts and meet the needs of the specific workers who need ancillary back support, wrist rests, specialized keyboards, trackballs, individually adjustable chairs, etc. Interaction among the professionals and the worker population, essential to the surveillance–cure cycle of ergonomics and the ability to respond to specific needs, is also critical to the legal process. If there is one overwhelmingly significant fundament to the decisions assessing punitive measures against employers, it is in the failure to communicate. The courts look for an exchange of information; they look for advice, cautions, and warnings to the workers from the professionals who are supposed to be watching out for them. Where there is proof of such interaction, it will be difficult for a court to find facts to support a punitive damages award. In *Cantrell*, the federal court of appeals was obviously influenced by the fact that the employer discharged its consultant, ignored his recommendations, and kept the plant population ignorant of the potential risks presented by the asbestos used in the manufacturing process. Recast the facts: The employer, now aware of a problem through interaction with its consultant, follows the recommendations about plant hygiene and surveillance, informs the workers of the risks it has discovered and the program it is establishing, and works with the population to minimize future risk. Will workers who develop asbestos-related problems after a latency period be compensated? Yes; that is precisely what the compensation system was designed to do. Will the workers have a case for an intentional tort and punitive damages? No. The employer responded to the emergent situation with appropriate safeguards and information. The employer acted responsibly. The employer did not conceal information critical to the health and safety of its worker population.

As a profession, ergonomics depends upon employer–worker interaction and worker awareness. Such an exchange, in the legal context, becomes a self-checking process. If ergonomists do their jobs with appropriate interaction, then the information exchange so critical to the legal process is a natural result of a job well done. Proof of that interaction may require some extra effort, for inherent in the nature of injuries that develop over time is the fact that memories fade over that same period of time. Contact records should be retained. Follow-up inquiries not only provide written confirmation of the resolution of a problem but also exist to show that the problem was addressed and resolved. Document all professional, supervisory, and worker involvement in the process. Keep track of announcements, postings, notices, meetings, and other publicity. Since documentation, again, is necessary to the overall ergonomic analysis of the process, the legal end of things is covered by simply doing the job in a professional manner. The standard of care for an ergonomics professional is to perform the job professionally.

## 35.6 OSHA ERGONOMIC STANDARD

OSHA's proposed rule on ergonomic safety and health management [6] has generated a significant amount of attention because it is intended to codify ergonomic concerns and establish an administrative process for review, comment, and ultimate enactment. When enacted, the standard will be valuable as a benchmark, or reference, standard for all industry. Employers with existing ergonomics programs will probably not be affected by the compliance requirements of the final rule, for such employers will already be familiar with the science, surveillance, and interaction involved in any ergonomics program. In addition, courts have held that ergonomic considerations are implicit in the OSHA General Duty clause [12], giving rise to the conclusion that specific ergonomic standards are not necessary to enforce ergonomic concerns in any event.

No standard can address every potential situation. It is important to keep in mind that, although compliance with an OSHA ergonomic standard will be mandatory, with fines and enforcement provisions for noncompliance, compliance alone will not guarantee a liability-free existence. The test is the standard of care expected of industry and the ergonomics professionals discussed earlier. That includes, of course, compliance with all statutory and administrative regulations. If a situation arises that, in the opinion of the professional, requires some action beyond that set forth in any administrative regulation, then the professional opinion, geared to the specific facts under analysis, becomes the standard of care. Consider the following example.

In warehousing operations, there is considerable manual handling of loads (see Chapter 28). If, for example, a rule were promulgated that limited the height from which heavy loads could be lifted to a minimum of 20 in. (50.8 cm), then the law must be followed in its prescriptive sense: it exists as a minimum requirement and if violated subjects the employer to sanctions and fines. As any properly trained ergonomist knows, or should know, the 20 in. minimum height for lifting heavy loads is based on ranges of "standing knuckle height" (see Chapter 28) for some ergonomically significant proportion of the population. However, if the population at your specific warehouse consists of individuals 6 ft 6 in. tall or taller, then mere "compliance" would not be in accord with the standard of care; standing knuckle height for such a tall population is probably something on the order of 32 in. Since a fundamental component of ergonomics consists of fitting the job to the person and "matching the capabilities of specific workers" [6], the 20 in. minimum lift-station height would be ergonomically inconsistent with your specific population. And the provision of a 20 in. lift station would not be in accord with the applicable standard of care. Because each situation has the potential for its own, individually analyzed standard of care, reference standards can only prescribe where to *begin* an analysis. An ergonomist provides custom services consistent with recommended guidelines.

## 35.7 AMERICANS WITH DISABILITIES ACT

The American justice system and the legal profession are frequently derided for the ponderous nature of their proceedings and their obvious preoccupation with economics and wealth redistribution. Bad jokes and egos aside, the Americans with Disabilities Act, discussed elsewhere in this book, will necessarily involve a significant amount of ergonomist, attorney, and court time. An act so fundamentally premised upon individual accommodation must be vague and therefore subject to interpretation, and case-specific analyses will be required for any disputed failure to comply with the act's provisions. Ergonomists obviously play a major role in the application of the act, for it will fall to them to consider, analyze, discuss, reject, and implement accommodations for the industry and individuals they serve. Ergonomists evaluate and design "facilities, environments, jobs, tasks, tools, equipment, processes and training methods to match the capabilities of specific workers" [6], and the ADA *mandates* such matching for the disabled.

The ADA is enforced by an individual's ability to place employers on trial for violations of civil rights. The enforcement provisions are the same as those used to enforce Title VII of the Civil Rights Act of 1964. The Equal Employment Opportunity Commission (EEOC) receives and investigates charges of discrimination, and if it is unable to mediate or resolve differences, the EEOC might elect to file suit, or it might issue a "right to sue" letter to the person making the charge of discrimination. Available relief includes hiring, reinstatement, accommodation, attorneys' fees, expenses, compensatory damages, and punitive damages according to a scale.

If an ergonomics professional or an employer is placed on trial for a violation of civil rights under the ADA, it is important to remember that neither will be judged by other ergonomists. Judges and juries will assess compliance, and it will be the obligation of the professionals to prove that their actions were reasonable under the circumstances. The greatest weakness—and strength—of the ADA is that it is vague and subjective; ergonomists are relatively free to use their best judgment and will be judged by what is reasonable—that is, whether they complied with the standard of care.

The recently publicized Rodney King cases—one set of facts, two disparate verdicts—illustrate the unpredictable nature of juries and trials. Jurors available for the jury pool will be similar individuals, selected from the voting population at random, who have the time and ability to serve without hardship. Consequently, a broad cross section of people will consider any ADA claim. The ergonomists' job, when they defend an employer's actions under the ADA, is to explain the complexities of the situation and the significance of the decisions made. The plaintiff, or claimant, has the affirmative burden of proof; the claimant has to present facts that, if believed, would indicate that there had been a violation of the provisions of the act. However, the jury will tend to give the claimant the benefit of the doubt and will perceive the ergonomist's role to be that of explaining just why the company doesn't think it owes this claimant a job or at least some money. Sympathy is generally thought to lie with the individual, not the employer.

Read the act: 29 CFR 1630. Read it all. Give it to the company attorney and to the management. Read the actual text of the law, the interpretive guidelines, and the EEOC technical assistance manual [13].

The act itself is designed to remove the barriers that prevent qualified individuals with disabilities from enjoying the same employment opportunities as are available to persons without disabilities. It is designed to ensure access to equal employment opportunity based on merit. It accomplishes these goals by legislating that where an individual's disability creates a barrier to employment, the employer must consider whether a reasonable accommodation could remove that barrier. Where an individual's functional limitation impedes job performance, an employer must take steps to reasonably accommodate, and thus overcome, the particular impediment, unless to do so would impose an undue hardship. This is not, however, something to be accomplished by relieving the job applicant of the obligation to perform the essential function of the job itself.

As an illustration of how the process is designed to work, consider the situation of a sight-impaired individual who has applied for a job as an all-purpose executive secretary/administrative assistant. The essential functions of this job are many and varied, including frequent movement about the entire office and plant; out-of-town, overnight travel; and the operation of a large number of office machines, such as a copier, computer, fax, dictaphone, shredder, telephone console, printer, and mail machine. Fully half of the job requires reading, document handling, sorting, abstracting, analyzing, composing, and keyboard entry.

It takes little analysis to conclude that a sight-impaired individual cannot perform many of these essential functions, especially when they are combined into one job. With such a disability, the more frequently considered "accommodation" is the provision of a pair of eyes—a reader or leader. With our theoretical applicant, the question arises, as a matter of law, whether it is a reasonable accommodation to provide a reader or leader to assist the disabled applicant with the job.

With this job description, the answer is undoubtedly no, for the reader, coach, or guide would actually have to perform and duplicate certain of the essential elements of the job: part of any clerical position is the nonlinear, simultaneous analysis of visual input from multiple sources, such as the letter to which the manager is responding, its attachments, the letter written by the manager, its attachments, the keyboard, screen, dictation machinery and controls, and the layout of the desk. A reader is limited to linear output; an executive secretary/administrative assistant requires simultaneous input from a variety of sources. A "reasonable" accommodation does not contemplate hiring an additional individual to actually do the job, or any "essential" portion of it, for the disabled applicant.

By contrast, consider another example. The position available is that of quality control inspector and sealer on the output line in a bulk potato chip operation. The process station is at that portion of the line where the chips are in an unsealed polyethylene bag that has been placed in a 5 gal can. The essential functions of this job are to look in; examine the contents for contamination, quantity, and breakage, sometimes lifting the bag out of the can to examine the sides and bottom; guide the bag back into the can; twist the bag; apply a wire tie to the bag; place a lid on the can; and go on to the next can. The cycle takes about 6 s. The applicant, otherwise qualified and experienced, has a diagnosed cumulative trauma syndrome, such as carpal tunnel syndrome or thoracic outlet syndrome, and cannot twist a wire tie 2000 times per shift. However, the applicant wants the job and asks for a reasonable accommodation.

The employer is now required to work with the applicant to determine potential reasonable accommodations. The essential functions of the job are inspection, bagging, moving the can, and sealing the bag. The applicant cannot seal the bag with the wire tie currently in use. The first accommodation proposal is a device, similar to that frequently seen in produce departments, that tapes bags shut with a straightforward push of the bag's neck through a slot in the device. Though the proposition sounds reasonable, the sales department quite correctly points out that the container used is the large economy-size can, not an individual portion that is opened only once. The company has to provide, in order to remain competitive, a reusable, simple-to-use closure device, and the tape used on the proposed device is a one-time, nonreusable closure. Through the dialogue that follows, a second accommodation is proposed: a reusable plastic tab, frequently used to seal bread packages, roughly square, with a serrated notch. The employer agrees, but the applicant points out that use of such a device requires a push-twist motion almost as aggravating as the wire-tie twist. Instead of the low-load, high-repetition twist required by the wire tie, there is an equally destructive high-load, low-repetition motion required by the plastic tab. What is the solution? Can the plastic tab or the wire tie be machine or tool applied? The plant ergonomist should be consulted to see if another accommodation is reasonable.

These simplistic examples illustrate the process required by the ADA to determine if an accommodation is possible and whether that accommodation is reasonable. The ergonomist will be an intimate part of this process, and if disagreement results, the ergonomist will be involved in its resolution as well. Whether involved in an accommodation effort or an actual dispute, the ergonomist will need to concentrate on defining the essential functions of a job, analyzing accommodations proposed for specific disabled individuals who are otherwise qualified, analyzing and proving the defense of substantial hardship, and, finally, designing and implementing workplace ergonomic controls to reduce and prevent injuries and to minimize aggravation of preexisting conditions. While this last task sounds like a general job description for a professional ergonomist, it is, in fact, an essential element of compliance under the ADA.

We have already discussed the *Village* decision where a court held that gradually developing injuries arising out of the performance of work are compensable. Courts have also held that a physical injury occasioned solely by mental or emotional stress incurred in the course and scope of employment is also compensable if the injury results from greater strain or tension than that to which all workers are occasionally subjected [14]. Finally, courts have also held that a worker who has proven an aggravation of a preexisting condition is not required to prove that the aggravation was substantial in order to receive compensation [15]. Such decisions are representative of the current evolution of workers' compensation schemes in general: cumulative trauma cases are compensable; physical or mental conditions that are not related to an *event* are compensable; a minor aggravation of preexisting conditions is compensable. However, under the ADA, an employer may not inquire into preexisting medical conditions or workers' compensation history before offering employment to an otherwise qualified employee. An employer may administer a medical examination or make inquiry into compensation history only if such examinations are required of all applicants in the same job category. An employer may not require a medical examination simply because the results from a medical inquiry disclose a previous job-related injury unless all applicants are required to have the same examination.

Employers will hire people, whether they like it or not, with preexisting conditions or a predisposition to mental or physical injury because the ADA prohibits prescreening on such bases. If the hiring process is conducted in accord with the ADA and the post-offer screening discloses preexisting conditions or predisposition, the applicant can ask for an accommodation. As the working population in a plant ages, or as employees succumb to the effects of job stresses, activities, or even gravity, productivity will diminish. However, increasingly, workers will respond to management's admonition with requests for accommodation. Consequently, the combined effect of the ADA and evolving workers' compensation law has created a serious need for professional ergonomists who will identify the essential elements of a job, design ADA accommodations, and redesign the work practices and stations so that jobs do not cause or aggravate injury. In all of these tasks, the ergonomist is subject to the "reasonableness" standard of care.

## 35.8 ADA GUIDELINES

The ergonomist should prepare written descriptions of the essential elements of the jobs for which an employer is hiring. Be reasonable. Write these up in advance. The ADA states that advance written descriptions are "relevant." That makes it easier for personnel to defend against a later claim that the description of the essential functions was contrived to rule out the specific applicant or specific disability. Essential functions of a job are those functions that the individual who holds the position must be able to perform unaided or with the assistance of a reasonable accommodation. The interpretive guidelines to Section 1630.2(a) of the ADA outline the factors to be considered in determining if a function is essential. The ADA also requires an employer to consider the terms of applicable collective bargaining agreements.

If a listed job function is not essential, an employer, as a reasonable accommodation, *must* (1) provide an accommodation, (2) transfer the function, or (3) exchange the function. The process of accommodation will require ergonomic supervision or intervention, and the ergonomist who fills this role must work with the otherwise qualified applicant to determine if a particular accommodation is appropriate. Ergonomists must be reasonable and uniform: They may not arbitrarily set production requirements unless those requirements are applied uniformly. The interpretive guidelines suggest a few examples of reasonable accommodations: making employer-provided transportation available; providing personal assistants, such as page turners, for an employee with no hands; providing a travel attendant to act as a sighted guide to assist a blind employee on an occasional business trip; permitting an employee who is blind to bring a guide dog to work, even though the animal is not furnished by the employer.

The benchmark for determining a reasonable accommodation is that it be without undue hardship to the employer. Any accommodation that would be unduly costly, extensive, substantial, or disruptive or that would fundamentally alter the nature or operation of the business could be deemed to be an undue hardship.

The process for determining the appropriate reasonable accommodation is clearly described in the interpretive guidelines relating to Section 1630.09(a). That process includes job analysis, individual consultation with the individual seeking the accommodation, and identification of potential accommodations and the preferences of the individual. If these steps do not result in an agreement, then the guidelines suggest more individualized assessment, including the employer's consultation with the EEOC, state or local rehabilitation agencies, or disability constituent organizations. The guidelines are not prescriptive and contain comments to that effect. However, they do exist as a checklist for items that will be considered as the minimum standard of care for the ergonomist and the employer in determining whether both or either of them have caused or committed any violation of an individual's civil rights under the ADA. What the guidelines really are is a list of the minimum efforts necessary to arrive at an accommodation. Professionals should follow the suggestions, for if they do not, someone will claim that this failure represents a lack of "reasonableness." The ADA was designed to favor the disabled applicant; therefore, to get into the best position possible to avoid a claim, the ergonomist must follow and document every suggestion contained in the act.

Ultimately, occupational ergonomists dealing with the ramifications of the ADA must follow the same standard of care that they apply to the normal performance of the profession: reasonableness. The "reasonableness" requirement suggests a negligence analysis: whether there is a duty, whether that duty was breached, and whether damages resulted. The duty is provided by the ADA statute itself, which also goes a long way to defining the minimum standard of care, that is, what is regarded as reasonable under the circumstances.

Ergonomists should not fear the ADA. It contains a set of nonspecific standards with which industry should comply, but compliance will ultimately be judged on a case-by-case analysis. The courts will look to the specifics of the job and the options, if any, for accommodation. The ergonomist who has acted reasonably has the best chance of prevailing when a claim is presented under the ADA.

## 35.9 CONCLUSION

No brief survey of current trends in workers' compensation law, the ADA, intentional torts, or OSHA-proposed rules can substitute for detailed knowledge of the actual cases and laws. The references cited in this survey provide the best available source, or starting point, for fluency in these areas. The actions of industry-employed ergonomists will be imputed to their employers, so it is critical that ergonomists comply with the standard of care for their profession; ergonomists must act in a reasonable and responsible manner and document contacts with management and workers. There is no prescribed standard for any particular situation; whether an ergonomist is reasonable will be judged by what is expected of other practicing ergonomists under similar circumstances.

> There ain't no answer. There ain't going to be any answer. There never has been an answer. That's the answer.
>
> GERTRUDE STEIN

## REFERENCES


1. *Village v. General Motors* (1984) 15 Ohio St. 3d 129, 472 N.E.2d 1079.
2. *Bowman v. National Graphics Corp.* (1978) 55 Ohio St. 2d 222, 378 N.E.2d 1056.
3. *Van Fossen v. Babcock & Wilcox Co.* (1988) 36 Ohio St. 3d 100, 522 N.E.2d 489.
4. *General Electric v. Industrial Commission* (1982) 89 Ill. 2d 432, 433 N.E.2d 671.
5. *Peoría County Bellwood Nursing Home v. Industrial Commission of Illinois* (1985) 138 Ill. App. 3d 880, 487 N.E.2d 356.
6. 29 CFR 1910; 57 FR 34192-01 (August 3, 1992).
7. See, generally, NIOSH HETA 88-180-1958, John Morrell & Co., Sioux Falls, SD (April 1989).
8. *Blankenship v. Cincinnati Milacron Chemicals* (1982) 69 Ohio St. 2d 608, 433 N.E.2d 572.
9. Section 4121.80(G) *Ohio Revised Code* (1986); *Jones v. VIP Development Co.* (1984) 69 Ohio St. 2d 608, 472 N.E.2d 1046; *Van Fossen v. Babcock & Wilcox Co.* (1988) 36 Ohio St. 3d 100, 522 N.E.2d 489; *Fyffe v. Jeno's Inc.* (1991) 59 Ohio St. 3d 115, 570 N.E.2d 1108.
10. *Fyffe v. Jeno's* (1991) 59 Ohio St. 115, 570 N.E.2d 1108; paragraph 1 of the syllabus.
11. *Cantrell v. GAF Corporation* (1993) 999 F.2d 1007.
12. *In the Matter of the Establishment Inspection of Kelly-Springfield Tire Co.* (1992) 808 F. Supp. 657, which held that even in the absence of specific regulations regarding ergonomics, the general duty provisions of the Occupational Safety and Health Act do cover ergonomic concerns.
13. EEOC, *Title I of the Americans with Disabilities Act*, Technical Assistance Manual, January 26, 1992.
14. *Ryan v. Connor* (1986) 28 Ohio St. 3d 406, 503 N.E.2d 1379.
15. *Schell v. Globe Trucking, Inc* (1990) 48 Ohio St. 3d 1, 548 N.E.2d 920.

CHAPTER 36

# Real-Time Exposure Assessment and Job Analysis Techniques to Solve Hazardous Workplace Exposures*

James D. McGlothlin and Sandra S. Cole

CONTENTS



* This chapter is largely condensed from a NIOSH technical report [3].

## 36.1 INTRODUCTION

This chapter discusses real-time exposure assessment techniques to solve workplace hazards. Three case studies illustrate useful equipment and techniques for controlling problems of dust, gas, and radiation exposure in industry. These case studies also show how integrating ergonomic and industrial hygiene principles pinpoint exposure sources and provide effective solutions.

Industrial hygienists often measure a worker's exposure to industrial air contaminants by sampling the air he or she breathes. A small, battery-powered pump draws a known flow rate of air through a filter or other collection medium for a measured period of time. The collection medium is analyzed to quantify the contaminant collected and to compute the average exposure for the sampling period. Although these results indicate the extent of exposure, integrated air sampling provides little insight into the specific causes of the worker's exposure.

In the case of personal radiation exposure, a worker's exposure to radiation in the workplace is often measured by the interaction of ionizing radiation with film in film badges or by the interaction of ionizing radiation with inorganic crystals, known as thermoluminescent dosimeters (TLDs). Film badges use photographic film, which darkens upon exposure to radiation, with a series of filters to measure the cumulative dose of radiation. Exposure is measured by developing the film and then the optical density is analyzed to quantify the cumulative dose. TLDs use inorganic crystals that have the property of trapping electrons/holes within the bandgap between the valence and conduction band when the crystals are irradiated. Then when the TLDs are heated in processing the TLDs, the trapped electrons/holes emit visible light, which is measured as they return to the ground state. The amount of light emitted depends on the cumulative dose of radiation that the individual was exposed to [1,2]. The use of film badges or TLDs gives the cumulative radiation exposure, but usually these dosimeters do not give immediate feedback and so correlation between cause and worker's exposure is unknown.

Recommendations for controlling air, dust, and radiation contaminant exposures are often based upon the industrial hygienist's judgment and can result in control measures that do not address the major sources of exposure to air and radiation contaminants. Direct-reading instruments and data-recording devices can overcome the problem by recording events and exposures in the workplace as a function of time. The data from such a system associates events and exposures and promotes more effective and focused recommendations for controlling the contaminant exposures.

Through studies conducted in a variety of industries, researchers with the National Institute for Occupational Safety and Health (NIOSH) have developed a systematic approach to identify the sources of workplace exposures to air and dust and to provide an effective means for communicating these results to workers and management [3–7]. Researchers in the academic and private sector have been involved in developing approaches and technology that can identify the sources of radiation exposure in the workplace. These systems employ

- Direct-reading instruments and data-recording devices to monitor and store data characterizing worker exposures
- Video cameras or web cameras and recorders to document worker activities
- Task analyses to evaluate work activities
- Statistical techniques to develop predictive models and to summarize the results
- Personal computers to perform analyses on the data and to combine the activity data and the exposure data into a presentable form

The present systems evolved from a series of studies, conducted either to evaluate the effectiveness of engineering controls or to identify characteristics of worker exposures. The early direct-reading instruments permitted researchers to monitor exposure changes over short intervals (on the order of seconds). The output from these instruments was stored in an electronic recording device, rather than on a strip chart recorder, so that the data would not require rekeying for statistical analysis. Some recent versions of some direct reading instruments can send the data directly to the computer where it is analyzed, stored, or transferred to other storage media. Workers' activities were documented by video-recording systems to determine whether exposures were the result of particular work practices. Work activity data were combined with the real-time exposure data by determining both the exposure and the activity at any given time. Time series analysis of the combined real-time and work activity data set resulted in a model to predict worker exposures. After several studies, however, it became apparent that time series analysis could become a prohibitive task because of processing the tremendous amount of data that can be collected over a very short period. To ease this problem, several simplified analysis techniques were developed. Although these techniques were not as powerful as the time series analysis, they identified those activities that contributed the most to the worker's contaminant exposures.

During the completion of the initial studies, a need became obvious: Communication of the study results to workers and management. The consensus among the studies' researchers was to provide the facility with a video recording of the work activity, combined with a display of the real-time exposure measurement. The exposure data could be presented in two forms on the video screen: numerically, with the value of the exposure measure displayed, or graphically. Both options were exploited by displaying both the numerical exposure concentration and a bar representing the relative magnitude of the exposure. To place the bar and number on the video screen, a computer program read the exposure data file and generated and updated the bar with time. The system required the use of consumer-quality video and ordinary personal computer equipment; the only specialized equipment required was a special graphics card for the personal computer. The result was a video recording that graphically showed how exposure to a particular substance was affected by activities of the worker.

## 36.2 VIDEO EQUIPMENT

Two types of video equipment were used in the evaluation of dust and air contaminants: conventional video equipment for documenting the worker's activities and infrared (IR) video equipment for visualizing specific air contaminant plumes. Conventional equipment is used for conducting video exposure monitoring (VEM); IR equipment is used with direct-reading instruments to characterize workplace contaminant concentrations. For the more recent developments in the evaluation of radiation exposure, web cameras can be used to record the worker's activities.

### 36.2.1 Conventional Video Equipment

The conventional video recording system consists of a video camera and a videotape recorder. A camcorder, having both capabilities, provides better portability. Mounting the video camera onto a tripod eliminates the need to hold the camera throughout the process. The tape format (Beta, VHS, 8 mm) is not important, and many consumer-quality video recording systems are suitable for VEM. There are, however, two important requirements. First, the video system must have a National Television System Committee (NTSC) standard video output signal—a signal used by the video overlay system described in Section 36.5 in this chapter. This standard is used by most home video equipment. Second, an on-screen clock or timer is needed—one that can be synchronized with the real-time clock of the data-recording device. Synchronizing the data-recording device with the video camera can be as simple as starting the timer in the camera at the same time the data logger is turned on. The clock or timer should have a resolution of at least 1 s. The on-screen clock permits an exposure to be coordinated with an associated activity. The video recording of the work cycle or process can then be reviewed while simultaneously tracking the worker's exposure from a printout or plot of the real-time exposure data.

### 36.2.2 Web Cameras

Recent developments in VEM systems consist of using web camera(s) with real-time detectors, which are synchronized to the clock on the computer. The web camera(s) connect to a computer via USB port(s) that allow video to be transmitted directly to a computer, where the video is viewed and stored. The video can then be transferred to other storage devices so that the large files do not take up hard drive space on the computer. One challenge with some detectors, software, and web cameras are finding commercial off-the-shelf computers that have Bluetooth installed as a default on the computer, instead of a separate dongle that has to be used to connect and allow the Bluetooth connection to occur between devices. Another challenge may be finding computers that can run real-time video reliably while sampling at a high rate such as at 1 sample/s. An advantage of web cameras is that the size makes them very portable and adjustable in a variety of situations.

### 36.2.3 Infrared Video Equipment

Effective control of air contaminants depends on understanding the characteristics of their release. It is important to know not only the concentration but also the source and path of the emission. Although some gases and vapors are visible, most are not. IR imaging is

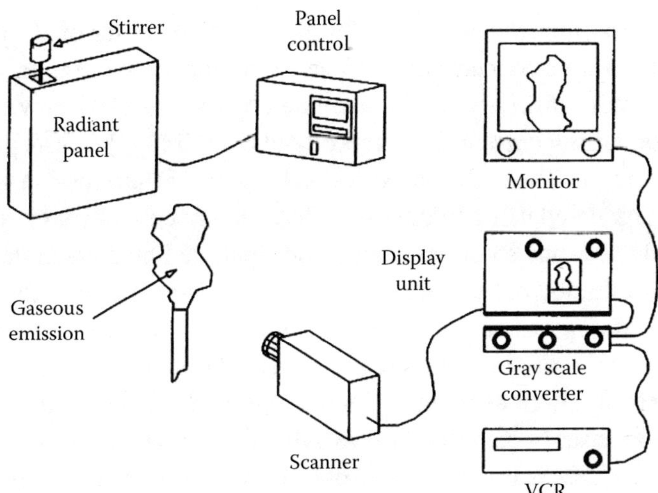


FIGURE 36.1 IR imaging system.

a technique that can provide a real-time picture of some otherwise invisible emissions. A schematic of such an IR imaging system is presented in Figure 36.1. An IR scanner (Thermovision 782) [8] detects changes in absorption of IR radiation by contaminant gases or vapors. Two versions of the scanner may be used, depending on the range in which the gases absorb IR radiation: a shortwave band (2–5.6 microns) and a longwave band (8–12 microns). The images received by the scanner are transmitted to a display unit and may be converted from the normal IR gray-scale image to a colored scale. This image is then simultaneously transmitted to a monitor and video recorder for real-time viewing and recording.

The system uses a flat, black panel as an IR radiator. The panel is a square, 2 in. thick, aluminum tank filled with water. A flat-sheet electrical heater is glued to the back surface of the tank; the front surface is painted black. An electronic temperature controller maintains the tank at a constant temperature (120°F). The water in the tank is circulated by a laboratory stirrer to inhibit the formation of a temperature gradient across the panel surface.

The radiant panel and the IR scanner are positioned so that the emission source is between them. The scanner sees the panel as a constant temperature source and displays it as a uniform image. As a contaminant gas passes between the scanner and the heat source, it absorbs some of the radiated IR energy. The scanner detects the gas as a lower temperature, which is then displayed as a different color or shade of gray and recorded.

This system is useful for detecting certain process emissions because it provides a real-time image that identifies both the source and path of the emissions. Medical processes, such as the release of nitrous oxide ($N_2O$) during dental surgery, and industrial processes can be monitored. Also, the IR imaging system, using tracer gas, can determine flow patterns around exhaust openings. This technique has the advantage that the effect of specific work activities or changes in control configuration can be determined immediately.

The most important limitation of this system is sensitivity. The absorption of the emission cloud is related directly to the concentration of the emission and the path length through the cloud. Thus, lower concentrations must be present in greater quantities to be visualized. For example, the sensitivity for $N_2O$ is on the order of 200 ppm meter, that is, a cloud of NO having a concentration of 200 ppm must be more than 1 m in diameter to be detected. System sensitivity can be increased by the use of narrow band pass filters that filter radiation falling outside the narrow band containing the absorption peak of the monitored contaminant. The high concentrations typical of the emission generation point can generally be visualized using this system. Detection of contaminants at lower concentrations typical of the range recommended by occupational health standards is, however, limited. Another limitation of the system is lack of portability. Because the radiant panel is a water-filled tank, it is quite heavy (~55 lbs) and not easily positioned. Although this is not a severe limitation for laboratory use, it does make field operation difficult.

Recent advances in thermal imaging technology address some of these limitations. A system that uses a laser in combination with the IR scanner to detect changes in energy is now available. The laser scans the viewed object, thus eliminating the need for a radiant panel, greatly increasing portability, and making the system much more convenient for field use. This system also has a sensitivity approximately one order of magnitude greater than the one previously described.

## 36.3 MONITORING EQUIPMENT

Any air contaminant–monitoring instrument that produces an output signal of the concentration measurements can predict real-time assessments of a worker's exposure to an air contaminant. Similarly, an instrument that produces an output signal of the radiation dose rate or radiation dose can assess in real time a worker's exposure to radiation. The usefulness of a specific instrument will vary with the situation. To evaluate the utility of an instrument, consider (1) the nature of the analog or serial output, (2) the response time of the instrument, (3) specificity for the contaminant of interest, (4) portability and size, and (5) the ease of storing data on data logger and the transference of data to a computer versus the direct transmission of data to the computer from the instrument.

### 36.3.1 Output

The collection of real-time concentration data to evaluate the relationship between events in the workplace and air contaminant concentrations generally must be recorded automatically. For a monitor to be useful, it should produce a digital or analog output, the latter often being voltage that is proportional to concentration or radiation dose. Techniques for recording analog data are given in Section 36.6. Some instruments also provide a digital output that is periodically updated. The frequency of these measurements is usually a function of the instrument and normally cannot be adjusted by the user. However, newer instruments allow the sampling rate to be adjusted by the user to fit the situation.

### 36.3.2 Response Time

The total system response time (for the monitor and the setting being evaluated) can be defined as the sum of (1) the time required for the contaminant to be transported to the worker and begin to accumulate and (2) the time required for the instrument to respond to a change in concentration in the worker's breathing zone or a change in the radiation dose rate or radiation dose in the worker's immediate work area. To conduct VEM studies of air contaminant concentrations or radiation dose, the total system response time must be less than that of the events of interest. As a result of the response time delays, the instrument output lags behind work events in the workplace.

Monitoring instruments often measure some general parameter that is proportional to concentration or dose. For example, aerosol photometers respond to any aerosol that scatters light. Such a limitation requires either that the monitor be calibrated for the specific contaminant being measured or that the results be reported as a relative concentration or relative dose.

### 36.3.3 Portability

To allow for worker acceptance, the monitoring equipment should not interfere with the worker's performance of his or her job and be light enough to be worn comfortably. The equipment should be battery-operated and should weigh as little as possible. In the case of assessment of air contaminant exposure, if workers cannot wear the equipment, tubing can transport the air contaminant from a worker's breathing zone to the instrument. This arrangement, however, adds some complications. The monitoring system's response time will increase because of the time needed to transport the air contaminant through the tube to the monitor. In addition, aerosols and contaminants can be lost to the tubing walls and other collecting surfaces. If the tubing is later struck or vibrated, these contaminants can be released and may contribute to the instruments' signal. Organic vapors can be adsorbed onto the tubing walls during periods of high concentration and desorbed during periods of low concentration.

Users need to consider the limitations and capabilities of the direct-reading instruments when designing and conducting studies to yield useful information about exposure sources. Background information on these instruments can be obtained from *Air Sampling Instruments*, by the American Conference of Governmental Industrial Hygienists (ACGIH) and the NIOSH Manual of Analytical Methods (NMAM) [9,10].

## 36.4 PERSONAL COMPUTER SOFTWARE

Several types of software collect and analyze real-time exposure data. Control software operates an analog-to-digital converter card, and communications software downloads portable data loggers. Spreadsheets are valuable for manipulating real-time exposure data, as well as for performing some simple analyses. For more sophisticated data analyses, full-function statistical analysis packages may be required. In addition, if the exposure data are to be combined with a work activity video recording, a customized computer program can be used to generate a graphical representation of the worker's exposure.

Control software operates the analog-to-digital converter that is either a card located in the computer or a stand-alone system with an interface to the computer. These software packages usually require special device drivers for the particular hardware system in use. Many of these control packages can process the real-time data during collection, and some packages provide limited data analysis capability. Besides collecting data from an analog source (e.g., a direct-reading instrument), control software can also instruct the computer to send out signals, although this function is beyond the scope of this chapter.

Configuring the analog-to-digital system is normally done by menu-driven software. Many of the software packages allow readings—graphical or tabular—to be displayed on the computer screen as the data are being collected. Once data collection is complete, the readings are stored in a data file. Some programs link directly with a spreadsheet program, making it possible to save the data in a spreadsheet file. For other programs, the data are stored in file formats that can be imported into the spreadsheet. There are several different control programs with many different functions and capabilities. Two specific packages are Labtech Notebook (Laboratory Technologies Corporation, Wilmington, MA) and ASYST (MacMillan Software Company, New York, NY). Both of these programs work with a variety of analog-to-digital converter cards.

If a portable data-recording device (data logger) records the real-time exposure data, control software is not needed. Instead, a program to download the data logger to a personal computer is required. Downloading software either comes complete with most data loggers or is available for an additional cost. After the data logger is downloaded, some programs allow simple data analysis to be performed. Many of these programs store the data in a file that can then be imported into a spreadsheet program. In addition to the programs supplied with the data loggers, there are communications programs such as Crosstalk (Crosstalk Communications, Roswell, GA) and Procomm (Datastorm Technologies Inc., Columbia, MO) that can download some data loggers through the computer's asynchronous communications port. Communications programs may require nonstandard use, since the format of the data from the data logger may vary with the device.

After the data have been collected and stored in a file, spreadsheet programs can manipulate data and do simple data analysis. Lotus 1-2-3 (Lotus Development Corporation, Cambridge, MA) and Microsoft Excel (Microsoft Corporation, Redmond, WA) are examples of two spreadsheet programs. If the data are to be analyzed by worker activity, a spreadsheet is useful for keying activities with the real-time exposure data. The researcher must first determine the time a particular reading was recorded and then observe the worker's activities for that time on the video recording of the work activity. Spreadsheets not only sort data and perform elementary statistical analysis but also format data sets for analysis in a statistical analysis program, or combine the work activities and the real-time exposure data onto videotape.

To combine the real-time exposure data with the video recording of the worker's activities, NIOSH researchers have written a program for IBM-compatible computers that generates a graphical representation of the worker's exposure [3,11,12]. This IBM-compatible program reads a real-time data file, generates a bar to represent the magnitude of the exposure, and then displays the bar on the screen. When this program is run through a video overlay system, a video recording graphically shows how a worker's exposure is influenced by the work activity. The video overlay system is discussed in Section 36.5. The bar is updated with each time interval of readings in the data set. The program allows either one or two bars to be displayed on the screen at one time. Two bars can be displayed if the exposures of two workers are to be compared, or if one worker is monitored with two different instruments. To use the program, the real-time exposure data must be stored in a properly formatted ASCII file. For the program to display one bar, the format of the data file must have three columns of data: two columns for the time the reading was recorded (minutes and seconds) and one column for the exposure measurements. For the program to display two bars, the data file format must have an additional column for the second exposure measurement. The first data set is displayed on the left side of the screen, whereas the second data set is displayed on the right. The time interval between the readings must be constant. The spreadsheet program arranges the data file into the proper format and generates a bar that is overlaid onto the work activity video recording by a video overlay system.

## 36.5 PERSONAL COMPUTER HARDWARE

The computer hardware required for collecting and presenting real-time data is basic. Specialized equipment is required only for combining the graphical exposure bars with the video recordings of the work activity or for running a computer-based analog-to-digital converter system. The basic computer system used by NIOSH researchers is an IBM PC-compatible personal computer. The computer should have sufficient memory (i.e., 1 GB RAM), several USB ports to plug in auxiliary equipment (i.e., real-time instruments, web cameras, data loggers, storage devices, etc.) and a hard disk drive. Additional memory (i.e., 1–2 GB) may be desirable to improve performance if unusually large data sets (i.e., 1200 readings) are to be manipulated in a spreadsheet. If video is to be directly transmitted to the computer from the cameras, additional memory will be needed for the video file storage. As video files even when compressed can be relatively large compared to other types of files, it is advisable to have an external hard drive (1–2 TB) to transfer the video files for additional analysis. If the real-time exposure data are not going to be combined with the video recording of the work activity, then the type of graphics card is not critical. If data loggers are to be downloaded to the computer, an asynchronous (serial) communications port is required (most computers are sold with this port as standard equipment).

Computer-based analog-to-digital converters are special cards that fit into an expansion slot of the computer. Special software drivers and control programs may be required to operate this board. Section 36.6 contains more detailed descriptions of the analog-to-digital converter systems.

To overlay the real-time exposure data with the video recording of the work activity, the computer will need both an enhanced graphics adapter (EGA) card and a video overlay board, or a variable graphics array (VGA) card with the overlay features built in. A monitor appropriate for the graphics card also is needed. Both VGA and EGA are high-resolution color graphics adapters, with VGA having slightly higher resolution.

If an EGA card is used, it must be combined with a video overlay system. One such system, the Video Charley (Progressive Image Technology, Folsom, CA), consists of a single computer card. The Video Charley requires the EGA card to have a standard features connector (most EGA cards do). The features connector links the video overlay board with the computer. The video overlay board converts the computer's graphics signal to an NTSC signal and overlays the graphics onto the activity video recording. Besides the features connector, most EGA cards also have a DB9-pin connector for the EGA monitor and two RCA-type connectors. Under normal circumstances (without the Video Charley), the two RCA connectors serve no function. With the Video Charley board installed, however, one RCA connector inputs the activity video signal, and the other outputs the video signal with computer graphics overlaid.

When overlaying computer graphics using the Video Charley, signal differences require the computer display system to operate at a resolution of 640 × 200 pixels, rather than at the typical EGA resolution of 640 × 350 pixels. To combine the activity video signal with the computer graphics signal, the two signals must have the same synchronization frequencies. In the case of the video signal, an NTSC signal, the horizontal sync frequency is 15.7 kHz and the vertical sync frequency is 60 Hz. In the 640 × 200 pixels mode, the horizontal and vertical sync frequencies are also 15.7 kHz and 60 Hz, respectively. In the 640 × 350 pixels mode, the vertical sync frequency is 60 Hz; however, the horizontal sync frequency is 21.8 kHz. To get both signals at the same horizontal sync frequency, the computer graphics card must operate at the lower resolution mode. Depending on the type of EGA card used, either software drivers or hardware switches can set the resolution.

If the VGA option is chosen, an appropriate VGA card is required. Two such cards are the USVideo VGA/NTSC Recordable® graphics card with the Genlock Overlay Module (USVideo, Stamford, CT) and the Willow Peripherals VGA-TV GE/O® (Willow Peripherals, Bronx, NY). These two systems allow computer graphics to be overlaid onto video images at a higher resolution than does the EGA system with the Video Charley. To overlay on VGA systems, only one setting needs to be changed to direct the card's output to the video monitor: On the USVideo card, this is done with a hardware switch; on the Willow Peripherals card, a software program is run. Both cards have two RCA-type ports, one for video-in and one for video-out. The cabling setup, shown in Figure 36.2, is the same for both the EGA/Video Charley and the VGA systems. Operation of the VGA overlay system is similar to the normal use of the computer, except that the video monitor (connected to the video-out–RCA-type port) is the primary monitor.

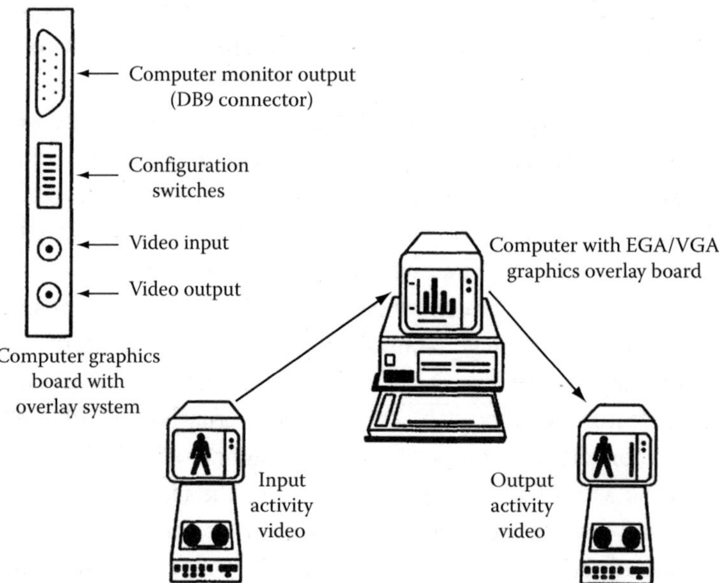


FIGURE 36.2 Diagram of personal computer system equipment and connections needed to overlay exposure data with work activities.

## 36.6 DATA ACQUISITION

Many direct-reading monitors have analog output capability, usually in the form of a DC voltage signal, typically on the order of 1–10 V, full scale. Before the proliferation of the personal computer, this analog output typically drove a strip chart recorder. To perform data analysis with a computer, the data from the strip chart was keyed into the computer—a tedious process. With advances in personal computers, the analog output from these monitors can now be stored digitally, allowing the data to be transferred to the computer in just a few easy steps.

Data-recording devices generally fall into two categories: portable data loggers and computer-based analog-to-digital (A/D) converter systems. Both types of devices have a limited resolution over their working voltage range. Depending on the type, the device has either a fixed voltage range (e.g., 0–2 V) or a working range that can be chosen with hardware switches or control software. This working range is then broken down into intervals. Resolution of a data-recording device is usually given in bits. For example, the working range of 0–2 V for an 8 bit data logger consists of 256 intervals. The number of intervals is determined as follows:

$$\text{Number of intervals} = 2^{\text{bits}} \tag{36.1}$$

The magnitude of these voltage intervals is calculated from

$$V_1 = \frac{V_U - V_L}{N} \tag{36.2}$$

where

$V_1$ is the interval, V
$V_U$ is the upper working range, V
$V_L$ is the lower working range, V
$N$ is the number of intervals

For the example of an 8 bit, 0–2 V working range, there must be a difference of at least 0.008 V for the data-recording device to detect a difference between two voltage readings. In most instances, an 8 bit device should be sufficient. Data loggers are typically 8 bit devices, whereas A/D converters range from 8 to 16 bits; 12 bit boards are very common.

Computer-based systems store the data directly onto the computer's hard drive or onto a disk drive. These systems require software programs to control the parameters. Depending on the program, the exposure measurements can be displayed on the computer screen as the data are being collected. Computer-based A/D converter systems are more flexible than portable data loggers. Computer-based systems are usually more expensive; A/D boards can cost $1000 or more and the control software can cost another $1000.

Portable data loggers store the data in a built-in bank of memory. After data collection, the data logger must be downloaded to the computer, typically through the computer's communication port. A general communications program or a program written specifically for downloading a particular data logger can control the downloading procedure. Most data loggers have parameter-setting programs built in and require no additional control software, and most display only limited amounts of data while recording. Since the data logger is likely to be fastened to a worker, observation of the data as it is being generated is not feasible. Portable data loggers, including downloading software, can be purchased for as little as $500.

A hybrid of the A/D systems and the portable data loggers is a telemetry system. Telemetry systems use a transmitter and receiver to transfer data from an instrument to a base unit for storage. The base unit may include a personal computer and may allow the data to be displayed as it is being generated. As with portable data loggers, telemetry systems do not require a worker to be tethered to a computer. Unfortunately, most commercially available telemetry systems tend to be expensive (i.e., >$10,000) [13].

## 36.7 WORKER ACTIVITY ANALYSIS

Activity analysis is an important step in VEM because such an analysis helps catalog work activities. This systematic method breaks a complex job into its elements, permitting study of and improvements in a specific task. More importantly, these elements can be sorted so those contributing most to a worker's contaminant exposure can be dealt with first.

The first phase of activity analysis is a time-and-motion study, which determines the work content of the job. Time-study and production records, as well as timed observations, provide the necessary interval data. Managers, supervisors, and workers can provide job descriptions and demonstrations from which to determine tasks. A job is described as a set of tasks, with each task consisting of a series of steps or elements [14], that is, the fundamental movements or acts (reaching, grasping, moving, positioning, using, etc.) required to perform a job. Groups of elements making up a task (or an activity) are usually performed in the same sequence to accomplish a common end. Examples of tasks might include the following: "turn on machine," "operate machine," and "cleanup." By observing the job or slow-motion video recordings of that job, the researcher identifies the elements composing a task. Gilbreth suggested that formal element definitions are arbitrary in that one can increase or decrease detail as necessary. For example, "get" adequately describes the process of "reach–grasp–move," and "put" works well for "move–position–release" [14].

The second phase of activity analysis is an actual review of the job for recognized occupational risk factors that may cause excess exposure to contaminants. If a trained investigator can record the risk factors as the worker is performing the job, this analysis can be done at the worksite. A more thorough analysis can be done, however, by viewing the video recording of the worker's activities. The clock or timer in the video camera, direct-measuring instrument, and/or computer documents the time it takes the worker to perform the various activities. The clock or timer also allows activities to be studied as changes occur in the contaminant exposure as measured by direct-reading instruments. When evaluating contaminant data, the researcher needs not analyze the job in more detail than what the real-time exposure data can reveal. For example, if the response time of the instrument measuring the contaminant exposure is longer than the time required to complete a set of tasks being video recorded, then analysis of those individual activities is of little value, and they should be combined into a "principal" activity. This principal activity can then be studied with regard to contaminant exposure.

## 36.8 DATA ANALYSIS

To perform data analysis, researchers must combine worker exposure measurements and descriptions of events in the workplace into a single data set. Descriptive statistics describe the contribution of workplace events to a worker's contaminant exposure. In addition, statistical analysis evaluates whether workplace events significantly affect exposure. The findings of the data analyses help to focus control measures upon actual sources of contaminant exposure to workers.

### 36.8.1 Transportation Lag and Autocorrelation

As a prologue to data analysis, an appreciation is needed of how events in the workplace affect the contaminant concentration measured by an instrument. For example, consider a worker standing at a workstation. Turbulence in front of the worker transports the air contaminant from a source at the workstation into the worker's breathing zone. If it takes 2 s for the air contaminants to travel from the source to the worker, the concentration in the worker's breathing zone does not start to change until 2 s after the event has occurred. In statistical terms, the concentration is said to lag behind the workplace events. This can be referred to as transportation lag. The actual magnitude of this lag can be estimated by observation and measurement, or it can be addressed in the selection of a statistical modeling and data analysis package.

After the air contaminant has been transported into the worker's breathing zone, the direct-reading monitor begins to respond to the changing concentration. A monitor with a time constant of 1 s would require 3 s to complete 95% of the change in response to an abrupt change in concentration. Because of the dynamics of the monitor's response, the measured concentration at any moment in time is a function of the concentration in the preceding time intervals. This phenomenon is called autocorrelation.

### 36.8.2 Assembling the Data Set

Concentration or radiation exposure measurements from a direct-reading instrument are recorded and stored by data-logging devices and/or by the computer directly. Because the software written for controlling or downloading data-logging devices has limited data analysis capabilities, real-time concentration data can be imported into a spreadsheet program for manipulation and data analysis. If the data is directly recorded and stored in the computer, the real-time data usually can be imported directly into a spreadsheet file for further analysis. Many of the downloading or control programs include utilities for storing the real-time con centration measurements in a *print file* that can be imported into the spreadsheet. (A print file is a text file in ASCII format that can be printed directly by the operating system's print command.) The interval between the concentration measurement readings is set either before the data are recorded by the data-logging device or when the data are stored in the print file.

If you are using an older command line–based system, the real-time exposure data are loaded into the spreadsheet using the "import" command. (The name of this command may vary from program to program, but the command loads a print file.) However, if you have a more user-friendly Windows or Macintosh system, the import of data into the spreadsheet will be menu-based and depend on the system. The print file loaded into the spreadsheet may contain several columns of numbers, depending on the type of data-logging device used. These columns may include several time columns (elapsed time, clock time, etc.), event markers, and concentration measurements. The data can be manipulated in the spreadsheet to create a data set that includes only two columns, one for the real-time concentration measurement and one for the time the readings were recorded. This time reading can be elapsed time or clock time, depending on how the data-logging device was synchronized with the video camera's clock or timer.

After the time and concentration readings have been isolated, work activity variables can be added to the spreadsheet. The video recording of the work activities can be viewed while tracking the worker's exposure in the data set. From this recording, the worker's activities can be defined in two different ways: so that only one activity can occur at any given time or so that any one of several activities can occur at any given time. For each concentration or radiation exposure measurement, the activity can be coded into the data set in one of several ways, depending on how the activities were defined and on the type of data analysis to be conducted. Two methods are frequently used: (1) to enter the activity as a single variable with a different value for each activity or (2) to enter each activity as a separate variable, with one value if the activity occurs, and another value if it does not occur (e.g., "1" and "0"). If the activities are defined such that only one activity can occur at a time, the single-variable method is usually more appropriate since it will result in a smaller data set than if each activity were to be entered as a separate variable. If, however, several activities can occur at a time or if data analysis involves using a spreadsheet program to perform multiple regressions, then each activity is usually entered as a separate variable. If the activities were entered using the single-variable method, a different value would be needed for every combination of activities.

As discussed earlier, the air contaminant concentration lags behind the causal activities because of the time required to transport the air contaminant from the source to the monitor. If the transportation lag is not addressed by a statistical analysis package, the air contaminant concentration measurements can be "slipped" with respect to the worker activity variables, after the researchers estimates the magnitude of the lag. The lag time matches the worker's activities with the measurements of associated air contaminant concentration.

### 36.8.3 Data Analysis Techniques

After the data set is assembled as a time series, it can be analyzed to determine the effect of workplace activities on changes in contaminant exposures to workers. Autocorrelation considerably complicates (1) statistical analysis for modeling worker exposures and (2) examination of whether the worker's activities are affecting the air contaminant exposures. When conducting statistical analysis, researchers should compare the extent of the changes in exposure attributing to a worker's activities with the variability of the exposure data. When the changes in exposure are large with respect to the exposure data variability, one can conclude that the activities significantly affect the exposure. Autocorrelation can cause the variability of the exposure data to be underestimated during regression analysis and analysis of variance. Thus, autocorrelation can cause these two data analysis techniques to overstate the level of confidence in the conclusion that workplace events affect the worker's exposure. Special techniques, called time-series analysis, have been devised to deal with autocorrelated data.

A variety of techniques are available to analyze real-time data and deal with autocorrelation, but because of the time and complexity required to deal with autocorrelation, descriptive statistics are commonly used instead. For a quantitative evaluation of whether activities are causing air contaminant exposures, autocorrelation in the data can be addressed either by censoring the data to remove autocorrelation or by performing time-series analysis. At times, too much information is lost when the data are censored to remove autocorrelation. Time-series analysis methods can evaluate the relationship between the worker's activities and contaminant concentrations or radiation dose without censoring the data set [15]. Because time-series analysis can be very complicated, the assistance of a statistician may be needed.

Descriptive statistics can aid exploratory data analysis. In such an analysis, the identity of workplace activities causing differences in the worker's exposure is investigated. If there are no differences or if the differences are greater than an order of magnitude, conclusions can usually be based on the findings of the descriptive statistics. However, when the observed differences in concentration are less than an order of magnitude, statistical analysis should be performed. In conducting statistical analysis, the effect of autocorrelation on the analysis must be evaluated.

Real-time data are frequently analyzed to evaluate whether specific workplace activities affect worker exposures. When a workplace activity occurs and the worker's exposure increases, one can conclude that the activity has contributed to the exposure. Because many activities can occur simultaneously in an industrial environment, the change in the worker's exposure may be due to some unrecognized activity. Thus, judgment must be exercised when interpreting the results of the data analysis. After one analyzes the real-time data, control measures can be focused on actual exposure sources.

## 36.9 PRACTICAL APPLICATIONS OF VIDEO EXPOSURE MONITORING

VEM is effective for identifying those specific activities that contribute most to a worker's exposure to an air contaminant. Integrated monitoring, such as sorbet tube or filter sampling, is normally conducted to determine the extent of the worker's exposure (averaged over the sampling period) before VEM. After determining the extent of the exposures, the researcher can apply the techniques for VEM. A typical VEM evaluation might proceed in the following manner:

1. With the worker activity and contaminant of concern identified, the appropriate direct-reading monitor must be chosen. The monitor should be appropriate for the contaminant, for example, an aerosol photometer to monitor for aerosols. It should have a minimal time constant so that activities of short duration can be evaluated, and it should be as portable as possible. The monitor should be set to zero and calibrated according to the manufacturer's instructions.

2. In addition to the direct-reading monitor, an IR video system may prove useful, depending on the contaminant being sampled. Such a video system can visualize air

contaminant plumes, identify contaminant sources, and identify work practices that may contribute to a worker's exposures.

3. The output of the direct-reading instrument should be recorded by a data acquisition system. Setup of this system consists either of programming the data logger or of running the control software of the analog-to-digital converter system. The clock on the video camera and the data-recording device should also be synchronized at this point.

4. Data collection begins by starting the data-recording device and the video camera, and continues for a period judged to be representative of the process being studied. After the data-collection period, the data must be stored in a data file. If a data logger has been used, the data must be downloaded to a computer for storage to a file.

5. After the data are collected and filed, they are imported into a spreadsheet program. Work activity is analyzed from the video recording of the work activities. The activity variables are entered into the spreadsheet to accompany the contaminant exposure data. Data analysis can be conducted with the spreadsheet or by statistical analysis programs. The spreadsheet analyses can consist of simple descriptive statistics or regression analysis. Statistical analysis programs are used for more sophisticated analyses, such as time-series analysis.

6. If the exposure data are to be overlaid onto the video recording of the work activities, the video overlay system must be assembled and the exposure data stored in an ASCII file specifically formatted for use by the bar-generating program. To overlay the exposure data onto the video recording of the work activities, the bar-generating program is set up (inputs entered); the work activity videotape then is played back. When the time on the video image reaches the time of the initial reading from the data file, the program's display is started. This synchronizes the exposure data with the video recording. The overlaid signal can be displayed on a video monitor and recorded on a second video recorder. Newer software associated with some direct-monitoring instrument systems will synchronize the video recordings with the exposure data, so this step is not required.

7. In some situations, the real-time concentration data is useful for evaluating ventilation systems for contaminant dilution and for determining the contaminant generation rate for the process. In these instances, the spreadsheet's regression function or a statistical analysis program can be used to determine the room's mixing factor. With this factor, the generation rate can be estimated, and ventilation systems can be further evaluated.

### 36.9.1 Case Study 1: Manual Material Weigh-Out

This plant manufactures a variety of plastic and rubber materials. At the operation studied, powdered acrylic copolymer was weighed into batch lots at a weigh-out booth as diagrammed in Figure 36.3 [4]. The final manufactured product from this operation was vinyl

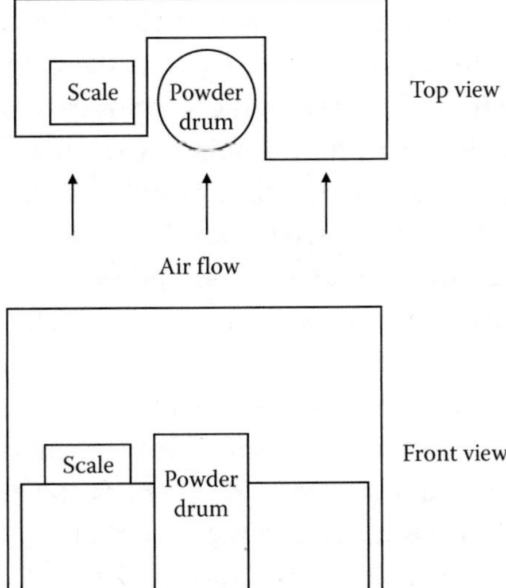


FIGURE 36.3 Diagram of a powder weigh-out workstation.

wall covering. In the weigh-out booth, a hinged segment of the work platform could be raised to allow a drum of raw material to be placed inside the booth. An exhaust plenum formed the back wall of the booth. At the booth, the worker emptied 22.7 kg (50 lb) bags of powder into a fiber drum measuring 84 cm (33 in.) high and 55 cm (21.5 in.) in diameter. Then, using a scoop, the worker transferred the powder from the drum to a small paper bag. The bag was placed on the scale and the weight of powder in the bag adjusted. Usually, two scoops of the powder were required to achieve the proper weight. Finally, the filled bag was closed and placed in a bin behind the worker. This process was repeated until the required number of batches was filled or the fiber drum was emptied.

**Methodology**

Direct-reading monitors measured the effect of depth of material in the drum and the elements of the job cycle on dust exposure. The worker began with a full drum and weighed the powder into paper bags. An aerosol photometer, the Hand-held Aerosol Monitor (HAM) (PPM Inc., Knoxville, TN), showed the dust concentration in the worker's breathing zone. Every 2 s, the HAM's analog output was recorded by an Apple II Plus computer, equipped with an AI 13 analog-to-digital converter (Interactive Structures Inc., Bala Cynwyd, PA). The evaluation ended when the drum was nearly empty (about 22 min).

The voltage output was statistically analyzed to determine if the amount of powder in the drum affected worker dust exposure, and if it did, which activities contributed the most to this increase. The strategy for this analysis was to fit a regression model involving the relation of the variable "worker" (a time-dependent measure of dust exposure) to the independent variables "bagcount," "scooping," "weighing," and "turning." "Worker" was the voltage output of the direct-reading monitor mounted on the worker. "Bagcount" was the cumulative number of bags that were weighed. "Scooping" was the cumulative time during each cycle spent scooping material from the drum and into the bag. "Weighing" was the cumulative time during each cycle spent weighing the bag on the scale and adjusting the amount of powder in the bag. "Turning" was the cumulative time during each cycle spent placing the bag in the bin. The worker's exposure was modeled closely enough to provide a fair representation of its relationship to the variables. There was no attempt to continue to add terms to the model until the lack of fit was not statistically significant.

A key assumption in the data analysis was the independence of measurements. Successive readings from the instrument were not independent. When a dust-generating event occurred, dust concentrations did not increase immediately; time was needed for the air to transport the dust cloud from the point of generation to the inlet of the instrument. In addition, the HAM was operated with a time constant of 1 s, and it required some time to respond to fluctuating concentrations. The total instrument response time appeared to be 2–5 s, meaning the instrument responded 2–5 s after a dust-generating activity occurred. As a result, autoregressive terms were used in the analysis.

The results of the regression analysis are shown in Figure 36.4, which shows that dust exposure during the scooping activity increased as the bag count increased. Bag count was a surrogate measure for the level of powder in the drum; a bag count of 0 corresponded to a full drum, and a bag count of 55 corresponded to an empty drum. During weighing and turning, the worker's dust exposure either remained constant or failed to increase as fast as the exposures during scooping.

Figure 36.5 illustrates the effect of job cycle upon dust exposure. During the scooping activity, the dust exposure increased. During the weighing and turning activities, the dust exposure decreased. This suggests that most of the worker's dust exposure was caused when scooping the powder from the drum. Dust exposures caused by weighing and turning were much smaller than the dust exposures caused by scooping and may have been controlled by the ventilation system. The weighing activity appeared to be associated with higher dust exposure than did the turning activity. This difference, however, may be an artifact caused by the delay of the HAM's response to the high dust exposures during scooping.

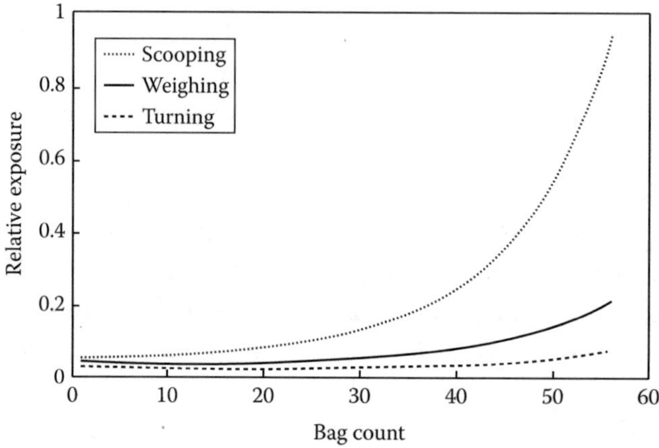


FIGURE 36.4 Modeled dust exposure of a worker as a function of bag count for scooping, weighing, and turning.

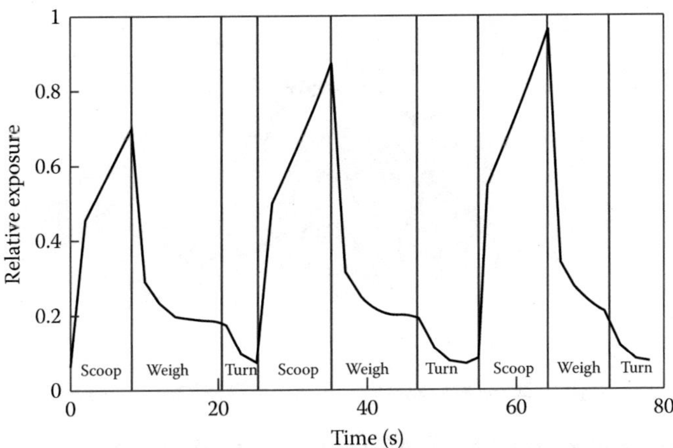


FIGURE 36.5 Modeled dust exposure of worker for filling bags 51 through 53.

**Findings**

Figure 36.4 shows that dust exposure increased with bag count, which is a surrogate variable for depth of scooping. The data were collected over ~20 min. This same conclusion was reached with the use of conventional short-term measurement of dust concentrations with pumps and filters. The filter data, which required three full shifts to collect, however, did not provide any insight into the relationship between job cycle and the worker's dust exposure. Knowledge of the specific task that elevated the worker's dust exposures was crucial to the redesign of the weigh-out booth.

**Recommendations**

Based upon the results presented in Figures 36.4 and 36.5, researchers recommend that the facility use shorter storage drums for bulk powder to reduce the dust generated by scooping. This case study clearly showed that direct-reading monitors can qualitatively and quantitatively measure sources of dust exposure during the work cycle—exposures too short with regard to time to be studied with integrated air sampling methods.

### 36.9.2 Case Study 2: Dental Administration of Nitrous Oxide

**Case Study 2a: Effectiveness of Scavenging System in Reduction of Occupational Exposure to Nitrous Oxide**

This study was conducted to evaluate how effectively scavenging systems reduce occupational exposure to waste $N_2O$ [16]. For more than 100 years, dentistry has used $N_2O$ as a general anesthetic agent, analgesic, and sedative [17]. Today, $N_2O$ is used primarily for psychosedation to reduce fear and anxiety in the conscious patient [17]. $N_2O$ scavenging systems typically have three principal components: an $N_2O$ and oxygen ($O_2$) gas delivery system, a nasal cone for the patient from which to inhale the gases, and an exhaust system that carries the respired gas from the patient out of the building. A schematic of the nasal cone is shown in Figure 36.6.

Occupationally exposed workers such as dentists or dental assistants have documented adverse effects from exposure to concentrations of $N_2O$ as low as 50 ppm. The effects

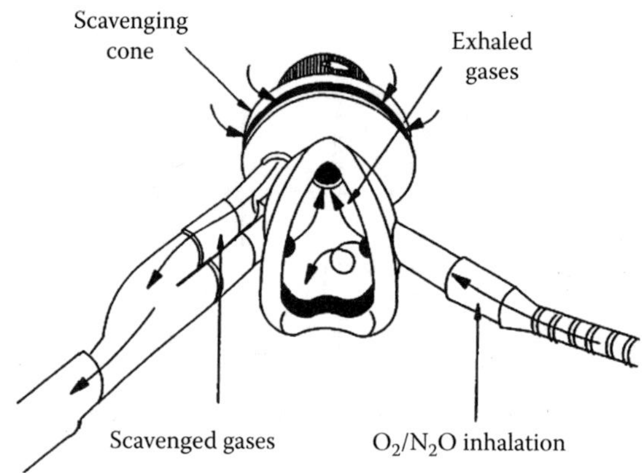


FIGURE 36.6 Diagram of a typical dental $N_2O$ scavenging system.

involve the reproductive, hematological, nervous, and audiovisual systems [7,18,19]. The ACGIH has established a threshold limit value of 50 ppm 8 h TWA [20]. Although the studies show that scavenging systems significantly reduce $N_2O$ concentrations, the systems do not reduce it to the NIOSH recommended exposure limit (REL) of 25 ppm during the time of administration [21]. In addition to evaluating the effectiveness of scavenging systems, this study was conducted also to determine why exposures exceeded 25 ppm.

**Methodology**

NIOSH researchers evaluated a dental facility that uses a commercial scavenging system during dental surgery. Ten dental operations (e.g., filling, extracting) were monitored by using a combination of sampling strategies: personal breathing zone sampling (dentist and dental assistant), general area sampling, and real-time sampling. A Miran 1A (Foxboro Instruments, Foxboro, MA) monitored the real-time $N_2O$ concentrations. A sampling probe connected to the Miran 1A was placed ~12 in. above the patient's head. In addition, video and IR scanning equipment recordings monitored the dental practices (activities). Because $N_2O$ is IR absorbing, it can be "visualized" by using IR thermography. Motion and time measurement techniques were used to document activities of the dentist, dental assistant, and patient during the operation [14]. These activities, listed in the following, were coded into a computer spreadsheet along with the associated $N_2O$ concentration data:

Injecting local anesthetic

Extracting a tooth

Filling a tooth

Using the aspirator

Using the water and air syringe

Using the rubber dam (small rubber sheet that isolates the operative site)

Using the curing light for restorative composite resin material

Talking, coughing, and yawning of patient

Turning on $N_2O$

Turning off $N_2O$

Adjusting $N_2O$ flow rate

Statistical analysis of the $N_2O$ concentration and changes in concentration were modeled as a function of these work elements from the spreadsheet [22].

**Findings**

Average real-time $N_2O$ concentrations for the 10 operations ranged from 206 to 770 ppm. The average real-time concentration over all 10 operations was 442 ppm. The average personal breathing zone (integrated sample) concentration over all 10 operations for the dentists was 487 ppm. There was no significant difference ($p < 0.68$) between the real-time and personal breathing zone concentrations for dentists. There was, however, a significant difference ($p < 0.014$) between the overall average real-time sampling concentration and the average personal breathing zone concentrations among dental assistants (150 ppm). The differences in dental assistant breathing zone concentrations and the real-time concentrations may have been because the sampling probe was placed closer to the patient's and the dentist's breathing zone than to the dental assistant's breathing zone. Thus, these real-time sampling results may be more representative of the dentists' exposure than that of the dental assistants. It also was determined that the dentists, by nature of the dental surgery, worked closer to the patient's breathing zone than did the dental assistants.

Real-time sampling results and work activities were combined to determine if changes in $N_2O$ concentrations were related to these activities. From the video recordings, several dental surgery activities were selected for analysis. For data analysis, the real-time concentrations were matched with the identified dental activities. A plot of this relationship is shown in Figure 36.6. Based on this analysis, the only activities that showed significant $N_2O$ concentration changes occurred (1) when the dentist turned on the $N_2O$ gas, (2) when the dentist adjusted the $N_2O$ concentration during the operation, and (3) when the dentist turned off the $N_2O$ gas, following the operation. Statistical analysis showed that 98% of the changes in $N_2O$ exposure could be accounted for by the $N_2O$ concentration of the gas delivered to the patient as opposed to the specific dental surgery activities (note the "saw-tooth" pattern in Figure 36.7). Thus, the primary source of $N_2O$ exposure was not from the work practices of the dentists but from $N_2O$ delivery and the inadequacy of the scavenging exhaust system.

During two of the ten dental operations, an IR video camera qualitatively evaluated scavenging mask leakage. The IR camera revealed $N_2O$ leakage between the mask and face seal, indicating that the scavenging mask did not fit the patient's face properly. The off-gassing of $N_2O$ during patient mouth breathing also affected exposure during these two operations. The IR video camera also revealed that a sudden increase in $N_2O$ exposure could be traced to the patient's expired breath. This increase was corroborated by real-time data. When the patient inspired, the $N_2O$ concentrations decreased. Synchronization of

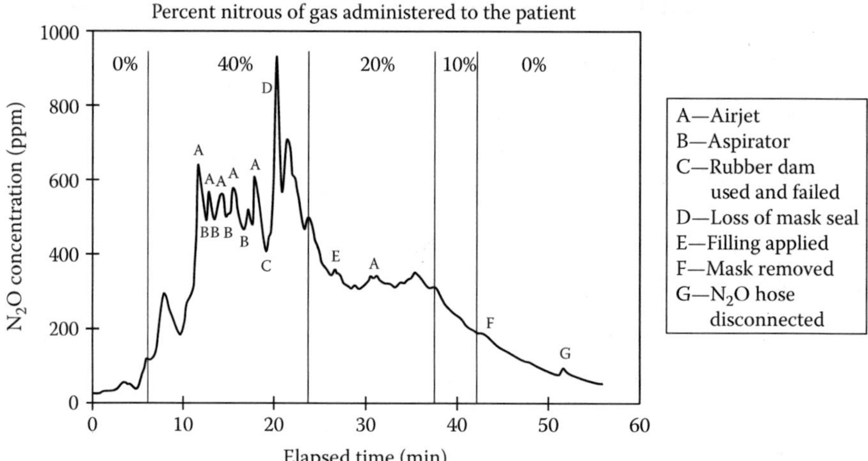


FIGURE 36.7 Plot of real-time $N_2O$ concentration with activities and supply concentrations.

the real-time data with the IR video camera helped to confirm that patient's mouth breathing was also a source of $N_2O$ exposure.

**Recommendations**

Scavenging mask leakage and an inadequate scavenging exhaust system caused most of the $N_2O$ exposure in this study. Patient mouth breathing was a secondary source of exposure. If the scavenging system had been more efficient, the work practices, such as use of the aspirator, air, and water syringes, and patient mouth breathing might have had a greater impact on the $N_2O$ exposures of the dentists.

The IR video camera proved to be a valuable tool for detecting $N_2O$ leakage from the patient's mask as well as from patient mouth breathing. By following the real-time data patterns, NIOSH researchers discerned when there was a mask leak, when the patient was mouth breathing, or both. This ability to determine these exposure sources helped provide recommendations for improving scavenging system mask design and work practices, and for reducing overall $N_2O$ exposures.

**Case Study 2b: Evaluation of Two Types of Scavenging Systems**

Due to reproductive, hematological, nervous, and audiovisual systems effects, waste gas scavenging systems are recommended to be used in dental offices to reduce exposure to $N_2O$ to dental office workers. A geometric mean of 25 ppm in the breathing zone of the dentist was obtained when using one type of scavenging system under optimal conditions with no actual patient involvement in this study. Realistically, factors such as engineering controls, work practices, and patient behavior may cause the levels of waste anesthetic gases to be increased under conditions of actual patient involvement during the evaluation of scavenging systems.

This study, which applied the techniques of video monitoring, real time–exposure monitoring of $N_2O$, and IR thermography to two types of scavenging gas systems in dental offices during oral surgery, had three main objectives. The first objective was to assess and compare the efficacy of two different designs of waste gas scavenging systems used in a dental office. The second objective was to compare the performance of the systems to the relevant occupational exposure limits. The third objective was to use one of the scavenger systems and evaluate the effect of work practices in two dental offices based on experience using the system, work practices, and $N_2O$ exposure.

**Methodology**

The overall design of this study involved three phases. All three phases were performed during exclusively wisdom teeth extraction procedures of at least two wisdom teeth for patients. Phase 1 evaluated a new scavenging system design, described as System 1 (Safe Sedate Dental Nasal Mask™) shown in Figures 36.8 and 36.9, in a dental office (Office 1) performing typical wisdom teeth extractions. The dentist in Office 1 was experienced with System 1. Phase 2 was performed in Office 1 with a different waste gas scavenging system, described as System 2 (PORTER scavenging System: PORTER Instrument Company, Inc., Hatfield, PA), shown in Figure 36.10. Phase 3 involved System 1 in a different office, Office 2, with a dentist (Dentist 2), who was a novice using System 1 [23].

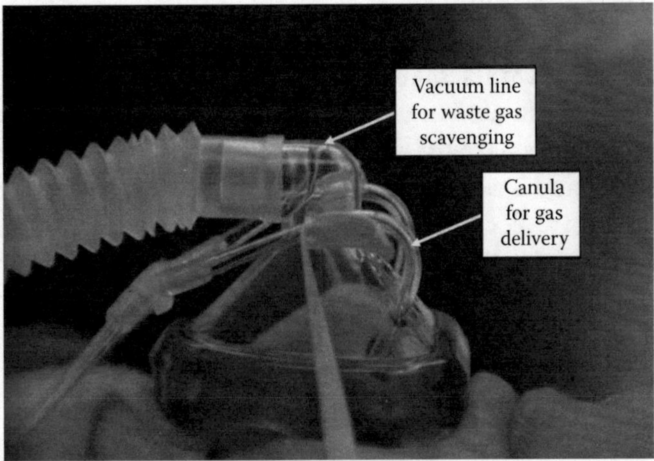


FIGURE 36.8 System 1 (Safe Sedate Dental Nasal Mask™).

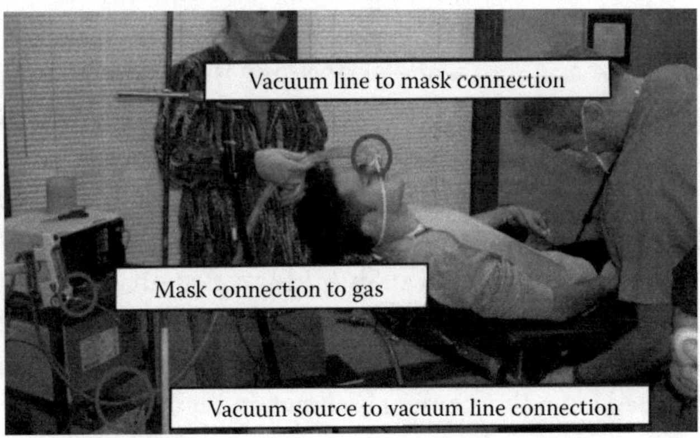


FIGURE 36.9 Typical setup for System 1.

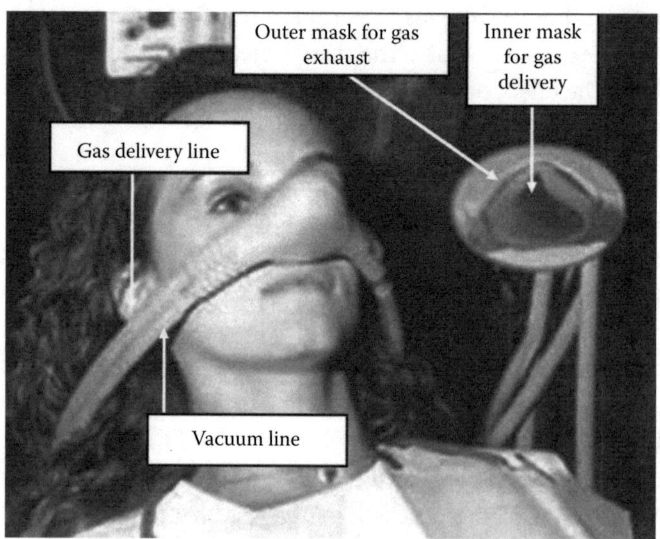


FIGURE 36.10 System 2 (PORTER™) design.

Phases 1 and 2 were used to assess and compare the efficiency of two different designs of waste gas scavenging systems used in a dental office. In addition, Phases 1 and 2 compared the performance of the systems to the relevant occupational exposure limits of the NIOSH REL. Finally, Phases 1 and 3 were used to compare the effects of work practices (MVTA used for comparison) in Office 1 versus Office 2 using System 1. In other words, part of the comparison involved the difference in actual work practices due to the differences in experience using System 1.

The main equipment required for the monitoring of exposure in each of these phases was a digital video camera, a waste gas scavenging system, an IR camera, and a real-time ambient air analyzer for $N_2O$. Other necessary equipment were external hard drives for data storage, a laptop computer for operation and image storage from the IR camera, and a digital camera for photographic records of events occurring during the study. The overall camera setup for the digital video camera and IR camera are shown in Figure 36.11.

The process followed in obtaining exposure data was $N_2O$ exposure data was obtained while simultaneously the digital video camera (SONY HandyCam® DCR-SR100) filmed the activities of the dentist and staff while performing the wisdom teeth procedure and IR thermography images were obtained. Then the video images saved to the camera were then uploaded to a computer where then ImageMixer software was used to edit the video and prepare it for synchronization to the IR thermography images and to the data concerning $N_2O$ exposure to the dentist and staff obtained in real time by the ambient air analyzer.

The IR [a Flir Systems Merlin® Mid InSb Mid-wavelength Infrared (MWIR)] camera was used to detect IR radiation in the 3–5 μm wavelength range. Flir Systems ThermaCAM Researcher software Pro 2.8 Sr-3 software was use to view the thermographic images. A typical image obtained from this camera is shown in Figure 36.12. This camera is able to detect $N_2O$ in the environment and convert it to an image visible by humans. The white circles show where $N_2O$ was detected around the subject's head as it was being administered

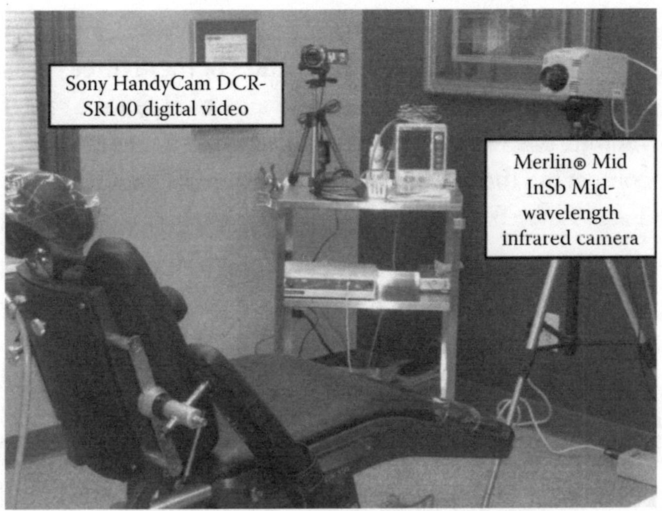


FIGURE 36.11 Example of camera setup.

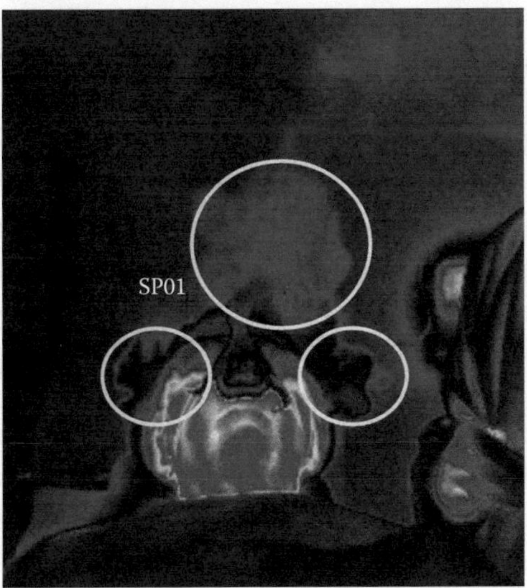


FIGURE 36.12 Typical image of $N_2O$ (shown by circles) from IR camera.

and the surgery was proceeding. The variations in color reflect the concentration of $N_2O$ with higher concentration (nearer the source) being more opaque.

The Miran SapphIRe ambient air analyzer (Thermofisher Scientific) was used to measure the air concentration of $N_2O$ in real time. During each subject's procedure, this instrument was attached to the dentist's shoulder of the dominant arm when he was in the room to measure the concentration within the dentist's breathing zone, otherwise when the dentist left the room, the instrument's probe was placed on a surrogate stand as near to the typical breathing zone for the dentist. This was to indicate the airborne exposure to $N_2O$ for an individual remaining in the area for the entire time of the $N_2O$ administration. The data obtained from these procedures were then uploaded to a computer for further data analysis.

MTVA software was used to combine the $N_2O$ air concentration data with the digital videos and IR thermography videos. The combination of these three types of data allowed the researchers to determine a relevant exposure value of $N_2O$ for a specific time point or interval of time. Therefore, an evaluation of the best work practices using System 1 could be performed by comparing the work practices and occupational exposures to $N_2O$ of the dentists in Phase 1 and Phase 3.

**Findings**

The results as shown in Table 36.1 [23] and ANOVA demonstrated that an experienced user of System 1 controlled occupational exposure to $N_2O$ significantly better than System 2. However, for an inexperienced user in Phase 3, System 1 did not achieve the same levels as the experienced user and was not significantly different from System 2. Neither system achieved the NIOSH REL of 25 ppm TWA for the time of exposure. Earlier studies have shown that it is possible for System 2 to achieve levels less than 50 ppm [24,25], but this study did not show this. Prior studies have evaluated the system in mock surgeries not using real patients. Thus, the prior studies have controlled variables, such as patient talking and breathing. However, in actual surgeries controlling these types of variables may be difficult. Hence, VEM with real-time detection of $N_2O$ allows the researcher to observe the effect of these variables.

While neither scavenging system demonstrated the ability to control occupational exposure to $N_2O$ below the NIOSH REL, System 1 performed better at this task than System 2. Combining consistent best work practices with System 1's usage achieves maximal efficiency. This study demonstrates that not only is control of occupational exposure to $N_2O$ in the oral surgery environment dependent on system design, it is also strongly dependent on experience and motivation for proper system utilization, work practices, and patient and surgical variables which are less easily controlled.

The combination of IR thermography, digital videos, and real-time ambient air monitoring of $N_2O$ concentration allows the comparison of occupational exposure to $N_2O$ between two scavenging systems for waste gas while examining the best work practices to use while performing wisdom teeth extractions such as ensuring that the mask of the scavenging system is fitted properly for either System 1 or System 2. The comparison between the two dentists' offices showed from the combination of these three data analysis that the use of best practices by the experienced dentist was significantly less than the less-experienced dentist.

TABLE 36.1 Summary of Descriptive Statistics for Each Study Phase of $N_2O$ Case Study

| | Phase 1: System 1—Experienced User | Phase 2: System 2—Experienced User | Phase 1: System 1—Inexperienced User |
|---|---|---|---|
| *n* | 16 | 14 | 13 |
| Mean air concentration of nitrous oxide/surgery (ppm) (S.D.) | 61.6 (35.8) | 225.6 (71.1) | 166.5 (68.9) |
| Minimum, maximum (ppm) | 25.5, 155.8 | 80.2, 363.8 | 64.5, 332.8 |
| Confidence interval (95%) | (42.5, 80.6) | (184.6, 266.6) | (124.9, 208.1) |

The evaluation of IR thermography and air concentration data using MTVA suggests that key work practices, patient, and surgical variables affect waste gas emissions.

**Recommendations**

VEM using a real-time detector of $N_2O$ together with digital videography and IR thermography allowed the evaluation of the efficacy of two different types of waste gas scavenging systems. In addition, it allows the examination of the work practices of two different dentists with different levels of experience with a specific scavenging system. These work practices were then able to be examined in detail using VEM and MTVA to determine specific times/intervals of time when exposure was highest and determine how the differences in work practices resulted in different exposures for the same task. These examinations then could lead to recommendations for improvements in work practices leading to lower occupational exposure to $N_2O$.

### 36.9.3 Case Study 3: University Laboratory Using Radioactive Materials

Diagnostic nuclear medicine procedures have increased 500%–600% in the past 40 years in the United States according to the National Council on Radiation Protection and Measurements (NCRP) Scientific Committee 6-2 medical subgroup [26]. In 1982, brain and thyroid procedures predominated in 1982, consisting of more than 56% of the total nuclear medicine procedures. While in 2005, higher dose cardiac procedures predominated, making up about 70% of the procedures [26], resulting in an increase of 550% in the estimated per capita effective dose from in vivo diagnostic nuclear medicine in the period of 1986–2005 [26]. Hence, patients receive a higher dose, but nuclear pharmacists and nuclear medicine technologists, who prepare, dispense, and administer the radiopharmaceuticals, are handling higher doses. Nuclear medicine and nuclear pharmacy workers have higher radiation exposure than other radiation workers in medicine due to preparation, dispensing, and administration of radiopharmaceuticals. The hands are of particular concern in a nuclear pharmacy. The NCRP states that the dose limit to the hands is 500 mSv in a year and stresses the concepts of "as low as reasonably achievable (ALARA)" [27]. While most workers in nuclear medicine and nuclear pharmacies record equivalent dose to the hands below the annual limits [28], the total exposure recorded may underestimate the actual dose, as a ring dosimeter worn at the base of the finger does not record the actual dose received by the fingertips. Several reports have shown [27–30] that the fingertips can receive a dose of 1.4, 5, and even 9 times the dose that the ring dosimeter records. Extrapolation of the exposure the fingertips receive has been shown to have great variability in the annual dose with the potential for the fingertips to exceed the annual dose limits if optimized work practices are not used [28–32].

Several factors may come into play in the exposure to workers' hands when handling radionuclides in the nuclear pharmacy. Factors include the type(s) of radiation emitted by the radioisotope, the energy of the radiation, the procedures and length of time required to prepare a specific radiopharmaceutical, the type of shielding used, the workstation layout/design, anthropometric variations in individuals, work experience, and gender of the nuclear pharmacists and nuclear medicine technologists.

Simultaneous to radiation exposure, individuals working in the nuclear pharmacy perform repetitive tasks, which place their bodies under musculoskeletal stresses. Nuclear pharmacy tasks are similar to those performed in other medical laboratories, which have been shown in earlier studies to have risk of musculoskeletal injury [33,34]. However, required shielding for radiation exposure protection places additional musculoskeletal stresses on the bodies of nuclear pharmacists. In particular, the upper extremities, hands, neck, and shoulders are at risk of musculoskeletal injury due to static loading, repetitive tasks, hand tool design, awkward positions, and postures.

Workstation design that minimizes radiation exposure using principles of radiation protection, namely, time, distance, and shielding, can be in conflict with the workstation design using principles of good ergonomic work practices. The interplay between anthropometric variations in individuals and minimizing radiation exposure are complex and involve interactions between workstation layout/design, principles of radiation protection, and ergonomics as shown in Figure 36.13.

Traditionally, in the nuclear pharmacy, film badges and TLDs have been used to monitor cumulative personal radiation exposure. These dosimeters are worn as chest badges and as a ring on the finger(s). These dosimeters must be processed before the dose received by the worker is known. Recently, real-time VEM systems that monitor the instantaneous dose rate and the total dose of gamma radiation of the fingers obtained by an individual have been developed [35–39]. The advantages of these systems are that a worker can view video and then correlate the activity being performed together with the dose received at the time, peak exposures can be detected, and the total dose received over an interval of time is known immediately. The disadvantages of some of these systems are that they only detect gamma radiation and rely on the worker to record his or her activities. Work performed in the United Kingdom using the early advanced exposure gamma instrument system (AEGIS) as well as a system developed in the United States by Wang and McGlothlin [35–40] shows the feasibility of VEM for the assessment of radiation exposure to individuals. The latest edition of AEGIS [38–41] uses two cameras together with two dosimeter

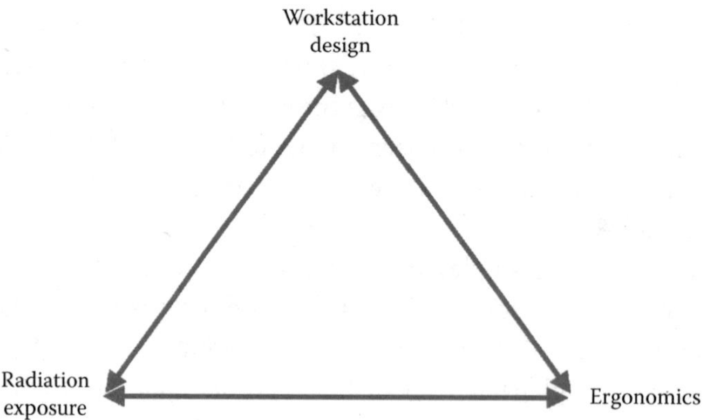


FIGURE 36.13 Interaction between ergonomics, radiation exposure, and workstation design.

probes worn on a finger on each hand to assess exposure and worker posture. AEGIS permits the assessment of radiation exposure and worker posture simultaneously in conjunction with tasks being performed as well as identifying unintended procedures or actions that may increase radiation exposure. Therefore, task performed can be correlated with the instantaneous dose rate and total dose. In addition, the researcher can identify work practices that may be causing higher (peak) exposures during certain tasks' elements rather than obtaining a just cumulative exposure dose and then having to reconstruct the events (tasks) that lead to the cumulative dose. VEM allows the researcher recommend ways to help the worker reduce exposure based on the work environment because the layout and design of the workplace is observed. The advanced extremity gamma instrumentation system (AEGIS) is the VEM system used in this study [38–40]. Observations of the video, recording the musculoskeletal stresses to the body as well as other conditions, such as glare that causes stresses to the body, provide an initial assessment of the ergonomics of the workstation design and layout. The initial assessment provides a perspective on elements of the task that are the most stressful and have the potential for musculoskeletal injuries.

**Methodology**

The main equipment and supplies used in this project were a small nuclear pharmacy teaching lab containing $^{99}$Mo-$^{99m}$Tc generator(s), laminar flow hoods, L-blocks, syringe shields, vials containing radioactive materials, lead containers, syringes, tongs, dose calibrators, Geiger-Muller counters, microscopes, materials for product quality assurance, a heat source, single channel analyzers for wipe tests, and waste containers. Some of the equipment is shown in Figure 36.14.

In addition, Figure 36.15 demonstrates the flow of data and the equipment used in this study. The equipment includes web cameras and the advanced extremity gamma instrumentation system (AEGIS), which consists of an ED2 personal extremity dosimeter, and software that can capture and synchronize video and radiation exposure measurements via Bluetooth technology to a computer.

The study involved recording tasks of a typical nuclear pharmacist's duties. The system used two webcams to record video on a laptop computer. One camera was used for

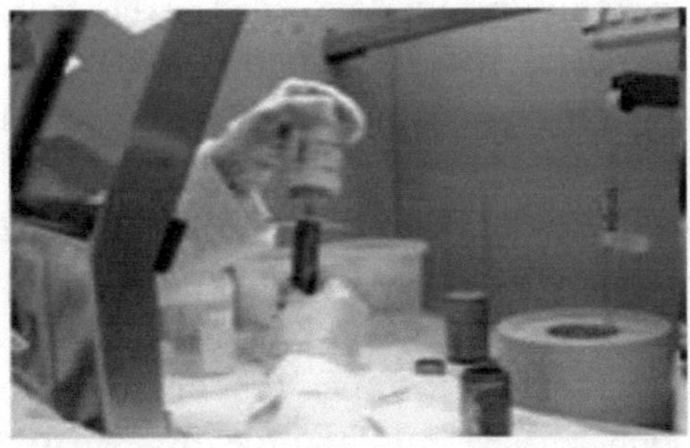

FIGURE 36.14 Typical equipment found in a nuclear pharmacy lab: L-block. Vials with radioactive materials, dose calibrator, syringe, and lead containers.

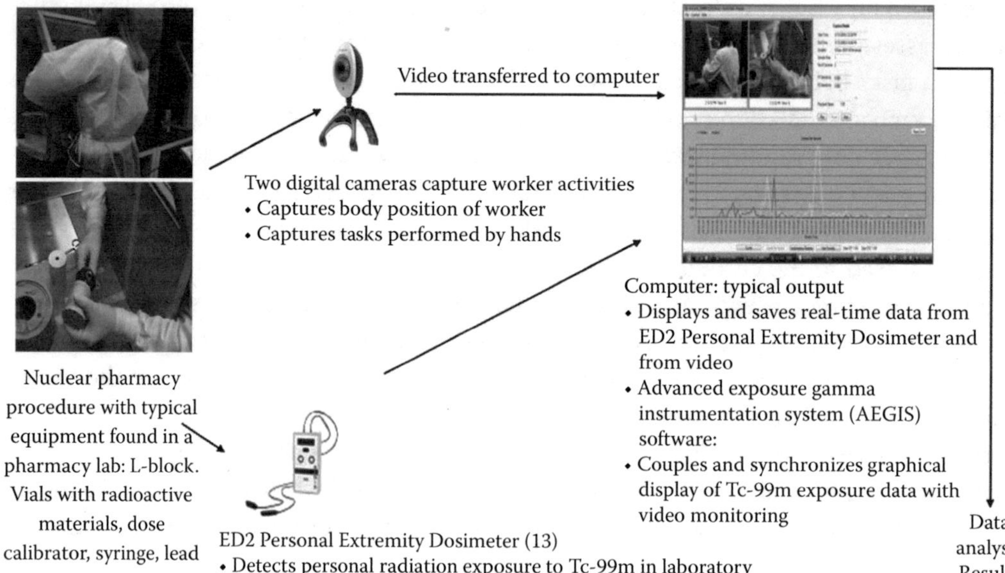


FIGURE 36.15 Research equipment and data stream of VEM of radiation exposure.

recording the worker's body position while the second camera was used to record the AEGIS dosimeter measurements and the worker's activities being performed by the hands. Still pictures were taken of the task of interest for radiation exposure and ergonomic concerns. Both web cameras were started simultaneously and were synchronized to the AEGIS dosimeter measurements.

The measurements of radiation exposure (mSv/h) to the hands and fingers were obtained in real time using an ED2 Dosimeter while simultaneously video recording the tasks performed by the hands. The ED2 Dosimeter has two probes that were attached to fingertips on each hand. Then the AEGIS Software, which connects to the dosimeter, permits wireless capture of radiation exposure data and sends it to a computer with/without video recording. The dosimeter is able to measure instantaneous dose rate and cumulative dose of exposure, and sets off an alarm if above a preset threshold level for accumulated dose and dose rate. The actual tasks observed and recorded by the (radiation exposure measurements recorded by AEGIS software at sampling rate of 1/s).

The ergonomic assessment involved observation of activities to see if the worker has an awkward posture with the shoulders abducted, whether the elbow is in neutral posture, static loading on the neck, glare/visual interferences from the screen, and the barrier height for different physical differences between the workers. More specifically, the ergonomic evaluations consisted of measurements of the height of the participants, the arm reach of participants, workstation height, dimensions of L-block used for radiation protection while preparing radiopharmaceuticals, and the distance to the dose calibrator from the front of the workbench. In addition to these measurements, participants were observed while in the lab and on video while preparing radiopharmaceuticals. The observations documented body positions and activities performed.

The overall design of the experiment was as follows:

1. Two cameras: one focused on the hands and one focused on the back and upper extremities
2. AEGIS with ED2 Extremity Dosimeter (taped to fingers, one on each hand) monitors radiation exposure
3. Measured participants' height, eye height, shoulder height, arm reach (thumb-tip reach), elbow height, waist height, shoulder breadth, and hand size (breadth and length)
4. Subject performed simulated radiopharmaceutical compounding/dose drawing exercise. Subjects randomly assigned to perform either Part A, then Part B or Part B, and then Part A in a cross-over design of the experiment
5. Reviewed video and correlated task elements with radiation exposure and musculoskeletal stresses to the individual

The simulated dose drawing sequence for Part A was (1) radionuclide (Tc-99m) drawn into syringe using syringe shield, (2) needle recapped while in the syringe shield, (3) syringe removed from syringe shield with tongs, (4) syringe placed in the dose calibrator using tongs, (5) dose measured, (6) syringe removed from dose calibrator, (7) syringe placed back into syringe shield, (8) needle cap removed, (9) inject radioactive material in syringe back into pig with radioactive material, (10) cap replaced, and (11) syringe placed in radioactive waste container. For Part B the sequence was as follows: (1) a new uncontaminated syringe is obtained and then (2) the process is repeated in Part A without the using syringe shield or tongs to manipulate syringe.

**Findings**

The study had 13 subjects; one trial was not included in results due to equipment problems. Typical results from a trial are shown in Figures 36.16 and 36.17. Overall, this specific trial allowed the specific elements of the task when peak exposures occurred to be identified (Figure 36.16). It also allowed the differences in using best practices leading to the lowest cumulative dose to be seen (Figure 36.17).

*Radiation Exposure Assessment*

The results from the assessment of radiation exposure are summarized in Table 36.2. The key findings in the table reveal that when Part B is compared to Part A that first Part B (not using tong or the syringe shield) was nearly twice as fast as Part A (use tongs and syringe shield). Therefore, the use of tongs and syringe shield may hinder the manipulations of syringes and other equipment, which may be due to individuals' anthropometric differences, experience, workstation layout/design, and equipment. However, making comparisons between peak and cumulative exposures for Parts A and B, it is found that that for the dominant hand, Part B has an average peak exposure of four times the average peak exposure of Part A. For the nondominant hand,

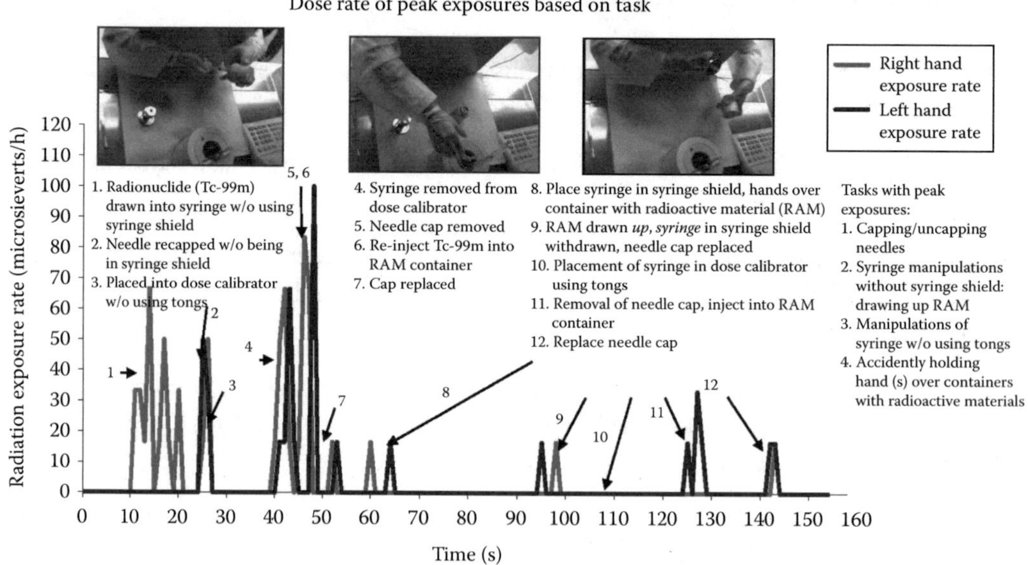


FIGURE 36.16 Typical output results of radiation exposure assessment based on task: dose rate of peak exposures based on task.

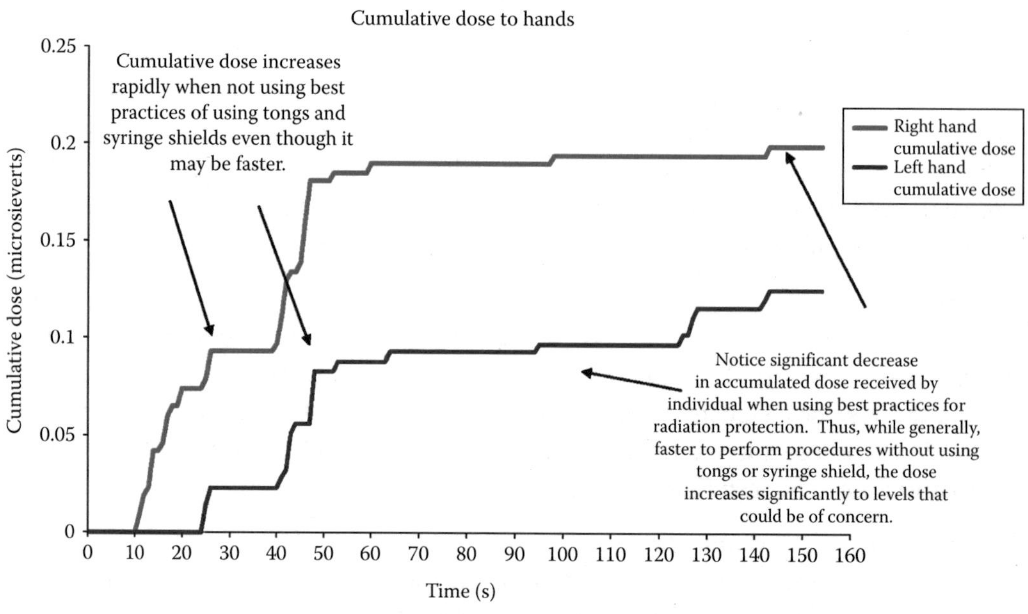


FIGURE 36.17 Typical output results of radiation exposure assessment based on task: cumulative dose to hands.

Part B has an average peak exposure of 1.4 times the average peak exposure of Part A. Comparison of the cumulative exposures shows that for the dominant hand, Part B has an average cumulative exposure of 14 times the average cumulative exposure of Part A. For the nondominant hand, Part B has an average peak exposure of twice the average peak exposure of Part A.

TABLE 36.2 Summary of the Results of Radiation Exposure to the Hands

| | Part A | | | | Part B | | | |
|---|---|---|---|---|---|---|---|---|
| | Peak Exposure | | Cumulative Exposure (Ave. Length of Part A: 96.833; Exposure Time during Part A: 13.833 s) | | Peak Exposure | | Cumulative Exposure (Ave. Length of Part B: 50.417; Exposure Time during Part B: 24.917 s) | |
| | Normalized Peak Exposure Dose Rate to Dominant Hand (μSv/h) | Normalized Peak Exposure Dose Rate to Nondominant Hand (μSv/h) | Normalized Dose to Dominant Hand in 1 h (μSv) | Normalized Dose to Nondominant Hand in 1 h (μSv) | Normalized Peak Exposure Dose Rate to Dominant Hand (μSv/h) | Normalized Peak Exposure Dose Rate to Nondominant Hand (μSv/h) | Normalized Dose to Dominant Hand in 1 h (μSv) | Normalized Dose to Nondominant Hand in 1 h (μSv) |
| Range | 3.3328–16.9552 | 0–40.6929 | 0.0783–0.6538 | 0.0704–1.9693 | 6.8609–77.8073 | 6.6656–32.3345 | 0.8429–10.7372 | 0.4246–3.3714 |
| Median | 5.1535 | 6.8774 | 0.2027 | 0.3068 | 17.8432 | 10.1830 | 3.197 | 00.9771 |
| Mean | 6.9497 | 9.4668 | 0.2566 | 0.5475 | 28.4300 | 13.7203 | 3.7230 | 1.1010 |
| Std. Dev | 4.7277 | 10.8002 | 0.1785 | 0.5634 | 21.2431 | 8.5911 | 2.6218 | 1.214 |

The relationship between anthropometric variables and radiation exposure was examined using SAS analysis. The generalized linear model (GLM) procedure in the SAS software with the categorical variables of gender, experience, height, elbow height, shoulder breadth, average handbreadth across the metacarpal and thumb, average hand length and average arm reach, and interactions against six radiation exposure measurements: Part A, the dominant hand cumulative dose; Part A, the nondominant hand cumulative dose; Part A, cumulative dose (cumulative dose was larger of dominant or nondominant hand); Part B, the dominant hand cumulative dose; Part B, the nondominant hand cumulative dose; and Part B, cumulative dose (cumulative dose was larger of dominant or nondominant hand). The significant variable (s) at 0.15 significance level (this may improve with large sample size) was for each part as follows:

Part A, the dominant hand cumulative dose: gender, experience, height, shoulder breadth, and average arm reach, with the model explaining 98.2% of the relationship between the variables and the exposure

- Part A, the nondominant hand cumulative dose: height, with the model explaining 91.9% of the relationship between the variables and the exposure
- Part A, the cumulative dose (for each subject, cumulative dose was larger of dominant or nondominant hand): height, with the model explaining 91.5% of the relationship between the variables and the exposure
- Part B, the dominant hand cumulative dose: no single significant variable, maybe complicated interaction between workstation design and anthropometric variables, with the model explaining 77.5% of the relationship between the anthropometric variables and the exposure
- Part B, the nondominant hand cumulative dose: average handbreadth across the metacarpals with the model explaining 84.9% of the relationship between the anthropometric variables and the exposure
- Part B, the cumulative dose (for each subject, cumulative dose was larger of dominant or nondominant hand): height, with the model explaining 77.1% of the relationship between the variables and the exposure

Significance level may improve if experience is not a factor, so work in facility with experienced workers may better define significant anthropometric variables.

*Ergonomic Assessment in a Nuclear Pharmacy*

Workers in the nuclear pharmacy face musculoskeletal stresses to the hands, wrists, arms, shoulders, neck and back as demonstrated in Figure 36.18 and Table 36.3. The most significant stresses appear to be awkward hand and arm positions from holding lead containers as well as standing at a non-adjustable L-block and reaching around it to perform tasks. The severity of musculoskeletal stresses may be a result of anthropometric

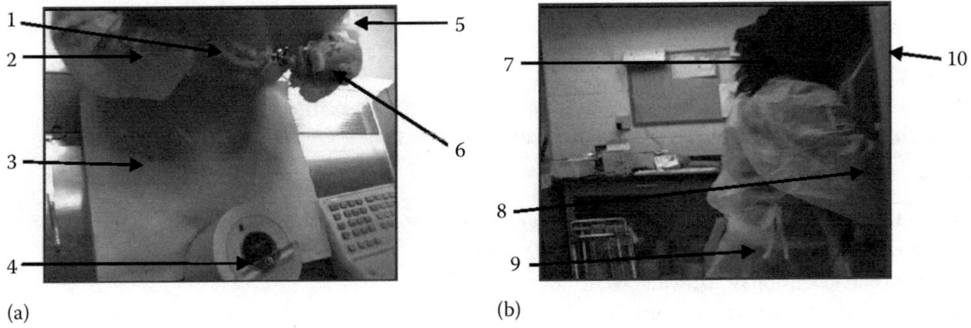


FIGURE 36.18 Example of stresses to the body while drawing up radionuclide at one instance in time.

variation of the participants, partially due to the layout/design of the workstation and the need to minimize radiation exposure.

**Recommendations**

Radiation exposure increases with specific tasks of making radiopharmaceuticals: capping and uncapping syringe needle, adding saline, drawing up final dose to be administered to patients, and measurements of radioactive material in dose calibrator. However, the radiation exposure can be minimized by using best practices. Hence, while it may be slower to use best practices to use tongs and syringe shield to perform the task, the reduction in radiation exposure is significant.

Anthropometric variations in individuals could significantly affect the dose received; variables that could significantly influence radiation exposure and musculoskeletal stress to the body that an individual receives are gender, experience, height, shoulder breadth, average arm reach, and average handbreadth across the metacarpals. Hence, designing a workstation (including the shield blocking the body from receiving radiation) that is adjustable to fit the nuclear pharmacist could minimize the dose received and the musculoskeletal stresses to the body. Figure 36.19 shows two commercially designed radiation shields that provide the required radiation shielding while having some adjustability in height, angle of inclination, and may lessen musculoskeletal disorders due to more ease in wrapping arms around shield [42].

Overall, the main recommendation is the recognition of workers and employers that radiation safety should be considered in all ergonomic recommendations for the redesign of a nuclear pharmacy workstation. In addition, ergonomic and anthropometric variations in individuals need to be considered in designing workstations, providing tools, and equipment that minimizes radiation exposure in a nuclear pharmacy facility.

VEM and a real-time radiation dosimeter has the potential to be used as a tool for training nuclear pharmacists and nuclear medicine technologists in correct work practices and techniques that can minimize radiation exposure while lessening stresses to body by using best ergonomic practices. In addition, VEM has the potential to be used as a real-time monitor of work practices in actual nuclear pharmacy aid in the maintenance of correct work practices by individuals.

TABLE 36.3 Summary of Findings and Recommendations of Observations of Ergonomic Risks to the Body at One Instance in Time Based on Figure 36.18

| Finding | Recommendation if Feasible |
|---|---|
| 1. Right hand and fingers in awkward position; repetitive motion of drawing up radioactive materials (RAM)—flexion and extension of hand and wrist<br>Gloves ill-fitting—may increase grip strength needed to hold equipment which may cause tendinitis or other musculoskeletal disorders (MSDs) | Provide well-fitting gloves |
| 2. Forearm unsupported at standing workstation doing precision work | Support forearms at standing workstation or perform task as either a sitting workstation of sit–stand workstation |
| 3. Width of L-block makes it difficult to reach around to compound radiopharmaceuticals. However, it protects the body from radiation exposure | Make shield easier to put arms around |
| 4. Placement of dose calibrator makes it difficult for individuals with smaller arm reaches to reach, may have higher radiation exposures and greater stress to back | Dose calibrator in designated place designed for it, set down in hole. Difficult to move it elsewhere at this time |
| 5. Upper arm horizontal and vertical adduction and abduction during compounding process<br>Static loading on shoulder, arm, and left hand from holding lead container with RAM in it while drawing out RAM or adding saline to radiopharmaceutical | Support forearms at standing workstation or perform task as either a sitting workstation of sit–stand workstation<br>Have shield angle be adjustable<br>Make L-block easier to put arms around L-block shield |
| 6. Awkward posture of hand (extension and flexion of hand and wrist) and static loading of hand while performing task repetitively | Train in the correct technique to hold lead container—use a power grip at end to minimize radiation exposure |
| 7. Static loading on neck from bending to look through leaded glass window | Make L-block adjustable in height and angle of inclination to relieve stress on neck and allow a more upright posture |
| 8. Right arm: static loading, arm in awkward posture around L-block, arm is at heart level so increases workload, arm is abducted and adducted during procedure | Shield needs to be adjustable in height<br>Shield for this participant needs to be narrow to fit body<br>Sitting or sit–stand workstation may be a preferable situation<br>Transparent shield may allow individual to lower arms since procedure will be visible at a lower height |
| 9. Static loading on back due to standing in one position at workstation for long period of time—may lead to awkward postures as the day goes on | Precision work performed—sit–stand or sitting workstation relieves static loading on back but there may not be enough clearance for legs at workstation |
| 10. L-block window: Limited view angles, so students must strain their necks to see what they are doing | Provide completely transparent, adjustable shields equivalent to solid L-block shielding in radiation protection |

FIGURE 36.19 Examples of commercially available shields that are adjustable and/or shaped to the body for easier maneuverability around the radiation shielding.

## 36.10 CRITICAL REVIEW OF CURRENT STATUS

VEM is a set of flexible techniques that can determine specific sources of a worker's exposure to contaminants. However, there are a number of practical considerations when integrating real-time exposures with the worker's activities. The researcher has to make sure the changes in exposures match worker activities. Problems challenging the researcher are (1) instrument transportation lag, such as natural diffusion of chemical from their source to the detectors; (2) response time, such as the instruments' sensors responding to a contaminant once it has arrived; (3) synchronization of video of worker's activities with exposure data; and (4) autocorrelation of data, which arises from instrument lag and response time. These challenges are equally matched by the difficulty of analyzing videotapes of worker activities and defining the work activity elements so that they correspond to changes in chemical or radiation exposure. For example, if the worker is performing several work activities in a short period of time, the changes in chemical exposure shown in the data may not match these activities. Therefore, the researcher has to take special precautions to collect enough data (i.e., videotape and real-time personal sampling) so that patterns of exposure emerge. This will guide the researcher to identify exposure sources and, from this, to devise effective controls. The key to successful integration of both real-time exposure assessment and job analysis techniques is to use both methods in such a way that patterns of exposure quickly emerge.

## 36.11 FUTURE CONCERNS

Federal and academic researchers [43–45] have conducted most of the integration of real-time direct-reading instruments with work analysis. For real-time exposure assessment and workplace job analysis techniques to succeed as a useful tool in industry, several things need to occur: (1) real-time instruments need to be more compact, portable, and easy to operate; (2) real-time instrument costs need to be in line with costs of traditional sampling instruments; (3) the availability and reliability of real-time sampling instruments need to be improved; (4) standards and guidelines need to be developed so that real-time data can be compared with traditional integrated sampling results; and

(5) industry needs to invest in educating and training its health and safety professionals to perform real-time sampling and work analysis assessments. The future of these sampling techniques may depend on a number of factors. The first is for government and academic researchers to continue applied research and development in this area and then to transfer this technology to industry to demonstrate the effectiveness of pinpointing exposure sources so that controls can be more effectively applied. The second is for industry to invest in this technology by training its own personnel to use these methods for cost-effective controls. We believe that if these factors occur, real-time sampling and work analysis will become a valued tool in evaluating and controlling hazardous chemical and physical agents in the workplace.

We are indebted to Michael Gressel, William Heitbrink, and Paul Jensen for their original contributions to the first edition of this chapter.

## REFERENCES

1. G. F. Knoll, *Radiation Detection and Measurement*, 3rd edn., John Wiley & Sons, New York, 2000.
2. H. F. Johns and J. R. Cunningham, *The Physics of Radiology*, 4th edn., Charles C. Thomas, Springfield, IL, 1983.
3. U.S. Department of Health and Human Services, Public Health Service, Centers for Disease Control and Prevention, National Institute for Occupational Safety and Health (NIOSH), *Analyzing Workplace Exposures Using Direct Reading Instruments and Video Direct Reading Instruments and Video Exposure Monitoring Techniques*, M. G. Gressel and W. A. Heitbrink, Eds., DHHS (NIOSH) Publ. No. 92-104, Cincinnati, OH, 1992.
4. J. D. McGlothlin, W. A. Heitbrink, M. G. Gressel, and T. J. Fischbach, Dust control by ergonomic design, *Proceedings of the IXth International Conference on Production Research*, Cincinnati, OH, 1987, pp. 687–694.
5. M. G. Gressel, W. A. Heitbrink, J. D. McGlothlin, and T. J. Fishbach, Advantages of real-time data acquisition for exposure assessment, *Appl. Ind. Hyg.* 3(11):316–320 (1988).
6. M. G. Gressel, G. J. Deye, W. A. Heitbrink, and J. D. McGlothlin, Graphical exposure demonstration: Computers and video link together, *Presented at the American Industrial Hygiene Conference*, St. Louis, MO, May 21–26, 1989.
7. U.S. Department of Health and Human Services, Public Health Service, Centers for Disease Control and Prevention, National Institute for Occupational Safety and Health (NIOSH), *Control of Nitrous Oxide in Dental Operatories*, J. D. McGlothlin, K. G. Crouch, and R. L. Mickelsen, Eds., DHHS (NIOSH) Publ. No. 94-129, Cincinnati, OH, 1994.
8. AGA Infrared Systems AB, *Operating Manual, Thermovision 782*, AGA Infrared Systems AB, Secaucus, NJ, 1984.
9. American Conference of Governmental Industrial Hygienist, *Air Sampling Instruments for Evaluation of Atmospheric Conditions*, 8th edn., ACGIH, Cincinnati, OH, 1995.
10. U.S. Department of Health and Human Services, Public Health Service, Centers for Disease Control and Prevention, National Institute for Occupational Safety and Health, *NIOSH Manual of Analytical Methods*, 4th edn., DHHS (NIOSH) Publ. No. 94-113, Cincinnati, OH, 1994.
11. M. G. Gressel, W. A. Heitbrink, and P. A. Jensen, Video exposure monitoring—A means of studying sources of occupational air contaminant exposure, Part I: Video exposure monitoring techniques, *Proceedings of the International Symposium on Air Sampling Instrument Performance, Appl. Occup. Environ. Hyg.* 8(4):334–338 (1993).

12. W. A. Heitbrink, M. G. Gressel, T. C. Cooper, T. J. Fischbach, D. M. O'Brien, and P. A. Jensen, Video exposure monitoring—A means of studying sources of occupational air contaminant exposure, Part 2: Data interpretation, *Proceedings of the International Symposium on Air Sampling Instrument Performance, Appl. Occup. Environ. Hyg. 8*(4):339–343 (1993).
13. D. Skoog and D. West, *Principles of Instrument Analysis*, Holt, Rinehart and Winston, New York, 1971.
14. R. M. Barnes, *Motion and Time Study: Design and Measurement of Work*, 7th edn., John Wiley & Sons, New York, 1980.
15. G. Box and G. Jenkins, *Time Series Analysis: Forecasting and Control*, Holding-Day, San Francisco, CA, 1981.
16. J. D. McGlothlin, P. A. Jensen, W. F. Todd, T. J. Fishbach, and C. L. Fairfield, Control of anesthetic gases in dental operatories at Children's Hospital Medical Center, Dental Facility, Cincinnati, Ohio, NTIS Publ. No. PB-90-155-946, National Technical Information Service, Springfield, VA, 1990.
17. E. I. Eger, Ed., *Nitrous Oxide/N$_2$O*, Elsevier, New York, 1985.
18. OSHA, Occupational and Safety Guidelines for Nitrous Oxide, Retrieved August 30, 2006, from http://www.osha.gov/SLTC/healthguidelines/nitrousoxide/recognition.html
19. M. S. Clark and A. L. Brunick, *Handbook of Nitrous Oxide and Oxygen Sedation*, 2nd edn., Mosby, St. Louis, MO, 2003.
20. ACGIH, TLVs and BEIs, ACGIH Worldwide, Cincinnati, OH, 1996.
21. U.S. Department of Health and Human Services, Public Health Service, Centers for Disease Control, National Institute for Occupational Safety and Health, Criteria for a recommended standard, occupational exposure to waste anesthetic gases and vapors, Publ. No. 77-140, NIOSH, Cincinnati, OH, 1977.
22. J. D. McGlothlin, P. A. Jensen, W. F. Todd, and T. J. Fischbach, Study protocol: Control of anesthetic gases in dental operatories, U.S. DHHS (NIOSH) Report No. ECTB 166-03, NIOSH, Cincinnati, OH, 1989.
23. A. Rademaker, Evaluation of two dental operatory scavenging systems using infrared thermography and real-time air analysis for nitrous oxide, MA Thesis, Purdue University, West Lafayette, IN, 2007, Print.
24. D. Donaldson and J. Orr, A comparison of the effectiveness of nitrous oxide scavenging devices, *J. Can. Dent. Assoc. 55*:535–537 (1989).
25. F. Certiosimo, M. Walton, D. Hartzell, and J. Farris, Clinical evaluation of the efficacy of three nitrous oxide scavenging units during dental treatment, *Dent. Assist. 72*:22–30 (2003).
26. F. A. Mettler et al., Nuclear medicine exposure in the United States, 2005–2007: Preliminary results, *Semin. Nucl. Med. 38*(5):384–391 (2008).
27. U.S. Nuclear Regulatory Commission, 10 CFR part 20. Standards for protection against radiation [online], 2007 [cited January 30, 2009], Available from: http://www.nrc.gov/reading-rm/doc-collections/cfr/part020/
28. F. Vanhavere et al., An overview on extremity dosimetry in medical applications, *Rad. Prot. Dosim. 129*(1–3):350–355 (2008).
29. W. Chruscielewski et al., Hand exposure in nuclear medicine workers, *Rad. Prot. Dosim. 101*(1–4):229–232 (2002).
30. J. Jankowski, J. Olszewski, and K. Kluska, Distribution of equivalent doses to skin of the hands of nuclear medicine personnel, *Rad. Prot. Dosim. 106*(2):177–180 (2003).
31. C. Martin and M. Whitby, Application of ALARP to extremity doses for hospital workers, *J. Radiol. Prot. 23*(4):405–421 (2003).
32. M. Wrzesien, J. Olszewski, and J. Jankowski, Hand exposure to ionizing radiation of nuclear medicine workers, *Rad. Prot. Dosim. 130*(3):325–330 (2008).

33. G. David and P. Buckle, A questionnaire survey of the ergonomic problems associated with pipettes and their usage with specific reference to work related upper limb disorders, *Appl. Ergon. 28*(4):257–262 (1997).
34. N. Kilroy and S. Dockrell, Ergonomic intervention: Its effect on working posture and musculoskeletal symptoms in female biomedical scientists, *Br. J. Biomed. Sci. 57*(3):199–206 (2000).
35. A. Montgomery et al., Application of a gamma extremity monitoring system in a radiopharmaceutical dispensary, *Nucl. Med. Commun. 18*(7):673–679 (1997).
36. W.-H. Wang et al., Evaluation of a radiation survey training video developed from a real-time video radiation detection system, *Health Phys. 90*(2):S33–S39 (2006).
37. W.-H. Wang, J. D. McGlothlin, and R. M. Traylor, Development of a real-time video radiation detection system and its application in radiation survey and radiation exposure assessment, *Radiat. Prot. Manage. 19*(6):25–31 (2002).
38. M. Whitby and C. Martin, Investigation using an advanced extremity gamma instrumentation system of options for shielding the hand during the preparation and injection of radiopharmaceuticals, *J. Radiol. Prot. 23*(1):79–96 (2003).
39. M.J. Guy, C.D. Greaves, and R.J. Morton, Development of a combined audiovisual and extremity dose monitoring software tool for use in nuclear medicine, *Nucl. Med. Commun. 26*(12):1147–1153 (2005).
40. M. Whitby and C. Martin, A multi-centre study of dispensing methods and hand doses in UK hospitals radiopharmacies, *Nucl. Med. Commun. 26*(1):49–60 (2005).
41. John Caunt Scientific Ltd., ED2 Extremity Dosimeter, 2008 [cited December 7, 2008], Available from: http://www.johncaunt.com/pages/dosimetry/ed2_2probe.html
42. Topac, Instrumentation for Research, Production and Quality Control, Beta and gamma shields, 2001, Retrieved November 30, 2008 from http://www.topac.com/shield.html
43. K. Willeke and P. A. Baron, Bridging science and application in aerosol measurement, in *Aerosol Measurement: Principles Techniques and Applications*, K. Willeke and P. A. Baron, Eds., Van Nostrand Reinhold, New York, 1993, pp. 3–7.
44. D. O'Brien, Data acquisition and analysis, in *Aerosol Measurement: Principles, Techniques, and Applications*, K. Willeke and P. A. Baron, Eds., Van Nostrand Reinhold, New York, 1993, pp. 521–532.
45. P. A. Jensen and D. O'Brien, Industrial hygiene, in *Aerosol Measurement: Principles, Techniques, and Applications*, K. Willeke and P. A. Baron, Eds., Van Nostrand Reinhold, New York, 1993, pp. 537–555.

CHAPTER 37

# Ergonomics and Concurrent Design

Peter M. Budnick, Donald S. Bloswick, and Don R. Brown

CONTENTS



## 37.1 INTRODUCTION

Many industries are adopting product development strategies intended to streamline the time and financial commitments during the conception-to-market phase of product life cycles. Termed "concurrent design," "design for manufacturing (DFM)," or "simultaneous engineering," for example, these methods strive to integrate traditionally separate engineering disciplines to achieve a more comprehensive design at a lower initial cost.

The time saved during this process can also provide significant advantage by beating competitors to the market. As related by Walsh et al. [1],

> In the 1990's … the rules of successful competition are changing. Companies not only have to develop and manufacture well-designed, high quality products at a price that offers the customer value for money, but this has to be done at least as fast as, or preferably faster than, the competition.

The later the design change made, the greater the cost burden. De Fazio et al. [2] estimate that 70% of a product's life-cycle costs are determined when it is designed. The choices made during design determine materials, fabrication, and assembly methods and to some extent the material handling, inspection, and other options available to the process designers. Decisions that affect assembly in particular also affect nearly every other aspect of the manufacture and use of a product [2]. Figure 37.1 indicates the relative relationships between cost, design change, and development time for a typical product from the point of product conception through production, comparing concurrent design methods with traditional design methods.

Production methods are, of course, a primary concern to ergonomics experts, because assembly, materials handling, and inspection tasks are major contributors to industrial ergonomic problems, including productivity, quality, and health and safety issues. With concurrent design, process design viewpoints are integrated into early product and process design stages. In this way, many potential problems can be eliminated or accommodated early in development. This saves time and money and can positively influence productivity and quality as well. Currently, far too many ergonomic recommendations are denied because they are presented or solicited late in the product life cycle when changes are financially prohibitive. We contend that a proactive ergonomics advocate is crucial in the design synthesis stages.

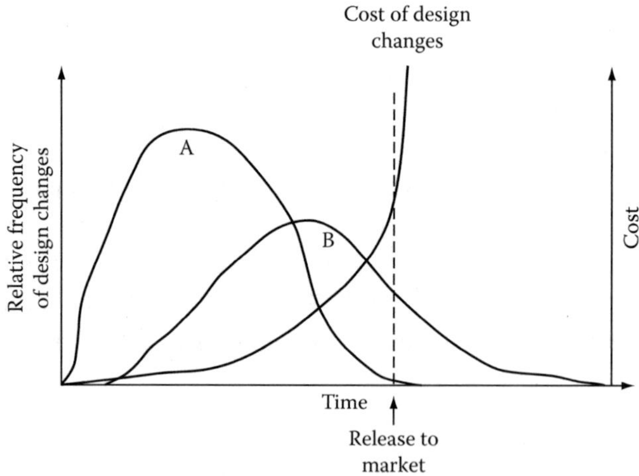


FIGURE 37.1 Relative relationships between cost, design changes, and development time for traditional design methods (curve B) and concurrent design methods (curve A).

## 37.2 BACKGROUND AND SIGNIFICANCE TO OCCUPATIONAL ERGONOMICS

### 37.2.1 Design Process

The product development process is, as stated by Rosenthal [3], "inherently a creative, interactive problem-solving process, not just a technological accomplishment." However, not all designs are innovative, and not all arise from R&D. In fact, the vast majority of new products involve variations on existing technology or incremental improvements to existing designs. Many design efforts are actually redesigns focusing on reliability, quality, and economic manufacturing [1].

Arriving at a final design involves narrowing a large set of possible choices, often representing conflicting views, until a satisfactory product can be developed. There is rarely one unique solution to a design quest. The ultimate choice depends upon the way the problem is defined or specified and the nature of the information available to solve it. The viewpoints and knowledge applied by the designers themselves probably have the strongest influence on final design choices. Based on these observations, Stoll [4] proposes the following: (1) in general, many different solutions to a design problem are possible; (2) the solution that is chosen may be selected for one or a combination of many right and wrong reasons; and (3) the design process provides no intrinsic guarantee that the selected solution is, in fact, the best solution or even the right solution. If this is true, then how can the "one best solution" be arrived at in the shortest period of time? This question is addressed in Section 37.2.1.2. First, traditional and concurrent design methods are reviewed and compared.

#### *37.2.1.1 Traditional Design Methods*

Traditional product life cycles followed an incremental and sometimes linear model, as depicted in Figure 37.2. In the worst cases, little communication occurred between the

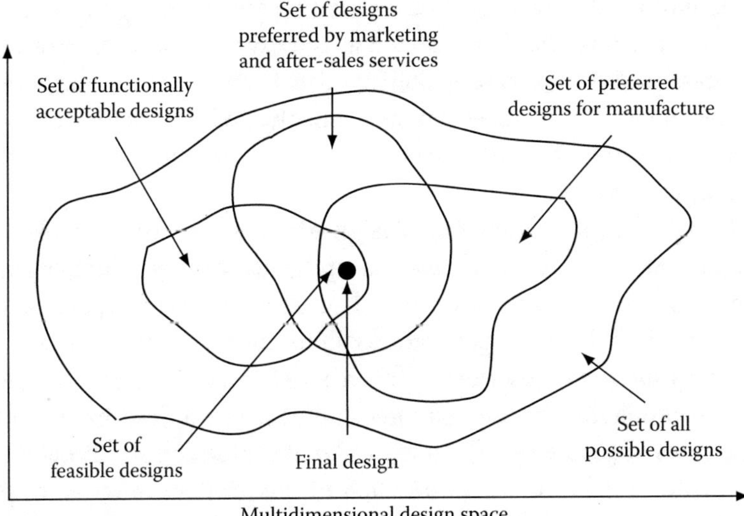


FIGURE 37.2 Representation of a multidimensional design space that narrows the final solutions set for a particular design problem.

traditionally separate departments of design, manufacturing, marketing, distribution, and after-sales service. Each department performed its duties and would then "throw it over the wall" to the next in line. Because decisions made in one area affect choices in others, such methods resulted in a particular product being "thrown" back and forth, losing valuable time and money before it was released to market. Significant compromises in product and process quality were also evident in this turbulent process.

It has become readily evident that traditional compartmental information hoarding and decision making is not competitive in the world market. There must be some sort of communication mechanism between the different concerns that encourages and accommodates varied viewpoints in a timely fashion. Corporate "culture" and organizational structure can either be a barrier or provide the potential for positive interaction among those with different perspectives (see, e.g., Rosenthal [3] for discussion relating to such issues).

#### *37.2.1.2 Concurrent Design Methods*

Recommended methodologies and case examples of successful concurrent design abound (see, e.g., Ettlie and Stoll [5], De Fazio et al. [2], Susman [6]). We do not propose or endorse any one method but conceptualize the underlying principles and identify specific methods and types of communication that we believe are important in the effort to infuse ergonomic information into the design process. Returning to the question previously posed: How can the "one best solution" be arrived at in the shortest period of time?

Concurrent design methods approach this problem by superimposing critical product and process life-cycle information on the problem definition. This approach, while using more information than has traditionally been handled during early development stages, quickly narrows the solution set to a manageable set of choices that can be optimized in a shorter period of time (see Figure 37.2). The greatest opportunity to "get the design right" exists in the earliest stages of design formulation. Thus, it is critical to get as comprehensive as possible a set of product knowledge in the planning and requirements definition design stage. The information must be clearly stated as requirements and constraints. Design is an iterative process, and the greatest flexibility exists in the planning stage. In concurrent design, the range of available choices narrows quickly, and changes become increasingly expensive and difficult [4]. The challenge for the field of ergonomics is to integrate ergonomic knowledge into this process.

As typically represented, concurrent design approaches can be broken down into three general and interrelated categories: concept generation, product engineering, and process engineering (in the following text, this is expanded by considering distribution and after-sales services). Concept generation often reflects the viewpoint of marketing, which provides a *functional* description in terms of target customer desires, expectations, needs, and problems. Product engineering develops a description of the *form* of the product in terms of geometry, tolerance, and material selections. Process engineering develops a description defining the *fabrication* of the product in terms of plant layout, tool and equipment design, work design, and so forth [4]. This presents three distinct viewpoints in which to infuse ergonomic knowledge, to which we will add distribution and after-sales services. This provides a more robust framework in which to advocate and expand ergonomic influences in product life cycles.

Functional knowledge may be viewed as more conceptual in nature, whereas form and fabrication information relies on more objective ergonomic knowledge. For example, a functional constraint may be that the product must be easy to grasp. This problem could be solved in a variety of ways, such as by providing traditional handles or by defining product geometry such that it "invites" a user to grasp in a certain location. This general functional information can be further refined into form criteria by imposing anthropometric, biomechanical, and psychophysical constraints relating to handles. Likewise, fabrication constraints defined by form can be positively influenced by industrial ergonomic principles and criteria. The key is that each design iteration and decision must be accompanied by specific and applicable ergonomic criteria. The type of ergonomic knowledge and its organizational structure is discussed in greater detail in the following text.

## 37.3 INFUSING ERGONOMICS INTO CONCURRENT DESIGN METHODS

The "best" methods for getting ergonomic knowledge included in concurrent design are a matter of debate. It would be ideal if an ergonomic expert were part of every design effort, as suggested by Mital and Morse [7], for instance, but for numerous reasons, this is rarely the case. These reasons include, for example, the following: (1) Ergonomists do not even agree among themselves on who qualifies as an expert; (2) ergonomists have been largely unsuccessful in the past at presenting their knowledge to nonexperts; (3) ergonomic knowledge is often ambiguous, heuristic, and less universally applicable than knowledge in other design domains; (4) ergonomists are generally unable to objectify the need for ergonomic design influences in terms of economics; and (5) ergonomists often have limited knowledge regarding design and engineering practices.

Alternatively, since a designer's knowledge, inspired by culture, world view, and life experiences, inevitably affects design choices, it would be advantageous if engineers received basic ergonomics training as part of their general education. Although efforts are under way to incorporate ergonomic and safety knowledge into the undergraduate engineering curriculum (see, e.g., Kavianian et al. [8]; Bloswick [9]; Talty [10]), ergonomists must not wait for the next generation of engineers. Product and process developers need tools that can be useful and instructive today. There are many viable ways to satisfy this need, including incorporating ergonomic knowledge into existing design tools or developing new tools specific to ergonomics. However, practicing engineers have warned that stand-alone products sit on the shelf [11]. Therefore, it is to the ergonomist's advantage to enhance existing tools by providing an ergonomic viewpoint. Since ergonomists must appeal to designers and engineers, it is to their benefit to work within the framework, language, and formats with which such professionals are already familiar. Computers are increasingly the vehicle for engineering tools and thus will play a major role in efforts to incorporate ergonomic knowledge into concurrent design practices. We believe that computerized methods will be most applicable in the future, especially for large organizations.

### 37.3.1 Product Life-Cycle Model

"Product life cycle" includes every stage of a product, from the time it is merely an idea in someone's mind to the time when it becomes inert (which may not necessarily be the time of disposal).

It has already been noted that traditionally compartmentalized and sequential product life cycles erect barriers to efficient concurrent development strategies. Therefore, rather than encouraging and prolonging the potential pitfalls associated with the traditional view depicted in Figure 37.3, we will consider the product life cycle from a higher-level mission- or goal-oriented viewpoint of function, form, and fabrication, to which distribution and after-sales services have been added, creating the model shown in Figure 37.4.

The product life cycle begins with a functional concept, which is then formed, fabricated, marketed, and serviced. Within this model, when, where, and how is ergonomic knowledge best presented? Expanding on Brown et al. [12], we want to encourage and work within a framework that represents the viewpoints of all concerns within one central design space. Most likely this is supported by a computerized framework within which decisions made by any participant are immediately available and evident to all other participants. It is this comprehensive and expedient flow of critical information that streamlines the portion of the product life cycle prior to release to market.

Figure 37.4 depicts the relationship between differing product viewpoints as each simultaneously monitors and contributes to the design space. Naturally, contributions and viewpoints will be prioritized based upon organizational and functional decisions, but the concept of multiple viewpoints should be clear. Ergonomic knowledge should be organized

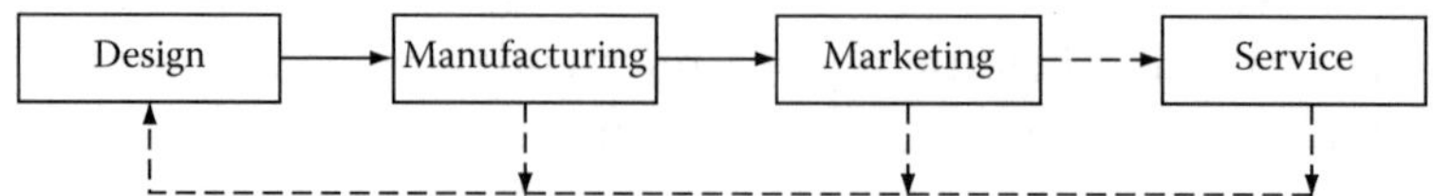


FIGURE 37.3 Traditional sequential, compartmentalized product life cycle. The broken lines denote the often-poor or nonexistent functions and lines of communication in older product development and life-cycle methods.

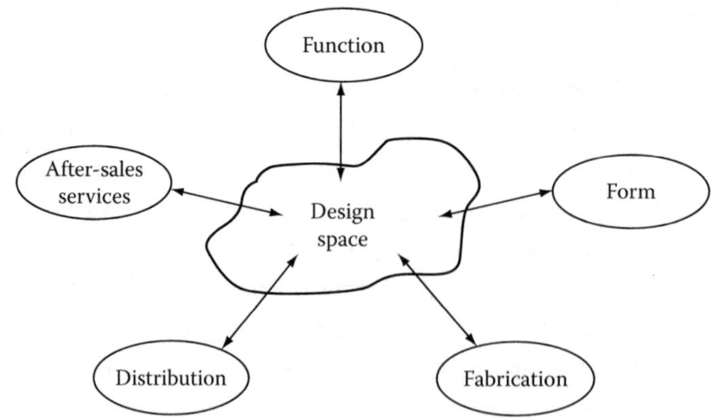


FIGURE 37.4 Model of all viewpoints having access to the design space.

in such a way that each viewpoint is afforded an additional human-oriented perspective. Furthermore, the knowledge and its presentation should act as an advocate for human interface perspectives, which may be in competition with purely technical or economic perspectives offered by other viewpoints. Our objectives must be to define and demonstrate the need for ergonomics, not to just supply solution-oriented data.

### 37.3.2 Ergonomic Knowledge Base

The knowledge base should consist of information taken from multidisciplinary ergonomics literature, design and analysis tools, expert knowledge from in-house and external experts, and experiential knowledge gleaned from existing processes and methods within a given organization (see Figure 37.5). (Nagamachi and Yamada [13] review participatory ergonomics, providing an example method that can be used as a platform to generate and capture experiential knowledge.)

"Ergonomic knowledge base" is used very broadly to describe a source of well-organized ergonomic concepts and data. The organization and structure of the knowledge should make it easily accessible, interpretable, and applicable. It could reside in an expert, or it could be a hard copy reference source, or it could be an electronically stored and manipulated source.

We put forth ideas and goals for such a knowledge base but leave the actual content and structures to those interested in further research and development in this area.

#### *37.3.2.1 Presentation Format*

For an audience of nonexperts, ergonomic knowledge must be presented in a way that advocates its inclusion into areas where it has traditionally been neglected. The knowledge base must also be able to defend ergonomic solutions in the context important to product development, which includes technical, organizational, and financial considerations—the

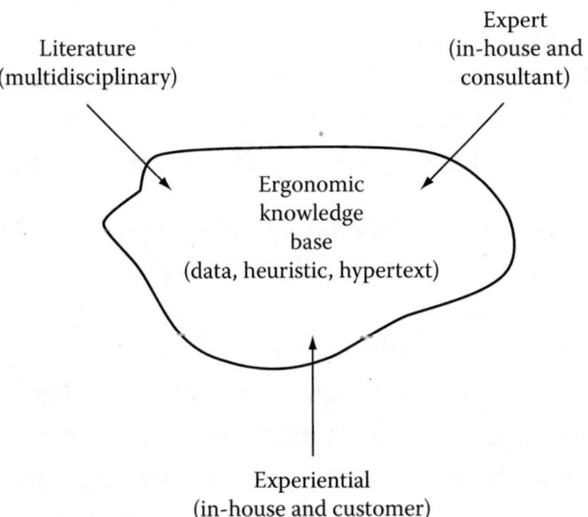


FIGURE 37.5 The ergonomic knowledge base should consist of knowledge elicited from multidisciplinary literature sources, in-house and out-of-house experts, and experiences gathered in-house and in the environment of use.

so-called bottom line. The nonexpert must not be burdened by a requirement to sift and search through volumes of extraneous materials, yet should be free to navigate easily when interested. In short, the format must present ergonomic knowledge so that it is respectful, useful, and easily applied by nonexperts of varying specializations and backgrounds. Nonexperts often do not understand the need for, let alone the context of, ergonomic information. Therefore, it is up to us to positively and proactively influence developers, which is no small task.

In conceptual stages, problem solutions are very uncertain. Minimizing the uncertainty surrounding design choices is central to concurrent design. As modeled by Stoll [4], this can be viewed as a "funnel process," where the set of choices is wide at first but quickly narrows as iterative decisions are made. Beginning with a set containing all ergonomic knowledge, the set must be quickly narrowed to only applicable information as choices are made. This quick elimination of unnecessary information and the bringing forward of applicable information implies a structured ergonomic knowledge base. (See Lehto [14], for example, for a thorough discussion relating to knowledge base "development with structured application to safety ergonomics.") In general, the ergonomic knowledge should be organized from top-level goal-oriented information that progressively funnels to specific solution-oriented data that can be applied as material and geometric choices are made.

An additional benefit of presenting ergonomics in a goal-oriented fashion is that it can appeal to the problem-solving creativities and abilities of engineers and developers. While the knowledge base must be able to provide specific solution-oriented data based on design choices, it must also be flexible enough to encourage, recognize, and accept new solutions to old problems.

The type of ergonomic knowledge that is applicable to each global perspective is now explored.

#### *37.3.2.2 Topical Knowledge Structure and Organization*

While each of the general product life-cycle categories is treated as a separate topical activity in this discussion, an effective concurrent design strategy dictates that they occur together and that much of the knowledge be applied during functional formulations.

*Function*: Functional ergonomic knowledge and its presentation format should focus on topics applicable early in product conceptualization and development. Product functional characteristics are defined during the conceptual stage for a new product and revised and updated during product redesign. This is a crucial point at which to advocate the human interface in any product or process design. It is important that experiential ergonomic knowledge gleaned from in-house sources and customer experiences monitored by after-sales services be added to the design space for product redesign. For a new product, it is important that the traditional technologically driven parameters be augmented with human-oriented goals. It must be impressed upon decision makers that all products will interface with humans in a variety of ways and that product success requires that ergonomic human interface criteria play a significant role in defining functional criteria.

During functional definition, human-oriented concepts such as product usability, maintainability, serviceability, and safety should be advocated and prioritized. Depending on the product and its intended application and user environment, some ergonomic topics may be given priority over others.

One question that should be answered early on is where will humans interfere with a product during its life cycle? Nearly all products will have some in-house human interface during design, manufacturing, and distribution. After market, depending on the nature of the product, it may have varying degrees of interface with people of different needs and skill levels. A consumer product, for example, might require considerable ergonomic knowledge relating to user operation and training, skilled or unskilled maintenance and service providers, and disposal. On the other hand, a product that serves as a component within a larger machine or system may require ergonomic knowledge geared primarily to assembly, maintenance, and service. By targeting the human interfaces that will occur during the product life cycle, one can determine the ergonomic knowledge that will be most applicable and eliminate the information extraneous to the project at hand.

Those responsible for market research can benefit from human-oriented research methods. Experimental design and statistical methods used in the field of ergonomics can help target user populations and more accurately determine potential user attributes, concerns, and desires. Just as statistically sound experimental and measurement methods have improved product quality control and reliability, ergonomists should advocate appropriate research methods to identify human responses, desires, capabilities, expectations, etc. Crucial topics such as the skill and product knowledge level of the target user population are often overlooked or given low priority by designers. Proactive ergonomic knowledge can and should positively influence designers to consider such parameters.

Another area of market research that is useful in this stage of design relates to market experience with similar products. Information concerning both successes and failures can be very instructive in early product development decisions. In particular, accident and injury data, consumer dissatisfaction, standards, laws, and related liability information can be beneficial in early stages of development. While the ergonomic knowledge base may not contain such specific information (unless it is gleaned from in-house experimental knowledge, which should be continually updated), it should advise as to how and why developers can and should conduct such research.

Unless market research has identified ergonomic parameters, it is up to the ergonomic knowledge base and presentation format to advocate them. At the top level, the advice might be quite heuristic, such as "portable consumer products must be easy to move, lift, or carry." Getting slightly more detailed, "if a product is meant to be lifted, it should be easy to grasp and hold." Thus, the knowledge first reminds developers to consider ergonomic issues and then immediately supplies applicable data when a choice is made. For instance, if "easy to grasp" survives as a design goal, then ergonomic data pertinent to grasps, such as handhold types and the advantages and disadvantages of each, must be immediately provided. If a particular handhold is selected, then location, orientation, shape, and clearance data should be immediately presented, and so on as the functional product decisions become product form decisions.

*Form*: As the functional parameters are set, the design quickly progresses toward form-oriented solutions. Ideally, the ergonomics advocate is able to influence design goals and parameters as functional determination progresses. Either way, it must continue to infuse applicable ergonomic knowledge into the development of form criteria. Choices made here will impact all aspects of the human interface throughout the product life cycle. Therefore, the knowledge base must offer industrial and consumer ergonomic expertise at varying levels of detail (conceptual and detailed) to an audience consisting mostly of design and manufacturing engineering experts.

For product redesign, experiential knowledge will be very valuable. For new products, it is important to learn from similar product and fabrication experiences. Many redesigns focus on product and process quality and reliability issues, driven primarily from a technical standpoint. The role of the human operator is often given the least consideration in fabrication processes, yet without ergonomic considerations, it can be the least controllable and least predictable component in the system. Not only must an ergonomic knowledge base impress this fact upon engineers, it must also demonstrate that ergonomics can minimize this uncertainty and improve productivity, quality, and reliability.

Geometric, tolerance, and material choices have direct effects on manufacturing processes and the people who interface with them. It is this relationship between form and fabrication that has received the greatest attention to date and has resulted in the proliferation of DFx (Design for *x*) tools, where *x* represents a particular interest or viewpoint. For example, there are DFManufacturability, DFAssembly, DFDisassembly, DFMaintainability, DFEnvironment, etc. Unfortunately, few of these tools are influenced by ergonomic knowledge.

Assembly operations are defined in part by the choices made while defining the form of a product and its components. The type of connection and the assembly procedures directly affect consumer concerns such as product quality, reliability, maintainability, and serviceability. Connection methods also have a dramatic effect on industrial ergonomic concerns, which can translate into productivity, product quality, reliability, and inspection problems. Many engineers are unaware of the role human capabilities and limitations can have on these important product characteristics.

Heuristic and specific ergonomic knowledge is applicable in the determination of form. For example, heuristic knowledge might identify to the product and process developers potential ergonomic hazards related to the use of threaded fasteners as component connectors. If threaded fasteners are nevertheless chosen, then methods to minimize potential hazards, such as recommended tools and assembly workstation design parameters, must be immediately provided. In this way, designers can make efficient and informed iterations and decisions.

In short, an effective ergonomic knowledge base must be able to specify geometric, tolerance, and material selections for both product and fabrication processes that will minimize ergonomic hazards and enhance usability, productivity, quality, and reliability.

Design of product support materials, which should begin in this design phase, can also significantly benefit from ergonomic perspectives. The importance of product user and maintenance manuals, product warnings, product packaging, and marketing materials is often neglected or poorly understood. These materials cannot only help consumers and users to better use a product, thereby increasing customer satisfaction, but they can also have significant impacts on safety and health issues and related liability.

*Fabrication*: In concurrent design, determination of design criteria that reflect form and fabrication considerations will occur in parallel. This relationship has received the most attention in industry, resulting in the variety of DFx tools mentioned earlier. Ergonomic knowledge must become a part of this critical process, because decisions made here will affect every aspect of a product. Industrial ergonomic knowledge, which is most applicable in fabrication, is undoubtedly one of the most complete sections in the body of ergonomic knowledge. To be useful in concurrent design, the knowledge must be structured and directed proactively, rather than reactively, as Corbett [15] notes is the typical mode of use for ergonomics design criteria.

If ergonomics is to assume a role in product and fabrication design, the ergonomics advocate must defend, with bottom-line data, the need for, and consequent benefits of, its application. Grossmith [16], for example, presents design-oriented ergonomic considerations from the viewpoint of reducing manufacturing costs, which is the type of bottom-line knowledge that will influence design decisions. Ergonomic methods that increase productivity or improve quality, reliability, and inspection processes will also receive serious attention from fabrication experts.

Engineers involved in the design of product fabrication processes typically plan assembly sequences and shop floor layouts, select appropriate fabrication machinery, design workstations, and determine the means of transport and the in-house materials handling, storage, and distribution schemes. Each of these activities can be enhanced by the inclusion of ergonomic principles.

Fabrication knowledge should again be organized from a conceptual, goal-oriented level down to specific applicable process design data. Conceptually, ergonomic input can help process designers distribute tasks between humans and machines, taking advantage of distinctly human talents. Processes and methods that have been known to contribute to ergonomic hazards and degradation in productivity, quality, or reliability can be flagged early, at which time they can be replaced with better alternatives or improved from an ergonomic standpoint.

*Distribution*: In-house employees, employees of distributors, and consumers are all impacted by marketing methods, packaging, layout, and product-handling methods. Commonly, such distribution characteristics are driven by operational and quality (minimizing transport damage) concerns but pay little attention to the human interface with this phase of the product life cycle.

From the perspective of in-house distribution, ergonomic knowledge should be applied to the design of shipping packaging (size, shape, weight, and handling characteristics) and storage and handling methods. Where applicable, packaging should be designed so that units can be handled manually with minimal biomechanical hazards. Where this is not possible, lifting assistance devices should be provided. When products will pass to second- and third-party distributors and outlets, specifying the ergonomic handling and training methods may be beneficial. However, since control over handling methods may not be possible once a product leaves a manufacturing facility, especially once it is passed to the consumer, we encourage that safe handling methods be achieved through initial product and packaging design. As with all perspectives in concurrent design, such principles must be presented and defended in the earliest stages of product functional formulation.

*After-sales service*: Depending on the type of after-sales services provided, varying degrees of ergonomic hazards may exist for in-house employees, consumers, and third-party service providers. Effective after-sales services ideally monitor and provide follow-up services for customers. There are two main roles for effective after-sales services: (1) product and customer support and (2) product and customer monitoring. In many organizations, the functions of product development and after-sales services are far removed from each other, which often negates the inclusion of experiential knowledge in future production and redesign efforts. An effective concurrent design program encourages and provides an infrastructure to bring this critical information to the functional development stage. Ergonomics can enhance this process by providing human-oriented product and consumer research and monitoring methods as well as by advocating design parameters that improve maintenance and serviceability parameters such as accessibility, visibility, tool selection, and biomechanical accommodations. For product maintenance and repair services, the design can either hinder or enhance maintainability and serviceability, which can have a significant impact on customer satisfaction and safety.

## 37.4 CRITICAL REVIEW OF CURRENT STATUS

Significant efforts are under way to integrate ergonomics into design, but no single example has demonstrated great success. Further, most ergonomic tools are evaluation-oriented rather than being synthesis methods that can be applied during concept formulation for new products and processes. Many tools that perform specific ergonomic functions have been proposed or developed, but none are known to have achieved widespread use within the design or engineering community to date. Young [17] suggests that "widespread use of [such tools] has been inhibited due to several factors: cost, complexity, ease of integration, functionality and validity." It must also be understood that until the need for ergonomics can be advocated and defended within the "bottom-line" parameters that drive development, this lack of use will continue.

### 37.4.1 Review of Related Research and Applications

#### *37.4.1.1 Useful References*

A number of useful discussions relating to ergonomics in design for manufacturability can be found in the volume edited by Helander and Nagamachi [7,13,16,19]. In particular, Mital and Morse [7] discuss the role of ergonomics in designing for manufacturing and identify some data sources that can be used in the process. Drury [18] discusses a systematic procedure for designing for inspectability. Grossmith [16] presents design-oriented ergonomic considerations from the viewpoint of reducing manufacturing costs, which is the type of bottom-line knowledge that will influence design decisions. He also presents some useful manufacturing ergonomic design parameters relating to materials, fasteners, and product handling. Seidel [19] reviews an assembly sequence planning system from a human factors perspective. Each of these example references represents a portion of ergonomic knowledge that could be valuable in a concurrent design effort; however, none appears to be part of a larger system to capture and disseminate ergonomic principles, concepts, and design data in a proactive manner.

#### *37.4.1.2 Ergonomic Design Tools*

Many ergonomic tools described as "design tools" are probably better called "design evaluation tools." That is, they do not provide design synthesis or conceptual development knowledge in the functional definition stages, which is necessary in the concurrent design environment. Further, many of these tools, though computerized, are stand-alone and have not been integrated into the tools and systems to which designers are accustomed (geometric modelers, material databases, etc.). Interested readers should refer to the Mattila Karwowski volume [20,21], for example, for a review of computerized ergonomic applications.

A specific type of computerized ergonomic tools are expert systems. One such system that we have under development is briefly described in the following text. The long-term value of expert systems, an outgrowth of artificial intelligence research, is debatable. However, because ergonomic experts are somewhat scarce, and because ergonomic knowledge is sometimes uncertain and heuristic, such knowledge is a prime candidate for expert systems. A number of ergonomic expert systems have been proposed and prototypes constructed. Brown et al. [23], Jarvinen and Karwowski [20], and Kern and Bauer [21], for instance, provide reviews and references for ergonomic expert systems. In addition to expert systems, we also suggest computerized hypertext methods to present ergonomic knowledge to nonexperts, which is discussed in the following text.

### 37.4.2 Review of CDEEPS

CDEEPS (concurrent design and engineering for the ergonomics of production systems) is a prototype system currently under development within a computational framework for concurrent engineering developed by Brown et al. [12]. This prototype system provides a common computing environment that uses an object-oriented programming language (Common Lisp). The computing environment is designed such that all critical design decision data can be viewed by all design participants. Currently, CDEEPS is geared toward product and process design and is limited to providing advice relating to industrial ergonomic concerns only. The architecture, however, is expandable to accommodate ergonomic concerns throughout the entire product life cycle. In essence, it captures the design space as depicted in Figure 37.4 and allows product designers and process designers to view and alter design criteria.

CDEEPS uses expert systems and hypertext technology, outgrowths from the field of artificial intelligence. Hypertext can be thought of as an on-line electronic library. Hypertext allows the user to browse through information by selecting keywords from the text, which then immediately opens other sources for review, just as one might use cross-referencing in a library. The hypertext tool is quite useful in the conceptual formulation stages, as it allows the nonexpert to easily peruse background ergonomic knowledge to any level of detail. In this sense, it may serve as an educational tool as well.

Currently, the expert system triggers the hypertext in the design formulation phase and performs more detailed task analysis as assembly processes are developed. The expert system prototype contains over 700 rules and is able to predict manual assembly process times using MTM [22] time measurement units (tmu's) and to predict a general hazard level for a given task.

Interested readers may find more detailed information concerning CDEEPS in Brown et al. [23] and Budnick et al. [24].

## 37.5 FUTURE CONCERNS

Due partially to the proliferation of computers and their ability to quickly process and pass on information, product development activities are becoming more organized and accelerated as competitors seek to introduce better products to market faster. Historically, it has been difficult to establish ergonomics in design practices, especially in the design synthesis and functional definition stages. Missed opportunities while setting design parameters translate to form definitions with inferior human interface characteristics throughout the product life cycle, including both workplace and consumer environments. To fully realize its benefits, the inclusion of ergonomics must be advocated and defended early in the design stages. To be useful to product and process developers, ergonomic knowledge must be organized in such a way that each design decision is immediately accompanied by appropriate and applicable concepts and data. This implies a highly organized and structured knowledge base with an interface that presents design information in a format consistent with that of other engineering and development data. Such a knowledge base must also be integrated with the environment and tools used by design experts. Short of an ergonomics expert on every design team, a worthy goal but unrealistic in the near future, we suggest hard copy and computerized tools targeted for proactive use rather than the reactive use that characterizes most ergonomic applications to date. It is strongly suggested that those interested in concurrent engineering applications study and work within the environments in which those applications will be used, which includes advanced computer hardware and software systems.

## REFERENCES


1. V. Walsh, R. Roy, M. Bruce, and S. Potter, *Winning by Design: Technology, Product Design and International Competitiveness*, Blackwell, Cambridge, MA, 1992, pp. 4–5, 10.
2. T. L. De Fazio, A. C. Edsall, R. E. Gustavson, R. E. Metzinger, and W. A. Dvorak, in *Concurrent Design of Products and Processes: A Strategy for the Next Generation in Manufacturing*, J. L. Nevins and D. E. Whitney, Eds., McGraw-Hill, New York, 1989, p. 2.
3. S. R. Rosenthal, Bridging the cultures of engineers: Challenges in organizing for manufacturing product design, in *Managing the Design-Manufacturing Process*, J. E. Ettlie and H. W. Stoll, Eds., McGraw-Hill, New York, 1990, pp. 21–52.
4. H. W. Stoll, Design for life-cycle manufacturing, in *Managing the Design-Manufacturing Process*, J. E. Ettlie and H. W. Stoll, Eds., McGraw-Hill, New York, 1990, pp. 79–113.
5. J. E. Ettlie and H. W. Stoll, Eds., *Managing the Design-Manufacturing Process*, McGraw-Hill, New York, 1990, p. 22.
6. G. I. Susman, Ed., *Integrating Design and Manufacturing for Competitive Advantage*, Oxford University Press, New York, 1992.

7. A. Mital and I. E. Morse, The role of ergonomics in designing for manufacturability, in *Design for Manufacturability: A Systems Approach to Concurrent Engineering and Ergonomics*, M. Helander and M. Nagamachi, Eds., Taylor & Francis, Washington, DC, 1992, pp. 147–159.
8. H. R. Kavianian, N. Meshkati, C. A. Wentz, and J. K. Rao, Should engineering schools address occupational and environmental safety and health issues?, *Prof. Safety 38*:48–49 (June 1993).
9. D. S. Bloswick, Developing and marketing safety and ergonomics courses in traditional engineering curricula: Process and content, *Proc. NW Workshop on Occup. Safety and Heath Engineering Curricula*, Corvallis, OR, April 27–28, 1992.
10. J. T. Talty, Project shape—Integration of safety and health into engineering school curricula, *Proc. NW Workshop Occup. Safety and Health in Engineering Curricula*, Corvallis, OR, April 27–28, 1992.
11. B. Joseph, Corporate Ergonomist, Ford Motor Company, Dearborn, MI, Personal communication, 1992.
12. D. R. Brown, M. R. Cutkosky, and J. M. Tenenbaum, Next-cut: A second generation framework for concurrent engineering, in *Computer Aided Cooperative Product Development*, D. Sririam and R. Logcher, Eds., Springer-Verlag, New York, 1991.
13. M. Nagamachi and Y. Yamada, Design for manufacturability through participatory ergonomics, in *Design for Manufacturability: A Systems Approach to Concurrent Engineering and Ergonomics*, M. Helander and M. Nagamachi, Eds., Taylor & Francis, Washington, DC, 1992, pp. 219–231.
14. M. R. Lehto, *A Structured Methodology for Expert System Development with Application to Safety Ergonomics*, UMI Dissertation Services, Ann Arbor, MI, 1985.
15. J. M. Corbett, Human centered advanced manufacturing systems: From rhetoric to reality, *Int. J. Ind. Ergon. 5*:83–90 (1990).
16. E. J. Grossmith, Product design considerations for the reduction of ergonomically related manufacturing costs, in *Design for Manufacturability: A Systems Approach to Concurrent Engineering and Ergonomics*, M. Helander and M. Nagamachi, Eds., Taylor & Francis, Washington, DC, 1992, pp. 232–243.
17. M. F. Young, Development of an integrated computer aided ergonomics toolbox for industrial workstation design, Masters Thesis, Department of Mechanical Engineering, University of Utah, Salt Lake City, UT, 1994.
18. C. G. Drury, Product design for inspectability, in *Design for Manufacturability: A Systems Approach to Concurrent Engineering and Ergonomics*, M. Helander and M. Nagamachi, Eds., Taylor & Francis, Washington, DC, 1992, pp. 204–216.
19. U. A. Seidel, MI system for assembly sequence planning: Human factors considerations, in *Design for Manufacturability: A Systems Approach to Concurrent Engineering and Ergonomics*, M. Helander and M. Nagamachi, Eds., Taylor & Francis, Washington, DC, 1992, pp. 189–203.
20. J. Jarvinen and W. Karwowski, Applications of knowledge-based expert systems in industrial ergonomics: A review and appraisal, in *Computer Applications in Ergonomics, Occupational Safety and Health*, M. Mattila and W. Karwowski, Eds., Elsevier, New York, 1992, pp. 45–54.
21. P. Kern and W. Bauer, Expert systems in ergonomics, perspective applications in industrial practice, in *Computer Applications in Ergonomics, Occupational Safety and Health*, M. Mattila and W. Karwowski, Eds., Elsevier, New York, 1992, pp. 37–44.
22. MTM Association for Standards and Research, 1111 East Touhy Ave. Des Plaines, IL 60018.
23. D. R. Brown, P. M. Budnick, D. S. Bloswick, and J. Zhou, Design for ergonomics in a concurrent environment, Department of Mechanical Engineering, University of Utah, Salt Lake City, UT, 1993.
24. P. M. Budnick, D. S. Bloswick, and D. R. Brown, Integrating industrial ergonomics into the design process: Accommodating the design engineer, in *Advances in Industrial Ergonomics and Safety*, Vol. IV, S. Kumar, Ed., Taylor & Francis, Washington, DC, 1992, pp. 11–18.

CHAPTER 38

# Economics of Ergonomics*

Tapas K. Ray, Thomas R. Waters, and Steve Hudock

CONTENTS



## 38.1 INTRODUCTION

Ergonomics is the science of designing safe and productive workplaces and jobs for workers. Studies have shown that workers whose jobs are physically demanding or are poorly designed from an ergonomics perspective are at significantly increased risk of developing work-related musculoskeletal disorders (WMSDs), such as low-back pain (LBP) and distal upper-extremity musculoskeletal disorders (MSDs) (NIOSH 1997; NRC and IOM 2001).

* The findings and conclusions in this chapter are those of the author(s) and do not necessarily represent the views of the National Institute for Occupational Safety and Health.

Workers in these high-risk jobs also have significantly more lost workdays (LWDs) than those whose jobs are considered to be ergonomically designed. Studies have shown that WMSDs and the associated reductions in productivity account for significant financial costs for companies.

If an existing job or workplace is thought to be of poor ergonomic design, then some type of ergonomic solution or intervention is needed to decrease the risk to the worker. The interventions typically include "retrofitting" or modifying the job or workplace and can involve either significant changes (such as in work processes) or simple changes (such as in layout of the workstation). Jobs that are still in the conceptual phase can benefit from ergonomic assessment, by which they are critically evaluated and designed from an ergonomic perspective before workers actually perform them. This approach is sometimes referred to as *design in ergonomics* or *prevention through design*. The cost of ergonomically designing a safe job is reportedly more than one-tenth that of retrofitting the job after a problem has been discovered.*

From an employer's perspective, it is important to consider how the design of the work, its processes, and the work site affect the financial success of the business. Rapidly changing economic conditions have a profound effect on a firm's decision-making as well. The safety manager, industrial hygienist, managers, and executive officers will base occupational-safety investment decisions on employee health, risk management, and the financial consequences of those investments. Decisions to implement ergonomic interventions may be driven either by the benefits of maintaining a healthy and safe workforce or by their contributions toward productivity and business profits. In either case, however, it is necessary to calculate the costs and benefits—in terms of health or money—resulting from such actions.

This chapter gives readers insight into the types of economic evaluations that can be performed and how they can be used to assist in making decisions about implementing specific ergonomic interventions. We first briefly introduce the principles of economics and ergonomics and their underlying philosophies. We then describe current types of economic analyses and provide simple examples of their application in real-world situations. The chapter also lays out the steps of an economic analysis, briefly discusses ergonomic intervention approaches, and introduces a host of available tools commonly used for cost-benefit analysis (CBA) of ergonomic interventions. To broaden the scope of this chapter, we discuss decision-making in the face of uncertainty, health-valuation approaches, and intervention-evaluation examples. The goal is to assist readers in subjective and objective valuation of ergonomic practices.

## 38.2 PRINCIPLES OF ECONOMICS

Resources such as land, labor, capital, and time are scarce. The goal of economics is to study how efficiently these resources can be allocated to maximize the welfare of respective economic agents.† For this chapter the term economic agent will imply the following: (1) the firm's owner, manager, employer, and share holders; (2) the employee, worker,

* This has been opined by Mr. D. Alexander, PE, CPE, in his address to National Advisory Committee on Ergonomics (NACE) meeting in 2004. Refer to http://www.osha.gov/SLTC/ergonomics/nace_mins_1_2004.html

† See Samuelson and Nordhaus (2001).

union member, safety engineer, manager, and industrial hygienist; and (3) the family members of the workers, insurers, health care workers, neighbors, regulators, and the government. Similarly, resources will imply the firm's capital, budget appropriated for workplace safety, business reputation and morals; workers' mental and physical health, work time, and leisure time. Efficient allocation of resources, within and outside a workplace, can be achieved only with consideration of the true costs and benefits associated with maintaining (1) worker safety and health, (2) high worker productivity, and (3) high product quality.

Identifying the principle economic agents involved and the resources to be analyzed is of primary concern and constitute important building blocks of any economic analysis of occupational safety and health (OSH). Oftentimes, managements need to be convinced that a recommended ergonomic solution is cost-effective (CE) before authorizing expenditures to implement the solution. To prove that an ergonomic implementation will be CE, an economic analysis must show that the benefits outweigh the costs. If the economic analysis is not properly conducted, it may fail to convince management and other related stakeholders and may deter the workplace and the society from reaping the true benefits (both in health and monetary terms) of ergonomic interventions. This will lead to misallocation of vital resources such as labor and capital.

## 38.3 PRINCIPLES OF ERGONOMICS

The primary goal of ergonomic design is to ensure that work processes and tasks are within the safe physical capabilities of the workers. This means that the required forces and weights required to perform the job should be kept sufficiently low so as to not exceed workers' strength or mechanical tolerance limits of their musculoskeletal tissue. It also means that the frequency of the task should be kept sufficiently low so that the work does not induce repeated mechanical insult to the musculoskeletal tissues or induce unhealthy physiological responses that can lead to cumulative trauma disorders (CTDs). In addition, the work should be designed so that the required posture of a worker's trunk and limbs is maintained in as near neutral position as possible (i.e., extreme or awkward postures are avoided) during the work processes. Postural stress can lead to increased risk of developing WMSDs, just as static body positions (those that must be held for more than a few minutes at a time) can increase the risk of local muscle fatigue and injury, as well as ligament strain.

When work is poorly designed from an ergonomics perspective, often some type of intervention is implemented to reduce employees' exposure to the risk factors. This may involve a number of different approaches, such as changing the design or layout of the workstation, adding assistive equipment or technology, adding workers or rotating workers to reduce exposures, and modifying the work itself. These approaches almost always have associated costs, which must be considered in choosing the best solution for a specific job.

## 38.4 ECONOMICS AND ERGONOMICS AND THEIR COMMON GOALS

As noted previously, the goal of ergonomics is to design safe and productive jobs and workplaces to protect workers from injury and illness. Similarly, one of the goals of economics is efficient allocation of resources. Clearly these goals are in alignment because safe and productive work environment not only protects workers but controls wastage of resources such as compensation expenses, lost time and effort, and workers' morale and health. Appropriate allocation of these vital resources also leads to productive efficiency by encouraging more output with fewer resources and at a lower cost. The problem is that it is not always easy to determine the payback that may result from implementation of an ergonomic intervention. The benefits of such a program are in terms of reduced future injuries, illnesses, and increased potential output, results which might not have occurred otherwise and thus are seldom visible. Properly designed economic modeling can incorporate such uncertainties and may estimate expected program benefits. It will also predict how long it will take for interventions to provide a return on the investment (ROI). Often, employers ask, "Will this intervention pay for itself, or will it just cost us a lot of money?" Management will be more likely to embrace interventions if they result in efficient utilization of the company's critical resources, namely, money and employees' time. In fact, an intervention may bring forth a positive change in the relationship between resources (labor and capital) and outcome (production output). Consider Figure 38.1. The bold curve, commonly known as a production function,* relates labor with output for specific capital inputs. Any point on the curve signals production efficiency. However, poor ergonomics may prevent the firm from reaching point A and force it down to point A′, thereby losing output A–A′ in the production process. A properly redesigned ergonomic workstation not only helps in attaining point A but also may generate a new production frontier by alternating the existing relationship between labor and output, shown by the dotted line in the picture. The investment may lead the firm to produce at point B. Most decision-makers focus on A–A′ while deciding to intervene or not. A properly designed economic study should focus on B–A′ in order to predict the real returns on investment.

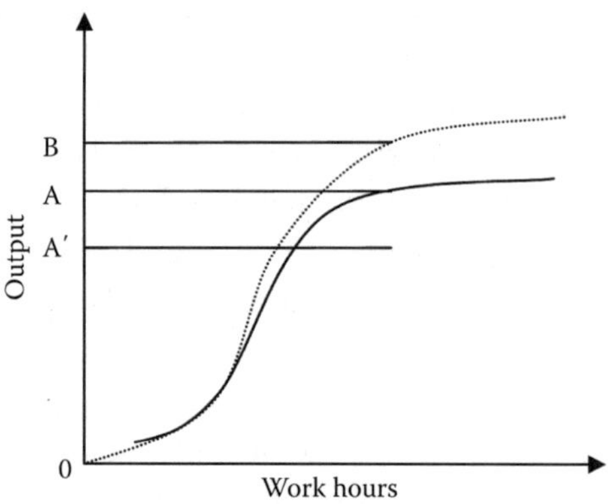


FIGURE 38.1 Production frontier and unrealized gains from intervention.

* A production function is a function that specifies the output of a firm, an industry, or an entire economy for all combinations of inputs. It is generally believed that Philip Wicksteed (1894) was the first economist to algebraically formulate the relationship between outputs and inputs as $P = f(x_1, x, \ldots, x_m)$, although there are some evidences suggesting that Johann von Thünen first formulated it in the 1840s. See Humphrey (1997).

Seabury et al. (2005) sum it up in this way: "If poor health makes the worker less productive and if employers are unable to replace unhealthy workers with healthy ones at no cost (or unhealthy trained workers with healthy untrained workers), then employers will also obtain some benefit from reducing poor health among workers."

## 38.5 TYPES OF ECONOMIC ANALYSES

Economic evaluation of ergonomic interventions can be defined as applying analytic methods to *identify*, *measure*, *value*, and *compare* the costs and consequences of alternative prevention strategies. The purpose of economic evaluation is to replace ad hoc decision-making with informed decision-making. Every decision regarding ergonomic intervention—be it primary, secondary, or tertiary—is associated with an opportunity cost. Decisions backed by economic evaluation help in reducing those costs. The three broad categories of economic evaluation of ergonomic hazards are (1) evaluating the economic burden of the hazard or exposure to that hazard, (2) measuring effectiveness of hazard controls or preventive interventions, and (3) comparing alternative preventive strategies to help choose the optimal one. The corresponding economic analyses are cost analysis (CA), cost-benefit analysis (CBA), and cost-effectiveness analysis (CEA).

### 38.5.1 Cost Analysis: Costs Associated with Poor Ergonomics

CA is a form of *economic evaluation* that involves the systematic collection, categorization, and analysis of data on the costs associated with poor ergonomic design of work or a workplace and the losses due to associated injuries and illness.* Oftentimes, injuries or illnesses are the end results of cumulative problems that have already been costly in terms of presenteeism† and increased absenteeism.‡ The economic burden to the workers, their families, and society in terms of endangered health and reduced household production might be significant even without injuries and illnesses; therefore, tracking numbers of injuries or illnesses may not reveal the already diminishing productivity§ due to declining labor health. Cumulative exposures to physical stressors may take time to manifest; however CA can be used as a problem identification process. Exposure assessment and CA can estimate production loss. Even when the number of injuries or illnesses is known, the associated costs and business impact may not be fully realized or estimated. CA can be used to reveal those losses to the company. Costs are measured in terms of *opportunity costs*—the output that could have been achieved if the resources had been properly used. Ergonomic-related injuries and illnesses will lead to deficient outcomes such as worker-health impairment; lost work time due to workers' non-comfort, illness, or injury; and decreased output due to work stoppage. In turn, these factors may lead to increased employer-sponsored or privately sponsored medical expenses and disability adjustments, reduced employee morale, pain and suffering of workers and their

* Cost of injury and/or illness hereafter signifies costs related to occupational injuries and illnesses.

† Presenteeism occurs when workers are present at work but are less than 100% efficient.

‡ Absenteeism occurs when workers are absent from work because of health conditions.

§ Productivity in simple terms is measured as output over input.

families, and negative social externalities. A true assessment of these costs will trigger intervention-seeking behavior by decision-makers. Traditionally, these costs are broadly categorized as direct or indirect:

1. Direct costs include the medical expenses associated with injuries and illnesses. Often this is combined with insurance administration costs* (the overhead for running an insurance company). For disabling injuries and illnesses, medical costs include worker compensation (WC) and immediate and long-term medical expenses associated with occupational injury or illness, as well as loss of income if an employee is not able to return to work, such as with disability (Leigh et al. 2004):

   $$\text{Direct cost} = \text{medical expenses} + \text{insurance administration cost}$$

   A majority of injuries and illnesses, although originating from the workplace, may not result in WC claims, and the associated medical expenses may differ from those in WC cases. The average cost of a WMSD case that is not reported through WC claims can still be estimated. Leigh et al. (2004) suggests that such costs can be estimated by multiplying WC claim costs by 89% (Baker and Krueger 1995).

   Direct costs are visible and hence termed direct. They are associated with mainly injuries and illnesses and may fail to reflect the majority of the costs associated with work hazards due to improper ergonomic designs, especially when the relationship between the hazard exposure and the injuries is not clear or may take time to build.

2. Indirect costs do not include costs covered by WC insurance but are expenses paid out by the company or the worker or the society. Although these are real costs, they are often thought of as hidden because they are not easily tracked or assigned a value and they are often ignored. These indirect costs include such things as absenteeism, presenteeism, hiring replacement workers, training new workers, loss of productivity and quality during the training and learning period, overtime payments, accident investigation, delayed return to work, counseling, and other costs that may be difficult to assess. Indirect injury costs have typically been reported to be two to five times greater than the direct costs. According to Leigh et al. (2004), they constitute 79% of total occupational injury costs, whereas the indirect costs of illnesses (such as Carpal Tunnel Syndrome [CTS]) constitute 34% of total occupational illness costs. Because many companies try to reduce direct costs by reassigning injured workers to modified or light-duty jobs, the cost of this modified work is often ignored or not included in a CBA:

* These are the costs incurred by insurers and are usually measured by the difference in their (insurers') total revenue (premium) and the WC benefit payments. For private insurers, this will include the abnormal profits they earn. See Rice et al. (1989) and Leigh et al. (2004) for further discussion. Leigh et al. (2004) assume administrative costs for WC claims to be 31% of the total medical costs, and Cutler (1994) assumes 15% for non-WC claims.

Indirect costs = lost work time of injured workers (absenteeism)
+ lost efficiency of injured worker while at work (presenteeism)
+ wastage of resources (idle capital and labor due to the absence of injured workers)
+ higher turnover costs (hiring and training)
+ worsened employee morale, resulting in low productivity (output per worker) as well as low quality of service or product
+ personal losses such as household services (social cost component)

Workplace hazard exposures that result in cumulative trauma and injuries/illnesses take time to become acute and to lead to workplace absences. Thus, before the injuries/illnesses get severe enough to be recognized and recorded, they may already have resulted in a significant loss of productivity. Efficiency is affected in the form of presenteeism (production below potential). Measuring presenteeism can be a challenge.

Absenteeism is measured by the numbers of days a worker is absent because of sickness (sick leave) and the resulting loss in potential output, usually estimated by the hourly wage. Most wage estimates are after-tax figures and should be adjusted to before-tax replacement costs. The literature includes various related conversion factors, such as described by Leigh et al. (2004). Further, a wage reflects the contribution of the worker at the margin and fails to reflect any surplus he or she generates. It would theoretically be more appealing to replace wages with the average output generated, assuming zero unemployment. Higher unemployment implies a lower opportunity cost for the replaced worker and at the extreme, costs due to absenteeism could be zero. For example, if the injured worker is unskilled, part-time, and fails to file a WC claim, then his or her absence may not affect the overall output if a replacement worker is readily available.

Absenteeism costs should also include fringe benefits. U.S. Chamber of Commerce (1993) estimated fringe benefits to be 54% of pretax compensation.

Presenteeism is difficult to value monetarily because it depends on the specific production process, intervention, and technical sophistication of the enumerator, data availability, and market condition. A worker may produce at less than optimal capability and are more prone to error, which seldom goes noticed if there is no one-to-one mapping between individuals' working hours and the output generated, if the production function is nonlinear, or if the market constraints are such that working at full potential may generate idle surplus.* Sometimes, employers do not share complete information about the production frontier that is technically efficient.

* Interested readers should look at the classic paper of Alchian and Demsetz (1972) on presence of moral hazard in team production.

Unlike other cost components, the burden of presenteeism is borne by employers alone, and the chances of directly shifting these costs to other parties are remote. Few studies have focused on quantifying the cost of presenteeism. Goetzel et al. (2004) estimated the annual costs at between \$22 and \$157 per employee. These costs were calculated by determining the number of unproductive hours attributable to presenteeism and multiplying that number by \$23.15, the average U.S. wage and benefit in 2001. Similar estimates can be found for several illness conditions in Pauly et al. (2008).

Turnover costs have three main components: (1) staffing (in addition to the recruiting and hiring costs for the initial worker, the organization must spend a similar amount to hire the replacement), (2) vacancy (the period during which an employee is not working results in lost productivity and potentially lost business), and (3) training (employees are not 100% productive from the moment they start working). It is important to understand how much money is spent on these activities.

CA not only highlights the importance of intervention by revealing the true damage caused by ergonomic-related hazards but also may be effective in finding out who bears the burden. The economic burden of poor ergonomics may be borne by the employers in terms of higher costs; by the employees in terms of the detrimental impact on health and wages; by insurance agents in terms of higher compensatory expenses; by the families in terms of suffering and a reduction in assistance from the injured workers; and by society in terms of higher prices, lower output, and less value. The burden may be shifted from one agent to another and then to the society as a whole subject to the market constraints.

In addition to estimating economic burden, CA is used to estimate the costs of intervention. The economic costs of an intervention program (workplace ergonomic redesign) are not just the dollar amount spent on it but also the value derived if the resources had been allocated to their next best use. The cost categories are similar to those outlined previously. The important difference, however, might be reflected in the time period considered. Typically, costs of ergonomic-related illnesses and injuries reflect longer time spans, accounting for the life span of the endangered cohort of workers. Intervention costs, on the other hand, are estimated on the basis of short-term investment and long-term maintenance costs. Although cost categories are intervention-specific, Table 38.1 suggests various intervention cost entries.

CA is an important component of all economic evaluation techniques. Once the costs associated with ergonomic hazards and the costs of eliminating those hazards are identified, the decision-maker must weigh the two and decide on the validity and effectiveness of intervening. This may also answer questions such as how much intervention is needed. The economic purpose of an intervention is to reduce overall opportunity costs. Cost minimization involves equalizing the marginal costs (MCs) before and after the intervention, as depicted in Figure 38.2. At I*, MC of intervention (ΔCost of Intervention/ΔIntervention) is equal to MC of ergonomic hazard (ΔCosts due to Hazard/ΔIntervention).

Unlike CA, CBA is the primary tool for weighing costs of an intervention/program against its benefits (reductions in hazard exposure and associated injuries/illnesses), and it is often regarded as the gold standard of economic analysis for public policy in the United

TABLE 38.1 Costs of an Ergonomic Intervention

| Factors | Metrics |
|---|---|
| | **Direct Costs of Equipment** |
| Purchase | Opportunity cost or market price |
| Fabrication | Market price |
| Replacement | Recycling cost for old equipment + installation costs |
| Maintenance | Time × wage + price of tools |
| Capital recovery factor | Market rate of interest + depreciation |
| | **Direct costs of personnel** |
| Training and consultancy | Payment to the trainer + time taken from work × wage rate |
| New hiring | Wages paid + benefits + search cost |
| | **Indirect costs** |
| Employee downtime | Time away from normal work and in reduced capacity × wage |
| Loss of production due to learning | Time away from work × wage + value of foregone output |
| Overhead costs | Utilities, facility maintenance, general administration, as calculated by accounting department as a percentage of direct costs |

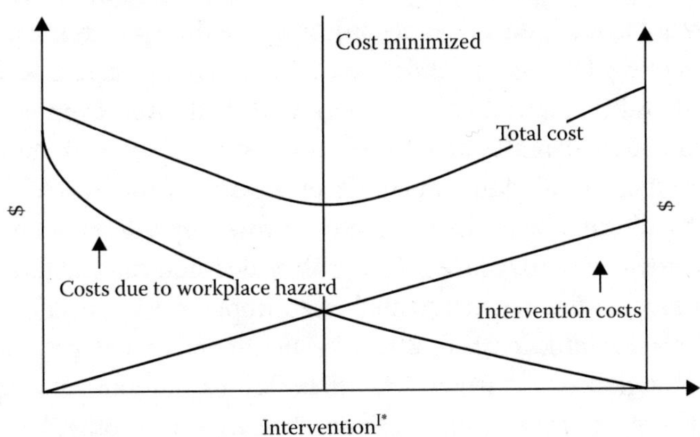


FIGURE 38.2 The economic factors involved in cost minimization CBA: ergonomic net benefits and intervention profitability.

States and Europe.* CBA is a form of economic evaluation that standardizes both costs and benefits in monetary terms (dollars) and provides a list of all costs and benefits accrued during a period. It is particularly helpful for deciding whether to implement a specific ergonomic program, deciding whether net benefits are positive, or for choosing the program option offering the highest ROI. If implementing a program results in a positive net benefit to society, then the program is efficient according to the Pareto improvement criterion.† This does not necessarily imply that the program may be beneficial from the company's

* See Dupuit (1952) and Rice (1966) for earlier utilization in public policy and health.

† An allocation is Pareto Efficient if there is no way to reallocate resources such that one individual is better off without making someone else worse off. However, if in principle the gainers can compensate the losers and still have a positive net gain, then the later situation is called Pareto Improvement over the former. See Stiglitz (1991).

perspective. Corporations may be interested in interventions that generate positive net benefits to the corporation. Incongruent cost and benefit timelines are adjusted, and *summary measures* for a CBA in monetary terms are typically presented as a single value, such as net present value (NPV), or as a ratio of benefit to cost:

$$\text{Net present value (NPV)} = \sum_{t=0}^{N} \frac{Benefit(B) - Cost(C)}{[1+r]^t},$$

where

$t$ is the time period normally in years 0, 1,……N
$N$ is the number of years being evaluated
$r$ is the discount rate

An intervention project is feasible if NPV > 0. Similarly, one can calculate a benefit-cost ratio (BCR) as BCR = $\sum_{t=0}^{N} \frac{Benefit}{(1+r)^t} \Big/ \sum_{t=0}^{N} \frac{Cost}{(1+r)^t}$. A BCR > 1 implies that benefits will outweigh the cost. Discounting makes it possible to compare benefits and costs that occur at different times by adjusting their values according to the time preference corresponding to the chosen perspective. CDC recommends that a 3% social discount rate be used in analyses.

The conceptual and theoretical framework of CBA is derived from *welfare economics,* the branch of economics that attempts to maximize overall social welfare by examining the economic activities of individual members of society. At the societal level, CBA helps in deciding how to allocate the limited funding approved by federal, state, or local legislature. At the company or industry level, CBA enables decision-makers to assess how much money they will save or the benefits gained from implementing a safety or health program at the workplace. Unlike costs of illness as mentioned earlier, *cost* here stands solely for expenditures, tangible or not (such as labor, buildings, equipment, supplies, time, and effort) associated with a particular intervention. *Benefits,* on the other hand, are the monetary values of desirable consequences attached to that intervention. In the previous CA discussion, benefits may be equal to the costs of illness that will not be incurred as a result of intervention. Together with intervention costs, they reflect the changes in a company's profits or changes in individual and social welfare that result from implementing alternative programs. In comparison with benefits, costs of ergonomic interventions are easier to measure because cost factors are often fewer and the necessary accounting data are easily available. Table 38.1 earlier briefly lists some of these cost factors. These costs vary according to how early an ergonomic program is put into place (Alexander, 1999).

Benefits are generally classified as *direct*, *indirect*, and *intangible*. In general terms, direct benefits are the values of desirable health and non-health outcomes directly related to the implementation of proposed interventions that can be estimated by using market-based data (e.g., reduction in injuries or illnesses of the workers). Indirect benefits are the averted costs and savings resulting from the interventions but not related directly to them (such as reduction in absenteeism and presenteeism). Intangible benefits include the values of positive outcomes (e.g., reductions in pain and suffering), which cannot be estimated from market data.

*Health factors*: Changes in health outcomes are the direct or immediate result of an intervention. In most cases, improvements in health or safety conditions are the primary aim of the intervention. The majority of health-related outcomes considered in CBAs in public health fall into one of the following categories: decreased mortality, decreased morbidity, increased life expectancy, or reduced disability. In the context of ergonomic interventions, health-related outcomes will be reduced injuries and illnesses and reduced discomfort associated with them. For example, in the CBA of an intervention to reduce occupational lower-back pain in a nursing home, health factors might include reduced discomfort, reduced disability, reduced medical claims, and reduced WC claims. Identifying the relevant health factors for an economic evaluation often involves close consideration of available clinical, epidemiologic, and cost data.

*Non-health factors*: Such outcomes typically include elements such as productivity, staffing and salary issues, and product quality. Often, non-health outcomes may be significantly impacted by the health outcomes. When this occurs, they are attributed to the impact of an OSH intervention or program. For example, in the CBA of an intervention to reduce occupational lower-back pain in a nursing home, non-health factors would include (1) less whole or partial days lost due to sickness-related absence, (2) higher productivity and hence more efficient work force leading to lower time loss, (3) affected quality of care leading to patient well-being, and (4) less overtime and increased staffing needs.

A CBA should account for the values of these non-health outcomes to arrive at a correct measure of the economic benefits associated with an intervention.

*Intangible factors*: Intangible health outcomes (e.g., improved company reputation and improved quality of life) might be major considerations in implementing an intervention or health program and should be included if possible. However, companies typically do not include intangible costs and benefits.

The following steps are important in conducting a CBA*:

1. Define the problem.
2. Identify the prevention intervention to be evaluated.
3. Identify and classify effects of intervention (health and non-health) as costs or benefits.
4. Assign dollar values to the outcome of the preceding step.
5. Determine their present value, taking into account discount rate and inflation.
6. Determine and calculate the NPV or the BCR.
7. Evaluate the results for robustness with a sensitivity analysis on key parameters.†
8. Prepare the results for presentation.

---

* See Messonnier and Metzler (2003).

† See Appendix, part I.

Identifying the economic variables and classifying them as benefits or costs are the most important steps in conducting an economic analysis. Metrics to consider include health, non-health, and intangible factors. An analyst must include all metrics relevant to the study perspective while considering practical feasibility issues. Each metric is classified as either a benefit or a cost, according to the adopted viewpoint. For further examples of such analyses for ergonomic interventions, see Oxenburg (2005), Loisel et al. (2002), Hendrick (2003), and Lahiri (2005). We provide an example at the end of this chapter from Seeley and Marklin (2003).*

### 38.5.2 Cost-Effectiveness Analysis: Intervention Effectiveness and Choosing between Alternatives

Similar to CBA, CEA is a method used to compare effectiveness of alternative interventions in terms of their contributions toward a common economic outcome, such as better health or fewer injuries. CEA is used to compare the opportunity costs of alternative intervention strategies that produce a common health effect. CEA is preferred over CBA in situations where monetization of outcome effects is difficult and socially unacceptable (DeRango and Franzini 2003; Haddix et al. 2003; Drummond et al. 2005). Such effect measures are often expressed in physical or natural health units and include final outcomes (e.g., number of injuries prevented) or are given in terms of healthy outcome variables such as reduced absenteeism. The primary step is to measure the CE ratio for various alternative intervention scenarios (e.g., intervention $A$, intervention $B$, no intervention). The numerator consists of the net costs of a typical intervention (costs of intervention—costs of adverse health outcomes averted), and the denominator measures the net health outcome (health outcome post-intervention—health outcome pre-intervention). The ratio (net cost over net health outcome) will measure the effectiveness of the expenses made and will make a comparison between intervention programs easier. The smaller the ratio, the higher is the program's worth for the expenses. The various types of CE ratios are average cost-effectiveness ratio (ACER), marginal cost-effectiveness ratio (MCER), and incremental cost-effectiveness ratio (ICER). Suppose, to alleviate MSD problems among a cohort of workers, two interventions ($A$ and $B$) are considered. Intervention A can be of large scale or small scale. The ACERs for the two intervention are $(C/E)_A$ and $(C/E)_B$:

$$\left(\frac{C}{E}\right)_A = \frac{(\text{intervention cost}_A - \text{injury/illness cost averted}_A)}{\text{total injuries/illnesses prevented}_A}$$

$$\left(\frac{C}{E}\right)_B = \frac{(\text{intervention cost}_B - \text{injury/illness cost averted}_B)}{\text{total injuries/illnesses prevented}_B}$$

The ACER provides useful information about the overall affordability of the intervention. If an intervention is cost-saving, then the corresponding ACER will be less than 0. In contrast to ACER, MCER provides information on the change in cost-effectiveness as

* See Appendix, part IV.

a result of expansion or contraction of an intervention and should be considered in conjunction with ACER. MCER is the ratio of additional costs to outcome obtained from one additional unit of intervention. For example, to solve back problems among desk workers, varying the number of workers being provided with back-supporting chairs will alter the cost-effectiveness ratio. The changes in net costs when divided by the extra reductions in lower-back pain incidences will give the MCER. Thus, by varying the degree of type $A$ intervention, MCER is defined as

$$\left(\frac{\Delta C}{\Delta E}\right)_A = \frac{(\text{extra intervention cost}_A - \text{extra injury/illness cost averted}_A)}{\text{extra injuries/illnesses prevented}_A}.$$

Whether or not intervention $A$ is CE depends on whether or not ACER is less than 0. However, to what degree intervention $A$ is prudent depends on the value of MCER. Thus, MCER in conjunction with ACER can provide the decision-maker with the threshold value of intervention.

Prevention-effectiveness studies often require comparison between two alternative and mutually exclusive intervention strategies. Such studies benefit from the utility of the incremental cost-effectiveness ratio, or ICER:

$$\left(\frac{C}{E}\right)_{A-B} = \frac{(\text{additional intervention cost} - \text{additional injury/illness cost averted})}{\text{additional injuries/illnesses prevented}}$$

$$= \frac{[\{\text{intervention cost}_B - \text{intervention cost}_A\} - \{\text{injury/illness cost}_A - \text{injury/illness cost}_B\}]}{\text{injury/illness}_A - \text{injury/illness}_B}$$

A negative ICER implies that the alternative program is dominated by the program in question. For detailed discussion of intervention selection and ranking by ACER and ICER, consult Gift et al. (2003).

### 38.5.3 Cost Utility Analysis (CUA): Ranking Interventions with Regard to Nonmonetary Outcomes

CUA is a type of CEA in which benefits are expressed as utility preferences (measured in cardinal numbers) over numbers of life years saved, adjusted to account for loss of quality from morbidity of the health outcome or side effects from the intervention (Haddix et al. 2003). CUA is a unique form of economic evaluation that attempts to capture timing and duration of injuries, illnesses, and disability by comparing the utility associated with different health outcomes. Utility represents a person's preference for a specific outcome (or health state). CUA typically measures outcomes in quality-adjusted life years (QALYs) or disability-adjusted life years (DALYs). CUA is measured by the cost-utility ratio and can be expressed as the dollar value per QALY or DALY saved.

There are various steps involved in undertaking an economic analysis of ergonomic interventions. Knowing these steps will enable development of an economic evaluation study. Studies dealing with economic evaluation of OSH interventions are often criticized for the lack of understanding of the different steps of an economic analysis.

## 38.6 STEPS INVOLVED IN FRAMING AN ECONOMIC ANALYSIS

The first and the foremost step in conducting an economic analysis is framing of the study, which determines the research framework. Framing involves the following stages:

1. *Defining the problem*: The study problem must be identified at the outset of any analysis. A clearly stated problem defines the objective of the study. At this initial stage, consider what questions need to be answered and which aspects of the problem need to be explained.

2. *Defining the options*: The scope of the study and the variety of outcomes to be included are determined to a large extent by the nature of the programs under consideration.

3. *Defining the audience*: Understanding what information the audience needs and how the study results will be used are the major factors that must be considered at this stage.

   These questions will help to identify the audience (1) who will use the results of the analysis, (2) what information the audience might need, and (3) how the results will be used.

4. *Defining the perspective* (e.g., *company, industry, employee, societal)*: When using any perspective, only the benefits and costs relevant to that specific perspective should be considered. The CBA can be conducted from any of these perspectives, depending on the audience.

5. *Defining the time frame and analytic horizon*: The time frame and the analytic horizon are largely determined by the interventions under consideration. Time frame refers to the period through which intervention is undertaken while analytic horizon refers to the period over which the benefits and costs of intervention are distributed. The analytic horizon is usually longer than the time frame because the majority of interventions or treatments produce multiple health and non-health outcomes for periods far exceeding the durations of the interventions or treatments.

6. *Defining the discount rate, adjustment for inflation, and time preference*: *Discounting* makes it possible to compare benefits and costs that occur at different times by adjusting their values according to the time preference corresponding to the chosen perspective. CDC recommends that a 3% *social discount rate* be used in analyses.

The discount rate is one parameter that can be varied in a sensitivity analysis to test its impact on the results and to make the results of studies based on different discount rates comparable. At the company level, the cost of capital (borrowing/interest rate) is the appropriate discount rate.

*Robustness of study results and treatment of uncertainty*: Uncertainty may arise due to the presence of imperfect information and/or incomplete information. Intervention outcome is a result of interaction among many factors. In situations involving imperfect information, the statistical uncertainty that exists while mapping intervention to injury outcomes can be tackled by using range values around the defined confidence intervals. The economic evaluation analysis will then provide a range of possible outcomes to assist the decision-maker. However, when there is incomplete information on the costs and consequences of intervention, proxy measures and assumptions about certain parameters are necessitated. Univariate and multiway sensitivity and threshold analyses are undertaken in such situations to ascertain the robustness of outcome measures and their dependence on key parameters. Thus decision analysis can be defined as the use of analytic methods to inform complex decisions under conditions of uncertainty. The decision-maker calculates the expected outcome against probability-weighted outcome.* A detailed discussion of sensitivity analysis is beyond the scope of the current chapter. Ambitious readers should consult Briggs et al. (2006).

## 38.7 ERGONOMIC INTERVENTION APPROACHES

When a job is found to present high risk for WMSDs, then some type of ergonomic intervention is required. Industrial hygiene practice has adopted a general hierarchy of controls for implementing interventions. The hierarchy of controls suggests that the first choice for interventions is to use engineering controls, followed by administrative controls, and last by the use of personal protective equipment (PPE), if applicable. This hierarchy of control is based on the premise that the best approach for intervention is to use engineering controls to reduce or eliminate exposure to the job risk factors through automation, the use of mechanical aids, or ergonomic redesign of the workplace. Ergonomic redesign of the workplace typically includes modifying the workstation or tools to eliminate risk factors by modifying the job layout and procedures to reduce bending, twisting, horizontal extensions, heavy lifting, forceful exertions, and repetitive motions. Administrative controls, the second choice for interventions in the hierarchy of controls, typically aim to mitigate exposures to the risk factors by assigning additional workers, reducing exposure times, or attempting to hire specific workers or increasing training for workers. These approaches, however, do not always modify the risk factors themselves but rather serve to reduce the "dose" of physical exposure to each individual worker. Finally, the concept behind the use of PPE as an intervention is to try to protect the worker from exposure to specific risk factors. In industrial hygiene practice, this typically means using gloves, respirators, or body suits. For physical exposures, however, there is little PPE available for the risk factors associated with development of MSDs. Therefore, this is usually used only as a last resort. A more detailed discussion of ergonomic interventions can be found in the chapter on manual material handling (MMH), and a good source of information about ergonomic interventions for MMH jobs can be found in the NIOSH publication *Guidelines for Manual Material Handling* (NIOSH 2007).

* This is based on the expected utility theory by von Neumann, Morgenstern, Savage, and others. See Raiffa (1968) and Von Neumann, J. and O. Morgenstern (1953).

Some of the economic considerations of interest in choosing an ergonomic intervention include the costs and benefits associated with factors such as (1) obtaining and implementing the intervention, (2) conducting training classes, (3) changes in personnel levels, (4) changes in productivity, and (5) changes in product quality.

## 38.8 SUMMARY OF COST-BENEFIT MODELING TOOLS

A number of tools have recently been created to assist in the development of cost-benefit models to quantify the effects of implementing workplace interventions, often of an ergonomic nature. The tools vary greatly in cost, sophistication, focus, and input information required. The tools and associated software, if available, range in price from free to several hundred dollars each. Some of the tools are relatively simple, utilizing default values and requiring limited user input, but others are fairly complicated, with specific geographic location and injury default data and requiring more user-supplied data. Several models focus on changes to productivity as the basis for analysis, whereas other tools focus on the potential savings from injury or illness avoidance. The examples provided are not intended to be all-inclusive but representative of the types of tools available.

The "$afety Pays" program of the U.S. Department of Labor Occupational Safety and Health Administration OSHA (2009) is a free, interactive, Internet-based system which assists users in estimating the cost of occupational injuries and illnesses and their impact on the profitability of a company. The program uses the company's profit margin (real or default value), the cost of an occupational injury or illness (actual or national average), and an indirect cost multiplier to compute an estimated amount of sales a company would have to produce to cover the associated costs of the injury or illness. The program is also available in a plain text version.

The "productivity assessment tool" described by Oxenburgh and Marlow (2005) focuses on the costs and benefits associated with the employment of workers, primarily through measures of their productivity. This tool is often used to measure the increase in cost-effectiveness of employees after an intervention by contrasting "before" and "after" findings. The primary measure of this tool is productivity, defined as the ratio of time paid for by the employer and the time the employee spends actively working, coined "the productive hours." This measure is typically taken for a 1 year period. Nonproductive hours paid by the employer include workplace absences for injuries, illnesses, vacations, holidays, training, etc. Shorter time periods are addressed by considering "reduced productivity" from an ideal or optimal condition. This tool, however, does not directly address either the quantity or quality of the output during productive time. Nor does it address presenteeism, in which an individual may be at work but not working up to expectations because of illness, distraction, or motivation. In general, there is no consideration of injury or illness costs with this tool.

The "return on health, safety, and environmental investments" (ROHSEI) tool and supporting software tool set described by Linhard (2005) was developed to provide occupational health, safety, and environmental staff a comprehensive look at investment decisions, including the implementation of ergonomic interventions. The ROHSEI system uses a four-step process to build a business case for each investment: (1) define the focus of the investment opportunity, (2) identify alternative solutions, (3) consider direct and hidden benefits, and (4) make recommendations. This system was designed for use by a team of analysts and may be too complicated for an individual to use effectively. This tool is also the most expensive of those presented.

The net-cost model reported by Lahiri et al. (2005) was developed to estimate the net economic costs of investments in ergonomic intervention to address LBP from a company or facility perspective. The model consists of a series of 23 equations that assist the user in determining the direct equipment costs, avoided medical care costs, lost productivity due to injury, improved productivity due to the intervention, and net economic cost of the intervention. Some of the variables used in the analysis, such as capital recovery factor, rate of return on private capital, and depreciation rate, may not be readily available to the OSH staff conducting the analysis.

However, it is often the case as Boff and Rouse (1997) recognizes that, in ergonomics and human factors arena, much of the traditional ROI data are simply not available. In the absence of detailed data and with lack of understanding of the available tools, economic analysis using those tools will largely affect the results obtained. It is always better to follow a simplified approach of CBA relying on the better understanding of the scope of the study which includes identifying stakeholders, defining benefits and costs, and evaluating alternative work practices.

## 38.9 CONCLUSIONS

Ergonomics and economics are in alignment: safe and productive work likely will cost less in the long run and increase profits for the company. Although the primary focus of occupational ergonomics is the design of safe and productive jobs, if the costs to implement ergonomic interventions are too great or the payback period for the investment is too long, then it may be difficult to convince management to invest in the most effective solutions. CBA can be an effective tool for making decisions about which intervention to implement when a high-risk job has been identified. A significant ROI and/or a short payback period can be effective in obtaining management approval for implementing ergonomic interventions. A conservative estimate is often preferred to ensure that future business case estimates are believable and achievable. To the extent feasible, assessments of costs and benefit should include factors associated with prevention of MSDs, work productivity, and quality. Oftentimes, the best benefit of an ergonomic intervention will be an increase in productivity rather than a savings due to reducing risk of MSDs. Also, a company may have high exposures to risk factors for MSDs but not have any significant reports of MSD injuries, for a number of reasons. There may be some incentive program for not reporting injuries, injuries may not be recorded as work-related, or workers may visit their own primary physician rather than the corporate doctor, so cases may be overlooked. In these instances, cost avoidance may be an important factor to consider when building a business case for an ergonomic intervention.

## 38.A APPENDIX A

### 38.A.1 General Health Valuation Methods

#### *38.A.1.1 Cost-of-Illness Method*

The *cost-of-illness (COI)* method is used to estimate the direct and indirect costs of an intervention. This method is based on the *human capital approach* and is discussed in detail in the CA module. *Intangible costs* of illness and the benefits of averted illness are estimated by means of *nonmarket valuation* methods.

#### *38.A.1.2 Nonmarket Valuation Methods*

The choices people make in the marketplace (i.e., their behaviors) result from their valuations of the goods they buy and the prices they are willing and able to pay. If a market exists, we can comfortably use what consumers are willing to pay as an estimate of how much they value the good. Therefore, existence of a market is relevant in this search. However, because the majority of health outcomes (e.g., health risks) do not have markets, valuing them presents analytical challenges.

Economists have developed theoretically sound and intuitively appealing ways of estimating the value of nonmarket goods. These methods and procedures are based on generally accepted economic theories and are broadly referred to as *nonmarket valuation* methods. Until recently, these methods were applied primarily to the valuation of environmental and natural resources. They are now a relevant and necessary part of contemporary safety and health research. In general, two main valuation methods are employed: *revealed preference* method (RPM) and *stated preference* method.

38.A.1.2.1 Revealed Preference Method *Hedonic pricing* and *averting behavior* are the two main types of *RPMs* used in health economics. By the use of these indirect methods, a person's valuation of a safety or health outcome is observed through actions in markets that already exist. This behavior in the market is used to make inferences about the person's valuation of the related safety or health outcome.

38.A.1.2.2 Hedonic Pricing (Wage) Most goods come in bundles. Each bundle is made up of various components (or characteristics) that are not separable but that might be presented in different ways and sold altogether at one price. Each of these components (or characteristics) might confer a different benefit on the buyer. Therefore, a buyer might pay more to have each additional characteristic or to have more of any one particular characteristic.

The *hedonic* model is based on this principle. The prices that consumers pay or receive are reflections of the different components of the goods in question. For example, the wage rate that underground miners receive is a price for their labor that is determined on the labor market. This price depends on characteristics of the miner that can be objectively measured (such as number of years on the job [experience] and educational level).

On the labor market (which is assumed to be competitive), employers are willing and able to pay more for each characteristic, in both quality (type) and quantity (amount).

A functional relationship therefore exists between the wage and all the measurable characteristics. This relationship can be expressed as follows:

$$W = f(q, e, ex, a, g),$$

where
W is the wage
q is the qualification
e is the education
ex is the experience
a is the age
g is the gender

If data were available, all of these characteristics and outcomes could be subjected to a rigorous empirical analysis (such as a regression analysis) to obtain how much more employers pay to have a unit increase in each of the characteristics, with controls for other factors. This amount is the implicit price received by the worker or paid by the employers for a unit of that particular characteristic. It is used to represent the employer's or worker's value of that characteristic, as determined by the existing conditions in the labor market.

38.A.1.2.3 Averting Behavior The value of a small change in safety and health status can be measured by the amount of money a person is willing and able to spend on some controlling or preventive device or defensive (averting) action. This amount of money may be used to represent the person's valuation of safety against a perceived risk.

Some examples of preventive devices on which persons might spend money to mitigate perceived risk include safety helmets, smoke detectors, seatbelts on tractors, larger cars, and non-mandated PPE.

38.A.1.2.4 Stated Preference Method This direct method uses primary surveys that ask individuals to place values on an intervention to attain a level of health outcome in question. The technique is generally known as the *contingent valuation method (CVM)*. The major advantage of the *CVM* over the *RPM* is that it estimates both *tangible* and *intangible values*, thereby providing a better estimate of the true value of the intervention or action.

Like any other economic evaluation, proper framing of a CBA is of major importance. To assess the effectiveness of an ergonomic intervention, one needs to understand and emphasize the consequences that are only specifically attributed to the intervention. An intervention to reduce stress may have different consequences on the study population, depending on whether or not the economy is in recession or not. Understanding these macro-level factors and similar other confounding factors is of primary importance. Equally important is the time period through which the costs are incurred and benefits are accrued. Intervention investments may be limited to a few years in the beginning, but the benefits may accrue over time. It is important to translate these long-term costs and benefits into their current monetary values for the decision-maker. Although CBA is used more often in other fields of decision-making, in health care research, it is often replaced by CEA and CUA.

## 38.B APPENDIX B

*Example of economic analyses of ergonomic interventions*: Business case for implementing two ergonomic interventions at an electric power utility (Seeley and Marklin 2003) is provided here. Ergonomic analysis of line workers (n = 370) in a large Midwestern U.S. electric utility company revealed 32 common line worker tasks that were rated at medium to high magnitude of risk factors MSDs. A university ergonomics team rated the potential interventions and came up with a list of 9 low cost–high value interventions out of which 2 were believed to have a significant improvement on occupational health of workers.

*Jobs needing ergonomic intervention: (1) manual tool for applying compression connectors to wire; (2) manual wire cutting.*

*Ergonomic interventions: (1) battery-operated tool for pressing connector and (2) battery-operated tool for cutting wire.*

Figures 38.A.1 and 38.A.2 depict the battery-operated tools. The risk factors with the manual tools are the source of the cost of the pre-intervention work practices and are the

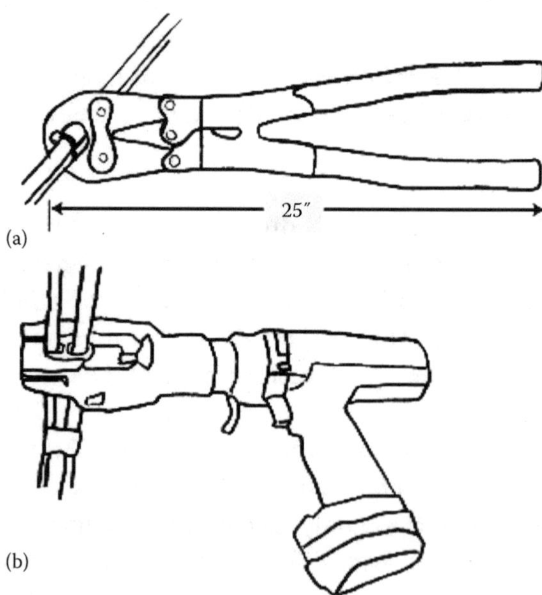


FIGURE 38.A.1 Intervention 1: a compression connector crimped with a manual operation (a) and battery-operated press (b). (EPRI 2001.)

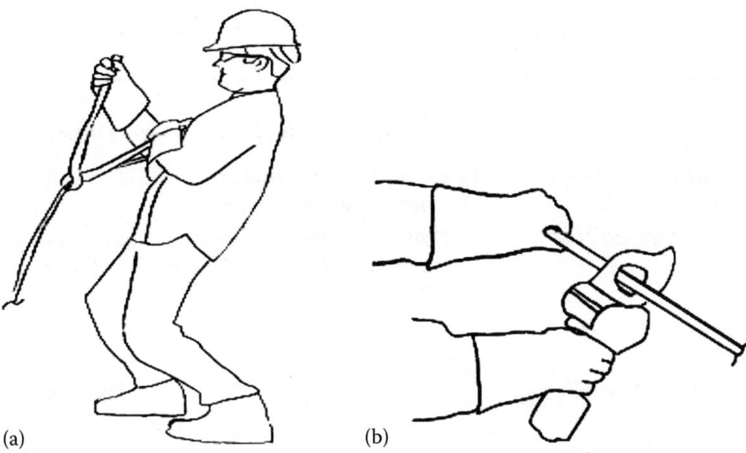


FIGURE 38.A.2 Intervention 2: A line worker cutting a large diameter wire with a manual cutter (a) and a battery-operated cutter (b). (EPRI 2001.)

benefits that are gained when manual tools are replaced with the battery-operated ones. For CA of pre-intervention work practices (manual tools), the analyst has to monetize the health problems and associated work place costs and define them as costs. While doing CBA for the intervention (battery-operated tools), these cost estimates will enter into the benefit side of the equation as avoided costs (Tables 38.A.1 through 38.A.6).

TABLE 38.A.1 Risk Factors of MSDs from Making Compression Connections with the Manual Press and Improvements with the Battery-Operated Press

| **Risk Factor** | **Manual Press Risks** | **Improvement with Battery-Operated Tool** |
|---|---|---|
| Amount of handle force required too empress a 1/0–12 who pair is approximately 311 N at the outer location of the manual press | Only 1% capable of the general population has the peak forces to make this connection | Nearly all workers capable |
| High shoulder force exertions working from a pole | Peak shoulder muscle EMG using the manual press was over 50% MVC (maximal voluntary contraction) | Reduced from over 50% MVC to 30% MVC (40% relative reduction) |
| High shoulder force exertions working from a bucket | Peak EMG 45% MVC | Reduced to less than 10% MVC (over 80% relative reduction) |
| Peak forces of flexor muscles in forearm from working on pole | Peak EMG of 90% MVC | Peak decreased from 90% MVC to 60% MVC (a 33% relative reduction) |
| Peak forces of flexor muscles in forearm from working in bucket | Peak EMG of 100% MVC | From 100% MVC to 60% MVC (a 40% relative decrease) |
| Jarring action of manual compression tool | Not measured, but substantial | Eliminated 100% virtually all workers capable |
| Twisted and awkward trunk postures | | Improved posture from the bucket |

*Source:* Reprinted from Seeley, P.A. and Marklin, R.W., *Appl. Ergonomics*, 34, 429, 2003.

TABLE 38.A.2 Risk Factors of MSDs from Cutting Wire with Manual Tools and Improvements with the Battery-Operated Cutter

| Risk Factor | Current Practice Risks | Improvement with Battery-Operated Cutter |
|---|---|---|
| High upper extremity force exertions to close cutter handles | Operating a cutter is similar to operating a manual press | Substantially reduced |
| Forceful lowering of arms (shoulder abduction) | High forces exerted by the upper extremity and shoulders although not measured | Substantially reduced with a finger pull on the trigger |
| Arms raised above shoulder level (shoulder abduction) | | Substantially reduced |
| Jarring action from quick drop in force when wire is cut | | Eliminated |
| Repetitive upper extremity exertions using the rather cutter | | Eliminated |
| Twisted and awkward trunk postures from bucket | | Improved posture |

*Source:* Reprinted from Seeley, P.A. and Marklin, R.W., *Appl. Ergonomics,* 34, 429, 2003.

TABLE 38.A.3 All Reported MSDs (January 1999–July 2001) of Line Workers ($n$ = 370) at the Utility by Body Part and Progression of Reporting

| | Progression by Reporting | | | | |
|---|---|---|---|---|---|
| Body Part or Injury Type | Early (First Aid or First Time Medical Treatment) | Mid (OSHA Recordable Injuries) | Late (Occupational Illness, Inflammation, RD, and LWDs) | Total | % Reported Late |
| Non-repetitive | 11 | 5 | 1 | 17 | 3 |
| Wrist | 5 | 1 | 1 | 7 | 14 |
| Elbow | 4 | 8 | 2 | 14 | 14 |
| Shoulder | 6 | 6 | 7 | 19 | 37 |
| Upper arm | 0 | 0 | 1 | 1 | 100 |
| Forearm | 1 | 1 | 1 | 3 | 33 |
| Buck | 25 | 12 | 20 | 57 | 35 |
| Leg | 1 | 0 | 0 | 1 | 0 |
| Knee | 15 | 8 | 2 | 25 | 8 |
| Neck | 0 | 0 | 3 | 3 | 100 |
| Inflammation | 1 | 2 | 3 | 6 | 50 |
| Carpal tunnel syndrome | 0 | 2 | 3 | 5 | 60 |
| Other occupational illness | 0 | 1 | 1 | 2 | 50 |
| Multiple body parts | 3 | 3 | 2 | 7 | 17 |
| Totals | 72 | 48 | 47 | 167 | 28 |

*Source:* Reprinted from Seeley, P.A. and Marklin, R.W., *Appl. Ergonomics,* 34, 429, 2003.

The bold-faced numbers reflected the body parts with high number of incidences. Early and mid injuries resulted in no LWDs or RD days; late injuries resulted in RD days or LWDs or were relatively severe, such as those requiring surgery. A thorough review of the table shows that approximately 73% of injuries were not reported. Also, it was found that costs of treating injuries are much lower at the early stage than at the later stage. Average costs for injury treatment were calculated based on only reported injury data and are given in Table 38.A.4.

TABLE 38.A.4 Utility's Medical and Workers' Compensation Costs for 47 MSD Injuries That Resulted in LWDs or RD Days (January 1999–July 2001)

| Type of Injury | Total Medical Costs, Indemnity, Reimbursements for Drugs, Parking, Transportation | Annualized | Annualized per Employee $N = 370$ |
|---|---|---|---|
| Upper extremity injuries | $179,296 | $71,718 | $194 |
| Neck injuries | $160,573 | $64,229 | $174 |
| Knee injuries | $46,207 | $18,522 | $50 |
| Back injuries | $92,429 | $36,971 | $99 |
| Totals | $478,605 | $191,440 | $317 |

*Source:* Reprinted from Seeley, P.A. and Marklin, R.W., *Appl. Ergonomics*, 34, 429, 2003.

The costs of replacing 43 of the 47 workers who incurred injuries resulting in restricted or lost work days are shown in Table 38.A.5. Based on this, the annualized cost per line worker for replacing skilled injured workers categorized by upper extremity and back are given in Table 38.A.6.

TABLE 38.A.5 Restricted Duty Days and Lost Workdays of MSDs Requiring Full-Time Replacement Workers Categorized by All Body Parts and Injury Types

| Body Part/ Injury | # of RD Cases | # of RD Days | # of LWD Cases | # of LW Days | Totals RD/LW Days Missed | Replacement Worker Costs Totals (Skilled Worker at $36/h) | Annualized per Line Worker ($n = 370$) |
|---|---|---|---|---|---|---|---|
| Shoulder | 7 | 185 | | | 185 | $56,448 | $61 |
| Back | 17 | 407 | 3 | 90 | 497 | $143,136 | $155 |
| Knee | 1 | 90 | 1 | 43 | 133 | $51,552 | $56 |
| Elbow | 2 | 2 | | | 2 | $576 | $1 |
| Inflammation | 3 | 60 | | | 60 | $17,280 | $19 |
| Carpal tunnel syndrome | 1 | 14 | 2 | 41 | 55 | $15,840 | $17 |
| Occupational illness | 1 | 56 | | | 56 | $16,128 | $17 |
| Multiple body parts | 1 | 20 | 1 | 11 | 31 | $5,760 | $6 |
| Wrist | 1 | 3 | | | 3 | $864 | $1 |
| Upper arm | 1 | 2 | | | 2 | $576 | $1 |
| Forearm | 1 | 46 | | | 46 | $0 | $0 |
| Totals | 36 | 885 | 7 | 185 | 1070 | $308,160 | 334 |

*Source:* Reprinted from Seeley, P.A. and Marklin, R.W., *Appl. Ergonomics*, 34, 429, 2003.

*Notes:* Total number of RD and LWD cases with full data was 43; four of the 47 did not have full data (January 1999–July 2001).

TABLE 38.A.6 Average Restricted Duty Days and Lost Workdays of MSDs Requiring Full-Time Replacement Workers Categorized by Upper Extremity and Back (January 1999–July 2001)

| Body Part Injury | # of RD Cases | # of RD Days | # of LWD Cases | # of LW Days | Totals RD/ LW Days Missed | Replacement Worker Costs Totals (Skilled Worker at $36/h) | Annualized per Line Worker ($n = 370$) |
|---|---|---|---|---|---|---|---|
| Upper extremity and neck only | 18 | 388 | 3 | 52 | 440 | $126,720 | $137 |
| Back only | 17 | 407 | 3 | 90 | 497 | $143,136 | $155 |

The following data from the utility during the January 1999–July 2001 period were reviewed:

1. Medical and workers' compensation costs for 47 cases of MSDs that resulted in time off work, either restricted duty (RD) days or LWDs.
2. The total number of MSDs ($n$ = 167) experienced by all the line workers at the utility not resulting in time off work.
3. Training and productivity costs for apprentice line workers. This is the replacement cost for skilled line workers experiencing permanent time off work.
4. Replacement costs for skilled line workers temporarily off work due to injury/illness.
5. Personnel records for line workers leaving their job. This helps in estimating turnover costs.
6. Personal information on hourly wages and benefits by job classification.

Estimating the costs associated with work injuries: Data for the following were obtained to calculate work injury costs:

1. Permanent partial or total disability costs
2. Lost time (hourly rate multiplied by number of hours missed per occurrence); includes operation downtime due to injury
3. Medical costs: doctors, tests, hospital, physical therapy, chiropractic care, company medical evaluations
4. Time spent discussing the injury or illness with workers, supervisors, health care professionals
5. Sick days' costs for injuries not reported as work-related
6. Clerical time handling claims and case investigation

### 38.B.1 Cost-Benefit Analysis

Seeley and Marklin (2003) assume that the intervention will reduce injuries and illnesses by 50%. This is a critical assumption about the most important parameter in the analysis. Usually, such parameter estimates rely on a suitable Bayesian prior where the base value comes from previous studies and reliable sources. In general terms, one can assume this as flexible and vary the value to do a sensitivity analysis. For example, we take the value as $\phi$, where $\phi \in [0,1]$. $\phi = 0$ implies that the interventions are 100% effective in controlling injuries.

For medical and workers' compensation cost prior and post-intervention, we follow Seeley and Marklin (2003) and consider only those injuries that are restricted to upper extremity and neck region. We also follow them and assume that medical costs increase by 8% annually. Table 38.A.7 gives the costs pre- and post-intervention on an annual basis.

TABLE 38.A.7 Annual Cost Pre and Post Intervention

| Factor | Manual Tools | Battery-Operated Tools | Difference = Avoided Costs = Benefits |
|---|---|---|---|
| Medical and WC | \$146,822 | \$(146,822 × ϕ) | \$146,822 (1 − ϕ) |
| Replacement workers | \$50,688 | \$(50,688 × ϕ) | \$50,688 (1 − ϕ) |
| Retraining costs | \$91,000 | \$(91,000 × ϕ) | \$91,000 (1 − ϕ) |
| Late injury reporting | \$80,000 | \$(80,000 × ϕ) | \$80,000 (1 − ϕ) |
| Total | | | \$368,510 (1 − ϕ) |

| ϕ = 0.0 | ϕ = 0.1 | ϕ = 0.2 | ϕ = 0.3 | ϕ = 0.4 | ϕ = 0.5 | ϕ = 0.6 | ϕ = 0.7 | ϕ = 0.8 | ϕ = 0.9 | ϕ = 1.0 |
|---|---|---|---|---|---|---|---|---|---|---|
| 308510.8 | 271659.7 | 234808.6 | 197957.5 | 161106.5 | 124255.4 | 87404.3 | 50553.23 | 13702.15 | −23148.9 | −60000 |

Annual costs to obtain equipment: Total costs to provide tools for 100 crews × \$3,000 per crew (\$2,000/battery-operated press; \$1,000/battery-operated cutter) divided by 5 years equals \$300,000/5 = \$60,000/year.

Net annualized benefits at the end of year 1 = \$[368,510 (1 − ϕ)−60,000]. Following are the values for respective ϕ values. It is evident that the net benefits will be positive if the interventions are 20% effective. Similarly, the benefit-cost ratio is greater than 1, when $\phi \leq 0.84$.

Payback period: (\$60,000/\$368,510 (1 − ϕ)) × 12 months. For ϕ = 0.5, the payback period becomes 3.9 months.

## REFERENCES


Alchian, A. and H. Demsetz (1972). Production, information costs, and economic organization. *Am Econ Rev* **62**:777–795.

Alexander, D. (1999). *The Cost Justification Process.* Auburn, AL, Auburn Engineers, Inc.

Baker, L. C. and A. B. Krueger (1995). Medical costs in workers' compensation insurance. *J Health Econ* **14**(5):531–549.

Boff, K. R. and W. B. Rouse (1997). Assessing cost/benefits of human factors. In: *Handbook of Human Factors and Ergonomics,* 2nd Edn. New York, Gavriel Salvendy, Wiley, pp. 1617–1633.

Briggs, A., K. Claxton et al. (2006). *Decision Modelling for Health Economic Evaluation.* New York, Oxford University Press.

Cutler, D. M. (1994). A guide to health care reform. *J Econ Perspect* **8**(3):13–29.

DeRango, K. and L. Franzini (2003). Economic evaluations of workplace health interventions: Theory and literature review. In J. C. Quick and L. E. Tetrick, (eds). *Handbook of Occupational Health Psychology.* American Psychological Association, Washington, DC.

Drummond, M., M. Sculpher et al. (2005). *Methods for the Economic Evaluation of Health Care Programmes.* Oxford, U.K., Oxford University Press.

Dupuit, J. (1952). On the measurement of utility of public works. *International Economic Papers* **2**:83–110.

Electric Power Research Institute (EPRI) (2001). Final report for ergonomics handbook for the electric power industry: Overhead distribution line workers interventions. EPRI Report Number 1005199, EPRI, Palo Alto, CA.

Gift, T. L., A. Haddix, and P. Corso (2003). Cost-effectiveness analysis. In: A. Haddix et al., eds. *Prevention Effectiveness: A Guide to Decision Analysis and Economic Evaluation,* 2nd Edn. Oxford, U.K., Oxford University Press, p. 156.

Goetzel, R. Z., S. R. Long, R. J. Ozminkowski, K. Hawkins, S. Wang, and W. Lynch (2004). Health, absence, disability, and presenteeism cost estimates of certain physical and mental health conditions affecting U.S. employers. *J Occup Environ Med* **46**(4):398–412.

Haddix, A., P. Corso, and R. Gorsky (2003). Costs. In: A. Haddix et al., eds. *Prevention Effectiveness: A Guide to Decision Analysis and Economic Evaluation,* 2nd Edn. Oxford, U.K., Oxford University Press, p. 53.

Hendrick, H. W. (2003). Determining the cost-benefit of ergonomics projects and factors that lead to their success. *Appl. Ergonomics* **34**:419–427.

Humphrey, T. M. (1997). Algebraic production functions and their uses before Cobb-Douglas. *Econ Q* **83**(1):51–83.

Lahiri, S., J. Gold, and C. Levenstein (2005). Net-cost model for workplace interventions. *J Safety Res* **36**:241–255.

Leigh, J. P., S. Markowitz et al. (2004). *Costs of Occupational Injuries and Illnesses.* Ann Arbor, MI, The University of Michigan Press.

Linhard, J. B. (2005). Understanding the return on health, safety and environmental investments. *J Safety Res* **36**:257–260.

Loisel, P., J. Lemaire, and S. Poitras et al. (2002). Cost-benefit and cost-effectiveness analysis of a disability prevention model for back pain management: A six year follow up study. *Occup Environ Med* **59**:807–815.

Messonnier, M. and M. Metzler (2003). Cost-benefit analysis. In: A. Haddix et al., eds. *Prevention Effectiveness: A Guide to Decision Analysis and Economic Evaluation,* 2nd Edn. Oxford, U.K., Oxford University Press, p. 127.

NIOSH (1997). *Musculoskeletal Disorders and Workplace Factors: A Critical Review of Epidemiologic Evidence for Work-Related Musculoskeletal Disorders of the Neck, Upper Extremity, and Low Back.* Cincinnati, OH, U.S. Department of Health and Human Services, Centers for Disease Control and Prevention, National Institute for Occupational Safety and Health.

NIOSH (2007). *Ergonomic Guidelines for Manual Material Handling.* Cincinnati, OH, Department of Health and Human Services, Centers for Disease Control and Prevention, National Institute for Occupational Safety and Health, DHHS (NIOSH) Publication No. 2007-13 1 [http://www.cdc.gov/niosh/docs/2007-13 111].

NRC and IOM (2001). *Musculoskeletal Disorders and the Workplace Low Back and Upper Extremities.* Washington, DC, National Academy Press.

OSHA (2009). "$afety Pays" Program [www.osha.gov/dcsp/smallbusiness/safetypays/index.html]. Washington, DC, U.S. Department of Labor, Occupational Safety and Health Administration. Accessed December 8, 2009.

Oxenburgh, M. and P. Marlow (2005). The productivity assessment tool: Computer-based cost benefit analysis model for the economic assessment of occupational health and safety interventions in the workplace. *J Safety Res* **36**:209–214.

Pauly, M. V., S. Nicholson et al. (2008). Valuing reductions in on-the-job illness: Presenteeism from managerial and economic perspectives. *Health Economics* **17**(4):469–485.

Raiffa, H. (1968). *Decision Analysis: Introductory Lectures on Choices under Uncertainty.* Reading, MA, Addison-Wesley.

Rice, D. (1966). *Estimating the Cost of Illness.* Washington DC: Department of Health, Education and Welfare, Public Health Service, Health Economic Series 6.

Rice, D. P., E. J. MacKenzie et al. (1989). *Cost of Injury in the United States: A Report to Congress.* San Francisco, CA, Institute of Health and Aging, University of California Injury Prevention Center, Johns Hopkins University.

Samuelson, P. A. and W. D. Nordhaus (2001). *Macroeconomics.* Boston, MA, McGraw-Hill/Irwin.

Seabury, S. A., D. Lakdawalla, R. T. Reville (2005). *The Economics of Integrating Injury and Illness Prevention and Health Promotion Programs*. RAND Institute for Civil Justice working paper series, WR-243 = ICJ. Santa Monica, CA, RAND, Institute for Civil Justice.

Seeley, P. A. and R. W. Marklin (2003). Business case for implementing two ergonomic interventions at an electric power utility. *Appl Ergonomics* **34**:429.

Stiglitz, J. E. (1991). *The Invisible Hand and Modern Welfare Economics in Information, Strategy and Public Policy*. Oxford and Cambridge, MA, Blackwell, pp. 12–50.

U.S. Chamber of Commerce (1993). Table 4. Employee Benefits, 1992. Washington, DC, U.S. Chamber of Commerce.

Von Neumann, J. and O. Morgenstern (1953). *Theory of Games and Economic Behavior*. Princeton, NJ, Princeton University Press.

CHAPTER 39

# Microergonomics

## *Healthy Workplace and Healthy Lifestyle in University Residence Halls*

Balmatee Bidassie

CONTENTS

## 39.1 INTRODUCTION

Many thousands worldwide conduct physically demanding work in university residence halls, especially those with a dining hall. Injuries in an occupational setting account for a significant amount of human suffering, a loss of productivity, an increase in workers' absenteeism rates, and an increase in workers' compensation (WC) cost [1].

To manage the workplace conditions through proactive application of ergonomic principles, an ergonomics training program can be used to identify the occupational risk factors that contribute to injuries in the workplace [2,3]. The risk factors identified from an ergonomics training program can provide direction to appropriate intervention(s) to improve workplace safety processes and reduce workplace injuries [4].

In addition to occupational risk factors, there are lifestyle risk factors that may contribute to injuries in the workplace. Therefore, it is important to understand the relationship of both occupational and lifestyle risk factors with workplace injuries to identify controllable risk factors. If the controllable risk factors are known, it will be possible for individuals and companies to empower themselves to take action to prevent and reduce injuries in the workplace [4].

The perception of risk is a necessary precursor to protective behaviors which has been included in preventive health models, such as the health belief model [5] and the protection motivation theory [6,7,8]. Therefore, safety professionals in university residence halls should not plan prevention strategies focused only on reducing workplace injuries. Instead, education and training on prevention strategies should be focused on both workplace and lifestyle (outside of the workplace) risk factors to help staff reduce injuries and illnesses in the workplace [4].

## 39.2 BACKGROUND AND SIGNIFICANCE TO OCCUPATIONAL ERGONOMICS

According to the United States Bureau of Labor Statistics [1], occupational safety continues to be a great concern. There is a need for appropriate safety training programs that provide interventions for occupational safety in the workplace. Managers and supervisors should work with staff to reduce risk factors that lead to injuries and illness in the workplace. The importance of organizational changes must be recognized by others in the workplace. The objective of this study is to implement a continuous improvement participatory ergonomic program with an emphasis on action, feedback, and individual learning to create a safety culture and climate. In addition, workplace risk factors (occupational risk factors) and lifestyle risk factors (nonoccupational risk factors) will be investigated to gain a comprehensive understanding into the risk factors that may contribute to injuries in the workplace. Understanding the early predictors of workplace injuries could help increase the knowledge to explain why some employees become injured whereas others do not. This knowledge may lead to more effective secondary prevention efforts focused on controllable risk factors.

### 39.2.1 Relevance to National Occupational Research Agenda

In 1996, the National Occupational Research Agenda (NORA) program was created to foster partnerships to conduct research that will support safer and healthier workplaces [9]. Musculoskeletal disorders in the workplace are often related to workplace ergonomic concerns and account for 33% of lost workdays due to injuries and illnesses [1]. Musculoskeletal disorders due to manual material handling, falls, and repetitive motion tend to result in longer periods away from the job than do more traumatic injuries. Low back musculoskeletal disorders is ranked in the top five reasons for injuries in university residence halls. Therefore, this research, also responds to the NORA [10] mission to identify the most critical existing musculoskeletal disorders that prevent safety and health in university residence halls.

## 39.3 RESEARCH PLAN

The goal of this pilot research project is to describe a process and provide data to encourage a sense of individual ownership and commitment toward supporting organizational initiatives for workplace injury reduction. Data trends will support the premise that employee involvement creates willingness on the worker's part to support an injury reduction program and to engage in the required cultural change process.

### 39.3.1 Research Objective and Purpose

The overall objective of this research is to document tangible and nontangible occupational and nonoccupational risk factors that contribute to workplace injuries, so as to guide the future development of university-wide safety and health programs. The purpose is to follow the proposed research model [4] and apply the research-to-practice initiative guidelines [11] to reduce injury occurrence, severity of injury, and WC cost in the workplace.

### 39.3.2 Research Goals and Hypotheses

There are four main goals for this project. First, to understand the relationship between workplace injuries in university residence halls and the following variables: the number of recordable cases (incident rate), days away from work and days on restricted work or job transfer (DART) rate, lost time case (LTC) rate, lost work days (LWDs) rate, severity rate, and incident cost. Second goal is to determine if there was a decrease in injuries and/or WC cost paid after the implementation of a safety program. Third is to determine if a person's overall health (physical, stress level, depression level, fatigue) and lifestyle outside of the workplace [4] are associated with the top three causes of workplace injuries. The final goal is to determine the occupational and lifestyle risk factors that are associated with low-back musculoskeletal disorders among university residence hall employees. The hypothesis is there may be a statistically significant difference between the selected occupational and lifestyle risk factors based on the associated cause of injuries.

### 39.3.3 Study Population and Sample Selection

The study consists of benefit eligible full-time university employees covered by WC insurance who worked in 14 university residence halls in a midwest university campus. Within a university residence hall, there are four classifications for job types: administrative, operational/technical staff, service, and clerical staff. Administrative staff includes individuals in charge of directing and assigning work to other employees, as well as those individuals with tasks requiring special knowledge. Operational/technical staff includes individuals in charge of managing departments and organizational units, as well as individuals with very advanced knowledge in a certain area [12]. A clerical staff member is an employee whose assignments typically are associated with clerical activities or are specifically of an administrative nature. A service staff member is an employee who performs duties that result in or contribute to the comfort, convenience, and hygiene of personnel and the student body, or who contribute to the maintenance and care of buildings, facilities, or grounds of the institutional property [13].

Two groups, each of seven university residence halls, were selected based on high WC claims paid for musculoskeletal disorders. University residence halls assigned to group 1 received an ergonomic evaluation and feedback; this group is also referred to as "case." University residence halls assigned to group 2 did not receive any ergonomic evaluation and feedback; this group is also referred to as "control."

## 39.4 SAFETY AND HEALTH MANAGEMENT PROGRAM OVERVIEW

The safety and health management program focuses on a positive safety culture (or climate), including the lifestyle of the employee, in order to minimize the number of preventable workers' injuries and overall WC cost. To reduce the high incidence of injuries and WC claim paid in the workplace, the intervention consists of three phases (safety, health, and sustainability). Phase 1 is an ergonomic evaluation which focuses on safety issues to address injuries in the workplace (assessment of occupational risk factors). The ergonomic evaluation and feedback is designed to engage and observe the university residence hall employees, with the intent of helping them analyze the ergonomic aspects of overall injuries in the workplace and to make ergonomics recommendations based on that analysis. Phase 2 is a work-life program which focuses on employees' health and promotes healthy lifestyles and healthy living. The intent is to determine the lifestyle risk factors and the effects of a safety culture that is beyond the organizational control. Phase 3 is sustainability which focuses on implementing an electronic system to sustain the results of the safety and health management program.

### 39.4.1 Phase 1: Safety (Ergonomic) Evaluation

Phase 1 consists of a comprehensive safety and ergonomics program to identify, control, improve, and sustain safe working conditions. It is proposed that occupational injuries can be controlled through a university-centered program that includes creation of a task force steering committee, thorough review of health and safety records, evaluation of employee work and lifestyle patterns, in-depth job analysis and observation, workers' training and proactive participation, and implementation of engineering controls for manual material handling (such as lift assist devices and workstation redesign). The specific objectives of this phase are (1) to determine ergonomic challenges for university residence hall workers that result in workplace injuries; (2) to provide quantitative measures to evaluate the challenges and job analyses methods, such as the revised NIOSH lifting guidelines [14]; (3) to develop, implement, and evaluate an ergonomics program for prevention and control of workplace injuries; and (4) to gradually integrate an ergonomics program within the university's safety and health program. During this initial phase, employee involvement includes identifying physically demanding tasks and considering possible solutions. Employees are involved in discussions about the design of equipment, tools, and methods in order to develop specifications for future purchases. In addition, employees' involvement in the identification and implementation of solutions will increase the effectiveness of solutions and reduce the resistance to change [17].

### 39.4.2 Phase 2: Health Work-Life Program

The university's health work-life program developed and sponsored a health improvement program with the purpose of determining the impact of a healthy lifestyle on employee's health in the workplace [15]. It is proposed that having a healthier lifestyle will reduce the risk of work-related injuries. It is important to determine how a person's lifestyle inside and outside the workplace has an impact on a person's risk of injury in the workplace. The health risk appraisal (HRA) dataset provides information into the investigation of (1) the personal risk factors associated with incidents in the workplace and (2) the relationship between a healthy lifestyle and the likelihood of workplace injuries. The findings of this investigation will provide information to develop, implement, and evaluate a pilot ergonomics program on high-risk occupations that considers both occupational and non-occupational risk factors for the prevention of the major causes of workplace injuries in university residence halls.

### 39.4.3 Phase 3: Sustainability

In the final year of the data collection, the initiative is to empower the employee to "own" the safety culture/climate with a focus on sustaining the results through a risk reduction program. One of the goals of sustainability is to continuously improve processes to reduce the risk of "X" injury while doing "Y" activity by substantially involving employees at all levels in developing the risk reduction strategies. As a result, workers' safety is improved, and the WC cost incurred is reduced. Another goal of sustainability is to reverse short-term financial results that can be achieved by driving closed loop, cyclical thinking rather than linear, goal-oriented thinking. As a result, sustainability actually advances whole-system thinking, as it requires the employees to consider the long-term unintended consequences of their decisions. Therefore, with regard to sustainability thinking, anything that prevents the ability of the employee to sustain a healthy workplace and/or healthy lifestyle should be reduced or eliminated. Hence, individual groups can intentionally redesign their processes so that their output from process become useful safety inputs into other processes. Metrics will be used to assess if the reality of the vision of the intervention can become real by transferring the inferences into practice. Figure 39.1 illustrates the proposed approach for risk reduction and sustainability toward a safe culture in the workplace.

Within the sustainability phase, there are four approaches which are outlined as follows:

1. "Reacting to a problem" approach
   On a daily basis, the employees (representative from each department/process) focus on safety activities rather than the processes. The potential safety issues are entered into the Safety Project Tracking Database, and the employees react to immediate needs and problems which are also documented in the Safety Project Tracking Database.
2. "Early systematic" approach (planning intervention)
   On a daily basis, regular and timely safety meetings (30 min to 1 h) are conducted with employee representatives attending from each department/process. During these meetings, the team reviews/discusses the potential safety risk(s) and proposes potential process improvement(s). All recommended actions are also documented in the Safety Project Tracking Database. On a monthly basis, achievable safety goals are defined among employees from all university residence halls. The Safety Project

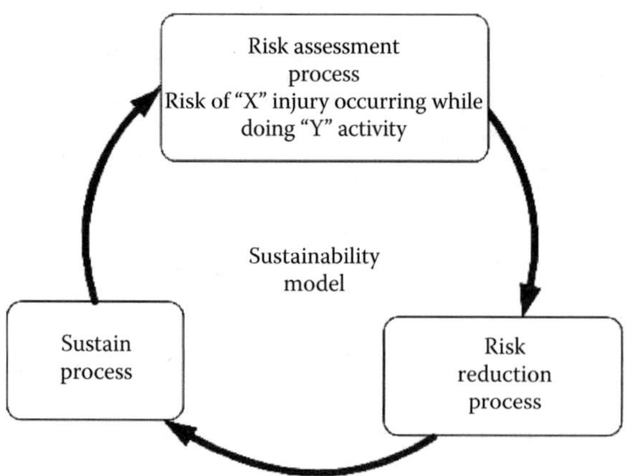


FIGURE 39.1 Sustainability model: safety culture.

Tracking Database is reviewed for similar projects that were completed or for existing projects currently in process, with an attempt to reuse processes through repeatability, evaluation, and improvement.

3. "Aligned project priorities" approach based on safety datasets (OSHA 200/300 logs and Workers' Compensation Claims)
   Safety projects address key safety goals for the university residence halls as they align with the injury and illness dataset. Operations are characterized by processes that are repeatable and regularly evaluated for improvement. Cross-functional learning is shared and coordinated among organizational units through the project documented in the Safety Project Tracking Database.
4. "Integration" approach
   A web-based, single repository Safety Project Tracking Database is used to share repeatable safety processes across all university residence halls. Operational processes are regularly evaluated for change and improvement in collaboration with other university residence halls. Efficiencies across departments are achieved through analysis, innovation, and sharing. Processes and measurements track progress on key strategic and operational safety goals.

The following section outlines the two key processes of the "early systematic" approach: (1) the risk reduction program structure which the daily safety meetings are based and (2) the description of the functionality of the Safety Project Tracking Database.

#### *39.4.3.1 Risk Reduction Program Structure*

In this program, the general message to all employees is that "employees are the leaders of safety"; the focus is on "what is the right way to reduce safety risks in the workplace" rather than "what may be wrong." People generally want to work in safe workplaces; therefore, they will be attracted to a company where the system supports and promotes "safety first." Specific, observable behaviors known to prevent injury eventually become a habit, so the goal of continuous improvement is achieved as the process repeats and refreshes itself. As a result, employee leaders emerge from every operational level of the university residence hall. There are scheduled team meetings where the employees identify safety hazards in the workplace by either looking at past injuries, listening to employee concerns, brainstorming with the committee, or using a rating process for existing projects in the Safety Project Tracking Database. The rating process assigns a risk score to represent the overall risk effect based on the following three variables: (1) frequency of injury, (2) the likelihood of injury for an employee exposed to the risk, and (3) the worst plausible injuries. Next, the team identifies processes to reduce the risk (by affecting at least one variable). Best practices are developed, and a training program is also developed to communicate the best practices to all affected employees. After training all affected employees, the safer methods are implemented. An observation system is used to track the degree to which employees use the recommended safety equipment and methods developed by the risk reduction project.

The web-based Safety Project Tracking Database serves as the single data repository where each project (planned, in progress, or completed) is assigned a risk score and documented in terms of the potential risk of "X" injury while doing "Y" activity. Any associated implementation plans and documentation generated or used for each project (completed or in progress) will be linked to the respective project and is accessible to authorized users. The web-based Safety Project Tracking Database was populated with the ergonomic evaluation observations (2005–2008) and with the employees' own safety evaluation (2009–2010), based on the following steps. First, a feedback report was produced from the ergonomic evaluation, which was presented to the university residence hall managers, including observed risk for injury, recommended interventions, discomfort surveys, biomechanics analysis, and photographs of the safety risks. Each observed risk was coded as "the risk of a potential injury (X) while doing task (Y)" and entered into the Safety Project Tracking Database. Each project is assigned a total risk score which was based on the product of the respective individual score for each of the three variables: frequency of risk exposure, likelihood of injury type(s), and severity of the risk. Risk reduction interventions related to a particular observed risk included changes to one or more of the following: equipment, process, training, and/or behavior.

#### *39.4.3.2 Web-Based Safety Project Tracking Database*

The risk observations from the ergonomic projects, the rating for each observation, and the other components can be accessed from authorized locations using a Web browser. The Safety Project Tracking Database start up menu contains the following options: project generator, top-down risk analysis, project matrix, and project archive.

39.4.3.2.1 Project Generator The university residence hall safety manager reviews the Safety Project Tracking Database containing all safety risk observations to ensure that the reporting format, "the risk of a potential injury (X) while doing task (Y)," is documented accurately. Based on the total risk score for each project and the impact of the safety risk to the employee as it relates to WC cost and the frequency of occurrence, projects are given priority and the appropriate risk reduction intervention(s) is(are) determined. The proposed interventions for the risk reduction project are entered into the Safety Project Tracking Database by a department representative or by an authorized database manager. The Safety Project Tracking Database also tracks details about project implementation, estimated timeline for implementation in days, planned start date for implementation, actual start date of implementation, planned end date for implementation, the actual end date of implementation, anticipated cost of implementation, and actual cost of project implementation. Also available is the option to follow up on the progress of complete projects and assign a total risk score for sustainability efforts.

39.4.3.2.2 Project Matrix The project matrix provides a link for each completed project or project in process based on "cause of injury" categories. All information is displayed in an easy-to-read tabulated summary.

39.4.3.2.3 Project Archive An archive of the project implementation is documented and stored in the database, which is linked to the project observation results. After the project has been implemented for a reasonable period of time, a new risk score is calculated as a subjective indicator of project success, and more importantly, to identify remaining residual risk that may be considered for further reduction. This also serves as a management system to generate a top-down view of projects' results to ensure that projects are being sustained.

## 39.5 DATA COLLECTION

Data collection is grouped into two areas: the safety datasets and the health dataset. Variables for the safety datasets were collected from Occupational Safety and Health Administration (OSHA) 300 logs [16] and associated WC cost. According to OSHA [16], the primary indicator of safety performance is the incident rate. OSHA incident rates provide information on the frequency of incidents occurrence and the severity of the incidents based on the following five metrics: total recordable incident rate, DART rate, LTC rate, LWD rate, and severity rate. Recordable incidents include "all work-related deaths, illnesses, and injuries which result in a loss of consciousness, restriction of work or motion, permanent transfer to another job within the company, or requirement of some type of medical treatment or first-aid." DART describes the number of recordable injuries and illnesses, per 100 full-time employee (FTE), that resulted in days away from work, restricted work activity, and/or job transfer, as experienced by a company in any given time frame. LTC rate describes the amount of lost time, per 100 FTE, in any given time frame. For the LTC, the LWD rate is based on the number of days an employee lost work because of an incident, and it is capped at 180 days for incidents that result in lost time over a longer period of time. Severity rate calculates the average number of lost days per recordable incident. WC cost was used as a metric because it was considered to be a more stable indicator of the severity of injury claims. However, only a portion of observed WC cost savings can be regarded as savings potentially attributable to the intervention since WC cost tends to fluctuate due to differing claims management practices. Disability payments, time loss payments, awards, and other extraneous factors are not included. According to Bidassie et al. [17], the incident cost, a measurement for year-to-year cost per work hour comparisons, was used to calculate the average WC claims paid (the ratio of the WC claims paid to the total number of recordable injuries).

The university's work-life program promoted lifestyle changes in employees to improve their health [15]. An HRA questionnaire [4] was administered to provide feedback to employees on how their lifestyle affects their overall health and wellness. Wellness coaches work with each employee on strategies to develop healthier habits. The following variables for the health dataset were collected from the HRA questionnaire: demographics (gender, age, job status, work experience, shift, ethnicity, marital status, number of dependents), biometrics (cholesterol, systolic and diastolic blood pressure, weight, height, body mass index (BMI)), personal lifestyle (amount of physical activity, tobacco and smoking habits, alcohol use, depression, stress factors, hours of sleep, driving habits, eating habits, personal care), current health, health history, and family health history.

## 39.6 DATASET CONSTRUCT

Six years of baseline data were grouped for the time period January 1, 1999 through December 31, 2004, and 4 years of intervention data were collected from January 1, 2005 through December 31, 2008. The final dataset constructed was a combined dataset of both safety dataset and HRA dataset (employees who reported an injury *and* employees who participated in the work-life program) to create a single dataset containing both workplace (occupational) and lifestyle risk factors. Employees who reported an injury and who also completed an HRA questionnaire were profiled to determine a relationship between workplace risk factors and health risk. From January 1, 2005 through December 31, 2008, there were 1382 incidents of injuries among university residence halls. Fifty-nine percent (n = 813) of the injuries were reported by employees from university residence halls in group 1, and 41.2% (n = 569) were from employees from the university residence halls in group 2. There was no statistically significant difference between the average age, the annual wages, and the average years of work experience between the case and control group.

## 39.7 STATISTICAL ANALYSIS

The datasets were analyzed using SPSS 16.0.1. A t-test is used to determine if there is any statistically significant difference between the means of the baseline data compared to the intervention data for both case and control group. A one-way ANOVA is used to determine if there is a relationship among the top three causes of workplace injuries with occupational and lifestyle risk factors. A profile consisting of demographic, occupational, and lifestyle risk factors will be summarized for each of the three major workplace risk factors in university residence halls. Finally, a logistic regression model for low-back musculoskeletal disorders in university residence halls will be determined based on the factors outlined in Bidassie doctoral dissertation [4]. For the logistic equation $Y = \boldsymbol{\beta}_0 + \boldsymbol{\beta}_1\mathbf{X}_1 + \boldsymbol{\beta}_2\mathbf{X}_2 + \cdots + \boldsymbol{\beta}_n\mathbf{X}_n$, the logistic regression coefficient (β) predicts the log odds of the low-back MSDs, where Y = logit(π) is the log odds (logit) of low-back musculoskeletal disorders $\boldsymbol{\beta}_0$ is the constant **n** are predictor risk factors ($X_i$), some of which may be interaction terms **β** are logistic regression coefficients, also called parameter estimates.

## 39.8 RESULTS

### 39.8.1 Incident Rate Metrics

There were no statistically significant difference between case and control during the intervention for the number of recordable cases, DART rate, LTC rate, LWD rate, severity rate, and incident cost. During the baseline stage (Table 39.1), there was no statistically significant difference between case and control for the number of lost days. However, there was statistically significant difference between case and control for the number of restricted days from work and average of WC claims paid. University residence hall group 1 (case) had more restricted days (mean = 2.27, SD = 12.3, $p < .013$) and a higher average of WC claims paid (mean = \$1418.9, SD = \$6182, $p < .000$) compared to the control group.

TABLE 39.1 Comparison between Group 1 (Case) and Group 2 (Control) Metrics

| | Metrics | University Residence Halls | N | Mean | Std. Deviation | F | p-Value |
|---|---|---|---|---|---|---|---|
| *Baseline* | Number of lost days | Case | 920 | 1.18 | 8.38 | 1.48 | ns |
| | | Control | 816 | 1.49 | 11.38 | | |
| | Number of restricted days | Case | 920 | 2.27 | 12.26 | 6.16 | * |
| | | Control | 817 | 1.53 | 8.32 | | |
| | Actual WC paid | Case | 813 | $1,419 | $6,182 | 20.11 | *** |
| | | Control | 707 | $751 | $3,782 | | |
| *Intervention* | Number of lost days | Case | 734 | 1.55 | 10.73 | 7.01 | ** |
| | | Control | 537 | 2.67 | 16.68 | | |
| | Number of restricted days | Case | 734 | 1.75 | 9.99 | 4.30 | * |
| | | Control | 537 | 2.55 | 15.77 | | |
| | Actual WC paid | Case | 534 | $1,346 | $5,740 | 9.35 | ** |
| | | Control | 394 | $2,213 | $10,257 | | |

*p < .05 **p < .01 ***p < .001.
*ns:* not statistically significant.

During the intervention (Table 39.1), there was a statistically significant difference for all three metrics. The university residence halls that received ergonomic evaluation and feedback (case) had fewer lost days (mean = 1.55, SD = 10.7, p < .008), fewer restricted days (mean = 1.75, SD = 10, p < .038), and a lower average of WC claims paid (mean = $1346, SD = $5740, p < .002) compared with the university residence halls that did not receive any ergonomic evaluations (control).

## 39.8.2 Workplace Risk Factors

Based on the analysis of the baseline data for group 1 (case), the top four workplace risk factors (lifting/twisting/bending, slips/trips/falls, struck by something or someone, and repetitive motion) accounted for 73% (n = 718) of the total injuries and contributed to 72% ($1,109,392) of the total WC claims paid. The risk assessment indicated that risk reduction projects need to focus on FTEs between the ages of 30 and 40 years, with an average of 5–10 years of work experience and why the injuries tend to be reported during the winter months. During the intervention, the set "target goals" were achieved, resulting in a reduction in injuries among FTEs between the ages of 30 and 40 years, from 21.2% (n = 208) to 8.5% (n = 69), and there was also a reduction in possible risk factors during the winter months, from 26.5% (n – 259) to 21.2% (n = 172). There was an overall decrease in the percentage of reported injuries and a decrease in the percentage of WC cost claims paid. For injuries due to lifting/twisting/bending, there was a 12% decrease in the number of reported injuries, however, there was an increase in the actual WC claims paid, mostly due to some costly back injuries. For injuries due to slips/trips/falls, there was a 1% decrease in the number of reported injuries and also a decrease in WC cost. For injuries due to struck by something or someone, overall, there was a 5% increase in the number of reported injuries and an

TABLE 39.2 Summary of Workplace Risk Factors

| Workplace Top Risk Factors | Phase | Injuries Reported | | Workers' Compensation Claims Paid | | | |
|---|---|---|---|---|---|---|---|
| | | n | % | n | % | $$ | % Paid |
| Lifting/twisting/ bending | Baseline | 241 | 25 | 224 | 28 | $216,858 | 19 |
| | Intervention | 102 | 13 | 86 | 16 | $296,169 | 41 |
| Slip/trip/fall | Baseline | 213 | 22 | 182 | 22 | $640,734 | 56 |
| | Intervention | 169 | 21 | 125 | 23 | $198,996 | 28 |
| Stuck by something or someone | Baseline | 198 | 20 | 131 | 16 | $122,552 | 11 |
| | Intervention | 200 | 25 | 139 | 26 | $80,800 | 11 |
| Repetitive motion | Baseline | 66 | 7 | 51 | 6 | $129,250 | 11 |
| | Intervention | 17 | 2 | 10 | 2 | $22,452 | 3 |
| Cut/puncture (knife, syringe, etc.) | Baseline | 109 | 11 | 86 | 11 | $14,821 | 1 |
| | Intervention | 152 | 19 | 51 | 10 | $42,743 | 6 |
| Infection/burn | Baseline | 85 | 9 | 77 | 9 | $9,229 | 1 |
| | Intervention | 97 | 12 | 55 | 10 | $10,409 | 1 |
| Other | Baseline | 67 | 7 | 62 | 8 | $20,118 | 2 |
| | Intervention | 76 | 9 | 68 | 13 | $67,322 | 9 |
| *Total* | Baseline | 979 | 100 | 813 | 100 | $1,153,562 | 100 |
| | Intervention | 813 | 100 | 534 | 100 | $718,891 | 100 |

increase in WC cost. For repetitive motion, there was a 5% decrease in both the number of injuries and the WC claims paid (Table 39.2).

### 39.8.3 Occupational and Nonoccupational Risk Factors (Top Three Injuries in the Workplace)

During 2006–2008, there were 468 university residence hall employees from the 14 university residence halls who reported an injury in the workplace and who also participated in the work-life program.

#### *39.8.3.1 Occupational Risk Factors*

Sixty-six percent (n = 309) of the causes for the injuries were based on the following three risk factors: lifting/twisting/bending (n = 66, 14%, $228,933), slips/trips/falls (n = 120, 25.6%, $326,708), and struck by something or someone (n = 123, 26.3%, $93,551). Of the 468 injuries, 9% (n = 41) were low-back musculoskeletal disorders, which accounted for 38% (WC cost = $314,282), where 73% (n = 30) were associated with lifting/twisting/bending activities and 24% (n = 10) of the injuries were due to slips/trips/falls.

#### *39.8.3.2 Demographic Profile*

The demographic profile of employees from university residence halls (Table 39.3) who were injured due to the three major workplace risk factors (lifting/twisting/bending, slips/trips/falls, struck by something or someone) indicates that 78% (n = 241) were female

TABLE 39.3 Demographics Description for Top Three Workplace Injuries

| Demographic Variables | LTB (n = 66) | | STF (n = 120) | | SBS (n = 123) | | Total (n = 309) | | | | |
|---|---|---|---|---|---|---|---|---|---|---|---|
| | n | % | n | % | n | % | n | % | $\chi^2$ | df | p-Value |
| *Gender* | | | | | | | | | | | |
| Female | 42 | 63.6 | 103 | 85.8 | 96 | 78.0 | 241 | 78 | 12.2 | 2 | ** |
| Male | 24 | 36.4 | 17 | 14.2 | 27 | 22.0 | 68 | 22 | | | |
| *Marital status* | | | | | | | | | | | |
| Single | 16 | 24.2 | 33 | 27.5 | 51 | 41.5 | 100 | 32 | **10.1** | 3 | * |
| Married | 39 | 59.1 | 63 | 52.5 | 50 | 40.7 | 152 | 49 | | | |
| Div/sep/other | 8 | 12.1 | 11 | 9.2 | 8 | 6.5 | 27 | 9 | | | |

*p < .05 **p < .01.

($\chi^2$ = 12.2, df = 2, p < .002), 58% (n = 179) were married or divorced/separated ($\chi^2$ = 10.1, df = 3, p < .039) and they tend to have an average age of 48 years old (SD = 11.3).

#### 39.8.3.3 Lifestyle Risk Factors

There were no statistically significant differences among the three workplace risk factors for social support network, eating habits, feeling depressed, and their personal care. However, employees who had a workplace injury due to lifting/twisting/bending tended to have three or more drinks on the days that they consume alcohol, compared with the employees who had an injury due to slips/trips/falls or struck by something or someone, who tended to have one to two drinks per day on the days that they consumed alcohol ($\chi^2$ = 6.0, df = 2, p < .049).

#### 39.8.3.4 Financial, Occupational, and Nonoccupational Stressors

There was a statistically significant difference between the type of workplace injury and the total number of major stressors reported by the employee (F = 5.4, df = 2, p < .005). For each of the three workplace injuries, the employees reported the same top four stressors and had a differing stressor for the fifth. The common four major stressors are financial (finances, loan, mortgage), job responsibilities, relationships at work, and family problems. The fifth major stressor differed for each group: people who had an injury due to lifting/twisting/bending reported their fifth stressor as "personal illness and injury," those with an injury due to slips/trips/falls reported "illness or injury of a loved one," and those with a struck by something or someone injury reported "death of a spouse, life partner, or other loved one."

#### 39.8.3.5 Perceived Stress Level

There was a statistically significant difference between the type of workplace injury and the stress level reported by the employee ($\chi^2$ = 11.6, df = 4, p < .02). Employees who had an injury due to lifting/twisting/bending reported feeling quite stressed to extremely stressed (36.4%, n = 24) from minor annoyances, major pressures, problems or difficulties, compared with employees who had an injury due to slips/trips/falls (22.5%, n = 27) or an injury due to struck by something or someone (18.7%, n = 23).

#### 39.8.3.6 Current Health and Biometric Status

There is a statistically significant difference for BMI among the three workplace injury group ($F = 571.3$, $df = 2$, $p < .022$). Employees who had an injury due to slips/trips/falls (mean = 32.3, SD = 7.9) tended to have a higher BMI compared with the employees who had an injury due to lifting/twisting/bending (mean = 30.1, SD = 10.2) or struck by something or someone (mean = 29.4, SD = 8.3).

#### 39.8.3.7 Fatigue

There is a relationship between the numbers of hours of sleep employees got on a daily basis and how often they felt tired during their waking hours. Fewer hours of sleep resulted in employees feeling tired quite often or almost always during their waking hours ($\chi^2 = 28.9$, $df = 15$, $p < .017$). People who had a workplace injury due to lifting/twisting/bending or slips/trips/falls reported being more tired quite often or almost often during the waking

TABLE 39.4 Occupational and Lifestyle Risk Factors That Predict Low-Back Musculoskeletal Disorders among University Residence Hall Employees

| Predictors Variables for Low-Back Musculoskeletal Disorders | n | B | Exp(B) = Odds | 95.0% C.I.EXP(B) | | p-Value |
|---|---|---|---|---|---|---|
| | | | | Lower | Upper | |
| *Demographics* | | | | | | |
| Age 30–40 years | 37 | 1.64 | 5.16 | 1.21 | 21.98 | * |
| Autumn (September–November) months | 89 | 1.63 | 5.13 | 1.93 | 13.62 | ** |
| Have one or more dependents | 161 | 1.09 | 2.98 | 1.19 | 7.49 | * |
| *Workplace risk factors* | | | | | | |
| Lifting/twisting/bending | 58 | 5.68 | 291.50 | 55.08 | 1542.78 | *** |
| Slipped/trips/fall | 103 | 2.99 | 19.80 | 4.06 | 96.64 | *** |
| *Social support network* | | | | | | |
| Have contact with close relatives: one per month or less | 59 | 1.37 | 3.93 | 1.07 | 14.35 | * |
| *Lifestyle* | | | | | | |
| Does not smoke either cigarettes, cigar, pipe, smokeless or chewing tobacco | 257 | 1.24 | 3.44 | 1.15 | 10.26 | * |
| Sleep less than 8 h | 293 | 1.14 | 3.14 | 1.00 | 9.87 | * |
| Slipping up under stress a major barrier to making lifestyle change | 171 | 0.83 | 2.30 | 0.88 | 6.03 | 0.089 |
| *Physical and emotional* | | | | | | |
| Ongoing problem with back pain that is serious enough to interfere with daily activities | 71 | 1.06 | 2.87 | 0.98 | 8.41 | * |
| Does not feel stressed from minor annoyances, major pressures, problems, or difficulties | 32 | 3.16 | 23.50 | 4.75 | 116.19 | *** |
| *Constant* | | −9.34 | 0.00 | | | *** |

*p < .05 **p < .01 ***p < .001.

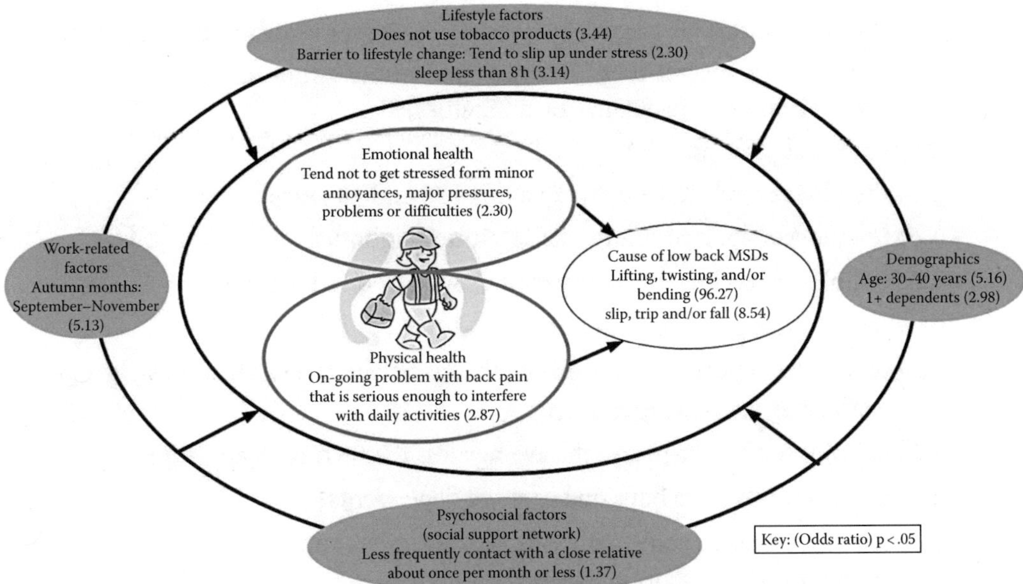


FIGURE 39.2 Predictive risk factors for low-back musculoskeletal disorders among university residence hall employees.

hours compared to those employees who had a workplace injury due to struck by something or someone ($\chi^2 = 13.2$, df = 4, $p < .010$).

### 39.8.4 Predictive Risk Factors for Low-Back Musculoskeletal Disorders among University Residence Hall Employees

Based on the conceptual framework proposed by Bidassie [4], the predictive model for low back musculoskeletal disorder is illustrated in Figure 39.2. A logistic regression model (see Table 39.4) was developed for university residence hall employees who reported low-back musculoskeletal disorders. The model ($\chi^2 = 144.44$, df = 11, $p < 0.000$) adequately fits the data since the Hosmer–Lemeshow statistical significance value is greater than 0.05 [18]. The odds ratio, Exp (B), greater than 1 indicates that the odds of a low-back musculoskeletal disorder occurring increases when the independent variable increases. An odds ratio less than 1 indicates that low-back musculoskeletal disorders are less likely to occur than it is to not occur. The sample size is based on employees reporting the occurrence of low-back musculoskeletal disorders. Typically, the average percentage of annual reporting of low-back musculoskeletal disorders varied from 8% to 10%. Logistic regression tends to systematically overestimate odds ratios or beta coefficients when the sample size is small. The magnitude of overestimation diminishes, and the estimated odds ratio asymptotically approaches the true population value when the sample size increases.

Equation 39.1 outlines the relationship between prevalence of low-back musculoskeletal disorder with occupational and lifestyle risk factors ($\beta$ values sorted descending chorological order).

$$
\begin{aligned}
\text{Logit(low-back musculoskeletal disorder)} = &-9.34 + 5.68 \text{ (cause: lifting/twisting/bending)} \\
&+ 3.16 \text{ (employee tends } \textit{not} \text{ to get stressed from minor annoyances,} \\
&\quad \text{major pressures, problems, or difficulties)} \\
&+ 2.99 \text{ (cause: slips/trips/falls)} \\
&+ 1.64 \text{ (employee tends to be between ages of 30 and 40 years)} \\
&+ 1.63 \text{ (employee reported injury during autumn months)} \\
&+ 1.37 \text{ (employee has less frequently contact with a close relative} \\
&\quad \text{about once per month or less)} \\
&+ 1.24 \text{ (employee reported to } \textit{not} \text{ using tobacco products, i.e., cigarettes/cigar/pipe/} \\
&\quad \text{smokeless or chewing tobacco)} \\
&+ 1.14 \text{ (employee tends to get on the average less than 8 h of sleep daily)} \\
&+ 1.09 \text{ (employee tends to have one or more dependents)} \\
&+ 1.06 \text{ (employee has ongoing back pain that is serious enough} \\
&\quad \text{to interfere with daily activities)} \\
&+ 0.83 \text{ (employee tends to report ``to slipping up when under stress'')}
\end{aligned}
\tag{39.1}
$$

Interpretation of the model (assuming all other things being equal):

*Demographics*:

- An employee who is between the age 30 and 40 years old is **5.16 (95% CI: 1.21, 21.98)** times more likely to report low-back musculoskeletal disorders compared with an employee who is younger than 30 years or older than 40 years.
- An employee is **5.13 (95% CI: 1.93; 13.62)** times more likely to report low-back musculoskeletal disorders during the autumn months (September–November) compared with spring, summer, and winter (December–August).
- An employee who has one or more dependents is **2.98 (95% CI: 1.19, 7.49)** times more likely to report low-back musculoskeletal disorders compared with an employee who does not have dependents.

*Workplace risk factors*:

- An employee whose job involves lifting, twisting, and/or bending is **291.50 (95% CI: 55.08, 1542.78)** times more likely to report low-back musculoskeletal disorders compared with an employee whose job does not involve lifting, twisting, and/or bending.
- An employee who slips, trips, and/or falls is **19.80 (95% CI: 4.06, 96.64)** times more likely to report low-back musculoskeletal disorders compared with an employee who did not slip, trip, and/or fall.

*Social support network*:

- An employee who has contact with a close relative less frequently (about once a month or less) is **3.93 (95% CI: 1.07, 14.35)** times more likely to report low-back musculoskeletal disorders compared with an employee who has contact with a close relative more frequently (once a week or more to two to three times a month)

*Lifestyle risk factors*:

- An employee who does not use tobacco products (cigarettes, cigar, pipe, smokeless or chewing tobacco) is **3.44 (95% CI: 1.15, 10.26)** times more likely to report low-back musculoskeletal disorders compared with an employee who uses tobacco products.
- An employee who gets on the average less than 8 h of sleep daily is **3.14 (95% CI: 1.00, 9.87)** times more likely to report low-back musculoskeletal disorders compared with an employee who gets more than 8 h of sleep.
- An employee who reports "slip up when under stress as a barrier for lifestyle changes" is **2.30 (95% CI: 0.88, 6.03)** times more likely to report low-back musculoskeletal disorders compared with an employee who does not "slip up when under stress as a barrier for lifestyle changes."

*Physical and emotional risk factors*:

- An employee who reports "ongoing back pain that is serious enough to interfere with their daily activities" is **2.87 (95% CI: 0.98, 8.41)** times more likely to report low-back musculoskeletal disorders compared with an employee who does not have "an ongoing back pain that is serious enough to interfere with their daily activities."
- An employee who does *not* get stressed from minor annoyances, major pressures, problems, or difficulties is **23.50 (95% CI: 4.75, 116.19)** times more likely to report low-back musculoskeletal disorders compared with an employee who tends to get stressed from minor annoyances, major pressures, problems, or difficulties.

## 39.9 LIFESTYLE CHANGES

The university contracted an independent wellness management company to provide coaching and services to employees to improve the following lifestyles: (1) eating a healthier diet, (2) losing weight, (3) getting more exercise, (4) quitting smoking or tobacco use, (5) handling stress better, (6) acting to reduce back pain, and (7) lowering blood pressure and cholesterol levels. Along with sharing the results of the HRA survey with an employee, a wellness representative also provides health and wellness coaching strategies based on a suggested roadmap and services that the university will make available to facilitate possible lifestyle changes based on the individual's HRA results.

TABLE 39.5 Lifestyle Changes Important to Make for Top Three Workplace Injuries

| Lifestyle | LTB (n = 66) | | STF (n = 120) | | SBS (n = 123) | | Total | | | | |
|---|---|---|---|---|---|---|---|---|---|---|---|
| Important to Make | n | % | n | % | n | % | n | % | $\chi^2$ | df | p-Value |
| *Eat a healthier diet* | | | | | | | | | | | |
| No | 15 | 22.7 | 23 | 19.2 | 43 | 35.0 | 81 | 26.2 | 8.36 | 2 | * |
| Yes | 51 | 77.3 | 97 | **80.8** | 80 | 65.0 | 228 | **73.8** | | | |
| *Get more exercise* | | | | | | | | | | | |
| No | 16 | 24.2 | 23 | 19.2 | 43 | 35.0 | 82 | 26.5 | 8.00 | 2 | * |
| Yes | 50 | 75.8 | 97 | **80.8** | 80 | 65.0 | 227 | **73.5** | | | |
| *Handle stress better* | | | | | | | | | | | |
| No | 25 | 37.9 | 68 | 56.7 | 68 | 55.3 | 161 | 52.1 | 6.85 | 2 | * |
| Yes | 41 | **62.1** | 52 | 43.3 | 55 | 44.7 | 148 | 47.9 | | | |

*p < .05.

## 39.9.1 Lifestyle Changes That Injured Employees Feel Are Important to Make

When asked which of the seven lifestyle changes employees are very confident that they can make, there were three lifestyle changes that were statistically significant within the three major workplace risk factors: eat healthier diet, get more exercise, and handle stress better (Table 39.5).

Employees who had an injury due to slips/trips/falls tend to want to change their lifestyle so that they could eat a healthier diet ($\chi^2 = 8.36$, df = 2, $p < .015$) and get more exercise ($\chi^2 = 8.00$, df = 2, $p < .015$). Employees who had an injury due to lifting/twisting/bending tend to want to change their lifestyle so that they could better handle their stress from minor annoyances, major pressures, problems, or difficulties ($\chi^2 = 6.85$, df = 2, $p < .033$).

## 39.9.2 Lifestyle Changes That Injured Employees Are Confident to Make

For the lifestyles that employees are very confident that they can make, there were two lifestyle changes that were statistically significant within the three major workplace risk factors: eat healthier diet and lose more weight (Table 39.6).

TABLE 39.6 Lifestyle Confident to Make for Top Three Workplace Injuries

| Lifestyle | LTB (n = 66) | | STF (n = 120) | | SBS (n = 123) | | Total | | | | |
|---|---|---|---|---|---|---|---|---|---|---|---|
| Confident to Make | n | % | n | % | n | % | n | % | $\chi^2$ | df | p-Value |
| *Eat a healthier diet* | | | | | | | | | | | |
| No | 29 | 43.9 | 33 | 27.5 | 48 | 39.0 | 110 | 35.6 | 6.07 | 2 | * |
| Yes | 37 | 56.1 | 87 | **72.5** | 75 | 61.0 | 199 | 64.4 | | | |
| *Lose weight* | | | | | | | | | | | |
| No | 36 | 54.5 | 38 | 31.7 | 54 | 43.9 | 128 | 41.4 | 9.70 | 2 | ** |
| Yes | 30 | 45.5 | 82 | **68.3** | 69 | 56.1 | 181 | 58.6 | | | |

*p < .05 **p < .01.

Employees who had an injury due to slips/trips/falls tend to be confident that they could change their lifestyle to eat a healthier diet ($\chi^2 = 6.07$, df = 2, $p < .048$) and get more exercise ($\chi^2 = 9.70$, df = 2, $p < .008$).

### 39.9.3 Lifestyle Changes That Injured Employees Plan to Make/Made

According to Prochaska and DiClemente [19,20] and El-Bassel et al. [21], people tend to go through five distinct stages of change: precontemplation, contemplation, determination, action, and maintenance. Employees' responses on their planned action items based on the suggested roadmap from the health coaches' recommendations were categorized into two groups. Group A includes people in the precontemplation, contemplation, and determination stages, while group B includes people in the action and maintenance phase. Group A is considered "no plans to change or may start in future," and group B is considered "started or completed." There was a statistically significant difference between group A and group B for lifestyle changes when it comes to handling stress better ($\chi^2 = 10.25$, df = 2, $p < .006$) and taking action to reduce back pain ($\chi^2 = 7.55$, df = 2, $p < .023$). Employees who had workplace injuries due to lifting, twisting, and/or bending; slips, trips, and/or fall; and struck by something tend to report that they had no plans to change their life or may start a lifestyle change in future to handle stress and take action to reduce back pain (Table 39.7).

### 39.9.4 Barriers to Lifestyle Changes

Seventy-five percent (n = 232) of the employees reported to have one to three major barriers to making lifestyle changes. The top four barriers were (1) not having enough time (63.4%, n = 196); (2) difficulty of giving up temptations (52.4%, n = 162); (3) difficulty of staying committed (46.3%, n = 143); and (4) slipping up under stress (43%, n = 134). There was a statistically significant difference (Table 39.8) among the three major workplace risk factors for two major barriers to lifestyle changes "not having enough time" ($\chi^2 = 8.47$, df = 2, $p < .014$) and "slipping up under stress" ($\chi^2 = 5.89$, df = 2, $p < .053$). Fifty-two percent (n = 34) of employees who had an injury due to lifting/twisting/bending reported

TABLE 39.7 Lifestyle Changes Plan to Make for Top Three Workplace Injuries

| Lifestyle Changes Plan to Make | LTB (n = 66) | | STF (n = 120) | | SBS (n = 123) | | Total | | | | |
|---|---|---|---|---|---|---|---|---|---|---|---|
| | n | % | n | % | n | % | n | % | $\chi^2$ | df | p-Value |
| *Handle stress better* | | | | | | | | | | | |
| No plans to change or may start in future | 49 | 76.6 | 74 | 64.3 | 95 | 82.6 | 218 | 74.1 | 10.25 | 2 | ** |
| Started or completed | 15 | 23.4 | 41 | **35.7** | 20 | 17.4 | 76 | 25.9 | | | |
| *Act to reduce back pain* | | | | | | | | | | | |
| No plans to change or may start in future | 44 | 75.9 | 78 | 69.6 | 92 | 85.2 | 214 | 77.0 | 7.55 | 2 | * |
| Started or completed | 14 | 24.1 | 34 | **30.4** | 16 | 14.8 | 64 | 23.0 | | | |

*p < .05 **p < .01.

TABLE 39.8 Barriers to Lifestyle Changes for Top Three Workplace Injuries

| Major Barriers | LTB (n = 66) | | STF (n = 120) | | SBS (n = 123) | | Total | | | | |
|---|---|---|---|---|---|---|---|---|---|---|---|
| | n | % | n | % | n | % | n | % | $\chi^2$ | df | p-Value |
| *Do not have enough time* | | | | | | | | | | | |
| No | 32 | 48.5 | 84 | 70.0 | 78 | 63.4 | 194 | 62.8 | 8.47 | 2 | * |
| Yes | 34 | 51.5 | 36 | 30.0 | 45 | 36.6 | 115 | 37.2 | | | |
| *Slipping up under stress* | | | | | | | | | | | |
| No | 34 | 51.5 | 61 | 50.8 | 80 | 65.0 | 175 | 56.6 | 5.89 | 2 | * |
| Yes | 32 | 48.5 | 59 | 49.2 | 43 | 35.0 | 134 | 43.4 | | | |

*p < .05.

"not having enough time" as a major barrier to lifestyle changes compared to employees who had an injury due to slips/trips/falls (30%, n = 36) or struck by something or someone (36.6%, n = 45). Also, employees who had an injury due to slips/trips/falls (49.2%, n = 59) reported that "slipping up under stress" was a major barrier to lifestyle changes compared to employees who had an injury due to lifting/twisting/bending (48.5%, n = 44) or struck by something or someone (35.0%, n = 43).

## 39.10 LIMITATIONS

There are numerous constraints with regard to the collection of different datasets and the correlation of the information from the different datasets, as well as the standardization of the different methods used by recordkeeping personnel, who had changed over the years. For the baseline data the amount of information provides allow for a conservative conclusion to be made with respect to the constraints of the datasets.

## 39.11 CONCLUSION

Based on the results, it was confirmed that the university residence halls that received ergonomic evaluations did show a lower WC cost in each of the 4 years when compared with the university residence halls which did not receive any ergonomic evaluations. In university residence halls, preventative strategies need focus on reducing the cause of injuries due work task that require lifting/twisting/bending and workplace condition and activities that may lead to slips/trips/falls or being struck by someone or something; especially those activities or conditions which may lead to a low back musculoskeletal disorder. The ergonomics evaluation in the seven selected university residence halls increased the awareness of a safety culture and ergonomic practices throughout the other university residence halls. This awareness led to an increase in the number of university employees proactively employing good ergonomic practices and improving safety and health procedures in their work environment. Hence, the safety culture focus increasingly promotes prevention of workplace injuries rather than seeking intervention after an injury has occurred. In order to increase the ownership of the ergonomic program, to increase awareness of ergonomics, and to proactively increase the maintenance of a safe and healthy work environment, managing one's lifestyle outside of the workplace appears to be an integral part of an overall healthy workplace.

Employees feedback reported that being a member of a team that is developing and implementing an ergonomics training program is motivating, rewarding, and beneficial to both the individual and the organization. In addition, a major contributing factor to the success of any safety and health program, including the work-life program, is the voluntary nature of completing an HRA questionnaire and using the results to determine lifestyle changes. This voluntary and participatory approach ensures that an individual has ownership in his/her own health and leads to an increase of proactive initiative, which creates and maintains a safe and healthy work environment.

An important aspect of any of these safety and health programs is the proper documentation of all costs and benefits associated with a program. A number of limitations in the available datasets restricted or prevented a more comprehensive form of study. Also, it is recommended that prior to any ergonomics program, a list of associated costs and benefits should be drawn up and followed up on throughout the course of the program. The major categories of costs associated with ergonomics projects are personnel, equipment and materials, reduced productivity, and overhead. Economic benefits typically fall into three general classes: personnel (reduced accidents, injuries and illnesses, reduced absenteeism, etc.), materials and equipment (reduced equipment damage and rework), and less tangible benefits (increased employee commitment and improved university image). The evaluation process should be ongoing and part of an effective ergonomics program.

## 39.12 DISCUSSION

This study provided a better understanding and knowledge of injuries and illnesses in the workplace by developing an improved safety culture to reduce injuries (errors) in the workplace due to unacceptable errors or shortcomings in human performance, with the intent to reduce avoidable injuries and reduce WC claims paid. Total WC costs were broken down into two categories: adverse events and preventable adverse events. Hence, the team was able to focus on the reduction of injuries for the three major preventable adverse workplace injuries, which were activities related to lifting/twisting/bending, slips/trips/falls, and incidents where the employee was struck by something or someone.

The implementation of a participatory ergonomics program had been used to promote the safety and physical well-being of employees with cooperation between experts and nonexperts (e.g., employees) and consideration and respect for employees' opinions, as the program addressed ergonomics and psychosocial risk factors in the workplace. Employee involvement in the ergonomics process helps to address psychosocial issues and increases the likelihood of success in reducing injuries in the workplace. While language barriers can add to the challenges of involving employees, an honest attempt to seek their opinions will encourage employees to feel that their opinions are valuable and valued, thereby increasing the likelihood of the program's success. Employees who are given an active role in the identification and analysis of ergonomic risk factors will have a better understanding of the risk factors associated with their work tasks and that can affect their behavior at work, their health, and their safety. With the combination of the workplace risk factors, workers' knowledge, and skills, employees can have a significant input into the design of ergonomic solutions (e.g., job redesign, training, work organization).

This study is significant because workplace injuries and illnesses have been and continue to be a difficult and unmanageable phenomenon that harms employees and leads to excessive WC cost. Over the past 10 years, the severity of injuries has caught the attention of the university leadership, resulting in some reduction driven by awareness. However, there are complex human factors issues and lifestyle that contribute significantly to the majority of injuries in the workplace [22]. Therefore, it is important to understand that injuries and illness involve diverse human performances and are situation-specific; consequently, the selected appropriate preventative and corrective techniques are critical. The success of the practical application depends on the safety manager's ability to understand concepts, identify needs, and undertake appropriate actions to reduce injuries and illness in the workplace. Consideration of the resources needed to accomplish this goal will depend on the university system's size, its budget, its special needs, the location of its facilities, the conceptual flexibility of staff and directors, the personalities and disciplines involved, and the overall policy and rules under which the university operates.

Even in the best organizations with all the available services, humans are fallible, and unhealthy lifestyles are to be expected. The questions that this study set out to answer were as follows: (1) "Can understanding lifestyle risk factors and taking advantage of available services to promote a healthier lifestyle have an impact on the risk of injuries and illnesses in the workplace?" and (2) "can a 'blameless' workplace culture reduce unwanted variability in human behavior?" During the intervention, it was observed that (1) in a "blameless" environment, the number of injuries reported will increase (as more people are reporting injuries); (2) even the best people (regardless of age or experience) are capable of faulty judgment that contributes to workplace injuries; (3) the three major workplace injuries are more likely to be lifting/twisting/bending, slips/trips/falls, struck by something or someone; and (4) management should expect its employees to make errors and to train employees to recognize and recover from these errors. Also, the work-life program provides knowledge, coaching, and training needed to reduce workers' discomfort and pain. As a result, the work-life program's services may have a positive effect on the health and well-being of employees. The goal to study lifestyles that leads to injuries and illness in the workplace is not to apportion blame but to ask "why?" Understanding causality enables prevention of unsafe acts that can lead to injuries and illness in the workplace. Some observed human behavior that contributes to unsafe acts (variability) is forgetfulness, inattention, poor motivation, carelessness, negligence, and recklessness. During the baseline period, it was assumed that the basic human conditions would not change. Also, there was a culture of "blame"—asking "who made the mistake?" rather than "why was it made?" and "how it can be avoided?" The safety and health management program served as an effective risk management system that could be depended on for establishing a reporting culture. It was discovered that the risk of blame was a potential setback that was preventing people from reporting injuries and illness.

The result of this study confirms the findings of Bidassie et al. [17] that when a successful ergonomics intervention program is implemented, one of many results will be an increase in the worker's ability to change his/her work environment, which in turn will lead to enhanced individual and organizational effectiveness across multiple categories.

## ACKNOWLEDGMENTS

The author would like to give a special thanks to Mr. John Meyers, Project Manager of Housing and Food Services, for his assistance with the third phase (Sustainability) of the project. Without his valuable contribution and insightful comments, this important phase of the project would not have been possible. Thanks to Mindy Paulet and Dr. Joseph Thomas III from the University's Worklife Program. Thanks to Dr. James D. McGlothlin and the students in HSCI 580: Occupational Ergonomics at Purdue University for conducting the ergonomic evaluations at the university residence hall and his mentorship in the development of this research which brought this chapter to fruition.

## REFERENCES

1. Bureau of Labor Statistics. *Workplace Injuries and Illness in 2006*. United States Department of Labor, Washington, DC, 2007.
2. J.D. McGlothlin. An ergonomics program to control work-related cumulative trauma disorders of the upper extremities. PhD dissertation, University of Michigan, Ann Arbor, MI, 1998.
3. A.L. Cohen, C.C. Gjessing, L.J. Fine, B.P. Bernard, and J.D. McGlothlin. *Elements of Ergonomics Programs: A Primer Based on Workplace Evaluations of Musculoskeletal Disorders*. U.S. Department of Health and Human Services, National Institute for Occupational Safety and Health, Washington, DC, 1997.
4. B. Bidassie. Development of a predictive model for low back musculoskeletal disorders based on occupational and lifestyle risk factors. PhD dissertation, Purdue University, West Lafayette, IN, 2010.
5. P. Banyard. *Psychology in Practice: Health*. Hodder & Stoughton, London, U.K., 2002.
6. S. Prentice-Dunn, B. Mcmath, and R. Cramer. Protection motivation theory and stages of change in sun protective behavior. *Journal of Health Psychology*, 14(2): 297–305, (2009).
7. J.E. Maddux and R.W. Rogers. Protection motivation theory and self-efficacy: A revised theory of fear appeals and attitude change. *Journal of Experimental Social Psychology*, 19(5): 469–479, (1983).
8. R.W. Rogers. A protection motivation theory of fear appeals and attitude change. *Journal of Psychology*, 91(1): 93–114, (1975).
9. National Occupational Research Agenda (NORA). National Institute of Occupational Safety and Health [NIOSH]. About NORA … Partnerships, Research and Practice. http://www.cdc.gov/niosh/nora/ (accessed January 28, 2010).
10. National Occupational Research Agenda (NORA). National Institute of Occupational Safety and Health [NIOSH]. National Occupational Research Agenda for Musculoskeletal disorders research topics for the next decade. A Report by the NORA Musculoskeletal Disorders Team. NIOSH Publication No. 2001-117 (January 2001). http://www.cdc.gov/niosh/docs/2001-117/agenda.html (accessed March 14, 2010).
11. r2p: Research to Practice at NIOSH. National Institute for Occupational Safety and Health [NIOSH]. http://www.cdc.gov/niosh/r2p/default.html (accessed January 29, 2010).
12. Classification of Exempt and Non-Exempt Administrative and Professional Employees. Purdue University. http://www.purdue.edu/policies/pages/human_resources/iv_4_1.html#classification (accessed December 1, 2009).
13. Classification of Clerical and Service Staff. Purdue University. http://www.purdue.edu/policies/pages/human_resources/iv_4_1.html#classification (accessed December 1, 2009).
14. T.R. Waters, V. Putz-Anderson, A. Garg et al. Revised NIOSH equation for the design and evaluation of manual lifting tasks. *Ergonomics*, 36(7): 749–776, (1993).

15. Mission of Work-life Program. Purdue University, Human Resources Services. http://www.purdue.edu/hr/WorkLife (accessed June 26, 2006).
16. Occupational Safety and Health Act (OSHA), Voluntary Protection Program (VPP) Policies and procedures manual (TED.8.4)—Appendix A. OSHA (2003). www.**osha**.gov/Publications/**OSHA**3317first-aid.pdf (accessed January 29, 2010).
17. B. Bidassie, J. McGlothlin, A. Goh, R. Feyen, and J. Barany. Limited economic evaluation to assess the effectiveness of a University-Wide Office Ergonomics Program. *Applied Ergonomics,* 41: 417–427, (2010).
18. A. Agresti. *Categorical Data Analysis*. Wiley, New York, 2002.
19. O.J. Prochaska and C.C. DiClemente. Toward a comprehensive model of change. In *Treating Addictive Behaviors*, eds. W.R. Miller and N. Heather, pp. 3–27. Plenum, New York, 1986.
20. O.J. Prochaska and C.C. DiClemente. The transtheoretical approach: Towards a systematic electric framework. In *Handbook of Electric Psychotherapy*, ed. J.C. Norcross. Brunner/Mazel, New York, 1986.
21. N. El-Bassel, R.F. Schilling, A. Ivanoff, D. Chen, M. Hanson, and B. Bidassie. Stages of change profile among incarcerated drug-using women. *Applied Behaviors,* 23: 389–394, (1998).
22. B. Bidassie, J. McGlothlin, I. Mena, V. Duffy, and J. Barany. Evaluation of lifestyle risk factors and job status associated with back injuries among employees at a Mid-Western University. *Applied Ergonomics,* 41: 106–114, (2010).

CHAPTER 40

# Research to Practice in Solving Ergonomic Problems*

Janice Huy, Heidi Hudson, Elizabeth Dalsey, John Howard, and R. DeLon Hull

CONTENTS



* The findings and conclusions in this report are those of the author(s) and do not necessarily represent the official position of the National Institute for Occupational Safety and Health, Centers for Disease Control and Prevention.

## 40.1 INTRODUCTION

For research to benefit society, it needs to be shared, communicated, and translated into policy and practice. Moving occupational ergonomic research effectively into practical application is a process that combines the generation of knowledge with the adoption of that knowledge in the workplace to enhance worker health and prevent injury and illness. Research that is not used cannot fulfill the twin goals of ergonomics: health and productivity.

Although still relatively new, the strategy of *research to practice (r2p)* is applied in various contexts within public health, medicine, and community action. It is increasingly called for in agency mandates and in organizational missions, strategic plans, priorities, and initiatives. This growing demand demonstrates the significance of ensuring that what is known is practiced to advance societal welfare. *r2p* is a strategy to ensure that research has an impact or makes a difference.

Not all research immediately results in changes in workplace practice. Often it takes incremental steps over time to change work practices to impact the safety and health of the entire workforce. Surveillance research leads to problem identification; basic research ultimately leads to intervention. Over time, our fund of knowledge increases, which can result in changes in workplace policies, programs, and practices. Being successful in getting one workplace at a time to implement change may influence further changes over the entire workforce. A broad understanding of how research can be applied to the workforce will help researchers design more research with greater impact.

The purpose of this chapter is to introduce the concept of *r2p* as it is conceived and implemented in the National Institute for Occupational Safety and Health (NIOSH), to describe r2p components, to share alternative models of research to practice, to present examples of how r2p concepts are applied to physical hazards by NIOSH scientists, and to provide a list of references.

The information presented in this chapter is primarily written from the NIOSH perspective of r2p. However, concepts of *r2p* from other science research agencies are also provided. When the term *r2p* is used in this chapter it represents the NIOSH view; when the term *research to practice* is used, it represents a view other than that of NIOSH.

### 40.1.1 Introduction to Research to Practice at NIOSH

In 2004, NIOSH, a part of the Centers for Disease Control and Prevention (CDC) in the U.S. Department of Health and Human Services (HHS), established an initiative that focused on the translation of research findings, technologies, and information into effective occupational safety and health policies, programs, practices, and products [1]. This initiative was referred to as *Research to Practice,* or *r2p*. NIOSH defines *r2p* as an interactive process in which occupational safety and health communities—research organizations, academic institutions, industry, trade associations, employers, and employees—work collaboratively to (1) prioritize research needs; (2) design, plan, and conduct research; (3) transfer research findings into the practice of occupational safety and health; and (4) evaluate results to determine the impact on the risk of injury and illness in the workplace. This process was designed with the intent to ensure NIOSH research has relevance, quality, and impact.

Since its inception at NIOSH, r2p has evolved and become well-integrated into the mission, intramural and extramural research programs, strategic plans, and National Occupational Research Agenda, called NORA [2].

The foundation for all r2p efforts is *collaboration with partners*. In occupational safety and health, a partner is an organization or person who is actively involved in efforts to help promote the movement of research into practice to prevent work-related injury or illness. A checklist of r2p components is provided in Table 40.1. It includes the major elements researchers may want to consider when planning for or describing their *r2p* efforts.

### 40.1.2 Alternative Models of Research to Practice

As many research institutions begin to transform the way they conduct and use their research, it is important to recognize how research to practice is conceptualized and operationalized in other research organizations.

At the Canadian Institutes for Health Research (CIHR), research to practice is conceptualized by the term *knowledge translation* (KT). Integrated in the CIHR mandate and strategic plan, KT is broadly conceived and is widely used in their 13 institutes to encompass all steps between the creation of new knowledge and its application to yield beneficial outcomes to society. KT is defined as the exchange, synthesis, and ethically sound application of knowledge, within a complex set of interactions among researchers and users [3]. The definition emphasizes the active exchange between creators of knowledge and users of knowledge during all stages of the research cycle. Effective and meaningful partnerships are emphasized in every KT activity. The seven stages of KT are research priority setting, research, knowledge priority setting, knowledge synthesis, knowledge distribution and application, use, and evaluation of uptake.

At the National Institutes of Health (NIH), the movement of research into practice is conceptualized by the term *translational research*. Translational research, designed around the biomedical research model, is infused as a priority in the *roadmap to medical research* [4], an NIH initiative to address barriers to research and to transform the way biomedical research is conducted by overcoming specific hurdles or filling defined knowledge gaps. Although each stage appears to work harmoniously, each phase differs in scientific context, partners, institutional support, goals, settings, study designs, and investigator skills [5]. As one of the 46 NIH Centers for Clinical and Translational Science, The Ohio State University Medical Center (OSUMC) envisions translational research as having up to four different yet interrelated phases. Phase 1 is often referred to as "bench to bedside" and seeks to move a basic discovery of disease mechanisms (i.e., basic science) into initial testing in humans (i.e., bedside). Phase 2 assesses the value of phase 1 applications for health practice, leading to the development of evidence-based guidelines. Phase 3 attempts to move evidence-based guidelines or practice-based research into practice through delivery, dissemination, and diffusion of research. Phase 4, translational research, seeks to evaluate the real-world health outcomes of a phase 1 application in practice [6].

Table 40.2 illustrates the core components of each of these three frameworks. Although not easily depicted, each framework conceptualizes research to practice as a nonlinear,

TABLE 40.1 Research to Practice Checklist: Components of Planning and Reporting

Partners

- Partners for the research process (*may include different partners for each step*)
  - Have a philosophy, a mission, and goals complementary to yours
  - Are recognized as opinion leaders in their area of expertise
  - Represent a diversity of thought and opinions
  - Are trustworthy
  - Will influence others
  - Will contribute resources, either direct or in-kind
  - Will benefit from the partnership

Funding source (*check one or more*)

- Employing agency or organization
- Partner(s), partially or in total
  - No conflicts of interest related to partner funding
  - Resolve existing conflicts of interest

---

Priority research needs (*check one or more*)

- Address an organizational priority need
- Address recognized gap in knowledge or practice
- Are based on surveillance or other data
- Are determined in collaboration with partners (*preferable*)

---

Research collaborators

- Involve in conduct of the research
- Determined to be scientifically competent and ethical

---

Transfer of knowledge and technology (*select the path to practice appropriate for research findings*)

Basic and applied research knowledge translation (*when pooled with other research knowledge and/or ready for workplace implementation*)

- Target audience
  - Identify
  - Needs, desires, and values
  - Motivations for change
  - Sources of influence
- Partners to assist
- Desired change
- Research translation methods
  - Diffusion
  - Dissemination of results
    - Submit to peer-reviewed scientific journal
    - Make available to other researchers via conference presentations and related mechanisms
    - Make available to standard-setting organizations, as appropriate
    - Engage partners in process

Research technology

- Report discovery or invention to employing organization
- Assess discovery or invention as to patentability and marketability
- Develop invention without assistance from external partners

*or*

- Seek partner(s) to develop invention
  - Acquire partners' signature on confidential disclosure agreement
  - Secure partnership intermediary
- Determine whether to patent
  - Is product or process unique?
  - Will patent advance goals of organization?
  - Market analysis
  - Economic analysis
  - Industry investigation
- Develop prototype or working model
- Identify target companies who may be interested in commercializing product

TABLE 40.1 (continued) Research to Practice Checklist: Components of Planning and Reporting

| | |
|---|---|
| Social networks (e.g., blogs, Wikipedia) | Seek assistance from partnership intermediary |
| Multichannel communication | Prepare "sales" plan |
| Health communication/marketing | Authorize personnel to negotiate licensing agreements |
| Integrated knowledge translation | Collect royalties |
| Implementation science | |
| Unified strategic plan: develop and communicate to all parties | |
| Challenges and opportunities | |
| Goals | |
| Measureable outcomes | |
| Communication strategy | |
| Evaluation measures | |
| Resources | |
| Impact | |
| Involve partners in evaluation effort | |
| Determine and implement evaluation measures | |

TABLE 40.2 Frameworks of Three Health-Research Agencies, Designed for Moving Research into Practice

| | Framework of Agency Concept | | |
|---|---|---|---|
| **Phase** | **NIOSH [1] Research to Practice** | **CIHR [3] Knowledge Translation** | **NIH [4] Translational Research** |
| 1 | Prioritization of research needs | Research prioritization | Bench to bedside |
| 2 | Designing, planning, and conducting research | Research | Health application to evidence-based practice guidelines |
| 3 | Transfer of knowledge and technology | Knowledge prioritization | Practice guidelines to health practices |
| 4 | Evaluation of impact | Knowledge synthesis | Practice to population impact |
| 5 | None | Knowledge distribution and application | None |
| 6 | None | Use | None |
| 7 | None | Evaluation of uptake | None |

iterative process. Each framework also emphasizes partnerships and the need to bring together creators and users of knowledge at each phase within the process.

All phases, in both r2p and KT, are cyclical and can be applied to both basic sciences and applied sciences. In translational research, the phases differ according to scientific context and are specific to the type of research being conducted. In addition, the types of research, users, and products are distinctive. For example, phase 1 involves basic research, phase 2 involves applied science, and phase 3 involves implementation science.

In the frameworks of both NIOSH and CIHR, the process starts with prioritizing research needs and the second phase involves conducting research. It is in the subsequent phases that the two frameworks start to move in different directions. NIOSH moves to knowledge transfer, but CIHR cycles back to prioritizing research-based knowledge and then moves to knowledge synthesis. CIHR begins to disseminate or distribute the information to users in phase 5, which is similar to the NIOSH phase 3 (knowledge transfer). The transfer mechanism in NIH translational research is not explicit and appears to be inherent in the process. Evaluation consistently is the last component; in the NIOSH framework, this is the fourth and final phase, involving evaluation of impact. In the CIHR framework, it is the sixth phase, involving evaluation of the uptake of research. And in the NIH framework, it is the fourth phase in which evaluation of health outcomes actively involves another phase of research (implementation research).

Other notable differences between the three programs may be related to their origins and ideology and to their financial, legal, and organizational support for *r2p* initiatives. These and other combined elements affect the extent to which a researcher, a program manager, or an institution can have an impact on health outcomes. As the previous comparisons illustrate, the concept of research to practice has evolved within and throughout health science research organizations. These large, health-related organizations continue to build upon their institutional frameworks to transform the way research is conducted.

## 40.2 PARTNERSHIPS

Collaboration with partners is a basic tenet of NIOSH r2p. Partnerships support, enhance, and advance research. Forming long-lasting, productive partnerships is essential in demonstrating and sustaining impact in the workplace [7,8].

A partnership is a collaboration, either formal or informal, between two or more individuals or organizations to work together toward mutually beneficial goal(s). Partners can contribute to the goals by sharing people, funds, property, or knowledge. There are many types of collaborations for mutual benefit, for example, partnerships, coalitions, or alliances. Unless it is a legal partnership, the label is not as important as how the collaboration is defined by the partners. Formal partnerships generally are bound by a written agreement in which the roles and responsibilities of each partner are delineated. Informal partnerships involve no signed agreement, and the roles and responsibilities of each party are more flexible.

The ultimate goal of partnerships is to identify and advance research and/or implementation objectives. The most fruitful partnerships are those in which all partners have the same vision and work together to accomplish more than any party could on its own. Every successful partnership takes time to develop and may be difficult to maintain [7]. Certain questions should arise when individuals and groups are considering a partnership [7,9]:

1. Are the philosophy, mission, and goals of all the partners compatible?
2. Are the partners recognized as thought leaders in their areas of expertise? Will they bring credibility to the project? Can they contribute intellectually? At what stage of research are the potential partners most needed?
3. Do the partners represent a diversity of thought and opinion? Have partners from government, academia, industry, labor, and professional practice associations been considered? Is there added value in including international partners?
4. Can you rely on partners to maintain confidentiality, protect proprietary information, and contribute fairly to the contributions, risks, and rewards of the activity?

5. Can the partners bring other necessary collaborators to the partnership? Will all partners advocate for and support change in the way people think and act as related to worker safety and health?
6. Are the partners potential diffusers of knowledge?
7. Do the partners have the necessary resources, either direct or in-kind, to contribute? Will the partners support scientific independence if they also provide funding? [10,11]

Having a diverse cadre of partners is a valuable asset to any research organization. Although it is tempting to rely on people who are "friends" of the scientist or research institution, the added value of bringing in new thoughts and perspectives in solving a problem from outside the scientists' circle cannot be overstated.

For r2p efforts to be most successful in protecting workers, partners should be involved throughout the research process. The right partners must be matched to each stage of research. For example, a partner who helps define the scope of the research may not be qualified to be involved in the actual conduct of the research. Similarly, a collaborating scientist who helps conduct the research may not be the best partner to assist in developing the dissemination strategy. Partnering with subject matter experts—who may not be scientists, but might be the workers themselves—always helps to ensure a successful r2p venture. Table 40.3 shows the different roles of partners and describes some common concerns with the involvement of partners in the research cycle.

TABLE 40.3 Involving Partners in Each Stage of Research to Practice: Roles, Concerns, and Strategies

| Stage of Research to Practice | Role of Partner | Concern and Strategy |
|---|---|---|
| Identifying priority research needs: identify the most important occupational health and injury issues facing workers | Identify the problem to be addressed; help to demonstrate need for research; or help articulate a specific research or practice gap | Partner may focus on personal interests and needs rather than research priorities; address this risk by including a range of partners with varying interests |
| Designing and conducting research | Assist in planning the research protocol and/or serve as a member of the research team | Scientific integrity may be compromised or research ideas may be taken; carefully select and monitor partners to prevent this occurrence |
| Transferring knowledge and technology: use communication science (for knowledge) and technology transfer principles (for inventions) to guide the movement of research into the workplace | Synthesize and convert the research findings into regulatory requirements, consensus standards, professional practices, other workplace recommendations, or equipment or processes; distribute research findings to specific target audiences | Partner may overemphasize or underemphasize certain research findings; address this risk by including a range of partners with varying interests |
| Evaluating impact: build data collection into each research effort to determine effectiveness in preventing workplace injury and illness | Assess the research to determine workplace impact | Partner may try to sabotage, distort, or discredit research findings; avoid this by resolving conflicts as they occur |

## 40.3 IDENTIFY PRIORITY RESEARCH NEEDS

Involving partners in the research process begins by identifying a research topic in cooperation with those who will be affected by or will use the research findings. Traditionally, new research ideas emerge from a variety of sources, including personal or organizational interests, surveillance data, recognized knowledge or practice gaps, and literature reviews. Following the r2p concepts, researchers work in collaboration with partners to use traditional methods in identifying research topics. Jointly identifying and agreeing on a research topic and research approach with partners can help to further refine the topic and increase its relevance to those who will benefit from or use the findings.

Obtaining or combining research funds from outside partners is an obvious way to expand research dollars. Researchers must use caution when accepting funds from commercial partners, to avoid or minimize conflicts of interest. Full disclosure of the role of each funding party must be made at the time of publication.

**r2p IN ACTION**

*Physical hazard*: Nurses and other health care workers have reported thumb and wrist issues related to opening pill packages (personal communication)

*r2p Component addressed*: Identify priority research needs

*Description*: A NIOSH partner, an occupational health supervisor, reported to NIOSH that many nurses and pharmacists in her health care system had described thumb and wrist issues resulting from opening pill packages and containers. These issues were especially noted by workers who are required to open many pill packages each day. The nurses use force to push the pills through the packaging because they are not allowed to use a knife or other implement to open the packages. In addition, the lids on pill containers often are tight and require extreme effort to open. The partner wondered whether NIOSH would be interested in researching this issue to identify preventive strategies.

Following the r2p concepts, NIOSH investigated whether this issue was a priority research need. The investigation was accomplished by reviewing national surveillance data to determine whether such injuries were commonly reported. No such surveillance data were found. NIOSH also contacted other partners, such as a national nurses association and a national association representing occupational health professionals, to learn if their constituents had reported this problem; it was not a known concern for either association. The Food and Drug Administration (FDA) was consulted and referred NIOSH to pharmaceutical manufacturers, because it does not deal with packaging. An Internet search identified two patents for devices to open pill (compressed tablet) containers and two inexpensive devices that were marketed as solutions for individuals taking medications. In conversations with pharmaceutical companies, NIOSH learned that developing a tool to assist with opening the containers would not be easy, because of variations in sizes from different vendors. NIOSH also learned that research was already being conducted elsewhere to address this problem.

NIOSH, in collaboration with many partners, determined that thumb and wrist issues related to opening pill packages were not a priority research topic at the time.

## 40.4 DESIGN AND CONDUCT RESEARCH

Involving partners in the conduct of the research can lend credibility among the target audience, which may help when research results are translated into practical solutions. Another benefit of involving partners in this stage of the research is that they may facilitate access to the study cohort. Maintaining scientific integrity is essential during this phase, and many scientists are hesitant to allow outside partners to participate in the actual research process so as not to taint the research findings. In addition, researchers do not want to take the chance of having their work or ideas stolen. Carefully selecting partners and monitoring their involvement can alleviate the concern.

**r2p IN ACTION**

*Physical hazard*: Occupational lower-back disorders are among the most common health problems facing society [12].

*r2p Component addressed*: Design and conduct research

*Description:* NIOSH formed a research team to develop a revised lifting equation as a practical analysis tool for evaluating the physical demands of two-handed manual lifting tasks. The research team included NIOSH scientists, academicians, and industry professionals, including national and international subject matter experts in the fields of biomechanics, epidemiology, and ergonomics. The research team collaboratively reviewed scientific literature, developed the lifting equation, and disseminated the tool.

The research team decided on the format and scope of the equation and chose the critical factors to be included. In addition, a series of evaluations and a peer review were conducted to examine the reliability and accuracy of the equation.

The result of this collaboration was the revised NIOSH lifting equation [13], an analysis tool consisting of two equations for estimating the recommended weight limit for a specified manual lifting task and a lifting index for estimating the physical demands of the job.

## 40.5 TRANSFER KNOWLEDGE AND TECHNOLOGY

To produce high-impact outcomes, it is necessary to carefully plan the transfer of research-based knowledge and technology. The transfer of knowledge or technology involves sharing and exchanging information among all relevant parties. However, there are distinct differences between knowledge transfer and technology transfer. The following introduces these two areas of research transfer and the necessary steps to plan for effective outcomes.

### 40.5.1 Knowledge Translation

As described earlier in the chapter, knowledge translation and research to practice may be perceived as having the same meaning, but they may have slight differences according to their organizational processes and scientific context. Knowledge translation is broadly conceived as a comprehensive process that includes dissemination, communication, technology transfer, knowledge utilization, implementation research, and exchanges between researchers and those who apply the knowledge [3]. The term *knowledge translation* is widely applied within multiple disciplines, such as medicine, public and global health, health services, and disability and rehabilitation.

As the research to practice process grows to engage multiple disciplines, the terminology used to describe the process and the associated activities evolves to become more refined and precise. The processes are often tailored to the areas of specialty, which can lead to confusion in thinking and speaking about the concepts. Terms frequently used in scientific literature and in academia to describe research to practice activities include diffusion, knowledge transfer, communication, research utilization, knowledge exchange, and dissemination. These terms are interrelated and may often be interpreted as having the same meaning; however, there are some distinctions between the terms and in how they are used in describing research to practice activities. These terms, defined in Table 40.4, will become more useful and relevant during the planning and discussing activities related to research to practice.

The goal of all occupational research to practice efforts is to improve worker safety and health. Achieving this goal will require incremental steps, such as first getting one company—and then many—to implement a workplace intervention. To achieve either the incremental or ultimate goals, social change principles can be used. Adapted from Siegel and Doner [14], the remainder of this section provides guidance on how to plan the transfer research knowledge using the seven social change principles.

#### *40.5.1.1 Craft an Intervention That Will Impact Behavior*

When research is created without considering its relevance to all users at each stage of the research process, a gap can be created between what is known in the research community

TABLE 40.4 Research to Practice Terminology

| Terminology | Definition |
|---|---|
| *Communication* | A process in which participants create and share information with one another in order to reach a mutual understanding |
| *Diffusion* | A process by which an innovation is communicated through certain channels over time [15] |
| *Dissemination* | A function of communication; the process through which tailored information about research-based knowledge, interventions, or technologies is transferred to target audiences through preidentified communication channels |
| *Knowledge exchange* | A collaborative problem-solving between researchers and decision-makers that happens through linkage and exchange. Effective knowledge exchange involves interaction between decision-makers and researchers and results in mutual learning through the process of planning, producing, disseminating, and applying existing or new research in decision-making [3] |
| *Knowledge transfer* | A broad, overarching term describing the exchange of research results, skills, or good ideas between research organizations, industry, and other relevant parties to enable the development of innovations aimed to improve occupational safety and health for all workers [1] |
| *Research utilization or knowledge utilization* | The uptake of research findings; a complex process in which research-based knowledge is transformed from the findings of one or more studies into instrumental, conceptual, or persuasive utilization |

and what is practiced in the workplace. The challenge for researchers is to think beyond the end points of the originating research stage (e.g., the technology development). Another challenge researchers may face is the balance between promoting research (selling what you can make) and marketing research (making what you can sell). Moving research through each stage, from basic to applied research and then to organizational practice, requires futuristic thinking. It involves understanding the relevance of the research to all potential users. Comprehensive planning of research translation requires

- Iterative planning
- Effective partners at each stage
- An understanding of the environment in which the innovation and intervention will operate
- An understanding of the innovation
- Identifying and understanding the target audiences

#### *40.5.1.2 Be Theory Driven*

Theories and frameworks can play important roles in shaping how occupational safety and health knowledge, interventions, and technologies are used. Theories are helpful for explaining how a worker, work organization, or industry may respond to a new technology, recommended work practice, or recently identified hazard. Frameworks are helpful for developing strategies for improving worker health, selecting the optimal target audience, setting reasonable objectives, and tailoring interventions to the unique needs of the target population. Both theories and frameworks can be used to effectively transfer research at each stage of the research process. Two of the most popular and widely used theories are described hereafter.

40.5.1.2.1 Diffusion of Innovations Developed by Everett Rogers in 1964, the diffusion of innovations theory [15] has been widely studied in a variety of settings and disciplines. It consists of four main elements: an innovation, communication channels, time, and a social system. Combined, these elements explain a process by which an innovation is communicated through certain channels over time among members of a social system. When innovations are preventive (e.g., industry work practices to prevent back injury), the rate of adoption is expected to be slow. Variables determining the rate of an innovation's adoption include the following: perceived attributes (i.e., relative advantage, compatibility, complexity, trialability, and observability); type of innovation decision (e.g., optional, collective, or authority); communication channels; nature of social system; and extent of the change agents' promotion efforts.

The innovation-decision process involves the time component and explains at which of these five time-ordered steps individuals or decision-making units occur: knowledge, persuasion, decision, implementation, or confirmation. Populations (individuals or decision-making units) are divided into six categories based upon their rate of adoption: innovators, earlier adopters, early majority, later majority, late adopters, and laggards.

A researcher may consider using Rogers' diffusion of innovations theory in the planning stages to help identify the innovation's rate of adoption, where the target audiences are in terms of adopter categories and in the innovation-decision process, and what type of diffusion network exists for seeking potential partners.

40.5.1.2.2 Social Marketing Social marketing provides an audience-focused and theory-driven approach to developing and implementing a research program, technology, or intervention. On the basis of commercial marketing principles, social marketing can be used to transfer occupational safety and health research. Although there are several extensive, comprehensive frameworks for applying social marketing theory, only the key concepts of social marketing are discussed here. These are the five key principles [16]:

1. Understand everything from your audience's perspective—including your partners'. This is the consumer orientation of social marketing. Understand how consumers think, what they do, what they value, and what they like and dislike about the innovation.
2. Define a specific, simple action that you want your audience to take.
3. Understand the exchange. What is it that you want your audience to give up in order to adopt your research or technology? What are the benefits that matter to them? What is the competition for the behavior you are encouraging?
4. Define the marketing mix, otherwise known as the 4 Ps:
   a. Product: the behavior, idea, good, or service that is exchanged with the target audience for a price and may be tangible or intangible.
   b. Price: the target audience's cost for making the behavioral change.
   c. Promotion: the selected media of communication for behavioral change.
   d. Place: where the target audience members will be when they engage with the product.
5. Monitor and evaluate the progress of your efforts. This will help you determine whether the directions and decisions you made are moving in the expected direction.

*40.5.1.3 Know Your Target Audience*

A target audience is a small part of the population that shares similar characteristics, such as demographics, perceptions, behaviors, wants, readiness to change, and channels [17]. Target audiences inform the researcher on how the innovation or research is perceived in the desired setting and may provide insight into how likely the research is to be adopted or received. Segments are subgroups of the target audience that have similarities in a variety of dimensions (e.g., geography, lifestyle, and behavioral characteristics).

At each stage in the research cycle it is important to understand and select the target audiences that will likely be impacted by the research. These are some questions to ask: (1) Who are the potential target audiences? (2) what are the sizes of these audiences?

TABLE 40.5 Potential and Existing Target Audiences of Occupational Safety and Health Research and Rationales for Engaging Each Group

| Target Audience | Rationale |
|---|---|
| Academia | Potential collaborators |
| Employers | Key decision-makers in the workplace |
| General public | Advocates to decision-makers |
| Government agencies | Potential collaborators; regulators |
| Industry associations | Potential collaborators, adding credibility and prestige; key influencers of the workplace |
| International health organizations | Potential collaborators, adding credibility and prestige |
| Labor organizations | Potential collaborators, adding credibility and prestige; key influencers of the workplace; provide access to new audiences |
| News media | Communication networks |
| Nongovernment organizations (private and not for profit) | Potential collaborators; have established communication networks with decision-makers |
| Policy-making officials and organizations conducting similar research | Potential collaborators; have established communication networks with decision-makers |
| Professional organizations | Potential collaborators, adding credibility and prestige; provide access to new audiences |
| Worker advocacy groups | Potential collaborators, adding credibility to audiences; key influencers of the workplace; have established communication networks with decision-makers; provide access to new audiences |
| Workers | Key influencers in the workplace |

*Source:* Canadian Population Health Initiative, An environmental scan of research transfer strategies, http://secure.cihi.ca/cihiweb/products/EnviroScan.pdf, (accessed January 30, 2011), 2001; National Institute for Occupational Safety and Health, Research to practice at NIOSH, http://www.cdc.gov/niosh/r2p, 2009.

(3) why does each target audience engage in its respective behavior? (4) what are their needs, desires, and values? (5) what is the cost to the audiences for exchanging their current behavior for the desired behavior? and (6) who is most likely to influence the audiences to accept research-based knowledge or adopt recommended behavior?

Table 40.5 shows potential and existing target audiences of occupational safety and health research and a brief rationale for engaging the groups. This list was modified on the basis of findings from the Canadian Population Health Initiative [18] and NIOSH [1].

#### *40.5.1.4 Set Realistic Expectations*

Establishing occupation-based social change takes time, especially for innovations providing longer-term benefits (e.g., use of a computer keyboard). When developing a plan for change or implementing research into practice, researchers need to consider three factors that influence the uptake of research: the innovation, the potential adopter(s), and the setting or social context [19]. These factors, combined with external, uncontrollable factors, may affect the success or failure of the effort.

#### *40.5.1.5 Create a Unified Strategic Plan*

A strategic plan describes how the innovation, intervention, or research-based knowledge is expected to work. It identifies who it should reach, what action people are expected to take, and how the strategies are expected to bring about action [14]. A strategic plan helps all parties involved understand their role and what is planned. This is meant to change over time to reflect current knowledge. The components of a comprehensive strategic plan include the executive summary, background and mission challenges and opportunities, goals, measurable outcomes, communication strategy, and components for implementing and evaluating the strategy. When developing a strategic plan with partners, it is critical to (1) ensure agreement of all players on the need for change; (2) define the nature of the evidence for change; (3) describe the methods, including evaluation of the impact of the change; and (4) determine who is responsible for each of the components.

A communication strategy is a subcomponent of the strategic plan and describes how the research-based knowledge, technology, or innovation will be framed or positioned for target audiences. Communication specialists may help researchers develop effective communication strategies for transferring their innovations.

TABLE 40.6 Hierarchy of Knowledge Translation Methods, Based upon Normative Standards at NIOSH

| Knowledge Translation Method | Rationale |
|---|---|
| Diffusion | A one-way, passive form of communication. Focusing on reach and access, it is used to ensure research is accessible and available for wider, undefined audiences. Content is informative and broad in context. Examples include press releases, television, and Web site postings |
| Dissemination | A one-way, targeted, active form of communication. Also focusing on reach and access, it is used to ensure research is available and accessible for audiences that are more defined. Examples include trade publications, professional presentations, and peer-reviewed publications |
| Social media | A newer and advancing means of communication, emphasizing a two-way exchange of information that focuses on depth and credibility. Audiences are defined according to channel style and access to information; thus, they are segmented by their likelihood to seek, receive, and interact with information initially sent. Examples include blogs, Wikipedia, Flickr®, Facebook, and Twitter |
| Multichannel communication | A combination of the previous methodologies, with the intent to address all or most factors of communication (access, reach, target audience, depth, credibility, and agenda setting). It is known to have more success in reaching a broad audience |
| Health marketing and communication | Creating, communicating, and delivering health information and interventions by means of customer-centered and science-based strategies to protect and promote the health of diverse populations [34]. It involves a planned application of behavioral and/or organizational change theories |
| Integrated knowledge translation | The shaping of the exchange of research-based knowledge by senders and users. Stakeholders or potential research users are engaged in the entire research process [35] |
| Implementation science | The scientific study of methods to promote the systematic uptake of research findings and other evidence-based practices into routine practice, hence to improve the relevance and impact of occupational safety and health research. It assesses and explains questions pertaining to the design, administration, operations, and outcomes of social programs: What is happening? Is this happening expected or desired? And why is it happening as it is?[a] |

[a] Unpublished material.

Table 40.6 provides a description of knowledge translation methods. These general communication strategies provide a hierarchical approach, where the complexity of communication exchange (to educate or to persuade) increases between the senders and receivers of information in each consecutive stage. More planning and technical skills are also needed as the stages progress.

#### *40.5.1.6 Leverage Resources and Relationships*

Many scientists agree that involving partners at the transfer stage is crucial. Fundamental to understanding and enhancing the process of transferring research to practice is a comprehensive understanding of how all stakeholders at each stage of the research cycle receive and interpret research-based knowledge. Partnerships and collaborations with stakeholders are critical to help bring about change. Partners can provide additional resources, reach audience members, and add credibility and expertise that your organization may not possess [14]. Effective partnerships take time to form and build trust. Refer to the partnership section for questions to consider when selecting and planning for partners.

#### *40.5.1.7 Build in Evaluation from the Start*

Research programs are increasingly called upon to demonstrate their value or impact. When developing an r2p plan at the start of the research effort, it is important to build in evaluation. It is also important to involve partners in evaluation planning to help ensure shared goals and successes. During this planning stage, determine what you will measure, how you will measure, and when you will measure your success.

**r2p IN ACTION**

*Physical hazard*: Patient handling and movement are recognized as the primary causes of musculoskeletal disorders in nurses [12].

*r2p Component addressed*: Transfer knowledge

*Description*: NIOSH partnered with a national nursing association, external researchers, and schools of nursing to develop, evaluate, and transfer an evidence-based training program targeted to educators at schools of nursing. NIOSH researchers, in collaboration with partners, designed the *safe patient handling and movement curriculum* [20] to teach nursing students about the risks involved in manually lifting patients and to educate students on safe patient handling and movement. The new curriculum moved away from the traditional curriculum of body mechanics and manual lifting to an evidence-based curriculum that focused on equipment-assisted safe patient lifting.

The research products were theory driven, guided by the theory of planned behavior [21]. Products were targeted to two specific audiences: nurse educators and students at schools of nursing throughout the United States. The knowledge transfer occurred through peer-reviewed journal articles [22,23], presentations at the Safe Patient Handling and Movement Conference, and curriculum modules posted on the Web sites of NIOSH and various partners.

### 40.5.2 Technology Transfer

The following section relies heavily on information contained in the *FLC technology transfer desk reference*, as prepared by the Federal Laboratory Consortium for Technology Transfer (FLC) [24].

One important tool for moving research findings into practice is technology transfer. As defined by the FLC, technology transfer is a process by which an invention or innovation developed in one place or for one purpose is applied and used in another place. The U.S. government has actively supported and encouraged the transfer of federally generated technologies, with the ultimate purpose of fulfilling public and private needs. The phrase *technology transfer* most often refers to transfers occurring between federal laboratories and any nonfederal organization, including private industry, academia, and state and local governments. The technology transfer process is not solely limited to the transfer of federal research but encompasses the general transfer of innovations and discoveries from inventors to the marketplace.

Technology transfer programs help to maximize the impact of research activities and to increase the return on investment of the research and development budget. For federal research laboratories, technology transfer programs assist in meeting agency mission requirements and enhance the competitiveness of the United States in the world economy.

The technology transfer process involves a research laboratory, a user of technology, and an interface that connects the two to facilitate movement of the technology from one organization to the other. Sharing among federal laboratories, private industry, and academia includes not only technologies, but personnel, facilities, methods, expertise, and technical information. This process has been made easier through the use of the Internet. Many individuals and organizations seek assistance in promoting and sharing inventions through Web sites dedicated to matching inventors with organizations that can produce and market the product.

#### *40.5.2.1 Technology Transfer Benefits*

According to the FLC, technology transfer has had an impact on fostering mutually beneficial partnerships between industry and federal laboratories. Government, industry, and academia all benefit from a strong technology transfer program:

*Government*: Technology transfer activities can be used to assist in accomplishing mission-related activities by moving technology out of the laboratories and moving technical expertise into the laboratories.

*Industry*: The business partner can leverage research and development (R&D) costs by building on the relevant R&D already done in federal laboratories.

*Universities and nonprofit organizations*: These entities can benefit financially from technology transfer partnerships. Individual researchers may benefit intellectually from the close contact with leading researchers in both government and industry.

#### *40.5.2.2 r2p Technology Transfer Procedure*

An invention can be either a product or a process. Every research laboratory and organization should have an established process for bringing inventions to market. What follows is a simplified, generic description of the technology transfer procedure for an invention, as implemented at NIOSH. The step-by-step process in the example is only for illustration; in actuality, the steps may be combined or followed in a different order.

*Step 1: Report the discovery or invention*

A scientist believes he or she has discovered a new product or process and reports it to the employer. An assessment is made of the patentability and marketability of the invention. A comprehensive market and economic analysis is conducted. The market analysis will determine whether the product or process is already in the market, whether it can significantly improve what is available, and whether the market will welcome the invention. The economic analysis will determine whether the cost of producing the product or process is comparable to what it might yield in income [25].

*Step 2: Seek partners*

After the product or process has been identified as novel, it can now be developed in-house or in collaboration with external partners. If seeking external partners, the scientist will look for partners with common research goals and, to protect the invention, will require potential partners to sign a confidential disclosure agreement. In such an agreement, each party agrees to recognize the confidential information and employ all reasonable efforts to maintain the confidential information of the other parties. A confidential disclosure agreement is critical to preserving U.S. and international patent rights [26].

*Step 3: Decide whether to seek patent protection*

A patent for an invention is the grant of a property right to the inventor, issued by the U.S. Patent and Trademark Office (USPTO) [27]. Generally, the term of a new patent is 20 years from the date on which the application for the patent was filed in the United States or, in special cases, from the date an earlier related application was filed, subject to the payment of maintenance fees. U.S. patent grants are effective only within the United States, U.S. territories, and U.S. possessions. Under certain circumstances, patent term extensions or adjustments may be available. A U.S. patent does not protect an invention internationally; additional procedures and costs are associated with an international patent [27].

The inventor and the employer may seek assistance from public or private individuals or companies who will conduct patent searches and provide assistance in the patent process.

Not all inventions can or should be patented. The patent process is long and costly. Many more products or processes are invented than can be supported in the market. When deciding whether or not to patent, consider the following [25]:

1. Is the product or process unique? To be patentable, a product or process must be novel. Also determine the potential life expectancy of the product or process. If the market for the invention is volatile or quickly advancing, it may not be cost-effective to patent the invention.

2. Has a market analysis been conducted? If there will be insurmountable hurdles to getting the product to market, it may not be feasible to seek patent protection.

3. Has an economic analysis been completed?
4. Will the patent be of value to a commercial partner?

*Step 4: Commercialize the product or process*

To begin the commercialization process, the inventor develops a prototype (usually functional) to show to companies to solicit interest. The prototype might not be the version that becomes commercialized, if its architecture is cost prohibitive [28].

If appropriate patent protection of the discovery has not been filed, company representatives should be asked to sign a confidentiality agreement before the scientist divulges the full details. Generally, it is necessary to discuss the invention in broad terms to solicit interest but to save detailed information until after the nondisclosure form is signed. If patent protection has been filed, the confidentiality agreement is not necessary [26].

When a private company decides to commercialize an invention, a complex series of negotiations follow. These negotiations are generally handled by those experienced in them. The negotiations form the foundation for developing the licensing terms and include the plans for commercializing the invention [26].

*Step 5: Collect royalties*

The company sends royalty payments as agreed upon in the licensing agreement, which may include payments before the product is commercially available.

**r2p IN ACTION**

*Physical hazard*: Musculoskeletal stress during hand-intensive work

*r2p Component addressed*: Transfer technology

*Description*: NIOSH scientists invented an ergonomically designed handle to maximize the power of the hand yet minimize musculoskeletal stress during hand-intensive work. The handle was originally designed to reduce the physical stressors associated with terminal insertion for factory workers, but it was found to have broader application for gardening hand tools, knives, box cutters, screwdrivers, and other handled tools for pushing, pulling, or turning.

The inventors, working without external partners, reported the design innovation to the CDC Technology Transfer Office, who conducted a patentability search and marketability analysis. Patent protection was pursued because the innovation was unique and the commercial applications were broad. Both U.S. and international patent protections were obtained (US 6,094,780, issued August 1, 2000; Australian patent 698779, issued February 18, 1999).

The NIOSH inventors developed a prototype and promotional material and sought commercial vendors. The potential vendors were identified through an Internet search, perusal of trade publications, and personal contacts at trade shows. Although many industry candidates expressed interest in this discovery and agreed on its need, none committed to manufacturing it. Interest in commercializing this invention waned when the Occupational Safety and Health Administration (OSHA) rescinded the ergonomics standard. If OSHA revisits the ergonomics standard, NIOSH may consider engaging an external organization to further explore the commercial viability of this innovation.

## 40.6 EVALUATE IMPACT

Improving the health and productivity of workers is what occupational ergonomists strive to accomplish. Without evaluation, one cannot determine whether the efforts are effective or perhaps even harmful [29]. According to the CDC [30], "effective program evaluation is a systematic way to improve and account for public health actions by involving procedures that are useful, feasible, ethical, and accurate." Evaluation generally answers questions related to the quality, cost-effectiveness, and importance of the effort.

In 2004, NIOSH contracted with the National Academies to conduct a series of independent evaluations on eight NIOSH research programs, focusing on the relevance and impact of those programs on reducing work-related injuries, illnesses, and hazardous exposures. NIOSH used this as an instructive activity in which it judged each program's activities and outputs on the following criteria: (1) its contribution, if any, to scientific knowledge and (2) the extent to which it was making a significant and important contribution to improving worker safety and health [31]. These evaluations yielded a number of recommendations to improve the NIOSH programs.

The CDC *framework for evaluation in public health* [30] is directed toward evaluating large public health programs; however, the concepts can be applied in evaluating the effectiveness of r2p efforts. CDC notes that the six steps it mandates for any evaluation are interdependent and may be taken in a nonlinear sequence. The steps are outlined in Table 40.7.

The NIOSH experience with the national academies and the CDC framework are two examples of large evaluation efforts. Evaluations on a much smaller scale or in combination with other evaluation efforts will still be effective in determining whether what you set out to do has been accomplished.

## 40.7 CHAPTER SUMMARY

This chapter outlined the four basic components of r2p at NIOSH: identification of research needs, design and conduct of research, transfer of knowledge and technology, and evaluation of impact. These components are seen and interpreted differently in other institutional models. Partnerships are at the heart of the NIOSH r2p framework. Careful selection of partners for each stage of research is essential. Partners may help with all of these basic components while also lending credibility to the appropriate audiences and providing resources. Comprehensive planning for the transfer of research knowledge or technology aids both the researcher and the practitioner and increases the likelihood that the research findings will be adopted for the benefit of workers. Evaluation ensures accountability by focusing on measurable health and productivity outcomes.

In occupational ergonomic research, the application and understanding of transferring research into practice is just beginning. The gap between "what is known" and "what is practiced" will narrow as the knowledge gained from research to practice continues to mature. Further understanding of how research moves into practice to benefit society is needed. Following the principles outlined in this chapter will help ensure ergonomic research contributes to the improvement of health and productivity.

TABLE 40.7 Steps in Program Evaluation, as Outlined in the Framework for Program Evaluation in Public Health

| Step | Description |
|---|---|
| Engage stakeholders | Stakeholders (the persons or organizations having an investment in what will be learned from an evaluation and what will be done with the knowledge) must be engaged in the inquiry to ensure that their perspectives are understood. When stakeholders are not engaged, an evaluation might not address important elements of a program's objectives, operations, and outcomes. Therefore, evaluation findings might be ignored, criticized, or resisted because the evaluation did not address the stakeholders' concerns or values |
| Describe the program | Program descriptions convey the mission and objectives of the program being evaluated. Descriptions should be sufficiently detailed to ensure understanding of program goals and strategies. Aspects to include in a program description are need, expected effects, activities, resources, stages of development, context, and logic model |
| Focus the evaluation design | The evaluation must be focused to assess the issues of greatest concern to stakeholders yet use time and resources as efficiently as possible. A thorough plan anticipates intended uses and creates an evaluation strategy with the greatest chance of being useful, feasible, ethical, and accurate. Among the items to consider when focusing an evaluation are purpose, users, uses, questions, methods, and agreements |
| Gather credible evidence | An evaluation should strive to collect information that will convey a well-rounded picture of the program so that the information is seen as credible by the evaluation's primary users |
| Justify conclusions | The evaluation conclusions are justified when they are linked to the evidence gathered and judged against agreed-upon values or standards set by the stakeholders. Stakeholders must agree that conclusions are justified before they will use the evaluation results with confidence. Justifying conclusions on the basis of evidence includes the use of standards, analysis and synthesis, interpretation, judgment, and recommendations |
| Ensure use and share lessons learned | Deliberate effort is needed to ensure that the evaluation processes and findings are used and disseminated appropriately. Preparing for use involves strategic thinking and continued vigilance, both of which begin in the earliest stages of stakeholder engagement and continue throughout the evaluation process. Five elements are critical for ensuring use of an evaluation: design, preparation, feedback, follow-up, and dissemination |

*Source:* Centers for Disease Control and Prevention, Framework for program evaluation in public health, *MMWR Recommendations and Reports* 48(RR11), 1–40, 1999.

**r2p IN ACTION**

*Physical hazard*: Previous work with California wine grape vineyard workers has indicated high rates of musculoskeletal disorders in this population [32].

*r2p Component addressed*: Evaluate impact

*Description*: Researchers partnered with vineyard owners/operators, vineyard workers, and external ergonomics researchers to compare the use of a small picking tub with the use of known alternative technologies for handling cut grapes during harvest. The evaluation compared preintervention and postintervention ergonomic stress data, musculoskeletal disorder–related pain, symptom data, and adoptability data.

Surveillance data collected subsequent to the evaluation project demonstrated that management and workers in wine grape production adopted the smaller, lighter picking tubs (>3000 tubs in 2002; 3400 in 2003). The only incentive for adoption was improved working conditions [33].

## REFERENCES


1. National Institute for Occupational Safety and Health. *Research to Practice at NIOSH.* http://www.cdc.gov/niosh/r2p (accessed October 6, 2009), 2009.
2. J. Howard. Informing public health policy and practice: The strategic management of research processes and organizations. *Governance: An International Journal of Policy, Administration, and Institutions* 22(2): 203–216 (2009).
3. Canadian Institutes of Health Research. *Knowledge Translation Strategy 2005–2009: Innovation in Action.* http://www.cihr-irsc.gc.ca/e/26574.html#blueprint (accessed October 6, 2009), 2009.
4. National Institutes of Health. *Re-engineering the Clinical Research Enterprise: Translational Research.* http://commonfund.nih.gov/aboutroadmap.aspx (accessed October 6, 2009), 2011.
5. S.H. Woolf. The meaning of translational research and why it matters. *JAMA* 299: 211–213 (2008).
6. The Ohio State University Medical Center. *Translational Research.* http://medicalcenter.osu.edu/research/translational_research/Pages/index.aspx (accessed October 6, 2009), 2009.
7. Centers for Disease Control and Prevention. *CDC Partnership Tool Kit, Draft Version 1.1.* http://intranet.cdc.gov/nchm/dpsa/ptk/full_version.pdf (accessed November 8, 2009), 2006.
8. K. Schmeer. *Guidelines for Conducting a Stakeholder Analysis.* Partnerships for Health Reform, ABT Associates, Inc., Bethesda, MD, 1999.
9. Employers' Organisation. *Making the Most of Partnership Working.* http://www.lgpartnerships.com/ (accessed November 8, 2009).
10. D. Michaels and W. Wagner. Disclosure in regulatory science. *Science* 302: 2073 (2003).
11. National Institutes of Health. Developing sponsored research agreements: Considerations for recipients of NIH research grants and contracts, *NIH Guide for Grants and Contracts* 23(25). http://grants.nih.gov/grants/guide/notice-files/not94-213.html (accessed October 6, 2009), July 1, 1994.
12. Bureau of Labor Statistics. *Incidence Rates for Nonfatal Occupational Injuries and Illnesses Involving Days Away from Work Per 10,000 Full-Time Workers by Industry and Selected Events Or Exposures Leading to Injury Or Illness.* http://www.bls.gov/data/home.htm (accessed October 8, 2009), 2006.
13. T. Waters, V. Putz-Anderson, A. Garg et al. Revised NIOSH equation for the design and evaluation of manual lifting tasks. *Ergonomics* 36: 749–776 (1993).
14. M. Siegel and L. Doner. *Marketing Public Health: Strategies to Promote Social Change,* Aspen Publishers, Gaithersburg, MD, 1998.
15. E. Rogers. *Diffusion of Innovations.* Free Press, New York, 2003.
16. P. Kotler, N. Roberto, and N. Lee. *Social Marketing: Improving the Quality of Life.* SAGE Publications, Thousand Oaks, CA, 2002.
17. W. Smith and J. Strand. *Social Marketing Behavior: Practical Resources for Social Change Professionals.* Academy for Educational Development, Washington, D.C., 2008.

18. Canadian Population Health Initiative. *An Environmental Scan of Research Transfer Strategies.* http://secure.cihi.ca/cihiweb/products/EnviroScan.pdf (accessed January 30, 2011), 2001.
19. K. Graham and J. Logan. Using the Ottawa model of research use to implement a skin care program. *Journal of Nursing Care Quality,* 19: 18–24 (2004).
20. National Institute for Occupational Safety and Health. *Safe Patient Handling Training for Schools of Nursing.* http://www.cdc.gov/niosh/docs/2009-127/ (accessed January 31, 2011), 2009.
21. I. Ajzen. From intentions to actions: A theory of planned behavior, in *Action-Control: From Cognition to Behavior.* J. Kuhl and J. Beckmann, (eds.), Springer, Heidelberg, Germany, 1985.
22. N.N. Menzel, N.L. Hughes, T. Waters, L. Shores, and A. Nelson. Preventing musculoskeletal disorders in nurses: A safe patient handling curriculum module for nursing schools. *Nursing Educator* 3: 130–135 (2007).
23. A.L. Nelson, T.R. Waters, N.N. Menzel et al. Effectiveness of an evidence-based curriculum module in nursing schools targeting safe patient handling and movement. *International Journal of Nursing Education Scholarship* 4: 1–19 (2007).
24. Federal Laboratory Consortium for Technology Transfer. *FLC Technology Transfer Desk Reference, 2006.* http://www.federallabs.org/pdf/T2_Desk_Reference.pdf (accessed October 1, 2009), 2006.
25. D. Lancaster. When to patent, *Midnight Engineering.* http://www.tinaja.com/glib/when2pat.pdf (accessed November 8, 2009), 1994.
26. Centers for Disease Control and Prevention. *Technology Transfer.* http://www.cdc.gov/osels/laboratory_science/tech_tran/index.htm (accessed February 15, 2011), 2010.
27. U.S. Patent and Trademark Office. *General Information Regarding Patents.* http://www.uspto.gov/patents/basics/index.html (accessed October 19, 2009), 2005.
28. D.M. Anderson. *Commercialization.* http://www.halfcostproducts.com/commercialization.htm (accessed February 15, 2011), 2010.
29. R. Vaughan. Evaluation and public health. *American Journal of Public Health* 93(3): 360 (2004).
30. Centers for Disease Control and Prevention. Framework for program evaluation in public health. *MMWR Recommendations and Reports* 48(RR11): 1–40 (1999).
31. Institute of Medicine and National Research Council. Committee on the Review of NIOSH Research Programs. *Evaluating Occupational Health and Safety Research Programs: Framework and Next Steps.* http://www.nap.edu/catalog/12639.html (accessed November 16, 2009), 2009.
32. National Institute for Occupational Safety and Health. *Vineyard Harvest Ergonomics Intervention Partnership.* http://www.cdc.gov/niosh/pgms/commpart/vinyard.html (accessed February 10, 2011), 2003.
33. J.A. Miles, F.A. Fathallah, J. Faucett, J.M. Meyers, and I. Janowitz. *Ergonomic Evaluation of Vineyard Systems.* http://www.cdc.gov/niosh/test/n2-const/article-2340.html (accessed November 8, 2009), 2004.
34. Centers for Disease Control and Prevention. *What is Health Marketing?* http://www.cdc.gov/healthmarketing/whatishm.htm (accessed October 9, 2009), 2005.
35. J. Tetroe. Knowledge translation at the Canadian Institutes of Health Research: A primer. *Focus.* Technical Brief No. 18. http://www.ncddr.org/kt/products/focus/focus18/. (accessed October 9, 2009), 2007.

# VII

# International Perspective on Ergonomics

CHAPTER 41

# Overview of Ergonomic Research and Some Practical Applications in Sweden

Lennart Dimberg

CONTENTS

## 41.1 INTRODUCTION

The Scandinavian countries, especially Sweden, have a long history of concern for the health of their workers, Sweden being one of the first countries in the world to adopt regulations on the ergonomics of the work environment, The Swedish Ordinance Concerning Work Postures and Working Movements, in 1983 [1].

Sweden has 8.7 million inhabitants and one of the Western world's highest proportion of female workers: 86% of women 15–64 years of age with children less than 10 years old had employment in 1988. It has 2.2 physicians per 1000 inhabitants, and in 1990, about 75% of all workers were covered by an occupational health program. The retirement age is 65 years of age, and in 1993, the unemployment rate was 8.8%. The Swedes have one of the world's highest average life expectancies—78 years for women and 73 for men in 1992—and the second lowest infant mortality rate in the world (second to Japan).

Volvo has used ergonomic designs for the interior of its cars for many years, and a study by Kelsey and Hardy [2] suggested that its low back support seats, developed in cooperation with orthopedic surgeons, reduced the frequency of lumbar disk surgery. Over the years, Volvo has transformed the worker-oriented approach into practical manufacturing examples in the car industry as well, in both the Kalmar (1974–1994) and Uddevalla (1985–1993) plants. However, in spite of all the work in the field of ergonomics, the number of reported work-related musculoskeletal injuries as well as overall sickness absenteeism continued to grow during the 1980s, and the Volvo workers were equally affected.

## 41.2 HISTORICAL BACKGROUND AND SIGNIFICANCE TO OCCUPATIONAL ERGONOMICS

### 41.2.1 General Aspects

In Sweden, the first law on workers' protection was adopted in 1912. In those days, the main problem was long working hours (12–14 h/day), which was thought to be the main reason for work-related accidents. In 1920, the 8 h workday was stipulated by law. A new law was adopted in 1949 that called for the introduction of mandatory health and safety committees in companies with more than 50 employees and worker safety representatives in companies with more than five employees.

The National Board of Occupational Health and Safety was also formed. The current Work Environmental Act was adopted in 1977 and emphasizes the importance of prevention in both physical and psychological terms. It applies to all areas of occupational life, private and public sectors alike. Through special ordinances, this board instructs employers as to how the work environment law should be followed.

From 1951 to 1992, agreements between management and the unions regulated the development of occupational health care services. Since 1993, these agreements have been under renegotiation. Occupational health care is not required by law.

#### *41.2.1.1 Labor Inspectorate*

The National Board of Occupational Safety and Health is the central administrative authority in Sweden and is also the authority guiding the field organization, the Labor Inspectorate. The Labor Inspectorate is a law enforcement agency whose inspectors regularly monitor workplaces. In recent years, its actions have been directed more toward system inspection, looking at the structure of how health and safety work is organized and implemented. Traditionally, it has focused on accidents, and the implementation of ergonomic regulations has, relative to all the reported musculoskeletal work injuries, only rarely been enforced.

#### *41.2.1.2 Work Environment Fund*

The work environment fund was founded in 1972 and grants research funding of about US $100 million per year (1990) for work environment-related research.

#### *41.2.1.3 Fund of Working Life*

As of 1989, a special fee (1.5% of salary costs) has been paid to a work environment fund, which distributes grants totaling about US $300 million annually to employers for special projects on rehabilitation and improvement of the work environment. To receive funds, an employer must present a written program describing existing problems and showing plans to deal with them. The fund will pay part of the costs of projects that meet its criteria.

#### *41.2.1.4 Occupational Health*

There are, in principle, three types of occupational health services in Sweden: in-plant service, out-of-plant group service, and branch service. In-plant service is the most common in the larger companies (with more than 1000 employees); out-of-plant service is most common in smaller companies; and branch service is representative of, for instance, the construction industry (Bygg-hälsan). In general, in the larger companies, there are 1 physician, 1 safety engineer, 1/2 physiotherapist, and 2 nurses per 2000 employees. Together, these professionals usually form a team to work with ergonomic problems.

### 41.2.2 Time Trends: Only Being Best Is Good Enough

Sweden has, like all other highly industrialized countries, a high cost of labor. Therefore, Swedish industry has found other ways of competing with low-cost countries, manufacturing high-value products with a focus on high technology, robotization, and a highly competent workforce. Swedish managers realize the high potential of a motivated worker, and a good work environment is an important motivator. However, now that high technology is readily available in low-cost countries as well and new computer programs make it easier to learn and use computers, the employers' demand for the best possible team is understandable. This also constitutes a risk that sick or disabled workers will no longer be taken care of by employers. It is important to find a humane solution to this problem.

#### *41.2.2.1 Decentralized Organization and Increased Worker Involvement*

Swedish companies like Sandvik, Asea Brown Boveri (ABB), Ericsson, and Volvo all have programs to decentralize their organizations. The hierarchy is flattened (fewer bosses), meaning that the number of decision levels is reduced to hasten the decision-making process. Work groups are being introduced, and supervisory responsibilities for planning, economy, quality, personnel, and production are shared among the workers. The goals of a group are set, and it is up to the group to decide on the means to achieve them. This has meant that health and safety matters are also delegated, which constitutes both an opportunity and a risk. If a worker needs an ergonomic chair or a lifting device, that can be decided on the floor, but there is also a risk that the specialized knowledge of rules and regulations is lost because the responsibility is now shared by several people.

The Japanese have shown that worker participation in safety work is crucial to good results [3]. However, individuals with certain handicaps may need special consideration, which, in such groups, may not necessarily exist. Leif Wallin [4] developed an interesting method to have the group analyze their work situation by means of a questionnaire and subsequent discussion of the results.

#### *41.2.2.2 Job Enlargement and Job Enrichment*

Since repetitive work, especially in fixed body positions, has been associated with muscular strain problems [5,6], occupational physicians have applauded when jobs have been enlarged. In the beginning, this often meant that a worker rotated between different, but similarly stressful, production operations—for example, from assembly to inspection of parts. It was clear that this was not enough to eliminate the repetitive strain problems. Also, work technique and its consequences for musculoskeletal disorders were evaluated by Kilbom and Persson [7]. Anyone who has played the guitar knows what the right technique means and the muscular fatigue of an incorrect technique.

The recent ideas of enriching the work by introducing completely different types of work such as planning, economics, and quality may be a better way to prevent these problems, but this still remains to be shown.

#### *41.2.2.3 Automation, Robots, and Computers*

The fast growth of computer technology has meant that the next generation of industrial robots will be run by workers without extensive assistance from programmers and computer personnel. This clearly represents a breakthrough in the prevention of muscular pain problems. For example, a very high frequency of neck and arm problems was reported on a production line at Volvo Aero Corp., where connecting rods were made (Figure 41.1). An automatic process using a line of robots was introduced (Figure 41.2), and now the workers only have to watch and serve the machines, and orthopedic problems are rare. At the Matsushita Company in Tokyo that makes radios for Volvo, robots have now relieved women of tasks that gave them pains in the assembly of radios.

#### *41.2.2.4 ISO 9000 and Total Quality Management*

Quality control is a fashionable term now in Sweden. The ISO 9000 quality standard calls for the documentation of every process and routine. Total quality management is an instrument of wider scope used to define, describe, and measure all processes and their relation to a business.

FIGURE 41.1 Manual work with connecting rods.

FIGURE 41.2 A numerically controlled (NC) machine.

The quality of work in occupational health services also needs to be looked at. For example, in the field of ergonomics, there is often, out at the plants, a lack of prognosis and little or no follow-up of implemented ergonomic solutions to work-related health problems.

#### *41.2.2.5 Process Organization Analysis*

Volvo's main process is the manufacture of transport equipment. The role of occupational health services in this process concerns the prevention of injuries and diseases, the treatment of medical problems, and rehabilitation supporting the core process. In some companies, many of the peripheral processes such as cleaning and guarding are contracted outside the company. In-plant occupational health services have also been questioned by some companies, but experience of outside contracting is that quality deteriorates, and that is especially true of ergonomic work.

#### *41.2.2.6 Kaizen*

It is clear that the closer one is to the workplace, the better one understands the problems. Many workers have suggestions not only on how to improve the quality of work but also on ergonomic solutions. The Japanese term *kaizen* means daily improvement, and this must also include the ergonomics of the workplace. It is important to act before the problems get out of hand. In Sweden, there has been a tendency to let the worker health and safety representatives (*skyddsombud*) deal with these problems through their organization. However, the law has recently been changed to emphasize that the employer alone is responsible, which has meant that these issues are now dealt with together with other production and quality issues. It is hoped that this will lead to the greater involvement of all workers in ergonomics.

#### *41.2.2.7 Worker Selection Process*

Over the years, much hope has been placed on preventing musculoskeletal problems by selecting the right workers for the job. Unfortunately, the conclusion reached by Kilbom and Hagberg [8] in a major literature survey is that no good screening program exists for the prevention of repetitive strain injuries. The idea that strength and fitness are of importance in the prevention of back problems [9] is contradicted by Battié's prospective study at Boeing in Seattle [10], where only the number of previous back problems was a predictive variable.

#### *41.2.2.8 Health Promotion and Worker Fitness*

Saab has developed a questionnaire and physical test that are presented to each participant to obtain a health profile. The general idea is that this profile should be an instrument for behavior change to a sounder lifestyle. Volvo has a similar program called the Life Line. In an evaluation, this has been shown to increase general well-being, but no preventive effect against muscular stain problems has been documented. A proven prophylactic program, however, has been the Volvo Back School, which is a prevention program that was used in a prospective, controlled study of patients with chronic back problems [11].

#### *41.2.2.9 Flexible Working Hours*

A system of flexible hours for blue-collar workers has also been implemented in some Volvo plants. It has the advantage of reducing stress associated with rigid shift hours, and workers with rheumatoid problems and morning stiffness may, for example, benefit from starting the workday somewhat later.

#### *41.2.2.10 From Blue- and White-Collar Worker to Fellow Worker*

ABB and Volvo are presently discussing the problems associated with the separation of workers into blue- and white-collar collectives. From the perspective of rehabilitation, it has been shown to be very difficult, especially in a downsizing situation, to transfer, for example, a welder (blue collar) with a chronic neck problem to a suitable office (white collar) job. Here, the unions sometimes cannot agree.

### 41.2.3 Examples at Volvo

#### *41.2.3.1 Kalmar Plant*

Attempts to facilitate the production by preassembly of, for example, doors were introduced at the original Torslanda factory, as shown in Figure 41.3. The vision of the Kalmar plant was to break the traditional line system. Involving the workers in more varied assembly work was expected to improve worker satisfaction, avoid repetitive strain injuries, and thereby increase quality. In the Kalmar plant, which was started in 1974, the mechanical conveyor belt was replaced by self-propelled trolley carriers (Figure 41.4) on which the car bodies were moved. The carriers were also equipped with tilting equipment so that work on the underside of the car could be performed in a comfortable posture. The work team became a factor in an endeavor to break down the impersonality of a big operation into sections that were easier to assimilate. The car was still moving between different workstations, but the trolley did not move during the time a given team was working on the car.

FIGURE 41.3 Traditional car assembly line at Volvo Torslanda.

FIGURE 41.4 The moving line was broken at Kalmar and assembly performed on a stationary platform.

However, even though the rate of reported muscular strain problems was lower at this plant compared to traditional line work, it was still unacceptably high.

#### *41.2.3.2 Uddevalla Plant*

In 1985, the opening of the Uddevalla plant signaled a revolution in the process of making cars (Figure 41.5). The Uddevalla plant's new methods were based on small work teams. Since each work team had a number of different tasks to perform and it was possible to lengthen the work cycles greatly, the individual team member's work could be more varied.

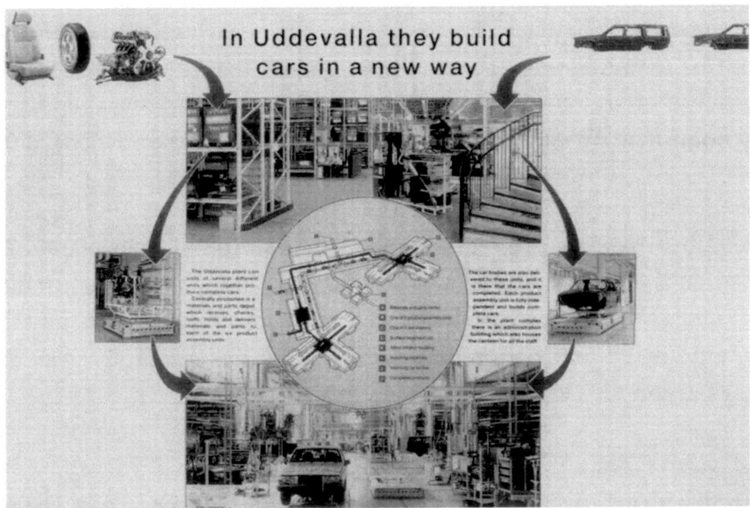


FIGURE 41.5 The concept of the Uddevalla plant.

The idea was that involving the team in planning, assembly, and quality work would provide so much variation that it would be mentally stimulating to the workers, and with this variation, one would also avoid all repetitive strain problems. In Uddevalla, the whole car was built at one station by a team of eight persons. Owing to a number of interesting ergonomic solutions, 80% of the time, the car could be manufactured in an upright ergonomically optimum body posture, as shown in Figure 41.6, by rotating it by 90°, compared to 20% in the old line system. Unfortunately, this did not prove to be the final solution for repetitive strain problems. In fact, the frequency soon increased to parallel the reported work accident frequency of the Kalmar plant. One reason was the high percentage of female

FIGURE 41.6 At the Uddevalla plant, 80% of the time the car was built in an upright economically optimum body posture.

assembly operators whose higher sickness absenteeism (40% higher than for male workers) was a contributing factor. Perhaps the main reason was that the cars were still being built almost completely by hand. Even if each particular assembly task was repeated only a few times per day, some assembly tasks required working with force in awkward positions.

#### *41.2.3.3 Skövde Plant*

The moving platform system, originally implemented at the Kalmar plant, has come into widespread use. In the new engine line at Skövde, the platforms have individually adjustable fixtures (Figure 41.7), which clearly afford less musculoskeletal strain. In this plant, an almost totally automated production line for gearboxes has shown the way to completely avoid manual assembly and thereby repetitive strain problems.

#### *41.2.3.4 Volvo Aero Plant*

Manual deburring of turbine blades for jet engines (Figure 41.8) has, over the years, caused large numbers of muscular strain problems in the neck and shoulders. The final solution to this problem turned out to be the introduction of a deburring robot (Figure 41.9).

FIGURE 41.7 The engines at the Skövde plant are built on platforms with individually adjustable fixtures.

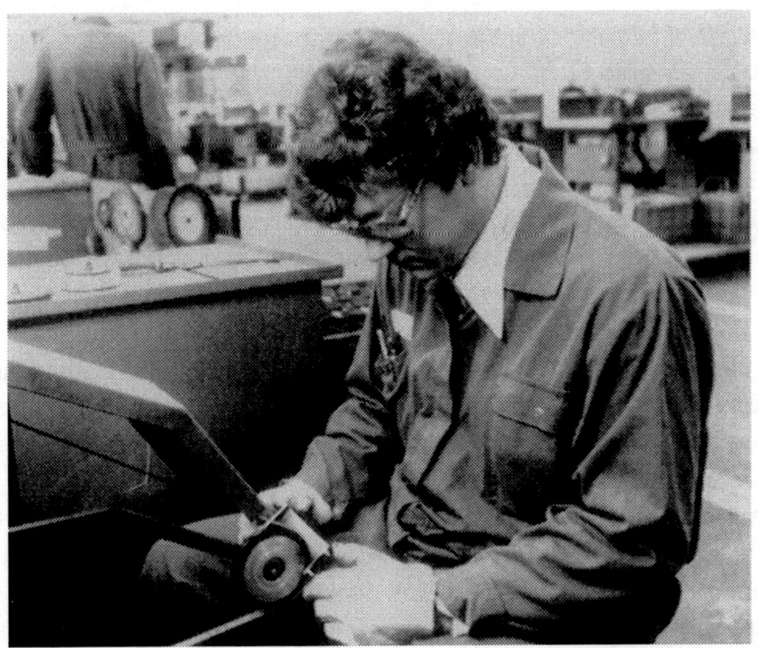

FIGURE 41.8 Manual polishing of turbine blades.

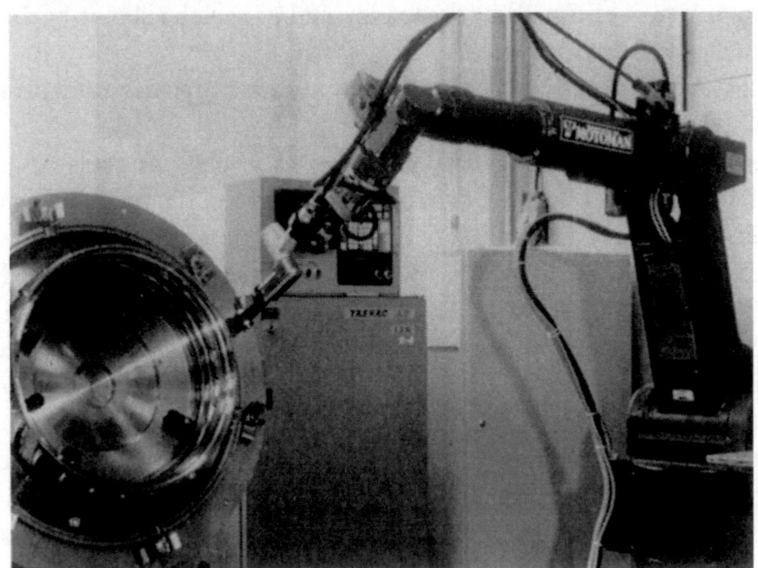


FIGURE 41.9 A deburring robot.

## 41.3 WHAT WENT WRONG?

### 41.3.1 Missing Link

Given the fact that improved ergonomics in general has not resolved the problem of repetitive strain injuries, one may speculate as to whether or not the injury mechanism really is clearly understood. Is there a missing link?

#### *41.3.1.1 Theoretical Injury Mechanisms of Repetitive Strain Disorders*

In a review article, Hagberg [12] summarizes injury mechanism. The following paragraphs draw strongly on that article.

Degenerative joint disease—osteoarthrosis or osteoarthritis—may be caused by increased stress on the cartilage, such as repetitive impulsive loading. According to some authors, this is sometimes preceded by trabecular microfractures in the subchondral bone caused by trauma. Other authors point to clinical evidence of polyarticular disease and suggest a metabolic abnormality of the articular cartilage. It is claimed that insertion disorders in tendons, ligaments, and articular capsules are caused by local ischemia leading to degeneration and producing inflammation and pain. In particular, the tendons of the supraspinatus, the biceps brachii, and the upper part of the infraspinatus muscles have a zone of avascularity. This has been found to be the site of microruptures and degeneration that may be accelerated by aging.

Impairment of the venous circulation may occur when the humeral head compresses the tendons (elevated arm) but also when there is increased tension in the tendon. Tendon inflammation has been provoked by repetitive contractions in rabbits.

Degenerative tendinitis in the shoulder girdle aroused by exertion, for example, may trigger a foreign body response inflammation.

Tenosynovitis is an inflammation of the tendon sheath and its synovia. In the long biceps tendon, this may be caused by the tendon and its sheath rubbing against the lesser tuberosity during overhead movements.

Postinfective arthritis as well as tendinitis may presumably predispose a person exposed to shoulder stress to a more severe reaction.

Muscle tenderness, myofascial syndrome, trapezius myalgia, and related disorders are obscure conditions because the pain does not originate from the contractile muscle fibers themselves. It may possibly derive from pain fibers within the blood vessels or the connective tissue. Hagberg [12] points to three pathophysiological routes. The first is mechanical failure with ruptures of Z-disks probably caused by temporary high local stress. The second is local ischemia due to the impairment of the circulation by continuous muscular performance, which may already occur at 10%–20% of the maximum voluntary contraction. This leads to a fall in pH and reversible enzyme inhibition. It is postulated that the tissue irritation causes extravasation of blood, edema, and fibrositis in some individuals. Highly repetitive work may then possibly cause cumulative trauma to the muscle cell [thence cumulative trauma disorders (CTDs)], affecting both morphology and energy metabolism.

The third pathophysiological route would be energy metabolism disturbance. Energy depletion in the muscle cell has been suggested as one factor in muscle pain. Defects in the energy metabolism are often associated with painful disorders of the muscle. Laboratory experiments involving repetitive shoulder flexions have produced energy depletion as indicated by an increasing serum creatine kinase and accompanying pain. It is hypothesized that this may also be important for static loads. The possibility of certain primary metabolic disturbances in certain individuals has also been proposed.

The carpal tunnel syndrome is a textbook example where the injury mechanism is clearly understood. Friction-caused synovitis of the tendon sheaths in the carpal tunnel causes pressure on the median nerve [13]. Another example is supraspinous tendinitis, where the supraspinous tendon is pressed against the acromion in a space that has been limited by inflammation [14].

Fibrositis/fibromyalgia and generalized muscular pain is a common condition but with a poorly understood pathogenetic mechanism. The theory that static muscular work causes ischemia in the muscle, creating morphological changes in the muscle fibers, has been suggested but remains to be proved [15]. It is easy to prove muscular fatigue and pain after minor static load [16,17], but the pathogenic link from repetitive chronic muscular fatigue to permanent damage and chronic pain remains to be shown.

Microfractures have been suggested by Hansson et al. [18] to be the reason for back pain in certain individuals, but no one knows how common this is. The lumbar disk and its degeneration have been connected to low back pain, and the intradiscal pressure as measured by Nachemson and Elfström [19] was for a long time the theoretical mechanism for ergonomic advice on lifting but has since been abandoned by Nachemson [20].

Spinal shrinkage as measured from height before and after loading provides a method for measuring mechanical load on the spine [21]. The shrinking is dependent on the elasticity of the intervertebral disks. However, no one knows if the shrinkage leads to permanent disk problems or whether this may be a pathogenetic mechanism for back pain.

#### *41.3.1.2 Recent Research*

*Overview*: Recent Swedish research on ergonomics has broadened the concept from physical workplace adjustment to psychological factors [22] and sociological aspects of work groups and leadership [23].

A comprehensive literature survey and analysis of health risks at work has been performed by the National Swedish Commission for the Work Environment [8]. Its conclusion is that the prevalence of and risk factors for many disorders in the locomotor system are still not satisfactorily known.

*Case studies*: Evaluations of the use of an ergonomic chair, the Ullman chair, as a secondary prophylaxis for low back problems has shown significantly less pain and fewer sick days for low back pain in this group than in a randomized control group using a traditional office chair in a 1-year prospective study [24]. Also, prophylactic inversion therapy (autotraction) by the Swedish Mastercare Inversion System (Svenska Hälsobänken) with self-training for 10 min daily has been shown to reduce the pain score significantly in chronic low back pain patients in a 1-year study at Volvo Aero Corp. [25].

### 41.3.2 Rules and Regulations

#### *41.3.2.1 Workers' Compensation System*

The Swedish workers' compensation system introduced in 1977 has been one of the most liberal in the world. This may be one explanation why Swedish industry, in spite of substantial improvements in the physical work environment and ergonomic advances, has had about the highest rate of work injuries in the world.

The acceptance of a work injury is based on two steps, the first being that a disorder should be known to be caused by work. If that was the case, in the next step, even a very low probability would allow a claim to be recognized as a work injury. This caused an explosion of filed claims for most pain problems involving the locomotor system, and over 90% were accepted as work injuries.

The acceptance of a work injury meant full salary during sick leave, and, in addition to this, compensation of pain and suffering and medical disability would provide the claimant with considerably more money than the regular salary and there was no incentive to return to work. There are several persons in Sweden with claimed muscular pain in the neck and shoulders that, at less than 30 years of age, have gone into early retirement in spite of few objective findings! In July 1993, the law was changed so that it now takes a high probability for work to be considered the causal factor of medical complaints.

#### *41.3.2.2 General Welfare System*

Sweden has for many years had a very liberal and generous health insurance paid by the government through equal fees from all employers. Although the Swedes have one of the highest life expectancy rates in the Western world, we also have one of the highest rates of sickness absenteeism. Although the Volvo plants had an average sickness rate of 12% among their blue-collar workers in 1991, it is important to know that two-thirds of these workers take sick leave less than 1 week/year. In 1991, the law was changed so that employers pay directly for only the first 2 weeks of sick leave, and the first 3 days of sickness compensation was reduced from 90% to 75% of the worker's salary. In 1992, the sickness rate decreased by 1.3% for blue-collar workers. When, in April 1993, a 1-day waiting period

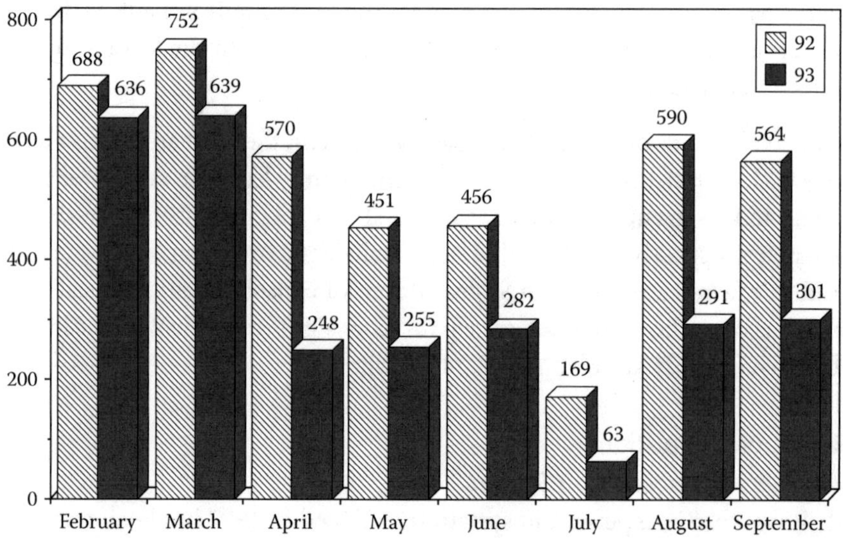


FIGURE 41.10 Sick reports by Volvo Flygmotor (3000 employees) during the time of a change in the compensation system. One day without pay was introduced in April 1993.

(with no pay) was introduced for health insurance, the number of sick reports per day at Volvo Aero Corp. was reduced by about 50% (see Figure 41.10). Against this background, it is probable that the real benefit of ergonomic applications in the work environment has been shadowed by effects of the health insurance system.

## 41.4 CRITICAL REVIEW OF CURRENT STATUS

### 41.4.1 Nachemson Back Review

In a major survey of international studies on low back pain, Nachemson [20] claims that very few prospective and controlled studies on work relatedness exist. He concludes that even if the effect of physical factors cannot be denied, work satisfaction seems to be the most important predictor of future work-limiting back problems. The importance of early and active rehabilitation is emphasized.

### 41.4.2 Svanborg Study of 70-Year-Old People

In one of the largest cross-sectional studies published to date, Svanborg [26] interviewed in great detail and physically examined some 700 70-year-old men and women in Göteborg. He concludes that there was no difference in the prevalence of pain problems in the locomotor system between the workers who had performed heavy manual labor and those who had been office workers. The only difference noted was that the former were physically stronger and had higher bone density. The group has since been followed for 10 years. At age 79, previously sedentary workers were more disabled in activities of daily living than those whose work had been physically strenuous.

### 41.4.3 Methodological Problems in the Evaluation of Implemented Solutions

Hagberg et al. [27] pointed out some methodological aspects that call for brief comment. The evaluation of doses of exposure to repetition injuries and the dose–response correlation is important. The diagnostic criteria should be stringent. Unfortunately, only a few clinical tests have been validated, and these clinical tests may be helpful for confirming a diagnosis but not for excluding a disease [28]. Ideally, an age-, gender-, and education-matched control group should be used to evaluate suggested ergonomic changes because over a period of time, job content may change, new work processes may be implemented, some workers may be transferred to other departments, and so on, which may be the real reason for improvements in health rather than the ergonomic solution. Also, time could be a healing factor that coincides with ergonomic change. Other factors such as tobacco use and alcohol consumption should also be checked [29]. The use of visits to physicians as an endpoint for measuring the prevalence of certain conditions such as tennis elbow at the workplace is not a good method. Workers with tennis elbow are more prone to see a doctor for their problem if they have a heavy job even if the real prevalence of tennis elbow is the same among workers in heavy and light jobs [30].

Swärd [31] studied top athletes and found that 85% of male gymnasts experience some low back pain and 54% have severe low back pain. Still they can operate at the top international level. Sickness absenteeism as a measure of the problem is thus seen to be affected by many psychological and motivational factors.

## 41.5 FUTURE CONCERNS

### 41.5.1 Finding the Answers

The criticism from Nachemson, the Svanborg study, and my own experience indicated that the irreversible chronic low back pain syndromes, neck/shoulder pain, fibrositis, etc., in many cases have a causative background (metabolic, infectious?) other than work. Work may aggravate aches and pains, but the aches and pains probably had another origin. It is important to distinguish between the pathological process and the provoking factor that produces pain. In such cases, ergonomic solutions may be of some help to the individuals but will usually not solve the problem.

The typical muscle strain work injury should be improved if working conditions are improved by, for instance, ergonomic changes. To be a candidate for ergonomic intervention, an injury should comply with most of the nine criteria stated for causation by Sir Bradford Hill [32]:

1. *Strength of association*: High odds ratios mean a strong association.
2. *Consistency*: Repeatedly observed by different investigators in different countries.
3. *Specificity of the association*: The injury is limited to the exposed persons.
4. *Temporal relationship*: Effects appear *after* the cause.
5. *A dose–response curve*: The higher the dose, the greater the number of cases.
6. *Biological plausibility*: The biological mechanism should be understood.
7. *Coherence of evidence*: Cause and effect should not seriously conflict with the generally known facts of natural history and biology.
8. *Reversible association*: As an example, when smokers stop smoking cigarettes, the rate at which they develop lung cancer should fall.
9. *Analogy*: After having discovered a drug effect of thalidomide on fetuses, we are more ready to accept fetal effects by other drugs.

The missing link of a pathologic process may suddenly appear. The etiologic process is better understood when, as in the case of rheumatoid arthritis, the immunologic mechanism is discovered. New imaging techniques such as magnetic resonance imaging (MRI) make it possible to visualize soft tissue and verify, for instance, changes in the carpal tunnel that explain the pain. The ultimate prerequisite for an ergonomic solution is that the pathogenetic mechanism is understood in detail.

The Swedish National Commission for the Work Environment (Arbetsmiljökommissionen, 1989 [8]) suggests that a systematic follow-up of implemented ergonomic solutions should be performed. Many suggested modifications are based more on individual ideas than on a solid base of knowledge.

### 41.5.2 Implementing the Solutions

Compiling and using collected examples of evaluated ergonomic modifications such as those published by Oxenburgh [33] is one way of implementing ergonomic solutions. A possible vision is that the next generation of intelligent, seeing, and feeling industrial robots will take over all the highly repetitive tasks that continue to provoke discomfort and pain in human workers.

## ACKNOWLEDGMENTS

I particularly wish to thank Ulf Jeverstam, AM Volvo, for valuable advice on the manuscript; Mats Ericson, Royal Swedish School of Technology, for critical comments; Dennis Savage for reviewing the language; and Sture Hård, Gothenburg University Medical Library, for help with the references.

## REFERENCES

1. Arbetarskyddsstyrelsens författningssamling (Swedish), The Swedish Ordinance Concerning Work Postures and Working Movements, 1983:6.
2. J. L. Kelsey and R. J. Hardy, Driving of motor vehicles as a risk factor for acute herniated lumbar intervertebral disc, *Am. J. Epidemiol. 102*(1):63–73 (1975).
3. K. Noro, *Participatory Ergonomics: Concept, Advantages and Japanese Cases in Human Factors in Organizational Design and Management*, North-Holland, Amsterdam, the Netherlands, pp. 83–86 (1990).
4. L. Wallin, Modification of work organization, *Ergonomics 30*(2):343–349 (1987).
5. S. Kvarnström, Occurrence of musculoskeletal disorders in a manufacturing industry with special attention to occupational shoulder disorders, *Scand. J. Rehabil. Med. Suppl. 8*:1–144 (1983).
6. L. Dimberg, A. Olafsson, E. Stefansson et al., The correlation between the work environment and the occurrence of cervicobrachial symptoms, *J. Occup. Med. 31*(5):447–453 (1989).
7. Å. Kilbom and J. Persson, Work technique and its consequences for musculoskeletal disorders, *Ergonomics 30*(2):273–279 (1987).
8. Å. Kilbom and M. Hagberg, Arbeten utsatta för särskilda hälsorisker, Arbetsmiljökommissionen, Swedish, 1989.
9. S. H. Snook, Approaches to preplacement testing and selection of workers, *Ergonomics 30*(2):241–247 (1987).
10. M. C. Battié, The reliability of physical factors as predictors of the occurrence of back pain reports. A prospective study within industry, Thesis, University of Göteborg, Gothenburg, Sweden, 1989.
11. A. Nachemson, I. Lindström, C. Öhlund et al., Symposium om "ryggproblemens lösning," *Läkartidningen 85*(50):4437 (1988) (in Swedish).
12. M. Hagberg, Occupational musculoskeletal stress and disorders of the neck and shoulder: A review of possible pathophysiology, *Int. Arch. Occup. Environ. Health 53*(3):269–278 (1984).
13. G. Lunborg, *Nerve Injury and Repair*, Churchill Livingstone, New York, 1988.
14. P. Herberts, R. Kadefors, C. Hogfors et al., Shoulder pain and heavy manual labor, *Clin. Orthop. 191*:166–178 (1984).
15. K. G. Henriksson, Exercise induced myopathies in man, *Opuscula Med. 28*(2):34–37 (1983).
16. M. Hagberg, On evaluation of local muscular load and fatigue by electromyography, *Arb. Hälsa 24*:1–53 (1981).

17. B. G. Jonsson, J. Persson, and Å. Kilbom, Disorders of the cervicobrachial region among female workers in the electronics industry. A two-year follow up, *Int. J. Ind. Ergon. 3*:1–12 (1988).
18. T. Hansson, T. Keller, and R. Jonsson, Fatigue fracture morphology in human lumbar motion segments, *J. Spinal Disord. 1*(1):33–38 (1988).
19. A. Nachemson and G. Elfström, Intravital dynamic pressure measurements in lumbar discs. A study of common movements, maneuvers and exercises, *Scand. J. Rehabil. Med. (Suppl.) 1*:1–40 (1970).
20. A. Nachemson, Ont i ryggen, orsaker, diagnostik och behandling, SBU, 1991, (in Swedish).
21. M. Ericson and I. Goldie, Spinal shrinkage with three different types of chair whilst performing video display unit work, *Int. J. Ind. Ergon. 3*:177–183 (1989).
22. T. Sivik, Diagnosis and treatment of patients with idiopathic low back pain, Thesis, University of Göteborg, Gothenburg, Sweden, 1992.
23. L. Dimberg, L. Wallin, and B. Eriksson, Unpleasant atmosphere at work increases the risk of musculoskeletal disorders, *Läkartidningen 88*(11):981–985 (1991) (in Swedish).
24. L. Dimberg, M. Ericson, and J. Ullman, The effects of low back pain and disorders by sitting in non-traditional chairs with forward sloping seats compared to traditional chairs with horizontal seats, *Proc. Swedish Med. Assoc. (Riksstämman) 46*:121 (1992) (in Swedish).
25. L. Dimberg, L. G. Josefsson, and B. Eriksson, Prophylactic inversion therapy in chronic low-back pain patients in the workplace—A prospective, randomized controlled study, *Proc. Swedish Med. Assoc. (Riksstämman) 9P*:114 (1993).
26. A. Svanborg, 70-Year-old people in Gothenburg, Sweden: A population study in an industralized Swedish city. Practical and functional consequences of aging, *Gerontology 34*(Suppl. 1): 11–15 (1988).
27. M. Hagberg, L. Jorulf, and Å. Kilbom, *Methodologic Problems in Muscular Strain Epidemiology*, Natl. Inst. of Occupational Health, Sweden, Report 19, 1988 (in Swedish).
28. E. Viikari-Juntura, Interexaminer reliability of observations in physical examinations of the neck, *Phys. Ther. 67*:1526–1532 (1987).
29. L. Dimberg, A. Olafsson, E. Stefansson et al., Sickness absenteeism in an engineering industry—An analysis with special reference to absence for neck and upper extremity symptoms, *Scand. J. Soc. Med. 17*(1):77–84 (1989).
30. L. Dimberg, The prevalence and causation of tennis elbow (lateral humeral epicondylitis) in a population of workers in an engineering industry, *Ergonomics 30*(3):573–579 (1987).
31. L. Swärd, The back of the young top athlete. Symptoms, muscle strength, mobility, anthropometric and radiologic findings, Thesis, University of Gothenburg, Gothenburg, Sweden, 1990.
32. B. Hill, Association is not causation, in *Principles and Practice of Research for Surgical Intervention*, H. Troidl, W. O. Spitzer, B. McPeek et al., Eds., Springer-Verlag, New York, 1986.
33. M. S. Oxenburgh, Increasing production and profit by health and safety, Sydney and Chicago' CCH International, 1991.

CHAPTER 42

# Current Status of the Ergonomics Research in China

He Lihua, Sheng Wang, and Bingshi Wang

CONTENTS



## 42.1 PREAMBLE

Although the formation and development of ergonomics has had a history of more than a century, it was only in the last 20 years that ergonomics assumed greater importance in China, which has accelerated growth in every related field.

The United Kingdom is one of the pioneering countries that initiated ergonomics research. A British scholar formed the word "ergonomics" with the Greek words "ergo" (work or effort) and "nomics" (law or normalization), that is, "ergonomics" means "work normalization" or "natural law of work." In the development of ergonomics, due to different research foci, scientists from different countries and regions have provided varied designations to this science. For example, in the United States, it is called "human factors" or "human engineering"; in Japan, it is called "人間工学 (にんげんこうがく)." The universal designation for this science is now "ergonomics," and China adopts this designation. Ergonomics, as a comprehensive application discipline, has a broad scope for research. It involves every aspect of human work and general living environment. The research in relation to work is called "occupational ergonomics." Occupational ergonomics is the study of relationships among humans, machinery, equipment, and their environment in order for people to stay healthy and work in a safe, comfortable, and effective manner.

## 42.2 DEVELOPMENT OF ERGONOMICS IN CHINA

### 42.2.1 Development of Research Disciplines and Organizational Societies

The study of ergonomics in China started very late. Although some research work in ergonomics started in the 1930s, it did not become an independent research field until the 1980s. In 1980, universities and research institutes in China started to set up ergonomics labs. In the 1980s, the first Chinese treatise on ergonomics was published and the occupational hygiene textbooks in colleges began to have chapters on occupational ergonomics. In 1980, the China Technical Committee of Ergonomics Standardization Administration was founded in Beijing. It is responsible for centralized planning, conducting research, and reviewing the establishment of the national ergonomics standards. In 1984, the State Commission of Science and Technology for National Defense Industry established the National Military Human-Machine-Environment System Engineering Standardization Technical Committee. The establishment of these two commissions was a big catalyst to the development of ergonomics research in China. In 1989, the Chinese Ergonomics Society (CES) was founded. In November 1989, the Safety and Environment Ergonomics Subcommittee of CES was founded which was later followed by the establishment of Ergonomics Management Subcommittee of CES and Human-Machine Engineering. Currently, CES has seven branches: cognition commission, human-machine engineering commission, biomechanics commission, safety and environment commission, ergonomics standardization commission, ergonomics management commission, and transportation ergonomics commission. Professor Walter Rohmert, a famous international ergonomics scholar and the former director of the Labor Science Research Institute of Technische Universität Darmstadt, had a great interest in the promotion of ergonomics in China and was very friendly to the Chinese ergonomics researchers. He always said that if ergonomics were not promoted in a large country like China, it would have been a very big deficiency for the world. With his drive and influence, at the meeting of the board of directors of the International Ergonomics Association (IEA) held in Paris in July 1991, with majority votes from the directors, CES was accepted as the member of the IEA. This was a milestone in the development of Chinese ergonomics. CES irregularly publishes newsletters, providing the latest news about CES, other affiliate societies, international academic trends, ergonomics research publication information, etc. [1–7] are some publications during that period. In September 1995, CES formally published the academic magazine *Ergonomics*.

In the early 1990s, the Beijing University of Aeronautics and Astronautics established the first program for a doctoral degree in ergonomics. Since then, the Psychology Research Institute of the Chinese Academy of Sciences (CAS), Nanjing University of Aeronautics and Astronautics, Beijing Institute of Technology, Peking University, Tsinghua University, Zhejiang University, and approximately 100 other universities and research institutes have successively established ergonomics majors and research areas. In 2005, CES established its website for the convenience of research and information exchange and for the promotion of the development of ergonomics.

In August 2009, CES successfully hosted the 17th International Symposium on Ergonomics in Beijing.

### 42.2.2 Academic Progress

The study of ergonomics in China started late. In the 1930s, the beginning stage of the Chinese ergonomics, in order to meet the application needs in the industry, Chinese psychologists introduced the concept of industrial psychology from western countries into China. Since then, ergonomics has been a research field in China. Some scholars conducted research projects on engineering psychology related to work fatigue, working environment, worker selection test, etc. In the 1950s, psychologists from the Institute of Psychology, CAS, and Hangzhou University conducted research on work psychology in the areas of operation optimization, technological innovation, accident analysis, employee training, etc. These research studies played an active role in increasing work efficiency and productivity. In addition, some scholars conducted research on the standardization of the design of gas masks, gloves, clothes, and airplane cabins. In the 1960s, some scholars shifted the focus of their research to human–machine interface relationships, such as engineering psychology covering areas such as railway signal display, power station control signal display, instrument display panel design, aviation illumination, aircraft cabin instrument display, etc. These research efforts established a firm foundation for the development of the ergonomics in China. In the late 1970s, the research on industrial psychology developed very rapidly in China.

These research efforts on engineering psychology mainly focused on fundamental applications of ergonomic principles. At present, there are many research projects dealing with driving safety and human–machine interaction [13,18–20,23,30]. The researches on driving behavior have found that the verbal communication behavior and the automobile-driving behavior have mutual impact on each other. The driver who is actively involved in conversation with another person is likely to be more affected. Furthermore, research studies also showed that when evaluating braking reaction, speech speed and speech repetition frequency are effective indices for evaluating the effect of conversation on driving behavior. Based on these findings, researchers at present mainly focus on measuring a driver's mental load, situational awareness, and fatigue in relation to electrophysiology. These types of research studies have generated some beneficial outcomes. In recent years, other research studies in addition to driving safety research work have included spatial cognition, stimulation–reaction compatibility, etc. Those research results provide design principles for manual control interface design that conforms to the human psychological processing.

As a result of rapid growth in computer science and technology, scientists from China have made significant contributions to the technology for Chinese character recognition and input into computer. These research efforts provided psychological evaluation principle of compatibility for a variety of Chinese character input methods. Based on the principles of cognition psychology, these research studies brought forward the design principles for stroke and chord keyboard Chinese character input methods. While the development of Chinese character handwriting recognition technology was just beginning, scientists discovered through mathematical modeling and experimentation the functional relationships among handwriting recognition time and recognition rate and word database size.

Based on this research, they defined the optimal handwriting recognition time setting. Thus, they solved the problem of the user's subjective perception of being delayed caused by the conflict between the recognition waiting time and word base size.

In the late 1990s, many multinational corporations, led by Motorola, took interest in the Chinese market and set up R&D centers and product manufacturing bases in China. In order to meet the demands of Chinese consumers, they cooperated with Chinese researchers and carried out many research projects for product localization, such as the definition of the cognition styles of the Chinese users, analysis of the usability of the Chinese translation of their products' English menus, etc. At the beginning of the twenty-first century, usability measurement made significant progress in China. In the past 2 years, the usability research has not only focused on the evaluation of the usability of prototype products but also emphasized on the discovery of user psychological models as well as the cognition of the processing mechanism during the product use process. For example, research on the multimode word input interface has discovered the bottleneck of the cognition of the voice recognition technology in mobile communication application. This finding has defined the direction for the future development of the voice recognition technology.

At present, research on engineering psychology in China has expanded to all sectors and industries including manufacturing, agriculture, transportation, health care, aerospace medicine, etc. It has also promoted the integration of this science with other engineering technologies and related sciences. Therefore, without a doubt, engineering psychology has become a very important interdisciplinary subject in China.

Ergonomics standardization means defining the requirements, standards, or codes for the products, facilities, and environmental conditions designed for human according to the principles of ergonomics. The establishment of ergonomics standards has played a very important role in improving people's work condition and quality of life and promoting the development of production and technology for the society. The China Technical Committee of Ergonomics Standardization Administration (TC7) mainly focuses on the standardization of the terminology, methodology, and human factor data in ergonomics. See Table 42.1 for corresponding of TC159 to ISO.

Since 1980, through systematic planning, research, and review by the China Technical Committee of the Ergonomics Standardization Administration, China has established a

TABLE 42.1 Comparison between ISO and China Technical Committee of Ergonomics Standardization Administration

| Name | International Standardization Organization | China Technical Committee of Ergonomics Standardization Administration |
|---|---|---|
| No. | TC159 | TC7 |
| Setup of branch technical committees | SC1 (ergonomic guiding principles)<br>SC3 (anthropometry and biomechanics)<br>SC4 (ergonomics of human–system interaction)<br>SC5 (ergonomics of the physical environment) | SC1 (ergonomics guiding principles)<br>SC3 (anthropometry and biomechanics)<br>SC4 (signal, display, and control)<br>SC5 (physical environment)<br>SC8 (lighting)<br>SC9 (labor safety) |

set of relatively comprehensive ergonomic standard systems. By the end of 2007, this committee had issued 45 national ergonomics standards. In general, the Chinese ergonomics standards are divided into four levels: national standards, national military standards, industrial standards, and corporate standards. These standards form an organic entity, but each one as well focuses on different fields [17,28,35,42].

The current ergonomics standard system consists of three parts: ergonomic guiding principles, anthropometry and biomechanics, and ergonomics of human–system interaction and ergonomics of the physical environment. These standards almost encompass every aspect of ergonomics including ergonomics of lighting and ergonomics of labor safety. Recently, China established the human–machine system interaction national ergonomics standard system. It includes human–machine interaction system universal design standards, design standards for visual display terminal, multimedia user interface, mental load, and operating with flat visual display panel. These standards are mainly used for research and development, design of the products in the related fields, and technical support role.

Biomechanics is an interdisciplinary subject that explores the use of laws of mechanics to better understand the movement of biological systems. Its primary focus is to explain various phenomena of biological bodies with established principles and viewpoints of the physics of motion. It studies the mechanical aspects of biological systems, such as humans, animals, organs, and cells, and elucidates their mechanical properties and their relation with the functions of biological systems using traditional theories of mechanics, in order to solve the problems existing in biomedicine, sports, and work.

The scope for research in biomechanics is very broad, and its research contents are very rich. Occupational biomechanics is the science of studying the stress imposed on individuals as a result of physical tasks at workplace. The field of occupational biomechanics focuses on reducing workplace injuries, protecting the health of workers, and increasing their job competence. The main research contents of occupational biomechanics are the musculoskeletal system, movement characteristics, mechanical/structural characteristics, and biomechanical models in different occupation and work environment; mechanical issues under different postures; special measurement and analysis methods; optimization of machine and tool design to accommodate the worker anthropometry and their work capacity; and related standards and specifications.

In the mid-twentieth century, biomechanics research efforts in China were mainly focused on sports medicine and athletics. At the beginning of the 1970s, China started research on occupational biomechanics. In the 1980s, biomechanics research mainly focused on investigating the effect of work postures (such as standing, sitting, and carrying heavy loads) on musculoskeletal stress and their relation to musculoskeletal diseases [3–12,29,31,34]. Chinese scholars, based on the development of anthropometry, established mechanical models for the stressed body parts of sitting operation including neck, shoulder, wrist, waist, etc. [14–16,21–25,36,37]. Those models allowed the calculation of the stresses associated with everyday working postures. Sheng Wang et al. [8–10] established waist and neck biomechanical models based on the biomechanical analysis of the workers involved in the production process in the electronics industry. In the 1990s, computer-based automatic analysis system, mechanical analysis, and synchronous analysis of bioelectrical and physiological functions were used by researchers in China [20,21,25]. The proper operation postures for sitting work were also studied. As the research went further, methodology was improved greatly. Experimental study was carried out for both standing and sitting operations. In the twenty-first century, as modern sciences and technologies are developed and applied, more advanced methods and approaches are becoming available for occupational biomechanics research and promote its future development. For example, the three-dimensional (3D) motion data collection and processing system has significantly replaced the manual motion analysis method. This system can quickly and accurately collect and analyze the biomechanical parameters of human body movement in the 3D space. Additionally, the combination of 3D dynamic biomechanical modeling and field surveys can provide the quantitative relationship between occupational risk factors and disease occurrence rate.

In recent years, research on occupational biomechanics has started to deepen into micro-level analysis to study muscle, tendon, ligament, and skeleton tissues at cellular and molecular level [24,27,40]. Occupational biomechanics has made great progress in the research on the pathogenesis and preventive measures of chronic musculoskeletal disorder, a frequently encountered occupational disease.

In the medical field, ergonomics is closely related to occupational hygiene. In terms of protecting the health and safety of the workers, ergonomics and occupational hygiene have different research approaches but the same research objectives. Much research about the human body and environment in ergonomics utilizes the knowledge and methodology of occupational hygiene. On the other hand, research results in ergonomics can be used in industrial hygiene to improve workers' health. In China, in occupational textbooks, ergonomics is defined as the science that is human oriented, researching the relationships among people, machinery, and the environment with the objective of ensuring the health, safety, and comfort of workers while promoting work efficiency.

As early as the 1960s, Chinese researchers started to pay attention to musculoskeletal disease. For example, manual material handling workers frequently had flatfoot and low limb varix disorder. However, elective classes in ergonomics and biomechanics in academic institutions had not been offered until ergonomics became an independent research disciplines in China. In the 1980s, some scholars began to study the relationship between workload and musculoskeletal diseases. Their research focus was mainly on labor-intensive workers, such as stevedores, refinery workers, construction workers, etc. Meanwhile, biological models were developed based on the characteristics of body mechanics, and workload standards were established regarding how to apply force reasonably. The School of Public Health, Peking University, carried out epidemiological surveys of the worker population and studied the relationships among different work tasks, different body parts, and work-related diseases and musculoskeletal disorders [8–10]. They also conducted ergonomic assessments. Tables 42.2 and 42.3 show work-related musculoskeletal disorders (WRMDs) and their prevalence in different types of work.

As the use of computer became prevalent, production processes started to change and resulted in changes of workload at the workplaces. With increased use of computers, the occupational health problems research focus shifted to the use of VDT (video display

TABLE 42.2 Comparison of 12 Month Prevalence and Point Prevalence in Different Body Parts

| | Point Prevalence | | 12 Month Prevalence | |
|---|---|---|---|---|
| **Body Parts** | *n* | % | *n* | % |
| Neck | 125 | 7.6 | 278 | 16.9 |
| Shoulder | 228 | 13.9 | 585 | 35.8 |
| Wrist | 48 | 2.9 | 171 | 10.5 |

TABLE 42.3 Prevalence of WRMDs in Different Body Parts Categorized by Type of Work

| | | Prevalence (Five) | | |
|---|---|---|---|---|
| **Work Type** | **Number** | **Neck** | **Shoulder** | **Wrist** |
| Preparing | 81 | 20.9 | 46.9 | 7.4 |
| Molding | 114 | 17.5 | 37.7 | 13.2 |
| Plastering | 251 | 20.7 | 50.6 | 9.2 |
| Operating | 183 | 10.3 | 37.7 | 8.2 |
| Packaging | 788 | 18.4 | 31 | 11.3 |
| Testing | 219 | 12.3 | 28.8 | 7.8 |

terminal, or sometimes referred to as visual display terminal). Those topics are visual strain analysis, fatigue factors analysis, musculoskeletal disease such as carpal tunnel syndrome, impact of radiation on the female reproduction system, and visual display and control panel ergonomics design. In the 1990s, as science and technology developed further, assembly-line-style work became dominant in industries. During this time period, research studies were carried out in the area of assembly-line WRMDs as well as the effect of shift work on labor competence. Studies were also carried out dealing with WRMD associated with repetitive and monotonous work such as sewing, typesetting, etc.

In the twenty-first century, as the pace of people's life became faster, social psychological factors became prominent risk factors. As a result, many industries started to pay attention to the psychological factor of employees. For example, research studies on medical staff's mental workload and emergency response, placing more emphasis on improving the task arrangement, proper job assignment, and efficiency, became of great importance. In spite of all the efforts made so far, WRMD is still one of the major concerns of the current occupational ergonomics. Using foreign WRMD standards questionnaires as a reference, Chinese researchers have established WRMD questionnaire which is more suitable for China [38,41].

Work environment research topics include analysis of the effect of physical work performed with exposure to environmental factors such as heat, cold, or high altitude and its impact on the body and work efficiency. Such research efforts help define optimal working condition. For example, by studying the temperature at workplace, scientists can define the optimal temperature needed for the workplace; by studying the noise level, they can define the optimal sound levels that do not affect normal verbal communication nor cause distraction. They are also able to define the optimal illumination needed for performing visually demanding tasks. Also, research studies were carried out on ergonomic assessments of workplace and workstation design. Based on those research studies, corresponding standards have been established. By 2007, 25 standards related to occupational health had been established in China. Among these national standards, some are mandatory and others are used as guidelines. Among all the occupational ergonomics standards, the research on workplace safety is the most successful (such as the machine safety guard distance and personal protection clothing requirement national ergonomics standards). Other ergonomics standards also have generated noticeable social and economic benefits.

In China, ergonomics research emphasizes on applications more than theory. Many scholars devote themselves to the application of general theories and methods of ergonomics in the actual design cases. They have achieved excellent results in product modeling, workplace design, and interface design and have gained a lot of experiences. Some researchers have performed literature reviews of over 4000 scientific journals in search of ergonomic research works done by Chinese scholars from 1994 to 2003 [13]. Literature search analysis showed that there is an increasing trend in research dealing with applied ergonomics in China. Based on the data shown in Figure 42.1 [2,13], applied ergonomics has received progressively more attention from all industries, sectors, and academic fields.

Researchers also summarized and analyzed the treatises published during the period of rapid development in ergonomics in China (2001–2005) [26,32,33,39]. Figure 42.2 provides a summary of the human–machine ergonomics research fields in China during 2001–2005.

It can be seen from Figure 42.2 that in those 5 years, research in ergonomics management and administration constituted the largest proportion (15.97%) in ergonomics research work followed by research in occupation and employee's qualifications (14.78%). Based on these research results, employee's health standards and job selection criteria were established. The amount of research done on working environment (12.39%) and human–machine system (10.55%) was very close, which indicates that the human comfort during work and relationship between human and machine—human–machine interface—draw increasingly more attention. Human error and system safety (12.14%) also got some

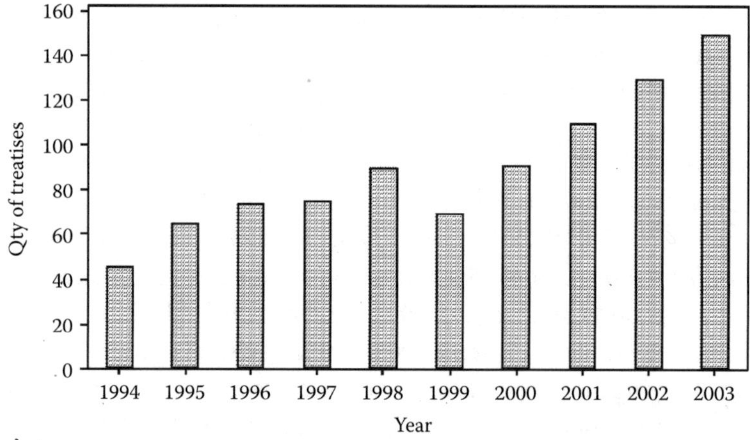


FIGURE 42.1 Cronicle summary of ergonomics application research treatises.

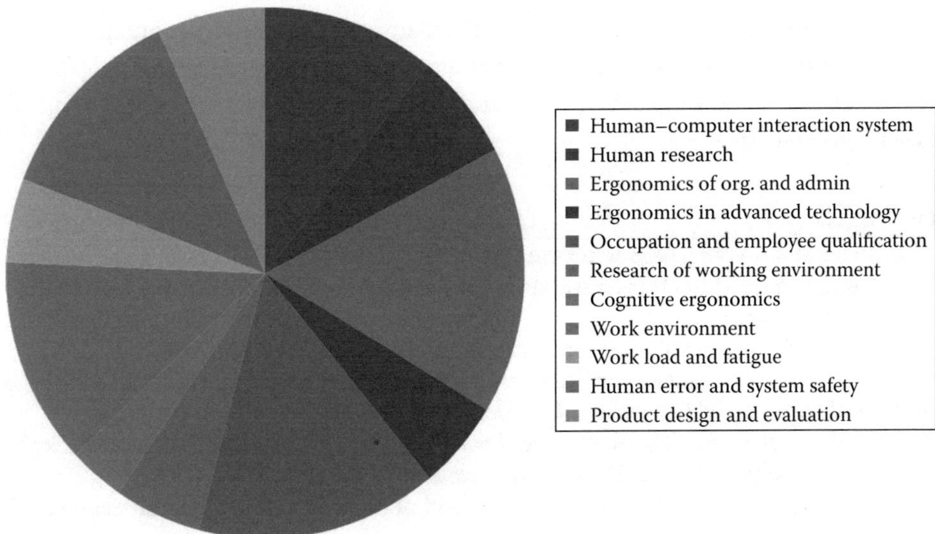


FIGURE 42.2 Proportions of the research fields of ergonomics in China.

attention especially in the area of traffic accidents and production system safety guards. Workload and fatigue, product design and evaluation, and work method and workplace improvement made up a small portion.

At present, among the existing ergonomics standards in China, most of them are qualitative guidelines. This situation cannot meet the practical needs of ergonomics for product design and production. For example, the existing ergonomic design requirement database defines only the general product design window. It does not define in detail the comfort zone, work zone, and endurance zone. In the future, design standards about the human body's work capability need to be created. This is an essential input for a product design. At present, in China, there are no ergonomic assessment standards. The gaps include a process to define the technical procedures and methods for ergonomic assessment and a method for selecting ergonomic assessment subjects. To fill those gaps, it is necessary to establish ergonomics assessment technical standards.

In terms of the actual application of occupational ergonomics in China, the research results from automobile, worker safety, and military material and equipment have made a significant impact on the development of science and technology as well as the advancement of society. Although China conducted fundamental occupational ergonomics theory research and also established and implemented standards according to the research findings, the majority of the contents of ergonomics have been limited to physical factors until recently. There is a need to expand the depth and breadth of ergonomics research. There is also a need to balance between theoretical research and actual application of occupational ergonomics.

Although ergonomics started late in China, it is developing fast. As science and technology and economy of China develop, expectation of good working conditions and high quality of living will gradually increase. People will pay increasingly more attention to the ergonomic aspects of products. At present, among the various products in the market, most products are designed to meet humanistic demands. Although ergonomics has very high application potential in China, it is not quite popular yet and many people still do not understand it. More work needs to be done for research, education, and promotion to increase its popularity.

## REFERENCES


1. S. Z. Zhang and S. Wang, Ergonomics in China, *Sino-Japanese Medical Conference: Occupational Hygiene Seminars*, China Preventive Medicine Association, Tianjin, 1990, pp. 120–122.
2. C. Qi, A brief history of development of human engineering, *Human Engineering*, Chen Du, Ed., Science-Technology Publishing House of Sichuan, Sichuan, China, 1988, p. 11.
3. S. Wang, L. Li, and J. J. Liu, Biomechanics analysis on waist of sitting work, *Chinese Journal of Industrial Hygiene and Occupational Diseases*, 13(1):29–31 (1994).
4. L. Li, S. Wang et al., Epidemiological study and ergonomical evaluation of lower back pain in sedentary workers, *Industrial Health and Occupational Diseases*, 21(1):4–7 (1995).
5. Z. Q. Ding, Current research status and prospect of biomechanics ergonomics and its effects, *Chinese Ergonomics*, 1(2):62–64 (1995).
6. L. H. He, S. Wang et al., Neck-shoulder-wrist injury of sedentary workers, *Chinese Journal of Industrial Hygiene and Occupational Diseases*, 8(6):337–339 (1995).
7. W. Sun Wei, D. P. Zhang, and S. Wang, Ergonomics study of manual material handling, *Railway Occupational Safety, Health, and Environment Protection*, 3(4):246–249 (1995).
8. L. H. He, S. Wang et al., Electromyographic study of neck, shoulder, and wrist of sitting working posture, *Railway Occupational Safety, Health, and Environment Protection*, 11(2):26–28 (1996).
9. Y. T. Niu, S. Wang et al., Field and simulation studies on low back fatigue in sedentary workers, *Chinese Journal of Industrial Medicine*, 10(3):145–148 (1997).
10. L. H. He, S. Wang, E. M. Huang et al., Epidemiological investigation on musculoskeletal injury in electronic industry, *Chinese Ergonomics*, 4(4):12–14 (1998).
11. J. J. Liu, D. X. Wu, Y. R. Xu, L. H. He, and S. Wang, An epidemiological investigation on musculoskeletal injury in sedentary workers, *Practical Preventive Medicine*, 5(6):331–332 (1998).
12. Q. Zheng, S. Wang et al., Epidemiological investigation of musculoskeletal injury in manufacturing workers, *Chinese Ergonomics*, 7(2):28–29 (1997).
13. P. Zheng, Z. H. Song, and Y. M. Zhou, Study and developing status and trend on detecting and evaluating techniques of motor driver fatigue, *Journal of China Agricultural University*, 6(6):101–105 (2001).
14. S. Wang and K. Zhang, *Occupational Ergonomics*, Tianjin Science and Technology Press, Tianjin, China, 2001.
15. L. H. He, S. Wang et al., The effects of static load on mitochondria function in rabbit muscle, *China Occupational Medicine*, 28(2):2–4 (2001).
16. S. Wang, L. H. He et al., Changes in energy metabolism of static load-induced muscle injury, *Chinese Journal of Industrial Hygiene and Occupational Diseases*, 19(3):172–174 (2001).
17. X. M. Xiao, G. H. Bi et al., Review of foreign ergonomics studies and standards, *Defense Science and Technology Industry Standardized Research Center GF Report*, Beijing, 6–32 (2002).
18. H. Yu, S. X. Hou, L. H. He, and S. Wang, Lower back pain in truck drivers working in plateau areas and its prevention, *Chinese Journal of Industrial Hygiene and Occupational Diseases*, 20(1):1–3 (2002).
19. S. X. Hou, H. Yu, L. H. He, and S. Wang, A epidemiological investigation on low back pain of truck driver in plateau, *Chinese Journal of Clinical Rehabilitation*, 6(10):1460–1462 (2002).
20. H. Y. Liang, W. W. Wu, S. Wang, and L. H. He, A monitoring study of electromyography median frequency on fatigue of erector spinalis in drivers working at high altitude, *Chinese Journal of Industrial Hygiene and Occupational Diseases*, 20(6):461–464 (2002).

21. S. Wang, L. H. He et al., An observation on infrared thermograph of lower back pain patients, *Industrial Health and Occupational Diseases*, 29(1):39–41 (2003).
22. Q. Y. Yu and J. F. Yu, Logistic regression analysis of nursing staff chronic MSDs' risk factors, *Acta Universitatis Medicinal Anhui*, 38(3):240–241 (2003).
23. L. H. He, S. Wang et al., Effect of lumbar protective belt on prevention of low back fatigue in personnel during simulated driving, *Chinese Journal of Industrial Hygiene and Occupational Diseases*, 4(4):18–20 (2004).
24. G. Z. Li, S. Wang, and X. Feng, Protective effect of vitamin C on muscle strain injury-induced peroxidative damage on rats, *Chinese Journal of Industrial Hygiene and Occupational Diseases*, 22(4):283–285 (2004).
25. J. Chen, Z. L. Wang, and L. Yang, Analysis of surface electromyography on repetitive lifting task induced fatigue of back muscles, *Chinese Journal of Industrial Hygiene and Occupational Diseases*, 22(6):402–405 (2004).
26. Z. W. Xie and C. Wu, Research developments on ergonomics application in China in recent ten years, *Industrial Safety and Environmental Protection*, 31(3):52–54 (2005).
27. L. H. He, S. Wang, and K. P. Ye, Levels of serum creatine kinase, lactate dehydrogenase and their iso-enzymes in low back pain patients, *Industrial Health and Occupational Diseases*, 31(1): 15–17 (2005).
28. H. C. He, The research progresses on human engineering and its standardization, *Technology Foundation of National Defence*, (9):5–8 (2005).
29. L. H. He and S. Wang, The investigation of work seat in some industry, *Chinese Ergonomics*, 11(6):37–39 (2005).
30. Z. H. Zhang, *A Research Method for Driving Fatigue Based on Physiological Signal and Its Application*, Harbin Industrial University, Heilongjiang, China, 2006.
31. L. Zhang, H. N. Zhang et al., Investigation on occupational muscular-skeletal disorder in three types of occupations, *Journal of Public Health and Preventive Medicine*, 17(2):74–75 (2006).
32. F. Guo, Y. L. Sun, and Q. H. Ye, Comparative analysis on domestic and overseas ergonomics research, *Industrial Engineering and Management*, (6):118–122 (2007).
33. P. P. Pau and X. W. Wu, Apply self-organizing map test mining algorithm to analyze Chinese ergonomics research area, *Chinese Ergonomics*, 13(1):17–20 (2007).
34. T. Y. Jin, *Occupational Health and Occupational Medicine*, (6th edn.), People's Medical Publishing House, Beijing, China, 2008, pp. 59–73.
35. Q. X. Zhou, H. Xiao, and X. Zhang, Review on China ergonomic standardization system research, *World Standardization & Quality Management*, 5(5):13–16 (2008).
36. L. H. He, G. Z. Li et al., Experimental study on the animal model of musculoskeletal strain injury, *Industrial Health and Occupational Diseases*, (2008).
37. W. D. Liu and Z. X. Wang, Musculoskeletal disorders and ergonomics, *Journal of Environmental and Occupational Medicine*, 25(6):605–607 (2008).
38. L. Yang, S. F. Yu et al., Occupational musculoskeletal disorder questionnaire, *Industrial Health and Occupational Diseases*, 35(1):25–31 (2009).
39. R. F. Shen, Reflections of ergonomics and Chinese ergonomics society, *Chinese Ergonomics*, 152:12 13 (2009).
40. Z. L. Wang, J. J. Li, and L. Yang, Experimental study on assessment of manual lifting techniques by surface electromyography and electrocardiography, *Industrial Health and Occupational Diseases*, 35(2):69–73 (2009).
41. S. S. Wu, L. H. He et al., Health survey of airport visual display terminal operators, *Industrial Health and Occupational Diseases*, (2010).
42. China National Institute of Standardization, *Fundamental Standardization*. http://www.cnis.gov.cn (accessed August, 2010).

CHAPTER 43

# Overview of Ergonomic Needs and Research in Taiwan

Chi-Yuang Yu and Eric Min-yang Wang

CONTENTS



## 43.1 HISTORICAL REVIEW

Ergonomics was first introduced to Taiwan in the early 1960s by industrial designers. In 1962, the Chinese Productivity Center (an affiliate agency of the Ministry of Economic Affairs) sponsored a series of introductory industrial design training courses that were aimed at the promotion of the design of industrial products. During these courses, "human engineering" was presented by Professor Ohara Zirou of Ciba University in Japan. A course in human engineering was taught for the first time 3 years later in the industrial design department of Ming-Chi Institute of Technology in Taipei. Within a decade, many universities, colleges, and junior colleges followed this new trend, offering human engineering courses in their industrial design departments. The term *human engineering* was increasingly heard by many people through commercials in the media. However, most people had only a vague impression of ergonomics, which was considered to be the knowledge of fitting products (e.g., cars and chairs) to the physical dimensions of human beings.

Ergonomics has grown rapidly in Taiwan since 1983, becoming increasingly important since the initiation of the course in industrial engineering departments. The industrial engineering department at National Tsing Hua University (NTHU) in Hsinchu was one of the pioneers in offering this course. The course was titled "human factors," as in the United States, where a majority of the faculty members had received their graduate degrees. The first ergonomics graduate program in Taiwan was initiated a couple of years later at NTHU. Many other universities and colleges followed suit. More than 20 schools currently offer at least one human factors course for undergraduate studies. Human factors has become an important branch in the field of industrial engineering and is increasingly known for its many applications, for example, workplace and protective equipment design, traffic and transportation system design, and control room design for nuclear power plants and chemical plants.

Two governmental agencies, the National Science Council (NSC) and the Council of Labor Affairs (CLA), have devoted a substantial amount of energy over the past several years to supporting the promotion of ergonomics. In 1985, NSC invited ergonomics professors from a number of disciplines, for example, industrial design, industrial engineering, and psychology, to discuss and plan the future development of ergonomics in Taiwan. Those meetings concluded with a directive plan, which was then adopted by the NSC as a guideline to grant research projects thereafter.

In a different approach, CLA, focusing its efforts on the welfare of laborers, has devoted much of its efforts to promoting ergonomics in the industrial sector for practical purposes. In recent years, CLA has sponsored many introductory seminars for its employees and industrial safety and health personnel. In 1992, the council established the Institute for Occupational Safety and Health (IOSH), one of whose primary missions involves promoting ergonomics research that would lead to the establishment of safety and hygiene standards for the entire nation. This mission is having a significant and positive impact for the future development of ergonomics in Taiwan.

## 43.2 ERGONOMICS EDUCATION IN TAIWAN

Ergonomics courses in Taiwan have primarily been taught in the departments of industrial design and industrial engineering in universities and colleges. For industrial design departments, 6 universities and 20 junior colleges offer ergonomics courses that stress design applications; therefore, anthropometry and biomechanics seem to be the core. Among these universities, two have master's programs in ergonomics. The educational backgrounds of those faculty members are industrial design, industrial engineering, psychology, or related areas. For industrial engineering departments, 5 universities and 20 junior colleges have general ergonomics courses, and of these, 3 universities currently offer doctoral and master's programs. The educational backgrounds of those faculty members are mainly ergonomics, industrial engineering, and psychology. Ergonomics education in the industrial engineering departments is more diversified and in depth

TABLE 43.1 Ergonomics Education Programs in Universities in Taiwan as of 1993

| | National Tsing Hua University | National Chiao Tung University | National Taiwan Institute of Technology |
|---|---|---|---|
| Faculty | 4 | 3 | 4 |
| Doctoral students | 5 | 4 | 9 |
| Master's students | 13 | 5 | 10 |
| Undergraduate courses | Psychology | Motion and time study | Industrial psychology |
| | Work study | Human factors | Human factors |
| | Human factors 1 | Industrial safety and hygiene | Industrial safety |
| | Human factors 2 | Industrial organization psychology | Consumer behavior |
| | Industrial safety | | Time and motion study |
| Postgraduate courses | Human performance | Human information process | Biomechanics |
| | Work physiology | Advanced human factors | Fundamental physiology |
| | Environmental analysis | Measurement and evaluation of human performance | Fundamental psychology |
| | Human-machine system | | Advanced human factors |
| | Human-computer interface | Human performance | Cognition |
| | Industrial safety | Human and computer | Human reliability |
| | Safety engineering | Human-machine system | Human-machine system |
| | Biomechanics | Problem solving | |
| | Vision and eye physiology | | |
| | Color science | | |
| | Design evaluation | | |

than that in industrial design departments. The numbers of faculty members and students along with the titles of the ergonomics courses of the three universities that offer doctoral programs are listed in Table 43.1.

## 43.3 ERGONOMICS SOCIETY OF TAIWAN ROC

The Ergonomics Society of Taiwan ROC (EST) was founded in February 1993, with the stated objectives of promoting indigenous research and technological applications as well as encouraging international cooperation and exchange in ergonomics. The EST's executive council tentatively planned to hold one national conference and publish at least one journal each year and also to occasionally sponsor international symposiums. The EST also coordinates research projects and provides training programs for interested parties, such as IOSH and those in the industrial sector. The EST is an independent organization. However, it closely coordinates with local institutes of industrial engineering and the Industrial Design Society for their common interests.

The society had roughly 180 individual members and seven organizational chapters by the end of 1993. In terms of educational background, industrial engineering is the largest group and industrial design the second largest (see Figure 43.1). In terms of occupations, university and college faculty are the largest group, with students being second (see Figure 43.2). Three of the seven organizational chapters are governmental institutions, that is, IOSH, Taiwan Power Company, and the Center for Industrial Safety and Health Technology at the Industrial Technology Research Institute; the remaining four are colleges. The officers

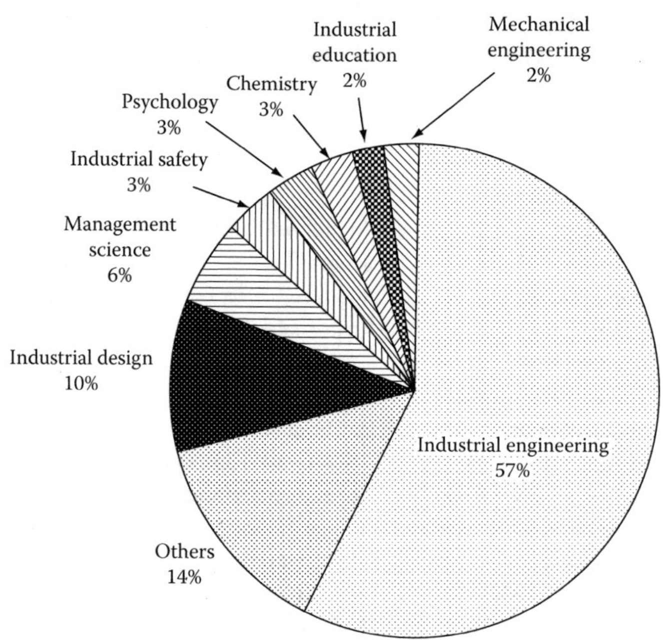


FIGURE 43.1 Educational backgrounds of EST's 180 members as of 1993.

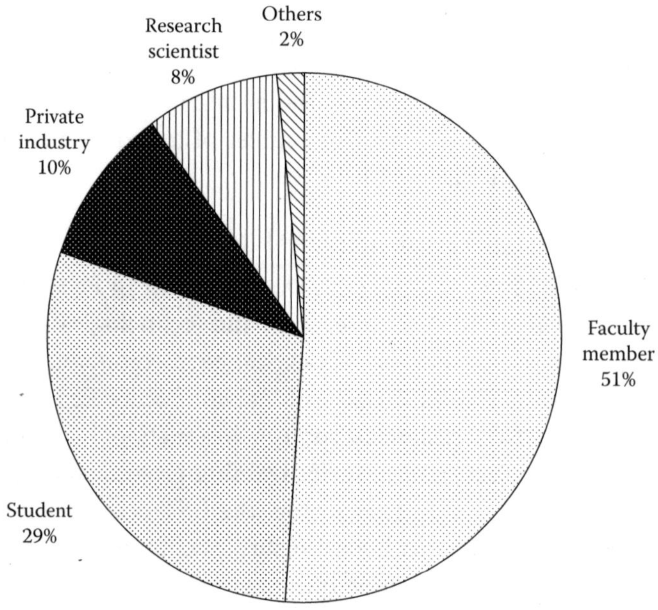


FIGURE 43.2 Classification of EST members' occupations as of 1993.

of EST are striving diligently to promote their society in all sectors of Taiwan. It is hoped that more individual members will be enrolled from among nonacademic professionals and that more organizational chapters will be formed in the private sector.

Less than 6 months after its formation, the society had already sponsored an industrial ergonomics and safety workshop and had also conducted a domestic ergonomics manpower survey. On its first anniversary, an international conference on ergonomics and occupational safety and health was held. Around 200 local researchers and practitioners as well as many from Japan and the United States participated. The society is currently coordinating an ergonomic industrial chair development and evaluation research project. The prototype of the chair is presently in the mass production stage and will be released to industry for a large-scale user test within months. Several other nationwide ergonomics projects—for example, anthropometric, biomechanical, craniofacial, and work physiological databases for the domestic population—are also in the planning stage. These databases are expected to be completed within the next 6 years.

In addition, the society is tentatively planning to place an emphasis on promoting ergonomics to the public in the coming years. Among others, one of the most effective approaches involves presenting introductory films and publishing articles for the mass media, for example, public television programs and newspapers.

## 43.4 RESEARCH PROJECTS AND FUTURE DEVELOPMENT

### 43.4.1 Research Projects Funded by NSC

Previous ergonomics research was primarily funded by NSC; however, several research projects have recently been funded by other governmental agencies, such as CLA and the Ministry of Transportation. Since 1985, NSC has given grants to more than 70 ergonomics-related research projects, and the total budget has exceeded US $8 million. As ergonomics-related research was comparatively new to NSC, the funded research projects were of small scale and not integrated. Among those projects, some important ones are listed in the following.

#### *43.4.1.1 Taiwanese Static Anthropometric Data Bank*

The primary purpose of this study focused on establishing a static anthropometric data bank for Chinese residents in Taiwan using a computerized photographic method [1]. Stratified random sampling was applied to determining the sampling site; in addition, a sample size was established by considering standard errors in a pilot study. A photographic method was next employed in addition to direct measurement of selected body dimensions. A cumulative total of 933 subjects were measured. The data were entered from photos via a digitizing tablet into a microcomputer for processing. The resulting anthropometric data bank was established, and recommendations were presented for future research efforts.

#### *43.4.1.2 Ergonomics Research for Chinese Computer Keyboard Operations*

The primary purpose of this research effort [2] involved developing ergonomic design guidelines for a Chinese computer keyboard. Some of the subproject titles translate as follows:

1. A questionnaire survey on Chinese VDT keyboard operation behaviors and the causes of discomfort
2. The anthropometry of hands for computer keyboard design

3. A study of the key grouping and layout for computer keyboard design
4. A study of finger movements in keyboard operation
5. A study of table-chair height and arm-shoulder postures in keyboard operation at a VDT workstation
6. A study on readability and legibility of Chinese VDT key legends
7. A study of keyboard design for the domestic population

#### *43.4.1.3 Driver Visibility and Safety for Motor Vehicle Design*

The primary purpose of this research effort focused on developing ergonomic design guidelines for cars and motorcycles by emphasizing driver visibility and safety. This project consisted of five subprojects:

1. A questionnaire survey on driver comfort and safety for heavy and light motorcycles
2. A study of the drivers' visual field, dark perception, and signal response time
3. An ergonomic study of motorcycle helmet design
4. The effects of tail light sizes on daytime braking reaction time for motorcycles
5. An evaluation of driving effectiveness on tail light configurations and the visual interference between dashboard displays and the steering wheel

#### *43.4.1.4 Ergonomics in a Nuclear Power Station*

This research effort was composed of a series of projects concerning supervisory control, error diagnosis, and problem solving for nuclear power stations. The research projects included

1. Application of queuing theory to quantify information workload in supervisory control systems [3]
2. Stochastic modeling of human errors on system reliability [4]
3. Dynamic hierarchical modeling of problem solving [5]
4. The prevention of human errors in nuclear power plant maintainability through root cause analysis [6]
5. The design strategies of Chinese information display in supervisory control systems [7]

### 43.4.2 Research Projects Funded by IOSH

Other than NSC, IOSH has funded several research projects during the past few years. Among these projects, two are especially important, the development of a high-mobility industrial chair and a computerized manual materials handling assessment system.

#### *43.4.2.1 Development of a High-Mobility Industrial Chair*

An anthropometric measurement procedure [8] was developed for investigating the design parameters for a low sit-stand posture, and the data on these parameters were applied to designing a chair for high-mobility industrial tasks.

The anthropometric procedure used ischial tuberosities as the seat reference point. Investigating the design parameters required tracing spinal curves of five postures and a posterior thigh curve via a three-dimensional (3D) curve-tracing device. The high-mobility chair was designed with a special seat pan and backrest profiles to accommodate the musculoskeletal geometrical configuration of a low sit-stand posture. The seat pan consists of pelvic support that supports the ischial tuberosities and the areas behind them and a thigh support that maintains the thighs at a 15° inclination angle, resulting in a 105° torso-to-thigh angle. The backrest consists of a lumbar support that preserves lumbar lordosis and a thoracic support that supports the upper back during backward leaning.

#### *43.4.2.2 Computerized Manual Materials Handling Assessment System*

The primary purpose of the second project [9] was to develop an automated computer system for evaluating the risk of low back injuries that occur in manual materials handling. The system employs a computer vision technique to assess postures while a worker is performing manual materials handling tasks. Based on one of the six biomechanical models available, for example, Chaffin's [10], the system can calculate the strain produced in the lower back and some joints on the basis of anthropometric data. The calculated strain values are then compared with the corresponding predicted injury load, indicating the worst posture to take during the job cycle. Recommendations for correct manual materials handling posture are finally provided.

### 43.4.3 Future Development

The NSC funded a research project in 1990 to plan ergonomics development for 5 years. The ergonomics research topics suggested by local researchers are summarized in the following text. This summary includes many of the most essential aspects in this field (excluding military applications), which indicates the enormous requirements and potential areas for ergonomics research.

1. Occupational safety and industrial ergonomics
    a. Design and evaluation of traffic signs and symbols
    b. Human factors applications in product design
    c. Measurement and standards for workplace environments
    d. Computerized biomechanical evaluation for manual materials handling
    e. The effects of workplace environment quality on work performance
2. Anthropometry, biomechanics, and work physiology
    a. Structural and functional anthropometric database of Taiwanese adults
    b. Physical strength data for Taiwanese

    c. Standards for physical work
    d. Epidemiological study of occupational musculoskeletal disorders
3. Human information processing and decision behaviors
    a. Error detecting and diagnostic behaviors
    b. Behavioral models of human errors
    c. Problem-solving behavior model and its applications in automated systems
    d. Human factors standards for the design of a knowledge-based information system
    e. The theory and applications of mental models
    f. The application of human decision-making behavior theory in a decision support system

In addition, IOSH has drawn up a 6 year ergonomics research plan that is aimed specifically at establishing a variety of work guidelines and standards for various tasks. The plan has been classified as a long-term directive plan in Executive Yuan (the highest public administration organization in Taiwan). The plan involves establishing indigenous anthropometric, strength, and work physiology databases for enhancing occupational safety and hygiene in the workplace. The main ideas are listed in the following.

*Anthropometric databases*: This project will consist of establishing static and dynamic databases for domestic workers. For static measurements, about 300 body dimensions will be measured from 3200 adult subjects. Stratified random sampling on gender, age, and other variables will be employed on the basis of the national demographic data. For dynamic measurement, more than 50 ranges of motion for specific postures will be measured from the same subjects.

This project, with its estimated budget of approximately US $1,200,000, will be initiated in 1994 and reach completion in 1996. A mobile measurement laboratory is proposed, and more than 30 investigators will participate in this project at three measurement centers stationed in major universities nationwide. Specific results of this project will consist of static and dynamic anthropometric databases. Three-dimensional (3D) and two-dimensional (2D) manikins will be made according to the actual data.

*Craniofacial anthropometric database*: The purpose of this project involves establishing a head and face anthropometric database for Taiwanese workers as well as constructing a set of 3D head and face models for various applications such as respiratory protective equipment. It is planned to have about 1000 subjects drawn by stratified random sampling on gender, occupation, age, and area based on national demographic data. The data will then be categorized, by considering the facial length, facial width, and standard deviations, into several groups, and a "typical model" will be recommended for each group.

This project started in 1993 and will be ended in 1995. The estimated budget for this project is about US $200,000. It is expected to yield a craniofacial anthropometric database and a complete set of typical 3D craniofacial models.

*Strength database*: The primary purpose of this project involves collecting basic strength data for various biomechanical applications as well as paving the way for establishing physical work standards. These data will be collected along with anthropometric measurement by the same group of researchers. The strength measurements will include lifting, pushing, pulling, carrying, gripping, and gripping torque. The data will be tabulated according to strength and torque moment with respect to specific conditions both isometrically and isotonically. Roughly 1000 subjects will be randomly selected from those of the anthropometric database project and measured for this purpose.

*Work physiology database*: The primary purpose of this project is to collect physiological work data for setting up physiological work standards. These data will include maximal aerobic power, oxygen uptake at various workload levels, respiratory rate, pulmonary ventilation capacity, and similar information. This project is still in its planning stage.

## 43.5 CONCLUSIONS

A significant amount of progress in ergonomics education and research has been achieved in Taiwan over recent decades. Ergonomics education is currently progressing at a good pace in terms of both quantity and quality. The EST, founded in 1993, will play an important role in promoting future research, education, and international exchange in the global village of ergonomics. Ergonomics research is gradually becoming solid and integrated, especially since the establishment of IOSH. Much faster growth is foreseeable in Taiwan over the next decade.

## REFERENCES

1. C.-C. Li, S.-L. Hwang, and M.-Y. Wang, Static anthropometry of civilian Chinese in Taiwan using computer-analyzed photography, *Hum. Factors 32*:359 (1990).
2. L. F. Chen, F. K. Wu, S. S. Lai, M. C. Lin, K. S. Chen, M. J. Suon, and S. S. Kuan, The human factors studies on computer keyboards (Proj. NSC75-0415-E006-02), 1987. (In Chinese, available from National Science Council, Taipei, Taiwan, ROC.)
3. C.-C. Her and S.-L. Hwang, Application of queuing theory to quantify information workload in supervisory control systems, *Int. J. Ind. Ergon. 4*:55 (1989).
4. C.-T. Hwang and S.-L. Hwang, A stochastic model of human errors on system reliability, *Reliab. Eng. Sys. Safe. 27*(2):139–153 (1990).
5. C.-H. Wang and S.-L. Hwang, The dynamic hierarchical model of problem solving, *IEEE Trans. Sys. Man. Cybern. 19*:946 (1989).
6. T.-M. Wu and S. L. Hwang, The prevention of human errors in nuclear power plant maintainability through root cause analysis, *Appl. Ergon. 20*:115 (1989).
7. S.-L. Hwang, M.-Y. Wang, C.-C. Her, D.-M. Wu, and C.-D. Hwang, The design strategies of Chinese information display in supervisory control systems, *Int. J. Hum. Comput. Interac. 2*:41 (1990).
8. C. Y. Yu, K. K. Li, and Y. P. Chen, Development of a high mobility industrial chair (Proj. IOSH82-H124), 1993. (In Chinese, available from Institute for Occupational Safety and Health, Council of Labor Affairs, Taipei, Taiwan, ROC.)
9. M. J. Wang and K. J. Hwang, Computerized manual material handling assessment system (Proj. IOSH82-H121), 1993. (In Chinese, available from Institute for Occupational Safety and Health, Council of Labor Affairs, Taipei, Taiwan, ROC.)
10. D. B. Chaffin, Biomechanical modeling of the low back during load lifting, *Ergonomics 32*: 685 (1988).

CHAPTER 44

# Overview of Ergonomics in Australia

Jean Mangharam

CONTENTS

## 44.1 INTRODUCTION

Interest in the field of ergonomics in Australia today is healthy, and it continues to develop within the realms of research, applied ergonomics, and public policy. There is activity within all three domains of ergonomics, including the physical, cognitive, and organizational domains. However, striking a balance between the application of ergonomics in all domains remains a challenge, as the interpretation of what "ergonomics" is by the Australian public remains inconsistent and the need to increase the awareness of its full potential and value is still needed (Caple 2008). Although acceptance of the application of "ergonomics" across industries has improved since the 1960s and there is evidence that there has been some consolidation of past and present work that has been translated into public policy by peak bodies, cost benefit analysis and rigorous field trials are still lacking in numbers and there remains a plea by ergonomists to maintain the quality of observation, especially for translation into public health policy, as it has been suggested that ergonomics owes much of its past success to this trait, as displayed by its pioneers (Welford 1976). Maintaining the level of specialization we have seen in Australia hitherto may also pose a challenge in the future as the number of universities that offer postgraduate diplomas and degrees in human factors and ergonomics in Australia has reduced over the past 10 years. Paralleling this are emerging issues faced by the Australian society, not dissimilar to those faced by other developed countries, all of which add further dynamics to the already present challenges.

## 44.2 HISTORICAL BACKGROUND AND SIGNIFICANCE OF OCCUPATIONAL ERGONOMICS

### 44.2.1 Demographics

Australia is a federation, with six states and two internal territories including the Australian Capital Territory (ACT), New South Wales (NSW), Victoria (Vic), Queensland (Qld), Western Australia (WA), South Australia (SA), Northern Territory (NT), and Tasmania (Tas). Much of the interior of the country is sparsely populated, and the bulk of the population lives on the east coast of the country (NSW, Vic, Qld, ACT). In 2009, the country had a population of approximately 22 million (Australian Bureau of Statistics 2009a). It is projected to increase to between 30.9 and 42.5 million people by 2056, and to between 33.7 and 62.2 million people by 2101. It has been projected that a large percentage of population growth will be owed to overseas migration. The ageing of Australia's population, already evident in the current age structure, is expected to continue. This is the result of sustained low levels of fertility combined with increasing life expectancy at birth (Australian Bureau of Statistics 2008).

In 2008–2009, 11 million people were employed across all industries. From an industry perspective, the retail trade industry employed the greatest number of people (1.2 million employed persons or 11% of total employment). Health care and social assistance employed 1.1 million people (just under 11% of total employment) followed by manufacturing and construction (both 9%), and education and professional, scientific, and technical services (both 7%). These industries were also the main employing industries in 1998–1999, although retail trade has displaced manufacturing as the largest employer (Australian Bureau of Statistics 2010). Of all employed persons in October 2009, 54% were made up of males (5,883,900 males employed), contributing to 932 million aggregated hours worked in the month of October 2009. Although almost half of the workforce (46%) is made up of female workers (4,924,000 female employed), females contribute to less hours worked at 38% of time worked by all employed Australians (587 million aggregated hours worked) (Australian Bureau of Statistics 2009b).

### 44.2.2 Ergonomics Profession

Activities that can be described as ergonomics in nature commenced in Australia in the late 1930s (Howie and Macdonald et al. 1988). The work was largely concerned with human factors in relation to flying. Following this, ergonomics research groups were established in government departments and universities. Ergonomics practice and its profile dramatically increased and altered during the 1980s as new occupational safety and health (OSH) legislation was introduced and an "epidemic" of work-related repetitive strain injury (RSI) was experienced in Australia. The profession has progressed in various realms since that period, and its involvement in the international ergonomics scene through formal engagement, such as through the International Ergonomics Association (IEA), and multinational collaborative projects has become significant.

Today, the Human Factors and Ergonomics Society of Australia (HFESA) is the professional organization for ergonomists and human factors specialists working in Australia, having representation from many disciplines. The HFESA was established in 1964 as the Ergonomics Society of Australia and New Zealand (ESANZ), arising from a conference held in Adelaide, South Australia (Welford 1976; Howie and Macdonald et al. 1988). In 1986, separate ergonomics societies for New Zealand and Australia (New Zealand Ergonomics Society and the Ergonomics Society of Australia) were formed. The society was renamed HFESA in 2003 to reflect its broad scope of interests and contributions (Bullock 2001).

The HFESA has branches in each state and territory which report to the national committee. The society's activities are directed by its strategic plan, the responsibilities for which are shared by branches. An annual national conference, regular newsletters, and state-run workshops and talks provide a means of communication for members. Several special interest groups exist to offer a focus for specialist fields including the Anthropometry Resource

Australia Special Interest Group (ARASIG), Computer-Human Interaction Special Interest Group (CHISIG), Rail Human Factors group (RAILSIG), and Health care Ergonomics group.

There are various levels of membership that may be obtained by an individual with HFESA, including certified professional ergonomist (CPE), full member, affiliate, and student membership. All CPEs and full members of the HFESA must follow the HFESA code of practice to retain membership (Human Factors and Ergonomics Society of Australia 2009a).

A national professional development committee was developed in 2007, and the HFESA annually appoints a national professional development officer to coordinate the program across the country. Professional development is considered to include activities which increase the human factors and ergonomics competencies of members, or otherwise enhance members' capacity to contribute to the aims of the society (Human Factors and Ergonomics Society of Australia 2009b).

The HFESA is a federated society of the IEA, having been admitted by the council for having met and continuing to fulfill the eligibility criteria provided by the IEA rules (International Ergonomics Association 2009).

Professional ergonomists from Australia have taken a leadership role at the international level including David Caple being the 17th President (2006–2009) and their contributions to this field have been recognised being recipients of IEA Liberty Mutual Research Award in 2008 (Andrew Shaw, Verna Blewett, Laurie Stiller, Christine Aickin, Drew Dawson, Sally Ferguson, Stephen Cox, Kaj Frick) and in 1999 (Shirley Ann Gibbs). Many Australian ergonomists are also active in the IEA technical committees.

### 44.2.3 Certification and Competencies

The need for a certification system and the criteria which could be used to certify professionally qualified ergonomists in Australia created debate when it was first raised in the society. In 1985, a proposal to proceed with developing a professional certification scheme was adopted by the society. In 1990, 21 society members were awarded professional certification status at the first ceremony of its kind in Australia. The Professional Affairs Board remains active in updating its criteria for membership.

In 1990, the Australian government moved to introduce competency based assessment in all occupations and professions. It was realized that a list of ergonomics competencies was vital for the certification and recertification procedures and as a resource in planning and accrediting education programs. Subsequently, the Ergonomics Society of Australia (ESA) defined and published its own outline of core competencies for an ergonomist. Professor Margaret Bullock had a major role to play in this development, as she was also leading an international IEA task force to outline international competency standards for a practicing ergonomist (Bullock 2001).

Currently to become a CPE in the HFESA, the applicant must demonstrate that they have the following:

- Been an active member of the HFESA or other IEA affiliated society for the past 2 years
- Completed an education program which provides a comprehensive set of ergonomics competencies

- Expertise in ergonomics demonstrated through the provision of at least one major work sample, supported by one or more work samples or products of smaller magnitude
- A minimum of 4 years of full-time practice in human factors and ergonomics or the part-time equivalent (Human Factors and Ergonomics Society of Australia 2009a)

## 44.3 LEGISLATION AND GOVERNMENT ACTIVITY

### 44.3.1 Occupational Safety and Health Legislation

The legislative powers of the Australian federal government are set out in the commonwealth constitution. The commonwealth constitution does not give the commonwealth a general power to legislate for occupational health and safety (OHS); hence, there are 10 OHS statutes (6 state acts, 2 territory acts, a commonwealth act covering commonwealth employees and employees of certain licensed corporations, and a commonwealth act covering the maritime industry). There are also specialist OHS statutes covering the mining industry in some states (National Research Centre for OSH Regulation 2009).

Historically each Australian state adopted most of the provisions of the nineteenth century British health and safety legislation (particularly the 1878 Factories Act, and later the 1901 Act) so that by 1970 each of the six states had an OHS statute implementing the traditional British model of OHS regulation. OHS laws in Australia were gradually reformed from the 1970s onward, much of which were influenced by the 1972 report chaired by Lord Robens (Robens 1972; Browne 1973). Each of the Australian OHS statutes today adopts the well-known three-tiered approach recommended by the Robens report—broad, overarching general duties and more-detailed provisions in regulations and codes of practice. Provisions in regulations have force of law (they are mandatory), whereas codes contain guidance material, which can be used as evidence in a prosecution for an alleged contravention of an applicable regulation or general duty provision.

The general duties generally cover employers, the self-employed, occupiers, designers, manufacturers and suppliers of plant and substances, employees, and some other duty holders in some jurisdictions. They require the duty holder to provide and maintain, as far as is reasonably practicable, a working environment that is safe and without risks to health—although the wording of these provisions differs from jurisdiction to jurisdiction and between duty holders (Johnstone 2004; National Research Centre for OSH Regulation 2009).

### 44.3.2 National Harmonization

Although, OSH legislation in Australia are currently separate for each state and territory, this system is in a state of transition as the Council of Australian Governments (COAG, the peak intergovernmental forum in Australia) initiates, develops, and monitors the implementation of policy reforms that are of national significance and which require cooperative action by Australian governments. All jurisdictions are currently working toward nationally consistent laws.

This nationalization process of OSH laws has been termed "harmonization." It takes on a consultative approach which considers input from multiple stakeholders, including representatives each of various jurisdictions, workers, and employers. In 2008, a review panel was established, and a report was released by the panel in 2009 with recommendations from which draft laws will be developed based on existing laws from the various jurisdictions. Nationally harmonized OHS laws and associated regulations are set to commence by December 2011 (Mayman 2009). The development of national OHS laws has been driven by Safe Work Australia, in partnership with the commonwealth, state, and territory governments. The Safe Work Australia Council held its inaugural meeting in June 2009 (Safe Work Australia 2009a).

### 44.3.3 Manual Tasks Legislation

One of the most significant ergonomics-related regulations that exist in all jurisdictions is those written primarily for the prevention of musculoskeletal disorders (MSDs). They have been termed "manual handling" or "manual tasks" regulations.

Throughout Australia certain terms are being phased out, and there has been greater acknowledgment of the wide-scope condition MSD. Sprains and strains, body stressing injuries, and occupational overuse syndrome terms are gradually being replaced by the umbrella term MSDs. Manual handling is being replaced by manual tasks, adding a wider scope to the activity so that it encompasses all risk factors associated with MSDs, not just those perceived to be related to handling loads or applying high forces.

In 2007, a national standard and a national code of practice for the prevention of MSDs from performing manual tasks at work (Australian Safety and Compensation Council 2007a,b) was developed to replace previous standards and codes which addressed manual handling and occupational overuse syndrome (National Occupational Health and Safety Commission 1994, 2005). All national standards and codes of practice are guidance and advisory documents only, and their implementation is dependent on legislation enacted by state/territory OHS authorities.

The code of practice provides practical guidance on the risk management process, particularly for employers. Employers have the duty to identify potential MSD hazards and understand the nature and sources of risk so that they can make informed decisions about what they need to do to eliminate or control them. To manage risk, they are required to

1. Identify hazardous manual tasks by screening work involving manual tasks to recognize those which have the potential to cause MSD
2. Assess risks of MSD that arise from these hazards
3. Eliminate tasks or parts of the task that have a potential to cause MSD, or if this is not reasonably practicable, implement risk controls to minimize the risk of MSD as far as is reasonably practicable
4. Monitor and review the effectiveness of the measures you have implemented

This practical guidance document requires duty holders to identify potentially hazardous manual tasks that present with physical risk factors related to adverse force, postures, movements, and vibration. As part of the risk assessment process, duty holders are required to identify the risk factors for each hazardous task and determine the source(s) of the risk. The risk assessment section does consider and require duty holders to assess some nonphysical risk factors associated with systems of work, work organization, and work practices (considered to be indirect risk factors). This category includes high workloads and tight deadlines, little latitude for workers to influence workload or work methods, and unsuitable or insufficient resources such as staffing levels, equipment, and guidance available to workers. The document states that if the risk cannot be eliminated by duty holders, the risks should be minimized by altering or redesigning the source of the risk. Physical and nonphysical aspects of the workplace are considered to be potential sources of risks. They include

- Design and layout of work area
- Nature of the load
- Nature of tool, equipment, or item
- Work environment
- Work organization, work system, or work practice

The development of this code and standard has not been without challenges. Debates have centered around multiple aspects of the practical guidance documents including the scope, definitions, duty holders that should be included and their responsibilities, risk factors that should be considered, risk assessment tools to be suggested, and size and complexity of the document. The process to update the national standards and codes relevant to the prevention of MSDs commenced in 2003, and the first draft for public comment was released in 2005. The process was revisited in 2006, and the final documents were endorsed by the Australian Safety and Compensation Council (ASCC) in 2007 for jurisdictions to implement in accordance with their legislation. The adoption of this standard and code by the various jurisdictions across Australia to support their regulations has been poor, with more states still referring to their own regulation and codes of practice (WorkCover Corporation of South Australia 1990; Commission of Occupational Safety and Health of Western Australia 2000; Victorian WorkCover Authority 2000; Workplace Health and Safety Queensland 2000). However, the process of developing a national code of practice will be revisited as the harmonization process progresses.

### 44.3.4 Safe Work Australia

Safe Work Australia is a national health and safety peak body that has seen name and role changes over time. The organization has been known as the National Occupational Health and Safety Council (NOHSC) and ASCC in the past. In November 2009, Safe Work Australia began operating as an independent statutory agency with primary responsibility to improve OHS and workers' compensation arrangements across Australia. The activities of the agency today include the coordination and development national laws, the publication of national standards and codes, and the coordination of research projects that may be applied to OHS policy.

Safe Work Australia has published several ergonomics-related research projects in the past few years including

- Hazard exposure research (Australian Safety and Compensation Council 2009) (described in the following text)
- MSDs—a series of publications to support the work toward the prevention of MSDs commenced in 2003 (Bryan Bottomley and Asscociates 2003a,b; David Caple and Associates Pty Ltd 2003; NOHSC Office Safe Design Team 2003; Uniquest Pty Ltd 2003), and more recent papers which collectively have provided a deeper understanding of the complexity of the risks associated and challenges for setting strategies and policy for the prevention of this condition (Safe Work Australia 2006a,d)
- Work-related mental disorders (Safe Work Australia 2006c)
- Safe design (Australian Safety and Compensation Council 2006)
- The manual handling of bariatric people (Safe Work Australia 2009c)
- Overweight and obesity: implications for workplace health and safety and workers' compensation (Australian Safety and Compensation Council 2008)
- Anthropometric data in Australia—refer to emerging issue section for details of this study (Safe Work Australia 2009b)
- Ageing workforce (Safe Work Australia 2005)
- Work-related fatigue (Safe Work Australia 2006b)

#### *44.3.4.1 Hazard Exposure Research*

The purpose of the National Hazard Exposure Worker Surveillance Survey was to gather information to guide decision makers in developing prevention initiatives that will ultimately lead to a reduction in occupational disease. The first publication presents research findings from a quantitative research study of 4500 telephone interviews with Australian workers. To find out more about MSD and mental disorders hazard exposures, biomechanical demands and psychosocial working conditions were asked of participants.

In relation to biomechanical risk factors, the study showed that across all industries, repetitive arm movement was the most common biomechanical demand, and that tiredness was the most common effect of the biomechanical demands placed on respondents at work. In relation to controls, the study showed that respondents from the manufacturing and mining industries were more likely to report that lifting equipment was provided, compared to respondents from other industries.

In relation to psychosocial risk factors, the study showed that across the industries, the most common time demand is the need to work quickly and the most common cognitive demand is work needing undivided attention. The majority of respondents reported to have a good working relationship with their work colleagues and felt that their fellow workers provided them with the help and support they needed and were willing to listen to their work-related problems. In terms of "bullying" in the workplace, 632 respondents reported they had experienced it. Most of those who experienced workplace bullying were bullied by their supervisor/manager or their coworkers. In relation to controls, the study showed that antistress and antibullying policies are the main strategies put in place to prevent workers from becoming too stressed; however, a third of workplaces have nothing in place (according to the workers).

### 44.3.5 Heads of Workplace Safety Authorities

The Heads of Workplace Safety Authorities (HWSA) is a group comprising the general managers (or their representatives) of the peak bodies responsible for the regulation and administration of OHS in Australia and New Zealand.

The HWSA coordinate national compliance campaigns targeted at specific industries across all jurisdictions. These campaign initiatives support the National OHS Strategy 2002–2012 and facilitate the development of consistent approaches to nationally recognized priorities.

Campaigns that have related to the field of ergonomics include the following:

#### *44.3.5.1 Design 4 Health Manual Handling*

In 2004, a campaign to address manual handling hazards in the Health and Community Services Industry Sector was conducted (Heads of Workplace Safety Authorities 2005). This project was one of the first collaborative campaigns conducted by the Australian OHS agencies. During the campaign, 643 randomly selected workplaces (171 hospitals and 472 aged care facilities) were audited across Australia. Information was obtained from a total of 8 focus groups involving 62 workplace health and safety inspectors and other relevant OHS staff. Findings showed that dramatic improvements occurred in the standard of manual handling risk management within the industry sector over the previous 5 years. The authors reported that these improvements were likely to have been a result of accreditation agency requirements, as well as the activities of the various OHS agencies and the widespread promotion of "no-lift" policies. The project showed that effective OHS management system for manual handling was more likely to be seen in organizations that had the presence of a person in the organization whose responsibilities include OHS. There was strong evidence that workplace-based manual handling safety culture, demonstrated by safety activity, does reduce the level of manual handling task risk at a workplace. Workplaces appeared to have controlled patient-handling risks reasonably well. However, risks remain uncontrolled in other general manual handling areas such as kitchens, food services, and laundries. It was identified that task risk assessments are less frequently undertaken in a systematic manner in nonclinical areas. Design issues relating to the design of buildings, furniture, and equipment which impacted on the uncontrolled manual handling risks were also identified during the task inspections. The design of wheeled equipment such as trolleys was identified as the most common design issue impacting on uncontrolled manual handling risk. Issues related to access and space was the second most common design issue, particularly problems relating to storage, space available within rooms, and the width of doors and corridors.

#### *44.3.5.2 Safe Steps: Manual Handling and Slips, Trips, and Falls in Hospitals*

As a follow-up, the 2004 Design 4 Health Campaign, the *Safe Steps—National Hospital Intervention and Compliance Campaign*, conducted in 2008, aimed to reduce the number of manual handling and slip and trip injuries sustained by nonclinical (kitchen, launderette, and clinical sterilization) hospital workers (Heads of Workplace Safety Authorities 2008a).

A total of 203 audits across the nation were undertaken. Most enforcement actions taken by inspectors (approximately 70%) were related with manual task activities, and the majority of improvement notices issued were associated with Central Sterilization and Supply Department (CSSD) work area. Most hospitals (approximately 70%) were rated as "above compliant" for hazard identification, risk assessment, risk control, purchasing, training, and management commitment.

The findings from the project showed that most of the time hospitals are controlling their manual tasks and slips/trips risks. However, it appears that there is some room for improvement for controlling risks in the nonclinical areas assessed, especially CSSDs. Although the audit approach does not allow direct data comparison with the Design 4 Health Campaign, risk identification and control in nonclinical areas were found to be better controlled in 2008 than 2004, and these findings suggest improvement has been made since the last campaign.

#### *44.3.5.3 National Falls from Heights in the Heavy Vehicle Sector Campaign Final Report*

In late 2003, an HWSA campaign was conducted in the transport industry focusing on falls from heights in the heavy vehicle sector (Heads of Workplace Safety Authorities 2006). The national strategy had two elements, information/advisory and compliance elements. The four primary sectors focused on during the program were car carriers, tankers, dry bulk haulage (pneumatic), and livestock transport. One of the controls that were seen to be implemented as part of the program was the modification of vehicles so that they were fitted with falls prevention systems. A market appeared to have been developed, and suppliers of falls systems retrofitted many vehicles. The true extent of the impact of the program could not be measured; however, a few anecdotal findings reflected positive outcomes.

#### *44.3.5.4 Prevention of Falls in Construction*

In 2007/2008, "Prevention of Falls in Construction" was conducted as a repeat of a 2003/2004 HWSA joint construction compliance project (Heads of Workplace Safety Authorities 2008b).

The two objectives of this campaign were to undertake a coordinated national compliance campaign focusing on falls prevention in the housing construction and the smaller general construction industry sectors, and compare the 2007/2008 level of fall prevention compliance with that found in the 2003/2004 campaign. A total of 1044 site inspections occurred and were distributed evenly between regional areas and the capital cities. Of the sites inspected, only 35% were compliant with falls prevention requirements. Compared with 2004, the proportion of enforcement action required was down by 24%. A significant sample of inspectors (104) from across Australia, who were to be involved in the 2008 campaign, undertook a voluntary online inspector survey prior to the field interventions commencing. This survey was used to gain an understanding of inspectors' perceptions of changes, if any, within the targeted construction sectors since 2004. Only 2% of inspectors believe that there had been deterioration in the use of fall prevention controls, while approximate 75% reported that there had been a slight or obvious improvement. Inspectors indicated that overall the housing subsector had seen the most obvious improvement since 2004.

#### *44.3.5.5 Manual Handling in Manufacturing*

The primary purpose of this campaign was to improve the capability of employers within targeted manufacturing sectors, to effectively manage manual handling risks in consultation with employees (Heads of Workplace Safety Authorities 2007). During the campaign, WorkSafe Victoria developed a guide to safety in the metal fabrication industry (WorkSafe Victoria 2007c). This document was used by a number of jurisdictions whose targeted manufacturing sector was sheet/structural fabricated metal manufacturing. As part of the project, a total of 334 workplaces were audited. Overall, the outcomes of the compliance audits indicated high levels of workplace manual handling compliance (workplaces performed a risk management process), which contrasted with the high injury rates of manual handling injuries within the manufacturing sector. One possible explanation for this discrepancy may be the nature in which risk management is expected to be triggered by workplaces. Tasks that are thought to be hazardous (i.e., present with risk factors) are expected to be identified, assessed, and managed. Unfortunately, many manual tasks performed by workers in this industry do not appear to be hazardous when looked at individually; however, the cumulative nature of exposing workers to multiple tasks that have several low-level risk factors poses more of a risk (e.g., grinding tasks that expose workers to sustained awkward postures, repetitive movement, and hand-arm vibration as shown in Figure 44.1). Some of the key drivers for effective management of manual handling risks were management commitment, positive safety culture, increases in productivity, and reduced workers compensation costs. Barriers included lack of management commitment, lack of awareness of OHS obligations, poor workplace culture, and costs of replacing or upgrading equipment.

### 44.3.6 Jurisdictional Activity

The numbers of scientific officers specializing in ergonomics that are employed by authority bodies across Australia vary greatly, and the roles of these specialists have changed over time. As an example, scientific officers of the human factors and ergonomics team at WorkSafe Western Australia (WA) have experience and educational background in an ergonomics-related discipline and are trained to be inspectors with legal powers to enforce the OSH laws where required. All inspectors at WorkSafe WA are expected to use a balance

FIGURE 44.1 Metal manufacturing worker performing a deburring task using a hand-held grinder.

of education and enforcement for the purpose of improving health and safety in the state. The role of the scientific officers/inspectors in WA includes conducting national and statewide proactive projects (targeted investigations of high-risk industries and specific hazard types); conducting reactive investigations (investigations of reportable injuries and high-risk incidents and public enquiries); contributing to the development of ergonomics-related publications, legislation, and policy; providing educational sessions at public forums and workshops; assisting in investigations and providing skills development courses for industry team inspectors; and conducting complex investigations of human factors and ergonomics-related hazards. Hazard types that the team has focused on over the past few years include those related to MSDs, slips, trips, and falls, and psychosocial and organizational issues especially associated with work-related stress, conflict in the workplace, aggression, and violence.

Most publications that are produced by jurisdictional authority bodies across Australia that have ergonomics-related principles have been written in plain English, with little technical jargon. Concepts have been simplified so that these publications may serve as practical guidance without the need for ergonomics specialist to interpret them. Some useful publications written by various jurisdictions include the codes of practice for manual handling and manual tasks (WorkCover Corporation of South Australia 1990; Commission of Occupational Safety and Health of Western Australia 2000; Victorian WorkCover Authority 2000; Workplace Health and Safety Queensland 2000); industry-specific manual task guidance material for the metal fabrication, saw milling, transport, textile, food and meat processing, mining, and council workers (Department of Health Safety and Welfare of Western Australia 1991, 1992; WorkSafe Victoria 2001, 2002, 2005, 2006a, 2007c; Department of Mines and Petroleum 2009); patient handling (Department of Health Safety and Welfare of Western Australia 1989; WorkSafe Victoria 2007b, 2009);

office ergonomics(WorkSafe Victoria 2006c); slips, trips and falls (Commission of Occupational Safety and Health of Western Australia 2004; WorkCover New South Wales 2006; WorkSafe Victoria 2006b; Workplace Health and Safety Queensland 2007); work-related stress (WorkSafe Victoria 2007a; Workplace Health and Safety Queensland 2009); and aggression and violence in the workplace (NSW Health 2003; WorkSafe Victoria 2008).

## 44.4 ERGONOMICS RESEARCH

Welford wrote in 1976 that with a population of only 13 million, Australia spends a far smaller proportion of its national income on research and development compared to other advanced countries, and that many of the best university students from that time have stayed overseas after leaving to gain wider research experience (Welford 1976). Although funding designated for human factors and ergonomics research in Australia remains limited, the proliferation of publications in the field of ergonomics since the 1970s shows that several key researchers have been innovative in obtaining necessary funds to conduct quality research in this field. A number of university departments in Australia have been successful in attracting researchers and students from overseas, as well as retaining their students to create productive research teams.

### 44.4.1 Early Research

Early ergonomics-related research focused on the aviation industry. Factors relating to visual standards, changes in atmospheric pressure with altitude, the problems of blackout in aircrew, and the problems of noise in aircraft were areas of particular interest (Bullock 2001). Dr. John Lane, Director of Aviation Medicine, published one of the early papers on ergonomics in 1953 entitled "Human Engineering: A new technology" (Lane 1953). A human engineering research group was established within the Aeronautical Research Laboratories of the Australian Defence Scientific Service in the Department of Supply, in about 1957. This represented the first formally constituted research group in ergonomics as such in Australia. Three of the principal researchers associated with this group, Dr. Colin Cameron, Prof. Ron Cumming, and Dr. John Lane, were to become instrumental in the later formation of the Ergonomics Society. The design of air traffic control systems, navigational aids to assist aircraft landing, and visual displays to eliminate irrelevant information and assist the operator to organize incoming data were also subjects of interest. Lane invited Dr. Margaret Bullock from the University of Queensland (UQ) to undertake functional anthropometric and force capabilities studies of relevance to design of small aircraft cockpits and standards for parachute rip cord release in the 1970s (Bullock 1973, 1974, 1978, 2001). Colin Cameron went on to study and write about the theory of fatigue (Cameron 1973) and how it presented in modern industry (Cameron 1971).

The effects of climate provided a special research interest. The Tropical and Fatigue Laboratory within the Department of Physiology at the UQ led major studies into the physical and psychological effects of tropical service, and the design of clothing for flying in the tropics and at low temperatures. Other studies in the School of Public Health and Tropical Medicine at the University of Sydney investigated the effects of climatic extremes on comfort and performance of people of all ages, whether healthy or sick. Dr. Provins, in South Australia, was concerned with studying the relationship between environmental conditions, body temperature, and the performance of skilled tasks (Provins and Cunliffe 1972a; Cumming 1973; Provins et al. 1973). He published several articles in the area of motor skill and handedness in the 1960s and 1970s (Provins and Clarke 1960; Provins 1967; Provins et al. 1968; Provins and Glencross 1968; Provins and Cunliffe 1972a,b,c). The results of these studies had application to work at Australia's station at Mawson, in Antarctica, and in the mining industry in Northern Australia. Engineers became actively involved in designing safety features in the design of load haul dump (LHD) vehicles and underground mining vehicles (Bullock 2001).

Collaborative research studies by the Departments of Mechanical Engineering and Physiotherapy at the UQ investigated the skills demanded of LHD vehicle drivers in coping with vibration. In an attempt to control the prevalent back injuries in industry and agriculture, extensive studies were carried out by Professor Margaret Bullock to determine the optimal worker-pedal relationship to minimize spinal movements. The physical stresses associated with manual sugarcane harvesting led engineers to develop an automated system of cane harvesting and bulk storage (Bullock 2001).

Demonstrating an early interest in rehabilitation ergonomics, engineers and medical practitioners collaborated in the 1960s to examine upper limb stresses in process work and the design of prosthesis which would enable disabled persons to become productive workers (Bullock 2001). Considerable research was also carried out during the 1970s and 1980s by Dr. Patkin (Patkin 1967, 1977, 1978, 1981), in the ergonomic design of surgical instruments,

The development of ergonomics practice in Australia has been closely associated with interests in OHS. The excessive amount of lost time from work because of musculoskeletal injury and the subsequent costs forced employers to introduce measures of control. Positive changes in OHS practices and also in management style were introduced into many work places in Australia during the 1980s, due in part to the major contributions of David Ferguson (Ferguson 1969, 1972, 1973).

In the early 1970s, the association between repetitive activity and posture and increased risk of musculoskeletal problems was established, especially in relation to process work and keyboard work (Ferguson 1971; Duncan and Ferguson 1974; Ferguson and Duncan 1974). In the 1980s, an influx of neck and upper limb disorders being reported as associated with keyboard work, termed RSI, was considered an "epidemic" in Australia (Oxenburgh et al. 1985; Hall and Morrow 1988; Quintner 1989; Ireland 1995, 1998; Gun and Jezukaitis 1999; Awerbuch 2004). Intense activity on the subject of "repetitive strain injury" coupled with new OSH legislation in Australia in the early 1980s led to an influx of ergonomics society members from the OH&S profession (Howie et al. 1988). Howie et al. describe that at that time, a simplistic biomechanical approach was taken, only to find after considerable moneys being spent on "ergonomic" furniture that problems were not solved. The authors also felt that although the reputation of ergonomics suffered as a result of this approach, it prompted the realization that a broader systems approach is necessary. The increasing use of computers as part of the new technology and the importance of developing effective user interfaces later led to the formation of a Computer-Human Interaction Special Interest Group of the ESA (Bullock 2001; Human Factors and Ergonomics Society of Australia 2009a).

In the 1970s, Welford published extensively in the area of cognitive ergonomics, including the influence of perception, mental load, and stress on performance, particularly reaction time (Welford 1970, 1971, 1973, 1978). One of Welford's early publications looked at the ergonomic approach to social behavior (Welford 1966).

### 44.4.2 Recent Research

Ergonomics research in Australia since the 1980s has covered many areas of interests including MSDs; health-care ergonomics; mining; transport; agriculture; slips, trips, and falls; anthropometry; human–computer interaction HCI; product design; human motor skill; job demands; and cost-benefit analysis of ergonomics intervention.

#### *44.4.2.1 Musculoskeletal Disorders*

MSDs have been a focal area of research in Australia since the 1980s. The risk factors of interest have included static and awkward postures, forceful exertion, repetitive movement, and psychosocial factors. These have been related to spinal pain, manual handling, computer interaction, prevalence in children, workplace assessments, and various psychosocial factors.

44.4.2.1.1 Spinal Pain Biomechanical studies conducted at the School of Physiotherapy at Curtin University of Technology have explored various patterns of muscular activity and sitting postures associated with nonspecific lower back pain (Dankaerts et al. 2006b; O'Sullivan et al. 2006). One of the studies (Dankaerts et al. 2006a) explored whether there were differences in trunk muscle activation during usual unsupported sitting between healthy controls and two subgroups of nonspecific chronic low back pain (LBP) patients. The authors concluded that subclassifying nonspecific chronic LBP patients into active extension pattern and flexion pattern patients revealed clear differences in surface EMG activity during sitting between pain-free subjects and subgroups of nonspecific chronic LBP patients. The flexion relaxation ratio of the back muscles was lower for nonspecific chronic LBP, suggesting a lack of flexion relaxation for those with nonspecific chronic LBP.

A number of reviews have been conducted on the prevalence and associated factors for thoracic pain in the working and general population (Briggs et al. 2009a,b; Briggs and Straker 2009).

44.4.2.1.2 Manual Handling The risks associated with various manual techniques and patterns of movement used during manual handling tasks have been explored by various methods including biomechanical, psychosocial, psychophysical, and physiological methods. Many of these studies were conducted at Curtin University of Technology and UQ (Burgess-Limerick et al. 1995; Burgess-Limerick and Abernethy 1997a,b, 1998; Straker et al. 1997a,b; Straker and Cain 1999; Straker and Duncan 2000; Burgess-Limerick 2003). Straker's studies in 1997 aimed to compare the risks assessed in single manual handling tasks with those in combination tasks (of pull, lift, carry, lower, and push), as most recommendations assume that a combination task can be split into its components for assessment (Straker et al. 1997a,b). Their studies found that in at least one of the twelve comparisons performed for each dependent variable, the combination-task value was significantly different to each single-task value. The differences occurred regardless of whether the most critical single-task value or an average of all single-task values was used. The authors concluded in their first study that the risk in combination manual tasks cannot be accurately assessed by using estimates from discomfort, exertion ratings, and heart rate measures of single tasks.

In their second study the maximum acceptable weight (MAW) of each combination task was compared to the MAWs of the single tasks. It was concluded that the current use of single-task MAWs to estimate the risk in combination tasks was unacceptable. Prediction models for combination-task MAWs based on single-task MAWs were also developed. It was argued that owing to their situation-specific nature, the prediction of combination-task risk using single-task MAWs was likely to result in unacceptable risk errors.

Burgess-Limerick's studies defined the stoop and squat postures and how they interrelated. In one of his studies, he found that two distinct patterns of coordination were evident during a cyclic lift and lowering task: a squat technique in which moderate range of hip, knee, and ankle movement was utilized and ankle plantar flexion occurred simultaneously with knee and hip extension; and a stoop technique in which the range of knee movement was reduced and knee and hip extension was accompanied by simultaneous ankle dorsiflexion (Burgess-Limerick et al. 2001). Abrupt transitions from stoop to squat techniques were observed during descending trials, and from squat to stoop during ascending trials. The authors of this study believed that the transitions may be a consequence of a trade-off between the biomechanical advantages of each technique and the influence of the lift height on this trade-off.

The effectiveness of participatory ergonomics for the reduction of the risk and severity of injuries from performing manual handling and manual tasks has been evaluated by several researchers (Straker et al. 2004; Carrivick et al. 2005; Burgess-Limerick et al. 2007).

Carrivick et al. (2005) evaluated the effectiveness of a participatory ergonomics risk assessment approach in reducing the rate and severity of injuries from manual and non-manual handling sustained by a cohort of 137 cleaners within a hospital setting. The date of injury and the workers' compensation claim cost and hours lost from work were obtained for each injury incurred during a 4 year preintervention and 3 year intervention period. Reductions of rate of injury by two-thirds, workers' compensation claim costs by 62%, and hours lost by 35% for manual handling injuries were found to be associated with the intervention period.

Straker et al. (2004) showed that participatory ergonomics can reduce the risks associated with performing manual tasks. The authors conducted a randomized controlled trial on 117 small to medium sized food, construction, and health companies, where participative ergonomics was the intervention. Workplaces were audited by government inspectors using a manual tasks risk assessment tool. The results showed a significant decrease in estimates of manual task risks and suggested better legal compliance in the experimental group.

44.4.2.1.3 Computer Interaction The risk of MSDs from performing computer work on laptops and standard desktops have been explored extensively, especially in relation to actions and postures associated with mouse and keyboard design and interaction, and screen design and position (Straker et al. 1997a,b; Burgess-Limerick et al. 1998, 1999a,b, 2000; Cooper and Straker 1998; Mon-Williams et al. 1999; Burgess-Limerick and Green 2000; Cook et al. 2000, 2004a,b; Evans and Patterson 2000; Straker and Mekhora 2000; Lindgaard and Caple 2001; Szeto et al. 2002, 2005a,b,c,d; Cook and Burgess-Limerick 2004; Straker et al. 2008, 2009b,e). Musculoskeletal discomfort among bank tellers and seated versus standing postures have also been studied (Roelefs and Straker 2002).

After years of research in this area and continued evidence of the rapid adoption of information technology in the workplace, it has been argued that many modern workers are at risk of insufficient physical workload and ergonomics will require a change in paradigm to tackle this issue (Straker and Mathiassen 2009). It is argued that the traditional physical ergonomics paradigm of reducing risk by reducing physical loads ("less is better") is not appropriate for many modern occupations. Straker and Mathiassen recently proposed that a new paradigm is required, where "more can be better." Their paper discusses the potential for work to be seen as an arena for improving physical health and capability and the challenges and responsibilities presented by this new paradigm for ergonomists, employers, health, and safety authorities and the community.

44.4.2.1.4 Prevalence in Children The published research on the current and potential deleterious effect and impact of information technology on children and adolescents continues to grow especially in relation to MSDs, postural changes, and muscular activity (Harris and Straker 2000; Zandvliet and Straker 2001; Straker et al. 2002, 2006, 2007, 2008a,b,c,d, 2009a,c; Briggs et al. 2004; Greig et al. 2005).

The collective studies of Straker and Pollock in the field of information technology and children have prompted the publication of a review of current exposure data and the evidence for positive and negative effects of computer use by children (Straker et al. 2009d). The authors described that the potential positive effects of computer use by children include enhanced cognitive development and school achievement, reduced barriers to social interaction, enhanced fine motor skills and visual processing, and effective rehabilitation. However, the negative effects are beyond physical issues related to MSDs and include threats to child safety, inappropriate content, and exposure to violence, bullying, Internet "addiction," displacement of moderate/vigorous physical activity, exposure to junk food advertising, sleep displacement, and vision problems. Progress toward child-specific guidelines is reported in their paper, with a set of guideline principles presented as the basis for more detailed guidelines on the physical, cognitive, and social impact of computer use by children.

Ranelli et al. (2008) explored the development of MSDs in musicians during childhood. This study investigated the risk factors and established the prevalence of playing-related musculoskeletal problems and their association with gender and age. This study found that females and older children were more likely to experience both symptoms and disorders. For children having reported the experience of a playing-related musculoskeletal symptom within the last month, 5% took medication to relieve the problem and 4% visited a health professional to seek advice for the problem.

44.4.2.1.5 Workplace Assessments Work-related assessments such as functional capacity evaluation and workplace assessments have been analyzed and written about in several publications by various authors especially Innes and Straker (Innes and Straker 1998a,b,c, 1999a,b, 2002a,b, 2003a,b; Tuckwell et al. 2002; Legge and Burgess-Limerick 2007).

Innes and Straker conducted two studies to determine the extent and quality of available evidence for the validity and reliability of work-related assessments. The study examined available literature and sources in order to review the extent to which validity and reliability have been established for 28 work-related assessments. Most work-related assessments were found to have limited evidence of validity and reliability. Of those that had adequate evidence of validity, the validity ranged from poor to good. There was no instrument that demonstrated moderate to good validity in all areas. Very few work-related assessments were able to demonstrate adequate validity in more than one area. For the limited number of work-related assessments with an adequate level of evidence on which to judge their reliability, most demonstrated a moderate to good level. Few assessments, however, demonstrated levels of reliability sufficient for clinical (and legal) purposes (Innes and Straker 1999a,b).

Innes and Straker researched further on this subject to understand the current beliefs of therapists in Australia and the strategies they use to address the issues of credibility, reliability, consistency, trustworthiness, validity, generalizability, and quality in conducting work-related assessments. The authors found that participants were aware of the issues of reliability and validity but believed it was not practical to establish these aspects formally in most work-related assessments (Innes and Straker 2003a).

44.4.2.1.6 Psychosocial Factors Significant reviews have been conducted looking at the association between psychosocial factors, workload, and stress on MSDs (Macdonald and Munk 2003; Macdonald 2004).

Wendy Macdonald and Owen Evans conducted a review of literature on the prevention of work-related musculoskeletal disorders (WMSDs) for Safe Work Australia (Safe Work Australia 2006a). Their paper showed that cumulative WMSDs can stem from a wide range of factors that together result in an inadequate margin between people's work demands and the coping resources available to them. They described the hazard as having both physical and psychosocial risk factors. The psychosocial risk factors included excessive amounts of work, long shifts, inadequate rest breaks, long weeks, time pressures, high responsibilities, inadequate time to cope with perceptual/cognitive task demands and excessive emotional demands of the work, inadequate personal control and autonomy, inadequate task variety and opportunities for skill utilization, and inadequate job security. They explained that besides exposure to certain risk factors, inadequate workplace support: poor materials/ information, poor supervisor support, poor social cohesion, low morale, and inadequate training provisions added to the risk.

The factors mentioned earlier, whether singly or in combination, can result in hazardous personal states such as high levels of fatigue (of varying types) and/or of psychological stress, which entail physiological responses that directly increase injury risk; these states can also induce behavioral changes which increase risk. In addition, hazardous task and job demands, particularly their physical components, can directly increase the risk of acute-onset WMSDs. There is also some evidence that at least some physical and psychosocial hazards can act synergistically in increasing WMSD risk, but such interactions are inadequately understood and there is a need for more research on this topic. It is concluded that it is no longer appropriate for psychosocial hazards to be seen as necessarily secondary or peripheral to physical hazards for WMSDs, particularly in light of the accumulating evidence of physiological mechanisms via which elements of the "stress response" can directly increase the risk of WMSDs. In many documented situations, WMSD risk has been shown to be highest when both physical and psychosocial hazard levels are high.

#### *44.4.2.2 Health-Care Ergonomics*

Patient handling and MSDs have been explored in various ways including the use of slings to handle people (Elford et al. 2000) and the implementation of no-lift policies as a high-level control (Engkvist 2006, 2007). The prevalence and impact of MSDs in relation to the nursing, occupational therapy, and physiotherapy professions have also been studied in Australia (Cromie et al. 2000, 2002, 2003; Retsas and Pinikahana 2000; King et al. 2001; Martin 2003).

The application of ergonomics in surgery has been studied and described in the areas of engineering technology. The studies are related to the use of robotics (Murphy et al. 2008), visual information, fine motor control, postures adopted, and instrument design (Patkin 1967, 1977, 1978, 1981; Patkin and Isabel 1995).

The ergonomics and human factors of visual and auditory patient monitoring display systems especially in the field of anesthesiology have been explored extensively by Penny Sanderson and her research team members (Watson et al. 2000, 2004; Seagull and Sanderson 2001; Sanderson et al. 2003, 2004, 2005a,b, 2006, 2008, 2009; Watson and Sanderson 2004, 2007; Sanderson 2006; Lacherez et al. 2007; Thompson and Sanderson 2008; Wee and Sanderson 2008; Anderson and Sanderson 2009; Grundgeiger and Sanderson 2009; Liu et al. 2009a,b,c). Sanderson et al. (2009) outlined and discussed the work on auditory displays, covering both auditory alarms that indicate technical or physiological threshold levels and informative auditory displays that provide a continuous awareness of a patient's well-being. The authors explained that auditory display in anesthesia can extend well beyond auditory alarms to displays that give the anesthesiologist a continuous peripheral awareness of patient well-being. However, they argue that much more rigorous approaches should be taken to evaluating auditory displays so they add information rather than noise.

#### *44.4.2.3 Mining*

A few studies have been conducted in the mining industry, including injury trends (Burgess-Limerick and Steiner 2006) and ergonomic intervention in underground coal mining industry of which some have been funded by the Australian Coal Association Research Program (Burgess-Limerick 2005; Burgess-Limerick et al. 2007; Zupanc et al. 2007). Burgess-Limerick has also been involved in publications on the application and value of ergonomics in U.S. mining industries (Burgess-Limerick and Steiner 2007;

Torma-Krajewski et al. 2007). Participatory ergonomics has been applied for the purpose of preventing MSDs, showing positive results (Burgess-Limerick et al. 2007).

McPhee has published a review, a handbook on ergonomics, and a handbook on whole-body vibration exposure specifically for the mining industry (McPhee 1993, 2004; Mcphee et al. 2001). McPhee reported that risks to health and safety in the mining industry posed by longer shift lengths, higher workloads, less task variation, and decision latitude have not really been well researched. Heavy physical workloads and stresses are still areas of concern, but are likely to be intermittent rather than constant. A report commissioned by the NSW minister for mineral resources confirmed that working hours were high and fatigue was experienced by a significant number of workers interviewed in the industry (Shaw Idea 2007). This study has been reported as Case Study 3.

McPhee points out that the contribution of slips, trips, and falls and increasing age of miners to manual handling injuries is still not clear. In some cases sedentary work and the operation of machinery have completely replaced heavy physical work. The issues of machinery design for operations and maintenance and whole-body vibration exposures when operating machines and vehicles are becoming more critical. The link between prolonged sitting, poor cab design, and vibration with back and neck pain is being recognized but has yet to be addressed in any systematic way by the mining industry (McPhee 2004).

#### *44.4.2.4 Transport*

Monash University Accident Research Centre (MUARC), formed in 1987, is Australia's largest multidisciplinary research center specializing in the study of injury prevention. The center undertakes applied research contracts for government and industry clients throughout Australia and internationally. More fundamental research is undertaken through research grants. A number of studies have centered around human factors including factors that may affect situational awareness (Salmon et al. 2009), human factors analysis from aviation crash (Lenné and Ashby 2008; Lenné et al. 2008a), railway crossings (Wigglesworth 2008), predicting pilot error (Stanton et al. 2009), the application of intelligent transport systems (Horberry et al. 2006b, 2007; Regan et al. 2007a,b; Salmon et al. 2007; Young and Regan 2007), driver distractions (Horberry et al. 2006a), including the use of mobile phones and text messaging (Hosking et al. 2006; Regan 2006), and detection of emergency vehicles (Lenné et al. 2008b). The MUARC has explored human error and road transport systems, developing a framework for an error-tolerant road (Salmon et al. 2006).

The ergonomics unit of WorkSafe Australia was engaged to identify and assess ergonomic problems in driver cabs of older trains on the Sydney city and suburban network (Stevenson et al. 2000).

Modeling and predicting workload in en route traffic control (Loft et al. 2007) and training systems in the field of aviation has also been explored (Naikar and Sanderson 1999).

Besides the research conducted at MUARC, road safety research has also been conducted extensively at the University of New South Wales. Fisher and Hall conducted many studies on night driving, road lighting, and car frontal design on pedestrian accident trauma in the 1970s and 1980s (Fisher and Hall 1972, 1973, 1976, 1978, 1982, 1985a,b, 1986; Hall and Fisher 1972, 1978, 1980; Hall 1976, 1979, 1980).

#### 44.4.2.5 *Agriculture*

Several studies have been conducted in the agriculture sector. A case-control study conducted in Victoria to identify risk factors for serious farm-work-related injury among men showed that the most common external causes of injury were machinery (26%), falls (19%), transport (18%), animals (17%), and being struck by an object (11%). Increased injury risk was observed for being an employee/contractor, not having attended farm training courses, absence of roll-over protective structures on all/almost all tractors, absence of personal protective equipment for chemical use, and a low average annual farm income (Day et al. 2009). The prevalence of headache and neck pain in farmers was explored in 1997 (Scutter et al. 1997). It was shown that 77.7% of farmers experienced neck pain and 79.2% experienced headaches. Driving a tractor was the activity which was most frequently described as increasing symptoms in both conditions. The contribution of factors related to driving a tractor, including being exposed to whole-body vibration and assuming a rotated neck posture, to the development of headache and neck pain were discussed by the authors.

Although most of the recent research related to sheep shearing has been published in New Zealand, some studies have been conducted in Australia. One of these studies includes the analysis of the forces required to drag sheep over various surfaces (Harvey et al. 2002). Results of this study showed that significant changes in mean dragging force occurred with changes in both surface texture and slope. The best floor tested was a floor sloped at 1:10 constructed of timber battens oriented parallel to the path of the drag.

Shearing in hot weather has been studied at the University of Adelaide (Gun RT and Budd GM 1995). Forty-three men were studied throughout 54 man-days of shearing sheep and pressing wool bales, in air temperatures 19°C–41°C and wet-bulb globe temperature index (WBGT) 16°C–29°C. Over the 10 h work day, the subjects sweated substantially, but they replaced their sweat losses so successfully that warmer weather and heavier sweating were not accompanied by significantly greater dehydration. It was shown that men with greater fat percentage felt cooler, and those who had drunk more alcohol the previous evening had lower core temperature readings and tended to be more productive. Age was not associated with any measured response. The findings highlighted the challenges in attempts to define safe limits for occupational heat stress, and they demonstrate the effectiveness of the behavioral responses that permit shearers to perform sustained strenuous work in a hot environment without excessive physiological strain.

The Accident Research Centre at Monash University researched on the implementation of published guidelines for retrofitting tractors with safe access platforms, which were developed to reduce the risk of serious injuries and deaths associated with mounting and dismounting (Day and Rechnitzer 2004). The results were based on farmer interviews and engineering-based inspections. It was shown that platform retrofitting had little effect on tractor operations and substantially improved ease of access. The authors recommended that mechanisms to increase adherence to the key criteria of bottom step positioning and rear wheel guarding should be included in future promotion.

Many of the recent publications on farmer's health in Australia have focused on mental health problems, mental illness, and suicide (Judd et al. 2006; McShane and Quirk 2009; Fraser et al. 2005; Stain et al. 2008).

Fraser et al. (2005) provided an overview of the literature examining mental health issues experienced by farming populations in the United Kingdom, Europe, Australia, Canada, and the United States and identifies areas for further research. This body of research studied male farmers, female farmers, farm workers, farming families, and young people living on farms. Research to date indicates that farmers, farm workers, and their respective families face an array of stressors related to the physical environment, structure of farming families, and the economic difficulties and uncertainties associated with farming which may be detrimental to their mental health. The authors concluded that while suicide rates in some groups of farmers are higher than the general population, conclusive data do not exist to indicate whether farmers and farming families experience higher rates of mental health problems compared with the non-farming community. It is clear, however, that farming is associated with a unique set of characteristics that is potentially hazardous to mental health and requires further research.

#### *44.4.2.6 Slips, Trips, and Falls*

There have been few studies conducted on occupational slips, trips, and falls in Australia. Many fall-related studies and peak body initiatives in Australia have been conducted for community-wide purposes, especially in relation to aged care (State Government Victoria 2007; Injury Control Council or Western Australia 2009; Mackenzie 2009; Comans et al. 2009; Hill et al. 2009; Mackenzie et al. 2009).

Having said that, a recent epidemiological study conducted by MUARC and commissioned by the Australian Building Codes Board (ABCB) was successful in placing research findings into the context of policy (Ozanne-Smith et al. 2008). Slip, trip, and fall injuries were examined across three levels of severity: emergency department presentations (without admission), hospital admissions, and deaths. Through a preliminary analysis of state, national, and international data and as identified in similar studies conducted internationally, the most prevalent hazards and harms related to slips, trips, and falls and the design and construction of buildings were identified. The major building structural and design components identified as being associated with fall injuries in this study were flooring surfaces, stairs, windows, balconies, verandas, and, indirectly, guttering and roofs in residential settings. Epidemiological data in this study show that many of the victims of fall injuries in buildings are from vulnerable populations, particularly the elderly, the sick, and children. The authors argued that these community members are equally entitled to a safe environment as the more physically robust members of society. Accordingly they should be protected by the Building Code of Australia and innovation in safe building design and construction.

Other studies related to occupational falls include issues related to wearing of fall-arrest harnesses in the construction (Zupanc and Burgess-Limerick 2003), horse-racing falls (Hitchens et al. 2009), portable ladder use (Shepherd et al. 2006), and injury trends of farmers (Day et al. 2009), miners (McPhee 2004), occupational spinal cord injuries (O'Connor 2001), and injuries from unpaid work at home (Driscoll et al. 2003).

#### *44.4.2.7 Anthropometry*

It has been shown that updated anthropometric data to be used for the Australian workforce has been lacking for some time. A report commissioned by the ASCC (Safe Work Australia 2009) showed that some anthropometric data targeted at working populations in Australia are proprietary or commercial-in-confidence. Of the publicly available data, many are out of date or of military origin, not civilian population-based, and thus, they are of limited value when applied to civilian populations. Some Australian standards contain anthropometric data, but this is out of date and therefore unreliable. There are international data available but they are not necessarily relevant to the Australian population. Designers, advisers to designers, and evaluators of products and spaces consulted during this project expressed their needs now, and into the future, for reliable, high-quality, accessible, and affordable anthropometric data that can inform their work. However, there is optimism in the future as the researches acknowledged that recent formation and rapid growth of HFESA's Anthropometry Resources Australia Special Interest Group (ARASIG) provides a forum for users and producers of anthropometric data in Australia that could be used to educate and inform people. A further emerging source of anthropometric data is the international, not-for-profit group, World Engineering Anthropometry Resource (WEAR). The database will contain some Australian data, and there is potential to produce further data and lodge it with WEAR for international use.

Further to the earlier findings, there has been recent activity in the gathering of anthropometrics data across Australia. The Australian Defense Anthropometric Personnel Testing project is a collaborative project of a number of high-profile national and interstate organizations, including the Australian Government Department of Defence, University of South Australia, Australian Government Australian Sports Commission, Sinclair Knight Merz (SKM), and University of Ballarat. These partners have joined together to undertake a project using Vitus Smart 3D whole-body scanner to gather and assess information about the size of potential aircrew recruits and their fit in the current inventory of ADF aircraft. The immediate aim of the project is to determine the appropriate body size and shape for aircrew flying a variety of aircraft. The longer-term aim is to develop a capability in a variety of new technologies for measuring human body size and shape and simulating the movements of aircrew in aircraft. Data was collected in 2009 from major cities of Australia, including Adelaide, Melbourne, Canberra, Brisbane, Sydney, and Perth. Most candidates were men and women aged 18–30, particularly those people who have technical or university qualifications, or who are currently enrolled courses (UniSA 2009).

Prior to the works described earlier, the development of normative data for hand strength and anthropometric dimensions in a population of automotive workers was conducted and published in 2007 (Kunelius et al. 2007).

#### *44.4.2.8 Human–Computer Interaction*

HCI research is conducted at various pockets of the country, especially at the UQ and Monash University. Sanderson's group at UQ have explored HCI models, theories, and frameworks and have applied them to various industries including defence and hydropower systems (Naikar and Sanderson 2001; Naikar et al. 2003; Sanderson 2003; Li et al. 2006). Monash University has conducted extensive amounts of research on cursor control devices(Phillips and Triggs 2000, 2001; Phillips et al. 2001, 2003a,b; Stevenson et al. 2004; Memisevic et al. 2007).

#### *44.4.2.9 Product Design*

Roger Hall, at the University of New South Wales, has written extensively on the ergonomics of product design and the prototyping (Hall 1998, 2001), and explored the usability of time functions on small electronic consumer products(Hall et al. 1998).

#### *44.4.2.10 Human Motor Skill*

Provins published newer studies in the area of manual asymmetry and motor skill in the 1990s (Provins and Magliaro 1993; Provins 1997). His publications show and discuss the activity-specific nature of motor skills. The findings of his studies show the importance of using the same criteria in making preference for performance comparisons and highlight the need to recognize and control the influence of previous training or experience.

Welford concentrated on cognitive ergonomics in the 1970s, but his more recent work focuses on the effects of ageing on work performance (Welford 1988a,b).

#### *44.4.2.11 Job Demands*

The ergonomics research team at Latrobe University, including Macdonald and Peterson, have contributed in the area of workload, job demands, job satisfaction, fatigue, stress, performance, health, and well-being (Peterson 1994; Macdonald and Bendak 2000; Weeks et al. 2000; Murphy et al. 2002; Macdonald 2003, 2006a,b; Peterson 2003).

#### *44.4.2.12 Cost-Benefit Analysis*

Oxenburgh has worked on and written about the concept of cost-benefit analysis and developed models in relation to ergonomics and other health and safety intervention (Oxenburgh and Guldberg 1993; Oxenburgh and Marlow 2005; Oxenburgh 2004) which have been applied independently in the health-care industry (Busse and Bridger 1997).

## 44.5 APPLIED ERGONOMICS AND CASE STUDIES

Ergonomics has been applied in most industry sectors in Australia. However, the evidence of implementation and uptake has been more apparent in some industries compared to others. For example, the health-care industry have applied widespread changes as No or Minimal Lift Policy programs were implemented across the nation, especially the state of Victoria, where the nursing board, in collaboration with other peak bodies, drove this intervention (State Government Victoria 2002). On the contrary, the construction industry, especially trade services, appears to have few publicized ergonomics intervention programs. WorkSafe WA produced a series of publications on reducing the risk of manual handling injuries in various trade services, including bricklayers, form workers, block layers and stonemasons, electricians, plasterers, and roof carpenters (WorkSafe WA 2009). These were formulated following stakeholder meetings with employer and employee representation. However, there have been no studies conducted to measure the level of uptake of the information.

Three case studies have been presented in the following. The first two cases are of specific interventions that were implemented in two companies that were working toward reducing the risk of MSDs in the workplace. The third case study is of a wider-scoped project that aimed to assess the impact of remuneration methods, fatigue management, and OHS management systems on OHS performance in the mining industry.

*Case Study* 1

An assay lab that conducts chemical analysis of soil samples had a job role that required the workers to grind soil samples to be analyzed using a semiautomated grinding machine. The job required a worker to retrieve soil samples from a trolley placed next to their workstation, place soil samples in the grinding machine, close the grinder door, switch the grinder on, place ground soil samples into labeled bags, and clean the grinder before the next process is conducted (as shown in Figure 44.2). Issues that presented were shoulder and wrist discomfort and exposure to dust (which have unknown levels of hazardous substances). The task required continuous hand and arm activities, exposing that part of the body to high biomechanical forces when lifting the metal rings of the machine and repetitive movement when bagging the soil samples. The company looked at various controls to tackle the source of the problem, which stemmed primarily from the design of the grinder machine and process. After evaluating what was available on the market and conducting a cost analysis, the executives decided to completely eliminate the semiautomated grinding process by using robotic technology. The supplier of the robotic machine tailored a design specifically for the company. The company has

FIGURE 44.2 Semiautomated grinding workstation: worker lifting metal rings before vacuuming the dust off the grinder.

FIGURE 44.3 Robotic system performing the grinding task equivalent of six semiautomated grinding machines.

since found that the robotic arm can perform the task of six semiautomated grinders at a faster pace than before (as shown in Figure 44.3).

*Case Study* 2

The next case study is of the simple interventions that have taken place in a metal manufacturing company that produces small metal parts, primarily for the construction industry. Many of the workers in this organization have physical and cognitive limitations. Processes, systems of work, work-area layout, workstations, and equipment throughout the workplace have been selected and designed for individual workers. To achieve health, safety, and productivity targets, collaboration and continuous consultation take place between the operations managers, engineers, occupational therapists, and workers within the organization.

Two examples of cost-effective ergonomics interventions include the following:

1. The design of a workstation for an individual with physical and cognitive limitations for a cardboard box assembling task. A bracket which helps form and stabilize the cardboard boxes while it is stapled together was developed, and a footrest was designed to suit the worker, to couple with the adjustable chair he was supplied with (as shown in Figure 44.4).

2. The intervention is of the design of a trolley tailored for the movement of metal sheeting. These were not commercially available and therefore were created in-house using suitable castors and brackets for metal sheets to be wheeled rather than manually lifted around the factory (as shown in Figure 44.5).

FIGURE 44.4 Metal manufacturing worker assembling boxes using a specially designed bracket and footrest.

FIGURE 44.5 Specially designed trolley for transporting metal sheets.

*Case Study* 3

Digging Deeper (Shaw Idea 2007) was a research project conducted in the mining industry and commissioned by the NSW minister for mineral resources aimed to assess the impact on OHS performance of

- Production bonus payments and safety-based incentive schemes
- Fatigue management and working hours in the mining industry
- OHS management systems and consultation

The researchers found that production bonus and safety incentive schemes that involve payment in exchange for achieving particular safety outcome targets have not proved themselves to consistently or reliably improve safety outcomes. Generally, sites reported that safety incentive schemes making payments as a result of achievement of outcome targets either made no difference at all or had negative effects on incident reporting. It was evident that hours of work in the NSW mining industry are high (average 49.8 per week) and far in excess of the hours worked in the mining industry on average around Australia (average 44.7 per week). The factors found to affect the hours worked were occupation (with management and professionals working longer hours), subsector, employment status (with contractors working longer hours), size of company (with large company workers working longer hours), and location. Many people interviewed reported that they were fatigued as a result of their hours of work and shift arrangements. Respondents reported statistically significant differences between fatigue according to shift. Night shift was reported to cause significantly worse effects on work performance and fatigue levels than either afternoon or day shift. The researchers found limited evidence that sites had a thorough understanding of the causes of fatigue, with most attitudes to fatigue focused around nonwork causes, rather than the contributions made by working arrangements. It was found that the importance of addressing the interrelated personal and organizational factors was not widely recognized. While the researchers found excellent examples of engineering risk controls, they also saw some reluctance on the part of the industry to manage risks at source. This was coupled with an increasing take-up of strategies that focus on worker behavior as the primary means of risk control. Twenty-five recommendations were provided to the NSW Department of Primary Industries, based on the findings of this project.

## 44.6 ERGONOMICS TRAINING PROGRAMS

Initially, education in ergonomics in Australia use to be offered within relevant professional programs, including engineering, psychology, physiology, architectural science, physiotherapy, occupational health, and applied arts and industrial engineering. Today, postgraduate qualifications in ergonomics (at postgraduate diploma or masters degree level) are offered within several tertiary institutions. No undergraduate program totally devoted to the preparation of an ergonomist is offered within Australia. In 2009, two universities were found to offer these courses including

1. *The UQ*—graduate certificate in ergonomics, diploma in ergonomics and masters of ergonomics (University of Queensland 2009)
2. *LaTrobe*—graduate certificate in occupational health, safety and ergonomics and masters of ergonomics, safety and health (La Trobe University 2009)

The IEA guidelines for accreditation of ergonomics education programs developed by an IEA committee chaired by Professor Margaret Bullock were accepted by the society, and these have also provided a resource for outlining curricula and philosophies for new ergonomics educational programs in Australia (Bullock 2001).

## 44.7 EMERGING ISSUES

The emerging issues apparent in Australia are very similar to those of other developed countries including greater use of computers by children and adults (Straker et al. 2006, 2007), increased sedentary work (Straker and Mathiassen 2009), increased social isolation at work, greater mobility of workers between jobs in a market-driven economy (Caple 2008), increased reports of work-related mental health disorders (Safe Work Australia 2006c), an ageing workforce (Safe Work Australia 2005), increased levels of obesity in the community, and changing anthropometrics (Safe Work Australia 2009b,c).

## 44.8 STUDY QUESTIONS

1. Are there any regulations in Australia that relate to the prevention of MSDs? If so, what are they called and in which states do they exist?

   Answer:

   a. Yes there are regulations that target MSDs.
   b. They are commonly termed "manual handling" or "manual task" regulations.
   c. They exist in all jurisdictions, including Australian Capital Territory, Northern Territory, Western Australia, Queensland, Victoria, South Australia, New South Wales, and Tasmania.

2. List 3 emerging issues in Australia and discuss them briefly, including whether they may be unique to Australia or be common in other developed countries, and how they may have an impact on future ergonomics work in the Australia.

   Answer:

   a. Emerging issues:
      - Use of computers by children and adults
      - Increased sedentary work

- Increased social isolation at work
- Greater mobility of workers between jobs in a market-driven economy
- Increased reports of work-related mental health disorders
- An ageing workforce
- Increased levels of obesity in the community
- Changing anthropometrics

## ACKNOWLEDGMENTS

Kathryn Jones, Rodney Powell, Christina Paterson, Rosalind Forward, and John Innes at WorkSafe WA for assistance and support.

David Caple, Robin Burgess-Limerick, and Leon Straker for names of leading researchers and topics of interest.

Margaret Bullock for her most recent contribution on the history of ergonomics in Australia to the society.

Effie Mangharam and Henri Rose for technical edits, encouragement, and support.

## REFERENCES

Anderson, J.E. and P. Sanderson, Sonification design for complex work domains: Dimensions and distractors, *J Exp Psychol: Appl*, **15**(3): 183–198 (2009).

Australia, S., Sizing up Australia: How contemporary is the anthropometric data Australian designers use? South Australia, Canberra, Australian Capital Territory, Australia, Commonwealth of Australia, 2009.

Australian Bureau of Statistics, Population projections Australia 2006 to 2101, Canberra, Australian Capital Territory, Australia, Commonwealth of Australia, 3222.0, 2008.

Australian Bureau of Statistics, Australian demographic statistics march quarter 2009, Canberra, Australian Capital Territory, Australia, Commonwealth of Australia, 3101.0, 2009.

Australian Bureau of Statistics, Labour force Australia October 2009, Canberra, Australian Capital Territory, Australia, Commonwealth of Australia, 6202.0, 2009.

Australian Bureau of Statistics, *Year Book Australia*, 2009–2010, Canberra, Australian Capital Territory, Australia, Commonwealth of Australia, 1301.0, 2010.

Australian Safety and Compensation Council, Guidance on the principle of safe design for work, Canberra, Australian Capital Territory, Australia, Commonwealth of Australia, 2006.

Australian Safety and Compensation Council, National code of practice for the prevention of musculoskeletal disorders from performing manual tasks at work, Canberra, Australian Capital Territory, Australia, Commonwealth of Australia, 2007a.

Australian Safety and Compensation Council, National standard for manual tasks, Australian Government, Canberra, Australian Capital Territory, Australia, Commonwealth of Australia, 2007b.

Australian Safety and Compensation Council, Overweight and obesity: Implications for workplace health and safety and workers' compensation, Australian Safety and Compensation Council, Canberra, Australian Capital Territory, Australia, Commonwealth of Australia, 2008.

Australian Safety and Compensation Council, National hazard exposure worker surveillance (NHEWS) survey: 2008 results, Canberra, Australian Capital Territory, Australia, Commonwealth of Australia, 2009.

Awerbuch, M., Repetitive strain injuries: Has the Australian epidemic burnt out? *Intern Med J*, July; **34**(7): 416–419 (2004).

Briggs, A.M. and P. Bragge et al., Prevalence and associated factors for thoracic spine pain in the adult working population, a literature review, *J Occup Health Saf Aust NZ*, **51**(3): 177–192 (2009a).

Briggs, A.M. and A.J. Smith et al., Thoracic spine pain in the general population: A systematic review of prevalence, incidence and associated factors in children, adolescents and adults, *BMC Musculoskelet Disord*, **10**: 77 (2009b).

Briggs, A. and L. Straker et al., Upper quadrant postural changes of school children in response to interaction with different information technologies, *Ergonomics*, **47**(7): 790–819 (2004).

Briggs, A.M. and L.M. Straker, Thoracic spine pain in youth: Should we be concerned, *Spine J*, **94**(4): 338–339 (2009).

Browne, R.C., Safety and health at work: The Robens report, *Br J Ind Med*, **30**: 87–91 (1973).

Bryan Bottomley and Associates, Identification of compliance requirements and tools for an effective risk management process for the national standard and code of practice for manual handling, National Occupational Health and Safety Commission, Canberra, Australian Capital Territory, Australia, Commonwealth of Australia, 2003a.

Bryan Bottomley and Associates, An assessment of current performance indicators as a measurement tool and the development of an adaptive model best practice for physical handling in the work environment, National Occupational Health and Safety Commission, Canberra, Australian Capital Territory, Australia, Commonwealth of Australia, February 2003b.

Bullock, M.I., Cockpit design—Pilot accommodation and accessibility to controls, *Aerosp Med*, November; **44**(11): 1295–1299 (1973).

Bullock, M.I., The determination of functional arm reach boundaries for operation of manual controls, *Ergonomics*, May; **17**(3): 375–388 (1974).

Bullock, M.I., Ripcord release capability of female parachutists, *Aviat Space Environ Med*, October; **49**(10): 1177–1183 (1978).

Bullock, M.I., Australia: Ergonomics society of Australia, *International Encyclopedia of Ergonomics and Human Factors*, W. Karwowski (ed.), Vol. 1, Taylor & Francis, London, U.K., and New York, 2001.

Burgess-Limerick, R., Stoop, squat, or something in-between, *Int J Ind Ergon*, **31**: 143–148 (2003).

Burgess-Limerick, R., Reducing injury risks associated with underground coal mining equipment, *Ergono Aust*, **19**(2): 14–20 (2005).

Burgess-Limerick, R. and B. Abernethy et al., Self-selected manual lifting technique: Functional consequences of interjoint coordination, *Hum Factors*, **37**: 395–411 (1995).

Burgess-Limerick, R. and B. Abernethy, Towards a quantitative definition of manual lifting postures, *Hum Factors*, **39**: 141–148 (1997a).

Burgess-Limerick, R. and B. Abernethy, Qualitatively different modes of manual lifting, *Int J Ind Ergon*, **19**: 413–417 (1997b).

Burgess-Limerick, R. and B. Abernethy, Effect of load distance on self-selected manual lifting technique, *Int J Ind Ergon*, **22**: 367–372 (1998).

Burgess-Limerick, R. and W. Green, Using multiple case studies in ergonomics: An example of pointing device use, *Int J Ind Ergon*, **26**: 381–388 (2000).

Burgess-Limerick, R. and M. Mon-Williams et al., Visual display height, *Hum Factors*, **42**: 140–150 (2000).

Burgess-Limerick, R. and A. Plooy et al., The effect of imposed and self-selected computer monitor height on posture and gaze angle, *Clin Biomech*, **13**: 584–592 (1998).

Burgess-Limerick, R. and A. Plooy et al., The Influence of computer monitor height on head and neck posture, *Int J Ind Ergon*, **23**: 171–179 (1999a).

Burgess-Limerick, R. and J. Shemmell et al., Wrist posture during pointing device use, *Clin Biomech*, **14**: 280–286 (1999b).

Burgess-Limerick, R. and J. Shemmell et al., Spontaneous transitions in lifting, *Hum Movement Sci*, **20**: 549–562 (2001).

Burgess-Limerick, R. and L. Steiner, Injuries associated with continuous miners, shuttle cars, load-haul-dump, and personnel transport in New South Wales underground coal mines, *Min Technol (TIMM A)*, **115**: 160–168 (2006).

Burgess-Limerick, R. and L. Steiner, Opportunities for preventing equipment related injuries in underground coal mines in the USA, *Min Eng*, October; 59(10): 20–32 (2007).

Burgess-Limerick, R. and L. Straker et al., Participative ergonomics for manual tasks in coal mining, *Int J Ind Ergon*, **37**: 145–155 (2007).

Busse, M. and B. Bridger, Cost benefits of ergonomic intervention in a hospital: A preliminary study using Oxenburgh's productivity model, *Curationis*, September; **20**(3): 54–58 (1997).

Cameron, C., Fatigue problems in modern industry, *Ergonomics*, November; **14**(6): 713–720 (1971).

Cameron, C., A theory of fatigue, *Ergonomics*, September; **16**(5): 633–648 (1973).

Caple, D., Emerging challenges to the ergonomics domain, *Ergonomics*, **51**(1): 49–54 (2008).

Carrivick, P.J. and A.H. Lee et al., Evaluating the effectiveness of a participatory ergonomics approach in reducing the risk and severity of injuries from manual handling, *Ergonomics*, June 22; **48**(8): 907–914 (2005).

Comans, T. and S. Brauer et al., A break-even analysis of a community rehabilitation falls prevention service, *Aust NZ J Public Health*, June; **33**(3): 240–245 (2009).

Commission of Occupational Safety and Health of Western Australia, Code of practice for manual handling, State Government of Western Australia, Perth, Western Australia, Australia, 2000.

Commission of Occupational Safety and Health of Western Australia, Code of practice for the prevention of falls at workplaces, Government of Western Australia, Perth, Western Australia, Australia, 2004.

Cook, C. and R. Burgess-Limerick et al., The prevalence of neck and upper extremity musculoskeletal symptoms in computer mouse users, *Int J Ind Ergon*, **26**: 347–356 (2000).

Cook, C. and R. Burgess-Limerick et al., The effect of upper extremity support on upper extremity posture and muscle activity during keyboard use, *Appl Ergon*, **35**: 285–292 (2004a).

Cook, C. and R. Burgess-Limerick et al., The effect of wrists rests during keyboard use with forearm support, *Int J Ind Ergon*, **33**: 463–472 (2004b).

Cook, C. and R. Burgess-Limerick, The effect of forearm support on musculoskeletal discomfort during call centre work, *Appl Ergon*, **35**: 337–342 (2004).

Cooper, A. and L. Straker, Mouse versus keyboard use: A comparison of shoulder muscle load, *Int J Ind Ergon*, **22**: 351–357 (1998).

Cromie, J.E. and V.J. Robertson et al., Work-related musculoskeletal disorders in physical therapists: Prevalence, severity, risks, and responses, *Phys Ther*, April; **80**(4): 336–351 (2000).

Cromie, J.E. and V.J. Robertson et al., Work-related musculoskeletal disorders and the culture of physical therapy, *Phys Ther*, May; **82**(5): 459–472 (2002).

Cromie, J.E. and V.J. Robertson et al., Physical therapists who claimed workers' compensation: A qualitative study, *Phys Ther*, December; **83**(12): 1080–1089 (2003).

Cumming, R.W., Human information processing under varying task demand, *Ergonomics*, September; **16**(5): 581–586 (1973).

Dankaerts, W. and P. O'Sullivan et al., Altered patterns of superficial trunk muscle activation during sitting in nonspecific chronic low back pain patients: Importance of subclassification, *Spine*, **31**(17): 2017–2023 (2006a).

Dankaerts, W. and P.B. O'Sullivan et al., Inter-examiner reliability of a classification method for non-specific chronic low back pain patients with motor control impairment, *Man Ther*, **11**: 28–39 (2006b).

David Caple and Associates Pty Ltd, Physical safety in small and medium-sized enterprises (SME's): Strategies for addressing the special needs of these organisations and their employees, National Occupational Health and Safety Commission, Canberra, Australian Capital Territory, Australia, Commonwealth of Australia, 2003.

Day, L. and G. Rechnitzer, Safe tractor access platforms: From guidance material to implementation, *J Agric Saf Health*, August; **10**(3): 197–209 (2004).

Day, L. and D. Voaklander et al., Risk factors for work related injury among male farmers, *Occup Environ Med*, May; **66**(5): 312–318 (2009).

Department of Health Safety and Welfare of Western Australia, Strategies to reduce the risk of back strain in nursing homes, Government of Western Australia, Perth, Western Australia, Australia, 1989.

Department of Health Safety and Welfare of Western Australia, Strain injuries to council workers: Reducing the risks, Government of Western Australia, Perth, Western Australia, Australia, 1991.

Department of Health Safety and Welfare of Western Australia, Manual handling in the meat industry: Reducing the risks, Government of Western Australia, Perth, Western Australia, Australia, 1992.

Department of Mines and Petroleum, Hazardous manual tasks [http://www.dmp.wa.gov.au/7221.aspx#7225]. Accessed October 2009.

Driscoll, T.R. and R.J. Mitchell et al., Unintentional fatal injuries arising from unpaid work at home, *Inj Prev*, March; **9**(1): 15–19 (2003).

Duncan, J. and D. Ferguson, Keyboard operating posture and symptoms in operating, *Ergonomics*, September; **17**(5): 651–662 (1974).

Elford, W. and L. Straker et al., Patient handling with and without slings: An analysis of the risk of injury to the lumbar spine, *Appl Ergon*, **31**: 185–200 (2000).

Engkvist, I.L., Evaluation of an intervention comprising a no lifting policy in Australian hospitals, *Appl Ergon*, March; **37**(2): 141–148 (2006).

Engkvist, I.L., Nurse's expectations, experiences and attitudes towards the intervention of a 'no lifting policy', *J Occup Health and Saf—Aust NZ*, **49**: 294–304 (2007).

Evans, O.M. and K. Patterson, Predictors of neck and shoulder pain in non-secretarial computer users, *Int J Ind Ergon*, **26**: 357–365 (2000).

Ferguson, D., Ergonomics in Australia, *Med J of Aust*, January 4; **1**(1): 30–33 (1969).

Ferguson, D., Repetition injuries in process workers, *Med J Aust*, August 21; **2**(28): 408–412 (1971).

Ferguson, D., Ergonomics and the quality of living, *Appl Ergon*, June; **3**(2): 70–74 (1972).

Ferguson, D., A study of occupational stress and health, *Ergonomics*, September; **16**(5): 648–664 (1973).

Ferguson, D. and J. Duncan, Keyboard design and operating posture, *Ergonomics*, November; **17**(6): 731–744 (1974).

Fisher, A.J. and R.R. Hall, The influence of car frontal design on pedestrian accident trauma, *Accid Anal Prevent*, **4**: 47–58 (1972).

Fisher, A.J. and R.R. Hall, The effect on vision of terminations and isolated lengths of fixed lighting, *Aust Road Res*, **5**: 34–39 (1973).

Fisher, A.J. and R.R. Hall, Road lighting levels based on detection of change of visual angle, *Light Res Technol*, **8**(4): 187–194 (1976).

Fisher, A.J. and R.R. Hall, The effect of presence lights on the detection of change of vehicle headway, *Aust Road Res*, **8**(2): 13–16 (1978).

Fisher, A.J. and R.R. Hall, Aids in night driving, *Aust J Optomet*, **65**(5): 194–195 (1982).

Fisher, A.J. and R.R. Hall, The lighting of curves on arterial roads, *Light Res Technol*, **17**(3): 129–137 (1985a).

Fisher, A.J. and R.R. Hall, The lighting of curves, *Providing Visibility and Visual Guidance to the Road User*, Transportation No. 297, pp. 27–28, 1985b.

Fisher, A.J. and R.R. Hall, Luminance, uniformity and glare in road lighting—Some more equal than others, *Aust Road Res*, **16**(3): 184–193 (1986).

Fraser, C.E. and K.B. Smith et al., Farming and mental health problems and mental illness, *Int J Soc Psychiatry*, December; **51**(4): 340–349 (2005).

Greig, A. and L. Straker et al., Cervical erector spinae and upper trapezius activity in children using desktop and laptop computers, *Physiotherapy*, **91**(2): 119–126 (2005).

Grundgeiger, T. and P. Sanderson, Interruptions in healthcare: Theoretical views, *Int J Med Inform*, **78**: 293–307 (2009).

Gun, R.T. and G.M. Budd, Effects of thermal, personal and behavioural factors on the physiological strain, thermal comfort and productivity of Australian shearers in hot weather, *Ergonomics*, July; **38**(7): 1368–1384 (1995).

Gun, R.T. and P.T. Jezukaitis, RSI: A perspective from its birthplace, *Occup Med*, January–March; **14**(1): 81–95, (1999).

Hall, R.R., CIE standard luminance program for the calculation of luminance in road lighting design—Users reference manual, Australian Road Research Board, Vermont, Victoria, Australia, 1976.

Hall, R.R., The delineation of rural roads and curve negotiation at night, Australian Road Research Board, Vermont South, Victoria, Australia, 1979.

Hall, R.R., The design and implementation of fixed lighting for arterial roads and freeways, Australian Road Research Board, Vermont South, Victoria, 1980.

Hall, R.R., Usability and product design, *Ergonomics, Usability and Product Development,* P. Jordan and W. Green (eds.), Taylor & Francis, London, U.K., pp. 85–91, 1998.

Hall, R.R., Prototyping for usability of new technology, *Int J Hum Comput Stud*, **55**(4): 485–501 (2001).

Hall, R.R. and A.J. Fisher, Some factors affecting the trauma of pedestrians involved in road accidents, *Med J Aust*, **1**(7): 313–317 (1972).

Hall, R.R. and A.J. Fisher, Measures of visibility and visual performance in road lighting, Australian Road Research Board, Vermont South, Victoria, Australia, 1978.

Hall, R.R. and A.J. Fisher, Freeway lighting: Analysis of installation performance and cost, Australian Road Research Board, Vermont South, Victoria, Australia, 1980.

Hall, W. and L. Morrow, Repetition strain injury: An Australian epidemic of upper limb pain, *Soc Sci Med*, **27**(6): 645–649 (1988).

Hall, R.R. and S. Zinser et al., The usability of time setting functions on small electronic consumer products—A test, *Int J Cogn Ergono*, **3**(2): 101–114 (1998).

Harris, C. and L. Straker, Survey of physical ergonomics issues associated with school children's use of laptop computers, *Int J Ind Ergon*, **26**: 337–347 (2000).

Harvey, J.T. and J. Culvenor et al., An analysis of the forces required to drag sheep over various surfaces, *Appl Ergon*, **33**(6): 523–531 (2002).

Heads of Workplace Safety Authorities, Design 4 health: National manual handling campaign 2004, 2005, [http://www.hwsa.org.au/activities/activities-campaign_final_reports.aspx]. Accessed October 2009.

Heads of Workplace Safety Authorities, National falls from heights in the heavy vehicle sector: Summary final report industry feedback, 2006, [http://www.hwsa.org.au/files/documents/Activities%20-%20Campaign%20Final%20Reports/afba065a-6f71-4871-91d9-be5c7616ce34.pdf]. Accessed October 2009.

Heads of Workplace Safety Authorities, Manual handling in manufacturing industry: HWSA final report, 2007, [http://www.hwsa.org.au/activities/activities-campaign_final_reports.aspx]. Accessed October 2009.

Heads of Workplace Safety Authorities, "Safe steps—Manual tasks, slips and trips in hospitals: Final report, 2008a, [http://www.hwsa.org.au/activities/activities-campaign_final_reports.aspx]. Accessed October 2009.

Heads of Workplace Safety Authorities, Prevention of falls in construction 2008: HWSA final report, 2008b, [http://www.hwsa.org.au/activities/activities-campaign_final_reports.aspx]. Accessed October 2009.

Hill, K.D. and D. LoGiudice et al., Effectiveness of balance training exercise in people with mild to moderate severity Alzheimer's disease: Protocol for a randomised trial, *BMC Geriatr*, July 16; **9**:29 (2009).

Hitchens, P.L. and C.L. Blizzard et al., The incidence of race-day jockey falls in Australia, 2002–2006, *Med J Aust*, January 19; **190**(2): 83–86 (2009).

Horberry, T. and J. Anderson et al., Driver distraction: The effects of concurrent in-vehicle tasks, road environment complexity and age on driving performance, *Accid Anal Prevent*, **38**(1): 185–191 (2006a).

Horberry, T. and A. Gunatilaka et al., Intelligent transport systems for industrial mobile safety equipment, *J Occup Health Saf—Aust NZ*, **22**(4): 323–334 (2006b).

Horberry, T. and M. Regan et al., Airport ramp safety and intelligent transport systems, *IET Intell Transp Sys*, **1**(4): 234–240 (2007).

Hosking, S.G. and K.L. Young et al., *The Effects of Text Messaging on Young Novice Driver Performance*, Monash University Accident Research Centre, Melbourne, Victoria, Australia, 2006.

Howie, A. and W. Macdonald et al., The ergonomics society of Australia, *Ergonomics*, **31**(5): 751–760 (1988).

Human Factors and Ergonomics Society of Australia, Membership, 2009a, [http://www.ergonomics.org.au/membership/cpe.aspx]. Accessed October 11, 2009.

Human Factors and Ergonomics Society of Australia, 2009b, Professional development, [http://www.ergonomics.org.au/events/professional_development.aspx]. Accessed October 11, 2009.

Injury Control Council or Western Australia, Falls prevention. http://www.iccwa.org.au/falls-prevention (accessed October 2009).

Innes, E. and L. Straker, A clinician's guide to work-related assessments: 1. Purposes and problems, *Work*, **11**: 183–189 (1998a).

Innes, E. and L. Straker, A clinician's guide to work-related assessments: 2. Design problems, *Work*, **11**: 191–206 (1998b).

Innes, E. and L. Straker, A clinician's guide to work-related assessments: 3. Administration and interpretation problems, *Work*, **11**: 207–219 (1998c).

Innes, E. and L. Straker, Reliability of work-related assessments, *Work*, **13**: 107–124 (1999a).

Innes, E. and L. Straker, Validity of work-related assessments, *Work*, **13**: 125–152 (1999b).

Innes, E. and L. Straker, Workplace assessments and functional capacity evaluations: Current practices of therapists in Australia, *Work*, **18**: 51–66 (2002a).

Innes, E. and L. Straker, Strategies used when conducting work related assessments, *Work*, **19**: 149–166 (2002b).

Innes, E. and L. Straker, Workplace assessments and functional capacity evaluations: Current beliefs of Australian therapists, *Work*, **20**(3): 225–236 (2003a).

Innes, E. and L. Straker, Attributes of excellence in work related assessments, *Work*, **20**(1): 63–76 (2003b).

International Ergonomics Association, IEA federated societies, [http://www.iea.cc/browse.php?contID=technical_committees_home]. Accessed November 20, 2009.

Ireland, D.C., Repetition strain injury: The Australian experience—1992 update, *J Hand Surg Am*, May; **20A**(3 Pt 2): S53–S56 (1995).

Ireland, D.C., Australian repetition strain injury phenomenon, *Clin Orthop Relat Res*, June; (351): 63–73 (1998).

Johnstone, R., Ed., *Occupational Health and Safety Law and Policy: Text and Materials*, Lawbook Co., Sydney, New South Wales, Australia, 2004.

Judd, F. and H. Jackson et al., Understanding suicide in Australian farmers, *Soc Psychiatry Psychiatr Epidemiol*, January; **41**(1): 1–10 (2006).

King, J. and P. Foreman et al., Back injuries among nurses: Prevalence, risk factors and prevention, *Aust J Rehabil Counsell*, **7**: 106–114 (2001).

Kunelius, A. and S. Darzins et al., Development of normative data for hand strength and anthropometric dimensions in a population of automotive workers, *Work: J Preven, Assess Rehabil*, **28**: 267–278 (2007).

La Trobe University, Postgraduate coursework programs in occupational health, safety and ergonomics—Distance mode, [http://www.latrobe.edu.au/ergonomics/PG_OHSE.html]. Accessed November 2009.

Lacherez, P. and E. Seah et al., Overlapping medical alarms are almost indiscriminable, *Hum Factors*, **49**(4): 637–645 (2007).

Lane, J.C., Human engineering: A new technology, *Med J Aust*, July 4; **2**(1): 7–10 (1953).

Legge, J. and R. Burgess-Limerick, Reliability of the jobfit system pre-employment functional assessment tool, *Work*, **28**: 299–312 (2007).

Lenné, M. and K. Ashby et al., Analysis of general aviation crashes in Australia using the human factors analysis and classification system, *Int J Aviat Psychol*, **18**: 340–352 (2008a).

Lenné, M. and K. Ashby, Characteristics on non-fatal general aviation crashes in Australia: Insights from insurance data, *Hum Factor Aero Saf*, **6**(4): 395–407 (2008).

Lenné, M. and C. Mulvihill et al., Detection of emergency vehicles: Driver responses to advanced warning in a driving simulator, *Hum Factors*, **50**(1): 135 (2008b).

Li, X. and P. Sanderson et al., Evaluating functional displays for hydropower system: Model-based guidance of scenario design, *Cogn Technol Work*, **8**(4): 269–282 (2006).

Lindgaard, G. and D. Caple, A case study in iterative keyboard design using participatory design techniques, *Appl Ergon*, February; **32**(1): 71–80 (2001).

Liu, D. and T. Grundgeiger et al., Interruptions and blood transfusion checks: Lessons from the simulated operating room, *Anesth Analg*, **107**(1): 219–222 (2009a).

Liu, D. and S. Jenkins et al., Monitoring with head-mounted displays: Performance and safety in a full-scale simulator and part-task trainer, *Anesth Analg*, **109**(4): 1135–1146 (2009b).

Liu, D. and S. Jenkins et al., Patient monitoring with head-mounted displays, *Curr Opin Anaesthesiol*, **22**(6): 796–803 (2009c).

Loft, S. and P. Sanderson et al., Modeling and predicting mental workload in en route air traffic control: Critical review and broader implications, *Hum Factors*, **49**(3): 376–399 (2007).

Ltd, U.P., Issues associated with force and weight limits and associated threshold limit values in the physical handling work environment, N. O. H. a. S. Commission, Canberra, Australian Capital Territory, Australia, Commonwealth of Australia, 2003.

Macdonald, W.A., The impact of job demands and workload on stress and fatigue, *Aust Psycholo*, **38**(2): 102–117 (2003).

Macdonald, W., Workload, stress and psychosocial factors as hazards for musculoskeletal disorders, *J Occup Health and Saf—Aust NZ*, **20**(1): 37–47 (2004).

Macdonald, W., Managing workloads to optimise performance, health and wellbeing, *International Encyclopedia of Ergonomics and Human Factors*, W. Karwowski (ed.), Taylor & Francis, Boca Raton, FL, pp. 2170–2174, 2006a.

Macdonald, W., Workload, performance, health and wellbeing: A conceptual framework, *International Encyclopedia of Ergonomics and Human Factors*, W. Karwowski (ed.), Taylor & Francis, Boca Raton, FL, pp. 2794–2799, 2006b.

Macdonald, W. and S. Bendak, Effects of workload level and 8- versus 12-h workday duration on test battery performance, *Int J Ind Ergon*, **26**(3): 399–416 (2000).

Macdonald, W. and K. Munk, Determinants of work-related stress in physical handling work environments: Issues Paper 6, Review of the National Manual Handling Standard, Codes of practice and associated guidance materials, N. O. H. S. Commission, Canberra, Australian Capital Territory, Australia, Commonwealth Government of Australia, p. 55, 2003.

Mackenzie, L., Perceptions of health professionals about effective practice in falls prevention, *Disabil Rehabil*, **31**(24): 2005–2012 (2009).

Mackenzie, L. and J. Byles et al., Longitudinal study of the home falls and accidents screening tool in identifying older people at increased risk of falls, *Australas J Ageing*, June; **28**(2): 64–69 (2009).

Martin, E., Work-related musculoskeletal disorders in Victorian occupational therapists, *22nd National Conference*, OT Australia, Melbourne, Victoria, Australia, 2003.

Mayman, S.C., Getting the dynamics right, [http://www.wairc.wa.gov.au/Files/DiscussionPapers/National%20Safety%20Summit%20(presentation).ppt]. Accessed October 11, 2009.

McPhee, B., *Ergonomics for the Control of Sprains and Strains in Mining*, Worksafe Australia, Canberra, Australian Capital Territory, Australia, 1993.

McPhee, B., Ergonomics in mining, *Occup Med* (*Lond*), August; **54**(5): 297–303 (2004).

Mcphee, B. and G. Foster et al., *Bad Vibrations: A Handbook on Whole-Body Vibration Exposure in Mining*, The Joint Coal Board Health and Safety Trust, Sydney, New South Wales, Australia, 2001.

McShane, C.J. and F. Quirk, Mediating and moderating effects of work-home interference upon farm stresses and psychological distress, *Aust J Rural Health*, October; **17**(5): 244–250 (2009).

Memisevic, R. and P. Sanderson et al., Investigating human-system interaction with an integrated hydropower and market system simulator, *IEEE Trans Power Syst*, **22**(2): 762–769 (2007).

Mon-Williams, M. and R. Burgess-Limerick et al., Vertical gaze direction and postural adjustment: An extension of the heuer model, *J Exp Psychol Appl*, **5**: 35–53 (1999).

Murphy, G. and J. Athanasou et al., Job satisfaction and organisational citizenship behaviour: A study of Australian human-service workers, *J Manag Psychol*, **17**: 287–297 (2002).

Murphy, D.G. and R. Hall et al., Robotic technology in surgery: Current status in 2008, *ANZ J Surg*, December; **78**(12): 1076–1081 (2008).

Naikar, N. and D. Drumm et al., Designing teams for first-of-a-kind, complex systems using the initial phases of cognitive work analysis: Case study, *Hum Factors*, **45**(2): 202–217 (2003).

Naikar, N. and P.M. Sanderson, Work domain analysis for training system definition and acquisition, *Int J Aviat Psychol*, **9**(3): 271–290 (1999).

Naikar, N. and P. Sanderson, Evaluating system design proposals with work domain analysis, *Hum Factors*, **43**(4): 529–542 (2001).

National Occupational Health and Safety Commission, National code of practice for the prevention of occupational overuse syndrome, Australian Government Publishing Service, Canberra, Australian Capital Territory, Australia, 1994.

National Occupational Health and Safety Commission, National code of practice for manual handling, Australian Government Publishing Agency, Canberra, Australian Capital Territory, Australia, 2005.

National Research Centre for OSH Regulation, About occupational health and safety regulation in Australia, [http://ohs.anu.edu.au/ohs/index.php]. Accessed November 2009.

NOHSC Office Safe Design Team, The role of design decision-makers in addressing the risk of manual handling injury in the workplace, National Occupational Health and Safety Commission, Canberra, Australian Capital Territory, Australia, Commonwealth of Australia, 2003.

NSW Health, Zero tolerance: Response to violence in the NSW health workplace, North Sydney, New South Wales, Australia, NSW Department of Health, 2003.

O'Connor, P., Work related spinal cord injury, Australia 1986–1997, *Inj Prev*, March; **7**(1): 29–34 (2001).

O'Sullivan, P. and W. Dankaerts et al., Lumbopelvic kinematics and trunk muscle activity during sitting on stable and unstable surfaces, *J Orthop Sports Phys Ther*, **36**: 19–25 (2006).

Oxenburgh, M.S., Ergonomics: Ron Cumming memorial lecture. Ergonomics, economics, and ethics, *J Occup Environ Hyg*, November; **1**(11): D126–D128 (2004).

Oxenburgh, M.S. and H.H. Guldberg, The economic and health effects on introducing a safe manual handling code of practice, *Int J Ind Ergon*, **12**: 241–253 (1993).

Oxenburgh, M. and P. Marlow, The productivity assessment tool: Computer-based cost benefit analysis model for the economic assessment of occupational health and safety interventions in the workplace, *J Saf Res*, **36**(3): 209–214 (2005).

Oxenburgh, M.S. and S.A. Rowe et al., Repetition strain injury in keyboard operators, *J Occup Health and Saf*, **1**(2): 106–112 (1985).

Ozanne-Smith, J. and J. Guy et al., *The Relationship between Slips, Trips and Falls and the Design and Construction of Buildings*, Monash University Accident Research Centre, Monash University, Melbourne, Victoria, Australia, 2008.

Patkin, M., Ergonomic aspects of surgical dexterity, *Med J Aust*, October 21; **2**(17): 775–777 (1967).

Patkin, M., Ergonomics applied to the practice of microsurgery, *Aust NZ J Surg*, June; **47**(3): 320–329 (1977).

Patkin, M., Ergonomics and the operating microscope, *Adv Ophthalmol*, **37**: 53–63 (1978).

Patkin, M., Ergonomics in microsurgery, *Aust NZ J Obstet Gynaecol*, August; **21**(3): 134–136 (1981).

Patkin, M. and L. Isabel, Ergonomics, engineering and surgery of endosurgical dissection, *J R Coll Surg Edinb*, April; **40**(2): 120–132 (1995).

Peterson, C.L., Work factors and stress: A critical review, *Int J Health Serv*, **24**(3): 495–519 (1994).

Peterson, C., Stress amongst the healthcare workforce, *J Occup Health Saf, Austr NZ*, **19**(1): 49–58 (2003).

Phillips, J.G. and J.W. Meehan et al., Effects of cursor orientation and required precision upon positioning movements on computer screens, *Int J Hum Comput Interact*, **15**(3): 379–389 (2003a).

Phillips, J.G. and T.J. Triggs, Cursor control device characteristics, *Aust J Inform Sys*, **7**(2): 115–119 (2000).

Phillips, J.G. and T.J. Triggs et al., Cursor orientation and computer screen positioning movements, *Hum Factors*, **43**: 435–441 (2001).

Phillips, J.G. and T.J. Triggs, Characteristics of cursor trajectories controlled by the computer mouse, *Ergonomics*, **44**: 527–536 (2001).

Phillips, J.G. and T.J. Triggs et al., Conflicting directional and locational cues afforded by arrowhead cursors in graphical user interfaces, *J Exp Psychol Appl*, **9**(2): 75–87 (2003b).

Provins, K.A., Handedness and motor skill, *Med J Aust*, September 2; **2**(10): 468–470 (1967).

Provins, K.A., The specificity of motor skill and manual asymmetry: A review of the evidence and its implications, *J Mot Behav*, June; **29**(2): 183–192 (1997).

Provins, K.A. and C.R. Bell et al., The cross-cultural measurement of perceptual and motor skills in relation to human adaptation, *Hum Biol*, December; **40**(4): 484–493 (1968).

Provins, K.A. and R.S. Clarke, The effect of cold on manual performance, *J Occup Med*, April; **2**: 169–176 (1960).

Provins, K.A. and P. Cunliffe, Motor performance tests of handedness and motivation, *Percept Mot Skills*, August; **35**(1): 143–150 (1972a).

Provins, K.A. and P. Cunliffe, The relationship between E.E.G. activity and handedness, *Cortex*, June; **8**(2): 136–146 (1972b).

Provins, K.A. and P. Cunliffe, The reliability of some motor performance tests of handedness, *Neuropsychologia*, July; **10**(2): 199–206 (1972c).

Provins, K.A. and D.J. Glencross, Handwriting, typewriting and handedness, *Q J Exp Psychol*, August; **20**(3): 282–289 (1968).

Provins, K.A. and D.J. Glencross et al., Thermal stress and arousal, *Ergonomics*, September; **16**(5): 623–631 (1973).

Provins, K.A. and J. Magliaro, The measurement of handedness by preference and performance tests, *Brain Cogn*, July; **22**(2): 171–181 (1993).

Quintner, J., The pain of 'RSI'. A historical perspective, *Aust Fam Physician*, August; **18**(8): 1003–1006, 1008–1009 (1989).

Ranelli, S. and L. Straker et al., Prevalence of playing-related musculoskeletal symptoms and disorders in children learning instrumental music, *Med Probl Perform Art*, **23**: 178–185 (2008).

Regan, M., Preventing traffic accidents by mobile phone users, *Med J Aust*, **185**: 11–12 (2006).

Regan, M. and K. Stephan et al., The effect on driver workload, attitudes and acceptability of in-vehicle intelligent transport systems: Selected final results from the TAC safecar project, *J Aust Coll Road Saf*, **18**(1): 30–36 (2007a).

Regan, M. and K. Young et al., Effects on driving performance of in-vehicle intelligent transport systems: Final results of the Australian TAC safecar project, *J Australas Coll Road Saf*, **18**(1): 23–30 (2007b).

Retsas, A. and J. Pinikahana, Manual handling activities and injuries among nurses: An Australian hospital study, *J Adv Nurs*, April; **31**(4): 875–883 (2000).

Robens, A., Safety and health at work, Vol. 1, London, U.K., Report of the Committee, pp. 1970–1972 (1972).

Roelofs, A. and L. Straker, The experience of musculoskeletal discomfort amongst bank tellers who just sit, just stand or sit and stand at work, *Ergonomics SA*, **14**(2): 11–29 (2002).

Safe Work Australia, OHS and ageing surveillance alert, Canberra, Australian Capital Territory, Australia, Commonwealth of Australia, 2005.

Safe Work Australia, Research on the prevention of work-related musculoskeletal disorders stage 1—Literature Review 2006, Canberra, Australian Capital Territory, Australia, Commonwealth of Australia, 2006a.

Safe Work Australia, Work-related fatigue- summary of recent indicative research, Canberra, Australian Capital Territory, Australia, Commonwealth of Australia, 2006b.

Safe Work Australia, Work-related mental health disorders in Australia, Canberra, Australian Capital Territory, Australia, Commonwealth of Australian, 2006c.

Safe Work Australia, Work-related musculoskeletal disorders in Australia, Canberra, Australian Capital Territory, Australia, Commonwealth of Australia, 2006d.

Safe work Australia, Safe work Australia council media release June 10, 2009a, [http://www.safeworkaustralia.gov.au/swa/NewsEvents/MediaReleases/SafeWorkAustraliaCouncilMediaRelease10June2009.htm]. Accessed October 2009.

Safe Work Australia, Sizing up Australia: How contemporary is the anthropometric data Australian designers use, Canberra, Australian Capital Territory, Australia, Commonwealth of Australia, 2009b.

Safe Work Australia, Manual handling risks associated with the care, treatment and transportation of bariatric (severely obese) patients in Australia, Canberra, Australian Capital Territory, Australia, Commonwealth of Australia, 2009c.

Salmon, P.M. and M.A. Regan et al., Human error and road transport: Phase two—A framework for an error tolerant road transport system, Monash University Accident Research Centre, Melbourne, Victoria, Australia, Vol. 257, p. 53, 2006.

Salmon, P. and M. Regan et al., Work domain analysis and intelligent transport systems: Implications for vehicle design, *Int J Vehi Des, Spec Iss Hum Factor Vehi Des*, **45**(3): 426–448 (2007).

Salmon, P.M. and N.A. Stanton et al., Measuring situation awareness in complex systems: Comparison of measures study, *Int J Ind Ergon*, **39**: 490–500 (2009).

Sanderson, P.M., Cognitive work analysis, *HCI Models, Theories, and Frameworks: Toward an Interdisciplinary Science*, J. Carroll (ed.), Morgan-Kaufmann, New York, 2003.

Sanderson, P., The multimodal world of medical monitoring displays, *Appl Ergon*, **37**(4): 501–512 (2006).

Sanderson, P. and J. Crawford et al., Visual and auditory attention in patient monitoring: A formative analysis, *Cogn Technol Work*, **6**(3): 172–185 (2004).

Sanderson, P.M. and D. Liu et al., Auditory displays in anaesthesiology, *Curr Opin Anaesthesiol*, **22**(6): 788–795 (2009).

Sanderson, P. and A. Pipingas et al., Process monitoring and configural display design: A neuroimaging study, *Theoretical Iss Ergon Sci*, **4**(2): 151–174 (2003).

Sanderson, P. and N. Tosh et al., Effects of ambient music on simulated anaesthesia monitoring with visual and auditory displays, *Anaesthesia*, **60**(11): 1073–1078 (2005a).

Sanderson, P. and M. Watson et al., Advanced patient monitoring displays: Tools for continuous informing, *Anesth Analg*, **101**(1): 161–168 (2005b).

Sanderson, P.M. and M.O. Watson et al., Advanced auditory displays and head mounted displays: Advantages and disadvantages for monitoring by the distracted anesthesiologist, *Anesth Analg*, **106**(6): 1787–1797 (2008).

Sanderson, P. and A. Wee et al., Learnability and discriminability of melodic medical equipment alarms, *Anaesthesia*, **61**: 142–147 (2006).

Scutter, S. and K.S. Türker et al., Headaches and neck pain in farmers, *Aust J Rural Health*, **5**(1): 2–5 (1997).

Seagull, F.J. and P.M. Sanderson, Anesthesia alarms in surgical context: An observational study, *Hum Factors*, **43**(1): 66–77 (2001).

Shaw Idea, Digging Deeper Wran Consultancy Project, NSW Department of Primary Industries, Orange, NSW, New South Wales Government, 2007.

Shepherd, G.W. and R.J. Kahler et al., Ergonomic design interventions—A case study involving portable ladders, *Ergonomics*, February 22; **49**(3): 221–234 (2006).

Stain, H.J. and B. Kelly et al., Social networks and mental health among a farming population, *Soc Psychiatry Psychiatr Epidemiol*, October; **43**(10): 843–849 (2008).

Stanton, N.A. and P.M. Salmon et al., Predicting pilot error: Testing a new method and a multi-methods and analysts approach, *Appl Ergon*, **40**(3): 464–471 (2009).

State Government Victoria, Victorian nurses back prevention programme: Evaluation report 2002, Department of Human Services, Victorian Government, Melbourne, Victoria, Australia, 2002.

State Government Victoria, Preventing falls in Victoria 2007–2012: Discussion paper. D. o. H. S. Aged Care Branch, State Government Victoria, Melbourne, Victoria, Australia, 2007.

Stevenson, M.G. and N. Coleman et al., Assessment, re-design and evaluation of changes to the driver's cab in a suburban electric train, *Appl Ergon*, October; **31**(5): 499–506 (2000).

Stevenson, R.S. and J.G. Phillips et al., Mouse and display tablets as cursor control devices, *Int J Pattern Recogn Artif Intell*, **18**: 1–12 (2004).

Straker, L. and A. Briggs et al., The effect of individually adjusted workstations on upper quadrant posture and muscle activity in school aged children, *Work*, **18**: 239–248 (2002).

Straker, L. and R. Burgess-Limerick et al., A randomized and controlled trial of a participative ergonomics intervention to reduce injuries associated with manual tasks: Physical risk and legislative compliance, *Ergonomics*, **47**(2): 166–188 (2004).

Straker, L. and R. Burgess-Limerick et al., Towards evidence based guidelines for wise use of computers by children, *Int J Ind Ergon*, **36**(12): 1045–1053 (2006).

Straker, L. and R. Burgess-Limerick et al., Children's head, neck and upper limb posture and muscle activity at different computer and paper display heights, *Hum Factors*, **50**: 49–61 (2008a).

Straker, L. and R. Burgess-Limerick et al., The impact of computer display height and desk design on 3D posture during information technology work by young adults, *J Electromyogr Kinesiol*, April; **18**(2): 336–349 (2008b).

Straker, L. and R. Burgess-Limerick et al., The effect of forearm support on children's head, neck and upper limb posture and muscle activity during computer use, *J Electromyo Kinesiol*, October; **19**(5): 965–974 (2009a).

Straker, L. and R. Burgess-Limerick et al., The influence of desk and display design on posture and muscle activity variability whilst performing information technology tasks, *Appl Ergon*, September; **40**(5): 852–859 (2009b).

Straker, L. and B.A. Maslen et al., Children have less variable postures and muscle activities when using new electronic information technology compared with old paper based information technology, *J Electromyogr Kinesiol*, **19**: e132–e143 (2009c).

Straker, L. and C. Pollock et al., Principles for the wise use of computers by children, *Ergonomics*, **52**(11): 1386–1401 (2009d).

Straker, L. and C. Cain, Psychophysical and physiological comparison of squat and semi-squat lifting by young males, *Occup Ergon*, **1**: 17–26 (1999).

Straker, L.M. and J. Coleman et al., A comparison of posture and muscle activity during tablet computer, desktop computer and paper use by young children, *Ergonomics*, **51**: 540–555 (2008c).

Straker, L. and P. Duncan, Psychophysical and psychological comparison of squat and stoop lifting by young females, *Aust J Physiother*, **46**: 27–32 (2000).

Straker, L. and K. Jones et al., A comparison of the postures assumed when using laptop computers and desktop computers, *Appl Ergon*, **28**: 263–268 (1997a).

Straker, L. and S.E. Mathiassen, Increased physical work loads in modern work–a necessity for better health and performance, *Ergonomics*, October; **52**(10): 1215–1225 (2009).

Straker, L. and K. Mekhora, An evaluation of visual display unit placement by electromyography, posture, discomfort and preference, *Int J Ind Ergon*, **26**: 335–336 (2000).

Straker, L. and P.B. O'Sullivan et al., Computer use and habitual spinal posture in Australian adolescents, *Public Health Rep*, **122**(5): 634–643 (2007).

Straker, L. and C. Pollock et al., The effect of shoulder posture on performance: Discomfort and muscle fatigue whilst working on a visual display unit, *Int J Ind Ergon*, **21**: 1–10 (1997b).

Straker, L. and C. Pollock et al., The impact of computer display height and desk design on muscle activity during information technology work by young adults, *J Electromyogr Kinesiol*, Aug; **18**(4): 606–617 (2008d).

Straker, L. and R. Skoss et al., Effect of visual display height on modelled upper and lower cervical gravitational moment, muscle capacity and relative strain, *Ergonomics*, February; **52**(2): 204–221 (2009e).

Straker, L. and M. Stevenson et al., A comparison of risk assessment of single and combination manual handling tasks: 2 Discomfort, rating of perceived exertion and heart rate measures, *Ergonomics*, **40**: 656–669 (1997a).

Straker, L. and M. Stevenson et al., A comparison of risk assessment of single and combination manual handling tasks: 3 Biomechanical measures, *Ergonomics*, **40**: 708–728 (1997b).

Szeto, G. and L. Straker et al., A field comparison of neck and shoulder postures in symptomatic and asymptomatic office workers, *Appl Ergon*, **33**: 75–84 (2002).

Szeto, G. and L. Straker et al., A comparison of symptomatic and asymptomatic office workers performing monotonous keyboard work—1: Neck and shoulder muscle recruitment patterns, *Man Ther*, **10**(4): 270–280 (2005a).

Szeto, G. and L. Straker et al., A comparison of symptomatic and asymptomatic office workers performing monotonous keyboard work—2: Neck and shoulder kinematics, *Man Ther*, **10**(4): 281–291 (2005b).

Szeto, G. and L. Straker et al., EMG median frequency changes in the neck-shoulder stabilizers of symptomatic office workers when challenged by different physical stressors, *J Electromyogr Kinesiol*, **15**(6): 544–555 (2005c).

Szeto, G.P.Y. and L.M. Straker et al., The effects of typing speed and force on motor control in symptomatic and asymptomatic office workers, *Int J Ind Ergon*, **35**(9): 429–444 (2005d).

Thompson, M. and P.M. Sanderson, Multisensory integration with a head-mounted display: Sound delivery and self-motion, *Hum Factors*, **50**(5): 789–800 (2008).

Torma-Krajewski, J. and C. Hipes et al., Ergonomics interventions at vulcan materials company, *Min Eng*, December; **59**(11): 54–58 (2007).

Tuckwell, N. and L. Straker et al., Test retest reliability on nine tasks of the physical work performance evaluation, *Work*, **19**: 243–253 (2002).

Uniquest Pty Ltd., Issues associated with force and weight limits and associated threshold limit values in the physical handling work environment, National Occupational Health and Safety Commission, Canberra, Australian Capital Territory, Australia, Commonwealth of Australia, 2003.

UniSA, ADAPT project (Australian defence anthropometric personnel testing), [http://www.unisa.edu.au/adapt/default.asp]. Accessed October 2009.

University of Queensland, Graduate certificate in ergonomics, [http://www.uq.edu.au/study/program.html?acad_prog=5281]. Accessed 2009.

Victorian WorkCover Authority, Code of practice for manual handling, State Government of Victoria, Melbourne, Victoria, Australia, 2000.

Watson, M. and W.J. Russell et al., Anaesthesia monitoring, alarm proliferation, and ecological interface design, *Aust J Inform Syst*, **7**(2): 109–114 (2000).

Watson, M. and P. Sanderson et al., Tailoring reveals information requirements: The case of anaesthesia alarms, *Interact Comput*, **16**: 271–293 (2004).

Watson, M. and P. Sanderson, Sonification helps eyes-free respiratory monitoring and task timesharing, *Hum Factors*, **46**(3): 497–517 (2004).

Watson, M. and P. Sanderson, Designing for attention with sound: Challenges and extensions to ecological interface design, *Hum Factors*, **49**(2): 331–346 (2007).

Wee, A. and P. Sanderson, Are melodic medical equipment alarms easily learned, *Anesth Analg*, **106**(2): 501–508 (2008).

Weeks, K. and C.L. Peterson et al., Stress and the workplace: The medical scientists' experience, *Lab Ind*, **11**(3): 95–120 (2000).

Welford, A.T., Perceptual selection and integration, *Ergonomics*, January; **13**(1): 5–23 (1970).

Welford, A.T., What is the basis of choice reaction-time, *Ergonomics*, November; **14**(6): 679–693 (1971).

Welford, A.T., Stress and performance, *Ergonomics*, September; **16**(5): 567–580 (September 1973).

Welford, A.T., Ergonomics: Where have we been and where are we going: I, *Ergonomics*, May; **19**(3): 275–286 (1976).

Welford, A.T., Mental work-load as a function of demand, capacity, strategy and skill, *Ergonomics*, March; **21**(3): 151–167 (1978).

Welford, A.T., Preventing adverse changes of work with age, *Int J Aging Hum Dev*, **27**(4): 283–291 (1988a).

Welford, A.T., Reaction time, speed of performance, and age, *Ann NY Acad Sci*, **515**: 1–17 (1988b).

Welford, A.T., The ergonomic approach to social behaviour, *Ergonomics*, September; **9**(5): 357–369 (1996).

Wigglesworth, E.C., Ergonomics, safety and railway level crossings, *Ergon Aust*, **21**(4): 4–13 (2008).

WorkCover Corporation of South Australia, Code of practice for manual handling adelaide, South Australian Government, Adelaide, South Australia, Australia, 1990.

WorkCover New South Wales, Preventing slips, trips and falls: Guide, WorkCover NSW, Gosford, New South Wales, Australia, 2006.

Workplace Health and Safety Queensland, Manual tasks code of practice 2000, DOJA Attorney-General, Queensland Government, Brisbane, Queensland, Australia, 2000.

Workplace Health and Safety Queensland, Occupational stress tip sheets: Risk assessment approach, [http://www.deir.qld.gov.au/workplace/resources/pdfs/occstress-riskmgt.pdf]. Accessed October 2009.

Workplace Health and Safety Queensland, Slips, trips, and falls prevention, D. o. E. a. I. Relations, Queensland Government, Brisbane, Queensland, Australia, 2007.

WorkSafe, W.A., Preventing manual handling injuries in the workplace, [http://www.commerce.wa.gov.au/WorkSafe/Content/Safety_Topics/Manual_handling/Preventing_manual_handling_inj.html]. Accessed October 2009.

WorkSafe, Victoria, Safety by design: Eliminating manual handling injuries in road transport, State Government Victoria, Melbourne, Victoria, Australia, 2001.

WorkSafe, Victoria, Manual handling solutions in the textile industry, State Government Victoria, Melbourne, Victoria, Australia, 2002.

WorkSafe, Victoria, Manual handling solutions in the sawmilling industry, State Government Victoria, Melbourne, Victoria, Australia, 2005.

WorkSafe, Victoria, A guide to manual handling in the food industry, State Government Victoria, Melbourne, Victoria, Australia, 2006a.

WorkSafe, Victoria, Slips, trips, and falls checklist, State Government Victoria, Melbourne, Victoria, Australia, 2006b.

Worksafe, Victoria, Stresswise-preventing work-related stress, State Government Victoria, Melbourne, Victoria, Australia, 2007a.

WorkSafe, Victoria, A guide to designing workplaces for safer handling of people, State Government Victoria, Melbourne, Victoria, Australia, 2007b.

WorkSafe, Victoria, A guide to safety in the metal fabrication industry, State Government Victoria, Melbourne, Victoria, Australia, 2007c.

WorkSafe, Victoria, Prevention and management of aggression in health services: A handbook for workplaces, State Government Victoria, Melbourne, Victoria, Australia, 2008.

WorkSafe, Victoria, Transferring people safely: A guide to handling patients, residents and clients in health, aged care, rehabilitation, and disability services, State Government Victoria, Melbourne, Victoria, Australia, 2009.

WorkSafe, Victoria Ergonomics Unit, Officewise: A guide to health and safety, State Government Victoria, Melbourne, Victoria, Australia, 2006c.

Young, K. and M. Regan, Use of manual speed alerting and cruise control devices by car drivers, *Saf Sci*, **45**(4): 473–485 (2007).

Zandvliet, D. and L. Straker, Physical and psychosocial aspects of the learning environment in information technology rich classrooms, *Ergonomics*, **9**: 838–857 (2001).

Zupanc, C. and R. Burgess-Limerick, Issues related to the wearing of fall-arrest harnesses in the construction industry, *Ergon Aust*, **17**(3): 18–24 (2003).

Zupanc, C. and R. Burgess-Limerick et al., Performance as a consequence of alternating control-response compatibility: Evidence from a coal mine shuttle car simulator, *Hum Factors*, **49**: 628–636 (2007).

CHAPTER 45

# Ergonomics in South Korea

Soo-Jin Lee, Kyung-Suk Lee, Yong-Ku Kong, Myung-Chul Jung, and Kermit G. Davis

CONTENTS

## 45.1 INTRODUCTION

The Korean economy was based on agriculture until the 1900s, when industrialization, especially in the areas of shipbuilding and steel and petrochemical production, began to play a larger role. Ergonomic improvements contributed to this process of economic development by maximizing work efficiency and productivity, improving the working environment, and advancing product safety and design. Ergonomic initiatives in Korea have concentrated on worker health and safety by focusing on work-related musculoskeletal disorders (WMSDs) and diseases that can result from workplace conditions. In Korea, ergonomic analysis began during the Korean War in the early 1950s, when the United States Air Force introduced ergonomic concepts through human research studies in aerial medicine [1]. In the 1970s, industrial ergonomics in Korea grew, particularly in the automobile and ship manufacturing sectors. This transformation continued in the 1980s, when computer- and office-based ergonomics became prevalent [1]. In the 1990s, WMSDs were finally recognized as medical disorders caused by occupational stress by the Workmen's Accident Compensation Insurance Act [2]. A constructive revision of that act resulted in the Industrial Safety and Health Law of 2000, which requires employers to enact measures to prevent WMSDs [3]. The Industrial Safety and Health Law regulates that companies must take an active role in controlling WMSDs. Ergonomic improvements for farm workers also took place in the 1990s but were minor relative to the ergonomic advancements in other industries. In recent years, efforts to understand WMSDs have concentrated on identifying risk factors for WMSDs in agriculture.

## 45.2 PROFESSIONAL DEVELOPMENT OF ERGONOMICS OF KOREA

### 45.2.1 Ergonomics Society of Korea

In 1982, the Ergonomics Society of Korea (ESK) was established by 40 founders, for "developing, spreading and applying ergonomics-related studies and technology for the profit of [the] general public [and] to contribute to advancement in ergonomics-related technology" [1]. By 1988, the society had expanded to 250 members and currently includes 591 members. Since 1982, the society has published the peer-reviewed *Journal of the Ergonomics Society of Korea*. Another major activity of the ESK is academic and research exchange during the ergonomics conference hosted every spring and fall. As a result of the expansion of ergonomics in Korea, 50 universities now have graduate programs in ergonomics or related fields, with approximately 550 graduates annually.

### 45.2.2 Research Areas of the ESK

The ESK has not only increased its membership over time but has also experienced qualitative and quantitative growth in research publications, with several industrial sectors now recognizing the importance and service of ergonomics in workplaces and product design. Publication of *Journal of the Ergonomics Society of Korea* increased from twice a year from 1982 to 1996, to three times a year from 1997 to 2001, to four times a year from 2002 to the present, demonstrating the revitalization of ergonomics research. Table 45.1 shows that the number of published articles has risen every year, increasing 2.43-fold during the 1990s.

TABLE 45.1 Number of Articles Focused on Ergonomics Published in the Journal of the Ergonomics Society of Korea

| Year | Vol. 1 | Vol. 2 | Vol. 3 | Vol. 4 | Total | Year | Vol. 1 | Vol. 2 | Vol. 3 | Vol. 4 | Total |
|---|---|---|---|---|---|---|---|---|---|---|---|
| 1982 | 4 | 4 | — | — | 8 | 1996 | 10 | 11 | — | — | 21 |
| 1983 | 2 | 2 | — | — | 4 | 1997 | 9 | 8 | 8 | — | 25 |
| 1984 | 3 | 2 | — | — | 5 | 1998 | 11 | 8 | 10 | — | 29 |
| 1985 | 4 | 4 | — | — | 8 | 1999 | 10 | 11 | 13 | — | 34 |
| 1986 | 5 | 6 | — | — | 11 | 2000 | 10 | 7 | 8 | — | 25 |
| 1987 | 5 | 4 | — | — | 9 | 2001 | 7 | 7 | 7 | — | 21 |
| 1988 | 6 | 4 | — | — | 10 | 2002 | 8 | 9 | 6 | 10 | 33 |
| 1989 | 4 | 5 | — | — | 9 | 2003 | 8 | 5 | 7 | 6 | 26 |
| 1990 | 5 | 8 | — | — | 13 | 2004 | 5 | 8 | 10 | 6 | 29 |
| 1991 | 4 | 6 | — | — | 10 | 2005 | 7 | 12 | 5 | 11 | 35 |
| 1992 | 11 | 6 | — | — | 17 | 2006 | 6 | 23 | 13 | 17 | 59 |
| 1993 | 6 | 6 | — | — | 12 | 2007 | 14 | 17 | 15 | 20 | 66 |
| 1994 | 6 | 8 | — | — | 14 | 2008 | 7 | 10 | 11 | 13 | 41 |
| 1995 | 10 | 9 | — | — | 19 | Total | | | | | 593 |

The annual average was 19.4 articles in this period, compared to an average of 8.0 from 1982 to 1999. From 2000 to 2008, the society published 37.2 articles per year, for a 4.65-fold increase from the initial phase of the journal and an approximate two-fold increase from the middle phase of the society in the 1990s.

Of the 593 articles published in *Journal of the Ergonomics Society of Korea*, most have concerned “physical ergonomics/WMSDs,” followed by articles on “product design/Kansei engineering,” including design of automobiles and cellular phones, “HCI/virtual environments,” and “cognitive engineering/human error” (Table 45.2). Other topics include anthropometry/standards, macroergonomics, environmental design, aging, and health care. Studies on physical ergonomics and WMSDs have consistently been published. Interestingly, more studies were conducted in the area of HCI than in product design in the 1980s. Nonetheless, more studies were published in the area of product design and Kansei engineering in the 1990s and 2000s than in the 1980s.

TABLE 45.2 Percentage of Articles Published in the Journal of the Ergonomics Society of Korea for the Different Research Areas of Ergonomics

| Research Area | 1980s | 1990s | 2000s | Mean |
|---|---|---|---|---|
| Physical ergonomics/WMSDs | 35.9 | 28.9 | 28.7 | 29.5 |
| Product design/Kansei engineering | 4.7 | 20.6 | 23.9 | 20.7 |
| HCI/virtual environments | 26.6 | 16.0 | 11.9 | 14.8 |
| Cognitive engineering/human error | 15.6 | 14.4 | 13.7 | 14.2 |
| Anthropometry/standards | 4.7 | 6.7 | 6.9 | 6.6 |
| Macroergonomics | 7.8 | 2.1 | 5.1 | 4.4 |
| Environmental design | 1.6 | 4.6 | 3.3 | 3.5 |
| Aging | 3.1 | 2.6 | 3.3 | 3.0 |
| Health care | — | 3.1 | 1.8 | 2.0 |
| Others | — | 1.0 | 1.5 | 1.2 |

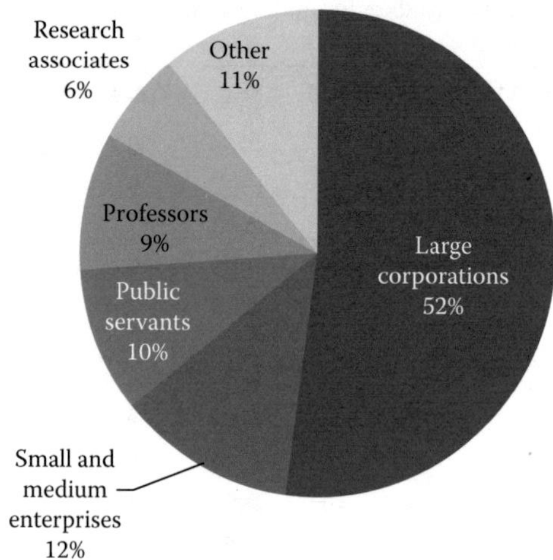


FIGURE 45.1 Distribution of ergonomic professionals in the Ergonomics Society of Korea.

### 45.2.3 Research Activities of the ESK

One of major activities of the ESK is academic exchange, exemplified by the annual spring and fall ergonomics conferences. Since the fall conference of 1992, the ESK has established a topic for each conference. The number of articles published and presented at the conference has increased every year, for a total of 49 articles from 1982 to 1989 and 768 articles from 1990 to 1999, and an annual average of 76.8 articles, representing a 12.6-fold increase. In the 2000s, the society published a total of 1205 articles, for an annual average of 133.9, demonstrating significant growth since the initial and middle phases of the society. A total of 2022 articles have been presented at the conference as of the end of 2008.

### 45.2.4 Activities of Ergonomists

Analysis of fields of activities for ESK ergonomics graduates from 50 universities with ergonomics or related programs shows that 281 of 545 total graduates entered large corporations and 27 went to small or medium enterprises, followed by 54 public servants, 51 professors, and 32 graduates working at national research institutes. Other graduates were involved in study-abroad programs, and 60 started new businesses. Figure 45.1 shows the distribution of ergonomics professionals in the ESK.

## 45.3 TYPICAL ERGONOMICS RESEARCH AREAS IN KOREA

### 45.3.1 Automobile Research

The automobile is one of the most complex human–machine interface systems. Designing an automobile requires technology that must consider performance, function, and appearance. Automobiles have been transformed from a simple transportation vehicle to a driver's living space, serving as mobile offices and entertainment centers. As a result, the current trend in automobile design applies not only to the automobile itself but also to the information systems and additional functional devices within the vehicle. The internal cab of the vehicle is a primary ergonomic focus area, specifically in the design of the seat, driving console, and accessory console with radio, heater, and air conditioner. The multiple themes that appear in many automobile-related ergonomics case studies include Kansei engineering methods, subjective preferences with respect to ergonomic stress, and objective quantitative methods.

Using Kansei engineering methods, Bahn et al. [4] developed complex models for automobile crash pads based on user perceptions. By manipulating the tactile and visible design characteristics of the crash pad, the Kansei engineering method was applied to incorporate luxuriousness. A similar Kansei engineering study developed perception models for the luxuriousness of interior materials [5]. Kansei engineering has also been applied to vehicle-driving simulators. Chung et al. [6] examined perception vocabularies in evaluating the motor sensation of a vehicle-driving simulator.

Choi et al. [7] determined the optimal position of the accelerator and brake pedals, leading to guidelines for accelerator and brake pedal positions, based on subjective lower body and extremity discomfort data for three different body sizes. Figure 45.2 shows the starting point and axle of the mock-up vehicle cab used in the research. Other studies in this area used a combination of subjective user satisfaction and objective experimental data. For example, one study examined driver satisfaction when entering and riding in a sport utility vehicle (Figure 45.3) and presented an ergonomic layout for the driver's seat that maintained an optimal driving position, generated using a three-dimensional motion analysis system to analyze driver movement [8]. Lim and associates [9] investigated three control types and two display menus to find the best choices for driver performance and preference. To reduce fatigue in lower leg muscles during repeated pedal pressing when driving, Kim and Seo [10] relied on objectively measured data rather than subjective user preference. They used electromyography (EMG) of the lower leg muscle to evaluate a heel rest during pedal activation. Park et al. [11] analyzed the relationship

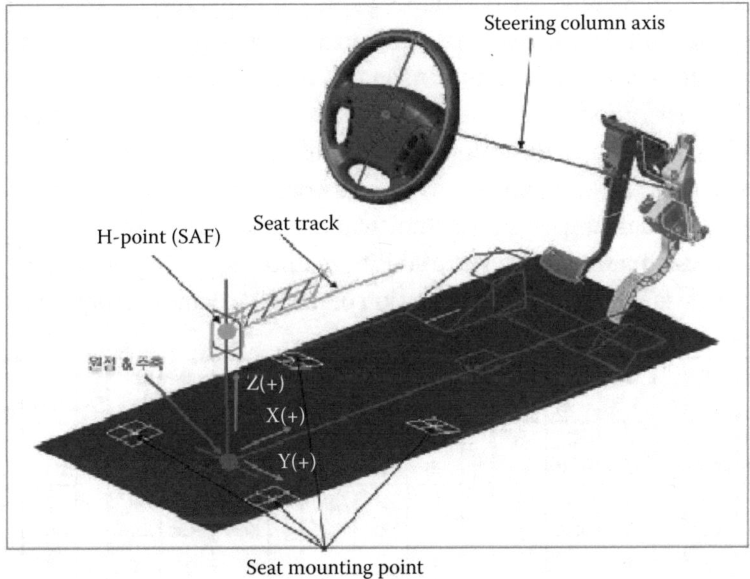


FIGURE 45.2 An example of a mock-up of a vehicle cab that includes starting point and axle used in ergonomic research.

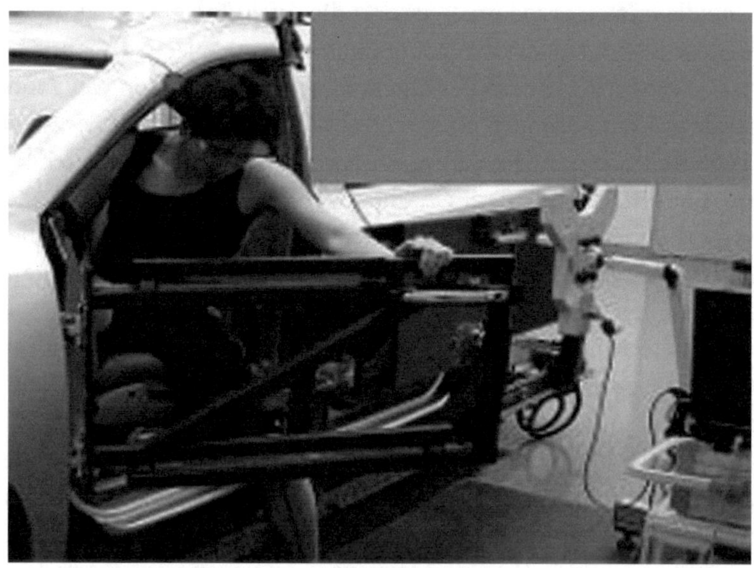

FIGURE 45.3 Simulated motor cab utilized in the analysis of satisfaction when entering an automobile.

between subjective body discomfort when seated and body pressure distribution of vehicle seats using a force-sensing resistor made of polymer.

### 45.3.2 Cellular Phone Research

Research on cellular phones in Korea has concentrated on usability, including universal design, sensible design for high-end cellular phones, and input functions of Hangul (the Korean alphabet) characters. Each cellular phone manufacturer uses different Hangul input methods, leading to the need for standardization. Figure 45.4 shows the different Hangul input methods, each with 19 consonants and 21 vowels on 12 buttons. The first input method (far left) appears to be superior based on the result of experiments measuring variables including user input time, input error, and satisfaction [12,13].

Jun and associates [14] developed a structural equation model to examine the perceptual factors that influence user satisfaction while using a cellular phone. Kim and colleagues [15] suggested a multicentered usability method involving developers, evaluators, and users. The goal is to prevent the production of many variable end products, an outcome

| 1<br>ㅣ | 2<br>· | 3<br>ㅡ |
|---|---|---|
| 4<br>ㄱㅋ | 5<br>ㄴㄹ | 6<br>ㄷㅌ |
| 7<br>ㅂㅍ | 8<br>ㅅㅎ | 9<br>ㅈㅊ |
| * | 0<br>ㅇㅁ | # |

| 1<br>ㄱㅋ | 2<br>ㄴㄹ | 3<br>ㄷㅌ |
|---|---|---|
| 4<br>ㅁ | 5<br>ㅂㅍ | 6<br>ㅅ |
| 7<br>ㅇ | 8<br>ㅈㅊ | 9<br>ㅎ |
| *<br>ㅣ | 0<br>· | #<br>ㅡ |

| 1<br>ㄱ | 2<br>ㄴ | 3<br>ㅏㅓ |
|---|---|---|
| 4<br>ㄹ | 5<br>ㅁ | 6<br>ㅗㅜ |
| 7<br>ㅅ | 8<br>ㅇ | 9<br>ㅣ |
| * | 0<br>ㅡ | # |

| 1<br>ㄱㅋ | 2<br>ㅣㅡ | 3<br>ㅏㅑ |
|---|---|---|
| 4<br>ㄷㅌ | 5<br>ㅂㄹ | 6<br>ㅓㅕ |
| 7<br>ㅁㅅ | 8<br>ㅂㅍ | 9<br>ㅗㅛ |
| *<br>ㅈㅊ | 0<br>ㅇㅎ | #<br>ㅜㅠ |

| 1<br>ㄱㅋ | 2<br>ㄴㅁ | 3<br>ㅏㅓ |
|---|---|---|
| 4<br>ㄷㅌ | 5<br>ㄹ | 6<br>ㅗㅜ |
| 7<br>ㅂㅍ | 8<br>ㅅ | 9<br>ㅣㅡ |
| *<br>ㅈㅊ | 0<br>ㅇㅎ | # |

FIGURE 45.4 Hangul input methods for cellular phones.

that occurs particularly when products are developed only by evaluators. Kim et al. [16] compared three approaches—quantification, neural network, and decision tree—for analyzing perceptions about cellular phone design. The strengths and weaknesses of each modeling method were discussed using a specific cellular phone example.

The concept of a universal design for cellular phones has particular ramifications for the elderly and people with disabilities. Hong [13] emphasized the need for less retentive functions and the use of simple keys when designing a cellular phone for the elderly. Kim et al. [17] developed a universal design index applicable to cellular phone evaluation.

### 45.3.3 Agricultural Research

Since the 1960s, policies that focus on increased food production have led to innovations in agricultural technologies and formed the basis of the green and white revolutions. However, these policies threatened the health of agricultural workers by leading to increases in the prevalence of WMSDs and increasing the number of safety concerns due to the use of agricultural machinery and agrichemicals. Among the key WMSD issues in farming are the high percentages of aging and female farmers [18].

The Korean government recently developed policies that protect agricultural workers while focusing on increasing agricultural production, balancing the need for worker health and safety, with meeting the demands for food in Korea. One specific initiative was the national action plans established in 2005 to promote the improvement and maintenance of the health of agricultural workers and fishermen. These action plans enabled the government to conduct specialized research programs to investigate the conditions under which accidents and injuries occurred during farm work, and provided an opportunity for the government to prioritize resources to address health outcomes [19].

#### *45.3.3.1 Aging Agricultural Community and Injuries from Agricultural Activities*

Rapid industrialization and economic development in Korea has significantly reduced the farming population. In 1945, 87% of the Korean population was in the agricultural sector, but by 2007, only 6.8% of the population made a living in agriculture. One of the major outcomes of this major shift in the farming workforce was the concentration of older individuals into a significant proportion of the agricultural workforce, of which 42% were 60 years of age and older [20]. This makes Korean agriculture one of the oldest working industry sectors in the developed world.

As a result of the workforce becoming older, degenerative and chronic diseases are on the rise among famers and pose a significant burden on these farmers and on Korean society. Lee [19] at the Rural Development Administration (RDA) reported prevalence rates of chronic diseases related to "farmer disease" as between 20% and 45%, as reported every 5 years since 1995. Many underlying conditions place these older farmers at increased risk for cumulative disorders and WMSDs, including high blood pressure, rheumatism, arteriosclerosis, and kidney disease.

The RDA has conducted studies on WMSD risk factors, nerve injuries from agrichemical poisoning, asthma, and immune deficiency since 2000 [21–23,25]. WMSDs vary among agricultural workers, depending on the type of work performed, the type of crop, and

TABLE 45.3 Surveys on WMSDs for Agricultural Workers Relative to Traditional Non-Agricultural Workers

| Survey[a] | Year | No. of People | Prevalence Rate | Remarks |
|---|---|---|---|---|
| RDA [21] | 2007 | 416 | Agricultural workers 75% | Medical examination, relation to agricultural work (41%) |
| RDA [22] | 1999 | 1700 | Agricultural workers/fishermen 62%<br>Nonagricultural workers 25% | National health and nutrition data at or above the age of 30 |
| RDA [23] | 2001 | 354 | Orchard 67%<br>Agriproducts 60%<br>Livestock farming 35%<br>Nonagricultural workers 31% | Medical examination, complete inventory as a function village unit |
| Paek [24] | 1999 | 1709 | Agricultural workers/fishermen 82%<br>Manufacturers 71% | Questionnaires for all males between the ages of 45 and 64 |

[a] RDA: Rural development administration.

whether livestock are involved (Table 45.3). Based on these studies, the predominant conclusion is that the prevalence of WMSDs in the agricultural sector is significantly higher than in the general public and nonagricultural workers (Table 45.3). In 2001, chronic diseases occurred in farmers 1.45 times more often than in nonagricultural general population workers, and WMSDs occurred 2.4 times more frequently [19].

According to workers' compensation data from the Ministry of Labor, farming had a higher injury rate than either the manufacturing or construction industries (Table 45.4) [25]. Specifically, the accident rate is a dangerous 34.7 per 1,000 workers from agricultural activities, as determined from compensation data from safety aid systems, which includes approximately 700,000 agricultural workers (Table 45.5) [26].

#### *45.3.3.2 Mid- to Long-Term Prevention Strategy for Agricultural Health and Safety*

As indicated earlier, an aging workforce is a concern of the agricultural industry, but it is not the only health and safety issue for the agricultural community. A specialized

TABLE 45.4 Injuries by Industrial Classification in the Workers' Compensation Data from the Ministry of Labor in 2006 [20]

| Type of Industry | No. of Workplaces | No. of Laborers | Industrial Disaster Victims | Injury Rate per 1000 People |
|---|---|---|---|---|
| Farming | 3,281 | 31,781 | 483 | 15.20 |
| Manufacturing | 236,429 | 3,032,667 | 35,914 | 11.84 |
| Construction | 149,874 | 2,547,754 | 17,955 | 7.05 |

TABLE 45.5 Injuries in Safety Aid System of Agricultural Workers and Mutual Aid System of Agricultural Machines in 2005 [24]

| Safety Aid System of Agricultural Workers | | | Mutual Aid System of Agricultural Machines | | |
|---|---|---|---|---|---|
| No. of Subscribers | No. of Accidents | Rate per 1000 People | No. of Subscribers | No. of Loss of Farm Machines | Rate per 1000 People |
| 702,000 | 13,930 | 34.7 | 8,742 | 731 | 83.6 |

TABLE 45.6 Mid to Long-Term Research Plan on the Safety and Health for Agricultural Work in the Rural Development Administration

| Component | Contents |
|---|---|
| Ergonomic work improvement technology development by agricultural process | • Research on working environment improvements that reduce worker burden of labor<br>• Development of work management guidelines<br>• Development of equipment and interventions to protect agricultural workers<br>• Development of ergonomic agricultural injury tracking system |
| Investigation of the causes of agricultural work-related injuries | • Research on the status of agricultural injuries<br>• Research to evaluate the influence of agricultural work risk factors on worker health<br>• Monitoring agricultural injury prevalence and development of management systems<br>• Implementation of a national statistical system for agricultural injuries |
| Research on management of exposure to primary risk factors in the agricultural working environment | • Investigating the prevalence and exposure to biological risk factors from agricultural livestock work<br>• Analysis of risk of exposing agricultural risk factors by agricultural species<br>• Establishing risk exposure level (REL) by risk factors |
| Research on implementation of a system tracking compensation of agricultural injuries | • Implementation of a prevention and management system for agricultural injuries<br>• Development of agricultural health and safety educational programs<br>• Development of regulations for agricultural worker injury compensation and development of a tracking system for compensated injuries |

system that identifies and tracks agricultural activities that create risks for farm workers, and determines the level of compensation for injuries is needed. Since researchers actively adopt strategies to understand the types and causes of injuries, the relevant research institutes and RDA specialists are conducting mid- to long-term studies in cooperation with academic institutions to examine the status of agricultural worker injuries and association with the type of work performed. They are also developing improvement strategies for farm working environments [27]. Table 45.6 provides a detailed outline of the plan that is currently being implemented [27].

#### *45.3.3.3 Research Achievements in Agricultural Safety and Health*

45.3.3.3.1 Developing Equipment to Improve Agricultural Working Postures To control farm work-related injuries, several interventions have been developed and implemented to improve agricultural working environments. Some general examples are equipment and devices to eliminate the carrying of unusually shaped or heavy loads, and protection guards on machinery. Many farm tasks require awkward postures that place stress on the

TABLE 45.7 Assistive Technology for Improving Working Posture

| Working Posture | Crop Type | Type of Work | Improvement |
|---|---|---|---|
| Bending [30] | Grapes | Berry thinning | Slope-regulated cycle work chairs |
| | Outdoor peppers | Harvesting | Small and large platform carriers |
| Squatting [31,32] | Strawberries under structure | Growing seedling, hydroponic, harvesting | Cultivation vehicle at the height of cultivated crops |
| | Cultivation under structure | Management, harvesting | Put-on chairs |
| Sitting on the ground [30] | Agarics | Grading, shipping | Grading and packaging bench |
| | Strawberries | Grading, shipping | Strawberry grading bench |
| | Cucumbers | Grading, shipping | Shipping work of cucumbers |

shoulders or back and result in fatigue and WMSDs [28]. Recent initiatives in farm mechanization have reduced much of the manual lifting, but awkward body posture, including bending of the back, squatting of the legs, or pulling back of the head, can lead to degeneration of the musculoskeletal system [29].

Researchers have developed various devices to improve working postures (Table 45.7), and some equipment is produced as part of government support programs to improve agricultural working environments. The RDA has conducted several demonstration projects for interventions and devices that reduce poor posture when performing agricultural work, including strawberry-harvesting vehicles, slope-regulated cycle work chairs, platform work chairs, and benches for grading and shipping mushrooms (Figure 45.5).

45.3.3.3.2 Reducing Heavy Loads in Farm Work Handling heavy loads in agricultural activities includes not only lifting, moving, and lowering but also pushing and pulling. Many agricultural tasks require long periods of carrying, moving, or lifting, sometimes with irregularly shaped loads. Loads carried include fertilizer for harvesting and regulating crops, heat insulators, and crops transferred into different harvesting containers. Recently, platform carriers that adapt to loads of various shapes have been developed and used to transport harvested crops [30]. However, more studies are needed to ensure safety when transporting heavy loads, especially for older and female workers. When possible,

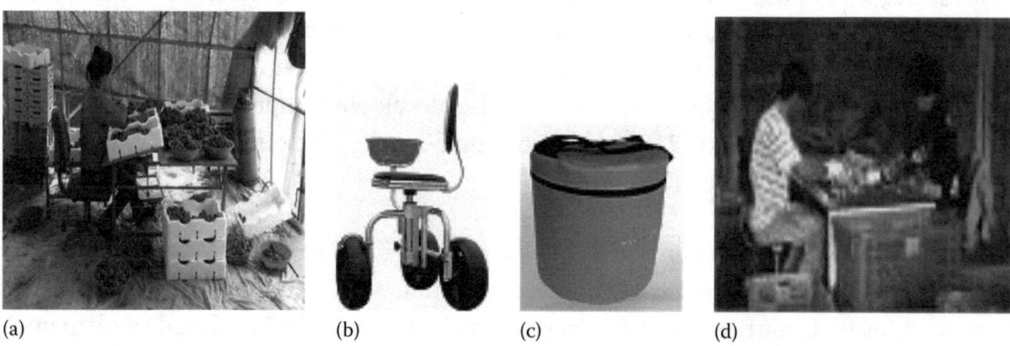

(a) (b) (c) (d)

FIGURE 45.5 Assistive technology for removing awkward working posture: (a) strawberry grading bench, (b) slope regulated cycle work chairs, (c) platform work chair, and (d) grading and shipping bench.

handles should be included on all objects being lifted or carried. Jung et al. [28] report that the maximum weight allowance of a box without handles is 7.2%–16% lower than that for a box with handles. Another study determined that container handles improve work efficiency and comfort level [33].

Interventions have been developed to assist in the transport of harvested peppers, cucumbers, strawberries, tomatoes, Chinese matrimony vines, fruit trees, squashes, chrysanthemums, mushrooms, and roses (Figure 45.6). In addition, a multipurpose vehicle has been developed to move seedling cases. Studies of platform carriers that evaluate work efficiency and comfort level by physiological, biomechanical, and subjective evaluation are in Table 45.8.

The agricultural work environment is designed for crops rather than workers. For example, the ridges along which crops are grown are awkward to walk on, and greenhouse aisles are often too narrow. In addition, fields become muddy and uneven over the planting and growing seasons, resulting in increased muscular demand and potential fatigue. Circumstances are even more burdensome when farm workers are required to step over fruit and vegetables on the ground such as watermelons, yellow melons, strawberries, and lettuce. To avoid crops, the farm worker may adopt unsafe postures that require sudden loading, an unstable gait, or unsafe posture. Another task that may induce WMSDs is extended reaching. Pruning and harvesting of apples, pears, and grapes, where workers

(a)

(b)

(c)

(d)

FIGURE 45.6 Assistive technology for transporting agricultural product: (a) platform carrier for outdoor peppers, (b) platform carrier for strawberries, (c) platform carrier for fruit vegetables, and (d) rail platform carrier.

TABLE 45.8 Assistive Technology for Transporting Agricultural Products

| Crops | Carriers | Evaluation Methods |
|---|---|---|
| Watermelons, lettuce, scallions [34,35] | Chairs with wheels, carriers | Physiological reaction, etc. |
| Cucumbers, tomatoes [30] | Human power and power-driven harvest platform carriers | Energy consumption, work efficiency |
| Strawberries [30] | Platform carriers for strawberries | Energy consumption, work efficiency |
| Yellow melons, strawberries, melons [36] | Rail platform carriers | Biomechanical evaluation, work efficiency |
| Greenhouse cultures [37] | Handy autonomous vehicles for greenhouse | Performance |
| Outdoor peppers [30] | Small and large platform carriers | Energy consumption, work efficiency |
| Scallions [35] | Chairs with wheels, handcarts | Physiological reaction, subjective symptoms |

TABLE 45.9 Assistive Technology for Working Condition

| Risk Factors | Characteristics | Improvement | Evaluation Methods |
|---|---|---|---|
| Ground conditions [34] | Improving and evaluating uneven and slanted ground | Flat ground | Heart rate, fatigue |
| Working area [38] | Ridges, furrows | Appropriate widths of ridges and furrows | Principles of motion economy |
| Height of bed culture [31] | Improving bed culture of strawberries | Elevated bed culture | Work hours, postures, and yields |

have trouble reaching the fruit, may produce excessive force on the lower back, shoulders, and upper extremities. Finally, poor posture may result from cleaning and sorting crops where a squatting position is usually adopted, which may place extreme stress on the lower extremities and lower back. To improve such work, Yoon et al. [31] raised the bed culture of strawberries to waist level, which improved working posture, decreasing the risk of WMSDs and increasing work efficiency (Table 45.9).

*45.3.3.4 Development of Protective Clothing for Workers in Cold and Hot Environments*
Agricultural workers are frequently exposed to extreme cold or hot temperatures. Since most work is outdoors, exposed to the elements, the primary environmental worry in the summer is heat exhaustion or, more seriously, heat stroke. In winter, the main issue is frostbite [29]. In extremely cold weather, loss of life can result from prolonged temperature loss. To combat extreme environmental temperatures, farmers often wear protective clothing to reduce the physiological burden. The RDA has developed and supplied various protective

TABLE 45.10 Assistive Technology for Agricultural Working Cloth

| Risk Factors | Characteristics | Improvement | Evaluation Methods |
|---|---|---|---|
| Ultraviolet rays [39] | Blocking ultraviolet rays, thermal comfort | Working clothing for a greenhouse | Physiological reaction, subjective sensation |
| High temperature [40] | Removability, cooling time, reduction in weight | Cooling vest | Physiological reaction, subjective sensation |
| High temperature [41] | — | Working clothing for outdoor peppers | Ergonomic workload test |
| Cold weather [42] | Waterproof, cold protection | Working clothing for dropwort | Physiological reaction, subjective sensation |
| Discomfort in body movement [43] | Convenience, functional improvement | Working clothing for fruit trees (pears) | Field test, functional test |
| Heat stress [44] | Materials, shapes and height of heels | Work shoes on a field | Physiological reaction, subjective sensation |
| Heat stress [30] | Waterproof coating, ventilation, wearability | Sprout and bud-picking gloves | Satisfaction, subjective sensation |
| Ultraviolet rays, solar radiation [45,46] | Blocking sunlight, ventilation, portability | Blocking sunlight, functionality | Subjective sensation, human physiological experiment |
| Exposure to agrichemicals, heat [47,48] | Waterproof, water permeability, water repellent | Enhanced material protecting clothing from agrichemicals | Physiological burden, subjective sensation, efficiency of materials |

clothing for working in greenhouses, interacting with livestock, handling agrichemicals, and providing protection from water and cold during the rainy season (Table 45.10 and Figure 45.7). Additional examples of protective clothing for farmers are specially designed fatigue-resistant work shoes for field work and antifatigue gloves for picking and pruning the sprouts and buds of fruit vegetables.

Advanced virtual human modeling has been used to determine the best method for completing specific farm tasks or using certain types of equipment (Figure 45.8) [36]. The premise is that problems in farming components can be improved and demonstrated through virtual human modeling, saving significant expense and time. In an industry with limited resources, these are always major concerns in developing new interventions.

The RDA also recently applied anthropometric principles to investigate unique characteristics of individuals and their relationship to farm work performance [49,50]. In anthropometric measurements of 2000 Korean agricultural workers, a high prevalence of "bow-shaped" legs was found in female agricultural workers (Figure 45.9) [49] and found to be related to a squatting working posture [49]. The RDA also reported that the waist size of rural women of all ages tended to be larger than those of urban women [50]. These data must be considered

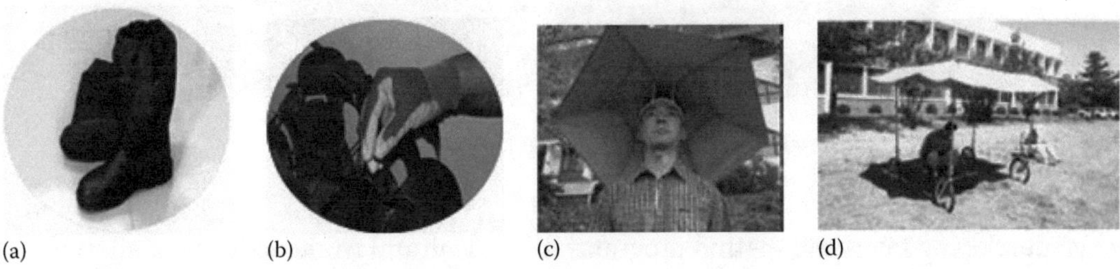


FIGURE 45.7 Assistive technology of protective clothing equipment: (a) agricultural work shoes, (b) sprout and bud picking glove, (c) heat protective hat, and (d) sun-blocking cart.

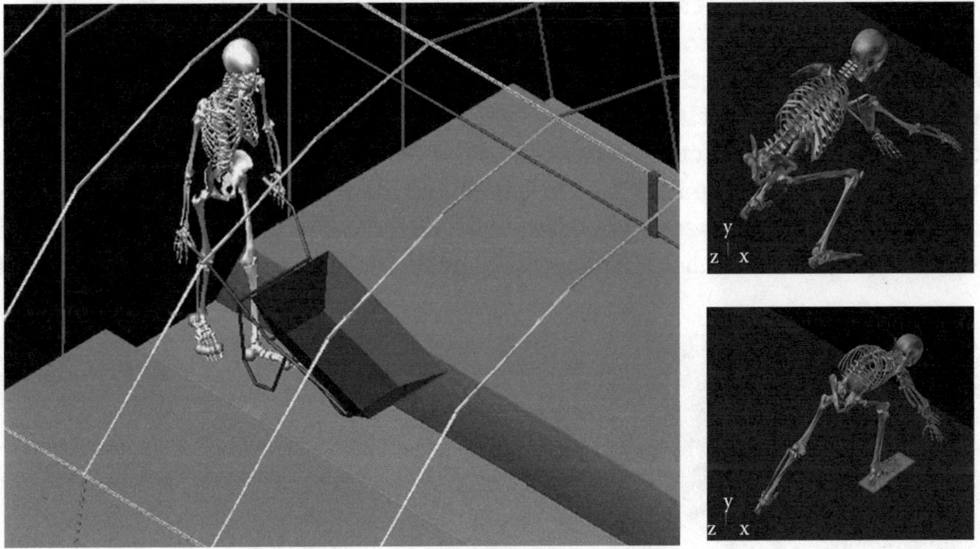


FIGURE 45.8 Evaluation of farmer while moving farm equipment using a human model.

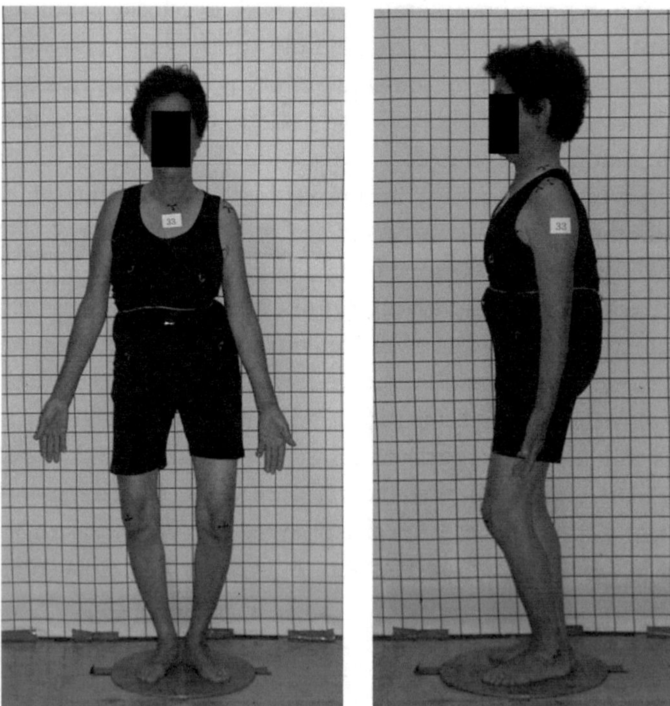

FIGURE 45.9 Anthropometric measurement of a agricultural worker.

when developing interventions and programs for agricultural workers since not all workers will be a standardized size. Equipment, machinery, clothing, and tools will not necessarily be one size fits all and will need to be adjusted for different working populations.

*45.3.3.5 Development of Educational Materials for Agricultural Work-Related Injuries*

Starting with an agricultural worker health management class model in 1995, the Korea Information Center of Agricultural Safety and Health developed educational programs for agricultural worker health and safety, including an exercise program to reduce worker fatigue (1997), an occupational disease-related risk factor management system (2000), and implementation of programs for the prevention and management of agricultural injuries (2006). Furthermore, the center provides summaries of the programs to the workers and the general public as handbooks (Figure 45.10) and through an online service (Figure 45.11).

FIGURE 45.10 Agricultural safety and management handbooks.

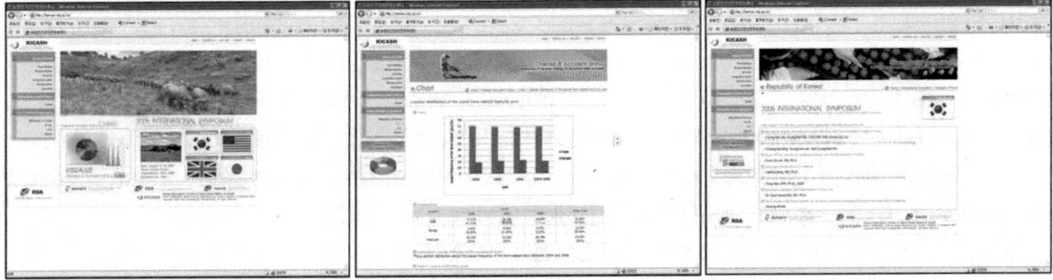

FIGURE 45.11 Korea Information Center of Agricultural Safety and Health (http://farmer.rda.go.kr).

## 45.4 MANAGEMENT OF WORK-RELATED MUSCULOSKELETAL DISORDERS

### 45.4.1 WMSDs in Industrial Workers

The first case of compensation for an occupational disease in Korea occurred in 1986, involving a broadcasting station typist. This served as a starting point for injury management, recording, and compensation for laborers [51].

Since 1980, the Ministry of Labor has announced the official status of occupational diseases each year in accordance with the Industrial Accident Compensation Law. According to these data, no cases of WMSDs from cumulative trauma were reported until 1992. In 1993, two cases were reported (Table 45.11). These cases increased the interest of labor unions, progressive nongovernment organizations, and industrial public health experts. Since then, the number of reported WMSD cases increased rapidly until 1997. In that year, the foreign currency crisis stagnated the Korean economy, with many companies laying off massive numbers of workers. As a result, laborers emphasized survival at their workplaces, giving priority to work over compensation for occupational injuries, and the number of WMSD cases decreased.

During the economic recovery, work intensity increased because of sustained productivity with a smaller workforce, leading to working environment that was more prone to WMSDs.

TABLE 45.11 Workplace Musculoskeletal Disorder (MSDs) Cases in South Korea (The Ministry of Labor, Labor Statistics 2008)

| | | Occupational Disease | |
|---|---|---|---|
| **Year** | **No. of Total Workers** | **No. of Total Diseases** | **No. of WMSDs (%)** |
| 1998 | 7,582,479 | 1,838 | 124(6.7) |
| 1999 | 7,441,160 | 2,732 | 344(12.6) |
| 2000 | 9,485,557 | 4,051 | 1,009(24.9) |
| 2001 | 10,581,186 | 5,653 | 1,634(28.9) |
| 2002 | 10,571,279 | 5,417 | 1,827(33.7) |
| 2003 | 10,599,345 | 9,130 | 4,532(49.6) |
| 2004 | 10,473,090 | 9,183 | 4,112(44.8) |
| 2005 | 11,059,193 | 7,495 | 2,901(38.7) |
| 2006 | 11,688,797 | 10,235 | 6,233(60.9) |
| 2007 | 12,528,879 | 11,472 | 7,723(67.3) |
| Total | 102,010,965 | 67,206 | 30,739(45.7) |

In 1998–1999, 200–300 cases of WMSDs were reported each year, including an increase from 1000 to 2000 from 2000 to 2003 and a further rise after 2003–2004, with approximately 4000 reported cases [52]. By 2007, 7723 people reported a WMSD, which accounted for 62.7% of the total number of laborers with an occupational disease [52]. The number of WMSD cases in 2007 was 166% of the 2005 number. One potential contribution to this sudden increase was a change in the reporting of low back pain (Table 45.11) [52].

Of all the WMSDs, 75% occurred in the manufacturing industry, specifically the automobile and shipbuilding industries [53]. Recently, more cases of WMSDs are being reported in other industries such as the restaurant, lodging, health, social welfare, banking, and insurance industries [53]. The average compensation cost per WMSD case is estimated to be $35,000–$45,000 [54]. The rapid increase in WMSDs not only threatens worker health and welfare but also imposes a heavy financial burden on the Korean economy as a whole.

### 45.4.2 WMSDs in Agricultural Workers

Korean farms are typically characterized as small-scale independent farms, often run by a family unit [55]. Even with an increase in mechanized farming equipment, many agricultural activities remain heavily dependent on human labor [55], so agricultural workers are exposed to long work hours and ergonomic risk factors for WMSDs.

While WMSDs are one of the most serious problems threatening agricultural worker health and welfare in Korea, researchers have only recently conducted systematic studies on work-related WMSDs. In 1233 agricultural workers examined by the RDA in 2006, prevalence rate for WMSD symptoms in the previous 12 months that lasted at least 1 week or occurred at least once a month was 81%. Prevalence rates by body part were 52% in the back, 51% in the legs/knees, and 38% in the shoulders [56]. In addition, 416 Korean agricultural workers were randomly selected and examined by physicians and radiological tests, including x-ray, CT, and MRI, revealing a WMSD prevalence rate of 75%. Injury types for the agricultural workers were 94 cases of "osteoarthritis of the knee," followed by 65 cases of "non-specific chronic back pain," 62 cases of "myofascial pain syndrome," and 49 cases of "radiculopathy of the back" [21]. Based on the collected data, farm work significantly contributed to the physical conditions in approximately 41% of the diseases [21].

Korean agricultural workers are exposed to different WMSD risk factors than laborers in the general manufacturing industries, and the risks are dependent upon the crop produced and livestock raised [57–59]. Pomiculture workers frequently report WMSDs of the neck, shoulders, arms, and elbows due to repetitive work with the neck flexed or bent backward or working with their arms above their shoulders [57,58]. Another example of crop-specific WMSDs is seen in greenhouse workers, or workers who harvest peppers, strawberries, yellow melons, and watermelons. This work, which occurs below knee level [58–60], requires the worker to bend forward significantly or sit in a squatting posture, potentially resulting in hip and knee WMSDs [58–60].

Lee and associates [21] examined ergonomic risk factors and WMSD symptoms in agricultural workers, finding higher prevalence rates of 45% in the hips and waist and 42%

in the lower bodies than in other parts of the body. Various postures were observed in agricultural workers, including a squatting posture, a bending posture in which workers bent their knees 30°–120°, and sitting cross-legged [21]. These postures may cause changes in body shapes [50,58]. Moreover, some have pointed out that such working postures may have led to frequent cases of "osteoarthritis of the knee" and "bow leg" [56,59].

The rapid entire body assessment (REBA) and Ovako working posture assessment system (OWAS) are frequently used as ergonomic assessment tools in agricultural work [61]. However, they oversimplify postures for the legs and knees or do not reflect the frequent working postures used in Korean agricultural work and thereby underestimate the WMSD risk for these body parts [61]. Based on this research, the RDA and other researchers [62] compared physiological burdens (e.g., heart rate), muscle activation levels (via EMG), and subjective discomfort rating in various lower body postures to develop a Korean ergonomic assessment tool specialized for leg and knee risk factors.

### 45.4.3 Korean Industrial Health and Safety Laws

In Korea, the enforcement regulation (Number 2, Paragraph 1, Article 16) in the Industrial Accident Compensation Law first recognized "cervicobrachial syndrome" as a type of occupational disease in 1994. In the late 1990s, the Korean government established "Work Management Guidelines [for] Laborers Handling Visual Display Terminals" ("VDT", hereafter; Ministry of Labor Announcement No. 1997-8), "Work Management Guidelines [for] Laborers Performing Simple Repetitive Work" (Ministry of Labor Announcement No. 1998-15), and "Work Management Guidelines [for] Prevention Of Occupational Lumbago" (Korea Occupational Safety and Health Agency (KOSHA) CODE H-05-1998) to initiate work environment management and prevention of WMSDs [2]. All regulations aimed to document and control WMSDs in the workplace.

In the last decade, WMSDs have become a major issue in Korean industrial settings, leading to the conclusion that previous government recommendations have not been effective in preventing WMSDs. As a result, the government revised the Industrial Safety and Health Law in 2003 and established employer obligations to prevent WMSDs. Based on this new regulation, Korean companies improved work environments and the employee health and safety budgets [63,64]. Large corporations hired WMSD experts or fostered in-house experts to investigate and improve risk factors in the workplace. However, many small businesses cannot carry out WMSD prevention and management measures at the level required by the revised law because of budgetary concerns.

The government has begun to enforce regulations for the most strenuous workplaces and jobs. Detailed management of the risk factors is regulated so that employers are required to provide notification of risk, provide education to employees, implement a medical management and prevention program, and reduce employee exposure to heavy load tasks. In this new regulation, the government suggested 11 items that should be considered strenuous and risky for WMSDs. The regulation was similar in content to the Washington State Ergonomic Rules—WAC-296-62-05174, Caution Zone Job—in the United States, which is used to define ergonomic requirements for risk factor investigations and working environment improvements [3]. The following is a list of the 11 key items that must be controlled to reduce WMSDs in the workplace:

1. Performing intensive tasks using a keyboard or mouse for more than 4h total per day
2. Repeating the same motion with the neck, shoulders, elbows, wrists, or hands (excluding keying activities) more than 2h total per day
3. Working with the hand(s) above the head, the elbow(s) above the shoulders, or the elbow(s) far away from or behind the body more than 2h total per day
4. Working with the neck or the back bent or twisted without support and without the ability to vary posture more than 2h total per day
5. Squatting or kneeling more than 2h total per day
6. Pinching an unsupported object(s) weighing 1 or more kg per hand, or pinching with a force of 2 or more kg per hand, more than 2h total per day
7. Gripping an unsupported objects(s) weighing 4.5 or more kg per hand, or gripping with a force of 4.5 or more kg per hand, more than 2h total per day
8. Lifting objects weighing more than 25kg more than 10 times per day
9. Lifting objects weighing more than 10kg above the shoulders, below the knees, or at arm's length more than 25 times per day
10. Lifting objects weighing more than 4.5kg if done more than twice per minute, more than 2h total per day
11. Using the hand (heel/base of palm) or knee as a hammer more than 10 times per hour, more than 2h total per day

Employers must perform risk factor investigations, covering facility and production line conditions, workload, task speed, cycle time, working posture, work method, and symptoms and signs of WMSDs in employees. A risk factor investigation must be conducted once every 3 years or once a WMSD is reported, when new jobs and equipment are introduced or when workload and working environment are modified. The law also states that representatives of the workers must participate in the investigation procedure.

Furthermore, employers are obligated to conduct a WMSD prevention program when 10 or more employees suffer from WMSDs, or 5 or more employees suffer from WMSDs, if that is equal to or greater than 10% of the total number of employees. The law defines specific guidelines for manual material handling jobs. It recommends limiting weight and considering the frequency, transfer distance and speed, and work/rest ratio in lifting tasks. The regulation also recommends that employers notify the employee of the exact weight of the material and the proper handling method for objects weighing more than 5kg.

## 45.5 FUTURE CONCERNS

The biggest problem facing the industrial health and safety system in Korea is that small business laborers and independent agricultural workers are extremely vulnerable to industrial injuries and occupational diseases. Since the Industrial Safety and Health Law applies only to workplaces with five or more employees, about 99% of small business workers and most agricultural workers on family-run farms are not covered. As a result, these workers do not benefit from many of the regulations on work environment management, surveillance of occupational diseases, or compensation and rehabilitation for industrial injuries. Overall improvement to these small businesses needs to be a high priority in the near future.

For a sustainable Korean agriculture, workers and their family members must be supported in adopting safe and healthy practices for performing farm work. Korea should reinforce research projects and education programs to prevent injuries on the job. Korea should also establish a compensation system to protect agricultural workers and their family members as well as to expand the social safety net, so quality of life is maintained after an injury has occurred.

## REFERENCES

1. N. S. Lee, C. M. Lee, K. Nah, B. Park, D. H. Lee, K. S. Kwon, D. C. Lee, S. J. Park, and H. S. Park, 20 Year's history of the ergonomics society of Korea, *J. Ergon. Soc. Korea* 22(2):69–92 (2003).
2. S. J. Lee and Y. S. Kim, Current trends of work-related musculoskeletal disorders (WMSDs), *Hanyang J. Med.* 19(2):31–39 (1999).
3. Korean Occupational Safety and Health Agency, Regulation and technical guideline for prevention for WMSDs, Incheon, Republic of Korea, 2003.
4. S. W. Bahn, C. Lee, J. H. Lee, and M. H. Yun, Development of luxuriousness models for automobile crash pad based on subjective and objective material characteristics, *J. Ergon. Soc. Korea* 25(2):187–195 (2006).
5. H. C. You, T. B. Ryu, K. H. Oh, M. H. Yun, and K. J. Kim, Development of satisfaction models for passenger car interior materials considering statistical, technical, and practical aspects of design variables, *IE Interfaces* 17(4):482–489 (2004).
6. Y. H. Chung, S. S. Eom, K. Sohn, and K. H. Choi, A human sensibility ergonomics method for vehicle driving simulator and verbal expressions collected, *J. Ergon. Soc. Korea* 19(2):1–14 (2000).
7. J. P. Choi, E. S. Jung, S. J. Park, and S. W. Jeong, Optimal layout of vehicle pedals depending on the types and positions of vehicle pedals, *J. Ergon. Soc. Korea* 26(4):91–100 (2007).
8. S. J. Park, Y. K. Yim, H. J. Lee, and C. J. Kim, The study on the satisfaction and a motion analysis when entering a car through change of vehicle conditions, *J. Ergon. Soc. Korea* 25(3):25–32 (2006).
9. H. U. Lim, E. S. Jung, S. J. Park, and S. W. Jeong, Comparisons of driver performance with control types of the driver information system, *J. Ergon. Soc. Korea* 26(1):1–100 (2007).
10. J. Y. Kim and K. B. Seo, The effect of the heel rest on the lower leg muscle activity and fatigue during repetitive pedaling, *J. Ergon. Soc. Korea* 24(4):55–62 (2005).
11. S. J. Park, N. S. Lee, and S. Lee, An investigation of the relationship between subjective comfort evaluation and body pressure distribution of vehicle seats, *J. Ergon. Soc. Korea* 12(1):3–15 (1993).
12. D. Kee, Evaluation for performance and preference of Hangul entry methods using real mobile phones, *J. Ergon. Soc. Korea* 25(3):33–41 (2006).

13. S. K. Hong, Universal design of Hangul input method for mobile phones, *J. Ergon. Soc. Korea* 26(3):117–124 (2007).
14. Y. H. Jun, I. K. Baek, J. I. Kim, and K. H. Son, A study on causal relationships among sensibility satisfaction factors for mobile phone, *J. Ergon. Soc. Korea* 22(2):1–13 (2003).
15. J. Y. Kim, H. Y. Lee, and Y. C. Choi, The structuring process of multi-centered usability evaluation method, *J. Ergon. Soc. Korea* 24(2):25–33 (2005).
16. I. K. Kim, C. Lee, and M. H. Yun, A comparison of modeling methods for a luxuriousness model of mobile phones, *J. Ergon. Soc. Korea* 25(2):161–171 (2006).
17. M. Y. Kim, S. Park, and E. S. Jung, Development process of a universal design evaluation index with a case study for mobile phone evaluation, *J. Ergon. Soc. Korea* 25(4):51–62 (2006).
18. K. R. Kim, A study on the farmers' health status and musculoskeletal workload, PhD thesis, Seoul National University, Seoul, Korea, 2008.
19. K. S. Lee, A study on the management system for occupational disease and injury of farmers, PhD thesis, Seoul National University, Seoul, Korea, 2006.
20. Korea National Statistical Office, Basic agricultural statistics survey, 2008.
21. Rural Development Administration, Development of the standardized diagnostic tool for farmer's work-related musculo-skeletal disorders and evaluation of the work-relatedness of musculo-skeletal disorders, 2009.
22. RDA, Report on comparison and analysis for main disease and health status in farmer and fisherman, National Academy of Agricultural Science, Suwon, Republic of Korea, 2004.
23. RDA, Hazard assessment for risk factors in farmer work-related disease and development of total management system, 2002.
24. D. M. Paek, In-depth and link analysis—The annual report of the National Health and Nutrition Survey in 1998, 1999.
25. Ministry of Labor, 2006 Annual report of work injury insurance, 2007.
26. Ministry of Agriculture, A study on patterns and characteristics of agriculture-related injury by analysis of safety aid system of farm workers, 2007.
27. Rural Development Administration, Mid-term and long-term plan for prevention strategy of agricultural accidents and management of farmer's health, 2007.
28. K. G. Davis and S. E. Kotowski, Understanding the ergonomic risk for musculoskeletal disorders in the United States agricultural sector, *Am. J. Ind. Med.* 50(7):501–511 (2007).
29. D. S. Lim, H. C. Kim, H. S. Chae, K. R. Kim, S. J. Lee, and K. S. Lee, A study on agricultural safety technology for ergonomic intervention in farm-work, *J. Ergon. Soc. Korea* 29(2):225–239, 2010.
30. Rural Development Administration, A handbook of the assistive technology for enhancement of farm-work: Tools and personal protective equipments, 2008.
31. H. S. Yoon, Y. H. Hwang, C. K. An, H. J. Hwang, and C. W. Rho, Growth and fruit yield of strawberry grown in raised bed culture using growing media with lower cost, *Kor. J. Hort. Sci. Technol.* 22(3):266–269 (2004).
32. H. S. Jung and H. S. Jung, Evaluation of proper height for squatting stool and the ergonomic design of wearable stool, *IE Interface* 18(3):288–296 (2005).
33. H. S. Jung, A. S. Park, and H. S. Jung, Ergonomic design and evaluation of carrying handles for bag, *IE Interface* 17(1):46–55 (2004).
34. M. J. Kim and J. W. Choi, A Study for farmers to reduce workload on the different working conditions (part 1)—Cultivating lettuce in the winter greenhouse, *Korean J. Commun. Living Sci.* 8(2):111–117 (1997).
35. M. J. Kim and J. W. Choi, A study for farmers to reduce workload on the different working conditions (part 2)—Cultivating welsh onion in the summer ground, *Korean J. Commun. Living Sci.* 8(2):119–124 (1997).
36. Rural Development Administration, Development of an electronic vehicle for chamwoae of protected cultivation using agricultural ergonomics, No.11-1390000-002058-01 (2008).

37. J. H. Lee and K. H. Ryu, Development of a simple autonomous vehicle for greenhouse works, *J. Korean Soc. Agric. Machinery* 21(4):422–428 (1996).
38. Rural Development Administration, Safety management booklets for elderly farmer (2006).
39. J. W. Choi and K. S. Hwang, Study in the effect of a cooling vest under hot environment, *J. Korean Soc. Cloth. Text.* 25(1):83–90 (2001).
40. J. W. Choi, M. J. Kim, and J. Y. Lee, Efficacy of cooling vests for alleviating heat strain of farm workers in summer, *J. Korean Soc. Cloth. Text.* 29(8):1176–1187 (2005).
41. J. W. Choi, M. J. Kim, and J. Y. Lee, Ergonomic investigation of the workload of red pepper harvest workers in summer: A pilot study for developing personal protective clothing, *J. Korean. Soc. Living. Environ. Sys* 14(1):9–19 (2007).
42. J. W. Choi and Y. J. Baek, Development the protective clothing for dropwort cropping workers, *J. Korean Soc. Cloth. Text.* 24(8):1266–1275 (2000).
43. M. S. Choi and S. P. Ashdown, The design and testing of work clothing for female pear farmers, *Cloth. Text. Res. J.* 20(4):253–263 (2002).
44. K. S. Lee and J. W. Choi, A study of agricultural fatigue shoes, *Korean J. Commun. Living Sci.* 7(2):99–108 (1996).
45. G. S. Kim and J. H. Choe, Development of movable sunshade tent, *J. Korean Home Econ. Assoc.* 42:1–15 (2004).
46. M. J. Kim and J. W. Choi, Thermal and subjective responses by sun hats for farmer in a hot climatic chamber, *J. Korean Soc. Cloth. Text.* 28:713–722 (2004).
47. J. W. Choi, M. J. Kim, and J. Y. Lee, Evaluation of the farmer's workload and thermal environments during cucumber harvest in the greenhouse, *Korean J. Commun. Living Sci.* 9(3):245–253 (2002).
48. K. S. Hwang, K. R. Kim, K. S. Lee, H. C. Kim, K. S. Kim, and Y. J. Baek, The textiles and the performance level in developing the pesticide proof clothing, *J. Korean Soc. Cloth. Text.* 31(11):1611–1620 (2007).
49. Y. J. Baek and K. S. Lee, Characteristics of the body shape of Korean woman farmers by crops, *J. Korean Soc. Cloth. Text.* 32(11):1584–1594 (2008).
50. Y. J. Baek, K. S. Lee, K. S. Hwang, K. R. Kim, and H. C. Kim, Characteristics of the somatotypes of Korean farmer, *J. Korean Soc. Cloth. Text.* 32(3):431–442 (2008).
51. B. K. Kim and J. Y. Kim, Current status of work-related musculoskeletal disorders in Korea, *Proc. 30th Conf. Korean Occupational and Environmental Medicine*, Seoul, Korea, pp. 7–12 (2003).
52. Ministry of Labor, Labor Statistics, 2008.
53. B. Y. Kwon, K. H. Park, Y. Kim, Y. M. Kim, and H. K. Choi, Analysis of incidence trend of work-related musculoskeletal disorders (1998–2007) in Korea, *Proc. Spring Conf. the Ergonomics Society of Korea*, Goomi, Korea, pp. 28–33 (2008).
54. C. H. Kim, S. Kim, J. Lee, D. Yoon, and S. Cho, A study of work-related musculoskeletal disorders at automobile industries in Korea, *Proc. Sixth Annual International Conf. on Industrial Engineering-Theory, Application, and Practice*, San Francisco, CA, pp. 18–20 (November 2001).
55. O. S. Ahn, Work environment and workload of farmers in four crop cultivation, PhD thesis, Seoul National University, Seoul, Korea, 2001.
56. Rural Development Administration, Agricultural workers' health and management of agricultural tasks, 2008.
57. Y. H. Lee, J. H. Lee, K. S. Lee, K. R. Kim, and S. J. Lee, Ergonomic risk factors related to musculoskeletal symptoms in the vineyard workers, *J. Korean Soc. Occup. Environ. Hyg.* 18(2):122–132 (2008).
58. Y. H. Lee, J. H. Lee, K. S. Lee, K. S. Kim, Y. K. Lee, Y. K. Kong, and S. J. Lee, Characteristics of lower-body posture of agricultural works in Korea, *Proc. 41th Conf. the Korean Society of Occupational & Environmental Medicine*, Busan, Korea, pp. 199–200 (November 2008).

59. H. C. Ryu, S. J. Lee, J. H. Lee, and K. S. Lee, Exposure rate of ergonomic risk factors related agricultural work, *Proc. 41th Conf. the Korean Society of Occupational & Environmental Medicine*, Busan, Korea, pp. 402–403 (November 2008).
60. K. S. Kim, K. R. Kim, H. C. Kim, and K. S. Lee, Risk assessment and symptoms of musculoskeletal disorders in melon farm workers, *J. Korean Soc. Occup. Environ. Hyg.* 16(4):385–397 (2006).
61. Y. H. Lee, J. H. Lee, K. S. Lee, K. S. Kim, Y. K. Kong, S. T. Shon, and S. J. Lee, Comparative analysis of ergonomic checklists by lower-body postures in Korean agricultural works, *Proc. 41th Conf. the Korean Society of Occupational & Environmental Medicine*, Busan, Korea, pp. 201–202 (November 2008).
62. Y. K. Kong, D. M. Kim, S. J. Lee, J. H. Lee, Y. H. Lee, K. S. Lee, and S. T. Shon, Evaluation of the effects of lower-limb postures on the subjective discomfort, heart rate, and EMGs of lower extremity muscles, *J. Ergon. Soc Korea* 28(1):9–19 (2008).
63. J. H. Kim, K. S. Lee, and S. K. Bae, The case study of the corporate ergonomics program at the automobile company, *2007 Spring Conf. of Ergonomics Society of Korea*, 2007.
64. Y. Pyo and B. Y. Jeong, An implementation case of ergonomics program at a Shipbuilding company, *J. Ergon. Soc. Korea* 26(3):45–52 (2007).

CHAPTER 46

# Overview of Ergonomic Needs and Research in India

Rabindra Nath Sen

CONTENTS

## 46.1 INTRODUCTION

Ergonomics has been defined as the science, technology, and art of man at work [1]. Its aim is mainly to optimize human–machine–environment interactions to get the maximum efficiency, productivity, and improvement of the working conditions and working life of the workers. Though the basic principles of ergonomics have been practiced by workers in India from prehistoric times without the specialized knowledge of ergonomics, their formal application started only a few decades ago.

## 46.2 HISTORICAL BACKGROUND AND SIGNIFICANCE OF OCCUPATIONAL ERGONOMICS

India is the second most populous country of the world, with about 920 million people, and its unemployment rate is huge even at the present time. A 3% annual population growth rate amounts to a 19-fold increase in a century. India was formerly a predominantly agricultural country, with about 80% of the population directly or indirectly involved in agriculture. Only since its independence from British rule and the initiation of the second five-year plan has it experienced industrial development. Hence, the application of occupational ergonomics is becoming more and more necessary.

Among the many constraints in industrial development, financial constraints are the most important. Hence, in industrially developing countries (IDCs) such as India, the application of ergonomics must in the initial stages consider the implementation of improvements whose cost will be low or negligible. The significance of the use of occupational ergonomics in the progress and development of the country as a whole, not only in industry but also in agriculture and other areas, is unquestionable.

## 46.3 SOME ASPECTS OF ERGONOMICS STUDIES IN INDIA

The work done in India regarding several ergonomics studies may also be applicable to other IDCs. Sen [1,2] and Kogi and Sen [3] reviewed the research work done in India in the following areas:

1. Energy expenditure of the workers in different types of jobs in various factories, and classification of heaviness of jobs on the basis of physiological responses
2. Anthropometric measurements in relation to the design of machines, tools, implements, and consumer products [1,2]
3. Problems of agricultural and other unorganized workers with special reference to manual materials handling [4–15]
4. Ergonomic solutions of problems in small-scale and cottage industries [16]
5. Low-cost improvements of working conditions, safety, health, and welfare of the workers [17]

6. Postural analysis [18]
7. Training programs for workers, supervisors, trade union officials, managers, government executives, etc., to bring awareness of ergonomics to people at all levels including the general populace

Because of the huge population of India, any slight ergonomic improvement at the individual level would yield in total a very significant quantitative effect. In the unorganized sectors such as in agriculture and among manual laborers in various vocations—for example, masons, carpenters, blacksmiths, construction workers, household workers, and service workers such as laundry workers, barbers, handloom workers [19], cobblers, and about 8 million handicraft workers in India [20]—the use and application of ergonomics are much less common than in the organized sectors. This is due mainly to a lack of awareness of the basic principles, poor economic conditions, and reluctance to change existing and traditional work methods and tools.

### 46.3.1 Agricultural Ergonomics

#### *46.3.1.1 Use of Tools and Implements and Their Design*

Several studies have been conducted for the improvement of the existing designs of tools, implements, and aids such as the sickle [5], plow [6,7], shovel [8,9], spade [10], combined shovel and hoe [11], and "float seat" [12] used in agricultural work.

#### *46.3.1.2 Use of Pesticides*

In agriculture, the increased use of pesticides, mainly spraying without adequate protection, has increased the health hazards to agricultural workers by several folds.

#### *46.3.1.3 Use of Work Methods*

Sen and coworkers [13–15,21,22] have studied the various traditional methods used in agricultural work.

#### *46.3.1.4 Use of Modern Sophisticated Machines*

Traditional agricultural workers are unfamiliar with modern agricultural mechanized systems such as the use of the combine harvester and thresher, and modern work methods [13]. Similarly, traditional industrial workers are not acquainted with modern sophisticated industrial machinery, complex tools, and work methods. Many work-related diseases occur very frequently. There are very few statistical data on accidents in either agriculture or industry, especially from small-scale and medium sized organizations [23]. Cultural, social, and economic differences make the problems of IDCs more difficult to solve than those of the developed countries. This poses a great challenge to professional ergonomists.

### 46.3.2 Industrial Ergonomics

Like most other IDCs, India is traditionally agriculture-based and has evolved into an IDC only in the last few decades.

#### *46.3.2.1 Ergonomic Design of Factory Buildings*

There are several monographs on the design of factory building in cold climates. However, there have been practically no studies on the ergonomic design of factory buildings suitable for a tropical climate characterized by high ambient temperatures and high relative humidity.

In any industry—big, medium-sized, or small—it is very important to plan much in advance the layout, design, and orientation of factory buildings. Sen [24] considered in detail the ergonomic design of factories suitable for the tropical climate of India. He stressed the importance of orienting the building to provide good natural ventilation and daylight, to reduce solar radiation by sun shades and double roofs, to circulate cooler underground air, and other factors. He emphasized the usefulness of a correct layout of the shop floor with respect to the positions of windows, fans, floor fans, materials, machines, workstations for different operations, storage, shipping, etc., so that efficient movement of personnel and a quicker flow of materials and products could be effected. It is imprudent to spend a huge sum of money to change a badly designed factory building at a later stage. It always pays to plan and design factory buildings ergonomically well in advance of the start of production.

There is an acute shortage of housing for industrial workers. Whereas most of our public sector employers have tried to provide some housing for workers, the records of private sectors are not at all encouraging.

#### *46.3.2.2 Reduction of Heat Stress*

Engineering control for the reduction of heat stress is very expensive in most cases. Hence, ergonomic designs of low-cost thermal barriers and low-cost personal protective clothing are of considerable importance.

*Thermal barriers*: In improving working conditions, various types of thermal barriers, personal protective devices, floor fans, “man coolers” (big, high-speed air circulating fans), etc., have been designed to reduce the effects of high heat stress, especially for workers in the steel industry, engineering work, glass factories, and the like.

To protect against thermal radiation from furnaces, molten metal, slag, etc., in various factories, low-cost barriers against thermal radiation have been fabricated from two oxidized iron sheets, each about 1 $m^2$ in size with a gap of 25 cm between them. The two sheets are held together by horizontal iron rods at four corners and at the middle. These portable thermal barriers are put on low-cost stands with different points of suspension to permit placing them between the sources of high thermal radiation and the workers. The hot air rises in the gap between the two sheets to significantly reduce the effects of thermal radiation [17].

*Molten metal or slag channel cover*: To reduce the thermal radiation from red hot molten metal or slag flowing in specific channels, small metal covers hinged on one side have been fabricated for the channels; these covers can be raised during maintenance and cleaning of the channels. The channel covers also act as guards against splashes of metal, sparks, etc., to effectively reduce the risks of burns and other accidents from spatters of molten metal [17].

*Personal protective devices*: The acceptance of personal protective devices, equipment, and protective measures for technical devices depends substantially on their effective functionality and ergonomic design.

Ergonomic designs of low-cost, washable, ventilated (by a "man cooler" at the back), vapor-permeable, flame-retardant, radiant-heat-reflecting type of special work clothing have been made for hot process workers such as furnace tenders and helpers in glass factories, and furnace tenders, teemers, chargers, cleaners, launderers, and bull-ladle operators in hot metal factories to reduce heat stress and physiological costs of work. The new special clothing was found to be 3–4 times more comfortable than the existing workwear, as revealed by worker polls. The use of this work clothing increased productivity in quality and quantity, especially during the summer months [3,25–27]. An ergonomically designed low-cost transparent face shield and fire-resistant, hard-toe safety boots with a hinged wooden sole are also suitable for workers working on the top of a coke oven battery and for furnacemen to protect them from thermal radiation and from injuries from metal splashes or sparks and to increase efficiency and productivity. Further research is necessary for ergonomic designs of similar low-cost personal protective devices for work situations having much more intense heat stress.

Sen and Das [28] ergonomically designed a manual metal arc (MMA) welders' screen and the protective work clothing for reducing the workload and risks of damage by ultraviolet radiation, shielding also against toxic fumes generated in the welding process, and reducing static local muscular fatigue. The improvement was due to the elimination of hand movements by the hand-held screen of the old design because the lower-jaw-controlled screen window protecting against ultraviolet rays was fixed in the helmet in the new design.

The welder's electro-optical protection filters and the welder's curtain must conform to the needs of labor safety, comfort of wearing, and good ergonomic layout of the workplace.

#### *46.3.2.3 Manual Materials Handling*

From time immemorial, one of the main problem areas in industry has been manual materials handling. The use of human labor is very extensive in IDCs like India. The use of machinery reduces the workload considerably for a worker and saves a lot of time, but it involves very high financial costs. It is often so expensive to mechanize and automate a manufacturing system that human adaptability makes it cheaper to use people for the awkward jobs such as materials handling and inspection. In a country with very cheap labor, a medium-sized or small-scale industry always uses workers in manual materials handling in the traditional manner.

#### *46.3.2.4 Use of Ergonomic Designs of Tools and Implements*

Another way to bring about improvement is to use ergonomically modified designs of existing tools and implements so as to enhance their efficiency. Sen and coworkers [2,8–11] made attempts to improve the designs of the traditional shovel (Figure 46.1), spade, and shovel cum hoe to increase productivity and to reduce physiological costs to the workers by minimizing the degree of stooping required and enhancing leverage.

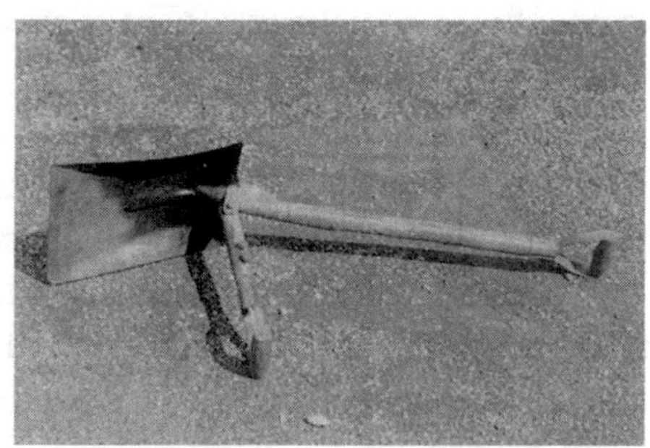

FIGURE 46.1 An additional handle fitted at the base of a traditional shovel significantly reduces the bending of the back and simultaneously increases the throwing distance by about 25%. A removable clip designed to temporarily hold the second handle when the shovel is used is not shown.

During the use of a standard shovel in lifting and handling materials such as sand and stone chips, the protracted bending or stooping posture causes backache and fatigue. An additional handle fitted at the base of the shovel significantly reduced the bending of the back and simultaneously increased the throwing distance by about 25% due to the improvement in leverage.

Sen [29] similarly improved the existing designs of the beater and the ballast rake used for the manual maintenance of railway tracks.

Efforts were also made to modify the designs of bullock-cart [30] and hand-pulled carts [31] and to use an ergonomically designed special harness [32] for the pullers.

##### *46.3.2.5 Mechanical Materials Handling*

*Special cart for moving heavy materials*: In a company manufacturing ceramic bricks, an ergonomically designed special cart for transporting the bricks to the drying chamber on a metal plate was fabricated from low-cost bicycle parts and iron angles, which reduced the efforts of the workers by about 25% and increased productivity by about 15% [17].

*Crane design*: In mechanical materials handling, improvements were also made in the ergonomic design of a chain pulley system, forklift truck, platform truck, and overhead transport crane [28,33]. In many of the factories, there was simultaneous use of different designs of overhead transport cranes that had different types of controls and displays according to their positions, purpose of use, distance from the operator, etc., and there were many control–display incompatibilities. The need for the same worker to use different types of cranes on the same day resulted in an increase in the risk of accidents due to the greater chance of making a mistake in operating the controls. Sen and coworkers [28,33] suggested standardizing the positions of controls and displays and changing their locations so that the normal motion stereotypes are maintained and there are no risks of accident. There is scope for similar improvements in the design of tools, implements, trucks, crane cabins, etc.

### 46.3.3 Application of Ergonomics to the Improvement of the Quality of Working Life

The issue of the quality of working life will become the major concern of the people of India in the coming few decades. A more participative and humanizing approach to labor relations will be helpful only in the presence of a cooperative attitude of top management who should foresee the benefits of such work redesign.

Ergonomists should undertake some studies relevant to improving the quality of working life. The challenge of change is faced by all big and small undertakings and enterprises.

Much of the information collected and knowledge gained should encourage enterprises who embark on innovations to give due prominence to the human aspects of the proposed changes so as to ensure both the advances that are sought and continued improvement in the quality of working life of the people involved.

#### *46.3.3.1 Consumer Ergonomics*

In most of the industries manufacturing products and consumer goods, much stress has been laid on the economic and aesthetic aspects, with very little or no attention paid to the functional and ergonomic aspects. A change in this attitude is desirable.

#### *46.3.3.2 Participatory Ergonomics*

In industry, there are many problems that could easily be solved by collaborative, cooperative, and participatory efforts among ergonomists, factory managers, other officials, supervisors, foremen, and workers. Typical examples are the various low-cost modifications and improvements of existing work conditions, methods, and tools suggested by various experts in IDCs such as India, Sri Lanka, Indonesia, Singapore, Thailand, Malaysia, and the Philippines, for a project supported by the International Labour Organization [17].

Studies on participation of workers in the management of problems concerning work methods, machines, tools, and the work environment including the effects of noise [34,35] have shown that, as in Japan, job satisfaction and high morale are possible only in small working units. This is encouraging a growing number of industrial companies to redesign their larger organizations to form a collection of interlocking small groups.

Considerable attention has been paid in recent years, by those interested, to change the attitude to work and restructure with innovations. Ergonomics is being applied to eliminate boring repetitive jobs [4], to create more interesting work, and thus to avoid physical and mental fatigue.

#### *46.3.3.3 Ergonomics Application in Information Technology*

Though the development of modern management with radical improvements in information technology and instant communication has provided effective and dynamic corporate functioning and growth, yet it has also added to stresses and strains due to the operation of VDUs and other components of office automation. The stress includes all manners of pressure—physical, physiological, sociological, and/or a combination of all these and other relevant factors. It is well accepted that only a relaxed and healthy mind in a healthy body can cope with demanding responsibilities. With explosive use of information technology in IDCs like India, it is essential that ergonomically bad design aspects are excluded during the technology transfer from industrially developed countries.

No system can claim to be the best; no organization possesses the one best form. There is a continuing interaction between all parts of the system. All parts of the system must therefore be developed together. The origin of a new philosophy of enterprise is to be found in the dramatic changes in India; these changes are evident in human relations in all aspects of living.

Employers and/or governments must realize that though workers may lack scientific training, knowledge, or familiarity with technical jargon, it is they who experience the hazards at first hand. They must also recognize that it is often the pressure to produce more, and not recklessness or carelessness, that causes accidents and exposes workers to ill health. Impractical goals and unrelenting drives force the workers to work faster than the safe limits and thus experience accidents.

Often trade unions too do not bother very much about shortcomings in enforcement of health and safety regulations. They are busy negotiating for higher wages and other monetary benefits. The trade union leaders should be put through special orientation and educational programs to make them aware of the importance and intricacies of safety and health issues. The message of the safety movement will have to reach every factory, farm, and even the workers on roadways and railways. Of late, there has been an increase in the number of accidents in the railway and road transport industries. Most of them could probably be avoided with better preventive action. Even work on farms, which used to be relatively safe from accidents, is becoming more and more hazardous because of the use of poisonous chemicals and pesticides and unsafe machines.

#### *46.3.3.4 Ergonomics Application for Low-Cost Improvements*

In the industrial sector, Sen [17] has suggested low-cost improvements in the work process, work organization, and working conditions of various industries such as textile, soap, food processing, and engineering.

Today many industrial managers believe that in order to achieve optimum efficiency in a plant, it should be kept rather smaller than what has generally been thought to be the optimum size.

#### *46.3.3.5 Ergonomics Application in Transport Systems*

*Railways transport*: In railway transport, Sen and Ganguli [36,37] considered improvements in the design of both diesel and electric railway locomotive drivers' cabins—control display aspects, the positions of primary and secondary controls and displays—the seating arrangements of the drivers and the assistants, and provision of their routine requirements. Sen [2,29] also stressed the need for ergonomic modifications of tools for railway track maintenance, existing ticket counters for punching tickets, the ergonomic design of signaling and signal control rooms, and arrangements of non-air-conditioned and air-conditioned passenger coaches.

*Motor vehicle transport system*: Sen and Nag [38] considered ergonomic improvements for double-decker public buses that were blind copies of London buses and consequently unsuitable for a tropical climate, having minimum facilities for ventilation and very high thermal radiation due to the metal body being heated by solar radiation. Sen and Nag [39] suggested several improvements such as higher ventilation, greater ease of passenger flow inside and out of the bus, better seating arrangements for the passengers as well as the driver, improved control and display arrangements for the driver, and greater access of the driver and the passengers into the bus.

*Public transport*: Similarly, application of ergonomics is needed for better designs of cars, minibuses, trucks, and nonmotorized transport for rural areas such as cycle rickshaws, bullock carts, handcarts, cycle vans, and bicycles. Future research is also required on the design of roads and road signs to reduce the high rate of traffic accidents in India.

*Nonmotorized transport systems*: Particularly in rural areas and in some urban and suburban areas in India, most passengers, goods, patients, etc., are transported in three-wheel cycle rickshaws manually operated by pullers [40]. Sen et al. [40] carried out studies on improving the design of hand-pulled and cycle rickshaws.

Cycle rickshaws form one of the most widely used transport systems in both urban and rural areas throughout the country for short journeys. Cycle-rickshaw pulling is an important occupation for millions of people. The main advantages of using cycle rickshaws are as follows:

1. It causes almost no environmental pollution.
2. It is the only transport available in some instances and at some places.
3. It is not very costly.
4. It provides self-employment to a large number of people.

In the years to come, research into improvements in the design of cycle rickshaws and the extent of such improvements should be ergonomically evaluated for implementing critical improvements from the anthropometric, physiological, and other relevant points of view. Design improvements in the line of new regenerative brakes, energy-storing devices, noncircular or elliptical or multisprocket drives, directional and positional stability, reduced cycle-rickshaw weight, lower center of gravity, and similar aspects constitute a few of the areas of possible improvement.

## 46.4 INTERNATIONAL SYMPOSIA AND MEETINGS ON ERGONOMICS

In the international symposia on work physiology and ergonomics and on applied physiology and ergonomics held in Calcutta in 1974 and 1983, respectively, various problems faced by the developing countries were discussed [41]. A review of papers from a number of IDCs confirmed that the main ergonomics effects are directed toward the analysis of work problems and the implementation of cost-efficient workplace improvement.

## 46.5 ERGONOMICS EDUCATION AND TRAINING PROGRAM

In IDCs, ergonomics education at the school, college, and university levels is as important as bringing awareness to the individual and the public by using mass media such as radio, television, newspapers, and magazines. It is very important to make the contents of each educational program suitable for those for which it is intended. Ergonomic ways of using audiovisual aids, suitable examples through practical demonstrations, and the organization of popular exhibitions would go a long way toward bringing the principles of ergonomics to the persons concerned and elucidating the scope and benefits of its applications in day-to-day activities in all spheres of life.

Invariably, the workers are the worst sufferers in all types of accidents. There has to be health and safety education for industrial workers as well as for supervisors and managers, both separately and jointly, particularly in accident-prone industries, such as the jute [42], mining, chemicals, fertilizer, engineering, and construction industries. There is as yet no such sustained educational program.

Social attitudes and practices cannot be forced to change through strict legislation. What is needed is a genuine change of heart and a scientific approach through awareness and training. These take both time and a lot of effort and cannot be achieved overnight.

## 46.6 ERGONOMIC ORGANIZATIONS IN INDIA AND THEIR ROLES

The pioneering research work on various aspects of work physiology and ergonomics such as anthropometry and energy costs of different activities including basal metabolic rate (BMR) of industrial workers was conducted in the Department of Physiology, Presidency College, Calcutta, beginning in 1953. Similarly, the Section of Physiological and Industrial Hygiene in the All India Institute of Hygiene and Public Health and the Department of Physiology, University Colleges of Science, Technology, and Agriculture, University of Calcutta started research on work physiology in the 1960s. Various institutions including the Defence Institutes of Physiology and Allied Sciences, Delhi; Industrial Physiology Division of Central Labour Institute, IIT and NITIE, Bombay; and the National Institute of Occupational Health, and NID, Ahmedabad, under the Government of India and other institutions supported by the Indian Council of Medical Research (ICMR), Indian Council of Social Research (ICSR), Council of Scientific and Industrial Research (CSIR), Department of Science and Technology (DST), and others carried out projects along similar lines. In 1970–1971, the University of Calcutta started the first postgraduate course in India on ergonomics and work physiology with limited resources. It is still the only university where this postgraduate specialty is taught, though several organizations run short-term orientation and refresher courses on work physiology and ergonomics.

More centers of research and training should be established in the rural and urban sectors to apply the multidisciplinary approach of ergonomics to solve various problems of both the organized and unorganized sectors by optimizing the human–machine–environment interactions to utilize human resources most efficiently without adverse effects.

Scientific and professional societies like the Indian Society of Ergonomics (ISE) and academic institutions should thus play a very important role in exploring science and technology applications to generate awareness of multidisciplinary ergonomics, occupational health, safety, and the environment.

As the interactions and the collaborative activities among the various institutions, industries, and agricultural institutes are very meager, steps must be taken to increase these interactions to have mutual benefits in solving pressing problems.

Ergonomics should be introduced in a very simple form even in the school curricula. Students are capable of enthusiastically applying ergonomics in project work to produce successful designs for various purposes.

An ergonomics survey of India similar in scope and function to the anthropological, zoological, and botanical surveys of India should be constituted to undertake comprehensive and large-scale surveys at the national level for the establishment of ergonomic norms of the Indian people. These should consider the different ethnic groups and habitual activities, static and dynamic body dimensions, abilities, performance, and environmental standards, for various working conditions.

## 46.7 ERGONOMIC DESIGN OF HOSPITALS AND HEALTH CARE UNITS

The ergonomic design of hospitals and health care units for a large number of patients, in both rural and urban areas, is one of the most important areas that need the attention of ergonomists to improve the efficiency and productivity of health care personnel and the quality of patient care. It is similarly important to consider ergonomic designs for children and elderly people and for the physically and mentally handicapped and functionally impaired population of developing countries such as India. This includes design of buildings and rooms and of products, such as crockery and cutlery, and crutches and other mobility aids [43].

It is apparent that in the years to come the importance of ergonomics will increase in all spheres of life in the IDCs, depending on present approaches in the application, training, and research needs of ergonomics.

Maintaining a balance between a few very expensive, high technology-oriented enterprises and a large number of low-cost small enterprises—all intending to bring benefits to the people—would be the key point in achieving improvement in the quality of working life in IDCs. The government of India in collaboration with the state governments should assume a supervisory and regulatory role in such matters.

## REFERENCES

1. R. N. Sen, Application of ergonomics to industrially developing countries, *Ergonomics 27*:1021–1032 (1984).
2. R. N. Sen, Ergonomics—Science and technology of man at work—Its role in our national development, *Proceedings of Indian Science Congress, 66th Session*, Part II, Hyderabad, India, 1979, pp. 53–77.
3. K. Kogi and R. N. Sen, Third world ergonomics, *Int. Rev. Ergon. 1*:77–118 (1987).
4. R. N. Sen and P. K. Nag, Optimal work load for Indians performing different repetitive heavy manual work, *Ind. J. Physiol. Allied Sci. 33*:18–25 (1979).
5. R. N. Sen and D. Chakraborti, An ergonomic study on sickle designs for a reaping task in Indian agriculture, *Proceedings of Ergonomics Society Conference, UK*, published as *Contemporary Ergonomics*, E. D. Megaw, Ed., Taylor & Francis, London, U.K., 1989, pp. 313–317.

6. R. N. Sen and D. Chakraborti, A new ergonomic design of a "desi" plough, *Ind. J. Physiol. Allied Sci. 38*:91–105 (1984).
7. A. De and R. N. Sen, Ergonomic evaluation of ploughing process of paddy cultivation in India, *J. Hum. Ergol. 15*:103–112 (1986).
8. R. N. Sen and S. N. Bhattacharya, Evaluation of an ergonomically designed shovel, *J. Physiol. Allied Sci. 38*:150–154 (1984).
9. R. N. Sen and S. Bhattacharya, Development of an ergonomic design of a shovel from the viewpoint of increasing productivity in manual material handling in India, Project Report No. 23, Ergonomics Lab., Dept. of Physiology, Univ. Calcutta, Kolkata, India, 1976, pp. 1–14.
10. R. N. Sen, C. K. Pradhan, S. Basu, and A. K. Ganguli, An ergonomic evaluation of design of spades ("kodal"), *Proceedings of Indian Science Congress, 66th Session,* Part III, Hyderabad, India, 1979, pp. 6–7.
11. U. R. N. Sen and S. Sahu, An ergonomic design of a multi-purpose shovel-cum-"kodal" for manual material handling, *Adv. Ind. Ergon. Safety 6:* 561–566 (1994).
12. R. N. Sen and D. Chakraborti, A new ergonomic "float-seat" for improvement of paddy cultivation in India, in *Ergonomics in Developing Countries*, OSH Ser. No. 58, ILO, Geneva, Switzerland, 1987, pp. 419–427.
13. R. N. Sen and A. De, Ergonomics study of sheafing in paddy cultivation in India, *Ind. J. Physiol. Allied Sci. 38*:85–94 (1984).
14. A. De and R. N. Sen, A work measurement method for application in Indian agriculture, *Int. J. Ind. Ergon. 10*:285–292 (1992).
15. R. N. Sen, A. K. Ganguli, G. G. Ray, A. De, and D. Chakraborti, Tea leaf plucking—Workloads and environmental studies, *Ergonomics 26*:887–893 (1983).
16. R. N. Sen, Case studies to improve working conditions through technological choices in small and medium-sized (food processing, textiles and metal working) enterprises in West Bengal, in *Improving Working Conditions in Small Enterprises in Developing Asia*, K. Kogi, Ed., International Labour Office, Geneva, Switzerland, 1985, p. 158.
17. R. N. Sen, Examples of low-cost improvements in working conditions and environment, work methods, occupational health and safety at the enterprise level in West Bengal, in *Low-Cost Ways of Improving Working Conditions—100 Examples from Asia*, K. Kogi, W. O. Phoon, and J. E. Thurman, Eds., International Labour Office, Geneva, Switzerland, 1988, p. 179.
18. R. N. Sen and D. Pal, A new ergonomic method for postural analysis, in *Biomechanics*, K. B. Sahay and R. K. Saxena, Eds., Wiley, New Delhi, India, 1989, pp. 226–270.
19. R. N. Sen and D. Ghoshthakur, Ergonomic study on work-rest cycle, physiological responses and production rate of hand loom weavers, *Ind. J. Physiol. Allied Sci. 38*:47–53 (1984).
20. R. N. Sen and A. Kar, An ergonomic study on bamboo handicraft workers, *Ind. J. Physiol. Allied Sci. 38*:69–77 (1984).
21. R. N. Sen, A. K. Ganguli, G. G. Roy, A. De, and D. Chakraborti, Ergonomics studies of tea-leaf plucking operations—Criteria for selection and categorization, *Appl. Ergon. 12*:83–85 (1981).
22. R. N. Sen, A. K. Ganguli, G. G. Roy, A. De, and D. Chakraborti, A preliminary ergonomic study on tea plucking operations: Two leaves and a bud, *TRA J. 27*:74–80 (1980).
23. R. N. Sen and S. Gangopadhyay, Ergonomics study of accidents during manual material handling in factories of West Bengal, *Proceedings of the 11th Asian Conference on Occupational Health*, Bombay, India, 1989, p. 95.
24. R. N. Sen, Certain ergonomic principles in the design of factories in hot climates, *Proceedings of the International Symposium on Occupational Safety, Health and Working Conditions*, International Labour Office, Geneva, Switzerland, 1982, pp. 123–147.
25. R. N. Sen and N. C. Dutta, An ergonomic design of low-cost protective work clothing for furnace workers, in *Ergonomics in Developing Countries*, Occupational Safety and Health Series No. 58, International Labour Office, Geneva, Switzerland, 1987, pp. 222–223.

26. R. N. Sen, Personal protective devices for workers working in radiant and hot environments in India, *Proceedings of the 2nd International Symposium on Clothing Comfort Studies, in Mount Fuji*, Japan Research Association for Textile End-Uses, Osaka, Japan, 1991, pp. 123–142.
27. R. N. Sen, Special work-wear and protective clothing for Indian furnace workers, *Proceedings of the International Conference on Human Environment System*, Pergamon, Tokyo, Japan, 1991, pp. 411–415; *J. Therm. Biol. 18*:677–681, 1993.
28. R. N. Sen and S. Das, Ergonomics, occupational health and safety improvements for the manual metal arc welders, *Proceedings of the 1st International Symposium on Ergonomics, Occupational Health, Safety and Environment*, Bombay, India, 1991, Indian Society of Ergonomics (ISE), Calcutta, India, 1993, pp. 97–102.
29. R. N. Sen, Ergonomics design of some tools for manual maintenance of railway tracks in India, *Proceedings of the 10th International Ergonomics Congress*, Sydney, Australia, 1988: *Ergonomics in Developing Countries*, Vol. 1, A. S. Adams, R. R. Hall, B. J. MePhee, and M. S. Oxenburgh, Eds., Ergonomics Society of Australia Inc., Baulkham Hills, New South Wales, Australia, 1988, pp. 227–229.
30. R. N. Sen and P. Chatterjee, An ergonomic approach to the design of Indian bullock-cart used for carrying load in the city of Calcutta, Project Rep. No. 140, Ergonomics Laboratory, Dept. of Physiology, Univ. Calcutta, Kolkata, India, 1986, p. 18.
31. S. R. Dutta and S. Ganguli, An ergonomic approach to the design of Indian hand-pulled carts ("Thelas"), *Ind. J. Physiol. Allied Sci. 33*:102–106 (1979).
32. R. N. Sen and S. Gangopadhyay, *New Ergonomic Harness Design for Hand-Cart Pullers*, Project Rep. No. 120, Ergonomics Laboratory, Univ. Calcutta, Kolkata, India, 1985, p. 20.
33. R. N. Sen, A. K. Ganguli, and D. Chakraborti, Some anthropometric considerations related to Indian railways locomotive drivers, *Ind. J. Physiol. Allied Sci. 38*:106–113 (1984).
34. R. N. Sen, Some studies on the physiological effects of noise, *Proceedings of the Seminar on Higher Productivity through Building Insulation*, Bombay, India, 1967, pp. 53–69.
35. S. K. Chatterjee, R. N. Sen, and P. N. Sana, Determination of the level of noise originating from room air conditioners, *J. Heating Vent. J. Air Cond. 38*:429–433 (1965).
36. R. N. Sen and A. K. Ganguli, An ergonomics analysis of railway locomotive driver functions in India, *J. Human Ergol. 11*:187–202 (1982).
37. R. N. Sen and A. K. Ganguli, Preliminary investigation into the loco-man factor on the Indian railways, *Appl. Ergon. 13*:107–117 (1982).
38. R. N. Sen and P. K. Nag, Are Calcutta public buses ergonomically designed? *Ind. J. Physiol. Allied Sci. 27*:156–157(1957).
39. R. N. Sen and P. K. Nag, Design ergonomics of some public buses in Calcutta, *Proceedings of Indian Science Congress, 61st Session*, Part IV, Nagpur, India, 1974, pp. 126–127.
40. R. N. Sen, S. Basu, and A. Goswami, An ergonomic design of hand-pulled rickshaw, *Proceedings of Indian Science Congress, 66th Session*, Hyderabad, India, 1979, pp. 8–9.
41. R. N. Sen and S. Bannerjee, Eds., *Proceedings of the International Satellite Symposium on Work Physiology and Ergonomics*, Calcutta, India, November, 1–3, 1974 (during the 24th World Congress of Physiological Sciences held at New Delhi). *Ind. J. Physiol. Allied Sci. 33*:1–145 (1979).
42. R. N. Sen and D. Majumdar, Ergonomics in relation to occupational safety and health in jute industries in eastern India, *Proceedings of the 10th Asian Conference on Occupational Health*, Singapore, 1982, pp. 289–298.
43. R. N. Sen and A. Dutta Gupta, Ergonomics study on rehabilitation of a bilateral below-elbow amputee, in *Advances in Industrial Ergonomics and Safety*, Vol. 6, F. Aghazadeh, Ed., Taylor & Francis, London, U.K., 1994, pp. 437–440.

# Appendix A: Biomechanical Modeling of Carpet Installation Task

Amit Bhattacharya

CONTENTS



First, a 3-D model was developed to estimate the joint stresses associated with static postures commonly assumed by the carpet layers. Second, a 2-D model was developed for analyzing the dynamic carpet-stretching task. The results emphasize joint reaction forces and ground reaction forces.

## A.1 DESCRIPTION OF THE MODELS

### A.1.1 Static 3-D Biomechanical Model

A static link segment model was developed to aid in the estimation of joint reaction forces and muscle moments at six joints on each side of the body. This model has 12 element link segments. These links are foot, shank, thigh, upper arm, and lower arm of each side of the body, plus one for torso and one for head and neck. During the carpet-stretching activity, three points of the body touch the ground [1]. These anatomical sites for a right-legged person are leg foot, knee, and palm of the hand. In addition, the knee kicker tool used for the carpet-stretching task makes ground contact at its front and back ends (Figure A.1). For the determination of ground reaction forces at these five contact points, the following equations apply:

The static equations of equilibrium are

$$\text{Sum of z-direction forces} = 0 \tag{A.1}$$

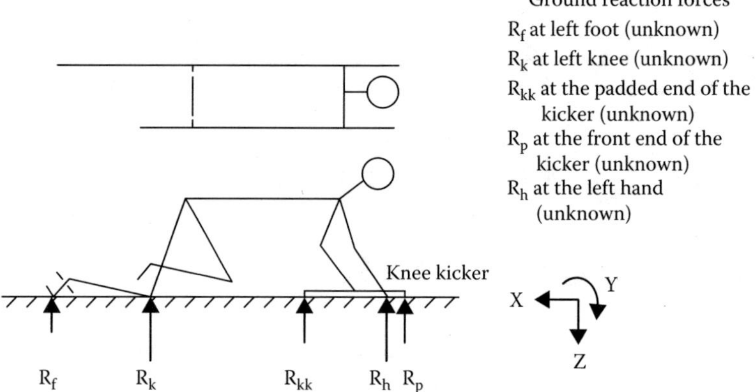


FIGURE A.1 Static free body diagram for carpet installation task.

$$\text{Sum of moments about x-axis} = 0 \tag{A.2}$$

$$\text{Sum of moments about y-axis} = 0 \tag{A.3}$$

$$R_f(x_f - x_k) = W_1(x_1 - x_k) + W_2(x_2 - x_k) \tag{A.4}$$

where

$R_f$ is the ground reaction force at the left foot (N)
$x_f$, $y_f$ are x and y coordinates of $R_f$ (m)
$W_1$ is the weight of the left foot segment (N)
$W_2$ is the weight of the left shank (N)
$x_1 \rightarrow x_{12}$ are x coordinates of segments 1–12 (m)
$y_1 \rightarrow y_{12}$ are y coordinates of segments 1–12 (m)
$x_k$, $y_k$ are x and y coordinates of left knee (m)

Equation A.4 is solved for $R_f$. The remaining equations are

$$\sum F_z = \sum_{I=1}^{12} W_I + (W_{kk} - R_p - R_f) - R_k - R_h - R_{kk} = 0 \tag{A.5}$$

$$\sum M_x = \sum_{I=1}^{12} W_I y_I + W_{kk} y_{kk} - R_p y_p - R_f y_f - R_k y_k - R_h y_h - R_{kk} y_{kk} = 0 \tag{A.6}$$

$$\sum M_y = \sum_{I=1}^{12} W_I x_I + W_{kk} x_{kk} - R_p x_p - R_f x_f - R_k x_k - R_h x_h - R_{kk} x_{kk} = 0 \tag{A.7}$$

Equations A.5 through A.7 are solved simultaneously for $R_k$, $R_h$, and $R_{kk}$

where

$R_p$ is the ground reaction force at the front end of the kicker (measured by the force platform) (N)

$R_k$ is the ground reaction force at the left knee (N) (unknown)

$R_k$, $R_{kk}$, are ground reaction forces at the left hand and at the padded end of the kicker (N) (unknown)

$W_{kk}$ is the weight of the kicker (N)

$W_1 \rightarrow W_{12}$ is the weight of segments 1–12 (N)

$x_p$, $y_p$ are x and y coordinates of the front end of the kicker (m)

$x_h$, $y_h$ are x and y coordinates of the left hand (m)

$x_{kk}$, $y_{kk}$ are x and y coordinates of the padded end of the kicker (m)

Once the ground reaction forces are calculated, the equations of static equilibrium are written for each of the 12 body segments, and the knee kicker and the head/neck segments. These are

$$\text{Sum of Z or vertical direction forces} = 0 \tag{A.8}$$

$$\text{Sum of moments about y-axis} = 0 \tag{A.9}$$

An example of static free body diagram for the right thigh is shown in Figure A.2A. Similar static free body diagrams for the other segments are drawn for the equations of static equilibrium. The static free body diagrams for the knee kicker and the head/neck segments are shown in Figure A.2B and C.

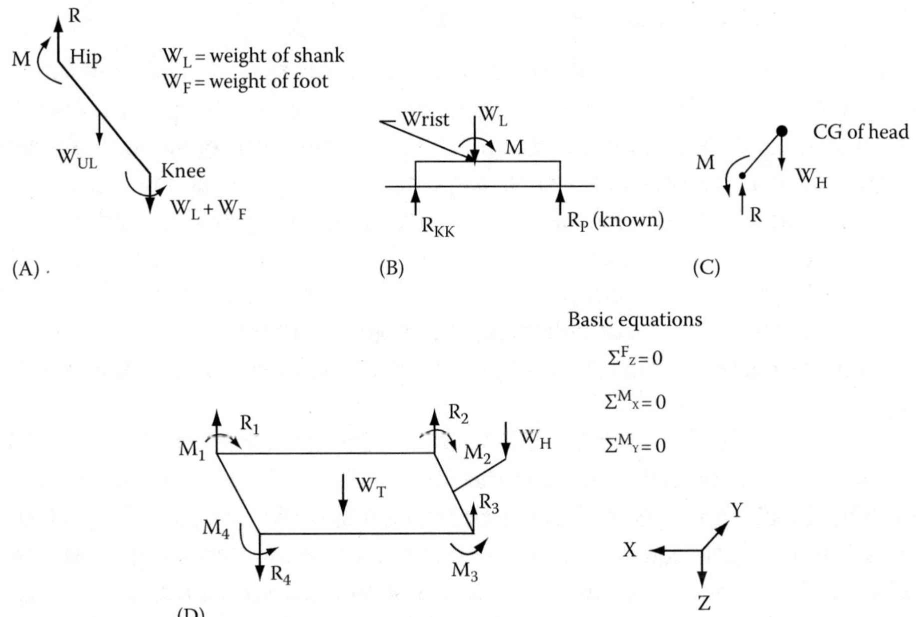


FIGURE A.2 Static free body diagram for the individual link segment. (A) Right thigh. (B) Knee kicker. (C) Head. (D) Trunk.

TABLE A.1 Weight of Body Segments[a]

| Segment | Percentage of Body Weight[b] | Weight (N) | Weight (lb) |
|---|---|---|---|
| Weight of foot = WA | 0.014 | 11.20 | 2.52 |
| Weight of calf = WK | 0.046 | 36.81 | 8.28 |
| Weight of thigh = WH | 0.097 | 77.61 | 17.46 |
| Weight of trunk = WS | 0.486 | 388.87 | 87.48 |
| Weight of upper arm = WE | 0.027 | 21.60 | 4.86 |
| Weight of lower arm = WW | 0.014 | 11.20 | 2.52 |
| Weight of hand = WN | 0.006 | 4.80 | 1.08 |

[a] For a subject of BW = 180 lb (800 N).
[b] Ref. [2].

TABLE A.2 Location of Segment Center of Gravity as a Function of Segment Length

| Segment | Segment Length from Proximal Joint (m) | Definition of Segment |
|---|---|---|
| Foot | 0.50 | Lateral malleolus to head metatarsal II |
| Calf | 0.433 | Femoral condyles to medial malleolus |
| Thigh | 0.433 | Greater trochanter to femoral condyles |
| Trunk | 0.50 | Greater trochanter to glenohumeral joint |
| Upper arm | 0.436 | Glenohumeral axis to elbow axis |
| Lower arm | 0.430 | Elbow axis to ulnar styloid |
| Hand | 0.506 | Ulnar styloid to knuckle II middle finger |

*Source:* Dempster, W., Space requirements of the seated operator, WADS-TR-55-150, Wright Patterson Air Force Base, OH, 1955.

The aforementioned generalized equations are used for the calculation of joint reaction force and the muscle moment at each body joint. A computer program written in IBM-PC BASIC for Equations A.4 through A.9 has built-in anthropometric tabular data (Tables A.1 and A.2) for segment weights as percent of body weight, and the location of segment center of gravity (CG) as percent of segment length (proximal and distal). This program first calculates the joint reaction forces and moments from the left ankle joint to the left hip joint. Then it calculates these values from the right ankle joint to right hip joint. After performing these calculations, the program determines the values of joint forces and moments from the right wrist joint to right shoulder joint and then for the left side of the body.

Finally, for the checking of force ($R_1$, $R_2$, $R_3$, $R_4$) and moment ($M_1$, $M_2$, $M_3$, $M_4$) balance at the four corners (right shoulder and hip joint and left shoulder and hip joint) of the trunk in association with the head/neck segment, a static free body diagram is used (Figure A.2D). For static equilibrium at the trunk and the head/neck segments, the program checks whether or not the sum of forces at all four joints equals the sum of weights of the trunk and the head/neck. Next, it checks whether all clockwise moments about the CG of the trunk equal the corresponding counterclockwise moments. For static equilibrium, these values are equal to each other.

*Model inputs*: These are body weight, $R_p$, $W_1 \rightarrow W_{12}$, $W_{kk}$, X, $\rightarrow X_{12}$, $Y_1 \rightarrow Y_{12}$, $Z_1$, $Z_{12}$, $X_k$, $Y_k$, $X_p$, $Y_p$, $X_h$, $Y_h$, $X_{kk}$, and $Y_{kk}$.

*Model outputs*: $R_f$; $R_k$; $R_h$; $R_{kk}$; X, Y, Z coordinates of segment CG; X, Y, Z coordinates of whole body CG; joint reaction forces (vertical) and moments ($M_y$) at all body joints; and force and moment check at the trunk segment.

### A.1.2 Two-Dimensional Dynamic Model for Carpet-Stretching Task

This phase describes a model to outline the dynamic aspects of the carpet-stretching task, which requires the use of a knee kicker [3]. This is a six-segment, or link, model of the kicking side of the body. The segments are foot, shank, thigh, torso, upper arm, and lower arm. The head and the hand are not included in this model at this stage of the model development. This model assumes that the major body motion is in the sagittal plane (xy) and in the longitudinal direction (along the x-axis) (Figure A.3), and the inertial forces in the

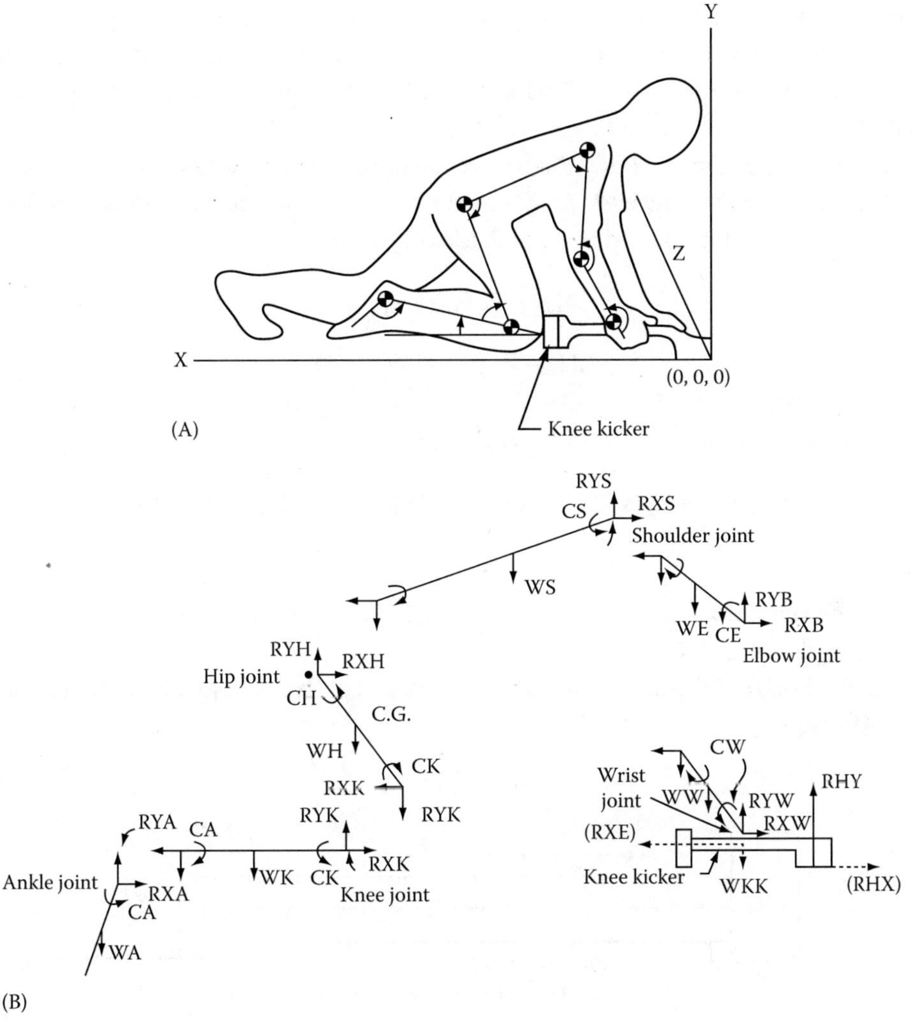


FIGURE A.3 (A) Schematic of coordinate system for this study. (B) Link segment model.

lateral direction (z-axis) are minimal. During the knee-kicking task, the kicking side of the body generally remains airborne until the impact phase, when the suprapatellar region of the knee makes contact with the padded end of the knee kicker. During the entire kicking cycle, the only point on the right side of the body that is in contact with the ground is the front end of the kicker. During dynamic knee-kicking experiments, we found that the padded end produces no (or negligible) ground reaction forces. Therefore, the ground reaction force at the front end of the kicker essentially provides an integrated effect of whole body weight shifts associated with knee kicking. This information is an important input for the determination of joint stresses and muscle moments for the right, or the kicking, side of the body. The lateral forces in the z-direction under the front end of the kicker are assumed insignificant in comparison to those in the vertical (y) and the horizontal (x) directions. It is realized that the body weight is also supported at the left or stationary knee and the left or stationary hand. However, it is assumed that the inertial effects along the lateral direction are minimal; therefore, ground reaction forces at these body contact points will affect mainly the joint stress calculations of the stationary, or left, side of the body.

As a first attempt in the modeling of dynamic aspects of knee kicking, the previous assumptions are reasonable. During the experiment phase, these assumptions were validated.

Figure A.4 is a static free body diagram showing the wrist joint–knee kicker segment. In the following, the hand weight is assumed negligible, and the wrist joint and the knee kicker are assumed stationary during each kicking cycle:

$$RH_x - R_x(w) + R_{xe} = 0 \tag{A.10}$$

$$RH_y - R_y(w) - W(R) = 0 \tag{A.11}$$

Taking moments around the CG of the kicker gives

$$-C(w) + R_y(w) \times 1 + RH_y \times X_1 + RH_x \times c + R_x(w) \times b = 0 \tag{A.12}$$

where

$RH_y$ is the vertical ground reaction force at the front end of the kicker (measured with a force platform) (N)

$RH_x$ is the horizontal friction force at the front end of the kicker (measured with a force platform) (N)

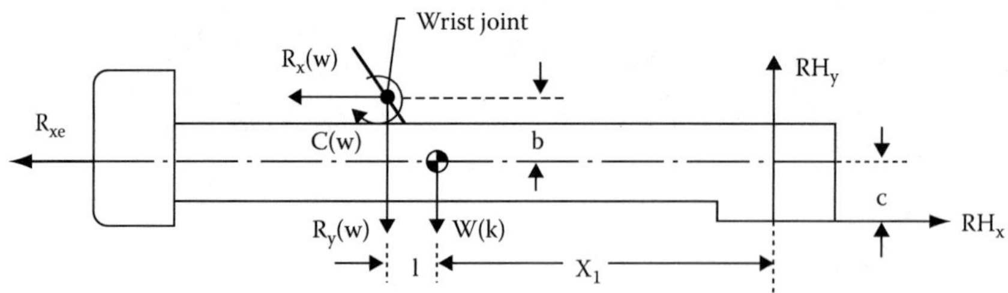


FIGURE A.4 Static free body diagram for the wrist joint–knee kicker segment.

$R_{xe}$ is the impact knee force applied at the padded end of the kicker (in Equation A.12, the direction of force applied is assumed to be toward the right) (N)
$R_x(w)$, $R_y(w)$ are horizontal and vertical joint reaction forces at the wrist joint (N)
W(k) is the weight of the kicker (N)
C(w) is the muscle moment at the wrist joint (N-m)
l is the horizontal distance between the CG of the kicker and the wrist joint (constant value, measured manually) (m)
$X_1$ is the horizontal distance between the CG of the kicker and the middle of the front end of the kicker (point 0) (constant value, measured manually) (m)
c is the vertical distance between the bottom of the front end of the kicker and the CG of the kicker (constant value, measured manually) (m)
b is the vertical distance between the wrist joint and the CG of the kicker (constant value, measured manually) (m)

For the calculations of joint reaction forces and muscle moments, the following generalized equations are used [4]:

$$R_x(j) - R_x(j-1) + R_{xe}(j) = M(j) \times A_x(j) \tag{A.13}$$

$$R_y(j) - R_y(j-1) - W(j) + R_{ye}(j) = M(j) \times A_y(j) \tag{A.14}$$

$$\begin{aligned} C(j) = {} & C(j-1) - R_x(j-1) \times \{L(j) - S(j)\} \times \sin\theta(j) + R_y(j) \\ & \times \{L(j) - S(j)\} \times \cos\theta(j) + R_y(j) \times S(j) \times \cos\theta(j) \\ & + R_{ye}(j) \times S(j) \times \cos\theta(j) - R_x(j) \times S(j) \times \sin\theta(j) \\ & - R_{xe}(j) \times S(j) \times \sin\theta(j) + IG(j) \times \ddot{\theta}(j) \end{aligned} \tag{A.15}$$

where
$R_x$, $R_y$ are joint reaction forces in the x and y directions, respectively, at (j)th and (j – 1)th joints (N)
$R_{xe}$, $R_{ye}$ are externally applied forces in the x and y directions, respectively, at (j)th and (j – 1)th joints (N)
$A_x$, $A_y$ is the linear acceleration in the x and y direction of the CG of the segment with respect to a fixed reference point (m/s²)
$\ddot{\theta}(j)$ is the angular acceleration of CG of the segment (rad/s²)
θ(j) is the angular location of segment with respect to right horizontal, degrees
W(j) and M(j) are weight (N) and mass (kg) of the segment
S(j) is the location of CG of the segment with respect to the joint under study (m)
L(j) is the segment length (m)
IG(j) is the mass moment of inertia of the segment about its CG (N m/s²)

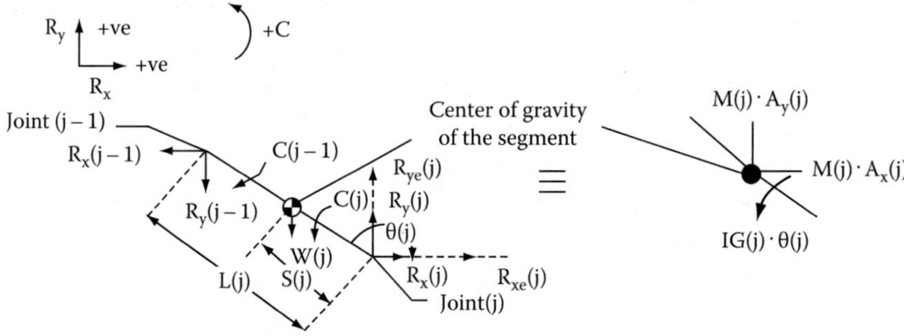


FIGURE A.5 Schematic of generalized segment for force and moment equilibrium.

j = 1 to n = number of joints

$$R_x(0) = 0, \quad R_y(0) = 0, \quad C(0) = 0$$

For the static case

$$A_x(j) = 0, \quad A_y(j) = 0 \quad \text{and} \quad \ddot{\theta}(j) = 0$$

Figure A.5 presents a generalized segment for force and moment equilibrium.

The linear acceleration of CG of the segment with respect to the jth joint is given by the following generalized equation:

$$AG_x(j) = S(j)\,\dot{\theta}^2(j)\cos\theta(j) + (m)\,S(j)\ddot{\theta}(j)\,\sin\theta(j) \tag{A.16}$$

$$AG_y(j) = S(j)\,\dot{\theta}^2(j)\,\sin\theta(j) - (m)\,S(j)\,\ddot{\theta}(j)\,\cos\theta(j) \tag{A.17}$$

The linear acceleration equation of CG of the generalized segment (in Figure A.6) with respect to a stationary reference point (usually a point of attachment) on the body is given by

$$A_x(j) = AG_x(j) + \text{linear x} - \text{acceleration of the jth joint with respect to the point of attachment} \tag{A.18}$$

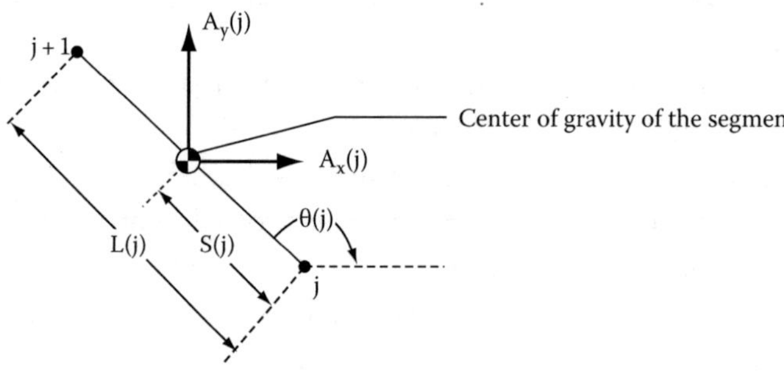


FIGURE A.6 Schematic of generalized segment for linear acceleration calculation.

$$A_y(j) = AG_y + y - \text{acceleration of the jth joint with respect to the point of attachment} \quad (A.19)$$

In the carpet installation modeling study, the wrist joint is considered to be a stationary reference point or point of attachment. The second terms in the preceding two equations are therefore equal to zero for the first segment

where

$j = 1$ to $n$ = number of joints
$m = +1$ for counterclockwise rotation
$m = -1$ for clockwise rotation
$\ddot{\theta}(j)$ is the angular acceleration of the segment CG; it can have either +ve or −ve values (rad/s$^2$)
$\dot{\theta}(j)$ is the angular velocity (rad/s)

The remaining parameters have been explained previously.

Equations A.10 through A.19 for the model were programmed in Microsoft FORTRAN for the IBM-PC laboratory computer. This program first starts calculating the values of $A_x(j)$ and $A_y(j)$ from the point of attachment, the wrist joint, and continues on to the foot segment. Next, the program calculates joint reaction forces at the ankle, knee, and hip joints. And, finally, it calculates these values at the wrist, elbow, and shoulder joints.

*Model input*: X and Y coordinates of body joints as a function of time, $RH_x$, $RH_y$, $R_{xe}$, $R_{ye}$, W(k), l, $X_1$, c, b, M(j), subject height, and body weight.

*Model output*: $R_x(j)$, $R_y(j)$, $A_x(j)$, $A_y(j)$, $\theta(j)$, $\ddot{\theta}(j)$, and C(j).

Developed by Amit Bhattacharya and his staff at the Biomechanics-Ergonomics Laboratory, University of Cincinnati Medical School, through a cooperative agreement with the National Institute for Occupational Safety and Health.

## REFERENCES


1. A. Bhattacharya, M. Mueller, and V. Putz-Anderson, Quantification of traumatogenic factors affecting the knee: A worksite study of carpet installation, *Applied Ergonomics 16*(4):243–250 (1985).
2. W. Dempster, Space requirements of the seated operator, WADS-TR-55-150, Wright-Patterson Air Force Base, OH, 1955.
3. A. Bhattacharya, Biomechanical analysis of carpet installation task, *ASME Conference*, Cincinnati, OH, 1987.
4. D. Winter, *Biomechanics of Human Movement*, John Wiley & Sons, New York, 1979.

# Appendix B: Ergonomics Checklists

James D. McGlothlin and Amit Bhattacharya

CONTENTS

## B.1 GENERAL WALKTHROUGH ERGONOMICS CHECKLISTS

### B.1.1 General Workstation Design Principles

The use of the following general workstation design principles will help achieve an optimum match between the work requirements and operator capabilities. This, in turn, will maximize the performance of the total system while maintaining human comfort, well-being, efficiency, and safety:

☐ 1. Make the workstation adjustable, enabling both large and small persons to fit comfortably and reach materials easily.

☐ 2. Locate all materials and tools in front of the worker to reduce twisting motions. Provide sufficient work space for the whole body to turn.

☐ 3. Avoid static loads, fixed work postures, and job requirements in which operators must frequently or for long periods
   a. Lean to the front or the sides.
   b. Hold a limb in a bent or extended position.
   c. Bend the torso or head forward/backward more than 15°.
   d. Support the body's weight with one leg.

☐ 4. Set the work height at 2 in. below the elbows for tasks requiring downward forces and heavy physical effort and 6 in. above the elbows for precision work.

☐ 5. Provide adjustable, properly designed chairs with adjustable seat height, adjustable up and down backrest, including a lumbar (lower-back) support, padding that will not compress more than an inch under the weight of a seated individual, and a chair that is stable to floor at all times (five-leg base).

☐ 6. Allow the workers, at their discretion, to alternate between sitting and standing. Provide floor mats/padded surfaces for prolonged standing.

☐ 7. Support the limbs. Provide elbow, wrist, arm, foot, and backrests as needed and feasible.

☐ 8. Use gravity to move materials.

☐ 9. Design the workstation so that arm movements are continuous and curved. Avoid straight-line, jerking arm movements.

☐ 10. Design so arm movements pivot about the elbow rather than around the shoulder to avoid stress on shoulder, neck, and upper back.

☐ 11. Design the primary work area so that arm movements or extension of more than 15 in. is minimized.

☐ 12. Provide dials and displays that are simple, logical, and easy to read, reach, and operate.

☐ 13. Eliminate or minimize the effect of undesirable environmental conditions such as excessive noise, heat, humidity, cold, and poor illumination.

The checklists in Section B.1 were developed by Dave Ridyard, CIH, CSP, CPE, Applied Ergonomics Technology, Jenkintown, Pennsylvania.

## B.1.2 Lifting and Lowering Tasks: Ergonomic Design

The following checklist should be used to eliminate the need to manually lift heavy or bulky materials, and reduce unnecessary bending, twisting, and reaching when lifting materials:

1. *Optimize material flow through the workplace*: ☐
   - ☐ Reduce manual handling of materials to a minimum.
   - ☐ Establish adequate receiving, storage, and shipping areas.
   - ☐ Maintain adequate aisle and access areas.
2. *Eliminate the need to lift or lower manually*:
   - ☐ Life tables and platforms
   - ☐ Life trucks
   - ☐ Cranes and hoists
   - ☐ Drum and barrel dumpers
   - ☐ Elevated pallets
   - ☐ Gravity dump systems
   - ☐ Elevating conveyors
   - ☐ Vacuum system
3. *Reduce the weight of the object*:
   - ☐ Reduce the weight and capacity of the container.
   - ☐ Reduce the load in the container.
   - ☐ Specify the quantity per container to suppliers.
4. *Reduce the hand distance from the body*:
   - ☐ Change the shape of the object or container.
   - ☐ Provide grips or handles.
5. *Convert lift/lower combined with a carry to a push or pull*:
   - ☐ Conveyors, hand trucks, carts
   - ☐ Ball caster tables

## B.1.3 Pushing and Pulling Tasks: Ergonomic Design

The following checklist should be used to eliminate manually pushing or pulling materials or to reduce the exertion hazard when materials are pushed or pulled:

1. *Eliminate the need to push or pull*:
   - ☐ Conveyors
   - ☐ Slides or chutes
   - ☐ Lift tables
   - ☐ Powered trucks
2. *Reduce force required to push or pull*:
   - ☐ Reduce size and/or weight of load.
   - ☐ Utilize four-wheel trucks or dollies.
   - ☐ Utilize non-powered conveyors.
   - ☐ Require that wheels and casters on hand trucks and dollies have the following:
     - Periodic lubrications of bearings
     - Adequate maintenance
     - Proper sizing (provide larger-diameter wheels and casters)
   - ☐ Maintain floors to eliminate holes and bumps.
   - ☐ Require surface treatment of floors to reduce friction.
3. *Reduce the distance of the push or pull*:
   - ☐ Relocate receiving, storage, production, or shipping areas.
   - ☐ Improve production to eliminate unnecessary material handling.

4. *Optimize technique of the push or pull*:
   - ☐ Replace pull with a push whenever possible.
   - ☐ Use ramps with slope less than 10%.
   - ☐ Provide variable-height handles so that both short and tall employees can maintain an elbow bend of 80°–100°.

## B.1.4 Repetitive Hand Tasks: Ergonomic Design

Workers may experience pain, discomfort, and disabling hand and wrist disorders such as carpal tunnel syndrome if the high number of repetitions is combined with abnormal wrist postures and excessive forces. The following checklist should be used as a guide for designing safe hand and wrist activities:

☐ 1. Reduce the number of repetitions per shift. Full or semiautomated systems should be used whenever possible.

☐ 2. Maintain neutral (handshake) wrist positions. Design jobs and select tools to reduce extreme flexion or deviation of the wrist. Avoid inward and outward rotation of the forearm when the wrist is bent to minimize stress to the elbow disorders (i.e., tennis elbow).

☐ 3. Reduce the force or pressure on the wrists and hands. Reduce the weight and size of objects that must be handled repetitively. Avoid tools that create pressure on the base of the palm which can obstruct blood flow and nerve function. Avoid repeated pounding with the base of the palm. Avoid repetitive, forceful pressing with the fingertips.

☐ 4. Design tasks so that a power grip rather than a finger pinch grip can be used to grasp materials. Note that a pinch grip is five times more stressful than a power grip.

☐ 5. Avoid reaching more than 15 in. in front of the body for materials:

   a. Avoid reaching above shoulder height, below waist level, or behind the body to minimize shoulder disorders.

   b. Avoid repetitive work that requires full arm extension (i.e., the elbow held straight and the arm extended).

☐ 6. Provide support devices where awkward body postures (elevated hands or elbows and extended arms) must be maintained. Use fixtures to relieve stressful hand/arm positions.

☐ 7. Select power tools and equipment with features designed to control or limit vibration transmissions to the hands, or alternatively design work methods to reduce time or need to hold vibrating tools.

☐ 8. Provide for protection of the hands if working in a cold or hot environment. Furnish a selection of glove sizes and sensitize users to problems of forceful over gripping when worn.

☐ 9. Select and use properly designed hand tools (e.g., grip size of tool handles should accommodate majority of workers).

## B.1.5 Selection of Ergonomic Hand Tools

Poorly designed hand tools that combine repetitive forceful grip exertions with bent wrist postures can cause carpal tunnel syndrome and other cumulative trauma disorders. The following checklist can be used as a guide for selecting hand tools:

☐ 1. Maintain straight wrists. Avoid bending or rotating the wrists. Remember, bend the tool, not the wrist. A variety of bent-handle tools are commercially available.

☐ 2. Avoid static muscle loading. Reduce both the weight and size of the tool. Do not raise or extend elbows when working with heavy tools. Provide counterbalanced support devices for heavier tools.

☐ 3. Avoid stress on soft tissues. Stresses result from poorly designed tools that exert pressure on the palms or fingers. Examples include short-handled pliers and tools with finger grooves.

☐ 4. Reduce grip force requirements. A compressible gripping surface rather than hard plastic is best.

☐ 5. Whenever possible, select tools that utilize a full-hand power grip rather than a precision finger grip.

☐ 6. Maintain optimal grip span. Optimal grip spans for pliers, scissors, or tongs, measured from the fingers to the base of the thumb, range from 6 to 9 cm. Recommended handle diameters for circular-handle tools such as screwdrivers are 3–5 cm when a power grip is required and 0.75–1.5 cm when a precision finger grip is needed.

☐ 7. Avoid sharp edges and pinch points. Select tools that will not cut or pinch the hands even when gloves are not worn.

☐ 8. Avoid repetitive trigger-finger actions. Select tools with large switches that can be operated with all four fingers. Also, the use of the thumb is preferred to using a single finger for trigger action. Proximity switches are the most desirable triggering mechanism.

☐ 9. Protect hands from excessive heat, cold, and vibration.

☐ 10. Wear gloves that fit properly to enhance strength and dexterity.

## B.1.6 Carrying Tasks: Ergonomic Design

1. Eliminate the need to carry by rearranging the workplace to eliminate unnecessary materials movement and using the following mechanical handling aids, when applicable:
   a. Conveyors (all kinds)
   b. Lift trucks and hand trucks
   c. Tables or slides between workstations
   d. Four-wheel carts or dollies
   e. Air or gravity press ejection systems
2. Reduce the weight that is carried by
   a. Reducing the weight of the object
   b. Reducing the weight of the container
   c. Reducing the load in the container
   d. Reducing the quantity per container to suppliers

3. Reduce the bulk of the materials that are carried by
   a. Reducing the size or shape of the object or container
   b. Providing handles or handgrips that allow materials to be held close to the body
   c. Assigning the job to two or more persons
4. Reduce the carrying distance by
   a. Moving receiving, storage, or shipping areas closer to production areas
   b. Using powered and non-powered conveyors
5. Convert carry to push or pull by
   a. Using non-powered conveyors
   b. Using hand trucks and push carts

### B.1.7 Computer Workstation: Ergonomic Design

The use of the following set of computer workstation design guidelines will maximize the performance of the computer operator while maintaining human comfort, well-being, efficiency, and safety. Note that *adjustability* of the workstation is the key:

- ☐ The *keyboard and mouse surface* should be height adjustable between 23 and 28 in. Both keyboard and mouse should be positioned at elbow height.
- ☐ *Workstation width* should be at least 30 in. (76 cm). *Workstation height* should adjust 23–30 in. (58.4–76.2 cm).
- ☐ *Depth of the workstation* should allow for screen, keyboard, and approximately 3 in. to serve as a wrist rest area.
- ☐ *Edges of the work surface* must be rounded and at least 1 in. (2.5 cm) thick to prevent stress on the arms and wrists.
- ☐ A *wrist rest* should be provided to enhance workstation adjustability. Π *Knee room design considerations*:
  - Knee room height should be a minimum of 26.2 in. (66.5 cm) for nonadjustable surface and 24 in. (70 cm) for adjustable surface.
  - Knee room width should be a minimum of 27 in. (76.2 cm).
  - Knee room depth should be a minimum of 23.5 in. (59.7 cm) knee level and 31.5 in. (80 cm) toe level.
- ☐ *Chair design considerations*:
  - Chair base should be five-point, on casters. Chair should swivel.
  - Chair should have an adjustable seat height adjustable to 16–20.5 in. (40–52.1 cm).
  - Seat size between 15 and 17 in. (38.1–43.2 cm) deep and 17.7 in. (45 cm) and 20 in. (51 cm) wide with a "waterfall" front edge.
  - Seat slope should be adjustable 0°–24° backward slope.
  - Seat pan and backrest should be upholstered. When seated, the seat pan and backrest should not compress more than 3/4 in.

- Backrest size should be 20 in. or higher (50.8 cm) and 13 in. wide (33 cm), backrest height should be adjustable between 3 and 6 in. (8–15 cm), and backrest tilt should be adjustable 30°.
- Removable, height-adjustable armrests should be incorporated.
- Backrest, if adjustable, should have a locking mechanism.

☐ *Screens* should be located directly in front of the operator at a distance of 16–22 in. (40.6–55.8 cm) for focusing at close range. The topmost line of the display should not be higher than the user's eyes. Screens should be located at right angle to windows.

☐ *Task lighting*, rather than overhead lighting, should be provided to minimize glare. Work surfaces and walls should be furnished with nonglare (matte) finish.

## B.1.8 Ergonomic Guidelines for Computer Operators

The simple adjustments outlined in the following text may increase the comfort of your computer workstation. Consider the following to prevent musculoskeletal and visual fatigue:

1. *Adjust the height of your work surface and the height of your chair* so that your keyboard is at elbow height and your feet are flat on the floor. If the work surface is not adjustable, a footrest should be used when the feet do not rest flat on the floor.
2. *Adjust the backrest of your chair* so that it provides support to your lower back. Do not sit on the edges of the chair: rest your back against the backrest.
3. *Position the screen* directly in front of you. The distance between your eyes and the screen should be approximately 16–22 in.
4. *Adjust the height of the screen* so that your eyes are level with the top of the screen.
5. *Tilt the screen to minimize glare.* Tilting the screen will help reduce glare caused by bright overhead lights.
6. *Draw drapes or shades and utilize task lighting* rather than bright overhead lighting when working at the computer to reduce glare.
7. *Use a document holder.* Documents placed flat on the desk will cause you to lean forward and flex your neck, leading to fatigue and discomfort. The document and screen should be located at approximately the same distance to eliminate constant eye refocusing at varying viewing distances.
8. *Keep the area under your desk clear* for adequate leg and knee room.
9. *When keying, rest your wrists and/or forearms, and keep the upper arms nearly vertical* to prevent fatigue. Use a wrist rest, if necessary, to maintain your wrists, hands, and arms in a straight horizontal line.
10. *Take frequent micro-breaks and stretch periodically* to reduce the soreness and stiffness related to fixed, static work postures.

### B.1.9 Ergonomic Guidelines for Laboratory Employees

1. Position equipment and tools directly in front of your body to reduce twisting motions and reach distances.
2. Eliminate excessive reaches over 16 in.
3. Set the work height (hand height) at 2 in. below the elbows to minimize static muscle fatigue of the shoulders, neck, and back.
4. Maintain good seated posture. Adjust the backrest so that it provides firm support to your lower back. Do not sit on the edge of the chair.
5. Minimize static loads and fixed work postures. Avoid:
   a. Leaning to the front or sides
   b. Holding an extremity in a bent or extended position
   c. Tilting your head forward more than 15°
   d. Bending your body forward or backward more than 15°
   e. Supporting your weight with one leg
6. Support the limbs. Use elbow, wrist, arm, and backrests when needed.
7. Reduce the number of repetitive hand, wrist, and finger motions. The use of the thumb is preferred to the use of a single finger for trigger action. Motorized pipettes eliminate repetitive, forceful trigger-finger motions.
8. Maintain neutral (handshake) wrist postures. Design experiments so that the wrist does not need to be flexed forward, extended backward, or bent from side to side.
9. Reduce grip force requirements. Select pipettes with a compressible rather than hard plastic-gripping surface. Use tools that utilize a full-hand power grip rather than a more forceful precision finger grip.
10. Avoid pounding with the pipette to pick up tips.
11. Select tools which minimize stress on soft tissues. Stress concentrations result from tool handles that exert pressure on the palms or fingers. Examples include short-handle pliers and tools with finger grooves that do not fit the specific employee's hand.

### B.1.10 Audit of Materials-Handling Risk Factors

The following items should be considered when evaluating new or existing materials-handling equipment, processes, work activities, or workstations:

1. *Does the task require any of the following activities?*
   - ☐ Lifting, lowering, or carrying more than 25 lb
   - ☐ Lifting, lowering, or carrying objects that are too bulky to easily grip and hold close to the body
   - ☐ Lifting, lowering, or carrying materials more than 50 times per shift
   - ☐ Lifting above shoulder height or below waist level
   - ☐ Lifting in cramped areas resulting in bending, reaching, and twisting
   - ☐ Pushing/pulling carts, etc., that require large forces to get moving
   - ☐ Maintaining a fixed or awkward work posture (e.g., overhead work, twisted or bent back, kneeling, stooping, or squatting)

2. *Do any of the following unsafe practices or conditions exist?*
   - ☐ Employees not following procedures (taking shortcuts, etc.)
   - ☐ Employees not using appropriate materials-handling equipment
   - ☐ Materials-handling equipment inadequate or damaged
   - ☐ Materials improperly stacked, loaded, or banded on fork trucks
3. *Do any of the following unsafe work area conditions exist?*
   - ☐ Unsafe (cracked or broken pavement, etc.) walking surfaces
   - ☐ Poor housekeeping (wet, oily floors; debris; clutter)
   - ☐ Poor layout of work area—crowding, congestion, excessive traffic
   - ☐ Excessive noise, heat, humidity, cold, or poor illumination

**Area audited:** ____________________

**Audited by:** ____________________

## B.1.11 Audit of Repetitive Hand Tasks

The following items should be evaluated when evaluating new or existing work activities requiring highly repetitive motions with the hands or wrists (i.e., manual packaging or inspection activities, and the use of hand tools):

1. *Does the task require any of the following activities?*
   - ☐ Performing the same motion every few seconds with the hands or wrists.
   - ☐ Repetitively shaking cartons, bottles, or other materials (repetitive bending of the wrists).
   - ☐ Bending the wrists when working. The wrists should be maintained in a neutral (handshake) position. The job task should not require the wrists to be flexed forward, extended backward, or bent from side to side.
   - ☐ Working for extended periods of time with awkward body postures (elevated hands and elbows, extended arms, reaching behind body).
   - ☐ Exerting high grip forces with the hands.
   - ☐ Using a pinch grip rather than a power (curled-finger) grip.
   - ☐ Using the hand as a "hammer."
   - ☐ Repetitively reaching more than 16 in. in front of the body.
2. *Do any of the following unsafe hand tool practices or conditions exist?*
   - ☐ Using vibrating or impact tools or equipment
   - ☐ Using tools that require bending or rotating the wrists
   - ☐ Raising or extending the elbows when working with heavy tools
   - ☐ Using tools requiring repetitive trigger-finger actions, excessive grip forces, or excessive forceful exertions

**Area audited:** ____________________

**Audited by:** ____________________

## B.2 WALKTHROUGH ERGONOMICS CHECKLIST FOR CARPENTRY TASKS

Site: ______________________ ID code: ____________

Date: ______________________

Contact: ______________________ Phone: ____________

Ergonomist: ______________________

Job specialty observed: ______________________ Job code: ____________

Task description: ______________________

Total time of observation: ______________________ (h) Start: ____________

Stop: ____________

### B.2.1 Worker Information

Subject name: ______________________ SS# ____________

Address: ______________________

Phone: ______________________

Age: ________ years Sex: M/F Ethnic code: ________

Height: ________ ft Weight: ________ lb.

Union local: ________ Years employed: ________

Tool belt: Yes/No Tools used: ________

### B.2.2 Work Experience (Yes = 2, No = 1) Score

1. High level of background noise? Yes No : ____________
2. Frequent, loud impact noise? Yes No : ____________
3. Exposure to vibration? Yes No : ____________

Body parts exposed ______________________ Rank 4 = : ____________

4. Weather conditions: Temperature : ________ °F

Cloudy/sunny wet/dry Working surface: : ____________

Comments /observations: ______________________

______________________

### B.2.3 Repetitive Motion

The following frequency periods and ratings are based on the analysis made during the observation period on the specific day of the ergonomic walkthrough:

| **Torso** | **Repetition/Minute** | | **Rating** | | **Score** |
|---|---|---|---|---|---|
| 1. Bending at waist (not lifting) | ______ | × | ______ | = | ______ |
| Rating: 1. For less than nine repetitions/minute | | | | + | |
| 2. For nine or more repetitions/minute | | | | | |
| 2. Stooping (bending with legs straight) | ______ | × | ______ | = | ______ |
| Rating: 1. For less than six repetitions/minute | | | | | |
| 2. For six or more repetitions/minute | | | | + | |
| 3. Turning/twisting of upper torso | ______ | × | ______ | = | ______ |
| Rating: 1. For less than nine repetitions/minute | | | | | |
| 2. For nine and more repetitions/minute | | | Rank 1 | = | ______ |

| **Lower Extremities** | **Repetition/Minute** | | **Rating** | | **Score** |
|---|---|---|---|---|---|
| 1. Squatting | ______ | × | ______ | = | ______ |
| Rating: 1. For less than six repetitions/minute | | | | + | |
| 2. For six or more repetitions/minute | | | | | |
| 2: Kneeling | ______ | × | ______ | = | ______ |
| Rating: 1. For less than six repetitions/minute | | | | | |
| 2. For six or more repetitions/minute | | | Rank 2 | = | ______ |

| **Upper Extremities/Wrist** | **Rating** | | **Score** |
|---|---|---|---|
| 1. Cycle time | | | |
| Rating: 1. For cycle time ≥30 s | | | |
| 2. For cycle time <30 s | ______ | = | ______ |
| 2. Deviation of wrist | | + | |
| Rating: 1. For deviation <30° | | | |
| 2. For deviation ≥30° | ______ | = | ______ |
| | Rank 3 | = | ______ |

## B.2.4 Postures

Neck–Shoulders

| | | |
|---|---|---|
| 1 | Free and relaxed<br>Time spent in this posture: ______ |  |
| 2 | In a natural posture, but limited by the work<br>Time spent in this posture: ______ |  |
| 3 | Tense due to the work<br>Time spent in this posture: ______ |  |
| 4 | Neck twisted or bent and/or the upper part of the arms at or above shoulder level<br>Time spent in this posture: ______ |  |
| 5 | Neck bent backward, great demand for strength in the arms<br>Time spent in this posture: ______ |  |

Back

| | | |
|---|---|---|
| 1 | In a natural posture and/or well supported in seated and standing position<br>Time spent in this posture: _____ |  |
| 2 | In a good posture, but limited by the work<br>Time spent in this posture: _____ |  |
| 3 | Bent and/or poorly supported<br>Time spent in this posture: _____ | 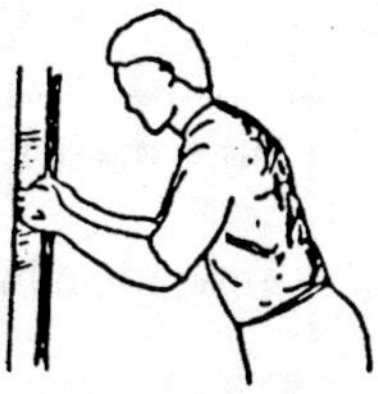 |
| 4 | Bent and twisted without support<br>Time spent in this posture: _____ |  |
| 5 | In a poor posture during heavy work<br>Time spent in this posture: _____ |  |

Elbow

| | | |
|---|---|---|
| 1 | Free in a posture of choice, demand for strength small<br>Time spent in this posture: ______ |  |
| 2 | Arms in a position required by the work, slightly tense at times<br>Time spent in this posture: ______ | 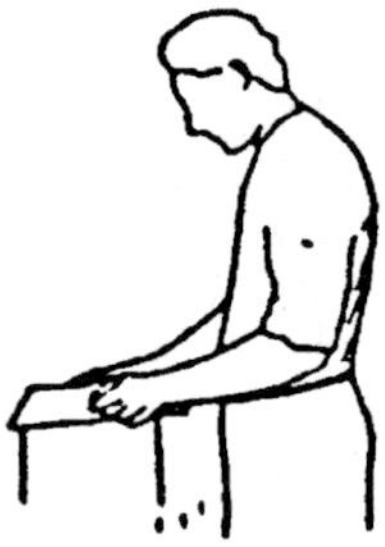 |
| 3 | Arms tense and/or the joints in an extreme posture<br>Time spent in this posture: ______ |  |
| 4 | Arms maintain static contraction with elbows elevated at an angle >30°<br>Time spent in this posture: ______ |  |
| 5 | Great demand for strength in the arms<br>Time spent in this posture: ______ |  |

Hips–Legs

| | | |
|---|---|---|
| 1 | In a free position that can be changed at will<br>Time spent in this posture: ______ |  |
| 2 | In a good posture, but limited by the work<br>Time spent in this posture: ______ |  |
| 3 | Poorly supported or unstable/soft footing; feet at two different levels<br>Time spent in this posture: ______ | 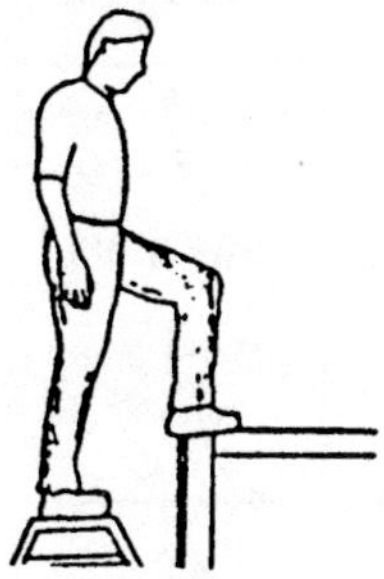 |
| 4 | Standing on one foot or in a kneeling or stooping position<br>Time spent in this posture: ______ |  |
| 5 | In a poor posture during heavy work<br>Time spent in this posture: ______ |  |

Wrist (Not Using Tools)

| | | |
|---|---|---|
| 1 | Wrist relaxed and straight, no bending, little movement, no stress on joint<br>Time spent in this posture: ______ | Neutral |
| 2 | Wrist bent side to side radial or ulnar deviation. Slight force on joint<br>Time spent in this posture: ______ | Radial<br>Ulnar |
| 3 | Wrist bent in up/down motion in flexion or extension. Slight force on joint<br>Time spent in this posture: ______ | Flexion<br>Extension |
| 4 | Wrist bent and used to grasp object (other than a tool) involving thumb as a support<br>Time spent in this posture: ______ | Pinch grip |
| 5 | Wrist in extreme/awkward posture requiring excessive force on joint<br>Time spent in this posture: ______ | |

Tool Design

| | | |
|---|---|---|
| 1 | Tool is lightweight and fits hand comfortably, wrist not deviated<br>Time spent in this posture: ______ | |
| 2 | Tool is powered, requires little exertion to use; little to no vibration<br>Time spent in this posture: ______ | |
| 3 | Tool handle pinches or puts pressure on palm; tool is hard to grasp and/or blows cool air currents on hand from tool motor<br>Time spent in this posture: ______ | |

| 4 | Tool is heavy and awkward to use and/or causes vibration/ impact<br>Time spent in this posture: ______ | |
|---|---|---|
| 5 | Tool design causes deviation of wrist more than 30° from neutral position and/or causes the use of pinch grip<br>Time spent in this posture: ______ | |

## B.2.5 Lifting Task

Physical Stress Job Analysis Sheet

Department______________________ Date______________

Job title ______________________ Analyst's name______________

| **Task Description** | **Object Weight** | | **Hand Location: Origin Destination** | | **Task Freq.** | **ASYM** | **Coupling** | **RWL** |
|---|---|---|---|---|---|---|---|---|
| | **Ave.** | **Max.** | **H V cm cm** | **H V cm cm** | | | | |
| | | | | | | | | |
| | | | | | | | | |
| | | | | | | | | |
| | | | | | | | | |
| | | | | | | | | |
| | | | | | | | | |
| | | | | | | | | |
| | | | | | | | | |
| | | | | | | | | |
| | | | | | | | | |
| | | | | | | | | |
| | | | | | | | | |
| | | | | | | | | |
| | | | | | | | | |
| | | | | | | | | |
| | | | | | | | | |
| | | | | | | | | |
| | | | | | | | | |

| | | | | | | | |
|---|---|---|---|---|---|---|---|
| | | | | | | | |
| | | | | | | | |
| | | | | | | | |
| | | | | | | | |

### B.2.6 Job/Task Requirement Data

Start time: ________________

Stop time: ________________

Resting heart rate: ________________________________________ beats/minute

Maximum heart rate: ________________________________________ beats/minute = (220 – age)

| **Time** | **Task** | **Heart Rate** | **Type of Force**[a] | **Measured Force** |
|---|---|---|---|---|
| | | | | |
| | | | | |
| | | | | |
| | | | | |
| | | | | |
| | | | | |
| | | | | |
| | | | | |
| | | | | |
| | | | | |
| | | | | |
| | | | | |
| | | | | |
| | | | | |
| | | | | |
| | | | | |
| | | | | |
| | | | | |
| | | | | |
| | | | | |

| | | | | |
|---|---|---|---|---|
| | | | | |
| | | | | |

*Note:* The times used in the earlier analysis should be rounded to the nearest second or minute, whichever the case may be.

[a] Impact, grip, push, pull, and lift.

## B.3 SYMPTOMS SURVEY: ERGONOMICS PROGRAM CHECKLIST

### B.3.1 Symptoms Survey Checklist

**Symptoms Survey: *Ergonomics Program***

Date_____/_____/_____

______ ______ ______ ______________________

Plant  Dept #  Job #  Job name

______ ______________ ______________ _____Years_____Months

Shift  Supervisor  Hours worked/week  Time on THIS job

*Other jobs you have done in the last year (for more than 2 weeks)*

______ ______ ______ ______________ _____Months_____Weeks

Plant  Dept #  Job #  Job name  Time on THIS job

______ ______ ______ ______________ _____Months_____Weeks

Plant  Dept #  Job #  Job name  Time on THIS job

(If more than two jobs, include those you worked on the most)

Have you had any pain or discomfort during the last year?

☐ Yes  ☐ No (If No, stop here)

If Yes, carefully shade in the area of the drawing which bothers you the most.

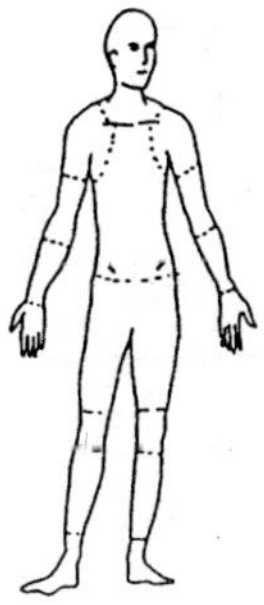

Front 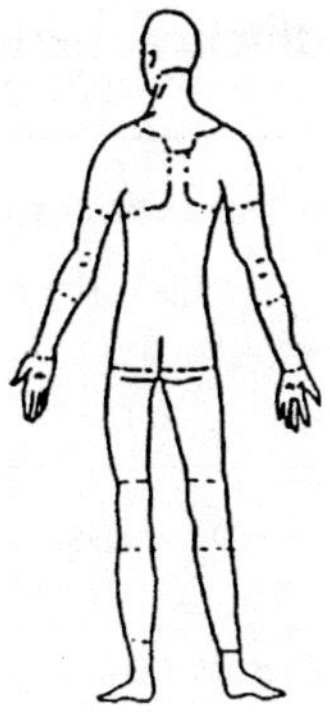  Back

Developed by Thomas R. Hales, of the National Institute for Occupational Safety and Health, and Patricia K. Bertsche, of the Occupational Safety and Health Administration. (Reprinted with permission from the *American Association of Occupational Health Nursing Journal 40*(3): 1992.)

## B.3.2 Physical Examination Recording Form for Health Care Providers

*(Complete a separate page for each area that bothers you)*

Check area: ☐ Neck ☐ Shoulder ☐ Elbow/Forearm

☐ Hand/Wrist ☐ Fingers ☐ Upper Back ☐ Low Back ☐ Thigh/Knee
☐ Low Leg ☐ Ankle/Foot

1. Please put a check by the word(s) that best describe your problem:
   - ☐ Aching
   - ☐ Burning
   - ☐ Numbness (asleep)
   - ☐ Tingling
   - ☐ Cramping
   - ☐ Pain
   - ☐ Weakness
   - ☐ Loss of color
   - ☐ Swelling
   - ☐ Other
   - ☐ Stiffness
2. When did you first notice the problem? ________ (Month) ________ (Year)
3. How long does each episode last? (Mark an X along the line)

   ______/______/______/______/______
   1 h 1 day 1 week 1 month 6 months

4. How many separate episodes have you had in the last year? ________
5. What do you think caused the problem? ______________________________

   ______________________________________________
6. Have you had this problem in the last 7 days? ☐ Yes ☐ No
7. How would you rate this problem? (Mark an X on the line)

   WOW

   ______________________________________________

   None Unbearable

   *When it was the worst*

   ______________________________________________

   None Unbearable

8. Have you had medical treatment for this problem? ☐ Yes ☐ No
   8a. If No, why not? ______________________________
   8b. If Yes, where did you receive treatment? ______________________________
      1. Company medical ☐ Times in past year _____
      2. Personal doctor ☐ Times in past year _____
      3. Other ☐ Times in past year _____
   8c. If Yes, did the treatment help? ☐ Yes ☐ No
9. How much time have you lost in the last year because of this problem? _____ days
10. How many days in the last year were you on restricted or light duty because of this problem? _____ days
11. Please comment on what you think would improve your symptoms.

______________________________

______________________________

______________________________

______________________________

Name: ______________________________ Current job: ______________

Examiner: ______________________________ Date:____/____/____

Discomfort scale: 1 = no discomfort, 2 = mild, 3 = moderate, 4 = severe, 5 = worst ever

*Neck*

Inspection: Inflammation (red, swollen, warm) ____Yes ____No

| Palpation: | *Right* | *Left* |
|---|---|---|
| Trapezius trigger point | ____ | ____ |
| Trapezius spasm | ____ | ____ |
| Maneuvers: | | |
| Resisted flexion | ____ | ____ |
| Resisted extension | ____ | ____ |
| Resisted rotation | ____ | ____ |

*Shoulder*

Inspection: Acromium inflammation ____Yes (R or L) ____No

| Maneuvers: | *Right* | *Left* |
|---|---|---|
| Passive abduction | ____ | ____ |
| Active abduction | ____ | ____ |
| Resisted abduction | ____ | ____ |
| Deltoid palpation | ____ | ____ |

*Elbow*

Inspection: Olecranon inflammation ——Yes (R or L) ——No

| Palpation: | *Right* | *Left* |
|---|---|---|
| Medial epicondyle | —— | —— |
| Lateral epicondyle | —— | —— |

*Forearm*

Inspection: Forearm inflammation ——Yes (R or L) ——No

| Maneuvers: | *Right* | *Left* |
|---|---|---|
| Passive wrist flexion | —— | —— |
| Passive wrist extension | —— | —— |
| Resisted wrist flexion | —— | —— |
| Resisted wrist extension | —— | —— |
| Resisted finger flexion | —— | —— |
| Resisted finger extension | —— | —— |
| Third digit resisted extension | —— | —— |

*Wrist*

Inspection: Inflammation ——Yes (R or L) ——No

Extensor ganglion cyst ——Yes (R or L) ——No

Flexor ganglion cyst ——Yes (R or L) ——No

| Maneuvers: | | *Right* | *Left* |
|---|---|---|---|
| Guyon Tinel's | —— | —— | |
| Carpal Tinel's | —— | —— | |
| Phalen's | | —— | —— |

*Hands and Fingers*

Inspection: Inflammation ——Yes (R or L) ——No

| Maneuvers: *Right* | *Left* | |
|---|---|---|
| Trigger finger | —— | —— |
| Finkelstein's | —— | —— |

# Appendix C: Electronic Sources of Information

James D. McGlothlin and Amit Bhattacharya

CONTENTS



## C.1 BIOMECHANICS YELLOW PAGES

This is an electronic database of products and services related to the field maintained by Pierre Baudin at the following WWW address:
http://www.orst.edu/~bowenk/byp.html

## C.2 ERGOWEB

This is a WWW site for useful information and services related to the field. It provides a comprehensive source of information, software, references, and simple software programs for practicing ergonomists and biomechanists:
http://www.ergoweb.com/

## C.3 BIOMECHANICS WORLDWIDE (BWW)

Another comprehensive and updated source maintained by Dr. Pierre Baudin, for navigating the WWW sites dealing with biomechanics and ergonomics, is the following:
http://dragon.acadiau.ca/~pbaudin/biomch.html

## C.4 BIOMECHANICS-ERGONOMICS RESOURCE TANK

This is a source of information related to biomechanics and ergonomics, with useful links to WWW sites serving a parallel purpose, maintained by the authors:
http://www.eh.uc.edu/bert/

## C.4.1 Electronic Sources

*International*:

| | |
|---|---|
| http://www.ace-ergocanada.ca/ | Association of Canadian Ergonomists |
| http://www.anzsb.asn.au/ | Australian and New Zealand Society of Biomechanics |
| http://www.bfe.org/ | Biofeedback Foundation of Europe |
| http://www.health.uottawa.ca/biomech/csb/ | Canadian Society for Biomechanics |
| http://www.ergonomics.org.au/ | Ergonomics Society of Australia |
| http://www.uq.edu.au/eaol/ | Ergonomics Australia On-Line |
| http://www.ergonomics.org.au/ | Ergonomics Society of the United Kingdom |
| http://osha.europa.eu/en/ | European Agency for Safety and Health at Work |
| http://www.esbiomech.org/ | European Society of Biomechanics |
| http://www.ise.org.in/ | The Indian Society of Ergonomics |
| http://www.ergonomics.com.au/ | Ergonomics in Australia |
| http://isbweb.org/ | International Society for Biomechanics |
| http://www.ccohs.ca/ | Canadian Centre for Occupational Health and Safety |

## C.4.2 United States

*Associations*:

| | |
|---|---|
| http://www.hfes.org/web/Default.aspx | The Human Factors and Ergonomics Society |
| http://www.aiha.org/Pages/default.aspx | American Industrial Hygiene Association |
| http://www.aota.org/ | American Occupational Therapy Association (AOTA) |
| http://www.apta.org//AM/Template.cfm?Section=Home | American Physical Therapy Association (APTA) |
| http://www.bcpe.org/ | Board of Certification in Professional Ergonomics |
| http://www.asse.org/ | American Society of Safety Engineers |

*Government*:

| | |
|---|---|
| http://www.os.dhhs.gov/ | Department of Health and Human Services |
| http://www.cdc.gov/ | Centers for Disease Control and Prevention |
| http://www.who.ch/ | World Health Organization (WHO) |
| http://www.hhs.gov/ | Department of Health and Human Services |
| http://thomas.loc.gov | Thomas: legislative information on the Internet |
| http://www.cdc.gov/niosh/ | NIOSH |
| http://www.dir.ca.gov/ | Department of Industrial Relations, California |
| https://www2.hf.faa.gov/HFPortalnew/index.aspx?AspxAutoDetectCookieSupport=1 | Federal Aviation Administration |
| http://www.osha.gov/SLTC/ergonomics/index.html | OSHA Ergonomics Resources |

*University*:

| | |
|---|---|
| http://www.engin.umich.edu/dept/ioe/C4E/ | Center for Ergonomics, University of Michigan |
| http://ergo.human.cornell.edu/ | Cornell University Ergonomics Web Site |
| http://www.hitl.washington.edu/home/ | Human Interface Technology Lab |
| http://ergo.berkeley.edu/ | UCSF/UCB Ergonomics Program |
| http://ergo.engr.wisc.edu/ | University of Wisconsin–Madison Occupational Ergonomics and Biomechanics Laboratory |

| | |
|---|---|
| http://www.stanford.edu/dept/EHS/prod/general/ergo/computer_ergo.html | Stanford University Ergonomics |
| http://safetyservices.ucdavis.edu/ | UC Davis Office Ergonomics |
| http://ergonomics.ucla.edu/oldergo/default.htm | UCLA Ergonomics |
| http://www2.ncsu.edu/CIL/NCERC/index.html | North Carolina Ergonomics Resource Center |

*Labor*:

| | |
|---|---|
| http://www.bls.gov/ | Bureau of Labor Statistics |
| http://www.uaw.org/ | United Auto Workers |

# Appendix D: Ergonomics Software Sources

H. Onan Demirel and Vincent Duffy

CONTENT



**AnyBody™**
Provider: AnyBody Technology
URL: http://www.anybodytech.com/
Description:

> The AnyBody Modeling System™ is a software system for simulating the mechanics of the live human body working in concert with its environment. The environment is defined in terms of external forces and boundary conditions, and the user may impose any kind of posture or motion for the human body—either from scratch or from a set recorded motion data. AnyBody then runs a simulation and calculates the mechanical properties for the body-environment system. From AnyBody the user can obtain results on individual muscle forces, joint forces and moments, metabolism, elastic energy in tendons, antagonistic muscle actions and much more.
>
> AnyBody can also scale the models to fit to any population from anthropometric data or to any individual. Or, you can parameterize your studies in AnyBody to match product design trade-offs, finding the optimum combination of parameters to fulfill a given purpose [1].

**Madymo™**
Provider: TNO
URL: http://www.tass-safe.com/en/products/madymo
Description:

> MADYMO is the worldwide standard software for analysing and optimizing occupant safety designs. It is without rival for its accuracy, efficiency and versatility.

Using MADYMO, researchers and engineers can model, thoroughly analyse and optimize safety designs early in the development process. This reduces the expense and time involved in building and testing prototypes. Adopting MADYMO also minimises the risk of making design changes late in development. So, for new or improved vehicle models and components, MADYMO ensures substantial cost savings and shortens the time-to-market [2].

**MIDAS™**
Provider: NASA
URL: http://hsi.arc.nasa.gov/groups/midas/
Description:

The Man-machine Integration Design and Analysis System (MIDAS) is a 3-D rapid prototyping human performance modeling and simulation environment that facilitates the design, visualization, and computational evaluation of complex man-machine system concepts in simulated operational environments. MIDAS combines graphical equipment prototyping, dynamic simulation, and human performance modeling with the aim to reduce design cycle time, support quantitative predictions of human-system effectiveness, and improve the design of crew stations and their associated operating procedures.

MIDAS links a virtual human, comprised of a physical anthropometric character, to a computational cognitive structure that represents human capabilities and limitations. The cognitive component is made up of a perceptual mechanism (visual and auditory), memory, a decision maker and a response selection architecture (Micro Saint Sharp). The complex interplay among bottom-up and top-down processes enables the emergence of unforeseen, and non programmed behaviors [3].

**RAMSIS™**
Provider: Human Solutions GmbH
URL: http://www.human-solutions.com/automotive/products_r_auto_en.php
Description:

The 3D CAD manikin RAMSIS is a highly accurate simulation software program for a wide range of design and construction analyses. RAMSIS addresses demands on ergonomics, comfort and safety as early as the planning stage. Latter-stage improvements can be dispensed with and the need for physical test benches is significant reduced.

RAMSIS analysis quality contributes considerably to series production readiness—and consequently to the economic success of the vehicle, since the system uses current, international body dimension databases to supply accurate representative results about product requirements for complex international target markets—roughly 3 to 5 times faster than conventional analysis methods. This reduces costs by up to 50% (Source: independent calculations by vehicle manufacturers). Analyses with RAMSIS can be applied to completely different vehicle types. Time and costs for time-consuming space-saving tests for new vehicle concepts can be dispensed with [4].

**SANTOS™**
Provider: Santos Human Inc.
URL: http://www.santoshumaninc.com/
Description:

The Santos™ virtual human modeling software represents a new generation of virtual humans that are highly realistic in terms of appearance, movement, and feedback (evaluation of the human body during task execution). Santos™ resides in a virtual environment with a set of variables that the user can manipulate to effect changes in the way the avatar and the environment interact. With two different operating modes, posture prediction and advanced inverse kinematics (IK), the software provides a broad range of simulation tools for testing equipment design and assessing task performance. An optimization-based approach to kinematic and dynamic motion analysis allows the avatar to operate with nearly complete autonomy, as opposed to depending on stored animation and data. This autonomy enables the user to conduct human-factors analysis, including, among other things, posture prediction, motion prediction, gate analysis, reach envelope analysis, and ergonomics studies [5].

**Siemens Jack™**
Provider: Siemens Inc.
URL: http://www.plm.automation.siemens.com/en_us/products/tecnomatix/
Description:

Manufacturing companies in a variety of industries are addressing the human element as a key component of the design, assembly and maintenance of products. Tecnomatix human products, Jack and Process Simulate Human, enable you to improve the ergonomics of your product designs and refine industrial tasks from the earliest stages of the engineering process. You can populate your digital environments with virtual people and size them to match worker populations. You can test your designs for a wide variety of human factors, including injury risk, timing and efficiency, user comfort, reachability, lines of sight, energy expenditure, fatigue limits and other important parameters. These products facilitate quality improvements, as well as cost and time savings by providing guidance for more user friendly designs throughout the engineering process [6].

**Delmia Human™**
Provider: Dessault Systems
URL: http://www.3ds.com/products/delmia/welcome/
Description:

> DELMIA's Virtual Ergonomics suite of digital human modeling tools are fully integrated into the DELMIA Digital Process for Manufacture (DPM) solutions. It provides such organizations with a comprehensive array of human simulation and Ergonomics tools specifically designed for understanding and optimizing the relationship between humans and the products they manufacture, install, operate, and maintain [7].

## REFERENCES

1. AnyBody Technology. AnyBody Technology: The AnyBody Modeling System™ [Internet]. [Cited February 10, 2011]; Available from: http://www.anybodytech.com/index.php?id=26
2. MADYMO. Cost-effective safety design [Internet]. [Cited February 10, 2011]; Available from: http://www.tass-safe.com/en/products/madymo
3. MIDAS-Man Machine Integration Design and Analysis System [Internet]. [Cited February 10, 2011]; Available from: http://hsi.arc.nasa.gov/groups/midas/
4. HUMAN SOLUTIONS [Internet]. [Cited February 10, 2011]; Available from: http://www.human-solutions.com/automotive/products_r_auto_en.php
5. SantosHuman™ Inc.-The Science of Virtual Human Motion Simulation [Internet]. [Cited February 10, 2011]; Available from: http://www.santoshumaninc.com/
6. Siemens Jack™ Inc.-Jack and Process Simulate Human [Internet]. [Cited February 10, 2011]; Available from: http://www.plm.automation.siemens.com/en_us/products/tecnomatix/assembly_planning/jack/index.shtml
7. Delmia Human™ Inc.-Delmia Digital Manufacturing and Production [Internet]. [Cited February 10, 2011]; Available from: http://www.3ds.com/fileadmin/PRODUCTS/DELMIA/OFFERS/Virtual-Ergonomics-Solutions/PDF/DELMIA-Virtual-Ergonomics-brochure.pdf

# Appendix E: Information Sources from the National Institute for Occupational Safety and Health

James D. McGlothlin and Amit Bhattacharya

CONTENTS



E-mail: cdcinfo@cdc.gov
Telephone: 1-880-CDC-INFO (24 h/day, every day). Technical information service number that provides access to NIOSH and its information resources.
(513) 533-8347 (fax)
(513) 533-8328 (outside United States)

For further information, write to

NIOSH
Education & Information Division
4676 Columbia Parkway
Cincinnati, OH 45226-1998
http://www.cdc.gov/niosh/ (NIOSH internet address)
pubstaft@niosdtl.em.cdc.gov (e-mail address request for NIOSH publications)

## E.1 SELECTED NIOSH PUBLICATIONS IN ERGONOMICS

The following site offers a complete list of NIOSH publications: http://www.cdc.gov/niosh/pubs/all_date_desc.html. The following are selected NIOSH publications in Ergonomics:

2011-117 Methods and analysis of economic impact of workplace injury
2011-116 Effects of extended work hours on intern health and safety

2007-131 Ergonomic guidelines for manual material handling
2005-112 A compendium of NIOSH economic research: 2002–2003
2004-164 Easy ergonomics: a guide to selecting non-powered hand tools
2001-117 National occupational research agenda for musculoskeletal disorders: next decade of research
2001-108 NIOSH testimony to OSHA—comments on the proposed ergonomics program—29 CFR Part 1910 Docket No. S-177
2000-116 Worker deaths by falls: a summary of surveillance findings and investigative case reports
1999-156 Investigation of a slip/fall accident
1997-141 Musculoskeletal disorders and workplace factors—a critical review of epidemiologic evidence for work-related musculoskeletal disorders of the neck, upper extremity, and low back
1997-142 The science of occupational musculoskeletal disorders (testimony from L. Rosenstock)
1997-117 Elements of ergonomics programs: a primer based on workplace evaluations of musculoskeletal disorders
1997-109 NIOSH research projects: musculoskeletal disorders
1995-119 Cumulative trauma disorders in the workplace: bibliography
1995-102 Effective interventions to reduce ergonomic injuries in meatpacking
1995-114 A strategy for industrial power hand tool ergonomic research design, selection, installation, and use in automotive manufacturing: proceedings of a NIOSH workshop
1994-124 Participatory ergonomic interventions in meatpacking plants

Bibliography: Carpal Tunnel Syndrome: Selected References reprint revised NIOSH equation for the design and evaluation of manual lifting tasks

1994-127 Back belts: do they prevent injury?
1994-122 Workplace use of back belts
1991-100 Selected topics in surface electromyography for use in the occupational setting: expert perspectives
1989-106 Occupational exposure to hand-arm vibration
1983-110 Vibration syndrome

## E.2 NIOSH EDUCATIONAL RESOURCE CENTERS OFFERING ERGONOMICS COURSES

Alabama Education and Research Center
University of Alabama at Birmingham
School of Public Health
1665 University Blvd.
Birmingham, AL 35294-0022
(205) 934-6208
Fax: (205) 975-6341

R. Kent Oestenstad, PhD, Director
E-mail: oestk@uab.edu

California Education and Research Center—Northern
University of California, Berkeley
School of Public Health
140 Warren
Berkeley, CA 94720-7360
(510) 643-4702
Fax: (510) 642-5815
John R. Balmes, MD, Director
E-mail: john.balmes@ucsf.edu

California Education and Research Center—Southern
University of California, Los Angeles
School of Public Health
650 Charles Young Drive South
Los Angeles, CA 90095-1772
(310) 206-6141
Fax: (310) 206-9903
John R. Froines, PhD, Interim Director
E-mail: jfroines@ucla.edu

Cincinnati Education and Research Center
University of Cincinnati
Department of Environmental Health ML. Box 670056
3223 Eden Avenue
Cincinnati, Ohio 45267-0056
(513) 558-1751
Fax: (513) 558-2772
Carol Rice, PhD, CIH, Director
E-mail: alerdilr@ucmail.uc.edu

Colorado Education and Research Center
University of Colorado School of Public Health
13001 E. 17th Place, B-119
Aurora, CO 80045
(303) 315-0880
Fax: (303) 315-7642
Lee S. Newman, MD, MA, FCCP, FACOEM, Director
E-mail: lee.newman@uchsc.edu

Harvard Education and Research Center
Harvard School of Public Health
Department of Environmental Health
665 Huntington Avenue

Bldg. 1, Rm. 1407
Boston, MA 02115
(617) 432-3323
Fax: (617) 432-3441
David C. Christiani, MD, Director
E-mail: dchris@hohp.harvard.edu

Illinois Education and Research Center
University of Illinois at Chicago
School of Public Health
2121 West Taylor St.
Chicago, IL 60612
(312) 996-7469
Fax: (312) 413-9898
Lorraine M. Conroy, ScD, CIH, Director
E-mail: lconroy@uic.edu

Iowa Education and Research Center
University of Iowa
College of Public Health
Department of Occupational and Environmental Health
UI Research Park, 137 IREH
100 Oakdale Campus-126 IREH
Iowa City, IA 52242-5000
(319) 335-4428
Fax: (319) 335-4225
Patrick O'Shaughnessy, PhD, Director
E-mail: patrick-oshaughnessy@uiowa.edu

Johns Hopkins Education and Research Center
Johns Hopkins University
Bloomberg School of Public Health
615 North Wolfe Street Rm 7503
Baltimore, MD 21205
(410) 955-4037
Fax: (410) 614-4986
Jacqueline Agnew, PhD, Director
E-mail: jagnew@jhsph.edu

Michigan Education and Research Center
University of Michigan
School of Public Health
1420 Washington Heights

Ann Arbor, MI 48109-2029
(734) 936-0757
Fax: (734) 763-8095
Thomas G. Robins, MD, Director
E-mail: trobins@umich.edu

Minnesota Education and Research Center
University of Minnesota
Prevention Research Training Program Division of Environmental Health Sciences
School of Public Health
Minnesota Mayo Mail Code 807
420 Delaware Street, S.E.
Minneapolis, MN 55455
(612) 625-5934
Fax: (612) 626-0650
Susan G. Gerberich, PhD, MSPH, Director
E-mail: gerbe001@umn.edu

New York/New Jersey Education and Research Center
Mount Sinai School of Medicine
Department of Community and Preventive Medicine
P.O. Box 1057
One Gustave L. Levy Pl.
New York, NY 10029-6574
(212) 824-7018
Fax: (212) 996-0407
Philip J. Landrigan, MD, MSc, Interim Director
E-mail: phil.landrigan@mssm.edu

North Carolina Education and Research Center
University of North Carolina at Chapel Hill
School of Public Health
1700 Airport Rd., CB 7502
Chapel Hill, NC 27599-7502
(919) 966-1765
Fax: (919) 966-8999
Bonnie Rogers, DrPH, COHN-S, FAAN, LNCC, Director
E-mail: rogersb@email.unc.edu

South Florida Education and Research Center
University of South Florida
College of Public Health
13201 Bruce B. Downs Blvd., MDC Box 56

Tampa, FL 33612-3805
(813) 974-6629
Fax: (813) 974-4718
Thomas E. Bernard, PhD, Director
E-mail: tbernard@ health.usf.edu

Texas Education and Research Center
University of Texas Health Science Center at Houston
School of Public Health
P.O. Box 20186
Houston, TX 77225-0186
(713) 500-9464
Fax: (713) 500-9442
Sarah A. Felknor, PhD, Director
E-mail: sarah.a.felknor@uth.tmc.edu

Utah Education and Research Center
University of Utah
Rocky Mountain Center for Occupational and Environmental Health
391 Chipeta Way, Suite C
Salt Lake City, UT 84108
(801) 581-4800
Fax: (801) 581-7224
Kurt Hegmann, MD, MPH, Director
E-mail: kurt.hegmann@hsc.utah.edu

Washington Education and Research Center
University of Washington
Department of Environmental Health and Occupational Health Sciences
P.O. Box 354695
Seattle, WA 98105
(206) 685-7189
Fax: (206) 616-6240
Noah S. Seixas, PhD, Director
E-mail: nseixas@u.washington.edu

### E.2.1 NIOSH Training Project Grants

Jerry Dzugan, MS
**Alaska Marine Safety Educ. Assoc. (AMSEA) [OT-Marine Safety]**
2924 Halibut Point Rd.
Sitka, AK 99835
(907) 747-3287
Fax: (907) 747-3259
E-mail: director@amsea.org

Jefferey L. Burgess, MD, MPH
College of Public Health
**University of Arizona [IH]**
1295 N. Martin Ave.
Tucson, AZ 85724
(520) 626-4918
Fax: (520) 626-8009
E-mail: jburgess@u.arizona.edu

Robert J. Harrison, MD, MPH
**Association of Occupational and Environmental Clinics [OSH]**
1010 Vermont Ave.
Washington, DC 20005
(510) 620-5769
Fax: (510) 620-5743
E-mail: rharriso.dhs.ca.gov

Vicki J. Magley, PhD
Department of Psychology
**University of Connecticut School [OT-OPSY]**
406 Babbidge Road
Storrs, CT 06030
(860) 486-9010
Fax: (860) 486-2760
E-mail: vicki.magley@uconn.edu

Paige Tolbert, PHD
Rollins School of Public Health
**Emory University [OEPI]**
1518 Clifton Road, N.E.
Atlanta, GA 30322
(404) 727-3801
Fax: (404) 727-8744
E-mail: ptolbert@sph.emory.edu

Patrick J. Morrison, Program Director
**International Association of Firefighters**
1750 New York Avenue, NW
Washington, DC 20006
(202) 737-8484
Fax: (202) 747-3259
E-mail: pmorrison@iaff.org

Robert H. McKnight, MPH, ScD
College of Public Health

**University of Kentucky [OT-AGR]**
1141 Red Mile Road
Lexington, KY 40504
(859) 323-6836
Fax: (859) 254-3760
E-mail: rmcknig@uky.edu

David Kriebel, Sc.D.
Department of Work Environment
**University of Massachusetts, Lowell [IH/OS/OEPI/WEP]**
Office of Research Administration
600 Suffolk St., 2nd Floor,
South Lowell, MA 01854
(978) 934-3270
Fax: (978) 452-5711
E-mail: David_Kriebel@uml.edu

Sangita Chakrabarty, MD
Department of Family and Community Medicine
**Meharry Medical College [OM]**
1005 Dr. D.B. Todd Blvd.
Nashville, TN 37208
(615) 321-2976
Fax: (615) 321-2979
E-mail: schakrabarty@mmc.edu

Daniel H. Anna, PhD
Occupational Safety and Environmental Health Program
**Millersville University [OSH]**
40 E. Frederick St.
Osborne Hall
P.O. Box 1002
Millersville, PA 17551
(717) 872-3334
Fax: (717) 872-3318
E-mail: dan.anna@millersville.edu

Roger C. Jensen, PhD
Safety, Health, and Industrial Hygiene
**Montana Tech of the University of Montana [OSH/IH]**
1300 W. Park Street
Butte, MT 59701-8997
(406) 496-4111
Fax: (406) 496-4650
E-mail: rjensen@mtech.edu

David G. Kraemer, PhD
Department of Occupational Safety & Health
**Murray State University [OSH]**
157 Industry & Technology Center
Murray, KY 42071
(270) 809-6654
Fax: (270) 809-3630
E-mail: david.kraemer@murraystate.edu

Crescente E. Figueroa, PhD
**University of North Alabama [IH]**
Faculty Box 5211
Florence, Alabama 35632
(256) 765-4236
Fax: (256) 765-4958
E-mail: cefigueroa@una.edu

Carolyn M. Sommerich, PhD
Department of Industrial, Welding and Systems Engineering
**Ohio State University [OS]**
1971 Neil Ave., 210 Baker Systems
Columbus, OH 43210
Phone: 614-292-9965
Fax: 614-292-7852
E-mail: sommerich.1@osu.edu

Margaret L. Phillips, PhD
Occupational & Environmental Health
**University of Oklahoma Health Sciences Center [IH]**
801 N.E. 13th Street
Oklahoma City, OK 73190
(405) 271-2070
Fax: (405) 271-1971
E-mail: margaret-phillips@ouhsc.edu

Joseph I. Boullata, PharmD
Associate Professor of Pharmacology & Therapeutics, Biobehavioral & Health Sciences (Faculty)
School of Nursing
**University of Pennsylvania [OHN]**
School of Nursing
418 Curie Blvd.
P221 Franklin Bldg.
Philadelphia, PA 19104-4217
215-573-7387

Fax: (215) 573-7387
E-mail: boullata@nursing.upenn.edu

Edward A. Emmett, MB, BS
Occupational Medicine, School of Medicine
**University of Pennsylvania [OM]**
3400 Spruce St.
Philadelphia, PA 19104
(215) 349-5708
Fax: (215) 662-4430
E-mail: emmetted@mail.med.upenn.edu

Leslie B. Hammer, PhD
Psychology Department
**Portland State University [OT/OPSY]**
317 Cramer Hall
P.O. Box 751
Portland, OR 97207
(503) 725-3971
Fax: (503) 725-3904
E-mail: hammerl@pdx.edu

Jesus Gonzalez, PhD
School of Public Health
**University of Puerto Rico [IH]**
Medical Sciences Campus
P.O. Box 365067
San Juan, PR 00936
(787) 758-2525, ext. 1437
Fax: (787) 758-6719
E-mail: jesus.gonzalez9@upr.edu

Frank S. Rosenthal, PhD
School of Health Sciences
**Purdue University [IH]**
Civil Engineering Bldg.
550 Stadium Mall Dr.
West Lafayette, IN 47907
(765) 494-0812
Fax: (317) 765-1377
E-mail: frank@purdue.edu

Victor L. Paquet, ScD
Department of Industrial Engineering
**SUNY Buffalo [OSH]**

Buffalo, NY 14260
(716) 645-2357
Fax: (716) 645-3302
E-mail: vpaquet@eng.buffalo.edu

Jerome J. Congleton, PhD, PE, CPE
School of Rural Public Health
**Texas A&M University [OSH]**
1266 TAMU
College Station, TX 77843
(979) 845-5574
Fax: (979) 690-0022
E-mail: jjcongleton@srph.tamhsc.edu

Jeffrey L. Levin, MD, MSPH
Occupational Health Sciences
**University of Texas Health Center at Tyler [OM]**
11937 U.S. Hwy. 271
Tyler, TX 75708
(903) 877-5900
Fax: (903) 877-7982
E-mail: jeffrey.levin@uthct.edu

Farhang Akbar, PhD, CIH, CSP
Department of Public Health
**University of Toledo [IH]**
3015 Arlington Avenue
Toledo, OH 43614
(419) 383-4543
Fax: (419) 383-5880
E-mail: farhang.akbar@utoledo.edu

Faye Grimsley, PhD
Environmental Health Sciences
**Tulane Univ. Health Sciences Center [IH]**
1440 Canal St.
New Orleans, LA 70112
(504) 988-8262
Fax: (504) 584-1726
E-mail: grimsley@tulane.edu

Maury Nussbaum, PhD
Grado Department of Industrial and Systems Engineering
**Virginia Polytechnic Institute and State University [OS]**
250 Durham Hall (0118)
Blacksburg, VA 24061-0118

(540) 231-6053
Fax: (540) 231-3322
E-mail: nussbaum@vt.edu

Chris Martin, MD, MS
Institute for Occup. & Envir. Health
**West Virginia University [OM/OS/IH]**
P.O. Box 9190
Morgantown, WV 26506
(304) 293-3693
Fax: (304) 293-2629
E-mail: cmartin@hsc.wvu.edu

Elbert Sorrell, EdD
College of Tech., Engineering & Mgmt.
**University of Wisconsin-Stout [OS]**
712 S. Broadway
P.O. Box 790
Menomonie, WI 54751
(715) 232-2630
Fax: (715) 232-5236
E-mail: sorrelle@uwstout.edu

Oyebode A. Taiwo, MD
Occup. & Environ. Medicine Program
**Yale University [OM]**
135 College St., 3rd Floor, Rm. 366
New Haven, CT 06510
(203) 785-7231
Fax: (203) 785-7391
E-mail: oyebode.taiwo@yale.edu

# Appendix F: Information Sources for Ergonomics Literature

James D. McGlothlin and Amit Bhattacharya

CONTENTS



## F.1 JOURNALS IN WHICH ARTICLES RELATED TO ERGONOMICS ARE LIKELY TO APPEAR

### F.1.1 Aging Titles

*Age and Aging*
http://ageing.oxfordjournals.org/

*Journal of Gerontology*
http://biomedgerontology.oxfordjournals.org/

### F.1.2 Biomechanics Titles

*ASME Transaction: Journal of Biomechanics*
http://scitation.aip.org/Biomechanical/

*Clinical Biomechanics*
http://www.elsevier.com/wps/find/journaldescription.cws_home/30397/description

*Journal of Biomechanics*
http://www.jbiomech.com/

### F.1.3 Engineering and Physics Titles

*American Journal of Physics*
http://ajp.aapt.org/

*Automatica*
http://www.elsevier.com/wps/find/journaldescription.cws_home/270/description#description

*Color Research and Application*
http://onlinelibrary.wiley.com/journal/10.1002/%28ISSN%291520-6378

*Computers and Industrial Engineering*
http://www.elsevier.com/wps/find/journaldescription.cws_home/399/description#description

*IEEE Transaction on Systems, Man, and Cybernetics*
http://ieeexplore.ieee.org/xpl/RecentIssue.jsp?reload=true&punumber=3477

*Journal of the Acoustical Society of America*
http://scitation.aip.org/JASA

*Journal of the Illuminating Engineering Society*
http://www.iesna.org/leukos/introduction.cfm

*Journal of Navigation*
http://journals.cambridge.org/action/displayJournal?jid=NAV

*Lighting Research and Technology*
http://lrt.sagepub.com/

*Noise Control Engineering Journal*
http://www.inceusa.org/publications/ncej

### F.1.4 Ergonomics Titles

*Applied Ergonomics*
http://www.elsevier.com/wps/find/journaldescription.cws_home/30389/description#description

*Human Factors*
http://www.hfes.org/web/Default.aspx

*Human Movement Science*
http://www.elsevier.com/wps/find/journaldescription.cws_home/505584/description#description

*International Journal of Industrial Ergonomics*
http://www.elsevier.com/wps/find/journaldescription.cws_home/505654/description#description

*International Journal of Man-Machine Studies*
http://www.sciencedirect.com/science/journal/00207373

*Japanese Journal of Ergonomics*
http://www.jstage.jst.go.jp/browse/jje

*Occupational Ergonomics*
adminrecords.ucsd.edu/ppm/docs/516–20.pdf

*Office Ergonomics*
www.lni.wa.gov/IPUB/417–133–000.pdf

*Office Ergonomics Review*
www.haworth.com/.../**Office**%20**Ergonomics**%20A%20**Review**%20of%20Pertinent.pdf

*Scandinavian Journal of Work, Environmental and Health*
http://www.sjweh.fi/

*Work and Occupations*
http://wox.sagepub.com/

*Work and Stress*
http://www.tandf.co.uk/journals/titles/02678373.asp

### F.1.5 Psychology Titles

*Applied Psychology: An International Review*
http://www.blackwellpublishing.com/journal.asp?ref=0269–994x

*Behavior and Information Technology*
http://www.tandf.co.uk/journals/tf/0144929X.html

*Cognitive Science*
http://www.cognitivesciencesociety.org/journal_csj.html

*European Journal of Applied Physiology and Occupational Physiology*
http://www.springerlink.com/content/108306/

*International Review of Applied Psychology*
http://www.wiley.com/bw/journal.asp?ref=0269–994X

*Journal of Applied Psychology*
http://jap.physiology.org/

*Journal of General Psychology*
http://www.tandf.co.uk/journals/titles/00221309.asp

*Journal of Managerial Psychology*
http://www.emeraldinsight.com/products/journals/journals.htm?id=jmp

*Quarterly Journal of Experimental Physiology*
http://www.researchgate.net/journal/0144-8757_Quarterly_journal_of_experimental_physiology_Cambridge_England

*Quarterly Journal of Experimental Physiology and Cognate Medical Science*
http://www.researchgate.net/journal/0033-5541_Quarterly_journal_of_experimental_physiology_and_cognate_medical_sciences

*Perceptual and Motor Skills*
http://ammons.ammonsscientific.com/php/browse_latest_issue_PMS.php

*Perception and Psychophysics*
http://www.psychonomic.org/psp/publications-resources.html

### F.1.6 Safety and Health Titles

*Accident Analysis and Prevention*
http://www.sciencedirect.com/science/journal/00014575

*American Journal of Industrial Medicine*
http://onlinelibrary.wiley.com/journal/10.1002/(ISSN)1097-0274/issues

*American Journal of Public Health and the Nation's Health*
http://www.apha.org/

*Annals of Occupational Hygiene*
http://annhyg.oxfordjournals.org/

*Applied Occupational & Environmental Hygiene*
http://www.informaworld.com/smpp/title~content=t713768777

*Canadian Journal of Public Health*
http://www.cpha.ca/en/cjph.aspx

*Health and Safety at Work*
http://www.hse.gov.uk/index.htm

*Health Laboratory Science*
http://www.ncbi.nlm.nih.gov/pmc/articles/PMC1254908/

*Health Physics*
http://journals.lww.com/health-physics/pages/default.aspx

*Industrial Health*
http://www.jstage.jst.go.jp/browse/indhealth

*Journal of Occupational Accidents*
http://www.sciencedirect.com/science/journal/03766349

*Journal of Occupational and Environmental Hygiene (Formerly known as: American Industrial Hygiene Association Journal and AIHA Journal)*
http://www.aiha.com/

*Occupational Health and Safety*
http://ohsonline.com/Home.aspx

*Occupational Health in Ontario*
http://www.labour.gov.on.ca/english/hs/

*Professional Safety*
http://www.asse.org/professionalsafety/

### F.1.7 Science and Medical Titles

*Acta Physiologica: (Formerly known as Acta Physiologica Scandinavica)*
http://www.wiley.com/bw/journal.asp?ref=1748-1708

*Advances in Biological and Medical Physics*
http://www.researchgate.net/journal/0065-2245_Advances_in_biological_and_medical_physics

*American Journal of Physiology*
http://www.physiology.org/

*American Journal of Science*
http://www.ajsonline.org/

*American Journal of the Medical Science*
http://journals.lww.com/amjmedsci/pages/default.aspx

*Annual Review of Physiology*
http://www.annualreviews.org/journal/physiol

*Aviation Space and Environmental Medicine*
http://www.asma.org/journal/

*Canadian Journal of Research (Section E-Medical Sciences)*
http://pubs.nrc-cnrc.gc.ca/rp-ps/journalDetail.jsp?jcode=cjr&lang=eng

*Canadian Medical Association Journal*
http://www.cmaj.ca/

*International Journal of Rehabilitation*
http://journals.lww.com/intjrehabilres/pages/default.aspx

*JAMA—the Journal of the American Medical Association*
http://jama.ama-assn.org/

*Journal of Applied Physiology*
http://jap.physiology.org/

*Journal of Hand Surgery: American Volume*
http://www.jhandsurg.org/

*Journal of Neurosurgery*
http://thejns.org/

*Journal of Occupational Medicine and Toxicology*
http://www.occup-med.com/

*Journal of Physiology*
http://jp.physoc.org/

*Journal of Labor Research*
http://www.springer.com/social+sciences/journal/12122

*Journal of Sport Science and Medicine*
http://www.jssm.org/

*Lancet*
http://www.thelancet.com/

*Life Sciences*
http://www.elsevier.com/wps/find/journaldescription.cws_home/525477/description#description

*Metabolism, Clinical and Experimental*
http://www.metabolismjournal.com/

*Muscle and Nerve*
http://onlinelibrary.wiley.com/journal/10.1002/%28ISSN%291097-4598/issues

*Neurology*
http://www.neurology.org/

*New England Journal of Medicine*
http://www.nejm.org/

*Orthopedics*
http://www.orthosupersite.com/

*Physiological Reviews*
http://physrev.physiology.org/

*Respiration Physiology*
http://www.sciencedirect.com/science/journal/00345687

*Rheumatology*
http://rheumatology.oxfordjournals.org/

*Science*
http://www.sciencemag.org/

*Science and Medicine in Sports*
http://www.jsams.org/

*Spine*
http://journals.lww.com/spinejournal/pages/default.aspx

*Sports Medicine*
http://emedicine.medscape.com/sports_medicine

### F.1.8 Other Titles

*Human Relations*
http://hum.sagepub.com/

*IIE Transactions*
http://www.tandf.co.uk/journals/titles/0740817x.asp

*New Technology, Work and Employment*
http://www.wiley.com/bw/journal.asp?ref=0268-1072

*Personnel Management*
http://www.opm.gov/

*Research Quarterly for Exercise and Sports*
http://www.aahperd.org/rc/publications/rqes/

# Appendix G: Ergonomics-Related Problems with Solutions

James D. McGlothlin and Amit Bhattacharya

1. In the following photos (A, B, C, and D; photos adapted from DHHS (NIOSH) Publication No. 2004-164), label the photos according to GOOD or POOR tool design based on ergonomic principles. Provide *justification* for your answers.

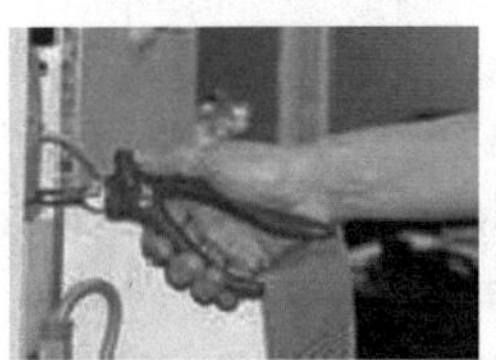
(A)

(B)

(C)

(D)

*Answers: A: Good (neutral wrist posture); B: Poor (wrist deviated); C: Poor (wrist deviated); D: Good (neutral wrist posture)*

2. In the following photos (A and B; photos adapted from DHHS (NIOSH) Publication No. 2004-164), identify which photo shows a better handle design for the bucket. Provide *ergonomic justifications* for your answer.

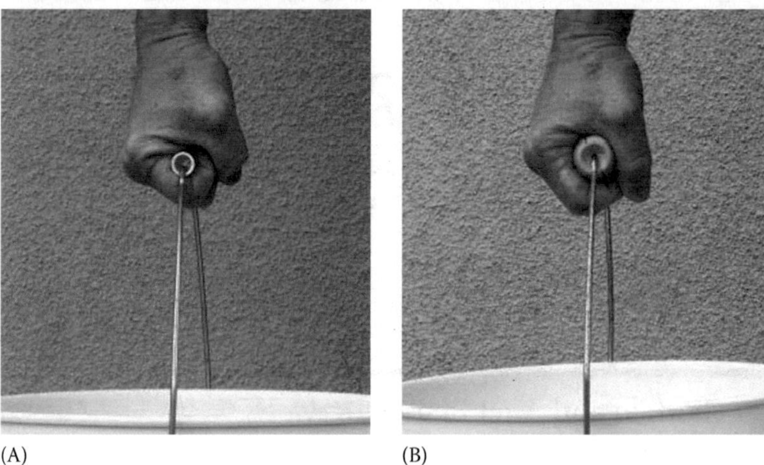

(A) (B)

*Answer: B has a better handle.*

3. In a 2010 article (Occupational Ergonomics Vol. 119–126), researchers from South Africa reported energy demands associated with forestry-related harvesting tasks, chainsaw activities (tasks of felling and crosscutting), and stacking of logs. They collected the following information for this project: (1) daily (24 h) energy consumption as dietary intake (i.e., fats, proteins, and carbohydrates) by the chainsaw workers and stackers; (2) energy cost of chainsaw job and stacker job over a typical shift period, 4.8 h; (3) energy intake as food consumption by the workers during the 4.8 h shift. As part of energy cost estimates associated with their jobs, both $VO_2$ consumption and heart rate responses were collected (shown in the below figure).

Based on the following data, provide answers to the following: (1) calculate energy deficit (kJ/shift) for chainsaw workers and stackers during task performance, (2) compare the daily energy consumption (kJ) based on recommended daily dietary intake (RDA) and actual

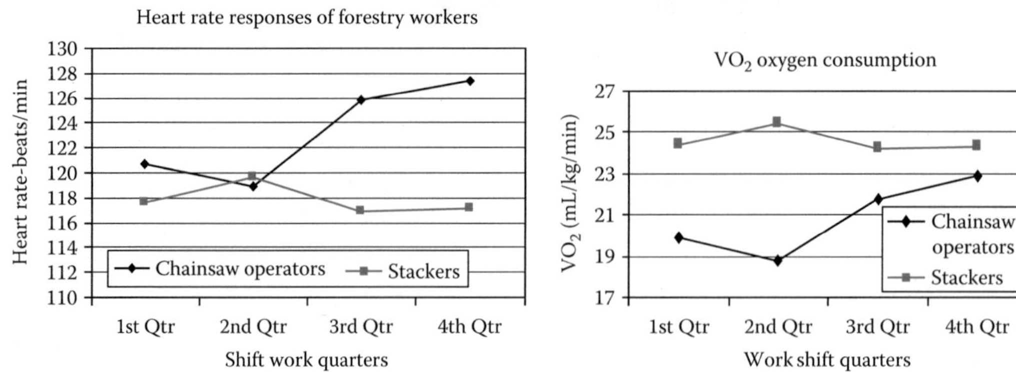

intake (in kJ) of the chainsaw workers and stackers over 24h, and (3) which worker group was more fatigued than the other and why? Justify your answers using the data given below.

| | RDA | Chainsaw Operators | Stackers |
|---|---|---|---|
| Protein (g) | 65 | 40.5 | 56.2 |
| Fat (g) | 30–80 | 26.7 | 23.1 |
| Carbohydrate (g) | 100 | 110 | 186.8 |
| *Total energy (kJ) | 8000 | 3949 | 5346 |

| | Chainsaw Operators | Stackers |
|---|---|---|
| Energy cost (kJ/shift) | 9481.2 | 8947.2 |
| Energy intake (kJ/shift) | 819.4 | 143 |

*Answers:*

1. *Energy deficit in chainsaw operators: 9481 – 819 = 8662 kJ/shift; energy deficit in stackers: 8947 – 143 = 8804 kJ/shift.*
2. *Chainsaw operators: consuming only 49% (8000 – 3949) of RDA; stackers: consuming only 67% (8000 – 5346) of RDA.*
3. *As per the oxygen demand and the heart rate plots given for these workers, the oxygen demand of stackers' job is much higher than that for the chainsaw operators' job. However, chainsaw operators' heart rate responses are markedly higher (than stackers') during majority of the work shift, implying potential fatigue setting in this group.*
4. Which one(s) of the following three jig workstation configurations is (are) ergonomically poor and therefore will cause the hand-wrist to assume awkward posture(s)?

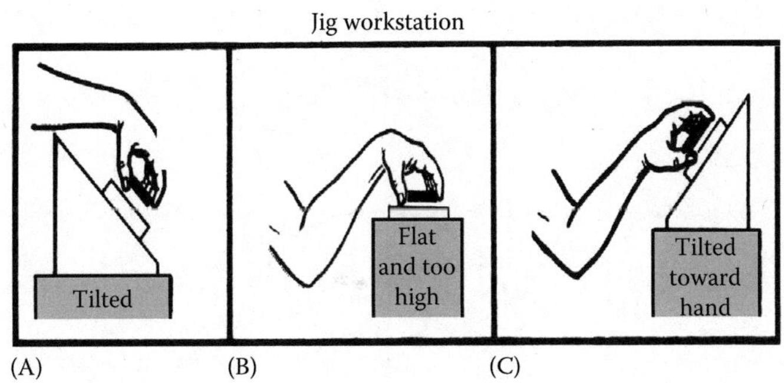


*Answers:*

*A: Poor; B: Poor; C: Good*

5. Calculate the zones of convenient reach ($d_{ZCR}$) for an older worker in the age range of 55–64 years standing 50 cm ($D$) from a vertically placed panel. Pick the appropriate anthropometric variable from the following table which gives mean values of a variety of anthropometric variables for a 24 year old person and percentage change in the variables for a 55–64 year old person.

Anthropometric Data

| | **Age (Years)** | |
|---|---|---|
| **Variables (cm)** | **≤24** | **55–64** |
| Biacromial breadth | 40.0 | −1% |
| Buttock–knee length | 58.9 | −1% |
| Butt–popliteal length | 49.2 | −2% |
| Functional reach | 75.1 | −3% |
| Knee height | 54.2 | −2% |
| Sitting height | 91.4 | −3% |
| Stature | 174.3 | −1% |

*Answer:*

*Anthropometric variable of interest is functional reach ($a_F$).*

*As per the previous tabler, the functional reach decreases by 3% (from 75.1 cm at age 24 years) for an older worker in the age range of 55–64 years.*

*Therefore, functional reach for an older worker in the age range of 55–64 years is computed as follows:*

$$a_F = 0.97 \times 75.1 = 72.84\ cm$$

*Use the formula for zone of convenient reach $d_{ZCR}$:*

$$d_{ZCR} = \sqrt{[(a_F)^2 - D^2]}$$
$$d_{ZCR} = 52.97\ cm$$

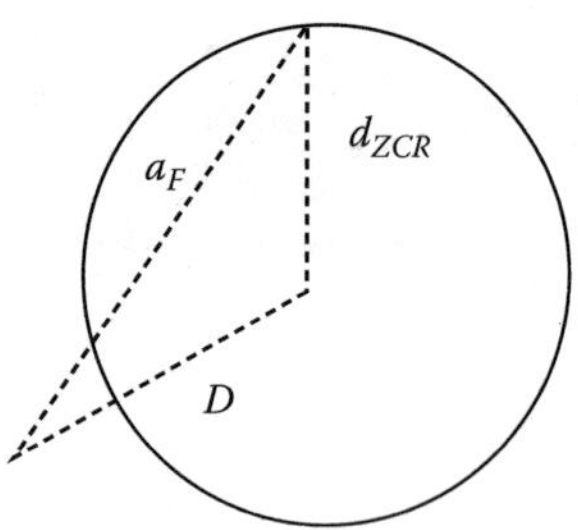

6. Lifting heavy objects from the floor usually requires the use of squatting posture as shown below (right). Squatting posture places biomechanical loading on the knee joint. In the picture below (adapted from DHHS (NIOSH) Publication No. 2007-131), the worker is holding a box weighing 25 lb in a squatting posture. The body weight carried by one of the supporting limbs is 94 lb (i.e., total body weight minus the weight of foot and leg).

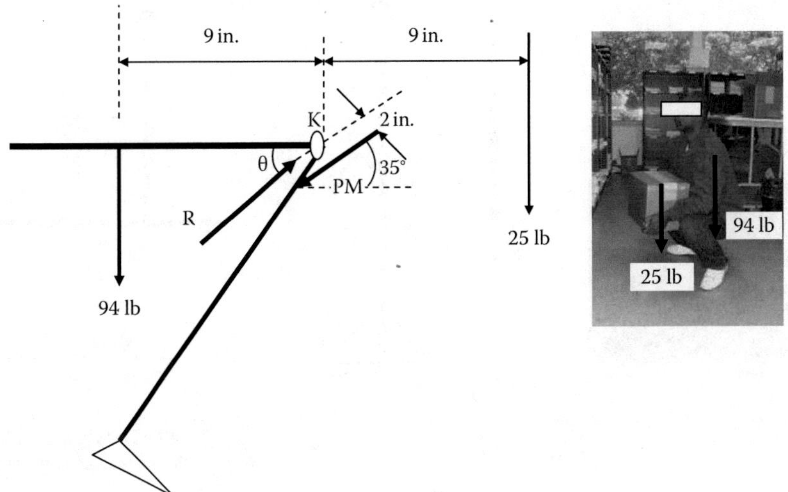


Using the earlier-mentioned free body diagram, determine the tibiofemoral joint reaction force, R, and its direction of action (θ) and quadriceps muscle force or pull force on the patellar tendon, PM, for the following cases: (a) without the 25 lb weight in hand and (b) with the 25 lb weight in hand. The moment arm for the patellar tendon, PM, is 2 in. The PM makes an angle of 35° with respect to the horizontal.

*Answer:*

*Applying the law of static equilibrium: Σ moment around the knee joint (K) = 0:*

a. *Without 25 lb weight in hand:* $-94 \times 9 + PM \times 2 = 0$; $PM = 423$ *lb*

b. *With 25 lb weight in hand:* $-94 \times 9 + PM \times 2 + 25 \times 9 = 0$; $PM = 310.5$ *lb*

*For the calculations of value of joint reaction force, R, and its direction, θ, use the static equilibrium law for the forces:*

*Σ All forces acting on the free body diagram shown = 0.*

*R with weight in hand = 390.9 lb and* θ = 49.2° with respect to the horizontal.

7. In the following picture (A), workers' arms are supported (with an overhead sling) to minimize biomechanical loading of shoulder joint. Assume that the length of upper arm + forearm + hand is 70 cm and it weighs 40 N, acting at its center of gravity location of 25 cm from the shoulder joint. (a) Draw a free body diagram of the arm–shoulder

joint system (C) for one side (e.g., right side), (b) calculate the torque about the shoulder (S) without the overhead support, and (c) calculate the location of sling to be placed (on the arm) if we wish to support 70% of the weight of the arm.

*Answer:*

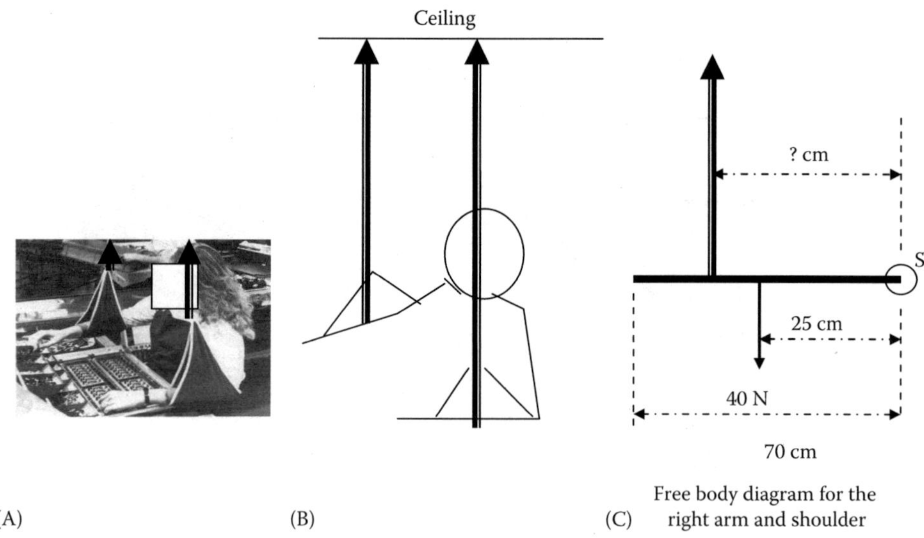


(A) (B) (C) Free body diagram for the right arm and shoulder

*Using the free body diagram shown in* figure C above:

*Applying the law of static equilibrium:* $\Sigma$ *moment around the shoulder joint* (*S*) = *0*:

*b. Torque about the shoulder* (*TS*) *without the overhead support*

$= 40 \times 25 = 1000$ *N.cm*

c. Location (y) of sling to be placed (on the arm) if we wish to support 70% of the weight of the arm

$$= 0.7 \times 40 \times y = 40 \times 25$$

*Location of sling to be placed* = *y* = *1000/28* = *35.7 cm from S*

8. In the following, physiological data [heart rate (b/min) and core body temperature (degrees F)] from a live fire training session for a firefighter (age 30 years) are presented. The live fire training consisted of the following activities: scenario 1, scenario 2, and scenario 3. For each scenario, the following tasks were performed: search and rescue, hose advancement, and backup (i.e., firefighter was serving as a backup person). Time-dependent changes in heart rate (HR) and core body temperature (CT) were obtained with a wireless POLAR chest belt–type sensor and an FDA-approved ingestible/wireless radio-pill sensor (*Musolin, K and Mani, A et al., Early Criteria for Evaluating the Cardiovascular and Psychophysical Effects of Heat Stress on Firefighters. Presented at NORA Symposium, Cincinnati, OH, July 2011*).

In the following, HR and CT data are given. Assuming a baseline (resting) heart rate of 70 b/min and baseline CT of 98.5°F, calculate (a) % max oxygen demand ($VO_2$) for all scenarios when the HR reached a peak value (shown by arrows in the following figure); (b) based on the resting HR data achieved immediately after each scenario, determine if the firefighter was experiencing fatigue and give your rationale for your answer.

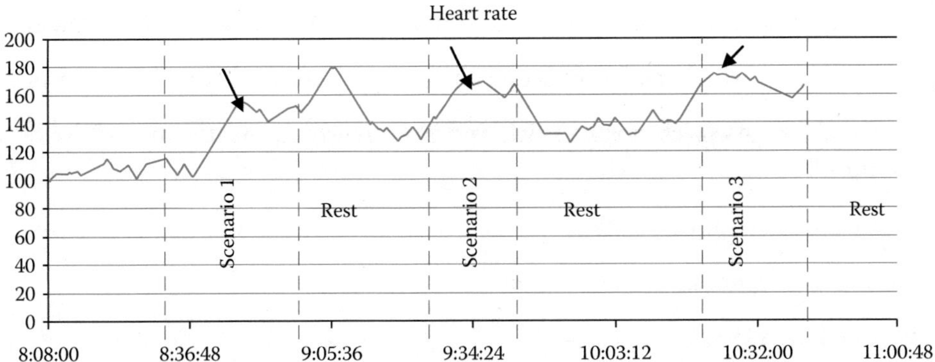


Using the baseline CT value of 98.5 and CT response data shown below, determine if the firefighter was potentially experiencing heat stress during the performance of live fire training.

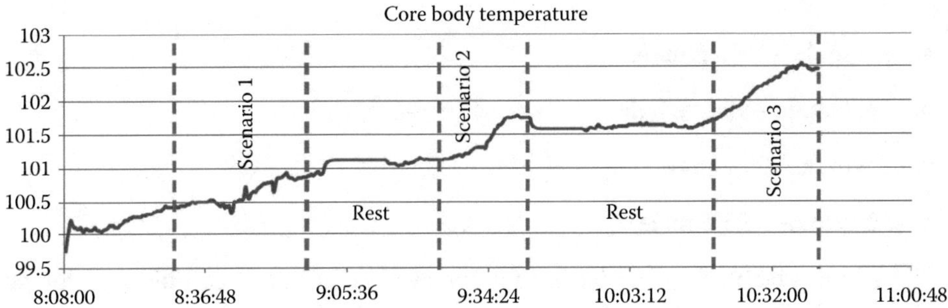


*Answer:*

$$Max\ HR = 220 - Age = 220 - 30 = 190\ b/min$$

*Use the following equation for calculating the values of* % max oxygen demand ($VO_2$) for all scenarios when the HR reached a peak value:

$$= 100 \times [(peak\ HR\ during\ the\ scenario\text{-}baseline\ resting\ HR)/(Max\ HR\text{-}baseline\ resting\ HR)]$$

*The firefighter is possibly becoming fatigued as his resting HR after each scenario is significantly higher than his baseline HR of 70 b/min.*

*The firefighter is possibly experiencing heat stress as his resting CT after each scenario is significantly higher than his baseline CT of 98.5°*F.

*Note*: In the following, answers are given in boldface.

9. Ergonomics is and means
   a. Latin: Ergo which means laws and Nomos which means work—"the laws of work"
   b. Greek: Ergo which means laws and Nomos which means work—"the work of laws"
   c. Latin: Ergo which means work and Nomos which means laws—"the work of laws"
   d. **Greek: Ergo which means work and Nomos which means laws—"the laws of work"**
   e. None of the above are true
10. The roots of ergonomics started with
    a. Newton in 1642
    b. **Ramazzini in 1700**
    c. Village vs. General Motors in 1981
    d. Ramazzini in 1532
11. When designing for human populations, it is a mistake to use the following anthropometric range:
    a. **5% male—95% female**
    b. 5% female—95% female
    c. 5% male—95% male
    d. 5% female—95% male
12. The major advantage of designing for the average is that it can be used for public places or facilities used by a large number of people and comfortably fit most people:
    a. True
    b. **False**
13. When thinking about human energy transformation as mechanical efficiency:
    a. 75%–80% is transformed into mechanical work performed with mechanical efficiency
    b. 50%–50% is transformed into mechanical work performed with mechanical efficiency

c. 10%–15% is transformed into mechanical work performed with mechanical efficiency

d. **20%–25% is transformed into mechanical work performed with mechanical efficiency**

14. A body part and a mechanical device that exhibit a class 2 lever system is raising up on the foot and the wheelbarrow:

a. **True**

b. False

15. The median vertical plane that divides the body into right and left parts is the

a. Coronal plane

b. Mid-coronal plane

c. Transverse plane

d. Sagittal plane

e. **Mid-sagittal plane**

16. How many bones are there in the human skeleton?

a. 602

b. 262

c. 226

d. 206

17. There are a total of _____ bones in the hand and _____ bones in the feet:

a. 32, 30

b. 30, 28

c. **28, 26**

d. 26, 24

e. None of the above are true

18. ***Aerobic glycolysis***: For most activities, glycogen (or fats or proteins) is broken down chemically with oxygen to generate ATP. The basic chemical reaction can be described as

$$\text{Glucose} + \text{Free fatty acids} + \text{P} + \text{ADP} + O_2 \rightarrow CO_2 + H_2O + \text{ATP}:$$

a. **True**

b. False

19. Women are generally 30% as strong as men. This is due to the fact that women have narrower (less cross-sectional area) muscles. However, the distributions of muscular strength overlap, so there are some women with greater strength values than some men:

a. True

b. **False**

20. **Reach** dimensions determine the maximum space allowable for the human being who operates equipment which are established by the smaller people in the expected user population (e.g., control height is determined by accommodating shorter people):

a. **True**

b. False

21. **Clearance** dimensions determine the minimum space required for the human being to perform his work activities in the workplace, such as operating and maintaining machines and so on. Clearances are established by the larger people from the expected user population (e.g., the size of a door frame is determined by the size of the largest expected user):

a. **True**

b. False

22. Heat exhaustion is characterized by

a. Hot, dry skin

b. **Clammy, moist skin**

c. Failure of the hypothalamus to properly regulate core temperature

d. A and C are true

e. B and C are true

23. Heat index combines _____ and _____ in an attempt to determine the human-perceived equivalent temperature:

a. Wind velocity, relative humidity

b. Wind velocity, air temperature

c. **Air temperature, relative humidity**

d. Evaporation rate, air temperature

24. Write the "lifting index (LI)" equation used in conjunction with the NIOSH-revised (1991) lifting equation. Define all variables—spell out the variables.

$$Answer\text{: } Lifting\ Index = \frac{Load\,Weight}{Recommended\,Weight\ Limit} = \frac{L}{RWL}$$

25. From the following diagram, calculate the muscle force required (F) to hold 5 lb in the palm of the hand. The following equation may be helpful:

**MA = force arm/resistance arm**

**Muscle force required = R/Ma**

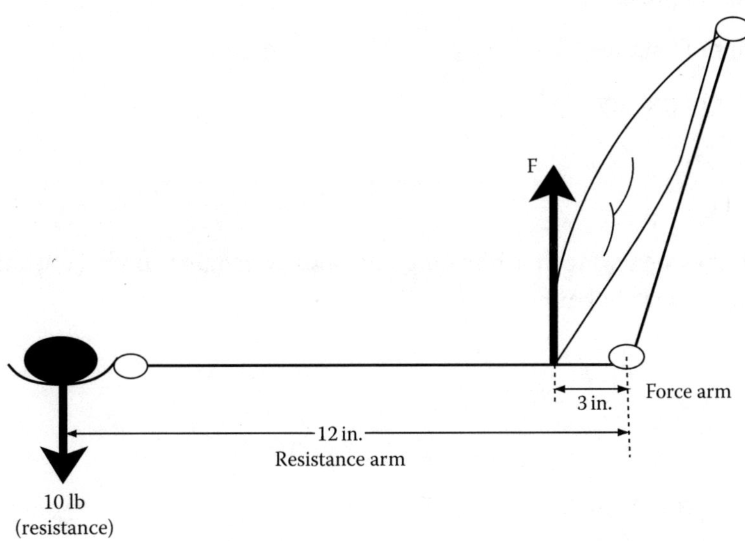


a. 18.4 kg

b. **9.07 kg**

c. 21.32 kg

d. 2.27 kg

**Please show your work.**
**Force arm = 3 in., resistance arm = 12 in., R = 5 lb**

**Answer:** $MA = \frac{force\,arm}{resistance\,arm} = \frac{3\,in.}{12\,in.} = 0.25$

$$Muscle\ force\,required = \frac{R}{MA} = \frac{5\,lb}{0.25} * \left(\frac{1\,lb}{0.454\,kg}\right) = 9.07\,lb$$

26. Slips, trips, and falls make up _____% of all accidental deaths with approximately _____% of reported injury claims each fiscal year:

    a. 5%, 10%

    b. **15%, 25%**

    c. 30%, 25%

    d. 15%, 40%

    e. None of the above

27. A trip, slip, or fall can occur when you are off your _____:

    a. Center of pressure

    b. Center of balance

    c. Center of gravity

    d. **A and B**

    e. B and C

28. This occurs when your foot hits an object and your upper body continues moving, throwing off your balance:

    a. Slip

    b. **Trip**

    c. Fall

    d. None of the above

29. Transitioning from one type of flooring to another with less traction without adjusting for the change may cause a

    a. **Slip**

    b. Trip or misstep

    c. Fall

    d. None of the above

30. CTDs most commonly appear as

    a. Carpal tunnel syndrome

    b. Tendinitis

    c. Tenosynovitis

    d. Bursitis

    e. **All of the above are true**

31. A ligament _____ occurs when it is pulled or pushed so that some of its fibers are torn. It can result from lifting a heavy object or holding up under an external force (usually when pulled by a traction force):

    a. Spasm

    b. **Sprain**

    c. Strain

    d. None of the above

32. Lateral epicondylitis is also known as

    a. Raynaud's syndrome

    b. Saturday-night palsy

    c. Golfer's elbow

    d. **Tennis elbow**

    e. Carpal tunnel syndrome

33. Thoracic outlet syndrome has symptoms of tingling and numbness in the _____ caused by _____:

    a. Thumb, index, and middle fingers; leaning on the elbows

    b. Fourth and little fingers; leaning on the elbows

    c. Fourth and little fingers; repetitive reaching behind to pick up objects

    d. **Thumb, index, and middle fingers; this can occur from repetitive reaching behind to pick up objects**

    e. None of the above

34. Tinel's test is performed to diagnose

    a. **Median nerve compression**

    b. Ulnar nerve compression

    c. Radial nerve compression

    d. Medial epicondylitis

    e. de Quervain's disease

35. Phalen's test: Patient is asked to put the backs of his/her two hands together and acutely flex his/her wrists for 1 minute. Development of tingling in the fingers is a positive sign of CTS:

    a. **True**

    b. False

36. During dynamic work,

    a. Blood needed is high, blood flow is low
    b. Blood needed is low, blood flow is high
    c. **Blood needed is high, blood flow is high**
    d. Blood needed is low, blood flow is low
    e. None of the above will work

37. Key considerations when designing workplaces using ergonomic principles:

    a. Anthropometric data
    b. User population
    c. Nature of the job
    d. Behavioral patterns of employees
    e. **All of the above**

38. A particular job requires reaching more than 41 cm (16 in.) forward or more than 15 cm (6 in.) above the work surface and has multiple tasks, some of which may require precision and some may require standing. What kind of workstation is recommended for this job?

    a. Seated workstation
    b. Standing workstation
    c. **Sit–stand workstation**
    d. Any of these would work equally well

39. The critical anthropometric measurement for determining the work-surface height of tables/desks (or workbenches and counters) is

    a. Stature (standing height)
    b. Waist height
    c. **Elbow height, sitting**
    d. Elbow height, standing
    e. Crotch height

# Appendix H: Americans with Disabilities Act: Reflections 10 Years Later

CONTENTS



## H.1 AMERICANS WITH DISABILITIES ACT: IMPLICATIONS FOR THE USE OF ERGONOMICS IN REHABILITATION, *REFLECTIONS 10 YEARS LATER*

*Robert E. Breslin*

The passage of the Americans with Disabilities Act (ADA) in 1991 was met with great anticipation by individuals with disabilities, their advocates, and professionals specializing in disability management, vocational rehabilitation, and ergonomics. This chapter was written soon after the passage of the ADA while the provisions of the law were in the early stages of implementation. The act was the result of a rare but broad bipartisan consensus regarding the under participation of individuals with disabilities in American society and, at the time this chapter was written, was widely viewed as the most significant expansion of civil rights under federal law since 1964. The findings section of the ADA estimated the number of Americans with disabilities to be approximately 45 million. The law was envisioned as a source of recourse and address for those individuals in that number who found that discrimination restricted their access to public accommodations, employment, transportation, telecommunication, and state and local government services.

Given the emphasis on employment discrimination found in Title I of the act and the emphasis on the elimination of environmental and administrative barriers to the employment of individuals with disabilities, the authors, and most other observers, anticipated that the implementation of the ADA would place a premium on the use of ergonomic and rehabilitation engineering interventions to accommodate the functional impairments of qualified individuals with disabilities in the workplace.

The majority of the provisions of the ADA took effect in 1992. As directed by the act, the Equal Employment Opportunity Commission (EEOC) issued regulations and guidelines to direct EEOC field investigators in their analysis of ADA-related complaints. Simultaneously, as ADA suits were brought to the federal courts, a body of case law was developed. The body of law related to the removal of barriers to employment for individuals with disabilities has been shaped by a series of court decisions, many at the level of the U.S. Supreme Court. These decisions have weakened the scope of the ADA by narrowing the definition of disability in a manner that has made it more difficult for individuals to claim protection under the act. As a result, employers have prevailed in the vast majority of ADA suits as the result of summary judgments (Hryniewicz, 2010).

The initial ADA cases heard by the U.S. Supreme Court appeared to follow the analytic approach common to civil rights law which resulted in a broad interpretation of the definition of disability and struck down the attempts of some employers to deny individuals protection under the ADA. In 1999, however, the court decided three cases, known collectively as the Sutton trilogy, which held that the effect of mitigating measures must be considered before an individual could be found to have a severe impairment under the law. These decisions had the perverse effect of not allowing qualified individuals to request reasonable accommodations on the job because the availability of such accommodations kept them from being considered disabled, even if the employer specifically forbade the individual from using the corrective measures in the workplace or when making a hiring or termination decision. In addition, in the *Toyota vs. Williams* decision, the court further narrowed the definition of disability by redefining major life activities in a manner that was expressly intended to reduce the number of Americans who would qualify for protection under ADA.

The continued narrowing of the scope of the ADA law resulted in the passage of the ADA Amendments Act (ADAAA) of 2008. This act, which was subtitled "An act to restore the intent and protections of the Americans with Disabilities Act of 1990," was passed unanimously by both houses in the 110th Congress and signed into law by President George W. Bush on September 25, 2008. It is intended to broaden the definition of disability and increase the number of individuals who are able to claim the protections of the ADA. It specifically reverses the interpretations of the Sutton trilogy and the *Toyota vs. Williams* case and directs the courts to interpret the definition of disability in a manner allowing a broad range of individuals with disabilities to claim protection.

As of this writing, there has not been sufficient case law to assess the impact of the ADAAA on the employment of individuals with disabilities. It is hoped that this amendment of the law will reverse the decade-long trend of individuals with disabilities losing employment discrimination actions in the federal courts. If this is indeed the case, the ADA and ADAAA may well fulfill the initial promise of promoting the accommodation of qualified individuals with disabilities in the workplace and increasing their level of employment (Graham and Davis 2010).

## H.2 FUNDAMENTAL LEGISLATIVE CHANGES BROUGHT WITH THE ADA AMENDMENT ACT OF 2008

*Michael White*

The laws set forth by the ADA of 1990 proved to be too vague and challenging for the courts to interpret; therefore, many employers won a large majority of litigation suits. The two most controversial Supreme Court cases being *Sutton vs. United Air Lines*, 527 U.S. 471 (1999) and *Toyota Motor Manufacturing, Kentucky, Inc. vs. Williams*, 534 U.S. 184 (2002). The major barrier being the statute's convoluted meaning of the term "disability."

As a result and nearly 20 years later, Congress amended the ADA of 1990 that became effective January 1, 2009, being called the ADAAA. The new act preserved the original definition of the term "disability" as being an impairment that substantially limits one or more major life activities, a record of such an impairment, or being regarded as having such an impairment. However, the act's purpose was to clarify and broaden the definition of the term "disability," while modifying the interpretation of the statutes by the courts. Instead of trying to decide if impairment falls within the umbrella of "disability," the courts were now to interpret a disability in favor of broad coverage to individuals. The act also overturned the Supreme Court's decisions in both the Sutton and Toyota cases, ultimately making it easier for impaired individuals to be covered (U.S. Equal Employment Opportunity Commission, 2009).

The act directed the Equal Employment Opportunity Commission (EEOC) to revise its previous ADA regulations to reflect the new changes. First, the definition of the term "substantially limits" was changed to include a lower degree of functional limitations and clarified that an impairment that is episodic or in remission is a disability if it would substantially limit a major life activity when active. Second, the ameliorative effects of mitigating measures (i.e., medication or hearing aids) other than "ordinary eyeglasses or contact lenses" should not be used to evaluate whether an impairment substantially limits one or more major life activities. Third, the definition of "a major life activity" was expanded to incorporate two indefinite lists: one including recognized and unrecognized physical activities, and the other including major bodily functions. Finally, and possibly the most significant revision, the definition of "regarded as" was changed to show that an employee is "regarded as" disabled if they are subject to employment action, which is prohibited by the ADA, based on an actual or perceived impairment, but one that is not transitory and minor. In addition, individuals covered only under the "regarded as" prong are not eligible for reasonable accommodation, but if they are covered under the "actual impairment" or "record of impairment" prongs they do qualify for reasonable accommodation. As of March 25, 2011, the EEOC's final revised regulations were approved and published in the Federal Register, and will go into effect on May 24, 2011 (U.S. Equal Employment Opportunity Commission, 2011).

To summarize according to the EEOC, the act

1. Directs EEOC to revise that portion of its regulations defining the term substantially limits

2. Expands the definition of "major life activities" by including two non-exhaustive lists

a. The first list includes many activities that the EEOC has recognized (e.g., walking) as well as activities that EEOC has not specifically recognized (e.g., reading, bending, and communicating)

b. The second list includes major bodily functions (e.g., "functions of the immune system, normal cell growth, digestive, bowel, bladder, neurological, brain, respiratory, circulatory, endocrine, and reproductive functions")

3. States that mitigating measures other than "ordinary eyeglasses or contact lenses" shall not be considered in assessing whether an individual has a disability

4. Clarifies that an impairment that is episodic or in remission is a disability if it would substantially limit a major life activity when active

5. Changes the definition of "regarded as" so that it no longer requires a showing that the employer perceived the individual to be substantially limited in a major life activity, and instead says that an applicant or employee is "regarded as" disabled if he or she is subject to an action prohibited by the ADA (e.g., failure to hire or termination) based on an impairment that is not transitory and minor

6. Provides that individuals covered only under the "regarded as" prong are not entitled to reasonable accommodation (U.S. Equal Employment Opportunity Commission, 2011)

Any accommodation is considered reasonable until it produces an "undue hardship" for the employer. According to the EEOC, "Undue hardship means significant difficulty or expense and focuses on the resources and circumstances of the particular employer in relationship to the cost or difficulty of providing a specific accommodation. Undue hardship refers not only to financial difficulty, but to reasonable accommodations that are unduly extensive, substantial, or disruptive, or those that would fundamentally alter the nature or operation of the business" (U.S. Equal Employment Opportunity Commission, 2002).

## REFERENCES

ADA Amendments Act of 2008, 42 U.S.C. Section 121010.

Malyina, J.H. (2010). The definition of major life activity under *Adams vs. Rice* is not substantially limiting, 20 *George Mason University Civil Rights Law Journal*, 417, 421, 427.

Graham, L.S. Jr. and Davis, M.E. Jr. Employment and the ADA/ADAAA, *The Journal of Forensic Vocational Analysis*, 13(1), 2010.

*Sutton vs. United Airlines*, 527 US 471 (1999).

*Toyota vs. Williams*, 535 US 184 (2002).

U.S. Equal Employment Opportunity Commission (2002). *Enforcement Guidance: Reasonable Accommodation and Undue Hardship under the Americans with Disabilities Act*. Notice, United States Federal Government.

U.S. Equal Employment Opportunity Commission (2009). *The Americans with Disabilities Act Amendments Act of 2008*. United States.

U.S. Equal Employment Opportunity Commission (2011). *Notice Concerning the Americans with Disabilities Act (ADA) Amendments Act of 2008*. United States.

Further Information can be found at the following:

ADA Act of 2008. (http://www.eeoc.gov/laws/statutes/adaaa_info.cfm).

EEOC's Final Regulations Published to the Federal Register as of March 25, 2011 (http://www.eeoc.gov/laws/regulations/adaaa_fact_sheet.cfm).

*Sutton vs. United Air Lines*, 527 US 471 (1999) (http://caselaw.lp.findlaw.com/scripts/getcase.pl?court=us&vol=000&invol=97-1943) (accessed July 21, 2011).

*Toyota Motor Manufacturing, Kentucky, Inc. vs. Williams*, 534 US 184 (2002) (http://caselaw.lp.findlaw.com/scripts/getcase.pl?court=us&vol=000&invol=00-1089).

# Index

## C

## D



## E

F



G



H

## I

## J



## K

## O

## T

## Z